A TEXTBOOK OF MINERALOGY

By the late JAMES D. DANA

SYSTEM OF MINERALOGY. *Seventh Edition.*

Rewritten and enlarged by Charles Palache,
the late Harry Berman, and Clifford Frondel.

VOL. I.   1944.
VOL. II.   1951.
VOL. III.   In preparation.

MANUAL OF MINERALOGY. *Sixteenth Edition.*
Revised by Cornelius S. Hurlbut, Jr.

By the late EDWARD S. DANA

A TEXTBOOK OF MINERALOGY. *Fourth Edition.*
Revised by the late William E. Ford.   1932.

MINERALS AND HOW TO STUDY THEM. *Third Edition*
Revised by Cornelius S. Hurlbut, Jr.   1949.

A TEXTBOOK

OF

MINERALOGY

WITH AN EXTENDED TREATISE ON

CRYSTALLOGRAPHY AND PHYSICAL MINERALOGY

\

BY

EDWARD SALISBURY DANA

Late Professor Emeritus of Physics, Yale University

FOURTH EDITION, REVISED AND ENLARGED

BY

WILLIAM E. FORD

*Late Professor of Mineralogy and Curator of the Mineral Collections,
Sheffield Scientific School of Yale University*

NEW YORK

JOHN WILEY & SONS, Inc.

London: CHAPMAN & HALL, Limited

PREFACE TO THE FOURTH EDITION

It is only a little over ten years since the third edition of this book was published. This period, however, has been one of such active mineralogical investigation that a new edition seemed desirable. The fourth edition, therefore, endeavors to present the important facts of the science as known on January 1, 1932. The changes in the book are essentially as follows: In Part I on Crystallography a section of some seventeen pages has been added on crystal structure and the methods of its investigation by means of X-rays. The remainder of this section remains substantially unchanged. Parts II and III on Physical and Chemical Mineralogy have been revised but have had only minor additions. Part IV on the Origin, Mode of Occurrence and Association of Minerals is new to this book. Part V has been entirely revised. Descriptions of about two hundred and twenty new species have been added. The attempt has been made to give a brief but complete statement of the important facts now known about all well-defined mineral species. In addition brief mention is made of doubtful species or those which have recently been discredited. It is realized that such a complete treatment of the subject would not normally have its place in a book of this character. At the present time, however, there is no book available that gives such a survey of the subject and until such a book is at hand it is thought that the inclusion here of brief descriptions of all known minerals will be of value. As far as possible simple statements of the results of the investigation of mineral structure by means of X-rays have been included in the mineral descriptions. The paragraphs on occurrence have been largely rewritten. In the case of the commoner minerals a careful selection of the localities of their occurrence has been made in order to give only those of most importance either from a scientific or economic point of view. In the case of rarer species all known localities have been mentioned. As far as possible the locality names have been checked, the spelling used by the Times Atlas having been followed in most instances. In the case of the south-central European countries, alternative names for the same place have been frequently given.

The mineral data presented have been gathered from the periodical literature, etc., and from many texts. Among the books which have been of particular value are the following: Doelter's Handbuch der Mineralchemie, Hintze's Handbuch der Mineralogie, Lehrbuch der Mineralogie by Niggli, Mineralogy by Miers-Bowman, and especially Winchell's Elements of Optical Mineralogy and Larsen's Microscopic Determination of the Nonopaque Minerals. The editor is also indebted to Professor Esper S. Larsen for the privilege of consulting the manuscript of the new edition of the last-named work. He also acknowledges gratefully many valuable suggestions made by Professor Charles Palache.

WILLIAM E. FORD

NEW HAVEN, CONN., *March* 1, 1932

PREFACE TO THE THIRD EDITION

The first edition of this book appeared in 1877 and approximately twenty years later (1898) the second and revised edition was published. Now, again after more than twenty years, comes the third edition. The changes involved in the present edition are chiefly those of addition, the general character and form of the book having been retained unchanged. In the section on Crystallography the important change consists in the introduction of the methods employed in the use of the stereographic and gnomonic projections. A considerable portion of the section on the Optical Characters of Minerals has been rewritten in the endeavor to make this portion of the book simpler and more readily understood by the student. In the section on Descriptive Mineralogy all species described since the previous edition have been briefly mentioned in their proper places. Numerous other changes and corrections have, of course, been made in order to embody the results of mineral investigation during the last two decades. Only minor changes have been made in the order of classification of the mineral species. It was felt that as this book is so closely related to the System of Mineralogy it was unwise to attempt any revision of the chemical classification until a new edition of that work should appear. The description of the methods of Crystal Drawing given in Appendix A has been largely rewritten. A new table has been added to Appendix B in which the minerals have been grouped into lists according to their important basic elements. Throughout the book the endeavor has been to present in a clear and concise way all the information needed by the elementary and advanced student of the science.

The editor of this edition is indebted especially to the published and unpublished writings of the late Professor Samuel L. Penfield for much material and many figures that have been used in the sections on Crystallography and The Optical Character of Minerals. He also acknowledges the cordial support and constant assistance given him by Professor Edward S. Dana.

WILLIAM E. FORD

NEW HAVEN, CONN., *Dec.* 1, 1921

PREFACE TO THE SECOND EDITION

The remarkable advance in the Science of Mineralogy, during the years that have elapsed since this Text-Book was first issued in 1877, has made it necessary, in the preparation of a new edition, to rewrite the whole as well as to add much new matter and many new illustrations.

The work being designed chiefly to meet the wants of class or private instruction, this object has at once determined the choice of topics discussed, the order and fullness of treatment and the method of presentation.

In the chapter on Crystallography, the different types of crystal forms are described under the now accepted thirty-two groups classed according to their symmetry. The names given to these groups are based, so far as possible, upon the characteristic form of each, and are intended also to suggest the terms formerly applied in accordance with the principles of hemihedrism. The order adopted is that which alone seems suited to the demands of the elementary student, the special and mathematically simple groups of the isometric system being described first. Especial prominence is given to the " normal group " under the successive systems, that is, to the group which is relatively of most common occurrence and which shows the highest degree of symmetry. The methods of Miller are followed as regards the indices of the different forms and the mathematical calculations.

In the chapters on Physical and Chemical Mineralogy, the plan of the former edition is retained of presenting somewhat fully the elementary principles of the science upon which the mineral characters depend; this is particularly true in the department of Optics. The effort has been made to give the student the means of becoming practically familiar with all the modern methods of investigation now commonly applied. Especial attention is, therefore, given to the optical properties of crystals as revealed by the microscope. Further, frequent references are introduced to important papers on the different subjects discussed, in order to direct the student's attention to the original literature.

The Descriptive part of the volume is essentially an abridgment of the Sixth Edition of Dana's System of Mineralogy, prepared by the author (1892). To this work (and future Appendices) the student is, therefore, referred for fuller descriptions of the crystallographic and optical properties of species, for analyses, lists of localities, etc.; also for the authorities for data here quoted. In certain directions, however, the work has been expanded when the interests of the student have seemed to demand it; for example, in the statement of the characters of the various isomorphous groups. Attention is also called to the paragraph headed " **Diff.**," in the description of each common species, in which are given the distinguishing characters, particularly those which serve to separate it from other species with which it might be easily confounded.

The list of American localities of minerals, which appeared as an Appendix in the earlier edition, has been omitted, since in its present expanded form it requires more space than could well be given to it; further, its reproduction here is unnecessary since it is accessible to all interested not only in the System of Mineralogy but also in separate form. A full topical Index has been added, besides the usual Index of Species.

The obligations of the present volume to well-known works of other authors — particularly to those of Groth and Rosenbusch — are too obvious to require special mention. The author must, however, express his gratitude to his colleague, Prof. L. V. Pirsson, who has given him material aid in the part of the work dealing with the optical properties of minerals as examined under the microscope. He is also indebted to Prof. S. L. Penfield of New Haven and to Prof. H. A. Miers of Oxford, England, for various valuable suggestions.

EDWARD SALISBURY DANA

NEW HAVEN, CONN., *Aug.* 1, 1898

CONTENTS

ix

PART II. PHYSICAL MINERALOGY

PART III. CHEMICAL MINERALOGY

PART IV

PART V. DESCRIPTIVE MINERALOGY

APPENDIX A

APPENDIX B

INTRODUCTION

1. The Science of Mineralogy treats of those inorganic species called *minerals*, which together in rock masses or in isolated form make up the material of the crust of the earth, and of other bodies in the universe so far as it is possible to study them in the form of meteorites.

2. Definition of a Mineral. — *A Mineral is a body produced by the processes of inorganic nature, having usually a definite chemical composition and, if formed under favorable conditions, a certain characteristic atomic structure which is expressed in its crystalline form and other physical properties.*

This definition calls for some further explanation.

First of all, a mineral must be a *homogeneous* substance, even when minutely examined by the microscope; further, it must have a *chemical composition* which commonly is definite and can be expressed by a chemical formula. In some cases the chemical composition is variable but only within certain limits and then usually according to a definite law. Thus, much basalt appears to be homogeneous to the eye, but when examined under the microscope in thin sections it is seen to be made up of different substances, each having characters of its own. Again, obsidian, or volcanic glass, though it may be essentially homogeneous, has not a sufficiently definite composition to be classed as a mineral.

Again, a mineral has in most cases a *definite atomic structure.* This atomic structure, as will be shown later, manifests itself in the physical characters and especially in the external crystalline form. A mineral, in the majority of cases, possesses both general properties, such as composition, specific gravity, melting point, etc., and directional properties, such as its atomic structure, crystal symmetry, optical characters, etc. The combination of these two kinds of characters serves to define a mineral species.

It is customary, as a matter of convenience, to limit the name mineral to those compounds which have been formed by the processes of nature alone, whereas compounds made in the laboratory or the smelting-furnace are at best called artificial minerals. Further, mineral substances which have been produced through the agency of organic life are not included among minerals, as the pearl of an oyster, etc. Finally, mineral species are, as a rule, limited to *solid substances*, the only liquids included being metallic mercury and water. Petroleum is not properly a homogeneous substance, consisting rather of several hydrocarbon compounds; it is hence not a mineral species.

It is obvious from the above that minerals, in the somewhat restricted sense usually adopted, constitute only a part of what is often called the Mineral Kingdom.

3. Scope of Mineralogy. — In the following pages, the general subject of mineralogy is treated under the following heads:

(1) *Crystallography.* — This comprises a discussion of crystals in general and especially of the crystalline forms of mineral species.

1

(2) *Physical Mineralogy.* — This includes a discussion of the physical characters of minerals, that is, those depending upon cohesion and elasticity, density, light, heat, electricity, and so on.

(3) *Chemical Mineralogy.* — Under this head are presented briefly the general principles of chemistry as applied to mineral species; their characters as chemical compounds are described, also the methods of investigating them from the chemical side by the blowpipe and other means.

(4) *Occurrence of Minerals.* — This section includes a brief description of the different modes of mineral occurrence, characteristic mineral associations, etc.

(5) *Descriptive Mineralogy.* — This includes the classification of minerals and the description of each species with its varieties, especially in its relations to closely allied species, as regards crystalline form, physical and chemical characters, occurrence in nature, and other points.

4. Literature. — Reference is made to the Introduction to the Sixth Edition of Dana's System of Mineralogy, pp. xlv–lxi, for an extended list of independent works on Mineralogy up to 1892 and to its Appendices I, II and III for works published up to 1915; the names are also given of the many scientific periodicals which contain original memoirs on mineralogical subjects. For the convenience of the student the titles of a few works, mostly of a general character, are given here. Further references to the literature of Mineralogy are introduced through the first half of this work, particularly at the end of the sections dealing with special subjects.

Crystallography and Physical Mineralogy

EARLY WORKS* include those of Romé de l'Isle, 1772; Haüy, 1822; Neumann, Krystallonomie, 1823, and Krystallographie, 1825; Kupffer, 1825; Grassmann, Krystallonomie, 1829; Naumann, 1829 and later; Quenstedt, 1846 (also 1873); Miller, 1839 and 1863; Grailich, 1856; Kopp, 1862; von Lang, 1866; Bravais, Études Crist., Paris, 1866 (1849); Schrauf, 1866–68; Rose-Sadebeck, 1873.

RECENT WORKS include the following:

Barker. Graphical and Tabular Methods in Crystallography, 1922; Systematic Crystallography, 1930.

Bayley. Elementary Crystallography, 1910.

Beale. Introduction to Crystallography, 1915.

Beckenkamp. Statische und kinetische Kristalltheorien, 1913–1915.

Bruhns. Elemente der Krystallographie, 1902.

Goldschmidt. Index der Krystallformen der Mineralien; 3 vols., 1886–91. Also Anwendung der Linearprojection zum Berechnen der Krystalle, 1887. Krystallographische Winkeltabellen, 1897. Atlas der Krystallformen, 1913–1923.

Goldschmidt and Gordon. Crystallographic Tables for the Determination of Minerals, 1928.

Gossner. Kristallberechnung und Kristallzeichnung, 1914.

Groth. Physikalische Krystallographie und Einleitung in die krystallographische Kenntniss der wichtigeren Substanzen, 1905; Elemente der Physikalischen und Chemischen Krystallographie, 1921.

Honess. Etch Figures on Crystals, 1927.

Klein. Einleitung in die Krystallberechnung, 1876.

Lewis. Crystallography, 1899.

Liebisch. Geometrische Krystallographie, 1881. Physikalische Krystallographie, 1891.

Mallard. Traité de Cristallographie géométrique et physique; vol. 1, 1879; vol. 2, 1884.

---

* The full titles of many of these are given in pp. li-lxi of Dana's System of Mineralogy, 1892.

Moses. Characters of Crystals, 1899.
Niggli. Lehrbuch der Mineralogie, Vol. I., 1924; Vol. II, 1926; Krystallographische und Strukturtheoretische Grundbegriffe, 1928.
Parker. Kristallzeichen, 1929.
Reeks. Hints for Crystal Drawing, 1908.
Reinhard. Universal Drehtischmethoden, 1931.
Sadebeck. Angewandte Krystallographie (Rose's Krystallographie, II. Band), 1876.
Schiebold. Über eine neue Herleitung und Nomenklatur der 230 kristallographischen Raumgruppen, 1929.
Schleede and Schneider. Röntgenspektroskopie und Krystallstrukturanalyse, 1929.
Schoenflies. Theorie der Kristallstruktur, 1923.
Sohnke. Entwickelung einer Theorie der Krystallstruktur, 1879.
Sommerfeldt. Geometrische Kristallographie, 1906; Physikalische Kristallographie, 1907; Die Kristallgruppe, 1911.
Story-Maskelyne. Crystallography: the Morphology of Crystals, 1895.
Tutton. Crystalline Structure and Chemical Constitution, 1926; Crystallography and Practical Crystal Measurement, 1922; The Natural History of Crystals, 1924.
Viola. Grundzüge der Kristallographie, 1904.
Walker. Crystallography, 1914.
Wallerant. Cristallographie, 1909.
Websky. Anwendung der Linearprojection zum Berechnen der Krystalle (Rose's Krystallographie III. Band), 1887.
Williams. Elements of Crystallography, 1890.
Wülfing. Die 32 krystallographischen Symmetrieklassen und ihre einfachen Formen, 1911.

In PHYSICAL MINERALOGY the most important general works are those of Schrauf (1868), Mallard (1884), Liebisch (1891), mentioned in the above list; also Rosenbusch, Mikr. Physiographie, etc. (1892). Important later works include the following.

Davy-Farnham. Microscopic Examination of the Ore Minerals, 1920.
Duparc and Pearce. Traité de Technique Minéralogique et Pétrographique, 1907.
Farnham. Determination of the Opaque Minerals, 1931.
Groth. Physikalische Krystallographie, 1905.
Groth Jackson. Optical Properties of Crystals, 1910.
Johannsen. Determination of Rock-Forming Minerals, 1908. Manual of Petrographic Methods, 1914. Essentials for the Microscopic Determination of Rock-forming Minerals and Rocks in Thin Sections, 1922.
Larsen. Microscopic Determination of the Nonopaque Minerals, 1921.
Murdoch. Microscopical Determination of the Opaque Minerals, 1916.
Nikitin, translated into French by **Duparc and de Dervies.** La Méthode Universelle de Fedoroff, 1914.
Schneiderhöhn and Ramdohr. Lehrbuch der Erzmikroskopie, 1931. Erzmikroskopische Bestimmungstafeln, 1931.
Short. Microscopic Determination of the Ore Minerals. U. S. Geol. Sur. Bull. 825, 1931.
Winchell. Elements of Optical Mineralogy, 1922–1931.
Wright. The Methods of Petrographic-Microscopic Research, 1911.

General Mineralogy

Of the many works, a knowledge of which is needed by one who wishes a full acquaintance with the historical development of Mineralogy, the following are particularly important. Very early works include those of Theophrastus, Pliny, Linnæus, Wallerius, Cronstedt, Werner, Bergmann, Klaproth.

Within the nineteenth century: Haüy's Treatise, 1801, 1822; Jameson, 1816, 1820; Werner's Letztes Mineral-System, 1817; Cleaveland's Mineralogy, 1816, 1822; Leonhard's Handbuch, 1821, 1826; Mohs's Min., 1822; Haidinger's translation of Mohs, 1824; Breithaupt's Charakteristik, 1820, 1823, 1832; Beudant's Treatise, 1824, 1832; Phillips' Min., 1823, 1837; Shepard's Min., 1832–35, and later editions; von Kobell's Grundzüge, 1838; Mohs' Min., 1839; Breithaupt's Min., 1836–1847; Haidinger's Handbuch, 1845; Naumann's Min., 1846 and later; Hausmann's Handbuch, 1847; Dufrénoy's Min., 1844–1847 (also 1856–1859); Brooke & Miller, 1852; J. D. Dana's System of 1837, 1844, 1850, 1854, 1868.

More RECENT WORKS are the following:

Bauer. Lehrbuch der Mineralogie, 1904.
Bauerman. Text-Book of Descriptive Mineralogy, 1884.
Baumhauer. Das Reich der Krystalle, 1889.
Bayley. Descriptive Mineralogy, 1917.
Blum. Lehrbuch der Mineralogie, 4th ed., 1873–1874.
Brauns. Das Mineralreich, 1903. English translation by **Spencer,** 1912.
Clarke. The Data of Geochemistry, 1924.
Dana, E. S. Dana's System of Mineralogy, 6th ed., New York, 1892. Appendix I, 1899; II, 1909; III, 1915. Also (elementary) Minerals and How to study them, New York, 1895.
Dana-Ford. Manual of Mineralogy, 1929.
Des Cloizeaux. Manuel de Minéralogie; vol. 1, 1862; vol. 2, 1er Fasc., 1874; 2me. 1893.
Groth and **Mieleitner.** Mineralogische Tabellen, 1921.
Hintze. Handbuch der Mineralogie, 1889–1931.
Iddings. Rock Minerals, 1906.
Kraus and **Hunt.** Mineralogy, 1928.
Lacroix. Minéralogie de la France et de ses Colonies, 5 vols., 1893–1913.
Miers-Bowman. Mineralogy, 1929.
Moses and **Parsons.** Mineralogy, Crystallography and Blowpipe Analysis, 1916.
Merrill. The Non-metallic Minerals, 1904.
Niggli. Lehrbuch der Mineralogie, 1924–1926.
Phillips. Mineralogy, 1912.
Rogers. Study of Minerals, 1921.
Schrauf. Atlas der Krystall-Formen des Mineralreiches, 4to, vol. 1, A–C, 1865–1877.
Spencer. The World's Minerals, 1916.
Tschermak. Lehrbuch der Mineralogie, 1884; 9th ed., 1923.
Weisbach. Synopsis Mineralogica, systematische Uebersicht des Mineralreiches, 1875.
Zirkel. 13th edition of Naumann's Mineralogy, Leipzig, 1897.
Wülfing. Die Meteoriten in Sammlungen, etc., 1897 (earlier works on related subjects, see Dana's System, p. 32).

For a catalogue of localities of minerals in the United States and Canada see the volume (51 pp.) reprinted from Dana's System, 6th ed. See also the volumes on the Mineral Resources of the United States published (since 1882) under the auspices of the U. S. Geological Survey and U. S. Bureau of Mines.

Chemical and Determinative Mineralogy

Bischoff. Lehrbuch der chemischen und physikalischen Geologie, 1847–54; 2d ed., 1863–66. (Also an English edition.)
Blum. Die Pseudomorphosen des Mineralreichs, 1843. With 4 Nachträge, 1847–1879.
Brush-Penfield. Manual of Determinative Mineralogy, with an Introduction on Blowpipe Analysis, 1896.
Doelter. Allgemeine chemische Mineralogie, Leipzig, 1890. Handbuch der Mineralchemie, 1912–1931.
Duparc and **Monnier.** Traité de Technique Minéralogique et Pétrographique, 1913.
Eakle. Mineral Tables for the Determination of Minerals by their Physical Properties, 1904.
Endlich. Manual of Qualitative Blowpipe Analysis, New York, 1892.
Kobell, F. von. Tafeln zur Bestimmung der Mineralien mitteist einfacher chemischer Versuche auf trockenem und nassem Wege, 11te Auflage, 1878.
Kraus and **Hunt.** Tables for the Determination of Minerals, 1930.
Lewis and **Hawkins.** A Manual of Determinative Mineralogy, 1931.
Putnam. A chart showing the Chemical Relationships in the Mineral Kingdom, 1925.
Rammelsberg. Handbuch der krystallographisch-physikalischen Chemie, Leipzig, 1881–82. Handbuch der Mineralchemie, 2d ed., 1875. Ergänzungsheft, 1, 1886; 2, 1895.
Rosenholtz and **Smith.** Tables and Charts of Specific Gravity and Hardness for use in the Determination of Minerals, 1931.
Roth. Allgemeine und chemische Geologie; vol. 1, Bildung u. Umbildung der Mineralien, etc., 1879; 2, Petrographie, 1887–1890.
Websky. Die Mineral Species nach den für das specifische Gewicht derselben angenommenen und gefundenen Werthen, Breslau, 1868.

Weisbach. Tabellen zur Bestimmung der Mineralien nach äusseren Kennzeichen, 3te Auflage, 1886. Also founded on Weisbach's work, **Frazer's** Tables for the determination of minerals, 4th ed., 1897.

Artificial Formation of Minerals

Dittler. Mineralsynthetisches Praktikum, 1915.
Gurlt. Uebersicht der pyrogeneten künstlichen Mineralien, namentlich der krystallisirten Hüttenerzeugnisse, 1857.
Fuchs. Die künstlich dargestellten Mineralien, 1872.
Daubrée. Études synthetique de Géologie expérimentale, Paris, 1879.
Fouqué and **M. Lévy.** Synthèse des Minéraux et des Roches, 1882.
Bourgeois. Réproduction artificielle des Minéraux, 1884.
Meunier. Les méthodes de synthèse en Minéralogie.
Vogt. Die Silikatschmelzlözungen, 1903–1904.

Mineralogical Journals

The following Journals are largely devoted to original papers on Mineralogy:

Amer. Min. American Mineralogist, 1916–.
Bull. Soc. Min. Bulletin de la Société Française de Minéralogie, 1878–.
Centralbl. Min. Centralblatt für Mineralogie, Geologie und Palæontologie, 1900–.
Chem. Erde. Chemie der Erde, 1914–.
Fortschr. Min. Fortschritte der Mineralogie, Kristallographie und Petrographie, 1911–.
Jb. Min. Neues Jahrbuch für Mineralogie, Geologie und Palæontologie, etc., from 1833.
Min. Mag. The Mineralogical Magazine and Journal of the Mineralogical Society of Gt. Britain, 1876–.
Min. Mitth. Mineralogische und petrographische Mittheilungen, 1878 ; Earlier, from 1871, Mineralogische Mittheilungen gesammelt von G. Tschermak.
Riv. Min. Rivista di Mineralogia e Crystallografia, 1887–1918.
Zs. Kr. Zeitschrift für Krystallographie und Mineralogie. 1877–.

ABBREVIATIONS

Ax. pl.	Plane of the optic axes.	H.	Hardness.
Bx, Bv$_a$	Acute bisectrix (p. 303).	Obs.	Observations on occurrence, etc.
Bx$_o$.	Obtuse bisectrix (p. 303).	O.F.	Oxidizing Flame (p. 362).
B.B.	Before the Blowpipe (p. 361).	Pyr.	Pyrognostics or blowpipe and
Comp.	Composition.		allied characters.
Diff.	Differences, or distinctive char	R.F.	Reducing Flame (p. 362).
	acters.	Var.	Varieties.
G.	Specific Gravity.		

The sign $\wedge$ is used to indicate the angle between two faces of a crystal, as *am* $(100 \wedge 110)$ = $11° 30'$.

PART I. CRYSTALLOGRAPHY

GENERAL MORPHOLOGICAL RELATIONS OF CRYSTALS

5. Crystallography. — The subject of Crystallography includes the description of the characters of crystals in general; of their internal or atomic structure; of the various forms of crystals and their division into classes and systems; of the methods of studying crystals, including the determination of the mathematical relations of their faces, and the measurement of the angles between them; finally, a description of compound or twin crystals, of irregularities in crystals, of crystalline aggregates, and of pseudomorphous crystals.

6. Definition of a Crystal. — *A crystal* is the regular polyhedral form, bounded by smooth surfaces, which is assumed by a chemical compound, under the action of its interatomic forces, when passing, under suitable conditions, from the state of a liquid or gas to that of a solid.*

As expressed in the foregoing definition, a crystal is characterized, first, by its definite internal structure, and, second, by its external form. A crystal is the *normal* form of a mineral species, as of all solid chemical compounds; but the conditions suitable for the formation of a crystal of ideal perfection in symmetry of form and smoothness of surface are never fully realized. Further, many species usually occur not in distinct crystals, but in massive form, and in some exceptional cases the definite internal structure is absent.

Three terms are sometimes used to indicate the different states in which a crystalline substance may appear. If a crystal is developed with all its proper plane surfaces (or crystal faces) it is said to be *euhedral*; if circumstances have permitted the formation of only a portion of the crystal it is termed *subhedral*; if no faces appear it is *anhedral*.

7. Molecular Structure in General. — By definite molecular structure is meant the special arrangement which the structural units assume under the action of the forces exerted between them during the formation of the solid. Some remarks are given in a later article (p. 23 *et seq.*) in regard to the kinds of structural arrangement theoretically possible, and their relation to the symmetry of the different systems and classes of crystals.

The definite internal structure is the essential character of a crystal, and the external form is only one of the ways, although the most important, in which this structure is manifested. Thus it is found that all similar directions in a crystal, or a fragment of a crystal, have like physical characters,[†]

---

* In its original signification the term *crystal* was applied only to crystals of quartz, which the ancient philosophers believed to be *water* congealed by intense cold. Hence the term, from κρύσταλλος, *ice*.

[†] This subject is further elucidated in the chapter devoted to Physical Mineralogy, where it is also shown that, with respect to many, but not all, of the physical characters the converse of this proposition is true, viz., that unlike directions in a crystal have in general unlike properties.

7

as of elasticity, cohesion, action on light, etc. This is clearly shown by the cleavage, or natural tendency to fracture in certain directions, yielding more or less smooth surfaces; as the cubic cleavage of galena, or the rhombohedral cleavage of calcite. It is evident, therefore, that a small crystal differs from a large one only in size, and that a fragment of a crystal is itself essentially a crystal in all its physical relations, though showing no crystalline faces.

Further, the external form without the corresponding molecular structure does not make a crystal of a solid. A model of glass or wood is obviously not a crystal, though having its external form, because there is no relation between form and structure. Also, an octahedron of malachite, having the form of the crystal of cuprite from which it has been derived by chemical alteration, is not a crystal of malachite, but what is known as a pseudomorph (see Art. **491**) of malachite after cuprite.

On the other hand, if the natural external faces are wanting, the solid is not called a crystal. A cleavage octahedron of fluorite and a cleavage rhombohedron of calcite are not properly crystals, because the surfaces have been yielded by fracture and not by the natural molecular growth of the crystal.

8. Crystalline and Amorphous. — When a mineral shows no external crystalline form, it is said to be *massive*. It may, however, have a definite molecular structure, and then it is said to be *crystalline*. If this structure, as shown by the cleavage, or by optical means, is the same in all parallel directions through the mass, it is described as a single individual. If it varies from grain to grain, or fiber to fiber, it is said to be a *crystalline aggregate,** since it is in fact made up of a multitude of individuals.

Thus in a granular mass of galena or calcite, it may be possible to separate the fragments from one another, each with its characteristic cubic, or rhombohedral, cleavage. Even if the individuals are so small that they cannot be separated, yet the cleavage, and hence the crystalline structure, may be evident from the spangling of a freshly broken surface, as with fine-grained statuary marble. Or, again, this aggregate structure may be so fine that the crystalline structure can only be resolved by optical methods with the aid of the microscope. In all these cases, the structure is said to be *crystalline*.

Statistical study has shown that more than 98 per cent of minerals show definite crystalline structure.

If optical means show a more or less distinct crystalline structure, which, however, cannot be resolved into individuals, the mass is said to be *cryptocrystalline;* this is true of some massive varieties of quartz.

If the definite molecular structure is entirely wanting, and all directions in the mass are sensibly the same, the substance is said to be *amorphous*. This is true of a piece of glass, and nearly so of opal. The amorphous state is rare among minerals.

A piece of feldspar which has been fused and cooled suddenly may be in the glass-like amorphous condition as regards absence of definite molecular structure. But even in such cases there is a tendency to go over into the crystalline condition by molecular rearrangement. A transparent amorphous mass of arsenic trioxide (As_2O_3), formed by fusion, becomes opaque and crystalline after a time.

---

* The consideration of the various forms of crystalline aggregates is postponed to the end of the present chapter.

The microscopic study of rocks reveals many cases in which an analogous change in molecular structure has taken place in a solid mass, as caused, for example, by great pressure.

9. External Form. — A crystal is bounded by smooth plane surfaces, called faces or planes,* showing in their arrangement a certain characteristic symmetry, and related to each other by definite mathematical laws.

Thus, without inquiring, at the moment, into the exact meaning of the term symmetry as applied to crystals, and the kinds of symmetry possible, which will be explained in detail later, it is apparent that the accompanying figures, 1–3, show the external form spoken of. They represent, therefore, certain definite types.

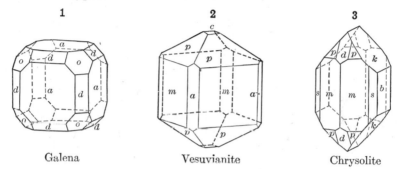

| 1 | 2 | 3 |
| Galena | Vesuvianite | Chrysolite |

10. Variation of Form and Surface. — Actual crystals deviate, within certain limits, from the ideal forms.

First, there may be variation in the size of like faces, thus producing what are defined later as *distorted forms*. In the second place, the faces are rarely absolutely smooth and brilliant; commonly they lack perfect polish, and they may even be rough or more or less covered with fine parallel lines (called striations), or show minute elevations, depressions or other peculiarities. Both the above subjects are discussed in detail in another place.

It may be noted in passing that the characters of natural faces, just alluded to, in general make it easy to distinguish between them and a face artificially ground, on the one hand, like the facet of a cut gem; or, on the other hand, the splintery uneven surface commonly yielded by cleavage.

11. Constancy of the Interfacial Angles in the Same Species. — The angles of inclination between like faces on the crystals of any species are essentially constant, wherever they are found, and whether products of nature or of the laboratory. These angles, therefore, form one of the important distinguishing characters of a species.

Thus, in Fig. 4, of apatite, the angle between the adjacent faces x and $m(130° 18')$ is the same for any two like faces, similarly situated with reference to each other. Further, this angle is constant for the species no matter what the size of the crystal may be or from what locality it may come. Moreover, the angles between all the faces on

Apatite

---

* This latter word is usually limited to cases where the direction, rather than the definite surface itself, is designated.

crystals of the same species (cf. Figs. 5–8 of zircon below) are more or less closely connected by certain definite mathematical laws.

12. Diversity of Form, or Habit. — While in the crystals of a given species there is constancy of angle between like faces, the forms of the crystals may be exceedingly diverse. The accompanying figures (5–8) are examples of a few of the forms of the species zircon. There is hardly any limit to the number of faces which may occur, and as their relative size changes, the *habit*, as it is called, may vary indefinitely. The variation in crystal habit of a given species is undoubtedly due to significant variations in the conditions under which it crystallized. It has been proved experimentally that even foreign material present in the crystallizing solution can affect the habit of the crystals formed. For instance, lead nitrate will crystallize in octahedrons from a pure water solution but when the solution is saturated with methylene-blue the crystals will have a cubic habit.

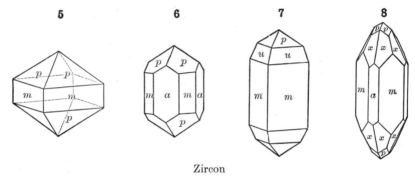

Zircon

13. Diversity of Size. — Crystals occur of all sizes, from the merest microscopic point to a yard or more in diameter. It is important to understand, however, that in a minute crystal the development is as complete as with a large one. Indeed the highest perfection of form and transparency is found only in crystals of small size.

A single crystal of quartz, now at Milan, is $3\frac{1}{4}$ feet long and $5\frac{1}{2}$ in circumference, and its weight is estimated at 870 pounds. A single cavity in a vein of quartz near the Tiefen Glacier, in Switzerland, discovered in 1867, afforded smoky quartz crystals, a considerable number of which had a weight of 200 to 250 pounds. A gigantic beryl from Acworth, New Hampshire, measured 4 feet in length and $2\frac{1}{2}$ in circumference; another, from Grafton, was over 4 feet long, and 32 inches in one of its diameters, and weighed about $2\frac{1}{2}$ tons; the largest beryls, however, have been found at Albany, Oxford Co., Maine, where one crystal was 18 feet in length, 4 feet in diameter, and weighed 18 tons.

14. Symmetry in General. — The faces of a crystal are arranged according to certain laws of symmetry, and this symmetry is the natural basis of the division of crystals into systems and classes. The symmetry may be defined in relation to (1) *a plane of symmetry*, (2) *an axis of symmetry*, and (3) *a center of symmetry*.

These different kinds of symmetry may, or may not, be combined in the same crystal. It will be shown later that there is one class, the crystals of which have neither center, axis, nor plane of symmetry; another where there is only a center of symmetry. On the other hand, some classes have all these elements of symmetry represented.

15. Planes of Symmetry. — A solid is said to be geometrically* symmetrical with reference to a plane of symmetry when for each face, edge, or solid angle there is another similar face, edge, or angle which has a like position on the opposite side of this plane.   Thus it is obvious that the crystal of amphibole, shown in Fig. 9, is symmetrical with reference to the central plane of symmetry indicated by the shading.

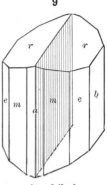

9

In the ideal crystal this symmetry is *right symmetry* in the geometrical sense, where every point on the one side of the plane of symmetry has a corresponding point at equal distances on the other side, measured on a line normal to it.   In other words, in the ideal geometrical symmetry, one half of the crystal is the exact *mirror-image* of the other half.

A crystal may have as many as nine planes of symmetry, three of one set and six of another, as is illustrated by the cube† (Fig. 16).   Here the planes of the first set pass through the crystal parallel to the cubic faces; they are shown in Fig. 10.   The planes of the second set join the opposite cubic edges; they are shown in Fig. 11.   A

Amphibole

plane of symmetry is always a possible crystal face and its normal is always parallel to a possible intersection between two crystal faces.

16. Axes of Symmetry. — If a solid can be revolved through a certain number of degrees about some line as an axis, with the result that it again occupies precisely the same position in space as at first, that axis is said to be an axis of symmetry.   There are four different kinds of axes of symmetry among crystals; they are defined according to the number of times which the crystal repeats itself in appearance during a complete revolution of 360°.   An axis of symmetry is always normal to a possible crystal face and parallel to the edge of intersection of two crystal faces.

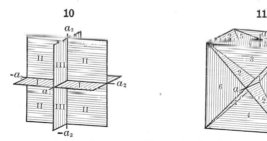

Symmetry Planes in the Cube

(*a*) A crystal is said to have an axis of *binary, digonal,* or *twofold,* symmetry when a revolution of 180° produces the result named above; in other words, when it occupies the same position twice in a complete revolution.   This is true of the crystal shown in Fig. 12 with respect to the vertical axis (and indeed each of the horizontal axes also).

---

* The relation between the ideal geometrical symmetry and the actual crystallographic symmetry is discussed in Art. **18.**
† This is the cube of the normal class of the isometric system.

(*b*) A crystal has an axis of *trigonal*, or *threefold*, symmetry when a revolution of 120° is needed; that is, when it occupies the same position three times in a complete revolution. The vertical axis of the crystal shown in Fig. 13 is an axis of trigonal symmetry.

(*c*) A crystal has an axis of *tetragonal*, or *fourfold*, symmetry when a revolution of 90° is called for; in other words, when it occupies the same position four times in a complete revolution. The vertical axis in the crystal shown in Fig. 14 is such an axis.

(*d*) Finally, a crystal has an axis of *hexagonal*, or *sixfold*, symmetry when a revolution of 60° is called for; in other words, when it occupies the same position six times in a complete revolution. This is illustrated by Fig. 15.

A symmetry axis which also forms the line of intersection of two or more symmetry planes is sometimes designated by prefixing *di* to its name, as didigonal, ditrigonal, etc. The different kinds of symmetry axes are sometimes known as *diad*, *triad*, *tetrad* and *hexad* axes.

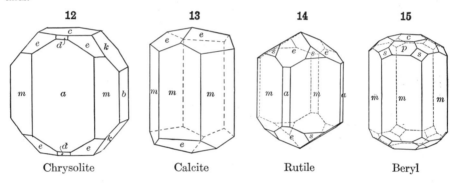

| 12 | 13 | 14 | 15 |

Chrysolite　　　　　Calcite　　　　　Rutile　　　　　Beryl

The cube * illustrates three of the four possible kinds of symmetry with respect to axes of symmetry. It has six axes of *binary* symmetry joining the middle points of opposite edges (Fig. 16). It has four axes of *trigonal* symmetry, joining the opposite solid angles (Fig. 17). It has, finally, three axes of *tetragonal* symmetry joining the middle points of opposite faces (Fig. 18).

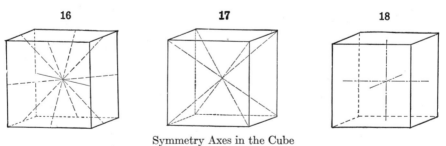

| 16 | 17 | 18 |

Symmetry Axes in the Cube

An axis of symmetry is termed an *alternating axis* when the faces at one end of the crystal can be obtained from the faces at the opposite end by a revolution (about the axis) through a certain arc and a simultaneous reflection across the plane normal to the axis. In Fig. 13 the vertical direction is such an axis of alternating symmetry, the *e* faces at the opposite

* This is again the cube of the normal class of the isometric system.

ends of the crystal having such positions that one set can be derived from the other only by a revolution about the vertical direction of 60° and a reflection across the horizontal plane.

17. Center of Symmetry. — Most crystals, besides planes and axes of symmetry, have also a center of symmetry. On the other hand, a crystal, though possessing neither plane nor axis of symmetry, may yet be symmetrical with reference to a point, its center. This last is true of the triclinic

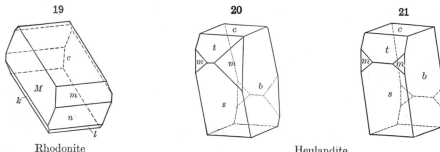

Rhodonite   Heulandite

crystal shown in Fig. 19, in which it follows that every face, edge, and solid angle has a face, edge, and angle similar to it in the opposite half of the crystal. In other words, a crystal has a center of symmetry if an imaginary line is passed from some point on its surface through its center, and a similar point is found on the line at an equal distance beyond the center. Another way of expressing such symmetry is to imagine the crystal turned about an axis for 60° or 180° and a simultaneous reflection of the faces over a plane normal to the axis of rotation.

18. Relation of Geometrical to Crystallographic Symmetry. — Since the symmetry in the arrangement of the faces of a crystal is an expression of the internal structure, which in general is alike in all parallel directions, the *relative size* of the faces and their *distance* from the plane or axis of symmetry are of no moment, their *angular position* alone is essential. The crystal represented in Fig. 20, although its faces show an unequal development, has in the crystallographic sense as truly a vertical plane of symmetry (parallel to the face b) as the ideally developed crystal shown in Fig. 21. The strict geometrical definition of symmetry would, however, apply only to the second crystal.*

Also in a normal cube (Fig. 22) the three central planes parallel to each pair of cubic faces are like planes of symmetry, as stated in Art. **15**. But a crystal is still crystallographically a cube, though deviating widely from the requirements of the strict geometrical definition, as shown in Figs. 23, 24, if only it can be proved, *e.g.*, by cleavage, by the physical nature of the faces, or by optical means, that the three pairs of faces are *like* faces, independently of their size, or, in other words, that the molecular structure is the same in the three directions normal to them.

Further, in the case of a normal cube, a face of an octahedron on any solid

---

* It is to be noted that the perspective figures of crystals usually show the geometrically ideal form, in which like faces, edges, and angles have the same shape, size, and position. In other words, the ideal crystal is uniformly represented as having the symmetry called for by the strict geometrical definition.

angle requires, as explained beyond, similar faces on the other angles. It is *not* necessary, however, that these eight faces should be of equal size, for in the crystallographic sense Fig. 25 is as truly symmetrical with reference to the planes named as Fig. 26.

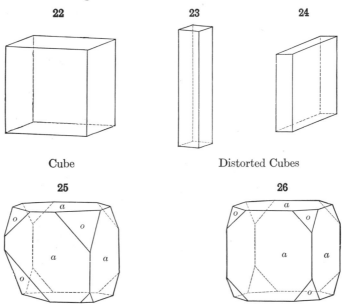

22

23

24

Cube   Distorted Cubes

25   26

Cube and Octahedron

19. On the other hand, the molecular and hence the crystallographic symmetry is not always that which the geometrical form would suggest. Thus, deferring for the moment the consideration of pseudo-symmetry, an illustration of the fact stated is afforded by the cube. It has already been implied and will be fully explained later that while the cube of the normal class of the isometric system has the symmetry described in Arts. **15, 16,** a cube of the same geometrical form but belonging molecularly, for example, to the tetrahedral class, has no planes of symmetry parallel to the faces but only the six diagonal planes; further, though the four axes shown in Fig. 17 are still axes of trigonal symmetry, the cubic axes (Fig. 18) are axes of binary symmetry only, and there are no axes of symmetry corresponding to those represented in Fig. 16. Other more complex cases will be described later.

Further, a crystal having interfacial angles of 90° is not necessarily a cube: in other words, the angular relations of the faces do not show in this case whether the figure is bounded by six like faces; or whether only four are alike and the other pair unlike; or, finally, whether there are three pairs of unlike faces. The question must be decided, in such cases, by the molecular structure as indicated by the physical nature of the surfaces, by the cleavage, or by other physical characters, as pyroelectricity, those connected with light phenomena, etc.

Still, again, the student will learn later that the decision reached in regard to the symmetry to which a crystal belongs, based upon the *distribution* of the faces, is only preliminary and approximate, and before being finally accepted it must be confirmed, first, by accurate measurements, and, second, by a minute study of the other physical characters. The best proof of the symmetry of a crystal is afforded by the analysis of its internal structure by the use of the X-rays (see Arts. **34–38**).

The method based upon the physical characters, which usually gives conclusive results and admits of wide application, is the skillful etching of the surface of the crystal by some appropriate solvent. By this means there are, in general, produced upon it minute depressions the shape of which conforms to the symmetry in the arrangement of the molecules. This process, which is in part essentially one involving the dissection of the molecular structure, is more particularly discussed in the chapter on Physical Mineralogy. However, doubt has arisen in several cases as to the exact symmetry class in which a given mineral belongs. Different lines of evidence have given different results. The development of small faces upon the crystals or the character of the artificial etching pits may have indicated a lower symmetry than the examination by X-rays or the evidence of pyroelectricity, etc., show. The word *amphisymmetry* has been suggested to indicate this group of substances with the tentative explanation that structurally they conform to the higher class of symmetry in each case but that they possess latent atomic tendencies toward a lower symmetry that may show themselves in etching, etc.

20. Pseudo-symmetry. — The crystals of certain species approximate closely in angle, and therefore in apparent symmetry, to the requirements of a system higher in symmetry than that to which they actually belong: they are then said to exhibit *pseudo-symmetry*. Numerous examples are given under the different systems. Thus the micas have been shown to be truly monoclinic in crystallization, though in angle they seem to be in some cases rhombohedral, in others orthorhombic.

It will be shown later that compound, or twin, crystals may also simulate by their regular grouping a higher grade of symmetry than that which belongs to the single crystal. Such crystals also exhibit pseudo-symmetry and are specifically called *mimetic*. Thus aragonite is an example of an orthorhombic species, whose crystals often imitate by twinning those of the hexagonal system.* Again, a highly complex twinned crystal of the monoclinic species, phillipsite, may have nearly the form of a rhombic dodecahedron of the isometric system. This kind of pseudo-symmetry also occurs among the classes of a single system, since a crystal belonging to a class of low symmetry may by twinning gain the geometrical symmetry of the corresponding form of the normal class. This is illustrated by a twinned crystal of scheelite like that figured (Fig. 442) in the chapter on twin crystals.

Pseudo-symmetry of still another kind, where there is an imitation of the symmetry of another system of lower grade, is particularly common in crystals of the isometric system (*e.g.*, gold, copper). The result is reached in such cases by an abnormal development of " distortion " in the direction of certain axes of symmetry. This subject is discussed and illustrated on a later page.

21. Possible Classes of Symmetry. — The theoretical consideration of the different kinds of symmetry possible among crystals built up of like structural units, as explained in Arts. **30–32,** has led to the conclusion that there are thirty-two (32) types in all, differing with respect to the combination of the different symmetry elements just described. Of these thirty-two natural classes among crystals based upon their symmetry, seven classes include by far the larger number of crystallized minerals. Besides these, some

---

* The terms *pseudo-hexagonal*, etc., used in this and similar cases explain themselves.

thirteen or fourteen others are distinctly represented, though several of these are of rare occurrence. The remaining classes, with possibly one or two exceptions, are known among the crystallized salts made in the laboratory. The characters of each of the thirty-two classes are given under the discussion of the several crystalline systems.

22. Crystallographic Axes. — In the description of a crystal, especially in regard to the position of its faces, it is found convenient to assume, after the methods of analytical geometry, certain lines passing through the center of the ideal crystal, as a basis of reference. (See further Art. **39** *et seq.*)

These lines are called the *crystallographic axes*. Their direction is to a greater or less extent fixed by the symmetry of the crystals, for an axis of symmetry is in almost all cases* a possible crystallographic axis. Further, the unit lengths assigned to these axes are fixed sometimes by the symmetry, sometimes by the position of the faces assumed as fundamental, *i.e.*, the unit forms in the sense defined later. The broken lines shown in Fig. 18 are the crystallographic axes to which the cubic faces are referred.

23. Systems of Crystallization and Symmetry Classes. — The thirty-two possible crystal classes which are distinguished from one another by their symmetry, are classified in this book under six systems. Each one of the six systems embraces several classes differing among themselves in their symmetry. One of these classes is conveniently called the *normal* class, since it is in general the common one, and since further it exhibits the highest degree of symmetry possible for the given system, while the others are lower in grade of symmetry.

It is important to note that the classes comprised within a given system are at once essentially connected together by their common optical characters, and in general separated† from those of the other systems in the same way.

Below is given a list of the six systems together with their subordinate classes, thirty-two in all. The order and the names given first are those that are used in this book while in the following parentheses are given other equivalent names that are also in common use. Under nearly all of the classes it is possible to give the name of a mineral or an artificial compound whose crystals serve to illustrate the characters of that particular class. There is some slight variation between different authors in the order in which the crystal systems and classes are considered but in the main essentials all modern discussions of crystallography are uniform.

ISOMETRIC SYSTEM

(*Regular, Cubic System*)

1. **Normal Class.** (Hexoctahedral. Holohedral. Ditesseral Central.) [Oh; Oi; Ia3d; 10]‡ Galena Type.
2. **Pyritohedral Class.** (Dyakisdodecahedral. Pentagonal Hemihedral. Diploidal. Tesseral Central.) [Th; Ti; Ia3; 7] Pyrite Type.

---

* Exceptions are found in the isometric system, where the axes must necessarily be the axes of tetragonal symmetry (Fig. 18), and cannot be those of binary or trigonal symmetry (Figs. 16, 17).

† Crystals of the tetragonal and hexagonal systems are alike in being optically uniaxial; but the crystals of all the other systems have distinguishing optical characters.

‡ For explanation of these symbols see Art. **33**.

3. TETRAHEDRAL CLASS. (Hextetrahedral. Tetrahedral Hemihedral. Ditesseral Polar.) [Td; Te; I43d; 6] Tetrahedrite Type.
4. PLAGIOHEDRAL CLASS. (Pentagonal Icositetrahedral. Plagiohedral Hemihedral. Gyroidal. Tesseral Holoaxial.) [O; O; I4₁3; 8] Cuprite Type.
5. TETARTOHEDRAL CLASS. (Tetrahedral Pentagonal Dodecahedral. Tesseral Polar.) [T; T; I2₁3; 5] Ullmannite Type.

TETRAGONAL SYSTEM

6. NORMAL CLASS. (Ditetragonal Dipyramidal. Holohedral. Ditetragonal Equatorial.) [D4h; 4Di; I4/acd; 20] Zircon Type.
7. HEMIMORPHIC CLASS. (Ditetragonal Pyramidal. Holohedral Hemimorphic. Ditetragonal Polar.) [C4v; 4e; I4cd; 12] Iodosuccinimide Type.
8. TRIPYRAMIDAL CLASS. (Tetragonal Dipyramidal. Pyramidal Hemihedral. Tetragonal Equatorial.) [C4h; 4Ci; I4₁/a; 6] Scheelite Type.
9. PYRAMIDAL-HEMIMORPHIC CLASS. (Tetragonal Pyramidal. Hemihedral Hemimorphic. Tetragonal Polar.) [C4; 4C; I4₁; 6] Wulfenite Type.
10. SPHENOIDAL CLASS. (Tetragonal Sphenoidal. Sphenoidal Hemihedral. Didigonal Scalenohedral. Ditetragonal Alternating.) [Vd (D2d); 4d; I42d; 12] Chalcopyrite Type.
11. TRAPEZOHEDRAL CLASS. (Tetragonal Trapezohedral. Trapezohedral Hemihedral. Tetragonal Holoaxial.) [D4; 4D; I4₁2; 10] Nickel Sulphate Type.
12. TETARTOHEDRAL CLASS. (Tetragonal Disphenoidal. Sphenoidal Tetartohedral. Tetragonal Alternating.) [S4 (C4); 4c; I4₁; 2] Anlif. 2 CaO.Al₂O₃.SiO₂ Type.

HEXAGONAL SYSTEM

A. HEXAGONAL DIVISION

13. NORMAL CLASS. (Dihexagonal Dipyramidal. Holohedral. Dihexagonal Equatorial.) [D6h; 6Di; C0/mmc, 4] Beryl Type.
14. HEMIMORPHIC CLASS. (Dihexagonal Pyramidal. Holohedral Hemimorphic. Dihexagonal Polar.) [C6v; 6e; C6mc; 4] Zincite Type.
15. TRIPYRAMIDAL CLASS. (Hexagonal Dipyramidal. Pyramidal Hemihedral. Hexagonal Equatorial.) [C6h; 6Ci; C6₃/m; 2] Apatite Type.
16. PYRAMIDAL-HEMIMORPHIC CLASS. (Hexagonal Pyramidal. Pyramidal Hemihedral Hemimorphic. Hexagonal Polar.) [C6; 6C; C6₃; 6] Nephelite Type.
17. TRAPEZOHEDRAL CLASS. (Hexagonal Trapezohedral. Trapezohedral Hemihedral. Hexagonal Holoaxial.) [D6; 6D; C6₃2; 6] β-Quartz Type.
18. TRIGONAL CLASS. (Ditrigonal Dipyramidal. Trigonal Hemihedral. Ditrigonal Equatorial.) [D3h; 6d; H6c; 4] Benitoite Type.
19. TRIGONAL TETARTOHEDRAL CLASS. (Trigonal Dipyramidal. Trigonal Equatorial.) [C3h; 6c; C6; 1] Disilverorthophosphate Type.

B. Trigonal or Rhombohedral Division

(*Trigonal System*)

20. **Rhombohedral Class.** (Ditrigonal Scalenohedral. Hexagonal Scalenohedral. Rhombohedral Hemihedral. Dihexagonal Alternating.) [D3d; 3Di; R̄3c; 6] Calcite Type.

21. **Rhombohedral Hemimorphic Class.** (Ditrigonal Pyramidal. Trigonal Hemihedral Hemimorphic. Ditrigonal Polar.) [C3v; 3e; R3c; 6] Tourmaline Type.

22. **Trirhombohedral Class.** (Rhombohedral. Trigonal Rhombohedral. Rhombohedral Tetartohedral. Hexagonal Alternating.) [C3i; 3Ci; R̄3; 2] Phenacite Type.

23. **Trapezohedral Class.** (Trigonal Trapezohedral. Trapezohedral Tetartohedral. Trigonal Holoaxial.) [D3; 3D; R32; 7] α-Quartz Type.

24. **Trigonal Tetartohedral Hemimorphic Class.** (Trigonal Pyramidal. Trigonal Polar.) [C3; 3C; R3; 4] Sodium Periodate Type.

ORTHORHOMBIC SYSTEM

(*Rhombic or Prismatic System*)

25. **Normal Class.** (Orthorhombic Dipyramidal. Holohedral. Didigonal Equatorial.) [Vh (D2h); 2Di; Imma; 28] Barite Type.

26. **Hemimorphic Class.** (Orthorhombic Pyramidal. Didigonal Polar.) [C2v; 2e; Ima; 22] Calamine Type.

27. **Sphenoidal Class.** (Orthorhombic Disphenoidal. Digonal Holoaxial.) [V (D2); 2D; I$2_12_12_1$; 9] Epsomite Type.

MONOCLINIC SYSTEM

(*Oblique or Monosymmetric System*)

28. **Normal Class.** (Prismatic. Holohedral. Digonal Equatorial.) [C2h; 2Ci; C2/c; 6] Gypsum Type.

29. **Hemimorphic Class.** (Sphenoidal. Digonal Polar.) [C2; 2C; C2; 3] Tartaric Acid Type.

30. **Clinohedral Class.** (Domatic. Hemihedral. Planar.) [C1h(Cs); 2c; Cc; 4] Clinohedrite Type.

TRICLINIC SYSTEM

(*Anorthic System*)

31. **Normal Class.** (Holohedral. Pinacoidal. Central.) [Ci (C$_2$ or S$_2$); 1Ci; P$\bar{1}$; 1] Axinite Type.

32. **Asymmetric Class.** (Hemihedral. Pedial.) [C1; 1C; P1; 1] Calcium Thiosulphate Type.

24. Systems of Crystallization; Crystal Axes and Symmetry. — In the paragraphs immediately following a statement is given describing the crystallographic axes of each of the crystal systems and also a synopsis of the symmetry of the *normal class* of each of the different systems. The symmetry is also given of one subordinate class of the hexagonal system, which is of so great importance that it is often conveniently treated as a sub-system even when, as in this work, the forms are referred to the same axes as those of the strictly hexagonal type — a usage not adopted by all authors.

I. ISOMETRIC SYSTEM. Three crystal axes at right angles to each other and of equal lengths. Three like axial* planes of symmetry (principal planes) parallel to the cubic faces, and fixing by their intersection the crystallographic axes; six like diagonal planes of symmetry, passing through each opposite pair of cubic edges, and hence parallel to the faces of the rhombic dodecahedron.

Further, three like axes of tetragonal symmetry coincident with the crystallographic axes and normal to the faces of the cube; four like diagonal axes of trigonal symmetry, normal to the faces of the octahedron; and six like diagonal axes of binary symmetry, normal to the faces of the dodecahedron. There is also obviously a center of symmetry.† These relations are illustrated by Fig. 27 also by Fig. 35; further by Figs. 110 to 143.

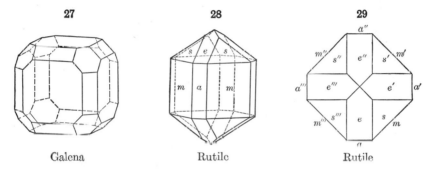

| 27 | 28 | 29 |
| Galena | Rutile | Rutile |

II. TETRAGONAL SYSTEM. Three crystal axes at right angles to each other; two of them — taken as the horizontal axes — being equal; the third — the vertical axis — being longer or shorter than the other two. Three axial planes of symmetry: of these, two are like planes intersecting at 90° in a line which is the vertical crystallographic axis, and the third plane (a principal plane) is normal to them and hence contains the horizontal axes. There are also two diagonal planes of symmetry, intersecting in the vertical axis and meeting the two axial planes at angles of 45°.

Further, there is one axis of tetragonal symmetry, a principal axis; this is the vertical crystallographic axis. There are also in a plane normal to this four axes of binary symmetry — like two and two — those of each pair at right angles to each other. Fig. 28 shows a typical tetragonal crystal, and Fig. 29 a basal projection of it, that is, a projection on the principal plane of symmetry normal to the vertical axis. See also Fig. 36 and Figs. 188–210.

---

* Two planes of symmetry are said to be *like* when they divide the ideal crystal into halves which are identical to each other; otherwise, they are said to be *unlike*. Axes of symmetry are also like or unlike. If a plane of symmetry includes two of the crystallographic axes, it is called an *axial plane* of symmetry. If the plane includes two or more like axes of symmetry, it is called a *principal plane* of symmetry; also an axis of symmetry in which two or more like planes of symmetry meet is a *principal axis* of symmetry.

† In describing the symmetry of the different classes, here and later, the center of symmetry is ordinarily not mentioned when its presence or absence is obvious.

III. HEXAGONAL SYSTEM. Four crystal axes, three of which are equal and lie in the horizontal plane making angles of 60° and 120° with each other, while the fourth axis is vertical and has a length different (longer or shorter) from that of the horizontal axes. In the *Hexagonal Division* there are four axial planes of symmetry; of these three are like planes meeting at angles of 60°, their intersection-line being the vertical crystallographic axis; the fourth plane (a principal plane) is at right angles to these. There are also three other diagonal planes of symmetry meeting the three of the first set in the vertical axis, and making with them angles of 30°.

Furtner, there is one principal axis of hexagonal symmetry; this is the vertical crystallographic axis; at right angles to it there are also six binary axes. The last are in two sets of three each. Fig. 30 shows a typical hexagonal crystal, with a basal projection of the same. See also Fig. 37 and Figs. 238–245.

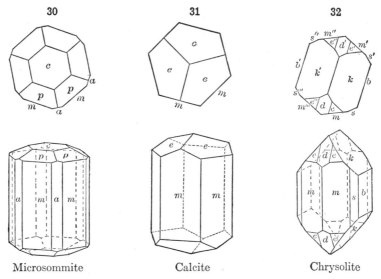

Microsommite   Calcite   Chrysolite

In the *Trigonal* or *Rhombohedral Division* of this system there are three like planes of symmetry intersecting at angles of 60° in the vertical axis. Further, the forms belonging here have a vertical principal axis of trigonal symmetry, and three horizontal axes of binary symmetry, coinciding with the horizontal crystallographic axes. Fig. 31 shows a typical rhombohedral crystal, with its basal projection. See also Figs. 261–287.

IV. ORTHORHOMBIC SYSTEM. Three crystal axes at right angles to each other, all of different lengths. Three unlike planes of symmetry meeting at 90°, and fixing by their intersection-lines the position of the crystallographic axes. Further, three unlike axes of binary symmetry coinciding with the last-named axes. Fig. 32 shows a typical orthorhombic crystal, with its basal projection. See also Fig. 38 and Figs. 316–338.

V. MONOCLINIC SYSTEM. Three crystal axes of unequal lengths, having one of their intersections oblique, the other two intersections being at 90°. One plane of symmetry which contains the two crystallographic axes that have the oblique intersection. Also one axis of binary symmetry, normal to this plane and coinciding with the third crystallographic axis.   See Fig. 33; also Fig. 39 and Figs. 353–367.

VI. TRICLINIC SYSTEM.   Three unequal crystal axes with mutually oblique intersections.   No plane and no axis of symmetry, but symmetry solely with respect to the central point.   Figs. 34 and 40 show typical triclinic crystals.   See also Figs. 379–386.

A statistical study has shown that more than one half of crystallized minerals belong in either the orthorhombic or monoclinic systems.   The majority of minerals crystallize in the class of highest symmetry in the respective systems.

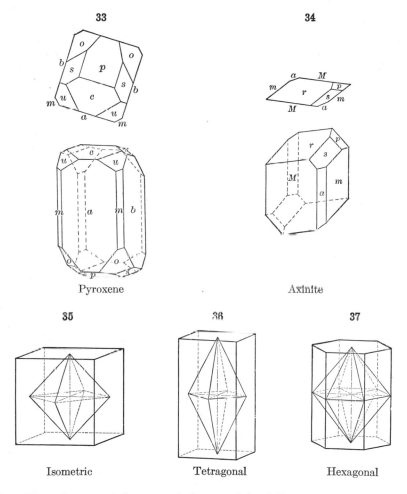

33

34

Pyroxene

Axinite

35

36

37

Isometric

Tetragonal

Hexagonal

25. The relations of the normal classes of the different systems are further illustrated both regarding the crystallographic axes and symmetry by the accompanying figures, 35–40.   The exterior form is here that bounded by faces each of which is parallel to a plane through two of the crystallographic axes indicated by the central broken lines.   Further, there is shown, within this, the combination of faces each of which joins the extremities of the unit lengths of the axes.

The full understanding of the subject will not be gained until after a study of the forms of each system in detail. Nevertheless the student will do well to make himself familiar at the outset with the fundamental relations here illustrated.

It will be shown later that the symmetry of the different classes can be most clearly and easily exhibited by the use of the different projections explained in Art. **44**, *et seq.*

26. Models. — Glass (or transparent celluloid) models illustrating the different systems, having the forms shown in Figs. 35–40, will be very useful to the student, especially in learning the fundamental relations regarding symmetry. They should show within, the crystallographic axes, and by colored threads or wires, the outlines of one or more simple forms. Models of wood are also made in great variety and perfection of form; these are indispensable to the student in mastering the principles of crystallography.

38

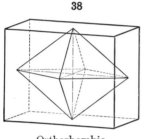

39

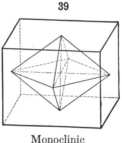

40

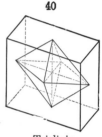

Orthorhombic   Monoclinic   Triclinic

27. So-called Holohedral and Hemihedral Forms. — It will appear later that each crystal form* of the normal class in a given system embraces *all* the faces which have a like geometrical position with reference to the crystallographic axes; such a form is said to be *holohedral* (from ὅλος, complete, and "δρα, face). On the other hand, under the classes of lower symmetry, a certain form, while necessarily having all the faces which the symmetry allows, may yet have but *half* as many as the corresponding form of the normal class; these half-faced forms are sometimes called on this account *hemihedral.* Furthermore, it will be seen that, in such cases, to the given holohedral form there correspond two similar and complementary hemihedral forms, called respectively positive and negative (or right and left), which together embrace all of its faces.

A single example will help to make the above statement intelligible. In the normal class of the isometric system, the octahedron (Fig. 41) is a "holohedral" form with all the possible faces — eight in number — which are alike in that they meet the axes at equal distances. In the tetrahedral class of the same system, the forms are referred to the same crystallographic axes, but the symmetry defined in Art. **19** (and more fully later) calls for but four similar faces having the position described. These yield a four-faced, or "hemihedral," form, the tetrahedron. Figs. 42 and 43 show the positive and negative tetrahedron, which together, it will be seen, embrace all the faces of the octahedron, Fig. 41.

In certain classes of still lower symmetry a given crystal form may have but *one-quarter* of the faces belonging to the corresponding normal form, and, after the same method, such a form is sometimes called *tetartohedral.*

---

* The use of the word *form* is defined in Art. **42**.

The development of the various possible kinds of hemihedral (and tetarto-hedral) forms under a given system has played a prominent part in the crystal-lography of the past, but it leads to much complexity and is distinctly less simple than the direct statement of the symmetry in each case.   The latter method is systematically followed in this work, and the subject of hemihe-drism is dismissed with the brief (and incomplete) statements of this and the following paragraphs.

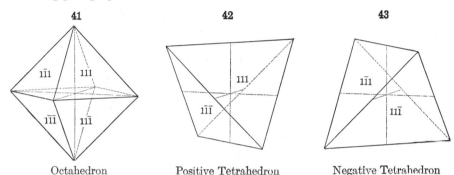

41	42	43
Octahedron	Positive Tetrahedron	Negative Tetrahedron

28. Hemimorphic Forms. — In several of the systems, forms occur under the classes of lower symmetry than that of the normal class which are characterized by this: that the faces present are only those belonging to one extremity of an axis of symmetry (and crystallographic axis). Such forms are conveniently called *hemimorphic* (*half-form*). A simple example under the hexagonal system is given in Fig. 44. It is obvious that hemi-morphic forms have no center of symmetry.

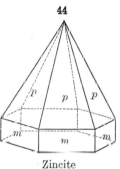

Zincite

29. Enantiomorphous Forms. — Crystal forms are said to be enantiomorphous when, possessing neither a plane nor center of symmetry, they may occur in two positions which are mirror images of each other. The two types cannot be converted into each other by any rotation, but are related to each other as are the right and left hand, and are commonly, therefore, designated as right- or left-handed. For an example, see under the quartz crystal class, Art. **143**.

30. Space-lattices. — Much light has recently been thrown upon the relations existing between the different types of crystals, on the one hand, and of these to the physical properties of crystals, on the other, by the consideration of the various possible methods of grouping of the structural units of which the crystals are built.   This subject, very early treated by Haüy and others (including J. D. Dana), was discussed at length by Frankenheim and later by Bravais.   More recently it has been extended and elaborated by Sohnke, Wulff, Schoenflies, Fedorov, Barlow, and others.

When a body passes from the state of a liquid or a gas to that of a solid, under such conditions as to allow perfectly free action to the interatomic forces, the result is a crystal of some definite type regarding symmetry. The simplest hypothesis which can be made assumes that the form of the crystal is determined by the way in which the atoms group themselves in a position of equilibrium under the action of the interatomic forces.

As, however, the crystallizing forces vary in magnitude and direction from one type of crystal to another, the resultant grouping of the crystal units must also vary, particularly regarding the distance between them and the angles between the planes in which they lie. This may be represented by a series of geometrical diagrams, showing the hypothetical groupings of a series of points. Such points may represent the positions of the chemical atoms, or the centers of similar groups of atoms. Such an arrangement of points is named a *network, point-system*, or a *space-lattice*. A space-lattice may be defined as a network of points arranged in such a manner that a straight line drawn through any two points and continued will pass at equal intervals through a succession of similar points and the like condition will hold true for any parallel line drawn through any other similar point.

45   **46**

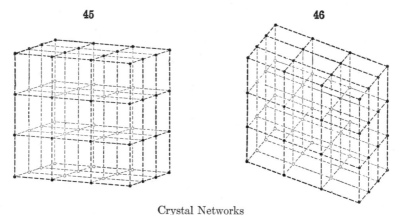

Crystal Networks

The subject may be illustrated by Figs. 45, 46 for two typical cases, which are easily understood. In Fig. 45, the most special case is represented where the points are grouped at equal distances, in planes at right angles to each other. The structure in this case obviously corresponds in symmetry to the cube described in Arts. **15** and **16**, or, in other words, to the normal class of the isometric system. Again, in Fig. 46, the general case is shown where the molecules are unequally grouped in the three directions, and further, these directions are oblique. The symmetry is here that of the normal class of the triclinic system.

If, in each of these cases, the figure be bounded by the simplest possible arrangement of eight points, the result is an *elementary parallelepiped*, which obviously defines the structure of the whole. In the grouping of these parallelepipeds together, as described, it is obvious that in whatever direction a line be drawn through them, the points will be spaced alike along it, and the grouping about any one of these points will be the same as about any other.

31. Certain important conclusions can be deduced from a consideration of such regular molecular networks as have been spoken of, which will be enumerated here though it is impossible to attempt a full explanation.

(1) The prominent crystalline faces must be such as include the largest number of points, that is, those in which the points are nearest together.

Thus in Fig. 47, which represents a section of a network conforming in symmetry to the structure of a normal orthorhombic crystal, the common crystalline faces would be expected to be those having the position *bb*, *aa*, *mm*, then *ll*, *nn*, and so on. This is found to be true in the study of crystals, for the common forms are, in nearly all cases, those whose position bears some simple relation to the assumed axes; forms whose position is complex are usually present only as small faces on the simple predominating forms, that is, as modifications of them. So-called *vicinal* forms, that is, forms taking the place of the simple fundamental forms to which they approximate very closely in angular position, are exceptional.

47

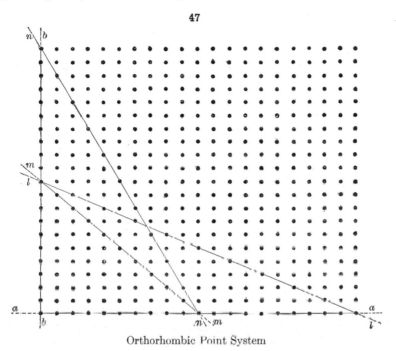

Orthorhombic Point System

(2) When a variety of faces occur on the same crystal, the numerical relation existing between them (that which fixes their position) must be rational and, as stated in (1), a simple numerical ratio is to be expected in the common cases. This, as explained later, is found by experience to be a fundamental law of crystals. Thus, in Fig. 47, starting with a face meeting the section in *mm*, *ll* would be a common face, and for it the ratio is 1 : 2 in the directions *b* and *a*; *nn* would be also common with the ratio 2 : 1.

(3) If a crystal shows the natural easy fracture, called cleavage, due to a minimum of cohesion, the cleavage surface must be a surface of relatively great molecular crowding, that is, one of the common or fundamental faces. This follows (and thus gives a partial, though not complete, explanation of cleavage) since it admits of easy proof that that plane in which the points are closest together is farthest separated from the next molecular plane. Thus in Fig. 47 compare the distance separating two adjoining planes parallel to *bb* or *aa*, then two parallel to *mm*, *ll*, *nn*, etc. Illustrations of the above will be found under the special discussion of the subject of cleavage.

32. Kinds of Space-lattices. — A theoretical study of the arrangements of points in space so that the conditions stated above under the definition of a space-lattice would be fulfilled showed that there were possible only fourteen such networks. These agree as to their symmetry with the seven classes defined in Art. **24** as representing respectively the normal classes of the six systems with also that of the trigonal (or the rhombohedral) division of the hexagonal system. Of the fourteen, three groupings belong to the isometric system; two to the tetragonal; one each to the hexagonal and the rhombohedral; four to the orthorhombic system; two to the monoclinic, and one to the triclinic. These fourteen different lattices are represented in Fig. 48; I represents the simple cubic lattice; II, the body-centered cubic lattice

48

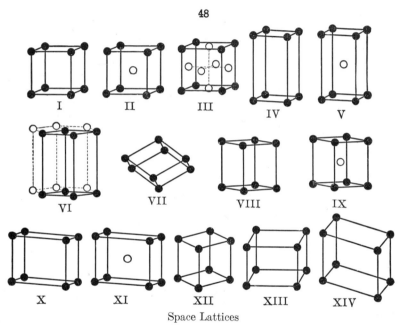

Space Lattices

which consists of two cubic lattices interpenetrating in such a way that the points of the second lattice lie at the centers of the unit cells of the first lattice; III, the face-centered cubic lattice in which four cubic lattices interpenetrate; IV, the tetragonal or square prism lattice; V, the body-centered tetragonal or square prism lattice; VI, the 120° prism lattice, three of whose unit cells together form the hexagonal prism; VII, the rhombohedron lattice; VIII, the orthorhombic prism lattice; IX, the body-centered orthorhombic prism lattice; X, the rectangular parallelepiped lattice; XI, the body-centered rectangular parallelepiped lattice; XII, the monoclinic prism lattice; XIII, the monoclinic parallelepiped lattice; XIV, the triclinic lattice.

33. Space-groups or Point-systems. — It will be noted that the fourteen space-lattices described above will yield only the normal or holohedral classes of the various systems. They cannot, therefore, account for the classes of lower symmetry. It is necessary to extend the theory in order to include these other classes. The points of a given space-lattice represent the structural units of a crystal, usually described as molecules or groups of molecules, and it is this structure that determines the crystal system, the axial ratios, etc. But the space-lattice of a crystal does not necessarily define its symmetry class. If, however, the positions of the individual atoms of the molecule are considered, other networks, that do not conform to the definition of a space-lattice as given above, can be formulated that will account for all possible symmetry classes. Sohnke showed that by assuming two or more identical lattices to interpenetrate each other in such a way that one system of points could be derived from another by some definite movement, a total of sixty-five different arrangements, or *point-systems*, could be derived. The movements, known as " coincidence movements," by which one set of points can be derived from another are of various kinds. The sixty-five point-systems of Sohnke may be obtained by assuming movements, (1) through a definite distance in some definite direction (" translation "), (2) by a rotation of a given number of degrees about a definite axis, (3) by a combination of (1) and (2) in which there is a rotation about some axis accompanied by a movement along the direction of the axis (such an axis is commonly called a screw-axis). These sixty-five point-systems account for all but eleven of the thirty-two symmetry classes. The missing classes are those that show enantiomorphous forms, or, in other words, those forms that have right- and left-handed relations to each other. In order to account for these classes it is necessary to assume that we may have two interpenetrating point-systems composed of different types of particles which have such characters that they show enantiomorphous relations to each other. These two interpenetrating point-systems are to be derived from each other by movements which involve mirror-like reflections, such as a reflection over a plane, a reflection over a plane accompanied by a movement (translation) of a definite amount and direction, a rotation about an axis with reflection over a plane normal to the axis, or repetition about a center of symmetry. By using such methods of derivation, Schoenflies, Fedorov, and Barlow independently and almost simultaneously between 1890 and 1894 extended the theory of point systems and determined that there were in all two hundred and thirty possible types of homogeneous structure. In all these groups it holds true that the arrangement of points about any single point is always identical or possesses a mirror-like similarity. Groth has on the basis of this theory of point-systems defined a crystal as follows: " a crystal — considered as indefinitely extended — consists of n interpenetrating regular point-systems, each of which is formed from similar atoms; each of these point-systems is built up from n interpenetrating space-lattices, each of the latter being formed from similar atoms occupying parallel positions. All the space-lattices of the combined system are geometrically identical or are characterized by the same elementary parallelepipedon."

These two hundred and thirty types of structure can all be grouped under the thirty-two symmetry classes of crystals. From the point of view of the geometrical crystallographer they are largely of only theoretical interest, but for the investigation of the atomic structure of crystals by use of the X-ray methods they are of the utmost practical importance. A complete exposition of their characters and nomenclature would be too complicated and extended to permit of its inclusion here. In the table (Art. 23) in which are listed the thirty-two symmetry classes of crystals the point-group symbols of these classes are given in brackets. The first symbol is that used by Schoenflies:

the second, the symbol proposed by Hilton and modified by Wyckoff; the third, the symbol adopted by an international conference held in Zürich, August, 1930; the numbers following these symbols in the brackets show how many different space-groups occur in the various symmetry classes. The particular space-group in question is indicated by using the proper exponent to modify the group symbol. For instance, the fourth space-group under the normal class of the orthorhombic system would be given the symbol Vh^4 or 2 Di–4.

The following books either concern themselves wholly with the theory of space-groups, or contain chapters giving detailed descriptions. To them the interested reader must be referred.

Hilton. Mathematical Crystallography, 1903.
Niggli. Krystallographische und Strukturtheoretische Grundbegriffe, 1928.
Schiebold. Über eine neue Herleitung und Nomenklatur der 230 kristallographischen Raumgruppen, 1929.
Schoenflies. Krystallsysteme und Krystallstruktur, 1891. Theorie der Kristallstruktur, 1923.
Sommerfeld. Physikalische Kristallographie, 1907.
Tutton. Crystallography and Practical Crystal Measurement, Vol. I, Chapters 30, 31, 1922.
Wyckoff. The Analytical Expression of the Results of the Theory of Space-Groups, 1922. The Structure of Crystals, 1931.
In Zeitschrift für Kristallographie, Vol. 79, p. 525, 1931 is given in tabular form a comparison of the various symbols that are in use for space-groups.

34. Analysis of Crystal Structure by Means of the X-rays. — In 1912, about twenty years after the first formulation of the two hundred and thirty space-groups, the theory of crystal structure outlined above received a definite proof when it was discovered that a beam of X-rays would suffer reflection and diffraction by the atomic planes of a crystal. X-rays, discovered in 1896 by Röntgen, are produced by the rays emanating from the cathode in a vacuum tube when they fall upon a metallic anti-cathode. The quantity and wave-length of the X-rays depend upon the metal of which the anti-cathode is made. X-rays travel outward from the point of origin in straight lines and because of their very short wave-lengths are able to penetrate all substances to a greater or less degree. Their effect upon a photographic film is like that of light; certain substances fluoresce when exposed to them; and gases are ionized and become conductors of electricity when X-rays are passed through them. Physicists had come to believe that X-rays were electromagnetic vibrations similar to those of light but with much shorter wave-lengths. It has since been shown that the wave-lengths of X-rays are approximately 10,000 times shorter than those of light. Because of this short wave-length it had been impossible to reflect or diffract the X-rays and so to prove their similarity to light vibrations. In 1912, Dr. Laue, then at Munich, made the suggestion that the supposed atomic structure of crystals might have interatomic distances comparable to the wave-length of X-rays and could therefore be used as a three-dimensional diffraction grating. Aided by Friedrich and Knipping, the experiment was tried of passing a beam of X-rays through a crystal of sphalerite behind which a photographic plate had been placed. When developed, the plate not only showed a dark spot in its center where the direct pencil of the X-rays had hit it but it also showed a large number of smaller spots arranged around the center in a regular geometrical pattern. Each of these spots was due to the reflection of the

X-rays by a very large series of parallel planes in the atomic structure. By this and subsequent similar experiments Laue showed the similarity between the vibrations of X-rays and those of light and at the same time indicated a method for the precise study of the atomic structure of crystals.

35. Laue Diffraction Patterns. — In Fig. 49 is given a diagrammatic representation of the spot pattern produced when X-rays are passed through a crystal of potassium chloride, sylvite, in a direction normal to a face of the cube. Each spot of the pattern indicates an atomic plane that corresponds to a possible crystal face. The arrangement of the spots is controlled by the symmetry of the crystal, although it should be noted that the spot patterns

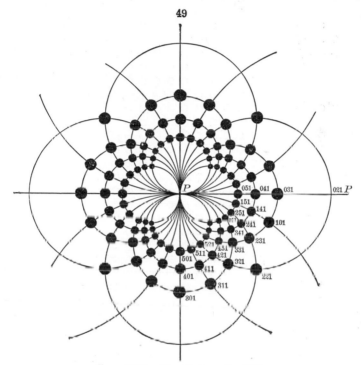

Laue Diffraction Pattern for KCl

may indicate a higher symmetry than that actually possessed by the crystal. The spots vary in their intensity, and this is due to a variation in the number of atoms present in a given area of the reflecting plane or to a variation in the reflecting power of the different kinds of atoms to be found in the given crystal. As the commoner crystal faces are those in which the larger number of atoms lie, such planes are indicated by the darker spots. On the photograph the spots lie at the intersections of ellipses, each of which passes through the center of the pattern. The spots from all planes lying in one crystal zone will be found on one ellipse. In Fig. 49 these ellipses have been reduced to circles. The central portion of the picture shows no spots. This is due to the fact that the distances between planes that have steep inclinations to the path of the X-rays are less than the wave-length of the X-rays. Faces with low inclinations are also not represented since the X-rays would be reflected at too great an angle to permit them to reach the photographic plate. Fig. 50 will aid in visualizing these facts. It shows a vertical plane through the cubic network of potassium chloride that includes the b (O–B) and c (O–C) crystal axes. The atomic planes having the indices (011),

50

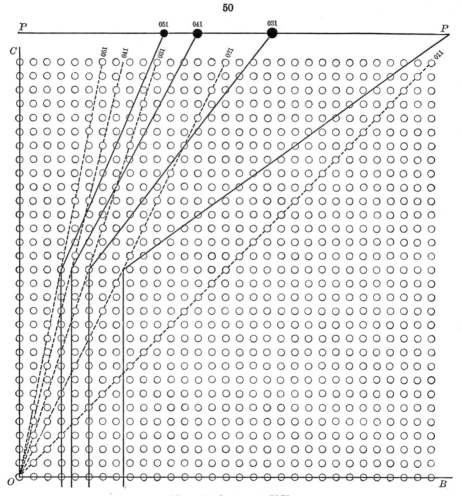

X-ray Reflection in KCl

(021), (031), (041), (051) are shown by the broken lines. The line P–P would represent a section of the photographic plate corresponding in position to line P–P in Fig. 49. The solid lines within the figure represent the paths of the X-rays before and after reflection from individual atomic planes. Reflections from (011) and (021) would not meet the photographic plate and the reflections from (031), (041), and (051) would show a decrease in intensity due to the decreasing numbers of atoms in the successive planes. The variation in intensity of the various reflections is indicated in Fig. 49, which is much idealized, by variation in the size of the spots.

The spots of the Laue patterns are due to reflection of the X-rays not from a single atomic plane but from a very great number of parallel planes. It is necessary therefore that the succession of reflected waves all fall together in the same phase of wave motion and so be able to reinforce each other, or else being in different phases they would interfere with each other and the wave motion in that particular direction would be destroyed. The law for the reflection of X-rays is given by the equation $n\lambda = 2d \sin \theta$, in which n equals some integer, λ the wave-length of the X-ray, d the distance between the parallel reflecting planes, and θ the angle between the path of the X-ray and the reflecting plane. This law is illustrated in Fig. 51 where the series of

parallel X-rays A, B, C, D, are reflected from the series of parallel planes p–p at the points E, F, G, H, in the direction I. In the reflection of X-rays according to the Laue method the distance d for any given series of atomic planes is fixed and the value of θ is also definite. In order, therefore, to satisfy the equation and have reflection taking place from any atomic plane it is necessary to be able to vary the value of λ. In making the Laue patterns it is there-

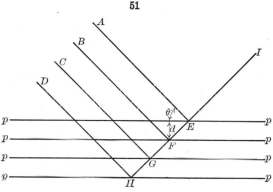

51

Reflection of X-rays

fore necessary to use the so-called "white radiation" of X-rays which contains a spectrum of varying wave-lengths, one of which will satisfy the conditions of the equation and yield a series of reflected waves, all agreeing in phase. All anti-cathodes provide some of this general radiation in addition to their own peculiar wave-length but in the case of platinum or tungsten the amount of such radiation is at a maximum and these are the metals commonly used as anti-cathodes in making Laue pictures.

Fig. 52 gives a diagrammatic representation of the Laue photograph obtained from a crystal of halite, sodium chloride. Comparison of this figure with that representing the spot photograph of potassium chloride shows marked differences. In Fig. 52 there are no spots at the intersections representing the planes (501), (521), (541), (341), etc. Further, the intensities of the spots do not vary in a regular manner. The explanation of these differences lies in the fact that in potassium chloride, the potassium and chlorine atoms have nearly the same atomic weights and nearly equal powers of reflection of X-rays. Consequently the structure of potassium chloride as far as X-rays are concerned can be considered to consist of only one kind of atom, as is represented in Fig. 50. In sodium chloride, on the other hand, the atomic weights of sodium and chlorine are quite different and as a result they reflect the X-rays with different intensities. The structure of sodium chloride must be represented as composed of two different kinds of atoms, as shown in Fig. 53. Some of the reflecting planes would contain only atoms of one kind, as for instance planes parallel to (011), (031), (051) are either all light or all dark spots in Fig. 53, while other planes (such as (021), (041)) would show an alternating arrangement of the two elements. The series of reflecting planes, therefore, in this structure are of two different types and should show differences in their reflections such as are seen in Fig. 52. From these observations it is concluded that the structure of potassium chloride can be represented by the simple cubic lattice, while that of sodium chloride belongs to the face-centered cubic lattice; I and III, Fig. 48.

The two Laue diagrams considered above were made normal to a principal symmetry axis of the crystal. Other pictures could be made normal

52

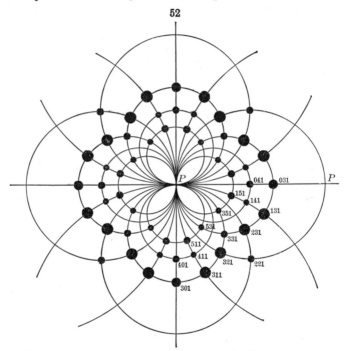

Laue Diffraction Pattern for NaCl

to other axes with corresponding differences in the patterns. Together these various patterns would enable one to learn much concerning the atomic structure of the material studied. To be of use the Laue pictures should have a definite and known crystal orientation.

36. The X-ray Spectrometer. — Other methods for using X-rays in the investigation of crystal structure were soon devised. The first of these was the X-ray spectrometer, developed by W. H. Bragg and W. L. Bragg. With this instrument the crystal is mounted on a central post which can be revolved about its axis. The arrangement is closely similar to that used in mounting and measuring crystals on the ordinary reflection goniometer. In place of a beam of light, a beam of X-rays is directed at the crystal. In this case X-rays are used that are predominantly of one known wave-length. Under certain conditions the X-rays are reflected from a series of parallel atomic planes of the crystal, the angles of incidence and reflection being equal. The position of the reflected beam is determined by revolving an ionization chamber about the central axis of the instrument until the electroscope connected with it shows that the gas in the chamber has become ionized. The angular position of the ionization chamber when this happens

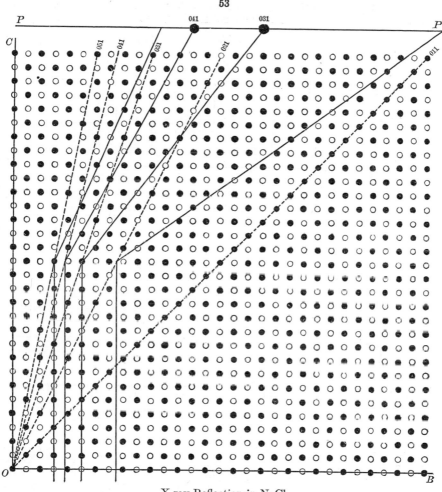

X-ray Reflection in NaCl

gives the angle of reflection of the beam of X-rays. The law of reflection of X-rays has been given in Art. **35** and the conditions of reflection are illustrated in Fig. 51. In this case, however, the wave-length of the X-ray has a definite value. Therefore in order to have a series of rays reflected from the successive atomic planes it is necessary to have a certain value of the angle θ. The length of the path of the ray represented by $BFEI$, for instance, must be one whole wave-length greater than that of ray AEI, in order that when

BFEI joins *AEI* at *E*, they will both be in the same phase of wave motion. The same holds true, of course, for the entire series of reflected rays, each in turn differing in phase by a whole wave-length. Under any other circumstances the rays would interfere with each other and there would be no vibration along *EI*. With λ known and θ measured, it is possible to calculate from the equation the value of *d*, the distance between the reflecting atomic layers. A similar reflection of the beam of X-rays may take place at such other angles as to make the difference in phase between the succession of

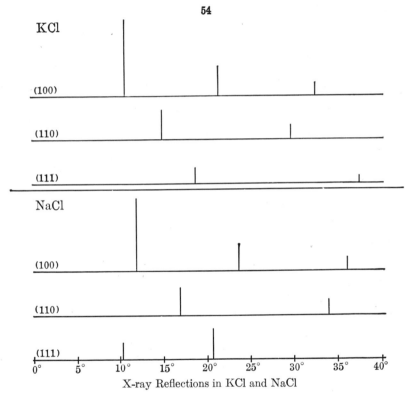

X-ray Reflections in KCl and NaCl

reflected rays 2, 4, 6, etc., wave-lengths, or even 3, 6, 9, etc., wave-lengths. These different reflections are known as reflections of the first, second, third, etc., orders. The orientation of the crystal planes from which the X-rays are reflected must be known. By changing the position of the crystal the spacing and characters of the atomic layers parallel to any crystal plane may be studied.

In Fig. 54 is shown diagrammatically the result of the investigation of the potassium and sodium chlorides by the X-ray spectrometer. Reflections from the atomic planes parallel to the cube (100), dodecahedron (110), and octahedron (111), were studied in each case. In the figure the vertical lines show the angles (2θ) at which the X-ray having the predominant wavelength of the palladium spectrum were reflected in each case, and the lengths of these lines indicate the relative intensity of the reflections. Two or three orders of reflection are shown.   From the observations with potassium chloride it was possible to calculate the value of d for the three sets of planes and it was found that the ratio was $\dfrac{1}{d_{(100)}} : \dfrac{1}{d_{(110)}} : \dfrac{1}{d_{(111)}} = 1 : \sqrt{2} : \sqrt{3}$. This ratio satisfies the conditions found in the simple cubic lattice.   Fig. 55A, represents eight unit cubes of this lattice.   The diagonal planes, A, B, and C

55

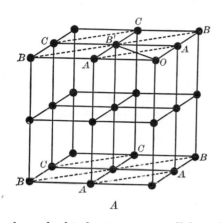
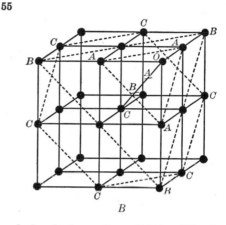

A   B

through this lattice are parallel to the dodecahedron plane (110).   The distance $d_{(110)}$ between the planes A and B is one-half the face diagonal $B'\!-\!O$ of the unit cubic cell; therefore $d_{(110)} = \dfrac{d_{(100)}\sqrt{2}}{2}$.   In Fig. 55B, the octahedral planes A, B, and C are indicated in the same cubic lattice.   The planes A and B divide the body diagonal $O\!-\!C$ of the unit cube into thirds as indicated at the points A and B.   Therefore $d_{(111)} = \dfrac{O\!-\!C}{3}$, or $\dfrac{d_{(100)}\sqrt{3}}{3}$.   If we let $d_{(100)} = 1$, then $d_{(110)} = \dfrac{\sqrt{2}}{2}$ or $\dfrac{1}{\sqrt{2}}$, and $d_{(111)} = \dfrac{\sqrt{3}}{3}$ or $\dfrac{1}{\sqrt{3}}$.   Therefore for these three interatomic distances we may derive the ratio $\dfrac{1}{d_{(100)}} : \dfrac{1}{d_{(110)}} : \dfrac{1}{d_{(111)}} = 1 : \sqrt{2} : \sqrt{3}$, agreeing with the ratio obtained from the X-ray measurements.   In this way it was shown that the atomic arrangement of potassium chloride was that of the simple cubic lattice.

Fig. 56A shows eight unit cells of the face-centered cubic lattice.   Inspection of the figure will show that $d_{(100)}$ and $d_{(110)}$ are the same as in the cubic lattice, Fig. 55A.   But of the octahedral planes shown in Fig. 55B, only the plane B appears in Fig. 56A, and the spacing between such planes must be twice as great as in the first case.   Therefore the ratio of the three interatomic distances for the face-centered cubic lattice is $\dfrac{1}{d_{(100)}} : \dfrac{1}{d_{(110)}} : \dfrac{1}{d_{(111)}} = 1 : \sqrt{2} : \dfrac{\sqrt{3}}{2}$.   Fig. 56B shows the body-centered cubic lattice and for this the ratio will become $\dfrac{1}{d_{(100)}} : \dfrac{1}{d_{(110)}} : \dfrac{1}{d_{(111)}} = 1 : \dfrac{1}{\sqrt{2}} : \sqrt{3}$.

The X-ray reflections obtained from sodium chloride, indicated in the lower half of Fig. 54, show certain interesting differences from those of potassium chloride. In the case of the (100) and (110) planes the reflections are similar but at a slightly greater angle in the case of NaCl, indicating that in

56

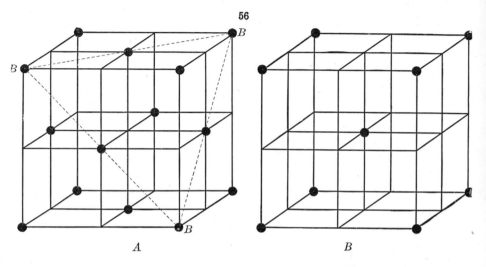

A   B

its structure the distances $d_{(100)}$ and $d_{(110)}$ are a little smaller than with KCl. In the case of the reflections from planes parallel to (111), while there is a reflection at an angle similar to the first reflection in the case of KCl, there is also another reflection in the case of NaCl at one-half the angle. This in-

57

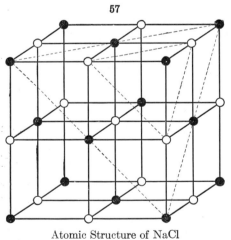

Atomic Structure of NaCl

dicates that the spacing of the octahedral planes corresponds to that of the face-centered cubic lattice instead of the cubic lattice. The first reflection from the octahedral planes in NaCl is of less intensity than the second reflection. This is contrary to the rule of regularly diminishing intensities when the reflecting planes are all of the same character. In order to account for these facts a structure for NaCl is assumed like that shown in Fig. 57, the solid dots representing atoms of sodium, while the circles indicate atoms of chlorine or vice versa. It will be noticed that the sodium and chlorine atoms considered separately are placed in face-centered cubic lattices, the two lattices interpenetrating each other to form a simple cubic lattice. Note that planes parallel to either (100) or (110) show both Na and Cl atoms alternating with each other, therefore these two sets of planes have uniform powers of reflection. On the other hand, the octahedral planes, two of which are outlined by the broken lines, alternately contain all Na or all Cl atoms. In the case of NaCl the atomic weights are different (Na = 23, Cl = 35.5), and the reflecting power of the alternating octahedral planes will be different. For this reason it is possible to obtain reflections from alternating planes at a small angle under conditions which would produce complete interference if all planes had the same reflecting power. But because of the partial interference by the reflections from the intervening planes the intensity of this small angle reflection is diminished. In the case of KCl the two elements have nearly the same atomic weights and therefore the same reflecting power. Here it would be impossible to obtain reflections from alternating planes because of the complete interference of the reflections from the intervening planes. As far as the X-ray is concerned KCl might be composed of uniform atoms arranged on a simple cubic lattice. Fig. 57, however, represents the structure of both KCl and NaCl.

These differences in the reflections between KCl and NaCl prove that it is the atoms that are arranged at the points of the lattice and not the molecules. If the latter were true, all reflections from the two substances would be similar.

From measurements of this kind it is possible to calculate the dimensions and volume of the unit cell of the atomic structure. In the case of the salts considered above, the volume would be $d_{(100)}^3$. From the known mass of the molecule and the specific gravity of the mineral it is possible to calculate the mass per unit volume and so determine how many unit cells one molecule must occupy. In the case of KCl and NaCl, this method yields two unit cells for each molecule. Each corner of the unit cube of the structure is occupied by one atom, but each corner is shared equally by eight unit cubes so that one eighth of each atom belongs to the enclosed unit cube or one atom to each unit. It will therefore take two unit cubes to contain the molecule of two atoms.

37. Other Methods of X-ray Investigation. I. *The Powder Method.* — In the methods previously described it is necessary to have a crystal sufficiently large and perfect in order to enable one to orientate it with accuracy and so make certain the crystallographic position of the reflecting atomic planes, etc. The so called powder method developed independently by Debye and Scherrer and by Hull does not require such a crystal. A small cylindrical tube of celluloid or glass is filled with fine crystalline powder of the material to be studied. In some cases the powder is pressed or cemented into a slender rod and the tube can be dispensed with. The tube or rod is placed in the path of a pencil of X-rays of uniform wave-length. The powder will contain crystalline particles in all possible crystallographic orientations. Therefore, there will always be enough particles properly orientated to give reflections from any series of parallel crystallographic planes. For instance, in the case of an isometric substance, crystal particles having certain orientations will possess octahedral planes at the proper angle to the path of X-rays in order to give rise to a series of reflections. The angular position of the octahedral planes to the path of the incident X-rays and also the angle of reflection depend upon the wave-length of the X-ray and upon the spacing of the octahedral planes. These angles would have to be different for any other series of crystallographic planes since the spacing of the planes would be different in each case. These various reflections will spread out from the powder tube in the form of cones and would meet a photographic plate placed normal to the incident beam of X-rays in a series of concentric circles. As in the case of the Laue photographs only reflections from steeply inclined planes can strike the plate. Consequently it is customary to surround the tube of powder with a strip of film arranged in a circle upon which reflections at large angles can be obtained. Such a film picture is represented diagrammatically in Fig. 58. Usually only the portion of the film along the medial

58

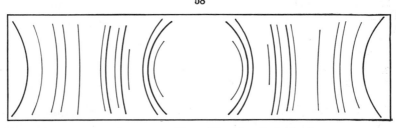

Diagram of a Powder X-ray Photograph of Aluminum

horizontal line is used and upon this the reflections appear practically as straight lines. The radius of the circle that the film makes about the powder tube being known, it is possible to determine the angles made by the reflected rays and from these angles and the varying intensities of the reflections the structure of the substance is determined. For instance the reflections obtained from the three different cubic lattices would have the relations shown

59

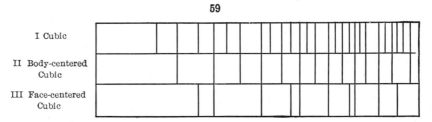

Diagrams of Powder X-ray Spectra for Isometric Structures

in Fig. 59. The reflections show different groupings in each case and a study of the powder photograph of an isometric substance would show to which lattice it belongs.

II. *Method of Rotation Photograph.* — In this method a crystal is mounted so that it can be revolved about a vertical axis which is some known crystallographic direction, usually a crystallographic or symmetry axis. A horizontal pencil of X-rays of uniform wave-length is directed upon the crystal and the crystal is slowly revolved on its axis. A photographic plate is placed behind the crystal normal to the path of the beam of X-rays or the crystal is surrounded by a strip of photographic film placed in a concentric circle about it. As the crystal is revolved various series of atomic planes come into the proper position for the reflection of the X-rays. The accumulation of the effects of these reflected rays produces spots upon the plate or film. Reflections from all planes that are parallel to the axis of rotation will lie in the horizontal plane and their spots will lie on the central horizontal line of the plate or film. Each series of vertical planes will be represented by two spots, equally spaced to the right and left of the center of the photograph. All planes that are inclined to the axis of rotation will have their reflections lying on the surfaces of cones whose axis coincides with the axis of rotation of the crystal. Each set of reflecting planes will be represented by four spots, two above and two below the central horizontal line and equally spaced to the right and left of a central vertical line. All planes having the indices $(hk1)$ will have their reflections lying on the same cone. This cone will intersect the plate in a flat hyperbola and the circular film in a straight line. The points from the series of $(hk1)$ planes will lie on the first line above or below the central horizontal line. Those having indices $(hk2)$ will lie on the

second line, etc. The spots in the picture will also have approximately a vertical arrangement over each other. If a spot on the central horizontal line is a reflection from $(hk0)$, the spots above it on the successive lines would be $(hk1)$, $(hk2)$, etc. From the distances between the lines of $(hk0)$, $(hk1)$, etc., and the distance from the crystal to the plate or film, can be calculated one dimension of the unit cell of the atomic structure. If the same crystal is orientated with other axes of rotation, data can be accumulated from which the details of the structure can be derived. Fig. 60 shows diagrammatically the spots of a rotation-photograph of NaCl, a crystallographic axis being used as the axis of rotation.

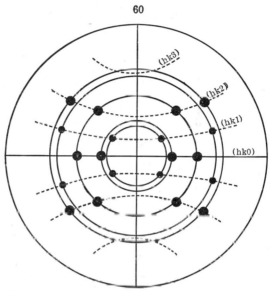

Diagram of a Rotation X-ray Photograph of NaCl
(Axis of Rotation parallel to Crystal Axis)

III. *The Oscillation Method.* — This is a modification of the rotation method. Instead of turning the crystal on an axis through an angle of 360° it is turned back and forth through an angle of about 30° from a known position. A beam of X-rays with uniform wave-length is used and the spectra are recorded on a photographic plate which is placed behind the crystal, parallel to the axis of oscillation and normal to the incident X-rays. The resultant photograph will be like the right or left half of a rotation photograph and the same methods of calculation are used in the two cases.

38. Examples of Crystal Structure as Determined by the Use of X-ray Methods. — Certain interesting examples of crystal structure that have been worked out by the use of X-rays are given below as a further illustration of the great importance of these methods. Lack of space will permit only the results of the investigations to be given.

1. *Sphalerite.* — This was the mineral first examined by X-rays by Laue. Its structure is represented in Fig. 61. The solid circles might indicate the zinc atoms. They are arranged on a face-centered cubic lattice. The hollow

circles, representing sulphur atoms, lie on a similar lattice, interpenetrating the first and which can be derived from it by a movement along the body diagonal of the cube for a distance equal to one-fourth of its length. The direction and amount of movement in order to derive one lattice from the other is indicated by the arrows.

61

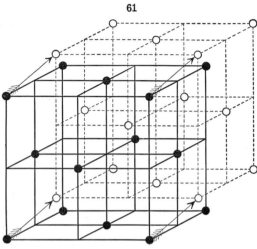

Structure of Sphalerite

2. *Diamond.* — The diamond structure was also one of the earlier ones studied. The carbon atoms are arranged on two interpenetrating face-centered cubic lattices in exactly the same way as the zinc and sulphur atoms of sphalerite. If in Fig. 61 all the lattice points were represented by solid circles we would have an illustration of the diamond structure. Fig. 62 represents the diamond structure in a different way. The cube shown in broken lines gives the crystal orientation of the structure. If it is imagined that this figure is indefinitely extended it will be seen that the structure is such that each carbon atom lies at the center of a group of four other atoms, which taken together form the points of a tetrahedron. Further, the carbon atoms are arranged in hexagonal rings. This is a very solid and compact atomic grouping and presumably accounts for some of the striking physical properties of the diamond.

3. *Graphite.* — It is interesting to compare the structure of graphite with that of diamond. In graphite the carbon atoms are arranged in parallel layers, those of each layer forming interlocking hexagons. Fig. 63 represents the projection of one such layer. The atoms in these rings do not all lie in the same plane but are either slightly above or below the central plane.

62

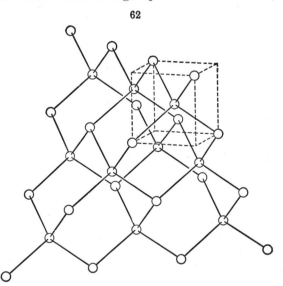

Structure of Diamond

The distances between the atoms in these hexagons is even smaller than the interatomic distances in the diamond structure.   On the other hand the distance between successive layers is unusually large, about two and one-half times the spacing in the layer itself.   The wide spacing of the layers undoubtedly accounts for the foliated structure and basal cleavage of graphite, while the close-knit structure of the individual layer accounts for the fact that graphite cannot be readily powdered but separates rather into minute flakes.   This is the property which makes it such a good lubricant.   It is interesting also to note that in both diamond and graphite the carbon molecules are arranged in hexagonal rings, a kind of atomic grouping long recognized as characteristic of carbon in organic compounds.   Another interesting fact is that if the distances between the warped hexagonal rings of the graphite structure be shortened a structure very similar to that of the diamond can be derived.

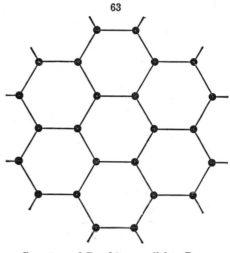

Structure of Graphite parallel to Base

4. *Fluorite.* — Fig. 64 represents the structure of fluorite, in which the calcium atoms lie at the points of a face-centered cubic lattice, while the fluorine atoms lie at the centers of the eight small cubes that compose one face-centered cube.

5. *Pyrite.* — The structure of pyrite is similar to that of fluorite, the iron atoms lying on a face-centered cubic lattice, but the sulphur atoms instead of lying at the center of the eight small cubes, lie on a body diagonal of each of the small cubes at a point one-fifth of the length of the diagonal from one of its ends.   Fig. 65 attempts to show this arrangement. The top view shows the distribution of the atoms and clearly indicates why pyrite belongs to a crystal class of lower symmetry than fluorite.

6. *Calcite.* — Fig. 66 attempts to show the atomic arrangement of calcite.   The calcium and carbon atoms lie on interpenetrating rhombohedral lattices and their positions are analogous to those of the sodium and chlorine atoms in the structure of halite (compare Fig. 57).   Except for the fact that the unit cell for calcite is a rhombohedron instead of a cube, the two structures are identical.   It has been possible with calcite to place the oxygen atoms in the structure.   They are grouped in sets of three about the carbon atoms, and they all lie in the

Structure of Fluorite; Ca = ●, F = ○

planes that contain the carbon atoms. Passing through the structure parallel to the vertical crystallographic axis, planes containing all calcium atoms alternate with planes containing only the groups of carbon and oxygen. The arrangement of the atoms in the three uppermost planes of the figure (designated by ′, ″, and ′ ″) is shown in the horizontal projection.

The above examples illustrate some of the earlier and simpler structures that were studied by X-ray methods. Many other minerals

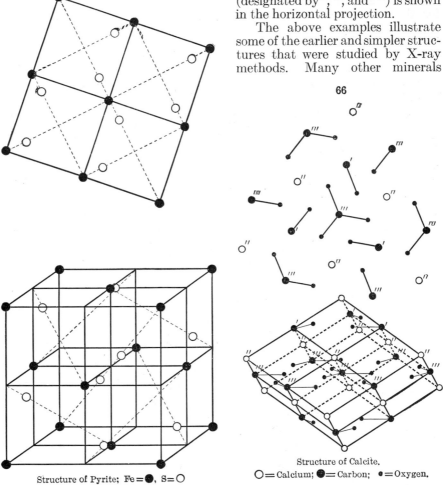

Structure of Pyrite; Fe = ●, S = O

Structure of Calcite.
O = Calcium; ◉ = Carbon; ● = Oxygen.

have been examined, some of them, as in the case of various silicates, of considerable complexity. It is impracticable to adequately treat such cases here, and the interested student must be referred to the original papers. The subject is more fully developed in the following books.

W. H. Bragg and W. L. Bragg. X-rays and Crystal Structure, 1924.
W. H. Bragg. An Introduction to Crystal Analysis, 1929.
G. L. Clarke. Applied X-rays, 1927.
R. W. G. Wyckoff. The Structure of Crystals, 1931.

GENERAL MATHEMATICAL RELATIONS OF CRYSTALS

39. Axial Ratio, Axial Plane. — The crystallographic axes have been defined (Art. 22) as certain lines, the position of which is usually determined by the symmetry of the crystal and which are used in the description of the faces of crystals, and in the determination of their position and angular inclination. With these objects in view, certain lengths of these axes are assumed as units to which the occurring faces are referred.

The axes are, in general, lettered a, b, c, to correspond to the scheme in Fig. 67. If two of the axes are equal, they are designated a, a, c; if the three are equal, a, a, a. In one system, the hexagonal, there are four axes, lettered a, a, a, c.

Further, in the systems other than the isometric, one of the horizontal axes is taken as the *unit* to which the other axes are referred; hence the lengths of the axes express strictly the *axial ratio*. Thus for sulphur (orthorhombic, see Fig. 67) the axial ratio is

$$a : b : c = 0\cdot8131 : 1 : 1\cdot9034.$$

For rutile (tetragonal) it is

$$a : c = 1 : 0\cdot64415, \quad \text{or, simply,} \quad c = 0\cdot64415.$$

67

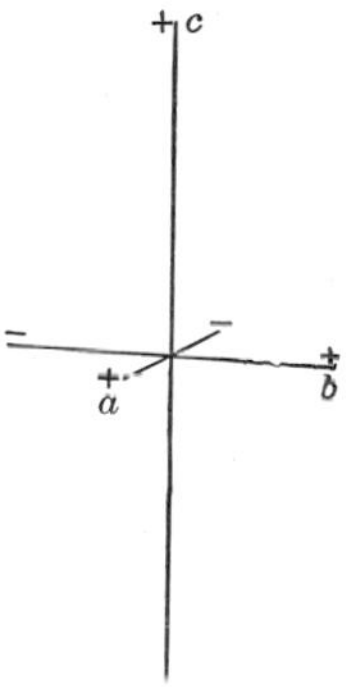

Orthorhombic Crystal Axes

The plane of any two of the axes is called an *axial plane*, and the space included by the three axial planes is an *octant*, since the total space about the center is thus divided by the three axes into eight parts. In the hexagonal system, however, where there are three horizontal axes, the space about the center is divided into twelve parts, or *sectants*.

40. Parameters, Indices, Symbol. — *Parameters.* The parameters of a plane consist of a series of numbers which express the relative intercepts of that plane upon the crystallographic axes. They are given in terms of the established unit lengths of those axes. For example, in Fig. 68 let the lines OX, OY, OZ be taken as the directions of the crystallographic axes, and let OA, OB, OC represent their unit lengths, designated (always in the same order) by the letters a, b, c. Then the intercepts for the plane (1) HKL are OH, OK, OL; for the plane (2) ANM they are OA, ON, OM. But in terms of the unit lengths of the axes these give the following parameters,

(1) $\frac{1}{4}a : \frac{1}{3}b : \frac{1}{2}c$

and

(2) $1a : \frac{4}{3}b : 2c.$

It is to be noted that since the two planes HKL and MNA are parallel to each other and hence crystallographically the same, these two sets of parameters are considered to be identical. Obviously each of them may be changed into the other by multiplying (or dividing) by 4.

Indices and Symbol. — Simplified and abbreviated expressions which have been derived from the parameters of a crystal form are commonly used to give its relations to the crystallographic axes. These are known as indices.

A number of different methods of deriving indices have been devised and several are in use at present. The so-called Miller indices are most widely employed and will be exclusively used in this work.* Below, a description of the other important systems of indices is given together with the necessary directions for transforming one type into another.

The Miller indices may be derived from the parameters of any form by taking their reciprocals and clearing of fractions if necessary. For instance take the two sets of parameters as given above.

$$(1) \quad \tfrac{1}{4}a : \tfrac{1}{3}b : \tfrac{1}{2}c, \quad \text{and} \quad (2) \quad 1a : \tfrac{4}{3}b : 2c.$$

By inversion of these expressions we obtain

$$(1) \quad 4a : 3b : 2c, \quad \text{and} \quad (2) \quad 1a : \tfrac{3}{4}b : \tfrac{1}{2}c.$$

In the case of (2) it is necessary to clear of fractions, giving

$$(2) \quad 4a : 3b : 2c.$$

The indices of this form then are $4a : 3b : 2c$. The letters indicating the different axes are commonly dropped and the indices in this case would be written simply as 432, the intercepts on the different axes being indicated by the order in which the numbers are given.

A general expression frequently used for the indices of a form belonging to any crystal system which has three crystallographic axes is hkl. In the hexagonal system, which has four axes, this becomes $hkil$. If the parameters of a form be written so that they are fractions with the numerators always unity then the denominators will become the same as the corresponding indices. The general expression in this case would therefore be, parameters $= \dfrac{1}{h}\dfrac{1}{k}\dfrac{1}{l}$.

The *symbol* of a given form is the indices of the face of that form which has the simplest relations to the crystallographic axes. The symbol is commonly used to designate the whole form.

Various examples are given below illustrating the relations between parameters and indices.

---

* In the hexagonal system the indices used are those adapted by Bravais after the method of Miller.

Parameters		Miller's Symbol
$\frac{1}{2}a : \frac{1}{2}b : 1c$ $1a : 1b : 2c$	$= \frac{1}{2}a : \frac{1}{2}b : \frac{1}{4}c =$	221
$\frac{1}{2}a : 1b : \frac{1}{2}c$ $1a : 2b : 1c$	$= \frac{1}{2}a : \frac{1}{4}b : \frac{1}{2}c =$	212
$\frac{1}{2}a : \infty b : 1c$ $1a : \infty b : 2c$	$= \frac{1}{2}a : \frac{1}{0}b : \frac{1}{4}c =$	201
$\frac{1}{2}a : 1b : \infty c$ $1a : 2b : \infty c$	$= \frac{1}{2}a : \frac{1}{4}b : \frac{1}{0}c =$	210
$1a : \infty b : \infty c$	$= \frac{1}{4}a : \frac{1}{0}b : \frac{1}{0}c =$	100

If the axial intercepts are measured in behind on the a axis, or to the left on the b axis, or below on the c axis, they are called negative, and a minus sign is placed over the corresponding number of the indices; as

Parameters	Indices
$-\frac{1}{2}a : -\frac{1}{2}b : \frac{1}{4}c =$	$\overline{2}\overline{2}1$
$-\frac{1}{2}a : \frac{1}{0}b : \frac{1}{4}c =$	$\overline{2}01$

Different Systems of Indices. — The *Weiss* indices are the same as the parameters described above. The different axes are represented by the letters a, b and c, each being preceded by a number indicating the relative intercept of the face in question upon that particular axis. For instance, a possible orthorhombic pyramid face might be represented as $1a : 2b : \frac{3}{4}c$. The Weiss indices may be converted into the Miller indices by inversion and clearing of fractions, the above symbol becoming then 213. In the *Naumann* indices the unit pyramidal form is indicated by O in the isometric system where the three crystal axes all have the same unit lengths or by P where the axes differ in their unit lengths. For other forms the indices become mPn (or mOn) in which m gives the intercept upon the vertical axis, c, and n the intercept upon one of the horizontal axes (a or b), the intercept upon the other horizontal axis being always at unity. To which particular horizontal axis this number refers may be indicated by a mark over it as $\bar{n}$ for the b axis, $\acute{n}$ or n' for the a axis. If the intercept m or n is unity it is omitted from the indices. The pyramid face used as an example above would therefore in the Naumann notation be represented by $\frac{3}{4}P2$. Other examples are given in the table below. *J. D. Dana* modified the Naumann indices by substituting a hyphen for the letter P or O and i for the infinity sign, ∞. He designated the fundamental pyramid form simply by 1. When the only parameter differing from unity was that one which referred to the intercept upon the vertical axis, it was written alone; for example the pyramid face having the parameter relations of $1a : 1b : 2c$ would be indicated by 2. The Naumann and Dana indices are easily converted into the Miller indices by arranging them in the proper order, inverting and then clearing of fractions. *Goldschmidt* has proposed another method of deriving indices. This has the ad-

EXAMPLES OF INDICES ACCORDING TO VARIOUS SYSTEMS OF NOTATION

Weiss	Naumann	Dana	Goldschmidt	Miller
$1a : 1b : 2c$...............	$2P$	2	2	221
$1a : 2b : 1c$...............	$0P2$	1–2	$1\frac{1}{2}$	212
$1a : \infty b : 2c$...............	$2P\infty$	2–i	20	201
$1a : 2b : \infty c$...............	$\infty P2$	i–2	2∞	210
$1a : \infty b : \infty c$...............	$\infty P\infty$	i–i	$\infty 0$	100

vantage that the indices for any particular face can be derived directly from the position of its pole on the gnomonic projection. The first number gives the linear position of the pole in respect to the left to right medial line of the projection and in terms of the unit pace distance of the projection (see Art. **88**). The second figure similarly gives the position of the pole in reference to the front to back medial line. These two figures constitute the Goldschmidt indices of the face. If the two numbers should be the same the second is omitted. The Goldschmidt indices are easily converted into the Miller indices by adding 1 as the third figure and clearing of fractions and eliminating any ∞ sign.

The relations between the Miller and the Miller-Bravais indices for the hexagonal system are given in Art. **173**.

41. Law of Rational Indices. — The study of crystals has established the general law that the ratios between the intercepts on the axes for the different faces on a crystal can always be expressed by rational numbers. These ratios may be $1 : 2$, $2 : 1$, $2 : 3$, $1 : \infty$, etc., but never $1 : \sqrt{2}$, etc. Hence the values of hkl in the Miller symbols must always be either *whole numbers or zero*.

If the form whose intercepts on the axes a, b, c determine their assumed unit lengths — the *unit form* as it is called — is well chosen, these numerical values of the indices are in most cases very simple. In the Miller symbols, 0 and the numbers from 1 to 6 are most common.

69

The above law, which has been established as the result of experience, in fact follows from the consideration of the molecular structure as hinted at in an earlier paragraph (Art. **31**).

42. Form. — A *form* in crystallography includes *all* the faces which have a like position relative to the planes, or axes, of symmetry. The full meaning of this will be appreciated after a study of the several systems. It will be seen that in the most general case, that of a form having the symbol (hkl), whose planes meet the assumed unit axes at unequal lengths, there must be forty-eight like faces in the isometric system* (see Fig. 139), twenty-four in the hexagonal (Fig. 244), sixteen in the tetragonal (Fig. 205), eight in the orthorhombic (Fig. 69), four in the monoclinic, and two in the triclinic. In the first four systems the faces named yield an enclosed solid, and hence the form is called a *closed form*; in the remaining two systems this is not true, and such forms in these and other cases are

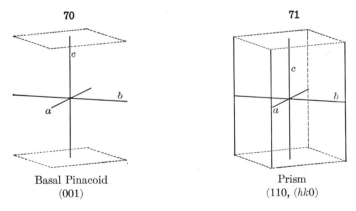

70	71
Basal Pinacoid (001)	Prism (110, $(hk0)$)

called *open forms*. Fig. 316 shows a crystal bounded by three pairs of unlike faces; each pair is hence an open form. Figs. 70–73 show open forms.

The *unit or fundamental form* is one where the parameters correspond to the assumed unit lengths of the axes. Fig. 69 shows the unit pyramid of

---

* The *normal* class is referred to in each case.

sulphur whose symbol is (111); it has eight similar faces, the position of which determines the ratio of the axes given in Art. **39**.

The forms in the isometric system have special individual names, given later. In the other systems certain general names are employed in this book which may be briefly mentioned here. A form whose faces are parallel to two of the axes* is called a *pinacoid* (from

72

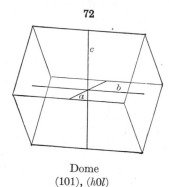

Dome
(101), (*h0l*)

73

Dome
(011), (*0kl*)

πῖναξ, a board). It is shown in Fig. 70. One whose faces are parallel to the vertical axis but meet both the horizontal axes is called a *prism*, as Fig. 71. If the faces are parallel to one horizontal axis only, it is a *dome* (Figs. 72, 73). If the faces meet all the axes, the form is a *pyramid* (Fig. 69); this name is given even if there is only one face belonging to the form.

In Fig. 74, *b*(010) is a pinacoid; *m*(110), *s*(120) are prisms; *d*(101) and *k*(021) are domes; all these are open forms. Finally, *e*(111) is a pyramid, this being a closed form. The relation existing in each of these cases between the symbol and the position of the faces to the axes should be carefully studied.

As shown in the above cases, the symbol of a *form* is usually included in parentheses, as (111), (100); or it may be in brackets [111] or {111}.

74

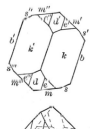

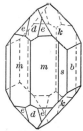

Chrysolite

43. Zone. — A zone includes a series of faces on a crystal whose intersection-lines are mutually parallel to each other and to a common line drawn through the center of the crystal, called the *zone-axis*. It follows that all edges between the faces that lie in the same zone are mutually parallel to each other and to the zone-axis. Some simple numerical relation exists, in every case, between all the faces in a zone, which is expressed by the *zonal equation* (see Art. **50**). The faces *m*, *s*, *b* (Fig. 74) are in a zone; also, *b* and *k*.

If a face of a crystal falls simultaneously in two zones, it follows that its symbol is fixed and can be determined from the two zonal equations, without the measurement of angles. Further, it can be proved that the face corresponding to the intersection of two zones is always a possible crystal face, that is, one having rational values for the indices which define its position.

---

* In the tetragonal system the form (100) is, however, called a prism and (101) a pyramid.

In many cases the zonal relation is obvious at sight, but it can always be determined, as shown in Arts. **50, 51** by an easy calculation.

Illustrations will be given after the methods of representing a crystal by the various projections have been explained.

44. Horizontal Projections. — In addition to the usual perspective figures of crystals, projections on the basal plane (or more generally the plane normal to the prismatic zone) are very conveniently used. These give in fact a map of the crystal as viewed from above looking in the direction of the axis of the prismatic zone. Figs. 30–33 give simple examples. In these the successive faces may be indicated by accents, as in Fig. 74, passing around in the

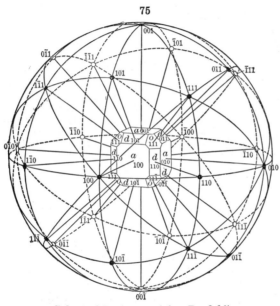

75

Spherical Projection (after Penfield)

direction of the axes a, b, $\bar{a}$, that is, counter-clockwise. On the construction of these projections see Appendix A.

45. Spherical Projection. — The study of actual crystals, particularly regarding the angular and zonal relations of their faces, is much facilitated by the use of various projections. The simplest of these and the one from which the others may be derived is known as the *spherical projection*.

In making a spherical projection of a crystal it is assumed that the crystal lies within a sphere, the center of which coincides with the center of the crystal (*i.e.* the point of intersection of its crystallographic axes). From this common center normals are drawn to the successive faces of the crystal and continued until they meet the surface of the sphere. The points at which these normals touch that surface locate the poles of the respective faces and together form the spherical projection of the crystal. The method of formation and the character of the spherical projection are shown in Fig. 75.

It is to be noted that all the poles of faces which lie in the same zone on the crystal, *i.e.* faces whose intersection lines are mutually parallel, fall upon the same great circle on the sphere. This is illustrated in the figure in the case of the zones a–d–a and a–o–d. Conversely, of course, all faces whose poles fall on the same great circle of the spherical projection must lie in the same zone. A face whose pole falls at the intersection of two or more great circles lies in two or more independent zones, as for instance $o(111)$, in Fig. 75. The angular relations between the faces on the crystal are of course preserved in the angles existing between their respective poles on the spherical projection. The angles between the poles, however, are the supplementary angles to those between the faces on the crystal, as shown in Fig. 76. The supplementary angles are those which are commonly measured and recorded when studying a crystal (see Art. 235).

The spherical projection is very useful in getting a mental picture of the relations existing between the various faces and zones upon a crystal but because of its nature does not permit of the close study and accurate measurements that may be made on the other projections described below which are made on plane surfaces.

46. The Stereographic Projection. — The stereographic projection may be best considered as derived from the spherical projection in the following manner. The plane of the projection is commonly taken as the equatorial plane of the sphere. Imaginary lines are drawn from the poles of the spherical projection to the south pole of the sphere. The points in which these lines pierce the plane of the equator locate the poles in the stereographic projection. The relation between the two projections is shown in Fig. 77. Fig. 78 shows the same stereographic projection without the foreshortening of Fig. 77. Commonly only the poles that lie in the northern hemisphere, including those on the equator, are transferred to the stereographic projection.

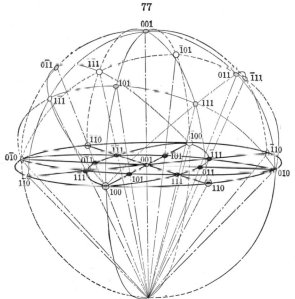

77

Relation between Spherical and Stereographic Projections

Certain facts concerning the stereographic projection need to be noted. Its most important character is that all circles or circular arcs on the spherical projection are projected as arcs of true circles on the stereographic projection.* The poles of all crystal faces that are parallel to the vertical crystallographic axis fall on the equator of the spherical projection and occupy the same positions in the stereographic projection. The pole of a horizontal face will fall at the center of the stereographic projection. All north and south meridians of the spherical projection will appear as straight radial lines in the stereographic projection

(*i.e.* as arcs of circles having infinite radii). Other great circles on the spherical projection, as already stated, will be transferred to the stereographic as circular arcs. Examples of all these are shown in Fig. 78.

---

* For proof of this statement see Penfield, Am. Jour. Sci., **11**, 10, 1901.

The angular relations between the poles of the various faces are preserved in the stereographic projection but the linear distance corresponding to a degree of arc naturally increases from the center of the projection toward its

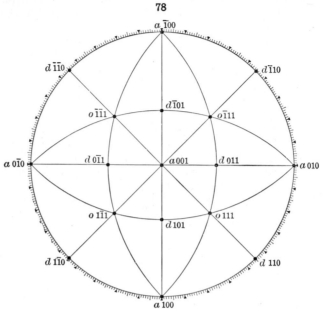

78

Stereographic Projection of the Isometric Forms, Cube, Octahedron, and Dodecahedron

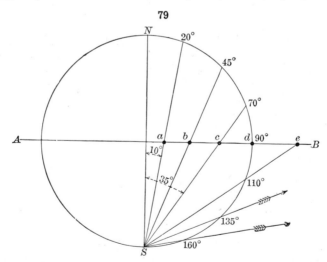

79

circumference. This is illustrated in Fig. 79 where the circle represents a vertical section through the spherical projection and the line $A-B$ represents the trace of the horizontal plane of the stereographic projection. A point

20° from N on the sphere is projected to the point a on the stereographic projection, a point 45° from N is projected to b, etc. In this way a protractor can be made by means of which angular distances from the center of the stereographic projection can be readily determined. Fig. 80 represents such a protractor which was devised by Penfield.* The mathematical relation between the linear distance from the center of the projection and its angular value is seen by study of Fig. 79. If the radius of the circle of the projection is taken as unity the distance from its center to any desired point is equal to the tangent of one half of the angle represented. For instance the distance from the center to the point a is equivalent to the tangent of 10°, to point c the tangent of 35°, etc.

Fig. 81 represents a chart used by Penfield for making stereographic projections. The circle has a diameter of 14 cm. and is graduated to degrees. With it go certain scales that are very useful in locating the desired points and zonal circles. These will be briefly described later.

For detailed descriptions of the principles of the stereographic projection and the methods of its use the reader is referred to the various books and articles, the titles of which are given beyond. It is possible here to give only a brief outline of the more important methods of construction used.

80

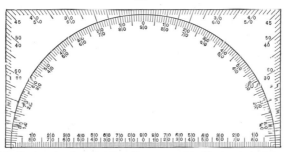

Stereographic Protractor for plotting Stereographic Projections (after Penfield; reduced one-half)

(1) *To locate the pole of a face lying on a known north and south great circle, its angular distance from the center or a point on the circumference of the projection being given.* — The stereographic protractor, Fig. 80, or the tangent relation as stated above, gives the proper distance. The poles labeled o (isometric octahedron), Fig. 78, may be located in this way.

(2) *To locate the projection of the arc of a great circle which is not a north and south meridian or the equator.* — The projections of three points on the arc must be known. Then, since the projection of the circle will be still a circular arc, its position can be determined by the usual geometric construction for a circle with three points on its arc given. If, as is commonly the case, the points where the great circle crosses the equator and the angle it makes with the equator are known it is possible to get the radius of the projected arc directly from Scale No. 1, Fig. 81. The location of such a desired arc is shown in Fig. 82. The arcs shown in Fig. 78 were also located in this way.

---

* This protractor and the other protractors and scales used by Penfield can be obtained from the Mineralogical Laboratory of the Sheffield Scientific School of Yale University, New Haven, Connecticut.

81

PENFIELD'S DIVIDED CIRCLE AND SCALES FOR PLOTTING STEREOGRAPHIC PROJECTIONS.

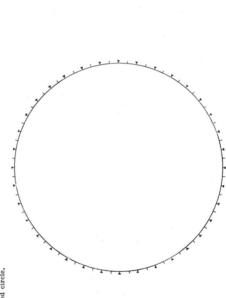

No. 1   Radii of Stereographically projected arcs of great circles, degrees measured from the divided circle.
No. 2   Radii of Stereographically projected arcs of small vertical circles, degrees measured from the divided circle.
No. 3   Degrees of a vertical great circle stereographically projected on a diameter.
No. 4   Decimal parts of the radius of the divided circle.

On the original engine-divided plate the circle has a diameter of 14 cm. and is divided into degrees.   The scales likewise are subdivided so as to give desired parts to degrees

(3) *To locate the position of the pole of a face lying on a known great circle, which is not a north and south meridian, its angle from a point on the circumference of the projection being known.* — The projected arc of a small vertical circle, whose radius is the known angle, is drawn about the point on the circumference of the projection and since all points on this arc must have the required angular distance from the given point, the intersection of this circle with the known great circle will give the desired point. The radius of the projected arc of the small vertical circle can be determined by finding the position of three points on the projection which have the required angular distance from the point given on the circumference of the projection and then obtaining the center of the required circle in the usual way. Or by the

82

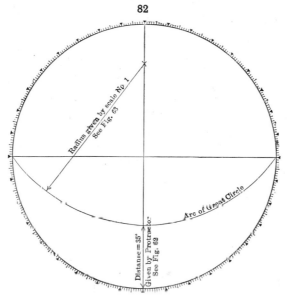

Location of the arc of a great circle in the Stereographic Projection at a given angle above the equator

use of Scale No. 2, Fig. 81, the required radius is obtained directly. It is to be noted that the known point on the circumference of the projection, while the stereographic center of the small circle, is not the actual center of the projected arc. The center will lie outside the circumference on a continuation of the radial line that joins the given point with the center of the projection. Therefore, even if the radius of the required arc is taken from Scale No. 2, it will be necessary to establish at least one point on the required circle in order to find its center. These methods of construction are illustrated in Fig. 83, in which the position is determined of the pole n (isometric trapezohedron) which lies on the great circle passing through the poles a (isometric cube) and o (isometric octahedron), and makes a known angle ($35\frac{1}{4}°$) with a.

(4) *To locate the position of the pole of a face given the angles between it and two other faces whose poles lie within the divided circle.* — Circumscribe about the poles of the two known points small circles with the proper radii and

the desired point will be located at their intersection.   The two small circles
may touch at only a single point or they may intersect in two points.   In
the latter case both points will meet the required conditions.   The positions
of the projected small circles are readily found by drawing radii from the

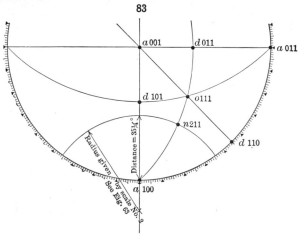

Location of pole of trapezohedron, $n(211)$, in Stereographic Projection

center of the projection through the two known poles and then laying off on
these radii points on either side of the known poles with the required angular
distances.   The center is then found between these two points in each case
and a circle drawn through them.   The line of this circle will then be everywhere the required number of degrees away from the known pole.   The re-

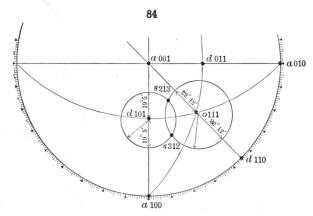

Location of two poles of hexoctahedron, s, in Stereographic Projection

quired points may be found readily by means of the Stereographic Protractor, Fig. 80, remembering that the zero point on the protractor must always
lie at the center of the projection.   This construction is illustrated in Fig. 84,
in which the points s (isometric hexoctahedron), are 22° 12′ and 19° 5′ from

the points o (isometric octahedron), and d (isometric dodecahedron). It is to be noted here, also, that while the points o and d are the stereographic centers of the circles about them, the actual centers are points which are somewhat farther out from the center of the projection.

(5) *To measure the angle between two given points on the stereographic projection.* — If the two points lie on the circumference of the projection the

85

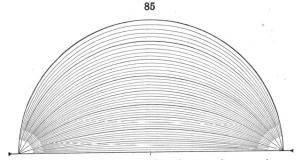

Stereographic Protractor, giving the great circles of every alternate degree (second, fourth. etc.). (After Penfield, reduced one-half)

angle between them is read directly from the divisions of the circle. If they lie on the same radial line in the projection, the angle is given by the use of the Stereographic Protractor, Fig. 80. In other cases it is necessary first to find the arc of a great circle upon which the two points lie. This is most easily accomplished by the use of a transparent celluloid protractor upon which the arcs of great circles are given, Fig. 85. Place this protractor over the projection with its center coinciding with the center of the projection and

86

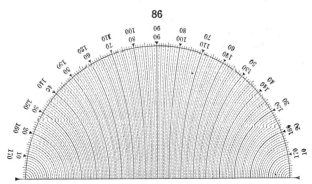

Stereographic Protractor, giving small circles for every degree measured from a given point on the circumference. (After Penfield, reduced one-half)

turn it about until the required great circle is found. Note the points where this circle intersects the circumference of the projection. Then place a second transparent protractor on which small vertical circles are given, Fig. 86, over the projection with its ends on the points of the circumference just determined. Now note the angular distance between the two given points. The whole operation may also be done by use of a third transparent protractor, on which the arcs of both great and small circles are given.

(6) *To measure the angle between the arcs of two great circles on the stereographic projection.* — This is most conveniently accomplished by constructing the arc of a great circle which shall have a 90° radius about the point at which the two arcs in question cross each other and then measuring the angular distance between the two points at which they intersect this great circle. Fig. 87, after Penfield, will serve to illustrate the method. First, if it is wished to measure the angle between the divided circle and the arc of the great circle that crosses it at *C* it is only necessary to draw a straight line through the center of the projection, *N*, which shall intersect the divided circle at points 90° distant from *C*.

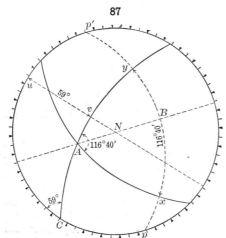

This line will be the projection of the arc of a great circle about the sphere at 90° distant from *C*. The angle at *C* is then determined by measuring with the stereographic protractor the angle between *u* and *v*.

In the case of the angle between two great circles that meet at some point within the divided circle as at *A*, Fig. 87, it is necessary to construct the projected arc of the great circle 90° distant from this point. This is done by drawing the radial line through *N* and *A* and measuring with the stereographic protractor an angle of 90° from *A* to the point *B*. The required arc will pass through this point and the points *p* and *p'* which are each 90° away from the points at which the line *A–N–B* crosses the divided circle. The angle between *x* and *y* measured on this great circle gives the value of the required angle at *A*. This is most readily measured by the use of the transparent protractor showing small circles, Fig. 86. This is placed across the projection from *p* to *p'* and the angle between *x* and *y* read directly from it.

Wülfing has described a stereographic net, which gives both great and small circles for every two degrees. Over this is placed a sheet of tracing paper upon which the stereographic projection is made. If the paper is fastened at the center of the drawing so that it can be turned into various positions in respect to the stereographic net below, the various great and small circles needed can be sketched directly upon the drawing. Or the required points can be transferred from the net to a separate drawing by means of three point dividers.

Examples of the use of the stereographic projection will be given later under each crystal system.

47. The Gnomonic Projection. — The characters of the gnomonic projection can best be understood by considering it to be derived from the spherical projection (see Art. **45**). In the case of the gnomonic projection the plane of the projection is usually taken as the horizontal plane which lies tangent to the north pole of the sphere of the spherical projection. Imaginary lines are then taken from the center of the sphere through the poles of the crystal faces that lie on its surface and extended until they touch the plane of the projection. The points in which these lines touch that plane constitute the gnomonic projection of the forms represented. Fig. 88 shows the relations between the spherical and gnomonic projections, using the same isometric crystal forms (cube, octahedron and dodecahedron) as were em-

88

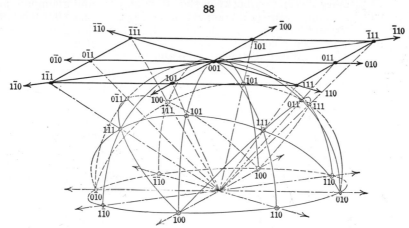

Relation between Spherical and Gnomonic Projections

89

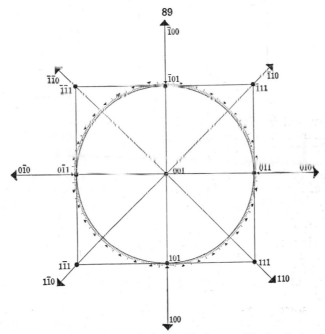

Gnomonic Projection of Cube, Octahedron and Dodecahedron

ployed to illustrate the principles of the Stereographic Projection (Art. **46**). Fig. 89 shows the gnomonic projection of the same set of forms.

The following features of the gnomonic projection are important. All great circles on the spherical projection become straight lines when transferred to the gnomonic. The poles of a series of crystal faces which belong in the same zone will, therefore, on the gnomonic projection, lie on a straight line. This primary distinction between the stereographic and gnomonic projections will be readily seen by a comparison of Figs. 78 and 89. The pole of a horizontal crystal face (like the top face of the cube) will fall at the center of the projection. The poles of vertical crystal faces will lie on the plane of projection only at infinite distances from the center. This is shown by a consideration of Fig. 88. Such faces are commonly indicated on the projection by the use of radial lines or arrows which indicate the directions in which their poles lie. This is illustrated in the case of the vertical cube and dodecahedron faces in Fig. 89. Crystal faces having a steep inclination with the horizontal plane must frequently be indicated in the same way.

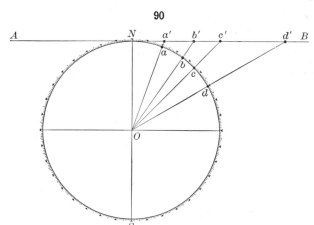

90

A simple relation exists between the linear distance from the center of the projection to a given point and the angular distance represented. This is shown in Fig. 90 where the circle is assumed to be a vertical cross section of the sphere of the spherical projection and the line A–B represents the trace of the plane of the gnomonic projection. It is evident from this figure that if the radius of the circle is taken as unity the linear distances N–a', N–b', etc., are the tangents of the angles 20°, 35°, etc. Consequently in the gnomonic projection the distance of a given pole from the center of the projection, considering the fundamental distance O–N, Fig. 90, to be unity, is equal to the tangent of the angle represented. In the case of the stereographic projection this distance is equal to the tangent of one half the angle (see Art. 46). The stereographic scale, used in the stereographic protractor, Fig. 80, can therefore be adapted for use in the gnomonic projection by taking the point on it reading at twice the desired angle. The simplest method of plotting, however, is to make a direct use of the tangent relation. The distance O–N, Fig. 90, is taken at some convenient length and then by multiplying this distance by the natural tangent of the angle desired the linear distance of the pole in question from the center of the projection is obtained. Frequently the distance O–N is taken as 5 cm. In making a gnomonic projection a circle is commonly drawn about the center of the projection, known as the fundamental circle, with a radius equal to this chosen distance. Points that have an angular distance of 45° with the center point of the projection will lie on the circumference of this circle. Commonly also the gnomonic projection is surrounded by a square border of two parallel lines on which are indicated the directions in which lie the poles that cannot appear on the projection because of the vertical or steeply inclined position of their faces.   These characters are shown in Fig. 89.

To measure the angle between two poles on the gnomonic projection. — In Fig. 91 let A_1 and A_2 be any two points the angle between which is desired. First draw a straight line through them or, in other words, find the direction of the zonal line upon which they lie.   Next erect the line $O–A$ perpendicular to this zonal line and passing through the center O of the projection.   On this line establish the point N, the distance $A–N$ being equal to the hypothenuse of the right triangle AOP or the distance $A–P$.   The point N is known as the *angle-point* of the zone $A_1–A_2$. The angle A_1NA_2 is equal to the desired angle between the points A_1 and A_2.   In the case of zonal lines that pass through the center of the projection this angle-point will lie on the circumference of the fundamental circle at the terminus

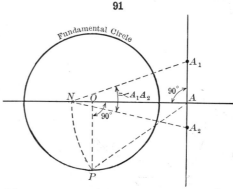

Measurement of angle between any two poles (A_1, A_2) on the Gnomonic Projection

of a radius which is at right angles to the zonal line in question.   In the case of vertical crystal faces whose poles lie at an infinite distance the center of the projection is itself the angle-point.

The explanation of the above method may be given as follows.   In Fig. 92 let the circle represent a vertical section through the sphere of the spherical projection and the line $N–A$ the trace of the plane of the gnomonic projection.   Let the line $A–C$ represent the intersection of a zonal plane lying at right angles to the plane of the drawing.   The zonal line representing the intersection of this zonal plane with the plane of the gnomonic projection would therefore be a straight line through point A which would be perpendicular to the plane of the drawing.   The angle be-

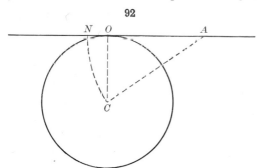

tween any two poles lying on this zonal line would be determined by the angle formed by the lines drawn from these poles to the point C.   If we consider this zonal line which passes through A perpendicular to the drawing as an axis around which we may revolve its zonal plane, the point C may be moved so that it will lie in the plane of the gnomonic projection and fall at N, the distance $A–N$ being equal to $A–C$.   The character of the point C has not been changed by this transfer and the point N becomes the angle-point of the zonal line running through A and the angle between any two poles on this line may be determined by running lines from them to this point and measuring the included angle. The point N lies on the line running through O (center of the gnomonic projection) and the distance $A–N$ is equal to the hypothenuse, $A–C$, of the right triangle one side of which is equal to $A–O$ and the other to $O–C$ (the radius of the fundamental circle).

To measure the angle between parallel zonal lines on the gnomonic projection. — In Fig. 93 let the two lines Zone 1 and Zone 2 represent two parallel zonal lines the angle between which is desired.   Draw the radial line from the center of the projection, O, at right angles to these zonal lines intersecting them at the points A_1 and A_2. Make $O-P$ at right angles to $O-A_1A_2$. The angle A_1PA_2 will give the angle between the two zones. The construction will be readily understood if the figure is supposed to be turned on the line $O-A_1A_2$ as on an axis until the point P becomes the center of the spherical projection. The broken arc now represents a vertical cross section of the sphere of the spherical projection and the points a_1 and a_2 the points where the two zonal lines cross it. The angle at P is obviously the angle between the two zones.

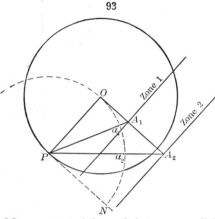

93

Measurement of the angle between parallel zones on the Gnomonic Projection

The angle between Zone 2 and the prism zone, the line of which lies at infinity on the gnomonic projection, is given in Fig. 93 by the angle A_2PN which is the same as A_1A_2P.

A gnomonic net, similar in character to the stereographic net described in Art. **46,** is useful in plotting the points of a projection or in making measurements upon it. The straight lines upon it represent the projection of the arcs of great circles of the spherical projection, while the hyperbola curves represent those of the small vertical circles.

The gnomonic projection is most commonly used in connection with the measurement of crystal angles by means of the two-circle goniometer. This use will be explained later, see Art. **237.** For more detailed descriptions of the principles and uses of the gnomonic projection the reader is referred to the literature listed below.

References on the Stereographic and Gnomonic Projections

In addition to the descriptions of these projections that are given in many general crystallographic texts the following books and papers are of value.

Barker, T. V. Graphical and Tabular Methods in Crystallography, 1922.

Boecke, H. E. Die Anwendung der stereographischen Projektion bei kristallographischen Untersuchungen, 1911. Die gnomonische Projektion in ihrer Anwendung auf kristallographische Aufgaben, 1913.

Evans, J. W. Gnomonic Projections in two planes. Min. Mag., **14,** 149, 1905.

Goldschmidt, V. Über Projektion und graphische Kristallberechnung, 1887.

Gossner, B. Kristallberechnung und Kristallzeichnung, 1914.

Hilton, H. The Gnomonic Net, Min. Mag., **14,** 18–20, 1904. The Construction of Crystallographic Projections, Min. Mag., **14,** 99–103, 1905. Some Applications of the Gnomonic Projection to Crystallography, Min. Mag., **14,** 104–108, 1905.

Hutchinson, A. On a protractor for use in constructing stereographic and gnomonic projections of the sphere, Min. Mag., **15,** 94–112, 1908.

Palache, Charles. The Gnomonic Projection. Amer. Min., **5,** 67, 1920.

Penfield, S. L. The Stereographic Projection and Its Possibilities from a Graphical Standpoint, Am. J. Sci., **9,** 1–24, 115–144, 1901. On the Solution of Problems in Crystallography by Means of Graphical Methods based upon Spherical and Plane Trigonometry, Am. J. Sci., **14,** 249–284, 1902. On the Drawing of Crystals from Stereographic and Gnomonic Projections, Am. J. Sci., **21,** 206–215, 1906.

Smith, G. H. H. On the Advantages of the Gnomonic Projection and its use in the Drawing of Crystals, Min. Mag., **13,** 309–321, 1903.

48. Angles between Faces. — The angles most conveniently used with the Miller symbols, and those given in this work, are the *normal angles*, that is, the angles between the poles or normals to the faces, measured on arcs of great circles joining the poles as shown on the stereographic projection. These normal angles are the supplements of the actual interfacial angles, as has been explained.

94

Chrysolite

The relations between these normal angles, for example in a given zone, is much simpler than those existing between the actual interfacial angles.   Thus it is always true that, for a series of faces in the same zone, the normal angle between two end faces is equal to the sum of the angles of faces falling between.   Thus (Figs. 94, 95) the normal angle of $ab(100 \wedge 010)$ is the sum of $am(100 \wedge 110)$, $ms(110 \wedge 120)$, and $sb(120 \wedge 010)$.   This relation holds true in all the systems.

Furthermore, it will be seen that, supposing aca' (Fig. 95) is a plane of symmetry as in the orthorhombic system, the angle $100 \wedge 110$, or am (Fig. 94), is half the angle $110 \wedge 1\overline{1}0(mm''')$.   Similarly $010 \wedge 120(bs)$ is half the angle $120 \wedge \overline{1}20(ss')$; again, $100 \wedge 111(ae)$ is the complement of half the angle $111 \wedge \overline{1}11(ee')$ and $010 \wedge 111(be)$ the complement of half the angle $111 \wedge 1\overline{1}1(ee''')$.

Here, as throughout this work, the sign $\wedge$ is used to represent the angle between two faces, usually designated by letters.

49. Use of the Stereographic Projection to Exhibit the Symmetry. — The symmetry of any one of the crystalline classes may be readily exhibited by the help of the stereographic projection.

The axes of binary, trigonal, tetragonal and hexagonal symmetry are represented respectively by the following signs:

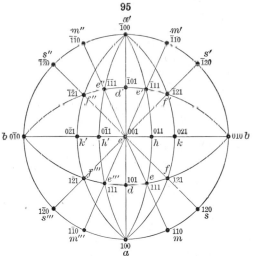

Stereographic Projection of Faces on Chrysolite Crystal, Fig. 94

95

Further, a plane of symmetry is represented by a full line (zone-circle), while a dotted line indicates that the plane of symmetry is wanting.   The position of the crystallographic axes is shown by arrows at the extremities of the lines.   The pole of a face in the upper half of the crystal (above the plane of projection) is represented by a cross; one below by a circle.   If two like faces fall in a vertical zone a double sign is used, a cross within the circle. Figs. 109, 146, 158, etc., give illustrations.

50. General Relations between Planes in the Same Zone. — Certain important relations exist between the indices of faces that lie in the same zone.   If the indices of two faces lying in the same zone are added to each other, the sum will be the indices of a face lying between them, or in other words, a face that truncates the edge between them. Note in Fig. 94 that the indices of e can be obtained by adding the indices of d and f (101 + 121 = 222 or 111); also by adding the indices of m and c (110 + 001 = 111); etc. All faces to belong to the same zone, *tautozonal faces* as they are called, must have their mutual intersections parallel to a given direction, see Art. **43.** This direction is known as the axis of the zone. The position of this zonal axis can be expressed by what is known as the zonal symbol. Consider

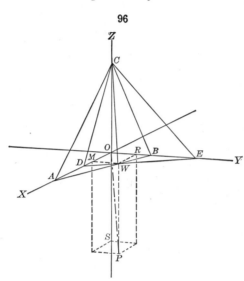

96

Fig. 96, where are represented two crystal faces, ABC, and CDE, intersecting the crystallographic axes X, Y and Z. In the illustration, for simplicity, both faces have been assumed to pass through the point C on the axis Z. This, of course, is possible since any crystal plane may be moved parallel to itself without altering its relative intercepts on the crystal axes. These two planes intersect in the line C–W, which then becomes the direction of the zonal axis for the zone in which they lie. Let the line O–P which has been drawn parallel to this direction represent that axis. In the parallelepiped of which it is a diagonal the length of the edge O–S and its parallel edges have been taken as equal to the distance O–C. The point P on the zonal axis and therefore the direction of the axis itself is fixed by the three coördinates, O–M, O–R, and O–S. By means of the consideration of similar triangles it is possible to prove that the values of these coördinates may be expressed by,

$$O\text{–}M = (kr - lq)a; \quad O\text{–}R = (lp - hr)b; \quad O\text{–}S = (hq - kp)c,$$

where a, b, c represent the unit lengths of the three crystallographic axes, X, Y, Z and (hkl) and (pqr) represent the indices of the two faces ABC and CDE. These expressions are usually simplified by substituting $u = kr - lq$, $v = lp - hr$, $w = hq - kp$, giving $O\text{–}M = ua$, $O\text{–}R = vb$ and $O\text{–}S = wc$. The three figures $[uvw]$ are said to be the symbol of the zone in question. They represent the reciprocals of the values of the three coördinates, or in other words are the indices of a point, P, on the zonal axis. They may most readily be obtained by a system of cross-multiplication and subtraction according to the following scheme. Write the indices of one face twice in their proper order and directly under them the corresponding indices of the second face. Cross off the first and last number of each series. Then multiply the figures joined by the cross lines, see below, and subtract the product of the two joined by light lines from that of those joined by heavy lines, working from left to right. The three numbers obtained will in their order correspond to u, v, and w.

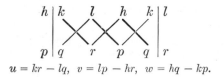

$$u = kr - lq, \quad v = lp - hr, \quad w = hq - kp.$$

Since the zonal symbol for a given zone may be obtained from the indices of any two faces lying in that zone it follows that the indices of every possible face in that zone must have definite relations to the zonal symbol. For a given face with indices (xyz), in a zone having the symbol $[uvw]$ the following equation, known as the zonal equation, must hold true.

$$ux + vy + wz = 0.$$

In this way it can be readily shown whether or not a given face can lie in a certain zone.

Further if $[uvw]$ be the symbol of one zone and $[efg]$ that of another intersecting it, then the point of intersection will always be the pole of a possible crystal face. Its indices (hkl) must satisfy the equations of both zones and may be obtained by combining them or the same result may be had by taking the symbols of the two zones and subjecting them to the same sort of cross-multiplication by which they were themselves originally derived.

51. Examples of Zones and Zonal Relations. — The following are cases in which the zonal equation is seen at once. In Figs. 94 and 95 the faces $a(100)$, $m(110)$, $s(120)$, $b(010)$ form a vertical zone with mutually parallel intersections, since they are all parallel to the vertical axis; that is, for all faces in this zone it must be true that $l = 0$.

Again, the faces $a(100)$, $d(101)$, $c(001)$ are in a zone, all being parallel to the horizontal axis b, hence for them and all others in this zone $k = 0$. Also $b(010)$, $k(021)$, $h(011)$, $c(001)$ are in a zone, all being parallel to the axis a, so that $h = 0$.

Also the faces $f(121)$, $e(111)$, $d(1\ 1)$, $e'''(1\bar{1}1)$, $f'''(1\bar{2}1)$ are in a zone, since they have a common ratio for the axes $a : c$. With them, obviously, $h = l$.

The faces $c(001)$, $e(111)$, $m(110)$ are also in a zone, and again $c(001)$, $f(121)$, $s(120)$, f in the other. For each of these zones it is true that there is a common ratio of the horizontal axes, that is, of h to k in the indices. For the first it may be shown that $h = k$; for the second, that $2h = k$.

All the relations named may be obtained at once from the above scheme. For example, for the faces $s(120)$ and $f(121)$ the scheme gives

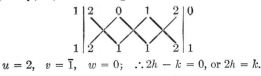

$$u = 2, \quad v = \bar{1}, \quad w = 0; \quad \therefore 2h - k = 0, \text{ or } 2h = k.$$

The symbol of a face lying at once in two zones, as stated above, must satisfy the zonal equation of each; these symbols are hence easily obtained either by combining the equations or by a scheme of multiplication like that given above.

For example, in Fig. 97, of sulphur, the face lettered x is in the Zone (1) with $b(010)$ and $s(113)$, also in Zone (2) with $p(111)$ and $n(011)$. These zones give, respectively:

(1)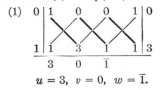

$$u = 3, \quad v = 0, \quad w = \bar{1}.$$

(2)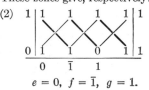

$$e = 0, \quad f = \bar{1}, \quad g = 1.$$

Hence for (1) the zonal equation is $3h = l$; for (2) $k = l$. Combining these, we obtain $h = 1, k = 3, l = 3$.

The symbol of the face x is, therefore, 133.

The same result is given by multiplying the zonal symbols $0\bar{1}1$, $30\bar{1}$, together after the same method, thus:

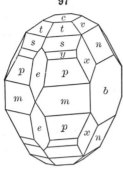

97

Sulphur

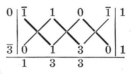

Hence, again, $x = 133$.

This method of calculation belongs to all the different systems. In the hexagonal system, in which there are four indices, one of the three referring to the horizontal axes (usually the third) is omitted when the zonal relations are applied. See Art. 170.

52. Methods of Calculation. — In general the angles between the poles can be calculated by the methods of spherical trigonometry from the triangles shown in the spherical projection — which for the most part are right-angled. Certain fundamental relations connect the axes with the elemental angles of the projection; the most important of these are given under the individual systems. Some general relations only are explained here.

53. Relations between the Indices of a Plane and the Angle made by it with the Axes. — In Fig. 98 let the three lines, X, Y, and Z represent three crystallographic axes making any angles with each other and let a, b and c represent their unit lengths. Assume any face HKL cutting these axes with the intercepts $O–H$, $O–K$ and $O–L$. Let $O–p–P$ be a normal to the plane HKL intersecting the plane at p and the enveloping surface of the spherical projection at P. Let hkl represent the indices of the given form. Since the line $O–p$ is normal to the plane HKL the triangles HOp, KOp and LOp are right angles and the following relations hold true.

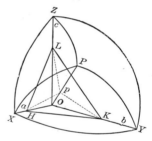

98

$$\frac{Op}{OH} = \cos HOp; \quad \frac{Op}{OK} = \cos KOp; \quad \frac{Op}{OL} = \cos LOp.$$

The angles HOp, KOp, and LOp are equal, respectively, to the angles represented on the spherical projection by the arcs PX, PY, and PZ and $OH = \dfrac{a}{h}$, $OK = \dfrac{b}{k}$, $OL = \dfrac{c}{l}$. By substituting we have,

$$Op = \frac{a}{h} \cos PX = \frac{b}{h} \cos PY = \frac{c}{l} \cos PZ.$$

This equation is fundamental, and several of the relations given beyond are deduced from it.

The most useful application is that when the axial angles are 90°, as represented in Fig. 99, then X, Y, Z are the normals to 100, 010, 001, respectively. Also if the plane HKL is taken as a face of the unit pyramid, that is, if its intercepts on the axes are taken as the unit lengths

$$OH = a, \qquad OK = b, \qquad OL = c.$$

Then the lines HK, HL, KL give also the intersections of the planes 110, 101, 011 on the three axial planes, and their poles are hence at the points fixed by normals to these lines drawn from O. It will be obvious from this figure, then, that the following relations hold true:

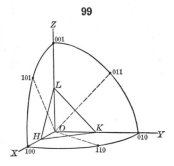

99

$$\tan (100 \wedge 110) = \frac{a}{b};$$

$$\tan (001 \wedge 101) = \frac{c}{a};$$

$$\tan (001 \wedge 011) = \frac{c}{b}.$$

These values are often used later.

54. Cotangent and Tangent Relations. — In the case of four faces in a zone concerning which we know either the angles between all the faces and the indices of three of them, or the angles between three faces and all the indices, it is possible by either a simple graphical method of plotting or by calculation to determine the missing angle or indices.

To illustrate the graphic method first let Fig. 100 represent a cross section perpendicular to the prism zone of a rhodonite crystal. The traces upon the

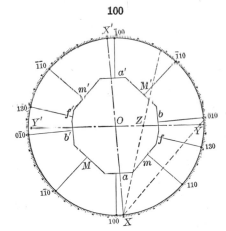

100

plane of the drawing of the faces $a(100)$ and $b(010)$ provide the direction of the lines of reference X and Y. It is assumed that the position of the third face $m(110)$ is known and a line drawn parallel to its trace upon the plane of the drawing from the point X will give its relative intercepts upon the two lines of reference. These intercepts do not correspond to the unit lengths of the axes a and b since, rhodonite being triclinic, these axes do not lie in the plane of the drawing but they represent rather the unit lengths of these axes as foreshortened by projection upon that plane. This makes no difference, however, since it will still be true that all faces lying in the prism zone of rhodonite must intercept these two lines in distances which will have rational relations to the lengths of the intercepts of $m(110)$. It is now assumed that a fourth face f has the indices (130) and its angular position in respect to the other faces in the zone is required. From its indices it must intercept the two lines of reference $X-X'$ and $Y-Y'$ in the ratio of 1 to $\frac{1}{3}$. Let OX equal 1 on $X-X'$ and OZ equal $\frac{1}{3}$ on $Y-Y'$. Then a line joining these two points will give the direction of the trace of f upon the plane of the drawing and so determine the angles it will make with the other faces in the zone.

If, on the other hand, the angles between f and the other faces in the zone were known, the position of the trace of f upon the plane of the drawing could be found, and so its relative intercepts (and indices) upon the two lines of reference be determined.

101

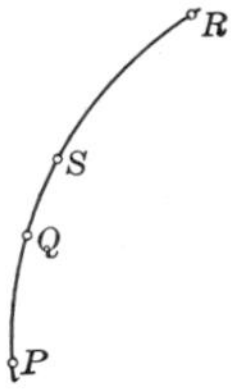

If the method of calculation is used let P, Q, S and R be the poles of four faces in a zone (Fig. 101) taken in such an order* that $PQ < PR$ and let the indices of these faces be respectively

P	Q	R	S
hkl	pqr	uvw	xyz

Then it may be proved that

$$\frac{\cot PS - \cot PR}{\cot PQ - \cot PR} = \frac{(P.Q)}{(Q.R)} \times \frac{(S.R)}{(P.S)}$$

where

$$\frac{(P.Q)}{(Q.R)} \begin{bmatrix} P, hkl \\ Q, pqr \\ Q, pqr \\ R, uvw \end{bmatrix} \overset{123}{=} \frac{hq - kp}{pv - qu} \overset{1 \times 2}{=} \frac{kr - lq}{qw - rv} \overset{2 \times 3}{=} \frac{lp - hr}{ru - pw} \overset{3 \times 1}{}$$

$$\frac{(S.R)}{(P.S)} \begin{bmatrix} S, xyz \\ R, uvw \\ P, hkl \\ S, xyz \end{bmatrix} \overset{123}{=} \frac{xv - yu}{hy - kx} \overset{1 \times 2}{=} \frac{yw - zv}{kz - ly} \overset{2 \times 3}{=} \frac{zu - xw}{lx - hz} \overset{3 \times 1}{}$$

If one of these fractions reduces to an indeterminate form, $\frac{0}{0}$, then one of the others must be taken in its place.

This formula is chiefly used in the monoclinic and triclinic systems; and some special cases are referred to under these systems.

The cotangent relation becomes much simplified for a rectangular zone, that is, a zone between a pinacoid and a face lying in a zone at right angles to it so that the angle PR becomes 90°. In Fig. 101 let $P(hkl)$ and $Q(pqr)$ be two faces lying in the zone between $a(100)$ and $d(011)$ with the angle $a \wedge d = 90°$. Let Pa and Qa represent the angles between the two faces and the pinacoid a. Then the following holds true,

$$\frac{h}{p} \times \frac{\tan Pa}{\tan Qa} = \frac{k}{q} = \frac{l}{r},$$

or if the faces P and Q lie in zones with the other pinacoids $b(010)$ or $c(001)$ the expression becomes

$$\frac{h}{p} = \frac{k}{q} \times \frac{\tan Pb}{\tan Qb} = \frac{l}{r},$$

$$\frac{h}{p} = \frac{k}{q} = \frac{l}{r} \times \frac{\tan Pc}{\tan Qc}.$$

---

* In the application of this principle it is essential that the planes should be taken in the proper order, as shown above; to accomplish this it is often necessary to use the indices and corresponding angles, not of (hkl), but the face opposite ($\bar{h}\bar{k}\bar{l}$), etc.

If the zone in question lies between two pinacoids which are at right angles to each other so that the indices of the faces P and Q become either $hk0$ and $pq0$, $h0l$ and $p0r$ or $0kl$ and $0qr$, we have

$$\frac{\tan (100 \wedge hk0)}{\tan (100 \wedge pq0)} = \frac{k}{h} \cdot \frac{p}{q};$$

$$\frac{\tan (001 \wedge h0l)}{\tan (001 \wedge p0r)} = \frac{h}{l} \cdot \frac{r}{p};$$

$$\frac{\tan (001 \wedge 0kl)}{\tan (001 \wedge 0qr)} = \frac{k}{l} \cdot \frac{r}{q}.$$

These equations are the ones ordinarily employed to determine the symbol of any prismatic plane or dome.

The most common and important application of this tangent principle is where the positions of the unit faces 110, 101, 011 are known, then the relation becomes

$$\frac{\tan (100 \wedge hk0)}{\tan (100 \wedge 110)} = \frac{k}{h}, \quad \text{or} \quad \frac{\tan (010 \wedge hk0)}{\tan (010 \wedge 110)} = \frac{h}{k}.$$

Also,

$$\frac{\tan (001 \wedge h0l)}{\tan (001 \wedge 101)} = \frac{h}{l}, \quad \frac{\tan (001 \wedge 0kl)}{\tan (001 \wedge 011)} = \frac{k}{l}.$$

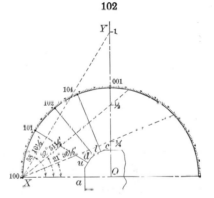

102

Thus the tangents of angles between the base, 001, and 102, 203, 302, 201, etc., are respectively $\frac{1}{2}$, $\frac{2}{3}$, $\frac{3}{2}$, 2 times the tangent of the angle between 001 and 101. Again, the tangent of the angle 100 $\wedge$ 120 is twice the tangent of 100 $\wedge$ 110 $\left(\text{here } \frac{k}{h} = 2\right)$, and one-half the tangent of 010 $\wedge$ 110.

These last relations are shown clearly in Fig. 102 which represents a cross section of a barite crystal showing the macrodome zone between $a(100)$ and $c(001)$. It is assumed that the angles between the faces a, u, d, l and c have been measured and the positions of their poles determined as indicated in the figure. The broken lines drawn from a point X on the line representing the a crystallographic axis show the direction of the traces of these faces upon the plane of the a and c axes. If the face u is assumed to be the unit dome (101) it will intersect the two axes at distances proportional to their unit lengths, namely $O-X$ and $O-Y$. The other faces d and l are seen to intersect the c axis at $\frac{1}{2}$ and $\frac{1}{4}$ the distance $O-Y$, giving them the indices (102) and (104). But the intercepts on $O-Y$ for the three faces u, d and l are proportional to the tangents of the angles between their poles and that of $c(001)$ as shown below.

$$\tan 58° \ 10\tfrac{1}{2}' = 1.6112 = 1$$
$$\tan 38° \ 51\tfrac{1}{2}' = .8056 = \tfrac{1}{2}$$
$$\tan 21° \ 56\tfrac{1}{2}' = .4028 = \tfrac{1}{4}$$

Note:— For a simple demonstration of the so-called law of Miller, see Cesàro, Min Mag. **17**, 324, 1916.

I. ISOMETRIC SYSTEM

(*Regular or Cubic System*)

55. The isometric system embraces all the forms which are referred to three axes of equal lengths and at right angles to each other. Since these axes are mutually interchangeable it is customary to designate them all by

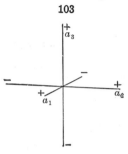

103

the letter *a*. When properly orientated (*i.e.* placed in the commonly accepted position for study) one of these axes has a vertical position and of the two which lie in the horizontal plane, one is perpendicular and the other parallel to the observer. The order in which the axes are referred to in giving the relations of any face to them is indicated in Fig. 103 by lettering them a_1, a_2, and a_3. The positive and negative ends of each axis are also shown.

Isometric Axes

There are five classes here included; of these the normal class,* which possesses the highest degree of symmetry for the system and, indeed, for all crystals, is by far the most important. Two of the other classes, the pyritohedral and tetrahedral, also have numerous representatives among minerals.

1. NORMAL CLASS (1). GALENA TYPE

(*Hexoctahedral, Holohedral or Ditesseral Central Class*)

56. Symmetry. 3 xl. Ax.-4; 4 diag. Ax.-3; 6 diag. Ax.-2; 3 xl. P.; 6 diag. P.; C.† — The symmetry of each of the types of solids enumerated in the following table, as belonging to this class, and of all their combinations, is as follows.

Axes of Symmetry. — There are three principal axes of tetragonal symmetry which are coincident with the crystallographic axes and are sometimes known as the cubic axes since they are perpendicular to the faces of the cube. There are four diagonal axes of trigonal symmetry which emerge in the middle of the octants formed by the cubic axes. These are known as the octahedral axes since they are perpendicular to the faces of the octahedron. Lastly there are six diagonal axes of binary symmetry which bisect the plane

---

* It is called *normal*, as before stated, since it is the most common and hence by far the most important class under the system; also, more fundamentally, because the forms here included possess the highest grade of symmetry possible in the system. The cube is a possible form in each of the five classes of this system, but although these forms are alike geometrically, it is only the cube of the normal class that has the full symmetry as regards molecular structure which its geometrical shape suggests. If a crystal is said to belong to the isometric system, without further qualification, it is to be understood that it is included here. Similar remarks apply to the normal classes of the other systems.

† In the symbolic descriptions of the symmetry of the various classes: Ax. = axis of symmetry, the preceding numeral indicating the number of like axes and the numeral following and joined to Ax. by a hyphen showing the degree of symmetry of the axis; P. = plane of symmetry; C. = center of symmetry; xl. indicates that the symmetry axis (or axes) referred to is also a crystallographic axis or that a plane contains crystallographic axes; diag. = diagonal to crystallographic axes or axial planes; vert. and hor. indicate vertical or horizontal respectively.

104

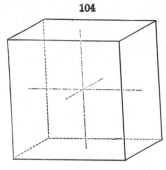

Axes of Tetragonal Symmetry
(Cubic Axes)

105

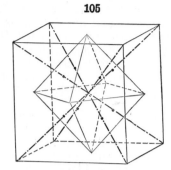

Axes of Trigonal Symmetry
(Octahedral Axes)

106

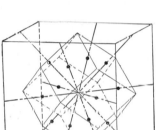

Axes of Binary Symmetry
(Dodecahedral Axes)

107

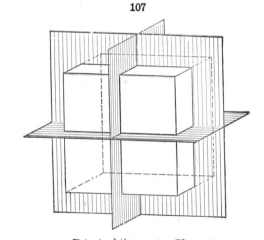

Principal Symmetry Planes

108

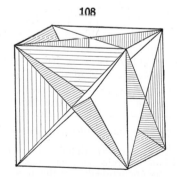

Diagonal Symmetry Planes

109

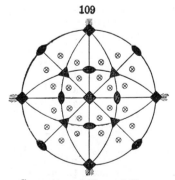

Symmetry of Normal Class,
Isometric System

angles made by the cubic axes. These are perpendicular to the faces of the dodecahedron and are known as the dodecahedral axes. These symmetry axes are shown in the Figs. 104–106.

Planes of Symmetry. — There are three principal planes of symmetry which are at right angles to each other and whose intersections fix the position of the crystallographic axes, Fig. 107. In addition there are six diagonal planes of symmetry which bisect the angles between the principal planes, Fig. 108.

The accompanying stereographic projection (Fig. 109), constructed in accordance with the principles explained in Art. **49**, shows the distribution of the faces of the general form, *hkl* (hexoctahedron) and hence represents clearly the symmetry of the class. Compare also the projections given later.

57. Forms. — The various possible forms belonging to this class, and possessing the symmetry defined, may be grouped under seven types of solids. These are enumerated in the following table, commencing with the simplest.

Indices

1. Cube......................(100)
2. Octahedron................(111)
3. Dodecahedron..............(110)
4. Tetrahexahedron..........(*hk0*) as, (310); (210); (320), etc.
5. Trisoctahedron...........(*hhl*) as, (331); (221); (332), etc.
6. Trapezohedron............(*hll*) as, (311); (211); (322), etc.
7. Hexoctahedron............(*hkl*) as, (421); (321), etc.

Attention is called to the letters uniformly used in this work and in Dana's System of Mineralogy (1892) to designate certain of the isometric forms.* They are:

Cube: *a*.
Octahedron: *o*.
Dodecahedron: *d*.
Tetrahexahedrons: $e = 210$; $f = 310$; $g = 320$; $h = 410$.
Trisoctahedrons: $p = 221$; $q = 331$; $r = 332$; $\rho = 441$.
Trapezohedrons: $m = 311$; $n = 211$; $\beta = 322$.
Hexoctahedrons: $s = 321$; $t = 421$.

58. Cube. — The *cube*, whose general symbol is (100), is shown in Fig. 110. It is bounded by six similar faces, each parallel to two of the axes. Each face is a square, and the interfacial angles are all 90°. The faces of the cube are parallel to the principal or axial planes of symmetry.

59. Octahedron. — The *octahedron*, shown in Fig. 111, has the general symbol (111). It is bounded by eight similar faces, each meeting the three axes at equal distances. Each face is an equilateral triangle with plane angles of 60°. The normal interfacial angle, $(111 \wedge 1\bar{1}1)$, is 70° 31′ 44″.

60. Dodecahedron. — The *rhombic dodecahedron*,† shown in Fig. 112, has the general symbol (110). It is bounded by twelve faces, each of which meets two of the axes at equal distances and is parallel to the third axis.

---

* The usage followed here (as also in the other systems) is in most cases that of Miller (1852).

† The *dodecahedron* of the crystallographer is this form with rhombic shaped faces commonly found on crystals of garnet. Geometricians recognize various solids bounded by twelve similar faces; of these the regular (pentagonal) dodecahedron is the most important. In crystallography this solid is impossible though the *pyritohedron* approximates to it. (See Art. **72**.)

Each face is a rhomb with plane angles of $70\frac{1}{2}°$ and $109\frac{1}{2}°$. The normal interfacial angle is $60°$. The faces of the dodecahedron are parallel to the six auxiliary, or diagonal, planes of symmetry.

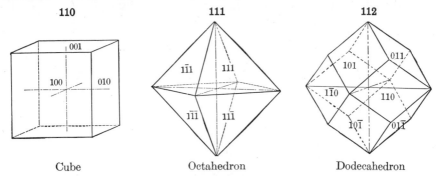

110	111	112
Cube	Octahedron	Dodecahedron

It will be remembered that, while the forms described are designated respectively by the symbols (100), (111), and (110), each face of any one of the forms has its own indices. Thus for the *cube* the six faces have the indices

$$100, \quad 010, \quad 001, \quad \bar{1}00, \quad 0\bar{1}0, \quad 00\bar{1}.$$

For the *octahedron* the indices of the eight faces are:

Above 111, $\bar{1}11$, $\bar{1}\bar{1}1$, $1\bar{1}1$;
Below $11\bar{1}$, $\bar{1}1\bar{1}$, $\bar{1}\bar{1}\bar{1}$, $1\bar{1}\bar{1}$.

For the *dodecahedron* the indices of the twelve faces are:

$$110, \quad \bar{1}10, \quad \bar{1}\bar{1}0, \quad 1\bar{1}0,$$
$$101, \quad \bar{1}0\bar{1}, \quad \bar{1}0\bar{1}, \quad 10\bar{1},$$
$$011, \quad 0\bar{1}1, \quad 0\bar{1}\bar{1}, \quad 01\bar{1}.$$

These should be carefully studied with reference to the figures (and to models), and also to the projections (Figs. 143, 144). The student should become thoroughly familiar with these individual indices and the relations to the axes which they express, so that he can give at once the indices of any face required.

61. Combinations of the Cube, Octahedron, and Dodecahedron. — Figs. 113–115 represent combinations of the cube and octahedron; Figs.

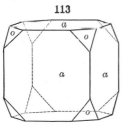
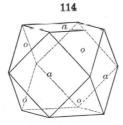
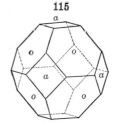

113	114	115
Cube and Octahedron	Cube and Octahedron	Octahedron and Cube

116, 119 of the cube and dodecahedron; Figs. 117, 118 of the octahedron and dodecahedron; finally, Figs. 120, 121 show combinations of the three forms. The predominating form, as the cube in Fig. 113, the octahedron in Fig. 115, etc., is usually said to be *modified* by the faces of the other forms. In Fig. 114 the cube and octahedron are sometimes said to be " in equilibrium," since the faces of the octahedron meet at the middle points of the edges of the cube.

It should be carefully noticed, further, that the octahedral faces replace the solid angles of the cube, as regular triangles equally inclined to the adja-

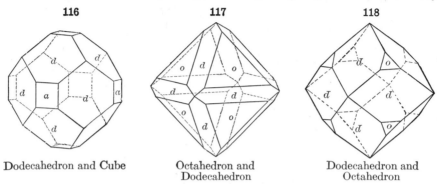

116	117	118
Dodecahedron and Cube	Octahedron and Dodecahedron	Dodecahedron and Octahedron

cent cubic faces, as shown in Fig. 113. Again, the square cubic faces replace the six solid angles of the octahedron, being equally inclined to the adjacent octahedral faces (Fig. 115). The faces of the dodecahedron *truncate** the twelve similar edges of the cube, as shown in Fig. 119. They also truncate the twelve edges of the octahedron (Fig. 117). Further, in Fig. 116 the cubic

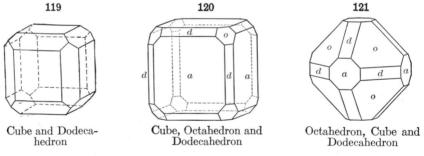

119	120	121
Cube and Dodeca-hedron	Cube, Octahedron and Dodecahedron	Octahedron, Cube and Dodecahedron

faces replace the six tetrahedral solid angles of the dodecahedron, while the octahedral faces replace its eight trihedral solid angles (Fig. 118).

The normal interfacial angles for adjacent faces are as follows:

Cube on octahedron, ao, 100 $\wedge$ 111 = 54° 44′ 8″.
Cube on dodecahedron, ad, 100 $\wedge$ 110 = 45° 0′ 0″.
Octahedron on dodecahedron, od, 111 $\wedge$ 110 = 35° 15′ 52″.

62. As explained in Art. **18** actual crystals always deviate more or less widely from the ideal solids figured, in consequence of the unequal development of like faces. Such crystals, therefore, do not satisfy the *geometrical* definition of right symmetry relatively to the three principal and the six auxiliary planes mentioned on p. 70 but they do conform to the conditions of crystallographic symmetry, requiring like angular position for similar faces. Again. it will be noted that in a combination form many of the faces do not actually meet

---

* The words *truncate, truncation,* are used only when the modifying face makes equal angles with the adjacent similar faces.

the axes within the crystal, as, for example, the octahedral face *o* in Fig. 113. It is still true, however, that this face would meet the axes at equal distances if produced; and since the *axial ratio* is the essential point in the case of each form, and the *actual lengths* of the axes are of no importance, it is not necessary that the faces of the different forms in a crystal should be referred to the same actual axial lengths. The above remarks will be seen to apply also to all the other forms and combinations of forms described in the pages following.

63. Tetrahexahedron. — The *tetrahexahedron* (*tetrakishexahedron*) (Figs. 122–124) is bounded by twenty-four faces, each of which is an isosceles triangle. Four of these faces together occupy the position of one face of the cube (hexahedron) whence the name commonly applied to this form. The general

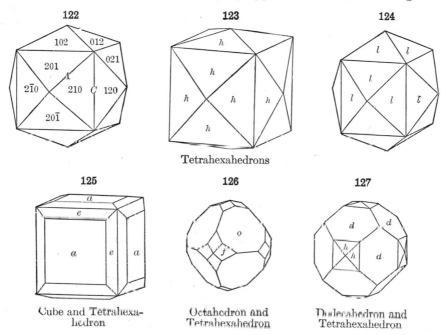

Tetrahexahedrons

Cube and Tetrahexahedron

Octahedron and Tetrahexahedron

Dodecahedron and Tetrahexahedron

symbol is (*hk*0), hence each face is parallel to one of the axes while it meets the other two axes at unequal distances which are definite multiples of each other. There are two kinds of edges, lettered *A* and *C* in Fig. 122; the interfacial angle of either edge is sufficient to determine the symbol of a given form (see below). The angles of some of the common forms are given on a later page (p. 79).

There may be a large number of tetrahexahedrons, as the ratio of the intercepts of the two axes, and hence of *h* to *k* varies; for example, (410), (310), (210), (320), etc. The form (210) is shown in Fig. 122; (410) in Fig. 123, and (530) in Fig. 124. All the tetrahexahedrons fall in a zone with a cubic face and a dodecahedral face. As *h* increases relatively to *k* the form approaches the cube (in which $h : k = \infty : 1$ or $1 : 0$), while as it diminishes and becomes more and more nearly equal to *k* in value it approaches the dodecahedron; for which $h = k$. Compare Fig. 123 and Fig. 124; also Figs. 107, 108. The special symbols belonging to each face of the tetrahexahedron should be carefully noted.

The faces of the tetrahexahedron bevel* the twelve similar edges of the cube, as in Fig. 125; they replace the solid angles of the octahedron by four faces inclined on the edges (Fig. 126; $f = 310$), and also the tetrahedral solid angles of the dodecahedron by four faces inclined on the faces (Fig. 127; $h = 410$).

64. Trisoctahedron. — The *trisoctahedron* (*triakisoctahedron*) (Fig. 128) is bounded by twenty-four similar faces; each of these is an isosceles triangle, and three together occupy the position of an octahedral face, whence the common name. Further, to distinguish it from the trapezohedron (or tetragonal trisoctahedron), it is sometimes called the *trigonal trisoctahedron*. There are two kinds of edges, lettered A and B in Fig. 128, and the interfacial angle corresponding to either is sufficient for the determination of the special symbol.

The general symbol is (*hhl*); common forms are (221), (331), etc. Each face of the trisoctahedron meets two of the axes at a distance less than unity

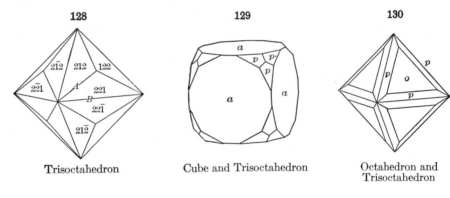

128	129	130
Trisoctahedron	Cube and Trisoctahedron	Octahedron and Trisoctahedron

and the third at the unit length, or (which is an identical expression†) it meets two of the axes at the unit length and the third at a distance greater than unity. The indices belonging to each face should be carefully noted. The normal interfacial angles for some of the more common forms are given on a later page.

65. Trapezohedron. — The *trapezohedron‡* (*icositetrahedron*) (Figs. 131, 132) is bounded by twenty-four similar faces, each of them a quadrilateral or trapezium. It also bears in appearance a certain relation to the octahedron, whence the name, sometimes employed, of *tetragonal trisoctahedron*. There are two kinds of edges, lettered B and C, in Fig. 131. The general symbol is *hll*; common forms are (311), (211), (322), etc. Of the faces, each cuts an axis at a distance less than unity, and the other two at the unit length, or (again, an identical expression) one of them intersects an axis at the unit length and the other two at equal distances greater than unity. The indices

---

* The word *bevel* is used when two like faces replace the edge of a form and hence are inclined at equal angles to its adjacent similar faces.

† Since $\frac{1}{2}a : \frac{1}{2}b : \frac{1}{4}c = 1a : 1b : 2c$. The student should read again carefully the explanations in Art. **40**.

‡ It will be seen later that the name trapezohedron is also given to other solids whose faces are trapeziums; conspicuously to the tetragonal trapezohedron and the trigonal trapezohedron.

belonging to each face should be carefully noted. The normal interfacial angles for some of the common forms are given on a later page.

66. The combinations of these forms with the cube, octahedron, etc., should be carefully studied. It will be seen (Fig. 129) that the faces of the trisoctahedron replace the solid angles of the cube as three faces equally

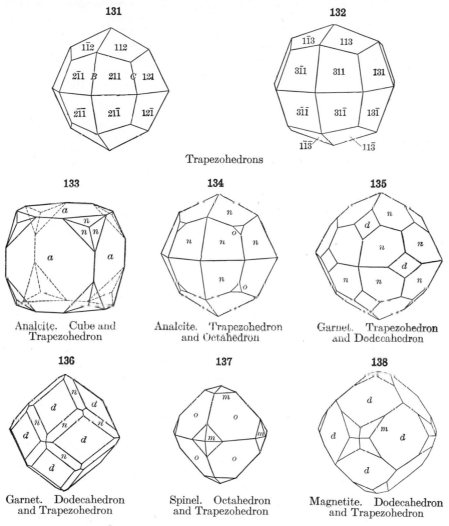

131

132

Trapezohedrons

133

Analcite. Cube and Trapezohedron

134

Analcite. Trapezohedron and Octahedron

135

Garnet. Trapezohedron and Dodecahedron

136

Garnet. Dodecahedron and Trapezohedron

137

Spinel. Octahedron and Trapezohedron

138

Magnetite. Dodecahedron and Trapezohedron

inclined on the *edges*; this is a combination which has not been observed on crystals. The faces of the trapezohedron appear as three equal triangles equally inclined to the cubic faces (Fig. 133).

Again, the faces of the trisoctahedron bevel the edges of the octahedron, Fig. 130, while those of the trapezohedron are triangles inclined to the faces at the extremities of the cubic axes, Fig. 137, $m(311)$. Still again, the faces of the trapezohedron n(211) truncate the edges of the dodecahedron (110), as shown in Fig. 136; this can be proved to follow at once from the zonal relations (Arts. **50, 51**), cf. also Figs. 143, 144. The position of the faces of the form m(311), in combination with o, is shown in Fig. 137; with d in Fig. 138.

It should be added that the trapezohedron n(211) is a common form both alone and in combination; m(311) is common in combination. The trisoctahedron alone is rarely met with, though in combination (Fig. 130) it is not uncommon.

67. Hexoctahedron. — The *hexoctahedron (hexakisoctahedron)*, Figs. 139, 140, is the general form in this system; it is bounded by forty-eight similar faces, each of which is a scalene triangle, and each intersects the three axes

139

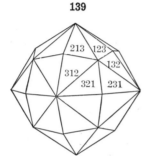

140

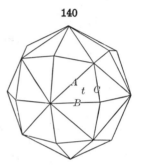

at unequal distances. The general symbol is (hkl); common forms are s(321), shown in Fig. 139, and t(421), in Fig. 140. The indices of the individual faces, as shown in Fig. 139 and more fully in the projections (Figs. 143, 144), should be carefully studied.

The hexoctahedron has three kinds of edges lettered A, B, C (longer, middle, shorter) in Fig. 140; the angles of two of these edges are needed to

141

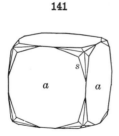

Fluorite. Cube and
Hexoctahedron

142

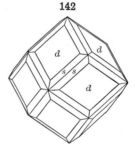

Garnet. Dodecahedron
and Hexoctahedron

fix the symbol unless the zonal relation can be made use of. In Fig. 142 the faces of the hexoctahedron bevel the dodecahedral edges, and hence for this form $h = k + l$; the form s has the special symbol (321). The hexoctahedron alone is a very rare form, but it is seen in combination with the cube (Fig. 141, fluorite) as six small faces replacing each solid angle. Fig. 142 is common with garnet.

68. Pseudo-symmetry in the Isometric System. — Isometric forms, by development in the direction of one of the cubic axes, simulate tetragonal forms. More common, and of greater interest, are forms simulating those of rhombohedral symmetry by extension, or by flattening, in the direction of an octahedral axis. Both these cases are illustrated later. Conversely, certain rhombohedral forms resemble an isometric octahedron in angle.

69. Stereographic and Gnomonic Projections. — The stereographic projection, Fig. 143, and gnomonic projection, Fig. 144, show the positions

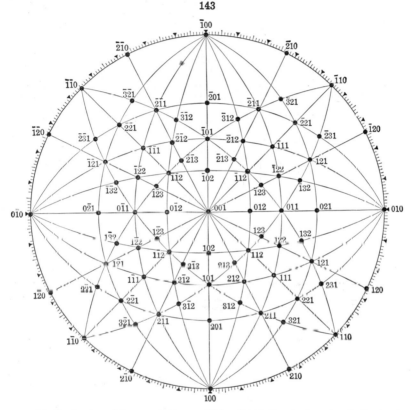

143

Stereographic Projection of Isometric Forms (Cube (100), Octahedron (111), Dodecahedron (110), Tetrahexahedron (210), Trisoctahedron (221), Trapezohedron (211), Hexoctahedron (321))

of the poles of the faces of the cube (100), octahedron (111), and dodecahedron (110); also the tetrahexahedron (210), the trisoctahedron (221), the trapezohedron (211), and the hexoctahedron (321). The isometric system is unique in that the angular relations of its forms with each other are constant, no matter upon what substance they may occur. Therefore the following projections and lists of angles will apply to all isometric crystals.

The student should study this projection carefully, noting the symmetry marked by the zones 100, 001, $\bar{1}$00, and 100, 010, $\bar{1}$00; also by 1$\bar{1}$0, 001, $\bar{1}$10; 1$\bar{1}$0, 00$\bar{1}$, $\bar{1}$10; 010, 10$\bar{1}$,

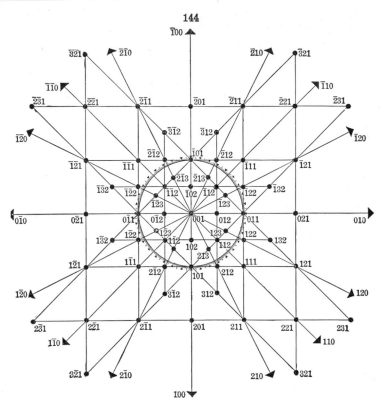

Gnomonic Projection of Isometric Forms (Cube (100), Octahedron (111), Dodecahedron (110), Tetrahexahedron (210), Trisoctahedron (221), Trapezohedron (211), Hexoctahedron (321))

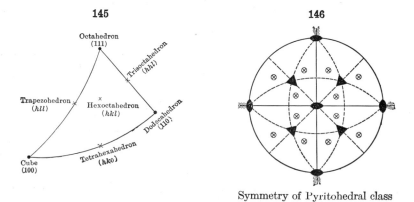

Symmetry of Pyritohedral class

0$\bar{1}$0; 010, $\bar{1}$01, 0$\bar{1}$0. Note further that the faces of a given form are symmetrically distributed about a cubic face, as 001; a dodecahedral face, as 101; an octahedral face, as 111.

Note further the symbols that belong to the individual faces of each form, comparing the projections with the figures which precede.

Finally, note the prominent *zones of planes*; for example, the zone between two cubic faces including a dodecahedral face and the faces of all possible tetrahexahedrons. Again, the zones from a cubic face (as 100) through an octahedral face (as 111) to a dodecahedral face (as 011) passing through the trapezohedrons, as 311, 211, 322, and the trisoctahedrons 233, 122, 133, etc. Also the zone from one dodecahedral face, as 110, to another, as 101, passing through 321, 211, 312, etc. At the same time compare these zones with the same zones shown on the figures already described. A study of the relations illustrated in Fig. 145 will be found useful. From it is seen that any crystal face falling in the zone between the cube and dodecahedron must belong to a tetrahexahedron; any face falling in the zone between the cube and octahedron must belong to a trapezohedron; and any face falling in the zone between the octahedron and dodecahedron must belong to a trisoctahedron, further, any face falling outside these three zones must belong to a hexoctahedron.

70. Angles of Common Isometric Forms.*

TETRAHEXAHEDRONS.

Cf. Fig. 122.	Edge A 210 ∧ 201, etc.	Edge C 210 ∧ 120, etc.	Angle on a(100)	Angle on o(111)
410	19° 45′	61° 55$\frac{3}{4}$′	14° 2$\frac{1}{4}$′	45° 33$\frac{3}{4}$′
310	25 50$\frac{1}{2}$	53 17$\frac{3}{4}$	18 26	43 5$\frac{1}{4}$
520	30 27	46 23$\frac{3}{4}$	21 48	41 22
210	36 52$\frac{1}{4}$	36 52$\frac{1}{4}$	26 34	39 14
530	42 40	28 4$\frac{1}{4}$	30 57$\frac{3}{4}$	37 37
320	46 11$\frac{1}{4}$	22 37$\frac{1}{4}$	33 41$\frac{1}{2}$	36 48$\frac{1}{2}$
430	50 12$\frac{1}{2}$	16 15$\frac{1}{2}$	36 52$\frac{1}{4}$	36 4$\frac{1}{4}$
540	52 25$\frac{3}{4}$	12 40$\frac{3}{4}$	38 39$\frac{1}{2}$	35 45$\frac{1}{2}$

TRISOCTAHEDRONS.

Cf. Fig 128.	Edge A 221 ∧ 212, etc.	Edge B 221 ∧ 221, etc.	Angle on a(100)	Angle on o(111)
332	17° 20$\frac{1}{2}$′	50° 28$\frac{3}{4}$′	50° 14$\frac{1}{4}$′	10° 1$\frac{1}{2}$′
221	27 16	38 56$\frac{1}{2}$	48 11	15 17$\frac{1}{2}$
552	33 33$\frac{1}{2}$	31 35$\frac{1}{4}$	47 7$\frac{1}{2}$	19 28$\frac{1}{4}$
331	37 51$\frac{1}{4}$	26 31$\frac{1}{4}$	46 30$\frac{1}{2}$	22 0
772	40 59	22 50$\frac{2}{3}$	46 7$\frac{1}{2}$	23 50$\frac{1}{2}$
441	43 20$\frac{1}{2}$	20 2$\frac{9}{4}$	45 52	25 14$\frac{1}{2}$

TRAPEZOHEDRONS.

Cf. Fig. 131.	Edge B 211 ∧ 2$\bar{1}$1, etc.	Edge C 211 ∧ 121, etc.	Angle on a(100)	Angle on o(111)
411	27° 16′	60° 0′	19° 28$\frac{1}{4}$′	35° 15$\frac{3}{4}$′
722	30 43$\frac{1}{2}$	55 50$\frac{3}{4}$	22 0	32 44
311	35 5$\frac{1}{4}$	50 28$\frac{3}{4}$	25 14$\frac{1}{4}$	29 29$\frac{3}{4}$
522	40 45	43 20$\frac{1}{2}$	29 29$\frac{3}{4}$	25 14$\frac{1}{4}$
211	48 11$\frac{1}{2}$	33 33$\frac{1}{2}$	35 15$\frac{3}{4}$	19 28$\frac{1}{4}$
322	58 2	19 45	43 18$\frac{3}{4}$	11 25$\frac{1}{4}$

HEXOCTAHEDRONS.

Cf. Fig. 139.	Edge A 321 ∧ 312, etc.	Edge B 321 ∧ 3$\bar{2}$1, etc.	Edge C 321 ∧ 231, etc.	Angle on a(100)	Angle on o(111)
421	17° 45$\frac{1}{4}$′	25° 12$\frac{1}{2}$′	35° 57′	29° 12$\frac{1}{4}$′	28° 6$\frac{1}{4}$′
531	27 39$\frac{1}{2}$	19 27$\frac{3}{4}$	27 39$\frac{1}{2}$	32 18$\frac{3}{4}$	28 33$\frac{3}{4}$
321	21 47$\frac{1}{4}$	31 0$\frac{1}{4}$	21 47$\frac{1}{4}$	36 42	22 12$\frac{1}{2}$
432	15 5$\frac{1}{2}$	43 36$\frac{1}{4}$	15 5$\frac{1}{2}$	42 1$\frac{3}{4}$	15 13$\frac{1}{2}$
431	32 12$\frac{1}{4}$	22 37$\frac{1}{4}$	15 56$\frac{1}{2}$	38 19$\frac{3}{4}$	25 4

* A fuller list is given in the Introduction to Dana's System of Mineralogy, 1892, pp. xx–xxiii.

2. PYRITOHEDRAL CLASS (2). PYRITE TYPE

(*Dyakisdodecahedral, Pentagonal Hemihedral, Diploidal, or Tesseral Central Class*)

71. Typical Forms and Symmetry. 3 xl. Ax.-2 ; 4 diag. Ax.-3 ; 3 xl. P. ; C. — The typical forms of the pyritohedral class are the *pyritohedron*, or pentagonal dodecahedron, Figs. 147, 148, and the *diploid*, or dyakisdodeca-hedron, Fig. 153. The symmetry of these forms, as of the class as a whole, is as follows: The three crystallographic axes are axes of binary symmetry only; there are also four diagonal axes of trigonal symmetry coinciding with the octahedral axes. There are but three planes of symmetry; these coincide with the planes of the crystallographic axes and are parallel to the faces of the cube.

The stereographic projection in Fig. 146 shows the distribution of the faces of the general form (*hkl*), diploid, and thus exhibits the symmetry of the class. This should be carefully compared with the corresponding pro-jection (Fig. 109) for the normal class, so that the lower grade of symmetry here present may be thoroughly understood. In studying the forms described and illustrated in the following pages, this matter of symmetry, especially in relation to that of the normal class, should be continually before the mind.

It will be observed that the faces of both the pyritohedron (Fig. 147) and the diploid (Fig. 153) are arranged in parallel pairs, and on this account these forms have been sometimes called *parallel hemihedrons*. Further, those authors who prefer to describe these forms as cases of hemihedrism call this type parallel-faced hemihedrism or pentagonal hemihedrism.

72. Pyritohedron. — The *pyritohedron* (Fig. 147) is so named because it is a typical form with the common species, pyrite. It is a solid bounded

147   148   149

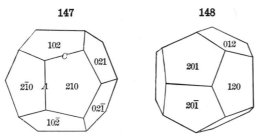
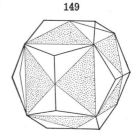

Pyritohedrons   Showing Relation between Pyrito-hedron and Tetrahexahedron

by twelve faces, each of which is a pentagon, but with one edge (*A*, Fig. 147) longer than the other four similar edges (*C*). It is often called a *pentagonal dodecahedron*, and indeed it resembles closely the regular dodecahedron of geometry, in which the faces are regular pentagons. This latter form is, however, an impossible form in crystallography.

The general symbol is (*hk*0) or like that of the tetrahexahedron of the normal class. Hence each face is parallel to one of the axes and meets the other two axes at unequal distances. Common forms are (410), (310), (210), (320), etc. Besides the positive pyritohedron, as (210), there is also the complementary negative form* shown in Fig. 148; the symbol is here (120). Other common forms are (250), (230), (130), etc.

The positive and negative pyritohedrons together embrace twenty-four faces, having the same position as the twenty-four like faces of the tetrahexahedron of the normal class. The relation between the tetrahexahedron and the pyritohedron is shown in Fig. 149, where the alternate faces of the tetrahexahedron (indicated by shading) are extended to form the faces of the pyritohedron.

73. Combinations. — The faces of the pyritohedron replace the edges of the cube as shown in Fig. 150; this resembles Fig. 119 but here the faces

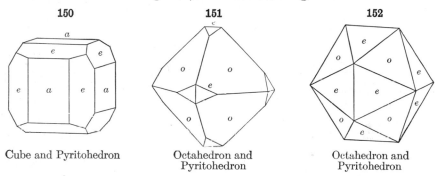

150	151	152
Cube and Pyritohedron	Octahedron and Pyritohedron	Octahedron and Pyritohedron

make unequal angles with the two adjacent cubic faces. On the other hand, when the pyritohedron is modified by the cube, the faces of the latter truncate the longer edges of the pentagons.

Fig. 151 shows the combination of the pyritohedron and octahedron, and in Fig. 152 these two forms are equally developed. The resulting combination bears a close similarity to the *icosahedron*, or regular twenty-faced solid, of geometry. Here, however, of the twenty faces, the eight octahedral are equilateral triangles; the twelve others belonging to the pyritohedron are isosceles triangles.

153

74. Diploid. — The *diploid* is bounded by twenty-four similar faces, each meeting the axes at unequal distances; its general symbol is hence (hkl), and common forms are $s(321)$, $t(421)$, etc. The form (321) is shown in Fig. 153; the symbols of its faces, as given, should be carefully studied. As seen in the figure, the faces are quadrilaterals or trapeziums; moreover, they are grouped in pairs, hence the common name diploid. It is also sometimes called a *dyakisdodecahedron*.

The complementary negative form bears to the positive form of Fig. 153 the same relation as the negative to the positive pyritohedron. Its faces have the symbols 312, 231, 123, in the front octant, and similarly with the proper negative signs in the others. The positive and negative forms together obviously embrace all the faces of the hexoctahedron of the normal class. The diploid can be considered to be

Diploid

---

* The negative forms in this and similar cases have sometimes distinct letters, sometimes the same as the positive form, but are then distinguished by a subscript accent, as $e(210)$ and $e_1(120)$.

derived from the hexoctahedron by the extension of the alternate faces of the latter and the omission of the remaining faces, exactly as in the case of the pyritohedron and tetrahexahedron (Art. **72**).

In Fig. 154 the positive diploid is shown in combination with the cube. Here the three faces replace each of its solid angles. This combination form

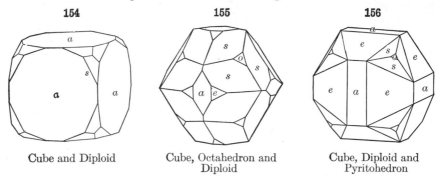

154	**155**	**156**
Cube and Diploid	Cube, Octahedron and Diploid	Cube, Diploid and Pyritohedron

resembles that of Fig. 129, but the three faces are here unequally inclined upon two adjacent cubic faces. Other combinations of the diploid with the cube, octahedron, and pyritohedron are given in Figs. 155 and 156.

75. Other Forms. — If the pyritohedral type of symmetry be applied to planes each parallel to two of the axes, it is seen that this symmetry calls for six of these, and the resulting form is obviously a cube. This cube cannot be distinguished geometrically from the cube of the normal class, but it has its own characteristic molecular symmetry. Corresponding to this it is common to find cubes of pyrite with fine lines (striations) parallel to the alternate edges, as indicated in Fig. 157. These are due to the partial development of pyritohedral faces (210). On a normal cube similar striations, if present, must be parallel to both sets of edges on each cubic face.

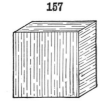

157

Pyrite.   Striated Cube

Similarly to the cube, the remaining forms of this pyritohedral class, namely, (111), (110), (hhl), (hll), have the same geometrical form, respectively, as the octahedron, dodecahedron, the trisoctahedrons and trapezohedrons of the normal class. In molecular structure, however, these forms are distinct, each having the symmetry described in Art. **71**.

76. Angles. — The following tables contain the angles of some common forms.

PYRITOHEDRONS.

Cf. Fig. 147.	Edge A 210 ∧ 2̄10, etc.	Edge C 210 ∧ 102, etc.	Angle on $a(100)$	Angle on $o(111)$
410	28° 4¼′	76° 23½′	14° 2¼′	45° 33¾′
310	36 52¼	72 32½	18 26	43 5¼
520	43 36½	69 49¾	21 48	41 22
210	53 7¾	66 25¼	26 34	39 14
530	61 55¾	63 49¼	30 57¾	37 37
320	67 22¾	62 30¾	33 41½	36 48½
430	73 44½	61 19	36 52¼	36 4¼
540	77 19¼	60 48¼	38 39½	35 45½
650	79 36¾	60 32½	39 48¼	35 35¾

DIPLOIDS.

Cf. Fig. 153.	Edge A 321 ∧ 3̄21, etc.	Edge B 321 ∧ 32̄1, etc.	Edge C 321 ∧ 213, etc.	Angle on a(100)	Angle on o(111)
421	51° 45¼′	25° 12½′	48° 11½′	29° 12¼′	28° 6½′
532	58 14½	37 51¾	35 20	35 47¾	20 30¾
531	60 56½	19 27¾	19 27¾	32 18¾	28 33¾
851	63 36¾	12 6	53 55¼	32 30¾	31 34
321	64 37½	31 0¼	38 12¼	36 42	22 12½
432	67 42½	43 36¼	26 17½	42 1¾	15 13½
431	72 4¾	22 37¼	43 3	38 19¾	25 4

3. TETRAHEDRAL CLASS (3). TETRAHEDRITE TYPE

(Hextetrahedral, Tetrahedral Hemihedral, or Ditesseral Polar Class)

77. Typical Forms and Symmetry. 3 xl. Ax.-2; 4 diag. Ax.-3; 6 diag. P.
— The typical form of this class, and that from which it derives its name, is the *tetrahedron*, shown in Figs. 159, 160. There are also three other distinct forms, shown in Figs. 167–169.

The symmetry of this class is as follows: There are three axes of binary symmetry which coincide with the crystallographic axes. There are also four diagonal axes of trigonal symmetry which coincide with the octahedral axes. There are six diagonal planes of symmetry. There is no center of symmetry.

The stereographic projection, (Fig. 158), shows the distribution of the faces of the general form (*hkl*), hextetrahedron, and thus exhibits the symmetry of the class. It will be seen at once that the like faces are all grouped in the *alternate octants*, and this will be seen to be characteristic of all the forms peculiar to this class. The relation between the symmetry here described and that of the normal class must be carefully studied.

In distinction from the pyritohedral forms whose faces were in parallel pairs, the faces of the tetrahedron and the analogous solids are inclined to each other, and hence they are some-

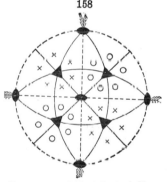

158

Symmetry of Tetrahedral Class

times spoken of as *inclined hemihedrons*, and the type of so-called hemihedrism here illustrated is then called inclined or tetrahedral hemihedrism.

78. Tetrahedron. — The tetrahedron,* as its name indicates, is a four-
faced solid, bounded by planes meeting the axes at equal distances. Its general symbol is (111), and the four faces of the positive form (Fig. 159) have the symbols 111, 1̄1̄1, 11̄1̄, 1̄11̄. These correspond to four of the faces of the octahedron of the normal class (Fig. 111). The relation between the two forms is shown in Fig. 161.

Each of the four faces of the tetrahedron is an equilateral triangle; the (normal) interfacial angle is 109° 28′ 16″. The tetrahedron is the regular

---

* This is one of the five regular solids of geometry, which include also the cube, octa-
hedron, the regular pentagonal dodecahedron, and the icosahedron; the last two, as already noted, are impossible forms among crystals.

triangular pyramid of geometry, but crystallographically it must be so placed that the axes join the middle points of opposite edges, and one axis is vertical.

There are two possible tetrahedrons: the positive tetrahedron (111). designated by the letter o, which has already been described, and the nega-

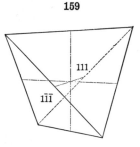

159

Positive Tetrahedron

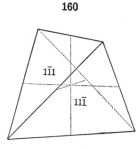

160

Negative Tetrahedron

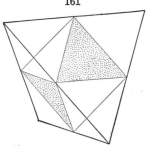

161

Showing Relation between
Octahedron and Tetrahedron

tive tetrahedron, having the same geometrical form and symmetry, but the indices of its four faces are $\bar{1}11$, $1\bar{1}1$, $11\bar{1}$, $\bar{1}\bar{1}\bar{1}$. This second form is shown in Fig. 160; it is usually designated by the letter $o_{\prime}$. These two forms are,

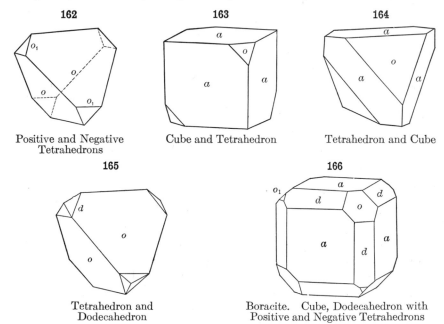

162

Positive and Negative
Tetrahedrons

163

Cube and Tetrahedron

164

Tetrahedron and Cube

165

Tetrahedron and
Dodecahedron

166

Boracite. Cube, Dodecahedron with
Positive and Negative Tetrahedrons

as stated above, identical in geometrical shape, but they may be distinguished in many cases by the tests which serve to reveal the molecular structure, particularly the etching-figures; also in many cases by pyroelectricity (see under boracite, p. 335), Art. **450.** It is probable that the positive and negative tetrahedrons of sphalerite (see that species) have a constant difference in this particular, which makes it possible to distinguish them on crystals from different localities and of different habit.

If both tetrahedrons are present together, the form in Fig. 162 results. This is geometrically an octahedron when the two forms are equally developed, but crystallographically it is always only a combination of two unlike forms, the positive and negative tetrahedrons, which can be distinguished as already noted.

The tetrahedron in combination with the cube replaces the alternate solid angles as in Fig. 163. The cube modifying the tetrahedron truncates its edges as shown in Fig. 164. The normal angle between adjacent cubic and tetrahedral faces is $54° 44'$. In Fig. 165 the dodecahedron is shown modifying the positive tetrahedron, while in Fig. 166 the cube is the predominating form with the positive and negative tetrahedrons and dodecahedron.

79. Other Typical Forms. — There are three other distinct types of solids in this class, having the general symbols (hhl), (hll), and (hkl). The first of these is shown in Fig. 167; here the symbol is (221). There are twelve

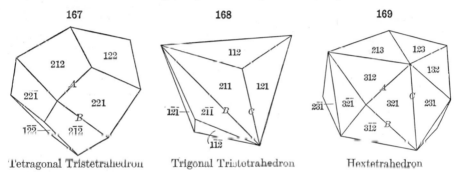

167	168	169

Tetragonal Tristetrahedron   Trigonal Tristetrahedron   Hextetrahedron

faces, each a quadrilateral, belonging to this form, distributed as determined by the tetrahedral type of symmetry. They correspond to twelve of the faces of the trisoctahedron, namely, all those falling in alternate octants. This type of solid is sometimes called a *tetragonal tristetrahedron*, or a *deltoid dodecahedron*. It does not occur alone among crystals, but its faces are observed modifying other forms.

There is also a complementary negative form, corresponding to the positive form, related to it in precisely the same way as the negative to the positive tetrahedron. Its twelve faces are those of the trisoctahedron which belong to the other set of alternate octants.

Another form, shown in Fig. 168, has the general symbol (hll), here (211); it is bounded by twelve like triangular faces, distributed after the type demanded by tetrahedral symmetry, and corresponding consequently to the faces of the alternate octants of the form (hll) — the trapezohedron — of the normal class. This type of solid is sometimes called a *trigonal tristetrahedron* or *triakistetrahedron*.* It is observed both alone and in combination, espe-

* It is to be noted that the tetragonal tristetrahedron has faces which resemble those of the trapezohedron (tetragonal trisoctahedron), although it is related not to this but to the trisoctahedron (trigonal trisoctahedron). On the other hand, the faces of the trigonal tristetrahedron resemble those of the trisoctahedron, though in fact related to the trapezohedron.

cially with the species tetrahedrite; it is much more common than the form
(*hhl*). There is here again a complementary negative form. Fig. 170
shows the positive form *n*(211) with the positive tetrahedron, and Fig. 171
the form *m*(311) with *a*(100), *o*(111), and *d*(110). In Fig. 172, the negative
form *n*,(2̄11) is present.

The fourth independent type of solids in this class is shown in Fig. 169.
It has the general symbol (*hkl*), here (321), and is bounded by twenty-four
faces distributed according to tetrahedral symmetry, that is, embracing all
the faces of the alternate octants of the forty-eight-faced hexoctahedron.
This form is sometimes called a *hextetrahedron* or *hexakistetrahedron*. The
complementary negative form (*h̄kl*) embraces the remaining faces of the
hexoctahedron. The positive hextetrahedron, *v*(531), is shown in Fig. 172
with the cube, octahedron, and dodecahedron, also the negative trigonal
tristetrahedron *n*,(2̄11).

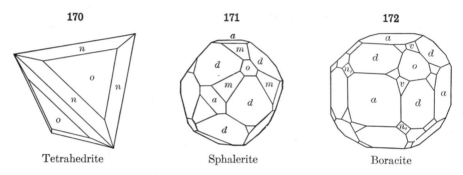

170   171   172

Tetrahedrite   Sphalerite   Boracite

80. If the tetrahedral symmetry be applied in the case of planes each
parallel to the two axes, it will be seen that there must be six such faces.
They form a *cube* similar geometrically to the cube both of the normal and
pyritohedral class but differing in its molecular structure, as can be readily
proved, for example, by pyroelectricity (Art. **450**). Similarly in the case
of the planes having the symbol (110), there must be twelve faces forming a
rhombic dodecahedron bearing the same relation to the like geometrical
form of the normal class. The same is true again of the planes having the
position expressed by the general symbol (*hk*0); there must be twenty-four
of them and they together form a tetrahexahedron.

In this class, therefore, there are also seven types of forms, but only four
of them are geometrically distinct from the corresponding forms of the
normal class.

81. Angles. — The following tables contain the angles of some common forms:

TETRAGONAL TRISTETRAHEDRONS.

Cf. Fig. 167.	Edge A 221 ∧ 2̄12, etc.	Edge B 221 ∧ 2̄1̄2, etc.	Angle on *a*(100)	Angle on *o*(111)
332	17° 20½′	97° 50¼′	50° 14¼′	10° 1½′
221	27 16	90 0	48 11½	15 47½
552	33 33½	84 41	47 7½	19 28⅔
331	37 51¾	80 55	46 30½	22 0

Trigonal Tristetrahedrons.

Cf. Fig. 168.	Edge B 211 ∧ 2$\bar{1}\bar{1}$, etc.	Edge C 211 ∧ $\bar{1}$21, etc.	Angle on a(100)	Angle on o(111)
411	38° 56$\frac{1}{2}$′	60° 0′	19° 28$\frac{1}{4}$′	35° 15$\frac{3}{4}$′
722	44 0$\frac{1}{4}$	55 50$\frac{3}{4}$	22 0	32 44
311	50 28$\frac{3}{4}$	50 28$\frac{3}{4}$	25 14$\frac{1}{4}$	29 29$\frac{3}{4}$
522	58 59$\frac{1}{4}$	43 20$\frac{1}{4}$	29 29$\frac{3}{4}$	25 14$\frac{1}{4}$
211	70 31$\frac{3}{4}$	33 33$\frac{1}{2}$	35 15$\frac{3}{4}$	19 28$\frac{3}{4}$
322	86 37$\frac{3}{4}$	19 45	43 18$\frac{3}{4}$	11 25$\frac{1}{4}$

Hextetrahedrons.

Cf. Fig. 169.	Edge A 321 ∧ 3$\bar{1}$2, etc.	Edge B 321 ∧ 3$\bar{1}\bar{2}$, etc.	Edge C 321 ∧ 231, etc.	Angle on a(100)	Angle on o(111)
531	27° 39$\frac{2}{3}$′	57° 7$\frac{1}{3}$′	27° 39$\frac{2}{3}$′	32° 18$\frac{2}{3}$′	28° 33$\frac{2}{3}$′
321	21 47$\frac{1}{4}$	69 4$\frac{1}{2}$	21 47$\frac{1}{4}$	36 42	22 12$\frac{1}{2}$
432	15 5$\frac{1}{2}$	82 4$\frac{1}{3}$	15 5$\frac{1}{2}$	42 1$\frac{3}{4}$	15 13$\frac{1}{2}$
431	32 12$\frac{1}{4}$	67 22$\frac{3}{4}$	15 56$\frac{1}{2}$	38 19$\frac{3}{4}$	25 4

4. PLAGIOHEDRAL CLASS (4). CUPRITE TYPE

(Pentagonal Icositetrahedral, Plagiohedral Hemihedral, Gyroidal, or Tesseral Holoaxial Class)

82. Typical Forms and Symmetry. 3 xl. Ax.-4; 4 diag. Ax.-3; 6 diag. Ax.-2. — The fourth class under the isometric system is called the plagiohedral or gyroidal class because the faces of the general form (hkl) are arranged in spiral order. This is shown on the stereographic projection, Fig. 173, and also in Figs. 174, 175, which represent the single typical form of the class. These two complementary solids together embrace all the faces of the hexoctahedron. They are distinguished from one another by being called respectively right-handed and left-handed *pentagonal icositetrahedrons*. The other forms of the class are geometrically like those of the normal class.

173

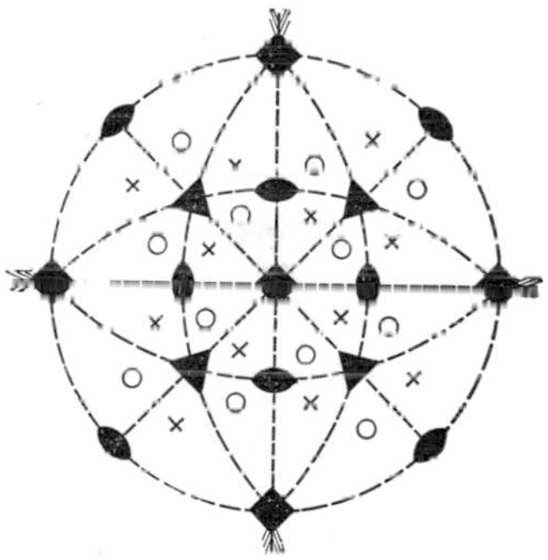

Symmetry of Plagiohedral Class

The symmetry characteristic of the class in general is as follows:

There are no planes of symmetry and no center of symmetry. There are, however, three axes of tetragonal symmetry normal to the cubic faces, four axes of trigonal symmetry normal to the octahedral faces, and six axes of binary symmetry normal to the faces of the dodecahedron. In other words, it has all the axes of symmetry of the normal class while without planes or center of symmetry.

83. It is to be noted that the two forms shown in Figs. 174, 175 are alike geometrically, but are not superposable; in other words, they are related to one another as is a right- to a left-hand glove. They are hence said to be *enantiomorphous*, and, as explained elsewhere, the crystals belonging here may be expected to show circular polarization of light. It will be seen that the complementary positive and negative forms of the preceding classes, unlike those here, may be superposed by being rotated 90° about one of the crystallographic axes. This distinction between positive and negative forms, and between right- and left-handed enantiomorphous forms, exists also in the case of the classes of several of the other systems.

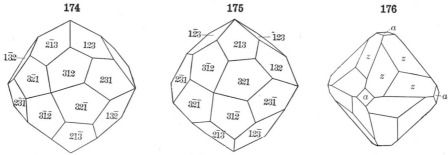

Right- and Left-handed Pentagonal Icositetrahedrons   Cuprite

This class is rare among minerals; it is represented by cuprite and sal ammoniac. It is usually shown by the distribution of the small modifying faces, or by the form of the etching figures. Fig. 176 shows a crystal of cuprite from Cornwall (Pratt) with the form $z(13 \cdot 10 \cdot 12)$.

5. TETARTOHEDRAL CLASS (5). ULLMANNITE TYPE

(*Tetrahedral-Pentagonal Dodecahedral or Tesseral Polar Class*)

84. Symmetry and Typical Forms. 3 xl. Ax.-2; 4 diag. Ax.-3. — The fifth remaining possible class under the isometric system is illustrated by Fig. 178, which represents the twelve-faced solid corresponding to the general

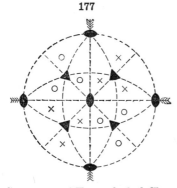

Symmetry of Tetartohedral Class

symbol (hkl). The distribution of its faces is shown in the projection, Fig. 177. This form is sometimes called a *tetrahedral-pentagonal dodecahedron*. It is seen to have one fourth as many faces as the form (hkl) in the normal class, hence there are four similar solids which together embrace all the faces of the hexoctahedron. These four solids, which are distinguished as right-handed (positive and negative) and left-handed (positive and negative), are enantiomorphous, like those of Figs. 174 and 175, and hence the salts crystallizing here may be expected to also show circular polarization. The remaining forms of the class are (besides the cube and rhombic dodecahedron) the tetrahedrons, the pyritohedrons, the tetragonal and trigonal tristetrahedrons; geometrically they are like the solids of the same names already described. This class has no plane of symmetry and no center of symmetry. There are three axes of binary symmetry normal to the cubic faces, and four axes of trigonal symmetry normal to the faces of the tetrahedron.

This group is illustrated by artificial crystals of barium nitrate, strontium nitrate, sodium chlorate, etc. Further, the species ullmannite, which shows sometimes pyritohedral and again tetrahedral forms, both having the same composition, must be regarded as belonging here.

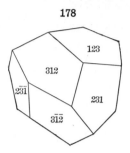

178

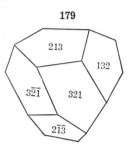

179

Mathematical Relations of the Isometric System

85. Most of the problems arising in the isometric system can be solved at once by the right-angled triangles in the sphere of projection (Fig. 143) without the use of any special formulas.

It will be remembered that the angles between a cubic face, as 100, and the adjacent face of a tetrahexahedron, 310, 210, 320, etc., can be obtained at once, since the tangent of this angle is equal to $\frac{1}{3}, \frac{1}{2}, \frac{2}{3}$, or in general $\frac{k}{h}$:

$$\tan (hk0 \wedge 100) = \frac{k}{h}$$

$$ac = k = 1$$
$$bc = h = 2$$
$$\angle\, adc = 90°$$

$$\tan \angle abc = \frac{ac}{bc} = \frac{k}{h} = \frac{1}{2}$$

$$\left.\begin{array}{c}\angle\, abc \\ (100) \wedge (210)\end{array}\right\} = 26°\,34'$$

This relation is illustrated in Fig. 180, which also shows the method of graphically determining the indices of a tetrahexahedron, the angle between one of its faces and an adjacent cube face being given.

Since all the forms of a given symbol under different species have the same angles, the tables of angles already given are very useful.

These and similar angles may be calculated immediately from the sphere, or often more simply by the formulas given in the following article.

86. Formulas. — (1) The distance of the pole of any face $P(hkl)$ from the cubic faces is given by the following equations. Here Pa is the distance between (hkl) and (100); Pb is the distance between (hkl) and (010); and Pc that between (hkl) and (001).

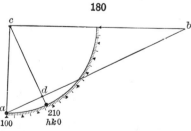

180

These equations admit of much simplification in the various special cases, for $(hk0)$, (hhl), etc.:

$$\cos^2 Pa = \frac{h^2}{h^2 + k^2 + l^2}; \quad \cos^2 Pb = \frac{k^2}{h^2 + k^2 + l^2}; \quad \cos^2 Pc = \frac{l^2}{h^2 + k^2 + l^2}.$$

(2) The distance between the poles of any two faces $P(hkl)$ and $Q(pqr)$ is given by the following equation, which in special cases may also be more or less simplified:

$$\cos PQ = \frac{hp + kq + lr}{\sqrt{(h^2 + k^2 + l^2)(p^2 + q^2 + r^2)}}.$$

(3) The calculation of the supplement interfacial or normal angles for the several forms may be accomplished as follows:

Trisoctahedron. — The angles A and B are, as before, the supplements of the interfacial angles of the edges lettered as in Fig. 128.

$$\cos A = \frac{h^2 + 2hl}{2h^2 + l^2}; \qquad \cos B = \frac{2h^2 - l^2}{2h^2 + l^2}.$$

For the *tetragonal-tristetrahedron* (Fig. 167), $\cos B = \dfrac{h^2 - 2hl}{2h^2 + l^2}.$

Trapezohedron (Fig. 131). B and C are the supplement angles of the edges as lettered in the figure.

$$\cos B = \frac{h^2}{h^2 + 2l^2}; \qquad \cos C = \frac{2hl + l^2}{h^2 + 2l^2}.$$

For the *trigonal-tristetrahedron* (Fig. 168), $\cos B = \dfrac{h^2 - 2l^2}{h^2 + 2l^2}.$

Tetrahexahedron (Fig. 122).

$$\cos A = \frac{h^2}{h^2 + k^2}; \qquad \cos C = \frac{2hk}{h^2 + k^2}.$$

For the *pyritohedron* (Fig. 147), $\cos A = \dfrac{h^2 - k^2}{h^2 + k^2};$ $\qquad \cos C = \dfrac{hk}{h^2 + k^2}.$

Hexoctahedron (Fig. 140).

$$\cos A = \frac{h^2 + 2kl}{h^2 + k^2 + l^2}; \qquad \cos B = \frac{h^2 + k^2 - l^2}{h^2 + k^2 + l^2}; \quad \cos C = \frac{2hk + l^2}{h^2 + k^2 + l^2};$$

For the *diploid* (Fig. 153), $\cos A = \dfrac{h^2 - k^2 + l^2}{h^2 + k^2 + l^2};$ $\quad \cos C = \dfrac{kl + lh + hk}{h^2 + k^2 + l^2}.$

For the *hextetrahedron* (Fig. 169), $\cos B = \dfrac{h^2 - 2kl}{h^2 + k^2 + l^2};$

87. To determine the indices of any face (hkl) of an isometric form, given the position of its pole on the stereographic projection. — As an illustrative example of this problem the hexoctahedron (321) has been taken.   It is assumed that the angles 100 $\wedge$ 321 = 36° 42′ and 111 $\wedge$ 321 = 22° 12′ have been given.   The methods by which the desired pole is located from these measurements have been described on page 53 and are illustrated in Fig. 181. Having located the pole (hkl) a line is drawn through it from the center O of the projection.   This line O–P represents the intersection with the horizontal plane (which is the plane of the horizontal crystal axes, a and b) of a plane which is normal to the crystal face (hkl).   Since two planes which are at right angles to each other will intersect a third plane in lines that are at right angles to each other, it follows that the plane of the hexoctahedral face will intersect the plane of the horizontal axes in a line at right angles to O–P.   If, therefore, the distance O–M be taken as representing unity on the a axis and the line M–P–N be drawn at right angles to O–P the distance O–N will represent the intercept of the face in question upon the b axis. O–N is found in this case to be $\frac{3}{2}O$–M in value. The intercepts upon the two horizontal axes are, therefore, $1a$, $\frac{3}{2}b$.   The plotting of the intercept upon the c axis is shown in the upper left-hand quadrant of the figure.   The angular distance from O to the pole (hkl) is measured by the stereographic protractor as 74° 30′.   This angle is then laid off from the line representing the c axis and the line representing the pole (hkl) is drawn.   The distance O–P is transferred from the lower part of the figure.   Then we can construct the right triangle, the vertical side of which is the c axis, the horizontal side is this line O–P (the intersection of the plane which is normal to the crystal face with the horizontal plane) and the hypothenuse is a line lying in the face and therefore at right angles to the pole of the face.   This line would intersect the c axis at a distance equal to $3O–M$. The same relation may be shown by starting this last line from a point on the c axis which is at a distance from the center of the figure equal to $O–M$. In this case the intercept on the horizontal line $O–P$ would be at one third its total length. By these constructions the parameters of the face in question are shown to be $1a$, $\frac{3}{2}b$, $3c$, giving (321) as its indices.

181

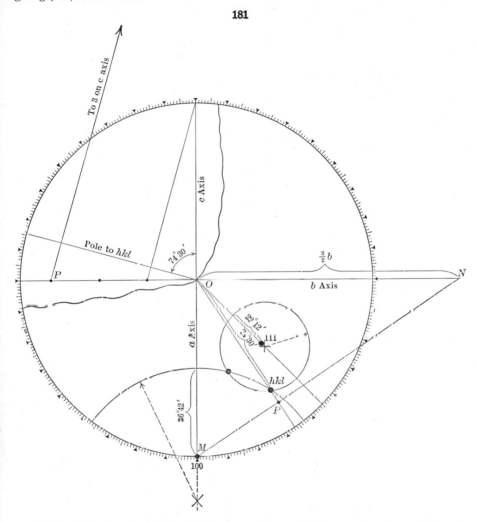

88. To determine the indices of the faces of isometric forms, given the positions of their poles on the gnomonic projection. — As an illustrative example of this problem the lower right-hand quadrant of the gnomonic projection of isometric forms, Fig. 144, has been taken and reproduced in Fig. 182. The lines $O–M$ and $O–N$ are at right angles to each other and may represent the horizontal crystallographic axes a_1 and a_2. If from each pole of the projection lines are drawn perpendicular to these two axial directions it will be seen that the intercepts made upon these lines have rational relations to each other. And since we are dealing with the isometric system in which the crystallographic axes are all alike and interchangeable with each other, it follows that the different intercepts upon $O–M$ and $O–N$ are identical. The distance $O–R$ (*i.e.* the distance from the center to the

45° point of the projection) must equal the unit length of the axes.   That this is true is readily seen by the consideration of Fig. 183.   The intercepts of the lines drawn from the different poles to the lines O–M and O–N are found to be $\frac{1}{3}$, $\frac{1}{2}$, $\frac{2}{3}$, 1, $\frac{3}{2}$, 2 and 3 times this unit distance.   To find the Miller indices of any face represented, it is only necessary to

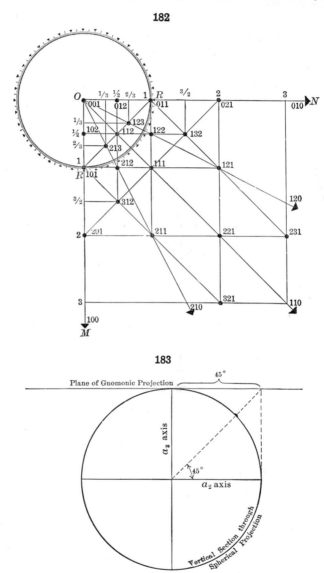

182

183

take the intercepts of the two lines drawn from its pole upon the two axes a_1 and a_2, place these numbers in their proper order and add a 1 as a third figure and then if necessary clear of fractions.   Take for example the hexoctahedron face with indices 312.   The lines drawn from its pole intercept the axes at $\frac{2}{3}a_1$ and $\frac{1}{2}a_2$, which gives the expression $\frac{2}{3}\,\frac{1}{2}\,1$, which, again, on clearing of fractions, yields 312, the indices of the face in question.   In the case of a face parallel to the vertical axis, the pole of which lies at infinity on the gnomonic projection, the indices may be obtained by taking any point on the radial line that points to the position of the pole and dropping perpendiculars to the lines representing the two horizontal axes. The relative intercepts formed upon these axes will give the first two numbers of the required indices while the third number will necessarily be 0.

II. TETRAGONAL SYSTEM

89. The tetragonal system includes all the forms which are referred to three axes at right angles to each other of which the two horizontal axes are equal to each other in length and interchangeable and the third, the vertical axis, is either shorter or longer. The horizontal axes are designated by the letter a; the vertical axis by c (see Fig. 184). The length of the vertical axis expresses properly the axial ratio of $a : c$, a being uniformly taken as equal to unity. The axes are orientated and their opposite ends designated by plus and minus signs exactly as in the case of the isometric system.

Seven classes are embraced in this system. Of these the normal class is common and important among minerals; two others have several representatives, and another a single one only. It may be noted that in four of the classes the vertical axis is an axis of tetragonal symmetry; in the remaining three it is an axis of binary symmetry only.

1. NORMAL CLASS (6). ZIRCON TYPE

(Ditetragonal Dipyramidal, Holohedral, or Ditetragonal Equatorial Class)

90. Symmetry. Vert. Ax.-4; 4 hor. Ax.-2; hor. P.; 4 vert. P.; C.—
The forms belonging to the normal class of the tetragonal system (cf. Figs. 188 to 210) have one principal axis of tetragonal symmetry (whence name of

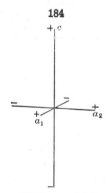

184

Axes of Tetragonal Mineral,
Octahedrite $a : c = 1 : 1.78$

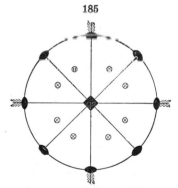

185

Symmetry of Normal Class
Tetragonal System

the system) which coincides with the vertical crystallographic axis, c. There are also four horizontal axes of binary symmetry, two of which coincide with the horizontal crystallographic axes while the other two are diagonal axes bisecting the angles between the first two.

Further, they have one principal plane of symmetry, the plane of the horizontal crystallographic axes. There are also four vertical planes of symmetry which pass through the vertical crystallographic axis c and make angles of 45° with each other. Two of these latter planes include the horizontal crystallographic axes and are known as axial planes of symmetry. The other two are known as diagonal planes of symmetry.

The axes and planes of symmetry are shown in Figs. 186 and 187.

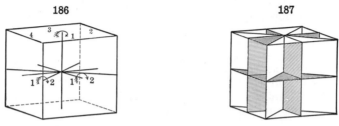

Symmetry of Normal Class, Tetragonal System

The symmetry and the distribution of the faces of the general form, hkl, is shown in the stereographic projection, Fig. 185.

91. Forms. — The various possible forms under the normal class of this system are as follows:

Symbols

1. Base or basal pinacoid............(001)
2. Prism of the first order...........(110)
3. Prism of the second order.........(100)
4. Ditetragonal prism................($hk0$) as, (310); (210); (320), etc.
5. Pyramid of the first order.........(hhl) as, (223); (111); (221), etc.
6. Pyramid of the second order......($h0l$) as, (203); (101); (201), etc.
7. Ditetragonal pyramid.............(hkl) as, (421); (321); (122), etc.

92. Base or Basal Pinacoid. — The *base* is that form which includes the two similar faces which are parallel to the plane of the horizontal axes. These faces have the indices 001 and 00$\bar{1}$ respectively; it is an " open form," as they do not inclose a space, consequently this form can occur only in combination with other forms. Cf. Figs. 188–191, etc. This form is always lettered c in this work.

93. Prisms. — Prisms, in systems other than the isometric, have been defined to be forms whose faces are parallel to the vertical axis (c) of the crystal, while they meet the two horizontal axes; in this system the four-faced form whose planes are parallel both to the vertical and one horizontal axis is also called a prism. There are hence three types of prisms here included.

94. Prism of First Order. — The *prism of the first order* includes the four faces which, while parallel to the vertical axis, meet the horizontal axes at equal distances; its general symbol is consequently (110). It is a *square prism*, with interfacial angles of 90°. It is shown in combination with the base in Fig. 188. It is uniformly designated by the letter m. The indices of its faces, taken in order, are 110, $\bar{1}$10, $\bar{1}\bar{1}$0, 1$\bar{1}$0.

95. Prism of Second Order. — The *prism of the second order* shown*
in combination with the base in Fig. 189 includes the four faces which are
parallel at once to the vertical and to a horizontal axis; it has, therefore, the
general symbol (100). It is a *square prism* with an angle between any two
adjacent faces of 90°. It is uniformly designated by the letter *a*, and its
faces, taken in order, have the indices 100, 010, $\bar{1}$00, 0$\bar{1}$0.

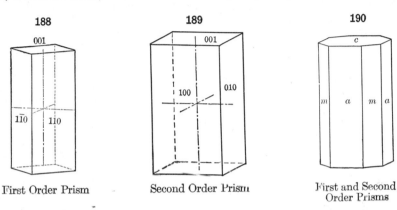

| 188 | 189 | 190 |

First Order Prism   Second Order Prism   First and Second
Order Prisms

It will be seen that the combination of this form with the base is the
analogue of the cube of the isometric system.

The faces of the prism of the first order truncate the edges of the prism
of the second order and *vice versa*. When both are equally developed, as in
Fig. 190, the result is a regular eight-sided prism, which, however, it must
be remembered, is a combination of *two* distinct forms.

It is evident that the two prisms described do not differ geometrically
from one another, and furthermore, in a given case, the symmetry of this
class allows either to be made the first order, and the other
the second order prism according to the position assumed
for the horizontal axes. If on crystals of a given species
both forms occur together equally developed (or, on the
other hand, separately on different crystals) and without
other faces than the base, there is no means of telling them
apart unless by minor characteristics, such as striations or
other markings on the surface, etchings, etc.

96. Ditetragonal Prism. — The *ditetragonal prism* is the
form which is bounded by eight similar faces, each one of
which is parallel to the vertical axis while meeting the two
horizontal axes at unequal distances. It has the general
symbol (*hk*0). It is shown in Fig. 191, where (*hk*0) = (210). Ditetragonal Prism

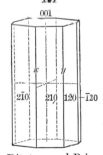

191

---

* In Figs. 188–191 the dimensions of the form are made to correspond to the assumed
length of the vertical axis (here $c = 1.78$ as in octahedrite) used in Fig. 204. It must be
noted, however, that in the case of actual crystals of these forms, while the tetragonal
symmetry is usually indicated by the unlike physical character of the face *c* as compared
with the faces *a*, *m*, etc., in the vertical prismatic zone, no inference can be drawn as to the
relative length of the vertical axis. This last can be determined only when a pyramid is
present; it is fixed for the species when a particular pyramid is chosen as the fundamental
or unit form, as explained later.

The successive faces have here the indices 210, 120, $\bar{1}20$, $\bar{2}10$, $\bar{2}\bar{1}0$, $\bar{1}\bar{2}0$, $1\bar{2}0$, $2\bar{1}0$.

In Fig. 203 a combination is shown of this form ($y = 310$) with the second order prism, the edges of which it bevels. In Fig. 207 ($h = 210$) it bevels the edges of the first order prism m. In Fig. 208 ($l = 310$) it is combined with both orders of prisms.

97. Pyramids. — There are three types of pyramids in this class, corresponding, respectively, to the three prisms which have just been described. As already stated, the name *pyramid* is given (in systems other than the isometric) to a form whose planes meet all three of the axes; in this system the form whose planes meet the axis c and one horizontal axis while parallel to the other is also called a pyramid. The pyramids of this class are strictly double pyramids (*bipyramids* or *dipyramids* of some authors).

98. Pyramid of First Order. — A *pyramid of the first order*, is a form whose eight similar faces intersect the two horizontal axes at equal distances and also intersect the vertical axis. It has the general symbol (hhl). It is a *square pyramid* with equal interfacial angles over the terminal edges, and the faces replace the horizontal, or basal, edges of the first order prism and the solid angles of the second order prism. If the ratio of the vertical to the horizontal axis for a given first order pyramid is the assumed axial ratio for the species, the form is called the *fundamental form*, and it has the symbol (111) as in Fig. 192. The indices of its faces given in order are: Above 111, $\bar{1}11$, $\bar{1}\bar{1}1$, $1\bar{1}1$; below $11\bar{1}$, $\bar{1}1\bar{1}$, $\bar{1}\bar{1}\bar{1}$, $1\bar{1}\bar{1}$.

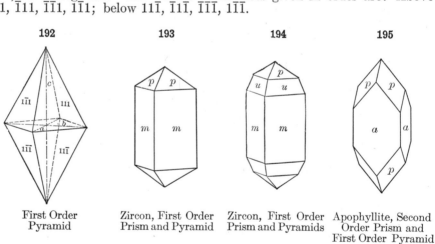

| 192 | 193 | 194 | 195 |

| First Order Pyramid | Zircon, First Order Prism and Pyramid | Zircon, First Order Prism and Pyramids | Apophyllite, Second Order Prism and First Order Pyramid |

Obviously the angles of the first order pyramid, and hence its geometrical aspect, vary widely with the length of the vertical axis. In Figs. 192 and 200 the pyramids shown have in both cases the symbol (111) but in the first case (octahedrite) $c = 1.78$, while in the second (vesuvianite), $c = 0.64$.

For a given species there may be a number of first order pyramids, varying in position according to the ratio of the intercepts upon the vertical and horizontal axes. Their symbols, passing from the base (001) to the unit prism (110), may thus be (115), (113), (223), (111), (332), (221), (441), etc. In the general symbol of these forms (hhl), as h diminishes, the form approximates more and more nearly to the base (001), for which $h = 0$;

as h increases, the form passes toward the first order prism. In Fig. 194 two pyramids of this order are shown, $p(111)$ and $u(331)$.

99. Pyramid of Second Order. — The *pyramid of the second order* is the form, Fig. 196, whose faces are parallel to one of the horizontal axes, while meeting the other two axes. The general symbol is $(h0l)$. These faces replace the basal edges of the second order prism (Fig. 197), and the solid angles of the first order prism (cf. Fig. 198). It is a *square pyramid* since its basal section is a square, and the interfacial angles over the four terminal edges, above and below, are equal. The successive faces of the form (101) are as follows: Above 101, 011, $\bar{1}$01, 0$\bar{1}$1; below 10$\bar{1}$, 01$\bar{1}$, $\bar{1}$0$\bar{1}$, 0$\bar{1}\bar{1}$.

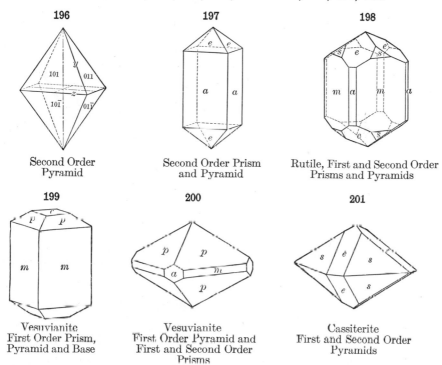

<table>
<tr><td>196</td><td>197</td><td>198</td></tr>
<tr><td>Second Order Pyramid</td><td>Second Order Prism and Pyramid</td><td>Rutile, First and Second Order Prisms and Pyramids</td></tr>
<tr><td>199</td><td>200</td><td>201</td></tr>
<tr><td>Vesuvianite First Order Prism, Pyramid and Base</td><td>Vesuvianite First Order Pyramid and First and Second Order Prisms</td><td>Cassiterite First and Second Order Pyramids</td></tr>
</table>

If the ratio of the intercepts on the horizontal and vertical axes is the assumed axial ratio of the species, the symbol is (101), and the form is designated by the letter e. This ratio can be deduced from the measurement of either one of the interfacial angles (y or z, Fig. 196) over the terminal or basal edges, as explained later. In the case of a given species, a number of second order pyramids may occur, varying in the ratio of the intercepts on the axes a and c. Hence there is possible a large number of such forms whose symbols may be, for example, (104), (103), (102), (101), (302), (201), (301), etc. Those mentioned first come nearest to the base (001), those last to the second order prism (100); the base is therefore the limit of these pyramids $(h0l)$ when $h = 0$, and the second order prism (100) when $h = 1$ and $l = 0$. Fig. 204 shows the three second order pyramids $u(105)$, $e(101)$, $q(201)$.

A second order pyramid truncating the pyramidal edges of a given first order pyramid as in Fig. 201 has the *same* ratio as it has for h to l. Thus (101) truncates the terminal edge of (111); (201) of (221), etc. This is obvious because each face has the same position as the corresponding edge of the other form (see Fig. 201, when $s = 111$ and $e = 101$; also, Figs. 204, 209, where $r = 115$, $u = 105$). Again, if a first order pyramid truncates the

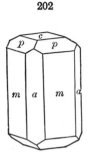

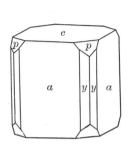

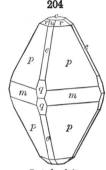

202	203	204
Vesuvianite	Apophyllite	Octahedrite
First and Second Order Prisms, First Order Pyramid and Base	Second Order Prism, Ditetragonal Prism, First Order Pyramid and Base	Two First Order Pyramids, First Order Prism, Three Second Order Pyramids and Base

pyramidal edges of a given second order pyramid, its ratio for h to l is *half* that of the other form; that is, (112) truncates the pyramidal edges of (101); (111) of (201), etc. This relation is exhibited by Fig. 204, where p(111) truncates the edges of q(201). In both cases the zonal equations prove the relations stated.

100. Ditetragonal Pyramid. — The *ditetragonal pyramid*, or double eight-sided pyramid, is the form each of whose sixteen similar faces meets the

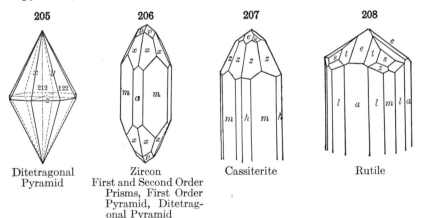

205	206	207	208
Ditetragonal Pyramid	Zircon	Cassiterite	Rutile
	First and Second Order Prisms, First Order Pyramid, Ditetragonal Pyramid		

three axes at unequal distances. This is the most general case of the symbol (hkl), where h, k, l are all unequal and no one is equal to 0. That there are sixteen faces in a single form is evident. Thus, for example, for the form (212) the face 212 is similar to 122, the two lateral axes being equal (not, however, to 221). Hence there are two like faces in each octant. Similarly the indices of all the faces in the successive octants are, therefore, as follows:

$$\begin{array}{ccccccccc}
\text{Above} & 212 & 122 & \bar{1}22 & \bar{2}12 & \bar{2}\bar{1}2 & \bar{1}\bar{2}2 & 1\bar{2}2 & 2\bar{1}2 \\
\text{Below} & 21\bar{2} & 12\bar{2} & \bar{1}2\bar{2} & \bar{2}1\bar{2} & \bar{2}\bar{1}\bar{2} & \bar{1}\bar{2}\bar{2} & 1\bar{2}\bar{2} & 2\bar{1}\bar{2}
\end{array}$$

The ditetragonal pyramid (212) is given in the above example instead of the (211) form because it lies in the *unit* pyramid zone between the first and second order unit pyramids, (111) and (101), and its indices are to be obtained by the addition of the indices of these two forms, as (111) + (101) = (212).

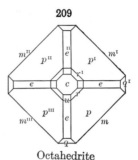

209

This form is common with the species zircon, and is hence often called a *zirconoid*. It is shown in Fig. 205. It is not observed alone, though sometimes, as in Figs. 206 (x = 311) and 207 (z = 321), it is the predominating form. In Fig. 208 two ditetragonal pyramids occur, namely, t(313) and z(321).

101. In addition to the perspective figures already given, a basal projection (Fig. 209) is added

Octahedrite

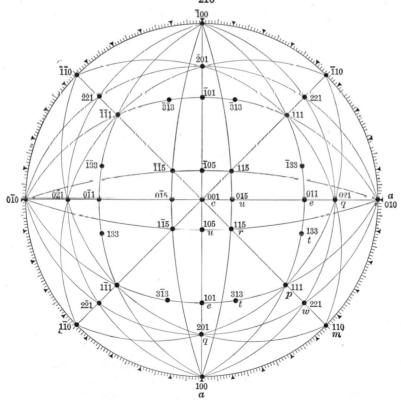

210

Stereographic Projection of Octahedrite

of the crystal of octahedrite already referred to (Fig. 204); also stereographic (Fig. 210) and gnomonic (Fig. 211) projections of the same with the faces of the forms $w(221)$ and $t(313)$ added. These exhibit well the general relations of the normal class of the tetragonal system. The symmetry here is to be noted, first, with respect to the similar zones 100, 001, $\bar{1}00$ and 010, 001, $00\bar{1}$; also, to the other pair of similar zones, 110, 001, $\bar{1}\bar{1}0$, and $1\bar{1}0$, 001, $\bar{1}10$.

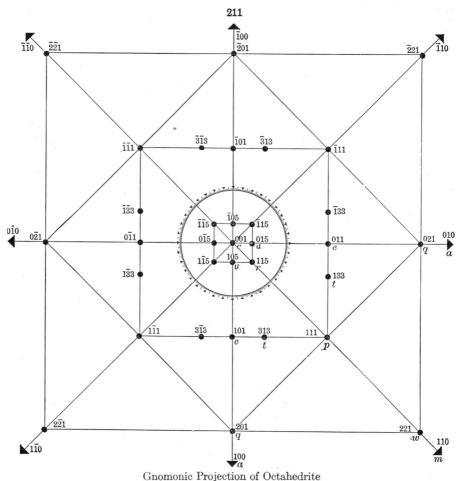

Gnomonic Projection of Octahedrite

2. HEMIMORPHIC CLASS (7). IODOSUCCINIMIDE TYPE

(Ditetragonal Pyramidal, Holohedral Hemimorphic, or Ditetragonal Polar Class)

102. Symmetry. Vert. Ax.-4; 4 vert. P. — This class differs from the normal class in having no horizontal axes or plane of symmetry; hence the forms are hemimorphic as defined in Art. **28**. It is not known to be represented among minerals, but is shown on the crystals of iodosuccinimide.

Its symmetry is illustrated by the stereographic projection (Fig. 212). Here the two basal planes (or pedions) are distinct forms, 001 and 00$\bar{1}$; the prisms do not differ geometrically from those of the normal class, though distinguished by their molecular structure; further, the pyramids are no longer double pyramids, but each form is represented by one half of Figs. 192, 196, 205 (cf. Fig. 44, p. 23). There are hence six distinct pyramidal forms, corresponding to the upper and lower halves of the first and second order pyramids and the ditetragonal pyramid.

212

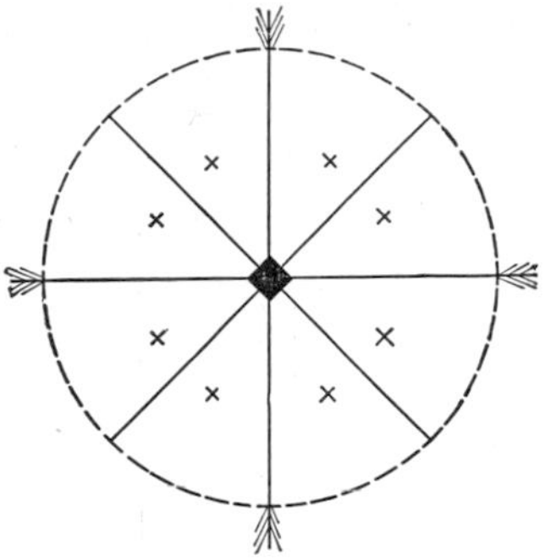

Symmetry of Hemimorphic Class

3. TRIPYRAMIDAL CLASS (8). SCHEELITE TYPE

(Tetragonal Dipyramidal, Pyramidal Hemihedral, or Tetragonal Equatorial Class)

103. Typical Forms and Symmetry. Vert. Ax.-4; hor. P.; C. — The forms here included have one plane of symmetry only, that of the horizontal crystallographic axes, and one axis of tetragonal symmetry (the vertical crystallographic axis) normal to it. The distinctive forms are the tetragonal prism (*hk*0) and pyramid (*hkl*) of the *third order*, shown in Figs. 214, 215.

The stereographic projection, Fig. 213, exhibits the symmetry of the class and the distribution of the faces of the general form (*hkl*). Comparing this, as well as the figures immediately following, with those of the normal class, it is seen that this class differs from it in the absence of the vertical planes of symmetry and the horizontal axes of symmetry.

213

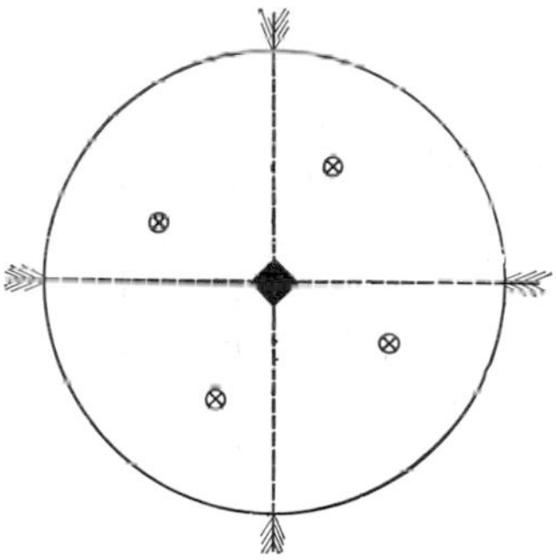

Symmetry of Tripyramidal Class

104. Prism and Pyramid of the Third Order. — The typical forms of the class, as above stated, are a square prism and a square pyramid, which are distinguished respectively from the square prisms *a*(100) and *m*(110), shown in Figs. 188 and 189, and from the square pyramids (*h0l*) and (*hhl*) of Figs. 192 and 196 by the name "*third order*."

The third order prism and pyramid may be considered as derived from the ditetragonal forms of the normal class by taking only one half the faces of the latter and the omission of the remaining faces. There are therefore two complementary forms in each case, designated *left* and *right*, which together include all the faces of the ditetragonal prism (Fig. 191) and ditetragonal pyramid (Fig. 205) of the normal class.

The indices of the faces of the two complementary prisms, as (210), are:

Left: 210, $\bar{1}$20, $\bar{2}\bar{1}$0, 1$\bar{2}$0.
Right: 120, $\bar{2}$10, $\bar{1}\bar{2}$0, 2$\bar{1}$0.

The indices of the faces of the corresponding pyramids, as (212), are:

Left: above 212, $\bar{1}22$, $\bar{2}\bar{1}2$, $1\bar{2}2$; below $21\bar{2}$, $\bar{1}2\bar{2}$, $\bar{2}\bar{1}\bar{2}$, $1\bar{2}\bar{2}$.
Right: above 122, $\bar{2}12$, $\bar{1}\bar{2}2$, $2\bar{1}2$; below $12\bar{2}$, $\bar{2}1\bar{2}$, $\bar{1}\bar{2}\bar{2}$, $2\bar{1}\bar{2}$.

Fig. 216 gives a transverse section of the prisms $a(100)$ and $m(110)$, also the prism of the third order (120). Figs. 214, 215 show the right prism (120) and pyramid (122) of the third order.

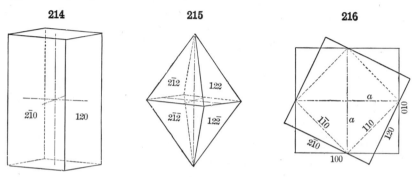

214 **215** **216**

Third Order Prism   Third Order Pyramid

105. Other Forms. — The other forms of this class, that is, the base $c(001)$; the other square prisms, $a(100)$ and $m(110)$; also the square pyramids $(h0l)$ and (hhl) are geometrically like the corresponding forms of the normal class already described. The class shows therefore three types of square pyramids and hence is called the *tripyramidal class.*

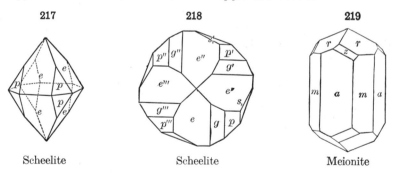

217 **218** **219**

Scheelite   Scheelite   Meionite

106. To this class belongs the important species scheelite; also the isomorphous species stolzite and powellite, unless it be that they are rather to be classed with wulfenite (p. 103). Fig. 217 shows a typical crystal of scheelite, and Fig. 218 a basal section of one similar; these illustrate well the characteristics of the class. Here the forms are $e(101)$, $p(111)$, and the third order pyramids $g(212)$, $s_{,}(131)$. Fig. 219 represents a meionite crystal with $r(111)$, and the third order pyramid $z(311)$. See also Figs. 221, 222, in which the third order prism is shown.

The forms of this class are sometimes described (see Art. **27**) as showing *pyramidal hemihedrism.*

4. PYRAMIDAL–HEMIMORPHIC CLASS (9). WULFENITE TYPE

(Tetragonal Pyramidal, Hemihedral Hemimorphic, or Tetragonal Polar Class)

107. Symmetry. Vert. Ax.-4. — The fourth class of the tetragonal system is closely related to the class just described. It has the same vertical axis of tetragonal symmetry, but there is no horizontal plane of symmetry. The forms are, therefore, hemimorphic in the distribution of the faces (cf. Fig. 220). The species wulfenite of the Scheelite Group among mineral species probably belongs here, although the crystals do not always show the difference between the pyramidal faces, above and below, which would characterize distinct complementary forms. Figs. 221, 222, could, therefore serve as illustrations of the preceding class, but in Fig. 223 a characteristic distinction is exhibited. In these figures the forms are $u(102)$, $e(101)$, $n(111)$; also $f(230)$, $k(210)$, $z(432)$, $x(311)$.

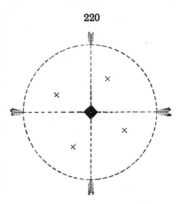

220

Symmetry of Pyramidal-
Hemimorphic Class

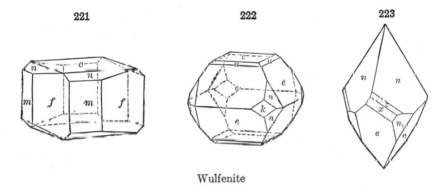

221   222   223

Wulfenite

5. SPHENOIDAL CLASS (10). CHALCOPYRITE TYPE

(Tetragonal Sphenoidal, Sphenoidal Hemihedral, Didigonal Scalenohedral, or Ditetragonal Alternating Class)

108. Typical Forms and Symmetry. 3 xl. Ax.-2 ; 2 vert. diag. P. — The typical forms of this class are the sphenoid (Fig. 225) and the tetragonal scalenohedron (Fig. 226). They and all the combinations of this class show the following symmetry. The three crystallographic axes are axes of binary symmetry and there are two vertical diagonal planes of symmetry.

This symmetry is exhibited in the stereographic projection (Fig. 224), which shows also the distribution of the faces of the general form (hkl). It is seen here that the faces are present in the alternate octants only, and it will be remembered that this same statement was made of the tetrahedral class under the isometric system. There is hence a close analogy between these two classes. The symmetry of this class should be carefully compared with that of the first and third classes of this system already described.

109. Sphenoid. — The *sphenoid*, shown in Fig. 225, is a four-faced solid, resembling a tetrahedron, but each face is an isosceles (not an equilateral) triangle. It may be considered as derived from the first order pyramid of the normal class by the development of only the alternate faces of the latter. There are therefore possible two complementary forms known as the positive and negative sphenoids. The general symbol of the positive unit sphenoid is (111), and its faces have the indices: 111, $\bar{1}\bar{1}1$, $\bar{1}1\bar{1}$, $1\bar{1}\bar{1}$, while the negative sphenoid has the symbol ($1\bar{1}1$). When the complementary forms occur together, if equally developed, the resulting solid, though having two unlike sets of faces, cannot be distinguished geometrically from the first order pyramid (111).

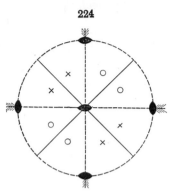

224

Symmetry of Spnenoida₁ Class

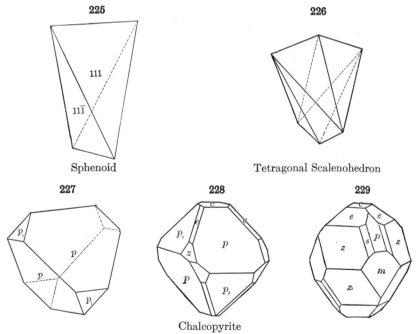

225

111

$11\bar{1}$

Sphenoid

226

Tetragonal Scalenohedron

227

$p_,$
p
p
$p_,$

228

$p_,$
p
p
$p_,$
z
c

229

c
e e
s p
z z
m
$z_,$

Chalcopyrite

In the species chalcopyrite, which belongs to this class, the deviation in angle and in axial ratio from the isometric system is very small, and hence the unit sphenoid cannot by the eye be distinguished from a tetrahedron (compare Fig. 227 with Fig. 162, p. 84). For this species $c = 0.985$ (instead of 1, as in the isometric system), and the normal sphenoidal angle is 108° 40′, instead of 109° 28′, the angle of the tetrahedron. Hence a crystal of chalcopyrite with both the positive and negative sphenoids equally developed closely resembles a regular octahedron.

In Fig. 228 the second order pyramids $e(101)$ and $z(201)$ and base $c(001)$ are also present.

110. Tetragonal Scalenohedron. — The sphenoidal symmetry yields another distinct type of form, that shown in Fig. 226. It is bounded by eight similar scalene triangles, and hence is called a *tetragonal scalenohedron*; the general symbol is (hkl). It may be considered as derived from the ditetragonal pyramid of the normal class by taking the alternate *pairs* of faces of the latter form. The faces of the complementary positive and negative forms therefore embrace all the faces of the ditetragonal pyramid. This form appears in combination in chalcopyrite, but is not observed independently. In Fig. 229 the form $s(531)$ is the positive tetragonal scalenohedron.

111. Other Forms. — The other forms of the class, namely, the first and second order prisms, the ditetragonal prism, and the first and second order pyramids (hhl) and $(h0l)$, are geometrically like those of the normal class. The lower symmetry in the molecular structure is only revealed by special investigation, as by etching.

6. TRAPEZOHEDRAL CLASS (11). NICKEL SULPHATE TYPE

(Tetragonal Trapezohedral, Trapezohedral Hemihedral, or Tetragonal Holoaxial Class)

112. Vert. Ax.-4; 4 hor. Ax.-2. — The trapezohedral class is analogous to the plagiohedral class of the isometric system, it is characterized by the absence of any plane or center of symmetry; the vertical axis, however, is

230

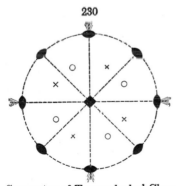

231

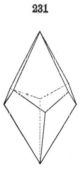

Symmetry of Trapezohedral Class   Tetragonal Trapezohedron

an axis of tetragonal symmetry, and perpendicular to this there are four axes of binary symmetry. This symmetry and the distribution of the faces of the general form (hkl) are shown in the stereographic projection, Fig. 230, and Fig. 231 gives the resulting solid, a *tetragonal trapezohedron*. It may be derived from the ditetragonal pyramid of the normal class by the extension of the alternate faces of that form. There are two complementary forms called right- and left-handed which embrace all the faces of the ditetragonal pyramid of the normal class. These two forms are enantiomorphous, and the salts belonging to this class show circular polarization of light.

Nickel sulphate and a few other artificial salts belong in this class.

7. TETARTOHEDRAL CLASS (12)

(*Tetragonal Disphenoidal, Sphenoidal Tetartohedral, or Tetragonal Alternating Class*)

113. Symmetry. Vert. Ax.-2. — The seventh and last possible class under this system has no plane or center of symmetry, but the vertical axis

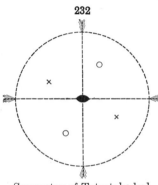

232

Symmetry of Tetartohedral Class

is an axis of binary symmetry. The symmetry and the distribution of the faces of the general form (*hkl*) are shown in the stereographic projection (Fig. 232), and the resulting solid is known as a *sphenoid of the third order*. It can be derived from the ditetragonal pyramid of the normal class by taking only one quarter of the faces of that form. There are therefore four complementary forms which are respectively distinguished as right (+ and −) and left (+ and −). These four together embrace all the sixteen faces of the ditetragonal pyramid. The other characteristic forms of this class are the prism of the third order (*hk*0), the positive and negative sphenoids of the first order (111), and also those of the second order (101). It is said that an artificial compound, $2CaO.Al_2O_3.SiO_2$, crystallizes in this class.

MATHEMATICAL RELATIONS OF THE TETRAGONAL SYSTEM

114. Choice of Axes. — It appears from the discussion of the symmetry of the seven classes of this system that with all of them the position of the vertical axis is fixed. In classes 1, 2, however, where there are two sets of vertical planes of symmetry, either set may be made the axial planes and the other the diagonal planes. The choice between these two possible positions of the horizontal axes is guided particularly by the habit of the occurring crystals and the relations of the given species to others of similar form. With a species whose crystal characters have been described it is customary to follow the orientation given in the original description.

115. Determination of the Axial Ratio, etc. — The following relations serve to connect the axial ratio, that is, the length of the vertical axis c, when $a = 1$, with the fundamental angles (001 $\wedge$ 101) and (001 $\wedge$ 111):

$$\tan (001 \wedge 101) = c; \qquad \tan (001 \wedge 111) \times \tfrac{1}{2}\sqrt{2} = c.$$

For faces in the same rectangular zone the tangent principle applies. The most important cases are:

$$\frac{\tan (001 \wedge h0l)}{\tan (001 \wedge 101)} = \frac{h}{l};$$

$$\frac{\tan (001 \wedge 0kl)}{\tan (001 \wedge 011)} = \frac{k}{l};$$

$$\frac{\tan (001 \wedge hhl)}{\tan (001 \wedge 111)} = \frac{h}{l}.$$

For the prisms

$$\tan (010 \wedge hk0) = \frac{h}{k}, \qquad \text{or} \qquad \tan (100 \wedge hk0) = \frac{k}{h}.$$

116. Other Calculations. — It will be noted that in the stereographic projection (Fig. 210) all those spherical triangles are right-angled which are formed by great circles (diameters) which meet the prismatic zone-circle 100, 010, $\bar{1}$00, 0$\bar{1}$0. Again, all those formed by great circles drawn between 100 and $\bar{1}$00, or 010 and 0$\bar{1}$0, and crossing respectively the zone-circles 100, 001, $\bar{1}$00, or 010, 001, 0$\bar{1}$0. Also, all those formed by great circles drawn between 110 and $\bar{1}\bar{1}$0 and crossing the zone-circle $\bar{1}$10, 001, 1$\bar{1}$0, or between $\bar{1}$10 and 1$\bar{1}$0 and crossing the zone-circle 110, 0$\bar{0}$1, $\bar{1}\bar{1}$0.

These spherical triangles may hence be readily used to calculate any angles desired; for example, the angles between the pole of any face, as hkl (say 321), and the pinacoids 100, 010, 001. The terminal angles (x and y, Fig. 205) of the ditetragonal pyramid, 212 $\wedge$ 2$\bar{1}$2 (or 313 $\wedge$ 3$\bar{1}$3, etc.), and 212 $\wedge$ 122 (or 313 $\wedge$ 133, etc.), can also be obtained in the same way. The zonal relations give the symbols of the poles on the zones 001, 100 and 001, 110 for the given case. For example, the zone-circle $\bar{1}$10, 313, 133, $\bar{1}$10 meets $\bar{1}$10, 001, 110 at the pole 223, and the calculated angle 313 $\wedge$ 223 is half the angle 313 $\wedge$ 133. If a large number of similar angles are to be calculated, it is more convenient to use a formula, as that given below.

117. Formulas. — It is sometimes convenient to have the normal interfacial angles expressed directly in terms of the axis c and the indices h, k, and l. Thus:

(1) The distances of the pole of any face $P(hkl)$ from the pinacoids $a(100) = Pa$, $b(010) = Pb$, $c(001) = Pc$ are given by the following equations:

$$\cos^2 Pa = \frac{h^2 c^2}{h^2 c^2 + k^2 c^2 + l^2} ; \quad \cos^2 Pb = \frac{k^2 c^2}{h^2 c^2 + k^2 c^2 + l^2} ; \quad \cos^2 Pc = \frac{l^2}{h^2 c^2 + k^2 c^2 + l^2} .$$

These may also be expressed in the form

$$\tan^2 Pa = \frac{k^2 c^2 + l^2}{h^2 c^2} ; \quad \tan^2 Pb = \frac{h^2 c^2 + l^2}{k^2 c^2} ; \quad \tan^2 Pc = \frac{h^2 c^2 + k^2 c^2}{l^2} .$$

(2) For the distance between the poles of any two faces (hkl), (pqr), we have in general

$$\cos PQ = \frac{hpc^2 + kqc^2 + lr}{\sqrt{[(h^2 + k^2)c^2 + l^2][(p^2 + q^2)c^2 + r^2]}}$$

The above equations take a simpler form for special cases often occurring; for example, for hkl and the angle of the edge y of Fig. 205.

118. Prismatic Angles. — The angles for the commonly occurring ditetragonal prisms are as follows:

	Angle on a(100)	Angle on m(110)		Angle on a(100)	Angle on m(110)
410	14° 2$\frac{1}{4}$′	30° 57$\frac{3}{4}$′	530	30° 57$\frac{3}{4}$′	14° 2$\frac{1}{4}$′
310	18 26	26 34	320	33 41$\frac{1}{4}$	11 18$\frac{3}{4}$
210	26 34	18 26	430	36 52$\frac{1}{4}$	8 7$\frac{1}{4}$

119. To determine, by plotting, the axial ratio, $a : c$, of a tetragonal mineral from the stereographic projection of its crystal forms. — As an illustrative example it has been assumed that the angles between the faces on the crystal of rutile, represented in Fig. 198, have been measured and from these measurements the poles of the faces in one octant located on the stereographic projection, see Fig. 233. In determining the axial ratio of a tetragonal crystal (or what is the same thing, the length of the c axis, since the length of the a axes are always taken as equal to 1) it is necessary to assume the indices of some pyramidal form. It is customary to take a pyramid which is prominent upon the crystals of the mineral and assume that it is the fundamental or unit pyramid of either the first or second order and has as its symbol either (111) or (101). In the example chosen both a first order and a second order pyramid are present and from their zonal relations it is evident that if the symbol assigned to the first order form be (111) that of the second order form must be (101). In order to determine the relative length of the c axis in respect to the length of the a axis for rutile therefore, it is only necessary to plot the intercept of either of these forms upon the axes. In the case of the second order pyramid it is only necessary to construct a right angle triangle (see upper left-hand quadrant of Fig. 233) in which the horizontal side shall equal the length of the a axis, (1), the vertical side shall represent the c axis and the hypothenuse shall show the proper angle of slope of the face. The angle between the center of the projection and the pole $e(101)$ is measured by the stereographic protractor and a line drawn making that angle with the line representing the

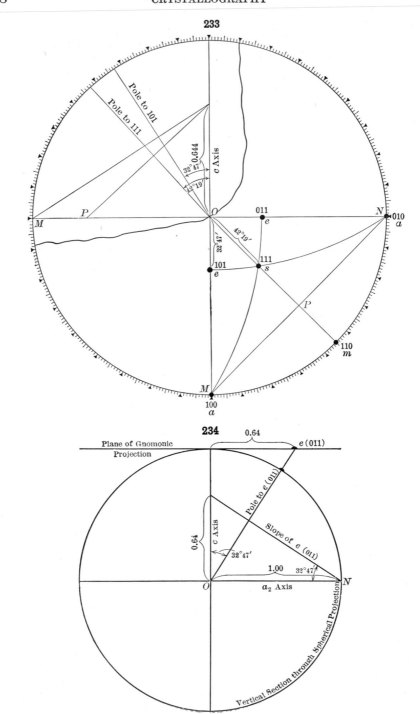

233

Pole to 10ī
Pole to 11ī
c Axis
32° 47′ 0.644
42° 19′
O
M
P
011
e
N
010
a
32° 47′
42° 19′
101
e
111
s
P
110
m
M
100
a

234

Plane of Gnomonic Projection

0.64
e (011)
Pole to e (011)
c Axis
0.64
Slope of e (011)
32° 47′
1.00
32° 47′
O
a₂ Axis
N
Vertical Section through Spherical Projection

c axis. The hypothenuse of the triangle must then be at right angles to this pole. Its intercept upon the vertical side of the triangle, when expressed in relation to the distance $(O-M)$ which was chosen as representing unity on the a axis, will therefore give the length of the c axis. In rutile this is found to be 0.644.

The same value is obtained when the position of the pyramid of the first order $s(111)$ is used. In this case the line $M-P-N$ is first drawn at right angles to the radial line $O-P$ drawn through the pole $s(111)$. The triangle to be plotted in this case has the distance $O-P$ as the length of its horizontal side. Its hypothenuse must be at right angles to the line representing the pole to (111). The intercept on the c axis is the same as in the first case.

120. To determine, by plotting, the indices of any face (hkl) of a tetragonal form from the position of its pole on the stereographic projection. — The solution of this problem is like that given in a similar case under the isometric system, see p. 90, except that the intercept of the face on the vertical axis must be referred to the established unit length of that axis and not to the length of the a axis. The method is exactly the reverse of the one used in the problem discussed directly above.

121. To determine, by plotting, the axial ratio, $a : c$, of a tetragonal mineral from the gnomonic projection of its crystal forms. — As an illustrative example consider the crystal of rutile, Fig. 198, the poles to the faces of which are shown plotted in gnomonic projection in Fig. 235. The pyramids of the first and second order present are taken as the unit forms with the symbols, $s(111)$ and $e(101)$. The lines $O-M$ and $O-N$ represent the two horizontal axes a_1 and a_2 and the distance from the center O to the circumference of the fundamental circle is equal to unity on these axes. The intercepts on $O-M$ and $O-N$ made by the poles of $e(101)$ or the perpendiculars drawn from the poles of $s(111)$ give the unit length of the vertical axis, c. In this case this distance, when expressed in terms of the assumed length of the horizontal axes (which in the tetragonal system always equals 1) is equal to 0.64.

That the above relation is true is obvious from a consideration of Fig. 234. This represents a vertical section through the spherical and gnomonic projection including the horizontal axis, a_2. The slope of the face $e(011)$ is plotted with its intercepts on the a_2 and c axes and the position of its pole in both the spherical and gnomonic projections is shown. It is seen through the two similar triangles in the figure that the distance from the center to the pole $e(011)$ in the gnomonic projection must be the same as the intercept of the face e upon the vertical axis c. And as e is a unit form this must represent unity on c.

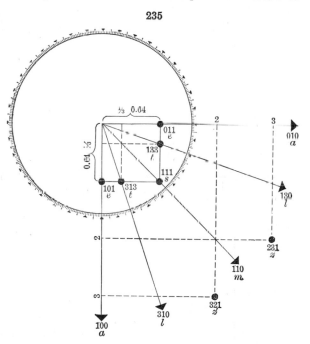

235

122. To determine, by plotting, the indices of any face of a tetragonal form from the position of its pole on the gnomonic projection. — It is assumed that in this case a mineral is being considered whose axial ratio is known. Under these conditions draw perpendiculars from the pole in question to the lines representing the two horizontal axes. Then space off on these lines distances equivalent to the length of the c axis, remembering that it must be expressed in terms of the length of the horizontal axes which in turn is equal to the distance from the center of the projection to the circumference of the fundamental circle. Give the intercepts of the lines drawn from the pole of the face to the axes a_1 and a_2 in terms of the length of the vertical axis, add a 1 as the third figure and if necessary clear of fractions and the required indices are the result. This is illustrated in Fig. 235, which is the lower right-hand quadrant of the gnomonic projection of the forms shown on the rutile crystal, Fig. 208. Consider first the ditetragonal pyramid $z(321)$. Perpendiculars drawn from its pole intersect the lines representing the horizontal axes in distances which are equal to 3 and 2 times the unit length of the c axis, 0.64. The indices of the face will therefore be 321. In the case of the ditetragonal pyramid $t(313)$, the intercepts are $1a_1$ and $\frac{1}{3}a_2$. This gives the expression $1.\frac{1}{3}.1$ which when cleared of the fraction yields 313, the indices of the face in question. The indices of a prism face like $l(310)$ can be readily obtained in exactly the same manner as described under the isometric system, Art. **88**, p. 91.

III. HEXAGONAL SYSTEM

123. The hexagonal system includes all the forms which are referred to four axes, three equal horizontal axes in a common plane intersecting at angles of 60°, and a fourth, vertical axis, at right angles to them.

Two sections are here included, each embracing a number of distinct classes related among themselves. They are called the *Hexagonal Division* and the *Trigonal* (or *Rhombohedral*) *Division*. The symmetry of the former, about the vertical axis, belongs to the hexagonal type, that of the latter to the trigonal type.

Miller (1852) referred all the forms of the hexagonal system to three equal axes parallel to the faces of the fundamental rhombohedron, and hence intersecting at equal angles, not 90°. This method (further explained in Art. **173**) had the disadvantage of failing to bring out the relationship between the normal hexagonal and tetragonal types, both characterized by a principal axis of symmetry, which (on the system adopted in this book) is the vertical crystallographic axis. It further gave different symbols to faces which are crystallographically identical. It is more natural to employ the three rhombohedral axes for trigonal forms only, as done by Groth (1905), who includes these groups in a *Trigonal System*; but this also has some disadvantages. The indices commonly used in describing hexagonal forms are known as the Miller-Bravais indices, since they were adopted by Bravais for use with the four axes from the scheme used by Miller in the other crystal systems.

124. Symmetry Classes. — There are seven possible classes in the Hexagonal Division. Of these the normal class is much the most important, and two others are also of importance among crystallized minerals.

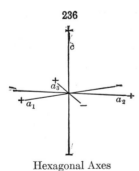

236

Hexagonal Axes

In the Trigonal Division there are five classes; of these the rhombohedral class or that of the Calcite Type, is by far the most common, and three others are also of importance.

125. Axes and Symbols. — The position of the four axes taken is shown in Fig. 236; the three horizontal axes are called a, since they are equal and interchangeable, and the vertical axis is c, since it has a different length, being either longer or shorter than the horizontal axes. The length of the vertical axis is expressed in terms of that of the horizontal axes which in turn is always taken as unity. Further, when it is desirable to distinguish between the horizontal axes they may be designated a_1, a_2, a_3. When properly orientated one of the horizontal axes (a_2) is parallel to the observer and the other two make angles of 30° either side of the line perpendicular to him. The axis to the left is taken as a_1, the one to the right as a_3. The positive and negative ends of the axes are shown in Fig. 236. The general position of any plane may be expressed in a manner analogous to that applicable in the other systems, viz.:

$$\frac{1}{h}a_1 : \frac{1}{k}a_2 : \frac{1}{i}a_3 : \frac{1}{l}c.$$

The corresponding indices for a given plane are then h, k, i, l; these always refer to the axes named in the above scheme. Since it is found convenient to consider the axis a_3 as negative in front and positive behind, the general symbol becomes $hk\bar{\imath}l$. Further, as following from the angular relation of the three horizontal axes, it can be readily shown to be always true that the algebraic sum of the indices h, k, i, is equal to zero:

$$h + k + i = 0.$$

A. Hexagonal Division

1. NORMAL CLASS (13). BERYL TYPE

(Dihexagonal Dipyramidal, Holohedral, or Dihexagonal Equatorial Class)

126. Symmetry. Vert. Ax.-6; 6 hor. Ax.-2; 6 vert. P.; hor. P.; C. — Crystals belonging to the normal class of the Hexagonal Division have one principal axis of hexagonal, or sixfold, symmetry, which coincides with the vertical crystallographic axis; also six horizontal axes of binary symmetry; three of these coincide with the horizontal crystallographic axes, the others bisect the angles between them. There is one principal plane of symmetry which is the plane of the horizontal crystallographic axes and six vertical planes of symmetry which meet in the vertical crystallographic axis. Three of these vertical planes include the horizontal crystallographic axes and the other three bisect the angles between the first set.

237

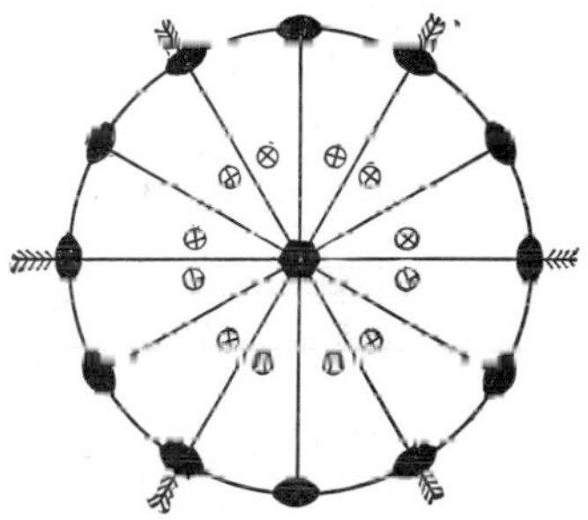

The symmetry of this class is exhibited in the accompanying stereographic projection, Fig. 237, and by the following crystal figures.

The analogy between this class and the normal class of the tetragonal system is obvious

Symmetry of Normal Class

at once and will be better appreciated as greater familiarity is gained with the individual forms and their combinations.

127. Forms. — The possible forms in this class are as follows:

		Miller-Bravais
1.	Base	(0001)
2.	Prism of the first order	(10$\bar{1}$0)
3.	Prism of the second order	(11$\bar{2}$0)
4.	Dihexagonal prism	($hk\bar{\imath}0$) as, (21$\bar{3}$0)
5.	Pyramid of the first order	($h0\bar{h}l$) as, (10$\bar{1}$1); (20$\bar{2}$1) etc.
6.	Pyramid of the second order	($h·h·2\bar{h}·l$) as, (11$\bar{2}$2)
7.	Dihexagonal pyramid	($hk\bar{\imath}l$) as, (21$\bar{3}$1)

In the above $h > k$, and $h + k = -i$.

128. Base. — The *base*, or *basal pinacoid*, includes the two faces, 0001 and 000$\bar{1}$, parallel to the plane of the horizontal axes. It is uniformly designated by the letter *c*; see Fig. 238 *et seq.*

129. Prism of the First Order. — There are three types of prisms, or forms in which the faces are parallel to the vertical axis.

The *prism of the first order*, Fig. 238, includes six faces, each one of which is parallel to the vertical axis and meets two adjacent horizontal axes at equal distances, while it is parallel to the third horizontal axis. It has hence the general symbol (10$\bar{1}$0) and is uniformly designated by the letter *m*; the indices of its six faces taken in order (see Figs. 238 and 247, 248) are:

$$10\bar{1}0, \quad 01\bar{1}0, \quad \bar{1}100, \quad \bar{1}010, \quad 0\bar{1}10, \quad 1\bar{1}00.$$

238

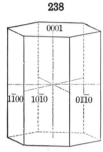

239

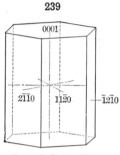

240

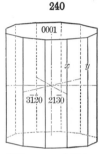

First Order Prism | Second Order Prism | Dihexagonal Prism

130. Prism of the Second Order. — The *prism of the second order*, Fig. 239, has six faces, each one of which is parallel to the vertical axis, and meets the three horizontal axes, two alternate axes at the unit distance, the intermediate axis at one half this distance; or, which is the same thing, it meets the last-named axis at the unit distance, the others at double this distance.* The general symbol is (11$\bar{2}$0) and it is uniformly designated by the letter *a*; the indices of the six faces (see Figs. 239 and 247, 248) in order are:

241

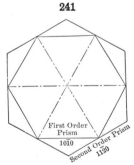

$$11\bar{2}0, \quad \bar{1}2\bar{1}0, \quad \bar{2}110, \quad \bar{1}\bar{1}20, \quad 1\bar{2}10, \quad 2\bar{1}\bar{1}0.$$

The first and second order prisms are not to be distinguished geometrically from each other since each is a regular hexagonal prism with normal interfacial angles of 60°. They are related to each other in the same way as the two prisms *m*(110) and *a*(100) of the tetragonal system.

The relation in position between the first order prism (and pyramids) on the one hand and the second order prism (and pyramids) on the other will be understood better from Fig. 241, representing a cross section of the two prisms parallel to the base *c*.

131. Dihexagonal Prism. — The *dihexagonal prism*, Fig. 240, is a twelve-sided prism bounded by twelve faces, each one of which is parallel to the

---

* Since $1a_1 : 1a_2 : -\frac{1}{2}a_3 : \infty c$ is equivalent to $2a_1 : 2a_2 : -1a_3 : \infty c$.

vertical axis, and also meets two adjacent horizontal axes at unequal distances, the ratio of which always lies between 1 : 1 and 1 : 2. This prism has two unlike edges, lettered x and y, as shown in Fig. 240. The general symbol is $(hki0)$ and the indices of the faces of a given form, as $(21\bar{3}0)$, are:

$$21\bar{3}0, \quad 12\bar{3}0, \quad \bar{1}3\bar{2}0, \quad 23\bar{1}0, \quad \bar{3}210, \quad \bar{3}120,$$
$$\bar{2}1\bar{3}0, \quad \bar{1}2\bar{3}0, \quad 1\bar{3}20, \quad 2\bar{3}10, \quad 3\bar{2}\bar{1}0, \quad 31\bar{2}0.$$

132. Pyramids of the First Order. — Corresponding to the three types of prisms just mentioned, there are three types of pyramids.

A *pyramid of the first order*, Fig. 242, is a double six-sided pyramid (or dipyramid) bounded by twelve similar triangular faces — six above and six below — which have the same position relative to the horizontal axes as the faces of the first order prism, while they also intersect the vertical axis above and below. The general symbol is hence $(h0\bar{h}l)$. The faces of a given form, as $(10\bar{1}1)$, are:

Above $\quad 10\bar{1}1, \quad 01\bar{1}1, \quad \bar{1}101, \quad \bar{1}011, \quad 0\bar{1}11, \quad 1\bar{1}01.$
Below $\quad 10\bar{1}\bar{1}, \quad 01\bar{1}\bar{1}, \quad \bar{1}10\bar{1}, \quad \bar{1}01\bar{1}, \quad 0\bar{1}1\bar{1}, \quad 1\bar{1}0\bar{1}.$

242

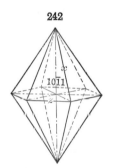

First Order Pyramid

243

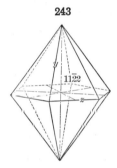

Second Order Pyramid

244

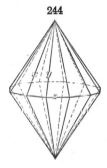

Dihexagonal Pyramid

On a given species there may be a number of pyramids of the first order, differing in the ratio of the intercepts on the horizontal to the vertical axis, and thus forming a zone between the base (0001) and the faces of the unit prism $(10\bar{1}0)$. Their symbols, passing from the base (0001) to the unit prism $(10\bar{1}0)$, would be, for example, $10\bar{1}4, 10\bar{1}2, 20\bar{2}3, 10\bar{1}1, 30\bar{3}2, 20\bar{2}1,$ etc. In Fig. 246 the faces p and u are first order pyramids and they have the symbols respectively $(10\bar{1}1)$ and $(20\bar{2}1)$, here $c = 0.4989$. As shown in these cases the faces of the first order pyramids replace the edges of the first order prism. On the other hand, they replace the solid angles of the second order prism $a(11\bar{2}0)$.

133. Pyramids of the Second Order. — The *pyramid of the second order* (Fig. 243), is a double six-sided pyramid including the twelve similar faces which have the same position relative to the horizontal axes as the faces of the second order prism, and which also intersect the vertical axis. They have the general symbol $(h \cdot h \cdot 2\bar{h} \cdot l)$. The indices of the faces of the form $(11\bar{2}2)$ are:

Above $\quad 11\bar{2}2, \quad \bar{1}2\bar{1}2, \quad \bar{2}112, \quad \bar{1}\bar{1}22, \quad 1\bar{2}12, \quad 2\bar{1}\bar{1}2.$
Below $\quad 11\bar{2}\bar{2}, \quad \bar{1}2\bar{1}\bar{2}, \quad \bar{2}11\bar{2}, \quad \bar{1}\bar{1}2\bar{2}, \quad 1\bar{2}1\bar{2}, \quad 2\bar{1}\bar{1}\bar{2}$

This form $(11\bar{2}2)$ is to be considered as the unit second order pyramid, rather than the form $(11\bar{2}1)$. This is seen when the parameters are noted, viz., $2a_1$, $2a_2$, $1a_3$, $1c$. Its indices are also to be obtained by adding those of the faces of the first order unit pyramid between which it lies, as $(10\bar{1}1) + (01\bar{1}1) = (11\bar{2}2)$.

The faces of the second order pyramid replace the edges between the faces of the second order prism and the base. Further, they replace the solid angles of the first order prism $m(10\bar{1}0)$. There may be on a single crystal a number of second order pyramids forming a zone between the base $c(0001)$ and the faces of the second order prism $a(11\bar{2}0)$, as, naming them in order: $11\bar{2}4$, $11\bar{2}2$, $22\bar{4}3$, $11\bar{2}1$, etc. In Fig. 245, s is the second order pyramid $(11\bar{2}1)$.

134. Dihexagonal Pyramid. — The *dihexagonal pyramid*, Fig. 244, is a double twelve-sided pyramid, having the twenty-four similar faces embraced under the general symbol $(hk\bar{\imath}l)$. It is bounded by twenty-four similar faces, each meeting the vertical axis, and also meeting two adjacent horizontal axes at unequal distances, the ratio of which always lies between $1:1$ and $1:2$. Thus the form $(21\bar{3}1)$ includes the following twelve faces in the upper half of the crystal:

$$2\bar{1}31, \quad 12\bar{3}1, \quad \bar{1}3\bar{2}1, \quad \bar{2}3\bar{1}1, \quad \bar{3}211, \quad \bar{3}121,$$
$$\bar{2}\bar{1}31, \quad \bar{1}\bar{2}31, \quad 1\bar{3}21, \quad 2\bar{3}11, \quad 3\bar{2}\bar{1}1, \quad 3\bar{1}\bar{2}1.$$

And similarly below with l (here 1) negative, $21\bar{3}\bar{1}$, etc. The dihexagonal pyramid is often called a *berylloid* because a common form with the species beryl. The dihexagonal pyramid $v(21\bar{3}1)$ is shown on Figs. 242, 243.

135. Combinations. — Fig. 245 of beryl shows a combination of the base $c(0001)$ and prism $m(10\bar{1}0)$ with the first order pyramids $p(10\bar{1}1)$ and

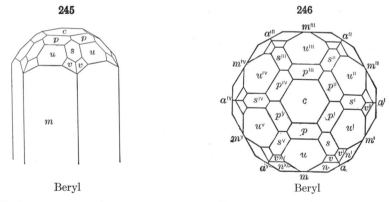

245

Beryl

246

Beryl

$u(20\bar{2}1)$; the second order pyramid $s(11\bar{2}1)$ and the dihexagonal pyramid $v(21\bar{3}1)$. The basal projection of a similar crystal shown in Fig. 246 is very instructive as exhibiting the symmetry of the normal hexagonal class. This is also true of the stereographic and gnomonic projections in Figs. 247 and 248 of a like crystal with the added form $o(11\bar{2}2)$.

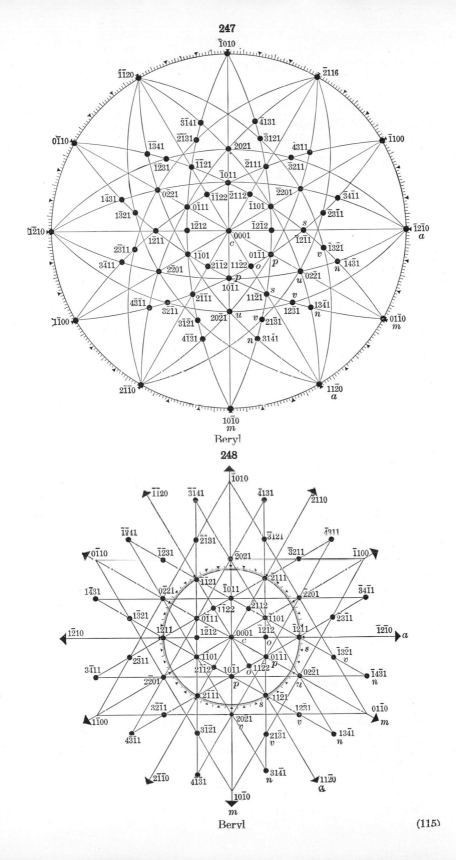

247

Beryl

248

Beryl

(115)

2. HEMIMORPHIC CLASS (14). ZINCITE TYPE

(Dihexagonal Pyramidal, Holohedral Hemimorphic, or Dihexagonal Polar Class)

136. Symmetry. Vert. Ax.-6; 6 vert. P. — This class differs from the normal class only in having no horizontal plane of principal symmetry and

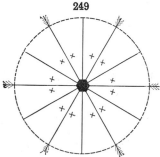

249

Symmetry of Hemimorphic Class

no horizontal axes of binary symmetry. It has, however, the same six vertical planes of symmetry meeting at angles of 30° in the vertical crystallographic axis which is an axis of hexagonal symmetry. There is no center of symmetry. The symmetry is exhibited in the stereographic projection, Fig. 249.

137. Forms. — The forms belonging to this class are the two basal planes (or pedions), 0001 and 000$\bar{1}$, here distinct forms, the positive (upper) and negative (lower) pyramids of each of the three types; also the three prisms, which last do not differ geometrically from the prisms of the normal class. An example of this class is found in zincite, Fig. 44, p. 23. Iodyrite, greenockite and wurtzite are also classed here.

3. TRIPYRAMIDAL CLASS (15). APATITE TYPE

(Hexagonal Dipyramidal, Pyramidal Hemihedral, or Hexagonal Equatorial Class)

138. Typical Forms and Symmetry. Vert. Ax.-6; hor. P.; C. — This class is important because it includes the common species of the Apatite Group, apatite, pyromorphite, mimetite, vanadinite. The typical form is the hexagonal prism ($hk\bar{i}0$) and the hexagonal pyramid ($hk\bar{i}l$), each designated as of the *third order*. These forms which are shown in Figs. 251 and 252 may be considered as derived from the corresponding dihexagonal forms of the normal class by the omission of one half of the faces of the latter. They and the other forms of the class have only one plane of symmetry, the plane of the horizontal axes, and also one axis of hexagonal symmetry (the vertical axis).

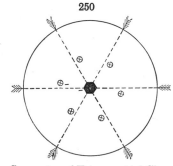

250

Symmetry of Tripyramidal Class

The symmetry is exhibited in the stereographic projection (Fig. 250). It is seen here, as in the figures of crystals given, that, like the tripyramidal class under the tetragonal system, the faces of the general form ($hk\bar{i}l$) present are half of the possible planes belonging to each sectant, and further that those above and below fall in the same vertical zone.

139. Prism and Pyramid of the Third Order. — The prism of the third order (Fig. 251) has six like faces embraced under the general symbol ($hk\bar{i}0$), and the form is a regular hexagonal prism with angles of 60°, not to be distinguished geometrically, if alone, from the other hexagonal prisms; cf. Figs. 238, 239, p. 112. The six faces of the right-handed form (21$\bar{3}$0) have the indices

$$21\bar{3}0, \quad \bar{1}3\bar{2}0, \quad \bar{3}210, \quad \bar{2}1\bar{3}0, \quad 1\bar{3}20, \quad 3\bar{2}\bar{1}0.$$

The faces of the complementary left-handed form have the indices:

$$12\bar{3}0, \quad \bar{2}3\bar{1}0, \quad \bar{3}120, \quad \bar{1}2\bar{3}0, \quad 2\bar{3}10, \quad 3\bar{1}\bar{2}0.$$

As already stated these two forms together embrace all the faces of the dihexagonal prism (Fig. 240).

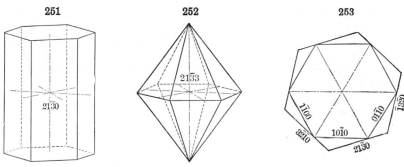

251   **252**   **253**

Third Order Prism   Third Order Pyramid

The pyramid is also a regular double hexagonal pyramid of the third order, and in its relations to the other hexagonal pyramids of the class (Figs. 242, 243) it is analogous to the square pyramid of the third order met with in the corresponding class of the tetragonal system (see Art. **104**). The faces of the right-handed form (21$\bar{3}$1) are:

Above   21$\bar{3}$1, $\bar{1}$3$\bar{2}$1, $\bar{3}$211, $\bar{2}$1$\bar{3}$1,   1$\bar{3}$21,   3$\bar{2}$$\bar{1}$1.
Below   21$\bar{3}$$\bar{1}$, $\bar{1}$3$\bar{2}$$\bar{1}$, $\bar{3}$21$\bar{1}$,   2131,   1$\bar{3}$2$\bar{1}$,   3$\bar{2}$$\bar{1}$$\bar{1}$.

There is also a complementary left-handed form, which with this embraces all the faces of the dihexagonal pyramid. The cross section of Fig. 253 shows in outline the position of the first order prism, and also that of the right-handed prism of the third order.

The prism and pyramid just described do not often appear on crystals as predominating forms, though this is sometimes the case, but commonly these faces are present modifying other fundamental forms.

140. Other Forms. — The remaining forms of the class are geometrically like those of the normal class, viz., the base (0001); the first order prism (10$\bar{1}$0); the second order prism (11$\bar{2}$0); the first order pyramids ($h0\bar{h}l$); and the second order pyramids ($h \cdot h \cdot 2\bar{h} \cdot l$). That their molecular structure, however, corresponds to the symmetry of this class is readily proved, for example, by etching. In this way it was shown that pyromorphite and mimetite belonged in the same group with apatite (Baumhauer), though crystals with the typical forms had not been observed. This class is given its name of *Tripyramidal* because its forms include three distinct types of pyramids.

141. A typical crystal of apatite is given in Fig. 254. It shows the third order prism $h(21\bar{3}0)$, and the third order pyramids, $\mu(21\bar{3}1)$, $n(31\bar{4}1)$; also the first order pyramids $r(10\bar{1}2)$, $x(10\bar{1}1)$, $y(20\bar{2}1)$, the second order pyramids $v(11\bar{2}2)$, $s(11\bar{2}1)$; finally, the prism $m(10\bar{1}0)$, and the base $c(0001)$.

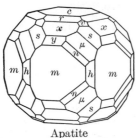

254

Apatite

4. PYRAMIDAL–HEMIMORPHIC CLASS (16). NEPHELITE TYPE

(*Hexagonal Pyramidal, Pyramidal Hemihedral Hemimorphic, or Hexagonal Polar Class*)

142. Symmetry. Vert. Ax.-6. — A fourth class under the hexagonal division, the *pyramidal-hemimorphic class*, is like that just described, except that the forms are hemimorphic. The single horizontal plane of symmetry is absent, but the vertical axis is still an axis of hexagonal symmetry. This symmetry is shown in the stereographic projection of Fig. 255. The typical

255

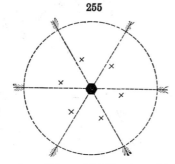

Symmetry of Pyramidal-Hemimorphic Class

256

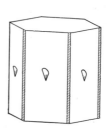

Nephelite

form would be like the upper half of Fig. 252 of the pyramid of the third order. The species nephelite is shown by the character of the etching-figures (Fig. 256, Groth after Baumhauer) to belong here.

5. TRAPEZOHEDRAL CLASS (17). β–QUARTZ TYPE

(*Hexagonal Trapezohedral, Trapezohedral Hemihedral, or Hexagonal Holoaxial Class*)

143. Symmetry. Vert. Ax.-6; 6 hor. Ax.-2. — The *trapezohedral class* has no plane of symmetry, but the vertical axis is an axis of hexagonal symmetry, and there are, further, six horizontal axes of binary symmetry. There is no center of symmetry. The symmetry and the distribution of the faces of the typical form ($hk\bar{i}l$) is shown in the stereographic projection (Fig. 257). The typical forms may be derived from the dihexagonal pyramid by the omission of the alternate faces of the latter. There are two possible types known as the right and left hexagonal trapezohedrons (see Fig. 258), which are enantiomorphous, and the few crystallized salts falling in this class show circular polarization. A modification of quartz known as β-quartz is also described as belonging here. The indices of the right form $(21\bar{3}1)$ are as follows:

Above $21\bar{3}1$, $\bar{1}3\bar{2}1$, $\bar{3}211$, $\bar{2}\bar{1}31$, $1\bar{3}21$, $3\bar{2}\bar{1}1$.
Below $12\bar{3}1$, $\bar{2}3\bar{1}1$, $312\bar{1}$, $\bar{1}2\bar{3}1$, $23\bar{1}1$, $3\bar{1}2\bar{1}$.

257

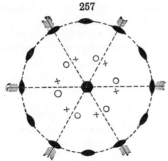

Symmetry of Trapezohedral Class

258

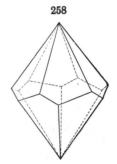

Hexagonal Trapezohedron

6. TRIGONAL CLASS (18). BENITOITE TYPE*

(Ditrigonal Dipyramidal, Trigonal Hemihedral, Trigonotype, or Ditrigonal Equatorial Class)

144. Typical Forms and Symmetry. Vert. Ax.-3; 3 hor. Ax.-2; 3 vert. P.; hor. P. — This class has, besides the vertical axis of trigonal symmetry, three horizontal axes of binary symmetry which are diagonal to the crystallographic axes. There are four planes of symmetry, one horizontal, and three

259

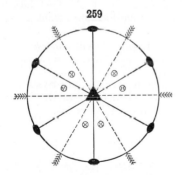

Symmetry of Trigonal Class

260

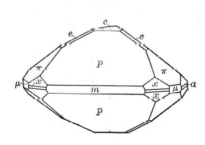

Benitoite (Palache)

vertical diagonal planes intersecting at angles of 60° in the vertical axis. The symmetry and the distribution of the faces of the positive ditrigonal pyramid are shown in Fig. 259. The characteristic forms are as follows: *Trigonal prism* consisting of three faces comprising one half the faces of the

---

* Symmetry Classes 18 and 19 were formerly grouped under the Rhombohedral Division because of their vertical axes of trigonal symmetry. They are now known to be structurally similar to the hexagonal classes and are more properly placed here.

hexagonal prism of the first order. They are of two types, called positive $(10\bar{1}0)$ and negative $(01\bar{1}0)$. *Trigonal pyramid*, a double three-faced pyramid, consisting of six faces corresponding to one half the faces of the hexagonal pyramid of the first order. The faces of the upper and lower halves fall in vertical zones with each other. There are two types, called positive $(10\bar{1}1)$ and negative $(01\bar{1}1)$. *Ditrigonal prism* consists of six vertical faces arranged in three similar sets of two faces and having therefore the alternate edges of differing character. It may be derived from the dihexagonal prism by taking alternating pairs of faces. *Ditrigonal pyramid* consists of twelve faces, six above and six below. It, like the prism, may be derived from the dihexagonal form by taking alternate pairs of faces of the latter. The faces of the upper and lower halves fall in vertical zones. The only representative of this class known is the rare mineral benitoite, a crystal of which is represented in Fig. 260. This crystal shows the trigonal prisms $m(10\bar{1}0)$ and $\mu(01\bar{1}0)$, the hexagonal prism of the second order, $a(11\bar{2}0)$, the trigonal pyramids, $p(10\bar{1}1)$ and $\pi(01\bar{1}1)$; $e(01\bar{1}2)$ and the hexagonal pyramid of the second order, $x(22\bar{4}1)$.

7. TRIGONAL TETARTOHEDRAL (19). DISILVERORTHO-PHOSPHATE TYPE

(Trigonal Dipyramidal, or Trigonal Equatorial Class)

145. Vert. Ax.-3 ; hor. P. — This class has one plane of symmetry — that of the horizontal axes, and one axis of trigonal symmetry — the vertical axis. There is no center of symmetry. Its characteristic forms are the three types of trigonal prisms and the three corresponding types of trigonal pyramids; cf. Fig. 261. This class has no known representation among minerals.

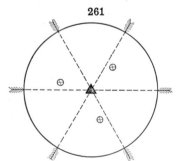

261

Symmetry of Trigonal Tetartohedral Class

B. Rhombohedral Division

Five classes are included in the Rhombohedral Division of the Hexagonal System, of which the Rhombohedral Class of the Calcite Type is by far the most important.

1. RHOMBOHEDRAL CLASS (20). CALCITE TYPE

(Ditrigonal Scalenohedral, Hexagonal Scalenohedral, Rhombohedral Hemihedral, or Dihexagonal Alternating Class)

146. Typical Forms and Symmetry. Vert. Ax.-3 ; 3 hor. xl. Ax.-2 ; 3 vert. diag. P. ; C. — The typical forms of the *rhombohedral class* are the rhombohedron (Fig. 263) and the scalenohedron (Fig. 278). These forms, with the projections, Figs. 262 and 288, illustrate the symmetry characteristic of the class. There are three planes of symmetry only; these are diagonal to the horizontal crystallographic axes and intersect at angles of 60° in the vertical crystallographic axis. This axis is with these forms an axis of trigonal symmetry; there are, further, three horizontal axes parallel to the crystallographic axes of binary symmetry; cf. Fig. 263, also Fig. 264 *et seq.*

By comparing Fig. 288 with Fig. 247, p. 115, it will be seen that all the faces in half the sectants are present. This group is hence analogous to the tetrahedral class of the isometric system, and the sphenoidal class of the tetragonal system.

147. Rhombohedron. — Geometrically described, the *rhombohedron* is a solid bounded by six like faces, each a rhomb. It has six like lateral edges forming a zigzag line about the crystal, and six like terminal edges, three above and three in alternate position below. The vertical axis joins the two trihedral solid angles, and the horizontal axes join the middle points of the opposite sides, as shown in Fig. 263.

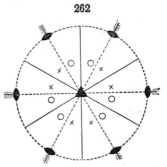

262

The general symbol of the rhombohedron is $(h0\bar{h}l)$, and the successive faces of the unit form $(10\bar{1}1)$ have the indices:

Above, $10\bar{1}1$, $\bar{1}101$, $0\bar{1}11$; below, $01\bar{1}\bar{1}$, $\bar{1}01\bar{1}$, $1\bar{1}0\bar{1}$.

Symmetry of Rhombohedral Class

The geometrical shape of the rhombohedron varies widely as the angles change, and consequently the relative length of the vertical axis c (expressed in terms of the horizontal axes, $a = 1$). As the vertical axis diminishes, the rhombohedrons become more and more obtuse or flattened; and as it increases they become more and more acute. A cube placed with an octahedral axis vertical is obviously the limiting case between the obtuse and acute forms where the interfacial angle is 90°. In Fig. 263 of calcite the normal rhombohedral angle is 74° 55' and $c = 0.854$, while for Fig. 265 of hematite this angle is 94° and $c = 1.366$. Further, Figs. 265–270 show other rhombohedrons of calcite, namely, $l(01\bar{1}2)$, $\phi(05\bar{5}4)$, $f(02\bar{2}1)$, $M(40\bar{4}1)$, and $\rho(16 \cdot 0 \cdot \bar{16} \cdot 1)$; here the vertical axes are in the ratio of $\frac{1}{2}$, $\frac{5}{4}$, 2, 4, 16, to that of the fundamental (cleavage) rhombohedron of Fig. 263, whose angle determines the value of c.

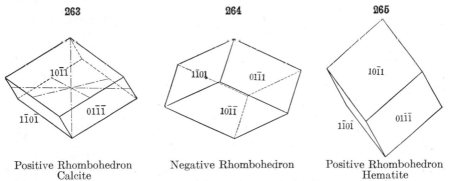

263

$10\bar{1}1$

$1\bar{1}0\bar{1}$ $01\bar{1}\bar{1}$

Positive Rhombohedron
Calcite

264

$\bar{1}10\bar{1}$ $0\bar{1}11$

$10\bar{1}\bar{1}$

Negative Rhombohedron

265

$10\bar{1}1$

$1\bar{1}0\bar{1}$ $01\bar{1}\bar{1}$

Positive Rhombohedron
Hematite

148. Positive and Negative Rhombohedrons. — To every positive rhombohedron there may be an inverse and complementary form, identical geometrically, but bounded by faces falling in the alternate sectants. Thus the negative form of the unit rhombohedron (01$\bar{1}$1) shown in Fig. 264 has the faces:

Above, 01$\bar{1}$1, $\bar{1}$01$\bar{1}$, 1$\bar{1}$01;   below, $\bar{1}$10$\bar{1}$, 0$\bar{1}$11, 101$\bar{1}$.

The position of these in the projections (Figs. 288, 289) should be carefully studied. Of the figures already referred to, Figs. 263, 265, 269 are positive, and Figs. 264, 266, 267, 268 negative, rhombohedrons; Fig. 270 shows both forms.

It will be seen that the two complementary positive and negative rhombohedrons of given axial length together embrace all the like faces of the double six-sided hexagonal pyramid of the first order. When these two rhombohedrons are equally developed the form is geometrically identical with this pyramid. This is illustrated by Fig. 273 of gmelinite r(10$\bar{1}$1), ρ(01$\bar{1}$1) and by Figs. 303, 304, p. 130, of quartz, r(10$\bar{1}$1), z(01$\bar{1}$1).* In each

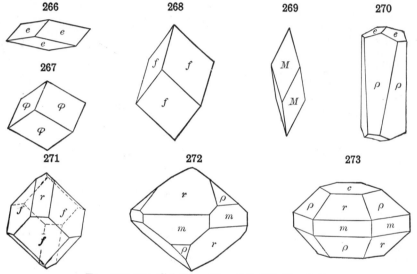

Figs. 266–271, Calcite   Figs. 272–273, Gmelinite

case the form, which is geometrically a double hexagonal pyramid (in Fig. 273 with c and m), is in fact a combination of the two unit rhombohedrons, positive and negative. Commonly a difference in size between the two forms may be observed, as in Figs. 272 and 305, where the form taken as the positive rhombohedron predominates. But even if this distinction cannot be established, the two rhombohedrons can always be distinguished by etching, or, as in the case of quartz, by pyroelectrical phenomena.

149. Of the two series, or zones, of rhombohedrons the faces of the *positive rhombohedrons* replace the edges between the base (0001) and the first order prism (10$\bar{1}$0). Also the faces of the *negative rhombohedrons* replace the alternate edges of the same forms, that is, the edges between (0001) and (01$\bar{1}$0) (compare Figs. 272, 273, etc.). Fig. 274 shows the rhombohedron in combination with the base. Fig. 275 the same with the prism a(11$\bar{2}$0).

---

* Quartz serves as a convenient illustration in this case, notwithstanding the fact that it belongs to the trapezohedral class of this division.

When the angle between the two forms happens to approximate to 70° 32′ the crystal simulates the aspect of a regular octahedron. This is illustrated by Fig. 276; here $co = 69°\ 42′$, also $oo = 71°\ 22′$, and the crystal resembles closely an octahedron with truncated edges (cf. Fig. 117, p. 72).

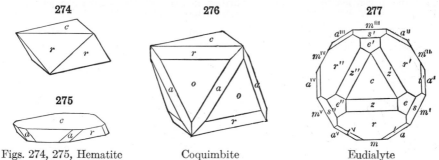

Figs. 274, 275, Hematite   Coquimbite   Eudialyte

150. There is a very simple relation between the positive and negative rhombohedrons which it is important to remember. The form of one series which truncates the terminal edges of a given form of the other will have one half the intercept on the vertical crystallographic axis of the latter. This ratio is expressed in the values of the indices of the two forms. Thus $(01\bar{1}2)$ truncates the terminal edges of the positive unit rhombohedron $(10\bar{1}1)$; $(10\bar{1}4)$ truncates the terminal edges of $(01\bar{1}2)$, $(10\bar{1}5)$ of $(20\bar{2}5)$. Again $(10\bar{1}1)$ truncates the edges of $(02\bar{2}1)$, $(40\bar{4}1)$ of $(02\bar{2}1)$, etc. This is illustrated by Fig. 271 with the forms $r(10\bar{1}1)$ and $f(02\bar{2}1)$. Also in Fig. 277, a basal projection, $z(10\bar{1}4)$ truncates the edges of $e(01\bar{1}2)$; $e(01\bar{1}2)$ of $r(10\bar{1}1)$; $r(10\bar{1}1)$ of $s(02\bar{2}1)$.

151. Scalenohedron. — The *scalenohedron*, shown in Fig. 278, is the general form for this class corresponding to the symbol $hk\bar{i}l$. It is a solid, bounded by twelve faces, each a scalene triangle. It has roughly the shape of a double six-sided pyramid, but there are two sets of terminal edges, one more obtuse than the other, and the lateral edges form a zig-zag edge around the form like that of the rhombohedron. It may be considered as derived from the dihexagonal pyramid by taking the alternating pairs of faces of that form. It is to be noted that the faces in the lower half of the form do not fall in vertical zones with those of the upper half. Like the rhombohedrons, the scalenohedrons may be either positive or negative. The positive forms correspond in position to the positive rhombohedrons and conversely.

278

Scalenohedron

The positive scalenohedron $(21\bar{3}1)$, Fig. 278, has the following indices for the several faces:

Above	$21\bar{3}1$,	$\bar{2}3\bar{1}1$,	$\bar{3}211$,	$\bar{1}\bar{2}31$,	$1\bar{3}21$,	$3\bar{1}\bar{2}1$.
Below	$12\bar{3}1$,	$\bar{1}3\bar{2}1$,	$\bar{3}12\bar{1}$,	$\bar{2}1\bar{3}1$,	$2\bar{3}1\bar{1}$,	$3\bar{2}1\bar{1}$.

For the complementary negative scalenohedron $(12\bar{3}1)$ the indices of the faces are:

Above	$12\bar{3}1$,	$\bar{1}3\bar{2}1$,	$\bar{3}121$,	$\bar{2}\bar{1}31$,	$2\bar{3}11$,	$3\bar{2}\bar{1}1$.
Below	$\bar{2}3\bar{1}1$,	$\bar{3}2\bar{1}1$,	$\bar{1}\bar{2}31$,	$1\bar{3}2\bar{1}$,	$3\bar{1}\bar{2}\bar{1}$,	$21\bar{3}\bar{1}$.

152. Relation of Scalenohedrons to Rhombohedrons. — It was noted above that the scalenohedron in general has a series of zigzag lateral edges like the rhombohedron. It is obvious, further, that for every rhombohedron there will be a series or zone of scalenohedrons having the *same* lateral edges. This is shown in Fig. 281, where the scalenohedron $v(21\bar{3}1)$ bevels the lateral edges of the fundamental rhombohedron $r(10\bar{1}1)$; the same would be true of the scalenohedron $(32\bar{5}1)$, etc. Further, in Fig. 282, the negative scalenohedron $x(13\bar{4}1)$ bevels the lateral edges of the negative rhombohedron $f(02\bar{2}1)$. The relation of

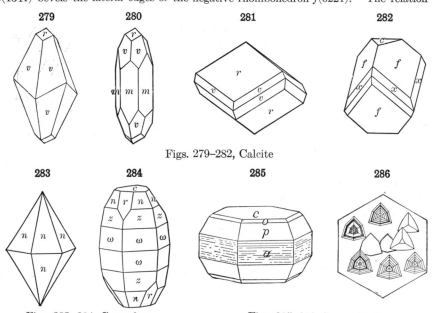

Figs. 279–282, Calcite

Figs. 283, 284, Corundum   Figs. 285, 286, Spangolite*

the indices which must exist in these cases may be shown to be, for example, for the rhombo-hedron $r(10\bar{1}1)$, $h = k + l$; again for $f(02\bar{2}1)$, $h + 2l = k$, etc. See also the projections, Figs. 288, 289. Further, the position of the scalenohedron may be defined with reference to its parent rhombohedron. For example, in Fig. 281 the scalenohedron $v(21\bar{3}1)$ has three times the vertical axis of the unit rhombohedron $r(10\bar{1}1)$. Again in Fig. 282 $x(13\bar{4}1)$ has twice the vertical axis of $f(02\bar{2}1)$.

153. Other Forms. — The remaining forms of this class which might be termed the normal class of the rhombohedral division are geometri-cally like those of the corresponding class of the hexagonal division — viz., the base $c(0001)$; the prisms $m(10\bar{1}0)$, $a(11\bar{2}0)$, $(hk\bar{i}0)$; also the second order pyramids, as $(11\bar{2}1)$. Some of these forms are shown in the accompanying figures. For further illustrations reference may be made to typical rhombohedral species, as calcite, hema-tite, etc.

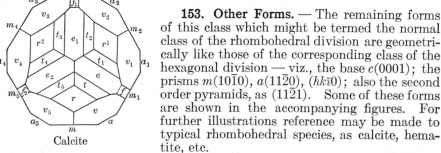

287

Calcite

With respect to the second order pyramid, it is interesting to note that if

---

* Spangolite belongs properly to the next (hemimorphic) group, but this fact does not destroy the value of the illustration.

it occurs alone (as in Fig. 283, $n = 22\bar{4}3$) it is impossible to say, on geometrical grounds, whether it has the trigonal symmetry of the rhombohedral type or the hexagonal symmetry of the hexagonal type. In the latter case, the form might be made a first order pyramid by exchanging the axial and diagonal planes of symmetry. The true symmetry, however, is often indicated, as with corundum, by the occurrence on other crystals of rhombohedral faces, as $r(10\bar{1}1)$ in Fig. 284 (here $z = 22\bar{4}1$, $\omega = 14\cdot14\cdot\bar{2}\bar{8}\cdot3$). Even if

288

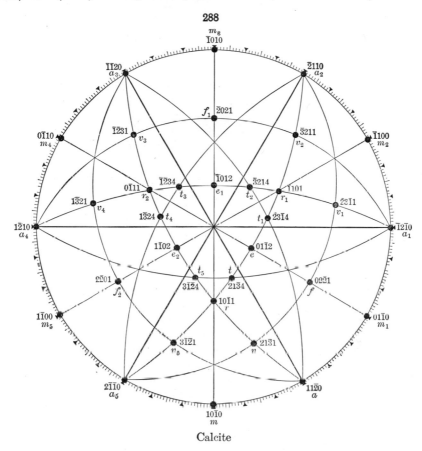

Calcite

rhombohedral faces are absent (Fig. 285), the etching-figures (Fig. 286) will often serve to reveal the true trigonal molecular symmetry; here $o = (11\bar{2}4)$, $p = (11\bar{2}2)$.

154. A basal projection of a somewhat complex crystal of calcite is given in Fig. 287, and stereographic and gnomonic projections of the same forms in Figs. 288 and 289; both show well the symmetry in the distribution of the faces. Here the forms are: prisms, $a(11\bar{2}0)$, $m(10\bar{1}0)$; rhombohedrons, positive, $r(10\bar{1}1)$, negative, $e(01\bar{1}2)$, $f(02\bar{2}1)$; scalenohedrons, positive, $v(21\bar{3}1)$, $t(21\bar{3}4)$.

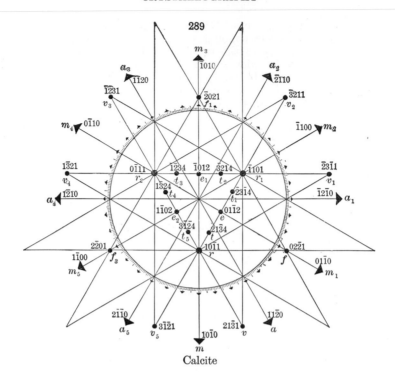

289

Calcite

2. RHOMBOHEDRAL–HEMIMORPHIC CLASS (21).
TOURMALINE TYPE

(Ditrigonal Pyramidal, Trigonal Hemihedral Hemimorphic, or Ditrigonal Polar Class)

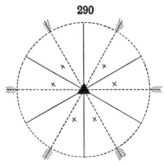

290

Symmetry of
Rhombohedral-Hemimorphic
Class

155. Symmetry. Vert. Ax.-3; 3 vert. diag. P. — A number of prominent rhombohedral species, as tourmaline, pyrargyrite, proustite, belong to a hemimorphic class under this division. For them the symmetry in the grouping of the faces differs at the two extremities of the vertical axis. The forms have the same three diagonal planes of symmetry meeting at angles of 60° in the vertical axis, which is an axis of trigonal symmetry. There are, however, no horizontal axes of symmetry, as in the rhombohedral class, and there is no center of symmetry; cf. Fig. 290.

156. Typical Forms. — In this class the basal planes (or pedions) (0001) and (000$\bar{1}$) are distinct forms. The other characteristic forms are the two trigonal prisms $m(10\bar{1}0)$ and $m_{/}(01\bar{1}0)$ of the first order series; also the four trigonal first order pyramids, corresponding respectively to the three upper or three lower faces of a positive rhombohedron, and the three upper or three lower faces of the negative rhombohedron; also the hemimorphic second order hexagonal pyramid; finally, the four ditrigonal pyramids, corresponding to the

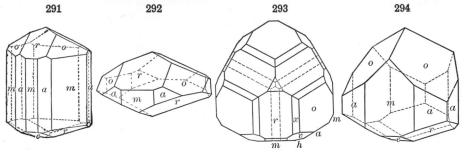

Figs. 291–294, Tourmaline

upper or lower faces respectively of the positive and negative scalenohedrons. Figs. 291–294 illustrate these forms. Fig. 293 is a basal section with $r_{,}(01\bar{1}1)$ and $e_{,}(10\bar{1}2)$ below.

3. TRI-RHOMBOHEDRAL CLASS (22). PHENACITE TYPE

(Rhombohedral, Trigonal Rhombohedral, Rhombohedral Tetartohedral, or Hexagonal Alternating Class)

157 Symmetry. Vert. Ax.-3; C. — This class, illustrated by the species dioptase, phenacite, willemite, dolomite, ilmenite, etc., is an important one. It is characterized by the absence of all planes of symmetry, but the vertical axis is still an axis of trigonal symmetry, and there is a center of symmetry; cf. Fig. 295.

158. Typical Forms. — The distinctive forms of the class are the rhombohedron of the second order and the hexagonal prism and rhombohedron, each of the third order. The class is thus characterized by three rhombohedrons of distinct types (each + and −), and hence the name given to it.

The *second order rhombohedron* may be derived by taking one half the faces of the normal hexagonal pyramid of the second order. There will be two complementary forms known as positive and negative. For example, in a given case the indices of the faces for the positive and negative forms are:

295

Symmetry of
Tri-Rhombohedral Class

Positive (above) $11\bar{2}2$, $\bar{2}112$, $1\bar{2}12$; (below) $\bar{1}2\bar{1}2$, $\bar{1}1\bar{2}2$, $2\bar{1}\bar{1}2$.
Negative (above) $\bar{1}2\bar{1}2$, $\bar{1}1\bar{2}2$, $2\bar{1}\bar{1}2$; (below) $\bar{2}11\bar{2}$, $1\bar{2}1\bar{2}$, $11\bar{2}\bar{2}$.

The *rhombohedron of the third order* has the general symbol $(hk\bar{i}l)$, and may be derived from the normal dihexagonal pyramid, Fig. 245, by taking one quarter of the faces of the latter.

There are therefore four complementary third order rhombohedrons, distinguished respectively as positive right-handed (21$\bar{3}$1), positive left-handed (3$\bar{1}$2$\bar{1}$), negative right-handed ($\bar{1}$3$\bar{2}$1), and negative left-handed (12$\bar{3}$1). The indices of the six like faces of the positive right-handed form (21$\bar{3}$1) are:

Above, 21$\bar{3}$1, $\bar{3}$211, 1$\bar{3}$21; below, $\bar{1}$3$\bar{2}\bar{1}$, $\bar{2}$1$\bar{3}\bar{1}$, 3$\bar{2}$1$\bar{1}$.

The *hexagonal prism of the third order* may be derived from the normal dihexagonal prism, Fig. 238, by taking one half the faces of the latter. There are two complementary forms known as right- and left-handed. The faces of these forms in a given case (21$\bar{3}$0) have the indices:

Right 21$\bar{3}$0, $\bar{1}$3$\bar{2}$0, $\bar{3}$210, $\bar{2}$1$\bar{3}$0, 1$\bar{3}$20, 3$\bar{2}\bar{1}$0.
Left   12$\bar{3}$0, $\bar{2}$3$\bar{1}$0, $\bar{3}$120, $\bar{1}$2$\bar{3}$0, 2$\bar{3}$10, 3$\bar{1}\bar{2}$0.

159. The remaining forms are geometrically like those of the rhombohedral class, viz.: Base c(0001); first order prism m(10$\bar{1}$0); second order prism a(11$\bar{2}$0); rhombohedrons of the first order, as (10$\bar{1}$1) and (01$\bar{1}$1), etc.

160. The forms of this group are illustrated by Figs. 296–298. Fig. 296 is of dioptase and shows the hexagonal prism of the second order a(11$\bar{2}$0) with a negative first order rhombohedron, s(02$\bar{2}$1) and the third order rhom-

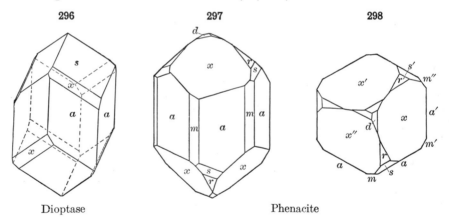

Dioptase   Phenacite

bohedron x(13$\bar{4}$1). Figs. 297 and 298 show the clinographic and horizontal projections of a crystal of phenacite with the following forms: first order prism, m(10$\bar{1}$0); second order prism, a(11$\bar{2}$0); third order rhombohedrons, x(12$\bar{3}$2) and s(21$\bar{3}$1); first order rhombohedrons, r(10$\bar{1}$1) and d(01$\bar{1}$2).

In order to make clearer the relation of the faces of the different types of forms under this class, Fig. 299 is added. Here the zones of the positive and negative rhombohedrons of the first order are indicated ($+R$ and $-R$) also the general positions of the four types of the third order rhombohedrons ($+r$, $-r$, $+l$, $-l$).

The following scheme may also be helpful in connection with Fig. 299. It shows the distribution of the faces of the four rhombohedrons of the third order ($+r$, $+l$, $-r$, $-l$) relatively to the faces of the unit hexagonal prism (10$\bar{1}$0).

PHENACITE TYPE

$+l$ $3\bar{1}\bar{2}1$	$+r$ $21\bar{3}1$	$-l$ $12\bar{3}1$	$-r$ $\bar{1}3\bar{2}1$	$+l$ $\bar{2}3\bar{1}1$	$+r$ $\bar{3}211$	$-l$ $\bar{3}121$	$-r$ $\bar{2}1\bar{3}1$	$+l$ $\bar{1}2\bar{3}1$	$+r$ $1\bar{3}21$	$-l$ $2\bar{3}11$	$-r$ $3\bar{2}\bar{1}1$
	$10\bar{1}0$		$01\bar{1}0$		$\bar{1}100$		$\bar{1}010$		$0\bar{1}10$		$1\bar{1}00$
$-l$ $3\bar{1}\bar{2}\bar{1}$	$-r$ $21\bar{3}\bar{1}$	$+l$ $12\bar{3}\bar{1}$	$+r$ $\bar{1}3\bar{2}\bar{1}$	$-l$ $\bar{2}3\bar{1}\bar{1}$	$-r$ $\bar{3}21\bar{1}$	$+l$ $\bar{3}12\bar{1}$	$+r$ $\bar{2}1\bar{3}\bar{1}$	$-l$ $\bar{1}2\bar{3}\bar{1}$	$-r$ $1\bar{3}2\bar{1}$	$+l$ $2\bar{3}1\bar{1}$	$+r$ $3\bar{2}\bar{1}\bar{1}$

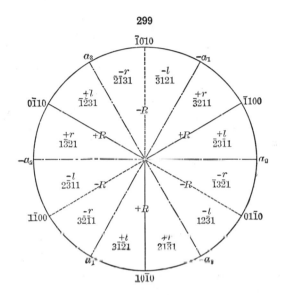

299

4. TRAPEZOHEDRAL CLASS (23). α–QUARTZ TYPE

(Trigonal Trapezohedral, Trapezohedral Tetartohedral, or Trigonal Holoaxial Class)

161. Symmetry. Vert. Ax.-3; 3 hor. Ax.-2. — This class includes, among minerals, the species quartz and cinnabar. The forms have no plane of symmetry and no center of symmetry; the vertical axis is, however, an axis of trigonal symmetry, and there are also three horizontal axes of binary symmetry, coinciding in direction with the crystallographic axes; cf. Fig. 300.

162. Typical Forms. — The characteristic form of the class is the trigonal trapezohedron shown in Fig. 301. This is the general form corresponding to the symbol $(hk\bar{i}l)$, the faces being distributed as indicated in the accompanying stereographic projection (Fig. 302). The faces of this form correspond to one quarter of the faces of the normal dihexagonal pyramid, Fig. 244. There are therefore four such trapezohedrons, two positive, called respectively right-handed (Fig 301) and left-handed (Fig. 302), and two similar negative forms, also right- and left-handed (see the scheme given in Art. **164**). It is obvious that the two forms of Figs. 301, 302 are enantiomorphous, and circular polarization is a striking character of the species belonging to the class as elsewhere discussed.

The indices of the six faces belonging to each of these will be evident on consulting Figs. 300, 247 and 248. The complementary positive form (*r* and *l*) of a given symbol include the twelve faces of a positive scalenohedron, while the faces of all four as already stated include the twenty-four faces of the dihexagonal pyramid.

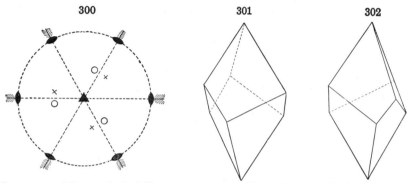

| 300 | 301 | 302 |

Symmetry of Trapezohedral Class   Trigonal Trapezohedrons

Corresponding to these trapezohedrons there are two *ditrigonal prisms*, respectively right- and left-handed, as $(21\bar{3}0)$ and $(3\bar{1}\bar{2}0)$.

The remaining characteristic forms are the right- and left-handed *trigonal prism* $(11\bar{2}0)$ and $(2\bar{1}\bar{1}0)$; also the right- and left-handed *trigonal pyramid*, as $(11\bar{2}2)$ and $(2\bar{1}\bar{1}2)$. They may be derived by taking respectively one half the faces of the hexagonal prism of the second order $(11\bar{2}0)$ or of the corresponding pyramid $(11\bar{2}2)$; these are shown in Figs. 239 and 243.

163. Other Forms. — The other forms of the class are geometrically like those of the normal rhombohedral class. They are the base $c(0001)$, the hexagonal first order prism $m(10\bar{1}0)$, and the positive and negative rhombohedrons as $(10\bar{1}1)$ and $(01\bar{1}1)$. These cannot be distinguished geometrically from the normal forms.

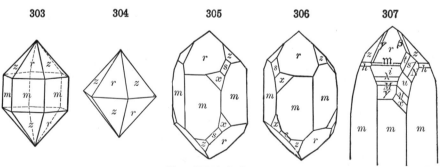

| 303 | 304 | 305 | 306 | 307 |

Figs. 303–307, Quartz

164. Illustrations. — The forms of this class are best shown in the species quartz. As already remarked (p. 122), simple crystals often appear to be of normal hexagonal symmetry, the rhombohedrons $r(10\bar{1}1)$ and $z(01\bar{1}1)$ being equally developed (Figs. 303, 304). In many cases, however, a difference in molecular character between them can be observed, and more commonly one rhombohedron, $r(10\bar{1}1)$, predominates in size; the distinction can always be made out by etching. Some crystals, like Fig. 305, show as modifying faces the *right* trigonal pyramid $s(11\bar{2}1)$, with a *right positive* trapezohedron, as $x(51\bar{6}1)$. Such crystals are called right-handed and rotate the plane of polarization of light transmitted in the direction of the vertical axis to the right. A crystal, like Fig. 306, with the *left* trigonal pyramid $s(2\bar{1}11)$ and one or more *left* trapezohedrons, as $x(6\bar{1}51)$, is called left-handed, and as regards light has the opposite character to the crystal of Fig. 305. Fig. 307 shows a more complex right-handed crystal with several positive and negative rhombohedrons, several positive right trapezohedrons and the negative left trapezohedron, N.

The following scheme shows the distribution of the faces of the four trapezohedrons $(+r, +l, -r, -l)$ relatively to the faces of the unit hexagonal prism $(10\bar{1}0)$; it is to be compared with the corresponding scheme, given in Art. **160**, of crystals of the phenacite type. In the case of the negative forms some authors prefer to make the faces $21\bar{3}1$, $12\bar{3}1$, etc., *right*, and $3\bar{1}2\bar{1}$, $\bar{1}3\bar{2}1$, etc., *left*.

QUARTZ TYPE

$+l$	$+r$	$-l$	$-r$	$+l$	$+r$	$-l$	$-r$	$+l$	$+r$	$-l$	$-r$
$3\bar{1}21$	$21\bar{3}1$	$12\bar{3}1$	$13\bar{2}1$	$2\bar{3}\bar{1}1$	$\bar{3}211$	3121	$\bar{2}\bar{1}31$	$\bar{1}231$	$13\bar{2}1$	$2\bar{3}11$	$3\bar{2}\bar{1}1$
	$10\bar{1}0$		$01\bar{1}0$		$\bar{1}100$		$\bar{1}010$		$0\bar{1}10$		$1\bar{1}00$
$-r$	$-l$	$+r$	$+l$	$-r$	$-l$	$\mid r$	$\vdash l$	$-r$	$-l$	$+r$	$+l$
$3\bar{1}2\bar{1}$	2131	$123\bar{1}$	$\bar{1}32\bar{1}$	$23\bar{1}\bar{1}$	$\bar{3}21\bar{1}$	$3\bar{1}2\bar{1}$	$\bar{2}1\bar{3}1$	$\bar{1}23\bar{1}$	$132\bar{1}$	$2\bar{3}1\bar{1}$	$3\bar{2}1\bar{1}$

5. TRIGONAL TETARTOHEDRAL HEMIMORPHIC (24). SODIUM PERIODATE TYPE

(Trigonal Pyramidal or Trigonal Polar Class)

165. The last class of this division has no plane of symmetry and no center of symmetry, but the vertical axis is an axis of trigonal symmetry. The forms are all hemimorphic, the prisms trigonal prisms, and the pyramids hemimorphic trigonal pyramids; cf. Fig. 308. No known mineral belongs to this class.

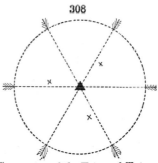

308

Symmetry of the Trigonal Tetartohedral Hemimorphic Class

MATHEMATICAL RELATIONS OF THE HEXAGONAL SYSTEM

166. Choice of Axis. — The position of the vertical crystallographic axis is fixed in all the classes of this system since it coincides with the axis of hexagonal or trigonal symmetry. The three horizontal axes are also fixed in direction except in the normal class and the subordinate hemimorphic class of the hexagonal division; in these there is a choice of two positions according to which of the two sets of vertical planes of symmetry is taken as the axial set.

167. Axial and Angular Elements. — The axial element is the length of the vertical axis, c, in terms of a horizontal axis, a; in other words, the axial ratio of $a : c$. A single measured angle (in any zone but the prismatic) may be taken as the fundamental angle from which the axial ratio can be obtained.

The angular element is usually taken as the angle between the base $c(0001)$ and the unit first order pyramid $(10\bar{1}1)$, that is, $0001 \wedge 10\bar{1}1$.

The relation between this angle and the axis c is given by the formula

$$\tan (0001 \wedge 10\bar{1}1) \times \frac{1}{2}\sqrt{3} = c.$$

The vertical axis is also easily obtained from the unit second order pyramid, since

$$\tan (0001 \wedge 11\bar{2}2) = c.$$

These relations become general by writing them as follows:

$$\tan (0001 \wedge h0\bar{h}l) \times \frac{1}{2}\sqrt{3} = \frac{h}{l} \times c;$$

$$\tan (0001 \wedge h{\cdot}h{\cdot}2\bar{h}{\cdot}l) = \frac{2h}{l} \times c.$$

In general it is easy to obtain any required angle between the poles of two faces on the spherical projection either by the use of the tangent (or cotangent) relation, or by the solution of spherical triangles, or by the application of both methods. In practice most of the triangles used in calculation are right-angled.

168. Tangent and Cotangent Relations. — The tangent relation holds good in any zone from $c(0001)$ to a face in the prismatic zone. For example:

$$\frac{\tan (0001 \wedge h0\bar{h}l)}{\tan (0001 \wedge 10\bar{1}1)} = \frac{h}{l}; \quad \frac{\tan (0001 \wedge h{\cdot}h{\cdot}2\bar{h}{\cdot}l)}{\tan (0001 \wedge 11\bar{2}2)} = \frac{2h}{l}.$$

In the prismatic zone, the cotangent formula takes a simplified form; for example, for a dihexagonal prism, $hk\bar{i}0$, as $(21\bar{3}0)$:

$$\cot (10\bar{1}0 \wedge hk\bar{i}0) = \frac{2h+k}{k}\sqrt{\frac{1}{3}};$$

$$\cot (11\bar{2}0 \wedge hk\bar{i}0) = \frac{h+k}{h-k}\sqrt{3}.$$

The sum of the angles $(10\bar{1}0 \wedge hk\bar{i}0)$ and $(11\bar{2}0 \wedge hk\bar{i}0)$ is equal to 30°.

Further, the last equations can be written in a more general form, applying to any pyramid $(hk\bar{i}l)$ in a zone, first between $10\bar{1}0$ and a face in the zone 0001 to $0\bar{1}10$, where the angle between $10\bar{1}0$ and this face is known; or again, for the same pyramid, in a zone between $11\bar{2}0$ and a face in the zone 0001 to $10\bar{1}0$, the angle between $11\bar{2}0$ and this face being given. For example (cf. Fig. 247, p. 115), if the first-mentioned zone is $10\bar{1}0{\cdot}hk\bar{i}l{\cdot}01\bar{1}1$ and the second is $11\bar{2}0{\cdot}hk\bar{i}l{\cdot}10\bar{1}1$, then

$$\cot (10\bar{1}0 \wedge hk\bar{i}l) = \cot (10\bar{1}0 \wedge 01\bar{1}1) \cdot \frac{2h+k}{k},$$

and

$$\cot (11\bar{2}0 \wedge hk\bar{i}l) = \cot (11\bar{2}0 \wedge 10\bar{1}1) \cdot \frac{h+k}{h-k}.$$

Also similarly for other zones,

$$\cot (10\bar{1}0 \wedge hk\bar{i}l) = \cot (10\bar{1}0 \wedge 02\bar{2}1) \cdot \frac{2h+k}{k}, \text{ etc.}$$

$$\cot (11\bar{2}0 \wedge hk\bar{i}l) = \cot (11\bar{2}0 \wedge 20\bar{2}1) \cdot \frac{h+k}{h-k}, \text{ etc.}$$

169. Other Angular Relations. — The following simple relations are of frequent use:
(1) For a *hexagonal pyramid of the first order*,

$$\tan \tfrac{1}{2} (10\bar{1}1 \wedge 01\bar{1}1) = \sin \xi \sqrt{\tfrac{1}{3}}, \quad \text{where } \tan \xi = c,$$

and in general

$$\tan \tfrac{1}{2} (h0\bar{h}l \wedge 0h\bar{h}l) = \sin \xi_{,} \sqrt{\tfrac{1}{3}}, \quad \text{where } \tan \xi_{,} = \frac{h}{l} c.$$

(2) For a *hexagonal pyramid of the second order*, as ($11\bar{2}2$),

$$2 \sin \tfrac{1}{2} (11\bar{2}2 \wedge \bar{1}2\bar{1}2) = \sin \xi, \quad \text{and} \quad \tan \xi = c.$$

(3) For a *rhombohedron*

$$\sin \tfrac{1}{2} (10\bar{1}1 \wedge \bar{1}101) = \sin \alpha \sqrt{\tfrac{3}{4}}, \quad \text{where } \alpha = (0001 \wedge 10\bar{1}1);$$

in general

$$\sin \tfrac{1}{2} (h0\bar{h}l \wedge \bar{h}h0l) = \sin \alpha_{,} \sqrt{\tfrac{3}{4}}, \quad \text{where } \alpha_{,} = (0001 \wedge h0\bar{h}l).$$

170. Zonal Relations. — The zonal equations, described in Arts **50**, **51**, apply here as in other systems, only that it is to be noted that one of the indices referring to the horizontal axes, preferably the third, i, is to be dropped in the calculations and only the other three employed. Thus the indices (u, v, w) of the zone in which the faces ($hk\bar{\imath}l$), ($pq\bar{r}t$) lie are given by the scheme

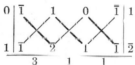

where
$$u = kt - lq, \quad v = lp - ht, \quad w = hq - kp.$$

For example (Fig. 244) the face n lies in the zone mv, $10\bar{1}0 \cdot 2\bar{1}\bar{3}1$ and also in the zone uu, $11\bar{2}0 \cdot 20\bar{2}1$. For the first zone the values obtained are: $u = 0$, $v = 1$, $w = 1$; for the second zone, $e = 1$, $f = \bar{1}$, $g = \bar{2}$. Combining these zone symbols according to the usual scheme

The face n has, therefore, the indices $31\bar{4}1$, since further $i = -(h + k)$.

171. Formulas. — The following formulas in which c equals the unit length of the vertical axis are sometimes useful:
(1) The distances (see Fig. 247) of the pole of any face ($hk\bar{\imath}l$) from the poles of the faces ($10\bar{1}0$), ($01\bar{1}0$), ($\bar{1}100$), and (0001) are given by the following equations:

$$\cos (hk\bar{\imath}l)(10\bar{1}0) = \frac{c(k + 2h)}{\sqrt{3l^2 + 4c^2 (h^2 + k^2 + hk)}}.$$

$$\cos (hk\bar{\imath}l)(01\bar{1}0) = \frac{c(2k + h)}{\sqrt{3l^2 + 4c^2 (h^2 + k^2 + hk)}}.$$

$$\cos (hk\bar{\imath}l)(\bar{1}100) = \frac{c(h - k)}{\sqrt{3l^2 + 4c^2 (h^2 + k^2 + hk)}}.$$

$$\cos (hk\bar{\imath}l)(0001) = \frac{l \sqrt{3}}{\sqrt{3l^2 + 4c^2 (h^2 + k^2 + hk)}}.$$

(2) The distance (PQ) between the poles of any two faces $P(hk\bar{\imath}l)$ and $Q(pq\bar{r}t)$ is given by the equation

$$\cos PQ = \frac{3lt + 2c^2\,(hq + pk + 2hp + 2kq)}{\sqrt{[3l^2 + 4c^2(h^2 + k^2 + hk)]\,[3t^2 + 4c^2(p^2 + q^2 + pq)]}}\,.$$

(3) For special cases the above formula becomes simplified; it serves to give the value of the normal angles for the several forms in the system. They are as follows:

(a) *Pyramid of First Order* $(h0hl)$, Fig. 242:

$$\cos X \text{ (terminal)} = \frac{3l^2 + 2h^2c^2}{3l^2 + 4h^2c^2}\,; \quad \cos Z \text{ (basal)} = \frac{4h^2c^2 - 3l^2}{3l^2 + 4h^2c^2}\,.$$

(b) *Pyramid of Second Order* $(h{\cdot}h{\cdot}2\bar{h}{\cdot}l)$, Fig. 243:

$$\cos Y \text{ (terminal)} = \frac{l^2 + 2c^2h^2}{l^2 + 4c^2h^2}\,; \quad \cos Z \text{ (basal)} = \frac{4c^2h^2 - l^2}{l^2 + 4c^2h^2}\,.$$

(c) *Dihexagonal Pyramid* $(hk\bar{\imath}l)$:

$$\cos X \text{ (see Fig. 244)} = \frac{3l^2 + 2c^2\,(h^2 + k^2 + 4hk)}{3l^2 + 4c^2\,(h^2 + k^2 + hk)}\,.$$

$$\cos Y \text{ (see Fig. 244)} = \frac{3l^2 + 2c^2\,(2h^2 + 2hk - k^2)}{3l^2 + 4c^2\,(h^2 + k^2 + hk)}\,.$$

$$\cos Z \text{ (basal)} \quad = \frac{4c^2\,(h^2 + k^2 + hk) - 3l^2}{3l^2 + 4c^2\,(h^2 + k^2 + hk)}\,.$$

(d) *Dihexagonal Prism* $(hk\bar{\imath}0)$, Fig. 240:

$$\cos X \text{ (axial)} = \frac{h^2 + k^2 + 4hk}{2\,(h^2 + k^2 + hk)}\,. \qquad \cos Y \text{ (diagonal)} = \frac{2h^2 + 2hk - k^2}{2\,(h^2 + k^2 + hk)}\,.$$

(e) *Rhombohedron* $(10\bar{1}1)$:

$$\cos X \text{ (terminal)} \quad = \frac{3l^2 - 2h^2c^2}{3l^2 + 4h^2c^2}\,.$$

(f) *Scalenohedron* $(hk\bar{\imath}l)$:

$$\cos X \text{ (see Fig. 277)} = \frac{3l^2 + 2c\,(2k^2 + 2hk - h^2)}{3l^2 + 4c^2\,(h^2 + k^2 + hk)}\,.$$

$$\cos Y \text{ (see Fig. 277)} = \frac{3l^2 + 2c^2\,(2h^2 + 2hk - k^2)}{3l^2 + 4c^2\,(h^2 + k^2 + hk)}\,.$$

$$\cos Z \text{ (basal)} \quad = \frac{2c^2\,(h^2 + k^2 + 4hk) - 3l^2}{3l^2 + 4c^2\,(h^2 + k^2 + hk)}\,.$$

172. Angles. — The angles for some commonly occurring dihexagonal prisms with the first and second order prisms are given in the following table:

	$m(10\bar{1}0)$	$a(11\bar{2}0)$
$51\bar{6}0$	8° 57′	21° 3′
$41\bar{5}0$	10 $53\frac{1}{2}$	19 $6\frac{1}{2}$
$31\bar{4}0$	13 54	16 6
$52\bar{7}0$	16 6	13 54
$21\bar{3}0$	19 $6\frac{1}{2}$	10 $53\frac{1}{2}$
$32\bar{5}0$	23 $24\frac{3}{4}$	6 $35\frac{1}{4}$
$54\bar{9}0$	26 $19\frac{1}{4}$	3 $40\frac{1}{4}$

173. The Miller Axes and Indices. — The forms of the hexagonal system were referred by Miller to a set of three equal oblique axes which were taken parallel to the edges of the unit positive rhombohedron of the species. Fig. 309 represents such a rhombohedron with the position of the Miller axes shown. This choice of axes for hexagonal forms has the grave objection that in several cases the faces of the same form are represented by two sets of different indices; for example the faces of the pyramid of the first order would have the indices, 100, $22\bar{1}$, 010, $\bar{1}22$, 001, $2\bar{1}2$. This objection, however, disappears if the Miller axes and indices are used only for forms in the Rhombohedral Division, that is, for forms belonging to classes which are characterized by a vertical axis of trigonal symmetry. It is believed, however, that the mutual relations of all the classes of both divisions of the hexagonal system among themselves (as also to the classes of the tetragonal system), both morphological and physical are best brought out by keeping throughout the same axes, namely those of Fig. 236, Art. **125**. The Miller method has, however, been adopted by a number of authors and consequently it is necessary to give the following brief description.

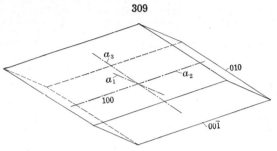

309

Fig. 310 shows in stereographic projection the common hexagonal-rhombohedral forms with their Miller indices and in parentheses the corresponding indices when the faces are referred to the four axial system. It will be noted that the faces of the

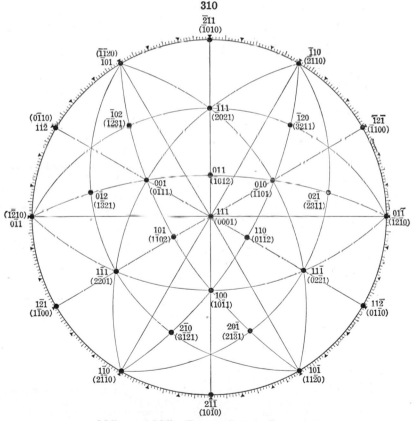

310

Miller and Miller-Bravais Indices Compared

unit positive rhombohedron have the indices 100, 010, and 001 and those of the negative one-half rhombohedron have 110, 011, 101. The hexagonal prism of the first order is represented by $2\bar{1}1$, etc., while the second order prism has $10\bar{1}$, etc. The dihexagonal pyramid has two sets of indices (hkl) and (efg); of these the symbol (hkl) belongs to the positive scalenohedron and (efg) to the negative form. In this as in other cases it is true that $e = 2h + 2k - l$, $f = 2h - k + 2l$, $g = -h + 2k + 2l$. For example, the faces of the form $20\bar{1}$, etc., belong in the Rhombohedral Division of this system to the scalenohedron $(21\bar{3}1)$ while the complementary negative form would have the indices $52\bar{4}$, etc.

The relation between the Miller-Bravais and the Miller indices for any form can be obtained from the following expressions, where $(hkil)$ represents the first and (pqr) the second.

$$p = 2h + k + l; \quad q = k - h + l; \quad r = -2k - h + l;$$

$$h = \frac{p - q}{3}; \quad k = \frac{q - r}{3}; \quad i = h + k; \quad l = \frac{p + q + r}{3}.$$

The relation between the Miller indices for hexagonal forms and those of isometric forms should be noted. If we conceive of the isometric cube as a rhombohedron with interfacial angles of 90° and change the orientation so that the normal to the octahedral face (111) becomes vertical we get a close correspondence between the two.

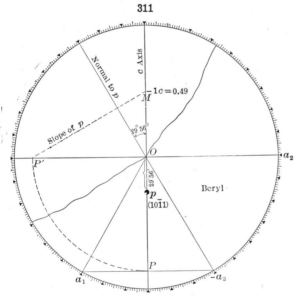

311

Determination of unit length of c axis, having given the position of $p(10\bar{1}1)$

174. To determine, by plotting, the length of the vertical axis of a hexagonal mineral, given the position on the stereographic projection of the pole of a face with known indices. To illustrate this problem it is assumed that the mineral in question is beryl and that the position of the pole $p(10\bar{1}1)$ is known, Fig. 311. Let the three lines a_1, a_2, a_3 represent the horizontal axes with their unit lengths equalling the radius of the circle. Draw a line from the center of the projection through the pole p. Draw another line (which will be at right angles to the first) joining the ends of a_1 and $-a_3$. This will be parallel to a_2 and will represent the intercept of $p(10\bar{1}1)$ upon the plane of the horizontal axes. In order to plot the intercept of p upon the vertical axis construct in the upper left-hand quadrant of the figure a right-angle triangle the base of which shall be equal to O–P, the vertical side of which shall represent the c axis and the hypothenuse shall show the slope of the face and give its intercept upon the c axis. The direction of the hypothenuse is determined by locating the normal to p from the angle measured from the center of the projection to its pole. Since the face has been assumed to have a unit intercept on the vertical axis the distance O–M, which equals 0·49 (in terms of the length of the horizontal axes, which equals 1·00), gives the unit length of the c axis for beryl.

175. To determine the indices of a face of a hexagonal form of a known mineral, given the position of its pole on the stereographic projection. — In Fig. 312 it is assumed that the position of the pole v of a crystal face on calcite is known. To determine its indices, first draw a radial line through the pole and then erect a perpendicular to it, starting the line from the end of one of the horizontal axes. This line will represent the direction of the intersection of the crystal face with the horizontal plane and its relative intercepts on the horizontal axes will give the first three numbers of the parameters of the face, namely $1a_1$, $2a_2$, $\frac{2}{3}-a_3$. To determine the relative intercept on the c axis transfer the distance O–P to the upper left-hand quadrant of the figure, then having measured the angular distance between the center of the projection and v by means of the stereographic protractor draw the pole to the face

312

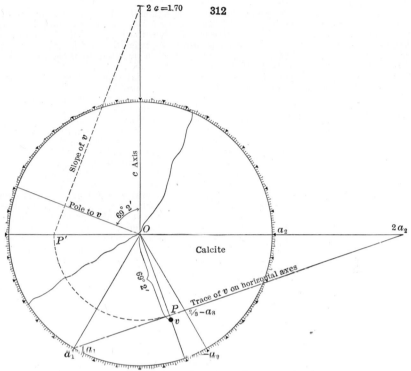

Determination of the indices for v on calcite

313

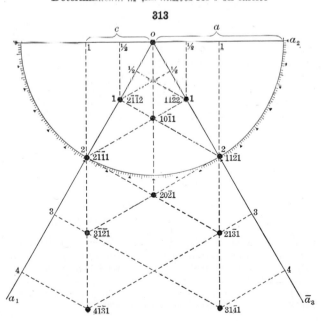

in the proper position.   Draw then a line at right angles to this pole starting from the point P'.   This line gives the intercept of the face upon the line representing the vertical axis.   In this case the intercept has a value of 1.7 when the length of the horizontal axes is taken as equal to 1.0.   This distance 1.7 is seen to be twice the unit length of the c axis for calcite, 0.85.   Therefore the parameters of the face in question upon the four axes are $1a_1, 2a_2, \frac{2}{3} - a_3, 2c$, which give $21\bar{3}1$ for the indices of the face v.

176. To determine, by plotting, the indices of hexagonal forms, given the position of their poles on the gnomonic projection. — To illustrate this problem one sectant of the gnomonic projection of the important forms of beryl, Fig. 246, is reproduced in Fig. 313.   The directions of the three horizontal axes, a_1, a_2 and a_3 are indicated by the heavy lines.   From the poles of the faces perpendiculars are drawn to these three axes.   It will be noted that the various intercepts made upon the axes by these lines have simple rational relations to each other.   One of these intercepts is chosen as having the length of 1 (this length will be equivalent to the unit length of the c crystallographic axis, see below) and the others are then given in terms of it.   The indices of each face are obtained directly by taking these intercepts upon the three horizontal axes in their proper order and by adding a 1 as the fourth figure. If necessary clear of fractions, as in the case of the second order pyramid, $11\bar{2}2$.

177. To determine the axial ratio of a hexagonal mineral from the gnomonic projection of its forms. — The gnomonic projection of the beryl forms, Fig. 313, may be used as an illustrative example.   The radius of the fundamental circle, a, is taken as equal to the length of the horizontal axes and is given a value of 1.   Then the length of the fundamental intercept of the lines dropped perpendicularly from the poles, i.e. the distance c, will equal the length of the c axis when expressed in terms of the length of a.   In the case of beryl this ratio is $a : c = 1.00 : 0.499$.   That this relationship is true can be proved in the same manner as in the case of the tetragonal system, see Art. **121, p. 109.**

IV. ORTHORHOMBIC SYSTEM

(*Rhombic or Prismatic System*)

178. Crystallographic Axes. — The *orthorhombic system* includes all the forms which are referred to three axes at right angles to each other, all of different lengths.

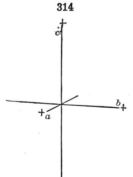

314

Orthorhombic Axes
(Barite)

Any one of the three axes may be taken as the vertical axis, c.   Of the two horizontal axes the longer is always taken as the b or macro-axis* and when orientated is parallel to the observer.   The a or brachy-axis is the shorter of the two horizontal axes and is perpendicular to the observer.   The length of the b axis is taken as unity and the lengths of the other axes are expressed in terms of it.   The axial ratio for barite, for instance, is $a : b : c = 0.815 : 1.00 : 1.31$.   Fig. 314 shows the crystallographic axes for barite.

1. NORMAL CLASS (25).   BARITE TYPE

(*Orthorhombic Dipyramidal, Holohedral, or Didigonal Equatorial Class*)

179. Symmetry.   3 xl. Ax.-2;   3 xl. P. ;   C. — The forms of the *normal class* of the orthorhombic system are characterized by three axes of binary symmetry, which directions are coincident with the crystallographic axes.

---

* The prefixes *brachy-* and *macro-* used in this system (and also in the triclinic system) are from the Greek words, βραχύs, *short*, and μακρόs, *long*.

There are also three unlike planes of symmetry at right angles to each other in which lie the crystallographic axes.

The symmetry of the class is exhibited in the accompanying stereographic projection, Fig. 315. This should be compared with Fig. 109 (p. 69) and Fig. 185 (p. 93), representing the symmetry of the normal classes of the isometric and tetragonal systems respectively. It will be seen that while normal isometric crystals are developed alike in the three axial directions, those of the tetragonal type have a like development only in the direction of the two horizontal axes, and those of the orthorhombic type are unlike in the three axial directions. Compare also Figs. 110 (p. 71), 189 (p. 95) and 316 (p. 140).

180. Forms. — The various forms possible in this class are as follows:

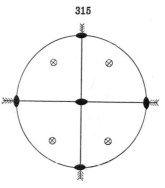

315

Symmetry of Normal Class
Orthorhombic System

		Indices
1.	Macropinacoid or *a*-pinacoid . . .	(100)
2.	Brachypinacoid or *b*-pinacoid. . .	(010)
3.	Base or *c*-pinacoid	(001)
4.	Prisms.	(*hk*0)
5.	Macrodomes.	(*h0l*)
6.	Brachydomes.	(0*kl*)
7.	Pyramids.	(*hkl*)

In general, as defined on p. 47, a *pinacoid* is a form whose faces are parallel to two of the axes, that is, to an axial plane; a *prism* is one whose faces are parallel to the vertical axis, but intersect the two horizontal axes; a *dome** (or *horizontal prism*) is one whose faces are parallel to one of the horizontal axes, but intersect the vertical axis. A pyramid is a form whose faces meet all the three axes.

These terms are used in the above sense not only in the orthorhombic system, but also in the monoclinic and triclinic systems; in the last each form consists of two planes only.

181. Pinacoids. — The *macropinacoid* includes two faces, each of which is parallel both to the macro-axis *b* and to the vertical axis *c*; their indices are respectively 100 and 100. This form is uniformly designated by the letter *a*, and is conveniently and briefly called the *a-face* or the *a-pinacoid.*

The *brachypinacoid* includes two faces, each of which is parallel both to the brachy-axis *a* and to the vertical axis *c*; they have the indices 010 and 0$\overline{1}$0. This form is designated by the letter *b*; it is called the *b-face* or the *b-pinacoid.*

The *base* or *basal pinacoid* includes the two faces parallel to the plane of the horizontal axes, and having the indices 001 and 00$\overline{1}$. This form is designated by the letter *c*; it is called the *c-face* or the *c-pinacoid.*

Each one of these three pinacoids is an open-form,† but together they make the so-called *diametral prism*, shown in Fig. 316, a solid which is the analogue of the cube of the isometric system. Geometrically it cannot be distinguished from the cube, but it differs in having the symmetry unlike in the three axial directions; this may be shown by the unlike physical character of the faces, *a*, *b*, *c*, for example as to luster, striations, etc.; or, again, by the cleavage. Further, it is proved at once by optical properties. This

---

* From the Latin *domus*, because resembling the roof of a house; cf. Figs. 319, 320.
† See p. 46.

diametral prism, as just stated, has three pairs of unlike faces. It has three kinds of edges, four in each set, parallel respectively to the axes a, b, and c; it has, further, eight similar solid angles. In Fig. 316 the dimensions are arbitrarily made to correspond to the relative lengths of the chosen axes, but the student will understand that a crystal of this shape gives no information as to these values.

182. Prisms. — The prisms proper include those forms whose faces are parallel to the vertical axis, while they intersect both the horizontal axes; their general symbol is, therefore, $(hk0)$. These all belong to one type of *rhombic prism*, in which the interfacial angles corresponding to the two unlike vertical edges have different values.

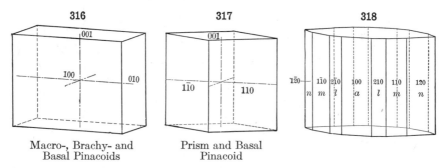

316   317   318

Macro-, Brachy- and   Prism and Basal
Basal Pinacoids   Pinacoid

The *unit prism*, (110), is that form whose faces intersect the horizontal axes in lengths having a ratio corresponding to the accepted axial ratio of $a : b$ for the given species; in other words, the angle of this unit prism fixes the unit lengths of the horizontal axes. This form is shown in combination with the basal pinacoid in Fig. 317; it is uniformly designated by the letter m. The four faces of the unit prism have the indices 110, $\bar{1}10$, $\bar{1}\bar{1}0$, $1\bar{1}0$.

There is, of course, a large number of other possible prisms whose intercepts upon the horizontal axes are not proportionate to their unit lengths. These may be divided into two classes as follows: *macroprisms*, whose faces lie between those of the macropinacoid and the unit prism, *brachyprisms* with faces between those of the brachypinacoid and the unit prism. A macroprism has the general symbol $(hk0)$ in which $h > k$ and is represented by the form $l(210)$, Fig. 318. A brachyprism has the general symbol $(hk0)$ with $h < k$ and is represented by $n(120)$, Fig. 318.

183. Macrodomes, Brachydomes. — The *macrodomes* are forms whose faces are parallel to the macro-axis b, while they intersect the vertical axis c and the horizontal axis a; hence the general symbol is $(h0l)$. The angle of the unit macrodome, (101), fixes the ratio of the axes $a : c$. This form is shown in Fig. 319 combined (since it is an open form) with the brachypinacoid.

In the macrodome zone between the base $c(001)$ and the macropinacoid $a(100)$ there may be a large number of macrodomes having the symbols, taken in the order named, (103), (102), (203), (101), (302), (201), (301), etc.; cf. Figs. 336 and 337 described later.

The *brachydomes* are forms whose faces are parallel to the brachy-axis, a, while they intersect the other axes c and b; their general symbol is $(0kl)$. The angle of the unit brachydome, (011), which is shown with $a(100)$ in Fig. 320, determines the ratio of the axes $b : c$.

The brachydome zone between $c(001)$ and $b(010)$ includes the forms (013), (012), (023), (011), (032), (021), (031), etc.; cf. Figs. 336 and 337.

Both sets of domes are often spoken of as *horizontal prisms*. The propriety of this expression is obvious, since they are in fact prisms in geometrical form; further, the choice of position for the axes which makes them domes, instead of prisms in the narrower sense, is more or less arbitrary, as already explained.

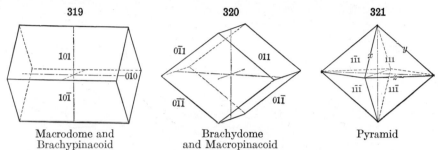

319	320	321
Macrodome and Brachypinacoid	Brachydome and Macropinacoid	Pyramid

184. Pyramids. — The pyramids in this system all belong to one type, the double *rhombic pyramid*, bounded by eight faces, each a scalene triangle. This form has three kinds of edges, x, y, z (Fig. 321), each set with a different interfacial angle; two of these angles suffice to determine the axial ratio. The symbol for this, the general form for the system, is (hkl).

The pyramids may be divided into three groups corresponding respectively to the three prisms just described, namely, unit pyramids, macropyramids, and brachypyramids.

The *unit pyramids* are characterized by the fact that their intercepts on the horizontal axes have the same ratio as those of the unit prism; that is, the assumed axial ratio ($a : b$) for the given species. For them, therefore, the general symbol becomes (hhl).

There may be different unit pyramids on crystals of the same species with different intercepts upon the vertical axis, and these form a *zone* of faces lying between the base $c(001)$ and the unit prism $m(110)$. This zone would include the forms, (119), (117), (115), (114), (113), (112), (111). In the symbol of all of the forms of this zone $h = k$, and the lengths of the vertical axes are hence, in the example given, $\frac{1}{9}, \frac{1}{7}, \frac{1}{5}, \frac{1}{4}, \frac{1}{3}, \frac{1}{2}$ of the vertical axis c of the unit pyramid.

The *macropyramids* and *brachypyramids* are related to each other and to the unit pyramids, as were the macroprisms and brachyprisms to themselves and to the unit prism. Further, each vertical zone of macropyramids (or brachypyramids), having a common ratio for the horizontal axes (or of $h : k$ in the symbol), belongs to a particular macroprism (or brachyprism) characterized by the same ratio. Thus the macropyramids (214), (213), (212), (421), etc., all belong in a common vertical zone between the base (001) and the prism (210). Similarly the brachypyramids (123), (122), (121), (241), etc., fall in a common vertical zone between (001) and (120).

185. Illustrations. — The following figures of barite (322–329) give excellent illustrations of crystals of a typical orthorhombic species, and show also how the habit of one and the same species may vary. The axial ratio for this species is $a : b : c = 0.815 : 1 : 1.314$. Here d is the macrodome

(102) and o the brachydome (011); m is, as always, the prism (110).　Figs. 322–325 and 327 are described as tabular ‖ c(001); Fig. 326 is prismatic in habit in the direction of the macro-axis (b), and Figs. 328, 329 prismatic in that of the brachy-axis (a).

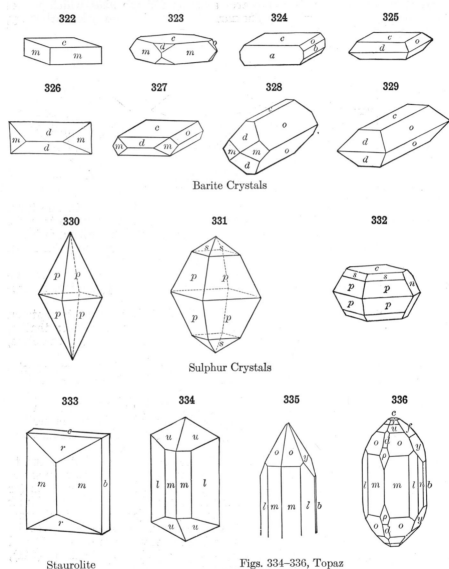

Barite Crystals

Sulphur Crystals

Staurolite　　　　　　　　　Figs. 334–336, Topaz

Figs. 330–332 of native sulphur show a series of crystals of pyramidal habit with the dome n(011), and the pyramids p(111), s(113).　Note n truncates the terminal edges of the fundamental pyramid p.　In general it should be remembered that a macrodome truncating the edge of a pyramid must have the same ratio of $h : l$; thus, (201) truncates the edge of (221), etc. Similarly of the brachydomes: (021) truncates the edge of (221), etc.; cf. Figs. 337–339.

Again, Fig. 333, of staurolite, shows the pinacoids $b(010)$, $c(001)$, the prism $m(110)$, and the macrodome $r(101)$.

Figs. 334–336 are prismatic crystals of topaz. Here m is the prism (110); l and n are the prisms (120), (140); d and ρ are the macrodomes (201) and (401); f and y are the brachydomes (021) and (041); i, u, and o are the pyramids (223), (111), (221).

186. Projections. — Basal, stereographic, and gnomonic projections are given in Figs. 337–339 for a crystal of the species topaz.

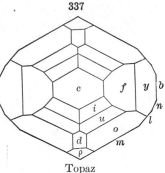

337

Topaz

Fig. 337 is the basal projection of the crystal shown in Fig. 336. Figs. 338 and 339 give the stereographic and gnomonic projections of the forms present upon it.

338

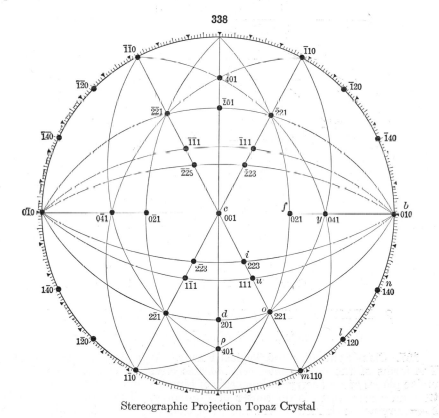

Stereographic Projection Topaz Crystal

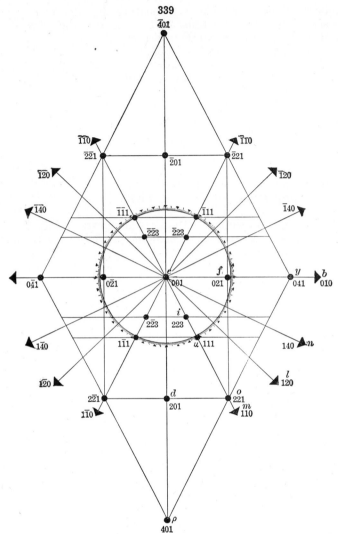

Gnomonic Projection Topaz Crystal

2. HEMIMORPHIC CLASS (26). CALAMINE TYPE

(Orthorhombic Pyramidal or Didigonal Polar Class)

187. Class Symmetry and Typical Forms. Vert. Ax.-2; 2 vert. xl. P. —
The forms of the *orthorhombic-hemimorphic* class are characterized by two
unlike planes of symmetry and one axis of binary symmetry, the line in which
they intersect; there is no center of symmetry. The forms are therefore
hemimorphic, as defined in Art. **28.** For example, if, as is usually the case,
the vertical axis is made the axis of symmetry, the two planes of symmetry

are parallel to the pinacoids $a(100)$ and $b(010)$. The prisms are then geometrically like those of the normal class, as are also the macropinacoid and brachypinacoid; but the two basal planes (or pedions) become independent forms, (001) and $(00\bar{1})$. There are also two macrodomes, (101) and $(10\bar{1})$, or in general $(h0l)$ and $(h0\bar{l})$; and similarly two sets, for a given symbol, of brachydomes and pyramids.

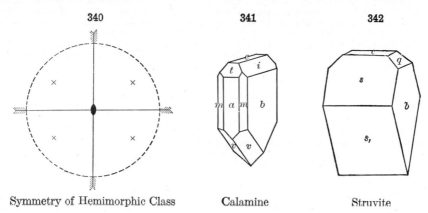

340   341   342

Symmetry of Hemimorphic Class   Calamine   Struvite

The general symmetry of the class is shown in the stereographic projection, Fig. 340. Further, Figs. 341, of calamine, and 342, of struvite, represent typical crystals of this class. In Fig. 341 the forms present are $t(301)$, $i(031)$, $v(12\bar{1})$; in Fig. 342 they are $s(101)$, $s_1(10\bar{1})$, $q(011)$.

3. SPHENOIDAL CLASS (27). EPSOMITE TYPE

(*Orthorhombic Disphenoidal or Digonal Holoaxial Class*)

188. Symmetry and Typical Forms. 3 xl. Ax.-2. — The forms of the remaining class of the system, the *orthorhombic-sphenoidal* class, are characterized by three unlike rectangular axes of binary symmetry which coincide with the crystallographic axes, but they have no plane and no center of symmetry (Fig. 343). The general form hkl here has four faces only, and the corresponding solid is a rhombic sphenoid, analogous to the sphenoid of the tetragonal system. The complementary positive and negative sphenoids are enantiomorphous. Fig. 344 represents a typical crystal of epsomite, with the positive sphenoid, $z(111)$. Other crystals of this species often show both positive and negative complementary forms, but usually unequally developed.

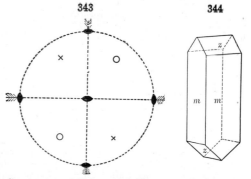

343   344

Symmetry of Sphenoidal Class   Epsomite

Mathematical Relations of the Orthorhombic System

189. Choice of Axes. — As explained in Art. **179**, the three crystallographic axes are fixed as regards direction in all orthorhombic crystals, but any one of them may be made the vertical axis, c; and of the two horizontal axes, which is the longer (b) and which the shorter (a) cannot be determined until it is decided which faces to assume as the fundamental, or unit, pyramid, prism, or domes.

The choice is generally so made, in a given case, as to best bring out the relation of the crystals of the species in hand to others allied to them in form or in chemical composition, or in both respects; or, so as to make the cleavage parallel to the fundamental form; or, as suggested by the common habit of the crystals, or other considerations.

190. Axial and Angular Elements. — The *axial elements* are given by the ratio of the lengths of the three axes in terms of the macro-axis, b, as unity. For example, with barite the axial ratio is

$$a : b : c = 0 \cdot 81520 : 1 : 1 \cdot 31359.$$

The *angular elements* are usually taken as the angles between the three pinacoids and the unit faces in the three zones between them. Thus, again for barite, these elements are

$$100 \wedge 110 = 39° \, 11' \, 13'', \quad 001 \wedge 101 = 58° \, 10' \, 36'', \quad 001 \wedge 011 = 52° \, 43' \, 8''.$$

Two of these angles obviously determine the third angle as well as the axial ratio. The degree of accuracy to be attempted in the statement of the axial ratio depends upon the character of the fundamental measurements from which this ratio has been deduced. There is no good reason for giving the values of a and c to many decimal places if the probable error of the measurements amounts to many minutes. In the above case the measurements (by Helmhacker) are supposed to be accurate within a few seconds. It is convenient, however, to have the angular elements correct, say, within $10''$, so that the calculated angles obtained from them will not vary from those derived direct from the measured angles by more than $30''$ to $1'$.

191. Calculation of the Axes. — The following simple relations (cf. Art. **53**) connect the axes with the angular elements:

$$\tan (100 \wedge 110) = a, \quad \tan (001 \wedge 011) = c, \quad \tan (001 \wedge 101) = \frac{c}{a}.$$

These equations serve to give either the axes from the angular elements, or the angular elements from the axes. It will be noted that the axes are not needed for simple purposes of calculation, but it is still important to have them, for example to use in comparing the morphological relations of allied species.

In practice it is easy to pass from the measured angles, assumed as the basis of calculation (or deduced from the observations by the method of least squares), to the angular elements, or from either to any other angles by the application of the tangent principle (Art. **54**) to the pinacoidal zones, and by the solution of the right-angled spherical triangles given on the sphere of projection.

Thus any face hkl lies in the three zones, 100 and $0kl$, 010 and $h0l$, 001 and $hk0$. For example, the position of the face 312 is fixed if the positions of two of the poles, 302, 012, 310, are known. These last are given, respectively, by the equations

$$\tan (001 \wedge 302) = \tfrac{3}{2} \times \tan (001 \wedge 101),$$
$$\tan (001 \wedge 012) = \tfrac{1}{2} \times \tan (001 \wedge 011) \quad \tan (100 \wedge 310) = \tfrac{1}{3} \times \tan (100 \wedge 110).$$

192. Example. — Fig. 345 represents a crystal of stibnite from Japan and Fig. 346 the stereographic projection of its forms, $p(111)$, $\tau(343)$, $\eta(353)$, $\omega_3(5 \cdot 10 \cdot 3)$, $m(110)$ and $b(010)$. On this the following measured angles were taken as fundamental:

$$\eta\eta' \quad (353 \wedge \bar{3}53) = 55° \, 1' \, 0'',$$
$$\eta\eta''' \quad (353 \wedge 3\bar{5}3) = 99° \, 39' \, 0''.$$

Hence, the angles $353 \wedge 010 = 40° \, 10\tfrac{1}{2}'$ and $353 \wedge 053 = 27° \, 30\tfrac{1}{2}'$ are known without calculation. The right-angled spherical triangle* $010 \cdot 053 \cdot 353$ yields the angle $(010 \wedge 053)$ and hence $(001 \wedge 053)$; also the angle at 010, which is equal to $(001 \wedge 101)$.

---

* The student in this as in every similar case should draw a projection, cf. Fig. 346 (not necessarily accurately constructed), to show, if only approximately, the relative position of the faces present.

But tan $(001 \wedge 011) = \frac{3}{5} \times$ tan $(001 \wedge 053)$, and tan $(001 \wedge 011) = c$. Also, since tan $(001 \wedge 101) = \dfrac{c}{a}$, the axial ratio is thus known, and two of the angular elements.

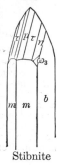

The third angular element $(001 \wedge 110)$ can be calculated independently, for the angle at $\overline{0}01$ in the triangle $001 \cdot 053 \cdot 353$ is equal to $(010 \wedge 350)$ and tan $(010 \wedge 350) \times \frac{5}{3} = (010 \wedge 110)$, the complement of $(100 \wedge 110)$.

Then since tan $(100 \wedge 110) = a$, this can be used to check the value of a already obtained. The further use of the tangent principle with the occasional solution of a right-angled triangle will serve to give any desired angle from either the fundamental angles direct, or from the angular elements.

Again, the symbol of any unknown face can be readily calculated if two measured angles of tolerable accuracy are at hand. For example, for the face ω, suppose the measured angles to be

$$b\omega \ (010 \wedge hkl) = 30° \ 15', \quad \omega\omega' \ (hkl \wedge \bar{h}kl) = 51° \ 32'.$$

The solution of the triangle $b \cdot \omega \cdot 0kl$ gives the angle $(010 \wedge 0kl) = 16° \ 25' \ 20''$, and

$$\frac{\tan (001 \wedge 0kl)}{\tan (001 \wedge 011)} = \frac{\tan 73° \ 34\frac{2}{3}'}{\tan 45° \ 30\frac{1}{2}'} = 3 \cdot 333+, = \frac{k}{l}.$$

But the ratio of $k : l$ must be rational and the number derived agrees most closely with $10 : 3$.

346

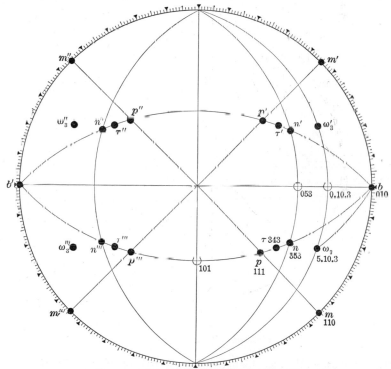

Stereographic Projection Stibnite Crystal

Again, the angle $(001 \wedge h0l)$ may now be calculated from the same triangle and the value $59° \ 38\frac{2}{3}'$ obtained. From this the ratio of h to l is derived since

$$\frac{\tan (001 \wedge h0l)}{\tan (001 \wedge 101)} = \frac{\tan 59° \ 38\frac{2}{3}'}{\tan 45° \ 43\frac{1}{4}'} = 1 \cdot 665 = \frac{h}{l}.$$

This ratio is nearly equal to 5 : 3, and the two values thus obtained give the symbol 5·10·3. If, however, from the triangle 001· 0kl·ω, the angle at 001 is calculated, the value 26° 42¾′ is obtained, which is also the angle (010 ∧ hk0). From this the ratio h : k is deduced, since

$$\frac{\tan\,(010 \wedge 110)}{\tan\,(010 \wedge hk0)} = \frac{\tan 45°\,12\frac{5}{6}′}{\tan 26°\,42\frac{3}{4}′} = 2·002 = \frac{k}{h}.$$

The value of $\frac{k}{h}$ is hence closely equal to 2; this combined with that first obtained $\left(\frac{k}{l} = \frac{10}{3}\right)$ gives the same symbol 5·10·3.

This symbol being more than usually complex calls for fairly accurate measurements. How accurate the symbol obtained is can best be judged by comparing the measured angles with those calculated from the symbol. For example, in the given case the calculated angles for ω(5·10·3) are bω(010 ∧ 5·10·3) = 30° 16′, ωω′ (5̄·10·3) = 51° 35′. The correctness of the value deduced is further established if it is found that the given face falls into prominent zones.

It will be understood further that the zonal relations, explained on pp. 61–64, play an important part in all calculations. For example, in Fig. 345, if the symbol of τ were unknown, it could be obtained from a single angle (as bτ), since for this zone h = l.

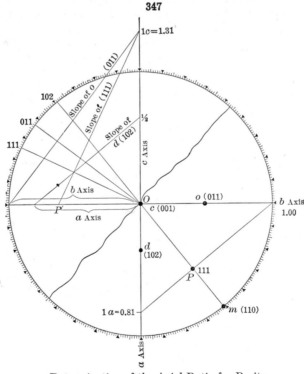

347

Determination of the Axial Ratio for Barite

193. Formulas. — Although it is not often necessary to employ formulas in calculations, a few are added here for sake of completeness. Here a and c in the formulas are the lengths of the two axes a and c.

(1) For the distance between the pole of any face P(hkl) and the pinacoids a, b, c, we have in general:

$$\cos^2 Pa = \cos^2 (hkl \wedge 100) = \frac{h^2 c^2}{h c^2 + k^2 a^2 c^2 + l^2 a^2};$$

$$\cos^2 Pb = \cos^2 (hkl \wedge 010) = \frac{k^2a^2c^2}{h^2c^2 + k^2a^2c^2 + l^2a^2};$$

$$\cos^2 Pc = \cos^2 (hkl \wedge 001) = \frac{l^2a^2}{h^2c^2 + k^2a^2c^2 + l^2a^2}.$$

(2) For the distance (PQ) between the poles of any two faces (hkl) and (pqr)

$$\cos PQ = \frac{hpc^2 + kqa^2c^2 + lra^2}{\sqrt{[h^2c^2 + k^2a^2c^2 + l^2a^2][p^2c^2 + q^2a^2c^2 + r^2a^2]}}.$$

194. To determine, by plotting, the axial ratio of an orthorhombic crystal, having given the stereographic projection of its forms. — In order to solve this problem it is necessary that the position of the pole of a pyramid face of known indices be given or the position of the faces of a prism and one dome or of both a macro- and a brachydome. For illustration it is assumed that a crystal of barite, such as represented in Fig. 323, has been measured on the goniometer and the poles of its faces plotted in the stereographic projection. The lower right-hand quadrant of this projection is shown in Fig. 347. The forms present are common ones on barite crystals and have been given the symbols, $m(110)$, $d(102)$, $o(011)$, $c(001)$. The ratio of $a : b$ can be determined readily from the position of the pole $m(110)$. A radial line is drawn to the pole of the face and then a perpendicular erected to it from the end of the line representing the b crystallographic axis. The intercept of this perpendicular on the line representing the a axis, when expressed in terms of the assumed unit length of the b axis, gives the length of a. It is to be noted that the fact that this line in the present case passes very nearly through the pole 111 is wholly accidental. The length of the vertical axis can be determined from the position of the pole of either $d(102)$ or $o(011)$. The construction used is given in the upper left-hand quadrant of the figure. If the brachydome, $o(011)$, is used the sloping line that gives the inclination of the face is started from a distance on the horizontal line equivalent to the length of the b axis, or 1, and its intercept on the c axis will equal the unit length of that axis. If, however, the position of $d(102)$ is used the base line of the triangle must be made equal to the unit length of the a axis as already established and the intercept on the c axis will equal $\frac{1}{2}$ of the latter's unit length.

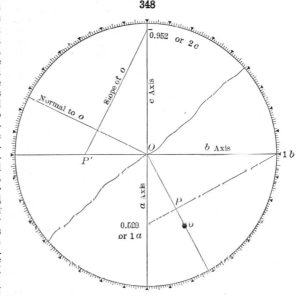

348

The problem could have been wholly solved from the position of the pyramid face, 111, if that form had been present on the crystal. The construction in this case is also illustrated.

195. To determine, by plotting, the indices of a face upon an orthorhombic crystal, given the position of its pole upon the stereographic projection and the axial ratio of the mineral. — To illustrate this problem it is assumed that the position of the pole in the stereographic projection of the face, o, Fig. 348, upon a topaz crystal is known. First draw a radial line through the pole o. Next erect a perpendicular to this line, starting it from the distance selected as representing 1 on the b crystallographic axis. The intercept of this line upon the line representing the a axis when expressed in terms of the unit length of the b axis is 0·53. This is equivalent to the established unit length of the a axis and therefore the parameters of the face o on the horizontal crystallographic axes are $1a$, $1b$. Next the distance $O–P$ is transferred into the upper left-hand quadrant of the figure. The position of the normal to the face is determined by measuring with a protractor the angular distance between O and o. The line giving the slope of the face is next drawn perpendicular to this normal and its intercept upon the line representing the vertical axis determined. This distance when expressed in terms of the

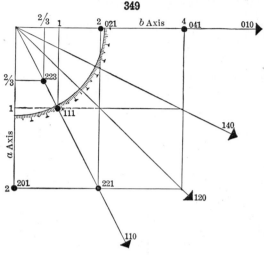

349

length of the b axis is 0·95. This is twice the established length of the c axis (0·476) and consequently the third parameter of the face o is $2c$. This gives the indices 221 for the face.

196. To determine, by plotting, the axial ratio of an orthorhombic crystal having given the gnomonic projection of its forms. — To illustrate this problem the gnomonic projection of the crystal of topaz already given in Fig. 339 will be used. In Fig. 349 one quadrant of this projection is reproduced. From each pole lines are drawn perpendicular to the two lines representing the a and b crystallographic axes. It will be found that the intercepts made in this way upon the a axis have rational relations to each other. The same is true of the intercepts upon the b axis. The intercepts upon the two axes, however, are irrational in respect to each other. A convenient intercept upon each axis is chosen as 1 and the other intercepts upon that axis are then expressed in terms of this length. Of course with a known mineral, whose forms have already had indices assigned to them, the intercept that shall be considered as 1 is fixed.

350

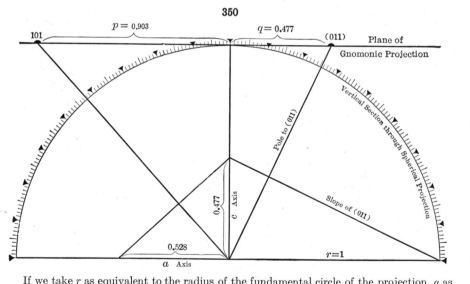

If we take r as equivalent to the radius of the fundamental circle of the projection, q as equal to the chosen intercept upon the b crystallographic axis and p that upon the a axis, then the axial ratio can be derived from the following expressions:

$$q = c; \quad \frac{a}{c} = \frac{r}{p},$$

The proof of these relationships is similar to that already given under the Tetragonal System, Art. **121**, p. 109, and is illustrated in Fig. 350 for the case of topaz. The position of the pole of the face (011) is shown on the right end of the tangential line representing the plane of the gnomonic projection. The line giving the slope of (011) is at right angles to the pole and by its intercept upon the line representing the vertical axis gives the unit length of that axis. The two right-angle triangles shown on the right-hand side of the figure are identical and q equals c. Similarly on the left-hand side of the figure the position of the pole to the face (101) is shown and the line giving the slope of that face starting from unity on the c axis will intercept the line representing the a axis at its unit length. The two right-angle triangles in this half of the figure are similar and therefore the values of a and c are proportional to r and p.

197. To determine, by plotting, the indices of a face upon an orthorhombic crystal, given the position of its pole upon the gnomonic projection and the axial ratio of the mineral. — The method of construction in this case is the reverse of that given in the problem above and is essentially the same as given under the Isometric and Tetragonal Systems, Arts. **88** and **122**. In the case of an orthorhombic mineral the intercepts of the perpendiculars drawn from the pole of the face to the a and b axes must be expressed in each case in terms of the unit intercept on that axis. These values, p and q, can be determined from the equations given in the preceding problem.

<hr />

V. MONOCLINIC SYSTEM

(*Oblique or Monosymmetric System*)

198. Crystallographic Axes. — The *monoclinic system* includes all the forms which are referred to three unequal axes, having one of their axial inclinations oblique.

The axes are designated as follows: the inclined or clino-axis is a; the ortho-axis is b, the vertical axis is c. The acute angle between the axes a and c is represented by the letter β; the angles between a and b and b and c are right angles. See Fig. 351. When properly orientated the inclined axis, a, slopes down toward the observer, the b axis is horizontal and parallel to the observer and the c axis vertical.

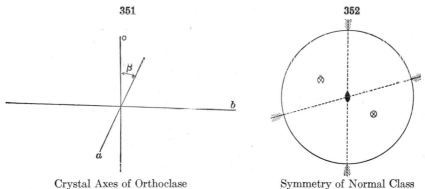

351

352

Crystal Axes of Orthoclase
$a:b:c = 0.66:1:0.55. \ \beta = 64°$

Symmetry of Normal Class

1. NORMAL CLASS (28). GYPSUM TYPE

(*Prismatic, Holohedral, or Digonal Equatorial Class*)

199. Symmetry. b xl. Ax.-2 ; a-c xl. P. ; C. — In the normal class of the monoclinic system there is one plane of symmetry and one axis of binary symmetry normal to it. The plane of symmetry is always the plane of the axes a and c, and the axis of symmetry coincides with the axis b, normal to this plane. The position of one axis (b) and that of the plane of the other two axes (a and c) is thus fixed by the symmetry; but the latter axes may occupy different positions in this plane. Fig. 352 shows the typical stereographic projection, projected on the plane of symmetry. Figs. 367, 368 are the projections of an actual crystal of epidote; here, as is usual, the plane of projection is normal to the prismatic zone.

200. Forms. — The various forms* belonging to this class, with their symbols, are given in the following table. As more particularly explained later, an orthodome includes two faces only, and a pyramid four only.

		Symbols
1.	Orthopinacoid or a-pinacoid.....................	(100)
2.	Clinopinacoid or b-pinacoid......................	(010)
3.	Base or c-pinacoid..............................	(001)
4.	Prisms..	$(hk0)$
5.	Orthodomes....................................	$\begin{cases} (h0l) \\ (\bar{h}0l) \end{cases}$
6.	Clinodomes....................................	$(0kl)$
7.	Pyramids......................................	$\begin{cases} (hkl) \\ (\bar{h}kl) \end{cases}$

201. Pinacoids. — The pinacoids are the orthopinacoid, clinopinacoid, and the basal plane.

The *orthopinacoid*, (100), includes the two faces parallel to the plane of the ortho-axis b and the vertical axis c. They have the indices 100 and $\bar{1}00$. This form is designated by the letter a, since it is situated at the extremity of the a axis; it is hence conveniently called the *a-face* or *a-pinacoid*.

The *clinopinacoid*, (010), includes the two faces parallel to the plane of symmetry, that is, the plane of the clino-axis a and the axis c. They have the indices 010 and $0\bar{1}0$. The clinopinacoid is designated by the letter b, and is called the *b-face* or *b-pinacoid*.

The *base* or *basal pinacoid*, (001), includes the two terminal faces, above and below, parallel to the plane of the axes a, b; they have the indices 001 and $00\bar{1}$. The base is designated by the letter c, and is often called the *c-face* or *c-pinacoid*. It is obviously inclined to the orthopinacoid, and the normal angle between the two faces (100 $\wedge$ 001) is the acute axial angle β.

The *diametral prism*, formed by these three pinacoids, taken together, Fig. 353, is the analogue of the cube in the isometric system. It is bounded by three sets of unlike faces; it has four similar vertical edges; also four similar edges parallel to the axis a, but the remaining edges, parallel to the axis b, are of two sets. Of its eight solid angles there are two sets of four each; the two above in front are similar to those below behind, and the two below in front to those above in behind.

202. Prisms. — The prisms are all of one type, the oblique rhombic prism. They may be divided into three classes as follows: the *unit prism*, (110), designated by the letter m, shown in Fig. 354; the *orthoprisms*, $(hk0)$, where $h > k$, lying between a(100) and m(110), and the *clinoprisms*, $(hk0)$ where $h < k$, lying between m(110) and b(010). The orthoprisms and clinoprisms correspond respectively to the macroprisms and brachyprisms of the

---

* On the general use of the terms pinacoid, prisms, domes, pyramids, see pp. 47, 139.

orthorhombic system, and the explanation on p. 140 will hence make their relation clear.   Common cases of these prisms are shown in the figures given later.

203. Orthodomes. — The four faces parallel to the ortho-axis b, and meeting the other two axes, fall into two sets of two each, having the general symbols $(h0l)$ and $\bar{h}0l)$.   These forms are called *orthodomes*; they are strictly hemiorthodomes.   For example, the unit orthodome (101) has the faces 101 and $\bar{1}0\bar{1}$; they would replace the two obtuse edges between $a(100)$ and $c(001)$ in Fig. 353.   The other unit orthodome ($\bar{1}01$) has the faces $\bar{1}01$ and $10\bar{1}$, and they would replace the acute edges between $a(100)$ and $c(001)$.   These two independent forms are shown together, with $b(010)$, in Fig. 355.

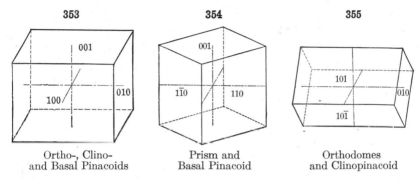

353	354	355

| Ortho-, Clino- and Basal Pinacoids | Prism and Basal Pinacoid | Orthodomes and Clinopinacoid |

Similarly the faces 201, $\bar{2}0\bar{1}$ belong to the form (201), and $\bar{2}01$, $20\bar{1}$ to the independent but complementary form ($\bar{2}01$).

204. Clinodomes. — The *clinodomes* are the forms whose faces are parallel to the inclined axis, a, while intersecting the other two axes.   Their general symbol is hence $(0kl)$ and they lie between the base (001) and the clinopinacoid (010).   Each form has four faces; thus for the unit clinodome these have the symbols, 011, $0\bar{1}1$, $0\bar{1}\bar{1}$, $01\bar{1}$.   The form $n(021)$ in Fig. 362 is a clinodome.

205. Pyramids. — The *pyramids* in the monoclinic system are all hemipyramids, embracing four faces only in each form, corresponding to the general symbol (hkl).   This obviously follows from the symmetry; it is shown, for example, in the fact already stated that the solid angles of the diametral prism (Fig. 353, see above), which are replaced by these pyramids, fall into two sets of four each.   Thus any general symbol, as (321), includes the two independent forms (321) and ($\bar{3}21$) with the faces

$$321, \quad 3\bar{2}1, \quad \bar{3}2\bar{1}, \quad \bar{3}2\bar{1}, \quad \text{and} \quad \bar{3}21, \quad \bar{3}\bar{2}1, \quad 32\bar{1}, \quad 3\bar{2}\bar{1}.$$

The pyramids may also be divided into three classes as *unit pyramids*, (hhl); *orthopyramids*, (hkl), when $h > k$; or *clinopyramids*, (hkl), when $h < k$. These correspond respectively to the three prisms already named.   They are analogous also to the unit pyramids, macropyramids, and brachypyramids of the orthorhombic system, and the explanation given on p. 141 should serve to make their relations clear.   But it must be remembered that each general symbol embraces two forms, (hhl) and $(\bar{h}kl)$ with four faces each, as above explained.

206. Illustrations. — Figs. 356–359 of pyroxene ($a : b : c = 1\cdot092 : 1 : 0\cdot589$, $\beta = 74° = a(100) \wedge c(001)$) show typical monoclinic forms.   Fig. 356

shows the diametral prism. Of the other forms, m is the unit prism (110); $p(\bar{1}01)$ is an orthodome; $u(111)$, $v(221)$, $s(\bar{1}11)$ are pyramids; for other figures see p. 556. Again, Figs. 360–362 represent common crystals of orthoclase ($a : b : c = 0.659 : 1 : 0.555$, $\beta = 64°$). Here $z(130)$ is a prism; $x(\bar{1}01)$ and $y(\bar{2}01)$ are orthodomes; $n(021)$ is a clinodome; $o(\bar{1}11)$ a pyramid. Since (Fig. 360) c and x happen to make nearly equal angles with the vertical

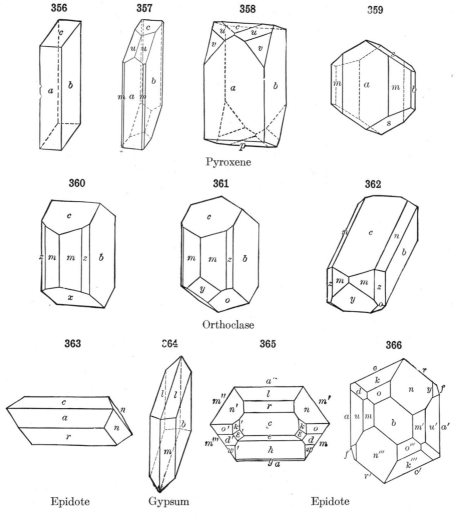

356   357   358   359

Pyroxene

360   361   362

Orthoclase

363   364   365   366

Epidote   Gypsum   Epidote

edge of the prism m, the combination often simulates an orthorhombic crystal.

Fig. 363 shows a monoclinic crystal, epidote, prismatic in the direction of the ortho-axis; the forms are $a(100)$, $c(001)$, $r(\bar{1}01)$ and $n(\bar{1}11)$. Fig. 364 of gypsum is flattened $\parallel b(010)$; it shows the unit pyramid $l(111)$ with the unit prism $m(110)$.

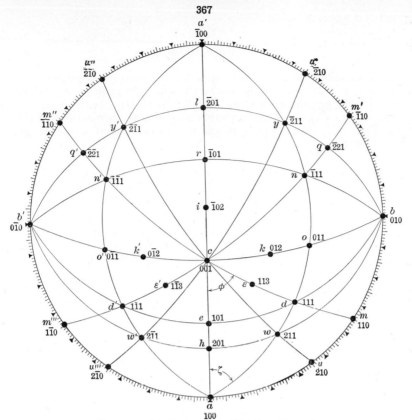

Stereographic Projection of Epidote Crystal

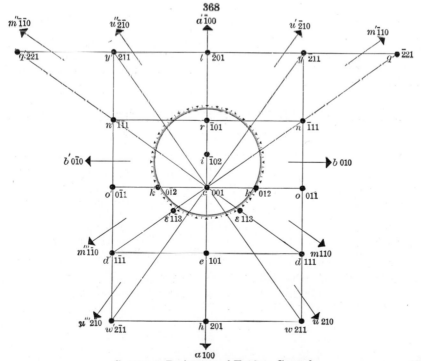

Gnomonic Projection of Epidote Crystal

(155)

207. Projections. — Fig. 365 shows a projection of a crystal of epidote (cf. Fig. 936, p. 623) on a plane normal to the prismatic zone, and Fig. 366 one of a similar crystal on a plane parallel to $b(010)$; both should be carefully studied, as also the stereographic and gnomonic projections of the same species, Figs. 367, 368. The symbols of the prominent faces are given in the latter figures.

2. HEMIMORPHIC CLASS (29). TARTARIC ACID TYPE

(*Sphenoidal or Digonal Polar Class*)

208. b xl. Ax.-2. — The *monoclinic-hemimorphic* class is characterized by a single axis of binary symmetry, the crystallographic axis b, but it has no plane of symmetry. It is illustrated by the stereographic projection (Fig. 369) made upon a plane parallel to $b(010)$. Fig. 370 shows a common form

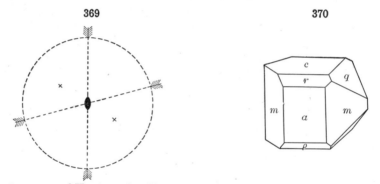

369   **370**

Symmetry of Hemimorphic Class   Tartaric Acid

of tartaric acid; sugar crystals also belong here. The hemimorphic character is distinctly shown in the distribution of the clinodomes and pyramids; corresponding to this the artificial salts belonging here often exhibit marked pyroelectrical phenomena.

3. CLINOHEDRAL CLASS (30). CLINOHEDRITE TYPE

(*Domatic, Hemihedral, or Planar Class*)

209. a-c xl. P. — The *monoclinic-clinohedral* class is characterized by a single plane of symmetry, parallel to the clinopinacoid, $b(010)$, but it has no axis of symmetry. This symmetry is shown in the stereographic projection made upon a plane parallel to $b(010)$, Fig. 371. In this class, therefore, the forms parallel to the b axis, viz., $c(001)$, $a(100)$, and the orthodomes, are represented by a single face only. The other forms have each two faces, but it is to be noted that, with the single exception of the clinopinacoid $b(010)$, the faces of a given form are never parallel to each other. The name given to the class is based on this fact.

Several artificial salts belong here, but there are few known representatives among minerals. One is the rare silicate, clinohedrite, a complex crystal of which is shown in two positions in Figs. 372, 373. As seen in these figures, the crystals of the group have a hemimorphic aspect with respect to their development in the direction of the vertical axis, although they cannot properly be called hemimorphic since this is not an axis of symmetry. The forms shown in Figs. 372, 373 are as follows: pinacoid, $b(010)$; prisms, $m(110)$, $m_1(\bar{1}10)$, $h(320)$, $n(120)$, $l(130)$; orthodomes, $e(101)$, $e_1(\bar{1}0\bar{1})$; pyramids, $p(111)$, $p_1(11\bar{1})$, $q(\bar{1}11)$, $r(\bar{3}31)$, $s(\bar{5}51)$, $t(\bar{7}71)$, $u(\bar{5}31)$, $o(\bar{1}31)$, $x(\bar{1}3\bar{1})$, $y(\bar{1}2\bar{1})$.

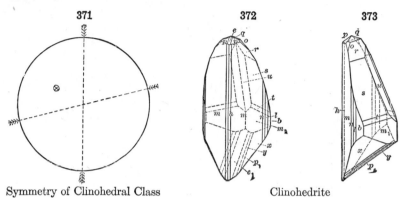

| 371 | 372 | 373 |

Symmetry of Clinohedral Class   Clinohedrite

MATHEMATICAL RELATIONS OF THE MONOCLINIC SYSTEM

210. Choice of Axes. — It is repeated here (Art. **199**) that the fixed position of the plane of symmetry establishes the direction of the plane of the a and c crystallographic axes and also of the axis b which is the symmetry axis and lies at right angles to this plane. The a and c axes, however, may have varying positions in the symmetry plane according to which faces are taken as the pinacoids $a(100)$ and $c(001)$, and which the unit pyramid, prism, or domes.

211. Axial and Angular Elements. — The *axial elements* are the *lengths* of the axes a and c in terms of the unit axis b, that is, the axial ratio, with also the acute angle of inclination of the axes a and c, called β. Thus for orthoclase the axial elements are:

$$a : b : c = 0.6585 : 1 : 0.5554 \quad \beta = 63°\,56\tfrac{3}{4}'.$$

The *angular elements* are usually taken as the angle $(100 \wedge 001)$ which is equal to the angle β; also the angles between the three pinacoids 100, 010, 001, respectively, and the unit prism 110, the unit orthodome (101 or $\bar{1}01$) and the unit clinodome 011. Thus, again, for orthoclase, the angular elements are:

$$001 \wedge 100 = 63°\,56\tfrac{3}{4}', \quad 100 \wedge 110 = 30°\,36\tfrac{1}{2}'.$$
$$001 \wedge \bar{1}01 = 50°\,16\tfrac{1}{2}', \quad 001 \wedge 011 = 26°\,31'.$$

212. The mathematical relations connecting axial and angular elements are given in the following equations in which a, b, and c represent the unit lengths of the respective crystallographic axes.

$$a = \frac{\tan(100 \wedge 110)}{\sin \beta} \quad \text{or} \quad \tan(100 \wedge 110) = a \cdot \sin \beta; \quad (1)$$

$$c = \frac{\tan(001 \wedge 011)}{\sin \beta} \quad \text{or} \quad \tan(001 \wedge 011) = c \cdot \sin \beta; \quad (2)$$

$$\left. \begin{aligned} c &= \frac{a \cdot \tan(001 \wedge 101)}{\sin \beta - \cos \beta \cdot \tan(001 \wedge 101)} \quad \text{or} \quad \tan(001 \wedge 101) = \frac{c \sin \beta}{a + c \cdot \cos \beta}, \\ c &= \frac{a \cdot \tan(001 \wedge \bar{1}01)}{\sin \beta + \cos \beta \cdot \tan(001 \wedge \bar{1}01)} \quad \text{or} \quad \tan(001 \wedge \bar{1}01) = \frac{c \sin \beta}{a - c \cdot \cos \beta}. \end{aligned} \right\} \quad (3)$$

These relations may be made more general by writing in the several cases —

in (1) $hk0$ for 110   and $\dfrac{k}{h} a$ for a;   in (2) $0kl$ for 011   and $\dfrac{k}{l} c$ for c;

in (3) $h0l$ for 101   and $\dfrac{h}{l} c$ for c.

Also

$$\frac{c}{a} = \frac{\sin\,(001 \wedge 101)}{\sin\,(100 \wedge 101)} = \frac{\sin\,(001 \wedge \bar{1}01)}{\sin\,(\bar{1}00 \wedge \bar{1}01)},$$

and more generally

$$\frac{h}{a} \cdot \frac{c}{l} = \frac{\sin\,(001 \wedge h0l)}{\sin\,(100 \wedge h0l)} = \frac{\sin\,(001 \wedge \bar{h}0l)}{\sin\,(\bar{1}00 \wedge \bar{h}0l)}.$$

Note also that

$$\tan \phi = a \quad \text{and} \quad \tan \zeta = c,$$

where ϕ is the angle (Fig. 367) between the zone-circles (001, 100) and (001, 110); also ζ is the angle between (100, 001) and (100, 011).

All the above relations are important and should be thoroughly understood.

213. The problems which usually arise have as their object either the deducing of the axial elements, *i.e.*, the angle β and the values of a and c in terms of $b(= 1)$, from three measured angles, or the finding of any required interfacial angles from these elements or from the fundamental angles.

The simple relations of the preceding article connect the angular and axial elements, and beyond this all ordinary problems can be solved* either by the solution of spherical triangles on the sphere of projection, or by the aid of the cotangent (and tangent) relation.

It is to be noted, in the first place, that all great circles on the sphere of projection (see the stereographic projection, Fig. 367) from 010 cut the zone circle 100, 001, $\bar{1}00$ at right angles, but those from 100 cut the zone circles 010, 001, $0\bar{1}0$ obliquely, as also those from 001 cutting the zone circle 100, 010, $\bar{1}00$.

214. Tangent and Cotangent Relations. — The simple *tangent relation* holds good for all zones from $0\bar{1}0$ to any pole on the zone circle 100, 001, $\bar{1}00$; in other words, for the prisms, clinodomes, and also zones of pyramids in which the ratio of $h : l$ is constant (from 001 to $h0l$ or to $\bar{h}0l$). Thus it is still true, as in the orthorhombic system, that the tangents of the angles of the prisms 210, 110, 120, 130 from 100 are in the ratio of $\frac{1}{2} : 1 : 2 : 3$, or, more generally, that

$$\frac{\tan\,(100 \wedge hk0)}{\tan\,(100 \wedge 110)} = \frac{k}{h} \quad \text{or} \quad \frac{\tan\,(010 \wedge hk0)}{\tan\,(010 \wedge 110)} = \frac{h}{k}.$$

Also for the clinodomes the tangents of the angles of 012, 011, 021 from 001 are in the ratio of $\frac{1}{2} : 1 : 2$, etc. A similar relation holds for the tangents of the angles of pyramids in the zones mentioned, as 121, 111, 212, etc.

For zones other than those mentioned as from 100 to a clinodome, or from 001 to a prism, the more general *cotangent formula* given in Art. **54** must be employed. This relation is simplified for certain common cases.

For any zone starting from 001, as the zone 001, 100, or 001, 110, or 001, 210, etc.: if two angles are known, viz., the angles between 001 and those two faces in the given zone which fall (1) in the zone 010, 101, and (2) in the prismatic zone 010, 100; then the angle between 001 and any other face in the given zone can be calculated.

Thus,

Let 001 $\wedge$ 101 = PQ and   001 $\wedge$ 100 = PR,
or " 001 $\wedge$ 111 = PQ "   001 $\wedge$ 110 = PR,
or " 001 $\wedge$ 212 = PQ "   001 $\wedge$ 210 = PR, etc.

Then for these, or any similar cases, the angle (PS) between 001 and any face in the given zone (as 201, or 221, or 421, etc., or in general $h0l$, hhl, etc.) is given by the equation

$$\frac{\cot PS - \cot PR}{\cot PQ - \cot PR} = \frac{l}{h}.$$

---

* The general formulas, from which it is possible to calculate directly the angles between any face and the pinacoids, or the angle between any two faces whatever, are so complex as to be of little value.

For the corresponding zones from 001 to $\bar{1}00$, to $\bar{1}10$, to $\bar{2}10$, etc., the expression has the same value; but here

$$PQ = 001 \wedge \bar{1}01, \quad PR = 001 \wedge \bar{1}00, \quad PS = 001 \wedge \bar{h}0l,$$
$$\text{or} \quad 001 \wedge \bar{1}11, \text{ etc.}, \quad 001 \wedge \bar{1}10, \text{ etc.}, \quad 001 \wedge \bar{h}hl, \text{ etc.}$$

If, however, 100 is the starting-point, and

$$100 \wedge 101 = PQ, \quad 100 \wedge 001 = PR,$$
$$\text{or } 100 \wedge 111 = PQ, \quad 100 \wedge 011 = PR, \text{ etc.},$$

then the relation becomes

$$\frac{\cot PS - \cot PR}{\cot PQ - \cot PR} = \frac{h}{l}.$$

215. To determine, by plotting, the axial elements of a monoclinic crystal, given the stereographic projection of its forms. — As an example of this problem it is assumed that an orthoclase crystal similar to the one shown in Fig. 361 has been measured and the poles

374

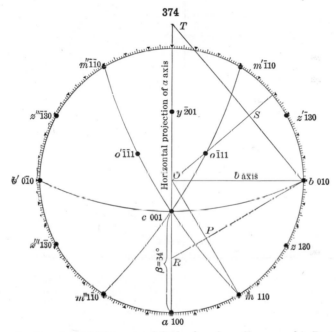

Determination of Axial Elements of Orthoclase from Stereographic Projection

of its faces located on the stereographic projection, Fig. 374. The inclination of the a axis or the angle β is given directly by measuring, by means of the stereographic protractor, the angular distance between the poles of $a(100)$ and $c(001)$. In the present case the $a(100)$ form does not actually occur on the crystal. β is measured as 64°. If the base is not present upon the crystal it will be possible usually to locate its position by means of some zone circle on which it must be. In the present case the great circle of the zone of $m'(\bar{1}10)$, $o(\bar{1}11)$, $m'''(1\bar{1}0)$ will cross the front to back line (zone of the orthodomes) at the point of the pole to the base.

The ratio between the lengths of the a and b axes can be readily determined from the position of the pole, $m(110)$. Draw the radial line O–P from the center of the projection to $m(110)$. From the end of the b axis draw a line at right angles to O–P. This represents the intersection of the prism face with the horizontal plane and the distance O–R gives the intercept of the prism upon the *horizontal projection of the a axis*. The distance O–R

therefore is not the unit length of the a axis but is that distance foreshortened somewhat because of the inclination of that axis. The construction by which the true length of the a axis is obtained is shown in Fig. 375. The line R–O–S–T represents the horizontal projection of the a axis upon which the distance O–R is transferred from Fig. 374. As the prism face is vertical its intercept upon the a axis can be found by dropping a perpendicular from R to intersect the line which represents the a axis. The inclination of this last line is found by use of the angle β, which has been already determined. The length of the a axis when expressed in terms of the b axis (1·00) was found to be 0·66.

The length of the c axis can be found best from the inclination of the $y(\bar{2}01)$ face. This face will intersect the negative end of the a axis and the upper end of the c axis at either $\frac{1}{2}a$, $1c$ or $1a$, $2c$. The angle between the center of the projection, O, Fig. 374, and the pole y

375

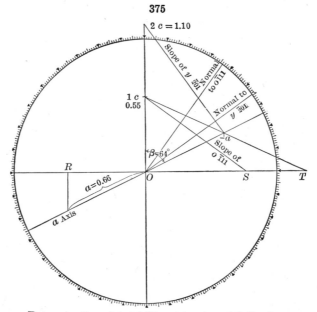

Determination of Axial Elements, etc., of Orthoclase

is measured by means of the stereographic protractor. From this angle the position of the normal to y, as shown in Fig. 375, is determined. The line representing the slope of the face is drawn at right angles to this normal, starting from the negative unit length of the inclined a axis. The intercept on the c axis was found to be equal to 1·11, which, as it is equal to $2c$, would give the unit length of the c axis as, 0·55.

The length of the c axis could also be determined from the inclination of the pyramid face, $o(\bar{1}11)$. The method of construction would be similar to that described in the problem below.

216. To determine the indices of a face upon a monoclinic crystal, having given the position of its pole upon the stereographic projection and the axial elements of the mineral. The pyramid face o on orthoclase will be used to illustrate the problem. First, see Fig. 374, a radial line is drawn through the pole o and a perpendicular S–T erected to it, starting from the unit length of the b axis. It is to be noted that the point T is the intersection of the face o with the horizontal projection of the a axis. Transfer the distance O–S to the horizontal line in Fig. 375 and locate the position of the normal to o by the angle, Fig. 374, between O and o. The line giving the slope of the face can then be drawn from the point S (Fig. 375) perpendicular to the normal. This line intersects the line representing the vertical axis at a distance equal to its unit length. Two points of intersection of the pyramid face with the plane of the a and c axes have now been determined, namely $1c$ and T. A line joining these two points will give the intersection of the two planes and the point where it crosses the line representing the a axis will therefore give the intercept of the pyramid upon that axis.   This is also found to be at the unit length and therefore the indices of *o* must be $\bar{1}11$.

217. To determine, by plotting, the axial elements of a monoclinic crystal, having given the gnomonic projection of its forms. — The construction by which this problem is solved is shown in Fig. 376.   The poles of the unit forms (101), (011), (001) and ($1\bar{1}1$) are located (in this case for pyroxene) and the zonal lines drawn.   The angle β is complementary to the angle from the center of the projection to 001.   This can be measured directly by means of the gnomonic tangent scale.   Then construct the triangles CST and XYZ.   The angles ρ and π, and μ and ν are measured.   This can most easily be done by means of the divided circle

376

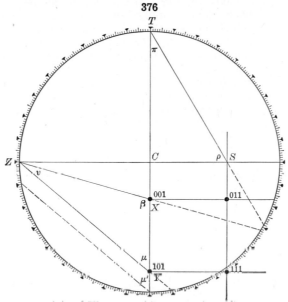

Determination of Axial Elements of Pyroxene from Gnomonic Projection

and the fact that an angle at the circumference of a circle is measured by one half its subtended arc.   The following relations will then yield the axial ratio

$$\frac{b}{c} = \frac{\sin \rho}{\sin \pi}; \quad \frac{a}{c} = \frac{\sin \mu}{\sin \nu}.$$

For the proof of these relations see the explanation of the more general case under the triclinic system, Art. **231**, p. 169.

218. To determine, by plotting, the indices of a face on a monoclinic crystal, having given the position of its pole upon the gnomonic projection. — There is no essential difference between the orthorhombic and monoclinic systems in the determination of indices from the gnomonic projection.   The intercepts of perpendiculars from the poles of the faces upon the front to back and left to right zonal lines running through the pole of c(001) give directly the first two numbers of the indices.   The gnomonic projection of the epidote crystal already given (Fig. 368) will serve to illustrate this problem.

VI. TRICLINIC SYSTEM

(*Anorthic System*)

219. Crystallographic Axes. — The *triclinic system* includes all the forms which are referred to three unequal axes with all their intersections oblique.

When orientated in the customary manner one axis has a vertical position and is called the *c* axis (cf. Fig. 377), a second axis lies in the front-to-back plane, sloping down toward the observer, and is called the *a* axis. The remaining axis is designated as the *b* axis. Usually the *a* and *b* axes are so chosen that the *a* axis is the shorter and, as in the orthorhombic system, is sometimes called the brachy-axis. In that case the *b* axis is longer and is known as the macro-axis. But this is not invariably true; thus with rhodonite the ratio of $a : b = 1\cdot073 : 1$. The angle between the axes *b* and *c* is called α, that between *a* and *c* is β, and that between *a* and *b* is γ (Fig. 377).

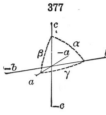

377

Triclinic Axes

It is to be noted that there is no necessary relation between the values of α, β, and γ, any one may be greater or less than 90°; this is determined by the choice of the fundamental forms.

1. NORMAL CLASS (31). AXINITE TYPE

(*Holohedral, Pinacoidal, or Central Class*)

220. Symmetry. C. — The normal class of the triclinic system is characterized by a center of symmetry, the point of intersection of the three axes, but there is no plane and no axis of symmetry. This symmetry is shown in the accompanying stereographic projection (Fig. 378).

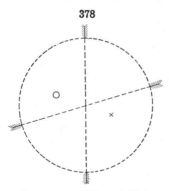

378

Symmetry of Normal Class

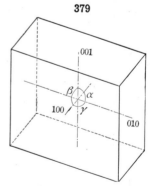

379

Triclinic Pinacoids

221. Forms. — Each form of the class includes two faces, parallel to one another and symmetrical with reference to the center of symmetry. This is true as well of the form with the general symbol (*hkl*) as of one of the special forms, as, for example, the *a*-pinacoid (100).

Hence, as shown in the following table, the four unit prismatic faces include two forms, namely, 110, $\bar{1}\bar{1}0$, and $\bar{1}10$, $1\bar{1}0$. The same is true of the domes. Further, any eight corresponding pyramidal faces belong to four distinct forms, namely, 111, $\bar{1}\bar{1}\bar{1}$; $\bar{1}11$, $1\bar{1}\bar{1}$; $\bar{1}\bar{1}1$, $11\bar{1}$; $1\bar{1}1$, $\bar{1}1\bar{1}$, and similarly in general.

The various types of forms are given in the following table:

	Indices
Macropinacoid or a-pinacoid	(100)
Brachypinacoid or b-pinacoid	(010)
Base or c-pinacoid	(001)
Prisms	$\begin{cases}(hk0)\\(h\bar{k}0)\end{cases}$
Macrodomes	$\begin{cases}(h0l)\\(\bar{h}0l)\end{cases}$
Brachydomes	$\begin{cases}(0kl)\\(0\bar{k}l)\end{cases}$
Pyramids	$\begin{cases}(hkl)\\(\bar{h}kl)\\(\bar{h}\bar{k}l)\\(h\bar{k}l)\end{cases}$

In the above table it is assumed that the axial ratio is such that $a < b$. If the opposite were true the names brachy- and macro- would be interchanged.

222. The explanations given under the two preceding systems make it unnecessary to discuss in detail the various forms individually, except as illustrated in the case of crystals belonging to certain typical triclinic species.

It may be mentioned, however, that Fig. 379 shows the *diametral prism*, which is bounded by three sets of unlike faces, the pinacoids a, b, and c. This is the analogue of the cube of the isometric system, but here the like faces, edges, and solid angles include only a given face, edge, and angle, and that opposite to it.

223. Illustrations. — A typical triclinic crystal is shown in Fig. 380 of axinite. Here $a(100)$ is the macropinacoid; $m(110)$ and $M(1\bar{1}0)$ the two

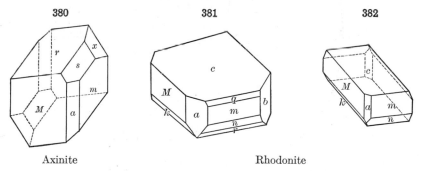

380   381   382

Axinite   Rhodonite

unit prisms; $s(201)$ a macrodome, and $x(111)$ and $r(1\bar{1}1)$ two unit pyramids. The axial ratio is as follows:

$$a : b : c = 0\cdot49 : 1 : 0\cdot48, \quad \alpha = 82° 54', \quad \beta = 91° 52', \quad \gamma = 131° 32'.$$

Figs. 381, 382 show two crystals of rhodonite, a species which is allied to pyroxene, and which approximates to it in angle and habit. Here the faces

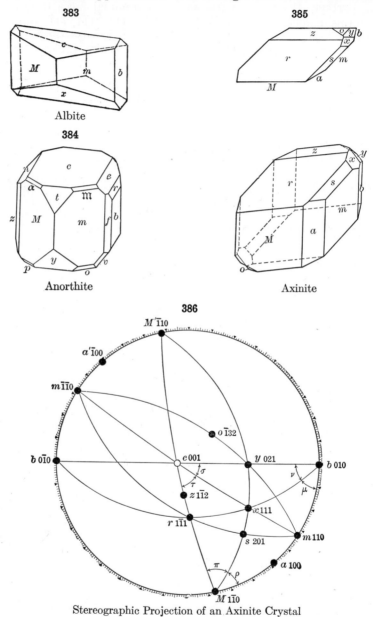

383

Albite

384

Anorthite

385

386

Axinite

Stereographic Projection of an Axinite Crystal

are: Pinacoids $a(100)$, $b(010)$, $c(001)$; prisms $m(110)$, $M(1\bar{1}0)$; pyramids $q(221)$, $k(\bar{2}21)$, $n(\bar{2}\bar{2}1)$, $r(\bar{1}\bar{1}1)$.

Further illustrations are given by Fig. 383 of albite and Fig. 384 of anorthite. The symbols of the faces, besides the pinacoids and the unit prisms, are as follows: Fig. 383, $x(\bar{1}01)$; Fig. 384, prisms $f(130)$, $z(1\bar{3}0)$; domes

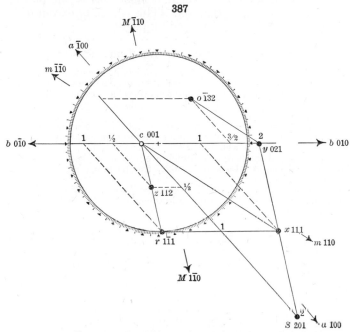

387

Gnomonic Projection of an Axinite Crystal

$t(207)$, $y(\bar{2}01)$, $e(021)$, $r(061)$, $n(0\bar{2}1)$; pyramids $m(111)$, $\alpha(1\bar{1}1)$, $v(\bar{1}\bar{1}1)$, $p(\bar{2}11)$. In Fig. 384 of anorthite the similarity of the crystal to one of orthoclase is evident on slight examination (cf. Figs. 360, 361), and careful study with the measurement of angles shows that the correspondence is very close. Hence in this case the choice of the fundamental planes is readily made.

Fig. 385 represents a crystal of axinite; Figs. 386 and 387 its stereographic and gnomonic projections.

2. ASYMMETRIC CLASS (32). CALCIUM THIOSULPHATE TYPE

(*Hemihedral or Pedial Class*)

224. No Symmetry. — Besides the normal class of the triclinic system there is another possible class, possessing symmetry neither with respect to a plane, axis nor center; in it a given form has *one face* only. This class finds examples among a number of artificial salts. One of these is calcium thiosulphate ($CaS_2O_3.6H_2O$); as yet no

388

Symmetry of Asymmetric Class

mineral species is known to be included here. This is the most general of all the thirty-two types of forms classified according to their symmetry and comes first, therefore, if the classes are arranged in order according to the degree of symmetry characterizing them. This class is one of those whose crystals may show circular polarization. This is true of eleven of the classes which have been described in the preceding pages.

MATHEMATICAL RELATIONS OF THE TRICLINIC SYSTEM

225. Choice of Axes. — It is obvious, from what has been said as to the symmetry of this system, that *any* three faces of a triclinic crystal may be chosen as the pinacoids, or the faces which fix the position of the axial planes and the directions of the axes; moreover, there is a like liberty in the choice of the unit prisms, domes or pyramids which further fix the lengths of the axes.

When the crystal in hand is allied in form or composition to other species, whether of the same or different systems, this fact simplifies the problem and makes the choice of the fundamental forms easy. This is well illustrated, as already noted, by the triclinic feldspars (*e.g.*, albite and anorthite, Figs. 383, 384) which are near in angle to the allied monoclinic species orthoclase. Rhodonite (Figs. 381, 382), the triclinic member of the pyroxene group, is another good example.

In other cases, where no such relationship exists, and where varied habit makes different orientations plausible, there is but little to guide the choice. This is illustrated in the case of axinite (Fig. 380), where at least ten distinct positions have been assumed by different authors.

226. Axial and Angular Elements. — The *axial elements* of a triclinic crystal are: (1) the axial ratio, which expresses the lengths of the axes a and c in terms of the third axis, b; and (2) the angles between the axes, α, β, γ (Fig. 377). There are here five quantities to be determined which obviously require the measurement of five independent angles between the faces.

The *angular elements* are usually taken as the angles between the pinacoids and, in addition, those between each pinacoid and the unit face lying in the zone of the other pinacoids; that is,

$$ab, \quad 100 \wedge 010, \quad ac, \quad 100 \wedge 001, \quad bc, \quad 010 \wedge 001;$$
also $\quad am, \quad 100 \wedge 110, \quad\quad 001 \wedge 101, \quad\quad 001 \wedge 011;$

or, instead, any one or all of these,

$$aM, \quad 100 \wedge 1\bar{1}0, \quad\quad 001 \wedge \bar{1}01, \quad\quad 001 \wedge 0\bar{1}1.$$

Of these six angles taken, one is determined when the others are known.

227. The mathematical relations existing between the axial angles and axial ratio, on the one hand, and the angles between the faces on the other, admit of being drawn out with great completeness, but they are necessarily complex and in general have little practical value. In fact, most of the problems likely to arise can be solved by means of the triangles of the spherical projection, together with the cotangent formula connecting four planes in the same zone (Art. **54**, p. 65); this will often be laborious and may require some ingenuity, but in general involves no serious difficulty. In connection with the use of the cotangent formula, it is to be noted that in certain commonly occurring cases its form is much simpli-

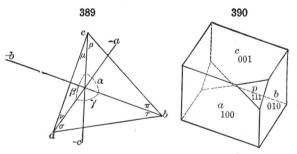

fied; some of these have already been explained under the monoclinic system (Art. **214**). The formulas given there are of course equally applicable here.

228. The first problem may be to find the axial elements from measured angles. Since these elements include five unknown quantities, viz., the three axial angles, α, β, γ and the lengths of the axes a and c in terms of b, five measured angles are required, as already stated.

Fig. 389 represents the crystallographic axes of the triclinic mineral rhodonite. The positive ends of the three axes are joined by lines forming three triangles the angles of which are very important. In the triangle, for instance, which has the *b* and *c* axes for two of its sides since the length of the *b* axis is taken as 1·0, it is only necessary to know the angle α and either ρ or π in order to determine the length of the *c* axis. In the triangle that has the *a* and *b* axes for two of its sides it is necessary to know the value of γ and either σ or τ in order to determine the length of the *a* axis. And lastly in the triangle formed between the *a* and *c* axes, if the length of either of these axes is known, the length of the other can be determined from the angle β and either μ or ν. It is assumed that a crystal of rhodonite showing the forms $a(100)$, $b(010)$, $c(001)$ and $p(111)$, see Fig. 390, has been measured and the poles of the faces plotted in the stereographic projection, Fig. 391. The angles between the great circles which connect these poles are the same as those shown in the triangles built upon the crystallographic axes, Fig. 389. With the angles between the different crystal faces known by measurement, it is easy, by the formulas of spherical trigonometry, to calculate the value of these other angles and from them obtain the axial ratio.

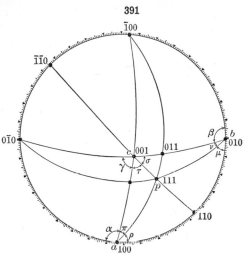

391

That the angles shown on the stereographic projection, Fig. 391, are identical with those in Fig. 389 may be proved as follows. Let Fig. 392 represent a vertical section cut through the spherical projection of rhodonite in such a way as to include the *b* and *c* crystallographic axes. The triangle, which has these axes as two sides and the three angles, α, π and ρ, lies therefore in the plane of the figure. The normals to all faces parallel to the *c* axis, *i.e.* the prism zone, would lie in a plane at right angles to that axis. This plane would intersect the sphere of the spherical projection in a great circle which is represented on the stereographic projection, Fig. 391, by the divided circle. On Fig. 392 this great circle would appear in orthographic projection as the line $C–C'$ lying at right angles to the *c* axis. In the same way all faces lying parallel to the *b* axis, *i.e.* the zone $(100)–(101)–(001)$, would have their normals in a plane which would be foreshortened to the line $B–B'$ in Fig. 392. Since the lines $C–C'$ and $B–B'$ are at right angles respectively to the *c* and *b* axes the angle between them must equal the axial angle, α. This same angle will appear therefore on the stereographic projection, Fig. 391, between the great circles of the two zones, the faces of which are parallel respectively to the *c* and *b* axes. Further, the normals to all faces which intersect the *b* and *c* axes at their unit lengths would lie in a plane at right angles to the line *b–c*, Fig. 392. This plane would appear in orthographic projection as the line $P–P'$. On the stereographic projection, Fig. 391, this would be represented as the zonal circle passing through (100), (111), (011), $(\bar{1}00)$. The angle between $B–B'$ and $P–P'$ will by construction equal π and that between $C–C'$ and $P–P'$ will equal ρ. These same angles will appear therefore in the stereographic projection between the corresponding zone circles. In the

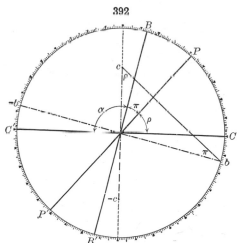

392

same way the identity of the angles γ, σ, τ, β, μ and ν in Figs. 389 and 391 can be proved.

With the necessary number of these angles given the formulas required for the calculation of the axial lengths are given below. The angles τ', σ', ν', μ', π' and ρ' are the corresponding angles to τ, σ, etc., in the adjacent quadrants, see Fig. 393.

$$\frac{\sin \tau}{\sin \sigma} = \frac{\sin \tau'}{\sin \sigma'} = \frac{a}{b}, \frac{\sin \nu}{\sin \mu} = \frac{\sin \nu'}{\sin \mu'} = \frac{c}{a}, \frac{\sin \pi}{\sin \rho} = \frac{\sin \pi'}{\sin \rho'} = \frac{c}{b}.$$

If the angles given are between the three pinacoids and the pyramid hkl (not the unit form) the relations are similar. That is, if for the face hkl the corresponding angles be represented by τ_0, σ_0, etc., where τ_0, σ_0 are the angles between the zone circles 100, 001 and 100, 010 respectively and the zone circle 001, $hk0$, these relations may be expressed in the general form

393

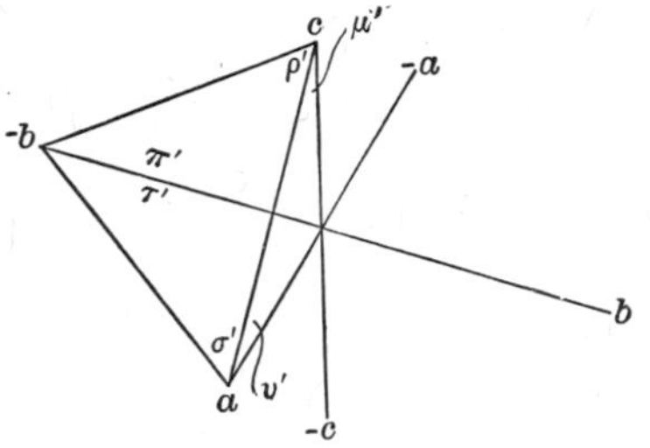

$$\frac{\sin \tau_0}{\sin \sigma_0} = \frac{\sin \tau_0'}{\sin \sigma_0'} = \frac{a}{\frac{h}{k}l} = \frac{k}{h} \cdot \frac{a}{b},$$

$$\frac{\sin \nu_0}{\sin \mu_0} = \frac{\sin \nu_0'}{\sin \mu_0'} = \frac{c}{\frac{l}{h}a} = \frac{h}{l} \cdot \frac{c}{a},$$

$$\frac{\sin \pi_0}{\sin \rho_0} = \frac{\sin \pi_0'}{\sin \rho_0'} = \frac{c}{\frac{l}{k}b} = \frac{k}{l} \cdot \frac{c}{b}.$$

Thus for the face 321 the formulas become

$$\frac{\sin \tau_0}{\sin \sigma_0} = \frac{a}{\frac{2}{3}b} = \frac{2a}{3b}, \frac{\sin \nu_0}{\sin \mu_0} = \frac{3c}{a}, \frac{\sin \pi_0}{\sin \rho_0} = \frac{2c}{b}.$$

It is also to be noted that

$$\alpha = 180° - A, \qquad \beta = 180° - B, \qquad \gamma = 180° - C,$$

where A, B, C are the angles in the pinacoidal spherical triangle 100·010·001 at these poles respectively. That is,

$$A = \pi + \rho = \pi_0 + \rho_0 = (180° - \alpha);$$
$$B = \nu + \mu = \nu_0 + \mu_0 = (180° - \beta);$$
$$C = \tau + \sigma = \tau_0 + \sigma_0 = (180° - \gamma).$$

Also

$$180° - A = \pi' + \rho' = \pi_0' + \rho_0' = \alpha.$$

Hence, having given, by measurement or calculation, the angles between the faces $ab(100 \wedge 010)$, $ac(100 \wedge 001)$ and $bc(010 \wedge 001)$, which are the sides of this triangle, the angles A, B, C are calculated and their supplements are the axial angles α, β, γ respectively.

Still another series of equations are those below, which give the relations of the angles μ, ν, ρ, etc., to the axes and axial angles. By means of them, with the sine formulas given above, the angular elements (and other angles) can be calculated from the axial elements.

$$\tan \mu = \frac{a \sin \beta}{c + a \cos \beta}; \quad \tan \nu = \frac{c \sin \beta}{a + c \cos \beta}.$$

$$\tan \rho = \frac{b \sin \alpha}{c + b \cos \alpha}; \quad \tan \pi = \frac{c \sin \alpha}{b + c \cos \alpha}.$$

$$\tan \tau = \frac{a \sin \gamma}{b + a \cos \gamma}; \quad \tan \sigma = \frac{b \sin \gamma}{a + b \cos \gamma}.$$

These equations apply when $\mu + \nu$, etc., is less than 90°; if their sum is greater than 90° the sign in the denominator is negative.

229. The following equations are also often useful.

$$\tan \alpha = \frac{2 \sin \rho \sin \rho'}{\sin (\rho - \rho')} = \frac{2 \sin \pi \sin \pi'}{\sin (\pi - \pi')}.$$

$$\tan \beta = \frac{2 \sin \mu \sin \mu'}{\sin (\mu - \mu')} = \frac{2 \sin \nu \sin \nu'}{\sin (\nu - \nu')}.$$

$$\tan \gamma = \frac{2 \sin \tau \sin \tau'}{\sin (\tau - \tau')} = \frac{2 \sin \sigma \sin \sigma'}{\sin (\sigma - \sigma')}.$$

Also,

$$\alpha + \pi + \rho = \beta + \mu + \nu = \gamma + \tau + \sigma = 180°.$$

The calculation, from the angular elements or from the assumed fundamental measured angles, either (1) of the angular position of any face whose symbol is given, or (2) of the symbol of an unknown face for which measured angles are at hand, requires no further explanation. The cotangent formula is all that is needed in a single zone, and the solution of spherical triangles on the projection (with the use of the sine formulas) will suffice in addition in all ordinary cases.

230. To determine, by plotting, the axial elements of a triclinic crystal, having given the stereographic projection of its forms. — In order to solve this problem it is necessary to have given the position of the poles of the unit forms (100), (010), (001), (111) or to be able to locate them by means of their zonal relations. Through these poles the various zonal circles are drawn as shown in the case of rhodonite, Fig. 391. The angles α, β, γ, π, ρ, etc., are then given upon the projection. These angles can be measured as described in Art. **46**, p. 49.

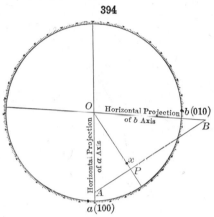

Taking next a certain line as representing the unit length of the b axis and knowing the angles α, π and ρ, the triangle that includes the b and c axes, see Fig. 389, can be drawn to scale and the unit length of the c axis determined. In a similar way the length of the a axis can be found.

231. To determine, by plotting, the indices of a face upon a triclinic crystal, having given the position of its pole in the stereographic projection and the axial elements of the mineral. — To illustrate this problem a possible pyramid face on rhodonite will be used. Its pole is located in the stereographic projection at x, Fig. 394. The position of the poles of the faces $a(100)$ and $b(010)$ must also be known. The directions of the intersections of the planes of the a–c and b–c axes with the plane of the projection can then be drawn. These lines will represent the horizontal projections of the a and b crystallographic axes. A radial line is then drawn from the center of the projection, O, through x. Another line, A–P–B, is drawn

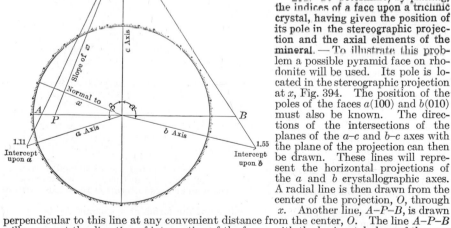

perpendicular to this line at any convenient distance from the center, O. The line A–P–B will represent the direction of intersection of the face x with the horizontal plane of the projection.   The intercept that the face will make upon the vertical axis can be found by the construction of a right triangle with $O-P$ as its base, a line representing the c axis as its vertical side and the angle between $O-x$ as the angle between the base and the hypothenuse, see Fig. 395.   Under the assumed conditions the face will intersect the c axis at a distance of 1·93, the radius of the circle in the figure being 1·0.   The face will also pass through the points A and B on the horizontal projections of the a and b axes.   With the known angles β and α it is possible to construct the a and b axes with their proper angular relations to the c axis. The intercepts of the face upon these two axes will be given by the extension of the lines from the point 1·93 on the c axis to the points A and B.   In this way the intercepts of the face upon the three axes were obtained as $1\cdot11a$, $1\cdot55b$, $1\cdot93c$.   By dividing these numbers by 1·55 we get the intercepts expressed in terms of the length of the b axis, considering that as

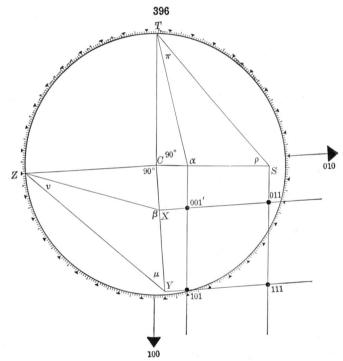

396

100

$1\cdot0$.   The intercepts then become $0\cdot71a$, $1b$, $1\cdot24c$.   When these are compared with the axial ratio of rhodonite, $a:b:c = 1\cdot114:1:0\cdot986$, the parameters of the face are found to be $\frac{2}{3}a$, $1b$, $2c$.   The indices of x are therefore 321.

232. To determine, by plotting, the axial elements of a triclinic crystal having given the gnomonic projection of its forms. — To illustrate this problem it is assumed that the positions of the poles of the faces, (100), (010), (001), (101), (011) and (111) on rhodonite are known, see Fig. 396.   If this figure is compared with the stereographic projection of the same forms given in Fig. 391, it will be seen that the angle between the zones (100)–(101)– (001) and (100)–(111)–(011) is equal to π, that between the zones (100)–(111)–(011) and (100)–(110)–(010) is equal to ρ, between (010)–(011)–(001) and (010)–(111)–(101) is equal to ν and between (010)–(111)–(101) and (010)–(110)–(100) is equal to μ.   The method by which the angles between these various zones may be measured was explained in Art. **47**, p. 56, and is illustrated by the construction of Fig. 396.   From these angles triangles can be readily constructed to give the lengths of the a and c axes in terms of the b axis, with its length taken as equal to 1·0.

233. To determine, by plotting, the indices of the forms of a triclinic crystal, having given the position of other poles upon the gnomonic projection. — The method for the solution of this problem is similar to that already described under the previous systems.   The difference lies in the fact that the lines of reference upon which are plotted the intercepts of the lines drawn to them from the poles of the faces make oblique angles with each other. These reference lines are taken as the zonal lines (001)–(101) and (001)–(011) and the intercepts from which the indices are determined are measured from the pole of (001). A study of the gnomonic projection of axinite, Fig. 387, will illustrate this problem.

MEASUREMENT OF THE ANGLES OF CRYSTALS

234. Contact-Goniometers. — The interfacial angles of crystals are measured by means of instruments which are called *goniometers*.

The simplest form is the contact- or hand-goniometer one form of which is represented in Fig. 397.

397

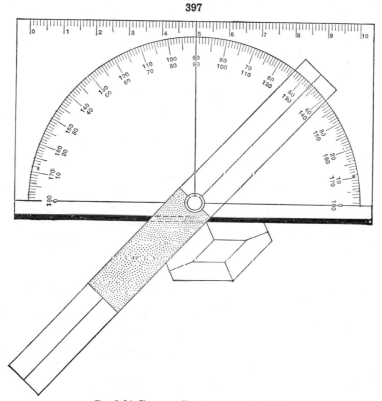

Penfield Contact Goniometer, Model B

This contact-goniometer consists of a card on which is printed a semi-circular arc graduated to half degrees, at the center of which is fastened a celluloid arm which may be turned to any desired position. The method of use of the goniometer is illustrated in Fig. 397. The bottom of the card and the blackened end of the celluloid arm are brought in as accurate contact as possible with the two crystal faces, the angle between which is desired. Care must be taken to see that the plane of the goniometer is at right angles to the edge of intersection between the two faces. Another model of the contact-goniometer, Fig. 398, has two arms swiveled together and separate from the graduated arc. The crystal angle is obtained by means of the arms and then the angle between them measured by placing them upon the graduated arc. This latter type is employed in cases where the crystal lies in such a position as to prevent the use of the former.*

The contact-goniometer is useful in the case of large crystals and those whose faces are not well polished; the measurements with it, however, are

398

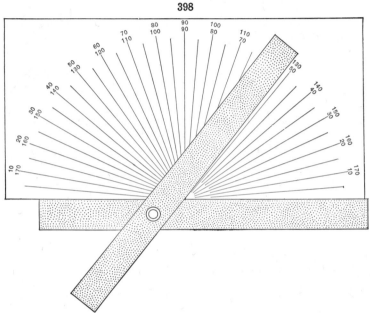

Penfield Contact Goniometer, Model A

seldom accurate within a quarter of a degree. In the finest specimens of crystals, where the faces are smooth and lustrous, results far more accurate may be obtained by means of a different instrument, called the reflecting goniometer.

Other more elaborate forms of contact goniometers have been described but it is doubtful if they can be used with any greater degree of accuracy than the simple ones described above.

235. Reflecting Goniometer. — This type of instrument was devised by Wollaston in 1809. It has undergone extensive modifications and improvements since that time. Only the perfected forms that are in common use today will be described.

The principle underlying the construction of the reflecting goniometer will be understood by reference to the figure (Fig. 399), which represents a

---

* These simple types of contact-goniometers were devised by S. L. Penfield and can be obtained by addressing the Mineralogical Laboratory of the Sheffield Scientific School of Yale University, New Haven, Connecticut.

section of a crystal, whose angle, *abc*, between the faces *ab*, *bc*, is required. Let the eye be placed at *p* and the point *m* be a source of light. The eye at *p*, looking at the face of the crystal, *bc*, will observe a reflected image of *m*, in the direction of *pn*. The crystal may now be so changed in its position that the same image is seen reflected by the next face and in the same direction, *pn*. To effect this, the crystal must be turned around until *abd* has the present direction of *bc*. The angle *dbc* measures, therefore, the number of degrees through which the crystal must be turned; it may be measured by attaching the crystal to a graduated circle, which turns with the crystal. This angle is the supplement of the interior angle between the two faces, or in other words is the *normal angle*, or angle between the two poles (see Art. **48**, p. 61). The reflecting goniometer hence gives directly the angle needed on the system of Miller here followed.

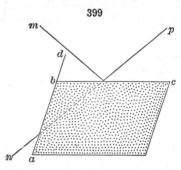

399

236. Horizontal Goniometer. — A form of reflecting goniometer well adapted for accurate measurements is shown in Fig. 400. The particular form of instrument here figured* is made by Fuess.

400

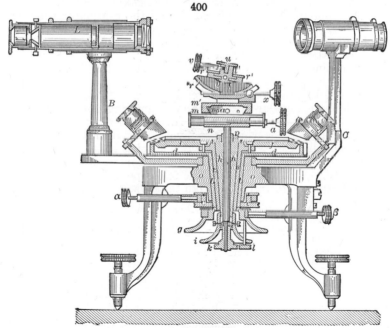

One-circle Reflection Goniometer

The instrument stands on a tripod with leveling screws. The central axis, *o*, has within it a hollow axis, *b*, with which the plate, *d*, turns, carrying

---

* The figure here used is from the catalogue of Fuess.

the verniers and also the observing telescope, the upright support of which is shown at B. Within b is a second hollow axis, e, which carries the graduated circle, f, above, and which is turned by the screw-head, g; the tangent screw, α, serves as a fine adjustment for the observing telescope, B, the screw, c, being for this purpose raised so as to bind b and e together. The tangent screw, β, is a fine adjustment for the graduated circle. Again, within e is the third axis, h, turned by the screw-head, i, and within h is the central rod, which carries the support for the crystal, with the adjusting and centering contrivances mentioned below. This rod can be raised or lowered by the screw, k, so as to bring the crystal to the proper height — that is, up to the axis of the telescope; when this has been accomplished, the clamp at p, turned by a set-key, binds s to the axis, h. The movement of h can take place independently of g, but after the crystal is ready for measurement these two axes are bound together by the set-screw, l. The signal telescope is supported at C, firmly attached to one of the legs of the tripod. The crystal is mounted on the plate, u, with wax, the plate is clamped by the screw, v. The *centering* apparatus consists of two slides at right angles to each other (one of these is shown in the figure) and the screw, a, which works it; the end of the other corresponding screw is seen at a'. The *adjusting* arrangement consists of two cylindrical sections, one of them, r, shown in the figure, the other at r'; the cylinders have a common center. The circle on f is graduated to degrees and quarter degrees, and the vernier gives the readings to $30''$.

A brilliant source of light is placed behind the collimator tube which is at the top of the support C. Openings of various size and character are provided at the rear end of this tube in order to modify the size and shape of the beam of light that is to be reflected from the crystal faces. The most commonly used opening is one made by placing two circular disks nearly in contact with each other leaving between them an hour-glass-shaped figure. The telescope tube L is provided with several removable telescopes with lenses which have different angular breadths and magnifying powers and hence are suitable for observing faces varying in size and degree of polish. At the front of the tube L there is a lens which is so pivoted that it may be thrown into or out of the axis of the telescope. When this lens lies in the axis of the tube it converts the telescope into a low-power microscope with which the crystal may be observed. Without this lens the telescope has a long-distance focus and only the beam of light reflected from the crystal face can be seen.

The method of use of the instrument is briefly as follows. The little plate u is removed and upon it is fastened by means of some wax the crystal to be measured. The faces of the zone that is to be measured should be placed as nearly as possible vertical to the surface of this plate. It will usually facilitate the subsequent adjustment if a prominent face in this zone be placed so that it is parallel to one of the edges of the plate u. This plate with the attached crystal is then fastened in place by the screw v. During the preliminary adjustments of the crystal the small lens in front of the tube L is placed in its axis and the crystal observed through the microscope thus formed. It is usually better also to make these first adjustments outside the dark room in daylight. By means of the screw-head k the central post is raised or lowered until the center of the crystal lies in the plane of the telescope. Next by means of the two sliding tables controlled by the screw-heads a and a' the crystal is adjusted so that the edge over which the angle is to be measured coincides with the axis of the instrument. This adjustment is most easily accomplished by turning the central post of the instrument until one of these sliding plates lies at right angles to the telescope and then by turning its screw-head bring the intersection in question to coincide with the vertical cross-hair of the telescope tube. Then turn the post until the other plate lies at right angles to the telescope and make a similar adjustment. Then in a similar manner by means of the tipping screws x and y bring the intersection between the faces to a position parallel with the vertical cross-hair of the telescope.   By a combination of these adjustments this edge should be made to coincide with the vertical cross-hair and to remain stationary while the crystal is revolved upon the central post of the instrument. Next the instrument is taken into the dark room and a light placed behind the collimator tube, and the crystal turned until one of the faces is seen through the tube L to be brightly illuminated.   Then the little lens in the front of this tube is raised and the reflection of the beam of light, or *signal* as it is called, should lie in the field.   If the preliminary adjustments were accurate the horizontal cross-hair will bisect this signal.   In the majority of cases, however, further slight adjustments will be necessary.   Before the angles between the faces can be measured their various signals must all be bisected by the horizontal cross-hair. When these conditions are fulfilled each signal in turn is brought into place so that it is bisected also by the vertical cross-hair and its angular position read by means of the graduated scale and vernier.   The difference between the angles for two faces gives the normal angle between them.   In making these readings care must be taken that the plate on which the graduated circle is engraved is turned with the central post.   In order to do this only the screw-head g must be used unless, as is wise, the two screw-heads i and g have been previously clamped together by means of l.   For the accurate adjustment of the signals on the vertical cross-hair the tangent screw β is used.   In making a record of the angles measured it is important to note accurately the face from which each signal is derived and the character of the signal.   It is frequently helpful to make a sketch of the outlines of the different faces and number or letter them.

237. Theodolite-Goniometer. — A form of goniometer* having many practical advantages and at present in wide use has two independent circles and is commonly known as the *two-circle goniometer*.   It is used in a manner analogous to that of the ordinary theodolite.   Instruments of this type were devised independently by Fedorow, Czapski and Goldschmidt.   Other models have been described since.   In addition to the usual graduated horizontal circle of Fig. 400, and the accompanying telescope and collimator, a second graduated circle is added which revolves in a plane at right angles to the first.   Fig. 401, after Goldschmidt, gives a cross-sectional view of one of the earlier machines devised by him.   It will serve to illustrate the essential features of the instrument.

The crystal to be measured is attached at the end of the axis (h) of the vertical circle and so adjusted by means of suitable centering and tipping devices that a given plane, called the polar plane, is normal to this axis and lies directly over the axis of the horizontal circle.   In using the instrument, instead of directly measuring the interfacial angles of the crystal, the position of each face is determined independently of the others by the measurement of its angular coördinates, or what might be called its latitude and longitude. These coördinates are the angles (φ and ρ of Goldschmidt) measured, respectively, in the vertical and horizontal circles from an assumed pole and meridian, which are fixed, in most cases, by the symmetry of the crystal.   In practice the crystal is usually so mounted that its prismatic zone is perpendicular to the vertical circle.   A plane at right angles to this zone, *i.e.*, the basal plane in the first four systems, is known as the polar plane and its position when reflecting the signal into the telescope establishes the zero position for the horizontal circle.   The position of a pinacoid, usually the 010 plane, in the prism zone establishes the zero position for the vertical circle.   For example, with an orthorhombic crystal, for the pyramid 111, the angle φ (measured on

* Fedorow, Universal or Theodolit-Goniometer, Zs. Kryst., **21**, 574, 1893; **22**, 229, 1893; Czapski, Zeitschr. f. Instrumentenkunde, 1, 1893; Goldschmidt, Zs. Kryst., **21**, 210, 1892; **24**, 610, 1895; **25**, 321, 538, 1896; **29**, 333, 589, 1898.   On the method of Goldschmidt, see Palache, Am. J. Sc., **2**, 279, 1896; Amer. Mineral., **5**, No. 2, *et seq.*, 1920.   A simplified form of the theodolite-goniometer is described by Stöber, Zs. Kryst., **29**, 25, 1897; **54**, 442.

the vertical circle) is equal to 010 $\wedge$ 110 and ρ (measured on the horizontal circle) is equal to 001 $\wedge$ 111.

Goldschmidt has shown that this instrument is directly applicable to the system of indices and methods of calculation and projection adopted by him, which admit the deducing of the elements and symbols of a given crystal with a minimum of labor and calculation.*   Fedorow has also shown that this in-

401

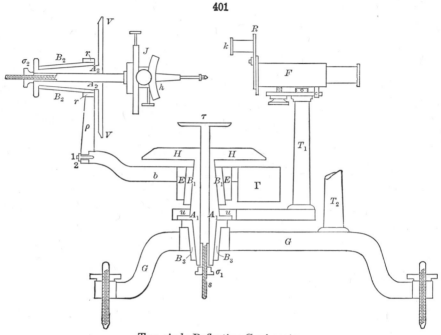

Two-circle Reflection Goniometer

strument, with the addition of the appliances devised by him, can be most conveniently used in the crystallographic and optical study of crystals.

The following hints as to the methods of using this instrument may prove helpful.

The telescope and collimator tube are placed at some convenient angle to each other (usually about 70°) and then clamped in position.   The next step is to find the polar position of the horizontal circle, *i.e.*, the position at which a crystal plane lying at right angles to the axis of the vertical circle will throw the reflected beam of light on to the cross-hairs of the telescope.   Obviously the plane under these conditions must be normal to the bisector of the angle between the axes of the collimator and telescope, the line *B–P*, Fig. 402.   The method by which this polar position is found is as follows: Some reflecting surface is mounted upon the end of the post *h*, Figs. 401, 402, making some small inclined angle to the plane normal to that post.   Then by turning the instrument in both the horizontal and vertical planes this surface is brought into the proper position to reflect the signal into the telescope, see position I, Fig. 402.   The horizontal angle of this position is noted.   Then the vertical

---

* See Goldschmidt's Krystallographische Winkeltabellen (432 pp., Berlin, 1897). This gives the angles required by his system for all known species.   See also Zs. Kryst., **29**, 361, 1898.   The same author's Atlas der Krystallformen, 1913 *et seq.*, is a monumental work giving all previously published crystal figures together with a discussion of the forms observed upon them.

circle is turned through an angle of 180°.   This brings the reflecting surface into the position indicated by the dotted lines in the figure.   In order to again bring this surface back to its reflecting position the vertical circle with the post h must be moved in the horizontal plane until the position II is reached.   The horizontal reading of this position is also noted.   The angle midway between these two readings is the polar position desired.   That is, when the post h lies in the direction of the broken line $P–B$ a plane normal to its axis would reflect a beam of light from the collimator into the telescope.   This position constitutes the zero position of the horizontal circle from which the ρ angles are measured.

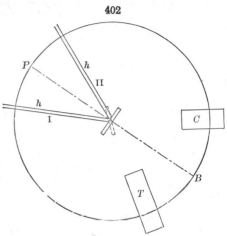

The method used to adjust a crystal upon the instrument so that it will occupy the proper position for measurement will vary with the character of the crystal.   A few illustrations follow.   1. *If the crystal has a basal plane at right angles to a prism zone.* The crystal is mounted upon the post h so that the faces of the prism zone lie as nearly as possible parallel to the axis of the post or the basal plane as nearly as possible normal to it.   Then the instrument is moved until the reading of the horizontal circle agrees with the polar position already determined. Then by means of the tipping screws the crystal is moved until the reflection from the basal plane is centered upon the cross-hairs of the telescope.   If the adjustments have been accurately made the signal will remain stationary while the vertical circle is revolved. Next the horizontal circle is moved through an angle of 90°.   This will bring the reflections from the faces of the prism zone into the telescope.   If the pinacoid 010 is present the vertical circle is turned until the reflected signal from this face falls on the horizontal cross-hair.   The reading of the vertical circle under these conditions establishes the position of the meridian from which the ϕ angles are measured.   If the pinacoid 010 is not present it is usually possible to determine its theoretical position from the position of other faces in the prism zone or in the zone between 010 and 100. 2. *If there is no basal plane present upon the crystal but a good prism zone.*   Under these circumstances the horizontal circle is turned until it is exactly 90° away from its determined polar angle and then the crystal adjusted by means of the tipping screws until the signals from the faces of the prism zone all fall on the vertical cross-hair as the vertical circle of the goniometer is turned.   3. *If neither basal plane nor prism zone is available but there are two or more faces present which are equally inclined to a theoretical basal plane.*   First adjust the crystal as nearly as possible in the

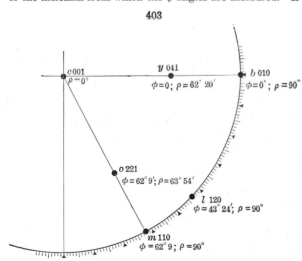

proper position and then obtaining reflections from these faces note the horizontal circle reading in each case.   Take an average of these readings and adding or subtracting this angle from the polar angle of the horizontal scale place the instrument in this position.   Then by tipping the crystal try to bring it into such a position that all of these faces will successively reflect the signal into the telescope as the vertical circle is turned.   The operation may have to be repeated two or three times before the final adjustment is made. If the angle between the inclined faces and the theoretical base is known the instrument can be set in the proper position at once and the crystal brought into adjustment very quickly. Other problems will arise in practice but their solution will be along similar lines to those suggested above. It may frequently happen that more than one method of adjustment may be used with a given crystal. In that case the faces giving the best reflections should be used. It should be emphasized that the preliminary adjustment of the crystal is of supreme importance since all measurements of the coördinates of the different faces depend upon it. It is wise to check the adjustment in all possible ways before making the measurements.

After these adjustments have been completed the crystal is turned about both the horizontal and vertical planes so that each face upon it successively reflects the signal into the telescope. The horizontal and vertical readings are made in each case. The forms present can then be readily plotted in either the stereographic or gnomonic projections. Fig. 403 shows how the forms of a simple crystal of topaz could be plotted in the stereographic projection from the ϕ and ρ angles obtained from it — the two-circle goniometer measurements. For each face the vertical circle angle, ϕ, is plotted on the divided circle, the position of $b(010)$ giving the zero point, while the horizontal circle angle is plotted on a radial line from the center of the projection, the position of $c(001)$ giving its zero point.

238. Goldschmidt's Tables of Angles (Winkeltabellen). — Goldschmidt's methods of crystal measurement, etc., are so widely used that it seems advisable to give an example of the tables of angles that he devised together with a brief explanation. The material has been taken from his Krystallographische Winkeltabellen, published in 1897. It is assumed that a crystal of topaz, possessing the faces shown in the horizontal plan of Fig. 337, has been measured on the two-circle goniometer. The basal plane $c(001)$ serves as the polar plane, and the position of the first meridian is determined by the face $b(010)$. Below are given the angles, etc., for these forms as listed in a typical Goldschmidt table.

TOPAZ

Orthorhombic

$a = 0.5285$ $a_0 = 0.5540$ $p_0 = 1.8049$
$c = 0.9539*$ $b_0 = 1.0483$ $q_0 = 0.9539$

Letter	Gdt. Symbol	Miller Symbol	ϕ	ρ	ξ_0	η_0	ξ	η	x	y	$d = \tan \rho$
c	0	001	..	0° 00	0° 00	0° 00	0° 00	0° 00	0	0	0
b	0∞	010	0° 00	"	90 00	90 00	"	90 00	"	∞	∞
m	∞	110	62 08	"	90 00	"	62 08	27 51	1.8922	"	"
l	∞ 2	120	43 25	"	"	"	43 25	46 35	0.9461	"	"
n	∞ 4	140	25 19	"	"	"	25 19	64 41	0.4730	"	"
f	01	011	0 00	43 39	0 00	43 39	0 00	43 39	0	0.9539	0.9539
y	02	021	"	62 20	"	62 20	"	62 20	"	1.9078	1.9078
o	1	111	62 08	63 54	61 00	43 39	52 33	24 48	1.8049	0.9593	2.0414
u	$\frac{1}{2}$	112	"	45 35	42 04	25 30	39 10	19 30	0.9024	0.4769	1.0207
i	$\frac{1}{3}$	113	"	34 14	31 02	17 38	29 49	15 14	0.6016	0.3179	0.6805

* Goldschmidt uses for the length of the vertical axis a value twice that of Dana. Consequently the indices of the domes and pyramids differ from those given under the description of the crystal represented (cf. Art. **186**).

Above the table are given after a and c the axial ratio assumed for topaz assuming that the length of b equals 1. The axial ratio with $c = 1$ is given after the symbols a_0 and b_0. The values p_0 and q_0 are the unit intercepts in the gnomonic projection of the normals from the poles of the faces upon the vertical and horizontal lines of reference; in Fig. 405 the x and y of the face $o(111)$ equal p_0 and q_0 of topaz. In the table itself the significance of the various angles and distances given is indicated in Figs. 404 and 405, where are shown, respectively, one quadrant of the stereographic and gnomonic projections of the face $o(111)$ of topaz.

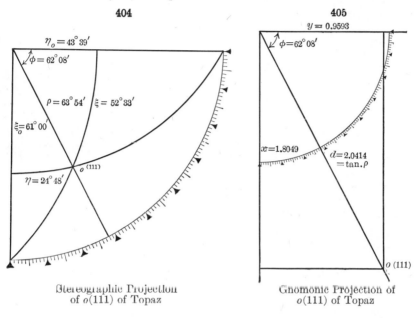

404

$\eta_0 = 43° 39'$

$\angle \phi = 62° 08'$

$\rho = 63° 54'$ $\xi = 52° 33'$

$\xi_0 = 61° 00'$

o (111)

$\eta = 24° 48'$

Stereographic Projection
of $o(111)$ of Topaz

405

$y = 0.9593$

$\angle \phi = 62° 08'$

$x = 1.8049$

$d = 2.0414$
$= \tan. \rho$

o (111)

Gnomonic Projection of
$o(111)$ of Topaz

COMPOUND OR TWIN CRYSTALS

239. Crystal Intergrowths. — Crystals commonly occur in groups, and the individuals composing any group may have various relations to each other. In the majority of cases these relations are irregular and depend upon the accidental conditions of growth, etc. In other cases the relations of the individual crystals to each other are definite and conform to some law. If all faces, edges, etc., of one individual are parallel to the similar elements of a second individual, the two are said to exhibit parallel growth or grouping. For a further discussion of this method of intergrowth see Art. **257.** When, however, two individuals grow together with only a part of their similar faces, edges, etc., in parallel positions they are said to form a twin crystal or group. The study of the types of twin crystals shown by different substances and the formulation of the laws governing their formation is an important part of crystallography.

240. Twin Crystals. — Twin crystals may be defined as the intergrowth of two or more individuals in such a way as to yield parallelism in the case of certain parts of the different individuals, and at the same time other parts of the different individuals are in reverse positions in respect to each other. They often appear externally to consist of two or more crystals symmetrically united, and sometimes have the form of a cross or star. They also exhibit their twinned structures in the reversed arrangement of part of the faces, in the striæ of the surface, and in reentrant angles; in certain cases the compound structure can only be surely detected by an examination in polarized light.   The figures below (Figs. 406–408) are examples of typical kinds of twin crystals, and many others are given on the pages following.

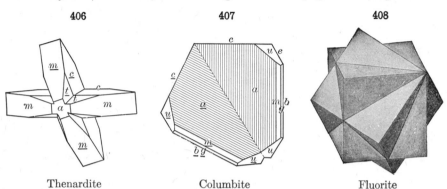

406	407	408
Thenardite	Columbite	Fluorite

241. Laws of Twinning. — In the majority of cases a twin crystal grows as such from the beginning.   Exceptions to this rule are discussed in Arts. **243** and **244.**   The two (or more) individuals comprising a twin crystal have different orientations of their atomic structures but the different positions of the crystal network must have certain planes or directions in common; they must fit together in some simple way, and it must be possible to derive one orientation from another by some simple movement.   Bearing in mind, therefore, that the formation of twin crystals is usually the result of a simultaneous and regular growth according to two interlocking orientations of the same atomic network, we may outline certain geometrical means of describing the results of such a compound growth.

The component parts of a twin crystal are geometrically related to each other, either as if one part was derived from the other by reflection over a plane common to both, or as if one part was derived from the other by a revolution of 180° about some crystal line common to both, or as if these two operations occurred simultaneously.   The plane and axis involved in the above supposed operations are known as the *twinning-plane* and *twinning-axis.* Such twinning-planes and -axes have in all ordinary cases simple and rational relations to the crystal axes.   In rare cases their relations to the crystal axes may not be rational, but then they have simple relations to other axes or planes which do possess such rational characters.   No plane that is a symmetry plane in the individual crystal can become a twinning-plane in its compound crystal.   Further, no axis of even symmetry (binary, tetragonal, or hexagonal) in the individual crystal can become a twinning-axis in the compound crystal.   It is obvious if such a plane or axis were assumed as twinning-plane or -axis the resulting crystal would be identical with the untwinned individual.   On the other hand, a symmetry plane or axis of a class of higher symmetry may become a twinning-plane or -axis of a crystal belonging in the same crystal system but to a class of lower symmetry. Such a twin crystal may therefore assume a symmetry higher than that belonging to its components.   Twin crystals of all crystal classes possessing a center of symmetry will have both a plane of twinning and an axis of twinning normal to the plane.   Where a center of symmetry is lacking a twinning-plane or -axis may occur independently.   These fundamental laws of twinning are illustrated in the following paragraphs.

Figs. 409 and 410 represent, respectively, a simple octahedron and its twinned form.   The twinning-plane is shown in Fig. 409 by the plane, *b–b*, indicated by the broken lines, this plane being parallel to a pair of octahedron faces.   In Fig. 410 is shown the twinned crystal; the faces of the front portion, which are marked *o*, lie in positions as if they had been reflected over the twinning-plane, which, therefore, though not a symmetry plane in the simple crystal, is one for the twin.   The same twin can be considered as derived by a revolution on the twinning-plane of 180° about an octahedral normal.   This direction, which is one of trigonal symmetry in the simple crystal, would

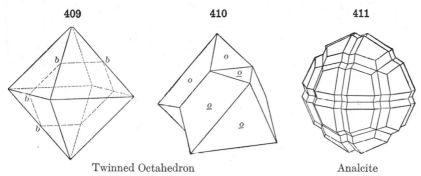

409   410   411

Twinned Octahedron   Analcite

therefore be the twinning-axis.   In this case, the crystal has a center of symmetry and therefore has both a twinning-plane and -axis, normal to each other; the twinning-plane is parallel to a common crystal face (111), and the twinning-axis is a simple crystallographic direction.   A twinning-plane is, with rare exceptions, parallel to a possible crystal face on the given species, and usually one of the more frequent or fundamental forms.   Exceptions to this rule occur only in the triclinic and monoclinic systems, where the twinning-axis is sometimes one of the oblique crystallographic axes, and then the plane of twinning normal to it is obviously not necessarily a crystallographic plane; this is conspicuously true in the case of the pericline law of twinning in the plagioclase feldspar group; see Art. **256**.

Crystals of quartz are enantiomorphous, *i.e.*, either right- or left-handed in their development.   Such crystals possess neither planes nor a center of symmetry.   It is impossible to derive one type of quartz crystal from the other by any revolution about a twinning-axis, and twin crystals which combine both right- and left-handed individuals therefore can possess only a plane of twinning.   The Brazilian law of twinning in quartz represents the case where the twin position is derived by reflection over a twinning-plane, namely, the second order prism $(11\bar{2}0)$, but where no twinning-axis exists.   Fig. 412A shows a left-handed quartz crystal; Fig. 412C, the corresponding right-handed crystal; and Fig. 412B, the combination of the two into a twinned individual with the plane $(11\bar{2}0)$, indicated by the broken lines, as the twinning-plane.

Fig. 434 represents a twin crystal of tetrahedrite.   It is composed of two interpenetrating tetrahedra with a normal to the tetrahedral face common to the two individuals as twinning-axis. In this case, however, there is no plane normal to the twinning-axis that could serve as a twinning-plane, *i.e.*, as a plane over which one twin component can be derived from the other by the process of reflection.

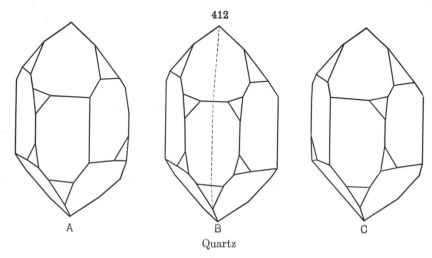

412

A   B   C

Quartz

Fig. 413 represents a twin crystal of calamine. Calamine is orthorhombic, hemimorphic. The twinning-plane here is the horizontal plane, the basal pinacoid, (001), the upper and lower halves of the twin being the reflections of each other over this plane. The vertical axis, normal to the twinning-plane, cannot be a twinning-axis since it is an axis of binary symmetry. If the development of the twin crystal is such that the prism and pinacoid faces of the two components coincide, thus eliminating the reentrant angles, the crystal will apparently possess normal orthorhombic symmetry. Such a possibility is shown in Fig. 414, where the vertical faces form composite faces joined together along the medial horizontal plane, indicated by the broken lines. Such twins, in which, owing to the twinning, all the faces are present that belong to a form of normal symmetry, are termed supplementary twins. Supplementary twins are also shown in Figs. 433–435, of pyrite, tetrahedrite and eulytite.

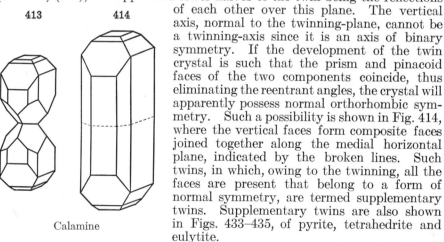

413   414

Calamine

242. Composition-Plane. — The plane by which the reversed crystals are united is the *composition-plane*. This and the twinning-plane very commonly coincide; this is true of the simple example of the twinned octahedron, described in Art. **241** and illustrated in Fig. 410. Here the plane about which the revolution may be conceived to take place (normal to the twinning-axis) and the plane by which the semi-individuals are united are identical. When not coinciding, the two planes are generally at right angles to each other — that is, the composition-plane is parallel to the axis of revolution. The micas furnish a case in which the composition-plane of the twin crystal is at right angles to the twinning-plane. Fig. 415 represents a mica twin, the base (001) being the composition-plane, the twinning-plane being the theoretical face (110) (not observed on mica crystals) and the twinning-axis being a horizontal direction. Still again, where the crystals are not regularly developed, and where they interpenetrate, the contact surface may be interrupted, or may be exceedingly irregular. In such cases the axis and plane of twinning have, as always, a definite position, but the composition-plane loses its significance.

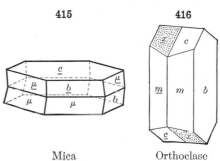

415   416

Mica   Orthoclase

Thus in quartz twins the interpenetrating parts have often no rectilinear boundary, but mingle in the most irregular manner throughout the mass, showing this composite irregularity by abrupt variations in the character of the surfaces. This irregular internal structure, found in many quartz crystals, even the common kinds, is well brought out by means of polarized light; also by etching with hydrofluoric acid.

The composition-plane has sometimes a more definite signification than the twinning-plane. This is due to the fact that in many cases, whereas the former is fixed, the twinning axis (and twinning-plane) may be exchanged for another line (and plane) at right angles to each, respectively, since a revolution about the second axis will also satisfy the conditions of producing the required form. An example of this is furnished by Fig. 416, of orthoclase; the composition-plane is here fixed — namely, parallel to the crystal face, b(010). But the axis of revolution may be either (1) parallel to this face and normal to a(100), though the axis does not coincide with the crystallographic axis; or (2) the twinning-axis may be taken as coinciding with the vertical axis, and then the plane normal to it is not a crystallographic face. In other simpler cases, also, the same principle holds good, generally in consequence of the possible mutual interchange of the planes of twinning and composition. In most cases the true twinning-plane is evident, since it is parallel to some face on the crystal of simple mathematical ratio.

243. An interesting example of the possible choice between two twinning-axes at right angles to each other is furnished by the species staurolite. Fig. 465 shows a prismatic twin from Fannin Co., Georgia. The measured angle for bb was 70° 30′. The twinning-axis deduced from this may be normal to the face (230), which would then be the twinning-plane. Or, instead of this axis, its complementary axis at right angles to it may be taken, which would equally well produce the observed form. Now in this species it happens that the faces, 130 and 230 (over 100), are almost exactly at right angles with each other, and, according to the latter supposition, 130 becomes the twinning-plane, and the axis of revolution is normal to it. Hence, either 230 or 130 may be the twinning-plane, either supposition agreeing closely with the measured angle (which could not be obtained with great accuracy). The former method of twinning (tw. pl. 230) conforms to the other twins observed on the species, and hence it may be accepted. What is true in this case, however, is not always true, for it will seldom happen that of the two complementary axes each is so nearly normal to a face of the crystal. In most cases one of the two axes conforms to the law in being a normal to a possible face, and the other does not, and hence there is no doubt as to which is the true twinning-axis.

Another interesting case is that furnished by columbite. The common twins of the species are similar to Fig. 407, p. 180, and have $e(021)$ as the twinning-plane; but twins also occur like Fig. 460, p. 191, where the twinning-plane is $q(023)$. The two faces, 021 and 0$\bar{2}$3, are nearly at right angles to each other, but the measured angles are in this case sufficiently exact to prove that the two kinds cannot be referred to one and the same law.

244. Contact- and Penetration-Twins. — In *contact-twins*, when normally formed, the two halves are simple connate, being united to each other by the composition-plane; they are illustrated by Figs. 407, 410, etc. In actual crystals the two parts are seldom symmetrical, as demanded by theory, but one may preponderate to a greater or less extent over the other; in some cases only a small portion of the second individual in the reversed position may exist. Very great irregularities are observed in nature in this respect. Moreover, the reentrant angles are often obliterated by the abnormal developments of one or other of the parts, and often only an indistinct line on some of the faces marks the division between the two individuals.

Penetration-twins are those in which two or more complete crystals interpenetrate, as it were crossing through each other. Normally, the crystals have a common center, which is the center of the axial system for both; practically, however, as in contact-twins, great irregularities occur.

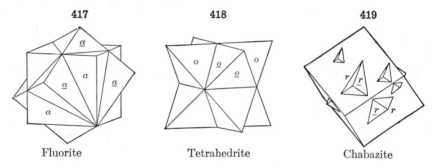

417 418 419

Fluorite   Tetrahedrite   Chabazite

Examples of twins of this second kind are given in the annexed figures, Figs. 408 and 417 of fluorite, Fig. 418 of tetrahedrite, and Fig. 419 of chabazite. Other examples occur in the pages following, as, for instance, of the species staurolite (Figs. 464–467), the crystals of which sometimes occur in nature with almost the perfect symmetry demanded by theory. It is obvious that the distinction between contact- and penetration-twins is not of great importance, and the line cannot always be clearly drawn between them.

245. Paragenetic and Metagenetic Twins. — The distinction of paragenetic and metagenetic twins belongs rather to crystallogeny than crystallography. Yet the forms are often so obviously distinct that a brief notice of the distinction is important.

In ordinary twins, the compound structure had its beginning in a nucleal compound molecule, or was compound in its very origin; and whatever inequalities in the result, these are only irregularities in the development from such a nucleus. But in others, the crystal was at first simple; and afterwards, through some change in itself or in the condition of the material supplied for its increase, received new layers, or a continuation, in a reversed position. This mode of twinning is *metagenetic*, or a result subsequent to the origin of the crystal; while the ordinary mode is *paragenetic*. One form of it is illustrated in Fig. 420. The middle portion had attained a length of half an inch or more, and then became geniculated simultaneously at either extremity. These geniculations are often repeated in rutile, and the ends of the crystal are thus bent into one another, and occasionally produce nearly regular prismatic forms.

This metagenetic twinning is sometimes presented by the successive layers of deposition in a crystal, as in some quartz crystals, especially amethyst, the inseparable layers, exceedingly thin, being of opposite kinds. In a similar manner, crystals of the triclinic feldspars, albite, etc., are often made up of thin plates parallel to $b(010)$, by oscillatory combination and the face $c(001)$, accordingly, is finely striated parallel to the edge, c/b.

246. Repeated Twinning, Polysynthetic and Symmetrical. — In the preceding paragraph one case of repeated twinning has been mentioned, that of the feldspars; it is a case of *parallel* repetition or parallel grouping in reversed position of successive crystalline lamellæ. This kind of twinning is often called *polysynthetic twinning*, the lamellæ in many cases being extremely thin, and giving rise to a series of parallel lines (striations) on a crystal face or a surface of cleavage. The triclinic feldspars show in many cases polysynthetic twinning and not infrequently on both $c(001)$ and $b(010)$, cf. p. 193. It is also observed with magnetite (Fig. 500), pyroxene, barite, etc.

420

Rutile

Another kind of repeated twinning is illustrated by Figs. 421–426, where the successively reversed individuals are not parallel. In these cases the axes

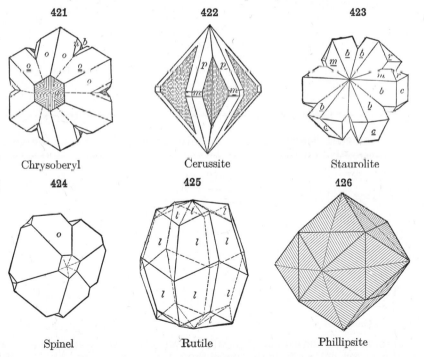

421	422	423
Chrysoberyl	Cerussite	Staurolite
424	425	426
Spinel	Rutile	Phillipsite

may, however, lie in a zone, as the prismatic twins of aragonite, or they may be inclined to each other, as in Fig. 423 of staurolite. In all such cases the repetition of the twinning tends to produce circular forms, when the angle between the two axial systems is an aliquot part of 360° (approximately). Thus six-rayed twinned crystals, consisting of three individuals (hence called *trillings*), occur with chrysoberyl (Fig. 421), or cerussite (Fig. 422), or staurolite (Fig. 423), since three times the angle of twinning in each case is not far from 360°. Again, five-fold twins, or *fivelings*, occur in the octahedrons of gold and spinel (Fig. 424), since $5 \times 70° 32' = 360°$ (approx.). Eight-fold twins, or *eightlings*, of rutile (Figs. 425, 439) occur, since the angle of the axes in twinned position goes approximately eight times in 360°.

Repeated twinning of the symmetrical type often serves to give the compound crystal an apparent symmetry of higher grade than that of the simple individual, and the result is often spoken of as a kind of pseudo-symmetry (Art. **20**), cf. Fig. 457 of aragonite, which represents a basal section of a *pseudo-hexagonal* crystal. Fig. 426 of phillipsite (cf. Figs. 478–480) is an interesting case, since it shows how a multiple twin of a monoclinic crystal may simulate an isometric crystal (dodecahedron).

Compound crystals in which twinning exists in accordance with two laws at once are not of common occurrence; an excellent example is afforded by staurolite, Fig. 467. They have also been observed with albite, orthoclase, and in other cases.

247. Secondary Twinning. — When there is reason to believe that the twinning has been produced subsequently to the original formation of the crystal, or crystalline mass, as, for example, by pressure, it is said to be *secondary*. Thus the calcite grains of a crystalline limestone often show such secondary twinning lamellæ. The same are occasionally observed ($\|c$, 001) in pyroxene crystals. Further, the polysynthetic twinning of the triclinic feldspars is often secondary in origin. This subject is further discussed on a later page, where it is also explained that in certain cases twinning may be produced artificially in a crystal individual — *e.g.*, in calcite (see Art. **287**).

EXAMPLES OF IMPORTANT METHODS OF TWINNING

248. Isometric System. — With few exceptions the twins of the normal class of this system are of one kind, the twinning-axis an octahedral axis, and

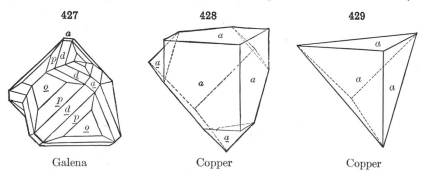

427   428   429

Galena   Copper   Copper

the twinning-plane consequently parallel to an *octahedral face*; in most cases, also, the latter coincides with the composition-plane. Fig. 410, p. 181,*

---

* It will be noted that here and elsewhere the letters used to designate the faces on the twinned parts of crystals are distinguished by a subscript line.

shows this kind as applied to the simple octahedron; it is especially common with the spinel group of minerals, and is hence called in general a *spinel-twin*. Fig. 427 is a similar more complex form; Fig. 428 shows a cube twinned by this method, and Fig. 429 represents the same form but shortened in the direction

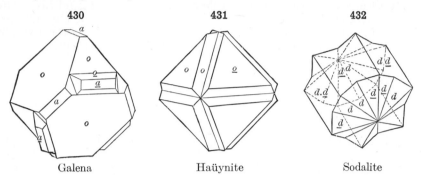

430	431	432
Galena	Haüynite	Sodalite

of the octahedral axis, and hence having the anomalous aspect of a triangular pyramid. All these cases are contact-twins.

Penetration-twins, following the same law, are also common. A simple case of fluorite is shown in Fig. 417, p. 184; Fig. 430 shows one of galena; Fig. 431 is a repeated octahedral twin of haüynite, and Fig. 432 a dodecahedral twin of sodalite.

249. In the *pyritohedral class* of the isometric system penetration-twins of the type shown in Fig. 433 are common (this form of pyrite is often called the *iron cross*). Here a dodecahedral plane serves as the twinning-plane and the normal to it as the twinning-axis. In this case both the plane and axis serve as symmetry elements in the normal class, and the twinned pyritohedron possesses all the planes of the normal tetrahexahedron. It is therefore a supplementary twin; cf. Art. 241.

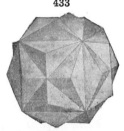

433

Pyrite

Figs. 434 and 435 show analogous forms with parallel axes for crystals belonging to the tetrahedral class. The peculiar development of Fig. 434 of

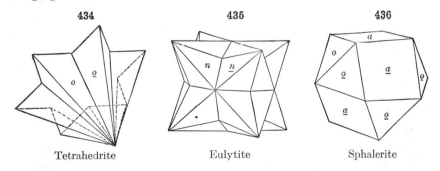

434	435	436
Tetrahedrite	Eulytite	Sphalerite

tetrahedrite is to be noted. Fig. 436 is a twin of the ordinary spinel type of another tetrahedral species, sphalerite; with it, complex forms with repeated twinning are not uncommon and sometimes polysynthetic twin lamellæ are noted.

250. Tetragonal System. — The most common method is that where the twinning-plane is parallel to a face of the pyramid, $e(101)$. It is especially characteristic of the species of the rutile group — viz., rutile and cassiterite: also similarly the allied species zircon. This is illustrated in Fig. 437, and again in Fig. 438. Fig. 439 shows a repeated twin of rutile, the twinning according to this law; the vertical axes of the successive six individuals lie in

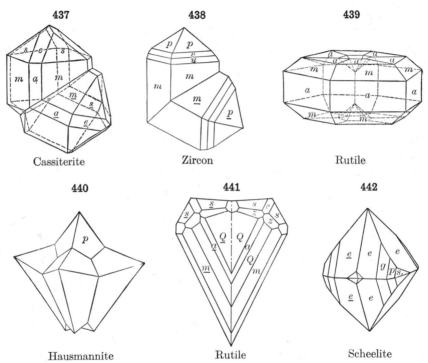

| 437 | 438 | 439 |
| Cassiterite | Zircon | Rutile |

| 440 | 441 | 442 |
| Hausmannite | Rutile | Scheelite |

a plane (the horizontal plane of the figure), and an inclosed circle is the result. Another repeated twin of rutile is shown in Fig. 425; here the successive vertical axes form a zigzag line; Fig. 440 shows an analogous twin of hausmannite.

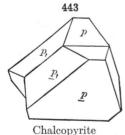

443

Chalcopyrite

Another kind of twinning with the twinning-plane parallel to a face of the pyramid (301) is shown in Fig. 441.

251. In the *pyramidal class* of the same system twins of the type of Fig. 442 are not rare. Here the basal pinacoid is the twinning-plane; such a crystal may simulate one of the normal class.

In chalcopyrite, of the *sphenoidal class*, twinning with a face of the unit sphenoid, $p(111)$, as the twinning-plane is common (Fig. 443). As the angles differ but a small fraction of a degree from those of a regular octahedron, such twins often resemble closely

spinel-twins.   The face e(101) may also be a twinning-plane and other rarer
types have been noted.

252. Hexagonal System. — In the *hexagonal* division of this system twins
are rare.   An example is furnished by pyrrhotite, Fig. 444, where the twinning-
plane is the pyramid (10$\bar{1}$1), the vertical axes of the individual crystals being nearly at right angles to each
other (since 0001 $\wedge$ 10$\bar{1}$1 = 45° 8′).

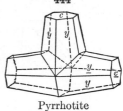

444

253. In the species belonging to the trigonal or
rhombohedral division, twins are common.   Thus the
twinning-axis may be the vertical axis, as in the contact-twins of Figs. 445 and 446, or the penetration-
twin of Fig. 419.   Or the twinning-plane may be the
obtuse rhombohedron e(01$\bar{1}$2), as in Fig. 447, the vertical axes crossing at angles of 127$\frac{1}{2}$° and 52$\frac{1}{2}$°.   Again,

Pyrrhotite

the twinning-plane may be r(10$\bar{1}$1), as in Figs. 448–451, the vertical axes nearly
at right angles (90$\frac{3}{4}$°); or (0$\bar{2}$21), as in Fig. 452, the axes inclined 53$\frac{3}{4}$°
and 126$\frac{1}{4}$°.

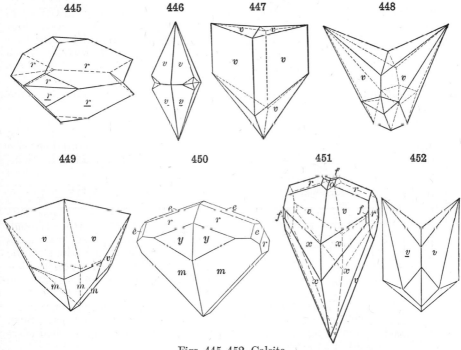

Figs. 445–452, Calcite

In the *trapezohedral class*, the species quartz shows several methods of
twinning.   In Fig. 453 the twinning-plane is the pyramid ξ(11$\bar{2}$2), the axes
crossing at angles of 84$\frac{1}{2}$° and 95$\frac{1}{2}$°.   In Fig. 454 the twinning-axis is c, the
axes hence parallel, the individuals both right- or both left-handed but unsymmetrical, r(10$\bar{1}$1) then parallel to and coinciding with z(01$\bar{1}$1).   The resulting forms, as in Fig. 454, are mostly penetration-twins, and the parts are

often very irregularly united, as shown by dull areas (*z*) on the plus rhombohedral face (*r*); otherwise these twins are recognized by pyroelectrical phenomena. In Fig. 455 the twinning-plane is $a(11\bar{2}0)$ — the *Brazil law* — the individuals respectively right- and left-handed and the twin symmetrical

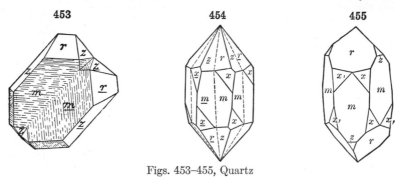

453   **454**   **455**

Figs. 453–455, Quartz

with reference to an *a*-face; these are usually irregular penetration-twins; in these twins *r* and *r*, also *z* and *z*, coincide. These twins often show, in converging polarized light, the phenomenon of Airy's spirals. It may be added that pseudo-twins of quartz are common — that is, groups of crystals which *nearly* conform to some more or less complex twinning law, but where the grouping is nevertheless only accidental.

254. Orthorhombic System. — In the orthorhombic system the commonest method of twinning is that where the twinning-plane is a face of a prism with the prism angle of 60°, or nearly 60°. This is well shown with the species of the aragonite group. In accordance with the principle stated in Art. **246,** the twinning after this law is often repeated, and thus forms with pseudo-hexagonal symmetry result. Fig. 456 shows a simple twin of aragonite; Fig. 457 shows a basal section of an aragonite triplet which although it resembles a hexagonal prism reveals its twinned character by the striations on the basal plane and by irregularities on its composite prism faces due to the fact that the prismatic angle is not exactly 60°. With

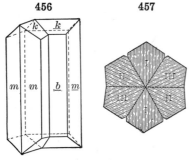

456   **457**

Aragonite

witherite (and bromlite), apparent hexagonal pyramids are common, but the true complex twinning is revealed in polarized light, as noted later. Such twins, which simulate the symmetry of another and more symmetrical crystal system, are often called *mimetic twins*.

Twinning of the same type, but where a dome of 60° is twinning-plane, is common with arsenopyrite (tw. pl. *e*(101)), as shown in Figs. 458, 459; also Fig. 460 of columbite, but compare Fig. 407 and remarks in Art. **243.** Another example is given in Fig. 421 of alexandrite (chrysoberyl). Chrysolite, manganite, humite, are other species with which this kind of twinning is common.

Another common method of twinning is that where the twinning is parallel

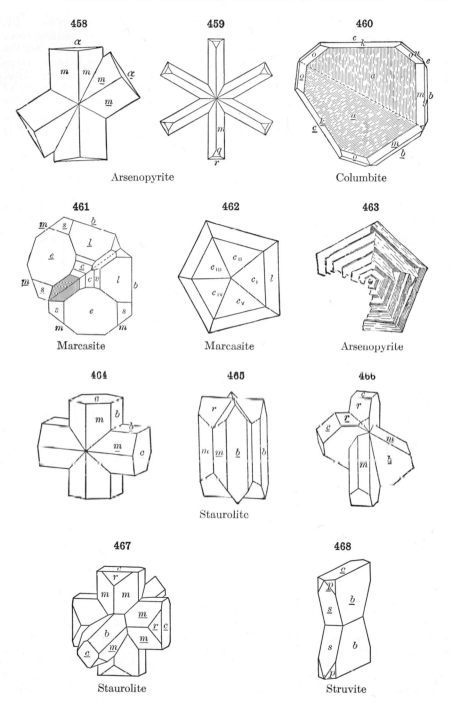

458

Arsenopyrite

459

Arsenopyrite

460

Columbite

461

Marcasite

462

Marcasite

463

Arsenopyrite

464

465

Staurolite

466

467

Staurolite

468

Struvite

to a face of a prism with a prism angle of about $70\frac{1}{2}°$, as shown in Fig. 461. With this method symmetrical fivelings not infrequently occur (Figs. 462, 463).

The species staurolite illustrates three kinds of twinning. In Fig. 464 the twinning-plane is (032), and since $(001 \wedge 032) = 45° 41'$, the crystals cross nearly at right angles. In Fig. 465 the twinning-plane is the prism (230). In Fig. 466 it is the pyramid (232), the crystals then crossing at angles of about 60°; stellate trillings occur (see Fig. 423), and indeed more complex forms. In Fig. 467 there is twinning according to both (032) and (232).

In the hemimorphic class, twins of the type shown in Fig. 468, with $c(001)$ as the twinning-plane, are to be noted.

469   470   471

Augite   Gypsum   Orthoclase

255. Monoclinic System. — In the monoclinic system, twins with the vertical axis as twinning-axis are common (see also Art. **242**); this is illustrated by Fig. 469 of augite (pyroxene), Fig. 470 of gypsum, and Fig. 471 of orthoclase (see also Fig. 416, p. 183). With the latter species these twins are called *Carlsbad twins* (because common in the trachyte of Carlsbad, Bohemia); they may be contact-twins (Fig. 416), or irregular penetration-twins (Fig. 471). In Fig. 416 it is to be noted that c and x fall nearly in the same plane.

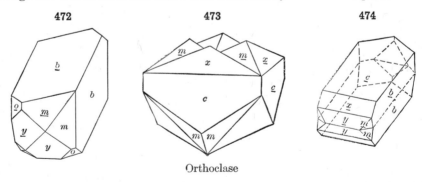

472   473   474

Orthoclase

In Fig. 472, also of orthoclase, the twinning-plane is the clinodome (021), and since $(001 \wedge 021) = 44° 56\frac{1}{2}'$, this method of twinning yields nearly square prisms. These twins are called *Baveno twins* (from a prominent locality at Baveno, Italy); they are often repeated (Fig. 473). In Fig. 474 a *Manebach twin* is shown; here the twinning-plane is $c(001)$. Other rarer types of twinning have been noted with orthoclase. Polysynthetic twinning with $c(001)$ as twinning-plane is common with pyroxene (cf. Fig. 487, p. 195).

Twins of the aragonite-chrysoberyl type are not uncommon with mono-clinic species, having a prominent 60° prism (or dome), as in Fig. 475. Stellate twins after this law are common with chondrodite and clinohumite. An analogous twin of pyroxene is shown in Fig. 476; here the pyramid ($\bar{1}22$) is the twinning-plane, and since ($010 \wedge \bar{1}22$) = 59° 21′, the crystals cross at angles of nearly 60°; further, the orthopinacoids fall nearly in a common zone, since

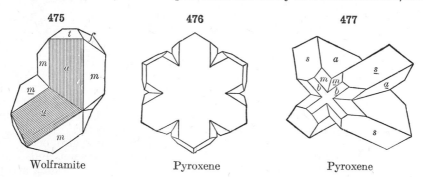

475	476	477
Wolframite	Pyroxene	Pyroxene

($100 \wedge \bar{1}22$) = 90° 9′. In Fig. 477 the twinning-plane is the orthodome (101). Phillipsite and harmotome exhibit multiple twinning, and the crystals often show pseudo-symmetry. Fig. 478 shows a cruciform fourling with c(001) as twinning-plane, the twinning shown by the striations on the side face. This is compounded in Fig. 479 with twinning-plane (011), making nearly square prisms, and this further repeated with m(110) as twinning-plane yields the form in Fig. 480, or even Fig. 426, p. 185, resembling an isometric dodecahedron, each face showing a fourfold striation.

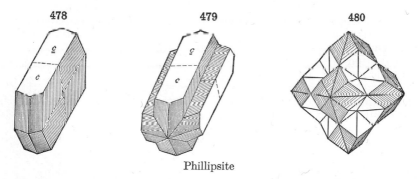

478	479	480

Phillipsite

256. Triclinic System. — The most interesting twins of the triclinic system are those shown by the feldspars. Twinning with b(010) as the twin-ning-plane is very common, especially polysynthetic twinning yielding thin parallel lamellæ, shown by the striations on the face c (or the correspond-ing cleavage-surface), and also clearly revealed in polarized light. This is known as the *albite law* (Figs. 481, 482). Another important method (Fig. 483) is that of the *pericline law*; the twinning-axis is the crystallographic axis b. Here the twins are united by a section (rhombic section) shown in the figure and further explained under the feldspars. Polysynthetic twinning after this law is common, and hence a cleavage-mass may show two sets of striations, one on the surface parallel to $c(001)$ and the other on that parallel to

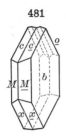

481

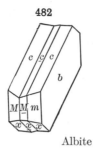

482

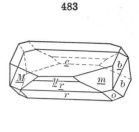

483

Albite

$b(010)$. The angle made by these last striations with the edge $001/010$ is characteristic of the particular triclinic species, as noted later.

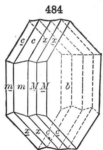

484

Albite

Twins of albite of other rarer types also occur, and further twins similar to the Carlsbad, Baveno, and Manebach twins of orthoclase. Fig. 484 shows twinning according to both the albite and Carlsbad types.

REGULAR GROUPING OF CRYSTALS

257. Parallel Grouping. — Connected with the subject of twin crystals is that of the parallel position of associated crystals of the same species, or of different species.

Crystals of the same species occurring together are very commonly in parallel position. In this way large crystals, as of calcite, quartz, fluorite, are sometimes built up of smaller individuals grouped together with corresponding faces parallel. This parallel grouping is often seen in crystals as they lie on the supporting rock. On glancing over a surface covered with crystals a reflection from one face will often be accompanied by reflections from the corresponding face in each of the other crystals, showing that the crystals are throughout similar in their positions.

With many species, complex crystalline forms result from the growth of parallel partial crystals in the direction of the crystallographic axes, or axes of symmetry. Thus *dendritic* forms, resembling branching vegetation, often of great delicacy, are seen with gold, copper, argentite, and other species,

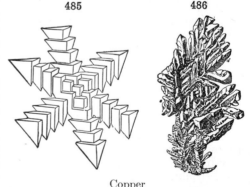

485   486

Copper

especially those of the isometric system. This is shown in Fig. 485 (ideal), and again in Fig. 486, where the twinned and flattened cubes (cf. Fig. 429, p. 186) are grouped in directions corresponding to the diagonals of an octahedral face.

258. Partial Parallel Grouping of the Same or Different Species. — Crystals of different species often show the same tendency to parallelism in mutual position. This is true most frequently of species which are more or less closely similar in form and composition. They will show a parallel position between certain (usually similar) planes in the two species, or the parallelism of certain directions. Such groupings of unlike species must be due to close similarities in crystal structure that control the mutual orientation of their crystals. The structure planes of the two minerals that are in contact with each other probably have atomic arrangements and spacings of almost identical character. Crystals of albite, implanted on a surface of orthoclase, are sometimes an example of this; crystals of amphibole and pyroxene (Fig. 487), of zircon and xenotime (Fig. 488), of various kinds of mica, are also at times observed associated in parallel position.

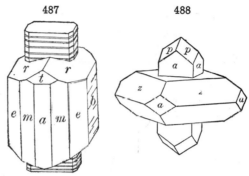

Amphibole enclosing pyroxene in parallel position

Xenotime enclosing zircon in parallel position

The same relation of position also occasionally occurs where there is no connection in composition, as the crystals of rutile on tabular crystals of hematite, the vertical axes of the former coinciding with the horizontal axes of the latter. Crystals of calcite have been observed whose rhombohedral faces had a series of quartz crystals upon them, all in parallel position; sometimes three such quartz crystals, one on each rhombohedral face, entirely envelop the calcite, and unite with reentering angles to form pseudo-twins (rather trillings) of quartz after calcite. Parallel growths of the sphenoidal chalcopyrite upon the tetrahedral sphalerite are common, the similarity in crystal structure of the two species controlling the position of the crystals of chalcopyrite. Cases have been described of similar grouping of the crystals of the same substance, in which a certain plane or direction is common to the different individuals but which the ordinary laws of twinning will not explain.

IRREGULARITIES OF CRYSTALS

259. The laws of crystallization, when unmodified by extrinsic causes, should produce forms of exact geometrical symmetry, not only the angles being equal, but also the homologous faces of crystals and the dimensions in the directions of like axes. This symmetry is, however, so uncommon that it can hardly be considered other than an ideal perfection. The various possible kinds of symmetry, and the relation of this ideal geometrical symmetry to the actual crystallographic symmetry, have been discussed in Arts. **14** and **18** *et seq.* Crystals are very generally distorted, and often the fundamental forms are so completely disguised that an intimate familiarity with the possible irregularities is required in order to unravel their complexities. Even the angles may occasionally vary rather widely.

The irregularities of crystals may be treated under several heads: **1,** *Variations of form and dimensions;* 2, *Imperfections of surface;* 3, *Variations of angles;* 4, *Internal imperfections and impurities.*

1. VARIATIONS IN THE FORMS AND DIMENSIONS OF CRYSTALS

260. Distortion in General. — The variations in the forms of crystals, or, in other words, their *distortion,* may be *irregular* in character, certain faces being larger and others smaller than in the ideal geometrical solid. On the other hand, it may be *symmetrical,* giving to the distorted form the symmetry of a group or system different from that to which it actually belongs. The former case is the common rule, but the latter is the more interesting.

261. Irregular Distortion. — As stated above and on p. 13, all crystals show to a greater or less extent an irregular or accidental variation from the ideal geometrical form. This distortion, if not accompanied by change in

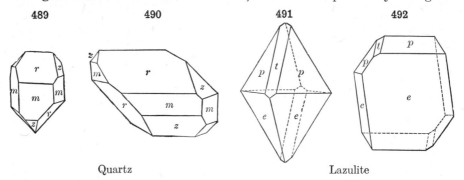

489   490   491   492

Quartz   Lazulite

the interfacial angles, has no particular significance, and does not involve any deviation from the laws of crystallographic symmetry. Figs. 489, 490 show distorted crystals of quartz; they may be compared with the ideal form, Fig. 303, p. 130. Fig. 491 is an ideal and Fig. 492 an actual crystal of lazulite.

The correct identification of the forms on a crystal is rendered much more difficult because of this prevailing distortion, especially when it results in the entire *obliteration* of certain faces by the enlargement of others. In deciphering the distorted crystalline forms it must be remembered that while the appearance of the crystals may be entirely altered, the interfacial angles remain the same; moreover, like faces are physically alike — that is, alike in degree of luster, in striations, and so on. Thus the prismatic faces of quartz show almost always characteristic horizontal striations.

In addition to the variations in form which have just been described, still greater irregularities are due to the fact that, in many cases, crystals in nature are attached either to other crystals or to some rock surface, and in consequence of this are only partially developed. Thus quartz crystals are generally attached by one extremity of the prism, and hence have only one set of terminal faces; perfectly formed crystals, having both ends complete, are rare.

262. Symmetrical Distortion. — The most interesting examples of the symmetrical distortion of crystalline forms are found among crystals of the isometric system. An elongation in the direction of one cubic axis may give the appearance of tetragonal symmetry, or that in the direction of two cubic axes of orthorhombic symmetry; while in the direction of an octahedral axis a lengthening or shortening gives rise to forms of apparent rhombohedral symmetry. Such cases are common with native gold, silver, and copper.

A *cube* lengthened or shortened along one axis becomes a right square prism, and if varied in the direction of two axes is changed to a rectangular prism. Cubes of pyrite, galena, fluorite, etc., are often thus distorted. It is very unusual to find a cubic crystal that is a true symmetrical cube. In some species the cube or octahedron (or other isometric form) is lengthened into a capillary crystal or needle, as happens in cuprite and pyrite.

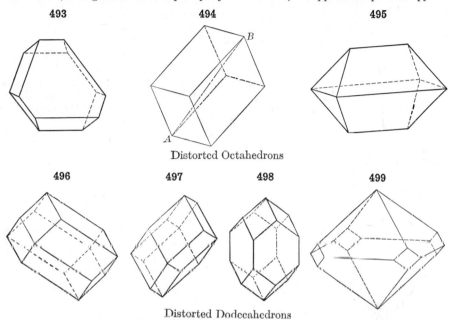

493 494 495

Distorted Octahedrons

496 497 498 499

Distorted Dodecahedrons

An octahedron *flattened* parallel to a face — that is, in the direction of a trigonal symmetry axis — is reduced to a tabular crystal resembling a rhombohedral crystal with basal plane (Fig. 493). If *lengthened* in the same direction (*i.e.* along line A–B, Fig. 494), to the obliteration of the terminal octahedral faces, it becomes an acute rhombohedron.

When an octahedron is extended in the direction of a line between two opposite edges, or that of a binary symmetry axis, it has the general form of a rectangular octahedron; and still further extended, as in Fig. 495, it resembles a combination of two orthorhombic domes (spinel, fluorite, magnetite).

The *dodecahedron* lengthened in the direction of a trigonal symmetry axis becomes a six-sided prism with three-sided summits, as in Fig. 496. If shortened in the same direction, it becomes a *short* prism of the same kind (Fig. 497). Both resemble rhombohedral forms and are common in garnet. When lengthened in the direction of one of the cubic axes, the dodecahedron becomes a square prism with pyramidal summits (Fig. 498), and shortened along the same axis it is reduced to a square octahedron, with truncated angles (Fig. 499).

The trapezohedron elongated in the direction of an octahedral (trigonal) axis assumes rhombohedral (trigonal) symmetry.

If the elongation of the trapezohedron takes place along a cubic axis, it becomes a double eight-sided pyramid with four-sided summits; or if these summit planes are obliterated by a farther extension, it becomes a complete eight-sided double pyramid.

Similarly the trisoctahedron, tetrahexahedron and hexoctahedron may show distortion of the same kind. Further examples are to be found in the other systems.

2. IMPERFECTIONS OF THE SURFACES OF CRYSTALS

263. Striations Due to Oscillatory Combinations. — The parallel lines or furrows on the surfaces of crystals are called *striæ* or *striations*, and such surfaces are said to be *striated*.

Each little ridge on a striated surface is inclosed by two narrow planes more or less regular. These planes often correspond in position to different faces of the crystal, and these ridges have been formed by a continued oscillation in the operation of the causes that give rise, when acting uninterruptedly, to enlarged faces. By this means, the surfaces of a crystal are marked in parallel lines, with a succession of narrow planes meeting at an angle and constituting the ridges referred to.

This combination of different planes in the formation of a surface has been termed *oscillatory combination*. The horizontal striations on prismatic crystals of quartz are examples of this combination, in which the oscillation has taken place between the prismatic and rhombohedral faces. Thus crystals of quartz are often tapered to a point, without the usual terminations.

Other examples are the striations on the cubic faces of pyrite parallel to the intersections of the cube with the faces of the pyritohedron; also the striations on magnetite due to the oscillation between the octahedron and dodecahedron. Prisms of tourmaline are very commonly bounded vertically

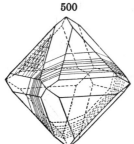

500

Magnetite

by three convex surfaces, owing to an oscillatory combination of the faces in the prismatic zone.

264. Striations Due to Repeated Twinning. — The striations of the basal plane of albite and other triclinic feldspars, also of the rhombohedral surfaces of some calcite, have been explained in Art. **246** as due to polysynthetic twinning. This is illustrated by Fig. 500 of magnetite from Port Henry, New York. (Kemp.)

265. Markings from Erosion and Other Causes. — The faces of crystals are often uneven, or have the crystalline structure developed as a consequence of etching by some chemical agent. Cubes of galena are frequently thus uneven, and crystals of lead sulphate (anglesite) or lead carbonate (cerussite) are sometimes present as evidence with regard to the cause. Crystals of numerous other species, even of corundum, spinel, quartz, etc., sometimes show the same result of partial change over the surface — often the incipient stage in a process tending to a final removal of the whole crystal. Interesting investigations have been made by various authors on the action of solvents on different minerals, the actual structure of the crystals being developed in this way. This method of etching is fully discussed, with illustrations, in another place (Art. **291**).

The markings on the surfaces of crystals are not, however, always to be ascribed to etching. In most cases such depressions, as well as the minute elevations upon the faces having the form of low pyramids (so-called *vicinal* prominences), are a part of the original molecular growth of the crystal, and often serve to show the successive stages in its history. They may be imperfections arising from an interrupted or disturbed development of the form, the perfectly smooth and even crystalline faces being the result of completed action free from disturbing causes. Examples of the markings referred to occur on the crystals of most minerals, and conspicuously so on the rhombo-hedral faces of quartz.

Faces of crystals are often marked with angular elevations more or less distinct, which are due to oscillatory combination.   Octahedrons of fluorite are common which have for each face a surface of minute cubes, proceeding from an oscillation between the cube and octahedron.   Sometimes an exami-nation of such a crystal shows that though the form is apparently octahedral, there are no octahedral faces present at all.   Other similar cases could be mentioned.

Whatever their cause, these minute markings are often of great importance as revealing the true molecular symmetry of the crystal.   For it follows from the symmetry of crystallization that like faces must be physically alike — that is, in regard to their surface character; it thus often happens that on all the crystals of a species from a given locality, or perhaps from all localities, the same planes are etched or roughened alike.   There is much uniformity on the faces of quartz crystals in this respect.

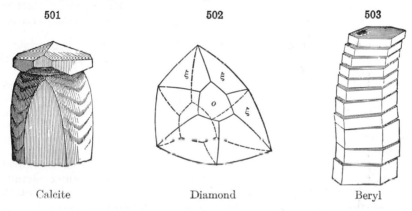

501	502	503
Calcite	Diamond	Beryl

266. Curved surfaces may result from (a) oscillatory combination; or (b) some independent molecular condition producing curvatures in the laminæ of the crystal; or (c) from a mechanical cause.

Curved surfaces of the *first* kind have been already mentioned (Art. 263). A singular curvature of this nature is seen in Fig. 501, of calcite; in the lower part traces of a scalenohedral form are apparent which was in oscillatory com-bination with the prismatic form.

Curvatures of the *second* kind sometimes have all the faces convex.   This is the case in crystals of diamond (Fig. 502), some of which are almost spheres. The mode of curvature, in which all the faces are equally convex, is less common than that in which a convex surface is opposite and parallel to a corresponding concave surface.   Rhombohedrons of dolomite and siderite are usually thus curved.   The feathery curves of frost on windows and the flagging-stones of pavements in winter are other examples.   The alabaster rosettes from the Mammoth Cave, Kentucky, are similar.   Stibnite crystals sometimes show very remarkable curved and twisted forms.

A *third* kind of curvature is of *mechanical* origin.   Sometimes crystals appear as if they had been broken transversely into many pieces, a slight displacement of which has given a curved form to the prism.   This is common in tourmaline and beryl. The beryls of Monroe, Connecticut, often present these interrupted curvatures, as represented in Fig. 503.

Crystals not infrequently occur with a deep pyramidal depression occupying the place of each plane, as is often observed in common salt, alum, and sulphur. This is due in part to their rapid growth.

3. VARIATIONS IN THE ANGLES OF CRYSTALS

267. The greater part of the distortions described in Arts. **261, 262** occasion no change in the interfacial angles of crystals. But those imperfections that produce convex, curved, or striated faces necessarily cause such variations. Furthermore, circumstances of heat or pressure under which the crystals were formed may sometimes have resulted not only in distortion of form, but also some variation in angle. The presence of impurities at the time of crystallization may also have a like effect.

Still more important is the change in the angles of completed crystals which is caused by subsequent pressure on the matrix in which they were formed, as, for example, the change which may take place during the more or less complete metamorphism of the inclosing rock.

The change of composition resulting in pseudomorphous crystals (see Art. **278**) is generally accompanied by an irregular change of angle, so that the pseudomorphs of a species vary much in angle.

In general it is safe to affirm that, with the exception of the irregularities arising from imperfections in the process of crystallization, or from the subsequent changes alluded to, variations in angles are rare, and the constancy of angle alluded to in Art. **11** is the universal law.

In cases where a greater or less variation in angle is observed in the crystals of the same species from different localities, the cause for this can usually be found in a difference of chemical composition. In the case of isomorphous compounds it is well known that an exchange of corresponding chemically equivalent elements may take place without a change of form, though usually accompanied with a slight variation in the fundamental angles.

The effect of heat upon the form of crystals is alluded to in Art. **445.**

4. INTERNAL IMPERFECTIONS AND INCLUSIONS

268. The transparency of crystals is often destroyed by disturbed crystallization; by impurities taken up from the solution during the process of crystallization; or, again, by the presence of foreign matter resulting from partial chemical alteration. The general name, *inclusion*, is given to any foreign body inclosed within the crystal, whatever its origin. These inclusions are extremely common; they may be gaseous, liquid, or solid; visible to the unaided eye or requiring the use of the microscope.

Rapid crystallization is a common explanation of inclusions. This is illustrated by quartz crystals containing large cavities full or nearly full of water (in the latter case, these showing a movable bubble); or, they may contain sand or iron oxide in large amount. In the case of calcite, crystallization from a liquid largely charged with a foreign material, as quartz sand, may result in the formation of crystals in which the impurity makes up as much as two thirds of the whole mass; this is seen in the famous Fontainebleau limestone, and similarly in that from other localities.

269. Liquid and Gas Inclusions. — Attention was early called by Brewster to the presence of fluids in cavities in certain minerals, as quartz, topaz, beryl, chrysolite, etc. In later years this subject has been thoroughly studied by Sorby, Zirkel, Vogelsang, Fischer, Rosenbusch, and others. The nature of the liquid can often be determined, by its refractive power, or by special physical test (*e.g.*, determination of the critical point in the case of CO_2), or by chemical examination. In the majority of cases the observed liquid is simply water; but it may be the salt solution in which the crystal was formed, and not infrequently, especially in the case of quartz, liquid carbon dioxide (CO_2), as first proved by Vogelsang. These liquid inclusions are marked as such, in many cases, by the presence in the cavity of a movable bubble of gas. Occasionally cavities contain two liquids, as water and liquid carbon dioxide, the latter then inclosing a bubble of the same substance as gas (cf. Fig. 504). Interesting experiments can be made with sections showing such inclusions (cf. literature, p. 203). The mixture of gases yielded by smoky quartz, meteoric iron, and other substances, on the application of heat, has been analyzed by Wright.

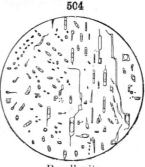

504

Beryllonite

In some cases the cavities appear to be empty; if they then have a regular form determined by the crystallization of the species, they are often called *negative crystals*. Such cavities are commonly of secondary origin, as remarked on a later page.

270. Solid Inclusions. — The solid inclusions are almost infinite in their variety. Sometimes they are large and distinct, and can be referred to known mineral species, as the scales of goethite or hematite, to which the peculiar character of aventurine feldspar is due. Magnetite is a very common impurity in many minerals, appearing, for example, in the Pennsbury mica; quartz is also often mechanically mixed, as in staurolite and gmelinite. On the other hand, quartz crystals very commonly inclose foreign material, such as chlorite, tourmaline, rutile, hematite, asbestos, and many other minerals. (Cf. also Arts. **271, 272.**)

The inclusions may consist of a heterogeneous mass of material, as the granitic matter seen in orthoclase crystals in a porphyritic granite; or the feldspar, quartz, etc., sometimes inclosed in large coarse crystals of beryl or spodumene, occurring in granite veins.

271. Microlites, Crystallites. — The microscopic crystals observed as inclusions may sometimes be referred to known species, but more generally their true nature is doubtful. The term *microlites*, proposed by Vogelsang, is often used to designate the minute inclosed crystals; they are generally of needlelike form, sometimes quite irregular, and often very remarkable in their arrangement and groupings; some of them are exhibited in Fig. 510 and Fig. 511, as explained below. Where the minute individuals belong to known species they are called, for example, feldspar microlites, etc.

Crystallites is an analogous term used by Vogelsang to cover those minute forms which have not the regular exterior form of crystals, but may be considered as intermediate between amorphous matter and true crystals. Some of the forms are shown in Figs. 505–509; they are often observed in glassy volcanic rocks, and also in furnace-slags. A series of names has been given to varieties of crystallites, such as globulites, margarites, etc. Trichite and belonite are names introduced by Zirkel; the former name is derived from θρίξ, *hair*; trichites, like that in Fig. 509, are common in obsidian.

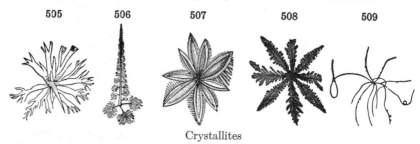

505   506   507   508   509

Crystallites

The microscopic inclusions may also be of an irregular glassy nature; this kind is often observed in crystals which have formed from a molten mass, as lava or the slag of an iron furnace.

272. Symmetrically Arranged Inclusions. — In general, while the solid inclusions sometimes occur quite irregularly in the crystals, they are more generally arranged with some evident reference to the symmetry of the form, or external faces of the crystals. Examples of this are shown in the following figures. Fig. 510 exhibits a crystal of augite, inclosing magnetite, feldspar

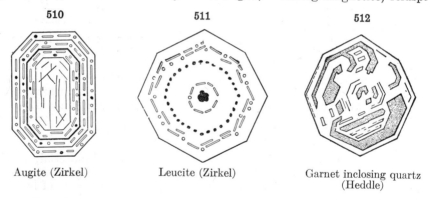

510   511   512

Augite (Zirkel)   Leucite (Zirkel)   Garnet inclosing quartz (Heddle)

and nephelite microlites, etc. Fig. 511 shows a crystal of leucite, a species whose crystals very commonly inclose foreign matter. Fig. 512 shows a section of a crystal of garnet, containing quartz.

513

Andalusite

Another striking example is afforded by andalusite (Fig. 513), in which the inclosed carbonaceous impurities are of considerable extent and remarkably arranged, so as to yield symmetrical figures of various forms. Staurolite occasionally shows analogous carbonaceous impurities symmetrically distributed.

The magnetite common as an inclusion in muscovite, alluded to above, is always symmetrically disposed, usually parallel to the directions of the percussion-figure (Fig. 517, p. 211). The asterism of phlogopite is explained by the presence of symmetrically arranged inclusions (cf. Art. 375).

514

Fig. 514 shows an interesting case of symmetrically arranged inclusions due to chemical alteration. The original mineral, spodumene, from Branchville, Connecticut, has been altered to a substance apparently homogeneous to the eye, but found under the microscope to have the structure shown in Fig. 514. Chemical analysis proves the base to be albite and the inclosed hexagonal mineral to be a lithium silicate ($LiAlSiO_4$) called eucryptite. It has not yet been identified except in this form.

Eucryptite in Albite

LITERATURE

Some of the most important works on the subject of microscopic inclusions are referred to here; for a fuller list of papers reference may be made to the work of Rosenbusch (1904); also that of Zirkel and others mentioned on pp. 3 and 4.

Brewster. Many papers, published mostly in the Philosophical Magazine, and the Edinburgh Phil. Journal, 1822–1856.

Blum, Leonhard, Seyfert, and Söchting. Die Einschlüsse von Mineralien in krystallisirten Mineralien. Haarlem, 1854. (Preisschrift.)

Sorby. On the microscopical structure of crystals, etc. Q. J. G. Soc., **14**, 453, 1858 (and other papers).

Sorby and Butler. On the structure of rubies, sapphires, diamonds, and some other minerals. Proc. Roy. Soc., No. 109, 1869.

Reusch. Labradorite. Pogg. Ann., **120**, 95, 1863.

Vogelsang. Labradorite. Arch. Néerland, **3**, 32, 1868.

Fischer. Kritische-microscopische mineralogische Studien. Freiburg in Br., 64 pp., 1860; 1te Fortsetzung, 64 pp., 1871; 2te Forts., 96 pp., 1873.

Kosmann. Hypersthene. Jahrb. Min., 532, 1869; 501, 1871.

Schrauf. Labradorite. Ber. Ak. Wien, **60** (1) 996, 1869.

Vogelsang. Die Krystalliten. 175 pp., Bonn, 1875.

Vogelsang and Geissler. Ueber die Natur der Flüssigkeitseinschlüsse in gewissen Mineralien. Pogg. Ann., **137**, 56, 257, 1869.

Hartley. Liquid CO_2 in cavities, etc. J. Chem. Soc., **1**, 137; **2**, 237, 1876; **1**, 241; **2**, 271, 1877; also, Proc. Roy. Soc., **26**, 137, 150, 1877.

Gümbel. Enhydros. Ber. Ak. München, **10**, 241, 1880, **11**, 321, 1881.

Hawes. Smoky quartz (CO_2). Am. J. Sc., **21**, 203, 1881.

A. W. Wright. Gases in smoky quartz. Am. J. Sc., **21**, 209, 1881.

Rutley. Notes on Crystallites. Min. Mag., **9**, 261, 1891.

Vater. Das Wesen der Krystalliten. Zs. Kr., **27**, 505, 1896.

CRYSTALLINE AGGREGATES

273. The greater part of the specimens or masses of minerals that occur may be described as aggregations of imperfect crystals. Many specimens whose structure appears to the eye quite homogeneous, and destitute internally of distinct crystallization, can be shown to be composed of crystalline grains. Under the above head, consequently, are included all the remaining varieties of structure among minerals.

The individuals composing imperfectly crystallized individuals may be:

1. *Columns*, or *fibres*, in which case the structure is *columnar* or *fibrous*.
2. *Thin laminæ*, producing a *lamellar* structure.
3. *Grains*, constituting a *granular* structure.

274. Columnar and Fibrous Structure. — A mineral possesses a *columnar* structure when it is made up of slender columns, as some amphibole. When the individuals are flattened like a knife-blade, as in kyanite, the structure is said to be *bladed*.

The structure again is called *fibrous* when the mineral is made up of fibres, as in asbestos, also the satin-spar variety of gypsum. The fibres may or may not be *separable*. There are many gradations between coarse columnar and fine fibrous structures. Fibrous minerals have often a silky luster.

The following are properly varieties of columnar or fibrous structure:

Reticulated: when the fibres or columns cross in various directions and produce an appearance having some resemblance to a net.

Stellated: when they radiate from a center in all directions and produce star-like forms. Ex. stilbite, wavellite.

Radiated, divergent: when the crystals radiate from a center without producing stellar forms. Ex. quartz, stibnite.

275. Lamellar Structure. — The structure of a mineral is *lamellar* when it consists of plates or leaves. The laminæ may be curved or straight, and thus give rise to the *curved* lamellar and *straight* lamellar structure. Ex. wollastonite (tabular spar), some varieties of gypsum, talc, etc. If the plates are approximately parallel about a common center the structure is said to be *concentric*. When the laminæ are thin and separable, the structure is said to be *foliaceous* or *foliated*. Mica is a striking example, and the term *micaceous* is often used to describe this kind of structure.

276. Granular Structure. — The particles in a granular structure differ much in size. When coarse, the mineral is described as *coarse-granular*; when fine, *fine-granular*; and if not distinguishable by the naked eye, the structure is termed *impalpable*. Examples of the first may be observed in granular crystalline limestone, sometimes called *saccharoidal*; of the second, in some varieties of hematite; of the last, in some kinds of sphalerite.

The above terms are indefinite, but from necessity, as there is every degree of fineness of structure among mineral species, from perfectly impalpable, through all possible shades, to the coarsest granular. The term *phanero-crystalline* has been used for varieties in which the grains are distinct, and *crypto-crystalline* for those in which they are not discernible, although an indistinct crystalline structure can be proved by the microscope.

Granular minerals, when easily crumbled in the fingers, are said to be *friable*.

277. Imitative Shapes. — The following are important terms used in describing the imitative forms of massive minerals.

Reniform: kidney-shaped. The structure may be radiating or concentric. Ex. hematite.

Botryoidal: consisting of a group of rounded prominences. The name is derived from the Greek βότρυς, a bunch of grapes. Ex. limonite, chalcedony, prehnite.

Mammillary: resembling the botryoidal, but composed of larger prominences. Ex. malachite.

Globular: spherical or nearly so; the globules may consist of radiating fibres or concentric coats. When attached, as they usually are, to the surface of a rock, they are described as *implanted globules*.

Nodular: in tuberose forms, or having irregular protuberances over the surface.

Amygdaloidal: almond-shaped, applied often to a rock (as diabase) containing almond-shaped or sub-globular nodules.

Coralloidal: like coral, or consisting of interlaced flexuous branchings of a white color, as in the variety of aragonite called *flos ferri*.

Dendritic: branching tree-like, as in crystallized gold. The term *dendrites* is used for similar forms even when not crystalline, as in the dendrites of manganese oxide, which form on surfaces of limestone or are inclosed in " moss-agates."

Mossy: like moss in form or appearance.

Filiform or *Capillary:* very slender and long, like a thread or hair; consists ordinarily of a succession of minute crystals. Ex. millerite.

Acicular: slender and rigid, like a needle. Ex. stibnite.

Reticulated: net-like. See Art. 274.

Drusy: closely covered with minute implanted crystals. Ex. quartz.

Stalactitic: when the mineral occurs in pendent columns, cylinders, or elongated cones. Stalactites are produced by the percolation of water, holding mineral matter in solution, through the rocky roofs of caverns. The evaporation of the water produces a deposit of the mineral matter, and gradually forms a long pendent cylinder or cone. The internal structure may be imperfectly crystalline and granular, or may consist of fibres radiating from the central column, or there may be a broad cross-cleavage. The most familiar example of stalactites is afforded by calcite. Chalcedony, gibbsite, limonite, and some other species, also present stalactitic forms.

The term *amorphous* is used when a mineral has not only no crystalline form or imitative shape, but does not polarize the light even in its minute particles, and thus appears to be destitute wholly of a crystalline structure internally, as most opal. Such a structure is also called *colloidal* or jelly-like, from the Greek κόλλα, for glue. The word amorphous is from ά *privative*, and μόρφη, *shape*.

---

278. Pseudomorphous Crystals. — Every mineral species has, when distinctly crystallized, a definite and characteristic form. Occasionally, however, crystals are found that have the form, both as to angles and general habit, of a certain species, and yet differ from it entirely in chemical composition. Moreover, it is often noted in such cases that, though in outward form complete crystals, in internal structure they are granular, or waxy, and have no regular cleavage. Even if they are crystalline in structure the optical characters do not conform to those required by the symmetry of the faces.

Such crystals are called *pseudomorphs*, and their existence is explained by the assumption, often admitting of direct proof, that the original mineral has been changed into the new compound; or it has disappeared through some agency, and its place been taken by another chemical compound to which the form does not belong. In all these cases the new substance is said to be a *pseudomorph after* the original mineral.

Common illustrations of pseudomorphous crystals are afforded by malachite in the form of cuprite, limonite in the form of pyrite, barite in the form of quartz, etc. This subject is further discussed in the chapter on Chemical Mineralogy.

PART II. PHYSICAL MINERALOGY

279. The physical characters of minerals fall under the following heads:

I. Characters depending upon *Cohesion* and *Elasticity* — viz., cleavage, fracture, tenacity, hardness, elasticity, etc.

II. *Specific Gravity*, or the *Density* compared with that of water.

III. Characters depending upon *Light* — viz., color, luster, degree of transparency, special optical properties, etc.

IV. Characters depending upon *Heat* — viz., heat-conductivity, change of form and of optical characters with change of temperature, fusibility, etc.

V. Characters depending upon *Electricity* and *Magnetism*.

VI. Characters depending upon the action of the senses — viz., taste, odor, feel.

280. General Relation of Physical Characters to Molecular Structure. — It has been shown (Arts. **30–32**) that the geometrical form of a crystallized mineral is the external expression of its internal molecular structure. It is also true that the internal structure controls many of the physical characters listed in the preceding paragraph, and their study is a help in the understanding of that structure. All the properties of a mineral must depend upon the character of the chemical elements of which it is composed and perhaps even more upon the way in which their atoms are arranged in the crystal structure.

Of these characters, the specific gravity merely gives indication of the atomic mass of the elements present, and further, of the state of molecular aggregation. The first of these points is illustrated by the high specific gravity of compounds of lead; the second, by the distinction observed, for example, between carbon in the form of the diamond, with a specific gravity of 3.5, and the same chemical substance as the mineral graphite, with a specific gravity of only 2.

All the other characters (except the relatively unimportant ones of Class VI) in general vary according to the direction in the crystal; in other words they have a definite orientation. For all of them it is true that *directions which are crystallographically identical have like physical characters.*

In regard to the converse proposition — viz., *that in all directions crystallographically dissimilar there may be a variation in the physical characters,* an important distinction is to be made. This proposition holds true for all crystals, so far as the characters of Class I are concerned; that is, those depending upon the cohesion and elasticity, as shown in the cleavage, hardness, the planes of molecular gliding, the etching-figures, etc. It is also true in the case of pyroelectricity and piezoelectricity.

It does *not* apply in the same way with respect to the characters which involve the propagation of light (and radiant heat), the change of volume with change of temperature; further, electric radiation, magnetic induction, etc.

Thus, although it will be shown that the optical characters of crystals are in agreement in general with the symmetry of their form, they do not show

207

all the variations in this symmetry. It is true, for example, that all directions are optically similar in a crystal belonging to any class under the isometric system; but this is obviously not true of its molecular cohesion, as may be shown by the cleavage. Again, all directions in a tetragonal crystal at right angles to the vertical axis are optically similar; but this again is not true of the cohesion. These points are further elucidated under the description of the special characters of each group.

I. CHARACTERS DEPENDING UPON COHESION AND ELASTICITY

281. Cohesion, Elasticity. — The name *cohesion* is given to the force of attraction existing between the molecules of one and the same body, in consequence of which they offer resistance to any influence tending to separate them, as in the breaking of a solid body or the scratching of its surface.

Elasticity is the force which tends to restore the molecules of a body back into their original position, from which they have been disturbed, as when a body has suffered change of shape or of volume under pressure.

The varying degrees of cohesion and elasticity for crystals of different minerals, or for different directions in the same crystal, are shown in the prominent characters: cleavage, fracture, tenacity, hardness; also in the gliding-planes, percussion-figures or pressure-figures, and the etching-figures.

282. Cleavage. — Cleavage is the tendency of a crystallized mineral to break in certain definite directions, yielding more or less smooth surfaces. It obviously indicates a minimum value of cohesion in the direction of easy fracture — that is, normal to the cleavage-plane itself. The cleavage parallel to the cubic faces of a crystal of galena is a familiar illustration. An amorphous body (p. 8) necessarily can show no cleavage.

Planes of cleavage are always planes of the crystal structure and therefore are parallel to possible crystal faces. These crystal planes have simple relations to the crystallographic axes and are usually commonly occurring forms on the crystal in question. That this is not always true is shown in the cases of fluorite and calcite, where the cleavage forms, though simple in their crystallographic relations, seldom occur as natural forms upon their crystals. Further, cleavage is the same in all directions in a crystal which are crystallographically identical; *i.e.*, if a cleavage exists parallel to one octahedral plane in an isometric substance it must occur also and with equal ease parallel to the three other octahedral planes. Cleavage planes have commonly been assumed to be those planes of the atomic structure in which the atoms are most closely packed together, while the distance between the successive planes is relatively large. Conditions of this sort are undoubtedly important in determining the existence of cleavage but they cannot be the only controlling factors. For instance, it has been shown that sphalerite and the diamond have the same atomic structure, but in one case the cleavage is dodecahedral and in the other octahedral. Cleavage apparently depends not only upon the geometrical position of the constituent atoms but also upon their electrical charges. The electrical forces existing between the different layers in the atomic structure are of great importance and cleavage takes place when the attractive forces are at a minimum.

Since cleavage commonly takes place parallel to some fundamental crystal form, it may be used, in cases where the choice of the position of the crystal axes is more or less arbitrary, as an aid to the proper orientation of the crystal in question.

Cleavage is defined, (1) according to its direction, as cubic, octahedral, rhombohedral, basal, prismatic, etc.   Also, (2) according to the ease with which it is obtained, and the smoothness of the surface yielded.   It is said to be *perfect* or *eminent* when it is obtained with great ease, affording smooth, lustrous surfaces, as in mica, topaz, calcite.   Inferior degrees of cleavage are spoken of as distinct, indistinct or imperfect, interrupted, in traces, difficult. These terms are sufficiently intelligible without further explanation.   It may be noticed that the cleavage of a species is sometimes better developed in some of its varieties than in others.

283. Cleavage in the Different Systems. — (1) In the ISOMETRIC SYSTEM, cleavage is *cubic*, when parallel to the faces of the cube; this is the common case, as illustrated by galena and halite.   It is also often *octahedral* — that is, parallel to the octahedral faces, as with fluorite and the diamond.   Less frequently it is *dodecahedral*, or parallel to the faces of the rhombic dodecahedron, as with sphalerite.

In the TETRAGONAL SYSTEM, cleavage is often *basal*, or parallel to the basal plane, as with apophyllite; also *prismatic*, or parallel to one (or both) of the square prisms, as with rutile and wernerite; less frequently it is *pyramidal*, or parallel to the faces of the square pyramid, as with scheelite.

In the HEXAGONAL SYSTEM, cleavage is usually either *basal*, as with beryl, or *prismatic*, parallel to one of the six-sided prisms, as with nephelite; *pyramidal* cleavage, as with pyromorphite, is rare and imperfect.

In the RHOMBOHEDRAL DIVISION, besides the basal and prismatic cleavages, *rhombohedral* cleavage, parallel to the faces of a rhombohedron, is also common, as with calcite and the allied species.

In the ORTHORHOMBIC SYSTEM, cleavage parallel to one or more of the pinacoids is common.   Thus it is *basal* wtih topaz, and in all three pinacoidal directions with anhydrite *Prismatic* cleavage is also common, as with barite; in this case the arbitrary position assumed in describing the crystal may make this cleavage parallel to a "horizontal prism," or dome.

In the MONOCLINIC SYSTEM, cleavage parallel to the clinopinacoid, is common, as with orthoclase, gypsum, heulandite and euclase; also *basal*, as with the micas and orthoclase, or parallel to the orthopinacoid; also *prismatic*, as with amphibole.   Less frequently cleavage is parallel to a hemi-pyramid, as with gypsum.

In the TRICLINIC SYSTEM, it is usual and proper to so select the fundamental form as to make the cleavage directions correspond with the pinacoids.

284.   In some cases cleavage which is ordinarily not observed may be developed by a sharp blow or by sudden change of temperature.   Thus, quartz is usually conspicuously free from cleavage, but a quartz crystal heated and plunged into cold water often shows planes of separation* parallel to both the + and − rhombohedrons and to the prism as well.   Similarly, the prismatic cleavage of pyroxene is observed with great distinctness in thin sections, made by grinding, while not so readily noted in large crystals.

When the cleavage is parallel to a closed form — that is, when it is cubic, octahedral, dodecahedral, or rhombohedral (also pyramidal in the tetragonal, hexagonal, and orthorhombic systems) — solids resembling crystals may often be broken out from a single crystalline individual, and all the fragments have the same angles.   It is, in general, easy to distinguish such a cleavage form, as a cleavage octahedron of fluorite, from a true crystal by the splintery character of the faces of the former.

285. Cleavage and Luster. — The face of a crystal parallel to which there is perfect cleavage often shows a pearly luster (see p. 273), due to the partial separation of the crystal into parallel plates.   This is illustrated by the basal plane of apophyllite, the clinopinacoid of stilbite and heulandite.   An iridescent play of colors is also often seen, as with calcite, when the separation has been sufficient to produce the prismatic colors by interference.

---

* Lehmann (Zs. Kr., **11**, 608, 1886) and Judd (Min. Mag., **8**, 7, 1888) regard these as gliding-planes (see Art. **286**).

286. Gliding-planes. — Closely related to the cleavage directions in their connection with the cohesion of the molecules of a crystal are the *gliding-planes*, or directions parallel to which a slipping of the molecules may take place under the application of mechanical force, as by pressure.

This may have the result of simply producing a separation into layers in the given direction, or, on the other hand, and more commonly, there may

515

Biotite

be a revolution of the molecules into a new twinning-position, so that *secondary twinning-lamellæ* are formed.

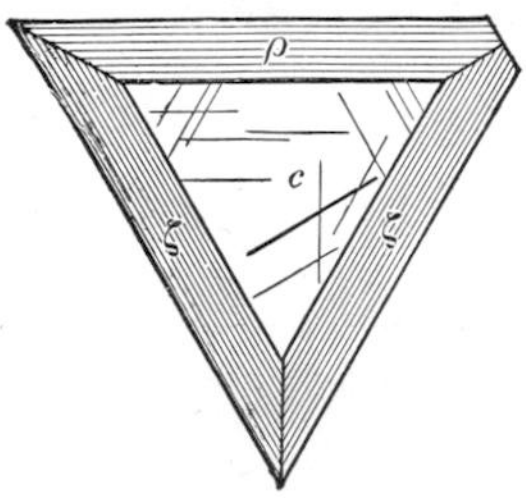

Thus, if a crystal of halite, or rock salt, be subjected to gradual pressure in the direction of a dodecahedral face, a plane of separation is developed normal to this and hence in the direction of another face of the same form. There are six such directions of molecular slipping and separation in a crystal of this substance. Certain kinds of mica of the biotite class often show pseudo-crystalline faces, which are undoubtedly secondary in origin — that is, have been developed by pressure exerted subsequently to the growth of the crystal (cf. Fig. 515).

In stibnite, the base, $c(001)$, normal to the plane of perfect cleavage, is a gliding-plane. Thus a slipping of the molecules without their separation may be made to take place by pressure in a plane ($\parallel c$) normal to the direction of perfect cleavage ($\parallel b$). A slender prismatic crystal supported near the ends and pressed downward by a dull edge is readily bent, or nicked, in this direction without the parts beyond the support being affected.

287. Secondary Twinning. — The other case mentioned in the preceding article, where molecular slipping is accompanied by a half-revolution (180°) of the molecules into a new twinning-position (see p. 179 *et seq.*), is well illustrated by calcite. Pressure upon a cleavage-fragment may result in the formation of a number of thin lamellæ in twinning-position to the parent mass, the twinning-plane being the obtuse negative rhombohedron, $e(01\bar{1}2)$. Secondary twinning-lamellæ similar to these are often observed in natural cleavage-masses of calcite, and particularly in the grains of a crystalline limestone, as observed in thin sections under the microscope.

516

Secondary twinning-lamellæ may also be produced (and are often noted in nature) in the case of the triclinic feldspars, pyroxene, barite, etc. A secondary lamellar structure in quartz has been observed by Judd, in which the lamellæ consisted of right-handed and left-handed portions.

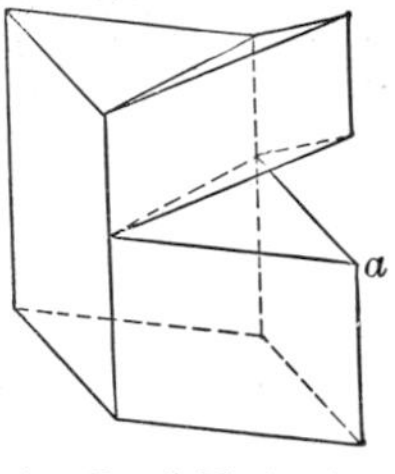

Artificial Twinning
in Calcite

By the proper means a complete calcite twin may be artificially produced by pressure. Thus, if a cleavage-fragment of prismatic form, say 6–8 mm. in length and 3–6 mm. in breadth, be placed with the obtuse edge on a firm horizontal support, and pressed by the blade of an ordinary tableknife on the other obtuse edge (at a, Fig. 516), the result is that a portion of the crystal is reversed in position, as if twinned parallel to the plane $(01\bar{1}2)$ which in the figure lies in a vertical position. If skillfully done, the twinning surface is perfectly smooth, and the reentrant angle corresponds exactly with that required by theory.

288. Parting. — The secondary twinning-planes described are often directions of an easy separation — conveniently called *parting* — which may be mistaken for cleavage.*   Parting may also occur along gliding-planes. The basal parting of pyroxene is a common example of such pseudo-cleavage; it was long mistaken for cleavage.   The basal and rhombohedral $(10\bar{1}1)$ and the less distinct prismatic $(11\bar{2}0)$ parting of corundum; the octahedral parting of magnetite (cf. Fig. 500, p. 198), are other examples.

An important distinction between cleavage and parting is this: parting can exist only in certain definite planes — that is, on the surface of a twinning-lamella — while the cleavage may take place in *any* plane having the given direction.

289. Percussion-figures. — Immediately connected with the gliding-planes are the figures — called *percussion-figures†* — produced upon a crystal section by a blow or pressure with a suitable point.   In such cases, the method described serves to develop more or less well-defined cracks whose orientation varies with the crystallographic direction of the surface.   Thus upon the cubic face of a crystal of halite a four-rayed, star-shaped figure is produced with arms parallel to the diagonals — that is, parallel to the dodecahedral faces.   On an octahedral face a three-rayed star is obtained.

The percussion-figures in the case of the micas have been often investigated, and, as remarked later, they form a means of fixing the true orientation of a cleavage-plate having no crystalline outlines.   The figure (Fig. 517) is here a six-rayed star one of whose branches is parallel to the clinopinacoid (b), the others approximately parallel to the intersection edges of the prism (m) and base (c).‡

Pressure upon a mica plate produces a less distinct six-rayed star, diagonal to that just named; this is called a *pressure-figure*.

290. Solution-planes. — In the case of many crystals, it is possible to prove the existence of certain directions, or structure-planes, in which chemical action takes place most readily — for example, when a crystal is under great pressure.   These directions of chemical weakness have been called *solution-planes*.   They often manifest themselves by the presence of a multitude of oriented cavities of crystalline outline (so-called negative crystals) in the given direction.

These solution-planes in certain cases, as shown by Judd, are the same as the directions of secondary lamellar twinning, as is illustrated by calcite.   Connected with this is the *schillerisation* (see Art. 376) observed in certain minerals in rocks (as diallage, schillerspar).

291. Etching-figures. — Intimately connected with the general subjects here considered, of cohesion in relation to crystals, are the figures produced by etching on crystalline faces; these are often called *etching-figures*. This method of investigation, developed particularly by Baumhauer, is of high importance as revealing the molecular structure of the crystal faces under examination, and therefore the symmetry of the crystal itself.

The etching is performed mostly by solvents, as by water in some cases, more generally the ordinary mineral acids, or caustic alkalies, also by steam at

---

* The lamellar structure of a massive mineral, without twinning, may also be the cause of a fracture which can be mistaken for cleavage.

† The percussion-figures are best obtained if the crystal plate under investigation be supported upon a hard cushion and a blow be struck with a light hammer upon a steel rod the slightly rounded point of which is held firmly against the surface.

‡ Cf. Walker, Am. J. Sc., **2**, 5, 1896, and G. Friedel, Bull. Soc. Min., **19**, 18, 1896. Walker found the angle opposite b(010) (χ in Fig. 517) to be 53° to 56° for muscovite, 59° for lepidolite, 60° for biotite, and 61° to 63° for phlogopite.

a high pressure and hydrofluoric acid; the last is especially powerful in its action, and is used frequently with the silicates. The figures produced are in the majority of cases angular depressions, such as low triangular or quadrilateral pyramids, whose outlines may run parallel to some of the crystalline edges. In some cases the planes produced can be referred to occurring crystallographic faces; more often they are of the nature of vicinal faces with com-

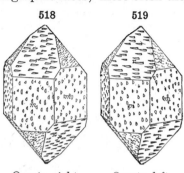

518   519

Quartz, right-   Quartz, left-
handed crystal   handed crystal

plicated indices or are in fact curved surfaces. They commonly lie, however, in definite and simple crystal zones. They appear alike on similar faces of crystals, and hence serve to distinguish different forms, perhaps in appearance identical, as the two sets of faces in the ordinary double pyramid of quartz; so, too, they reveal the compound twinning-structure common on some crystals, as quartz and aragonite. Further, their form in general corresponds to the symmetry of the group to which the given crystal belongs. They thus reveal the trapezohedral symmetry of quartz and the difference between a right-handed and left-handed crystal (Figs. 518, 519); the distinction between calcite and dolomite (Figs. 522, 523); the distinctive character of apatite, pyromorphite, etc.; the hemimorphic symmetry of calamine and nephelite (cf. Fig. 256, p. 118), etc.; they also prove by their form the monoclinic crystallization of muscovite and other micas (Fig. 521).

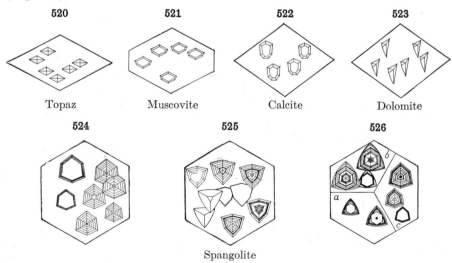

520   521   522   523

Topaz   Muscovite   Calcite   Dolomite

524   525   526

Spangolite

Fig. 520 shows the etching-figures formed on a basal plane (cleavage) of topaz by fused caustic potash; Fig. 521, those on a cleavage-plate of muscovite by hydrofluoric acid; Fig. 522, upon a rhombohedral face of calcite, and Fig. 523, on one of dolomite by dilute hydrochloric acid.

The shape of the etching-figures may vary with the same crystal with the nature of the solvent employed, though their symmetry remains constant. For example, Fig. 524 shows the figures obtained with spangolite by the action of sulphuric acid; Fig. 525, by the same diluted, and Fig. 526 by hydrochloric acid of different degrees of concentration.

Of the same nature as the etching-figures artificially produced, in their relation to the symmetry of the crystal, are the markings often observed on the natural faces of crystals. These are sometimes secondary, caused by a natural etching process, but are more often an irregularity in the crystalline development of the crystal. The inverted triangular depressions often seen on the octahedral faces of diamond crystals are an example. Fig. 527 shows natural depressions, rhombohedral in character, observed on corundum crystals from Montana (Pratt). Fig. 528 shows a twin crystal of fluorite with

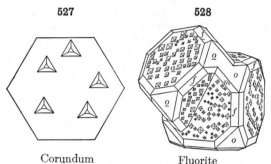

527

528

Corundum

Fluorite

natural etching-figures (Pirsson): these are minute pyramidal depressions whose sides are parallel to the faces of the trapezohedron (311).

292. Corrosion Forms. — If the etching process spoken of in the preceding article — whether natural or artificial — is continued, the result may be to destroy the original crystalline surface and to substitute for it perhaps a multitude of minute elevations, more or less distinct; or, further, new faces may be developed, the crystallographic position of which can often be determined, though the symbols may be complex. The mere loss of water in some cases produces certain corrosive forms.

Penfield subjected a sphere of quartz (from a simple right-handed individual) to the prolonged action of hydrofluoric acid. It was found that it was attacked rapidly in the direction of the vertical axis, but barely at all at the + extremities of the horizontal axes.

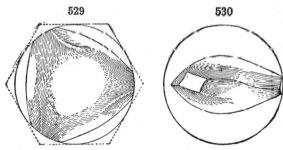

529

530

Etched Sphere of Quartz

Figs. 529, 530 show the form remaining after the sphere had been etched for seven weeks; Fig. 529 is a basal view; Fig. 530, a front view; the circle shows the original form of the sphere, the dotted hexagon the position of the axes.

293. Fracture. — The term *fracture* is used to define the form or kind of surface obtained by breaking in a direction other than that of cleavage in crystallized minerals, and in any direction in massive minerals. When the cleavage is highly perfect in several directions, as the rhombohedral cleavage of calcite, fracture is often not readily obtainable.

Fracture is defined as:

(*a*) *Conchoidal;* when a mineral breaks with curved concavities, more or less deep. It is so called from the resemblance of the concavity to the valve of a shell, from *concha, a shell*. This is well illustrated by obsidian, also by flint. If the resulting forms are small, the fracture is said to be *small-conchoidal;* if only partially distinct, it is *subconchoidal.*

(*b*) *Even;* when the surface of fracture, though rough with numerous small elevations and depressions, still approximates to a plane surface.

(*c*) *Uneven;* when the surface is rough and entirely irregular; this is true of most minerals.

(*d*) *Hackly;* when the elevations are sharp or jagged; broken iron.

Other terms also employed are *earthy, splintery,* etc.

294. Hardness. — The *hardness* of a mineral is measured by the resistance which a smooth surface offers to abrasion. The degree of hardness is determined by observing the comparative ease or difficulty with which one mineral is scratched by another, or by a file or knife.

In minerals there are all grades of hardness, from that of talc, impressible by the finger-nail, to that of the diamond. To give precision to the use of this character, a *scale of hardness* was introduced by Mohs. It is as follows:

1.	*Talc.*	6.	*Orthoclase.*
2.	*Gypsum.*	7.	*Quartz.*
3.	*Calcite.*	8.	*Topaz.*
4.	*Fluorite.*	9.	*Corundum.*
5.	*Apatite.*	10.	*Diamond.*

Crystalline varieties with smooth surfaces should be taken so far as possible.

If the mineral under examination is scratched by the knife-blade as easily as calcite its hardness is said to be 3; if less easily than calcite and more so than fluorite its hardness is 3·5. In the latter case the mineral in question would be scratched by fluorite but would itself scratch calcite. It need hardly be added that great accuracy is not attainable by the above methods, though, indeed, for purposes of the determination of minerals, exactness is quite unnecessary.

It should be noted that minerals of grade 1 have a greasy feel to the hand; those of grade 2 are easily scratched by the finger-nail; those of grade 3 are rather readily cut, as by a knife; of grade 4, scratched rather easily by the knife; grade 5, scratched with some difficulty; grade 6, barely scratched by a knife, but distinctly by a file — moreover, they also scratch ordinary glass. Minerals as hard as quartz (H. = 7), or harder, scratch glass readily but are little touched by a file; the few species belonging here are enumerated in Appendix B; they include all the gems.

295. Sclerometer. — Accurate determinations of the hardness of minerals can be made in various ways, one of the best being by use of an instrument called a *sclerometer.* The mineral is placed on a movable carriage, with the surface to be experimented upon horizontal; this is brought in contact with a steel point (or diamond point), fixed on a support above; the weight is then determined which is just sufficient to move the carriage and produce a scratch on the surface of the mineral.

By means of such an instrument the hardness of the different faces of a given crystal has been determined in a variety of cases. It has been found that different faces of a crystal (*e.g.,* kyanite) differ in hardness, and the same face may differ as it is scratched in different directions. The degree of ease with which a given mineral is scratched is a measure of its molecular cohesion, and in cases where the cohesion varies with the direction it is obvious that the hardness will likewise vary.   In general, differences in hardness are noted only with crystals which show distinct cleavage; the hardest face is that which is intersected by the plane of most complete cleavage.   Further, of a single face, which is intersected by cleavage-planes, the direction perpendicular to the cleavage-direction is the softer, those parallel to it the harder.

This subject has been investigated by Exner (p. 216), who has given the form of the *curves of hardness* for the different faces of many crystals.   These curves are obtained as follows: the least weight required to scratch a crystalline surface in different directions, for each 10° or 15°, from 0° to 180°, is determined with the sclerometer; these directions are laid off as radii from a center, and the length of each is made proportional to the weight fixed by experiment — that is, to the hardness thus determined; the line connecting the extremities of these radii is the curve of hardness for the given face.

The following table gives the results obtained* (see literature) in comparing the hardness of the minerals of the scale from corundum, No. 9, taken as 1000, to gypsum, No. 2.   Pfaff used the method of boring with a standard point, the hardness being determined by the number of rotations; Rosiwal used a standard powder to grind the surface, Jaggar employed his micro-sclerometer, the method being essentially a modification of that of Pfaff.   By means of this instrument he is able to test the hardness of the minerals present in a thin section under the microscope.   Measurements of absolute hardness have also been made by Auerbach.   Holmquist has recently made many hardness tests by the grinding method. His results with regard to the minerals of the scale of hardness agree fairly well with those of Rosiwal given below but show considerable discrepancies with the results obtained by the other methods.   He, like Rosiwal, finds that topaz is lower in the scale than quartz.

		Pfaff, 1884	Rosiwal, 1892	Jaggar, 1897
9.	Corundum.............	1000	1000	1000
8.	Topaz.................	459	138	152
7.	Quartz................	254	149	40
6.	Orthoclase............	191	28·7	25
5.	Apatite...............	53·5	6·20	1·23
4.	Fluorite..............	37·3	4·70	·75
3.	Calcite...............	15·3	2·68	·26
2.	Gypsum...............	12·03	·34	·04

296. Relation of Hardness to Chemical Composition.   Some general facts of importance can be stated† in regard to the connection between the hardness of a mineral and its chemical composition.

1.   Compounds of the heavy metals, as silver, copper, mercury, lead, etc., are *soft*, their hardness seldom exceeding 2·5 to 3.

Among the compounds of the common metals, the sulphides (arsenides) and oxides of iron (also of nickel and cobalt) are relatively *hard* (*e.g.*, for pyrite H. = 6 to 6·5; for hematite H. = 6, etc.); here belong also columbite, iron niobate; tantalite, iron tantalate; wolframite, iron tungstate.

2.   The sulphides are mostly relatively soft (except as noted in 1), also most of the carbonates, sulphates, and phosphates.

3.   Hydrous salts are relatively soft.   This is most distinctly shown among the silicates — *e.g.*, compare the feldspars and zeolites.

4.   The conspicuously hard minerals are found chiefly among the oxides and silicates; many of them are compounds containing aluminum — *e.g.*, corundum, diaspore, chrysoberyl, and many alumino-silicates.   Outside of these the borate, boracite, is hard (H. = 7); also iridosmine.

On the relation of hardness to specific gravity, see Art. **307**.

297. Practical Suggestions. — Several points should be regarded in the trials of hardness:

(1) If the mineral is slightly altered, as is often the case with corundum, garnet, etc., the surface may be readily scratched when this would be impossible with the mineral itself; a trial with an edge of the latter will often give a correct result in such a case.

---

* The numbers are here given as tabulated by Jaggar.
† See further in Appendix B.

(2) A mineral with a granular surface often appears to be scratched when the grains have been only torn apart or crushed.

(3) A relatively soft mineral may leave a faint white ridge on a surface, as of glass, which can be mistaken for a scratch if carelessly observed.

(4) A crystal, as of quartz, is often slightly scratched by the edge of another of the same species and like hardness.

(5) The scratch should be made in such a way as to disfigure the specimen as little as possible.

298. Tenacity. — Minerals may be either brittle, sectile, malleable, or flexible.

(a) *Brittle;* when parts of a mineral separate in powder or grains on attempting to cut it, as calcite.

(b) *Sectile;* when pieces may be cut off with a knife without falling to powder, but still the mineral pulverizes under a hammer. This character is intermediate between brittle and malleable, as gypsum.

(c) *Malleable;* when slices may be cut off, and these slices flattened out under a hammer; native gold, native silver.

(d) *Flexible;* when the mineral will bend without breaking, and remain bent after the bending force is removed, as talc.

The tenacity of a substance is properly a consequence of its elasticity.

299. Elasticity. — The elasticity of a solid body expresses at once the resistance which it makes to a change in shape or volume, and also its tendency to return to its original shape when the deforming force ceases to act. If the *limit of elasticity* is not passed, the change in molecular position is proportional to the force acting, and the former shape or volume is exactly resumed; if this limit is exceeded, the deformation becomes permanent, a new position of molecular equilibrium having been assumed; this is shown in the phenomena of gliding-planes and secondary twinning, already discussed. The magnitude of the elasticity of a given substance is measured by the *coefficient of elasticity*, or, better, the coefficient of restitution. This is defined as the relation, for example, between the elongation of a bar of unit section to the force acting to produce this effect; similarly of the bending or twisting of a bar. The subject was early investigated acoustically by Savart; in recent years, Voigt and others have made accurate measures of the elasticity of many substances and of the crystals of the same substance in different directions. The elasticity of an amorphous body is the same in all directions, but it changes in value with change of crystallographic direction in all crystals.

The distinction between *elastic* and *inelastic* is often made between the species of the mica group and allied minerals. Muscovite, for example, is described as " highly elastic," while phlogopite is much less so. In this case it is not true in the physical sense that muscovite has a high value for the coefficient of elasticity; its peculiarity lies rather in the fact that its elasticity is displayed through unusually wide limits.

LITERATURE

Hardness

Seebeck. Sklerometer. Programm d. Cöln Realgymnasiums, 1833.
Franz. Pogg., **80**, 37, 1850.
Grailich u. Pekárek. Ber. Ak. Wien, **13**, 410, 1854.
Pfaff. Mesosklerometer. Ber. Ak. München, **13**, 55, 1883.
Sohnke. Halite. Pogg., **137**, 177, 1869.
Exner. Ueber die Härte der Krystallflächen, 166 pp. Vienna, 1873 (Preisschrift Wiener. Akad.).

Auerbach. Wied. Ann., **43**, 61, 1891; **45**, 262, 277, 1892; **58**, 357, 1896.
Rosiwal. Verh. G. Reichs., 475, 1896. Verh. G. Reichs., 1916; Mitt. Wiener Min. Ges. No. 80, 69, 1917.
T. A. Jaggar, Jr. Microsclerometer. Am. J. Sc., **4**, 399, 1897.
Schroeder van der Kolk. Ueber Härte in Verland mit Spaltbarkeit, Verh. Ak. Amsterdam, **8**, 1901.
Holmquist. Ueber den Relativen Abnutzungswiderstand der Mineralien der Härteskala. Geol. För. Förh., **33**, 281, 1911. Die Schleifhärte der Feldspathe, ibid., **36**, 401, 1914. Die Härtestufe, 4–5, ibid., **38**, 501, 1916.

Etching-figures, etc.

Goldschmidt and Wright. Ueber Aetzfiguren, Lichtfiguren und Lösungskörper. With exhaustive references to the literature. N. Jb. Min. Beil-Bd., **17**, 355–390, 1903.
Honess. Etch Figures on Crystals, 1927.

Gliding-planes, Secondary Twinning, etc.

Reusch. "Körnerprobe," halite, calcite. Pogg. Ann., **132**, 441, 1867. Mica, ibid., **136**, 130, 632, 1869. Gypsum, ibid., p. 135. Ber. Ak., Berlin, 440, 1873.
Bauer. Mica, etc. Pogg. Ann., **138**, 337, 1869; Zs. G. Ges., **26**, 137, 1874. Galena, Jb. Min., **1**, 138, 1882.
Baumhauer. Calcite. Zs. Kr., **3**, 588, 1879.
Mügge. Calcite, augite, stibnite, etc. Jb. Min., **1**, 32, 1883; **2**, 13, 1883. Also ibid., **1**, 71, 1898.
J. W. Judd. Solution-planes, etc. Q. J. G. Soc., **41**, 374, 1885; Min. Mag., **7**, 81, 1887. Structure planes of corundum, Min. Mag., **11**, 49, 1895.
Voigt. See below.

Elasticity

Savart. Ann. Ch. Phys., **40**, 1, 113, 1829; also in Pogg. Ann., **16**, 206, 1829.
Neumann. Pogg. Ann., **31**, 177, 1834.
Ångström. Pogg. Ann., **86**, 206, 1852.
Baumgarten. Calcite. Pogg. Ann., **152**, 369, 1874.
Groth. Halite. Pogg. Ann., **157**, 115, 1876.
Coromilas. Gypsum, mica. Inaug. Diss., Tübingen, 1877 (Zs. Kr., **1**, 407, 1877).
Reusch. Ice. Wied. Ann., **9**, 329, 1880.
Klang. Fluorite. Wied. Ann., **12**, 321, 1881.
Koch. Halite, sylvite. Wied. Ann., **18**, 325, 1883.
Beckenkamp. Alum. Zs. Kr., **10**, 41, 1885.
Voigt. Pogg. Ann., Erg. Bd., **7**, 1, 177, 1876. Wied. Ann., **38**, 573, 1889. Calcite, **39**, 412, 1890. Dolomite, ibid., **40**, 642, 1890. Tourmaline, ibid., **41**, 712, 1890; **44**, 168, 1891. Also papers in Nachr. Ges. Wiss. Göttingen.
Tutton. The Elasmometer. Crystalline Structure and Chemical Constitution, 1926.

II. SPECIFIC GRAVITY OR RELATIVE DENSITY

300. Definition of Specific Gravity. — The specific gravity of a mineral is the ratio of its density[*] to that of water at 4° C. (39·2° F.). This relative density may be learned in any case by comparing the ratio of the weight of a certain volume of the given substance to that of an equal volume of water; hence the specific gravity is often defined as: *the weight of the body divided by the weight of an equal volume of water.*

---

[*] The *density* of a body is strictly *the mass of the unit volume.* Thus if a cubic centimeter of water (at its maximum density, 4° C. or 39·2° F.) is taken as the unit of mass, the density of any body — as gold — is given by the number of grams of mass (about 19) in a cubic centimeter; in this case the same number, 19, gives the relative density or specific gravity. If, however, a pound is taken as the unit of mass, and the cubic foot as the unit of volume, the mass of a cubic foot of water is 62·5 lbs., that of gold about 1188 lbs., and the specific gravity is the ratio of the second to the first, or, again, 19.

The statement that the specific gravity of graphite is 2, of corundum 4, of galena 7·5, etc., means that the densities of the minerals named are 2, 4, and 7·5, etc., times that of water; in other words, as familiarly expressed, any volume of them, a cubic inch for example, weighs 2 times, 4 times, 7·5 times, etc., as much as a like volume, a cubic inch, of water.

Strictly speaking, since the density of water varies with its expansion or contraction under change of temperature, the comparison should be made with water at a fixed temperature, namely 4° C. (39·2° F.), at which it has its maximum density. If made at a higher temperature, a suitable correction should be introduced by calculation. Practically, however, since a high degree of accuracy is not often called for, and, indeed, in many cases is impracticable to attain in consequence of the nature of the material at hand, in the ordinary work of obtaining the specific gravity of minerals the temperature at which the observation is made can safely be neglected. Common variations of temperature would seldom affect the value of the specific gravity to the extent of one unit in the third decimal place.

For the same reason, it is not necessary to take into consideration the fact that the observed weight of a fragment of a mineral is less than its true weight by the weight of air displaced.

Where the nature of the investigation calls for an *accurate* determination of the specific gravity (*e.g.*, to four decimal places), no one of the precautions in regard to the purity of material, exactness of weight-measurement, temperature, etc., can be neglected.* The accurate values spoken of are needed in the consideration of such problems as the specific volume, the relation of molecular volume to specific gravity, and many others.

301. Determination of the Specific Gravity by the Balance. — The direct comparison by weight of a certain volume of the given mineral with an equal volume of water is not often practicable. By making use, however, of a familiar principle in hydrostatics, viz., that a solid immersed in water, in consequence of the buoyancy of the latter, loses in weight an amount which is equal to the weight of an equal volume of the water (that is, the volume it displaces) — the determination of the specific gravity becomes a very simple process.

The weight of the solid in the air (w) is first determined in the usual manner; then the weight in water is found (w'); the difference between these weights — that is, the loss by immersion ($w - w'$) — is the weight of a volume of water equal to that of the solid; finally, the quotient of the first weight (w) by that of the equal volume of water as determined ($w - w'$) is the specific gravity (G).

Hence,

$$G = \frac{w}{w - w'}.$$

A common method of obtaining the specific gravity of a firm fragment of a mineral is as follows: First weigh the specimen accurately on a good chemical balance. Then suspend it from one pan of the balance by a horsehair, silk thread, or, better still, by a fine platinum wire, in a glass of water conveniently placed beneath, and take the weight again with the same care; then use the results as above directed. The platinum wire may be wound around the

---

* Cf. Earl of Berkeley in Min. Mag., **11**, 64, 1895.

specimen, or where the latter is small it may be made at one end into a little spiral support.

302. The Jolly Balance. — Instead of using an ordinary balance and determining the actual weight, the spiral balance of Jolly, shown in Fig. 531, may be conveniently employed; this is also suitable when the mineral is in the form of small grains.   The instrument consists of a spiral spring at the lower end of which are suspended two pans or wire baskets, c and d, Fig. 531. Upon the movable stand B rests a beaker filled with water.   When in adjustment for reading this stand has such a position that the pan d is immersed in the water while c hangs above it.   Upon the upright A there is a mirror upon which is marked a scale.   The position of the balance at any time is obtained by so placing the eye that the bead, m, and its reflection in the mirror coincide and then reading the position of the top of the bead upon the scale.   The first step in the operation consists in getting the position of the spring alone, having the pan d immersed in the water in the beaker.   Let this reading be represented by n.   The mineral whose specific gravity is to be determined is then placed on the pan or basket, c, and the platform B raised until d is properly immersed in the water.   The position of the bead m is again read.   Let this value be represented by N_1.   If from N_1 be subtracted the number n, expressing the amount to which the scale is stretched by the weight of spring and pans alone, the difference will be proportional to the weight of the mineral. Next, the mineral is placed in the lower pan, d, immersed in the water, and again the corresponding scale number, N_2, read.   The difference between these readings ($N_1 - N_2$) is a number proportional to the loss of weight in water. The specific gravity is then

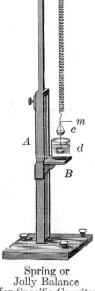

531

A m c d B

Spring or
Jolly Balance
for Specific Gravity

$$G = \frac{N_1 - n}{N_1 - N_2}.$$

It is obviously necessary to have the wires supporting the lower pan immersed to the same depth in the case of each of the three determinations.   If care is taken the specific gravity can be obtained accurately to two decimal places.   A much improved form of this balance has been devised by Kraus; see reference below.

303. The Beam Balance. — A beam balance described by Penfield is another very simple and quite accurate device for measuring the specific gravity.   It is illustrated in Fig. 532, which will make clear its essential parts. The beam is so balanced by a weight on its shorter end that it is very nearly in equilibrium when the lower pan is immersed in water.   An exact balance is then obtained by the small rider d.   When the beam is once balanced this rider is kept stationary and its position disregarded in the subsequent readings. The mineral is first placed in the upper pan and the beam balanced by another rider of such a weight that its position will be near the outer end of the beam. The position of this rider is then read from the scale engraved upon the beam. Let this value be equal to N_1.   The mineral is next transferred to the lower pan and the beam again brought into balance by moving this same rider back.

The second reading may be represented by N_2. The formula for obtaining the specific gravity is now:

$$G = \frac{N_1}{N_1 - N_2}.$$

532

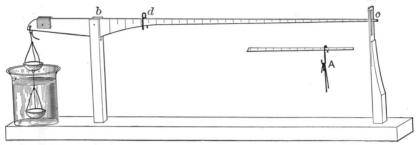

Beam Balance for Specific Gravity, ⅛th Natural Size (after Penfield)

304. Pycnometer. — If the mineral is in the form of grains or small fragments, the specific gravity may be obtained by use of the *pycnometer*. This is a small bottle (Fig. 533) having a stopper which fits tightly and ends in a tube with a very fine opening. A number of different types of bottles are used. The bottle is filled with distilled water, the stopper inserted, and the overflowing water carefully removed with a soft cloth and then weighed. The mineral whose density is to be determined is also weighed. Lastly the bottle is weighed with the mineral in it and filled with water as described

533

Pycnometer

above.* The weight of the water displaced by the mineral is obviously the difference between this last weight and that of the bottle filled with water plus the weight of the mineral. The specific gravity of the mineral is equal to its weight alone divided by the weight of the equal volume of water thus determined. Where this method is followed with sufficient care, especially avoiding any change of temperature in the water, the results may be highly accurate.

If the mineral forms a porous mass, it may be first reduced to powder, but it is to be noted that it has been shown by Rose that chemical precipitates have uniformly a higher density than belongs to the same substance in a less finely divided state. This increase of density also characterizes, though to less extent, a mineral in a fine state of mechanical subdivision. It is explained by the condensation of the water on the surface of the powder.

305. Use of Liquids of High Density. — It is often found convenient both in the determination of the specific gravity and in the mechanical separation of fragments of different specific gravities (*e.g.*, to obtain pure material for analysis, or again in the study of rocks) to use a liquid of high density —

---

* Care should be taken to prevent air-bubbles being included among the mineral particles. This may be accomplished by placing the bottle under an air-pump and exhausting the air or by suspending the bottle for a short time in a beaker filled with boiling water and then allowing it to cool again before weighing.

that is, a so-called *heavy solution*. One of these is the solution of mercuric iodide in potassium iodide, called the Sonstadt or Thoulet solution. When made with care it has a maximum density of nearly 3·2, which by dilution may be lowered at will.

A second solution, often employed, is the *Klein solution*, the borotungstate of cadmium, having a maximum density of 3·6. This again may be lowered at will by dilution, observing certain necessary precautions. Still a third solution of much practical value is that proposed by Brauns, methylene iodide, which has a specific gravity of 3·324. Clerici solution, composed of equal parts of thallium formate and malonate, has specific gravities above 4·00; see below under Literature. A number of other solutions, more or less practical, have also been suggested.* When one of these liquids is to be used for the determination of the specific gravity of fragments of a certain mineral it must be diluted until the fragments just float and the specific gravity then obtained, most conveniently by the Westphal balance (Art. **306**).

When, on the other hand, the liquid is to be used for the separation of the fragments of two or more minerals mixed together, the material is first reduced to the proper degree of fineness, the dust and smallest fragments being sifted out, then it is introduced into the solution and this diluted until one constituent after another sinks and is removed. For the convenient application of this method a suitable tube is called for and certain precautions must be observed; see the papers noted in the literature (p. 222), especially one by Penfield.

306. Westphal's Balance. — The Westphal balance is conveniently used to determine the specific gravity of a liquid, and hence of a mineral when a heavy solution is employed (Art. **305**). It consists essentially of a graduated steelyard arm, at one end of which a glass sinker is suspended by a fine wire. The arm is so weighted at the opposite end that the sinker is exactly balanced and the arm lies horizontal. A glass tube containing the liquid whose specific gravity is to be determined is placed so that the sinker is immersed in the liquid. In order to again balance the arm, weights in the form of riders are placed upon the graduated arm. These must be so adjusted that the sinker is freely suspended in the given liquid while the index at the opposite end of the arm points to the zero of the scale and shows that the arm is horizontal (cf. Johannsen, p. 533). The graduation usually allows of the specific gravity being read off directly without calculation.

307. Relation of Density to Hardness, Chemical Composition, etc. — The density, or specific gravity, of a solid depends, first, upon the nature of the chemical substances which it contains, and, second, upon the state of molecular aggregation.

Thus, as an illustration of the first point, all lead compounds have a high density (G. = about 6), since lead is a heavy metal, or, chemically expressed, has a high atomic weight (206·4). Similarly, barium sulphate, barite, has a specific gravity of 4·5, while for calcium sulphate or anhydrite the value is only 2·95 (atomic weight for barium 137, for calcium about 40).

On the other hand, while aluminum is a metal of low density (G. = 2·5 and atomic weight = 27), its oxide, corundum, has a remarkably *high* density (G. = 4) and is also very hard (H. = 9). Again, carbon (atomic weight = 12) has a high density in the diamond (G. = 3·5) and low in graphite (G. = 2); also, the first is hard (H. = 10), the second soft (H. = 1·5). In these and similar cases the high density signifies great molecular aggregation, and hence it is natural that it should be accompanied by great hardness and resistance to the attack of acids.

As bearing upon this point, it is to be noted that the density of many substances is altered by fusion. Again, the same mineral in different states of molecular aggregation may differ (but only slightly) in density. Furthermore, substances having the same chemical composition have sometimes different densities, corresponding to the different crystalline

* Johannsen, Manual of Petrographic Methods, p. 519 *et seq.*, gives in detail an account of the various solutions, the methods of their preparation, etc.

forms in which they appear. Thus in the case of calcium carbonate ($CaCO_3$), calcite has G. = 2·7, aragonite has G. = 2·9.

308. Average Specific Gravities. — It is to be noted that among minerals of NON-METALLIC LUSTER the *average* specific gravity ranges from 2·6 to 3. Here belong quartz (2·66), calcite (2·7), the feldspars (2·6–2·75), muscovite (2·8). A specific gravity of 2·5 or less is *low*, and is characteristic of soft minerals, and often those which are hydrous (*e.g.*, gypsum, G. = 2·3). The common species fluorite, tourmaline, apatite, vesuvianite, amphibole, pyroxene, and epidote lie just above the limit given, namely, 3·0 to 3·5. A specific gravity of 3·5 or above is relatively *high*, and belongs to hard minerals (as corundum, see Art. **307**), or to those containing a heavy metal, as compounds of strontium, barium, also iron, tungsten, copper, silver, lead, mercury, etc.

With minerals of METALLIC LUSTER, the average is about 5 (here belong pyrite, hematite, etc.), while if below 4 it is relatively low (graphite 2, stibnite 4·5); if 7 or above, relatively high (as galena, 7·5).

Tables of minerals arranged according to their specific gravity are given in Appendix B.

309. Constancy of Specific Gravity. — The specific gravity of a mineral species is a character of fundamental importance, and is highly constant for different specimens of the same species, if pure, free from cavities, solid inclusions, etc., and if essentially constant in composition. In the case of many species, however, a greater or less variation exists in the chemical composition, and this at once causes a variation in specific gravity. The different kinds of garnet illustrate this point; also the various minerals intermediate between the tantalate of iron (and manganese) and the niobate, varying from G. = 7·3 to G. = 5·3.

310. Practical Suggestions. — It should be noted that the determination of the specific gravity has little value unless the fragment taken is pure and is free from impurities, internal and external, and not porous. Care must be taken to exclude air-bubbles, and it will often be found well to moisten the surface of the specimen before inserting it in the water, and sometimes boiling (or the use of the air-pump) is necessary to free it from air. If it absorbs water this latter process must be allowed to go on till the substance is fully saturated. No *accurate* determinations can be made unless the changes of temperature are rigorously excluded and the actual temperature noted.

In a mechanical mixture of two constituents in known proportions, when the specific gravity of the whole and of one are known, that of the other can be readily obtained. This method is often important in the study of rocks.

It is to be noted that the hand may be soon trained to detect a difference of specific gravity, if like volumes are taken, even in a small fragment — thus the difference between calcite, or albite and barite, even the difference between a small diamond and a quartz crystal, can be detected.

LITERATURE

Specific Gravity

General:

   Beudant. Pogg. Ann., **14**, 474, 1828.
   Jenzsch. Pogg. Ann., **99**, 151, 1856.
   Jolly. Ber. Ak. München, 1864, 162.
   Gadolin. Pogg., **106**, 213, 1859.
   G. Rose. Pogg. Ann., **73**, 1; **75**, 403, 1848.
   Scheerer. Pogg. Ann., **67**, 120, 1846.
   Schröder. Pogg. Ann., **106**, 226, 1859. Jb. Min., 561, 932, 1873; 399, 1874, etc.
   Tschermak. Ber. Ak. Wien, **47** (1), 292, 1863.
   Websky. Die Mineralien nach den für das specifische Gewicht derselben angenommenen und gefundenen Werthen. 170 pp. Breslau, 1868.
   Kraus. Improved Jolly balance. Am. J. Sc., **31**, 561, 1911; Am. Min., **11**, 169, 1926.

Use of Heavy Solutions, etc.:

   Sonstadt. Chem. News, **29**, 127, 1874.
   Thoulet. Bull. Soc. Min., **2**, 17, 189, 1879.
   Bréon. Bull. Soc. Min., **3**, 46, 1880.
   Goldschmidt. Jb. Min., Beil.-Bd., **1**, 179, 1881.
   D. Klein. Bull. Soc. Min., **4**, 149, 1881.
   Rohrbach. Jb. Min., **2**, 186, 1883.
   Gisevius. Inaug. Diss., Bonn, 1883.
   Brauns. Jb. Min., **2**, 72, 1886; **1**, 213, 1888.

Retgers. Jb. Min., **2**, 185, 1889.
Salomon. Jb. Min., **2**, 214, 1891.
Penfield. Am. J. Sc., **50**, 446, 1895.
Merwin. Am. J. Sc., **32**, 425, 1911.
Vassar. Use and formation of Clerici solution. Am. Min., **10**, 123, 1925.

III. CHARACTERS DEPENDING UPON LIGHT
GENERAL PRINCIPLES OF OPTICS

311. Before considering the optical characters of minerals in general, and more particularly those that belong to the crystals of the different systems, it is desirable to review briefly some of the more important principles of optics upon which the phenomena in question depend.

For a fuller discussion of the optics of crystals, special reference is made to the works of Groth (translation by Jackson), Liebisch, Mallard, Duparc and Pearce, Rosenbusch (translation by Iddings), Iddings, Johannsen, Winchell, mentioned on p. 3, also to the various advanced text-books of Physics.

312. The Nature of Light. — Light is now considered to be an electromagnetic phenomenon due to a periodic variation in the energy given off by vibrating electrons. This energy is transmitted by a series of periodic changes that show all the characters of ordinary wave phenomena. The light waves, as they are commonly called, possess certain short wave-lengths that are of the correct magnitude to affect the optic nerves. Other similar waves with longer or shorter wave-lengths belong to the same class of phenomena. Immediately beyond the violet end of the visible spectrum come the so-called " ultra-violet " waves with still shorter wave-lengths and on beyond these we have the X-rays and the " gamma " rays produced by radium. Of the waves having greater lengths than those of light waves we have the waves that give rise to the sensation of heat and the Hertzian waves used in wireless. All of these vibrations, while varying enormously in their wave-lengths, belong to the same order of phenomena and obey the same laws. The proportion that the section of the series which produces the effect of light bears to the whole may be strikingly shown when we say that if ordinary white light is broken up into a spectrum a yard long and this then considered to be extended on either end so as to include all known electromagnetic waves the entire spectrum would be over five million miles in length.

The transmission of light through interstellar space, through liquids and transparent solids, has for some time been explained by the assumption that a medium, called the luminiferous ether, pervades all space, including the intermolecular space of material bodies. In this medium the vibrations of light waves are assumed to take place. For the purposes of the present work, however, it is unnecessary to consider closely the exact nature of light or the mode of its transmission. It will assist greatly, however, in obtaining a clear idea of the behavior of light in crystals if we assume that light waves are mechanical in nature and consist of periodic vibrations or oscillations in an all-prevading ether.

313. Wave-motion in General. — A familiar example of wave-motion is given by the series of concentric waves which on a surface of smooth water go out from a center of disturbance, as the point where a pebble has been dropped in. These surface-waves are propagated by a motion of the water-particles which is *transverse* to the direction in which the waves themselves travel; this motion is given from each particle to the next adjoining, and so on. Thus the particles of water at any one spot oscillate up and down,* while the wave moves on as a circular ridge of water of constantly increasing diameter, but of diminishing height. The ridge is followed by a valley, indeed both together properly constitute a wave in the physical sense. This compound wave is followed by another wave and another, until the original impulse has exhausted itself.

Another familiar kind of wave-motion is illustrated by the sound-waves which in the free air travel outward from a sonorous body in the form of concentric spheres. Here the actual motion of the layers of air is forward and back — that is, in the direction of propagation of the sound — and the effect of the transfer of this impulse from one layer to the next is to give rise alternately to a condensed and rarefied shell of air, which together constitute a sound-wave and which expand in spherical waves of constantly decreasing intensity (since the mass of air set in motion continually increases). Sound-waves, as of the voice, may be several feet in length, and they travel at a rate of 1120 feet per second at ordinary temperatures.

314. It is important to understand that in both the cases mentioned, as in every case of free wave-motion, each point on a given wave may be considered as a center of disturbance from which a system of new waves tend to go out. These individual wave-systems ordinarily destroy each other except so far as the onward progression of the wave as a whole is concerned. This is further discussed and illustrated in its application to light-waves (Art. **316** and Figs. **535, 536**).

In general, therefore, a given wave is to be considered as the resultant of all these minor wave-systems. If, however, a wave encounters an obstacle in its path, as a narrow opening (*i.e.*, one narrow in comparison with the length of the wave) or a sharp edge, then the fact just mentioned explains how the waves seem to bend about the obstacles, since new waves start from them as centers. This principle has an important application in the case of light-waves, explaining the phenomena of diffraction (Art. **337**).

315. Still another case of wave-motion may be mentioned, since it is particularly helpful in giving a correct apprehension of light-phenomena. If a long rope, attached at one end, be grasped at the other, a quick motion of the hand, up or down, will give rise to a half wave-form — in one case a crest, in the other a trough — which will travel quickly to the other end and be reflected back with a reversal in its position; that is, if it went forward as a hill-like wave, it will return as a trough. If, just as the wave has reached the end, a second like one be started, the two will meet and pass in the middle, but here for a brief interval the rope is sensibly at rest, since it feels two equal and opposite impulses. This will be seen later to be a case of the simple interference of two like waves opposed in phase.

Again, a double motion of the hand, up and down, will produce a complete wave, with crest and trough, as the result, and this again is reflected back as in the simpler case. Still again, if a series of like motions are continued rhythmically and so timed that each wave is an even part of the whole rope, the two systems of equal and opposite waves passing in the two directions will interfere and a system of so-called stationary waves will be the result, the rope seeming to vibrate in segments to and fro about the position of equilibrium.

Finally, if the end of the rope be made to describe a small circle at a rapid, uniform, rhythmical rate, a system of stationary waves will again result, but now the vibrations of the string will be sensibly in circles about the central line. This last case will be seen to indicate roughly the kind of transverse vibrations by which the waves of circularly polarized light are propagated, while the former case represents the vibrations of waves of what is called plane-polarized light.

---

* Strictly speaking, the path of each particle approximates closely to a circle.

All these cases of waves obtained with a rope deserve to be carefully considered and studied by experiment, for the sake of the assistance they give to an understanding of the complex phenomena of light-waves.

316. Light-waves. — In the discussion that follows, in order to make the explanations simpler and clearer, light-waves have been treated as if they consisted of mechanical disturbances in a material medium called the ether.

The vibrations in the ether caused by the transmission of a light-wave take place in directions transverse to the direction of the movement of the wave. These oscillations have the following characters. When an ether particle is set vibrating it moves from its original position with gradually decreasing velocity until the position of its maximum displacement is reached. Then with gradually increasing velocity it returns to its original position and since it is moving without friction it will continue in the same direction on past this point. Its velocity will then again diminish until it has reached a displacement equal but opposite in direction to its first swing, when it will start back on its course and repeat the oscillation. The varying velocity of such an oscillation would be the same as that shown by a particle moving around a circle with uniform speed if the particle was observed in a direction lying in the plane of the circle. Under these conditions the particle would appear to move forward and backward along a straight line with constantly changing velocity. Such a motion is called simple harmonic motion.

The motion of one ether particle is communicated to another and so on, each, in order, falling a little behind in the time of its oscillation. Conse-

quently, while the individual particles move only back and forth in the same line the wave disturbance moves forward. If, at a given instant of time, the positions of successive particles in their oscillations are plotted, a curve, such as shown in Fig. 534, will be formed. Such a curve is known as a harmonic

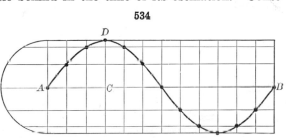

534

Harmonic Curve

curve. The oscillatory motion of the particles in a light-wave is called a *periodic motion* since it repeats itself at regular intervals. The maximum displacement of a particle from its original position of rest is called the *amplitude* of the wave (distance C–D, Fig. 534). The *phase* of a particle at a given instant is its position in the vibration and the direction in which it is moving.

The distance between any particle and the next which is in a like position — i.e., of like *phase*, as A and B — is the *wave-length;* and the time required for this completed movement is the *time* of vibration, or *vibration-period*. The wave-system therefore travels onward the distance of one wave-length in one vibration-period. The intensity of the light varies with the square of the amplitude of the vibration, and the color, as explained in a later article, depends upon the length of the waves; the length of the violet waves is about one half the length of the red waves.

In *ordinary light* the transverse vibrations are to be thought of as taking place in all planes about the line of propagation. In the above figure, vibrations in one plane only are represented; light that has only one direction of transverse vibration is said to be *plane-polarized*.

Light-waves have a very minute length, only 0·000023 of an inch for the yellow sodium flame, and they travel with enormous velocity, 186,000 miles per second in a vacuum; thus light passes from the sun to the earth in about eight minutes. The vibration-period, or time of one oscillation, is consequently extremely brief; it is given by dividing the distance traveled by light in one second by the number of waves included.*

317. Wave-front. — In an isotropic medium, as air, water, or glass — that is, one in which light would be propagated in all directions about a luminous point with the same velocity — the waves are spherical in form. The *wave-front* is the continuous surface, in this case spherical, which includes all particles that commence their vibration at the same moment of time. Obviously the curvature of the wave-front diminishes as the distance from the source of light increases, and when the light comes

535

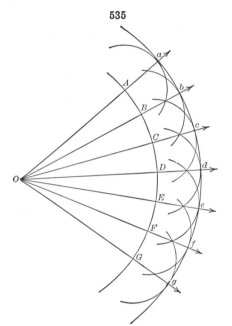

536

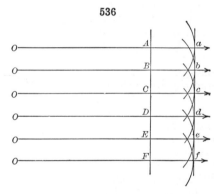

from an indefinitely great distance (as the sun) the wave-front becomes sensibly a plane surface. Such waves are usually called *plane waves*. These cases are illustrated by Figs. 535 and 536. In Fig. 535 the luminous point is supposed to be O, and the medium being isotropic, it is obvious that the wave-front, as $ABC \ldots G$, is spherical. It is also made clear by this figure how, as briefly stated in Art. **314,** the resultant of all the individual impulses which go out from the successive points, as A, B, C, etc., as centers, form a new wave-front, $abc \ldots g$, concentric with $ABC \ldots G$. In Fig. 536 the luminous body is supposed to be at a great distance, so that the wave-

---

* " On account of the tremendous speed at which light travels the rapidity of vibration, or " frequency " of light as it passes through a fixed point, is extremely great. About eight hundred trillion waves of violet light would pass through such a point in a second. The extreme brevity of the interval of time required for the passage of a single wave of this sort may perhaps be realized better when it is said that one eight-hundred-trillionth of a second is a vastly smaller part of a second than a second is of the whole of historic time." Comstock and Troland, " The Nature of Matter and Electricity," p. 157.

front $AB \ldots F$ is a plane surface. Here also the individual impulses from A, B, etc., unite to form the wave-front $ab \ldots f$ parallel to $AB \ldots F$.

318. Light-ray. — The study of light-phenomena is, in certain cases, facilitated by the conception of a *light-ray*, a line drawn from the luminous point to the wave-front, and whose direction is taken so as to represent that of the wave itself. In Fig. 535, OA, OB, etc., are diverging light-rays, and in Fig. 536, OA, OB, etc., are parallel light-rays. In both these cases, where the medium is assumed to be isotropic, the light-ray is normal to the wave-front. This is equivalent to saying that the light-wave moves onward in a direction normal to the wave-front.

It must be understood that the " light-ray " has no real existence and is to be taken only as a convenient method of representing the direction of motion of the light-waves under varying conditions. Thus when by appropriate means (*e.g.*, the use of lenses) the curvature of the wave-front is altered — for example, if from being a plane surface it is made sharply convex — then the light-rays, at first parallel, are said to be made to diverge. Again, if the convex wave-front is made plane, the diverging light-rays are then said to be made parallel.

319. Wave-length. Color. White Light. — Notwithstanding the very small length of the waves of light, they can be measured with great precision. The visual part of the waves going out from a brilliantly incandescent body, as the glowing carbons of an electric arc-light, may be shown to consist of waves of widely varying lengths. They include red waves whose length is

$$0.0007604 \text{ mm.} \left(\text{about } \frac{1}{39,000} \text{ of an inch} \right) \text{ and waves whose length constantly}$$

diminishes without break, through the orange, yellow, green, and blue to the violet, whose minimum length (0·0003968 mm.) is about half of that of the red. The symbol commonly used for the wave-length of light is lambda, λ.*
The color of light is commonly said to depend upon its wave-length and will be so spoken of here. This is not strictly true, however, because, since the velocity of light varies with the medium through which it is traveling while the vibration period remains constant under all conditions, it follows that the wave-length of light of the same color must be different in different media. It is, therefore, rather the frequency with which the light-waves reach the eye that determines the color sensation. Commonly a given color is produced by the combination of several different wave-lengths of light. It is strictly *monochromatic* only when it corresponds to one definite wave-length; this is nearly true of the bright-yellow sodium line, though strictly speaking this consists of two sets of waves of slightly different lengths.

The effect of *white light* is obtained if all the waves from the red to the violet come together to the eye simultaneously; for this reason a piece of platinum at a temperature of 1500° C. appears " white hot."

The radiation from the sources named, either the sun, the electric carbons, or the glowing platinum, includes also longer waves which do not affect the eye, but which, like the light-waves, produce the effect of sensible heat when received upon an absorbing surface, as one of lamp-black. There are also, particularly in the radiation from the sun, waves shorter than the violet which

---

* The wave-length of light is commonly given in millimicrons, *i.e.*, millionths of a millimeter, the symbol being $\mu\mu$. That is, for light λ varies from about 380 to 760 $\mu\mu$. The following wave-lengths occur in the middle of the colors of the spectrum, 424 $\mu\mu$ in the violet, 461 $\mu\mu$ blue, 505 $\mu\mu$ green, 558 $\mu\mu$ yellow, 623 $\mu\mu$ orange, 707 $\mu\mu$ red.

also do not affect the eye. The former are called *infra-red*, the latter *ultra-violet* waves.

The *brightness* of light depends upon the amplitude of its vibrations and varies directly as the square of this distance.

320. Complementary Colors. — The sensation of white light mentioned above is also obtained when to a given color — that is, light-waves of given wave-length — is combined a certain other so-called *complementary* color. Thus certain shades of pink and green combined, as by the rapid rotation of a card on which the colors form segments, produce the effect of white. Blue and yellow of certain shades are also complementary. For every shade of color in the spectrum there is another one complementary to it in the sense here defined. The most perfect illustration of complementary colors is given by the examination of sections of crystals in polarized light, as later explained.

537

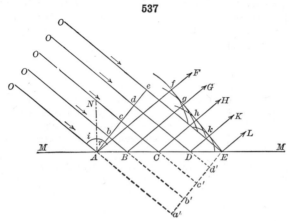

321. Reflection. — When light-waves come to the boundary which separates one medium from another, as a surface of water, or glass in air, they are, in general, in part *reflected* or returned back into the first medium.

The reflection of light-waves is illustrated by Figs. 537 and 538. In Fig. 537, MM is the reflecting surface — here a plane surface — and the light-waves have a plane wave-front ($Abcde$); in other words, the light-rays (OA, Ob, etc.) are parallel. It is obvious that the wave-front meets the surface first at A and successively from point to point to E. These points are to be regarded as the centers of new wave-systems which unimpeded would be propagated outward in all directions and at a given instant would have traveled through distances equal to the lines Aa, Bb', etc. Hence the common tangent $fghkE$ to the circular arcs drawn with these radii from A, B, etc., represents the direction of the new or reflected wave-front. But geometrically the angle eAE is equal to fEA, or the *incident and reflected wave-*

538

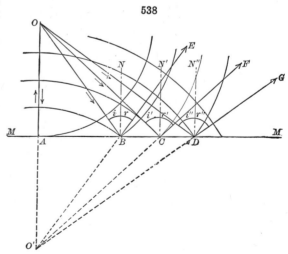

fronts make equal angles with the reflecting surface. If NA is a normal at A, the angle OAN — called the *angle of incidence* — is equal to NAF, the *angle of reflection.* Hence the familiar law:

The angle of incidence is equal to the angle of reflection.

Furthermore, the " incident and reflected rays " both lie in the same plane with the normal to the reflecting surface.

In Fig. 538, where the luminous point is at O, the waves going out from it will meet the plane mirror MM first at the point A and successively at points, as B, C, D, etc., farther away to the right (and left) of A. Here also it is easy to show that all the new impulses, which have their centers at A, B, C, etc., must together give rise to a series of reflected waves whose center is at O', at a distance equally great from MM measured on a normal to the surface ($OA = O'A$).

Now the lines OA, OB, etc., which are perpendicular to the wave-front, represent certain incident light-rays, and the eye placed in the direction BE, CF, etc., will see the luminous point as if at O'. It follows from the construction of the figure and can be proved by experiment that if BN, CN', etc., are normals to the mirror the angles of incidence, OBN, OCN', etc., are equal to the angles of reflection, NBE, $N'CF$, etc., respectively. Hence the above law applies to this case also.

If the reflecting surface is not plane, but, for example, a concave surface, as that of a spherical or parabolic mirror, there is a change in the curvature of the wave-front after reflection, but the same law still holds true.

The proportion of the reflected to the incident light increases with the smoothness of the surface and also as the angle of incidence diminishes. The intensity of the reflected light is a maximum for a given surface in the case of perpendicular incidence (OA, Fig. 538).

If the surface is not perfectly polished, *diffuse reflection* will take place, and there will be no distinct reflected ray. It is the diffusely reflected light which makes the reflected surface visible; if the surface of a mirror were *absolutely smooth* the eye would see the reflected body in it only, not the surface itself. Optically expressed, the surface is to be considered smooth if the distance between the scratches upon it is considerably less (say one fourth) than the wave-length of light.

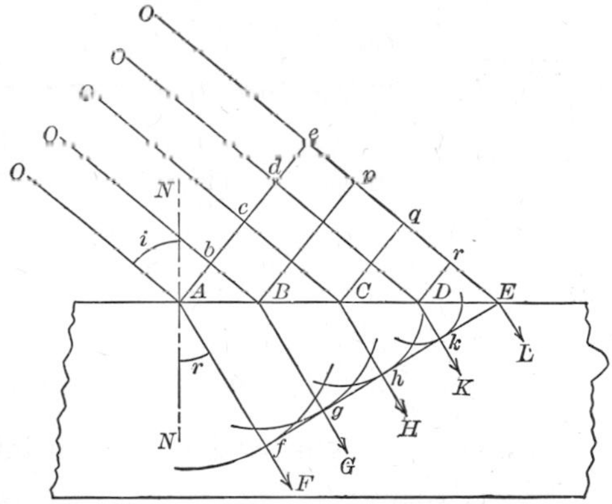

322. Refraction. — When light passes from one medium into another there is, in general, an increase or decrease in its velocity, and this commonly results in the phenomenon of *refraction* — that is, a change in the direction of propagation. The principles applicable here can be most easily shown in the case of light-waves with a plane wave-front, as shown in Fig. 539 — that is, where the light-rays OA, OB, etc., are paral-

539

lel. Suppose, for example, that a light-wave, part of whose wave-front is *Abcde*, passes from air obliquely into glass, in which its velocity is about two thirds as great as it was in the air and suppose the surface of the glass to be plane. At the moment that the ray $O–A$ enters the glass the ray $O–E$ has reached the point e. During the time that the latter ray travels from e

to E, the ray O–A will have advanced in the glass a distance equal to $\frac{2}{3}e$–E, or to some point on an arc having this distance as a radius (A–f). In the same way during the time ray O–E passes from the point p to E, ray O–B will have traveled in the glass the distance B–g, equal to $\frac{2}{3}p$–E. In this way arcs may be drawn about each one of the points A, B, C, etc., and the position of the new wave-front in the glass determined by their common tangent, *Ekhgf*. It is seen that there is a change of direction in the wave-front, or otherwise stated, in the light-ray, the magnitude of which depends on the ratio between the light-velocities in the two media, and, as discussed later, also upon the wave-length of the light. The light-ray is here said to be broken or *refracted*, and for a medium like glass, optically denser than air (*i.e.*, with a lower value of the light-velocity), the refraction is toward the perpendicular with the angle of refraction, r, smaller than the angle of incidence, i. In the opposite case — when light passes into an optically rarer medium — the refraction is away from the perpendicular and the angle of refraction is larger than that of incidence (Art. **328**).

323. Refractive Index. — It is obvious from the figure that whatever the direction of the wave-front — that is, of the light-rays — relatively to the given surface, the *ratio* of eE to Af, which determines the direction of the new wave-front (*i.e.*, the direction of a refracted ray, AF) is constant. This ratio is equal to $\dfrac{V}{v}$ where V is the value of the light-velocity for the first medium (here air) and v for the second (as glass). This constant ratio is commonly represented by n and is known as the *index of refraction*. Therefore

$$n = \frac{eE}{Af}$$

In Fig. 539, by construction,

$$\angle eAE = \angle i \quad \text{and} \quad \angle AEf = \angle r.$$

Also,

$$\frac{eE}{AE} = \sin i \quad \text{and} \quad \frac{Af}{AE} = \sin r.$$

Therefore,

$$\frac{\sin i}{\sin r} = \frac{\dfrac{eE}{AE}}{\dfrac{Af}{AE}} = \frac{eE}{Af} = n.$$

The law of refraction then is given by the expression, $n = \dfrac{\sin i}{\sin r}$, or may be formulated as follows:

The sine of the angle of incidence bears a constant ratio to the sine of the angle of refraction.

In the case of light passing from air into crown glass this ratio is found to be, $\dfrac{\sin i}{\sin r} = 1\cdot608$, and this number consequently gives the value of the refractive index, or n, for this kind of glass.*

---

* Strictly speaking, the index of refraction of the ether in a vacuum is taken as unity, but since air has practically the same power of refraction (n for air $= 1\cdot0003$), the index of any substance is obtained by comparison with that of air.

The above relation holds true for any wave-system of given wave-length in passing from one medium into another, whatever the wave-front or shape of the bounding surface. In Fig. 540 the luminous point is at O, and it can be readily shown that the new wave-front propagated in the second medium (of greater optical density) has a flattened curvature and corresponding to this a center at O' $\left(\text{where } \dfrac{O'A}{OA} = \dfrac{V}{v}\right)$. Here the incident rays OB, OC, are refracted at B and C, the corresponding refracted rays being BE and CF. For this case also the relation holds good,

$$n = \frac{\sin i}{\sin r} = \frac{\sin i'}{\sin r'}, \text{ etc.}$$

540

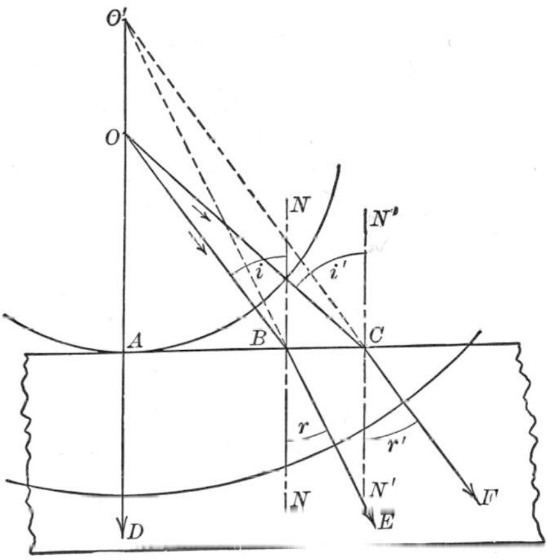

If the bounding surface is not plane but curved, as in lenses, there is a change in the curvature of the wave-front in the second medium, but the simple law, $n = \dfrac{\sin i}{\sin r}$, holds true here also, so long as the medium is isotropic.

The relation between wavelength and refractive index is spoken of in Art. **333**.

324. Relation of Refractive Index to Light-velocity. — The discussion of the preceding article shows that if n is the refractive index of a given substance for waves of a certain length, referred to air, V the velocity in air and v the velocity in the given medium, then

$$n = \frac{V}{v}.$$

For two media whose indices are n_1 and n_2 respectively, it consequently follows that

$$\frac{n_1}{n_2} = \frac{v_2}{v_1}.$$

Therefore: *The indices of refraction of two given media for a certain wavelength are inversely proportional to their relative light-velocities.*

In other words, if the velocity of light in air is taken as equal to 1 and the velocity of the same light is found to be one-half as great when passing through a given substance, the index of refraction, or n, of that substance when referred to air $(n - 1\cdot 0)$ will be equal to $2\cdot 0$.

325. Principal Refractive Indices. — The refractive index has, as stated, a constant value for every substance, referred, as is usual, to air (or it may be to a vacuum). In regard to solid media, it is evident from Art. **323** and will be further explained later that those which are *isotropic*, viz., amorphous substances and crystals of the isometric system, can have but a single value cf this index. Crystals of the tetragonal and hexagonal systems have, as later explained, *two* principal refractive indices, ω and ϵ, corresponding to the velocities of light having certain definite directions of vibration. Further, all orthorhombic, monoclinic, and triclinic crystals have similarly *three* principal indices, α, β, γ. In the latter cases of so-called anisotropic media, the mean refractive index is taken as the arithmetical mean, namely $\dfrac{2\omega + \epsilon}{3}$ and $\dfrac{\alpha + \beta + \gamma}{3}$.

326. Effect of Index of Refraction upon Luster, etc. — The luster and general appearance of a transparent substance depend largely upon its refractive index. For instance the peculiar aspect of the mineral cryolite, by means of which it is usually possible to readily identify the substance, is due to its low index of refraction. If cryolite is pulverized and the powder poured into a test tube of water it will disappear and apparently go into solution. It is quite insoluble, however, but becomes invisible in the water because its index of refraction (about 1·34) is near that of water (1·335). The light will travel with practically the same velocity through the cryolite as through the water and consequently suffer little reflection or refraction at the surfaces between the two. On the other hand powdered glass with a higher index of refraction than that of water appears white under the same conditions because of the reflection of light from the surfaces of the particles.

Substances having an unusually high index of refraction have an appearance which it is hard to define, and which is generally spoken of as an *adamantine luster*. This kind of luster may be best comprehended by examining specimens of diamond ($n = 2\cdot419$) or of cerussite ($n = 1\cdot98$). They have a flash and quality, sometimes almost a metallic appearance, which is not possessed by minerals of a low refractive index. Compare, for example, specimens of cerussite and fluorite ($n = 1\cdot434$). The usual index of refraction for minerals may be said to range not far from 1·55, and gives to minerals a luster which has been termed *vitreous*. Quartz, feldspar, and halite show good examples of vitreous luster.

Below is given a list of common minerals arranged according to their indices of refraction. For minerals other than those of the isometric system the average value (as defined in the preceding article) is given here.

Water	1·335	Kyanite	1·723
Fluorite	1·434	Epidote	1·750
Orthoclase	1·523	Corundum	1·765
Gypsum	1·524	Almandite	1·810
Quartz	1·547	Malachite	1·880
Muscovite	1·582	Anglesite	1·884
Beryl	1·582	Zircon	1·952
Calcite	1·601	Cerussite	1·986
Topaz	1·622	Cassiterite	2·029
Tremolite	1·622	Sulphur	2·077
Dolomite	1·626	Sphalerite	2·369
Aragonite	1·633	Diamond	2·419
Apatite	1·633	Rutile	2·711
Barite	1·640	Cuprite	2·849
Diopside	1·685	Cinnabar	2·969

327. Relations between Chemical Composition, Density, and Refractive Index. — That definite relations exist between the chemical composition of a substance, its specific gravity, and its index of refraction, has been conclusively shown in many cases. With the plagioclase feldspar group, for instance, the variation in composition which the different members show is accompanied by a direct variation in density and refractive index. Attempts have been made to express these relations in the form of mathematical statements. The two most satisfactory expressions are the one proposed by Gladstone and Dale,[*] $\dfrac{n-1}{d}$ = constant, and the one proposed independently by Lorentz[†] and Lorenz,[‡] $\dfrac{n^2-1}{n^2+2} \cdot \dfrac{1}{d}$ = constant. In these n is equal to the mean refractive index and d to the density. These were originally proposed for use with gases and solutions and for these bodies have been found to serve about equally well. When attempts are made, however, to apply them to crystalline solids the results are at the best only approximate.[§] This is probably because the formulas do not take into consideration the modifications that the crystal structure must introduce.

328. Total Reflection. Critical Angle. — In regard to the principle

stated in Art. **323** and expressed by the equation $n = \dfrac{\sin i}{\sin r}$, two points are to be noted. First, if the angle $i = 0°$, then $\sin i = 0$, and obviously also $r = 0$; in other words, when the ray of light (as OA, Fig. 540) coincides with the perpendicular, no change of direction takes place, the ray proceeds onward (AD) into the second medium without deviation, but with a change of velocity.

Again, if the angle $i = 90°$, then $\sin i = 1$, and the equation above becomes $n = \dfrac{1}{\sin r}$ or $\sin r = \dfrac{1}{n}$. As n has a fixed value for every substance, it is obvious that there will also be a corresponding value of the angle r for the case mentioned. From the above table it is seen that for water, $\sin r = \dfrac{1}{1\cdot335}$ and $r = 48° 31'$; for crown glass ($n = 1\cdot608$), $\sin r = \dfrac{1}{1\cdot608}$ and $r = 38° 27'$; for diamond, $\sin r = \dfrac{1}{2\cdot42}$ and $r = 24° 25'$.

This fact, that for each substance at a particular value of the angle r the angle i becomes equal to 90°, has an important bearing on the behavior of light when it is passing from an optically denser into an optically rarer medium. In Fig. 541 we may assume that light-rays coming from various directions meet the surface between a block of glass and the air at the point A. Light traveling along the path $O–A$ will pass out into the air without a change in its direction but with an increase in its velocity. If it emerges from the glass at any other angle than 90° the ray on entering the air will be bent away from the perpendicular and the angle of deviation will vary with the angle at which the ray touched the surface and with the index of refraction of the glass. The same law holds true in this case as in the case of a ray entering from the air, except that the formula now reads $n = \dfrac{\sin r}{\sin i}$, where $r =$ the angle the ray in air makes with the normal to the surface and $i =$ the angle that the ray makes within the glass to the same normal. In Fig. 541 the ray $C–A$ will pass

---

* Phil. Trans., **153**, 317, 1863.
† Wied. Ann., **9**, 641, 1880.
‡ Wied. Ann., **11**, 70, 1880.
§ E. S. Larsen, Am. Jour. Sci., **28**, 263, 1909. See also Cheneveau, Ann. Chem. Phys., **12**, 145, 289, 1907.

out into the air along the line $A–D$. But the angle i for the ray $E–A =$ 38° 27′ and, as shown in the preceding paragraph, for glass, where $n = 1·608$, the angle r in the air will be 90° and the ray will travel along the surface of the glass in the direction $A–F$. Consequently any ray, such as $G–A$, which meets

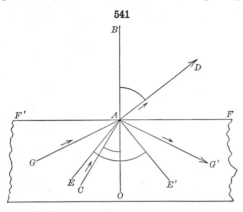

541

the surface of the glass at an angle greater than 38° 27′, will be unable to pass out into the air and will suffer *total reflection* at the surface, passing back into the glass in the direction $A–G'$, with angle $OAG =$ angle OAG'. The angle at which total reflection takes place for any substance is known as its *critical angle.*

The phenomenon of total reflection is taken advantage of in the cutting of gem stones. According to common practice such a stone is cut with a flat surface on top and with a number of inclined facets on the bottom. The light that enters the stone from above is in a large measure totally reflected from the sloping planes below and comes back to the eye through the stone. The amount of light reflected in this way and the consequent brilliancy of the gem increases with its index of refraction. Two stones cut exactly alike, one from diamond and the other, perhaps, from quartz, would have very different appearances due to this difference in the amount of light totally reflected from their lower facets. This principle is illustrated in Figs. 542 and 543. They represent cross sections of two hemispheres cut, one from fluorite and the other from diamond. It is assumed that light from all directions is focused on the center of the plane surface of each hemisphere. All the light that meets this surface at an angle greater

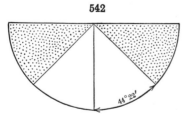

542

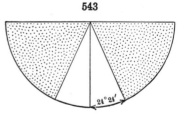

543

Total Reflection in Fluorite $n = 1.43$ Total Reflection in Diamond $n = 2.42$

than the critical angle for the mineral will be totally reflected back through the spherical surface. The shaded areas of the figures show the amount of light in each case that would be so reflected and clearly illustrate the optical difference between the two substances.

329. Effect of Index of Refraction upon Microscopic Phenomena. — In the study of minerals, especially in thin sections under the microscope, variations in the index of refraction give effects which are of importance. In Fig. 544 let it be assumed that L is the objective lens of a compound microscope, and that the instrument is exactly focused upon a point O, Fig. 544 A. If now we imagine that a section of some mineral of mean index of refraction is placed under the lens, Fig. 544 B, the point O' will now be in focus, or as in Fig. 544 C, where the mineral is supposed to have a high index of refraction, the focus will be at O''. Thus it is that with two sections of equal thickness and with the lens in the same position, one looks deeper into the mineral of higher index of refraction. Consequently, when there are two minerals in the same section, the one having a high and the other a low index of refraction (for example, a crystal of zircon, $n = 1.95$, embedded in quartz, $n = 1.55$), the one having the higher index of refraction will apparently have the greater thickness and will appear to stand up in relief above the surface of the mineral of lower index. The apparent relief is furthermore augmented by other properties to be explained below.

544

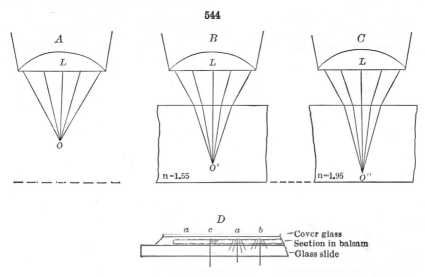

In preparing thin sections of minerals or rocks for study with the microscope the process, in brief, is to make first a flat surface upon the mineral or rock by grinding it upon a plate supplied with some abrasive. This flat surface is then cemented to a piece of glass by means of Canada balsam and the remainder of the mineral is ground away until only a thin film remains, which in the best rock sections is not over 0.03 mm. in thickness. The section is finally embedded in balsam, n about 1.54, and over it a thin cover glass is laid. In the preparation of a section the surfaces are not polished, hence, from the nature of the abrasive, they must be pitted and scratched and it may be assumed that in cross section such a preparation would be somewhat as represented in Fig. 544 D. When a thin section is examined under the microscope the light enters the section from below, having been reflected up into the microscope tube by an inclined mirror. Before it reaches the section it will have passed through a nicol prism and through a slightly converging lens. Let it be assumed that the mineral at a, Fig. 544 D, is one of mean refractive index. The convergent light entering the section will pass with little or no refraction from the mineral into the balsam because their refractive indices are nearly alike. Hence the roughness of the surface of the section is not apparent and the mineral appears as if polished.   If there is a crack, as at *b*, so much light penetrates it that it is scarcely visible when the convergent lens is close to the object, but when the latter is lowered, and especially when the light is restricted by the use of an iris diaphragm inserted into the microscope tube, the nearly parallel rays of light will suffer some total reflection along the line of the crack and so make it visible.   On the other hand, if the mineral has a high index of refraction there will be innumerable places all over the section where the surfaces are so inclined that the light will suffer total reflection in attempting to pass from the optically dense mineral into the rarer balsam.   Hence the uneven surface of the section due to its grinding is plainly visible.   This effect is more pronounced if the convergent lens is lowered. The cracks that may exist in a mineral of high index of refraction are for the same reasons much more distinct than in a mineral of low index.   Further, if a mineral of high index of refraction is embedded in one of low, *c*, Fig. 544 *D*, there will be places along its outer edge where total reflection will take place, thus causing its outline to be dark and distinct.   This effect combined with the roughened aspect of the surface and the apparent increase in thickness, as described in the preceding paragraph, all tend to make a mineral of high index of refraction stand out conspicuously in relief.   These effects are seen best with a lens of medium power, and as stated above, with the condenser lens lowered and the field partially darkened.

330. Determination of the Indices of Refraction of Mineral Grains under the Microscope. — The considerations of the preceding article suggest a means of determining the indices of refraction of mineral grains under the microscope.   If a grain is immersed in a liquid of known index of refraction it is possible to determine whether it has a higher or lower index of refraction than the liquid and by the use of a series of liquids of varying refractive indices it is possible to determine with considerable accuracy the index of refraction of the mineral.   A list of liquids* in common use for such purposes, with their indices of refraction is given below.

Petroleum distillates...................................	1·35 –1·45†
Mixtures of refined petroleum oils and turpentine..........	1·450–1·475
Turpentine and ethylene bromide or clove oil.............	1·480–1·535
Clove oil and α-monobromnaphthalene...................	1·540–1·635
Petroleum oils and α-monobromnaphthalene..............	1·475–1·650
α-monobromnaphthalene and methylene iodide............	1·650–1·740
Sulphur dissolved in methylene iodide...................	1·740–1·790
Mixtures of methylene iodide with iodides of antimony, arsenic and tin, also sulphur and iodoform (see Merwin)....	1·740–1·870
Methylene iodide and arsenic trisulphide (see Merwin).....	1·740–2·280
Resin-like substances formed from mixtures of piperine and the tri-iodides of arsenic and antimony. These fuse easily and mineral grains can be thus embedded in a thin film of the material......................................	1·680–2·100
Mixtures of the chloride, bromide and iodide of thallium (see Barth)...	2·24 –2·78

---

* Wright, Methods of Petrographic-Microscopic Research, p. 98; Merwin, Jour. Wash. Acad. Sc., **3**, 35, 1913; Larsen, Microscopic Determination of Nonopaque Minerals, 1921, 1932; Barth, Am. Min., **14**, 358, 1929.

† Harrington and Buerger, Am. Min., **16**, 45, 1931.

The indices of refraction of the test liquids can be determined either by the use of the total refractometer or by filling a hollow glass prism with the liquid and using the methods employed with ordinary mineral prisms, see Art. **332**.

A series of these liquids should be prepared which for most purposes might conveniently show differences in the indices of the different liquids of 0·01. For more exacting work smaller differences between the indices of the members of the series would be of advantage. If these are kept in well stoppered bottles and are protected from the light they will show very little change over considerable periods of time. It is advisable, however, to check their indices at least once a year.

The mineral to be studied should be broken down into uniform small grains (0·05 mm. is usually a good diameter) and then a few grains placed upon a glass slide. A drop of liquid with a known index of refraction is then placed upon the grains and the whole covered with a thin cover glass. When a mineral grain is immersed in a liquid of closely the same index of refraction it loses its sharpness of outline and if the mineral is colorless and the correspondence of the two indices exact it will quite disappear. Certain tests, however, are commonly used to determine the relative indices of the mineral and the liquid which with proper care can distinguish differences as small as 0·01 or with practice and especial care as small as 0·001.

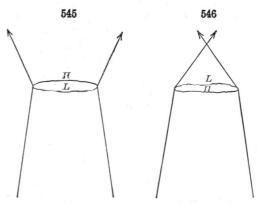

545   546

Grain with Low Refractive Index immersed in Liquid of High Refractive Index   Grain with High Refractive Index immersed in Liquid of Low Refractive Index

To make these tests the condenser below the microscope stage should be lowered and, if the instrument has a sub-stage iris diaphragm, this should be partly closed. Under these conditions the obliquity of the light is reduced and only a small pencil of light composed of nearly parallel rays enters the section. Such mineral grains will usually be thinner on their edges and act somewhat like lenses in their effect upon light. Let Fig. 545 represent a mineral grain illuminated in this way when immersed in a liquid of higher index of refraction. The light-rays as they pass from the mineral into the higher refracting liquid above will be bent away from the perpendicular. In the opposite case, Fig. 546, where the mineral has the higher index the reverse will be true and the light-rays will be bent toward the perpendicular. This will produce in one case a brighter illumination of the borders of the mineral grain and in the other a brighter illumination of its center. This difference in illumination is, however, commonly so slight as to be certainly detected only with difficulty. The so-called *Becke Test* is commonly used under these circumstances. This consists in focusing upon the grain with a high power objective and then slowly raising or lowering the microscope tube. The condenser lens beneath the microscope stage should be lowered so that the grains under observation lie above its focal point.   In the case illustrated by Fig. 545, when the tube is raised, a narrow line of light will be seen to move outward from the mineral, while when the tube is lowered this line will move inward.   In the case illustrated in Fig. 546 the opposite conditions will prevail.   A convenient rule to remember is that *when the microscope tube is raised the Becke line will move toward the material of higher refractive index and when the tube is lowered this line will move toward the material of lower index.* This makes a very satisfactory and quite delicate test for distinguishing differences in refractive indices.   Sometimes two lines will appear moving in opposite directions and it may be difficult to decide which is the Becke line.   This is usually obviated by lowering the condenser or decreasing the aperture in the iris diaphragm.   For the use of the Becke test in rock sections, see Art. **331.**

The test upon mineral grains immersed in a liquid may also be made by means of oblique illumination.   An oblique pencil of rays may be obtained

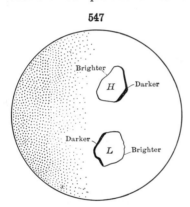

547

most conveniently by placing a pencil, a finger, or a piece of cardboard between the reflecting mirror and the polarizer in such a way as to darken one half of the field of vision.   The best results will be obtained by the use of an objective of medium magnifying power.   When a mineral grain is viewed under these conditions it will be noted that one of its edges is more brightly illuminated than the other.   With the condenser lens lowered and mineral with a lower index of refraction than the liquid, the bright edge of the mineral will be away from the shadow, while if the mineral has a higher index than the liquid the bright edge will be on the side toward the shadow.   These conditions are presented in Fig. 547, where L and H represent grains with indices respectively lower and higher than the liquid in which they are immersed.   If the condenser lens is raised so that the mineral grain will be below its focus, effects exactly opposite to those described above will be noted.   It is wise to test the apparatus used by observing mineral fragments of known indices and taking note of the effects produced.

Commonly the liquids used have a higher dispersion than the mineral to be tested.   In other words the liquid will have distinctly different indices of refraction for red and for blue light.   If the mineral should have an index intermediate between those for red and blue light in the liquid the grain when illuminated in oblique light will show colored borders.   With the condenser lens lowered the edge of the mineral next to the shadow will be colored an orange-red while the edge away from the shadow will be pale blue.   If the amount of the dispersion in the liquid (*i.e.*, the difference between the indices for blue and red light) is not too great this effect gives very closely the refractive index of the mineral.

It should be pointed out here that all minerals, except those of the isometric system, show different indices of refraction depending upon the crystal direction in which the light is vibrating while passing through the mineral.   Consequently unorientated grains of a mineral, unless it belongs to the isometric system, will show a variation in the refractive indices depending upon their position on the slide. Sometimes it is possible to determine the crystal orientation of a grain due to some significant cleavage or structure and so obtain the index for some particular crystal direction, but ordinarily all that can be determined is the mean index of refraction of the mineral.

331. The Becke Test in Rock Sections. — The Becke test can be often used in a rock section to determine the relative indices of refraction of two different minerals lying in contact with each other. Their contact plane should be nearly vertical in order to give clear results. The position of this plane can be determined by focusing on the surface of the section and then as the microscope tube is lowered note whether or not the position of the dividing line between the two minerals remains stationary or moves. If it remains stationary or moves only a little, the dividing plane is vertical or nearly so. Under these conditions assume that the cone of light entering from below is focused at point *O*, Fig. 548, lying on the dividing plane between *L* (mineral with lower index) and *H* (mineral with higher index). The light rays 1–6 passing as they do from a mineral of lower index into one of higher will suffer no total reflection and all emerge from the section on the side of *H*. On the other hand, rays 7–12 attempting to pass from *H* to *L* will only in part pass across the dividing plane while the others will be totally reflected and add themselves to rays 1–6 on the side of *H*. *H* will therefore show a brighter illumination than *L*. In this case also when the tube of the microscope is raised the Becke line will be seen moving toward the mineral of higher index or when the tube is lowered, toward that of lower index. The best results will be obtained by using an objective of high magnification and the condenser lens must be lowered.

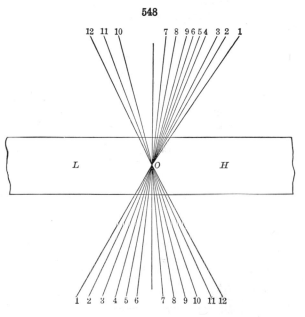

332. Dispersion Methods. — Various methods have been devised for the determination of indices of refraction with the microscope which depend upon the ability to vary the index of a liquid by a variation in the wavelength of the light or by a change in temperature. By this means it is possible to vary the index of the liquid until it exactly matches that of the mineral grain. It is possible to make determinations that are accurate to ±·001. The disadvantages of the methods are that they require more apparatus and greater skill in manipulation.

By using a monochromatic illuminator, the index of a given mineral may be determined for a certain wave-length of light when immersed in a certain liquid. The indices of refraction of that particular liquid for several wave-lengths of light being already known, it is possible to establish the index of refraction of the mineral for the wave-length of sodium light. In practice the same mineral could be immersed in two or three different liquids in order to check the results.*

Variation in the index of a liquid may also be produced by a change in its temperature. A cell through which water of known temperature can circulate is placed on the stage of the microscope, and the mineral grains and index liquid are placed on the thin glass plate that forms the upper surface of the cell. The temperature of the index liquid can in this way be varied until its index of refraction matches that of the mineral. The refractive indices of the liquid (its thermal dispersion) being known for varying temperatures, the index of the mineral can be determined for a given temperature.

A combination of these two methods, known as the double dispersion method, yields the best results.† The refractive index of the liquid decreases as the wave-length of the light used increases, and also as the temperature of the liquid rises. On the other hand, the variation in the index of the mineral grain with changes in temperature and wave-length of light will be so small as to be ordinarily negligible. By choosing liquids that show a maximum light and heat dispersion it is possible to cover the range of mineral indices with comparatively few liquids.

333. Determination of the Index of Refraction by Means of Prisms or Plates. — For the more accurate determination of the indices of refraction of minerals a natural or cut prism or plate of the mineral is used. In all cases, except minerals of the isometric system, the prism or plate used must have a certain crystallographic orientation. This matter, however, will be discussed when the optical characters of such minerals are given. For the present, we will assume that the mineral whose index of refraction is to be determined is isometric in its crystallization. There are two chief methods of determining the index of refraction by the use of a prism.

1. *The Method of Perpendicular Incidence.* — This method, although not the one most generally employed, is an excellent one to become acquainted with, as it may be used to advantage in some cases and from it the formula necessary for making the calculations is readily derived. It is necessary to have a prism of the mineral which has two plane surfaces meeting at a small angle.

This angle should be small enough so that the light may pass freely through the prism and not suffer any total reflection as it attempts to pass out into the air. For instance with fluorite in which $n = 1·434$, the prism angle must be less than $44° 12'$, for at this angle total reflection would take place. For a mineral of higher index the angle would have to be smaller still, as with diamond, $n = 2·419$, where total reflection would take place at $24° 24'$. On the other hand, more accurate results will be obtained if the prism angle is fairly near to the limit for the mineral being used.

Let Fig. 549 represent the cross section of such a prism. Let a–b represent a ray of light striking the face of this prism at $90°$ incidence. It will suffer

---

* See Posnjak and Merwin. J. Am. Chem. Soc. **44**, 1970, 1922. Tsuboi. Min. Mag., **20**, 108, 1923.
† Emmons. Am. Min., **13**, 504, 1928.

no deviation in its path on entering the prism but will proceed with somewhat diminished velocity until it reaches c. In passing out of the prism at this point, from a denser to a rarer medium, the light will be deflected away from the normal to the surface, P–P', making a deviation δ in the direction c–d. The data necessary for the calculation of the index of refraction under these conditions are the angle of the prism, α, and that of the deviation in the path of the light, δ. It is easy to see from the figure that α and α' are equal, for they are both parts of right-angled triangles having the angle $bP'c$ in common, and α'' is equal to α' because they are opposite angles. The angle of incidence, as defined in Art. **322**, is equal to $\alpha + \delta$ and the angle of refraction is equal to α. Therefore the usual formula $\dfrac{\sin i}{\sin r} = n$ becomes here $\dfrac{\sin \alpha + \delta}{\sin \alpha} = n$. In order to make

549

Refraction of Light through a Prism
Method of Perpendicular Incidence

a determination of the index of refraction, therefore, it is necessary to measure these two angles, α and δ.

The prism is mounted on a one-circle reflection goniometer and its angle α measured in the same way as an angle upon a crystal. The instrument is then adapted to the uses of a refractometer. For this purpose it is necessary to note that the telescope and vernier are both fastened to the outer rim of the instrument and move together. The graduated circle being clamped, the telescope tube is first moved to the position T', Fig. 550, so that the rays from the collimator tube, C, passing the edge of the prism, cause the light signal to fall on the vertical cross-hair of the telescope. The inner circle being clamped the telescope is next moved through an arc of exactly 60° to position T'' and then clamped. Next the

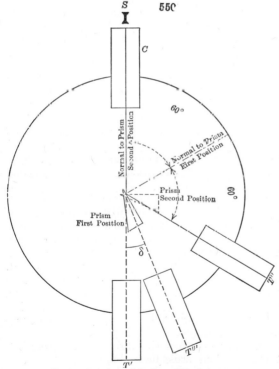

Determination of Index of Refraction
Method of Perpendicular Incidence

prism is turned to the first position so that the light from C is reflected from its right-hand face and the signal s falls on the cross-hair of T''. In this position the normal, N, to the prism face, must bisect the angle between the axes of C and T''. The prism is now turned through an angle of exactly $60°$ to its second position, which brings the normal N exactly in line with the axis of the collimator tube. When this has been accomplished the graduated circle is securely clamped. The telescope may now be unclamped and moved without altering the position of the prism, and somewhere between T' and T'' a position T''' will be found where the refracted ray falls on the cross-hair of the telescope. The movement of the telescope from the position T''' back to T' gives the angle of deviation, or δ, of the light-ray that has been refracted by the prism. In practice it is well to repeat the measurements both of α and δ several times and to go through all the operations of shifting the positions of the prism and telescope. If white light is used for illumination the refracted ray seen at T''' will appear as a narrow spectrum. To make an exact determination a monochromatic light (sodium light is best) must be employed.

2. *The Method of Minimum Deviation.* — This is the method that is most generally employed for determining indices of refraction by the use of prisms. It depends upon the principle that when a beam of light, *abcd*, Fig. 551, traverses a prism in such a way that the angles i and i' are equal, the beam suffers the minimum amount of deviation in its path of any possible course through the prism. This fact may be proven empirically by experimentation on the refractometer. In order to make a determination, the angle α of the prism is first measured on the goniometer. The angle of the prism with this method

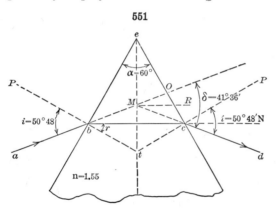

551

may be considerably larger than when the method of perpendicular incidence is used. The determination will increase in accuracy as the angle of the prism approaches its maximum possible size. The prism is then turned with its edge to the left about as in the position shown in Fig. 552, the telescope unclamped and moved until the refracted ray appears in it. Now, turn the central post with the prism on it toward the left and follow the signal with the telescope. The position of minimum deviation is soon reached, when, on turning the prism, the signal seems to remain stationary for a moment and then moves away to the right, no matter in which direction the prism is turned. A little practice is needed to determine exactly the position of minimum deviation and the measurement should be made in a monochromatic light. When the telescope is properly placed at this point the graduated circle is clamped and the telescope turned until the direct signal from the collimator tube is fixed upon the vertical cross-hair. The angle between these two positions of the telescope is the same as the angle of deviation, or δ. The formula for making the necessary calculation from these measurements follows very simply from a comparison of Figs. 551 and 549. It may be imagined that Fig. 551 is composed of two prisms like Fig. 549 placed back to back. This results in doubling the angles α and δ so that the formula now becomes

$$n = \frac{\sin \frac{1}{2}(\alpha + \delta)}{\sin \frac{1}{2}\alpha}.$$

3. *The Method of Total Reflection.* — This method is based upon the principle that light cannot always pass from an optically dense into an optically rarer medium but at a certain angle, known as the critical angle, will suffer total reflection. The critical angle for any substance varies with the index of refraction of that substance as explained in Art. **328**. Consequently if we can measure this critical angle we can calculate the index of refraction of the substance. This method is particularly useful because the measurement can be made upon a single polished surface, which may be quite small in area. This measurement is made by means of an instrument, known as the Total Refractometer, a description of which will be found in Art. **359**. The essential feature of this instrument is a hemisphere of glass with a known, high index of refraction. The upper surface of the hemisphere is plane and should be accurately adjusted in a horizontal position. The mineral to be tested may be of any shape provided that some surface upon it has been ground plane and polished. A drop of some liquid of high index of refraction is placed between the surface of the glass hemisphere and the flat surface of the mineral. This serves to unite the two substances and dispel the thin layer of air that would otherwise separate them. The liquid should have an index of refraction intermediate between that of the glass and that of the mineral. As the liquid lies between the two substances in the form of a thin film with parallel surfaces whatever optical effect it has upon the light as it enters will be balanced by the opposite effect as the light leaves the film. So the presence of the film of liquid can be ignored. Fig. 553 represents a cross section of such a hemisphere with a mineral plate resting upon it. Let it be now supposed that by

552

Colimator Tube

Incident Ray

Prism

Refracted Ray

Telescope

Determination of Index of Refraction
Method of Minimum Deviation

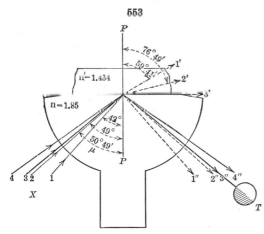

553

Determination of Index of Refraction
Method of Total Reflection, I

means of a mirror a beam of monochromatic light is thrown upon the apparatus from the direction of X. Rays 1 and 2 will suffer partial refraction at the dividing plane between the glass and the mineral to rays 1′ and 2′ and also partial reflection to rays 1″ and 2″. Ray 3 strikes the mineral at the critical angle for the combination of the glass and mineral and will in part be refracted at a 90° angle and emerge as ray 3′, just grazing the surface of the hemisphere. The greater part of ray 3 will however be reflected as ray 3″. Beyond this point, all the light must be totally reflected, thus 4 to 4″. If the optical axis of a telescope is now brought to the direction 3″, what appears to be a marked shadow will appear in the field of vision. One side will be illuminated by the total reflection of all rays beyond those of the critical angle while the other side will be distinctly darker since here a considerable amount of the light passed out into the mineral. The angle between the position of the shadow and the normal to the surface of the hemisphere, μ, Fig. 553, will be the critical angle for the combination of glass and mineral. As the index of refraction of the glass is known it is possible to calculate what the index of refraction of the mineral must be. If the mineral plate is transparent enough so that light may pass through it into the glass hemisphere another method of illumination may be used, as illustrated in Fig. 554. The reflecting mirror is so arranged that the light comes from the direction X. Rays 1 and 2 will be refracted to 1′ and 2′ and 3 which just grazes the surface to 3′. Beyond this point no light will pass into the hemisphere and a telescope placed with its axis along the line 3′ will show in its field a dark shadow. The contrast between the light and dark portions of the field, by this method of illumination, is much stronger than by the one first described. The telescope is so placed that the line of the shadow exactly divides the angle between the diagonal cross-hairs of the eyepiece. The telescope is attached to a graduated circle from which the angle μ can be directly read. With each of these instruments comes ordinarily a table giving the indices of refraction corresponding to the different possible values of μ. This table can easily be converted into a curve plotted on coördinate paper in such a way that the index of refraction for a particular angle can be read at a glance. Further, the calculation can be made having given the index of refraction of the glass of the hemisphere and the value of μ for a special mineral plate. Let $n′$ equal the index of refraction of the glass of the hemisphere and μ the critical angle measured; then the index of refraction of the mineral, $n = \sin \mu \times n′$.*

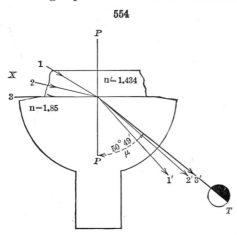

554

Determination of Index of Refraction
Method of Total Reflection, II

---

* The derivation of this formula follows. From the ordinary law for the index of refraction we have, $\dfrac{velocity\ of\ light\ in\ air}{velocity\ of\ light\ in\ mineral} = \dfrac{\sin i}{\sin r} = n$. But when the critical angle is

334. Dispersion. — Thus far the change in direction which light suffers in reflection and refraction has alone been considered. It is further true that the amount of refraction differs for light of different wave-lengths, being greater for blue than for red. In consequence of this fact, if ordinary light be passed through a prism, as in Fig. 551, it will not only be refracted, but it will also suffer *dispersion* or be separated into its component colors, thus forming the *prismatic spectrum*.

This variation for the different colors depends directly upon their wave-lengths; the red waves are longer, their transverse vibrations are slower, and it may be shown to follow from this that they suffer less change of velocity on entering the new medium than the violet waves, which are shorter and whose velocity of transverse vibration is greater. Hence the refractive index for a given substance is greater for blue than for red light. The following are values of the refractive indices for diamond determined by Schrauf:

> 2·40845 red (lithium flame).
> 2·41723 yellow (sodium flame).
> 2·42549 green (thallium flame).

335. Spectroscope. — The instrument most commonly used for the analysis of the light by dispersion is familiar to all as the *spectroscope*. There are a number of varieties of spectroscopes made, the simplest of which consists of a glass prism mounted at the center of the instrument with two tubes pointing away from it. The light from the given source is received through a narrow slit in the end of one tube and made to fall as a plane-wave (that is, as a " pencil of parallel rays ") upon one surface of a prism at the center. The light is dispersed by its passage through the prism and the spectrum produced is viewed through a suitable telescope at the end of the second tube.

If the light from an incandescent solid — which is " white hot " (Art. **319**) — is viewed through the spectroscope, the complete band of colors of the spectrum is seen from the red through the orange, yellow, green, blue, to the violet. If, however, the light from an incandescent vapor is examined, it is found to give a spectrum consisting of bright lines (or bands) only, and these in a definite position characteristic of it — as the yellow line (double line) of sodium vapor; the more complex series of lines and bands, red, yellow, and

---

reached $i = 90°$ and $\sin i = 1$. Therefore we may substitute and have

$$n = \frac{1}{velocity\ of\ light\ in\ mineral} \text{ or } velocity\ of\ light\ in\ mineral = \frac{1}{n}.$$ Further, we may derive

in the same way for the highly refracting glass of the hemisphere whose refractive index, n', is known, the expression, $velocity\ of\ light\ in\ glass = \frac{1}{n'}$. Further, we have in the case of the light attempting to pass from the glass (optically denser medium) into the mineral the expression, $\dfrac{velocity\ of\ light\ in\ mineral}{velocity\ of\ light\ in\ glass} = \dfrac{\sin 90°}{\sin \mu}$ (measured on instrument). By substituting this becomes

$$\frac{\frac{1}{n}}{\frac{1}{n'}} = \frac{\sin 90°}{\sin \mu} = \frac{1}{\sin \mu}$$

or

$$\frac{1}{\sin \mu} = \frac{n'}{n} \quad \text{or} \quad n = \sin \mu \times n'.$$

green, characteristic of barium; the multitude of bright lines due to iron vapor (in the intensely hot electric arc), and so on.

The monochromatic illuminator is an adaptation of the spectroscope so made that only a small portion of the spectrum, thus corresponding to practically one wave-length of light, is permitted to emerge through a narrow slit. In this way, light of any desired wave-length can be obtained.

336. Absorption. — Of the light incident upon the surface of a new medium, not only is part reflected (Art. **321**) and part transmitted and refracted (Art. **322**), but, in general, part is also *absorbed* at the surface and part also during the transmission. Physically expressed, absorption in this case means the transformation of the ether-waves into sensible heat, that is, into the motion of the molecules of the body itself.

The color of a body gives an evidence of this absorption. Thus a sheet of red glass appears red to the eye by *transmitted light*, because in the transmission of the light-waves through it, it absorbs all except those which together produce the effect of red. For the same reason a piece of jasper appears red by *reflected light*, because it absorbs part of the light-waves at the surface, or, in other words, it reflects only those which together give the effect of this particular shade of red.

Absorption in general is *selective* absorption; that is, a given body absorbs particular parts of the total radiation, or, more definitely, waves of a definite wave-length only. Thus, if transparent pieces of glass of different colors are held in succession in the path of the white light which is passing into the spectroscope, the spectrum viewed will be that due to the selective absorption of the substance in question. A layer of blood absorbs certain parts of the light so that its spectrum consists of a series of absorption bands. Certain rare substances, as the salts of didymium, etc., have the property of selective absorption in a high degree. In consequence of this, a section of a mineral containing them often gives a characteristic absorption spectrum.

This latter property may be made use of in testing certain minerals, more especially those that contain the rare earths or uranium. These give characteristic absorption bands in the spectrum. They may be tested by passing a strong white light through a thin section of the mineral and observing the resulting spectrum by means of a direct vision spectroscope. Often a better result will be obtained by illuminating the surface of the mineral and testing the reflected light for absorption bands. The light will have sufficiently penetrated the mineral before reflection to have had some of it absorbed. These tests can be made best by some sort of a microspectroscope, which will give a clear spectrum superimposed upon a scale of wave-lengths.*

The dark lines of the solar spectrum, of which the so-called Fraunhofer lines are the most prominent, are due to the selective absorption exerted by the solar atmosphere upon the waves emitted by the much hotter incandescent mass of the sun.

337. Diffraction. — When monochromatic light is made to pass through a narrow slit, or by the sharp edge of an opaque body, it suffers *diffraction*, and there arise, as may be observed upon an appropriately placed screen, a series of dark and light bands, growing fainter on the outer limits. Their presence is explained (see Arts. **341, 342**) as due to the interference, or mutual reaction,

---

* For details of this method of testing minerals see Wherry, Smithsonian Misc. Coll., **65**, No. 5, 1915.

of the adjoining systems of waves of light, that is, the initial light-waves, and further, those which have their origin at the edge or sides of the slit in question. It is essential that the opening in the slit should be small as compared with the wave-length of the light. If ordinary light is employed, the phenomena are the same, and for the same causes, except that the bands are successive colored spectra.

Diffraction spectra, explained on the principles alluded to, are obtained from diffraction gratings. These gratings consist of a series of extremely fine parallel lines (say, 15,000 or 20,000 to an inch) ruled with great regularity upon glass, or upon a polished surface of speculum metal. The glass grating is used with transmitted, and the speculum grating with reflected, light; the Rowland grating of the latter kind has a concave surface. Each grating gives a number of spectra, of the first, second, third order, etc. These spectra have the advantage, as compared with those given by prisms, that the dispersion of the different colors is strictly proportional to the wave-length.

338. Double Refraction. — As implied in Art. **325,** all crystallized substances may be divided into two principal optical classes, viz.: *isotropic*, in which light has the same velocity no matter what the direction of its vibration, and *anistropic*, in which the velocity of light in general varies with the direction of vibration. The anisotropic class is further divided into *uniaxial*, which includes crystals of the tetragonal and hexagonal systems, and *biaxial*, which includes crystals of the orthorhombic, monoclinic, and triclinic systems. The characters of these various optical classes will be explained in detail further on.

In the discussion of Art. **322,** applying to isotropic media, it was shown that light-waves passing from one medium into another, which is also isotropic, suffer simply a change in wave-front in consequence of their change in velocity. In anisotropic media, however, which include all crystals but those of the isometric system, there are, in general, two wave-systems propagated with different velocities and only in certain limited cases is it true that the light-ray is normal to the wave-front. This subject cannot be adequately explained until the optical properties of these media are fully discussed, but it must be alluded to here since it serves to explain the familiar fact that, while with glass, for example, there is only one refracted ray, many other substances give two refracted rays, or, in other words, show *double refraction.*

The most familiar example of this property is furnished by the mineral calcite, also called on account of this property " doubly-refracting spar." If *mnop* (Fig. 555) be a cleavage piece of calcite, and a ray of light meets it at *b*, it will, in passing through, be divided into two rays, *bc*, *bd*. For this reason, a dark spot or a line seen through a piece of calcite ordinarily appears double. As implied above, the same property is enjoyed by all crystallized minerals, except those of the isometric system. The wide separation of the two refracted rays by calcite, which makes the phenomenon so striking, is a consequence of the large difference in the values of its indices of refraction; in other words, as technically expressed, it is due to the *strength* of its double refraction, or its *birefringence.*

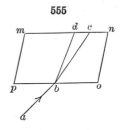

555

339. Double refraction also takes place in the anisotropic media just mentioned, in the majority of cases, even when the incident light is perpendicular to the surface. If the medium belongs to the uniaxial class (see p. 277, *et seq.*), one of the rays always retains its initial direction normal to the surface; but the other, except in certain special cases, is more or less deviated from it. With a biaxial substance, further, both rays are usually refracted and bent from their original direction. In the case of both uniaxial and biaxial media, however, it is still true that the normal to the wave-front remains unrefracted with perpendicular incidence.

340. Interference of Waves in General. — The subject of the interference of light-waves, alluded to in Art. **337,** requires detailed discussion. It is one of great importance, since it serves to explain many common and beautiful phenomena in the optical study of crystals.

Referring again to the water-waves spoken of in Art. **313**, it is easily understood that when two wave-systems, going out, for example, from two centers of disturbance near one another, come together, if at a given point they meet in the same phase (as crest to crest), the result is to give the particle in question a double amplitude of motion. On the other hand, if at any point the two wave-systems come together in opposite phases, that is, half a wave-length apart, the crest of one corresponding to the trough of the other, they interfere and the amplitude of motion is zero. Under certain conditions, therefore, two sets of waves may unite to form waves of double amplitude; on the other hand, they may mutually interfere and destroy each other. Obviously an indefinite number of intermediate cases lie between these extremes. What is true of the waves mentioned is true also of sound-waves and of wave-motion in general. A very simple case of interference was spoken of in connection with the discussion of the waves carried by a long rope (Art. **315**).

341. Interference of Light-waves. — Interference phenomena can be most satisfactorily studied in the case of light-waves. The extreme cases are as follows: If two waves of like length and intensity, and propagated in the same direction, meet in the same phase, they unite to form a wave of double intensity (double amplitude). This, as stated in Art. **316,** will cause an increase in the intensity of the light. If, however, the waves differ in phase by half a wave-length, or an odd multiple of this, they *interfere* and extinguish each other and no light results. For other relations of phase they are also said to interfere, forming a new resultant wave, differing in amplitude from each of the component waves. In the above cases monochromatic light-waves were assumed (that is, those of like length). If ordinary white light is used interference for certain wave-lengths may result with the consequent subtraction of the corresponding color from the white light and so give rise to various spectrum colors.

342. Illustrations of Interference. — A simple illustration is afforded by the bright colors of very thin films or plates, as a film of oil on water, a soap bubble, and like cases. To understand these, it is only necessary to remember that the incident light-waves are reflected in part from the upper and in part from the lower surface of the film or plate. The rays that are reflected from the under surface of the very thin film (see Fig. 556) having traveled a greater distance and with a different velocity will, when they unite with those rays reflected from the upper surface, show in general a different phase. For

556

some particular wave-length of light this difference is likely to be exactly a half wave-length or some odd multiple of this amount and so the corresponding color will be eliminated (assuming that ordinary white light is being used) and its complementary color will be seen. It is to be noted that the phenomena of interference by reflection are somewhat complicated by the fact that there is a reversal of phase (that is, a loss of half a wave-length) at the surface that separates the medium of greater optical density from the rarer one. Hence the actual relation in phase of the two reflected rays, as AC, BD (supposing them of the same wave-length) is that determined by the retardation due to the greater length of path traversed by BD, together with the loss of a half wave-length due to the reversal of phase spoken of. As shown in the figure, there are also two transmitted waves which also interfere in like manner.

A plano-convex lens of long curvature, resting on a plane glass surface (Fig. 557), and hence separated from it, except at the center, by a film of air of varying thickness, gives by reflected monochromatic light a dark center and about this a series of light and dark rings, called *Newton's rings*. The dark center is due to the interference of the

557

incident and reflected waves, the latter half a wave-length behind the former. The light rings correspond to the distances where the two sets of reflected waves meet in the *same phase*, that is (noting the explanation above) where the retardation of those having the longer path is a half wave-length or an odd multiple of this ($\frac{1}{2}\lambda$, $\frac{3}{2}\lambda$, $\frac{5}{2}\lambda$, etc.). Similarly the dark rings fall between these and correspond to the points where the two waves meet in opposite phase, the retardation being a wave length or an even multiple of this. The rings are closer together with blue than with red because of the smaller wave-length of blue light. In each of the cases described the ring is properly the intersection on the plane surface of the cone of rays of like retardation.

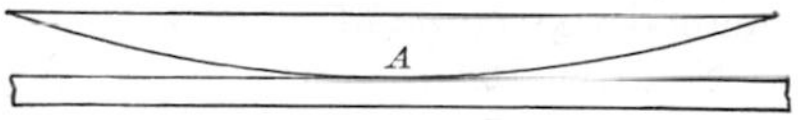

In ordinary white light we get, instead of the alternate light and dark rings described above, a series of colored bands. If the illumination was originally by sodium light the position of the dark rings indicates where light for that particular wave-length has been extinguished through interference. When white light is used the conditions in respect to its component having the yellow sodium-light wave-length have not changed and this light will still be eliminated at the same points, but now, instead of dark rings, we get rings having the complementary color blue. If our original illumination was by means of a red light the dark rings would have had different positions from those produced in sodium light. And again when white light is used red light is eliminated at those points and its complementary color shows. In this way we obtain a series of *colored rings*, each showing the successive colors of the spectrum. The series of the spectrum colors are repeated a number of times due to successive interferences produced by differences of phase of $\frac{1}{2}$, $1\frac{1}{2}$, $2\frac{1}{2}$, etc., wave-lengths. The different series are distinguished as of the first, second, third, etc., order; for a given color, as red, may be repeated a number of times. The interference rings for different colored lights are not evenly spaced, the rings shown in blue light being, for instance, closer together than for red. Consequently after three or four repetitions of the spectrum bands the different interference rings begin to overlap one another and the resulting colors become fainter and less pure. Ultimately this overlapping becomes so general that the effect of color is lost and white light, the so-called white of the higher orders, is shown.

Another most satisfactory illustration of the interference of light-waves is given by means of the diffraction gratings spoken of in Art. **337.**

Other cases of the composition of two systems of light-waves will be considered after some remarks on polarized light.

343. Polarization and Polarized Light. — Ordinary light is propagated by transverse vibrations of the ether which may take place in any direction as long as it is at right angles to the line of propagation. The direction of vibration is constantly changing and the resulting disturbance of the ether is a complex one. A ray of ordinary light will be symmetrical, therefore, only to the line of its propagation.

Plane-polarized light, on the other hand, as stated briefly in Art. **316,** is propagated by ether-vibrations which take place *in one plane only.* The change by which ordinary light is converted into a polarized light is called *polarization,* and the plane at right angles to the plane of transverse vibration is called the *plane of polarization.* *

Polarization may be accomplished (1) by reflection and by single refraction, and (2) by double refraction.

344. Polarization by Reflection and Single Refraction. — In general, light which has suffered reflection from a surface like that of polished glass is

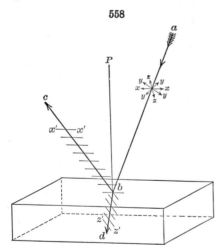

558

more or less completely polarized; that is, the reflected waves are propagated by vibrations to a large extent limited to a single plane, viz. (as assumed), the plane normal to the plane of incidence, which last is hence the plane of polarization. Furthermore, in this case, the light transmitted and refracted by the reflecting medium is also in like manner partially polarized; that is, the vibrations are more or less limited to a single plane, in this case a plane at right angles to the former and hence coinciding with the plane of incidence. For instance, in Fig. 558, let *a–b* represent an incident light ray in which the vibrations are taking place in all possible transverse directions as represented by the arrows, *x–x, y–y,* and *z–z.*

When this ray strikes the polished surface at *b* light with vibrations parallel to *x–x* will be reflected along *b–c* and other vibrations near to *x–x* in direction will be shifted to this direction so that the reflected ray will be largely polarized. In a similar manner the light having *z–z* vibrations will enter the transparent substance as the refracted ray *b–d* and other vibrations will be shifted to this direction so that the refracted ray is also largely polarized and in a plane at right angles to that of the reflected ray. Light reflected

---

* It is necessary to keep clear the distinction between the *plane of polarization* and the plane in which the vibrations take place. All ambiguity is avoided by speaking uniformly of the *vibration-plane* of the light.

from a polished and transparent surface is not completely polarized but there is an angle of incidence for every substance at which the amount of polarization will be at its maximum. This will happen, as illustrated in Fig. 559, when the angle between the reflected and refracted rays AB and AC equals 90°. It is evident from a consideration of the figure that the angle r is the complement of i; hence the formula $\dfrac{\sin i}{\sin r} = n$ becomes in this case

$$\frac{\sin i}{\cos i} = \tan i = n.$$

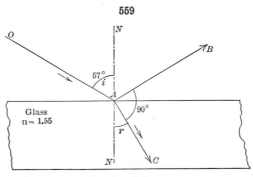

559

Brewster's Law

This law, established by Brewster, may be stated as follows:

The angle of incidence for maximum polarization is that angle whose tangent is the index of refraction of the reflecting substance. For crown glass this angle is about 57° (see Fig. 559). If light suffers repeated reflections from a series of thin glass plates, the polarization is more complete, though its intensity is weakened. Metallic surfaces polarize the light very slightly.

345. Polarization by Double Refraction. — When light in passing through a crystalline medium is doubly refracted (Art. **338**) or divided into two sets of waves, it is always true that both are completely polarized and in planes at right angles to each other. This subject can only be satisfactorily explained after a full discussion of the properties of anisotropic crystalline media, but it may be alluded to here since this principle gives the most satisfactory method of obtaining polarized light. For this end it is necessary that one of the two wave-systems should be extinguished, so that only that one due to a single set of vibrations is transmitted. This is accomplished by natural absorption in the case of tourmaline plates and by artificial means in the nicol prisms of calcite.

346. Polarized Light by Absorption. — Light passing through a strongly colored but transparent thin section of a tourmaline crystal — the section being cut parallel to the vertical crystallographic axis — will be almost completely polarized. This can be easily demonstrated in the following way. Select a polished floor surface, or a table top and stand in such a position that light from a window is reflected from the polished wood to the eye. Look at this reflected light through the tourmaline section, holding it first with the direction of the c crystal axis in a horizontal position and then turning the section until the c axis becomes vertical. The light passing into the tourmaline section is in considerable part polarized through its reflection from the wood surface and possesses a horizontal vibration direction. It will be noted that when the c axis of the tourmaline is horizontal the section readily transmits light but when this axis is vertical the section becomes practically opaque. The crystal structure of the tourmaline is such that light entering it is broken up into two rays (*i.e.*, it is doubly refracted), one of which has its vibrations parallel to the c axis, while the vibrations of the other lie in the plane of the horizontal crystal axes. From the foregoing experiment it is obvious that the light vibrating parallel to the c axis is readily transmitted by the crystal but that the other ray, vibrating in the horizontal axial plane, is almost completely absorbed. Under these conditions it is clear that the transmitted light belongs almost wholly to one ray, the vibrations of which take place in a single direction. In other words, the light transmitted by such a tourmaline section is *polarized*.

If two such sections of tourmaline are available it is instructive to make the following experiment with them. Place them together, first with their c axes parallel to each other, and then turn one section upon the other until these axes are at right angles to each other. In the first case, the light comes through the sections because the vibration planes of the transmitted rays in the two sections are parallel to each other. In the second case, all light is cut off because now these two vibration planes are at right angles to each other, the light that did get through the first section being wholly absorbed in the second.

347. Polarized Light by Double Refraction. — Calcite, as already stated in Art. **338,** possesses in an unusual degree the power to doubly refract light. If we take a cleavage block of clear calcite (Iceland spar) and look at an image through it, such as a dot or line drawn on a piece of paper, the image will appear double. If we take a card and make in it a pinhole, place the card upon one face of a cleavage rhombohedron and, looking through the calcite, hold it up against a source of light, we will observe two bright dots. Now if we look in the same way at the light reflected from a polished wooden surface, as described in the preceding article, we will find that when a line bisecting the acute angles of the rhombic face of the cleavage block is horizontal one of these images is bright while the other is almost invisible. If we then turn the block so that the line bisecting the obtuse angles of the rhombic face is horizontal the first image will fade while the second becomes bright. Remembering that the light reflected from the polished wooden surface is largely polarized with a horizontal vibration direction, it becomes evident from this experiment that the two rays into which the light is broken up in passing through the calcite are polarized and that their planes of vibration are at right angles to each other and respectively bisect the angles of the rhombic face of the cleavage block. As the double refraction of calcite is strong, it follows that the indices of refraction of the two rays show considerable differences. This fact is taken advantage of in constructing a prism from calcite in such a way as to wholly eliminate one of these rays and so, as only the other ray can come through the prism, effectively polarizing the light that emerges.

The prism referred to above is called the *Nicol Prism* or simply the *nicol*. A full explanation of the nicol cannot be made at this time, as there would be required a knowledge of the optical properties of hexagonal crystals, but a description may be given enabling one to understand its construction and uses. In Fig. 560 is represented a cleavage rhombohedron of calcite with its edges vertical. Let d represent a point of light underneath the rhombohedron. Light coming from d will be broken into two rays whose paths through the rhombohedron are shown by the lines o and e. As shown above, these two rays are polarized, with vibration directions as indicated by the double arrows in the top view in Fig. 560. In the construction of a nicol, the top and bottom surfaces of such a cleavage rhombohedron are ground and polished so that they make angles of 68° with the vertical edges. Then the block is cut in two along the diagonal a–f, as shown in Fig. 561. These two surfaces, after being polished, are cemented together by means of a thin layer of Canada balsam.   Let us assume that a ray of light enters the prism from below, as shown in Fig. 561.   It is broken up into the rays *o* and *e*. The ray *o* travels with the slower velocity, has therefore the higher index of refraction, and shows a greater deviation from the original path.   The Canada balsam has a lower index of refraction than ray *o*, which, therefore, when it strikes the layer of balsam, is attempting to pass from an optically dense into a rarer medium.   The construction of the prism is such that this ray meets the layer of balsam at an angle greater than the critical angle for this optical combination and suffers therefore total reflection toward the side of the prism, and will be absorbed by whatever fastening holds the nicol.

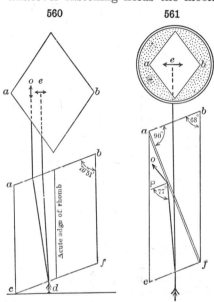

Nicol Prism

The second ray *e* passes through the prism with almost no deviation from its original course.   Its index of refraction and that of the Canada balsam are nearly the same, hence the ray suffers almost no deflection at this point and passes out of the upper face of the prism.   The light, therefore, that emerges from a nicol belongs wholly to one ray and is all vibrating parallel to the shorter diagonal of the rhombic end surface.   It should be noted, however, that some prisms are made in a different way and that the above statement concerning the plane of vibration of the light emerging from the prism may not always hold true.   It is always wise to test the plane of vibration of a nicol by looking through it at the floor or a table top as previously described.   The prism will show bright when its plane of vibration is horizontal, thus corresponding to the plane of vibration of the reflected light.

348. Polariscope.   Polarizer.   Analyzer. — The combination of two nicols, or other polarizing contrivances, between which transparent mineral sections may be examined in polarized light is called, in general, a *polariscope*, the common forms of which are described later.   In any polariscope the lower prism, or other contrivance, which polarizes the light given from the outside source is called the *polarizer;* the upper prism is the *analyzer*.   If these prisms have their vibration-planes at right angles to each other, they are said to be *crossed;* the incident light polarized by the polarizer will then be extinguished by the analyzer; briefly, under these conditions it is said to suffer *extinction*.

349. Interference of Plane-polarized Waves.   Interference Colors. — When sections of doubly refracting minerals are examined in polarized light certain interference effects are commonly obtained that are of great importance.   As shown in Art. **347,** calcite when it doubly refracts light also polarizes the two rays and in planes that are at right angles to each other.   In general, this is true of sections of doubly refracting minerals.   Consider, then, what takes place when a general section of a doubly refracting mineral is placed in a polariscope between the polarizer and analyzer the planes of vibration of which are at right angles to each other. In Fig. 562 let the rectangular outline represent such a section. The double arrows marked o and e show the two possible directions of vibration of light in the section. The direction $P-P'$ represents the plane of vibration of light which emerges from the polarizer below and $A-A'$ shows the direction in which light must vibrate when it emerges from the analyzer above. In the first case to be considered the directions o and e are taken as parallel to $P-P'$ and $A-A'$ respectively. The light that enters the section from below must all vibrate parallel to the direction $P-P'$. It enters the mineral section and must vibrate there as the ray labeled o. There will be no ray in the mineral vibrating parallel to the direction e, as a vibration parallel to o cannot be resolved into another at right angles to it. The light will leave the section, therefore, still vibrating parallel to $P-P'$ and enter the analyzer above. It will, however, be entirely reflected in the analyzer at the layer of balsam since only light vibrating parallel to

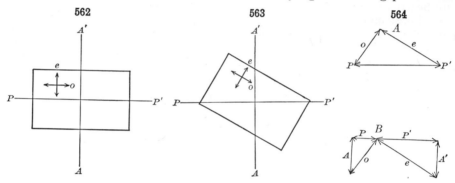

562   563   564

$A-A'$, which is at right angles to $P-P'$, can emerge from the analyzer. Consequently, when such a section has its planes of vibration parallel to those of the polarizer and analyzer, the section will appear dark. The same reasoning holds true when the section is turned to a position at 90° from the first. Consequently with such a section there are four positions at 90° to each other in which it appears dark during its complete rotation upon the stage of the polariscope. At such positions the section is said to be *extinguished*.

Next consider what happens when the vibration directions of the section are at oblique angles to those of the polarizer and analyzer. In Fig. 563 let o and e represent the directions of vibration in a section which makes some oblique angle with the directions $P-P'$ and $A-A'$. In Fig. 564 A let the line $P-P'$ represent the direction and amplitude of the vibration of the light entering the mineral section having come through the polarizer below. The light must vibrate in the mineral in directions parallel to o and e, Fig. 563. The vibration $P-P'$ will therefore be resolved into two vibrations at right angles to each other which will be parallel respectively to o and e. In Fig. 564 A the lines o and e representing the direction and amplitudes of such vibrations are found by the application of the principle of the parallelogram of forces. The two rays emerge from the mineral section vibrating in these two planes and enter the analyzer above. Since the planes of vibration in the analyzer are parallel to $A-A'$ and $P-P'$ these two rays o and e will resolve each into two new rays which will vibrate now parallel to $A-A'$ and $P-P'$. The two rays labeled P and P' in Fig. 564 B will be absorbed by the analyzer but the rays marked A and A' will emerge and meet the eye. The section in this position, therefore, will be illuminated. Consequently the section will be illuminated in all possible positions in which the directions of vibration of the light in the mineral make inclined angles with the directions of vibration of the polarizer and analyzer. It is easy to prove that this illumination will be at its maximum when the angle between the directions o and e and A–A' and P–P' is 45°. In addition to being illuminated, the section, if thin, will also be colored. This interference color, as it is called, of mineral sections when examined in a polariscope, now needs explanation.

The amount of refraction which any ray of light suffers on entering a mineral depends upon two things, namely, the angle of incidence at which the light enters and the index of refraction of the mineral. In the case of a doubly refracting mineral we have a light-ray entering the section at a given angle of incidence and then being broken up into two polarized rays which have different angles of refraction and so travel different paths. Consequently the

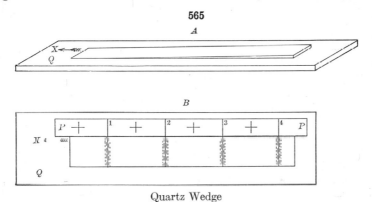

Quartz Wedge

indices of refraction for these rays must be different and from this it follows that the two rays must have different velocities and will therefore emerge from the mineral in different phases. Light-waves having different phases will in a greater or less degree interfere with each other and in case of light of certain wave-lengths, i.e., light of some particular color, the interference may lead to extinguishment of that particular wave-length. If one particular color is subtracted in this way from white light the result will be to produce the complementary color and under such conditions the section will no longer be white but colored. The color of thin sections of minerals when seen under the polariscope is known as their *interference color*. To develop this subject further use will be made of an accessory of the microscope known as the *Quartz Wedge*.

The quartz wedge consists simply of a very thin tapering wedge the faces of which are approximately parallel to the prism of a quartz crystal. It is mounted on a narrow glass plate, Fig. 565 A. The plate is generally marked with the letter Q (quartz) and with an arrow. If the wedge is cut, as is usually the case, with its longer direction at right angles to the vertical axis of a quartz crystal, the arrow is marked X (or $\mathbf{a}$), which indicates that of the two directions of vibration of light in the wedge the one which is parallel to this direction is that of the ray which is propagated with greater velocity. Some wedges are cut with their longer direction parallel to the vertical axis of quartz, and the arrow in this case would be marked Z (or c), which indicates that this is the direction of vibration of the slower ray. It is absolutely essential that the optical orientation of the wedge be known.

The quartz wedge furnishes a prismatic section of varying thickness and of known orientation and may be used to study the effects of polarized light on plates (short sections of the wedge) of different thicknesses. Take the simplest form of polariscope, a combination of polarizer and analyzer without lenses, and arrange it so that the vibration planes of the instrument are crossed. Illuminate with ordinary light and on the stage of the instrument place a quartz wedge with its X direction parallel to the plane of vibration of the polarizer. The light in entering the quartz will vibrate parallel to the X direction and without changing its plane of vibration will pass through the quartz and up into the upper nicol where it will suffer total reflection. Hence the wedge in this position will appear dark throughout its length. A similar result will be obtained when the X direction of the wedge is placed parallel to the vibration plane of the analyzer. But if the wedge is turned so that its X direction makes an angle of about 45° with the plane of vibration of the polarizer the wedge will exhibit a series of beautiful interference colors, arranged in transverse bands, the nature of which will be discussed in a later paragraph. If the wedge is turned from this 45° position the colors become less and less brilliant as the position of extinction is neared.

As preliminary to another experiment, paste a narrow strip of paper, $P-P$, Fig. 565 B, on the top, but to one side, of a quartz wedge. Place this on the stage of a polariscope (without lenses) and illuminate with diffused sodium light. When the wedge is examined under these conditions it will be found that it shows extinction when its vibration directions are parallel to those of the polariscope but at the 45° position it will show transverse dark bands upon a yellow field. The number of these bands will depend upon the thickness of the wedge; usually there will be two or three, although for this experiment it is interesting to have a longer and proportionally thicker wedge than those commonly supplied, so as to have more bands appearing. Mark on the strip of paper the position of each band, as illustrated in Fig. 565 B and number them, starting at the band nearest the thinner end of the wedge. The number 1 band marks the place where the faster of the two rays, into which the quartz breaks up the sodium light, has gained exactly one wave-length in its phase over the slower ray. At the point marked 2 the gain is two wave-lengths, etc.

In explaining the phenomenon just described, reference is made to Fig. 566 in which it is assumed that $P-P'$ is the plane of the polarizer and $A-A'$ is the plane of the analyzer, and a quartz wedge is between them at such an angle that the direction of the vertical crystal axis lies parallel to $C-C'$. If we explain the action of light in the wedge in a purely mechanical way we may say, let the amplitude of vibration of an ether particle before the light has entered the wedge be represented in the figure by the line $O-p$. The vibration may be likened to that of a pendulum, swinging back and forth from p to p'. If the impact, or disturbance, of an ether particle is communicated to the ether particles of the quartz when it is at O at the middle of an oscillation from p to p', there will result two disturbances, one to r parallel to $C-C'$ and the other to s at right angles thereto. The amplitude of the vibrations represented by $O–r$ and $O–s$ are determined by the parallelogram of forces, as indicated by the dotted lines in the figure.   During the passage of these two rays through the quartz the one whose vibrations are represented by $s–s'$ travels the faster and it is assumed that the thickness of the quartz wedge at the place under consideration is such that, on emerging, this ray is just one wave-length ahead of the one whose vibrations are parallel to $r–r'$.   Now, when one ray is exactly one wave-length ahead of another (it may be two, three or any exact number of wave-lengths) the conditions are such that, at the middle of the vibration, when an ether particle of the ray $s–s'$ is just starting from O to s, an ether particle of the ray $r–r'$ will be just starting from O to r.   Now consider the effects produced by the simultaneous impacts in the directions O to s and O to r upon the ether particles of the calcite constituting the analyzer.

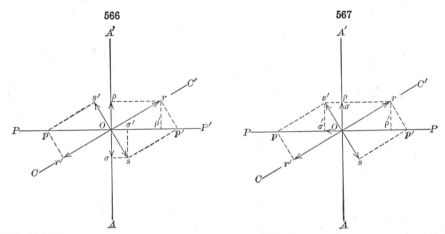

A vibration from s' to s acting at O will displace the ether particles of the calcite to σ and σ'.   Likewise a vibration from r' to r acting at O will displace the ether particles to ρ and ρ'.   Two of these resulting disturbances, namely $O–\sigma'$ and $O–\rho'$, are easily disposed of, for being in the plane $P–P'$ their effects cannot pass beyond the layer of Canada balsam in the nicol.   The other disturbances $O–\sigma$ and $O–\rho$ are both in the plane $A–A'$ and could emerge from the nicol, but since the ether particles at O are acted upon simultaneously by forces of equal magnitude acting in opposite directions no disturbance can take place and under these conditions the section is dark.   From the above it follows that, when a section of a doubly refracting mineral is observed between crossed nicols with its vibration planes making some oblique angle with the vibration planes of the nicols, complete interference will take place for some particular wave-length of light whenever the two polarized rays corresponding to this color emerge from the section in the same phase.

It is well to consider next the effects that result when, with the planes of vibration of the nicols crossed, light travels through such thicknesses of the quartz wedge that one ray gains $\frac{1}{2}$, $\frac{3}{2}$, or some other half wave-length over the second ray.   Let it be supposed, Fig. 567, that at O, the middle of an oscillation from p to p', the impact is communicated to the ether particles of a quartz section the vertical crystal axis of which lies parallel to the direction $C–C'$.   There will result two disturbances in the quartz, one from O to r and the other from O to s.   After traversing the section the phases of the two rays differ by one-half wave-length so that when the direction of the first oscillation is from O to r, that of the other will be from O to s'.   The impulse $O–r$ gives rise in the analyzer to two disturbances $O–\rho$ and $O–\rho'$.   The impulse $O–s'$ results in the two displacements $O–\sigma$ and $O–\sigma'$.   Of these disturbances $O–\rho'$ and $O–\sigma'$ do not extend beyond the layer of Canada balsam of the analyzer, while $O–\rho$ and $O–\sigma$, both of equal magnitude and vibrating in the plane $A–A'$, are additive and give rise to a disturbance and the sensation of light. Hence, in the experiment with the quartz wedge in sodium light, there are areas of light between the dark bands, Fig. 565 B.

An instructive experiment with the wedge should also be tried with sodium light illumination but with the planes of vibration of the polari-

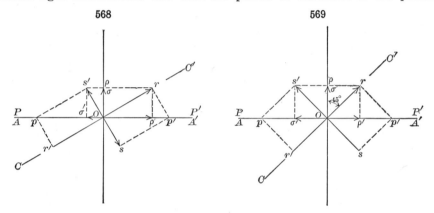

scope parallel to each other instead of crossed as in the previous cases.   If light traverses such a thickness of quartz that, on emerging, one ray has gained one-half of a wave-length over the other the conditions up to the time the vibrations enter the analyzer will be the same as in the previous case. The vibrations, however, which can now pass through the analyzer result, Fig. 568, from the disturbances $O–\rho'$ and $O–\sigma'$, and these acting on an ether particle in opposite directions but with unequal force would produce a disturbance in the direction $O–\rho'$ and, therefore, give rise to the sensation of light.   A wedge with the direction of the vertical crystal axis about parallel to $C–C'$ will appear yellow throughout its entire length.   This will not be the case, however, if the wedge is turned so that the vertical axis makes an angle of 45° with the plane of polarization, Fig. 569, for then the forces acting upon an ether particle at O are $O–\rho'$ and $O–\sigma$, which, being equal and in opposite directions, will neutralize each other and therefore will not produce any sensation of light.   A wedge in the 45° position will therefore show a series of dark bands, the first, starting from the thin end of the wedge, being where one ray has gained $\frac{1}{2}$ wave-length, the second where it has gained $\frac{3}{2}$ wave-lengths, etc., over the second ray.   In Fig. 565 B, the positions of the bands in this experiment are indicated by the crosses marked on the strip of paper pasted upon one side of the quartz wedge.   The lines and crosses on this paper strip indicating gains of whole and one-half wave-lengths for yellow light may now serve as starting points for further considerations.

For the next experiment use a microscope with crossed nicols, a number 3 or

4 objective, and illuminate with ordinary light; place the wedge in the 45° position and focus on that part of it opposite the first line drawn on the paper strip.   The field will show at its center a blue color, about at the point where it is beginning to merge into red.   A moment's consideration will indicate what this color really is.   It is a mixture of all colors of the spectrum except yellow.   That this is the case may be proved by analyzing the blue by means of a small direct-vision spectroscope.   This will show a spectrum through which runs a dark band between the red and green, that is, where the yellow would normally appear.   The blue of the wedge at this point is therefore the complement of yellow, which has been made to disappear by interference. Next focus the microscope on the wedge opposite the second line.   Here the color will be nearly a sky blue, with perhaps a tinge of green.   Upon analysis with the spectroscope again a dark band will be found in the yellow, this time due to interference brought about by a difference in phase of two wave-lengths for sodium light.   Proceeding next to opposite the third line the color will be found to be a light green, which on analysis shows a band where yellow should occur and a perceptible shortening of the spectrum, especially by cutting off the extreme blue and violet.   Opposite the fourth line the color would be a very pale green which upon analysis with the spectroscope would show two dark bands, one in the yellow and another in the blue.   The pale green color is therefore due to a mixture of red, green, and violet.   If, in the original experiment the wedge had been illuminated by a monochromatic blue light it would have been found at the thicker end of the wedge, where the fourth band for yellow light was located, there would have been a fifth band for the blue light.   Consequently the interference color at this point of the wedge is equivalent to white light from which both yellow and blue have been subtracted   If a wedge of extra length was available for study it would have been noted that opposite the eighth band for sodium light the color showing, when the wedge is studied in the polarizing microscope, was white.   This upon analysis would show a spectrum crossed by bands in the red, yellow, green, and blue.   In other words, in traversing the thickness of the quartz at this point, the faster ray has gained for red seven wave-lengths over the slower ray, for yellow eight, for green nine, for blue ten.   The white polarization effect seen when looking at this point with the microscope is known as white of the higher order.   It is a mixture of the several primary colors of the spectrum, some portions of all of which are present and combine to give the effect of white.

It is important to study carefully the polarization colors of the quartz wedge under the microscope, using Fig. 570 as a guide.   It will be noted that the colors occur in general in the following order as the thickness of the quartz increases: violet, blue, green, yellow, orange, red.   This sequence of colors is repeated quite distinctly three times and then as the thicker end of the wedge is approached the colors become fainter and not so clear.   This series of interference colors is divided into orders as indicated in Fig. 570.   It is to be noted that at the very thin end of the wedge before any interference can have taken place the color is white.   Also the thicker end of the wedge is white because here there is an overlapping of the various points of interference of the different colors.   The thickness of the wedge at the different points is given in millimeters in Fig. 570.

350. Sensitive Tint. — Among the accessories of the polarizing microscope is a thin plate of gypsum mounted between two plates of glass.   It is commonly marked S. T. and also with an arrow marked either X (a) or Z (c), indicating respectively the direction of vibration of the faster or slower ray. If this is placed on the microscope stage in the 45° position with the nicols crossed, the interference color shown is reddish violet, the same as that close to the red of the first order of the quartz wedge. It is an interesting experiment to first put a quartz wedge under the microscope and focus on the red-violet, just beyond the red of the first order and then cover it with the sensitive tint arranged in such a way that its X direction is at right angles to the

570

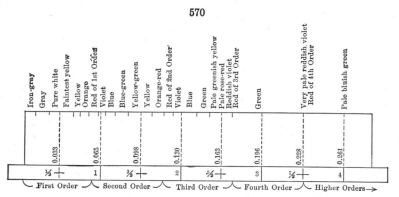

Interference Colors with Quartz Wedge

X direction of the quartz wedge. The resulting color will be gray. The explanation of this is simple. Whatever gain the faster ray had made over the slower in passing through the quartz has been overcome or neutralized by passing through a layer of gypsum of opposite optical orientation and of suitable thickness to produce the same interference as the quartz. The name *Sensitive Tint* is given to this gypsum plate because a slight increase of the double refraction which it shows will give a blue color while a corresponding slight decrease will change the color to yellow. Numerous uses of the sensitive tint will be given in subsequent articles.

351. Interference Colors of Mineral Sections. — The interference colors of mineral sections depend upon three things.

1. On the strength of the birefringence of the mineral, or in other words upon the amount of double refraction that the mineral shows. The greater the birefringence the higher the order of interference color, the other influencing factors remaining constant.

2. The thickness of the section. The thicker the section the greater will be the amount of double refraction and consequently the higher the order of the interference color.

3. The crystallographic orientation of the section. This will be explained later when the optical characters of the different crystal systems are described.

352. Determination of the Order of the Interference Color of a Mineral Section. — It is often important to determine to which order (see last paragraph of Art. **349**) the interference color of a given section belongs. If, as is often the case, the section has somewhere a tapering wedge-like edge, the successive bands of color shown there can be counted and the order of the color of the surface of the section determined. In other words the order of the color can be told in the same way as upon the quartz wedge itself. If such an edge cannot be found the quartz wedge is used as described below.

Suppose a certain mineral section showed an interference color of orange-red and it was desired to ascertain whether this color belonged to the first or second order. Under the microscope with crossed nicols find a position of extinction of the section and then turn it upon the stage of the microscope through an angle of 45°. By doing this the vibration directions of the section are brought into such a position that they make angles of 45° with the vibration directions of the polarizer and analyzer. Then insert above the section and below the analyzer a quartz wedge, the optical orientation of which is known. The thin end of the wedge should always be inserted first. A slot running through the microscope tube just above the objective and making an angle of 45° to the cross-hairs is provided for this purpose.

Under these conditions there are two possibilities. Either the optical orientations of the section and the quartz wedge agree; *i.e.*, the X direction of the section is parallel to the X direction of the wedge, or these two directions are at right angles to each other. The effect of the introduction of the wedge above the section will be either to increase or decrease the amount of double refraction of the light due to the mineral section. If the double refraction is increased, the optical effect will be as if the mineral section had been thickened and in this case its interference color will rise in its order. On the other hand, if the double refraction of the light is decreased by the introduction of the quartz wedge the effect will be as if the mineral section had been thinned and the interference color will fall in its order. In the first case the red interference color of the section would be changed as the wedge is pushed in, first to blue and then to green. In the second case it would change to orange, then to yellow and green. Arrange the section, therefore, so that upon the introduction of the quartz wedge the interference color will fall in its order. Then gradually continue to push in the wedge, noting the successive colors that occur as the amount of the double refraction is decreased. Finally the point will be reached where the thickness of the wedge will give practically the same amount of double refraction as the mineral section. The two having opposite optical orientations the result will be to eliminate all interference and a gray color of the first order will result. When this condition arises the quartz wedge is said to *compensate the mineral*. By noting the succession of colors that occurs until this point is reached the order of the original color of the section can be determined.

353. Determination of Strength of Birefringence. — The birefringence, or amount of double refraction, varies with different minerals. It is expressed numerically by a figure that is the difference between the greatest and least indices of refraction of a given mineral. In the case of calcite, for instance, the index of refraction for one ray is 1·486 and for the other is 1·658. The birefringence of calcite therefore equals 0·172. This is much higher than for most minerals, the strength of the birefringence of quartz being only 0·0091. An accurate estimation of the strength of the birefringence of a mineral is to be made only by determining the greatest and least indices of refraction. An approximate determination, however, can often be made in a thin section under the microscope. The order of the interference color of a section, as stated in Art. **351**, varies with the thickness of the mineral, its crystallographic orientation and the strength of its birefringence. If the first two factors are known the birefringence can be estimated by noting the interference color of the section. Fig. 571 will aid in this determination. The thickness of the section is shown in the column at the left. The strength of the birefringence is expressed along the top and right-hand side of the figure. Suppose that a given section was 0·03 mm. in thickness and showed an orange-red interference color of the first order. By following the diagonal line that crosses the horizontal line marked 0·03 mm. at a point lying in the middle of the orange-red of the first order it will be seen that the birefringence of the mineral must be about 0·015. This method of determining birefringence is most commonly used in the case of minerals observed in rock sections. In the case of the best

571

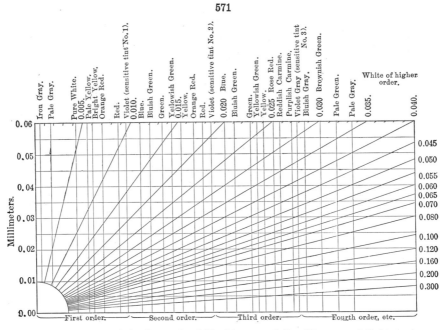

Determination of the Strength of Birefringence (after Pirsson and Robinson)

rock sections the thickness of the section is usually about 0·03 to 0·04 mm. The thickness of the section can also be judged from the interference color shown by some known mineral, like quartz or feldspar, which is to be observed in the section. As the strength of the birefringence of a mineral varies with its crystallographic orientation it is necessary always to look over the rock section and use in the observations that section of the mineral which shows the highest order of interference color. The birefringence of a mineral is always expressed as the maximum difference between the indices of refraction. Consequently, with a uniform thickness, such as is obtained in a rock section, that section of a mineral which shows the highest order of interference color most nearly approaches the proper orientation for the maximum birefringence.

The order of the interference color of a given section is to be determined by the method of compensation as explained in Art. **352**. Special quartz wedges are made with scales upon them giving the birefringence produced by the varying thicknesses of the wedge. Such a wedge, described by Wright, consists of a wedge of quartz placed on top of a plate of quartz, the two having opposite optical orientations, *i.e.*, the Z direction in the plate being parallel to the X direction in the wedge. The thickness of the plate is intermediate between the thicknesses of the two ends of the wedge. At the point, therefore, where the thickness of the wedge equals that of the plate, there will be a dark line showing compensation when the combination is placed between crossed nicols. This point is marked zero on the scale engraved on the wedge. When the wedge is placed above a mineral section, this dark line will be displaced by a distance proportional to the birefringence of the mineral. The latter can then be read directly from the scale engraved on the wedge. For a detailed description of the various wedges and compensators used for this purpose the reader must be referred to more special text-books.*

354. Determination of the Relative Optical Character of the Extinction Directions of any Section of a Doubly Refracting Mineral. — It frequently becomes important to determine which of the two rays of light in a doubly refracting mineral is being propagated with the greater or less velocity; in other words, to determine which of the two directions of vibration corresponds to the X and which to the Z direction. Place the given section under the microscope with the nicols crossed. Find a position of extinction and then turn the section through an arc of 45° so that its vibration directions make that angle with the planes of vibration of the nicols. If the section in this position shows a strong color or white of the higher order the quartz wedge is used. The optical orientation of the wedge must be known, *i.e.*, which are its X and Z directions. The wedge is then pushed through the slot above the objective lens, the thin end of the wedge being introduced first. The vibration directions of the wedge and the section will now coincide and the effect of the gradual introduction of the wedge above the mineral will be to slowly increase or decrease the birefringence due to the section. The result will be to either raise or lower the order of the interference color obtained. If the X directions of the wedge and the section coincide the effect will be additive in character and the color will rise in its order. If the optical directions of the two are opposed to each other the birefringence is decreased and the color will fall. By noting which effect takes place the X and Z directions of the section are determined.

In this use of the quartz wedge the following precaution must be observed. If the section originally showed a color of the first order and the wedge was introduced in the opposed position the effect would be to cause the color to fall rapidly to gray of the first order. The optical effect of the quartz wedge would thus quickly compensate that of the section. From this point on as the quartz wedge is pushed further in, the optical effect of the wedge will more and more preponderate over that of the section and the interference colors will now appear in ascending order. Under these conditions, if the first effect of the quartz wedge was overlooked, a wrong deduction would be made. It is always best to repeat the test with the section rotated 90° from the first position. The two results should be of opposite character and so serve to confirm each other.

Frequently a thick section of a mineral will show a tapering edge somewhere which will show bands of color. When the quartz wedge is introduced

---

* See Johannsen, Manual of Petrographic Methods; Wright, The Methods of Petrographic-Microscopic Research.

above the section these color bands will move, either toward the center of the section, or go off toward the edge. When the color bands move up on the section it means that the effect of the quartz wedge is such that a thicker part of the section is now showing the same interference as a thinner part did originally. In other words, the result is as if the section had been thinned. If this is so, then the X and Z directions of the section and the wedge must be opposed to each other. On the other hand, if the color bands move off the section it means that a thinner part of the section is showing the same interference effect that a thicker portion did originally. The introduction of the quartz wedge has in effect thickened the section and therefore the similar optical directions of the two coincide. This test is particularly useful for determining the X and Z directions of deeply colored minerals, as the natural color of the mineral may, over the thicker portion of the section, completely mask the interference color.

If a mineral section shows an interference color of white or gray of the first order the sensitive tint will give better results than the quartz wedge. If the similar optical directions of the section and the sensitive tint coincide the effect will be to raise the color of the sensitive tint (red of the first order) to blue. On the other hand, if the optical orientations of the two are opposed the color will fall to yellow. This test can be made to advantage only when the birefringent effect of the section is small enough to just raise or lower the color of the sensitive tint respectively to blue or yellow.

355. Circularly and Elliptically Polarized Light. — In the preceding articles the two interfering light-rays, after emerging from the second nicol, were assumed to be polarized in the same plane; for them the resulting phenomena as indicated are comparatively simple. If, however, two plane-polarized rays propagated in the same direction have their vibration-directions at right angles to each other, and if they differ one-quarter of a wave-length ($\frac{1}{4}\lambda$) in phase (assuming monochromatic light), then it may be shown that the composition of these two systems results in a ray of *circularly polarized* light. Briefly expressed, this is a ray that, looked at end-on, would seem to be propagated by ether-vibrations taking place in circles about the line of transmission. From the side, the onward motion would be like that of a screw, and either right-handed or left-handed.

If, again, two light-rays meet as above described, with a difference of phase differing from $\frac{1}{4}\lambda$ (but not equal to an even multiple of $\frac{1}{2}\lambda$), then the resulting composition gives rise to *elliptically polarized* light, that is, a light-ray propagated by ether-motions taking place in ellipses.

The above results are obtained most simply by passing plane-polarized light through a doubly refracting medium of the proper thickness (*e.g.*, a mica plate) which is placed with its vibration-planes inclined 45° to that of the polarizer. If the thickness is such as to give a difference in phase of $\frac{1}{4}\lambda$ or an odd multiple of this, the light which emerges is circularly polarized. If the phase differs from $\frac{1}{4}\lambda$ (but is not equal to $\frac{1}{2}\lambda$ or λ), the emergent light is elliptically polarized.

356. Rotation of Plane of Polarization. — In the case of certain doubly refracting crystallized media (as quartz), and also of certain solutions (as of sugar), it can be shown that the light is propagated by two sets of ether-vibrations which take place, not in definite transverse planes — as in plane-polarized light — but in circles; that is, each ray is circularly polarized, one being right-handed, the other left-handed. Further. of these rays, one will uniformly gain with reference to the other. The result is, that if a ray of plane-polarized light fall upon such a medium (assuming the simplest case, as of a section of quartz cut normal to the vertical crystal axis), it is found that the two rays circularly polarized within unite on emerging to a plane-polarized ray, but the plane of polarization has suffered an angular change or rotation, which may be either to the right (to one looking in the direction of the ray), when the substance is said to be *right-handed*, or to the left, when it is called *left-handed*.

This phenomenon is theoretically possible with all crystals of a given system belonging to any of the classes of lower symmetry than the normal class which show a plagiohedral development of the faces*; or, more simply, those in which the corresponding right and left (or + and −) typical forms are enantiomorphous (pp. 23, 87, 129), as noted in the chapter on crystallography. In mineralogy, this subject is most important with the common species of quartz, of the rhombohedral-trapezohedral class, and a further discussion of it is postponed to a later page (Art. **402**).

357. Abnormal Interference Colors. — Certain factors may at times modify the interference colors shown by a mineral section so as to produce abnormal effects. The most common of these is the natural color (or absorption) of the mineral section itself. If this is strong, the interference color may be considerably changed. Certain minerals may show quite different birefringence for different wave-lengths of light. In some cases, a mineral may have practically zero birefringence for a certain color and therefore that particular color will be lacking in its interference colors. Further, in some cases the vibration directions may be considerably divergent for light of varying wave-lengths. The section might therefore be at the position of extinction for one color when other colors could come through. In these various ways abnormal interference colors may be produced.

OPTICAL INSTRUMENTS AND METHODS

358. Measurement of Refractive Indices. Refractometer. — For the determination of the refractive indices of crystallized minerals various methods are employed. The most accurate results, when suitable material is at hand, may be obtained by the ordinary refractometer. This requires the observation of the angle of minimum deviation (δ) of a light-ray on passing through a prism of the given material, having a known angle (α), and with its edge cut in the proper direction. The measurements of α and δ can be made with an ordinary refractometer or with the horizontal goniometer described in Art. **236.** For the latter instrument, the collimator is made stationary, being fastened to a leg of the tripod support, but the observing telescope with the verniers moves freely. Further, for this object the graduated circle is clamped, and the screw attachments connected with the axis carrying the support, and the vernier circle and observing telescope are loosened. Light from a monochromatic source passes through an appropriate slit and an image of this is thrown by the collimator upon the prism. With a doubly refracting substance two images are yielded and the angle of minimum deviation must be measured

---

* Of the thirty-two possible classes among crystals, the following eleven may be characterized by circular polarization: Class 4, p. 87; 5, p. 88; 11 and 12, pp. 105, 106; 17, p. 118; 19, p. 120; 23, p. 129; 24, p. 131; 27, p. 145; 29, p. 156; 32, p. 165.

for each; the proper direction for the edge of the prism in this case is discussed later. When α and δ are known the formula in Art. **333** is used.

359. Total Refractometer. — The principle of total reflection (Art. **328**) may also be made use of to determine the refractive index. No prism is required, but only a small fragment having a single polished surface; this may have any direction with an isometric crystal, but in other cases must have a definite orientation, as described later. A number of different instruments have been devised by means of which indices of refraction may be measured

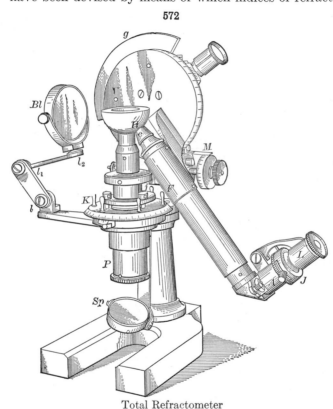

572

Total Refractometer

by the use of total reflection. A type widely used at present is represented in Fig. 572. This particular instrument was made by Leiss. It consists of a hemisphere of glass (H) having a high refractive index which is mounted upon a glass post through which light may be reflected from the mirror Sp. The tube P contains a nicol prism so that when a thin section of a mineral is placed upon the plane surface of the hemisphere it is possible to obtain its optical orientation in the same manner as with the polarizing microscope. The polished mineral surface is placed upon the plane surface of H with a film of some high refracting oil between them. Then a beam of light from some source of illumination, usually a monochromatic light, is reflected by means of the mirror Bl in such a way as to produce a total reflection shadow down on the opposite side of the hemisphere. For further details of the operation see Art. **333**. The telescope F is attached to the disk V which in turn carries a scale on its edge. The telescope is moved up or down until the line between the light and dark portions of the field lies on the cross-hairs. The angle which is read on the scale under these conditions is the desired critical angle for the combination of the glass of the hemisphere and the mineral plate. Knowing this angle and the index of refraction of the glass of the hemisphere it is possible to calculate the index of refraction of the mineral; see Art. **333**. Usually a table is furnished with the total refractometer by means of which the desired refractive index is obtained directly from the value of the measured critical angle. The post carrying the glass hemisphere may be revolved in the horizontal plane and the angle of rotation measured on the scale K. This permits the measurement of indices corresponding to different vibration directions in the mineral. L is an eye lens which in combination with the other lenses of the tube F makes a low power microscope, which is used in the preliminary operations in order to center the mineral plate, etc. In the tube A is an iris diaphragm and usually a small nicol prism that may be pushed in or out of the tube.

573

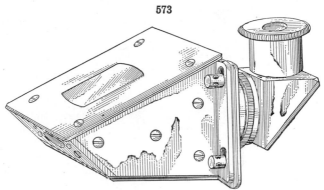

Fig. 573 represents a small total refractometer devised by G. F. H. Smith which depends upon the same principle. The mineral plate is placed upon the glass surface shown on the top of the instrument. The instru-

Smith Total Refractometer (Actual Size)

ment is so held that light enters at the forward end, and the totally reflected light is sent by means of an inclined mirror to the eyepiece. A scale is placed in the instrument in such a way that the boundary between the light and dark areas is seen superimposed upon it and so yields directly the value of the refractive index. For rapid and approximate determinations this instrument is very useful.

360. Tourmaline Tongs. — A very simple form of polariscope for converging light is shown in Fig. 574; it is convenient in use, but of limited application. Here the polarizer and analyzer are two tourmaline plates such as

574

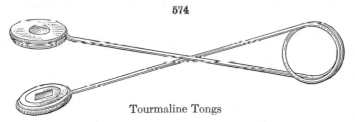

Tourmaline Tongs

were described in Art. **346.** They are mounted in pieces of cork and held in a kind of wire pincers. The object to be examined is placed between them and supported there by the spring in the wire. In use they are held close to the eye, and in this position the crystal section is viewed in *converging* polarized light, with the result of showing (under proper conditions) the axial interference-figures (Arts. **397** and **417**).

361. Polariscope. Conoscope. — The common forms of polariscopes employing nicol prisms are shown in Figs. 575 and 576.* Fig. 575 represents

---

* These figures are taken from the catalogue of Fuess.

the instrument arranged for converging light, which is often called a *conoscope*.

The essential parts are the mirror *S*, reflecting the light, which after passing through the lens *e* is polarized by the prism *p*. It is then rendered strongly converging by the system of lenses *nn*, before passing through the section under examination placed on a plate at *k*. This plate can be revolved through any angle desired, measured on its circumference. The upper tube contains the converging system *oo*, the lens *t*, and the analyzing prism *q*. The arrangements for lowering or raising the tubes need no explanation, nor indeed the special devices for setting the vibration-planes of the nicols at right angles to each other.

575 576 577

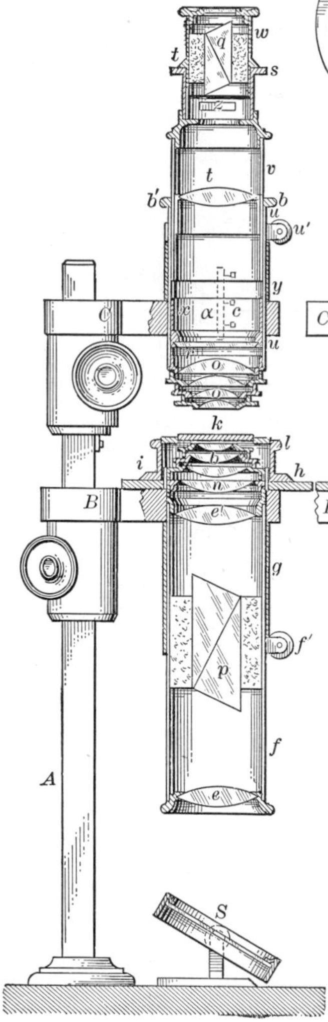

Conoscope   Polariscope   Polariscope ($\frac{1}{4}$ natural size)

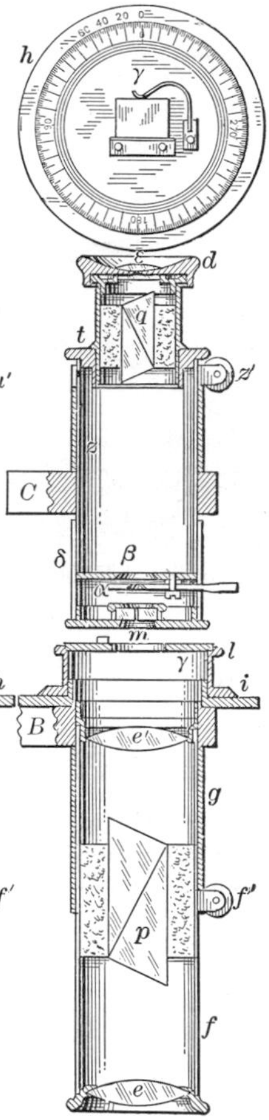

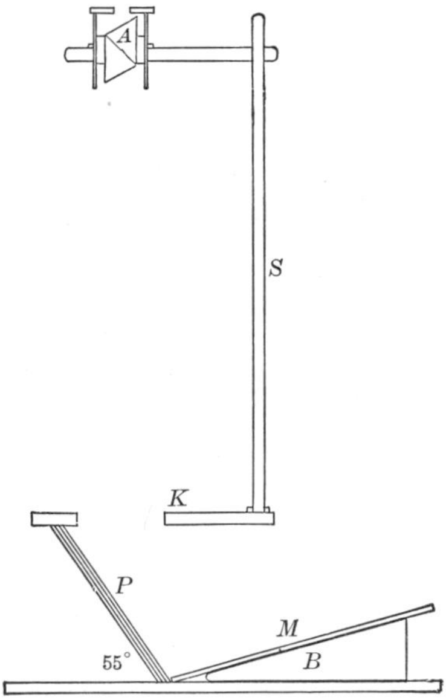

The accompanying tube (Fig. 576) shows the arrangement for observations in parallel light, the converging lenses having been removed.

Fig. 577 represents in cross section a simple, inexpensive but quite efficient form of polariscope. The polarizing device, *P*, is in the form of two or three thin glass sheets, the back of the bottom one being blackened. These glass plates are set at the appropriate angle to secure the maximum amount of polarization of the light reflected from them up through the opening in the stage *K*. *M* represents an adjustable mirror by means of which light is reflected upon *P*. The analyzer, *A*, is a small nicol prism which is held over the opening in the stage by means of the standard *S*. A double series of lenses may be placed upon the stage of the instrument and so convert it into a conoscope.

362. Polarization-Microscope. — The investigation of the form and optical properties of minerals when in microscopic form has been much facilitated by the use of microscopes* specially adapted for this purpose. First ar-

578

579

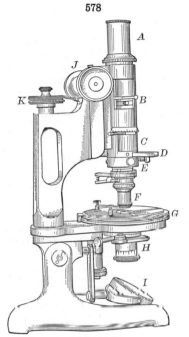
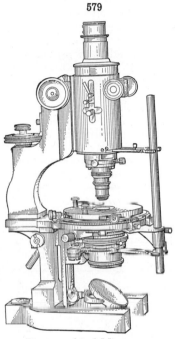

Pctrographical Microscope
(Laboratory Model, Bausch and
Lomb, $\frac{1}{4}$ actual size)

Pctrographical Microscope
(Research Model, Bausch and
Lomb, $\frac{1}{4}$ actual size)

ranged with reference to the special study of minerals as seen in thin sections of rocks, they have now been so elaborated as largely to take the place of the older optical instruments. They not only allow of the determination of the optical properties of minerals with greater facility, but are applicable to many cases where the crystals in hand are far too small for other means.

A highly serviceable microscope is the Laboratory Model made by the Bausch and Lomb Optical Co., and illustrated in Fig. 578. The essential arrangements of this instrument are as follows: The eyepiece at *A*, which is

* For detailed descriptions of the polarizing microscope and its accessories see Johannsen, Manual of Petrographic Methods; Wright, The Methods of Petrographic Research; etc.

removable, contains the cross-hairs with an eye lens adjustable for focusing upon them. At B is a Bertrand lens that slides in and out of the tube, with an iris diaphragm immediately above it. At C is the analyzer box which slides in and out of the body tube. This prism may be revolved through a quarter turn. D is a slot in the microscope tube with a dust-proof shutter for the introduction of various accessories, such as the quartz wedge, etc. At E is the nosepiece which can be centered by the two screws which work at right angles in the N and E positions. The objective F is held in place by a spring clamp and is quickly detached. The stage, G, revolves and carries a scale graduated into degrees, the attached vernier permitting the reading of angles to one-tenth degree. The substage at H carries condensing lenses, iris diaphragm and the polarizing prism. It can be moved upward and downward by means of a screw-head and when at its lowest point can be sprung to one side, out of the optical axis. The mirror at I is adjustable and has both a plane and a concave surface. The coarse focusing adjustment is at J, while the milled head at K provides a fine adjustment by means of which a vertical movement of 0·0005 mm. can be read.

363. The Research Model of the Bausch and Lomb microscope is illustrated in Fig. 579. This instrument is patterned after one described by Wright to whose papers reference is made for a more detailed account. The outstanding features of the instrument may be briefly summarized as follows: It has a large body-tube within which are always contained the analyzer and the Bertrand lens, both when they lie in or outside the optical axis of the microscope. The two nicols may be connected by means of the upright bar and rotated simultaneously through an arc of 90°. This enables the measurement of extinction angles, etc., to be made without the necessity of revolving the stage and the consequent difficulty in keeping the mineral grain under observation exactly centered in the field. This bar carries verniers that lie against the scale engraved upon the stage so that the angle of rotation of the nicols can be accurately measured. The polarizing prism can be entirely removed from the optical axis. A revolvable carrier for a sensitive plate is attached to the iris diaphragm mount of the substage.

It should be added that a number of different makes of petrographic microscopes, all excellent, are available. In the essential characters however, they do not differ materially from those described above.

GENERAL OPTICAL CHARACTERS OF MINERALS

364. There are certain characteristics belonging to all minerals alike, crystallized and non-crystallized, in their relation to light. These are:

1. DIAPHANEITY: depending on the relative quantity of light transmitted.
2. COLOR: depending on the kind of light reflected or transmitted, as determined by the selective absorption.
3. LUSTER: depending on the power and manner of reflecting light.

1. DIAPHANEITY

365. Degrees of Transparency. — The amount of light transmitted by a solid varies in intensity, or, in other words, more or less light may be *absorbed* in the passage through the given substance (see Art. **336**). The amount of absorption is a minimum in a transparent solid, as ice, while it is greatest in one which is opaque, as iron.   The following terms are adopted to express the different degrees in the power of transmitting light:

Transparent: when the outline of an object seen through the mineral is perfectly distinct.

Subtransparent, or *semi-transparent:* when objects are seen, but the outlines are not distinct.

Translucent: when light is transmitted, but objects are not seen.

Subtranslucent: when merely the edges transmit light or are translucent.

When no light is transmitted, even on the thin edges of small splinters, the mineral is said to be *opaque.*   This is properly only a *relative* term, since no substance fails to transmit some light, if made sufficiently thin.   Magnetite is translucent in the Pennsbury mica.   Even gold may be beaten out so thin as to be translucent, in which case it transmits a greenish light.

The property of diaphaneity occurs in the mineral kingdom, from nearly perfect opacity to transparency, and many minerals present, in their numerous varieties, nearly all the different degrees.

2. COLOR

366. Nature of Color. — As briefly explained in Art. **319,** the sensation of color depends, in the case of monochromatic light, solely upon the length of the waves of light which meet the eye.   If the light consists of various wave-lengths, it is to the combined effect of these that the sensation of color is due.

Further, since the light ordinarily employed is essentially white light, that is, consists of all the wave-lengths corresponding to the successive colors of the spectrum, the color of a body depends upon the selective absorption (see Art. **336**) which it exerts upon the light transmitted or reflected by it.   A yellow mineral, for instance, absorbs all the waves of the spectrum with the exception of those which together give the sensation of yellow.   In general, the color which the eye perceives is the result of the mixture of those waves which are not absorbed.

367. Streak. — The color of the powder of a mineral as obtained by scratching the surface of the mineral with a knife or file, or, still better, if the mineral is not too hard, by rubbing it on an unglazed porcelain surface, is called the *streak.*   The streak is often a very important quality in distinguishing minerals.   This is especially true with minerals having a metallic luster, as defined in Art. **370.**

368. Dichroism; Pleochroism. — The selective absorption, to which the color of a mineral is due, more especially by transmitted light, often varies with the crystallographic direction in which the light vibrates in the mineral. It is hence one of the special optical characters depending upon the crystallization, which are discussed later.   Here belong *dichroism* or *pleochroism,* the property of exhibiting different colors in different crystallographic directions by transmitted light.   This subject is explained further in Arts. **401** and **423.**

369. Varieties of Color. — The following eight colors were selected by Werner as fundamental, to facilitate the employment of this character in the description of minerals: *white, gray, black, blue, green, yellow, red,* and *brown.*

(a) The varieties of METALLIC COLORS recognized are as follows:
1. *Copper-red:* native copper. — 2. *Bronze-yellow:* pyrrhotite. — 3. *Brass-yellow:* chalcopyrite. — 4. *Gold-yellow:* native gold. — 5. *Silver-white:* native silver, less distinct in arsenopyrite. — 6. *Tin-white:* mercury; cobaltite. — 7. *Lead-gray:* galena, molybdenite. — 8. *Steel-gray:* nearly the color of fine-grained steel on a recent fracture; native platinum, and palladium.

(b) The following are the varieties of NON-METALLIC COLORS:
A. WHITE. 1. *Snow-white:* Carrara marble. — 2. *Reddish white*, 3. *Yellowish white* and 4. *Grayish-white:* all illustrated by some varieties of calcite and quartz. — 5. *Greenish white:* talc. — 6. *Milk-white:* white, slightly bluish; some chalcedony.

B. GRAY. 1. *Bluish gray:* gray, inclining to dirty blue. — 2. *Pearl-gray:* gray, mixed with red and blue; cerargyrite. — 3. *Smoke-gray:* gray, with some brown; flint. — 4. *Greenish gray:* gray, with some green; cat's-eye; some varieties of talc. — 5. *Yellowish gray:* some varieties of compact limestone. — 6. *Ash-gray:* the purest gray color; zoisite.

C. BLACK. 1. *Grayish black:* black, mixed with gray (without green, brown, or blue tints); basalt; Lydian stone. — 2. *Velvet-black:* pure black; obsidian, black tourmaline. — 3. *Greenish black:* augite. — 4. *Brownish black:* brown coal, lignite. — 5. *Bluish black:* black cobalt.

D. BLUE. 1. *Blackish blue:* dark varieties of azurite. — 2. *Azure-blue:* a clear shade of bright blue; pale varieties of azurite, bright varieties of lazulite. — 3. *Violet-blue:* blue, mixed with red; amethyst, fluorite. — 4. *Lavender-blue:* blue, with some red and much gray. — 5. *Prussian-blue,* or Berlin blue: pure blue; sapphire, kyanite. — 6. *Smalt-blue:* some varieties of gypsum. — 7. *Indigo-blue:* blue, with black and green; blue tourmaline. — 8. *Sky-blue:* pale blue, with a little green; it is called mountain-blue by painters.

E. GREEN. 1. *Verdigris-green:* green, inclining to blue; some feldspar (amazon-stone). — 2. *Celandine-green:* green, with blue and gray; some varieties of talc and beryl. It is the color of the leaves of the celandine. — 3. *Mountain-green:* green, with much blue; beryl. — 4. *Leek-green:* green, with some brown; the color of leaves of garlic; distinctly seen in prase, a variety of quartz. — 5. *Emerald-green:* pure deep green; emerald. — 6. *Apple-green:* light green with some yellow; chrysoprase. — 7. *Grass-green:* bright green, with more yellow; green diallage. — 8. *Pistachio-green:* yellowish green, with some brown; epidote. — 9. *Asparagus-green:* pale green, with much yellow; asparagus stone (apatite). — 10. *Blackish green:* serpentine. — 11. *Olive-green:* dark green, with much brown and yellow; chrysolite. — 12. *Oil-green:* the color of olive-oil; beryl, pitchstone. — 13. *Siskin-green:* light green, much inclining to yellow; uranite.

F. YELLOW. 1. *Sulphur-yellow:* sulphur. — 2. *Straw-yellow:* pale yellow; topaz. — 3. *Wax-yellow:* grayish yellow with some brown; sphalerite, opal. — 4. *Honey-yellow:* yellow, with some red and brown; calcite. — 5. *Lemon-yellow:* sulphur, orpiment. — 6. *Ocher-yellow:* yellow, with brown; yellow ocher. — 7. *Wine-yellow:* topaz and fluorite. — 8. *Cream-yellow:* some varieties of kaolinite. — 9. *Orange-yellow:* orpiment.

G. RED. 1. *Aurora-red:* red, with much yellow; some realgar. — 2. *Hyacinth-red:* red, with yellow and some brown; hyacinth garnet. — 3. *Brick-red:* polyhalite, some jasper. — 4. *Scarlet-red:* bright red, with a tinge of yellow; cinnabar. — 5. *Blood-red:* dark red, with some yellow; pyrope. — 6. *Flesh-red:* feldspar. — 7. *Carmine-red:* pure red; ruby sapphire. — 8. *Rose-red:* rose quartz. — 9. *Crimson-red:* ruby. — 10. *Peachblossom-red:* red, with white and gray; lepidolite. — 11. *Columbine-red:* deep red, with some blue; garnet. — 12. *Cherry-red:* dark red, with some blue and brown; spinel, some jasper. — 13. *Brownish-red:* jasper, limonite.

H. BROWN. 1. *Reddish brown:* garnet, zircon. — 2. *Clove-brown:* brown, with red and some blue; axinite. — 3. *Hair-brown:* wood-opal. — 4. *Broccoli-brown:* brown, with blue, red, and gray; zircon. — 5. *Chestnut-brown:* pure brown. — 6. *Yellowish brown:* jasper. — 7. *Pinchbeck-brown:* yellowish brown, with a metallic or metallic-pearly luster; several varieties of talc, bronzite. — 8. *Wood-brown:* color of old wood nearly rotten; some specimens of asbestos. — 9. *Liver-brown:* brown, with some gray and green; jasper. — 10. *Blackish brown:* bituminous coal, brown coal.

Various attempts have been made to classify colors more rigidly so that one color could be compared with another with a fair degree of accuracy. The most comprehensive and best attempt of this kind was made by Ridgeway, who gives over one thousand different tints arranged in a logical order.*

* Robert Ridgeway, Color Standards and Nomenclature. 1913.

3. LUSTER

370. Nature of Luster. — The luster of minerals varies with the nature of their surfaces. A variation in the quantity of light reflected produces different degrees of intensity of luster; a variation in the nature of the reflecting surface produces different kinds of luster.

371. Kinds of Luster. — The kinds of luster recognized are as follows:

1. METALLIC: the luster of the metals, as of gold, copper, iron, tin.

In general, a mineral is not said to have metallic luster unless it is opaque in the mineralogical sense, that is, it transmits no light on the edges of thin splinters. Some minerals have varieties with metallic and others with non-metallic luster; this is true of hematite.

Imperfect metallic luster is expressed by the term *sub-metallic*, as illustrated by columbite, wolframite. Other kinds of luster are described briefly as NON-METALLIC.

2. NON-METALLIC. *A. Adamantine:* the luster of the diamond. When also sub-metallic, it is termed *metallic-adamantine*, as cerussite, pyrargyrite.

Adamantine luster belongs to substances of high refractive index. This may be connected with their relatively great density (and hardness), as with the diamond, also corundum, etc., or because they contain heavy molecules; thus most compounds of lead, not metallic in luster, have a high refractive index and an adamantine luster.

B. Vitreous: the luster of broken glass. An imperfectly vitreous luster is termed *sub-vitreous.* The vitreous and sub-vitreous lusters are the most common in the mineral kingdom. Quartz possesses the former in an eminent degree; calcite, often the latter.

C. Resinous: luster of the yellow resins, as opal, and some yellow varieties of sphalerite.

D. Greasy: luster of oily glass. This is near resinous luster, but is often quite distinct, as nephelite.

E. Pearly: like pearl, as talc, brucite, stilbite, etc. When united with sub-metallic, as in hypersthene, the term *metallic-pearly* is used.

Pearly luster belongs to the light reflected from a pile of thin glass-plates; similarly it is exhibited by minerals, which, having a perfect cleavage, may be partially separated into successive plates, as on the basal plane of apophyllite. It is also shown for a like reason by foliated minerals, as talc and brucite.

F. Silky: like silk; it is the result of a fibrous structure. Ex. fibrous calcite, fibrous gypsum.

The different degrees and kinds of luster are often exhibited differently by unlike faces of the same crystal, but always similarly by like faces. For example, the basal plane of apophyllite has a pearly luster wanting in the prismatic faces, which have a vitreous luster.

As shown by Haidinger, only vitreous, adamantine, and metallic lusters belong to faces perfectly smooth and pure. In the first, the refractive index of the mineral is 1·3–1·8; in the second, 1·9–2·5; in the third, about 2·5. The true difference between metallic and vitreous luster is due to the effect which the different surfaces have upon the reflected light; in general, the luster is produced by the union of two simultaneous impressions made upon the eye. If the light reflected from a metallic surface be examined by a nicol prism (or the dichroscope of Haidinger, Art. **401**), it will be found that both rays, that vibrating in the plane of incidence and that whose vibrations are normal to it, are alike, each having the color of the material, only differing a little in brilliancy; on the contrary, of the light reflected by a vitreous substance, those rays whose vibrations are at right angles to the plane of incidence are more or less polarized, and are colorless, while those whose vibrations are in this plane, having penetrated somewhat into the medium and suffered some absorption, show the color of the substance itself. A plate of red glass thus examined will show a colorless and a red image. Adamantine luster occupies a position between the others.

372. Degrees of Luster. — The *degrees of intensity* of luster are classified as follows:

1. *Splendent:* reflecting with brilliancy and giving well-defined images, as hematite, cassiterite.

2. *Shining:* producing an image by reflection, but not one well-defined, as celestite.

3. *Glistening:* affording a general reflection from the surface, but no image, as talc, chalcopyrite.

4. *Glimmering:* affording imperfect reflection, and apparently from points over the surface, as flint, chalcedony.

A mineral is said to be *dull* when there is a total absence of luster, as chalk, the ochers, kaolin.

373. Play of Colors. Opalescence. Iridescence. — The term *play of colors* is used to describe the appearance of several prismatic colors in rapid succession on turning the mineral. This property belongs in perfection to the diamond, in which it is due to its high dispersive power. It is also observed in precious opal, where it is explained on the principle of interference; in this case it is most brilliant by candle-light.

The expression *change of colors* is used when each particular color appears to pervade a larger space than in the play of colors and the succession produced by turning the mineral is less rapid. This is shown in labradorite, as explained under that species.

Opalescence is a milky or pearly reflection from the interior of a specimen. Observed in some opal, and in cat's-eye.

Iridescence means the exhibition of prismatic colors in the interior or on the surface of a mineral. The phenomena of the play of colors, iridescence, etc., are sometimes to be explained by the presence of minute foreign crystals, in parallel positions; more generally, however, they are caused by the presence of fine cleavage-lamellæ, in the light reflected from which interference takes place, analogous to the well-known Newton's rings (see Art. **342**).

374. Tarnish. — A metallic surface is tarnished when its color differs from that obtained by fracture, as is the case with specimens of bornite. A surface possesses the *steel tarnish* when it presents the superficial blue color of tempered steel, as columbite. The tarnish is *irised* when it exhibits fixed prismatic colors, as is common with the hematite of Elba. These tarnish and iris colors of minerals are owing to a thin surface or film, proceeding from different sources, either from a change in the surface of the mineral or from foreign incrustation; hydrated iron oxide is one of the most common sources of it and produces the colors on anthracite and hematite.

375. Asterism. — This name is given to the peculiar star-like rays of light observed in certain directions in some minerals. This is seen by reflected light in the form of a six-rayed star in sapphire, and is also well shown by transmitted light (as of a small flame) with the phlogopite mica from South Burgess, Canada. In the former case it is explained by the presence of thin twinning-lamellæ symmetrically arranged. In the other case it is due to the presence of minute inclosed crystals, also symmetrically arranged, which are probably rutile or tourmaline in most cases. Crystalline faces which have been artificially etched also sometimes exhibit asterism. The peculiar light-figures sometimes observed in reflected light on the faces of crystals, either natural or etched, are of similar nature.

376. Schillerization. — The general term *schiller* is applied to the peculiar luster, sometimes nearly metallic, observed in definite directions in certain minerals, as conspicuously in schiller-spar (an altered variety of bronzite), also in diallage, hypersthene, sunstone, and others. It is explained by the reflection either from minute inclosed plates in parallel position or from the surfaces of minute cavities (negative crystals) having a common orientation. In many cases it is due to alteration which has developed these bodies (or the cavities) in the direction of solution-planes (see Art. **290**). The process by which it has been produced is then called *schillerization*.

377. Fluorescence. — The emission of light from within a substance while it is being exposed to direct radiation, or in certain cases to an electrical discharge in a vacuum tube, is called *fluorescence*. It is best exhibited by fluorite, from which the phenomenon gained its name. Thus, if a beam of white light be passed through a cube of colorless fluorite a delicate violet color is called out in its path. This effect is chiefly due to the action of the ultra-violet rays, and is connected with a change of refrangibility in the transmitted light.

The electrical discharge from the negative pole of a vacuum tube calls out a brilliant fluorescence not only with the diamond, the ruby, and many gems, but also with calcite and other minerals. Such substances may continue to emit light, or *phosphoresce*, after the discharge ceases.

378. Phosphorescence. — The *continued* emission of light by a substance (not incandescent) produced especially after heating, exposure to light or to an electrical discharge, is called *phosphorescence*.

Fluorite becomes highly phosphorescent after being heated to about 150° C. Different varieties give off light of different colors; the *chlorophane* variety, an emerald-green light; others purple, blue, and reddish tints. This phosphorescence may be observed in a dark place by subjecting the pulverized mineral to a heat below redness. It may even be produced by a sharp blow with a hammer. Some varieties of white limestone or marble, after slight heating, emit a yellow light; so also tremolite, danburite, and other species.

The X-ray and ultra-violet light will produce phosphorescence in willemite, kunzite, and some diamonds. The fact that willemite glows when exposed to ultra-violet light is made use of in testing the residues from a willemite ore to make certain the separation has been complete. Radium emanations cause certain minerals to phosphoresce, as willemite and wurtzite.

Exposure to the light of the sun produces very apparent phosphorescence with many diamonds, but some specimens seem to be destitute of this power. This property is most striking after exposure to the blue rays of the spectrum, while in the red rays it is rapidly lost. A mixture of calcium sulphide and bismuth will phosphoresce for a considerable period after being exposed to sunlight.

379. Triboluminescence and Crystalloluminescence. — Certain crystallized substances become luminous when rubbed or scratched. This property, known as *triboluminescence*, is exhibited by some diamond crystals. Some varieties of sphalerite give off light when scratched. Light is sometimes given off by certain substances in crystallizing from a solution. Arsenic oxide, As_2O_3, is an example. This property is called *crystalloluminescence*.

SPECIAL OPTICAL CHARACTERS BELONGING TO CRYSTALS OF THE DIFFERENT SYSTEMS

380. All crystallized minerals may be grouped into three grand classes, which are distinguished by their physical properties, as well as their geometrical form. These three classes are as follows:

A. *Isometric class*, embracing crystals of the isometric system, which are referred to three equal rectangular axes.

B. *Isodiametric class*, embracing crystals of the tetragonal and hexagonal systems, referred to two, or three, equal horizontal axes and a third, or fourth, axis unequal to them at right angles to their plane. Crystals of this class have a fixed principal axis of crystallographic symmetry.

C. *Anisometric class*, embracing the crystals of the orthorhombic, monoclinic, and triclinic systems, referred to three unequal axes.

381. Isotropic Crystals. — Of the three classes, the ISOMETRIC CLASS includes all crystals which, with respect to light and related phenomena involving the ether, are *isotropic* (from the Greek, signifying *equal turning*); that is, those which have like optical properties in all directions. Their distinguishing characteristic is that light travels through them with equal velocity in all directions, provided their molecular equilibrium is not disturbed by external pressure or internal strain. If it be imagined therefore that light starts from a point within an isotropic medium at a given moment of time the resulting wave surface will be a sphere.

It must be emphasized here, however, that such a crystal is *not* isotropic with reference to those characters which depend directly upon the molecular structure alone, as cohesion and elasticity. (See Art. **280**).

Further, amorphous bodies, as glass and opal, which are destitute of any orientated molecular structure — that is, those in which all directions are sensibly the same — are also isotropic, and not only with reference to light, but also as regards their strictly molecular properties.

382. Anisotropic Crystals; Uniaxial and Biaxial. — Crystals of the ISODIAMETRIC and ANISOMETRIC CLASSES, on the other hand, are in distinction *anisotropic* (from the Greek, signifying *unequal turning*). Their optical properties are in general unlike in different directions, or, more particularly, the velocity with which light is propagated varies with the direction of vibration.

Further, in crystals of the isodiametric class that variable property of the light-ether upon which the velocity of propagation depends remains constant for all directions which are normal to, or, again, for all those equally inclined to, the vertical crystallographic axis. In the direction of this axis there is no double refraction; it is hence called the *optic axis*, and the crystals of this class are said to be *uniaxial*.

Crystals of the third or anisometric class have more complex optical relations requiring special explanation, but in general it may be stated that in them there are always two directions analogous in character to the single optic axis spoken of above; hence, these crystals are said to be optically *biaxial*.

A. Isometric Crystals

383. It has been stated that crystals of the isometric system are optically *isotropic*, and hence light travels with the same velocity in every direction in them. Light can, therefore, suffer only single refraction in passing into an isotropic medium; or, in other words, there can be but one value of the refractive index for a given wave-length. If this be represented by n, while V is the velocity of light in air and v that in the given medium, then

$$n = \frac{V}{v}, \quad \text{or} \quad v = \frac{V}{n}.$$

The wave-front for light-waves propagated from any point within such an isotropic medium is, as already stated, a sphere. The sphere, therefore, may be taken to represent the optical properties of an isotropic medium. Sections of a sphere normal to any diameter will always be circles. These circular sections with like radii in all directions correspond to the fact that the optical character of an isotropic substance is the same in all directions normal to the line of light propagation. Or, in other words, light vibrations may take place in any direction normal to the direction of transmission; i.e., the light is not polarized. Further its velocity remains uniform no matter what may be the direction of its vibration.

This statement holds true of all the classes of isometric crystals. In other words, a crystal of maximum symmetry, as fluorite, and one having the restricted symmetry characteristic of the tetrahedral or pyritohedral divisions, have alike the same isotropic character. Two of the classes, however, namely, the plagiohedral and the tetartohedral classes, differ in this particular: that crystals belonging to them may exhibit what has already been defined (Art. **356**) as circular polarization.

384. Behavior of Sections of Isometric Crystals in Polarized Light. — In consequence of their isotropic character, isometric crystals exhibit no special phenomena in polarized light. As a section of an isotropic substance (isometric crystal or some amorphous material) has no polarizing or doubly refracting effect upon light it does not change at all the character of light that enters it from the polarizer of a polariscope. Therefore thin sections of isotropic media when examined in a polariscope or polarizing microscope with the nicols crossed will appear dark in all positions. In other words, they are always extinguished. Further, when a colored mineral is examined without the analyzer there will be no change in its color when the section is revolved with the stage of the microscope. Some anomalies are mentioned on a later page (Art. **441**).

The single refractive index of an isotropic substance may be determined by means of a prism (see Art. **333**) with its edge cut in any direction whatever.

B. UNIAXIAL CRYSTALS

General Optical Relations

385. The crystallographic and optical relations of crystals belonging to the tetragonal and hexagonal systems have already been briefly summarized (Art. **382**); it now remains to develop their optical characters more fully. This can be done most simply by making frequent use of the familiar conception of a light-ray to represent the character and motion of the light-wave.

386. Behavior of Light in Uniaxial Minerals. — Light entering a uniaxial mineral is in general broken up into two rays which are polarized in planes perpendicular to each other and which travel with different velocities and therefore have different indices of refraction. One of the two rays derived from a single incident ray always vibrates in the plane of the horizontal crystallographic axes. The other ray vibrates at right angles to the first and always in a vertical plane that includes the vertical crystallographic axis. The optical character of a uniaxial mineral is uniform for all directions lying in the horizontal crystallographic plane and therefore the ray whose vibrations lie in this plane will have uniform velocity no matter what its direction of transmission. This ray will therefore have a single and constant index of refraction, commonly designated by ω. Since this ray follows the usual law as to the constant ratio between the sines of the angles of incidence and refraction and in general behaves in an ordinary way it is called the *ordinary ray*. The ray which vibrates in a plane that includes the vertical crystallographic axis will have the direction of its vibration constantly changing as the direction of its path through the crystal changes and its velocity will correspondingly vary. Its index of refraction will therefore depend upon the direction of its propagation and it will not in general obey the usual sine law. This ray is therefore called the *extraordinary ray*.

When light travels in a uniaxial mineral in a direction parallel to the vertical crystallographic axis, since all its vibrations must take place in the horizontal plane, it behaves wholly as the ordinary ray with a single velocity and refractive index. There can be no double refraction of light, therefore, along this direction and in this case the mineral will behave like an isotropic substance. This direction of no double refraction, coincident with the vertical crystal axis, is known as the *optic axis* and as there is only one such direction in this optical group the latter is called *uniaxial*. As soon as the direction of transmission becomes inclined to the vertical crystal axis the light is doubly refracted and as the inclination increases the direction of vibration of the light of the extraordinary ray departs more and more from the plane of vibration of the ordinary ray with a corresponding change in its velocity and refractive index. The difference between the refractive indices of the two rays becomes a maximum when the light passes through the mineral in a horizontal direction with the direction of vibration of the extraordinary ray parallel to the vertical crystal axis — or in other words as divergent as possible from the horizontal plane. The value of the refractive index of the extraordinary ray when at its maximum difference from the constant index of the ordinary ray is the one always quoted and is indicated by ϵ. These two indices, ω and ϵ, are called the *principal indices* of a uniaxial crystal. A *principal section* of a uniaxial crystal is a section passing through the vertical axis.

387. Positive and Negative Crystals. — Uniaxial crystals are divided into two classes, depending upon whether the velocity of the extraordinary ray is greater or less than that of the ordinary ray. Those in which the refractive index of the ordinary ray, ω, is less than that of the extraordinary ray, ϵ ($\omega < \epsilon$), are called *positive*. This is illustrated by quartz for which (for yellow sodium light):

$$\omega = 1 \cdot 544. \qquad \epsilon = 1 \cdot 553.$$

On the other hand, if ϵ is less than ω ($\epsilon < \omega$), the crystal is said to be *negative*.* Calcite is an example for which (for sodium light)

$$\omega = 1 \cdot 658. \qquad \epsilon = 1 \cdot 486.$$

Other examples are given later (Art. **391**).

---

* It will assist in remembering these relations to note that the first vowel in the words positive and negative agrees with the symbol used for the smaller index of refraction in each case.

388. Determination of the Refractive Indices in Uniaxial Crystals. — The indices of refraction of uniaxial minerals are measured in much the same way as in the case of isotropic substances. With uniaxial crystals, however, the prism or plate used must have a definite crystallographic orientation. If a prism is employed its edge should be parallel to the optic axis, or in other words parallel to the vertical crystal axis of the mineral. When such a prism is examined on the refractometer two refracted rays are seen, the angles of refraction of which can be measured by either the method of minimum deviation or perpendicular incidence as described in Art. **333**. The two rays are polarized, the ordinary ray vibrating in the horizontal plane and the extraordinary ray vibrating in the vertical plane, *i.e.*, parallel to the edge of the prism. The plane of vibration of each ray must be determined by the use of a nicol prism held in front of the eyepiece of the refractometer. When the plane of the nicol is horizontal the image belonging to the ordinary ray will be visible and when the plane of the nicol is vertical only that of the extraordinary ray will appear. In this way the indices of the two rays are determined and the positive or negative character of the mineral is established. It is possible to obtain these measurements in prisms with different crystallographic orientation but the difficulties attending their preparation are so great that such prisms are very seldom used.

If the method of total reflection is used a single plate will suffice, provided it either lies in the prism zone of the crystal, or is parallel to the basal plane. In each case two shadows will be observed, corresponding in their position to the angles of total reflection of the two rays. When the plate is cut parallel to a face in the prism zone one of these shadows, that belonging to the ordinary ray, will remain stationary as the plate is revolved on the hemisphere of the total refractometer while the shadow of the extraordinary ray will vary from being coincident with that of the ordinary ray to a certain maximum divergence from that position. This maximum difference in position, which may yield a greater or less angle than that of the ordinary ray, depending upon the optical character of the mineral, is the angle corresponding to the true value of the refractive index of the extraordinary ray. There will be two positions at 180° apart during the complete revolution of the section at which this value may be measured. If the plate was cut parallel to the basal plane of the crystal the two shadows would both be stationary during such a revolution and the value of the angle for both rays can be measured in any position of the plate.

In case the method of immersion in liquids of known refractive indices is used, it is important, if possible, to obtain the true values for ω and ϵ. In general, each mineral grain will yield two indices corresponding to the two vibration directions indicated by the positions of extinction of the grain. One of these indices will belong to the ordinary ray and yield definitely the value of ω. The other index, while belonging to the extraordinary ray, will vary with the crystal orientation of the grain. In a series of variously orientated sections, the index having the maximum difference from ω will at least approximate to ϵ. If the grains are of uniform thickness, those that show the highest order of interference colors will give the closest approach to the true ϵ. On the other hand, grains that show no or only a little birefringence will give only the value for ω. If interference figures (see Art. **397**) can be obtained from the grains they will help in their orientation and so assist in finding the vibration direction of the extraordinary ray that will give the index ϵ.

389. Wave-surface. — Remembering that the velocity of light-propagation is always inversely proportional to the corresponding refractive index, it is obvious that the velocity of the ordinary ray for all directions in a uniaxial crystal must be the same, being uniformly proportional to $\frac{1}{\omega}$. In other words, supposing light originates at a point within a uniaxial crystal the ordinary ray would travel out in all directions with uniform velocity and its wave-front would form a sphere.

For the extraordinary ray, however, the velocity varies with the direction, being proportional to $\frac{1}{\epsilon}$ in a horizontal direction and becoming sensibly equal to $\frac{1}{\omega}$ when nearly coincident with the direction of the vertical axis. The law of the varying change of velocity between these values, $\frac{1}{\omega}$ and $\frac{1}{\epsilon}$, is given by an ellipse whose axes (OC, OA, Figs. 580, 581) are respectively proportional to the above values.

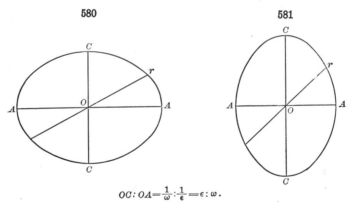

580   581

$$OC : OA = \tfrac{1}{\omega} : \tfrac{1}{\epsilon} = \epsilon : \omega.$$

The wave-front of the extraordinary ray is then a spheroid, or an ellipsoid of revolution whose axis coincides with the vertical crystallographic axis, that is, the optic axis. In the direction of the vertical axis it is obvious that the wave-fronts of the ordinary and extraordinary rays will coincide.

Figs. 582 and 583 represent vertical sections of the combined wave-surfaces for both rays. Fig. 582 gives that for a *negative* crystal like calcite ($\epsilon < \omega$), the ellipsoidal wave-surface of the extraordinary ray being outside the spherical surface of the ordinary ray; Fig. 583 that of a *positive* crystal like quartz ($\omega < \epsilon$) with the ellipsoidal surface within that of the sphere. Fig. 584 is an attempt to show the relations of the two wave-fronts of a negative crystal in perspective for a single octant. The constant value of the velocity of the ordinary ray $\left(\frac{1}{\omega}\right)$, whatever its direction in the plane of Figs. 582 and 583, is expressed by the radius of the circle (OC). On the other hand, the velocity of the extraordinary ray in the horizontal direction is given by $OA \left(\frac{1}{\epsilon}\right)$, while in any oblique direction, as Osr, Fig. 582 (Ors, Fig. 583), it is expressed by the length of this line, becoming more and more nearly equal to $OC\left(\dfrac{1}{\omega}\right)$ as its direction approaches that of the vertical axis.

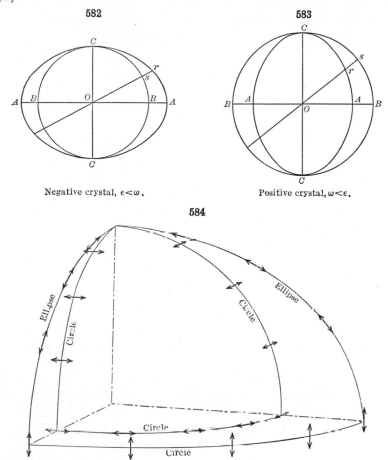

582

Negative crystal, $\epsilon < \omega$,

583

Positive crystal, $\omega < \epsilon$.

584

390. Uniaxial Indicatrix. — The optical structure of a uniaxial crystal can be represented by an ellipsoid of revolution, called the *indicatrix,** from which can be obtained the directions of vibration and indices of refraction of the ordinary and extraordinary rays derived from any single incident ray. Fig. 585 represents a principal section of such an ellipsoid for an optically negative crystal, the line $C\text{-}C$ being its axis of revolution. The axes of this ellipsoid are made inversely proportional to the indices of refraction of the two rays, ω and ϵ, as follows:

$$OC : OA = \frac{1}{\omega} : \frac{1}{\epsilon} \quad \text{or} \quad \epsilon : \omega.$$

* The Optical Indicatrix and the Transmission of Light in Crystals, by L. Fletcher London, 1892.

In this figure let Or be a direction of transmission of light. Let Vr and VR be tangents to the elliptical surface at the points r and R and OR be a radius vector parallel to the tangent Vr. Or and OR form then what are known as *conjugate radii*. From the geometrical properties of an ellipse it follows that the area of any parallelogram with conjugate radii forming two sides, such as $ORVr$ in Fig. 585, is constant and equal to the area of a parallelogram having OC and OA as two sides. Let RN be perpendicular to the extended line Or. Then the area of $ORVr$ will be equal to $RN{\cdot}Or$. It follows since $RN{\cdot}Or = OA{\cdot}OC$ = a constant, k, that

$$Or = \frac{OA{\cdot}OC}{RN} = \frac{k}{RN}; \quad \text{also} \quad OA = \frac{k}{OC}.$$

From the last expression we see that OA and OC are inversely proportional to each other, or, in other words, as OC represents the minimum index, OA will represent the corresponding velocity of light which will be the maximum for any transmission direction in the crystal. In the same way Or and RN are inversely proportional to each other, the distance Or representing the velocity of the extraordinary ray traveling along that direction while RN will represent its refractive index. The line RN will also give the direction of vibration of the extraordinary ray.

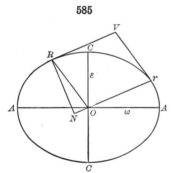

585

For the radius vector Or there will be another possible direction perpendicular to it and also normal to the ellipsoidal surface. This will be a line from O perpendicular to the principal section represented in Fig. 585. This line will lie in the horizontal circular section of the indicatrix ellipsoid with its length equal to OA which in turn is proportional to the index of the ordinary ray, ω. So for a given direction of transmission of light, such as Or, the two lines that are perpendicular to it and at the same time normal to the surface of the indicatrix yield both the indices of refraction of the two rays and the directions of their vibrations.

If, however, the light is passing parallel to the principal axis of the indicatrix, *i.e.*, C–C, Fig. 585, there will be an infinite number of lines which are perpendicular to this direction and at the same time normal to the surface of the indicatrix. These will lie in the horizontal circular section of the ellipsoid and consequently will be of a uniform length. From this it is evident that such a transmitted ray may vibrate in any transverse direction and will possess a single index of refraction and velocity. Along this direction, known as the optic axis, there will consequently be no double refraction of the light.

391. Examples of Positive and Negative Crystals. — The following lists give prominent positive and negative uniaxial crystals, with the values of the refractive indices, ω and ϵ for each, corresponding to yellow sodium light.* The difference between these, $\omega - \epsilon$ or $\epsilon - \omega$, is also given; this measures the birefringence or *strength* of the double refraction.

It may be remarked that in some species both $+$ and $-$ varieties have been observed. Certain crystals of apophyllite are positive for one end of the spectrum and negative for the other, and consequently for some color between the two extremes it has no double

---

* From tables by E. S. Larsen.

refraction. The same is true for some other species (*e.g.*, chabazite) of weak double refraction.

NEGATIVE CRYSTALS

	ω	ϵ	$\omega - \epsilon$
Proustite	2·979	2·711	0·268
Calcite	1·658	1·486	0·172
Tourmaline	1·638	1·620	0·018
Corundum	1·768	1·760	0·008
Beryl	1·584	1·578	0·006
Vesuvianite	1·720	1·715	0·005
Nephelite	1·542	1·538	0·004
Apatite	1·634	1·631	0·003

POSITIVE CRYSTALS

	ω	ϵ	$\epsilon - \omega$
Rutile	2·616	2·903	0·287
Cassiterite	1·997	2·093	0·096
Zircon	1·923	1·968	0·045
Brucite	1·559	1·580	0·021
Phenacite	1·654	1·670	0·016
Quartz	1·544	1·553	0·009
Leucite	1·508	1·509	0·001

Examination of Uniaxial Crystals in Polarized Light

392. Section Normal to the Axis in Parallel Polarized Light. — Suppose a section of a uniaxial crystal to be cut perpendicular to the vertical crystallographic axis. It has already been shown that light passing through the crystal in this direction suffers no double refraction; consequently, such a section examined in *parallel* polarized light behaves as a section of an isotropic substance. If the nicols are crossed it appears *dark*, or *extinguished*, and remains so when revolved.

393. Section Parallel to the Axis. — A section cut parallel to the vertical axis, as already explained, has two directions of light-vibration, one parallel to this axis, that of the extraordinary ray, and the other at right angles to it, that of the ordinary ray. A ray of light falling upon such a section with perpendicular incidence is divided into the two rays, ordinary and extraordinary, which travel on in the same path through the crystal,

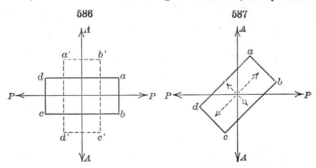

but one of them retarded relatively to the other. When such a section is examined in polarized light with crossed nicols it will appear dark, or be extinguished, when its vibration directions lie parallel to the vibration directions of the nicols. Assume that the section *abcd*, Fig. 586, lies with the direction of its vertical crystallographic axis parallel to *P–P*, which represents the vibration direction of the polarizer. The light entering the section under these conditions will be vibrating parallel to the vertical axis of the crystal and will therefore pass into the mineral wholly as the extraordinary ray, there being no vibration possible in the direction of the ordinary ray. The light will, therefore, leave the section with the same direction of vibration as when it entered and will be entirely lost by reflection in the analyzer. If the section is turned at an angle of 90°, as $a'b'c'd'$, Fig. 586, similar conditions prevail, although in this case the light will vibrate in the section as the ordinary ray. Therefore in such a section there will be four positions during its complete revolution on the stage of the polariscope or microscope when it will be extinguished.

If the section stand obliquely, as $abcd$ in Fig. 587, it will appear light to the eye (and usually colored), for the vibrations parallel to P–P that have passed through the polarizer have upon resolution a component in the direction of each of the vibration-planes of the section. Again, each of these components can be resolved along the direction of the vibration-plane of the upper nicol, A–A. Therefore, two rays will emerge from the analyzer, both having the same vibration-plane, but one more or less retarded with reference to the other, the amount of retardation increasing with the birefringence and the thickness of the section. In general, therefore, these rays will interfere, and if the thickness of the section is sufficient (and not too great) it will appear colored in white light and, supposing the thickness uniform, of the same color throughout.

394. Parallel Extinction. — When the vibration directions of a section coincide with those of the polarizer and analyzer, assuming them to be crossed, the section appears dark and it is said to be in the position of extinction. If a section extinguishes when its crystallographic axis or axial plane is parallel to one of the planes of vibration of the nicols it is said to show *parallel extinction*. If, on the other hand, no such parallelism exists between the crystallographic directions and the directions of vibration in the mineral the section is said to show *inclined extinction*.

In the case of uniaxial minerals, since the vibration directions always lie in some crystallographic axial plane, all sections of such minerals will show parallel extinction.

395. Determination of the Relative Character of the Extinction Directions of a Given Uniaxial Mineral. — The relative characters of the extinction directions of a section of a uniaxial mineral are to be determined by the use of the quartz wedge or the sensitive tint as described in Art. **354**. If the orientation of the section is known so that it can be told which of the directions of vibration belongs to the ordinary and which to the extraordinary ray the positive or negative character of the mineral can be determined. For instance, if the ordinary ray is proved to be the faster of the two (*i.e.*, the X direction) it follows that its index is the smaller, *i.e.*, $\omega < \epsilon$, and the mineral is positive.

396. Interference Colors of Uniaxial Minerals. Birefringence. — The interference color of any section of a uniaxial mineral depends upon the following: first upon the thickness of the section, second upon the strength of the double refraction of the mineral, *i.e.*, its birefringence, this being measured by the difference between the indices of refraction of the two rays in the section, and third upon the crystallographic orientation of the section. A section cut parallel to the basal plane shows no double refraction and therefore cannot exhibit any interference color. The strength of the birefringence, the other conditions remaining uniform, increases as the inclination of the section to the basal plane increases. The highest birefringence of a given mineral is therefore shown by its prismatic sections.

The following table* gives the thickness (in millimeters) of sections of a few uniaxial crystals which yield *red* of the first order:

	Birefringence $(\omega - \epsilon)$ or $(\epsilon - \omega)$	Thickness in Millimeters
Rutile	0·287	0·0019
Calcite	0·172	0·0032
Zircon	0·062	0·0089
Tourmaline	0·023	0·0240
Quartz	0·009	0·0612
Nephelite	0·004	0·1377
Leucite	0·001	0·5510

Again, as another example, it may be noted that with zircon ($\epsilon - \omega = 0·062$), a thickness of about 0·009 mm. gives red of the first order; of 0·017 red of the second order; of 0·026 red of the third order.

The methods ordinarily used to determine the birefringence of a section (not $\perp$ c axis) of a uniaxial crystal, as also to fix the relative value of its two vibration-directions, have already been discussed, see Arts. **353** and **354**.

397. Effects of Convergent Polarized Light upon Sections of Uniaxial Minerals. Uniaxial Interference Figures. — When certain sections of uniaxial minerals are observed in convergent polarized light they show what are known as *interference figures*. A symmetrical interference figure is obtained in uniaxial minerals by allowing converging polarized light to pass through a basal section of the crystal. Parallel polarized light entering such a section would suffer no double refraction and consequently give no interference. To convert the parallel polarized light that comes from the polarizer into convergent light a lens is placed between the polarizer and the section. Under these conditions a sharply converging cone of light rays enters the section. Another lens is placed above the section to change these oblique rays back again into a parallel position. Such an instrument is known as a *conoscope* and may be obtained by placing a pair of lenses between the polarizer and analyzer of a polariscope, or, in case the polarizing microscope is used, the small converging lens that lies above the polarizer is swung into position by a lever and at the same time a small lens known as the Bertrand lens is introduced into the microscope tube.

Under such conditions the light entering the section is composed of a converging system of rays polarized and vibrating in the plane P–P, Fig. 588. Let B–B (Fig. 588 A) be a vertical cross section of the mineral section along the line B–B, Fig. 588 B. Consider any ray, as a, entering the section. Since the ray enters the section obliquely it will be doubly refracted into the rays o and e. The mineral being taken as calcite the extraordinary ray (calcite being negative) will have the greater velocity and be less refracted. As the light enters the section in the form of a cone the traces of the two rays as they emerge from the section will be circles, Fig. 588 B. Now consider in a similar case the action of the two rays a and b or a' and b' (Fig. 589) upon each other. Ray a on entering the section is doubly refracted and polarized into the rays e and o which are considered as emerging from the section at the

---

* See further, Rosenbusch (Mikr. Phys. Min., 1904, p. 292), from whom these are taken.

points e and r. Ray b also on entering the section is doubly refracted and polarized. Suppose the extraordinary ray derived from b emerges from the section at the same point as the ordinary ray derived from a, that is at r. Since it travels with a greater velocity the extraordinary ray emerging at this point will have advanced in its phase over that of the ordinary ray. In that case they would be in a condition to interfere with each other except that they are vibrating in planes perpendicular to each other and so cannot. The two rays travel on, vibrating in planes at right angles to each other and maintaining this difference in phase until they reach the upper nicol; there they are each resolved into rays vibrating in the plane A–A, Fig. 588 B, and are now in condition to interfere with each other. Let it be assumed that the conditions are right for the extraordinary ray to emerge from the section just one wave-length ahead of the ordinary ray. Their components in the upper nicol will have opposite phases and therefore compensate each other, see Art. **341**. If the section is viewed in a monochromatic light (for instance,

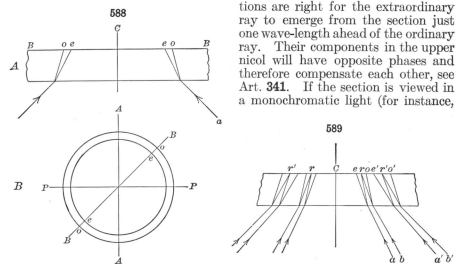

sodium light) this interference will result in a black point. But as these rays are converging in the form of a cone they will make, when they strike the section, a circular trace upon its surface and their interference will result in a dark ring. Going out from the center of the section there will be a succession of these rings corresponding to the interference of waves 1, 2, 3, 4, 5, etc., wave-lengths apart. As the distance from the center of the section is increased, the paths of the refracted rays in the section are lengthened and the points of interference are brought closer together. This will cause the interference rings to lie nearer together as the distance from the center of the figure increases.

Fig. 590 is a top view of the section without taking into consideration the effects of the upper nicol. Let the two circles represent the traces of the emergence of the two rays e and o into which one incident conical ray is divided; e, being the less refracted (for calcite), will be the inner one. The plane of vibration of e is always parallel to some plane passing through the vertical axis of the crystal, therefore the trace of its plane of vibration upon the surface of the section will always be in a radial direction. The plane of vibration of o is at right angles to that of the extraordinary ray and parallel to the horizontal axes of the crystal, therefore the trace of its plane of vibration upon the surface of the section will always be in a tangential direction, see Fig. 590. Along the line P–P, Fig. 590, only light vibrating in a radial plane or that of the extraordinary ray can come through the section, since the light entering the section cannot be resolved into the vibrations of the ordinary ray. The intensity and direction of vibration of the light that emerges from the section along the line P–P is represented by the double arrow on that line. Along the line A–A, since the light entering the section is still vibrating in the plane P–P, all the light passing through the section must vibrate as the ordinary ray. It is evident, therefore, that along these two directions, P–P and A–A the plane of vibration of the light is not changed by passage through the section and consequently such light will be completely absorbed in the upper nicol. In this way dark brushes will be formed along the lines P–P and A–A. These dark areas along which extinction occurs are known in both the uniaxial and biaxial interference figures, as *iso-gyres*. A dark spot will also be formed in the center of the field because any light entering the section at this point must enter in the direction of the optic axis and therefore will not be doubly refracted and consequently will also be absorbed in the analyzer.

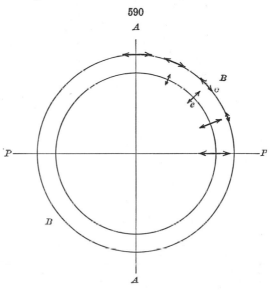

Now consider point B, Fig. 590, which lies 45° away from P and A. Here the directions of vibration of e and o would be equally inclined to the planes of vibration of the polariscope, A–A and P–P. Light striking the section at B would be vibrating in the plane P–P but by resolution a component vibrating in the direction B–B would come through the section as the ray e; in the same manner a component vibrating in a direction at right angles to B–B would emerge as o. The intensities and directions of vibration of these two rays at this point are represented by the double arrows. When these rays meet the analyzer above they would again each be resolved and their components which vibrate in the plane A–A would emerge from the analyzer. In this way it is seen that, except at the special points where complete interference takes place, light will result in the interference figure at all points away from the center of the figure and from the lines P–P and A–A. From the consideration of Fig. 590 it is evident that the greatest amount of light will come through the section at the 45° points, such as B. When viewed in monochromatic light, therefore, the interference figure consists of a series of concentric dark and light rings crossed by a vertical and a horizontal dark brush intersecting in the center of the field of the microscope, like Fig. 591.

If a basal section of a uniaxial mineral while in the conoscope is viewed in daylight colored rings will take the place of the light and dark rings observed in the monochromatic light. The change will be like that shown by the quartz wedge in the similar case described in Art. **349**. Where the first few dark rings near the center of the figure were formed by the interference of rays having the wave-length of sodium, light colored rings will result in the daylight illumination. These rings will be composed of all the components of white light with the yellow of sodium subtracted. The other colors are obtained in a similar manner by the elimination through interference of some particular wave-length of light. While the interference figure when illuminated in the monochromatic light showed a large number of distinct black rings, in daylight the corresponding colored rings are limited in number, and their colors, gradually becoming fainter as the distance from the center of the figure

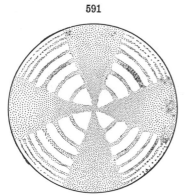

591

Uniaxial Interference Figure

increases, finally merge into the white of the higher order. This is due to the overlapping of the interference rings of the various colors in the same manner as observed in the quartz wedge, see Art. **349**. The interference figure viewed in daylight will of course retain the black cross and center since these are due to the cutting out of all the light by the analyzer and are not the result of interference.

The distance of each successive ring from the center of the interference figure obviously depends upon the birefringence, or the difference between the refractive indices, for the ordinary and extraordinary ray, and also upon the thickness of the plate. The stronger the double refraction and the thicker the plate, the smaller the angle of the light-cone which will give a certain amount of retardation, or, in other words, the nearer the circles will be to the center. Further, for the same section the circles will be nearer for blue light than for red, because of their shorter wave-length. When the plate is either quite thin or quite thick only the black brushes will be distinctly seen.

398. Determination of the Positive or Negative Character of the Birefringence of a Uniaxial Mineral from Its Interference Figure.

Use of the Mica Plate. — For the identification of a uniaxial mineral it is naturally important to determine whether the character of its birefringence is positive or negative. This can usually be best accomplished by tests made upon its interference figure. One of the common ways of making this test is by the use of a sheet of muscovite mica, cleaved so thin that, of the two rays of light passing through it, one has gained one quarter of a wave-length in phase over the other. The mica is usually mounted between long and narrow glass plates and is known as the one quarter wave-length mica plate. It is commonly marked *1/4M* with an arrow indicating the *Z* optical direction. In testing an interference figure by means of the mica plate the latter is inserted somewhere between the polarizer and analyzer (in the microscope commonly through the slot just above the objective) and is so orientated that the *Z* direction makes an angle of 45° with the planes of vibration of the nicols.

In Fig. 592 let *P–P* represent the plane of vibration of the polarizer and *A–A* the plane of vibration of the analyzer of a conoscope. Let *O* be the point of emergence of the optic axis of a positive uniaxial mineral.   Suppose a single conical ray of light enters the section.   It is broken up in the mineral into two rays, o and e, which emerge from the section along the arcs of the circles shown in Fig. 592.   The trace of the ordinary ray, o, will be within that of the extraordinary ray, e, because in a positive mineral the o ray travels the faster and is less refracted.   The directions of vibration of these two rays at the 45° points R and R' are represented by the double-headed arrows.   When these rays reach the analyzer they will be resolved into components vibrating parallel to A–A.   There are an infinite number of such rays entering and passing through the mineral section with varying angles of inclination and therefore varying lengths of path.   At some certain distance out from the center O two rays will emerge on the same circle with a difference of phase of one whole wave-length and when resolved in the upper nicol into rays vibrating in the same plane will interfere with each other and produce the first dark ring of the interference figure as it is viewed in monochromatic light.

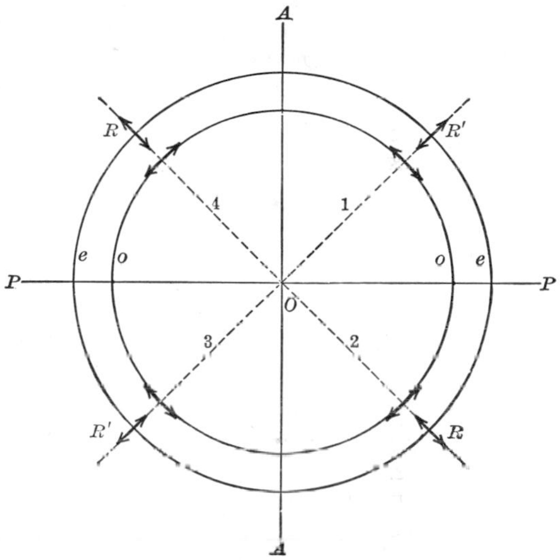

If the mica plate is introduced above the section a change in the interference figure is noted.   The optical character of the mica cannot be fully explained at this point.   It is sufficient for present purposes to know that it is a doubly refracting mineral which breaks light up into two rays which are polarized in planes at right angles to each other and which, traveling with different velocities through the mica, will emerge from it with different phases.   As stated above, the mica plate is cleaved to the requisite thickness so that the two rays emerge from it with a difference of phase of one quarter of a wavelength.   Consider what takes place when such a plate is introduced above the section represented in Fig. 592 in such a position that its vibration direction Z is parallel to the direction R–O–R of the figure.   Consider what takes place at the points R.   There the vibration direction of the e ray coincides with the vibration direction Z of the mica plate.   These vibration directions in each case are those of the rays traveling with the smaller velocity.   On the other hand, at the same point the vibration direction of the o ray in the mineral coincides with the vibration direction X in the plate, both of these being of the rays with the greater velocity.   So at this point the effect of the mica plate is to increase the difference of phase between o and e and to produce the same result as if the mineral section had been thickened.   Consequently the interference rings along the line R–O–R are increased in number and drawn toward the center of the figure.   At the points R' the opposite is true.   The vibration direction of e coincides now with that of X in the mica plate; the direction of less velocity in the mineral with that of the greater in the mica. Also the vibration direction of *o* coincides with that of *Z*; that of the greater velocity in the mineral with the less velocity in the mica. So at this point the mica will decrease the difference in phase between *o* and *e* and produce the effect of thinning the section and so spreading the interference rings farther apart along the line $R'-O-R'$. In quadrants 2 and 4, therefore, the rings will be drawn nearer the center, while in quadrants 1 and 3 they will be pushed farther out. Another effect caused by the introduction of the mica plate is even more pronounced. In quadrants 1 and 3, in the case illustrated in Fig. 592, black dots will appear near the center of the figure. In the interference figure, before the introduction of the mica plate, there were points in quadrants 1 and 3 at short distances from the center, *O*, where the two rays, *o* and *e*, emerged from the section with a difference of phase of one quarter wave-length. Under these conditions no interference could take place and these spots were light. The effect of the mica plate in these two quadrants is everywhere to reduce the birefringence due to the mineral by one quarter of a wave-length.

593

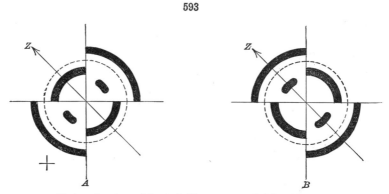

Determination of Optical Character with Mica Plate

Therefore at these two points the difference of phase caused by the birefringence of the mineral is annulled by the mica plate and consequently at these points interference will result and black dots appear. The mica plate produces still other effects. The brushes which were dark in the interference figure become light. Light coming from the crystal section along the lines of the brushes is vibrating only in the vibration direction of the polarizer and ordinarily is wholly cut out by the analyzer above. But with the mica plate intervening this light is broken up in the mica into two rays which vibrate in the vibration planes of the mica and as these are inclined to the plane of the analyzer a portion of the light will come through to the eye. As the light coming from the section along the lines of the brushes had only a single velocity (was entirely either the ordinary or extraordinary ray) there are only two rays emerging from the mica plate along these directions and their difference of phase is one quarter of a wave-length. Under these conditions there can be no interference and white brushes result. In the same way the dark center of the interference figure becomes light.

Fig. 593 *A*, is a diagrammatic representation of the interference figure of a positive mineral as affected by the insertion of the mica plate, the direction of the arrow indicating the direction Z of the mica, *i.e.*, the direction of vibration of the ray having the smaller velocity.   In the case of a negative mineral the conditions as described above will be completely reversed.   Fig. 593 B, represents the appearance of an interference figure of a negative mineral when the mica plate is used.

Therefore, to determine the optical character of a uniaxial mineral from its interference figure insert a mica plate above the section with its Z direction making 45° with the vibration planes of the nicols.   Then, if this direction Z is at right angles to a line joining the two black dots that appear near the center of the figure (*i.e.*, the two lines form a plus sign), the mineral is positive; if, on the other hand, these two directions coincide (form together a minus sign) the mineral is negative.

Use of the Sensitive Tint. — The sensitive tint, see Art. **350**, is used to determine the positive or negative character of a uniaxial mineral from its interference figure when the mineral section is so thin, or the mineral possesses such a low birefringence, as to show in the figure only a black cross without any rings. Under such conditions the mica plate would not give a decisive test. The sensitive tint is usually so mounted that its longer direction coincides with the direction of the vibration of the faster ray, *i.e.*, the direction X.

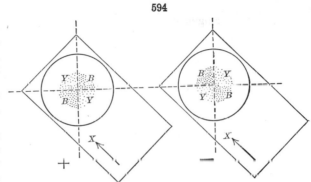

594

Determination of Optical Character with Sensitive Tint

The sensitive tint is introduced somewhere between the polarizer and analyzer in such a position that its vibration directions are at 45° with the planes of vibration of the nicols.   Let it be assumed that we have the interference figure from a positive mineral, such as is represented in Fig. 592.   If the sensitive tint is introduced in such a position that its X direction is parallel to the line R–O–R the X direction of the sensitive tint will be parallel to the direction of vibration of the e ray in the mineral.   Since the mineral is positive the e ray will have the smaller velocity and therefore in quadrants 2 and 4 the optical orientation of the mineral and the sensitive tint will be opposed to each other.   The sensitive tint alone would produce an interference color of red of the first order.   But if the effect of the birefringence of the mineral is such as to subtract from the birefringence of the sensitive tint the color will change to yellow.   Consequently in these quadrants yellow spots will appear near the center of the field at the points where the effect of the mineral has been sufficient to lower the interference color to that extent.   In the other quadrants, 1 and 3, the faster and slower rays of the mineral and sensitive tint coincide in their directions and the effect of the two substances is an additive one.   Consequently in these two quadrants the color will rise to blue.

In making the above test with the sensitive tint it is convenient to follow the rule that if the direction X of the sensitive tint crosses a line uniting the two blue dots (makes a plus sign) the mineral is positive; if, on the other hand, these two directions coincide (make together a minus sign) the mineral is negative. These conditions are illustrated in Fig. 594.

399. Interference Figures from Inclined Sections of Uniaxial Minerals. — It frequently happens that a mineral section under observation for an interference figure is not cut exactly parallel to the basal plane of the crystal. An interference figure obtained from such an inclined section will of course be eccentric to the microscope field. If the section is inclined only a little to the basal plane, the center of the figure (*i.e.*, the point of emergence of the optic axis) will still be within the field of vision and will move in a circle about the center of the field when the section is revolved upon the microscope stage. Fig. 595 *A*, shows the successive positions of such an interference figure during revolution. If the section is more sharply inclined the center of the interference figure may be quite outside the field. As the section is turned on the

595

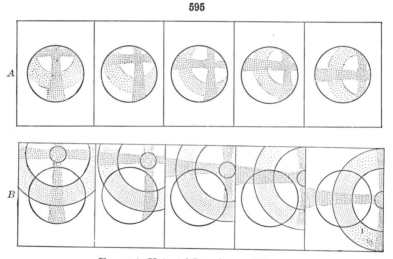

Eccentric Uniaxial Interference Figures

stage the four arms of the interference cross will traverse the field in succession. They will move across the field as straight bars and, provided the section has been cut not too highly inclined to the optic axis, will move across the field parallel to the cross-hairs of the microscope. This fact is of importance in order to distinguish such a uniaxial interference figure from certain biaxial figures. The latter will often show similar bars which, however, will usually curve as they cross the field of the microscope. If the first of these bars in the uniaxial figure moves from left to right across the field, the second will move from the top to the bottom, the third from right to left and the last from the bottom to the top, etc. Fig. 595 *B*, shows the different position of such a figure during one quarter of a revolution. If the section is considerably inclined to the optic axis, the black bars in the figure will show some curvature as they move across the field during the rotation of the section. They will cross the center of the field as straight bars, being parallel to either the horizontal or vertical cross-hairs of the microscope, but as they disappear from view the end away from the optic axis, which will be broader and more vague in its outline, will curve slightly, bending away from the cross-hair to which the bar as a whole is parallel.

The positive or negative character of the mineral can usually be determined from an eccentric figure if care is taken to make certain which quadrant is visible when the test is made. For instance, in Fig. 596 is shown how the test is made with the sensitive tint upon the eccentric interference figure of a positive mineral.

In examining unorientated sections of a mineral, such as the random sections found in a rock section or the small fragments of a mineral placed upon a glass slide, it is advisable always to hunt for that section that gives the lowest interference color. The amount of birefringence shown in various sections of a uniaxial mineral decreases as the section approaches the orientation of the basal plane. Consequently that section showing the lowest interference color will yield the most nearly symmetrical interference figure.

596

Blue

X

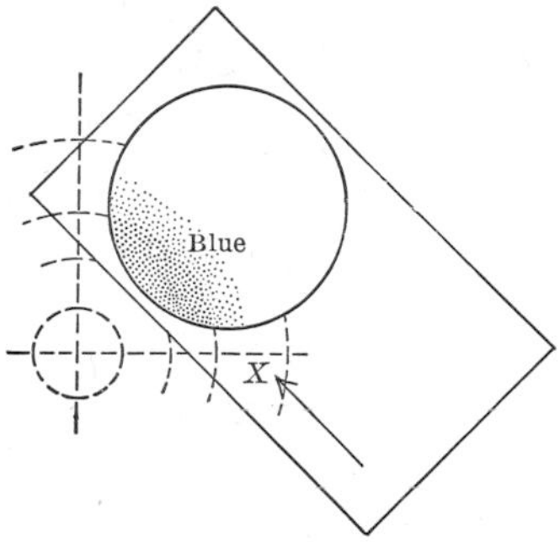

Sensitive Tint with
Eccentric Interference Figure

400. Interference Figure from a Prismatic Section of a Uniaxial Mineral.—When a prismatic section of a uniaxial mineral is examined for an interference figure the result is a figure which is analogous to one obtained in the case of biaxial crystals. The reasons for this resemblance will be pointed out in a later article. The two types of figures cannot be in this case easily differentiated. Two dark and usually indefinite hyperbolas approach each other as the section is turned on the micro-

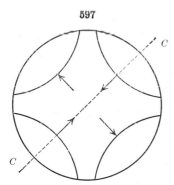

597

C

C

scope stage, form an indistinct cross, and rapidly separate and disappear. For this reason this figure is often called a "flash figure." These bars differ from those obtained in an ordinary biaxial interference figure in that they rapidly fade out as they move away from the crossed position. These vague bars always enter and leave the field in the quadrants which are bisected by the optic axis of the crystal. At the position of maximum illumination the field usually shows interference colors arranged in a pattern similar to that indicated in Fig. 597. The colors will rise in their order in the directions of the arrows shown in the figure. The optical axis — the direction $C-C$ — will bisect those opposite quadrants in which the colors fall in their order in passing outward from the center of the field. If it is possible to make definite observations of this kind, the direction of the optical axis can be determined and then by tests with plane polarized light the relative velocities of the ordinary and extraordinary vibrations can be found and so the optical sign of the mineral. It must be emphasized, however, that confusion can easily exist between this figure and a similar one obtained from certain sections of a biaxial crystal, see Art. **417**. This type of uniaxial figure can be easily obtained from the quartz wedge.

401. Absorption Phenomena of Uniaxial Crystals. Dichroism. — When light enters colored minerals as rays of white light, *i.e.*, containing vibrations of all wave-lengths from that of violet light at one end of the spectrum to that of red light at the other, certain wave-lengths will be absorbed during the passage of the light through the mineral, so that the light, as it emerges, has a definite color. It happens in certain deeply colored minerals that the amount and character of this absorption depends upon the direction of the light vibration. For instance in the case of uniaxial minerals, the ordinary and extraordinary rays may emerge from the section with distinctly different colors. Take, for instance, a prismatic section of a brown colored tourmaline and observe it in plane polarized light without the use of the upper nicol. As the section is revolved upon the stage of the polariscope the color may change from a dark brown to a light yellow-brown. The greatest difference in the

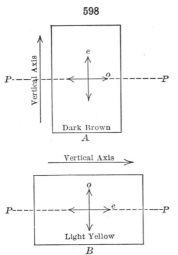

598

Dark Brown

A

Light Yellow

B

color occurs at positions 90° apart and when the crystallographic directions of the section, *i.e.*, the vertical crystallographic axis and the trace of the plane of the horizontal axes, are either parallel or perpendicular to the vibration plane of the polarizer. In other words, these extremes of color occur when the directions of the vibration of the ordinary and extraordinary rays in the section are parallel or perpendicular to the vibration plane of the light entering the section. In Fig. 598 *A*, let *P–P* represent the vibration direction of the light entering the section. The mineral section is so placed that the direction of the vertical crystal axis is perpendicular to *P–P*. The light on entering the section will therefore vibrate in the plane of the horizontal axes or as the ordinary ray, *o*. In this position the tourmaline section is dark colored and consequently it is seen that light vibrating in the mineral as the ordinary ray is

largely absorbed. Now turn the section through a 90° angle to the position shown in Fig. 598 *B*. In this position the light must vibrate in the section wholly as the extraordinary ray, *e*, and the color is a light yellow-brown. Therefore the extraordinary ray is only slightly absorbed. This difference in the absorption or the color of the two rays is known as *dichroism*. Either the ordinary or the extraordinary ray may be the most absorbed and the two cases are expressed as either $o > e$ ($\omega > \epsilon$) or $e > o$ ($\epsilon > \omega$). In uniaxial minerals dichroism is to be best observed in prismatic sections where it attains its full intensity. Basal sections show no dichroism, since light passing through the section parallel to the optic axis must all vibrate in the horizontal axial plane and belong wholly to the ordinary ray.

An instrument called a *dichroscope*, contrived by Haidinger, is sometimes used for examining this property of crystals. An oblong rhombohedron of Iceland spar is placed in a metallic cylindrical case, having a convex lens at one end, and a square hole at the other. On looking through it, the square hole appears double; one image belongs to the ordinary and the other to the extraordinary ray. When a pleochroic crystal is examined with it by transmitted light, on revolving it the two squares, at intervals of 90° in the revolution, have different colors, corresponding to the vibration-planes of the ordinary and extraordinary ray in calcite. Since the two images are situated side by side, a very slight difference of color is perceptible. A similar device is sometimes used as an ocular in the microscope.

402. Circular Polarization. — The subject of elliptically polarized light and circular polarization has already been briefly alluded to in Art. **356.** This phenomenon is most distinctly observed among minerals in the case of crystals belonging to the rhombohedral-trapezohedral class, that is, quartz and cinnabar.

It has been explained that a section of an ordinary uniaxial crystal cut normal to the vertical (optic) axis appears dark in parallel polarized light for every position between crossed nicols. If, however, a similar section of quartz, say 1 mm. in thickness, be examined under these conditions, it appears dark in monochromatic light only, and that not until the analyzer has been rotated so that its vibration-plane makes for sodium light an angle of 24° with that of the polarizer. In other words, this quartz section has rotated the plane of vibration some 24°, and here either to the right or to the left, looking in the direction of the light. The *amount* of this rotation increases with the thickness of the section, and, as the wave-length of the light diminishes (for red this angle of rotation for a section of 1 mm. is about 19°, for blue, 32°). The direction of the rotation is to the right or left, as defined above — according as the crystal is crystallographically right-handed or left-handed (p. 129).

If the same section of quartz (cut perpendicular to the axis) be viewed between crossed nicols in converging polarized light, it is found that the interference-figure differs from that of an ordinary uniaxial crystal. The central portion of the black cross has disappeared, and instead the space within the inner ring is brilliantly colored.* Furthermore, when the analyzing nicol is revolved, this color changes from blue to yellow to red, and it is found that in some cases this change is produced by revolving the nicol to the *right*, and in other cases to the *left*; the first is true with right-handed crystals, and the second with left-handed. If sections of a right-handed and left-handed crystal are placed together in the polariscope, the center of the interference-figure is occupied with a four-rayed spiral curve, called, from the discoverer, *Airy's spiral*. Twins of quartz crystals are not uncommon, consisting of the combination of right- and left-handed individuals (according to the Brazil law) which show these spirals of Airy. With cinnabar similar phenomena are observed. Twins of this species also not infrequently show Airy's spirals in the polariscope.

403. Summary of the Optical Characters of Uniaxial Crystals. — All sections of uniaxial crystals show double refraction except those that are cut parallel to the basal plane. All doubly refracting sections show parallel extinction. When viewed in convergent polarized light with crossed nicols all sections show a characteristic uniaxial interference figure except those that lie in the prism zone of the crystal or that are only slightly inclined to that zone. All doubly refracting sections have two refractive indices corresponding to the two extinction directions: one of these is *always* ω and the other has a value (ϵ') ranging from ω to ϵ, dependent on the inclination of the section to the optic axis. Dark colored minerals may show dichroism. Tetragonal and hexagonal substances cannot be distinguished from each other by optical tests. They may be at times told apart by characteristic cross sections of their crystals.

C. Biaxial Crystals

General Optical Relations

The crystals of the remaining systems, *i.e.*, the orthorhombic, monoclinic, and triclinic, belong optically to what is known as the Biaxial Group.

404. The Behavior of Light in Biaxial Crystals. — In biaxial crystals there are three especially important directions at right angles to each other which are designated as X, Y, and Z (also $\mathfrak{a}$, $\mathfrak{b}$, and $\mathfrak{c}$). These three directions are sometimes spoken of as *axes of elasticity* in reference to certain

---

* Very thin sections of quartz, however, show (*e.g.*, with the microscope) the dark cross of an ordinary uniaxial crystal.

assumed differences in the ether along them. The nature of these three directions is as follows. Light which results from vibrations parallel to X (axis of greatest elasticity) is propagated with the maximum velocity; that from vibrations parallel to Z (axis of least elasticity) with minimum velocity; and that from vibrations parallel to Y with an intermediate velocity. It is to be emphasized that these directions, X, Y, and Z refer to directions of vibration and not to directions of propagation. Corresponding to the maximum, intermediate, and minimum light velocities are three principal indices of refraction, designated respectively as α, β, and γ* (sometimes designated as N_p, N_m, and N_g). Of these α, belonging to light with the maximum velocity, will have the least value and γ belonging to light with the minimum velocity, will have the greatest value. The value of β will be intermediate between the other two, sometimes being nearer to α and at other times being nearer to γ; it is not the arithmetical mean between them. The various methods of determining the values of these three principal indices of refraction will be considered in a later article.

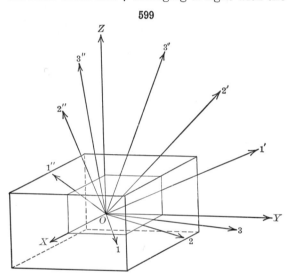

599

In studying the propagation of light within a biaxial crystal let it be assumed that Fig. 599 represents a rectangular parallelepiped in which the front to back axis is the direction X, the left to right axis is Y, and the vertical axis is Z. In connection with the figure and those which follow, it is helpful to make use of a model (a pasteboard box would answer) orientated so that its longer edge runs from front to back, its intermediate edge from left to right and its shortest edge vertical, corresponding to the X, Y, and Z directions of the figure. In the development of the figures that follow it has been assumed that the three principal indices of refraction are $\alpha = 1{\cdot}5$, $\beta = 1{\cdot}6$, $\gamma = 2{\cdot}5$, a difference between α and γ far exceeding anything observed in actual crystals. In general, this difference does not exceed $0{\cdot}1$; hence it is necessary to exaggerate the actual values greatly in order that the phenomena may be distinctly shown by diagrams drawn on a small scale.

In the discussion that follows it will be assumed that light originates at the center of a crystal, O, Fig. 599, and the endeavor will be made to determine the character of the rays which radiate from O in all directions. The simplest directions, and the ones which in reality are the most important, are those that lie in the axial planes of the figure, XOY, YOZ, and XOZ. These will be considered first.

---

* Some authors use the symbols α, β, and γ not only to indicate the three principal indices of refraction but also the vibration directions of the light (*i.e.*, X, Y, and Z) that correspond to these indices.

Consider the plane of the X and Y directions, Fig. 599.   Light will radiate from O toward X and Y and in all intermediate directions with vibrations parallel to Z and hence travel-ing with a uniform and at the same time minimum velocity, $1/\gamma$. The distance such light will travel in a given moment of time may be plotted by drawing a circle about O with the radius, $1/\gamma$, Fig. 600.   In the direction OX there must also travel a second polarized ray resulting from vibrations parallel to OY, hence travel-ing with intermediate ve-locity $1/\beta$.   Likewise in the direction OY there will be a ray resulting from vibrations parallel to OX, hence travel-ing with the maximum veloc-ity, $1/\alpha$.   In all directions in-termediate between X and Y the light velocities will be

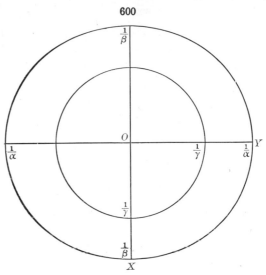

600

proportional to the radii of an ellipse having $1/\beta$ and $1/\alpha$ respectively as its semi-minor and semi-major diameters, Fig. 600.   In the plane of the X and Y directions, therefore, in a given moment of time light will radiate from the center as ordinary and extraordinary rays, the wave-fronts being represented by a circle within an ellipse.

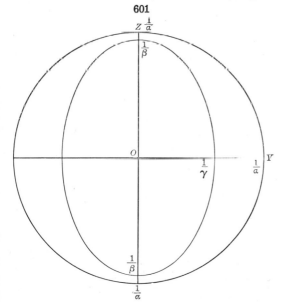

601

Consider next the plane of the Y and Z directions, Fig. 599.   Light will radiate from O toward Y and Z and in all intermediate directions re-sulting from vibrations paral-lel to OX.   It will therefore travel with uniform and the maximum velocity, $1/\alpha$. The distance traveled in a given moment of time may be plot-ted by drawing a circle about O with the radius $1/\alpha$, Fig. 601.   Likewise there will travel in the direction OY a second ray resulting from vi-brations parallel to OZ, hence moving with the minimum velocity, $1/\gamma$.   Also in the direction OZ there will be a ray resulting from vibrations parallel to OY with the velocity $1/\beta$.   In directions intermediate between Y and Z

the light velocities will be proportional to the radii of an ellipse having $1/\gamma$ and $1/\beta$ respectively as its semi-minor and semi-major diameters, Fig. 601.

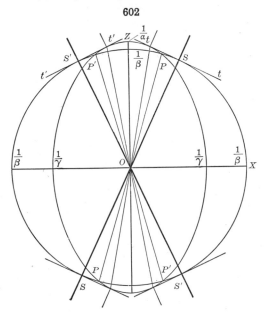

602

In the plane of the Y and Z directions, therefore, in a given moment of time, light will radiate from the center as ordinary and extraordinary rays, the wave-fronts being represented by an ellipse within a circle.

The last and most important plane to be considered is that of the X and Z directions, Fig. 602. Light will radiate from O toward X and Z and all intermediate directions with vibrations parallel to OY, hence traveling with a uniform and intermediate velocity, $1/\beta$. The distance traveled in a given moment of time is represented in Fig. 602 by the circle with the radius $1/\beta$. There will likewise travel in the direction OZ a ray resulting from vibrations parallel to OX, hence moving with the maximum velocity, $1/\alpha$.

Also a ray will travel in the direction OX with vibrations parallel to OZ, hence having the minimum velocity, $1/\gamma$. In intermediate positions the light velocity will be proportional to the radii of an ellipse with $1/\alpha$ and $1/\gamma$ respectively as its semi-major and semi-minor diameters, Fig. 602. In the plane of the X and Z directions, therefore, in a given moment of time, light will radiate from the center as ordinary and extraordinary rays, the wave-fronts represented by a circle intersecting an ellipse. It is to be noted that in this last plane there are four points where the two wave-fronts coincide. In other words, light traveling along the radial lines connecting these points

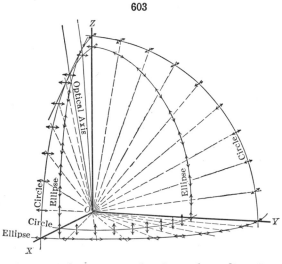

603

will be moving with uniform velocity and consequently along these directions there will be no double refraction. These directions are known as the optic axes of the crystal and since there are two of them the optical group is spoken of as biaxial. The character of these optic axes will be more fully developed in a later article.

In the above paragraphs the wave-fronts for light moving in the three principal optical planes of the crystal have been discussed. Fig. 603 represents the wave-fronts in these three planes as they appear when bounding one octant. The complete wave-surfaces for light propagated in all directions consist of warped figures which conform to the circular or elliptical wave-fronts already described in the three principal planes and have intermediate positions elsewhere. The only satisfactory way to represent these complete surfaces is by means of a model.

405. The Fresnel Ellipsoid. — The study of the behavior of light in biaxial crystals, especially from the mathematical point of view, has been greatly facilitated by the conceptions of two ellipsoidal solids, known as the Fresnel ellipsoid and the Fletcher indicatrix. The Fresnel ellipsoid — or vibration velocity ellipsoid — is an ellipsoidal solid, the three principal axes of which are made proportional to the velocities of light vibrating parallel to X, Y, and Z, respectively. From it can be derived the velocity and direction of vibration of any ray of light passing through the crystal in any direction. In general, a plane passed through the center of the ellipsoid will have an elliptical outline. The major and minor diameters of such an elliptical section will give the directions of vibration and their lengths will be proportional to the velocities of the two rays which can pass through the crystal in a direction normal to the chosen plane. Further, in such an ellipsoid there will be two sections that are circular in outline. Since all the radii of such sections are equal, light passing through the crystal in directions normal to them must have a uniform velocity and cannot be doubly refracted or polarized. The circular sections of the Fresnel ellipsoid are therefore perpendicular to what are known as the secondary optic axes of the crystal (see further Art. **407**).

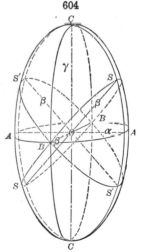

604

Biaxial Indicatrix

406. Biaxial Indicatrix.* — It is found further that the optical structure of a biaxial crystal can be represented by an ellipsoid, known as the *indicatrix*, having as its axes three lines which are at right angles to each other and proportional in length to the indices α, β, γ. This is analogous to the similar figure for uniaxial crystals described in Art. **390**.

This ellipsoid, whose axes represent in magnitude the three principal refractive indices, α, β, γ (where $\alpha < \beta < \gamma$) (see Fig. 604), not only exhibits the character of the optical symmetry, but from it may be derived the velocity and plane of vibration of any light-ray traversing the crystal.

In general, any section through the indicatrix will be an elliptical section. The major and minor diameters of such a section will yield the indices and planes of vibration of the two possible rays whose wave-fronts lie in this plane, or in other words, whose wave-normals coincide with the line perpendicular to the section. If

---

* L. Fletcher. The Optical Indicatrix and the Transmission of Light in Crystals. London, 1892.

this section happens to be one of the three principal sections of the indicatrix, $ABAB$, $ACAC$, or $BCBC$, Fig. 604, its major and minor diameters give the directions of vibration and their lengths the indices of refraction of the two rays. If the incident ray has some direction different from the directions of the three axes of the indicatrix ellipsoid the derivation of the character of the two refracted rays is not as simple. Let Fig. 605 represent such an elliptical section normal to $L-L$. In this case the major and minor diameters $R-O-R$ and $r-O-r$ of the elliptical section lie in the vibration planes of the two rays but the directions of vibration of the latter will be somewhat inclined to the elliptical section. These directions of vibration may be obtained by erecting normals to the surface of the indicatrix at the points R and r where the major and minor diameters of the elliptical section meet

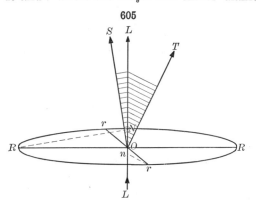

605

that surface. These normals RN and rn, when extended to the line $L-L$, yield the directions of vibration and the refractive indices of the two refracted rays. Their directions of transmission (the lines OS and OT) will be perpendicular to these normals and since neither of the latter lie in the elliptical section both rays will be refracted and behave as extraordinary rays.

There are two special sections of the indicatrix that require notice. The line $B-O-B$ (Fig. 604) is longer than the line $A-O-A$ but shorter than the line $C-O-C$. Obviously, in some position intermediate between $A-O-A$ and $C-O-C$ there will be a diameter of the ellipse $ACAC$ which will be equal in length to $B-O-B$. There are two such lines, as $S-O-S$ and $S'-O-S'$ in Fig. 604. The major and minor diameters of these sections of the indicatrix, $BSBS$ and $BS'BS'$, are equal and the sections therefore become circles. Consequently light passing through a section of a crystal cut parallel to either of these circular sections of its indicatrix will have a uniform velocity and may vibrate in any transverse direction. In other words, there will be no double refraction along the lines normal to these two sections. These lines constitute what are known as the primary optic axes of the crystal; see further in Art. **407**.

The major and minor diameters of any section of the indicatrix yield the traces upon that section of the planes of vibrations of the two rays whose wave-normals are perpendicular to the section. In other words, the major and minor diameters of the elliptical section of the indicatrix give the directions of extinction of a crystal section having this optical orientation. Further, these extinction directions bisect the angles made by the traces upon the section of two planes, each of which includes the pole of the section and one of the two optic axes. This may be demonstrated by aid of Fig. 606 which represents a general elliptical section of an indicatrix. $A-A$ and $B-B$ are the major and minor diameters of the ellipse and so represent the extinction directions of the mineral section. $C-C$ and $C'-C'$ represent the intersections of the two circular sections of the indicatrix with this elliptical section. As these lines are diameters of equal circles they must be equal in length and it therefore follows from the geometrical nature of an ellipse that the angles AOC and AOC' are equal.   Let the line P–P represent the intersection with this elliptical section of a plane in which lie the normal to the section and one of the optic axes. Since this plane includes an optic axis it must be perpendicular to the circular section of the indicatrix of which the line C'–C' is a diameter. Also since this plane includes the normal to the elliptical section under consideration it must be at right angles to the latter plane.   Under these conditions it is obvious that the lines P–P and C'–C' in Fig. 606 must be at right angles to each other.   In the same way it can be proved that the lines P'–P' and C–C are also at right angles to each other.   Since the angles AOC and AOC' are equal and the angles POC' and $P'OC$ are also equal it follows that the angles AOP and AOP' are likewise equal. In other words the lines A–A and B–B representing the directions of extinction of the section bisect the angles made by the traces upon the section of the two planes which respectively pass through each optic axis and the normal to the section.   This fact will be made use of later, see Art. **417**, in explaining the characters of the biaxial interference figure.

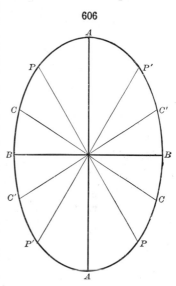

606

407. Primary and Secondary Optic Axes.   It has already been stated (Art. **406**) that there are two directions, namely, those normal to the circular cross sections of the indicatrix (SS, $S'S'$, Fig. 604) which are of such a character that all light having its wave-normals parallel to them travels in the crystal with uniform velocity.   These two directions bear so close an analogy to the optic axes of a uniaxial crystal that they are also called *optic axes*, and the crystals here considered are hence named *biaxial*.   In Fig. 602, which represents a cross section of the wave-surfaces in the plane of the X and Z directions, these optic axes have the direction SS, $S'S'$ normal to the tangent planes tt, $t't'$, and the direction of the external wave is given by the normal $S\sigma$ (Fig. 607).

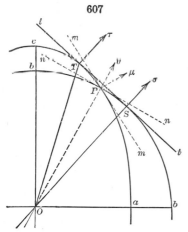

607

Properly speaking the directions mentioned are those of the *primary optic axes,* for there are also two other somewhat analogous directions, PP, $P'P'$, of Fig. 602, called for the sake of distinction the *secondary optic axes.*   The properties of the latter directions are obvious from the following considerations.

In the section of the wave-surface shown in Fig. 602 (also enlarged, in Fig. 607), corresponding to the axial plane XZ, it is seen that the circle with radius

$\dfrac{1}{\beta}$ intersects the ellipse whose major and minor axes are $\dfrac{1}{\alpha}$ and $\dfrac{1}{\gamma}$ in the four points P, P, P', P'. Corresponding to these directions the velocity of propagation is obviously the same for both rays. Hence within the crystal these rays travel together without double refraction. Since, however, there is no common wave-front for these two rays (for the tangent for one ray is represented by mm and for the other by nn, Fig. 607) they do suffer double refraction on emerging; in fact, two external light-waves are formed whose directions are given by the normals $P\mu$ and $P\nu$. These directions, PP, $P'P'$, therefore, have a relatively minor interest, and whenever, in the pages following, optic axes are spoken of, they are always the *primary* optic axes, that is, those having the directions SS, $S'S'$ (Fig. 602), or OS, Fig. 607. In practice, however, as remarked in the next article, the angular variation between the two sets of axes is usually very small, perhaps 1° or less.

408. Interior and Exterior Conical Refraction. — The tangent plane to the wave-surface drawn normal to the line OS through the point S (Fig. 607) may be shown to meet it in a small circle on whose circumference lie the points S and T. This circle is the base of the interior cone of rays SOT, whose remarkable properties will be briefly hinted at. If a section of a biaxial crystal be cut with its faces normal to OS, those parallel rays belonging to a cylinder having this circle as its base, incident upon it from without, will be propagated within as the cone SOT. Conversely, rays from within corresponding in position to the surface of this cone will emerge *parallel* and form a circular cylinder. This phenomenon is called *interior conical refraction.*

On the other hand, if a section be cut with its faces normal to OP, those rays having the direction of the surface of a cone formed by perpendiculars to mm and nn will be propagated within parallel to OP, and emerging on the other surface form without a similar cone on the other side. This phenomenon is called *exterior conical refraction.*

In the various figures given (600–607) the relations are much exaggerated for the sake of clearness; in practice the relatively small difference between the indices of refraction α and γ makes this cone of small angular size, rarely over 2°. The effects of conical refraction are negligible in all ordinary interference figures. Its existence does, however, explain the fact that sections of biaxial minerals cut normal to an optic axis show slightly less perfect extinction between crossed nicols than do similar sections of uniaxial crystals.

409. Optic Axial Angle. Bisectrices. Positive and Negative Biaxial Crystals. — The optic axes always lie in the plane of the X and Z optical directions; this plane is called the *optic axial plane* (or, briefly, ax. pl.). The Y direction is always normal to the axial plane and is therefore sometimes known as the *optic normal.* It is obvious from a consideration of the indicatrix ellipsoid that the position of its circular sections and consequently of the optic axes normal to them, will vary with a variation in the relative values of the indices of refraction. As already stated the index β is not an arithmetical mean between α and γ but may at times be nearer to α than to γ or the reverse. As these relations change, the shape of the indicatrix and the position of its circular sections and the angle between the optic axes will also change. The mathematical relations between the optic axial angle and the principal refractive indices are given in the next article. From the above it is obvious that for certain relative values of the refractive indices, the optic angle must be 90°.* Such a case, however, is rarely observed and when it

---

* The axial angle will equal 90° when the indices satisfy the following equation:

$$\frac{1}{\alpha^2} - \frac{1}{\beta^2} = \frac{1}{\beta^2} - \frac{1}{\gamma^2}.$$

occurs it is true for light of a certain color* (wave-length) only and not for others.

The X and Z optical directions bisect the angles between the optic axes and are therefore known as *bisectrices*. The one that bisects the acute axial angle is called the *acute bisectrix* (or Bx_a) while the one bisecting the obtuse angle is the *obtuse bisectrix* (or Bx_o). If the word *bisectrix* is used alone without special qualification it is always to be understood as referring to the acute bisectrix.

Either X or Z may be the acute bisectrix. If X is the acute bisectrix the substance is said to be *optically negative,* while if Z is the acute bisectrix it is *optically positive.*

Roughly expressed, the optic axes will lie nearer to Z than to X — that is, Z will be the bisectrix — when the value of the intermediate index, β, is nearer to that of α than to that of γ. It is obvious (cf. Fig. 602) that in this case, as the angle diminishes and becomes nearly equal to zero, the form of the ellipsoid then approaches that of the prolate spheroid of the positive uniaxial crystal as its limit (Fig. 583, p. 281); this shows the appropriateness of the + sign here used.

On the other hand, the optic axes will lie nearer to X than to Z — that is, X will be the bisectrix — if the value of the mean index β is nearer to that of γ than to that of α. Such a crystal, for which $Bx_a = X$, is called *optically negative*. In this case the smaller the angle the more the ellipsoid approaches the oblate spheroid of the negative uniaxial crystal (Fig. 582, p. 281).

The following are a few examples of positive and negative biaxial crystals:

Positive (+).	Negative (−).
Sulphur.	Aragonite.
Enstatite.	Hypersthene.
Topaz.	Muscovite.
Barite.	Orthoclase.
Chrysolite.	Epidote.
Albite.	Axinite.

110. Relation of the Axial Angle to the Refractive Indices. — If in a given case the values of α, β, and γ are known, the value of the interior optic axial angle known as $2V$ (see also Art. **418**) can be calculated from them by the following formulas:

$$\cos^2 V = \frac{\dfrac{1}{\beta^2} - \dfrac{1}{\gamma^2}}{\dfrac{1}{\alpha^2} - \dfrac{1}{\gamma^2}} \quad \text{or} \quad \tan^2 V = \frac{\dfrac{1}{\alpha^2} - \dfrac{1}{\beta^2}}{\dfrac{1}{\beta^2} - \dfrac{1}{\gamma^2}}.$$

In the majority of cases, the difference between $\gamma - \beta$ and $\beta - \alpha$ is small, and then a close approximation to the value of $2V$ can be obtained from the formula $\tan V = \sqrt{\dfrac{\beta - \alpha}{\gamma - \beta}}$. But in any case, the results of such calculations are usually not highly accurate, since a slight variation in the values of the indices of refraction will yield a disproportionate change in the value of the calculated angle.

---

* For danburite axial angle = 89° 14′ for green (thallium) and 90° 14′ for blue ($CuSO_4$).

Examination of Biaxial Crystals in Polarized Light

411. Sections in Parallel Polarized Light with Crossed Nicols.
Interference Colors. — Thin sections of biaxial crystals when examined between crossed nicols in general show some interference color. This color will depend upon the following factors: *the thickness of the section,* — the thicker the section the higher the order of color; *the birefringence of the substance,* — the higher the birefringence (*i.e.,* the greater the difference between the values of α and γ) the higher the order of color; *the optical orientation of the section,* — in general, the nearer the section comes to being parallel to the optic axial plane, in which lie the vibration directions of the fastest and slowest rays, the higher will be its birefringence and the order of its interference color.

Extinction Directions. — A section which, in general, is colored will show during a complete revolution on the microscope stage four positions at 90° intervals in which it appears dark. These are the positions of extinction, or are those positions in which the vibration planes of the section coincide with those of the nicols. When the directions of extinction of a section are parallel or at right angles to a crystallographic axis or to the trace, upon the section, of a crystallographic axial plane it is said to show *parallel extinction.* If the extinction directions are not parallel to these crystallographic directions the extinction is said to be *inclined.*

608

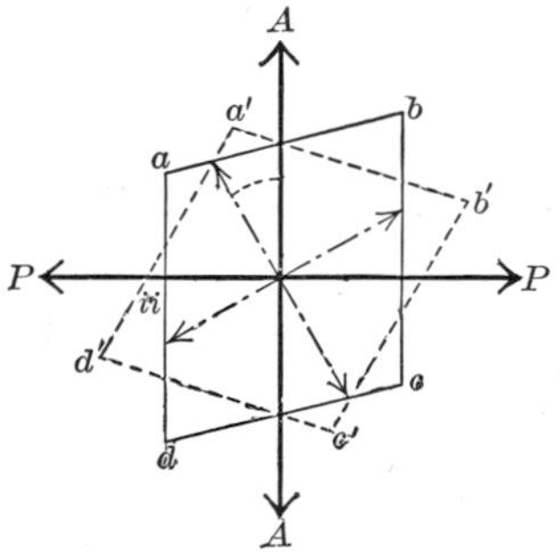

For example, in Fig. 608, let the two larger rectangular arrows represent the vibration directions for the two nicols, and between which suppose a section of a biaxial crystal, *abcd,* to be placed so that one edge of a known crystallographic plane coincides with the direction of one of these lines. The vibration directions of the section are indicated by the dotted arrows and as in this position of the section these directions do not coincide with the vibration directions of the nicols the section will appear light. The section will have to be turned to the position *a'b'c'd'* in order to achieve this coincidence and so bring about extinction. The angle (indicated in the figure) which it has been necessary to revolve the plate to obtain the effect described, is the angle which one of the vibration directions in the given plate makes with the given crystallographic edge *ad*; it is called the *extinction angle.* In some biaxial minerals the vibration directions for light of different colors may be sufficiently inclined to each other to permit complete extinction only when monochromatic light is used.

412. Measurement of the Extinction Angle. — It frequently becomes important to measure as accurately as possible the extinction angle of a section. This is most commonly done with a microscope which is provided with a revolving stage having a graduated circle for measuring angles of rotation. In order to measure an extinction angle it is of course necessary to be able to locate in the section some definite crystallographic direction. This is usually provided by some crystal outline or cleavage crack. This crystallographic direction is brought parallel to one of the cross-hairs of the microscope and the angular position of the microscope stage noted. Then the stage is rotated until the section shows its maximum darkness. The angle between these two positions is the angle of extinction desired. The difficulty in the measurement lies in the accurate determination of the position of maximum extinction. Frequently it is possible to rotate the microscope stage through an arc of one to two degrees without any appreciable brightening of the field. It will help in determining the point of maximum extinction if the plate is turned beyond the point of extinction until the first faint illumination is observed and then back in the other direction until the same strength of illumination occurs. The point half way between these two positions should be very close to the point desired. The measurements should be repeated a number of times and the average taken. It is also advisable to make the measurements on both sides of the position of the crystallographic direction. The illumination in most cases had better be in the monochromatic sodium-light.

Various devices are used at times in order to increase the accuracy with which the position of maximum extinction can be determined.* The sensitive tint is sometimes used for this purpose. If this is inserted in the diagonal slot of the microscope tube below the analyzer the field will be uniformly colored red of the first order when the section on the microscope stage is at the position of extinction. But if the section is turned, even very slightly, from this position it will also affect the light and change the interference color observed. The sensitive tint in specially favorable cases can be used in this way to advantage but it has been shown that in the majority of cases its use does not materially increase the accuracy of the measurements.

The power of quartz plates cut normal to the vertical crystallographic axis to rotate the plane of polarization of light (see Art. **402**) is used in other devices to increase the accuracy of the measurement of the angle of extinction. The *Bertrand ocular* contains four such sectors of quartz; two of these placed diagonally opposite to each other are from a right-handed quartz crystal while the other two are from a left-handed crystal. This ocular is inserted in the microscope tube in place of the regular ocular; the analyzer is pushed out of the microscope tube and a nicol prism mounted in an appropriate holder is placed over the ocular. If this upper nicol is turned about in various positions it will be noted that, in general, opposite quadrants of the field are colored alike but differ in color from the adjacent quadrants, see Fig. 609. But when the plane of the cap nicol is exactly at right angles to the plane of the polarizer below all four quadrants show the same color. If a double refracting mineral be placed on the stage of the microscope with its vibration directions parallel to those of the nicols, since in this position it has no birefringent effect upon the light, the field will still remain uniformly colored. But if the section is turned from its position of extinction its birefringent effect is added to that of two opposite quadrants of the ocular and subtracted from that of the remaining two. Consequently adjacent quadrants become differently colored. A very slight rotation of the section is sufficient to produce an appreciable effect.

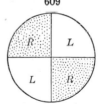

609

Bertrand Ocular

Another microscope accessory using the same principle as the Bertrand ocular is the so-called *bi-quartz wedge plate* described by Wright. This consists of two adjacent plates of quartz cut normal to the *c* crystal axis, one

---

* Detailed descriptions of these various devices with comment on their accuracy are given by F. E. Wright in The Methods of Petrographic-Microscopic Research.

from a left-handed and the other from a right-handed crystal. Above these are placed two wedges of quartz, a right-handed wedge above the left-handed plate, etc. At the point where the wedge is equal in thickness to the plate beneath there will be zero rotation of the light and between crossed nicols this will produce a dark line across the field. As the distance increases from this point the amount of rotation of the light increases equally but in opposite directions on either side of the central dividing line of the plate. Both halves of the plate will be equally illuminated if the mineral section is in the position of extinction, but if the latter is turned so that it adds or subtracts its birefringent effect to or from that of the quartz plate the two halves become differently illuminated. By moving the plate in or out a position can be found where this change in illumination is most marked. This quartz plate is used with a special ocular provided with a slot in such a position that the quartz plate may be introduced into the microscope tube at the focal plane of the ocular and with the medial line of the plate parallel to the plane of vibration of the polarizer. A cap nicol is used above the ocular.

413. Determination of the Birefringence with the Microscope. — The value of the maximum birefringence ($\gamma - \alpha$) is obviously given at once when the refractive indices are known. It can be approximately estimated for a section of proper orientation and of measured thickness by noting the interference-color as described in Art. **353**.

414. Determination of the Relative Refractive Power. — The relative refractive power of the two vibration-directions in a thin section is readily determined with the microscope (in parallel polarized light) by the method of compensation. This is applicable to any section, whatever its orientation and whether uniaxial or biaxial. The methods employed have already been described in Art. **354**.

A crystal-section is said to have *positive elongation* if its direction of extension approximately coincides with the ether-axis Z; if with X the elongation is *negative*. The same terms are also used, in general, according to the relative refractive power of the two directions.

415. Determination of the Indices of Refraction of a Biaxial Mineral. — The indices of refraction of a biaxial mineral are determined by the same methods as outlined previously, see Art. **333**, the only modification introduced being necessitated by the fact that three principal indices, α, β, and γ, are to be determined.

Measurement of the Angles of Refraction by Means of Prisms. — Two or three prisms must be used to determine the three indices. If three prisms are used they are cut so that their edges are parallel respectively to the X, Y, and Z directions of the mineral. In the case of an orthorhombic mineral, in which these directions are parallel to the directions of the three crystallographic axes, the prism edges would have to be respectively parallel to the a, b, and c crystal axes. In crystals of the monoclinic and triclinic systems the proper orientation of the three prisms is a matter of considerable difficulty. Each such prism will yield two refracted and polarized rays but only the one whose light has its vibrations parallel to the edge of the prism (to be determined by the use of a nicol) is considered. In certain cases all three indices may be obtained from two prisms. If one prism is cut so that not only is its edge parallel to one of the directions X, Y, and Z but so that its medial plane contains not only this direction but one other, then by the use of the method of minimum deviation an index may be determined from each of the two refracted rays. Or with a small angle prism cut so that one of its faces contains two of these directions the corresponding two indices may be determined when the method of perpendicular incidence is used upon this face. In making these measurements it is important to note the crystallographic directions parallel to which the different rays vibrate. In this way the optical orientation in respect to the crystallographic directions can be determined.

Method of Total Reflection. — The method of total reflection for determining the indices of refraction of a biaxial mineral has the obvious advantage that only polished plates of the mineral are required instead of carefully orientated prisms. In general, the plane surface of a plate will give with the total refractometer two boundaries of total reflection. Both of these shadows move when the section is rotated. Four readings should be taken corresponding to the maximum and minimum positions of each boundary. The largest and smallest angles read will give on calculation the values for the greatest and least indices of refraction, *i.e.*, γ and α. The mean index of refraction, β, can be derived from one of the other measurements. There are certain more or less complicated methods by which these two intermediate readings can be tested in order to prove which is the correct one for the index β. It is commonly simpler to make use of another plate having a different crystallographic orientation. It will be found that in the second plate one of the intermediate angles corresponds with one already observed on the first plate while the second angle shows no such correspondence. The angle that is common to the two plates is the one desired. If the plate is orientated so that its plane contains two of the three optical directions, X, Y, and Z, all three indices can be obtained easily from the single plate. In this case one of the boundaries of total reflection is stationary for different positions of the plate. This corresponds to the ray whose vibrations are normal to the surface of the plate. The other boundary will vary its position as the plate is rotated and yield at its maximum and minimum positions the angles corresponding to the other two indices of refraction.

416. By Immersion in Liquids of Known Refractive Index. — The methods employed are similar to those described in Art. **330**. In the case of biaxial minerals, however, it is important, if possible, to determine to what vibration direction the index determined belongs. If the mineral breaks up into irregular grains so that in a number of grains studied practically every optical orientation should be represented, the highest and lowest index values determined in a series of observations should at least approximate to the values for α and γ. Further, if it is possible to obtain interference figures (see Art. **417**) from the individual grains, those giving centered figures will yield the index β (from the vibration direction normal to the trace of the axial plane) and either α or γ. The index β will be given by any grain which lies perpendicular to the axial plane, whether its interference figure is centered or not. Such a grain can be recognized by the fact that in the interference figure a line joining the points of emergence of the optical axes (whether they lie within or outside the field) will in all cases pass through the center of the field. In case the mineral being studied has one or more good cleavages, the individual grains will in general lie on the cleavage planes. This limits the number of orientations available for study, but on the other hand a good cleavage plane will often lie parallel to one of the principal sections of the optical structure and therefore give definitely the values of two of the indices desired.

417. Sections of Biaxial Crystals in Convergent Polarized Light. — In general, sections of biaxial crystals when examined in convergent polarized light show interference figures. The best and most symmetrical figures are to be observed when the section has been cut perpendicular to a bisectrix, and preferably to the acute bisectrix. If such a section is examined under the conditions described in the case of uniaxial crystals, see Art. **397**, figures similar to those shown in Fig. 610 will be observed. When the axial plane, *i.e.*, the plane including the two optic axes, lies parallel to the direction of vibration of the polarizer the figure is similar to that of Fig. 610 *A*. When these two directions are inclined at a 45° angle the figure is like that shown in Fig. 610 *B*.

First consider the interference figure in the parallel position, Fig. 610 *A*, and when viewed in monochromatic light. It consists of two black bars that

610

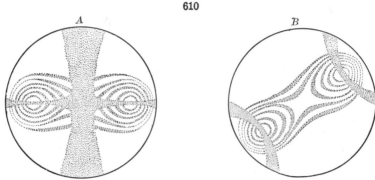

Biaxial Interference Figures

form a cross somewhat similar to the cross of a uniaxial figure. The horizontal bar is thinner and better defined than the vertical one. About two points on the horizontal bar, there will be observed a concentric series of dark elliptical curves which, as they enlarge, coalesce, forming first a figure eight and then a double curve. As the section is rotated on the microscope or polariscope stage, the black bars forming the cross separate at the center and curve across the field pivoting on these points until at the 45° position, Fig. 610 *B*, they form the two arms of a hyperbola.

A biaxial mineral has two directions, the directions of the optic axes, along which light travels with essentially no double refraction. At these points there would be no birefringence and consequently dark spots would result. As the paths of the light rays become inclined to the directions of the optic axes the light suffers double refraction and in increasing degree as the amount of inclination becomes greater. Consequently at short distances away from these points the light must be refracted into two rays which have a difference of phase of one wave-length for a certain colored light, the yellow of the sodium flame in this case. The result will be extinguishment at such points. The assemblage of all points where the difference of phase equals one wave-length yields the first dark elliptical curve, called a lemniscate, shown in the figure. Further out will be found curves embracing the points where the difference of phase is two wave-lengths, three wave-lengths, etc.

If the interference figure is viewed in daylight instead of the monochromatic light the black curves will be replaced by colored ones.   Each colored curve is produced by the elimination from the white light of some particular wave-length of light on account of the interference explained above.

The convergent bundle of light rays that pass through the section will each have its own particular plane of vibration.   The directions of the planes of vibration for light emerging from the section at any given point can be approximately found, as explained in Art. **406**, by bisecting the angles made by two lines connecting this point with the two points of emergence of the optic axes.   Fig. 611 shows how the direction of vibration of the two rays emerging from given points can be obtained in this way.   These directions of vibration vary over the field and consequently some of them must always be parallel or very nearly so to the planes of vibration of the nicol prisms.   When this happens the light is extinguished and darkness results.   This explains

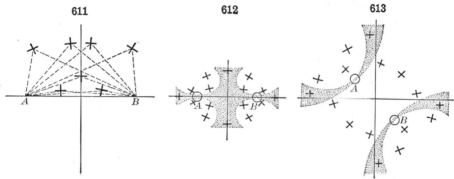

the formation of the black bars of the interference figure.   Fig. 612 shows the bars in the crossed position and Fig. 613 when separated into the hyperbola arms.   As the section is turned the vibration directions of new points successively become parallel to the planes of the nicols and so the dark bars sweep and curve across the field.

With a thick section or one of a mineral of high birefringence, the number of colored curves (when the figure is viewed in daylight) is greater than with a thinner section or one with low birefringence.   An instructive experiment can be made by noting the changes in the interference figure obtained from a section of muscovite as the mineral is cleaved into thinner and thinner sheets. In most rock sections the minerals are ground so thin that their interference figures do not show any colored curves but rather only the dark hyperbola bars.

The biaxial interference figure varies in appearance with the change in the angle between the optic axes.   Where this angle is very small the figure becomes practically the same as that of a uniaxial crystal.   Where this angle becomes greater than 60° the points of the emergence of the optic axes will commonly lie outside the microscope field.   In the latter case the hyperbola arms will appear as the section is brought into the parallel position, form a cross, and then as the section is further revolved will curve out of the field again.   The larger the axial angle the more rapidly will the bars disappear from the field.

A symmetrical interference figure may also be obtained from a section cut perpendicular to the obtuse bisectrix.   In general, the obtuse axial angle is considerably larger than the acute angle and the interference figure will differ therefore in this respect from that obtained from the section cut perpendicular to the acute bisectrix.   If, however, the axial angle approaches 90°, the distinction between the acute and obtuse interference figures is difficult to make. If the center of the figure coincides closely with the center of the field, it is possible sometimes to use the angle of rotation at which the hyperbolas leave the field as an indication as to whether the section is normal to the acute or obtuse bisectrix.   In case it is normal to the acute bisectrix, the angle of rotation necessary to take the hyperbolas out of the field is greater than in the case of the obtuse bisectrix.   It is reasonably certain that, if this angle (*i.e.*, the angle between the position where the hyperbolas form a cross and the point where they disappear from the field) is greater than 30°, the section is normal to the acute bisectrix; if the angle is less than 15°, it is normal to

614

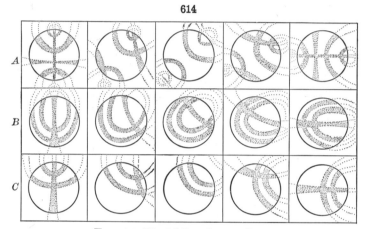

Eccentric Biaxial Interference Figures

the obtuse bisectrix.   For intermediate angles, unless the two sections of the same mineral are available for comparison, no positive decision can be made.

It is important to be able to recognize the biaxial interference figures which are obtained from inclined sections.   They are chiefly characterized by the fact that the hyperbola bars curve as they cross the field.   In the case where the section is normal to a plane of symmetry of the optical structure the bar will cross the center of the field in a straight line parallel to one or the other cross-hair.   In all other sections the bars are always curves.   This characteristic distinguishes the figure from an eccentric uniaxial figure in which the bars of the cross move in straight lines as the section is turned. Fig. 614 shows in the row *A* a series illustrating the appearance in different positions of the figure when the section is slightly inclined to the bisectrix; in row *B*, a series where the section is cut perpendicular to an optic axis and the hyperbola bar revolves in the field as upon a pivot.   In this case the bar curves with its convex side toward the acute bisectrix.   If the axial angle was 90° there would be no distinction between acute and obtuse bisectrices and the bar would then revolve as a straight line.   Therefore such a figure indicates by the amount of the curvature of the bar the size of the axial angle.   The figures given by planes cut nearly normal to an optic axis are often of great use in the optical examination of a mineral. Sections which will furnish them are easily found by noting those sections of the mineral that remain dark or nearly so during their rotation between crossed nicols. If the single bar shown in such a figure exhibits a decided curvature it indicates that the direction of the acute bisectrix is not very much inclined to the plane of the section and consequently its character, whether X or Z, can be determined by noting the character of that extinction direction which symmetrically bisects the curve (see Art. **422**). From this observation the positive or negative character of the mineral can be determined. In row C, Fig. 614, is shown a series of figures where the section has a still greater inclination. A section cut parallel to the axial plane does not give a decisive interference figure. Often it is difficult to distinguish it from the figure obtained from a section cut parallel to the optic axis of a uniaxial mineral, see Art. **400**. This similarity increases as the axial angle of the biaxial crystal decreases. When such an interference figure is rotated, broad and ill-defined hyperbolas pass rapidly into and out of the field. In the parallel position they may occupy practically all the field. When turned from this position the hyperbolas break and disappear in opposite quadrants. One pair of arms will usually disappear too quickly to be seen, but careful observation may establish into what opposite quadrants the other two arms disappear. The larger the optical angle of the crystal the more slowly they will leave the field. The line joining the two quadrants will be the direction of the acute bisectrix of the crystal. Also, if the figure is colored the line joining the quadrants which show colors of a lower order at equal distances from the center of the figure will be the direction of the acute bisectrix. Such an interference figure is often called a " flash figure." It must, however, be emphasized that cases are frequent where it is extremely difficult, if not impossible, to distinguish between the flash figures of uniaxial and biaxial crystals (see also Art. **400**).

418. Measurement of the Axial Angle. — The determination of the angle made by the optic axes is most accurately accomplished by use of the instrument shown in Fig. 615. The section of the crystal, cut at right angles to the bisectrix, is held in the pincers at p, with the plane of the axes *horizontal*, and making an angle of 45° with the vibration-plane of the nicols. There is a cross-wire in the focus of the eyepiece, and as the pincers holding the section are turned by the screw at the top (here omitted) one of the axes, that is, one black hyperbola, is brought in coincidence with the vertical cross-wire, and then, by a further revolution, the second. The angle which the section has been turned from one axis to the second, as read off at the vernier on the graduated circle above, is the *apparent* angle for the axes of the given crystal as seen in the air ($aca = 2E$, Fig. 616). It is only the *apparent* angle, for, on passing from the section of the crystal to the air, the true axial angle is more or less increased, according to the refractive power of the given crystal. The relation between the real interior angle and the measured angle is given below.

If the axial angle is large, the axes may suffer total reflection. In this case some oil or liquid with a high refractive power is interposed so that the axes will no longer be totally reflected but emerge into the liquid and thence into the air. In the instrument described a small receptacle holding the oil is brought between the tubes, as seen in the figure, and the pincers holding the section are immersed in this and the angle measured as before.

In the majority of cases it is only the acute axial angle that it is practicable to measure; but sometimes, especially when oil (or other liquid) is made use of, the obtuse angle can also be determined from a second section normal to the obtuse bisectrix.

615

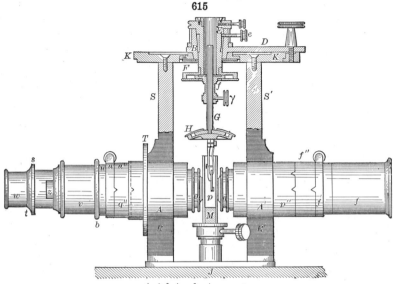

Axial Angle Apparatus

If E = the apparent semi-acute axial angle in air (Fig. 616),
H_a =   "   "   "   "   "   " in oil,
H_o =   "   "   semi-obtuse angle in oil,
V_a = the (real or interior) semi-acute angle,
V_o = the (real or interior) semi-obtuse angle,
n = refractive index for the oil or other medium,
β = the intermediate refractive index for the given crystallized substance, the following simple relations connect the various quantities mentioned:

616

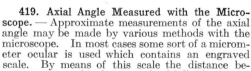

$$\sin E = \beta \sin V_a; \quad \sin E = n \sin H_a;$$

$$\sin V_a = \frac{n}{\beta} \sin H_a; \quad \sin V_o = \frac{n}{\beta} \sin H_o.$$

These formulas give the true interior angle $(2V)$ from the measured apparent angle in air $(2E)$ or in oil $(2H)$ when the refractive index (β) is known.

419. Axial Angle Measured with the Microscope. — Approximate measurements of the axial angle may be made by various methods with the microscope. In most cases some sort of a micrometer ocular is used which contains an engraved scale. By means of this scale the distance between the points of emergence of the optic axes can be determined. Mallard[*] showed that the distance of any point from the center of the interference figure as observed in

---

[*] Bull. Soc. Min., **5**, 77, 1882.

the microscope is very closely the same as the sine of the angle which the ray emerging at this point makes with the axis of the microscope. The Mallard equation for the derivation of the axial angle is $D = K \sin E$, in which D equals one half the measured distance between the optic axes and K a constant which varies with the microscope and the system of lenses used. K for a given set of lenses may be determined by observing the interference figures derived from plates of minerals with known axial angles and then substituting the values for D and E in the above equation. The angular values of the divisions on the micrometer scale of the ocular may also be determined directly by the use of an instrument known as the apertometer. The measurement of an axial angle by means of the microscope is naturally most easily accomplished when the points of emergence of both optic axes are visible in the field. It is possible, however, by various ingenious methods to determine its value when only one optic axis is in view, especially if the section is normal, or nearly so, to the optic axis. These methods are too complicated and too seldom used to be explained here and the reader is referred to the text-books on the methods of petrographic investigation for their details.*

420. The Universal Stage. — The universal stage, chiefly developed by Federov, is a substitute for the ordinary stage of the microscope, or it may be placed on top of it. It has four graduated circles which can be revolved on axes at right angles to them. By their use a given section may theoretically be turned to any desired optical and crystallographic position. The section is placed on a glass plate in the center of the stage or it may be mounted between two glass hemispheres with films of a high refracting liquid between the hemispheres and the section. This latter arrangement enables light to pass through the combination without any disturbing refractions. The universal stage is used to determine the optical orientation of a given section, to determine the position of the X, Y, and Z directions, the optical sign and the axial angle. Only low magnifications can be used with it, and the methods of investigation depend upon locating X, Y, and Z, by finding directions that show uniform extinction while the section is being rotated on a horizontal axis. After these three directions have been found and their positions noted on the various circular scales it is easy to establish the position of the axial plane. Then by rotating the section on to the Y direction as an axis the directions of the optical axes can be made vertical. When an optic axis is vertical the section will remain uniformly extinguished during rotation around a vertical axis. The optical sign and the axial angle can then be determined. It is impossible by this method to obtain interference figures, and it is necessary to have a thin section of a mineral of reasonable size. For the details of the method of investigation the reader is referred to the literature given below. †

421. Determination of the Optical Character of a Biaxial Mineral from Its Interference Figure. Use of the Quartz Wedge. — If the section is turned until its interference figure is in the 45° position and then the quartz wedge inserted above the section through the 45° slot in the microscope tube the vibration directions of the section along a line that joins the optical axes and a line at right angles to this through the center of the figure will be parallel to the vibration directions of the quartz wedge. Under these circumstances the effect of the introduction of the quartz wedge will be to gradually increase or diminish along these lines the birefringence due to the section alone. If the directions of vibration of the faster and slower rays in the quartz coincide with the vibration directions of the similar rays in the section, the total birefringence will be increased and the effect upon the interference figure will be as if the section had been thickened. Complete interference will take place with rays of less obliquity and the colored curves will be drawn closer together. They will move, as the quartz wedge is pushed in over the section, as indicated

---

* See especially Wright, The Methods of Petrographic Microscopic Research, and in Am. J. Sc., **24**, 317, 1907; Johannsen, Manual of Petrographic Methods; Phemister, J. Geol., **32**, 100, 1924.

† Johannsen. Manual of Petrographic Methods, p. 303. List of references to the original papers.

Wright. The measurement of the optic axial angle of minerals in thin section. Am. J. Sc., **24**, 317, 1907.

Winchell. Elements of Optical Mineralogy, p. 209, 1928.

by the arrows shown in Fig. 617. On the other hand, if the quartz wedge is so placed that its optical orientation is opposed to that of the section, the effect will be the same as if the section was being gradually thinned. The colored rings about the points of the optic axes will expand until they meet in the center as a figure eight and then grow outwards as a continuous curve. The directions of their movements are shown by the arrows in Fig. 618. Therefore, by knowing the optical orientation of the quartz wedge and noting the effect of its introduction over a section upon the interference figure, it is possible to determine the relative character of the two important extinction directions of the sections; that is, to determine whether the ray vibrating in

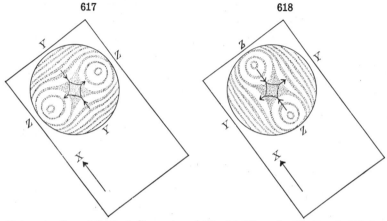

Determination of Optical Character of Biaxial Mineral with Quartz Wedge

the plane which includes the optic axes is faster or slower than the one which vibrates in the plane at right angles to this direction.

In the case of a positive mineral the acute bisectrix, which in a symmetrical interference figure is the direction normal to the section, is the direction Z. Consequently the direction of the line in the section which passes through the points of emergence of the two optic axes is the direction of the obtuse bisectrix, or in this case the direction X. The direction Y then will lie in the plane of the section and at right angles to the line joining the points of emergence of the optic axes. In the case, therefore, of a positive mineral, the faster ray has its vibrations lying in the optical axial plane. With a negative mineral the direction X becomes the acute bisectrix and will be normal to the section, while the direction Z will lie in the section along the line connecting the points of emergence of the optic axes. With a negative mineral, therefore, the vibration direction which lies in the optical axial plane is of the slower ray. By finding, therefore, the relative character of these two vibration directions the optical character of the mineral is determined. The effects produced by an interference figure which is perpendicular to an obtuse bisectrix would be exactly opposite to those described above. It is imperative, therefore, that the positions of the two bisectrices be definitely known. With sections that are very thin or with minerals of low birefringence the interference figure may show only the black hyperbolas without any colored rings. In such cases, frequently the introduction of the quartz wedge in such a position that its optical orientation is parallel to that of the section will suffice to so thicken the section in effect as to cause the appearance of colored rings. Further, with such sections it is possible to establish the directions in the section that are parallel and at right angles to the trace upon the section of the optical axial plane. Then, by use of the sensitive tint, when the convergent lens has been removed the character of the vibrations parallel to these two directions is easily determined.

422. Determination of the Optical Character of a Biaxial Mineral from the Interference Figure Obtained from a Section Normal to an Optic Axis. — The optical character may also be determined from the interference figure obtained from a section normal to an optic axis. As stated in Art. **417**, such a figure consists of a single bar, more or less curved, that rotates as on a pivot in the center of the field when the section is turned. The curve of the bar is always convex toward the acute bisectrix and concave toward the obtuse bisectrix. If the axial angle is 90° (or nearly), the bar becomes straight. In such a case the determination of the positive or negative character of the mineral is impossible, but fortunately such cases are rare. In this interference figure (Fig. 619) the trace of the axial plane bisects the curve of the bar and the optical direction Y is tangent to it.

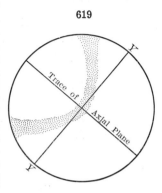

619

Light emerging at points beyond the center of the field in the direction of the acute bisectrix (*i.e.*, on the convex side of the curve) would have one of its vibrations parallel to Y and the other inclined toward the direction of the obtuse bisectrix. On the other side of the bar, in the direction of the obtuse bisectrix, light would have one of its vibrations still parallel to Y and the other inclined toward the acute bisectrix. These facts may be more clearly understood by a consideration of Fig. 620. It represents the principal elliptical section of the indicatrix of a biaxial mineral parallel to the optical axial plane. The section under consideration is normal to an optic axis, as P–P, and is represented in the figure by the line C–C, which lies in the circular section of the ellipsoid. The direction Y is normal to the plane of the figure at its center, O, and is always one of the diameters of any elliptical section of the indicatrix that is normal to the axial plane. Light is passing through this section as a cone of rays with varying inclinations. Therefore, for light passing in directions that lie between the lines P–P and Z–Z, the variable semi-diameter of the elliptical sections of the indicatrix that are normal to these rays will gradually decrease from the length of O–C (which is equal to Y) to that of O–X; that is, in all cases it will be less than Y. All such sections therefore will have the ray which vibrates in the axial plane partake more of the nature of the vibrations parallel to the obtuse bisectrix. In the same way, the elliptical sections normal to the

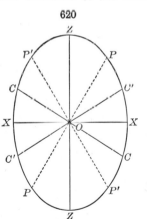

620

rays passing in directions between the lines $P–P$ and $X–X$ will have their variable semi-diameter increasing from the length $O–C$ (i.e., Y) to the length $O–Z$. In other words, light with such directions will have the ray which vibrates in the axial plane more nearly of the same velocity as the vibrations along the acute bisectrix. Therefore it is possible to determine the character of the acute and obtuse bisectrices from the figure and so the optical sign. The comparative velocities of the two vibrations on either side of the bar may be determined in various cases by the use of the quartz wedge, mica plate, or sensitive tint. If the sensitive tint can be used, the portions of the figure along the trace of the axial plane and on either side of the hyperbola bar will be colored yellow or blue. Knowing the orientation of the sensitive tint it is easy to determine whether the ray vibrating in the axial plane is faster or slower than the ray vibrating parallel to Y, and so establish the character of the two bisectrices. It is only necessary to remember that the color on the convex side of the hyperbola toward the acute bisectrix will establish the character of the obtuse bisectrix in relation to Y, and *vice versa*. With a mica plate, a dark spot will appear on one side or the other of the hyperbola. In the portion of the figure in which it appears, the vibrations in the mineral section and the mica plate are opposed to each other. When the quartz wedge is used, color bands will move toward the point of emergence of the optic axis in one half of the figure and away from it in the other. The movement of color bands toward the optic axis means correspondence between the optical orientation of the quartz wedge and the mineral section, and movement away from the optic axis indicates opposition.

423. Absorption Phenomena of Biaxial Crystals. Pleochroism. — Colored biaxial crystals like similar uniaxial crystals may show different degrees or kinds of absorption of the light passing through them depending upon the direction of vibration of the light. In biaxial crystals there may be three different degrees of absorption corresponding to three different directions of vibration lying at right angles to each other. In general, these directions coincide with the optical directions X, Y, and Z. Variations from this parallelism may be observed, however, in crystals of the monoclinic and triclinic systems. It is customary, however, to describe the absorption as it is observed parallel to the directions X, Y, and Z. If light vibrating parallel to X is the most absorbed and light vibrating parallel to Z is the least absorbed these facts are expressed as $X > Y > Z$. There are various other possibilities, such as $X > Y = Z, Z > X > Y$, etc. Further, according to the kind of selective absorption, the crystal may show distinctly different colors for light vibrating in the different directions, or in general show *pleochroism*. The character of the pleochroism is stated by giving the colors corresponding to the vibrations parallel to X, Y, and Z. For instance, in the case of *riebeckite*, X = deep blue, Y = light blue, Z = yellow-green. In order to investigate the absorption properties of a biaxial crystal at least two sections must be obtained in which will lie the directions X, Y, and Z. These sections are examined on the stage of the polariscope or microscope without the upper nicol. They will show as they are rotated upon the stage variations in absorption and in color as the light passing through them vibrates parallel to first one and then the other of their vibration directions. See the discussion of dichroism in uniaxial minerals, Art. **401.**

When a section cut normal to an optic axis of a crystal characterized by a high degree of color-absorption is examined by the eye alone (or with the microscope) in strongly converging light, it often shows the so-called *epoptic figures, polarization-brushes,* or *houppes* somewhat resembling the ordinary axial interference-figures. This is true of andalusite, epidote, iolite, also tourmaline, etc. A cleavage section of epidote $\|c(001)$ held close to the eye and looked through to a bright sky shows the polarization-brushes, here brown on a green ground. These figures are caused by the light being differently absorbed as it passes through the section with different degrees of inclination.

In certain minerals small circular or elliptical spots may be observed in which the color is deeper and the pleochroism stronger than in the surrounding mineral. These are commonly spoken of as *pleochroic haloes.* They are found to surround minute inclusions of some other mineral. There have been many diverse theories to account for these " haloes " but recently it has been shown that they are due to the ionization effect of the disintegration products of radioactive elements contained in the inclosed crystal. Pleochroic haloes have been observed in biotite, iolite, andalusite, pyroxene, hornblende, tourmaline, etc., while the included crystals belong to allanite, rutile, titanite, zircon, apatite, etc.

Special Optical Characters of Orthorhombic Crystals

424. Position of the Ether-axis. — In the ORTHORHOMBIC SYSTEM, in accordance with the symmetry of the crystallization, the three axes of the indicatrix, that is, the directions X, Y, and Z, *coincide* with the three crystallographic axes, and the three crystallographic axial planes of symmetry correspond to the planes of symmetry of the ellipsoid. Further than this, there is no immediate relation between the two sets of axes in respect to magnitude, for the reason that, as has been stated, the choice of the crystallographic axes is arbitrary so far as relative length and position are concerned, and hence made, in most cases, without reference to the optical character.

Sections of an orthorhombic crystal parallel to a pinacoid plane ($a(100)$, $b(010)$, or $c(001)$) appear dark between crossed nicols, when the axial directions coincide with the vibration-planes of the nicols; in other words, such sections show parallel extinction.

The same is true of all sections that are parallel to one of the three crystallographic axes, *i.e.,* sections lying in the prism, macrodome and brachydome zones. Sections, however, that are inclined to all three crystallographic axes, *i.e.,* pyramidal sections, will show inclined extinction.

425. Determination of the Plane of the Optic Axes. — The plane of the *optic axes,* that is, the plane including the directions X and Z, must be parallel to one of the three pinacoids. Further, the acute bisectrix must be normal to one of the two pinacoids that are at right angles to the optic axial plane while the obtuse bisectrix is normal to the other such pinacoid. The optical orientation, *i.e.,* the relation between the principal optical and crystallographic directions, can be easily determined by the examination of sections of a crystal which are cut parallel to the three pinacoids. To illustrate by an example, let it be assumed that such sections of the mineral aragonite are available. These are represented in Fig. 621 A, B, and C. If the relative characters of the vibration directions of each section are determined it will be found that light vibrating parallel to the c axis in sections parallel to (100) and (010) is in both cases moving with the greater velocity, that light vibrating parallel to the b axis in (100) and (001) is in both cases the slower ray, and that light vibrating parallel to the a axis is the faster ray in (001) but the slower ray in (010). From this it is seen that the a axis must coincide with the direction of vibration of the ray having the intermediate velocity, or be the same as the optical direction Y. Also it follows that the c axis $= X$ and the b axis $= Z$. The optic axial plane, therefore, since it must include X and Z, lies parallel to (100). If the sections parallel to (001) and (010) are examined in convergent light both will show biaxial interference figures with the points of emergence of the optic axes lying as illustrated in B and C, Fig. 621. The axial angle observed with the section parallel to (001) is much smaller than that obtained from (010). Consequently the acute bisectrix is normal to the

621

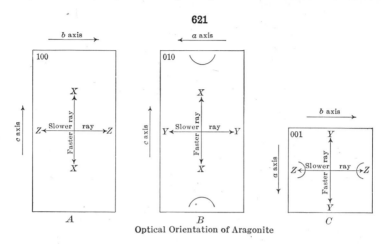

Optical Orientation of Aragonite

base (001) and since it is the direction X the mineral is optically negative. These facts of optical orientation may be summarized in the statements: optically $-$, Ax. pl. $\|$ a(100), Bx$_a$ $\perp$ c(001).

426. Dispersion of the Optic Axes in Orthorhombic Crystals. — In determining the indices of refraction of a crystal by means of the prism method it is to be noted that when the incident ray is of white light the refracted ray will in general show this white light dispersed into its primary colors. The

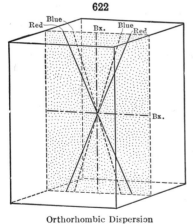

622

Orthorhombic Dispersion

amount of this dispersion is usually small but in certain substances becomes considerable. Obviously since the angle of refraction varies in this way with the different wave-lengths of light the indices of refraction will also vary. In biaxial minerals, as already stated, the optic axial angle is directly dependent upon the relative values of the three indices of refraction, α, β, and γ. As these indices may show considerable differences, depending upon the wave-length of the refracted ray, it follows that the optic axial angle will also vary with the color of the light used. In other words, the optic axes may be dispersed. Fig. 622 represents such a case in which the angle between the optic axes for red light is greater than that for blue. The opposite condition may hold, in which the angle for blue is greater than for red. From this it follows that the interference figure when observed in blue light will not exactly coincide with that produced by red light. The bisectrices of both figures will be the same but the position of the points where the optic axes emerge will be different and consequently the positions of the hyperbolas and lemniscate curves will also be different.   In the case of orthorhombic crystals the dispersion will always be symmetrical to the two symmetry planes of the indicatrix that pass through the acute bisectrix, *i.e.*, the directions *M–M* and

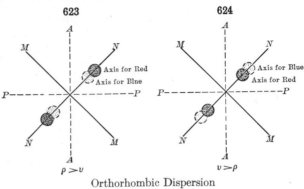

Orthorhombic Dispersion

N–N in Figs. 623 and 624.   This particular type of dispersion is said to be *orthorhombic dispersion*, in order to distinguish it from that observed in biaxial crystals of other systems. The two possible cases of orthorhombic dispersion are shown in Figs. 623 and 624.   In expressing these two cases the Greek letters ρ (for red) and υ (for violet) are used.   When the axes for red light are more dispersed than those for blue that fact is expressed as $\rho > \upsilon$ or in the reverse case it is $\rho < \upsilon$.

In the majority of cases the effect produced upon the interference figure by the dispersion of the optic axes is too slight to be noted.   In exceptional cases where the amount of dispersion is large the effects are clearly seen.   The hyperbola bars, which are ordinarily black throughout, will, when the figure is observed in white light, be seen, near the center, to be bordered on one side by a red fringe and on the other by a blue one.   The first one or two of the colored lemniscates will also be broadened out along the line joining the two optic axes.   As already stated these changes in the appearance of the figure will always be symmetrical in respect to the traces of the two symmetry planes lying at right angles to each other.   In the case, Fig. 623, where the axes for red light are farther apart than those for blue ($\rho > \upsilon$), the hyperbolas in the interference figure for the two different wavelengths of light will not coincide and the ones where the red light is extinguished will be farther out than

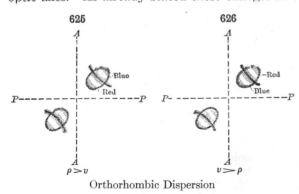

Orthorhombic Dispersion

those for blue light.   When red light is taken out of the white light, blue remains, and conversely when blue is subtracted the resultant color is red. Consequently in this case the hyperbola bars will be bordered on their concave sides by blue and on their convex sides by red, Fig. 625.   In the other case, where $\rho < \upsilon$, the hyperbolas will be bordered on their concave sides by red and on their convex sides by blue, Fig. 626.   In other words, if blue light shows at the larger angle it means that red light has been eliminated from these positions and the optic axes for red are more dispersed than those for blue, etc.

Special Optical Characters of Monoclinic Crystals

427. Optical Orientation of Monoclinic Crystals. — In monoclinic crystals there is one axis of symmetry, the b crystallographic axis, and one plane of symmetry, the plane of the a and c crystallographic axes. These are the only crystallographic elements that are definitely fixed in position. One of the three chief optical directions, X, Y, or Z, is coincident with the b crystallographic axis, while the other two lie in the symmetry plane, (010), but not parallel to any crystal direction. There are obviously three possible cases.

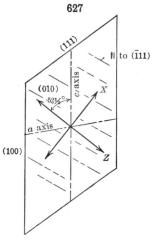

627

Optical Orientation of Gypsum

If Y coincides with the axis b (and this is apparently the most common case) the directions X and Z will lie in the crystal symmetry plane, which therefore becomes the optic axial plane. If X or Z coincides with the b axis the optic axial plane will be at right angles to (010) and either the acute or obtuse bisectrix will be normal to that plane. This clino-pinacoid of a monoclinic crystal is usually the best plane upon which to study its optical orientation. Fig. 627 represents such a section cleaved from an ordinary crystal of gypsum. The cleavages parallel to (100) and ($\bar{1}11$) will serve to give its crystallographic orientation. Examination of the section in convergent light fails to show a distinct interference figure, consequently it is to be assumed that the section itself is parallel to the optic axial plane and that the direction Y is normal to the section. When the section is rotated on the microscope stage between crossed nicols its extinction directions are seen to be inclined to the direction of the c crystallographic axis, the angle of inclination being measured as $52\frac{1}{2}°$. The relative character of the two extinction directions can be easily determined by the use of the quartz wedge and so the position of X and Z established. In this way the orientation of the X, Y, and Z directions can be determined. It is also possible from this section to determine whether the mineral is optically positive or negative. If the section is viewed in convergent light a somewhat vague interference figure is observed. When the section is turned from its position of extinction it will be noted that faint dark hyperbolas move rapidly out of the field. Careful observation will show that they disappear more slowly into one set of quadrants than into the other. The line bisecting the opposite quadrants into which the hyperbola bars disappear more slowly is the direction of the acute bisectrix. The X or Z character of this direction can be determined and from this the positive or negative character of the mineral. In a similar way the clino-pinacoid section of crystals belonging to the two other possible classes would yield data concerning their optical orientations.

428. Extinction in Monoclinic Crystals. — Since only one of the three principal optical directions, X, Y, or Z, of a monoclinic crystal coincides with a crystallographic axis, namely the symmetry axis b, it follows that only sections that are parallel to this axis, *i.e.*, sections in the orthodome zone, will show parallel extinction. All other sections will exhibit inclined extinction.

429. Dispersion in Monoclinic Crystals. — As previously stated there are three possible optical orientations of a monoclinic crystal. In the first case the vibration direction Y coincides with that of the symmetry axis b and the optic axial plane coincides with the symmetry plane (010). In the other cases either the vibration direction X or Z

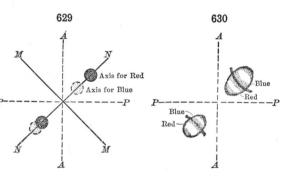

Inclined Dispersion   Inclined Dispersion $\rho > v$

coincides with the crystallographic axis b and the optic axial plane is at right angles to the crystallographic symmetry plane. Under these conditions either the acute or obtuse bisectrix may coincide with the axis b. Each of these three possibilities may produce a different kind of dispersion. It should be emphasized that the phenomenon of dispersion is seldom to be clearly observed and then commonly only in unusually thick mineral sections.

Case 1. *Inclined Dispersion.* — Inclined dispersion is observed in the case where the direction Y coincides with the axis b. This is illustrated in Fig. 628. In this case not only may the axial angles vary for light of different wavelengths but the bisectrices of these angles may lie along different lines. So, here, both the optical axes and the bisectrices may be dispersed. In Fig. 628 with $\rho > v$ the angle between the optic axes for red light is greater than that for blue. But because of the dispersion of the bisectrices it follows that on one side the point of emergence of the optic axis for red light lies beyond that for blue, while on the other side the conditions are reversed. Also the optic axes for red

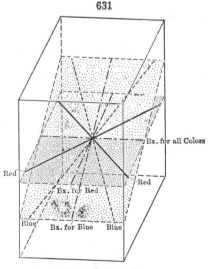

Horizontal Dispersion

and blue will be farther apart on one side of the interference figure than on the other side. With this sort of dispersion the interference figure will be symmetrical only in respect to the line which is the trace upon the section

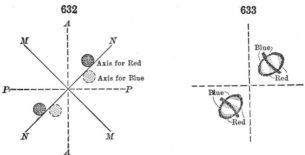

Horizontal Dispersion $\rho > v$

of the optic axial plane, $N-N$, Fig. 629, but is unsymmetrical to the line at right angles to it, $M-M$.

Inclined dispersion is shown in the interference figure by the fact that the colored borders to the hyperbola bars are reversed in the two cases, *i.e.*, if blue is on the concave side of one, red will be on the concave side of the other. Further, the amount of dispersion shown is much greater with one bar than with the other. Fig. 630 represents a case of inclined dispersion.

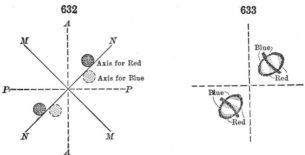

Crossed Dispersion

Case 2. *Horizontal Dispersion.* — In this case the crystallographic axis b coincides with the obtuse bisectrix which may be either the X or Z direction, depending upon whether the crystal is optically positive or negative in character. In this case the direction of the obtuse bisectrix is fixed for light of all wave-lengths. The angle between the optic axes may vary and further the position of the acute bisectrix may vary as long as it lies in the crystallographic symmetry plane. In other words, the axial planes may be dispersed, see Fig. 631. The points of emergence of the optic axes, when $\rho > v$, for blue and red light, might therefore be like that shown in Fig. 632. It will be noted that in this case the interference figure (obtained of course from a section approximately perpendicular to the acute bisectrix) is symmetrical to the line $M-M$ but unsymmetrical in respect to the line $N-N$. Fig. 633 shows the effect of horizontal dispersion upon the interference figure.

Case 3. *Crossed Dispersion.* — In this case the crystallographic axis b coincides with the acute bisectrix, which may be either the X or Z direction depending upon the optical character of the crystal. In this case the direction of the acute bisectrix is fixed for light of all wave-lengths. The angle between the optic axes may vary and further the position of the axial planes for different wave-lengths may vary as long as they remain perpendicular to the crystallographic symmetry plane. A case of this sort is shown in Fig. 634. The points of emergence of the optic axes when $\rho > v$ for blue and red light might therefore be like that shown in Fig. 635. It will be seen that in this case the figure is symmetrical to neither the line $M-M$ nor $N-N$ but only to the central

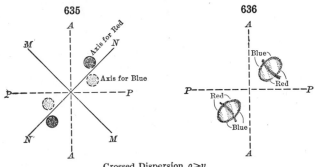

Crossed Dispersion $\rho > v$

point of the figure, *i.e.*, the point of emergence of the acute bisectrix. Fig. 636 shows the effect of crossed dispersion upon the interference figure.

Special Optical Characters of Triclinic Crystals

430. Optical Orientation of Triclinic Crystals. — The center of the optical ellipsoidal figure coincides with the center of the system of crystallographic axes but there is no further correspondence between optical and crystallographic directions.

431. Extinction in Triclinic Crystals. — Since there is no parallel relation existing between optical and crystallographic directions in triclinic crystals all sections will show inclined extinction.

432. Dispersion in Triclinic Crystals. — Because of the lack of coincidence between any optical and crystallographic direction in triclinic crystals it follows that the optic axes and bisectrices for different wave-lengths of light may be dispersed in any direction. Consequently the dispersion shown in an interference-figure obtained from a triclinic crystal is irregular and without symmetry.

433. Suggestions as to Methods and Order of Optical Tests upon an Unknown Mineral. — *Preparation of Material.* — The size and character of the fragments or section of a mineral to be studied will depend upon various circumstances. In the majority of cases it will probably be most convenient to crush the mineral into small uniform sized fragments. In other cases a cleavage flake of the mineral will serve, and under still other conditions it may be preferable to cut an unorientated or, better, an orientated section. For at least the preliminary examination small irregular fragments of varying orientation will most often be used. Take a few of these mineral grains and place them upon an object glass and immerse them either in Canada balsam or in some oil with known refractive index and cover with a piece of thin cover glass. In the majority of cases it will prove more expeditious and convenient to place the fragments in an oil.

Order of Optical Tests. — Below is given a brief outline of the natural order of observations and tests to be made upon the mineral.

1. *Observations in plane polarized light without the upper nicol.*

   a. Note color of mineral, whether uniform or not.
   b. By rotating slide on microscope stage test for possible pleochroism. If the mineral exhibits pleochroism it cannot be isotropic. Connect as far as possible the directions of absorption with crystallographic directions.
   c. Note crystal outline, if any; cleavage cracks, etc.
   d. Note any inclusions, their shape and arrangement.
   e. Index of refraction. Determine approximately the refractive index. Note character of relief and determine whether mineral has a higher or lower index than the medium in which it is immersed (see Art. **330**).

2. *Observations in plane polarized light with crossed nicols.*

   a. If the section is dark between crossed nicols and remains so during the rotation of the stage the mineral is either isotropic or orientated perpendicular to an optic axis. In the latter case test as indicated below under 3a.
   b. If the section is alternately light and dark during the rotation of the stage the mineral is anisotropic.
   c. Note position of extinction directions. If they are inclined to some known crystallographic direction measure the angle of inclination.
   d. Determine the relative character of the two vibration directions of the section (*i.e.*, the two extinction directions), as to which corresponds to the faster and which to the slower ray. Test to be made with quartz wedge or sensitive tint, see Art. **354**.
   e. Find the grain showing the highest order of interference color and so approximately determine the strength of the mineral's birefringence.
   f. By immersion in oils of known refractive indices determine as accurately as possible the range of the refractive indices shown by the mineral. It may be possible in connection with tests made under 3 to determine the values for certain of the principal refractive indices.

3. *Observations in convergent polarized light with crossed nicols.*

   a. Note whether the mineral shows an interference figure, and if so whether it is uniaxial or biaxial.
   b. If mineral is uniaxial determine the position of the optic axis in respect to the plane of the given section and if possible determine the positive or negative character of the mineral.
   c. If the mineral is biaxial determine the position of the axial plane in respect to the section. Determine, if possible, the positive or negative character of the mineral. Obtain, if possible, an approximate idea as to the size of the axial angle. Note any evidences of dispersion.

Note. — In making the above tests it is helpful to keep, as far as possible, a graphic record of the results, something like that illustrated in Fig. 621.

434. Effect of Heat upon Optical Characters. — The general effects of heat upon crystals as regards expansion, etc., are spoken of later. It is convenient, however, to consider here, briefly, the changes produced by this means in the special optical characters. It is assumed that no alteration of the chemical composition takes place and no abnormal change in molecular structure. In general, the effect of a temperature change causes a change in the refractive indices. In the majority of cases the indices decrease in value with rise of temperature but in certain cases the reverse is true. It is consequently important in any exact statement of a refractive index to give the temperature at which it was determined. The particular facts for the different optical classes are as follows:

(1) *Isotropic* crystals remain isotropic at all temperatures. Crystals, however, which, like sodium chlorate ($NaClO_3$ of Class 5, p. 88), show circular polarization may have their rotatory power altered; in this substance it is increased by rise of temperature.

(2) *Uniaxial* crystals similarly remain uniaxial with rise or fall of temperature; the only change noted is a variation in the relative values of ω and ϵ, that is, in the strength of the double refraction. This increases, for example, with calcite and grows weaker with beryl and quartz. It is, further, interesting to note that the rotatory power of quartz increases with rise of temperature, but the relation for all parts of the spectrum remains sensibly the same.

(3) With *Biaxial crystals*, the effect of change of temperature varies with the system to which they belong.

The axial angle of biaxial crystals may be measured at any required temperature by the use of a metal air-bath. This is placed at P (Fig. 615) and extends beyond the instrument on either side, so as to allow of its being heated with gas-burners; a thermometer inserted in the bath makes it possible to regulate the temperature as may be desired. This bath has two openings, closed with glass plates, corresponding to the two tubes carrying the lenses, and the crystal-section, held as usual in the pincers, is seen through these glass windows. Suitable accessories to the refractometer also allow of the measurement of the refractive indices at different temperatures.

In the case of *orthorhombic* crystals, the position of the three rectangular ether-axes cannot alter, since they must always coincide with the crystallographic axes. The values of the refractive indices, however, may change, and hence with them also the optic axial angle; indeed a change of axial plane or of the optical character is thus possible.

With *monoclinic* crystals, one ether-axis must coincide at all temperatures with the axis of symmetry, but the position of the other two in the plane of symmetry may alter, and this, with the possible change in the value of the refractive indices, may cause a variation in the degree (or kind) of dispersion as well as in the axial angle.

With *triclinic* crystals, both the positions of the ether-axes and the values of the refractive indices may change. The observed optical characters may therefore vary widely.

A striking example of the change of optical characters with change of temperature is furnished by gypsum, as investigated by Des Cloizeaux and recently by others. At ordinary temperatures, the dispersion is inclined, the optic axial plane is $\parallel b(010)$ and $2E_r = 95°$. As the temperature rises this angle diminishes, and finally the axial angle becomes zero (for a given wave-length of light) at temperatures about 90° C. Above this temperature the axial plane becomes perpendicular to $b(010)$ and the dispersion becomes horizontal.

Another interesting case is that of glauberite. Its optical characters under normal conditions are described as follows: Optically $-$. Ax. pl. $\perp b(010)$, Bx$_{a\cdot r} \wedge c$ axis $= -31°\,3'$, Bx$_{a\cdot y} \wedge c$ axis $= -30°\,46'$, Bx$_{a\cdot bl} \wedge c$ axis $= -30°\,10'$. The optical character $(-)$ and the position of the axes of elasticity remain sensibly constant between $0°$ and $100°$. The ax. pl., however, at first $\perp b(010)$ with horizontal dispersion and $\rho > v$ becomes on rise of temperature $\| b$ with inclined dispersion and $\rho < v$. The axial angle accordingly diminishes to $0°$ at a temperature depending upon the wave-length and then increases in the new plane. The temperature for the uniaxial condition varies from $18°$ C. for blue light, to $43°$ for sodium light and $52°$ for lithium light. In white light, therefore, the interference-figures are abnormal and change with rise in temperature.

Des Cloizeaux found that the feldspars, when heated up to a certain point, suffer a change in the position of the axes, and if the heat becomes greater and is long continued they do not return again to their original position, but remain altered.

In addition to the typical cases referred to, it is to be noted that when elevation of temperature is connected with change of chemical composition wide changes in optical characters are possible. This is illustrated by the zeolites and related species, where the effect of loss of water has been particularly investigated.

Further, with some crystals, heat serves to bring about a change of molecular structure and with that a total change of optical characters. For example, the greenish-yellow (artificial) orthorhombic crystals of antimony iodide (SbI_3) on heating (to about $114°$) change to red uniaxial hexagonal crystals. Note also the remarks made later in regard to the effect of heat upon leucite and boracite (Art. **441**).

435. Some Peculiarities in Axial Interference-figures.* — In the case of uniaxial crystals, the characteristic interference-figure varies but little from one species to another, such variation as is observed being usually due to the thickness of the section and the birefringence. In some cases, however, peculiarities are noted. For example, the interference-figure of apophyllite is somewhat peculiar, since its birefringence is very weak, and it may be optically positive for one part of the spectrum and negative for the other.

In the case of biaxial crystals, peculiarities are more common. The following are some typical examples:

Brookite is optically $+$ and the acute bisectrix is always normal to $a(100)$. While, however, the axial plane is $\| c(001)$ for red and yellow, with $2E_r = 55°$, $2E_y = 30°$, it is commonly $\| b(010)$ for green and blue, with $2E_{gr} = 34°$. Hence a section $\| a(100)$ in the conoscope shows a figure somewhat resembling that of a uniaxial crystal but with four sets of hyperbolic bands.

Titanite also gives a peculiar interference-figure with colored hyperbolas because of the high color-dispersion, $\rho > v$, the variation between $2E$ for red and green light being approximately $10°$; the dispersion of the bisectrices is, however, very small.

The most striking cases of peculiar axial figures are afforded by twin crystals (Art. **437**).

436. Relation of Optical Properties to Chemical Composition. — The effect of varying chemical composition upon the optical characters has been minutely studied in the case of many series of isomorphous salts, and with important results. It is, indeed, only a part of the general subject of the relation between crystalline form and molecular structure on the one hand and chemical composition on the other, one part of which has been discussed in Art. **327**. It was shown there that the refractive index can often be approximately calculated from the chemical composition.

Among minerals, the most important examples of the relation between composition and optical characters are afforded by the triclinic feldspars of the albite-anorthite series. Here, as explained in detail in the descriptive part of this work, the relation is so close that

---

* Variations in the axial figures embraced under the head of optical anomalies are spoken of later (Art. **441**).

the composition of any intermediate member of this isomorphous group can be predicted from the position of its ether-axes, or more simply from the vibration directions on the fundamental cleavage-directions, $\parallel c(001)$ and $\parallel b(010)$.

The effect of varying amounts of iron protoxide (FeO) is illustrated in the case of the monoclinic pyroxenes, where, for example, the angle $Bx_a \wedge c$ axis is 38° in diopside (2·9 per cent FeO) and 47° in hedenbergite (26 per cent FeO).   This is also shown in the closely related orthorhombic species of the same group, enstatite, $MgSiO_3$ with little iron, and hypersthene, $(Mg,Fe)SiO_3$ with iron to nearly 30 per cent.   With both of these species the axial plane is parallel to $b(110)$, but the former is optically $+ (Bx_a = Z)$ and the dispersion $\rho < v$; the latter is optically $- (Bx_a = X)$ and dispersion $\rho > v$.   In other words, the optic axial angle changes rapidly with the FeO percentage, being about 90° for FeO = 10 per cent.   In the case of the chrysolites, the epidotes, the species triphylite and lithiophilite, and others, analogous relations have been made out.

437. Optical Properties of Twin Crystals. — The examination of sections of twin crystals of any other than the isometric system in polarized light serves to establish the compound character at once and also to show the relative orientation of the several parts.   This is most distinct in the case of contact-twins, but is also well shown with penetration-twins, though here the parts are usually not separated by a sharp line.

Thus the examination of a section parallel to $b(010)$ of a twin crystal of gypsum, of the type of Fig. 637, makes it easy not only to establish the fact of the twinning but also to fix the relative positions of the ether-axes in the two parts.   The measurement can in such cases be made between the extinction-directions in the two halves, instead of between one of these and some definite crystallographic line, as the vertical axis.

637

638

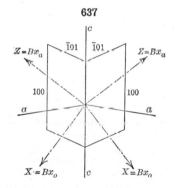
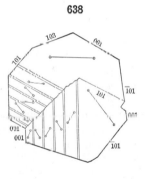

The polysynthetic twinning of certain species, as the triclinic feldspars, appears with great distinctness in polarized light.   For example, in the case of a section of albite, parallel to the basal cleavage, the alternate bands extinguish together and assume the same tint when the quartz section is inserted.   Hence the angle between these directions is easily measured, and this is obviously double the extinction-angle made with the edge $b(010) \wedge c(001)$.   A basal section of microcline in the same way shows its compound twinning according to both the albite and pericline laws, the characteristic grating structure being clearly revealed in polarized light.   Fig. 638 of a section of chondrodite (from Des Cloizeaux) shows how the compound structure is shown by optical examination; the position of the axial plane is indicated in the case of the successive polysynthetic lamellæ.   The complex penetration-twins of right- and left-handed crystals of quartz (see the description of that species) also have their character strikingly revealed in polarized light.

Still again, the true structure of complex multiple twins, exhibiting pseudo-symmetry in their external form, can only be fully made out in this way.   This is illustrated by Fig. 639, a basal section of an apparent hexagonal pyramid of witherite.   The analogous six-sided pyramid of bromlite (Fig. 640) has a still more complex structure, as shown in Fig.

641. Fig. 642 shows a simple crystal of stilbite; Fig. 643 is the common type of twin-crystal, and Fig. 644 illustrates how the complex structure ($\parallel b$ 010) is revealed in polarized light. Other illustrations are given in Art. **441.** It will be understood that the axial interference-figures of twin crystals, where the parts are superposed, often show many peculiarities; the Airy spirals of quartz (p. 295) will serve as an illustration.

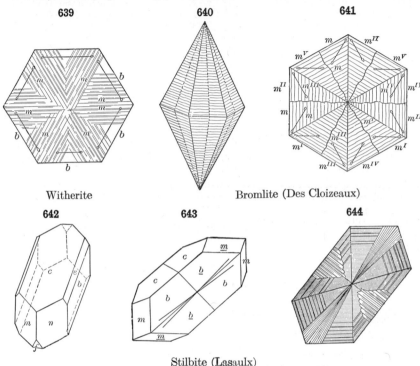

639   640   641

Witherite   Bromlite (Des Cloizeaux)

642   643   644

Stilbite (Lasaulx)

438. A particularly interesting case, related to the subject discussed in the preceding article, is that of the special properties of superposed cleavage-sections of mica. If three or more of these, say of rectangular form, be super-posed and so placed that the lines of the axial planes make equal angles of 60° (45°, etc.) with each other the effect is that polarized light which has passed through the center suffers circular polarization, with a rotation to right or left according to the way in which the sections are built up. The inter-ference-figure resembles that of a section of quartz cut normal to the axis.

If the sections are numerous and very thin the imitation of the phenomena of quartz is closer. These facts throw much light upon the ultimate molecu-lar structure of a crystallized medium showing circular polarization. Fur-ther, it is easy from this to understand how it is possible to have in sections of certain crystals (*e.g.*, of clinochlore) portions which are biaxial and others that are uniaxial, the latter being due to an intimate twinning after this method of biaxial portions.

439. Optical Properties of Crystalline Aggregates. — The special optical phenomena of the different kinds of crystalline aggregates described on pp. 204, 205, and the extent to which their optical characters can be determined, depend upon the distinctness in the development of the individuals and their relative orientation. The case of ordinary granular, fibrous, or columnar aggregates needs no special discussion. Where, however, the doubly refracting grains are extremely small, the microscope may hardly serve to do more than to show the *aggregate polarization* present.

A case of special interest is that of spherulites, that is, aggregates spherical in form and radiated or concentric in structure; such aggregates occur with calcite, various chlorites, feldspars, etc. If they are formed of a doubly refracting crystalline mineral, or of an amorphous substance which has birefringent characters due to internal tension, they commonly exhibit a dark cross in the microscope between crossed nicols; further, this cross, as the section is revolved on the stage, though actually stationary, seems to rotate backward.

A distinct and more special case is that of spherical aggregates of a mineral optically uniaxial (or biaxial with a small angle). Sections of these (not central) in parallel polarized light show more or less distinctly the interference-figure of a uniaxial crystal. The objective must be focussed on a point a little removed from the section itself, say on the surface of the sphere of which it is a part. In such cases the + or − character of the double refraction can be determined as usual.

440. Change of Optical Character Induced by Pressure. — As the difference between the optical phenomena exhibited by an isometric crystal on the one hand and a uniaxial or biaxial crystal on the other is referred to a difference in molecular structure modifying the properties of the ether, it would be inferred that if an amorphous substance were subjected to conditions tending to develop an analogous difference in its molecular structure it would also show doubly refracting properties.

This is found to be the case. Glass which has been suddenly cooled from a state of fusion, and which is therefore characterized by strong internal tension, usually shows marked double refraction. Further, glass plates subjected to great mechanical pressure in one direction show in polarized light more or less distinct interference-curves. Gelatine sections, also, under pressure exhibit like phenomena. Even the strain in a glass block developed under the influence of unlike charges of electricity of great difference of potential on its opposite sides is sufficient to make it doubly refracting.

In an analogous manner the double refraction of a crystal may be changed by the application of mechanical force. Pressure exerted normal to the vertical axis of a section of a tetragonal or hexagonal crystal which has been cut ⊥ c axis, changes the uniaxial interference-figure into a biaxial, and with substances optically positive, the plane of the optic axes is parallel, and with negative substances normal, to the direction of pressure.

The quartz crystals in rocks, which have been subjected to great pressure, are often found to be in an abnormal state of tension, showing an undulatory extinction in polarized light.

441. Optical Anomalies. — Since the early investigations of Brewster, Herschel, and others (1815 *et seq.*) it has been recognized that many crystals exhibit optical phenomena which are not in harmony with the apparent symmetry of their external form. Crystals of many isometric species, as analcite, alum, boracite, garnet, etc., often show more or less pronounced double refraction, and sometimes they are distinctly uniaxial or biaxial. A section examined in parallel polarized light may show more or less sharply defined doubly refracting areas, or parallel bands or lamellæ with varying extinction. Occasionally, as noted by Klein in the case of garnet, while most crystals are normally isotropic, others show optical characters which seem to be determined by the external bounding faces and edges; thus, a dodecahedron may appear to be made up of twelve rhombic pyramids (biaxial) whose apices are at the center; a hexoctahedron similarly may seem to be made up of forty-eight triangular pyramids, etc.

Similarly, crystals of many common tetragonal or hexagonal species, as vesuvianite, zircon, beryl, apatite, corundum, chabazite, etc., give interference-figures resembling those of biaxial crystals. Also, analogous contradictions between form and optical characters are noted with crystals of orthorhombic and monoclinic species, *e.g.*, topaz, natrolite, orthoclase, etc. All cases such as those mentioned are embraced under the common term of *optical anomalies.*

This subject has been minutely studied by many investigators in recent years and important additions have been made to it both on the practical and the theoretical side.   The result is that, though doubtful cases still remain, many of the typical ones have found a satisfactory explanation.   No single theory, however, can be universally applied.

The chief question involved has been whether the anomalies are to be considered as secondary and non-essential, or whether they belong to the inherent molecular structure of the crystals in question.   On the one hand, it has been urged that internal tension suffices (Art. **440**) to call out double refraction in an isotropic substance or to give a uniaxial crystal the typical optical structure of a biaxial crystal.   On the other hand, it is equally clear that twinning often produces pseudo-symmetry in external form, and at the same time conceals or changes the optical characters.   From the simplest case, as that of aragonite, we pass to more complex cases, as witherite (Fig. 639), bromlite (Figs. 640, 641), phillipsite (Figs. 426, 478–480), which last is sometimes pseudo-isometric in form though optical study shows the monoclinic character of the individuals.*   Reasoning from the analogy of these last cases, Mallard was led (1876) to the theory that the optical anomalies could in most cases be explained by the assumption of a similar but still more intimate grouping of molecules which themselves without this would unite to form crystals of a lower grade of symmetry than that which their complex twinned crystals actually simulate.

In regard to the two points of view mentioned, it seems probable that internal tension (due to pressure, sudden cooling, or rapidity of growth, etc.) can be safely appealed to to explain the anomalous optical character of many species, as diamond, halite, beryl, quartz, etc.   Again, it has been fully proved that the later growth of isomorphous layers of varying composition may produce optical anomalies, probably here also to be referred to tension.   Alum is a striking example.   The peculiarities of this species were early investigated by Biot and made by him the basis of his theory of " lamellar polarization," but the present explanation is doubtless the true one.   Fig. 645 (from Brauns)

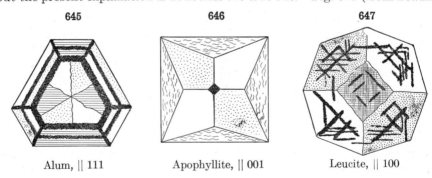

645	646	647
Alum, ‖ 111	Apophyllite, ‖ 001	Leucite, ‖ 100

shows the appearance in polarized light of a section ‖ $o(111)$ from a crystal in which the successive layers have different composition.   Further, according to Brauns, the optical peculiarities of many other species may be referred to this same cause.   He includes here, particularly, those cases (as with some garnets)

* Crystals showing pseudo-symmetry of highly complex type are called *mimetic* crystals by Tschermak.

in which the optical characters seem to depend upon the external form, as noted above. Here belongs also apophyllite, a section of which (from Golden, Colorado, by Klein) is shown in Fig. 646. The section has been cut $\parallel c(001)$ through the center of the crystal and is represented as it appears in parallel polarized light.

Another quite distinct but most important class is that including species such as boracite and leucite, which are *dimorphous*; that is, those species which at a certain elevation of temperature ($300°$ for boracite and $500°$ to $600°$ for leucite) become strictly isotropic. Under ordinary conditions, these species are anisotropic, but the fact stated makes it probable that originally their crystalline form and optical characters were in harmony. The relations for leucite deserve to be more minutely stated.

Leucite usually shows very feeble double refraction: $\omega = 1\cdot508$, $\epsilon = 1\cdot509$. This anomalous double refraction, early noted (Brewster, Biot), was variously explained. In 1873, Rath, on the basis of careful measurements, referred the seemingly isometric crystals to the tetragonal system, the trapezohedral face 112 being taken as 111 and 211, 121 as 421, 241, respectively; also 101, 011 as 201, 021. Later Weisbach (1880), on the same ground, made them orthorhombic; Mallard, however, referred them (1876), chiefly on optical grounds, to the monoclinic system, and Fouqué and Lévy (1879) to the triclinic. The true symmetry, corresponding to the molecular structure which they possess or tend to possess at ordinary temperatures, is in doubt, but it has been shown (Klein, Penfield) that at $500°$ to $600°$ sections become isotropic; and further (Rosenbusch) that the twinning striations disappear on heating, to reappear again in new positions on cooling. Sections ordinarily show twinning-lamellæ $\parallel d(110)$; in some cases a bisectrix ($+$) is normal to what corresponds to a cubic face, the axial angle being very small. The structure corresponds in general (Klein) to the interpenetration of three crystals, in twinning position $\parallel d$, which may be equally or unequally developed; or there may be one fundamental individual with inclosed twinning-lamellæ. Fig. 647 shows a section of a crystal ($\parallel a$, 100) which is apparently made up by the twinning of three individuals.

Still again, in a limited number of cases, it can be shown that the intergrowth of lamellæ having slightly different crystallographic orientation is the cause of the optical peculiarities. Prehnite is a conspicuous example of this class.

After all the various possible explanations have been applied there still remain, however, many species about which no certain conclusion can be reached. To many of these species the theory of Mallard may probably be applicable. Indeed it may be added that much difference of opinion still exists as to the cause of the " optical anomalies " in a considerable number of minerals.

LITERATURE

*Optical Anomalies**

Brewster. Many papers in Phil. Trans., 1814, 1815, and later; also in Ed. Trans., Ed. Phil. J., etc.

Biot. Recherches sur la polarisation lamellaire, etc. C. R., **12**, 967, 1841; **13**, 155, 391, 839, 1841; in full in Mem. de l'Institut, **18**, 539.

Volger. Monographie des Boracits. Hannover, 1857.

Marbach. Ueber die optischen Eigenschaften einiger Krystalle des tesseralen Systems. Pogg. Ann., **94**, 412, 1855.

Pfaff. Versuche über den Einfluss des Drucks auf die optischen Eigenschaften der Krystalle. Pogg. Ann., **107**, 333, 1859; **108**, 598, 1859.

* A complete bibliography is given in the memoir by Brauns (1891), see above.

Des Cloizeaux. Ann. Mines **11**, 261, 1857; **14**, 339, 1858; **6**, 557, 1864. Also Nouvelles Recherches, etc., 1867.

Reusch. Ueber die sogennante Lamellarpolarisation des Alauns. Pogg. Ann., **132**, 618, 1867.

Rumpf. Apophyllite. Min. petr. Mitth., **2**, 369, 1870.

Hirschwald. Leucite. Min. Mitth., 227, 1875.

Lasaulx. Tridymite. Zs. Kr., **2**, 253, 1878.

Mallard. Application des phénomènes optiques anomaux que présentent un grand nombre de substances cristallisées. Annales des Mines (Ann. Min.) **10**, pp. 60–196, 1876 (Abstract in Zs. Kr., **1**, 309–320). See also Bull. Soc. Min., **1**, 107, 1878. Sur les propriétés optiques des mélanges de substances isomorphes et sur les anomalies optiques des cristaux. Bull. Soc. Min., **3**, 3, 1880. Also ibid., **4**, 71, 1881; **5**, 144, 1882.

Bertrand. Numerous papers in Bull. Soc. Min., 1878–1882.

Becke. Chabazite. Min. petr. Mitth., **2**, 391, 1879.

Baumhauer. Perovskite. Zs. Kr., **4**, 187, 1879.

Tschermak. "Mimetische Formen." Zs. G. Ges., **31**, 637, 1879.

Jannettaz. Diamond. Bull. Soc. Min., **2**, 124, 1879; alum, ibid., **2**, 191; **3**, 20.

Bücking. Ueber durch Druck hervorgerufene optische Anomalien. Zs. G. Ges., **32**, 199, 1880. Also, Zs. Kr., **7**, 555, 1883.

Arzruni and **S. Kock.** Analcite. Zs. Kr., **5**, 483, 1881.

Klocke. Ueber Doppelbrechung regulärer Krystalle. Jb. Min., **1**, 53, 1880 (also **2**, 97, 13 ref.; **1**, 204, 1881, and Verh. nat. Ges. Freiburg, **8**, 31). Ueber einige optische Eigenschaften optisch anomaler Krystalle und deren Nachahmung durch gespannte und gepresste Colloide. Jb. Min., **2**, 249, 1881.

C. Klein. Boracite. Jb. Min., **2**, 209, 1880; **1**, 239, 1881; **1**, 235, 1884. Garnet. Nachr. Ges. Göttingen, 1882; Jb. Min., **1**, 87, 1883. Apophyllite (influence of heat). Jb. Min., **2**, 165, 1892. Garnet, vesuvianite, etc. Ibid., **2**, 68, 1895.

W. Klein. Beiträge zur Kenntniss der optischen Aenderungen in Krystallen unter dem Einflusse der Erwärmung. Zs. Kr., **9**, 38, 1884.

Brauns. Die optischen Anomalien der Krystalle. (Preisschrift), Leipzig, 1891. Also earlier papers: Jb. Min., **2**, 102, 1883; **1**, 96, 1885; **1**, 47, 1887.

Ben Saude. Beitrag zu einer Theorie der Optischen Anomalien der regulären Krystalle. Lisbon, 1894. Also earlier: Analcite, Jb. Min., **1**, 41, 1882. Perovskite (Preisschrift), Göttingen, 1882.

Wallerant. Théorie des anomales optiques, de l'isomorphisme et du polymorphisme. Bull. Soc. Min., **21**, 188, 1898.

IV. CHARACTERS DEPENDING UPON HEAT

442. The more important of the special properties of a mineral species with respect to heat include the following: Fusibility; conductivity and expansion, especially in their relation to crystalline structure; change in optical characters with change of temperature; specific heat; also diathermancy, or the power of transmitting heat radiation. The full discussion of these and other related subjects lies outside of the range of the present text-book. A few brief remarks are made upon them, and beyond these reference must be made to text-books on Physics and to special memoirs, some of which are mentioned in the literature (p. 334).

443. Fusibility. — The approximate relative fusibility of different minerals is an important character in distinguishing different species from one another by means of the blowpipe. For this purpose a scale is conveniently used for comparison, as explained in the articles later devoted to the blowpipe. Accurate determinations of the fusibility are difficult, and though of little importance for the above object, they are interesting from a theoretical standpoint. They have been attempted by various authors by the use of a number of different methods. The following are the approximate melting-point values for the minerals used in von Kobell's scale (Art. **504**): Stibnite, 525°; natrolite, 965°; almandite, 1200°; actinolite, 1296°; orthoclase, 1200°; bronzite, 1380°; also for quartz, about 1600°.

444. Conductivity. — The conducting power of different crystallized media was early investigated by Sénarmont. He covered the faces of the substance under investigation with wax and observed the form of the figure melted by a hot wire placed in contact with the surface at its middle point. Later investigations have been made by Röntgen (who modified the method of Sénarmont), by Jannettaz, and others. In general it is found that, as regards their thermal conductivity, crystals are to be divided into the three classes noted on p. 256. In other words, the conductivity for heat seems to follow the same general laws as the propagation of light. It is to be stated, however, that experiments by S. P. Thompson and O. J. Lodge have shown a different rate of conductivity in tourmaline in the opposite directions of the vertical axis.

445. Expansion. — Expansion, that is, increase in volume upon rise of temperature, is a nearly universal property for all solids. The increment of volume for the unit volume in passing from 0° to 1° C. is called the coefficient of expansion. This quantity has been determined for a number of species. Further, the relative expansion in different directions is found to obey the same laws as the light-propagation. Crystals, as regards heat-expansion, are thus divided into the same three classes mentioned on p. 276 and referred to in the preceding article.

The amount of expansion varies widely, and, as shown by Jannettaz, is influenced particularly by the cleavage. Mitscherlich found that in calcite there was a diminution of 8′ 37″ in the angle of the rhombohedron on passing from 0° to 100° C., the form thus approaching that of a cube as the temperature increased. The rhombohedron of dolomite, for the same range of temperature, diminishes 4′ 46″; and in aragonite, for a rise in temperature from 21° to 100°, the angle of the prism diminishes 2′ 46″. In some rhombohedrons, as of calcite, the vertical axis is lengthened (and the horizontal shortened), while in others, like quartz, the reverse is true. The variation is such in both cases that the birefringence is diminished with the increase of temperature, for calcite possesses negative double refraction, and quartz, positive.

It is to be noted that in general the expansion by heat, while it may serve to alter the angles of crystals, other than those of the isometric system, does not alter the zone-relations and the crystalline symmetry. In certain cases, however, the effect of heat may be to give rise to twinning-lamellæ (as in anhydrite) or to cause their disappearance (as in calcite). Rarely heat serves to develop a new molecular structure; thus, as explained in Art. **441**, boracite and leucite, which are anisotropic at ordinary temperatures, become isotropic when heated, the former to 300°, the latter to 500° or 600°. The change in the optical properties of crystals produced by heat has already been noticed (Art. **434**).

446. Specific Heat. — The specific heat of any substance is the amount of heat necessary to raise the temperature of one gram of the substance one degree Centigrade. The unit of specific heat is the heat necessary to raise one gram of water one degree Centigrade. Determinations of the specific heat of many minerals have been made by Joly, by Oeberg, and others. Some of the results reached are as follows:

	Joly	Oeberg		Joly	Oeberg
Galena, *cryst.*	0·0541	—	Orthoclase	0·1869	0·1877
Chalcopyrite	0·1271	0·1291	Albite	0·1983	0·1976
Pyrite	0·1306	—	Amphibole, *black*	0·1963 Augite 0·1830	
Hematite	0·1683	0·1645	Beryl	0·2066	0·1979
Garnet, *red cryst.* 0·1780	— 0·1793	0·1758	Calcite 0·2034	— 0·2044	0·2042
Epidote	0·1877	0·1861	Aragonite	0·2036	—

447. Diathermancy. — Besides the slow molecular propagation of heat in a body, measured by its thermal conductivity, there is also to be considered the rapid propagation of what is called radiant heat through it by the wave-motion of the ether which surrounds its molecules. This is merely a part of the general subject of light-propagation already fully discussed, since heat-waves, in the restricted sense, differ from light-waves only in their relatively greater length. The degree of absorption exerted by the body is measured by its diathermancy, which corresponds to transparency in light. In this sense halite, sylvite, and fluorite are highly *diathermanous*, since they absorb but little of the heat-waves passing through them; on the other hand, gypsum and, still more, alum are comparatively *athermanous*, since while transparent to the short light-waves they absorb the long heat-waves, transforming the energy into that of sensible heat. Measurements of the diathermancy were early made by Melloni, later by Tyndall, Langley, and others.

LITERATURE

Heat

Mitscherlich. Pogg. Ann., **1**, 125, 1824; **10**, 137, 1827.
F. E. Neumann. Gypsum. Pogg. Ann., **27**, 240, 1833.
Sénarmont. Ann. Ch. Phys., **21**, 457, 1847; **22**, 179, 1848; also in Pogg. Ann., **73**, 191; **74**, 190; **75**, 50, 482.
Ångström. Pogg. Ann., **86**, 206, 1852.
Grailich and **von Lang.** Ber. Ak. Wien, **33**, 369, 1858.
Fizeau. Thermal expansion. C. R., **58**, 923, 1864. Ann. Ch. Phys., **2**, 143, 1864; **8**, 335, 1866; also C. R., 1864–1867.
C. Neumann. Pogg. Ann., **114**, 492, 1868.
Pape. Thermic axes of blue vitriol. Wied. Ann., **1**, 126, 1877.
Röntgen. Pogg. Ann., **151**, 603, 1874; Zs. Kr., **3**, 17, 1878.
Jannettaz. Conductivity of crystals. Bull. Soc. Geol. (3), **1**, 117, 252; **2**, 264; **3**, 499; **4**, 116, 554; **9**, 196. Bull. Soc. Min., **1**, 19, 1879. C. R., 1848, **114**, 1352, 1892.
O. J. Lodge. Thermal conductivity. Phil. Mag., **5**, 110, 1878.
S. P. Thompson and **O. J. Lodge.** Conductivity of tourmaline. Phil. Mag., **8**, 18, 1879.
Arzruni. Effect of heat on refractive indices of barite, etc. Zs. Kr., **1**, 165, 1877.
Beckenkamp. Expansion of monoclinic and triclinic crystals. Zs. Kr., **5**, 436, 1881.
H. Dufet. Effect of heat on refractive indices of gypsum. Bull. Soc. Min., **4**, 113, 191, 1881.
A. Schrauf. Sulphur. Zs. Kr., **12**, 321, 1887; TiO_2, ibid., **9**, 433, 1884.
L. Fletcher. Expansion of crystals. Zs. Kr., **4**, 337, 1880.
Joly. Meldometer. Ch. News, **65**, 1, 16, 1892, and Proc. Roy. Irish Acad., **2**, 38, 1891. Specific heat. Proc. Roy. Soc., **41**, 250, 352, 1887.
Oeberg. Specific heat. Oefv. Ak. Stockh., No. 8, 43, 1885.
Doelter. For methods and results in fusing silicates, see Handbuch der Mineralchemie, **1**, 628 *et seq.*

V. CHARACTERS DEPENDING UPON ELECTRICITY AND MAGNETISM

1. ELECTRICITY

448. Electrical Conductivity. — The subject of the relative conducting power of different minerals is one of minor interest.* In general most minerals, except those having a metallic luster among the sulphides and oxides, are non-conductors. Only the non-conductors can show pyroelectrical phenomena, and only the conductors can give a thermoelectric current. Experiments have shown that electrical conductivity is alike in all directions in isometric crystals and that in crystals of other systems it varies with the crystal direction, conforming to the same laws that govern the transmission of light. In hematite, for instance, it has been shown that the conductivity in the horizontal symmetry plane is uniform in all directions and is about twice as great as in the direction of the vertical crystal axis.

449. Frictional Electricity. — The development of an electrical charge on many bodies *by friction* is a familiar subject. All minerals become electric by friction, although the degree to which this is manifested differs widely. There is no line of distinction among minerals, dividing them into *positively* electric and *negatively* electric; for both electrical states may be presented by different varieties of the same species, and by the same variety in different states. The gems are in general positively electric only when polished; the diamond, however, exhibits positive electricity whether polished or not. It is a familiar fact that the electrification of amber upon friction was early observed (600 b.c.), and indeed the Greek name (ἤλεκτρον) later gave rise to the word electricity.

450. Pyroelectricity. — The simultaneous development of positive and negative charges of electricity on different parts of the same crystal when its temperature is suitably changed is called pyroelectricity. If a crystal develops a positive charge in one portion during warming, a negative charge will be developed at the same point during cooling, etc. Crystals exhibiting such phenomena are said to be *pyroelectric*. This phenomenon was first observed in the case of tourmaline, which is rhombohedral-hemimorphic in crystallization, and it is particularly marked with crystals belonging to groups of relatively low symmetry, especially those of the hemimorphic type. It is possible, of course, only with non-conductors. This subject was early investigated by Riess and Rose (1843), later by Hankel, also by C. Friedel, Kundt, and others (see literature).

In all cases it is true that directions of like crystallographic symmetry show charges of like sign, while unlike directions may exhibit opposite charges. Substances not crystallized cannot show pyroelectricity. A few of the many possible examples will serve to bring out the most essential points.

Boracite (isometric-tetrahedral, p. 83) on heating exhibits + electricity on one set of tetrahedral faces and − electricity on the other. Cf. Fig. 648.

Tourmaline (rhombohedral-hemimorphic, p. 126) shows opposite charges at the opposite extremities of the vertical axis corresponding to its hemimorphic crystallization. In this and in other similar cases, the extremity which

---

* On the conductivity of minerals, see Beijerinck, Jb. Min., Beil.-Bd., **11**, 403, 1898.

becomes positive on heating has been called the *analogous* pole, and that which becomes negative has been called the *antilogous* pole.

Calamine and *struvite* (orthorhombic-hemimorphic, p. 144) exhibit phenomena analogous to those of tourmaline.

Quartz (rhombohedral-trapezohedral, p. 129) shows + electricity on heating at the three alternate prismatic edges and − electricity at the three remaining edges; the distribution for right-handed crystals is opposite to that of left-handed. Twins may exhibit a high degree of complexity. Cf. Figs. 649, 650.

Axinite (triclinic, p. 162), when heated to 120° or 130°, has an analogous pole (Riess and Rose) at the solid angle rxM'; the antilogous pole at the angle $mr'M'$ near plane n.

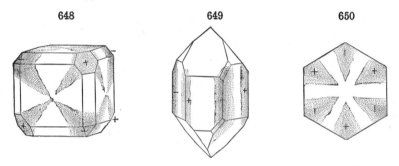

A very convenient and simple method for investigating the phenomena is the following, which is due to Kundt: First heat the crystal or section carefully in an air-bath; pass it several times through the flame of an alcohol lamp and then place it on a little upright cylinder of brass to cool. While cooling, a mixture of red lead and sulphur finely pulverized and previously agitated is dusted over it through a fine cloth from a suitable bellows. The positively electrified red lead collects on the parts having a negative charge, and the negatively electrified sulphur on those with a positive charge. This is illustrated by Figs. 648–650, and still better by the illustrations given by Kundt and others. (Cf. Plate III of Groth, Phys. Kryst., 1905.)

451. Piezoelectricity. — The heating (or cooling) of a crystal to induce pyroelectric effects involves a change in its volume. Further, if a change in volume can be brought about by other means, such as compression or tension, similar electric charges result. The name *piezoelectricity* has been given to the development of electrical charges on a crystallized body by pressure or by tension. Charges of opposite sign are produced by the two different operations. This phenomenon is most interesting where a relation can be established between the electrical excitement and the molecular structure, as is conspicuously true with quartz, tourmaline, and some other species.

This subject has been investigated by Hankel, Curie, and others, and discussed theoretically by Lord Kelvin (see literature). Hankel has also employed the term *actinoelectricity*, or, better, *photoelectricity*, for the phenomenon of producing an electrical condition by the influence of direct radiation; fluorite is a conspicuous example.

452. Thermoelectricity. — The contact of two unlike metals in general results in electrifying one of them positively and the other negatively. If.

further, the point of contact be heated while the other parts, connected with a wire, are kept cool, a continuous current of electricity — shown, for example, by a suitable galvanometer — is set up at the expense of the heat-energy supplied. If, on the other hand, the point of junction is cooled, a current is set up in the reverse direction. This phenomenon is called *thermoelectricity*, and two metals so connected constitute a thermoelectric couple. Further it is found that different conductors can be arranged in order in a table — a so-called thermoelectric series — according to the *direction* of the current set up on heating and according to the *electromotive force* of this current. Among the metals, bismuth (+) and antimony (−) stand at the opposite ends of the series; the current passes through the connecting wire from antimony to bismuth.

This subject is so far important for mineralogy, as it was shown by Bunsen that the natural metallic sulphides stand farther off in a series than bismuth and antimony, and consequently by them a higher electromotive force is produced. The thermoelectrical relations of a large number of minerals were determined by Flight.

It was early observed that some minerals have varieties which are both + and −. Rose attempted to establish a relation between the positive and negative pyritohedral forms of pyrite and cobaltite, and the positive or negative thermoelectrical character. Later investigations by Schrauf and Dana have shown, however, that the same peculiarity belongs also to glaucodot, tetradymite, skutterudite, danaite, and other minerals, and it is demonstrated by them that it cannot be dependent upon crystalline form, but rather upon chemical composition.

LITERATURE*

Pyroelectricity, etc.
> Rose. Tourmaline. Pogg. Ann., **39**, 285, 1830.
> Riess and Rose. Pogg. Ann., **59**, 353, 1843; **61**, 659, 1844.
> Kobell. Pogg Ann., **118**, 594, 1863.
> Hankel. Pogg. Ann., **49**, 493; **50**, 237, 1840; **61**, 281, 1844. Many important papers in Abhandl. K. Sächs. Ges., 1865 and later; also Wied. Ann., **2**, 66, 1877; **11**, 269, 1880, etc.
> J. and P. Curie. C. R., **91**, 294, 983, 1880; **92**, 186, 350, 1881; **93**, 204, 1882.
> Kundt. Ber. Ak. Berlin, 421, 1883; Wied. Ann., **20**, 592, 1893.
> Kolenko. Quartz. Zs. Kr., **9**, 1, 1884.
> C. Friedel. Sphalerite, etc. Bull. Soc. Min., **2**, 31, 1879.
> C. Friedel and Curie. Sphalerite, boracite. Bull. Soc. Min., **6**, 191, 1883.
> Mack. Boracite. Zs. Kr., **8**, 503, 1883.
> Voigt. Abhandl. Ges. Göttingen, **36**, 99, 1890.
> Kelvin. Phil. Mag., **36**, 331, 453, 1893.
> G. S. Schmidt. Photo-electricity of fluorite. Wied. Ann., **62**, 407, 1897.

Thermoelectricity
> Marbach. C. R., **45**, 705, 1857.
> Bunsen. Pogg. Ann., **123**, 505, 1864.
> Friedel. Ann. Ch. Phys., **17**, 79, 1869; C. R., **78**, 508, 1874.
> Rose. Pyrite and cobaltite. Pogg. Ann., **142**, 1, 1871.
> Schrauf and E. S. Dana. Ber. Ak. Wien, **69** (1), 142, 1874; Am. J. Sc., **8**, 255, 1874.

---

* See Liebisch, Phys. Krystallographie, 1891, for a full discussion of the topics briefly touched upon in the preceding pages, also for references to original articles; also Tutton, Crystallography and Practical Crystal Measurement, Vol. 2, Chapter 49, 1922.

2. MAGNETISM

453. Magnetic Minerals.   Natural Magnets. — A few minerals in their natural state are capable of being attracted by a strong steel magnet; they are said to be *magnetic*.   This is conspicuously true of magnetite, the magnetic oxide of iron; also of pyrrhotite or magnetic pyrites, and of some varieties of native platinum (especially the variety called iron-platinum).

A number of other minerals, as hematite, franklinite, etc., are in some cases attracted by a steel magnet, but probably in most if not all cases because of admixed magnetite (but see Art. **455**).   Occasional varieties of the three minerals mentioned above, as the lodestone variety of magnetite, exhibit themselves the attracting power and polarity of a true magnet.   They are then called *natural magnets*.   In such cases the magnetic polarity has probably been derived from the inductive action of the earth, which is itself a huge magnet.

454. Paramagnetism.   Diamagnetism. — In a very strong magnetic field, as that between the poles of a very powerful electromagnet, all minerals, as indeed all other substances, are influenced by the magnetic force.   According to their behavior they are divided into two classes, the *paramagnetic* and *diamagnetic*; those of the former appear to be attracted, those of the latter to be repelled.   For purposes of experiment the substance in question, in the form of a rod, is suspended on a horizontal axis between the poles of the magnet.   If paramagnetic, it takes a position parallel to the magnetic axis; if diamagnetic, transversely to it.   Iron, cobalt, nickel, manganese, platinum are paramagnetic; silver, copper, bismuth are diamagnetic.   Among minerals compounds of iron are paramagnetic, as siderite, also diopside; further, beryl, dioptase.   Diamagnetic species include calcite, zircon, wulfenite, etc.

By the use of a sphere it is possible to determine the relative amount of magnetic induction in different directions of the same substance.   Experiment has shown that in isometric crystals the magnetic induction is alike in all directions; that in those optically uniaxial, there is a direction of maximum and, normal to it, one of minimum magnetic induction; that in biaxial crystals, there are three unequal magnetic axes, the position of which may be determined.   In other words, the magnetic relations of the three classes of crystals are analogous to their optical relations.

455. Corresponding to the facts just stated, that all compounds of iron are paramagnetic, it is found that a sufficiently powerful electromagnet attracts all minerals containing iron, though, except in the cases given in Art. **453**, a bar magnet has no sensible influence upon them; hence the efficiency of the electromagnetic method of separating ores.

Plücker* determined the magnetic attraction of a number of substances compared with iron taken as 100,000.   For example, for magnetite he obtained 40,227; for hematite, crystallized, 533, massive, 134; limonite, 71; pyrite, 150.

LITERATURE

Magnetism
   Plücker.   Pogg. Ann., **72**, 315, 1847; **76**, 576, 1849; **77**, 447, 1849; **78**, 427, 1849; **86**, 1, 1852.
   Plücker and Beer.   Pogg. Ann., **81**, 115, 1850; **82**, 42, 1852.

---

* Pogg. Ann., **74**, 343, 1848.

Faraday. Phil. Trans., 1849–1857, and Experimental Researches, Series XXII, XXVI, XXX.

W. Thomson (Lord Kelvin). Theory of Magnetic Induction. Brit. Assoc., 1850, pt. 2, 23; Phil. Mag., **1**, 177, 1851, etc. Reprint of Papers on Electrostatics and Magnetism, 1872.

Tyndall. Phil. Mag., **2**, 165, 1851; **10**, 153, 257, 1855; **11**, 125, 1856; Phil. Trans., 1855, 1. Researches on diamagnetism and magne-crystallic action. London, 1870.

Knoblauch and **Tyndall.** Pogg. Ann., **79**, 233; **81**, 481, 1850 (Phil. Mag., **36, 37**, 1850).

Rowland and **Jacques.** Bismuth, Calcite. Am. J. Sc., **18**, 360, 1879.

Tumlirz. Quartz. Wied. Ann., **27**, 133, 1886.

Koenig. Wied. Ann., **31**, 273, 1887.

Stenger. Calcite. Wied. Ann., **20**, 304, 1883; **35**, 331, 1888.

VI. TASTE AND ODOR

In their action upon the senses a few minerals possess *taste*, and others under some circumstances give off *odor*.

456. Taste belongs only to soluble minerals. The different kinds of taste adopted for reference are as follows:

1. *Astringent:* the taste of vitriol.
2. *Sweetish astringent:* taste of alum.
3. *Saline:* taste of common salt.
4. *Alkaline:* taste of soda.
5. *Cooling:* taste of saltpeter.
6. *Bitter:* taste of Epsom salts.
7. *Sour:* taste of sulphuric acid.

457. Odor. — Excepting a few gaseous and soluble species, minerals in the dry unchanged state do not give off odor. By friction, moistening with the breath, and the elimination of some volatile ingredient by heat or acids, odors are sometimes obtained which are thus designated:

1. *Alliaceous:* the odor of garlic. Friction of arsenopyrite elicits this odor; it may also be obtained from arsenical compounds by means of heat.
2. *Horse-radish odor:* the odor of decaying horse-radish. This odor is strongly perceived when the ores of selenium are heated.
3. *Sulphurous:* friction elicits this odor from pyrite, and heat from many sulphides.
4. *Bituminous:* the odor of bitumen.
5. *Fetid:* the odor of sulphureted hydrogen or rotten eggs. It is elicited by friction from some varieties of quartz and limestone.
6. *Argillaceous:* the odor of moistened clay. It is obtained from serpentine and some allied minerals, after moistening them with the breath; others, as pyrargillite, afford it when heated.

458. Feel. — The FEEL is a character which is occasionally of some importance; it is said to be *smooth* (sepiolite), *greasy* (talc), *harsh*, or *meager*, etc. Some minerals, in consequence of their hygroscopic character, *adhere to the tongue* when brought in contact with it.

PART III CHEMICAL MINERALOGY

GENERAL PRINCIPLES OF CHEMISTRY AS APPLIED TO MINERALS

459. Minerals, as regards their chemical constitution, are either the uncombined elements in a native state, or definite compounds of these elements formed in accordance with chemical laws. It is the object of Chemical Mineralogy to determine the chemical composition of each species; to show the chemical relations of different species to each other where such exist; and also to explain the methods of distinguishing different minerals by chemical means. It thus embraces the most important part of Determinative Mineralogy.

In order to understand the chemical constitution of minerals, some knowledge of the fundamental principles of Chemical Philosophy is required; and these are here briefly recapitulated.

460. Chemical Elements. — Chemistry recognizes about ninety substances which cannot at will be decomposed, or divided into others, by any process of analysis at present known; these substances are called the chemical *elements*. A list of them is given in a later article (**464**); common examples are: oxygen, nitrogen, hydrogen, chlorine, gold, silver, sodium, etc.

461. Atom. Molecule. — The study of the chemical properties of substances and of the laws governing their formation led to the belief that there is for each element a definite, indivisible mass, which is the smallest particle which can play a part in chemical reactions; this indivisible unit is called the *atom*. The atom of a given element has a definite weight which is a major characteristic. The modern conception of an atom is that it is composed of positive and negative electrical charges. The nucleus, or central portion of the atom, which contributes the greater part of the mass of the atom, contains a number of positive electrical charges, called protons. The nucleus is surrounded by revolving negative charges that are called electrons. The number of electrons surrounding each nucleus is equal to the number of protons in the nucleus. The electrons revolve in successive concentric shells about the nucleus, the whole atom resembling a more or less complex solar system. The number of protons and electrons varies with each element. The number of such protons or electrons of each element gives what is called the *atomic number* of that element.

The maximum number of electrons belonging to the successive concentric shells about the nucleus are two in the first shell (as in helium), eight in the second (as in neon), eight in the third, eighteen in each of the fourth and fifth shells, and thirty-two in the sixth. Whenever an element has the maximum number of electrons in its outermost electronic shell it is found to be an inert substance that does not enter into chemical reactions. The chemically active elements are therefore those whose outer ring of electrons

340

has vacancies in it that might be filled by the electrons of other atoms in chemical combination.

With some rare exceptions, the atom cannot exist alone, but unites by the action of what is called chemical force, or chemical affinity, with other atoms of the same or different kind to form the *molecule*. The molecule, in the chemical sense, may be defined as the smallest particle into which a given kind of substance can be subdivided without undergoing chemical decomposition. For example, two *atoms* of hydrogen unite to form a *molecule* of hydrogen gas. Again, one atom of hydrogen and one of chlorine form a molecule of hydrochloric acid gas; two atoms of hydrogen and one of sulphur form a molecule of the gas hydrogen sulphide.

462. Atomic Weight. — The atomic weight of an element is the weight, or, better expressed, the mass of its atom compared with that of an atom of oxygen taken as 16. Of the methods by which the relation between the masses of the atoms is determined it is unnecessary here to speak; the results that have been obtained are given in the table on p. 342.

463. Symbol. Formula. — The symbol of an element is the initial letter, or letters, often of its Latin name, by which it is represented when expressing in chemical notation the constitution of substances into the composition of which it enters. Thus O is the symbol of oxygen, H of hydrogen, Cl of chlorine, Fe (from *ferrum*) of iron, Ag (from *argentum*) of silver, etc. Further, this symbol is always understood to indicate that definite amount of the given element expressed by its atomic weight; in other words, it represents one atom. If twice this quantity is involved, that is, two atoms, this is indicated by a small subscript number written immediately after the symbol. Thus, Sb_2S_3 means a compound consisting of two atoms of antimony and three of sulphur, or of 2×120 parts by weight of antimony and 3×32 of sulphur.

This expression, Sb_2S_3, is called the *formula* of the given compound, since it expresses in briefest form its composition. Similarly the formula of the mineral albite is $NaAlSi_3O_8$.

Strictly speaking, such formulas are merely *empirical formulas*, since they express only the actual result of analysis, as giving the relative number of atoms of each element present, and make no attempt to represent the actual constitution. A formula developed with the latter object in view is called a rational, structural, or constitutional formula (see Art. **481**).

464. Table of the Elements. — The following table gives a list of all the definitely established elements with their accepted symbols and also their atomic weights.[*]

Of the elements given in this list — more than eighty in all — only a very small number, say twelve, play an important part in making up the crust of the earth and the water and air surrounding it. The common elements concerned in the composition of minerals are: oxygen, sulphur, silicon, aluminum, iron, calcium, magnesium, sodium, potassium. These are estimated to form over 97 per cent of the material of the earth's crust with oxygen the most abundant (about 50 per cent). Besides these, hydrogen is present in water, nitrogen in the air, and carbon in all animal and vegetable substances. Only a very few of the elements occur as such in nature, as native gold, native silver, native sulphur, etc.

---

[*] These correspond in value to those commonly accepted, and are given accurate to one decimal place.

Of the elements, oxygen, hydrogen, nitrogen, chlorine, and fluorine are gases; bromine is a volatile liquid; mercury is also a liquid, but the others are solids under ordinary conditions.

	Symbol	O = 16 At. Weight		Symbol	O = 16 At. Weight
Aluminum	Al	26·9	Mercury (*Hydrargyrum*)	Hg	200·6
Antimony (*Stibium*)	Sb	121·7	Molybdenum	Mo	96·0
Argon	A	39·9	Neodymium	Nd	144·2
Arsenic	As	74·9	Neon	Ne	20·2
Barium	Ba	137·3	Nickel	Ni	58·6
Beryllium, Glucinum	Be (or Gl)	9·0	Niobium	Nb	93·1
Bismuth	Bi	209·0	Nitrogen	N	14·0
Boron	B	10·8	Osmium	Os	190·8
Bromine	Br	79·9	Oxygen	O	16·0
Cadmium	Cd	112·4	Palladium	Pd	106·7
Caesium	Cs	132·8	Phosphorus	P	31·0
Calcium	Ca	40·0	Platinum	Pt	195·2
Carbon	C	12·0	Potassium (*Kalium*)	K	39·0
Cerium	Ce	140·2	Praseodymium	Pr	140·9
Chlorine	Cl	35·4	Radium	Ra	225·9
Chromium	Cr	52·0	Radon, Niton	Rn	222·0
Cobalt	Co	58·9	Rhodium	Rh	102·9
Columbium, see *Niobium*			Rubidium	Rb	85·4
Copper (*Cuprum*)	Cu	63·5	Ruthenium	Ru	101·7
Dysprosium	Dy	162·5	Samarium	Sa	150·4
Erbium	Er	167·7	Scandium	Sc	45·1
Europium	Eu	152·0	Selenium	Se	79·2
Fluorine	F	19·0	Silicon	Si	28·0
Gadolinium	Gd	157·2	Silver (*Argentum*)	Ag	107·8
Gallium	Ga	69·7	Sodium (*Natrium*)	Na	22·9
Germanium	Ge	72·6	Strontium	Sr	87·6
Glucinum, see *Beryllium*			Sulphur	S	32·0
Gold (*Aurum*)	Au	197·2	Tantalum	Ta	181·5
Helium	He	4·0	Tellurium	Te	127·5
Holmium	Ho	163·4	Terbium	Tb	159·2
Hydrogen	H	1·0	Thallium	Tl	204·3
Indium	In	114·8	Thorium	Th	232·1
Iodine	I	126·9	Thulium	Tm	169·4
Iridium	Ir	193·1	Tin (*Stannum*)	Sn	118·7
Iron (*Ferrum*)	Fe	55·8	Titanium	Ti	48·1
Krypton	Kr	82·9	Tungsten (*Wolframium*)	W	184·0
Lanthanum	La	138·9	Uranium	U	238·1
Lead (*Plumbum*) (Ordinary)	Pb	207·2	Vanadium	V	50·9
Lead (*Uranium*)	Pb	206·0	Xenon	Xe	130·2
Lithium	Li	6·9	Ytterbium	Yb	173·6
Lutecium	Lu	175·0	Yttrium	Yt	88·9
Magnesium	Mg	24·3	Zinc	Zn	65·3
Manganese	Mn	54·9	Zirconium	Zr	91·0

465. Metals and Non-metals. — The elements may be divided into two more or less distinct classes, the metals and the non-metals. Between the two lie a number of elements sometimes called the semi-metals. The *metals*, as gold, silver, iron, sodium, are those elements which, *physically* described, possess to a more or less perfect degree the fundamental characters of the ideal metal, viz.: malleability, ductility, metallic luster (and opacity to light), conductivity for heat and electricity; moreover, *chemically* described, they commonly play the part of the positive or basic element in a simple compound, as later defined (Arts. **474–477**). The *non-metals*, as sulphur, carbon, silicon, etc., also the gases, as oxygen, chlorine, etc., have none of the physical characters alluded to: they are, if solids, brittle, often transparent to light-radiation, are poor conductors for heat and electricity. Chemically expressed, they usually play the negative or acid part in a simple compound.

The so-called *semi-metals*, or metalloids, include certain elements, as tellurium, arsenic, antimony, bismuth, which have the physical characters of a metal to a less perfect degree (*e.g.*, they are more or less brittle); and, more important than this they often play the part of the acidic element in the compound into which they enter. These points are illustrated later.

It is to be understood that the distinctions between the classes of the elements named cannot be very sharply applied. Thus the typical metallic characters mentioned are possessed to a very unequal degree by the different substances classed as metals; for example, by silver and tin. Corresponding to this a number of the true metals, as tin and manganese, play the part of an acid in numerous salts. Further, the mineral magnetite, $\overset{\text{II}}{\text{Fe}}\overset{\text{III}}{\text{Fe}}_2\text{O}_4$, is often described as an *iron ferrate*; so that in this compound the same element would play the part of both acid and base.

466. Positive and Negative Elements. — It is common to make a distinction between the *electro-positive* and *electro-negative* element in a compound. The passage of a sufficiently strong electrical current through a chemical compound in many cases results in its decomposition (or electrolysis) into its elements or parts. In such cases it is found that for each compound the atoms of one element collect at the negative pole (the cathode) and those of the other at the positive pole (the anode). The former is called the electro-positive element and the latter the electro-negative element. Thus in the electrolysis of water (H_2O) the hydrogen collects at the cathode and is hence called positive, and the oxygen at the anode and is called negative. Similarly, in hydrochloric acid (HCl) the hydrogen is thus shown to be positive, the chlorine negative. This distinction is also carried to complex compounds, as copper sulphate ($CuSO_4$), which by electrolysis is broken into Cu, which is found to be electro-positive, and SO_4 (the last separates into SO_3, forming H_2SO_4 and free oxygen).

For reasons which will be explained later, the positive element is said to play the basic part, the negative the acidic. The metals, as already stated, in most cases belong to the former class, the non-metals to the latter, while the semi-metals may play both parts.

It is common in writing the formula to put the positive or basic element first, thus H_2O, H_2S, HCl, H_2SO_4, Sb_2S_3, As_2O_3, AsH_3, $NiSb$, $FeAs_2$. Here it will be noted that antimony (Sb) and arsenic (As) are positive in some of the compounds named but negative in the others.

467. Periodic Law. — In order to understand the relations of the chief classes of chemical compounds represented among minerals, as still more their further subdivision, down finally to the many *isomorphous groups* — groups of species having analogous composition and closely similar form, as explained in Art. **483** — the fundamental relations and grouping of the elements must be understood, especially as shown in the so-called Periodic Law.

Although the subject can be only briefly touched upon, it will be useful to give here the general distribution of the elements into Groups and Periods, as presented in the Principles of Chemistry (Engl. Ed., 1891) of D. Mendeléeff, to whom is due more than any one else the development of the Periodic Law.

When the elements are arranged according to the values of their atomic weights, or better according to their atomic numbers, it is seen that they fall more or less into groups consisting of eight elements each, or double groups containing sixteen elements. The corresponding members of each group show similar chemical characters. The table given on p. 345 will illustrate these relationships. For the thorough explanation of this subject, more particularly as regards the periodic or progressive relation between the atomic weights and various properties of the elements, the reader is referred to the work above mentioned or to one of the many other excellent modern text-books of chemistry.

The relations of some of the elements of the first group are exhibited by the isomorphism (see Art. **483,** also the description of the various groups and species here referred to, which are given in Part V of this work) of $NaCl$, KCl, $AgCl$; or again of $LiMnPO_4$ and $NaMnPO_4$, etc. In the second group, reference may be made to the isomorphism of the carbonates and sulphates (p. 352) of calcium, barium, and strontium; while among the sulphides, ZnS, CdS, and HgS are doubly related. In the third group, we find boron and aluminum often replacing one another among silicates. In the fourth group, the relations of silicon and titanium are shown in the titano-silicates, while the compounds TiO_2, SnO_2, PbO_2, also $ZrSiO_4$ and $ThSiO_4$, have closely similar form. In the fifth group, many compounds of arsenic, antimony, and bismuth are isomorphous among metallic compounds, while the relations of phosphorous, vanadium, arsenic, also antimony, are shown among the phosphates, vanadates, arsenates, and antimonates; again the mutual relations of the niobates and tantalates are to be noted.

In the sixth group, the strongly acidic elements, sulphur, selenium, tellurium, are all closely related, as seen in many sulphides, selenides, tellurides; further, the relations of sulphur and chromium, and similarly of both of these to molybdenum and tungsten, are shown among many artificial sulphates, chromates, molybdates, and tungstates.

In the seventh group the relations of the halogens are too well understood to need special remark. In the eighth group, we have Fe, Co, Ni alloyed in meteoric iron, and their phosphates and sulphates are in several cases closely isomorphous; further, the relation of the iron series to that of the platinum series is exhibited in the isomorphism of FeS_2, $FeAsS$, $FeAs_2$, etc., with $PtAs_2$ and probably RuS_2.

When the atomic structure of the elements in any one group of the Periodic System is considered, it is found that in general it is similar for the group. For instance, the outer electronic shell for both sodium and potassium contains only one electron, etc. Therefore the elements in a single group in general show similar chemical characters. The elements that lie in the upper right-hand corner of the table are the acidic elements; the basic elements are placed to the left and bottom.

468. Combining Weight. — Chemical investigation proves that the mass of a given element entering into a compound is always proportional either to its atomic weight or to some simple multiple of this; the atomic weight is hence also called the *combining weight*. Thus in rock salt, sodium chloride, the masses involved of sodium and chlorine present are found by analysis to be equal to 39·4 and 60·6 in 100 parts, and these numbers are in proportion to 23 : 35·4, the atomic weights of sodium and chlorine; hence it is concluded that one atom of each is present in the compound. The formula is, therefore,

TABLE OF PERIODIC CLASSIFICATION OF THE ELEMENTS

Period	(0)	I	II	III	IV	V	VI	VII	VIII
1	2 He 4.0	1 H 1.0							
2	10 Ne 20.2	3 Li 6.9	4 Be 9.0	5 B 10.8	6 C 12.0	7 N 14.0	8 O 16.0	9 F 19.0	
3	18 A 39.9	11 Na 22.9	12 Mg 24.3	13 Al 26.9	14 Si 28.0	15 P 31.0	16 S 32.0	17 Cl 35.4	
4		19 K 39.0	20 Ca 40.0	21 Sc 45.1	22 Ti 48.1	23 V 50.9	24 Cr 52.0	25 Mn 54.9	26 Fe 55.8 27 Co 58.9 28 Ni 58.6
4	36 Kr 82.9	29 Cu 63.5	30 Zn 65.3	31 Ga 69.7	32 Ge 72.6	33 As 74.9	34 Se 79.2	35 Br 79.9	
5		37 Rb 85.4	38 Sr 87.6	39 Y 88.9	40 Zr 91.0	41 Nb 93.1	42 Mo 96.0		44 Ru 101.7 45 Rh 102.9 46 Pd 106.7
5	54 Xe 130.2	47 Ag 107.8	48 Cd 112.4	49 In 114.8	50 Sn 118.7	51 Sb 121.7	52 Te 127.5	53 I 126.9	
6		55 Cs 132.8	56 Ba 137.3	57 La 138.9	58 Ce 140.2				
6					72 Hf 180.8	73 Ta 181.5	74 W 184.0		76 Os 190.8 77 Ir 193.1 78 Pt 195.2
6		79 Au 197.2	80 Hg 200.3	81 Tl 204.4	82 Pb 207.2	83 Bi 209.0			
7	86 Rn 222.0		88 Ra 226	89 Ac 230?	90 Th 232.1		92 U 238.1		

NaCl. In calcium chloride, by the same method the masses present are found to be proportional to $39.9 : 70.8$, that is, to $39.9 : 2 \times 35.4$; hence the formula is $CaCl_2$.

Still again, a series of compounds of nitrogen with oxygen is known in which the ratios of the masses of the two elements are as follows: (1) 28 : 16, (2) 14 : 16, (3) 28 : 48, (4) 14 : 32, (5) 28 : 80. It is seen at once that these must have the formulas (1) N_2O, (2) NO, (3) N_2O_3, (4) NO_2, (5) N_2O_5. On the contrary, atmospheric air which contains these elements in about the ratio of 76.8 to 23.2 cannot be a chemical compound of these elements, since (aside from other considerations) these numbers are not in the ratio of $n \times 14 : m \times 16$ where n and m are simple whole numbers.

469. Molecular Weight. — The molecular weight is the weight of the molecule of the given substance, expressed in terms of the mass of the oxygen atom as unit. The molecular weight of hydrogen is 2 because the molecule can be shown to consist of two atoms. The molecular weight of hydrochloric acid (HCl) is 36.4, of water vapor (H_2O) 18, of hydrogen sulphide (H_2S) 34.

Since, according to the law of Avogadro, like volumes of different gases under like conditions as to temperature and pressure contain the same number of molecules, it is obvious that the molecular weight of substances in the form of gas can be derived directly from the relative density or specific gravity. If the density is referred to hydrogen, whose molecular weight is 2, it will be always true that the molecular weight is twice the density in the state of a gas and *vice versa*. Thus the observed density of carbon dioxide (CO_2) is 22, hence its molecular weight must be 44. It is this principle that makes it possible in the case of a gas to fix the constitution of the molecule when the ratio in number of the atoms entering into it has been determined by analysis. In the case of solids, where the constitution of the molecule in general cannot be fixed, it is best, as already stated, to write the molecular formula in its simplest form, as $NaAlSi_3O_8$ for albite. The sum of the weights of the atoms present is then taken as the molecular weight.

470. Valence. — The valence of an element is given by a number representing the capacity of its atoms to combine with the atoms of some unit element like hydrogen or chlorine. Thus, using the examples of Art. **468**, in NaCl, since one atom of sodium unites with one of chlorine, its valence is one; or, in other words, it is said to be *univalent*. Further, calcium (as in $CaCl_2$), also barium, etc., are bivalent; aluminum is trivalent; silicon is tetravalent, etc. The valence may be expressed by the number of bonds by which one element in a compound is united to another, thus:

$$Na - Cl, \qquad Ba = Cl_2, \qquad Au \equiv Cl_3, \qquad Sn \equiv Cl_4, \qquad \text{etc.}$$

The explanation of valence lies in the atomic structure. For instance, the atomic number of sodium is 11. It therefore has two electrons in its innermost ring, eight in the second shell, and only one in the third shell. On the other hand, the atomic number of chlorine is 17, that is, it lacks one electron in its third shell. When sodium and chlorine combine, the one electron in the third shell of the sodium atom fits into the vacancy in the third shell of the chlorine atom. Sodium can give up one of its electrons and chlorine can add to its atom one electron, or in other words, both elements have a valency of one.

A considerable number of the elements show a different valence in different compounds. Thus both Sb_2O_3 and Sb_2O_5 are known; also FeO and Fe_2O_3; $CuCl$ and $CuCl_2$. These possible variations are indicated in the following table which gives the valences for the common elements.

Univalent: H, Cl, Br, I, F; Li, Na, K, Rb, Cs, Ag, Hg, Cu, Au.

Bivalent: O, S, Se, Te; Be, Mg, Ca, Sr, Ba, Pb, Hg, Cu, Zn, Co, Ni, Fe, Mn, Cr, C, Sn.

Trivalent: B, Au, Al, Fe, Mn, Cr, Co, Ni, N, P, As, Sb, Bi.

Tetravalent: C, Si, Ti, Zr, Sn, Mn, Pb.

Pentavalent: N, P, As, Sb, V, Bi, Nb, Ta.

471. Chemical Reactions. — When solutions of two chemical substances are brought together, in many cases they react upon each other with the result of forming new compounds out of the elements present; this phenomenon is called a *chemical reaction.* One of the original substances may be a gas, and in many cases similar results are obtained from a liquid and a solid, or less often from two solids.

For example, solutions of sodium chloride (NaCl) and silver nitrate ($AgNO_3$) react on each other and yield silver chloride (AgCl) and sodium nitrate ($NaNO_3$). This is expressed in chemical language as follows:

$$NaCl + AgNO_3 = AgCl + NaNO_3.$$

This is a chemical equation, the sign of equality meaning that equal weights are involved both before and after the reaction.

Again, hydrochloric acid (HCl) and calcium carbonate ($CaCO_3$) yield calcium chloride ($CaCl_2$) and carbonic acid (H_2CO_3); which last breaks up into water (H_2O) and carbon dioxide (CO_2), the last going off as a gas with effervescence. Hence

$$CaCO_3 + 2HCl = CaCl_2 + H_2O + CO_2.$$

472. Radicals. — A compound of two or more elements according to their relative valence in which all their bonds are satisfied is said to be *saturated.* This is true of H_2O, or, as it may be written, H—O—H. If, however, one or more bonds is left unsatisfied, the resulting combination of elements is called a *radical.* Thus —O—H, called briefly hydroxyl, is a common radical, having a valence of one, or, in other words, univalent; NH_4 is again a univalent radical; so, too, (CaF), (MgF) or (AlO). Radicals often enter into a compound like a simple element; for example, in ammonium chloride, NH_4Cl, the univalent radical NH_4 plays the same part as the univalent element Na in NaCl. In the chemical composition of mineral species, the commonest radical is hydroxyl (—O—H) already defined. Other examples are (CaF) in apatite (see Art. **483**), (MgF) in wagnerite, (AlO) in many basic silicates, etc.

473. Chemical Compound. — A chemical compound is a combination of two or more elements united by the force of chemical attraction. It is always true of it, as before stated (Art. **468**), that the elements present are combined in the proportion of their atomic weights or some simple multiples of these. A substance which does not satisfy this condition is not a compound, but only a mechanical mixture.

Examples of the simpler class of compounds are afforded by the *oxides*, or compounds of oxygen with another element. Thus, among minerals we have Cu_2O, cuprous oxide (cuprite); ZnO, zinc oxide (zincite); Al_2O_3, alumina (corundum); SnO_2, tin dioxide (cassiterite); SiO_2, silicon dioxide (quartz); As_2O_3, arsenic trioxide (arsenolite).

Another simple class of compounds are the *sulphides* (with the selenides, tellurides, arsenides, antimonides, etc.), compounds in which sulphur (selenium, tellurium, arsenic, antimony, etc.) plays the same part as oxygen in the oxides. Here belong Cu_2S, cuprous sulphide (chalcocite); ZnS, zinc sulphide (sphalerite); $PbTe$, lead telluride (altaite); FeS_2, iron disulphide (pyrite); Sb_2S_3, antimony trisulphide (stibnite).

474. Acids. — The more complex chemical compounds, an understanding of which is needed in a study of minerals, are classed as acids, bases, and salts; the distinctions between them are important.

An *acid* is a compound of hydrogen, or hydroxyl, with a non-metallic element (as chlorine, sulphur, nitrogen, phosphorus, etc.), or a radical containing these elements. When dissolved in water they all give the positive hydrogen ion and a negative ionic substance such as Cl, SO_4, etc. The hydrogen atoms of an acid may be replaced by metallic atoms; the result being then the formation of a salt (see Art. **476**). Acids in general turn blue litmus paper red and have a sharp, sour taste. The following are familiar examples:

HCl, hydrochloric acid.
HNO_3, nitric acid.
H_2CO_3, carbonic acid.
H_2SO_4, sulphuric acid.
H_2SiO_3, metasilicic acid.
H_3PO_4, phosphoric acid.
H_4SiO_4, orthosilicic acid.

It is to be noted that with a given acid element several acids are possible. Thus normal, or orthosilicic, acid is H_4SiO_4, in which the bonds of the element silicon are all satisfied by the hydroxyl (HO). But the removal of one molecule of water, H_2O, from this gives the formula H_2SiO_3, or metasilicic acid.

Acids which, like HNO_3, contain one atom of hydrogen that may be replaced by a metallic atom (*e.g.*, in KNO_3) are called *monobasic*. If, as in H_2CO_3 and H_2SO_4, there are two atoms or a single bivalent atom (*e.g.*, in $CaCO_3$, $BaSO_4$), the acids are *dibasic*. Similarly, H_3PO_4 is *tribasic*, etc.

Most acids are liquids (or gases), and hence acids are represented very sparingly among minerals; $B(OH)_3$, boric acid (sassolite), is an illustration.

475. Bases. — The *bases*, or hydroxides, as they are also called, are compounds which may be regarded as formed of a metallic element (or radical) and the univalent radical hydroxyl, — (OH); or, in other words, of an oxide with water. They are compounds which in solution yield hydroxyl ions. Thus potash, K_2O, and water, H_2O, form $2K(OH)$, or potassium hydroxide; also $CaO + H_2O$ similarly give $Ca(OH)_2$, or calcium hydroxide. In general, when soluble in water, bases give an alkaline reaction with turmeric paper or red litmus paper, and they also neutralize an acid, as explained in the next article. Further, the bases yield water on ignition, that is, at a temperature sufficiently high to break up the compound.

Among minerals the bases are represented by the hydroxides, or hydrated oxides, as $Mg(OH)_2$, magnesium hydrate (brucite); $Al(OH)_2$, aluminum hydrate (gibbsite); also, $(AlO)(OH)$, diaspore, etc.

476. Salts. — A third class of compounds are the *salts*; these may be regarded as formed chemically by the reaction of a base upon an acid, or, in other words, by the neutralization of the acid. Thus calcium hydrate and sulphuric acid give calcium sulphate and water:

$$Ca(OH)_2 + H_2SO_4 = CaSO_4 + 2H_2O.$$

Here calcium sulphate is the salt, and in this case the acid, sulphuric acid, is said to be neutralized by the base, calcium hydroxide.   It is instructive to compare the formulas of a base, an acid, and the corresponding salt, as follows:

$$Base, \text{ Ca(OH)}_2; \qquad Acid, \text{ H}_2\text{SO}_4; \qquad Salt, \text{ CaSO}_4.$$

Here it is seen that a salt may be simply described as formed from an acid by the replacement of the hydrogen atom, or atoms, by a metallic element or radical.

477. Typical Salts. — The commonest types of salts represented among minerals are the following:

Chlorides: salts of hydrochloric acid, HCl; as $AgCl$, silver chloride (cerargyrite).

Nitrates: salts of nitric acid, HNO_3; as KNO_3, potassium nitrate (niter).

Carbonates: salts of carbonic acid, H_2CO_3; as $CaCO_3$, calcium carbonate (calcite and aragonite).

Sulphates: salts of sulphuric acid, H_2SO_4; as $CaSO_4$, calcium sulphate (anhydrite).

Phosphates: salts of phosphoric acid, H_3PO_4; as $Ca_3(PO_4)_2$, calcium phosphate.

Silicates: several classes of salts are here included.   The most common are the salts of metasilicic acid, H_2SiO_3; as $MnSiO_3$, manganese metasilicate (rhodonite).   Also salts of orthosilicic acid, H_4SiO_4; as Mn_2SiO_4, manganese orthosilicate (tephroite).

Numerous other classes of salts are also included among mineral species; their composition, as well as that of complex salts of the above types, is explained in the descriptive part of this work.

478. Normal, Acid, and Basic Salts. — A *neutral* or *normal salt* is one in which the basic and acidic elements have completely neutralized each other, or, in other words, one of the type already given as examples, in which *all* the hydrogen atoms of the acid have been replaced by metallic atoms or radicals, or the hydroxyl groups have been completely replaced by the acid radical. Thus, K_2SO_4 is normal potassium sulphate, but $HKSO_4$, on the other hand, is acid potassium sulphate, since in the acid H_2SO_4 only one of the bonds is taken by the basic element potassium.   Salts of this kind are called *acid salts.*   The formula in such cases may be written* as if the compound consisted of a normal salt and an acid; thus, for the example given, $K_2SO_4 . H_2SO_4$.

A *basic salt* is one in which the acid part of the compound is not sufficient to satisfy all the bonds of the base.   Thus malachite is a basic salt — basic carbonate of copper — its composition being expressed by the formula $Cu_2(OH)_2CO_3$.   This may be written $CuCO_3 . Cu(OH)_2$, or $(Cu_2) \genfrac{}{}{0pt}{}{= CO_3}{= (OH)_2}$.

The majority of minerals consist not of simple salts, as those noted above, but of more or less complex double salts in which several metallic elements are present.   Thus common grossular garnet is an orthosilicate containing both calcium and aluminum as bases; its formula is $Ca_3Al_2(SiO_4)_3$.

479. Sulpho-salts. — The salts thus far spoken of are all oxygen salts. There are also others, of analogous constitution, in which sulphur takes the place of the oxygen; they are hence called *sulpho-salts.*   Thus normal sulph-

---

* This early form of writing the composition explains the name often given to the compound, namely, in this case, " bisulphate of potash."

arsenious acid has the formula H_3AsS_3, and the corresponding silver salt is Ag_3AsS_3, the mineral proustite. Similarly the silver salt of the analogous antimony acid is Ag_3SbS_3, the mineral pyrargyrite. From the normal acids named, a series of other hypothetical acids may be derived, as $HAsS_2$, $H_4As_2S_5$, etc.; these acids are not known to exist, but their salts are important minerals. Thus zinkenite, $PbSb_2S_4$, is a salt of the acid $H_2Sb_2S_4$, and jamesonite, $Pb_2Sb_2S_5$, of the acid $H_4Sb_2S_5$, etc.

480. Water of Crystallization. — As stated in Art. **475**, the hydroxides, or bases, and further basic salts in general, yield water when ignited. Thus calcium hydroxide $Ca(OH)_2$ breaks up on heating into CaO and H_2O, as expressed in the chemical equation

$$2Ca(OH)_2 = 2CaO + H_2O.$$

So also the basic cupric carbonate, malachite, $Cu_2(OH)_2CO_3$, yields water on ignition; and the same is true of the complex basic orthosilicates, like zoisite, whose formula is $(HO)Ca_2Al_3(SiO_4)_3$. It is not to be understood, however, in these or similar cases, that water as such is present in the substance.

On the other hand, there is a large number of mineral compounds which yield water readily when heated, and in which the water molecules are regarded as present as so-called *water of crystallization*. Thus, the formula of gypsum is written

$$CaSO_4 + 2H_2O,$$

and the molecules of water ($2H_2O$) are considered as water of crystallization. So, too, in potash alum, $KAl(SO_4)_2 + 12H_2O$, the water is believed to play the same part.

481. Formulas of Minerals. — The strictly empirical formula expresses the kinds and numbers of atoms of the elements present in the given compound, without attempting to show the way in which it is believed that the atoms are combined. Thus, in the case of zoisite the empirical formula is $HCa_2Al_3Si_3O_{13}$. While not attempting to represent the structural formula (which will not be discussed here), it is convenient in certain cases to indicate the atoms which there is reason to believe play a peculiar relation to each other. Thus the same formula written $(HO)Ca_2Al_3(SiO_4)_3$ shows that it is regarded as a basic orthosilicate, in other words, a basic salt of orthosilicic acid, H_4SiO_4.

Again, the empirical formula of common apatite is $Ca_5FP_3O_{12}$; but if this is written $(CaF)Ca_4(PO_4)_3$, it shows that it is regarded as a phosphate of the acid H_3PO_4, that is, $H_9(PO_4)_3$, in which the nine hydrogen atoms are replaced by four Ca atoms together with the univalent radical (CaF). In another kind of apatite the radical (CaCl) enters in the same way. Similarly to this the formula of pyromorphite is $(PbCl)Pb_4(PO_4)_3$, of vanadinite $(PbCl)Pb_4(VO_4)_3$.

Further, it is often convenient to employ the method of writing the formulas in vogue under the old dualistic system. For example,

$$CaO . CO_2 \text{ for } CaCO_3,$$
$$3CaO . Al_2O_3 . 3SiO_2 \text{ for } Ca_3Al_2Si_3O_{12},$$
$$3Ag_2S . Sb_2S_3 \text{ for } Ag_3SbS_3, \text{ etc.}$$

It is no longer believed, however, that the molecular groups CaO, Al_2O_3, etc., actually exist in the molecule of the substance. But in part because these groups are what analysis of the substance affords directly, and in part because so easily retained in the memory, this method of writing is still often used.

482. Calculation of a Formula from an Analysis. — The result of an analysis gives the proportions, in a hundred parts of the mineral, of either the elements themselves, or of their oxides or other compounds obtained in the chemical analysis.   In order to obtain the atomic proportions of the elements:

Divide the percentages of the elements by the respective ATOMIC WEIGHTS; or, for those of the oxides: *Divide the percentage amounts of each by their* MOLECULAR WEIGHTS; *then find the simplest ratio in whole numbers for the numbers thus obtained.*

Example. — An analysis of bournonite from Wolfsberg gave C. Bromeis the results under (1) below.   These percentages divided by the respective atomic weights, as indicated, give the numbers under (2).   Finally the ratio of these numbers gives very nearly 1 : 3 : 1 : 1. Hence the formula derived is $CuPbSbS_3$.   The theoretical values called for by the formula are added under (4).

	(1)	(2)	(3)	(4)
Sb	$24 \cdot 34 \div 120 = 0 \cdot 203$		1	24·7
S	$19 \cdot 76 \div 32 = 0 \cdot 617$		3	19·8
Pb	$42 \cdot 88 \div 206 \cdot 4 = 0 \cdot 208$		1	42·5
Cu	$13 \cdot 06 \div 63 \cdot 2 = 0 \cdot 207$		1	13·0
	100·04			100·0

Second Example. — The mean of two analyses of a garnet from Alaska gave Kountze the results under (1) below.   Here, as usual, the percentage amounts of the several molecular groups (SiO_2, Al_2O_3, etc.) are given instead of those of the elements.   These amounts divided by the respective molecular weights give the numbers under (2).   In this case the amounts of the protoxides are taken together and the ratio thus obtained is 3·09 : 1 : 2·92, which corresponds approximately to the formula $3FeO.Al_2O_3.3SiO_2$, or $Fe_3Al_2(SiO_4)_3$.   The magnesium in this garnet would ordinarily be explained by the presence of the pyrope molecule ($Mg_3Al_2[SiO_4]_3$) together with the simple almandite molecule whose composition is given above.

	(1)	(2)	(3)
SiO_2	$39 \cdot 29 \div 60 = 0 \cdot 655$		3·09
Al_2O_3	$21 \cdot 70 \div 102 = 0 \cdot 212$		1
Fe_2O_3	*tr.*		
FeO	$30 \cdot 82 \div 71 \cdot 9 = 0 \cdot 429$		
MnO	$1 \cdot 51 \div 70 \cdot 8 = 0 \cdot 022$	0·619	2·92
MgO	$5 \cdot 26 \div 40 = 0 \cdot 132$		
CaO	$1 \cdot 99 \div 55 \cdot 0 = 0 \cdot 036$		
	100·57		

It is necessary, when very small quantities only of certain elements (as MnO, MgO, CaO, above) are present, to neglect them in the final formula, reckoning them in with the elements which they replace, that is, with those of the same quantivalence.   The degree of correspondence between the analysis and the formula deduced, if the latter is correctly assumed, depends entirely upon the accuracy of the former.

483. Isomorphism. — Chemical compounds which have an analogous composition and a closely related crystalline form are commonly said to be *isomorphous*.   This phenomenon, called ISOMORPHISM, was first clearly brought out by Mitscherlich.

Many examples of groups of isomorphous compounds will be found among the minerals described in the following pages.   Some examples are mentioned here in order to elucidate the subject.

In the brief discussion of the periodic classification of the chemical elements of Art. **467,** attention has been called to the prominent groups among the elements which form analogous compounds.   Thus calcium, barium, and strontium, and also lead, form the two series of analogous compounds,

Aragonite Group		Barite Group
$CaCO_3$, aragonite.	Also	$CaSO_4$, anhydrite.
$BaCO_3$, witherite.		$BaSO_4$, barite.
$SrCO_3$, strontianite.		$SrSO_4$, celestite.
$PbCO_3$, cerussite.		$PbSO_4$, anglesite.

Further, the members of each series crystallize in closely similar forms.   The carbonates are orthorhombic, with axial ratios not far from one another; thus the prismatic angle approximates to 60° and 120°, and corresponding to this they all exhibit pseudo-hexagonal forms due to twinning.   The sulphates also form a similar orthorhombic series, and though anhydrite deviates somewhat widely, the others are close together in angle and in cleavage.

Again, calcium, magnesium, iron, zinc, and manganese form a series of carbonates with analogous composition as shown in the list of the species of the *Calcite Group* given on p. 511.   This table brings out clearly the close relation in form between the species named.

Further, it is also generally true with an isomorphous series that the various molecules may enter in greater or less degree into the constitution of one of the members of the series without causing any marked change in the crystal characters.   For instance, in the Calcite group, calcite itself may contain small percentages of $MgCO_3$, $FeCO_3$ and $MnCO_3$.   These different molecules may assume in the crystal structure of the mineral the same functions as the corresponding amounts of $CaCO_3$ which they have replaced.   The molecules of magnesite and siderite, $MgCO_3$ and $FeCO_3$, may replace each other in any proportion and the same is true with siderite and rhodochrosite, $MnCO_3$. Various intermediate mixtures of these latter molecules have been described and given distinctive names to which definite formulas have been assigned. It is doubtful, however, if these compounds have any real existence but merely represent certain points in the complete isomorphous series that lies between the end members.   Dolomite, $CaMg(CO_3)_2$, on the other hand, is a definite compound and not an isomorphous mixture of $CaCO_3$ and $MgCO_3$.   It may, however, contain varying amounts of $FeCO_3$, $MnCO_3$ and also an excess of $CaCO_3$ or $MgCO_3$, all of which enter the regular molecule in the form of isomorphous replacements.

The *Apatite Group* forms another valuable illustration since in it are represented the analogous compounds, apatite and pyromorphite, both phosphates, but respectively phosphates of calcium and lead; also the analogous lead compounds pyromorphite, mimetite, and vanadinite, respectively lead phosphate, lead arsenate, and lead vanadate.   Further, in all these compounds the radical (RCl) or (RF) enters in the same way (see Art. **481**).   Thus the formulas for the two kinds of apatite and that for pyromorphite are as follows:

$$(CaF)Ca_4(PO_4)_3, \qquad (CaCl)Ca_4(PO_4)_3, \qquad (PbCl)Pb_4(PO_4)_3.$$

Some of the more important isomorphous groups are mentioned below.   For a discussion of them, as well as of many others that might be mentioned here, reference must be made to the descriptive part of this work.

Isometric System. — The Spinel group, including spinel, $MgAl_2O_4$; also magnetite, chromite, franklinite, gahnite, etc.   The Galena group, as galena, PbS; argentite, Ag_2S, etc.   The Garnet group, as grossularite, $Ca_3Al_2Si_3O_{12}$, etc.

Tetragonal System. — Rutile group, including rutile, TiO_2; cassiterite, SnO_2. The Scheelite group, including scheelite, $CaWO_4$; stolzite, $PbWO_4$; wulfenite, $PbMoO_4$.

Hexagonal System. — Apatite group, already mentioned, including apatite, pyromorphite, mimetite, and vanadinite. Corundum group, corundum, Al_2O_3; hematite, Fe_2O_3. Calcite group, already mentioned. Phenacite group, etc.

Orthorhombic System. — Aragonite group, and Barite group, both mentioned above. Chrysolite group, $(Mg,Fe)_2SiO_4$; Topaz group, etc.

Monoclinic System. — Copperas group, including melanterite, $FeSO_4 + 7H_2O$; bieberite, $CoSO_4 + 7H_2O$, etc. Pyroxene and Amphibole groups, and the Mica group.

Monoclinic and Triclinic Systems. — Feldspar group.

Isomorphism in the broader sense in which the word is commonly used expresses the relations that exist between a series of substances of analogous chemical formulas and analogous crystal structures. Analogous formulas possess an equal number of atoms and of positive and negative constituents. Analogous structures are composed of geometrically similar elementary parallelepipeds in which an equal number of atoms are arranged in a geometrically similar manner. Such isomorphism occurs when the relative size of the atoms and certain of their physical properties are nearly the same.* This theory, known sometimes as *volume isomorphism,* accounts for the fact that elements of distinctly different chemical character, and even of different valence, may substitute for each other without changing the crystal structure. It would explain not only why calcite, $CaCO_3$, and sodium nitrate, $NaNO_3$, crystallize in closely related forms but also why rhombohedrons of sodium nitrate will crystallize from a solution upon a cleavage rhombohedron of calcite with parallel orientation.

Isomorphism used in a narrower sense expresses the ability of qualitatively different substances to form together in a mixed crystal. Usually the elements that may substitute for each other are chemically closely related, as in the illustrations given above. But here also the volume of the atoms involved or, perhaps better, the volume of the atomic groups, must be nearly equal. The isomorphous feldspar series, described in the succeeding article, illustrates this point.

484. Isomorphous Mixtures. — It is important to note that the intermediate compounds in the case of an isomorphous series, such as those spoken of in the preceding article, often show a distinct gradation in crystalline form, and more particularly in physical characters (*e.g.,* specific gravity, optical properties, etc.). This is illustrated by the species of the calcite group already referred to; also still more strikingly by the group of the triclinic feldspars as fully discussed under the description of that group. See further Art. **436.**

The feldspars also illustrate two other important points in the subject, which must be briefly alluded to here. The triclinic feldspars have been shown by Tschermak to be isomorphous mixtures of the end compounds in varying proportions:

Albite, $NaAlSi_3O_8$.   Anorthite, $CaAl_2Si_2O_8$.

Here it is seen that these compounds have not an analogous composition in the narrow sense previously illustrated, and yet they are isomorphous and form an isomorphous series. Other examples of this are found among the pyroxenes, the scapolites, etc.

---

* For discussion of modern theories of crystal structure and chemical composition see V. M. Goldschmidt, Trans. Faraday Soc., **25**, p. 253, 1929.

Further, the Feldspar Group in the broader sense includes several other species, conspicuously the monoclinic orthoclase, $KAlSi_3O_8$, which, though belonging to a different system, still approximates closely in form to the triclinic species.

485. Variation in Composition of Minerals. Isomorphous Replacement and Solid Solution. — The idea that a mineral must rigidly conform in its chemical composition to a theoretical composition derived from its formula can no longer be strictly held. It is true that the majority of minerals do show a close correspondence to that theory, commonly within the limits of possible errors in the analyses. On the other hand, many minerals show slight, and certain ones considerable variations from their theoretical compositions. These variations can usually be explained by the principle of isomorphism. An instructive example is the case of sphalerite. Note in the analyses quoted below how the percentages of zinc diminish and those of iron correspondingly increase. It is evident from these analyses that iron, and in a much smaller degree other metals, may enter into the chemical compound and while replacing the zinc perform the same function as it, in the crystalline structure of the mineral. The iron is therefore spoken of as being isomorphous with the zinc or the iron sulphide molecule as isomorphous with the zinc sulphide molecule. There is no definite ratio between the amounts of the iron and zinc that may be present but there is a constant ratio (1 : 1) between the sum of the atoms of the metals and the atoms of sulphur. That is, although the composition may vary, the atomic ratios and the crystalline structure remain constant. In some cases this interchange between elements or radicals may be complete, in other cases there may be distinct limitations to the amount by which any element or radical may be replaced by another. For instance in sphalerite the maximum percentage of the isomorphous iron seems to be about 16 to 18 per cent.

Colorless Sphalerite		Brown Sphalerite		Black Sphalerite	
S	32.93	S	33.36	S	33.25
Zn	66.69	Zn	63.36	Zn	50.02
Fe	0.42	Fe	3.60	Fe	15.44
	100.04		100.32	Cd	0.30
				Pb	1.01
					100.02

Further, we have cases where a compound may, in a certain sense, dissolve another unrelated substance and form what is known as a *solid solution*. This kind of phenomenon is well recognized among artificial salts and has recently been definitely proved with certain minerals. For instance, it has been shown experimentally that the artificial iron sulphide, FeS, corresponding to pyrrhotite, can dissolve an excess of sulphur up to about 6 per cent. Natural pyrrhotite always contains an excess of sulphur over that required by the formula, FeS, and various formulas such as Fe_7S_8, Fe_nS_{n+1}, etc., have been assigned to the mineral. This extra sulphur in the mineral varies in amount but also has as its maximum about 6 per cent. In view of the experimental data there is no doubt but that pyrrhotite should be considered as the monosulphide of iron containing varying small amounts of excess sulphur in the form of a solid solution.

Another case of solid solution is undoubtedly shown by nephelite which commonly contains a small excess of SiO_2. It is very probable that further investigation will show that many minerals have this power of holding in solid solution small amounts of foreign substances and that many hitherto inexplicable discrepancies in their analyses may be explained in this way. Such an assumption should not be made, however, without convincing proof of its probability, since many analytical discrepancies are undoubtedly due either to faulty analyses or to impure material.

486. Colloidal Minerals or Mineral Gels.* — It has been recognized recently that our amorphous hydrated minerals frequently do not conform in their analyses to the usually accepted formulas and cannot be regarded in the strict sense as definite chemical compounds. They show rather the properties of solid colloids or as they are commonly called *mineral gels*. A colloidal solution may be conceived as being intermediate in its characters between a true solution in which the salt in solution exists in particles of molecular size and the case where the mineral material is definitely in suspension in a liquid, its particles being visible and removable by filtration. It is probable that all gradations between these two extremes may occur. The mineral gels, or hydrogels, as they are sometimes called, since water is the liquid involved, are apparently formed from such colloidal solutions by some process of coagulation. They are considered therefore to consist of a micro-heterogeneous mixture of excessively minute particles of mineral material and water. The size of the individual particles is sub-microscopic but usually much greater than that of the molecule. The particles might very well have a definite crystalline structure.

These mineral gels are formed at low temperatures and pressures and are characteristically found among the products of rock weathering and in the oxidized zone of ore deposits. Some of them also occur in hot spring deposits. These minerals ordinarily assume botryoidal, reniform or stalactitic shapes, although, when the conditions of formation do not permit free growth, they may be earthy or dendritic. Frequently a mineral originally colloidal may become more or less crystalline in character through a molecular rearrangement and develop a fibrous or foliated structure. These have been designated as *meta-colloids*.

One important character of the gel minerals is their power to adsorb foreign materials. If through some change in condition one of these hydrogels should lose a part of its water content the remaining material would have a finely divided and porous structure exactly adapted to exert a strong power of adsorption. Consequently, although in many cases the main mass of the mineral may have a composition closely similar to some definite crystallized mineral, it will commonly show a considerable range in composition due both to the non-molecular relations of the contained water and to this secondary adsorption. Common mineral gels or substances derived from them are opal, bauxite, psilomelane, various members of the phosphate and arsenate groups, etc. As suggested above, gel varieties of minerals that occur also in crystalline forms are thought to exist. For example some authors speak of bauxite as the gel form of hydrargillite, stilpnosiderite as the gel form of goethite, chrysocolla of dioptase, and further give new names, such as gelvariscite, gelpyrophyllite, etc., to the gel phases of the corresponding crystalline minerals.

---

* For a résumé of the subject of gel minerals and a complete bibliography reference is made to articles by Marc and Himmelbauer, Fortschritte Min. Krist. Pet., **3**, 11, 33, 1913.

487. Dimorphism. Isodimorphism. — A chemical compound, which crystallizes in two forms genetically distinct, is said to be *dimorphous*; if in three, *trimorphous*, or in general *polymorphous*. This phenomenon is called DIMORPHISM or POLYMORPHISM. The explanation of dimorphism lies in the different molecular structure that the two substances possess. This proves that the physical properties of a mineral depend not only upon the chemical character of its constituent atoms but quite as definitely upon the manner in which they are arranged in the crystal structure.

An example is given by the compound calcium carbonate ($CaCO_3$), which is dimorphous: appearing as calcite and as aragonite. As *calcite* it crystallizes in the rhombohedral class of the hexagonal system, and, unlike as its many crystalline forms are, they may be all referred to the same fundamental axes, and, what is more, they have all the same cleavage and the same specific gravity (2·7) and, of course, the same optical characters. As *aragonite*, calcium carbonate appears in orthorhombic crystals, whose optical characters are entirely different from those of calcite; moreover, the specific gravity of aragonite (2·9) is higher than that of calcite (2·7).

Many other examples might be given: Titanium dioxide (TiO_2) is trimorphous, the species being called *rutile*, tetragonal ($c = 0·6442$), G. = 4·25; *octahedrite*, tetragonal ($c = 1·778$), G. = 3·9; and *brookite*, orthorhombic, G. = 4·15. Carbon appears in two forms, in diamond and graphite. Other familiar examples are pyrite and marcasite (FeS_2), sphalerite and wurtzite (ZnS), etc.

When two or more analogous compounds are at the same time isomorphous and dimorphous, they are said to be *isodimorphous*, and the phenomenon is called ISODIMORPHISM. An example of this is given in the Pyrite and Marcasite groups described later. Thus we have in the isometric Pyrite group, pyrite, FeS_2, smaltite, $CoAs_2$; in the orthorhombic Marcasite group, marcasite, FeS_2, safflorite, $CoAs_2$, etc.

488. Chemical and Microchemical Analysis. — The analysis of minerals is a subject treated of in chemical works, and need not be touched upon here except so far as to note the convenient use of certain qualitative methods, as described in the later part of this chapter.

Microchemical methods of analysis are of importance. They are especially valuable when only a minute amount of material is available for testing. They are performed usually on a glass slide placed under a microscope. The mineral grain to be tested is dissolved in a very small amount of liquid and to this is added a drop of a suitable reagent. The tests involve usually the formation of characteristically shaped crystals. At times a color reaction constitutes the test. Microchemical methods have also been used in testing mineral grains that occur in uncovered rock sections. It is impossible to treat adequately here this very useful method of chemical analysis, and the reader must be referred to the various text-books listed below. The subject has been particularly developed by Boricky, Haushofer, Behrens, Streng, and others. Reference is made to the discussion by Rosenbusch (Mikr. Phys., 1904, p. 435, *et seq.*) and to Johannsen (Manual of Pet. Methods, p. 559, *et seq.*, including a bibliography). Microchemical methods have been applied to the study of the opaque minerals, particularly those that serve as ores. The reactions are performed upon uncovered polished surfaces of the minerals. Observation in polarized light (see Art. **489**) is often useful. Many important facts concerning the origin and interrelation of the ore minerals have been established by investigations of this character. The terms *mineragraphic*, *mineralographic*, and *chalcographic*, have been used to designate this method of study. For details see Murdock (Micro. Deter. Opaque Min., 1916), Davy-Farnham (Micro. Exam. of the Ore Min., 1920), Farnham (Deter. of the Opaque Min., 1931), Schneiderhöhn and Ramdohr (Lehrbuch der Erzmikroskopie, 1931), Short (Microscopic Deter. of the Ore Min., 1931).

489. Use of Polarized Light with Polished Opaque Minerals. — When a polished surface of an opaque mineral is illuminated by a beam of polarized light and then viewed through a nicol prism with its plane of vibration approximately perpendicular to the plane in which the reflected light is polarized, optical effects are to be observed similar to those given to transmitted polarized light by transparent minerals. In the case of isometric minerals, the illumination or color does not change when the mineral is rotated upon the stage of the microscope. But with anisotropic minerals there will be a distinct change in illumination with two or four positions during the rotation that correspond to the positions of extinction of transparent minerals. There may also be a change in color with the change in position of the mineral section. This method is very useful in distinguishing between isotropic and anisotropic minerals, and certain minerals show characteristic effects that aid in their determination, but the method cannot give the quantitative results that are to be obtained when transparent minerals are examined in polarized light. A discussion of this use of polarized light, together with the important references to the literature, is given by Short in Bull. **825**, U. S. Geol. Sur., pp. 39 *et seq.*, 1931.

490. Mineral Synthesis. — The occurrence of certain mineral compounds (*e.g.*, the chrysolites) among the products of metallurgical furnaces has long been noted. But it has only been in recent years that the formation of artificial minerals has been made the subject of minute systematic experimental study. In this direction the French chemists have been particularly successful and very important work has been done at the Geophysical Laboratory of the Carnegie Institution in Washington, and now it may be stated that the majority of common minerals — quartz, the feldspars, amphibole, mica, etc. — have been obtained in crystallized form. Even the diamond has been formed in minute crystals by Moissan. These studies are obviously of great importance, particularly as throwing light upon the method of formation of minerals in nature. The chief results of the work thus far done are given in the volumes mentioned in the Introduction, p. 5, and as far as possible the information available concerning the conditions of mineral formation has been summarized under the description of the individual minerals in Part V.

491. Alteration of Minerals. Pseudomorphs. — The chemical alteration of mineral species under the action of natural agencies is a subject of great importance and interest, particularly when it results in the change of the original composition into some other equally definite compound. A crystallized mineral which has thus suffered change so that its form no longer belongs to its chemical composition has already been defined (Art. **278**, p. 205) as a *pseudomorph*. It remains to describe more fully the different kinds of pseudomorphs. Pseudomorphs are classed under several heads:

1. Pseudomorphs by *substitution*.
2. Pseudomorphs by simple *deposition*, and either by (*a*) *incrustation* or (*b*) *infiltration*.

3. Pseudomorphs by *alteration;* and these may be altered

   (*a*) without a change of composition, by *paramorphism;*
   (*b*) by the loss of an ingredient;
   (*c*) by the assumption of a foreign substance;
   (*d*) by a partial exchange of constituents.

1. The first class of pseudomorphs, by *substitution,* embraces those cases where there has been a gradual removal of the original material and a corresponding and simultaneous replacement of it by another, without, however, any chemical reaction between the two. A common example of this is a piece of fossilized wood, where the original fiber has been replaced entirely by silica. The first step in the process was the filling of the pores and cavities by the silica in solution, and then as the woody fiber, by gradual decomposition, disappeared the silica further took its place. Other examples are quartz after fluorite, calcite, and many other species; cassiterite after orthoclase; native copper after aragonite, etc.

2. Pseudomorphs by *incrustation* form a less important class. Such are the crusts of quartz formed over fluorite. In most cases the removal of the original mineral has gone on simultaneously with the deposition of the second, so that the resulting pseudomorph is properly one of substitution. In pseudomorphs by *infiltration* a cavity made by the removal of a crystal has been filled by another mineral.

3. The third class of pseudomorphs, by *alteration,* includes a considerable proportion of the observed cases, of which the number is very large. Conclusive evidence of the change which has gone on is often furnished by a nucleus of the original mineral in the center of the altered crystal — *e.g.,* a kernel of cuprite in a pseudomorphous octahedron of malachite; also of chrysolite in a pseudomorphous crystal of serpentine, etc.

(*a*) An example of *paramorphism* — that is, of a change in molecular constitution without change of chemical substance — is furnished by the change of aragonite to calcite (both $CaCO_3$) at a certain temperature; also the *paramorphs* of rutile after brookite (both TiO_2) from Magnet Cove, Arkansas.

(*b*) An example of the pseudomorphs in which alteration is accompanied by a loss of ingredients is furnished by crystals of native copper in the form of cuprite.

(*c*) In the change of cuprite to malachite — *e.g.,* the familiar crystals from Chessy, France — an instance is afforded of the assumption of an ingredient — viz., carbon dioxide (and water). Pseudomorphs of gypsum after anhydrite occur where there has been an assumption of water alone.

(*d*) A partial exchange of constituents — in other words, a loss of one and gain of another — takes place in the change of feldspar to kaolin, in which the potash silicate disappears and water is taken up; pseudomorphs of limonite after pyrite or siderite, of chlorite after garnet, pyromorphite after galena, are other examples.

The chemical processes involved in such changes open a wide and important field for investigation. Their study has served to throw much light on the chemical constitution of mineral species and the conditions under which they have been formed. For the literature of the subject see the Introduction, p. 4 (Blum, Bischoff, Roth, etc.).

CHEMICAL EXAMINATION OF MINERALS

492. The complete investigation of the chemical composition of a mineral includes, first, the identification of the elements present by qualitative analysis, and, second, the determination of the relative amounts of each by quantitative analysis, from which last the formula can be calculated. Both processes carried out in full call for the equipment of a chemical laboratory. An approximate qualitative analysis, however, can, in many cases, be made quickly and simply with few conveniences. The methods employed involve either (a) the use of acids or other reagents " in the wet way," or (b) the use of the blowpipe, or of both methods combined. Some practical instructions will be given applying to both cases.

EXAMINATION IN THE WET WAY

493. Reagents, etc. — The most commonly employed chemical reagents are the three mineral acids, hydrochloric, nitric, and sulphuric. To these may be added ammonium hydroxide, also solutions of barium chloride, silver nitrate, ammonium molybdate, ammonium oxalate; finally, distilled water in a wash-bottle.

A few test-tubes are needed for the trials and sometimes a porcelain dish with a handle called a casserole; further, a glass funnel and filter-paper. The Bunsen gas-burner (p. 361) is the best source of heat, though an alcohol lamp may take its place.

In testing the powdered mineral with the acids, the important points to be noted are. (1) the degree of solubility, and (2) the phenomena attending entire or partial solution; that is, whether (a) a solution is obtained quietly, without effervescence and, if so, what its color is; or (b) a gas is evolved, producing effervescence; or (c) an insoluble constituent is separated out.

494. Solubility. — In testing the degree of solubility hydrochloric acid is most commonly used, though in the case of many metallic minerals, as the sulphides and compounds of lead and silver, nitric acid is required. Less often sulphuric acid and aqua regia (nitro-hydrochloric acid) are resorted to.

The trial is usually made in a test-tube, and in general the fragment of mineral to be examined should be first carefully pulverized in an agate mortar. In most cases the heat of the Bunsen burner must be employed.

(a) Many minerals are completely *soluble without effervescence;* among these are some of the oxides, as hematite, limonite, goethite, etc.; some sulphates, many phosphates and arsenates, etc. Gold and platinum are soluble only in aqua regia or nitro-hydrochloric acid.

A yellow solution is usually obtained if much iron is present; a blue or greenish blue solution (turning deep blue on the addition of ammonium hydroxide in excess) from compounds of copper; pink or pale rose from cobalt, etc.

(b) *Solubility with effervescence* takes place when the mineral loses a gaseous ingredient, or when one is generated by the mutual reaction of acid and mineral. Most conspicuous here are the *carbonates,* all of which dissolve with effervescence, giving off the odorless gas *carbon dioxide* (CO_2), though some of them only when pulverized, or, again, on the addition of heat. In applying this test dilute hydrochloric acid is employed.

Hydrogen sulphide (H_2S) is evolved by some sulphides when dissolved in hydrochloric acid: this is true of sphalerite, stibnite, etc. This gas is readily recognized by its offensive odor.

Chlorine is evolved by oxides of manganese and also chromic and vanadic acid salts when dissolved in hydrochloric acid.

Nitrogen dioxide (NO_2) is given off, in the form of red suffocating fumes, by many metallic minerals, and also some of the lower oxides (cuprite, etc.) when treated with nitric acid.

(c) The *separation of an insoluble ingredient* takes place: with many silicates, the *silica* separating sometimes as a fine powder, and again as a jelly; in the latter case the mineral is said to *gelatinize* (sodalite, analcite). In order to test this point the finely pulverized silicate is digested with strong hydrochloric acid, and the solution afterward slowly evaporated nearly to dryness. With a considerable number of silicates the gelatinization takes place only after the mineral has been previously fused; while some others, which ordinarily gelatinize, are rendered insoluble by ignition.

With many sulphides (as pyrite) a separation of *sulphur* takes place when they are treated with nitric acid.

Some compounds of titanium and tungsten are decomposed by hydrochloric acid with the separation of the oxides of the elements named (TiO_2, WO_3). The same is true of salts of molybdic and vanadic acids, only that here the oxides are soluble in an excess of the acid.

Compounds containing silver, lead, and mercury give with hydrochloric acid insoluble residues of the *chlorides*. These compounds are, however, soluble in nitric acid.

When compounds containing tin are treated with nitric acid, the *tin dioxide* (SnO_2) separates as a white powder. A corresponding reaction takes place under similar circumstances with minerals containing arsenic and antimony.

Insoluble Minerals. — A large number of minerals are not sensibly attacked by any of the acids. Among these may be named the following oxides: corundum, spinel, chromite, diaspore, rutile, cassiterite, quartz; also cerargyrite; many silicates, titanates, tantalates, and niobates; some of the sulphates, as barite, celestite; many phosphates, as xenotime, lazulite, childrenite, amblygonite; also the borate, boracite.

495. Examination of the Solution. — If the mineral is difficultly, or only partially, soluble, the question as to solubility or insolubility is not always settled at once. Partial solution is often shown by the color given to the liquid, or more generally by the precipitate yielded, for example, on the addition of ammonium hydroxide to the liquid filtered off from the remaining powder. The further examination of the solution yielded, whether from partial or complete solution, after the separation by filtration of any insoluble residue, requires the systematic laboratory methods of qualitative analysis.

It may be noted, however, that in the case of sulphates the presence of *sulphur* is shown by the precipitation of a heavy white powder of barium sulphate ($BaSO_4$) when barium chloride is added. The presence of *silver* in solution is shown by the separation of a white curdy precipitate of silver chloride ($AgCl$) upon the addition of any chlorine compound; conversely, the same precipitate shows the presence of *chlorine* when silver nitrate is added to the solution.

Again, *phosphorus* may be detected if present, even in small quantity, in a nitric acid solution of a mineral by the fine yellow powder which separates, sometimes after standing, when ammonium molybdate has been added.

EXAMINATION BY MEANS OF THE BLOWPIPE*

496. The use of the blowpipe, in skilled hands, gives a quick method of obtaining a partial knowledge of the qualitative composition of a mineral. The apparatus needed includes the following articles:

Blowpipe, lamp, forceps, preferably with platinum points, platinum wire, charcoal, glass tubes; also a small hammer with sharp edges, a steel anvil an inch or two long, a horseshoe magnet, a small agate mortar, a pair of cutting pliers, a three-cornered file.

651

Further, test-paper, both turmeric and blue litmus paper; a little pure tin-foil; also in small wooden boxes the fluxes: borax (sodium tetraborate), soda (anhydrous sodium carbonate), salt of phosphorus or microcosmic salt (sodium-ammonium phosphate), acid potassium sulphate ($HKSO_4$); also a solution of cobalt nitrate in a dropping bulb or bottle; further, the three acids mentioned in Art. **493.**

497. Blowpipe and Lamp. — A good form of *blowpipe* is shown in Fig. 651. The air-chamber, at *a*, is essential to stop the condensed moisture of the breath, the tip (*b*), which is removable, is usually of brass, (*c*) is a removable mouthpiece which may or may not be used as preferred.

The most convenient form of *lamp* is that furnished by an ordinary Bunsen gas-burner† (Fig. 652), provided with

652

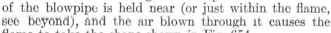

a tube, *b*, which when inserted cuts off the air supply at *a*; the gas then burns at the top with the usual yellow flame. This flame should be one to one and a half inches high. The tip of the blowpipe is held near (or just within the flame, see beyond), and the air blown through it causes the flame to take the shape shown in Fig. 654.

It is necessary to learn to blow *continuously*, that is, to keep up a blast of air from the compressed reservoir in the mouth-cavity while respiration is maintained through the nose. To accomplish this successfully and at the same time to produce a clear flame without unnecessary fatiguing effort calls for some practice.

When the tube, *b*, is removed, the gas burns with a colorless flame and is used for heating glass tubes, test-tubes, etc.

498. Forceps. Wire. — The *forceps* (Fig. 653) are made of steel, nickel-plated, and should have a spring strong enough to support firmly the small

---

* The subject of the blowpipe and its use is treated very briefly in this place. The student who wishes to be fully informed not only in regard to the use of the various instruments, but also as to all the valuable reactions practically useful in the identification of minerals, should consult a manual on the subject. The Brush-Penfield Manual of Determinative Mineralogy, with an introduction on Blowpipe Analysis, is particularly to be recommended.

† Instead of this, a good stearin candle will answer, or an oil flame with flat wick.

fragment of mineral between the platinum points at *d*. The steel points at the other end are used to pick up small pieces of minerals, but must not be inserted in the flame. Care must be taken not to injure the platinum by allowing it to come in contact with the fused mineral, especially if this contains antimony, arsenic, lead, etc. Cheaper forceps, made of steel wire, etc., while not so convenient, will also serve reasonably well.

A short length of fairly stout platinum wire to be used in the making of bead tests should be available. A similar length of finer wire for making flame tests is also desirable.

499. Charcoal. — The *charcoal* employed should not snap and should yield but little ash; the kinds made from basswood, pine or willow are best. It is most conveniently employed in rectangular pieces, say four inches long, an inch wide, and three-quarters of an inch in thickness. The surface must always be perfectly clean before each trial.

500. Glass Tubes. — The glass tubes should be preferably of two grades: a hard glass tubing with about 5 mm. interior diameter to be cut in five inch lengths and used in open tube tests and a soft glass tubing with about 3 mm. interior diameter to be in about six inch lengths, each length yielding two closed tubes.

653

501. Blowpipe Flame. — The blowpipe flame, shown in Fig. 654, consists of three cones: an inner of a blue color, *c*, a second pale violet cone, *b*, and an outer invisible cone, *a*. The cone *c* consists of unburned gas mixed with air from the blowpipe. There is no combustion in this cone and therefore no heat. The cone *b* is the one in which combustion is taking place. This cone contains carbon monoxide which is a strong reducing agent, see below. Cone *a* is merely a gas envelope composed of the final products of combustion, CO_2 and H_2O. The heat is most intense near the tip of the cone *b*, and the mineral is held at this point when its *fusibility* is to be tested.

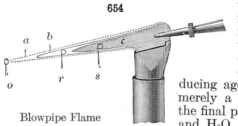

654

Blowpipe Flame

The point *o*, Fig. 654, is called the OXIDIZING FLAME (O.F.); it is characterized by the excess of the oxygen of the air and has hence an *oxidizing* effect upon the assay. This flame is best produced when the jet of the blowpipe is inserted a very little in the gas flame; it should be entirely non-luminous.

The cone *b* is called the REDUCING FLAME (R.F.); it is characterized by the excess of the carbon or hydrocarbons of the gas, which at the high temperature present tend to combine with the oxygen of the mineral brought into it (at *r*), or, in other words, to *reduce* it. The best reducing flame is produced when the blowpipe is held a little distance from the gas flame; it should retain the yellow color of the latter on its upper edge.

502. Methods of Examination. — The blowpipe investigation of minerals includes their examination, (1) in the forceps, (2) in the closed and the open tubes, (3) on charcoal or other support, and (4) with the fluxes on the platinum wire.

1. Examination in the Forceps

503. Use of the Forceps. — Forceps are employed to hold the fragment of the mineral while a test is made as to its fusibility; also when the presence of a volatile ingredient which may give the flame a characteristic color is tested for, etc.

The following practical points must be regarded: (1) Metallic minerals, especially those containing arsenic or antimony, which when fused might injure the platinum of the forceps, should first be examined on charcoal*; (2) the fragment taken should be thin, and as small as can conveniently be held, with its edge projecting well beyond the points; (3) when decrepitation takes place, the heat must be applied slowly, or, if this does not prevent it, the mineral may be powdered and a paste made with water, thick enough to be held in the forceps or on the platinum wire; or the paste may, with the same end in view, be heated on charcoal; (4) the fragment whose fusibility is to be tested must be held in the hottest part of the flame, just beyond the extremity of the blue cone.

504. Fusibility. — All grades of fusibility exist among minerals, from those which fuse in large fragments in the flame of the candle (stibnite, see below) to those which fuse only on the thinnest edges in the hottest blowpipe flame (bronzite); and still again there are a considerable number which are entirely infusible (*e.g.*, corundum).

The exact determination of the temperature of fusion is not easily accomplished (cf. Art. **443**, p. 332), and for purposes of determination of species it is unnecessary. The approximate *relative* degree of fusibility is readily fixed by referring the mineral to the following scale, suggested by von Kobell:

1. Stibnite.
2. Natrolite (or Chalcopyrite).
3. Almandite Garnet.
4. Actinolite.
5. Orthoclase.
6. Bronzite.

505. In connection with the trial of fusibility, the following phenomena may be observed: (*a*) *coloration* of the flame (see Art. **506**); (*b*) *swelling up* (stilbite), or *exfoliation* of the mineral (vermiculite); or (*c*) *glowing* without fusion (calcite); and (*d*) *intumescence*, or a spurting out of the mass as it fuses (scapolite).

The color of the mineral after ignition is to be noted; and the nature of the fused mass is also to be observed, whether a clear or blebby glass is obtained, or a black slag; also whether the bead or residue is magnetic or not (due to iron, less often nickel, cobalt), etc.

The ignited fragment, if nearly or quite infusible, may be moistened with the cobalt solution and again ignited, in which case, if it turns *blue*, this indicates the presence of aluminum (as with kyanite, topaz, etc.); but note that zinc silicate (calamine) also assumes a blue color. If it becomes *pink*, this indicates a compound of magnesium (as brucite).

Also, if not too fusible, it may, after treatment in the forceps, be placed upon a strip of moistened turmeric paper, in which case an *alkaline* reaction proves the presence of an alkali, sodium, potassium; or an alkaline earth, calcium, barium, strontium.

506. Flame Coloration. — The color often imparted to the outer blowpipe flame, while the mineral held in the forceps is being heated, makes possible the identification of a number of the elements.

---

* Arsenic, antimony, and easily reducible metals like lead, also copper, form more or less fusible alloys with platinum.

The colors which may be produced, and the substances to whose presence they are due, are as follows:

Color	Substance
Carmine-red	Lithium.
Purple-red	Strontium.
Orange-red	Calcium.
Yellow	Sodium.
Yellowish green	Barium.
Siskine-green	Boron.
Emerald-green	Oxide of copper.
Bluish green	Phosphoric acid (phosphates).
Greenish blue	Antimony.
Whitish blue	Arsenic.
Azure-blue	Chloride of copper; also selenium.
Violet	Potassium.

A yellowish green flame is also given by the oxide or sulphide of molybdenum; a bluish green flame (in streaks) by zinc; a pale bluish flame by tellurium; a blue flame by lead.

507. Notes. — The presence of soda, even in small quantities, produces a yellow flame, which (except in the spectroscope) more or less completely masks the coloration of the flame due to other substances, *e.g.*, potassium. A filter of blue glass held in front of the flame will shut out the monochromatic yellow of the sodium flame and allow the characteristic violet color of the potassium to be observed.* Silicates are often so difficultly decomposed that no distinct color is obtained even when the substance is present; in such cases (*e.g.*, potash feldspar) the powdered mineral may be fused on the platinum wire with an equal volume of gypsum, when the flame can be seen (at least through blue glass). Again, a silicate like tourmaline fused with a mixture of fluorite and acid potassium sulphate yields the characteristic green flame of boron. Phosphates and borates give the green flame in general best when they have been pulverized and moistened with sulphuric acid. Moistening with hydrochloric acid makes the coloration more distinct in many cases (as with the carbonates of calcium, barium, strontium).

2. Heating in the Closed and Open Tubes

508. The tubes are useful chiefly for examining minerals containing volatile ingredients, given off at the temperature of the gas flame.

In the case of the *closed tube*, the heating goes on practically uninfluenced by the air present, since this is driven out of the tube in the early stages of the process. In the *open tube*, on the other hand, a continual stream of hot air, that is, of hot oxygen, passes over the assay, tending to produce oxidation and hence often materially changing the result.

509. Closed Tube. — A small fragment is inserted, or a small amount of the powdered mineral — in this case with care not to soil the sides of the tube — and heat is applied by means of the ordinary Bunsen flame. The presence of a volatile ingredient is ordinarily shown by the deposit, or *sublimate*, upon the tube at some distance above the assay where the tube is relatively cool.

Independent of this, other phenomena may be noted, namely: *decrepitation*, as shown by fluorite, calcite, etc.; *glowing*, as exhibited by gadolinite; *phosphorescence*, of which fluorite is an example; *change of color* (limonite), and here the color of the mineral should be noted both when hot, and again after cooling; *fusion*; giving off *oxygen*, as mercuric oxide; yielding *acid* or *alkaline vapors*, which should be tested by inserting a strip of moistened litmus or turmeric paper in the tube.

Of the *sublimates* which form in the tube, the following are those with which it is most important to be familiar:

---

* A didymium glass may also be used as such a light filter.

Substance	Sublimate in the Closed Tube
Water (H_2O)	Colorless liquid drops.
Sulphur (S)	Red to deep yellow, liquid; pale yellow, solid.
Tellurium dioxide (TeO_2)	Pale yellow to colorless, liquid; colorless or white, solid.
Arsenic sulphide (As_2S_3)	Dark red, liquid; reddish yellow, solid.
Antimony oxysulphide (Sb_2S_2O)	Black to reddish brown on cooling, solid.
Arsenic (As)	Black, brilliant metallic to gray crystalline, solid
Mercury sulphide (HgS)	Deep black, red when rubbed very fine.
Mercury (Hg)	Gray metallic globules.

In addition to the above: Tellurium gives black fusible globules; selenium the same, but in part dark red when very small; the chloride of lead and oxides of arsenic and antimony give white solid sublimates.

510. Open Tube. — The small fragment is placed in the tube about an inch from the lower end, the tube being slightly inclined (say 20°), but not enough to cause the mineral to slip out, and heat applied beneath. The current of air passing upward through the tube during the heating process has an oxidizing effect. The special phenomena to be observed are the formation of a *sublimate* and the *odor* of the escaping gases. The acid or alkaline character of the vapors is tested for in the same way as with the closed tube. The most common gas to be obtained in this way is sulphur dioxide, SO_2, when sulphides are being oxidized. This gas is to be recognized by its irritating, pungent odor and its acid reaction upon moistened blue litmus paper.

The more important sublimates are as follows:

Substance	Sublimate in the Open Tube
Arsenic trioxide (As_2O_3)	White, crystalline, volatile.
Antimony antimonate (Sb_2O_4)	Straw-yellow, hot; white, cold. Infusible, non-volatile, amorphous, settling along bottom of tube. Obtained from compounds containing sulphur as stibnite, also the sulphantimonites (e.g., bournonite) as dense white fumes. Usually accompanied by the following:
Antimony trioxide (Sb_2O_3) . . .	White, crystalline, slowly volatile, forming as a ring on walls of tube.
Tellurium dioxide (TeO_2)	White to pale yellow globules.
Selenium dioxide (SeO_2)	White, crystalline, volatile.
Molybdenum trioxide (MoO_3)	Pale yellow, hot; white, cold.
Mercury (Hg)	Gray metallic globules, easily united by rubbing.

It is also to be noted that if the heating process is too rapid for full oxidation, sublimates, like those of the closed tubes, may be formed, especially with sulphur (yellow), arsenic (black), arsenic sulphide (orange), mercury sulphide (black), antimony oxysulphide (black to reddish brown).

3. HEATING ON CHARCOAL, ETC.

511. The fragment (or powder) to be examined is placed near one end of the piece and this so held that the flame passes along its length. If the mineral decrepitates, it may be powdered, mixed with water, and then the material employed as a paste.

The reducing flame is employed if it is desired to *reduce* a metal (*e.g.*, silver, copper) from its ores: this is the common case. If, however, the mineral is to be *roasted*, that is, heated in contact with the air so as to oxidize and volatilize, for example, the sulphur, arsenic, antimony present, the oxidizing flame is needed and the mineral should be in powder and spread out.

The points to be noted are as follows:

(*a*) The *odor* given off after short heating. In this way the presence of *sulphur, arsenic* (garlic or alliaceous odor), and *selenium* (odor of decayed horse-radish) may be recognized.

(b) *Fusion.* — In the case of the salts of the alkalies the fused mass is absorbed into the charcoal; this is also true, after long heating, of the carbonates and sulphates of barium and strontium.   (Art. **515**.)

(c) *The Sublimate.* — By this means the presence of many of the metals may be determined.   The color of the sublimate, both near the assay (N) and at a distance (D), as also when hot and when cold, is to be noted.

The important sublimates are the following:

Substance	Sublimate on Charcoal
Arsenic trioxide (As_2O_3)	White, very volatile, distant from the assay; also garlic fumes.
Antimony oxides (Sb_2O_3 and Sb_2O_4)	Dense white, volatile; forms near the assay.
Zinc oxide (ZnO)	Canary-yellow, hot; white, cold; moistened with cobalt nitrate and ignited (O.F.) becomes green.
Molybdenum trioxide (MoO_3)	Pale yellow, hot; yellow, cold; touched for a moment with the R.F. becomes azure-blue. Also a copper-red sublimate (MoO_2) near the assay.
Lead oxide (PbO)	Dark yellow, hot; pale yellow, cold. Also (from sulphides) dense white (resembling antimony), a mixture of oxide, sulphite, and sulphate of lead.
Bismuth trioxide (Bi_2O_3)	Dark orange-yellow (N), paler on cooling; also bluish white (D). See further, p. 370.
Cadmium oxide (CdO)	Nearly black to reddish brown (N) and orange-yellow (D); often iridescent.

To the above are also to be added the following:

Selenium dioxide, SeO_2, sublimate steel-gray (N) to white tinged with red (D); touched with R.F. gives an azure-blue flame; also an offensive selenium odor.

Tellurium dioxide, TeO_2, sublimate dense white (N) to gray (D); in R.F. volatilizes with green flame.

Tin dioxide, SnO_2, sublimate faint yellow hot to white cold; becomes bluish green when moistened with cobalt solution and ignited.

Silver (with lead and antimony), sublimate reddish.

(d) *The Infusible Residue.* — This may (1) glow brightly in the O.F., indicating the presence of calcium, strontium, magnesium, zirconium, zinc, or tin. (2) It may give an alkaline reaction after ignition: alkaline earths.   (3) It may be magnetic, showing the presence of iron (or nickel).   (4) It may yield a globule or mass of a metal (Art. **513**).

512. Heating on Plaster of Paris Tablets.* — In some cases it is preferable to collect sublimates on the surface of a plaster of Paris tablet rather than on charcoal.   Such tablets can be easily made by spreading a thin layer of the wet plaster upon a glass plate, the surface of which has been oiled.   While the plaster is still moist it should be cut into rectangular strips measuring about one and one-half by four inches.   After the plaster has hardened, these can be broken out into the desired tablets.   The material to be tested is placed in a small depression made near one end of the tablet and then heated before the blowpipe exactly as in the case of charcoal.   The iodide tests are especially marked on the plaster tablet.   These are obtained when either hydriodic acid or a flux composed of a mixture of potassium iodide and sulphur is added to the mineral powder before heating.   The important sublimates are listed below.

---

* A method by which sublimates are formed on sheets of mica has been described by Braly in Bull. Soc. Min., **44**, 8, 1921.

Substance	Sublimate on Plaster Tablet
Selenium oxide (SeO_2).....	Red to crimson, volatile, giving reddish fumes and characteristic odor.
Tellurium oxide (TeO_2)....	Dark brown, volatile.
Cadmium oxide (CdO)....	Greenish yellow with brown, both (N) and (D). Nonvolatile.
Lead iodide (PbI_2)........	Chrome-yellow.
Bismuth iodide (BiI_3).....	Chocolate-brown with underlying red. Subjected to ammonia fumes sublimate becomes first orange-yellow, then red.
Molybdenum iodide (MoI_4)	Deep ultramarine blue.
Antimony iodide (SbI_3)....	Orange to red. Disappears when subjected to ammonia fumes.

513. Reduction on Charcoal. — In many cases the reducing flame alone suffices on charcoal to separate the metal from the volatile element present, with the result of giving a globule or metallic mass. Thus silver is obtained from argentite (Ag_2S) and cerargyrite (AgCl); copper from chalcocite (Cu_2S) and cuprite (Cu_2O), etc. The process of reduction is always facilitated by the use of sodium carbonate or borax as a flux, and this is in many cases (sulpharsenites, etc.) essential.

The finely pulverized mineral is intimately mixed with two or three times its volume of soda, and a drop of water added to form a paste. This is placed in a cavity in the charcoal, and subjected to a strong reducing flame. More soda is added as that present sinks into the coal, and, after the process has been continued some time, a metallic globule is often visible, or a number of them, which can be removed and separately examined. If not distinct, the remainder of the flux, the assay, and the surrounding coal are cut out with a knife, and the whole ground up in a mortar, with the addition of a little water. The charcoal is carefully washed away and the metallic globules, flattened out by the process, remain behind. Some metallic oxides are very readily reduced, as lead, while others, as copper and tin, require considerable skill and care.

The metals obtained (in globules or as a metallic mass) may be: *copper*, color red; *bismuth*, lead-gray, brittle; *gold*, yellow, not soluble in nitric acid; *silver*, white, soluble in nitric acid, the solution giving a silver chloride precipitate (p. 372); *tin*, white, harder than silver, soluble in nitric acid with separation of white powder (SnO_2); *lead*, lead-gray (oxidizing), soft and fusible. The coatings (see the list of sublimates above) often serve to identify the metal present.

514. Detection of Sulphur in Sulphates. — By means of soda on charcoal the presence of sulphur in the sulphates may be shown, in the following manner. Fuse the powdered mineral with soda and charcoal dust. The latter acting as a strong reducing agent changes the sulphate to a sulphide with the formation of sodium sulphide. When the fused mass is placed with a drop of water upon a clean silver surface a black or yellow stain of silver sulphide will be formed. A similar reaction would of course be obtained from a sulphide. The latter can however be readily distinguished by roasting in the open tube or upon charcoal and noting the formation of SO_2.

4. Treatment on the Platinum Wire

515. Use of the Fluxes. — The three common fluxes are borax, salt of phosphorus, and carbonate of soda (p. 361). They are generally used with the platinum wire, less often on charcoal (see above). If the wire is employed it must have a small round loop at the end; this is heated to redness and dipped into the powdered flux, and the adhering particles fused to a bead; this operation is repeated until the loop is filled. Sometimes in the use of soda the wire may at first be moistened a little to cause it to adhere.

When the bead is ready, it is, while hot, brought in contact with the powdered mineral, some of which will adhere to it, and then the heating process may be continued. Very little of the mineral is in general required, and the experiment should be commenced with a minute quantity and more added if necessary. The bead must be heated successively first in the oxidizing flame (O.F.) and then in the reducing flame (R.F.), and in each case the color noted when *hot* and when *cold*. The phenomena connected with fusion, if it takes place, must also be observed.

Minerals containing sulphur or arsenic, or both, must be first *roasted* (see p. 365) till these substances have been volatilized. If too much of the mineral has been added and the bead is hence too opaque to show the color, it may, while hot, be flattened out with the hammer, or drawn out into a wire, or part of it may be removed and the remainder diluted with more of the flux.

With salt of phosphorus, the wire should be held above the flame so that the escaping gases may support the bead; this is continued till quiet fusion is attained.

It is to be noted that the colors vary much with the amount of material present; they are also modified by the presence of other metals.

516. Borax. — The following list enumerates the different colored beads obtained with borax, both in the oxidizing (O.F.) and reducing flames (R.F.), and also the metals to the presence of whose oxides the colors are due. Compare further the reactions given in the list of elements (Art. **518**).

Color in Borax Bead	Substance
1. OXIDIZING FLAME	
Colorless, or opaque white...	Silica, calcium, aluminum; also silver, zinc, etc.
	Iron, cold — (pale yellow, hot, if in small amount).
Red, red-brown to brown....	Chromium (CrO_3), hot — (yellowish green, cold).
	Manganese (Mn_2O_3), amethystine-red — (violet, hot).
	Iron (Fe_2O_3), hot — (yellow, cold) — if saturated.
	Nickel (NiO) red-brown to brown, cold — (violet, hot).
	Uranium (UO_3), hot — (yellow, cold).
Green.....................	Copper (CuO), hot — (blue, cold, or bluish green if highly saturated).
	Chromium (CrO_3), yellowish green, cold — (red, hot).
Yellow....................	Iron (Fe_2O_3), hot — (pale yellow to colorless, cold) — but red-brown and yellow if saturated.
	Uranium (UO_3), hot, if in small amount; paler on cooling.
	Chromium (CrO_3), hot and in small amount — (yellowish green, cold).
Blue......................	Cobalt (CoO), hot and cold.
	Copper (CuO), cold if highly saturated — (green, hot).
Violet....................	Nickel (NiO), hot — (red-brown, cold).
	Manganese (Mn_2O_3), hot — (amethystine-red, cold).
2. REDUCING FLAME (R.F.)	
Colorless.................	Manganese (MnO), or a faint rose color.
Red......................	Copper (Cu_2O, with Cu), opaque red.
Green....................	Iron (FeO), bottle-green.
	Chromium (Cr_2O_3), emerald-green.
	Uranium (U_2O_3), yellowish green if saturated.
Blue.......	Cobalt (CoO), hot and cold.
Gray, turbid..............	Nickel (Ni).

517. Salt of Phosphorus. — This flux gives for the most part reactions similar to those obtained with borax. The only cases enumerated here are those which are distinct, and hence those where the flux is a good test.

With *silicates* this flux forms a glass in which the bases of the silicate are dissolved, but the silica itself is left insoluble. It appears as a skeleton readily seen floating about in the melted bead.

The colors of the beads, and the metals to whose oxides these are due, are:

Color	Substance
Red............	Chromium in O.F., hot — (fine green when cold).
Green..........	Chromium in O.F. and R.F., when cold — (red in O.F., hot). Molybdenum in R.F., dirty green, hot; fine green, cold — (yellow-green in O.F.). Uranium in R.F., cold; yellow-green, hot. Vanadium, chrome-green in R.F., cold — (brownish red, hot). In O.F., dark yellow, hot, paler on cooling.
Yellow.........	Molybdenum, yellowish green in O.F., hot, paler on cooling — (in R.F., dirty, green, hot; fine green, cold). Uranium in O.F., hot; yellowish green, cold — (in R.F., yellowish green, hot; green, cold). Vanadium in O.F., dark yellow, hot, paler on cooling — (in R.F., brownish red, hot; chrome-green, cold).
Violet.........	Titanium (TiO$_2$) in R.F., yellow, hot. (Also in O.F., yellow, hot; colorless, cold.)

CHARACTERISTIC REACTIONS OF THE IMPORTANT ELEMENTS AND OF SOME OF THEIR COMPOUNDS

518. The following list contains the most characteristic reactions, chiefly before the blowpipe and in some cases also in the wet way, of the different elements and their oxides. It is desirable for every student to gain familiarity with them by trial with as many minerals as possible. Many of them have already been briefly mentioned in the preceding pages. For a thoroughly full description of these and other characteristic tests (blowpipe and otherwise) reference should be made to the volume by Brush and Penfield referred to on p. 361.

It is to be remembered that while the reaction of a single substance may be perfectly distinct if alone, the presence of other substances may more or less entirely obscure these reactions; it is consequently obvious that in the actual examination of minerals precautions have to be taken, and special methods have to be devised, to overcome the difficulty arising from this cause. These will be gathered from the "pyrognostic characters" (**Pyr.**) given in connection with the description of each species in Part V of this work.

Aluminum. — The presence of aluminum in most infusible minerals, containing a considerable amount, may be detected by the blue color which they assume when, after being heated, they are moistened with cobalt solution and again ignited (*e.g.*, kyanite, andalusite, etc.). Very hard minerals (as corundum) must be first finely pulverized. The test is not conclusive with fusible minerals since a glass colored blue by cobalt oxide may be formed. It is to be noted that the infusible calamine (zinc silicate) also assumes a blue color when treated with cobalt nitrate. From solutions aluminum will be precipitated as a flocculent white or colorless precipitate on the addition of ammonium hydroxide in excess.

Antimony. — Antimonial minerals roasted on charcoal give dense white odorless fumes; metallic antimony and its sulphur compounds give in the open tube a white sublimate of oxide of antimony (see p. 365). Antimony sulphide (stibnite), also many sulphantimonites, give in a strong heat in the closed tube a sublimate of antimony oxysulphide, black when hot, brown-red when cold. Gives iodide test on plaster, orange to red. Sublimate disappears when subjected to ammonia fumes. See also p. 367.

In nitric acid, compounds containing antimony deposit white insoluble metantimonic acid.

Arsenic. — Arsenides, sulpharsenites, etc., give off fumes when roasted on charcoal, usually easily recognized by their peculiar garlic odor. In the open tube they give a white, volatile, crystalline sublimate of arsenic trioxide. In the closed tube arsenic sulphide gives a sublimate dark brown-red when hot, and red or reddish yellow when cold; arsenic and some arsenides yield a black to gray mirror of metallic arsenic in the closed tube. In arsenates the arsenic can be detected by the garlic odor yielded when a mixture of the powdered mineral with charcoal dust and sodium carbonate is heated (R.F.) on charcoal; or same mixture heated in closed tube will yield sublimate of metallic arsenic.

Barium. — A yellowish green coloration of the flame is given by all barium salts, except the silicates; an alkaline reaction is usually obtained after intense ignition.

In solution the presence of barium is proved by the heavy white precipitate ($BaSO_4$) formed upon the addition of dilute sulphuric acid.

Bismuth. — On charcoal alone, or better with soda, bismuth gives a very characteristic orange-yellow sublimate; brittle globules of the reduced metal are also obtained (with soda). Also when treated with 3 or 4 times the volume of a mixture in equal parts of potassium iodide and sulphur, and fused on charcoal, a beautiful red sublimate of bismuth iodide is obtained; near the mineral the coating is yellow; on plaster the iodide sublimate is chocolate-brown with underlying red. Changes to orange-yellow and red on exposure to ammonia fumes.

Boron. — Many compounds containing boron (borates, also the silicates, datolite, danburite, etc.) tinge the flame intense yellowish green, especially if moistened with sulphuric acid. For some silicates (as tourmaline) the best method is to mix the powdered mineral with one part powdered fluorite and two parts potassium bisulphate. The mixture is moistened and placed on platinum wire. At the moment of fusion the green color appears, but lasts but an instant.

A dilute hydrochloric acid solution containing boron gives a reddish brown color to turmeric paper which has been moistened with it and then dried at 100°; the color changes to black when ammonia is poured on the paper.

Calcium. — Many calcium minerals (carbonates, sulphates, etc.) give an alkaline reaction on turmeric paper after being ignited. A yellowish red color is given to the flame by some compounds (*e.g.*, calcite after moistening with HCl); the strontium flame is a much deeper red.

In weakly acid or alkaline solutions calcium is precipitated as oxalate by the addition of ammonium oxalate. Calcium sulphate is precipitated in concentrated solutions only on addition of sulphuric acid.

Cadmium. — On charcoal with soda, compounds of cadmium give a characteristic sublimate of the reddish brown oxide; on plaster the coating is greenish yellow with brown.

Carbonates. — All carbonates effervesce with *dilute* hydrochloric acid, yielding the odorless gas CO_2 (*e.g.*, calcite); many require to be pulverized, and some need the additior of heat (dolomite, siderite). Carbonates of lead should be tested with nitric acid.

Chlorides. — If a small portion of a mineral containing chlorine (a chloride, also pyromorphite, etc.) is added to the bead of salt of phosphorus, saturated with copper oxide, the bead when heated is instantly surrounded with an intense purplish flame of copper chloride.

In solution chlorine gives with silver nitrate a white curdy precipitate of silver chloride which darkens in color on exposure to the light; it is insoluble in nitric acid, but entirely soluble in ammonia.

Chromium. — Chromium gives with borax a bead which (O.F.) is yellow to red (hot) and yellowish green (cold) and R.F. a fine emerald-green. With salt of phosphorus in O.F. the bead is dirty green (hot) and clear green (cold); in R.F. the same. Cf. vanadium beyond (also pp. 368, 369).

Cobalt. — A beautiful blue bead is obtained with borax in both flames from minerals containing cobalt. Where sulphur or arsenic is present the mineral should first be thoroughly roasted on charcoal.

Copper. — On charcoal, at least with soda, metallic copper can be reduced from most of its compounds. In the case of sulphides the powdered mineral should be roasted first in order to eliminate the major part of the sulphur before fusion with soda. With borax it gives (O.F.) a green bead when hot, becoming blue when cold; also (R.F.), if saturated, an opaque red bead containing Cu_2O and often Cu is obtained. Copper chloride, obtained by moistening the mineral with hydrochloric acid (in the case of sulphides the mineral should be previously roasted) yields a vivid azure-blue flame; copper oxide gives a green flame.

Most metallic compounds are soluble in nitric acid. Ammonia in excess produces an intense blue color in the solution.

Fluorine. — Heated in the closed tube with potassium bisulphate and powdered glass produces a white sublimate of SiO_2. This sublimate and the hydrofluosilicic acid present form a volatile combination. But if the lower end of the tube is broken off and the open tube then dipped in a test tube of water so that the acid is removed, the deposit of SiO_2 which will appear when the tube is dried will be found to be no longer volatile.

Heated gently in a platinum crucible with sulphuric acid, many compounds (*e.g.*, fluorite) give off hydrofluoric acid, which corrodes the exposed parts of a glass plate placed over it which has been coated with wax and then scratched.

Iron. — Minerals which contain even a small amount of iron yield a magnetic mass when heated in the reducing flame. With borax iron gives a bead (O.F.) which is yellow to brownish red (according to quantity) while hot, but is colorless to yellow on cooling; R.F. becomes bottle-green (see p. 368). Ferric iron is precipitated from a solution by the addition of ammonium hydroxide in excess as red-brown ferric hydroxide $Fe(OH)_3$. Ferric iron in hydrochloric acid solution gives deep blue precipitate when potassium ferrocyanide is added; a similar test is obtained for ferrous iron by the addition of potassium ferricyanide.

Lead. — With soda on charcoal a malleable globule of metallic lead is obtained from lead compounds; the coating has a yellow color near the assay; the sulphide gives also a white coating ($PbSO_3$) farther off (p. 366). On being touched with the reducing flame the coating disappears, tingeing the flame azure-blue. On plaster, lead iodide coating is chrome-yellow.

In solutions dilute sulphuric acid gives a white precipitate of lead sulphate; when delicacy is required an excess of the acid is added, the solution evaporated to dryness, and water added; the lead sulphate, if present, will then be left as a residue.

Lithium. — Lithium gives an intense carmine-red to the outer flame, the color somewhat resembling that of the strontium flame but deeper; in very small quantities it is evident in the spectroscope.

Magnesium. — Moistened, after heating, with cobalt nitrate and again ignited, a pink color is obtained from some infusible compounds of magnesium (*e.g.* brucite). In solution the addition of ammonium hydroxide in large excess and a little hydrogen sodium phosphate produces a white granular precipitate of NH_4MgPO_4. Elements precipitated by ammonium hydroxide or ammonium oxalate should be removed first.

Manganese. — With borax manganese gives a bead violet-red (O.F.), and colorless (R.F.). With soda (O.F.) it gives a bluish green bead; this reaction is very delicate and may be relied upon, even in presence of almost any other metal.

Mercury. — In the closed tube a sublimate of metallic mercury is yielded when the mineral is heated with dry sodium carbonate. In the open tube the sulphide gives a mirror of metallic mercury; in the closed tube a black lusterless sublimate of HgS, red when rubbed, is obtained.

Molybdenum. — On charcoal molybdenum sulphide gives near the assay a copper-red stain (O.F.), and beyond a white coating of the oxide; the former becomes azure-blue when for a moment touched with the R.F. The iodide coating on plaster is deep ultramarine blue. The salt of phosphorus bead (O.F.) is yellowish green (hot) and nearly colorless (cold); also (R.F.) a fine green. Molybdates are tested by putting powdered mineral in dry test tube with small scrap of paper, adding a few drops of water and an equal amount of concentrated sulphuric acid. Heat until acid fumes. Cool or add slowly drops of water and solution will assume deep blue color.

Nickel. — With borax, nickel oxide gives a bead which (O.F.) is violet when hot and red-brown on cooling; (R.F.) the glass becomes gray and turbid from the separation of metallic nickel. A solution of nickel becomes pale blue (cf. copper) on the addition of ammonia in excess. If an alcoholic solution of dimethylglyoxine is added to an ammoniacal solution a scarlet precipitate is formed.

Niobium (Columbium). — An acid solution boiled with metallic tin gives a blue color which slowly changes to brown on continued boiling and disappears on dilution. The reactions with the fluxes are not very satisfactory.

Nitrates. — These detonate when heated on charcoal. Heated in a tube with sulphuric acid they give off red fumes of nitrogen dioxide (NO_2).

Phosphorus. — Most phosphates impart a green color to the flame, especially after having been moistened with sulphuric acid, though this test may be rendered unsatisfactory by the presence of other coloring agents. If they are fused in the closed tube with a fragment of metallic magnesium or sodium, and afterward moistened with water, phosphureted hydrogen is given off, recognizable by its disagreeable odor.

A few drops of a nitric acid solution, containing phosphoric acid, produce in a solution of ammonium molybdate a pulverulent yellow precipitate of ammonium phosphomolybdate.

Potassium. — Potash imparts a violet color to the flame when alone. The flame is best observed through a blue glass filter which will eliminate the sodium flame color which will almost invariably be present. It is best detected in small quantities, or when soda or lithia is present, by the aid of the spectroscope. See also p. 364.

Selenium. — On charcoal selenium fuses easily, giving off brown fumes with a peculiar disagreeable organic odor; the sublimate on charcoal is volatile, and when heated (R.F.) gives a fine azure-blue flame. On plaster gives red to crimson volatile sublimate with characteristic odor.

Silicon. — A small fragment of a silicate in the salt of phosphorus bead leaves a skeleton of silica, the bases being dissolved.

Most silicates are insoluble in acids. The fine powder of such a silicate is fused with sodium carbonate and the mass then dissolved in hydrochloric acid and evaporated to dryness, the silica separates as a gelatinous mass and on evaporation to dryness is made insoluble. When strong hydrochloric acid and then water is added to the dry residue in the test tube, the bases are dissolved and the silica left behind.

Many silicates are directly soluble in hydrochloric acid, and the solution on evaporation will yield the characteristic silica jelly. Other silicates when treated with acid are decomposed, the bases present going into solution, but the silica separating and remaining suspended in the liquid. There are gradations between these three groups.

Silver. — On charcoal in O.F. silver gives a brown coating. A globule of metallic silver may generally be obtained by heating on charcoal in R.F., especially if soda is added. Under some circumstances it is desirable to have recourse to cupellation.

From a solution containing any salt of silver, the insoluble chloride is thrown down when hydrochloric acid is added. This precipitate is insoluble in acid or water, but entirely so in ammonia. It changes color on exposure to the light.

Strontium. — Compounds of strontium are usually recognized by the fine crimson-red which they give to the blowpipe flame; many yield an alkaline reaction after ignition (cf. barium). In concentrated and slightly diluted solutions strontium sulphate will be precipitated on the addition of sulphuric acid (cf. calcium and barium).

Sodium. — Compounds containing sodium give a strong and persistent yellow flame. Some compounds will give an alkaline reaction after ignition.

Sulphur, Sulphides, Sulphates. — In the closed tube some sulphides give off sulphur; in the open tube they yield sulphur dioxide, which has a characteristic odor and reddens a strip of moistened litmus paper. In small quantities, or in sulphates, sulphur is best detected by fusion on charcoal with soda and charcoal dust. The fused mass, when sodium sulphide has thus been formed, is placed on a clean silver coin and moistened; a distinct black stain on the silver is thus obtained (the precaution mentioned on p. 367 must be exercised).

A solution of a sulphate in hydrochloric acid gives with barium chloride a white insoluble precipitate of barium sulphate.

Tellurium. — Tellurides heated in the open tube give a white or grayish sublimate, fusible to colorless drops (p. 365). On charcoal they give a white coating and color the R.F. green. On plaster they give a dark brown volatile sublimate. A small amount dissolved in warm concentrated sulphuric acid yields a deep red solution. The color will disappear if acid is heated too hot or if on cooling, water is added.

Tin. — Minerals containing tin (*e.g.*, cassiterite), when heated on charcoal with soda or potassium cyanide, yield metallic tin in minute globules and a white and difficultly volatile oxide coating is formed; the globules are malleable, but harder than silver. Dissolved in nitric acid, white insoluble stannic oxide separates out.

Titanium. — Titanium gives in the R.F. with salt of phosphorus a bead which is violet when cold. Fused with sodium carbonate and dissolved in hydrochloric acid, and heated with a piece of metallic tin, the liquid takes a violet color, especially after partial evaporation. A yellow to amber color results when hydrogen peroxide is added to a sulphuric acid solution.

Tungsten. — Tungsten oxide gives a blue color to the salt of phosphorus bead (R.F.). Fused and treated as titanium (see above) with the addition of zinc instead of tin, gives a fine blue color. The color changes to brown on continued reduction. The color does not disappear when the solution is diluted (cf. niobium).

Uranium. — Uranium compounds give to the salt of phosphorus bead (O.F.) a greenish yellow bead when cool; also (R.F.) a fine green on cooling (p. 369).

Vanadium. — With borax (O.F.) vanadates give a bead yellow (hot) changing to yellowish green and nearly colorless (cold); also (R.F.) dirty green (hot), fine green (cold). With salt of phosphorus (O.F.) a yellow to amber color (thus differing from chromium); also (R.F.) fine green (cold).

Zinc. — On charcoal in the reducing flame compounds of zinc give a coating which is yellow while hot and white on cooling, and moistened by the cobalt solution and again heated becomes a fine green.   Note, however, that the zinc silicate (calamine) becomes blue when heated after moistening with cobalt solution.

Zirconium. — A dilute hydrochloric acid solution, containing zirconium, imparts an orange-yellow color to turmeric paper, moistened by the solution.

DETERMINATIVE MINERALOGY

519. Determinative Mineralogy may be properly considered under the general head of Chemical Mineralogy, since the determination of minerals depends mostly upon chemical tests.   But crystallographic and all the physical characters have also to be carefully observed.

There is but one exhaustive way in which the identity of an unknown mineral may in all cases be fixed beyond question, and that is by the use of a complete set of determinative tables.   By means of such tables the mineral in hand is referred successively from a general group into a more special one, until at last all other species have been eliminated, and the identity of the one given is beyond doubt.

A careful preliminary examination of the unknown mineral should, however, always be made before final recourse is had to the tables.   This examination will often suffice to show what the mineral in hand is, and in any case it should not be omitted, since it is only in this way that a practical familiarity with the appearance and characters of minerals can be gained.

The student will naturally take note first of those characters which are at once obvious to the senses, that is: *crystalline form,* if distinct; *general structure, cleavage, fracture, luster, color* (and *streak*), *feel;* also, if the specimen is not too small, the apparent weight will suggest something as to the *specific gravity.* The characters named are of very unequal importance.   Structure, if crystals are not present, and fracture are generally unessential except in distinguishing varieties; color and luster are essential with metallic, but generally very unimportant with nonmetallic minerals.   *Streak* is of importance only with colored minerals and those of metallic luster (p. 271).   Crystalline form and cleavage are of the highest importance, but may require careful study.

The first trial should be the determination of the *hardness* (for which end the pocket-knife is often sufficient in experienced hands).   The second trial should be the determination of the *specific gravity.*   Treatment of the powdered mineral with acids may come next; by this means (see pp. 359, 360) a carbonate is readily identified, and also other results obtained.   Then should follow blowpipe trials, to ascertain the *fusibility;* the *color* given to the flame, if any; the character of the *sublimate* given off in the tubes and on charcoal; the metal reduced on the latter; the reactions with the *fluxes,* and other points as explained in the preceding pages.

How much the observer learns in the above way, in regard to the nature of his mineral, depends upon his knowledge of the characters of minerals in general, and upon his familiarity with the chemical behavior of the various elementary substances with reagents and before the blowpipe (pp. 359 to 369). If the results of such a preliminary examination are sufficiently definite to suggest that the mineral in hand is one of a small number of species, reference may be made to their full description in Part V of this work for the final decision.

A number of tables, in which the minerals included are arranged according to their crystalline and physical characters, are added in Appendix B. They will in many cases aid the observer in reaching a conclusion in regard to a specimen in hand.

The first of these tables gives lists of minerals arranged primarily according to their principle basic elements and secondarily according to their acid radicals.

The second of these tables is intended to include all important species, grouped according to the crystalline system to which they belong and arranged under each system in the order of their specific gravities; the hardness is also added in each case. Following this are minor tables enumerating species characterized by some one of the prominent crystalline forms; that is, those crystallizing in cubes, octahedrons, rhombohedrons, etc. Other tables give the names of species prominent because of their cleavage; structure of different types; hardness; luster; the various colors, etc. The student is recommended to make frequent use of these tables, not simply for aid in the identification of specimens, but rather because they will help him in the difficult task of learning the prominent characters of the more important minerals.

PART IV. ORIGIN, MODE OF OCCURRENCE
AND ASSOCIATION OF MINERALS*

520. General. — The subject-matter of the first three parts of this book has been concerned chiefly with the individual mineral as it is found in nature. Its physical and chemical characters have been discussed. There are other important aspects of the subject that should, at least briefly, be indicated. These are the history of the mineral, how it happened to be formed and under what circumstances, what changes may have occurred since its formation, what are the characteristic conditions of its usual occurrence, with what other minerals it is commonly associated, etc. These are some of the questions that are concerned with the subject of the origin, mode of occurrence, and association of minerals. Information of this sort that has been noted about given minerals is briefly summarized under the descriptions of the individual species in Part V. This whole subject belongs more properly to Geophysics and Geochemistry, but brief and general statements of the more important facts are given below.

521. Chief Modes of Mineral Origin. (1) *From Fusion.* — The greater part of the minerals that form the crust of the earth have been formed by solidification from fused rock material. That is, they are, or were originally, constituents of igneous rocks.

(2) *From Solution.* — Many minerals have been formed by crystallization from a solution. This is obviously true of the group of saline minerals, such as halite, sylvite, borax, etc. It is also true of those minerals that occur lining or filling cracks and fissures in the rocks, *i.e.*, vein minerals. The conditions of their formation are often different from any that can be duplicated in the laboratory, but the evidence of origin from a solution is strong.

(3) *From a Vapor.* — Minerals may be formed by a direct crystallization from a gas. Such minerals are most commonly found in connection with volcanic fumeroles, etc., where the contents of the escaping gases may condense directly into solid forms. Vapors also play an important part in the complicated processes that are concerned in the formation of the so-called pneumatolytic minerals.

It should be emphasized that the boundary lines between the three modes of formation listed above are very indefinite. The fused rock material may be considered as only a form of very dense solution and a vapor on the other hand a very attenuated one.

522. The Formation of Minerals from Fusion. — An igneous rock has its origin in a fused mass of rock constituents known as a magma. An igneous magma, as stated above, may be considered as a heavy complex solution in

---

* Much of the material of this part has been taken with only minor additions and changes from the section on Occurrence and Association of Minerals in Dana's Manual of Mineralogy, 1929.

which the various elements present are more or less free to circulate and under the proper conditions to unite to form mineral molecules. The composition of the magma will determine in large part the character of the minerals that compose the resulting rock. By the study of thousands of analyses of igneous rocks it is shown that the following constitute about 99 per cent of the elements present: oxygen, silicon, aluminum, iron, magnesium, calcium, sodium, and potassium. These elements occur in varying proportions in igneous magmas, but the constituent minerals of igneous rocks are composed essentially of them. The conditions under which the various minerals are formed are complex. They crystallize out from the cooling magma in general in the order of their solubilities. Although this order is a fairly definite one, the variation of the chemical composition of the magma will vary the degree of solubility of the various mineral constituents and so may vary the order of their crystallization. The fluidity of the magma and the temperature at which various minerals will crystallize are influenced greatly by the common presence in a magma of small amounts of substances that are known as mineralizers. These are usually water vapor, carbon dioxide, fluorine, boric acid, sulphur, and chlorine. To some extent, these enter into the composition of various rock-making minerals, but the rôle they play in the formation of igneous rocks is more a physical one. It has, for instance, been shown experimentally that such common rock minerals as orthoclase, albite, and quartz can rarely be formed from a dry fusion — as a rule only uncrystallized glasses will result in such a case. It is only in the presence of water vapor that the temperature of crystallization is sufficiently lowered and the fluidity of the material increased so as to permit the ready growth of crystals of these minerals.

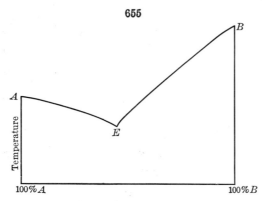

When two substances (A and B) are mixed together in various proportions and then fused, the temperatures of solidification of the mixtures are in general

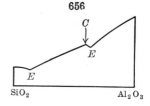

lower than the corresponding temperatures for either of the pure substances. In Fig. 655 the percentages of the mixtures of A and B are plotted as abscissas and the temperatures of solidification as ordinates. The point E represents the particular mixture of A and B that has the lowest fusing point of any mixture. Such a point is known as the *eutectic point*, and the mixture of A and B that corresponds to it is called the *eutectic mixture*. This simple case of two components illustrates the effect on the solidification temperature of one compound by the presence in the fused mass of a second compound. The diagram becomes more complicated if the two components A and B form for some certain mixture a definite compound, A_xB_y. Then the curves will show two eutectic points, one between A and A_xB_y and the other between

A_xB_y and B. Such a condition is represented in Fig. 656, which gives the fusing curves for mixtures of SiO_2 and Al_2O_3. The point C represents the mineral sillimanite, Al_2SiO_5, and the points E, the eutectic points on either side of it. If there are two or more possible different molecular combinations of A and B, the curves become increasingly complex. It can be readily seen that if in addition the fused mass (*i.e.*, the rock magma) contains, as is usually the case, several different chemical components, the problems involved in the crystallization of the constituent minerals of the rocks are enormously difficult of solution.

There has been in recent years, however, a great amount of experimental investigation concerning the behavior of various systems of oxides under conditions of high temperature and pressure. In this way our knowledge of the physical and chemical processes involved in the crystallization of minerals from rock magmas is being greatly extended.*

523. The Formation of Minerals from Solution. (1) *By the Evaporation of Saline Waters.* — Ocean waters and the waters of salt lakes contain various salts in solution; when such waters become concentrated through evaporation, certain minerals are deposited. The saline content of sea-water is composed of the following constituents; NaCl about 78 per cent; KCl, 2 per cent; $MgCl_2$, 9 per cent; $MgSO_4$, 6 per cent; $CaSO_4$, 4 per cent. When such waters are evaporated, the above salts, or certain combinations, such as carnallite, $KCl.MgCl_2.6H_2O$, crystallize from the solution in the order of their solubility. In general, the normal

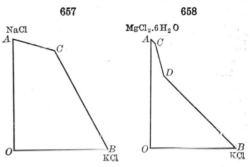

order of crystallization is as follows: carbonates of lime and magnesia, calcium sulphate, sodium chloride, magnesium chloride and sulphate, potassium chloride, etc. The factors of concentration, temperature, and proportions of the various constituents in the solution control the character of the minerals formed, their relative amounts, the order of the deposition, etc. For instance, in the case of NaCl and KCl, the amount of each that can be dissolved in a given solution is controlled by the amount of the other salt also present. In Fig. 657 the distance $O-A$ represents the amount of NaCl that can be dissolved in a given amount of water at a given temperature, and $O-B$ in the same way the amount of KCl. The line $A-C-B$ gives the varying proportions of the two salts that together form a saturated solution. The point C represents the maximum amount of the two salts that may be dissolved. The line of saturation for mixtures of two salts in a solution will be further complicated if they may form one or more double salts. Such a case

---

* The results of these investigations are to be found in many papers in the various journals. They are summarized in the following books:

Clarke. The Data of Geochemistry. 1916.
Behrend and Berg. Chemische Geologie. 1927.
Eitel. Physikalische Chemie der Silikate. 1929.

For an explanation of the physical chemical processes see any text-book on physical chemistry, and especially, The Phase Rule and its Applications, by Findlay, 1927.

is illustrated in Fig. 658. The two salts are bischofite, $MgCl_2.6H_2O$, and sylvite, KCl. Together these may unite to form carnallite, $MgCl_2.KCl.6H_2O$. Point D represents the condition of saturation for both carnallite and sylvite; C, that for carnallite and bischofite. The conditions governing the crystallization of saline minerals from bodies of salt water have been extensively studied by van't Hoff and his students, both experimentally and in the famous salt beds of Prussia.

Another less frequent type of saline deposit occurs when the waters of lakes that lie in regions of strong volcanic activity are evaporated. Their saline contents have been derived from the leaching of volcanic rocks, ashes, etc., from the material contained in the waters of hot springs, and from the gases of fumeroles, etc. The resulting salt beds contain the sulphate and carbonate of sodium as well as the chloride; often borax and other borates, glauberite, sometimes nitrates, etc.

(2) *By Precipitation from Ground Waters.* — The rocks of the earth's crust have many openings within them. These openings vary in size from microscopic cracks to cavities of considerable extent. They may be irregular and discontinuous, or they may be in the form of fissures which are continuous for greater or less distances. Below a certain inconsiderable depth, these openings are largely filled with water, commonly known as ground water. Much of this water has accumulated by a slow infiltration of surface waters; a small amount may have come from sea-waters; a still smaller amount has been derived from the original water of igneous rocks, given off in the form of vapor during the solidification of the rocks. The underground water slowly circulates through the rocks by means of the openings. Through a large part of its circulation, the water must exist at a high temperature and pressure, and under these circumstances it becomes a strong solvent and active chemical agent. Underground water in general descends slowly through the smaller openings in the rocks, and then, gradually finding its way into the bigger openings, will at last enter some larger fissure and changing its course will begin to ascend. On its passage through the rocks, it will have dissolved their more soluble constituents, and when it ultimately enters the larger fissures and commences to rise it will be carrying considerable amounts of dissolved mineral material. The igneous rocks in particular are important factors in furnishing underground waters with mineral constituents, partly because of the effect of their heat upon its activity, and partly because they give off in the form of vapors a large amount of mineral material which ultimately gets into underground circulation. When these mineral-laden waters commence to rise in the larger fissures, they slowly come into regions of lower pressure and temperature. Under these changing conditions the water will not be able to retain all its mineral constituents in solution, and various minerals, their points of saturation being reached, will begin to crystallize out and be deposited on the walls of the fissure. In time, if the process continues, the fissure may be completely filled from wall to wall with minerals deposited in this way. Such a filled fissure is known as a *mineral vein*. Evidence that the minerals of a vein have been deposited from solution is given by the following facts. Often a mineral vein shows a distinctly banded or ribboned structure. That is, the different minerals occur in more or less regular layers which lie parallel to the walls. This shows that the various minerals have not been deposited simultaneously, but in a definite order of succession. Again, frequently it will be observed that the vein material has not completely filled the fissure,

but that there are openings left along its central line.   These openings, termed *vugs*, are often lined with crystallized minerals.   These conditions cannot be easily explained except on the assumption that the contents of a mineral vein have been deposited from solution.   It must be remembered that the formation of a mineral vein may involve very long periods of time, a condition that precludes its duplication in any experimental way.

Ground waters often reach the surface in the form of springs.   Under rather exceptional conditions such spring waters may be highly charged with mineral materials which may be deposited around the spring opening at the surface.   The most common substance deposited from spring waters is calcium carbonate, often called travertine.   A siliceous material is also frequently a spring deposit, the material called geyserite being an example.   Iron oxides in the form of ocher may also occur.   Sodium salts, various sulphides, etc., are also found in such deposits.   Usually such spring waters are hot when they reach the surface.

In arid countries solutions ascending through the soil by capillary action will evaporate on reaching the surface and leave their saline contents behind in the form of efflorescences or crusts on the surface of the ground.   These are usually sodium or calcium salts.

(3) *The Formation of Minerals from Vapors.* — The direct formation of minerals from vapors is confined to such volcanic regions where mineral gases are discharged from fumeroles.   Such deposits are rare and usually of small extent.   Materials deposited in this way include sulphur, tellurium, arsenic sulphides, boric acid, various chlorides, etc.   Vapors play a much more important rôle in the formation of the so-called pneumatolytic minerals; see Art. **529**.

524. Occurrence and Association of Minerals.   *General.* — Clarke* has estimated the percentage composition of the earth's crust to be approximately as follows: SiO_2, 59.8; Al_2O_3, 14.9; Fe_2O_3, 2.7; FeO, 3.4; MgO, 3.7; CaO, 4.8; Na_2O, 3.2; K_2O, 3.0; H_2O, 2.0; TiO_2, 0.8; CO_2, 0.7; P_2O_5, 0.3; remainder, 0.7.   Over 99 per cent, in other words, of the constituents of minerals consists of twelve elements, and of the remaining 0.7 per cent no one element totals as much as 0.01 per cent.   Further, the greater part of these prevailing elements are to be found in a short list of minerals.   There is, therefore, an enormous difference in frequency of occurrence between minerals, only a few being properly termed " common."   The mineral beryl, for instance, is not considered an especially rare mineral, and yet Clarke does not include beryllium oxide in his list of oxides that form more than 0.01 per cent of the composition of the earth's crust.   The most common minerals are quartz, various aluminum-bearing silicates, such as the feldspars, and the so-called ferromagnesium silicates.   Even calcite, which as limestone occurs in such large masses, actually forms a small proportion of the total.

The association of minerals, or their *paragenesis*, is of great interest and significance.   Further, the order in which the various minerals of a deposit have crystallized can often be learned by noting that certain minerals are deposited upon other species and are therefore later in origin or that certain minerals have been enclosed by the crystals of other minerals subsequently formed.   By such studies, much can be learned concerning the conditions under which the various minerals of a deposit were formed.

---

* Data of Geochemistry, p. 32, 1916.

The following paragraphs describe briefly the more important modes of mineral occurrence and the more characteristic types of mineral association.

525. Occurrence in Rocks. — Although many minerals are found as rock constituents, those which can be termed common and characteristic rock-making minerals are comparatively few in number. The following list includes all such species, and some of these occur only in rare rock types: quartz, the feldspars, nephelite, sodalite, leucite, the micas, the pyroxenes, the amphiboles, chrysolite, kaolin, the chlorites, serpentine, talc, calcite, and dolomite. In addition to the more important and common rock-making minerals there is a group of minerals which are characteristically found as rock constituents but in a minor way. They occur usually only as small and scattered crystals in the rock and seldom become one of its prime constituents. These minerals are known as *accessory rock-making minerals.* A great many different minerals may at times occur as accessory rock constituents, but the following lists include all those that characteristically occur in this way. The most common are garnet, epidote, staurolite, kyanite, zircon, titanite, apatite. The following are of rarer occurrence: rutile, iolite, scapolite, andalusite, sillimanite. In addition, the iron minerals, magnetite, hematite, ilmenite, and pyrite, also may occur as accessory rock minerals.

526. Igneous Rocks. — Igneous rocks, as the name indicates, are those which have been formed by the cooling and consequent solidification of a once hot and fluid mass of rock material, which is known as a magma; see Art. **522.** In most igneous rocks a more or less definite order of crystallization of the mineral constituents can be determined. In general, the more basic minerals, or those which contain the smaller amounts of silica, which is the acid element in igneous rocks, are found to crystallize first and the more acid minerals last. Among the commoner rock-making minerals the following would be the usual order of crystallization: iron oxides like magnetite are first, then the ferromagnesian minerals like pyroxene, next the plagioclase feldspars, then orthoclase, and lastly quartz.

The type of minerals to be found in any igneous rock would depend chiefly upon the chemical composition of the original magma. If the magma was acid in character, *i.e.*, had a high percentage of silica, the resulting rock would contain the more acid minerals and an abundance of free quartz. It would usually be light in color. If, on the other hand, the magma had a low percentage of silica, or in other words was basic in character, the resulting rock would contain the more basic minerals and would not show free quartz. It would also in general be dark in color.

In addition to the wide variation in chemical and mineral composition shown by igneous rocks there is also a variation in their physical structure. This is dependent upon the mode of origin of the rock. If a rock has been formed from a magma buried at a considerable depth in the crust of the earth it must have cooled very slowly and taken a long period of time for its gradual crystallization and solidification. Under these conditions the mineral particles have had the opportunity, because of the slowness of crystallization, to grow to considerable size. A rock having such a deep-seated origin has, therefore, a coarse-grained structure and the various minerals that go to form the rock can in general be differentiated and recognized by the unaided eye. Such rocks are commonly termed *plutonic.* They have also been termed *grained rocks* or *phanerites.*

On the other hand, if, by volcanic forces, the magma has been extruded upon the surface of the earth or intruded in the form of dikes into the rocks lying close to the surface, its subsequent cooling and solidification go on quite rapidly. Under these conditions the mineral particles have little chance to grow to any size and the resulting rock is fine-grained in character. In some cases, indeed, the cooling has been too rapid to allow the separation of any minerals and the resulting rock is like a glass. Ordinarily the mineral constituents of such a rock are to be definitely recognized only by a microscopic examination of a thin section of the rock. Such igneous rocks are known as *volcanic* rocks and have also been called *aphanites.*

An igneous rock, because of the mode of its formation, consists of crystalline particles which may be said to interlock with each other. In other words, it is a solid mass, and each mineral particle is intimately and firmly embedded in the surrounding particles. This structure will enable one ordinarily to distinguish an igneous from a sedimentary rock, the latter being ocmposed of grains which do not interlock with each other but stand out, more or less, by themselves. A sedimentary rock is not so firm and coherent as an igneous rock. Further the texture of an igneous rock is the same in all directions and it forms a fairly uniform and homogeneous mass. This characteristic will enable one to distinguish an igneous from a metamorphic rock, since the latter shows a more or less definite parallel arrangement of its minerals and a banded structure.

Because of the almost infinite variation possible in the chemical composition of their magmas, and because of the various conditions under which they may form, igneous rocks show likewise a wide variation in character. The more common and important types, however, are very briefly described below.*

Plutonic, Coarse-grained Rocks

1. *Granite.* — A granite is a medium- to coarse-grained, light-colored rock having an even texture and consisting chiefly of quartz and a feldspar. Frequently both orthoclase and a plagioclase feldspar, and usually also small amounts of mica or hornblende are present. Granites in which plagioclase is abundant are called *quartz monzonites,* and those in which there is more plagioclase than orthoclase are called *granodiorites.* These distinctions, however, can usually be made only after a microscopic examination. Granite is a common rock type.

2. *Syenite.* — A syenite is a medium- to coarse-grained light-colored rock with an even texture and much like a granite in appearance. It is to be distinguished from granite, however, by the fact that it contains little or no quartz. Its chief minerals are the feldspars, with more or less hornblende, mica or pyroxene. A variety, known as *nephelite-syenite,* is characterized by the presence of considerable amounts of nephelite. Syenites are not very common.

3. *Diorite.* — Diorite is a medium- to coarse-grained dark gray or greenish colored rock having an even texture and consisting chiefly of plagioclase feldspar and a ferromagnesian mineral, such as biotite, hornblende, or pyroxene. The plagioclase equals or exceeds the dark minerals in amount. Some-

---

* For an adequate megascopic description of rock types, etc., the reader must be referred to some elementary petrologic text-book, such as Rocks and Rock Minerals, by Pirsson and Knopf, 1925.

times quartz is present, and if in notable amounts the rock is called *quartz-diorite*. It is a common rock type.

4. *Gabbro.* — Gabbro is a medium- to coarse-grained dark gray to greenish black rock with an even texture composed chiefly of a ferromagnesian mineral (pyroxene, hornblende, chrysolite) and plagioclase. It is closely similar to diorite, the distinction lying in the fact that the amount of the ferromagnesian minerals present is greater than that of the plagioclase. If the pyroxene present is the orthorhombic hypersthene the rock is called a *norite*. It is a common rock. A rock composed almost wholly of a plagioclase, frequently labradorite, usually with small amounts of pyroxene, is called an *anorthosite*. It is not common.

5. *Dolerite.* — This is a name often given to fine-grained varieties of diorite and gabbro, diabase, etc., that cannot be distinguished from each other megascopically.

6. *Peridotite.* — A peridotite is a medium- to coarse-grained dark green to black rock with an even texture which consists wholly of ferromagnesian minerals. These are chiefly olivine, pyroxene and hornblende. As one or the other of these minerals predominates, various variety names are used, such as *dunite* for an olivine rock and *pyroxenite* and *hornblendite* for respectively pyroxene and hornblende rocks. Common accessory minerals found in these rocks are ilmenite, chromite and garnet. The peridotites are not very common in their occurrence.

Volcanic, Fine-grained Igneous Rocks

Because of their very fine-grained structure volcanic rocks cannot in general be readily told apart. A number of different types are recognized, the distinction between them being based, however, chiefly upon microscopic study. In the field only an approximate classification, depending upon whether the rock is light or dark in color, can be made. A brief description of these two types of volcanic rocks follows.

1. *Felsite.* — This is a dense fine-grained rock type with a stony texture and includes all colors except dark gray, dark green or black. These rocks may, by the aid of a lens, still show a very fine-grained structure or their mineral constituents may occur in such small particles as to give them a dense and homogeneous, often a flinty, appearance. By microscopic study the felsites have been divided into the following groups: *rhyolite*, consisting chiefly of alkaline feldspars and quartz; *dacite*, lime-soda feldspars and quartz; *trachyte*, alkaline feldspars with little or no quartz; *andesite*, soda-lime feldspars with little or no quartz; *phonolite*, alkaline feldspars and nephelite. As a rule these varieties are not to be distinguished from each other in the field. The felsites are widespread in their occurrence, being found as dikes and sheets intruded into the upper part of the earth's crust or as lava flows which have been poured out upon the earth's surface.

2. *Basalt.* — The basalts are dense fine-grained rocks that are of very dark color, green or black. They are composed of microscopic grains of a soda-lime feldspar with pyroxene, iron ore, often more or less olivine and at times biotite or hornblende. These rocks are formed under the same conditions as the felsites and are to be found occurring in the same ways. *Trap* is a field name given to any dark-colored fine-grained rock, the mineral con-

stituents of which cannot be determined by inspection.   It would include basalts, dolerites, etc.

3. *Glassy Rocks.* — Some of the volcanic rocks have cooled so rapidly that they are wholly or in part made up of a glassy material in which the different elements have not had the necessary opportunity to group themselves into definite minerals.   If the entire rock is composed of glass it is called *obsidian,* when it has a bright and vitreous luster; *pitchstone* when its luster is dull and pitchy; *perlite* if it is made up of small spheroids; and *pumice* if it has a distinctly cellular structure.   These rocks may also have distinct crystals of various minerals embedded in the glass, in which case they are known as *glass porphyries* (see below for a definition of a porphyry) or *vitrophyres.*

Porphyries. — Igneous rocks at times show distinct crystals of certain minerals which lie embedded in a much finer-grained material.   These larger crystals are known as *phenocrysts,* and the finer-grained material as the *groundmass* of the rock.   Rocks exhibiting such a structure are known as *porphyries.* The phenocrysts may vary in size from crystals an inch or more across down to quite small individuals.   The groundmass may also be composed of fairly coarse-grained material or its grains may be microscopic in size.   It is the distinct difference in size existing between the phenocrysts and the particles of the groundmass that is the distinguishing feature of a porphyry.   This peculiar structure is due to certain conditions prevailing during the formation of the rock which permitted some crystals to grow to considerable size before the main mass of the rock consolidated into a finer- and uniform-grained material.   The explanation of the reasons why a certain rock should assume a porphyritic structure would involve a more detailed discussion than it is expedient to give in this place.   Any one of the above described types of igneous rocks may have a porphyritic variety, such as *granite-porphyry, diorite-porphyry, foloite-porphyry,* etc.   Porphyritic varieties are more liable to occur in connection with volcanic rocks, and they are also found most frequently in the case of the more acid types.

Magmatic Segregations. — During the process of solidification of an igneous rock certain heavy minerals will be the first to crystallize.   In cases where their constituents form a considerable percentage of the composition of the magma these minerals may gradually sink in the fluid material and slowly gather into large irregular masses, known as magmatic segregations.   Bodies of magnetite, ilmenite, chromite, and cassitorite are frequently formed in this way.   More rarely, the sulphides, pyrrhotite, chalcopyrite, arsenopyrite, etc., occur in such segregations.   Important ore bodies have been formed in this way.

527. Sedimentary Rocks. — Sedimentary rocks are secondary in their origin, the materials of which they are composed having been derived from the decay and disintegration of some previously existing rock mass.   They have been formed by a deposition of sediments in a body of water.   They may be divided into two classes, depending upon whether their origin has been mechanical or chemical in its nature.   In the case of the sedimentary rocks of a mechanical origin, their constituent particles have been derived from the disintegration of some rock mass, and have been transported by streams into a large body of quiet water, where they have been deposited in practically horizontal layers.   Sedimentary rocks of chemical origin have had the materials of which they are composed dissolved by waters circulating through the rocks and brought ultimately by these waters into a sea, where through some chemical change they are precipitated upon its floor, also in horizontal layers. These horizontal beds of sediments are ultimately consolidated into the masses known as sedimentary rocks.

Sedimentary rocks are therefore characterized by a parallel arrangement of their constituent particles into layers and beds which are to be distinguished from each other by differences in thickness, size of grain and often in color. It is to be noted, further, that sedimentary rocks in general are composed of an aggregate of individual mineral particles, each of which stands out in a way by itself and does not have that intimate interlocking relation with the surrounding particles which is to be seen in the minerals of an igneous rock. In all the coarser-grained sedimentary rocks there is some material which, acting as a cement, surrounds the individual mineral particles and binds them together. This cement is usually either silica, calcium carbonate or iron oxide. The chief minerals to be found in sedimentary rocks are quartz and a carbonate, calcite or dolomite. These give rise to the two chief types of sedimentary rocks, the sandstones and the limestones. A brief description of these rocks follows.

1. *Sandstone.* — Sandstones are mechanical in their origin, being formed by the consolidation into rock masses of beds of sand and gravel. Usually the constituent grains are rounded and water-worn, but at times they may be more or less angular in shape. With the variation in the size of the mineral particles the rocks themselves vary in their grain. Coarse-grained sandstones formed from water-worn gravels are known as *conglomerates*. A *breccia* is a similar rock in which the individual grains are angular and sharp edged. The cement which serves to bind the sand grains together may be deposited silica, a carbonate, usually calcite, an iron oxide, hematite or limonite, or fine-grained argillaceous or claylike material. The color of the rock will depend in large measure upon the character of the cement. The rocks which have silica or calcite as their binding material are light in color, usually pale yellow, buff, white to gray, while those that contain an iron oxide are red to reddish brown. It is to be noted that when a sandstone breaks it is usually the cement that is fractured, while the individual grains remain unbroken, so that the fresh surfaces of the rock have a granular appearance and feeling. The chief mineral of sandstones is quartz, but at times a rock may contain notable amounts of feldspar and is then termed an *arkose*. *Graywacke* is a sandstone, usually of a gray color, which in addition to quartz and feldspar contains particles of other rocks and minerals.

2. *Shale.* — The shales are very fine-grained sedimentary rocks which have been formed by the consolidation of beds of mud, clay or silt. They have usually a thinly laminated structure. Their color is commonly some tone of gray, although they may be white, yellow, brown, green to black. They are composed chiefly of kaolin, mica, etc., but are too fine-grained to permit the recognition of their mineral constituents by the eye alone. By the introduction of quartz and an increase in the size of grain they grade into the sandstones.

3. *Limestone.* — The limestones are carbonate rocks composed usually chiefly of calcite, although dolomite may also be at times an important constituent. The carbonate has in the great majority of cases been extracted from the sea water by the agency of minute organisms and then deposited in beds which ultimately are consolidated into rock. These rocks are usually fine- and even-grained in structure and sometimes quite dense. Some lime-

stones are quite pure calcite, while others contain claylike materials and various oxides as impurities.   The color of a limestone is usually gray, although it may be white, yellow, brown to almost black.   *Oölite* or *oölitic limestone* is a variety which consists of an aggregate of small spherical concretions. *Chalk* is a very fine-grained friable limestone composed of shells of minute sea animals known as foraminifera.   *Travertine* is a deposit of calcium carbonate formed by springs.   A fine example exists in the deposits formed by the Mammoth Hot Springs, Yellowstone Park.   *Marl* is a loose, earthy material composed of a carbonate mixed with clay in variable amount.

4. *Other Sedimentary Formations.* — Embedded in the sedimentary rocks and at times occurring in large masses are various minerals that have been precipitated from water; see also Art. **523**.   *Rock salt* occurs widely distributed as a part of the sedimentary series and at many different geological horizons.   Its beds vary in thickness from one to more than four hundred feet.   *Gypsum* also occurs in many places and at times in thick beds.   It commonly underlies salt beds and is especially associated with limestones and shales.   *Anhydrite*, though less frequent in its occurrence, is found under similar conditions.   Silica in various forms may occur as a sediment.   *Geyserite* is deposited from hot springs either by the evaporation of the water or by the action of algæ which secrete it from the water in which they live.   *Diatomaceous earth* is a fine-grained chalk-like material formed from the silica skeletons of minute organisms that live in water.   At times it occurs in beds of considerable extent.   The *iron ores*, limonite, hematite, and siderite, may occur in sedimentary beds (see further under the descriptions of the individual species).   In certain regions, beds of impure calcium phosphates, known as *phosphorite* or *phosphate rock*, occur (see under apatite, p. 704).

520. Metamorphic Rocks. — Metamorphic rocks are rocks which have undergone some chemical or physical change subsequent to their original formation.   This change has been brought about by means of high temperature and pressure aided by the action of water and other chemical agents. The changes involve the formation of new minerals, the adding or subtracting of chemical constituents and a physical readjustment of the mineral particles to conform to the existing pressure.   The original rock from which a metamorphic rock has been derived may be either igneous or sedimentary.   As these rocks become involved in movements of the earth's crust, they are subjected to extreme pressures accompanied usually by high temperatures. The result will be frequently to transform the existing minerals into others more stable under the new conditions.   Although many minerals that were constituents of the original rock may still exist in it after its metamorphism, there are certain other minerals that are characteristically developed during the process.   Some of these minerals peculiar to metamorphic rocks are tremolite, wollastonite, kyanite, zoisite, staurolite, talc, paragonite, grossularite.   The physical structure of the rock will also ordinarily be changed during the process.   Because of the pressure to which the rock is subjected the mineral particles will be more or less broken and flattened and rearranged in parallel layers.   This banded or laminated character given by the parallel arrangement of its minerals is the most striking peculiarity of a metamorphic rock.   Because of this structure a metamorphic rock can be distinguished from an igneous rock.   Further, in the great majority of cases a metamorphic rock has a crystalline structure which distinguishes it from a sedimentary rock.

There are, of course, all gradations from a typical metamorphic rock into an unaltered sedimentary rock on the one hand and into an unaltered igneous rock on the other. The most common types of metamorphic rocks are briefly described beyond.

1. *Gneiss.* — A gneiss is a coarse-grained, roughly foliated, metamorphic rock. When the word gneiss is used alone it usually refers to a rock composed essentially of quartz, feldspar and a mica. The quartz and feldspar occur together in layers which are separated from each other by thin drawn-out bands of mica. A gneiss has usually a light color, although this is not necessarily so. Various varieties of gneiss have received distinctive names, most of which are self-explanatory, like *banded-gneiss, lenticular-gneiss, biotite-gneiss, hornblende-gneiss, granite-gneiss, diorite-gneiss,* etc. Gneiss is a very common rock type, especially in regions in which the oldest rocks, those of the Archæan age, are found. Gneisses have been more commonly derived by the metamorphism of igneous rocks, mostly granites, but may have been formed from sedimentary rocks as well.

2. *Mica-schist.* — Mica-schist is a rock composed essentially of quartz and a mica, usually either muscovite or biotite. The mica is the prominent mineral, occurring in irregular leaves and in foliated masses. The mica plates all lie with their cleavage planes parallel to each other and give to the rock a striking laminated or " schistose " structure. They are related to both the gneisses and quartzites and intermediate rock types occur. The mica-schists frequently carry characteristic accessory minerals, such as garnet, staurolite, kyanite, epidote, hornblende, etc. They may have been derived from either an igneous or a sedimentary rock. Next to the gneisses, they are the most common metamorphic rocks.

3. *Quartzite.* — As its name indicates, a quartzite is a rock composed essentially of quartz. It is a firm, compact rock which breaks with an uneven, splintery or conchoidal fracture. It is usually light in color. Quartzite has been derived from a sandstone by intense metamorphism. It is a common and widely distributed rock.

4. *Slate.* — Slates are exceedingly fine-grained rocks which have a remarkable cleavage which permits them to be split into thin and broad sheets. Their color is commonly gray to black, but may be green, yellow, brown, red, etc. They have been formed commonly by the metamorphism of shales. Their characteristic slaty cleavage may or may not be parallel to the bedding planes of the original shales. They are quite common in occurrence.

5. *Various Schists.* — There are various other kinds of schistose rocks, which are chiefly derived by the metamorphism of the ferromagnesian igneous rocks. The most important types are *talc-schist, chlorite-schist, amphibolite* or *hornblende-schist.* They each are characterized, as their names indicate, by the preponderance of some metamorphic ferromagnesian mineral.

· 6. *Marble.* — A marble is a metamorphosed limestone. It is a crystalline rock composed of grains of calcite, or more rarely of dolomite. At times the individual grains are so small that they cannot be distinguished by the eye, and again they may be quite coarse and show clearly the characteristic cleavage of the mineral. Like limestone, a marble is characterized by its softness and its effervescence with acids. When pure, marble is white in color, but it may show a wide range of color, due to various impurities that it contains. It is a rock which is found in many localities and at times in thick and extensive beds.

529. Contact Metamorphic Minerals. — When an igneous rock magma is intruded into the earth's crust, it causes through the attendant heat and pressure a greater or less alteration in the surrounding rock. This alteration, or metamorphism, of the rocks lying next to an igneous intrusion usually consists partly in the development of new and characteristic mineral species. To a less degree certain characteristic modifications of the igneous rock may occur near the contact. These are chiefly due to the effects of gases that are given off by the igneous rock during its cooling and result in the formation of various pneumatolytic minerals (see below) in the igneous rock itself. The minerals that are formed under these conditions are known as *contact metamorphic minerals*, since they are produced by a metamorphic change and are to be found at or near the contact line between the rocks. Any rock into which an igneous mass is intruded will be affected in a greater or less degree, the amount and character of the change depending chiefly upon the size of the intruded mass and upon the chemical and physical character of the surrounding rock. The most striking and important contact metamorphic changes take place when the igneous rock is intruded into impure limestones. When a pure limestone is affected, it is recrystallized and converted into a marble, but without any development of new species. But, on the other hand, in the case of an impure limestone the heat and pressure caused by the igneous intrusion will serve to develop new and characteristic minerals in the rock. An impure limestone will ordinarily contain, besides the calcium carbonate of the rock, varying amounts of quartz, clay, iron oxide, etc. Under the influence of the heat and pressure these materials will combine with the calcium carbonate to form new minerals. For instance, the calcite and quartz may react together to form wollastonite, $CaSiO_3$. If the limestone contains dolomite, the reaction of this mineral with quartz may produce pyroxene, $(Ca,Mg)SiO_3$. If clay is present, aluminum will enter into the reaction and such minerals as spinel, $MgAl_2O_4$, and grossularite, $Ca_3Al_2Si_3O_{12}$, may result. If any carbonaceous materials are present, the effect of the metamorphism may convert them into graphite. The common contact metamorphic minerals found in limestone are as follows: *graphite, spinel, corundum, wollastonite, tremolite, pyroxene* and the lime garnets, *grossularite* and *andradite*.

As mentioned in Art. **523**, an igneous rock in cooling often gives off large amounts of mineralizing vapors. These consist largely of water vapor, but often include boron and fluorine gases. Under the influence of these vapors, other minerals are often formed in the contact zone of a limestone. These particular minerals are commonly spoken of as *pneumatolytic minerals*, since they are formed, partly at least, through the agency of mineral gases. They consist chiefly of calcium and aluminum silicates which contain hydroxyl, fluorine or boron. The most common of the pneumatolytic contact minerals are *chondrodite, vesuvianite, scapolite, phlogopite, tourmaline, topaz* and *fluorite*.

530. Pegmatite Dikes or Veins. — In connection with the deep-seated, coarse-grained igneous rocks, especially the granites, we frequently find mineral deposits which are known as pegmatite dikes or veins. These bodies have the general shape and character of an igneous dike or a broad mineral vein although in certain respects they differ markedly from either of these. They are to be found running through the main mass of the igneous rock or filling fissures in the other surrounding rocks. They are composed chiefly of the same minerals as occur in the igneous rock, but usually in very coarse crystallizations. A granite pegmatite is therefore made up principally of quartz, feldspar and mica. The quartz and feldspar crystals may be several feet in length and the mica plates are at times more than a foot across. In addition to the coarseness of the crystallization of the minerals, these veins possess other peculiar features. The minerals of a pegmatite vein, for instance, have not apparently been deposited in the definite order that prevailed in the igneous rock mass, but their crystals have grown more nearly simultaneously. These veins will also at times show a ribboned or banded structure where the different minerals occur in distinct layers which lie parallel to the walls of the deposit. Their minerals are also commonly quite irregularly distributed through the mass, so that at times the vein is composed chiefly of feldspar and again becomes nearly pure quartz. Frequently, along the central portion of the dike, cavities and openings will be observed into which crystals of the different minerals project. These characteristics point to a somewhat different origin for the pegmatite veins from that of the igneous rock with which they are associated.

No extended and detailed discussion of the theory of the origin of pegmatite veins can be given here, but it may be briefly summarized as follows. Pegmatite veins are formed during the last stages of the cooling and solidification of a plutonic igneous rock. As an igneous magma cools and slowly solidifies, it shrinks somewhat in volume and various cracks and fissures open up throughout the mass. The pressure due to the weight of the rock forces any still fluid material from the interior of the mass up through these cracks and also into any fissures that may exist in the surrounding rocks. The filling up of these fissures both in the igneous rock itself and in the neighboring rocks constitutes a pegmatite vein. As a magma cools and its minerals crystallize, large amounts of water vapor are frequently set free so that the residue of the still fused rock material must contain much higher percentages of water than the original magma. Consequently it becomes in its character and behavior more like a solution than a fused mass. This would account for the peculiar features observed in pegmatite veins which differentiate them from ordinary igneous deposits.

The minerals found in pegmatite veins may be divided into three general divisions. First come those minerals which form the main mass of the deposit and which, as stated above, are the same as the prominent minerals of the igneous rock with which the pegmatite dike is associated. These are commonly *quartz*, a feldspar which is usually either *orthoclase* or *microcline*, but may be *albite*, and a mica which may be either *muscovite* or *biotite*. *Garnet* is also at times in a smaller way a characteristic constituent. Second comes a series of rare minerals which are, however, quite commonly observed in pegmatite deposits, and which are characterized by the presence in them of fluorine, boron or hydroxyl. Their presence in the veins indicates also that gases under high pressures have been instrumental in their formation. The minerals of this type include *beryl, tourmaline, topaz, apatite* and *fluorite*. A third class of minerals found in pegmatite veins includes species containing rare elements such as lithium, molybdenum, tin, niobium and tantalum, the rare earths, etc. These are minerals which are rarer still in their occurrence, but when they do occur are usually to be found in pegmatite deposits. The most important members of this group are *molybdenite, lepidolite, spodumene, triphylite, columbite, cassiterite* and *monazite*.

Because of the frequent occurrence in pegmatite veins of the rare minerals mentioned above, some of which are often found finely colored and well crystallized, these deposits are of particular interest to students of mineralogy. Pegmatite veins are also of commercial importance, for it is from them that most of the feldspar and mica used in the arts are obtained. Many beautiful gem stones, such as beryl and tourmaline, are also found in them. Pegmatite veins are widely distributed in their occurrence, being almost universally found wherever plutonic igneous rocks are exposed. Important districts for pegmatite veins in the United States include the New England states, the Black Hills in South Dakota and Southern California.

531. Veins and Vein Minerals. — A discussion has already been given, see Art. **523**, of the origin of mineral veins. As most of the important mineral deposits, especially those that furnish the valuable metals, are found in veins, certain further facts should be mentioned.

The shape and general physical character of a vein depend upon the type of fissure in which its minerals have been deposited, and the type of fissure in turn depends upon the character of the rock in which it lies and the kind of force which originally caused its formation. In a firm homogeneous rock, like a granite, a fissure will be fairly regular and clean cut in character. It is liable to be comparatively narrow in respect to its horizontal and vertical extent and reasonably straight in its course. On the other hand, if a rock that is easily fractured and splintered, like a slate or a schist, is subjected to a breaking strain, we are more liable to have formed a zone of narrow and interlacing fissures, rather than one straight crack. In an easily soluble rock like a limestone, a fissure will often be extremely irregular in its shape and size due more or less to a solution of its walls by the waters that have flowed through it.

A typical vein consists of a mineral deposit which has filled a fissure solidly from wall to wall, and shows sharply defined boundaries. There are, however, many variations from this type. Frequently, as observed above, irregular openings termed vugs may occur among the vein minerals. It is from these vugs that we obtain many of our crystallized mineral specimens. Again, the walls of a vein may not be sharply defined. The mineralizing waters that filled the fissure may have acted upon the wall rocks and partially dissolving them may have replaced them with the vein minerals. Consequently we may have almost a complete gradation from the unaltered rock to the pure vein filling, and with no sharp line of division between. Some deposits have been largely formed by the deposition of vein minerals in the wall rocks. Such deposits are known as *replacement deposits*. They are more liable to be found in the soluble rocks like limestones. There is every gradation possible from a true vein with sharply defined walls to a replacement deposit with indefinite boundaries.

The mineral contents of a vein depend chiefly upon the chemical composition of the waters from which its minerals have crystallized. There are many different sorts of veins, and many different mineral associations are observed in them. There are, however, certain minerals and associations that are more frequent in their occurrence to which attention should be drawn. The sulphides form perhaps the most characteristic chemical group of minerals to be found in veins. The following minerals are very common vein minerals, *pyrite*, FeS_2, *chalcopyrite*, $CuFeS_2$, *galena*, PbS, *sphalerite*, ZnS, *chalcocite*, Cu_2S, *bornite*, Cu_5FeS_4, *marcasite*, FeS_2, *arsenopyrite*, $FeAsS$, *stibnite*, Sb_2S_3, *tetrahedrite*, $Cu_6Sb_2S_7$, etc. In addition to these, which in large part comprise our ore minerals, certain nonmetallic minerals are also commonly to be observed. These being of no particular commercial value are called *gangue minerals* (*gangue* is from *gang, a vein*). They include the following: *quartz*, SiO_2, *calcite*, $CaCO_3$, *dolomite*, $CaMg(CO_3)_2$, *siderite*, $FeCO_3$, *barite*, $BaSO_4$, *fluorite*, CaF_2, *rhodochrosite*, $MnCO_3$, etc.

Lindgren* has classified vein minerals according to the range of temperature that probably prevailed during their formation and, since in general temperature will increase with depth, also roughly according to the depth below the earth's surface at which deposition took place. The low temperature minerals (50°–150° C.) include pyrite, marcasite, stibnite, and cinnabar as ore minerals, and adularia, quartz, calcite, opal, fluorite, zeolites (but rarely) as gangue minerals. The deposits of intermediate temperature (150°–300° C.) show gold, galena, sphalerite, pyrite, chalcopyrite, bornite, arsenopyrite, tetrahedrite, enargite among the ore minerals, whereas the principal gangue minerals are quartz, calcite, dolomite, siderite and barite. The high temperature deposits (300°–500° C.) contain gold, gold tellurides, molybdenite, cassiterite, wolframite, magnetite, ilmenite, various micas, garnet, topaz, apatite, etc.

While comparatively few positive statements concerning the associations of vein minerals can be made, the following points are of interest.

1. *Gold-bearing Quartz Veins.* — Native gold is most commonly found in quartz veins. It may occur alone in the quartz either in nests or in finely disseminated particles, or it may occur in connection with certain sulphides in the veins. The most common sulphides found in such connections are pyrite, chalcopyrite and arsenopyrite.

2. *Gold- and Silver-bearing Copper Veins.* — The gold and silver content of these veins is associated with the various copper sulphides. Frequently the amount of the precious metals is quite small. The chief minerals are chalcopyrite, tetrahedrite, bornite, chalcocite, pyrite and various rarer silver minerals.

3. *Silver-bearing Lead Veins.* — Silver and lead minerals are very commonly associated with each other. These veins contain such minerals as galena, argentite, tetrahedrite, sphalerite, pyrite, calcite, dolomite, rhodochrosite, etc.

4. *Lead-zinc Veins.* — Lead and zinc minerals often occur together particularly in deposits that lie in limestones. The chief minerals of such deposits are galena, sphalerite, marcasite, chalcopyrite, smithsonite, calamine, cerussite, calcite, dolomite.

5. *Copper-iron Veins.* — Copper and iron sulphides are quite commonly associated with each other, the prominent minerals of such veins being pyrite, chalcopyrite, chalcocite, bornite, tetrahedrite, enargite, etc.

532. Primary and Secondary Vein Minerals. Secondary Enrichment. — In many mineral veins, it is obvious that certain minerals belong to the original vein deposit while certain others have been formed subsequently. These two classes of minerals are known respectively as *Primary* and *Secondary Minerals*. The primary vein minerals are those which were originally deposited by the ascending waters in the vein fissure. The primary metallic vein minerals are comparatively few in number, the more important being pyrite, chalcopyrite, galena and sphalerite. The secondary vein minerals have been formed from the primary minerals by some subsequent chemical

---

* Mineral Deposits, 1928.

reaction. This change is ordinarily brought about through the influence of oxidizing waters which, coming from the surface of the earth, descend through the upper portions of the vein. Under these conditions, various new minerals are formed, many of them being oxidized compounds. As the descending waters lose their oxygen content within a comparatively short distance of the earth's surface, the secondary minerals are only to be found in the upper part of a vein. Together with the formation of these secondary minerals, there is frequently a downward migration of the valuable metals in the vein. This is brought about by the solution of the minerals in the uppermost portion of the vein and a subsequent reprecipitation a little farther down. As the surface of the earth is gradually lowered by erosion, the upper part of a vein is continually being worn away. But the metallic content of the uppermost part of the vein is always being carried downward by the descending oxidizing waters. In this way, the metallic content of the upper part of many veins has been notably enriched since there is concentrated in this short space most of the original contents of hundreds, perhaps thousands, of feet of the vein which have been slowly worn away by the general erosion of the country. Consequently the zone of the secondary vein minerals is also frequently a zone of secondary enrichment. This is an important fact to be borne in mind since, because of it, the upper two or three hundred feet of a vein are ordinarily the richest portion of a deposit. The ore below that depth gradually reverts to its original unaltered and unenriched character and may frequently prove too low in value to warrant its being mined. The prevalent idea that the ore of a vein must increase in value with increasing depth is not true in the great majority of cases.

It will be of interest to consider the more important primary vein minerals and the secondary minerals that are commonly formed from them.

1. *Iron Minerals.* — The common primary vein iron mineral is pyrite, FeS_2. Marcasite, FeS_2, while not so common in occurrence is also a primary mineral. When oxidized, these minerals yield ordinarily the hydrated oxide limonite, $Fe_4O_3[OH]_6$. The upper portion of a vein that was originally rich in pyrite will often show a cellular and rusty mass of limonite. This limonite deposit near the surface is commonly termed *gossan*. The yellow rusty character of the outcrop of many veins enables one frequently to locate them and to trace them across the country.

2. *Copper Minerals.* — The one common primary copper mineral is chalcopyrite, $CuFeS_2$. At times, some of the other sulphides may be primary in their origin, but this is not generally the case. The secondary formation of bornite and chalcocite may be explained as follows. The copper sulphide existing in the original chalcopyrite is oxidized by the descending waters at the surface to copper sulphate which is then dissolved and carried farther down the vein. Here it comes in contact with unaltered chalcopyrite and a reaction takes place which enriches the sulphide, changing it to bornite, Cu_5FeS_4. Later, more copper sulphate in solution comes in contact with the bornite and a further enrichment takes place with the formation of chalcocite, Cu_2S. In each case, there is an interchange of metals, the iron in the original sulphide going into solution as a sulphate thus taking the place of the copper which has been precipitated. If the copper deposit lies in limestone rocks, we commonly find the various carbonates and oxides of copper also formed in the upper parts of the deposit. The secondary copper minerals therefore include *chalcocite*, Cu_2S, *bornite*, Cu_5FeS_4, *native copper*, Cu, *cuprite*, Cu_2O, *malachite*,

$(Cu.OH)_2CO_3$, *azurite*, $Cu(Cu.OH)_2(CO_3)_2$, *chrysocolla*, $CuSiO_3.2H_2O$, *chalcanthite*, $CuSO_4.5H_2O$.

3. *Lead Minerals.* — The one primary lead mineral is galena, PbS. The secondary minerals of lead are all oxidized compounds and include the following: *cerussite*, $PbCO_3$, *anglesite*, $PbSO_4$, *pyromorphite*, $Pb_4(PbCl)(PO_4)_3$, *wulfenite*, $PbMoO_4$.

4. *Zinc Minerals.* — Sphalerite, ZnS, is the only common primary zinc mineral. The chief secondary minerals are *smithsonite*, $ZnCO_3$, and *calamine*, $H_2(Zn_2O)SiO_4$.

5. *Silver Minerals.* — Probably most of the sulphide minerals of silver are primary in their origin. The following minerals are usually secondary, although native silver at times appears primary: *native silver*, Ag, *cerargyrite*, AgCl, *embolite*, Ag(Cl,Br), etc.

PART V. DESCRIPTIVE MINERALOGY

533. Scope of Descriptive Mineralogy. — It is the province of Descriptive Mineralogy to describe each mineral species, as regards: (1) form and structure; (2) physical characters; (3) chemical composition including blowpipe and chemical tests; (4) occurrence in nature with reference to geographical distribution and association with other species; also in connection with the above to show how it may be distinguished from other species. Further, it should classify mineral species into more or less comprehensive groups according to those characters regarded as most essential. Other points which may or may not be included are the investigation of the methods of origin of minerals; the changes that they undergo in nature and the results of such alteration; also the methods by which the same compounds may be made in the laboratory; finally, the uses of minerals as ores, for ornament and in the arts.

534. Scheme of Classification. — The method of classification adopted in this work, and the one which can alone claim to be thoroughly scientific, is that which places similar chemical compounds together in a common class and which further arranges the mineral species into groups according to the more minute relations existing between them in chemical composition, crystalline form and other physical properties.

Upon this basis there are recognized eight distinct chemical classes, beginning with the Native Elements; these are enumerated on the following page. Under each of these, sections of different grades are made, also based on chemical relationships. Finally, the mineral species themselves are arranged, as far as possible, in isomorphous groups, including those which have, at once, analogous chemical composition and similar crystallization (see Art. **483**). It is unnecessary to take the space here to develop the entire scheme of classification in detail, since a survey of the successive sub-classes under any one of the divisions will make the principles followed entirely clear. A few remarks, only, are added for sake of illustration.

Under the Oxides, for example, the classification is as follows: First, the Oxides of silicon (quartz, tridymite, opal). Second, the Oxides of the semi-metals, tellurium, arsenic, antimony, bismuth, also molybdenum, tungsten. Third, the Oxides of the metals, as copper, zinc, iron, manganese, tin, etc. The third section is then subdivided into the anhydrous and hydrous species. Further, the former fall into the four divisions: Protoxides, R_2O and RO; Sesquioxides, R_2O_3; Intermediate oxides, $RO.R_2O_3$; Dioxides, RO_2. Under each of these heads come finally the individual species, arranged so far as possible in isomorphous groups. Thus we have the Hematite group, the Rutile group, etc.

In regard to the various classes of salts it may be stated that, in general,

393

they are separated into anhydrous, acid, basic and hydrous sections; the special subdivisions called for, however, vary in the different cases.

For an explanation of the abbreviations used in the description of species, see p. 5.

SCHEME OF CLASSIFICATION

I. NATIVE ELEMENTS.
II. SULPHIDES, SELENIDES, TELLURIDES, ARSENIDES, ANTIMONIDES.
III. *Sulpho-salts.* — SULPHARSENITES, SULPHANTIMONITES, SULPHO-BISMUTHITES.
IV. *Haloids.* — CHLORIDES, BROMIDES, IODIDES; FLUORIDES.
V. OXIDES.
VI. *Oxygen Salts.*
    1. CARBONATES.
    2. SILICATES, TITANATES.
    3. NIOBATES, TANTALATES.
    4. PHOSPHATES, ARSENATES, VANADATES; ANTIMONATES; NITRATES.
    5. BORATES, URANATES.
    6. SULPHATES, CHROMATES, TELLURATES.
    7. TUNGSTATES, MOLYBDATES.
VII. *Salts of Organic Acids:* OXALATES, MELLATES, ETC.
VIII. HYDROCARBON COMPOUNDS.

I. NATIVE ELEMENTS

The NATIVE ELEMENTS are divided into the two distinct sections of the Metals and the Non-metals, and these are connected by the transition class of the Semi-metals. The distinction between them as regards physical characters and chemical relations has already been given (Art. **465**).

The only *non-metals* present among minerals are carbon, sulphur, and selenium; the last, in one of its allotropic forms, is closely related to the semi-metal tellurium.

The native *semi-metals* form a distinct group by themselves, since all crystallize in the rhombohedral class of the hexagonal system with a fundamental angle differing only a few degrees from 90°, as shown in the following list:

Tellurium, $rr' = 93°\ 3'$.   Arsenic, $rr' = 94°\ 54'$.
Antimony, $rr' = 92°\ 53'$.   Bismuth, $rr' = 92°\ 20'$.

An artificial form of selenium is known with metallic luster and rhombohedral crystallization, with $rr' = 93°$. Zinc (also only artif.) is rhombohedral ($rr' = 93°\ 46'$) and connects the semi-metals to the true metals. Metallic tantalum has been described in cubic crystals.

Among the *metals* the isometric GOLD GROUP is prominent, including gold, silver, copper, mercury, amalgam (AgHg), and lead.

Another related isometric group includes the metals platinum, iridium, palladium, and iron. An allotropic form of palladium and also iridosmine (IrOs) are both rhombohedral.

DIAMOND.

Isometric. The development of certain crystals as well as etching tests have suggested tetrahedral symmetry, but X-ray study (cf. p. 40) shows that the atomic structure conforms to normal symmetry. Commonly showing octahedral, hexoctahedral, and other forms; faces frequently rounded or striated and with triangular depressions (on $o(111)$). Twins common with tw. pl. $o(111)$ and often flattened parallel to o. Crystals often distorted. In spherical forms; massive.

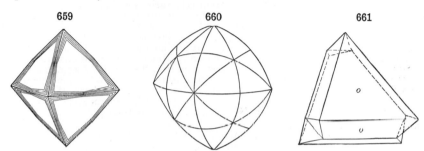

Cleavage: $o(111)$ highly perfect. Fracture conchoidal. Brittle. H. = 10. G. = 3·516–3·525 crystals. Luster adamantine to greasy. Color white or colorless; occasionally various pale shades of yellow, red, orange, green, blue, brown; rarely deeply colored; sometimes black. Usually transparent; also translucent, opaque. Refractive and dispersive power high; index n = 2·4195. (See Art. 334.) Usually isotropic but sometimes shows abnormal birefringence. Shows at times brilliant phosphorescence when rubbed or exposed to the electric discharge in a vacuum tube; to ultra-violet light, etc.

Var. — 1. *Ordinary.* In crystals usually with rounded faces and varying from those which are colorless and free from flaws (*first water*) through many faint shades of color, yellow being the most common; often full of flaws and hence of value only for cutting purposes.

2. *Bort, Boort* or *Bourt*; rounded forms with rough exterior and radiated or confused crystalline structure. Also applied to diamonds of inferior quality or small fragments.

3. *Carbonado* or *Carbon*; black diamond. Massive, crystalline, granular to compact, without cleavage. Color black or grayish black. Opaque. Obtained chiefly from Bahia, Brazil.

Comp. — Pure carbon; the variety carbonado yields on combustion a slight ash.

Pyr., etc. — Unaffected by heat except at very high temperatures, when (in an oxygen atmosphere) it burns to carbon dioxide (CO_2); out of contact with the air transformed into graphite at 1900° C. Not acted upon by acids or alkalies.

Diff. — Distinguished (*e.g.*, from quartz crystal) by its extreme hardness and brilliant adamantine luster; the form, cleavage, and high specific gravity are also distinctive characters; it is optically isotropic; transparent to X-rays, which distinguishes it from glass imitations.

Artif. — Various ways have been described in which minute diamonds have been formed artificially. It has been shown, however, that, at least in some cases, the supposed diamond crystals do not possess the requisite refractive index or structure.

Obs. — The diamond occurs chiefly in alluvial deposits of gravel, sand, or clay, associated with quartz, gold, platinum, zircon, octahedrite, rutile, brookite, hematite, ilmenite, and also andalusite, chrysoberyl, topaz, corundum, tourmaline, garnet, etc.; the associated minerals being those common in granitic rocks or granitic veins. Also found in quartzose conglomerates, and further in connection with the laminated granular quartz rock or quartzose hydromica schist, *itacolumite*, which in thin slabs is more or less flexible. This rock occurs at the mines of Brazil and the Ural Mts.; and also in Georgia and North Carolina, where a few diamonds have been found.

It has been reported as occurring *in situ* in a pegmatite vein in gneiss at Bellary in India. It occurs further in connection with an eruptive peridotite in South Africa and in a similar formation in Pike County, Arkansas. It has been noted as grayish particles forming one per cent of the meteorite which fell at Novo-Urei, Russia, Sept. 22, 1886; also in the form of black diamond (H. = 9) in the meteorite of Carcote, Chile; in the meteoric iron of Cañon Diablo, Arizona.

India was the chief source of diamonds from very early times down to the discovery of the Brazilian mines; the yield is now small. Of the localities, that in southern India, in the Madras presidency, included the famous " Golconda mines." The diamond deposits of Brazil have been worked since the early part of the 18th century, and have yielded very largely, although at the present time the amount obtained is small. The most important region was that near Diamantina in the province of Minas Geraes; also from Bahia, etc.

The discovery of diamonds in South Africa dates from 1867. They were first found in the gravel of the Vaal River; they occur from Potchefstroom down to the junction with the Orange River, and along the latter as far as Hope Town. More recently they have been found in gravels in the Somabula Forest, Rhodesia and at Lüderitzbucht, German Southwest Africa. These *river diggings* are now of much less importance than the *dry diggings*, discovered in 1871.

The latter are chiefly in Griqualand-West, south of the Vaal river, on the border of the Orange Free State. There are here near Kimberley a number of limited areas approximately spherical or oval in form, with an average diameter of some 200 to 300 yards, of which the Kimberley, De Beer's, Dutoitspan and Bultfontein mines are the most important. A circle $3\frac{1}{2}$ miles in diameter encloses these four principal mines. The general structure is similar: a wall of nearly horizontal black carbonaceous shale with upturned edges enclosing the diamantiferous area. The upper portion of the deposit consists of a friable mass of little coherence of a pale yellow color, called the " yellow ground." Below the reach of atmospheric influences, the rock is more firm and of a bluish green or greenish color; it is called the " blue ground " or simply " the blue." This consists essentially of a serpentinous breccia: a base of hydrated magnesian silicate penetrated by calcite and opaline silica and enclosing fragments of bronzite, diallage, also garnet, magnetite, and ilmenite, and less commonly smaragdite, pyrite, zircon, etc. The diamonds are rather abundantly disseminated through the mass, in some claims to the amount of 4 to 6 carats per cubic yard. The original rock seems to have been a peculiar type of peridotite. These areas are believed to be volcanic pipes, and the occurrence of the diamonds is obviously connected with the eruptive outflow, they having probably been brought up from underlying rocks. Other important mines, similar in character to those near Kimberley, are the Jagersfontein mine in Orange Free State and the Premier, near Pretoria, Transvaal.

Diamonds are also obtained in Borneo, associated with platinum, etc.; in Australia and the Ural Mts.

In the United States a few stones have been found in gravels in North Carolina, Georgia, Virginia, Colorado, California, and Wisconsin. Reported from Idaho and from Oregon with platinum. In 1906 diamonds were found near Murfreesboro in Pike County, Arkansas, both loose in the soil and enclosed in a peridotite rock associated with tuffs and breccias that occur in Carboniferous and Lower Cretaceous strata. Considerable exploration work has been done at this locality and probably between two and three thousand stones found. The stones have been of good color but usually small, although one weighing 21 carats was found in 1921.

Some of the famous diamonds of the world with their weights are as follows: the Kohinoor, which weighed when brought to England 186 carats, and as recut as a brilliant, 106 carats; the Orloff, 194 carats; the Regent or Pitt, 137 carats; the Florentine or Grand Duke of Tuscany, 133 carats. The " Star of the South " found in Brazil weighed before and after cutting respectively 254 and 125 carats. The orange-yellow Tiffany diamond weighs 125 carats. Also famous because of the rarity of their color are the green diamond of Dresden, 40 carats, and the deep blue Hope diamond from India, weighing 44 carats.

South Africa has yielded some very large stones. Among these may be mentioned the following: From the Jagersfontein mine the Excelsior weighing 969 carats; the Jubilee,

634 carats; and the Imperial, 457 carats. The largest diamond known was found in 1905 at the Premier mine. It was named the Cullinan or Star of Africa and was presented by the Transvaal Assembly to King Edward VII of England. When found it weighed 3025 carats or over 1¼ pounds. It has since been cut into 105 separate stones, the two largest weighing 516 and 309 carats, respectively, being the largest cut stones in existence. The history of the above stones and of others is given in many works on gems.

Use. — In addition to its use as a gem, the diamond is extensively used as an abrasive. Crystal fragments are used to cut glass. The fine powder is employed in grinding and polishing gem stones. The noncrystalline, opaque varieties, especially the carbonado, are used in the bits of diamond drills. The diamond is also used in wire drawing and in the making of tungsten filaments for electric lights.

The name has been derived by corruption from the ancient Greek name for the mineral, ἀδάμας (*the invincible*).

CLIFTONITE. — Carbon in minute cubic and cubo-octahedral crystals. H. = 2·5. G. = 2·12. Color and streak black; from the Youndegin, West Australia, meteoric iron, found in 1884, and other meteoric irons.

GRAPHITE. Plumbago. Black Lead.

Rhombohedral. In six-sided tabular crystals. Sometimes shows triangular markings on the basal plane. Commonly in embedded foliated masses, also columnar or radiated; scaly or slaty; granular to compact; earthy. See p. 40, for a discussion of atomic structure.

Cleavage: basal, perfect. Thin laminæ, flexible, inelastic. Feel greasy. H. = 1–2. G. = 2·09–2·23. Luster metallic, sometimes dull, earthy. Color iron-black to dark steel-gray. Opaque. A conductor of electricity.

Comp. — Carbon, like the diamond; often impure from the presence of ferric oxide, clay, etc.

It has been shown that graphite may contain some iron in solid solution.

Pyr., etc. At a high temperature some graphite burns more easily than diamond, other varieties less so. B.B. infusible. Unaltered by acids.

Diff. Characterized by its extreme softness (soapy feel); iron-black color, metallic luster; low specific gravity; also by infusibility. Its rubbed streak is black. Cf. molybdenite, p. 413.

Artif. — It is a common furnace product being formed from the fuel. It is produced extensively by heating coke in the electric furnace.

Obs. — Graphite occurs as beds, veins and embedded masses, as laminæ or scales in granite, gneiss, crystalline schist, quartzite, limestone, and also in basic eruptive rocks. It is formed in a number of different ways; in many cases the exact origin is obscure and has had various interpretations. It may occur as a separation from a silicate magma, the graphite being either an original constituent of the magma or derived by absorption of carbonaceous matter with which it has come into contact. It is found in the iron-bearing basalts of Greenland and is reported in a nephelite syenite from India. It occurs in meteoric iron. It is also the result of pneumatolytic action, occurring in pegmatite and similar veins, as in the case of the occurrence in Ceylon. Of a similar origin are many contact metamorphic deposits. It commonly occurs in schistose rocks and has been thought to have been either a constituent of the original rock or introduced during metamorphism through the agency of carbon-bearing gases. In some places it has resulted through the alteration of coal by heat. Common associated minerals include feldspar, kaolinite, rutile, titanite, pyrite, chlorite, micas, apatite.

Important localities are: In the Tunkinsk Mts., west of Irkutsk, Siberia, large quantities of fine graphite have been found in gneiss. The chief district for the production of graphite is the Island of Ceylon, where it occurs in veins in granulite. Crystals occur in the granular limestone at Pargas, Finland. Graphite schists are common in central Europe. A compact graphite ("black-lead") has been mined for a long time at Borrowdale, Cumberland, England. It occurs in quantity with schistose rocks in Madagascar. Sonora, Mexico, furnishes a fine-grained graphite suited for use in pencils. Graphite occurs plentifully in the gneisses of eastern Ontario and adjacent portions of Quebec.

The most productive district in the United States is in the eastern and southeastern Adirondack region, particularly at Ticonderoga, New York. It occurs there in graphitic quartzites, with quartz in small veins running through gneiss and in pegmatite veins. At

Rossie, near Grasse Lake, New York; found in metamorphosed Carboniferous rocks near Providence and Tiverton, Rhode Island; in granite and schists in Clay Co., and elsewhere, Alabama; in Chester Co., Pennsylvania; near Dillon, Montana; as amorphous graphite in Colfax Co., New Mexico, formed by the metamorphic action of igneous intrusions on coal beds.

Use. — Its chief uses are for making crucibles and other refractory products, in lubricants, paint, stove polish, "lead" pencils, electrodes, and for foundry facings.

The name *black lead*, applied to this species, is inappropriate, as it contains no lead. The name graphite, of Werner, is derived from γράφειν, *to write*, alluding to its use for pencils.

QUISQUEITE. — A black lustrous material composed chiefly of carbon and sulphur from the vanadium ores of Minasragra, Peru.

SULPHUR.

Orthorhombic.   Axes $a : b : c = 0.8131 : 1 : 1.9034$.

Crystals commonly acute pyramidal; sometimes thick tabular $\parallel c(001)$. See also Fig. 97, p. 64.   Crystals are sometimes sphenoidal in habit, and etch figures suggest such a symmetry, but study by X-rays indicates normal symmetry for the atomic structure. Rarely twinned.   Also massive, in reniform shapes, incrusting, stalactitic and stalagmitic; in powder.

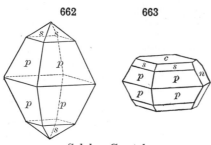

662   663

Sulphur Crystals

Cleavage: $c(001)$, $m(110)$, $p(111)$ imperfect.   Fracture conchoidal to uneven.   Rather brittle to imperfectly sectile.   H. = 1.5–2.5.   G. = 2.05–2.09.   Luster resinous.   Color sulphur-yellow, straw- and honey-yellow, yellowish brown, greenish, reddish to yellowish gray.   Streak white.   Transparent to translucent.   A non-conductor of electricity; by friction negatively electrified.   A poor conductor of heat. Optically +.   Double refraction strong.   Ax. pl. $\parallel$ (010).   Bx. $\perp$ (001). Dispersion $\rho < v$, weak. $2V = 68° 58'$. $\alpha = 1.958$, $\beta = 2.038$, $\gamma = 2.245$.

Comp. — Pure sulphur; often contaminated with clay, bitumen, and other impurities.

Sulphur is polymorphous.   The orthorhombic or α-sulphur is the common natural form. When sulphur is fused and again crystallized it is monoclinic, β-sulphur.   This is unstable and changes readily into the orthorhombic form.   There is another monoclinic form, γ-sulphur.   These three modifications have been described as occurring on the island of Vulcano. γ-sulphur has been found in minute crystals associated with pyrite near Letovice in Moravia and given the name of *rosickyite*.   Two or three other modifications are known. *Daiton-sulphur*, from northern Formosa, originally described as a distinct monoclinic type, has been shown to be ordinary orthorhombic sulphur.   Sulphur also at times appears to occur in an amorphous condition.

Pyr., etc. — Melts at 108° C., and at 270° burns with a bluish flame yielding sulphur dioxide.   Insoluble in water, and not acted on by the acids, but soluble in carbon disulphide.

Diff. — Readily distinguished by the color, fusibility and combustibility.

Obs. — Sulphur may be formed in various ways.   It is frequently the result of volcanic activity.   It occurs in the gases given off at fumeroles, at times being deposited as a direct sublimation product.   It is also formed by the incomplete oxidation by the oxygen of the air of hydrogen sulphide gas derived from volcanic sources.   Further, it is formed by the decomposition of the hydrogen sulphide that frequently occurs in thermal spring waters. This may come from volcanic sources, by the action of acid water on metallic sulphides, or by the reduction of sulphates, especially gypsum.   This last is aided by the action of certain micro-organisms.   Also the living processes of the so-called sulphur bacteria result in the separation of sulphur. The decomposition of metallic sulphides may produce sulphur, often in small crystals.

Sulphur is most commonly found in the Tertiary sedimentary rocks and is most frequently associated with gypsum and limestone; often in clay rocks; frequently associated with bituminous deposits. It occurs at times in fine crystals but more often in compact masses; at times in crusts and kidney or stalactitic forms; sometimes as a friable aggregate known as flour or meal sulphur. It may be in rounded nodules. It is often impure and variously colored.

The sulphur deposits on the island of Sicily are especially noteworthy because of their size and the fine crystals that occur in them. The chief associated minerals are celestite, calcite, aragonite, gypsum. The important localities are at Girgenti and the neighboring Cianciana, Racalmuto, and Cattolica. Sulphur occurs elsewhere in Italy as in fine crystals with asphalt at Perticara, south of Cesena, Romagna; near Bologna; at Carrara, Tuscany; in Spain, at Conil, near Cadiz. It is common in the volcanic regions of Iceland, Japan, Hawaii, Mexico, and western South America.

In the United States the most productive deposits are in Louisiana and Texas. Near Lake Charles in Calsasieu Parish, Louisiana, a bed of sulphur 100 feet thick is found at a depth of between 300 and 400 feet. It is underlain by beds of gypsum and salt. A similar deposit occurs near Freeport in Brazoria Co., Texas. In numerous other western localities; Utah, at Sulphurdale, Beaver Co., in a rhyolitic tuff; Wyoming, in limestones near Cody and Thermopolis and about the fumeroles in Yellowstone Park; Nevada, in Esmeralda Co., near Luning and Cuprite, in Humboldt Co., near Rosebud at times in crystals, and at Eureka, Eureka Co.; California, at the Sulphur Bank mercury mine on Clear Lake, Lake Co., at the geysers in Napa Valley, Sonoma Co., on Lassen Peak, Tehama Co., in Colusa, San Bernardino, and other counties; Colorado, at Vulcan, Gunnison Co., and in Mineral Co.

Use. — In manufacture of sulphuric acid, in the process of making paper from wood pulp, in making matches, gun powder, fireworks, insecticides, for vulcanizing rubber, for medicinal purposes, etc. Sulphuric acid is now largely derived from the oxidation of pyrite.

Selensulphur. Volcanite. — Contains sulphur and selenium; the amount of selenium is usually very small and in most cases at least the mineral could more properly be considered as a seleniferous variety of sulphur; orange-red or reddish brown; from the islands Vulcano and Lipari; from Hawaii, etc.

ARSENIC.

Rhombohedral. Rarely in cube-like rhombohedral crystals. Generally granular massive; sometimes reticulated, reniform, stalactitic.

Cleavage: $c(0001)$ highly perfect but owing to usual structure seldom conspicuous. Fracture uneven and fine granular. Brittle. H. = 3·5. G. = 5·7. Luster nearly metallic. Color and streak tin-white, tarnishing to dark gray.

Comp. — Arsenic, often with some antimony, and traces of iron, silver, gold, or bismuth.

Pyr. — B.B. on charcoal volatilizes without fusing, coats the coal with white arsenic trioxide, and affords a garlic odor; the coating treated in R.F. volatilizes, tingeing the flame blue. In the closed tube gives a volatile sublimate of arsenic.

Obs. — Arsenic frequently occurs in metallic veins but rarely in any quantity. It is associated most commonly with silver, cobalt, or nickel ores and with such other minerals as barite, arsenolite, cinnabar, realgar, orpiment, galena, sphalerite.

Arsenic is found in many places in Saxony, especially at Freiberg, Schneeberg, Marienberg, and Annaberg. Fine reniform masses occur at Andreasberg, Harz Mts. It is found in Bohemia at Joachimstal (Jáchymov); in Transylvania at Nagy-Ág; at Kapnikbánya (Capnic) in Rumania. Fine specimens come from the silver mines at Ste. Marie aux Mines in Alsace. Globular masses composed of small rhombohedral crystals are found at Akatani, Echizen Prov., Japan.

Arsenic is found infrequently in the United States. It has been found at Washington Camp, Santa Cruz Co., Arizona, in reniform masses, sometimes several pounds in weight. In Canada it occurs at Montreal and on Alden Island, Queen Charlotte Islands, British Columbia.

Use. — An ore of arsenic.

Allemontite. — Arsenical Antimony, $SbAs_3$. In reniform masses. G. = 6·203. Luster metallic. Color tin-white or reddish gray. A study of allemontite from Atlin, British Columbia, shows that it is an intergrowth of metallic arsenic and antimony. It is probable that all allemontite is of a similar nature. From Allemont, France; Příbram, Bohemia; Andreasberg in the Harz Mts., etc.

Tellurium. — Rhombohedral. In prismatic crystals; commonly columnar to fine-granular massive. Perfect prismatic cleavage. H. = 2–2·5. G. = 6·2. Luster metallic. Color and streak tin-white. B.B. wholly volatile. In warm concentrated sulphuric acid gives red solution. From Rumania, West Australia, and a number of places in Colorado.

ANTIMONY.

Rhombohedral. Generally massive, lamellar and distinctly cleavable; also radiated; granular. Often shows polysynthetic twinning on (01$\bar{1}$2).

Cleavage: c(0001) highly perfect; also other cleavages. Fracture uneven; brittle. H. = 3–3·5. G. = 6·7. Luster metallic. Color and streak tin-white.

Comp. — Antimony, containing sometimes silver, iron, or arsenic.

Pyr. — B.B. on charcoal fuses very easily and is wholly volatile giving a white coating. The white coating tinges the R.F. bluish green. Crystallizes readily from fusion.

Obs. — Antimony occurs in metal-bearing veins with silver, antimony, and arsenic ores, often with stibnite. Occurs near Sala in Sweden; Andreasberg in the Harz Mts., Germany; Allemont, Isère, France; Příbram, Bohemia; Chile; Borneo. In the United States, in Kern Co., and at South Riverside, California. At South Ham, Quebec; Prince William Parish, York Co., New Brunswick.

Use. — An ore of antimony.

BISMUTH.

Rhombohedral. Usually reticulated, arborescent; foliated or granular.

Cleavage: c(0001) perfect. Sectile. Brittle, but when heated somewhat malleable. H. = 2·5. G. = 9·8. Luster metallic. Streak and color silver-white, with a reddish hue; subject to tarnish. Opaque.

Comp. — Bismuth, with traces of arsenic, sulphur, tellurium, etc.

Pyr., etc. — B.B. on charcoal fuses very easily and entirely volatilizes, giving a coating orange-yellow while hot, lemon-yellow on cooling. With potassium iodide and sulphur B.B. on charcoal gives a brilliant red coating. Dissolves in nitric acid; subsequent dilution causes a white precipitate. Crystallizes readily from fusion.

Obs. — Occurs in veins in granite, gneiss, and other crystalline rocks, accompanying ores of cobalt, nickel, silver, and sometimes tin. Also noted in topaz-bearing quartz veins. Often associated with galena. Abundant in Saxony, at Schneeberg, etc.; in Bohemia at Joachimstal (Jáchymov). It occurs with various bismuth minerals at Meymac, Corrèze, France. Found in various places in Norway (Modum) and Sweden (Broddbo near Falun). In Cornwall and Devonshire. The most productive deposits occur in Bolivia at San Baldomero and elsewhere near La Paz. There are important deposits in Australia, at Bamford and Wolfram, North Queensland, and at Kingsgate, New England Range, N. S. W., where it has been found in masses weighing up to 30 pounds.

Bismuth occurs sparingly in the United States; in Monroe, Connecticut; in the Chesterfield district, South Carolina; in Colorado in the placers of French Creek, Summit Co., and at the Las Animas mine, Boulder Co. In Canada it occurs in fine specimens associated with the silver ores of the Cobalt district, Ontario.

Use. — An ore of bismuth.

Zinc. — Probably does not occur in the native state. In the laboratory it is obtained in hexagonal prisms with tapering pyramids; also in complex crystalline aggregates. It also appears to crystallize in the isometric system, at least in various alloys.

Tantalum. — Isometric. In cubic crystals and fine grains. Color grayish yellow. Found containing small amounts of niobium in the gold washings of the Ural and Altai Mts.

Gold Group

GOLD.

Isometric. X-ray study shows a face-centered cubic lattice (cf. p. 26). Distinct crystals rare, $o(111)$ most common, also $d(110)$ and $m(311)$; crystals often elongated in direction of an octahedral axis, giving rise to rhombohedral-like forms, and arborescent shapes; also in plates flattened $\parallel o(111)$, and branching at 60° parallel either to the edges or diagonals of an o face (see p. 194). Twins: tw. plane o. Skeleton crystals common; edges salient or rounded; in filiform, reticulated, dendritic shapes. Also massive and in thin laminæ; often in flattened grains or scales.

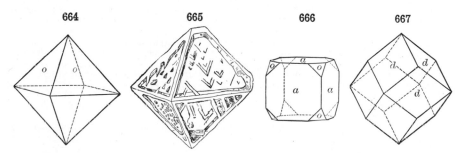

Cleavage none. Fracture hackly. Very malleable and ductile. H. = 2·5–3. G. = 15·6–19·3, 19·33 when pure. Luster metallic. Color and streak gold-yellow, sometimes inclining to silver-white and rarely to orange-red. Opaque.

Comp. — Gold, but usually alloyed with silver in varying amounts and sometimes containing also traces of copper or iron.

Var. — 1. *Ordinary.* Containing up to 16 per cent of silver. Color varying accordingly from deep gold-yellow to pale yellow, and specific gravity from 19·4 to 15·5. The ratio of gold to silver of 3 : 1 corresponds to 15·1 per cent silver. For G. = 17·6, Ag = 9 per cent; G. = 16·9, Ag = 13·2; G. – 14·6, Ag = 38·4. The purest gold which has been described is the so-called "sponge gold" from Kalgoorlie, West Australia, with 99.91 per cent gold, .09 per cent silver.

2. *Argentiferous; Electrum.* Color pale yellow to yellowish white; G. = 15·5–12·5. Ratio for the gold and silver of 1 : 1 corresponds to 36 per cent of silver; $1\frac{1}{2}$: 1, to 26 per cent; 2 : 1, to 21 per cent; $2\frac{1}{2}$: 1, to 18 per cent. The word in Greek means also *amber;* and its use for this alloy probably arose from the pale yellow color it has as compared with gold.

Varieties have also been described containing copper up to 20 per cent from the Ural Mts.; palladium to 10 per cent (*porpezite*), from Porpez, Brazil; bismuth, including the black gold of Australia (*maldonite*); also rhodium(?).

Pyr., etc. — B.B. fuses easily (at 1100° C.). Not acted on by fluxes. Insoluble in any single acid; soluble in aqua regia, the separation not complete if more than 20 per cent Ag is present.

Diff. — Readily recognized (*e.g.*, from other metallic minerals, also from scales of yellow mica) by its malleability and high specific gravity, which last makes it possible to separate it from the gangue by washing; distinguished from chalcopyrite and pyrite since both sulphides are brittle and soluble in nitric acid.

Obs. — Gold is widely distributed in the earth's crust. It has been found in various igneous rocks, more commonly in the acid types, and sometimes in visible particles. It occurs in sedimentary rocks and quite frequently in connection with metamorphic rocks. It is a constituent of sea water. It is most frequently found in notable amounts in quartz veins and in the various forms of placer deposits.

The gold, when occurring in quartz, is often irregularly distributed, in strings, scales, plates, and in masses which are sometimes an agglomeration of crystals. Frequently the scales are invisible to the naked eye. The associated minerals are: pyrite, which far exceeds in quantity all others, and is generally auriferous; next, chalcopyrite, galena, sphalerite, arsenopyrite, each frequently auriferous; often tetradymite and other tellurium ores, native bismuth, native arsenic, stibnite, cinnabar, magnetite, hematite; sometimes barite, scheelite, apatite, fluorite, siderite, chrysocolla. The quartz at the surface, or in the upper part of a vein, is usually cellular, and rusted from the more or less complete disappearance of the pyrite and other sulphides by decomposition; but below, it is commonly solid.

The gold of the world was early gathered, not directly from the quartz veins (the " quartz reefs " of Australia and Africa), but from the gravel and sand deposited in the valleys in auriferous regions, or on the slopes of the mountains or hills, whose rocks contain in some part, and generally not far distant, gold-bearing veins. Such deposits are known as placer deposits. The gold is obtained by some method involving the use of a current of water and the separation of the gold from the sand and gravel by means of its high specific gravity. These hydraulic methods have been very extensively used in California and Alaska and indeed most of the gold of the Ural Mts., Brazil, Australia, and many other gold regions has come from such alluvial washings. At the present time, however, placer deposits are much less depended upon and in many regions all the gold is obtained directly from the rock.

The alluvial gold is usually in flattened scales of different degrees of fineness, the size depending partly on the original condition in the quartz veins, and partly on the distance to which it has been transported and assorted by running water. The rolled masses when of some size are called *nuggets*; in rare cases these occur very large and of great value. The Australian gold region has yielded many large nuggets; one of these found in 1858 weighed 184 pounds, and another (1869) weighed 190 pounds. In the auriferous sands, crystals of zircon are very common; also garnet and kyanite in grains; often also monazite, diamond, topaz, corundum, iridosmine, platinum.

Besides the free gold of the quartz veins and gravels, much gold is also obtained from auriferous sulphides or the oxides produced by their alteration, especially pyrite, also arsenopyrite, chalcopyrite, sphalerite, marcasite, etc. The only minerals containing gold in combination are the rare tellurides (sylvanite, calaverite, etc.).

Gold is widely distributed over the earth. It occurs under many different conditions and with many different rocks, being, however, more commonly associated with the acid types. A brief summary of the more important districts follows.

Europe. The gold deposits of Europe are to be found chiefly in three great districts, namely the Ural Mts., the Balkan States and a less important Alpine district reaching from Carinthia through the Austrian Tyrol and the Italian Alps to the Pyrenees. There are two minor districts in Czechoslovakia, one in the central portion and one near the Galician frontier. The third district, which is the most important district in Europe, is in Transylvania, lying in the southeastern portion of the Bihar mountains. Its important centers are Offenbánya, Verespatak (crystals), Nagy-Ág (largely tellurides), Boicza, etc.

Asia. In Siberia gold is found on the eastern slope of the Ural Mts. for a distance of 500 miles. The important districts are Bogoslovsk, Nizhne Tagilsk, Beresovsk and other localities near Ekaterinburg, Sysertsk and Kyshtym, the Miask district including Zlatoust and Mt. Ilmen, Kotchkar and at the southern limit of the fields, Orsk. Siberia also has the important placer districts in Tomsk, which include Altai and Marinsk, and in Yeniseisk, the Achinsk, Minusinsk and the north and south Yenisei districts. Farther east there are deposits in Transbaikalia and the Lena district in Yakutsk. In India the chief districts are the Kolar field near Bangalore in Mysore and the Gadag and Hutti districts to the northwest. Gold has been mined in China in Chihli, Shantung, Weihaiwei, Szechwan and Fu-Kien. In Manchuria on the Liao-tung Peninsula. In Korea principally at Un San. Gold-quartz veins, many of which have been worked for a long time, occur on a number of the Japanese islands.

Australasia. The most important districts in New Zealand lie on the Hauraki Peninsula with the Waihi mine as the most famous. Other districts are the West Coast area on the western slopes of the Alps of the South Island and the Otago area. In Queensland the districts of Charters Towers and the Mount Morgan mine are important. There are many gold districts in New South Wales among which are Hillgrove, Mount Boppy and Hill End. Rich districts in Victoria are the Bendigo and Ballarat. The principal gold fields of Tasmania are Beaconsfield, Mathinna and the copper deposits at Mount Lyell. The chief gold field in West Australia is near Kalgoorlie where the ores are largely tellurides.

Africa. Gold is found in Egypt in the section between the Nile and the Red Sea. Some of these deposits were worked in very early days. Gold has been produced for a long time from the Gold Coast district on the Gulf of Guinea. Important deposits are found in Matabeleland and Mashonaland in Southern Rhodesia. The most important gold district in the world is that of the Witwatersrand in the Transvaal. The mines occur in an east and west belt, some sixty miles in length, near Johannesburg. The gold is found scattered in small amounts through a series of steeply dipping quartz conglomerate rocks.

South America. Colombia has in the past produced large amounts of gold. The chief districts today are in the states of Antioquia and Cauca. Comparatively small amounts are produced at the present time in the other northern countries. The important deposits of Brazil lie 200 miles to the north of Rio de Janeiro in Minas Geraes along the Sierra do Espinhaco. The gold deposits in Chile lie chiefly in the coast ranges in the northern and central parts of the country.

Mexico. While Mexico is chiefly noteworthy for its silver output it produces also considerable gold. Important districts are as follows: Altar, Magdalena and Arizpe in Sonora; various places in Chihuahua, especially about Parral, and the Dolores mine on the western border of the state; the El Oro mines in the state of Mexico; the Pachuca district in Hidalgo; also various places in Guanajuato and Zacatecas.

Canada. The three important placer districts of Canada are the Klondike in Yukon Territory and the Atlin and Cariboo in British Columbia. The most productive vein deposits are found in British Columbia in the West Kootenay and Yale districts. Gold is also found in Ontario and Nova Scotia.

United States. Gold occurs in the United States chiefly along the mountain ranges in the western states. Smaller amounts have been found along the Appalachians in the states of Virginia, North and South Carolina and Georgia. The more important localities in the western states are given below, the states being arranged approximately in the order of their importance. California. At the present time about two thirds of the state's output comes from the lode mines and one third from placer deposits. The quartz veins are chiefly found in what is known as the Mother-Lode belt that lies on the western slope of the Sierra Nevada and stretches from Mariposa County for more than 100 miles toward the north. The veins occur chiefly in a belt of slates. The lode mines are found chiefly in Amador, Calaveras, Eldorado, Kern, Nevada, Shasta, Sierra and Tuolumne Counties. Crystals have been noted at localities in Amador and Eldorado Counties. The important placer mines are located in Butte, Sacramento and Yuba Counties. About 90 per cent of the placer gold is obtained by the use of dredges. Colorado. Gold is mined in various districts in Gilpin County, from the Leadville district and others in Lake County, in the region of the San Juan mountains in the Sneffels, Silverton and Telluride districts, Cripple Creek district (telluride ores) in Teller County, placer deposits in the Breckenridge district in Summit County. Alaska. The most important lode mines are in the Juneau district, while the chief placer deposits are those of Fairbanks and Iditarod in the Yukon basin and the Nome district on the Seward Peninsula. Nevada. The most important districts are those of Goldfield in Esmeralda County and Tonopah in Nye County. South Dakota. The output is chiefly from the Homestake mine at Lead in Lawrence County. Montana. There are various producing districts, the more important being in Madison (largely placers), Deer Lodge and Silver Bow Counties. Arizona. The important counties are Mohave and Cochise. Utah. Gold is produced chiefly from the Bingham and Tintic districts in Salt Lake County and from Juab County.

Use. — The chief ore of gold.

SILVER.

Isometric. X-ray study shows that silver has a face-centered cubic lattice (cf. p. 26). Crystals commonly distorted, in acicular forms, reticulated or arborescent shapes; coarse to fine filiform; also massive, in plates or flattened scales.

Cleavage none. Ductile and malleable. Fracture hackly. H. = 2·5–3. G. = 10·1–11·1, pure 10·5. Luster metallic. Color and streak silver-white, often gray to black by tarnish.

Comp. — Silver, with some gold (up to 10 per cent), copper, and sometimes platinum, antimony, bismuth, mercury.

Pyr., etc. — B.B. on charcoal fuses easily to a silver-white globule, which in O.F. gives a faint dark red coating of silver oxide; crystallizes on cooling; fusibility about 1050° C. Soluble in nitric acid, and deposited again by a plate of copper. Precipitated from its solutions by hydrochloric acid in white curdy forms of silver chloride.

Diff. — Distinguished by its malleability, color (on the fresh surface), and specific gravity.

Obs. — Native silver is much rarer in occurrence than native gold but is widely distributed in small amounts. It may be primary in its origin but usually is clearly secondary and commonly only to be found in the upper portions of silver-bearing veins, associated with other silver minerals. When secondary it may have been formed in various ways: by the action of hot water vapors or of oxygen upon silver sulphides, or by the action of metallic sulphides or arsenides upon silver chloride, etc. It occurs in masses, or in arborescent and filiform shapes, in veins which most commonly traverse gneiss or other crystalline metamorphic rocks but it is also found associated with eruptive rocks and at times in sedimentary rocks. It also frequently occurs disseminated, but usually invisibly, in various metallic sulphides.

The mines of Kongsberg, Norway, have afforded magnificent specimens of native silver, sometimes in very large masses. It is common in Saxony, especially at Freiberg, Schneeberg, etc.; in Bohemia at Přibram and Joachimstal (Jáchymov). It is found at Andreasberg, Harz Mts.; at Wittichen in Baden; at Schemnitz (Selmeczbánya), Czechoslovakia. In the mines of Monte Narba, near Sarrabus, Sardinia. At Broken Hill, New South Wales; at Copiapo and Chañarcillo, Atacama, Chile. Mexico has been the most important country for the production of silver. The native metal is found especially at Batopilas, Chihuahua, and at Guanajuato, also at various places in the states of Sonora, Durango, and Zacatecas.

In the United States it occurs associated with the copper of Keweenaw peninsula, Michigan; in Montana at Butte and at Elkhorn, Jefferson Co.; in Idaho large amounts were obtained at the Poorman mine, Owyhee Co., and it is found with cerussite in the mines of Shoshone Co. In Colorado considerable amounts were found at Aspen and it also occurs in Boulder Co.; in Arizona at the Silver King mine, Pinal Co., and in the Globe district. In Ontario, Canada, silver is found in notable amounts in the Timiskaming district, especially at Cobalt, Coleman Township; also in the Thunder Bay district on the north shore of Lake Superior, at Silver Islet, etc.

Use. — An ore of silver.

COPPER.

Isometric. Atomic structure shows a face-centered cubic lattice. The tetrahexahedron a common form (Fig. 668); also in octahedral plates.

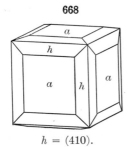

668

$h = (410).$

Distinct crystals rare. Frequently irregularly distorted and passing into twisted and wirelike forms; filiform and arborescent. Massive; as sand. Twins: tw. pl. $o(111)$, very common, often flattened or elongated to spear-shaped forms. Cf. p. 194. At times in pseudomorphs after cuprite, azurite, chalcopyrite, calcite, aragonite, etc.

Cleavage none. Fracture hackly. Highly ductile and malleable. H. = 2·5–3. G. = 8·8–8·9. Luster metallic. Color copper-red. Streak metallic, shining. Opaque. An excellent conductor for heat and electricity.

Comp. — Pure copper, sometimes containing small amounts of iron, silver, bismuth, tin, lead or antimony.

Pyr., etc. — B.B. fuses readily; on cooling becomes covered with a coating of black oxide. Dissolves readily in nitric acid, giving off red nitrous fumes, and produces a deep azure-blue solution with excess of ammonia. Fusibility 1080° C.

Obs. — Native copper is usually, if not always, secondary in its origin, and though found in numerous localities it occurs only seldom in commercial quantities. Various theories have been proposed to account for the reduction of copper minerals or solutions so as to produce native copper. It may be precipitated from a sulphate solution by the action of organic substances, as in the case of the action of wood on mine waters. Usually the reduction is apparently accomplished by the simultaneous oxidation of iron, which may occur in sulphide minerals or in the silicates of the neighboring rocks. Copper occurs in beds or veins associated with chalcopyrite, chalcocite, etc., or with cuprite, malachite, and azurite. It is found in sandstone, limestone, clay slate; it is frequently abundant in the vicinity of igneous rocks and may occur in the cavities of their amygdaloidal layers.   Cuprite, malachite, and azurite occur as pseudomorphs after copper.

In Russia, copper occurs in fine crystals in a limestone at Turnisk, near Bogolovsk, Perm, also near Nizhne-Tagilsk.   Crystallized copper occurs in various places in Cornwall.   In South Australia it occurs abundantly at Wallaroo and nearby on Yorke Peninsula; at Broken Hill, New South Wales.   In Bolivia there is a noted occurrence in sandstone at Corocoro, southwest of La Paz, where it is found in crystals and in pseudomorphs after aragonite.   In Mexico at Cananea, Sonora.

Copper is found in small amounts at various localities throughout the red sandstone region of the eastern United States, in Massachusetts, Connecticut, and more abundantly in New Jersey.   The Lake Superior copper district on Keweenaw Peninsula in northern Michigan is the most important locality in the world.   Here the copper is practically all as the native metal and is obtained over an area 200 miles in length.   Masses of great size have been found; the largest, discovered in 1857, weighed about 420 tons.   The copper is associated with prehnite, datolite, analcite, laumontite, pectolite, epidote, chlorite, wollastonite, calcite, etc.   Native silver is frequently found in small amounts.   The rocks of this district consist of a series of interbedded lava flows (basalts), sandstones, and conglomerates which dip steeply to the northwest.   The copper occurs as (1) a cement filling the interstices in the sandstone and conglomerate, sometimes replacing in large part the grains and pebbles themselves, (2) filling the amygdaloidal cavities in the trap, and (3) in veins that traverse all kinds of rocks.   Copper is also found in crystals at the Copper Queen mine, Bisbee, Arizona; also in the Globe district; in New Mexico at Georgetown, Grant Co.

Use. — An ore of copper.

MERCURY.   Quicksilver.

In small fluid globules scattered through its gangue.   Crystallizes at $-40°$ C. and then shows by X-ray study a rhombohedral structure.   G. = 13·6.   Luster metallic, brilliant.   Color tin-white.   Opaque.

Comp. — Pure mercury (Hg); with sometimes a little silver.

Pyr., etc. — B.B. entirely volatile, vaporizing at 350° C.   Dissolves in nitric acid.

Obs. — Mercury in the metallic state is a rare mineral.   In most cases it is secondary in origin and intimately associated with cinnabar.   It is commonly in small isolated drops but at times has been found in larger fluid masses occupying cavities in the rocks of the deposit. Mercury ores occur in rocks of various kinds and ages but usually in regions of volcanic activity.   Metallic mercury as well as cinnabar is found at times in connection with deposits from hot springs.

Found at Idria, in Gorizia, Italy (formerly in Carniola, Austria); at Mt. Avala near Belgrad, Yugoslavia; at Mochellandsberg (Landsberg near Ober-Moschel), Palatinate, Bavaria. See also under cinnabar.

LEAD.

Isometric.   Face-centered cubic lattice.   Crystals rare.   Usually in thin plates and small globular masses.   Very malleable, and somewhat ductile. H. = 1·5.   G. = 11·4.   Luster metallic.   Color lead-gray.   Opaque.

Comp. — Nearly pure lead; sometimes contains a little silver, also antimony.

Pyr. — B.B. fuses easily, coating the charcoal with a yellow to white oxide.   Fusibility 330° C.   Dissolves easily in dilute nitric acid.

Obs. — Native lead is extremely rare in its occurrence.   Many of the reported occurrences are of doubtful character.   The best-known localities are in Vermland, Sweden, at the Harstig mine, Pajsberg, at Långbanshyttan and Nordmark.   Crystals are known only from the Harstig mine.   Found in the gold placers of the Urals in the district of Ekaterinburg.   In Mexico near Jalapa, state of Vera Cruz.   In the United States has been found at Franklin, New Jersey, and at the Jay Gould mine, Wood River district, Blaine Co., Idaho.

AMALGAM.

Isometric.   Common habit dodecahedral.   Crystals often highly modified. Also massive in plates, coatings, and embedded grains.

Cleavage: dodecahedral in traces.   Fracture conchoidal, uneven.   Rather brittle to malleable. H. = 3–3·5. G. = 13·75–14·1. Luster metallic, brilliant. Color and streak silver-white. Opaque.

Comp. — (Ag,Hg), silver and mercury, varying from Ag_2Hg_3 to $Ag_{36}Hg$. Undoubtedly represents a solid solution of the two metals in each other.

Var. — *Ordinary amalgam*, Ag_2Hg_3 (silver 26·4 per cent) or AgHg (silver 35·0); also Ag_5Hg_3, etc. *Arquerite*, $Ag_{12}Hg$ (silver 86·6); G. = 10·8; malleable and soft. *Kongsbergite*, $Ag_{32}Hg$ or $Ag_{36}Hg$.

Pyr., etc. — B.B. on charcoal the mercury volatilizes and a globule of silver is left. In the closed tube the mercury sublimes and condenses on the cold part of the tube in minute globules. Dissolves in nitric acid. Rubbed on copper it gives a silvery luster.

Obs. — Amalgam is of rare occurrence, being found in either mercury or silver deposits, usually in small scattered grains or crystals. It is often associated with cerargyrite and therefore belongs to the oxidation zone of the deposits. Occurs in fine crystals at Moschellandsberg (Landsberg near Ober-Moschel), Palatinate, Bavaria; found at Friedrichssegen, near Ems, Nassau; at Szlana, near Rosenau (Rožňava), Czechoslovakia; at Březina near Plsen (Pilsen), Bohemia. Occurs in France at Les Chalantes near Allemont, Isère; at Almaden, Ciudad Real, Spain; in Norway at Kongsberg (*kongsbergite*) and at Sala, Sweden. In Chile it was found at Chañarcillo near Copiapo and at Arqueros (*arquerite*), Coquimbo. The variety *arquerite* occurs at Vitalle Creek, British Columbia.

Tin. — Tetragonal. Native tin has been reported from several localities. The only occurrence fairly above doubt is that from the washings at the headwaters of the Clarence river, near Oban, New South Wales. It has been found here in grayish white rounded grains, with platinum, iridosmine, gold, cassiterite, and corundum.

Platinum-Iron Group

PLATINUM.

Isometric. Crystals rare; usually in grains and scales.

Cleavage none. Fracture hackly. Malleable and ductile. H. = 4–4·5. G. = 14–19 native; 21–22 chem. pure. Luster metallic. Color and streak whitish steel-gray; shining. Sometimes magnetic and occasionally shows polarity.

Comp. — Platinum alloyed with iron, iridium, rhodium, palladium, osmium, and other metals.

Most platinum yields from 8 to 15 or even 18 per cent of iron, 0·5 to 2 per cent palladium, 1 to 3 per cent each of rhodium and iridium, a trace of osmium and finally 0·5 to 2 per cent or more of copper.

Var. — 1. *Ordinary.* *Non-magnetic* or only slightly magnetic. G. = 16·5–18·0 mostly. 2. *Magnetic.* G. about 14. Much platinum is magnetic, and occasionally it has polarity. The magnetic property seems to be connected with high percentage of iron (iron-platinum), although this distinction does not hold without exception.

Pyr., etc. — B.B. infusible. Not affected by borax or salt of phosphorus, except in the state of fine dust, when reactions for iron and copper may be obtained. Soluble only in heated aqua regia.

Diff. — Distinguished by its color, malleability, high specific gravity, infusibility and insolubility in ordinary acids.

Obs. — Native platinum is a rare mineral and has been found in paying quantities in only a few localities. The most important district is in the Ural Mts. Here the platinum had its original home in basic igneous rocks; these are in most cases the olivine rocks called dunites. Less frequently it is found in olivine-pyroxenites and rarely in gabbro. Its occurrence in these rocks is the result of the process of magmatic differentiation. In the Transvaal it has been found in basic derivatives of an original norite. In few instances does platinum occur in these rocks except in very small and scattered amounts. The productive deposits are the placers that are to be found in the neighborhood of the original platinum-bearing rocks. The associated minerals to be found with platinum are olivine, and the serpentine derived from its alteration, chromite, magnetite, zircon, corundum, and the various platinum metals.

Platinum was first found in the alluvial deposits of the river Pinto near Papayan in the Intendencia of Choco, Department of Cauca, Colombia, from whence it was taken to Europe in 1735. Here it first received its name from *platina*, silver-like (*platina del Pinto*).

The world's most important district is in the Ural Mts., Russia, where it was discovered in 1822. It is found here over a large area on the upper Tura River and especially on the Iss River, a tributary of the Tura. The region is in the province of Perm, east of the crest of the Urals. This district, of which Nizhne-Tagilsk is about the center, extends from Denezh-kin-Kamen, north of Bogoslovsk to south of Ekaterinburg. It has been found in the sand of the Ivalo River, northern Finland; sparingly in the sand of the Rhine; and in County Wicklow, Ireland. Platinum occurs also in Borneo; from New South Wales it is reported as occurring *in situ* in the Broken Hill district; also in a gold placer at Platina (Fifield), and elsewhere. In New Zealand from the Tayaka and Gorge Rivers in an area of dunite rocks. From placers on the east coast of Madagascar. From the Lydenburg and Waterburg districts, Transvaal.

Platinum has been reported as found in traces in the gold sands of Rutherford and Burke Counties, North Carolina. Occurs in the black sands of placers of California in a number of districts, especially in Trinity Co. and at Oroville, Butte Co.; similarly in Oregon at Cape Blanco (Port Orford), Curry Co., and elsewhere. In Canada it was found in the gold washings of Rivière-du-Loup and Rivière des Plantes, Beauce Co., Quebec. In British Columbia in the sands of a number of streams, especially in the Kamloops mining district along the Fraser and Tranquille Rivers; in the Similkameen mining district on Granite, Cedar and Slate Creeks, tributaries of the Tulameen River. It has also been found in this last section in fine grains in the peridotites of Olivine Mountain. In Alberta on the North Saskatchewan River near Edmonton.

Use. — Practically the only ore of platinum.

Iridium. Platiniridium. — Isometric. Iridium alloyed with platinum and other allied metals. Occurs usually in angular grains of a silver-white color. H. = 6–7. G. = 22·6–22·8. Very rare. With the platinum of the Ural Mts. and in Brazil; also in northern California. In the gold sands of Ava, near Mandalay, Burma.

IRIDOSMINE. Osmiridium.

Rhombohedral. Axis $c = 1·3823$. Usually in irregular flattened grains. Cleavage: $c(0001)$ perfect. Slightly malleable to nearly brittle. H. = 6–7. G. = 19·3–21·12. Luster metallic. Color tin-white to light steel-gray. Opaque.

Comp. — Iridium and osmium in different proportions. Some rhodium, platinum, ruthenium, and other metals are usually present.

Var. -- 1. *Nevyanskite.* II. = 7; G. = 18·8–19·5. In flat scales; color tin-white. Over 40 per cent of iridium. 2. *Siserskite.* In flat scales, often six-sided, color grayish white, steel-gray. G. = 20–21·2. Not over 30 per cent of iridium. Less common than the light-colored variety.

Diff. — Distinguished from platinum by greater hardness and by its lighter color.

Obs. — Occurs with platinum in the dependency of Choco, Colombia, and in the district of Ekaterinburg, Ural Mts. In New South Wales at Platina (Fifield) and elsewhere. In Brazil; rather frequently found in the gold sands of northern California and southern Oregon. From the Quesnel and Similkameen mining districts, British Columbia.

Palladium. — Isometric. Palladium, alloyed with a little platinum and iridium. Mostly in grains. H. = 4·5–5. G. = 11·3–11·8. Color whitish steel-gray. Occurs with platinum in Brazil; also from the Ural Mts. *Potarite* was described as a mercury-palladium amalgam from the diamond gravels of the Potaro River, Kangaruma district, British Guiana. Mineragraphic investigation shows that it consists of two materials, one isometric and the other anisotropic. Both are apparently palladium-mercury compounds. *Allopalladium* is an hexagonal modification of palladium from Tilkerode, Harz Mts. Shown by mineragraphic tests to be distinct from palladium and potarite.

Stibiopalladinite Pd_3Sb. Silver-white to steel-gray. H. = 4–5. G. = 9.5. In rounded grains with platinum ore near Potgietersrust, Transvaal.

IRON.

Isometric. Usually massive, rarely in crystals.

Cleavage: $a(100)$, perfect; also a lamellar structure $\parallel o(111)$ and $\parallel d(110)$. Fracture hackly. Malleable. H. = 4–5. G. = 7·3–7·8. Luster metallic. Color steel-gray to iron-black. Strongly magnetic.

Var. — 1. Terrestrial Iron. — Found in masses, occasionally of great size, as well as in small embedded particles, in basalt at Blaafjeld, Ovifak (or Uifak), Disco Island, West Greenland; also elsewhere on the same coast. This iron contains 1 to 2 per cent of Ni. In small grains with pyrrhotite in basalt from near Cassel, Hessen-Nassau, Germany. In minute spherules in feldspar from Cameron Township, Nipissing Dist., Ontario. Some other occurrences, usually classed as meteoric, may be in fact terrestrial.

A nickeliferous metallic iron (FeNi₂) called *awaruite* occurs in the drift of the Gorge River, which empties into Awarua Bay on the west coast of the south island of New Zealand; associated with gold, platinum, cassiterite, chromite; probably derived from a partially serpentinized peridotite. *Josephinite* is a nickel-iron (FeNi₃) from Oregon, occurring in stream gravel. Similar material from near Lillooet on the Fraser River, British Columbia, has been called *souesite*. Native iron also occurs sparingly in some basalts; reported from gold or platinum washings at various points.

2. Meteoric Iron. — Native iron also occurs in most meteorites, forming in some cases (*a*) the entire mass (*iron meteorites*); also (*b*) as a spongy, cellular matrix in which are embedded grains of chrysolite or other silicates (*siderolites*); (*c*) in grains or scales disseminated more or less freely throughout a stony matrix (*meteoric stones*). Rarely a meteorite consists of a single crystalline individual with numerous twinning lamellæ ∥ *o*(111). Cubic cleavage sometimes observed; also an octahedral, less often dodecahedral, lamellar structure. Etching with dilute nitric acid (or iodine) commonly develops a crystalline structure (called *Widmanstätten figures*) (Fig. 669); usually consisting of lines or bands crossing at various angles according to the direction of the section, at 60° if ∥ *o*(111), 90° ∥ *a*(100), etc. They are formed by the edges of crystalline plates, usually ∥ *o*, of the nickeliferous iron of different composition (*kamacite, tænite, plessite*), as shown by the fact that they are differently attacked by the acid. Irons with cubic structure and with twinning lamellæ have a series of fine lines corresponding to those developed by etching (*Neumann lines*). A damascene luster is also produced in some cases, due to quadrilateral depressions. Some irons show no distinct crystalline structure upon etching.

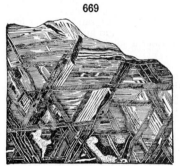

669

Glorieta Mt., New Mexico

The exterior of masses of meteoric iron is usually more or less deeply pitted with rounded thumblike depressions, and the surface at the time of fall is covered with a film of iron oxide in fine ridges showing lines of flow due to the melting caused by the heat developed by the resistance of the air; this film disappears when the iron is exposed to the weather.

Meteoric iron is always alloyed with nickel, which is usually present in amounts varying from 5 to 10 per cent, sometimes much more; small amounts of other metals, as cobalt, manganese, tin, copper, chromium, are also often present. Occluded gases can usually be detected. Graphite, in seams or nodules, also troilite (iron sulphide), schreibersite (iron-nickel phosphide) are common in masses of meteoric iron; diamond, daubreelite, etc., are rare. *Cohenite*, sometimes identified, is (Fe,Ni,Co)₃C in tin-white crystals.

Moissanite. — CSi. This material, originally produced artificially as *carborundum*, has been found occurring naturally as small green hexagonal plates in the meteoric iron of Cañon Diablo, Arizona.

II. SULPHIDES, SELENIDES, TELLURIDES, ARSENIDES, ANTIMONIDES

The sulphides, etc., fall into two groups according to the character of the positive element.

I. Sulphides, Selenides, Tellurides of the Semi-Metals.

II. Sulphides, Selenides, Tellurides, Arsenides, Antimonides of the Metals.

I. Sulphides, etc., of the Semi-Metals

This section includes one distinct group, the Stibnite Group, to which orpiment is related; the other species included stand alone.

REALGAR.

Monoclinic.   Axes $a : b : c = 0.7203 : 1 : 0.4858$; $\beta = 66° 15'$.

$$ll''', 110 \wedge 1\bar{1}0 = 66° 48'.$$

Crystals short prismatic; striated vertically.   Also granular, coarse or fine; compact; as an incrustation.

Cleavage: $b(010)$ and $c(001)$, fair.   Fracture small conchoidal.   Sectile. H. = 1·5–2.   G. = 3·56. $\alpha = 2.54, \beta = 2.68, \gamma = 2.70$.   Optically −.   Ax. pl. ‖ (010). $X \wedge c$ axis = +11°.   2V = 40°.   Strong dispersion, $\rho > v$. Luster resinous.   Color aurora-red or orange-yellow.   Streak varying from orange-red to aurora-red.   Transparent — translucent.   Disintegrates on exposure to light, changing to a mixture of As_2S_3 and As_2O_3.

670

Comp. — Arsenic monosulphide, AsS = Sulphur 29·9, arsenic 70·1 = 100.

Nagy-Ág

Pyr., etc. — In the closed tube melts and gives a dark red liquid when hot and a reddish yellow solid when cold; in the open tube (if heated very slowly) sulphurous fumes, and a white crystalline sublimate of arsenic trioxide.   B.B. on charcoal burns with a blue flame, emitting arsenical and sulphurous odors.   Soluble in caustic alkalies.

Artif. — Realgar is frequently noted as a sublimation product from furnaces roasting ores of arsenic.   Crystals are produced when arsenic sulphide is heated in a sealed tube with a solution of sodium bicarbonate.

Obs.   Realgar occurs commonly as a minor constituent of certain ore veins associated with orpiment and other arsenic minerals, with stibnite, and with lead, silver, and gold ores. It is also found in certain limestones or dolomites, in clay rocks, and as a volcanic sublimation product or a deposit from hot springs.

Realgar occurs with silver and lead ores at Felsöbánya (Baia Sprie), Kapnikbánya, and Nagy-Ág in Rumania; at Allchar near Rozsdan, northwest of Salonika, Macedonia.   It is found in fine crystals in the dolomite of the Binnental, Valais, Switzerland.   In Japan in the provinces of Rikuzen and Shimotsuke.   In the United States it is found in the Norris Geyser Basin, Yellowstone National Park, Wyoming, where it occurs with orpiment as a deposition from the hot waters.   From Mercur, Tooele Co., Utah, and at Manhattan, Nye Co., Nevada.   The name realgar is from the Arabic, Rahj al ghār, powder of the mine.

Use. — Was used in fireworks to give a brilliant white light when mixed with saltpeter and ignited.   The artificial material is now used for this purpose.

ORPIMENT.

Monoclinic.   Axes $a : b : c = 0.596 : 1 : 0.665$, $\beta = 89° 19'$.   Pseudo-orthorhombic.

Crystals small, rarely distinct.   Usually in foliated or columnar masses; sometimes with reniform surface.

Cleavage: $b(010)$ highly perfect, cleavage face vertically striated; $a(100)$ in traces; gliding-plane $c(001)$.   Sectile.   Cleavage laminæ flexible, inelastic. H. = 1·5–2.   G. = 3·4–3·5. $\alpha = 2.4, \beta = 2.8, \gamma = 3.0$.   Optically +.   Ax. pl. (001). $Z = a$ axis.   2E = 70°.   Strong dispersion, $\rho > v$.   Luster pearly on b (cleavage); elsewhere resinous.   Color lemon-yellow of several shades; streak the same, but paler.   Subtransparent — subtranslucent.

Comp. — Arsenic trisulphide, As_2S_3 = Sulphur 39·0, arsenic 61·0 = 100.

Pyr., etc. — Same as for realgar, p. 409.

Diff. — Distinguished by its fine yellow color, pearly luster, easy cleavage, and flexibility when in plates.

Artif. — Orpiment has been synthesized by heating solutions of arsenic with ammonium sulphocyanate in a sealed tube; also by the treatment under pressure of arsenic acid with hydrogen sulphide.

Obs. — Orpiment is very commonly associated with realgar, although somewhat rarer in its occurrence. Its results from the alteration of arsenic and silver minerals and also from realgar. It occurs in small crystals, embedded in clay at Tajowa near Neusohl (Ban Bystrica), Czechoslovakia. It is usually in foliated or fibrous masses, and in this form is found at Moldawa, in the Banat, Rumania; at Kapnikbánya and Felsöbánya (Baia Sprie), Rumania, it exists in metalliferous veins associated with realgar and native arsenic. Occurs at Allchar, near Rozsdan, northwest cf Salonika, Macedonia; from Balia, Asia Minor; a large deposit near Julamerk, Kurdistan. From Jozankei, Province of Ishikari, Island of Yezo, Japan. In the United States it is found in fine crystals at Mercur, Tooele Co., Utah, and also from Manhattan, Nye Co., Nevada. As a hot spring deposit at Steamboat Springs, Nevada, and with realgar in the Yellowstone National Park, Wyoming.

The name orpiment is a corruption of its Latin name auripigmentum, *golden paint*, given in allusion to the color, and also because the substance was supposed to contain gold.

Use. — For a pigment, in dyeing and in a preparation for the removal of hair from skins. The artificial material is largely used as a substitute for the mineral.

JEROMITE. A sulphide of arsenic containing some selenium. In globular masses occurring as a coating on rock fragments beneath iron hoods placed over vents from which sulphur dioxide gases were escaping from a portion of the United Verde copper mine at Jerome, Arizona, which was burning. Opaque. Isotropic. Black color. Cherry-red on thin edges.

Stibnite Group

		$a : b : c$
Stibnite	Sb_2S_3	0·9926 : 1 : 1·0179
Bismuthinite	Bi_2S_3	0·9679 : 1 : 0·9850
Guanajuatite	Bi_2Se_3	1 : 1 approx.

The species of the Stibnite Group crystallize in the orthorhombic system and have perfect brachypinacoidal cleavage, yielding flexible laminæ.

The species orpiment is in physical properties somewhat related to stibnite, but is monoclinic in crystallization. Groth notes that in a similar way, the oxide, As_2O_3, is monoclinic in claudetite, while the corresponding compound Sb_2O_3 (valentinite), is orthorhombic.

STIBNITE. Antimonite, Antimony Glance.

Orthorhombic. Axes $a : b : c = 0·9926 : 1 : 1·0179$.

mm''', 110 ∧ 1$\bar{1}$0 = 89° 34'.	bv, 010 ∧ 121 = 35° 8'.
pp', 111 ∧ $\bar{1}$11 = 71° 24$\frac{1}{2}$'.	$b\eta$, 010 ∧ 353 = 40° 10$\frac{1}{2}$'.
ss', 113 ∧ $\bar{1}$13 = 35° 52$\frac{1}{2}$'.	$b\tau$, 010 ∧ 343 = 46° 33'.
ss''', 113 ∧ 1$\bar{1}$3 = 35° 36'.	bp, 010 ∧ 111 = 54° 36'.

Crystals prismatic; striated or furrowed vertically; often curved or twisted (cf. p. 210). The better crystals are frequently very rich in faces. Common in confused aggregates or radiating groups of acicular crystals; massive, coarse or fine columnar, commonly bladed, less often granular to impalpable.

Cleavage: $b(010)$ highly perfect; less perfect cleavages parallel to (001), (101), (110), and (100). The planes of better cleavage are also gliding planes, movement on which yields at times curved and warped crystals. Slightly sectile. Fracture small sub-conchoidal. H. = 2. G. = 4·52–4·62. Luster metallic, highly splendent on cleavage or fresh crystalline surfaces. Color and streak lead-gray, inclining to steel-gray; subject to blackish tarnish, sometimes iridescent.

Comp. — Antimony trisulphide, Sb_2S_3 = Sulphur 28·3, antimony 71·7 = 100. Sometimes auriferous, also argentiferous.

Pyr., etc. — Fuses very easily (at 1), coloring the flame greenish blue. In the open tube sulphurous (SO_2) and antimonial (chiefly Sb_2O_4) fumes, the latter condensing as a white non-volatile sublimate on bottom of tube. On charcoal fuses, spreads out, gives sulphurous fumes, and coats the coal white with oxide of antimony; this coating treated in R.F. volatilizes and tinges the flame greenish blue. When pure, perfectly soluble in hydrochloric acid; in nitric acid decomposed with separation of antimony pentoxide.

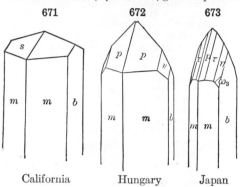

671   672   673

California   Hungary   Japan

Diff. — Distinguished (*e.g.*, from galena) by cleavage, color, softness; also by its fusibility and other blowpipe characters. It is harder than graphite. Resembles sometimes certain of the rarer sulphantimonites of lead, but yields no lead coating on charcoal.

Artif. — Stibnite, like orpiment, has been artificially produced by heating in a sealed tube, a solution of antimony with ammonium sulphocyanate; also by passing hydrogen sulphide at a red heat over compounds of antimony.

Obs. — Stibnite is the most common antimony mineral and the chief source of the metal. Although it is widely distributed, its occurrence in quantity is rare. It is of primary origin and found most commonly in veins with quartz. The contents of these veins were deposited from alkaline solutions at comparatively shallow depths. The veins frequently occur in granitic rocks. Sometimes found as beds in schists, as a replacement in limestone, in hot spring deposits, in deep-seated veins lying in or near intrusive rocks. Beside quartz, the associated minerals include other antimony compounds, galena, sphalerite, silver ores, pyrite, barite, calcite, cinnabar, realgar, gold, etc. Often alters to various antimony oxides.

Notable localities are: In Germany at Woltsberg in the Harz Mts., and near Arnsberg, Westphalia; in Rumania at Felsöbánya (Baia Sprie) and Kapnikbánya; in Czechoslovakia at Schemnitz (Selmeczbánya) and Kremnitz (Körmöczbánya); in Italy at Pereta and near Siena, Tuscany; in France at Lubilhac, Haute-Loire, and the neighboring locality of Massiac, Cantal and at La Lucette, Mayenne. Abundant in Province of Sarawak, Borneo; from Province of Puno, Peru. Magnificent groups of crystals were formerly found in the Province of Iyo, Island of Shikoku, Japan. The chief antimony deposits of the world are in the Province of Hunan and elsewhere in central and southern China, in Algeria (here largely oxidized ores), and in various portions of Mexico. Stibnite occurs sparingly in the United States, the only notable locality being at Hollister, Benito Co., California. Crystals occur at Manhattan, Nye Co., Nevada.

Use. — The most important ore of antimony.

Metastibnite. — An amorphous brick-red deposit of antimony trisulphide, Sb_2S_3, occurring with cinnabar and arsenic sulphide upon siliceous sinter at Steamboat Springs, Washoe Co., Nevada.

BISMUTHINITE. Bismuth Glance.

Orthorhombic. Rarely in acicular crystals. mm''', 110 $\wedge$ $1\overline{1}0$ = 88° 8′. Usually massive, foliated or fibrous.

Cleavage: $b(010)$ perfect. Somewhat sectile. H. = 2. G. = 6·4–6·5. Luster metallic. Streak and color lead-gray, inclining to tin-white, with a yellowish or iridescent tarnish. Opaque.

Comp. — Bismuth trisulphide, Bi_2S_3 = Sulphur 18·8, bismuth 81·2 = 100. Sometimes contains a little copper and iron.

Stibiobismuthinite is a bismuth sulphide from Nacozari, Sonora, Mexico, containing over 8 per cent Sb; *aurobismuthinite*, from the same locality, is a doubtful species containing gold.

Pyr., etc. — Fusibility = 1.   In the open tube sulphurous fumes, and a white sublimate which B.B. fuses into drops, brown while hot and opaque yellow on cooling.   On charcoal at first gives sulphurous fumes; then fuses with spurting, and coats the coal with yellow bismuth oxide; with potassium iodide and sulphur gives a yellow to bright red coating of bismuth iodide.   Dissolves readily in hot nitric acid, and a white precipitate of a basic salt falls on diluting with water.

Artif. — Bismuthinite has been produced artificially by treating the volatilized chloride of bismuth with hydrogen sulphide; in crystals by heating bismuth sulphide in a sealed tube with an alkaline sulphide.

Obs. — Bismuthinite is a comparatively rare mineral.   It is apparently confined in its occurrence to those deposits that have definite relations to igneous rocks and is usually strictly a primary mineral but at times it is found in the secondary sulphide zone.   In southern Norway it occurs with igneous rocks, associated with magnetite, garnet, calcite, pyrite, chalcopyrite, galena, sphalerite, etc.   Found also in tourmaline-bearing copper deposits and in tourmaline-quartz veins.   In Bolivia it is associated with tin and tungsten ores.

Bismuthinite is found in Saxony at Schneeberg and Altenberg; in Rumania at Rezbánya and Moravicza (Vaskö); in Sweden at the Bastnaës mine at Riddarhyttan, Vastmanland, and at Persberg, Vermland.   In England it occurs at various places in Cornwall, in Devonshire at Tavistock, and in Cumberland at Carrock Fell.   On Mount Shamrock, Queensland, and at Balhannon, Adelaide Co., South Australia.   The most important deposits of bismuthinite are in Bolivia, at San Baldamero, near Sorata, at Llallogua, and in the Huanina, Tazna, and Chorolque districts, Potosi.   In the United States it has been found at Haddam, Connecticut, with chrysoberyl; in Delaware Co., Pennsylvania; at Wickes. Jefferson Co., Montana; abundantly with garnet and barite in the Granite district, Beaver Co., Utah.

Use. — An ore of bismuth.

Guanajuatite.   Frenzelite.   Selenobismutite. — Bismuth selenide, Bi_2Se_3, sometimes with a small amount of sulphur replacing selenium.   In acicular crystals; also massive, granular, foliated or fibrous.   Cleavage: b(010) distinct.   H. = 2·5–3·5.   G. = 6·25–6·98. Luster metallic.   Color bluish gray.   From the Santa Catarina and La Industrial mines, near Guanajuato, Mexico.   Noted from near Salmon, Lemhi Co., Idaho.

TETRADYMITE.

Rhombohedral.   Crystals small, indistinct.   Commonly in bladed forms foliated to granular massive.

Cleavage: basal, perfect.   Laminæ flexible; not very sectile.   H. = 1·5–2; soils paper.   G. = 7·2–7·6.   Luster metallic, splendent.   Color pale steel-gray.

Comp. — Consists of bismuth and tellurium, with sometimes sulphur and a trace of selenium; the analyses for the most part afford the general formula $Bi_2(Te,S)_3$.

Var. — 1. *Free from sulphur.* Bi_2Te_3 = Tellurium 48·1, bismuth 51·9.   G. = 7·642 from Dahlonega.   2.   *Sulphurous.* $2Bi_2Te_3.Bi_2S_3$ = Tellurium 36·4, sulphur 4·6, bismuth 59·0 = 100.   This is the more common variety and includes the *tetradymite* in crystals from Schubkau.

Pyr. — In the open tube a white sublimate of tellurium dioxide, which B.B. fuses to colorless drops.   On charcoal fuses, gives white fumes, and entirely volatilizes; tinges the R.F. bluish green; coats the coal at first white (TeO_2), and finally orange-yellow (Bi_2O_3); some varieties give sulphurous and selenious odors.

Obs. — Bismuth tellurides are most commonly found in the gold-quartz veins, also in deposits formed near the surface by the action of hot waters on igneous rocks, rarely in contact metamorphic deposits.

Occurs at Schubkau northwest of Schemnitz (Selmeczbánya), Czechoslovakia; at Oravicza, Rumania.   Found in various places in Norway and at Bastnaës mine, near Riddarhyttan, Vastmanland, Sweden.   From Carrock Fell, Cumberland, and at Llanaber near Dolgelly, Merioneth, Wales.   At Mount Shamrock, Queensland; near Sorata, Bolivia.   In the United States, in Virginia in Spotsylvania, Stafford, and Fluvanna counties; at various points in North and South Carolina and in Georgia, near Dahlonega, Lumpkin Co.; near

Helena, Montana; at Goldhill, Boulder Co., Colorado. With hessite and altaite near Liddle Creek, Slocan district, British Columbia. Named from τετρ'δυμος, *fourfold*, in allusion to complex twin crystals sometimes observed.

Grünlingite. — Bi_4TeS_3. Massive. One distinct cleavage. Color, gray. G. = 7·321. From Carrock Fell, Cumberland, England. *Oruetite* is a similar mineral, Bi_8TeS_4, from Serrania de Ronda, Spain.

Joseïte. — A bismuth telluride (Te 80 per cent, also S and Se). G. = 7·9. San José, near Marianna, Minas Geraes, Brazil.

Wehrlite. — A foliated bismuth telluride (Te 30 per cent) of doubtful formula. G. = 8·4. From Börzsöny (Deutsch-Pilsen) near Esztergom, Hungary.

MOLYBDENITE.

Hexagonal. $c = 3·816$. Crystals hexagonal in form, tabular, or short prisms slightly tapering and horizontally striated. Commonly foliated, massive or in scales; also fine granular.

Cleavage: basal eminent. Laminæ very flexible, but not elastic. Sectile. H. = 1–1·5. G. = 4·7–4·8. Luster metallic. Color pure lead-gray; a bluish gray trace on paper, greenish gray on porcelain. Opaque. Feel greasy.

Comp. — Molybdenum disulphide, MoS_2 = Sulphur 40·0, molybdenum 60·0 = 100.

Pyr., etc. — In the open tube sulphurous fumes and a pale yellow crystalline sublimate of molybdenum trioxide (MoO_3). B.B. in the forceps infusible, imparts a yellowish green color to the flame; on charcoal the pulverized mineral gives in O.F. a strong odor of sulphur dioxide and coats the coal with crystals of molybdic oxide, yellow while hot, white on cooling; near the assay the coating is copper-red, and if the white coating be touched with an intermittent R.F., it assumes a beautiful azure-blue color. Decomposed by nitric acid, leaving a white or grayish residue.

Diff. — Much resembles graphite in softness and structure (see p. 397), but differs in color of streak on paper and porcelain and readily yields sulphur fumes on charcoal.

Artif. — Molybdenite has been made artificially by adding molybdic oxide to a fused mixture of potassium carbonate and sulphur; also by heating a mixture of molybdates and lime in an atmosphere of hydrochloric acid and hydrogen sulphide.

Obs. — Molybdenite is the most common mineral of molybdenum; it occurs widely, but never in large quantities. It is frequently in pneumatolytic contact deposits associated with cassiterite, scheelite, wolframite, fluorite, etc. Also found in pegmatite and quartz veins associated with granite, syenite, gneiss, etc. It is more rarely found in granular limestones and with garnet rocks.

Found with the tin ores of Schlaggenwald and Zinnwald, Bohemia, Czechoslovakia, and at Altenberg, Saxony. In Norway it occurs in considerable amount in Telemarken; also in fine specimens at Raade, near Moss, at Brevik in the Langesund district, and at Arendal. Occurs at Carrock Fell, Cumberland. It is found in Australia at Kingsgate and Deepwater in the New England Range, New South Wales; at Wolfram near Chillagoe, in Queensland. In South West Africa at Windhoek. In the United States it has been found in Maine, in large crystals at Blue Hill Bay and Camdage Farm; at Westmoreland, New Hampshire; at Haddam, Connecticut; in large crystals at Frankford, Philadelphia, Pennsylvania; in Okanogan and Chelan counties, Washington. In Ontario at Ross and elsewhere in Renfrew Co.; in Quebec at Aldfield, Pontiac Co., and at Wakefield, Hull Co.

Named from μό'λυβδος, *lead;* the name, first given to some substances containing lead, later included graphite and molybdenite, and even some compounds of antimony. The distinction between graphite and molybdenite was established by Scheele in 1778–79.

Use. — An important ore of molybdenum.

Tungstenite. — Probably WS_2. Earthy or foliated. Color and streak, dark lead-gray. H. = 2·5. G. = 7·4. Found at Emma mine, Salt Lake Co., Utah.

Patronite. Rizopatronite. — Complex composition, containing large amounts of a vanadium sulphide, perhaps VS_4. Amorphous. Color black. Occurs in a complex mixture of mineral substances among which are *quisqueite* and *bravoite*, at Minasragra, near Cerro de Pasco, Peru.

II. Sulphides, Selenides, Tellurides, Arsenides, Antimonides of the Metals

The sulphides of this second section fall into four divisions depending upon the proportion of the negative element present. These divisions with the groups belonging to them are as follows:

A. Basic Division

B. Monosulphides, Monotellurides, etc., R_2S, RS, etc.

1. *Galena Group.* Isometric-normal.
2. *Chalcocite Group.* Orthorhombic.
3. *Sphalerite Group.* Isometric-tetrahedral.
4. *Cinnabar — Wurtzite — Millerite Group.* Hexagonal and rhombohedral.

C. Intermediate Division

Embraces Bornite, $5Cu_2S.Fe_2S_3$; Linnæite, $CoS.Co_2S_3$; Chalcopyrite, $Cu_2S.Fe_2S_3$; etc.

D. Disulphides, Diarsenides, etc., RS_2, RAs_2, etc.

1. *Pyrite Group.* Isometric-pyritohedral.
2. *Marcasite Group.* Orthorhombic.

A. Basic Division

The basic division includes several rare basic compounds of silver, copper or nickel chiefly with antimony and arsenic. Of these the crystallization of dyscrasite and maucherite only is known.

DYSCRASITE. Antimonsilver.

Orthorhombic. Axes $a : b : c = 0.5775 : 1 : 0.6718$. Crystals rare, pseudo-hexagonal in angles $(mm''', 110 \wedge 1\bar{1}0 = 60° 1')$ and by twinning. Also massive. Cleavage, (001), (011), good. Fracture uneven. Sectile. H. = 3.5–4. G. = 9.44–10.02. Luster metallic. Color and streak silver-white, inclining to tin-white; sometimes tarnished yellow or blackish. Opaque.

Comp. — A silver antimonide Ag_3Sb = Antimony 27.1, silver 72.9 = 100.

X-ray study shows dyscrasite to be identical with artificial Ag_3Sb. Analyses show wide variations, owing perhaps to mixtures of dyscrasite with solid solutions of antimony and silver.

Pyr., etc. B.B. on charcoal fuses (1.5) to a globule, coating the coal with white antimony trioxide and finally giving a globule of almost pure silver. Soluble in nitric acid, leaving antimony trioxide.

Obs. — Dyscrasite occurs as the chief silver mineral at Wolfach on the Kinzig in the Black Forest, Baden, the associated minerals being galena, the ruby silvers, argentite, native silver, and barite. Crystals in calcite are found at Andreasberg in the Harz Mts. From Broken Hill, New South Wales; Chañarcillo near Copiapo, Atacama, Chile. Reported from the Reese River district, Nevada. Occurs with the silver ores at Cobalt, Ontario. Named from δυσκρασις, a bad alloy.

HUNTILITE, ANIMIKITE. The ores from Silver Islet, Lake Superior, apparently contain a silver arsenide (*huntilite*, Ag_3As?). *Animikite*, originally described as a silver antimonide, has been shown to be a mixture of silver, galena, and niccolite.

Horsfordite. — A silver-white, massive copper antimonide, probably Cu_6Sb (Sb 24 per cent). G. = 8·8. Asia Minor, near Mytilene.

Domeykite. — Copper arsenide, Cu_3As. X-ray study (Machatschki) of structures of domeykite and related compounds shows that natural domeykite has a different structure from the artificial material, and further the structure is not related to the structure of dyscrasite and is distinct from that of algodonite and whitneyite. Another study (Ramsdell) affirms definite structure for domeykite and algodonite but not for whitneyite. An earlier microscopic examination of polished specimens indicated that domeykite was a mixture of two constituents. Reniform and botryoidal; also massive, disseminated. G. = 7·2–7·75. Luster metallic. Color tin-white to steel-gray, readily tarnished. From several Chilian mines; also Zwickau, Saxony. In North America, in Michigan on Portage Lake, Houghton Co., and from the Mohawk mine, Keweenaw Co., with niccolite at Michipicoten Island, Lake Superior, Ontario.

MOHAWKITE. — Like domeykite, Cu_3As, with Ni and Co. Massive, fine granular to compact. Color gray with faint yellow tinge; tarnishes to dull purple. H. = 3·5. Brittle. G. = 8·07. Microscopic examination shows it to be a mixture. In Michigan from Mohawk mine, Keweenaw Co., and from Kearsage, Houghton Co. *Ledouxite* from the Mohawk mine said to be Cu_4As has been shown to be a mixture. *Dienerite* described as Ni_3As, found in a single cubic gray-white crystal near Radstadt, Salzburg, Austria.

Algodonite. — Copper arsenide, Cu_6As (As 16·5 per cent); G. = 7·62. Resembles domeykite. From Chile; also Lake Superior. Microscopic examination shows this mineral to be a mixture of two constituents. X-ray study (Machatschki) shows a mixture of two substances having respectively a cubic and a hexagonal lattice. Other investigators consider it a definite compound.

WHITNEYITE. — Copper arsenide, Cu_9As (As 11·6 per cent). G. = 8·4–8·6. Color pale reddish white. From Houghton Co., Michigan; Sonora, Lower California. Both microscopic and X-ray study shows it to be a mixture.

CHILENITE. Perhaps Ag_6Bi. Copiapo, Chile. Shown to be an intergrowth of native silver and cuprite.

COCINERITE. Copper, silver sulphide, Cu_4AgS. Massive. Color silver-gray, tarnishing black. H. = 2·5. G. = 6·1. From Cocinera mine, Ramos, San Luis Potosi, Mexico.

Stützite. — A rare silver telluride (Ag_4Te?). From Nagy-Ag, Rumania.

Rickardite. — Cu_4Te_3. Massive. H. = 3·5. G. = 7·5. Color deep purple, dulling on exposure. Fusible. Found at Vulcan, Colorado. Has also been noted in small amounts at the Empress Josephine mine, Bonanza, Colorado; in copper ores from Warren, Arizona, and at the San Sebastian mine, Salvador.

WEISSITE. Cu_5Te_3. Massive. Dark bluish black on fresh surface, tarnishes to deep black. Black streak. Luster shiny-metallic. H. = 3. G. = 6. Associated with pyrite, tellurium, sylvanite, petzite, rickardite, etc. in the Good Hope and Mammoth Chimney mines at Vulcan, Gunnison Co., Colorado.

Maucherite. — Ni_3As_2 or Ni_4As_3. Tetragonal. Habit, square tabular. H. = 5. G. = 7·83. Color reddish silver-white tarnishing to gray copper-red. Streak blackish gray. Easily fusible. From Eisleben, Thuringia. Also found in the mixture called *temiskamite* from Elk Lake, Ontario. The furnace product, *placodine*, is identical with *maucherite*.

B. Monosulphides, Monotellurides, etc., R_2S, RS, etc.

1. *Isometric*

Galena Group		Argentite Group	
Galena	PbS	Argentite	Ag_2S
Also,	$(Pb,Cu_2)S$, $(Cu_2,Pb)S$	Hessite	Ag_2Te
Altaite	PbTe	Aquilarite	Ag_2Se
Clausthalite	PbSe	Naumannite	$(Ag_2,Pb)Se$
		Eucairite	$Cu_2Se.Ag_2Se$

GALENA. GALENITE. Lead Glance.

Isometric. Structure according to a face-centered cubic lattice. Commonly in cubes, or cubo-octahedrons, less often octahedral. Also in skeleton crystals, reticulated, tabular. Twins: tw. pl. *o*(111), both contact- and penetration-twins (Figs. 427, 430, pp. 186, 187), sometimes repeated; twin crystals often tabular ‖ *o*. Also other tw. planes giving polysynthetic tw. lamellæ. Massive cleavable, coarse or fine granular, to impalpable; occasionally fibrous or plumose.

674   675   676   677

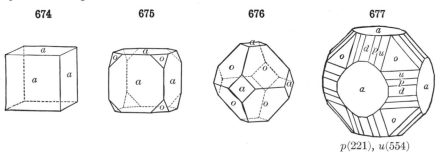

p(221), *u*(554)

Cleavage: cubic, highly perfect; less often an octahedral parting. Fracture flat subconchoidal or even. H. = 2·5–2·75. G. = 7·4–7·6. Luster metallic. Color and streak pure lead-gray. Opaque.

Comp. — Lead sulphide, PbS = Sulphur 13·4, lead 86·6 = 100. Often contains silver, and occasionally selenium, zinc, cadmium, antimony, bismuth, copper, as sulphides; besides, also, sometimes native silver and gold.

Var. — 1. *Ordinary.* (*a*) Crystallized; (*b*) somewhat fibrous and plumose; (*c*) cleavable, granular coarse or fine; (*d*) crypto-crystalline. The variety with octahedral parting is rare; in it the usual cubic cleavage is obtained readily after heating to 200° or 300°; the peculiar parting may be connected with the bismuth usually present. One variety showing octahedral parting contained a small amount of tellurium.

2. *Argentiferous.* All galena is more or less argentiferous, and no external characters serve to distinguish the kinds that are much so from those that are not. The silver is detected by cupellation, and may amount from a few thousandths of one per cent to one per cent or more; when mined for silver it ranks as a *silver ore.*

3. Containing arsenic, or antimony, or a compound of these metals, as impurity. Here belong *bleischweif*, an almost massive type from Klaustal, Harz Mts., with 0·22 Sb, and *steinmannite* from Přibram, Bohemia, with both arsenic and antimony.

Pyr. — In the open tube gives sulphurous fumes. B.B. on charcoal fuses, emits sulphurous fumes, coats the coal yellow near the assay (PbO) and white with a bluish border at a distance (PbSO₃, chiefly), and yields a globule of metallic lead. Decomposed by strong nitric acid with the separation of some sulphur and the formation of lead sulphate.

Diff. — Distinguished, except in very fine granular varieties, by its cubic cleavage; the color and the high specific gravity are characteristic; also the blowpipe reactions.

Alter. — Galena may alter, giving rise to various secondary lead minerals, as cerussite, anglesite, pyromorphite, mimetite, phosgenite, etc. On the other hand, it has been noted in pseudomorphs after cerussite, anglesite, pyromorphite, etc.

Artif. — Crystallized galena has been formed in numerous ways. In nature it is apparently commonly formed by hydrochemical reactions perhaps similar to the following laboratory methods: galena was produced by allowing a mixture of lead chloride, sodium bicarbonate and a solution of hydrogen sulphide to remain in a sealed tube for several months. Pyrite or marcasite heated with a solution of lead chloride will produce galena; a solution of lead nitrate when heated with ammonium sulphydrate will yield galena. Galena is frequently observed in furnace slags.

Obs. — Galena is one of the most widely distributed of the metallic sulphides. It is formed commonly by hydrochemical reactions, and it occurs in beds and veins, both in crystalline and non-crystalline rocks. The veins are commonly in eruptive rocks or closely associated with their occurrence. Associated minerals are most frequently sphalerite, chalcopyrite, pyrite, tetrahedrite, bournonite and other sulpho-salts, silver ores, quartz, calcite, dolomite, and other carbonates, barite, fluorite, etc. Galena often carries a notable amount of silver, and this together with its frequent association with silver minerals makes its deposits important sources of that metal. Galena also occurs in contact metamorphic deposits. A very important type of occurrence is that in which more or less irregular bodies occur as replacement deposits in limestone or dolomite rocks. The dolomitization of a limestone frequently accompanies the formation of such ore bodies. These may or may not show an association with intrusive rocks. The associated minerals include almost invariably sphalerite, and also commonly smithsonite, limonite, calcite, dolomite, etc. An unusual type of deposit occurs near Commern not far from Aix-la-Chapelle in Prussia where the galena occurs in concretions in the variegated sandstone. Various secondary lead minerals, in particular cerussite and anglesite, frequently accompany galena.

Only those occurrences of galena which are of importance because of their commercial value or because of the mineralogical interest of their specimens can be mentioned here. From Schemnitz (Selmeczbánya), Czechoslovakia; in the silver mines of Přibram and Mies, Bohemia; in Rumania at Kapnikbánya and O-Rodna; from Bleiberg, Carinthia, Austria. In Germany it occurs in fine crystals in the silver-lead veins of Freiberg, Saxony; from near Ems, Hessen-Nassau, and at Dillenburg; in Westphalia in unusual crystals from the Gonderbach mine, near Laaspe, also from Müsen; very common and often in fine specimens from the silver mines of the Harz Mts., especially at Andreasberg, Neudorf, Klaustal, and Zellerfeld, and at Pfaffen Berg near Bad-Sachsa. In Italy it has been found at Bottino near Seravezza, Tuscany, and also with the lavas of Vesuvius. Notable localities in France are at the mines of Pontgibaud, Puy-de-Dôme; those of Pont-Péan near Bruz, Ille-et-Vilaine, and at Poullaouen, Finistère. Fine specimens come from Truro and Liskeard in Cornwall, at Alston Moor in Cumberland, and at Weardale, Durham; in Scotland at Wanlockhead, Dumfries. Galena is found also in many places in Australia, Chile, Bolivia, Peru, etc.

Extensive deposits of galena in the United States are to be found in Missouri, Illinois, Iowa, and Wisconsin. The ore usually occurs filling cavities in stratified limestone, of different periods from Silurian to Carboniferous. It is associated with sphalerite, smithsonite, calcite, pyrite, etc. The Missouri mines are situated in three districts in the southern part of the state: (1) southeastern, chiefly in St. François, Washington, and Madison counties, (2) central, (3) southwestern of Joplin district, the latter producing chiefly zinc. Other districts in the upper Mississippi Valley are found in southwestern Wisconsin, at Mineral Point and Schullsburg, in northwestern Illinois at Galena, in eastern Iowa. Found in crystals with calcite and chalcopyrite at Rossie, St. Lawrence Co., New York, and also at Ellenville, Ulster Co.; in Pennsylvania at Phœnixville, and elsewhere. In fine crystals from Pitcher, Oklahoma. In Colorado, at Leadville and Aspen there are productive mines of argentiferous galena, also at Georgetown, the San Juan district, and elsewhere; fine specimens come from Breckenridge, Summit Co., and Lake City, Hinsdale Co. Mined for silver in the Cœur d'Alene region in Idaho; at the Park City and Tintic districts, Utah.

The name galena is from the Latin *galena* (Greek γαλῆνη), a name given to lead ore or the dross from melted lead.

Use. — The most important ore of lead and frequently a valuable ore of silver.

CUPROPLUMBITE. A massive mineral, from Chile, varying in characters from galena to those of chalcocite and covellite; composition, $Cu_2S.2PbS(?)$. Material classed here from Butte, Montana, gave formula, $5Cu_2S.PbS$. *Alisonite* is massive, deep indigo-blue quickly tarnishing; corresponds to $3Cu_2S.PbS$. From Mina Grande, Chile. Whether these and similar minerals represent definite homogeneous compounds, or only ill-defined alteration-products, is uncertain, and if so it is not clear whether they should be classed with isometric galena or with orthorhombic chalcocite.

Altaite. — Lead telluride, PbTe. Atomic structure like galena. Rarely in cubic or octahedral crystals, usually massive with cubic cleavage. H. = 3. G. = 8·16. Color tin-white, with yellowish tinge tarnishing to bronze-yellow. From near Ziryanovsk in the Altai Mts., Siberia, with hessite; Coquimbo, Chile; in North Carolina at King's Mountain mine, Gaston Co.; in Colorado at Goldhill, Boulder Co.; in the Organ Mts., near Las Cruces, Dona Ana Co., New Mexico; various localities in California.

Clausthalite. — Lead selenide, PbSe. Hg and Pt noted in material from Tilkerode. Atomic structure like galena. Commonly in fine granular masses resembling galena. Cleavage: cubic. G. = 7·6-8·8. Color lead-gray, somewhat bluish. Occurs in the Harz Mts., with hematite near the diabase rocks of Klaustal, Lerbach, Tilkerode, and Zorge. In the Cerro de Cacheuta, on the Mendoza River, Argentina. The following names have been given to impure clausthalite from the Harz Mts.: *tilkerodite*, a mixture of clausthalite, cobaltite, etc.; *lerbachite*, clausthalite and tiemannite; *zorgite*, clausthalite and umangite.

ARGENTITE. Silver Glance.

Isometric. Crystals often octahedral, also cubic; often distorted, frequently grouped in reticulated or arborescent forms; also filiform. Massive; embedded; as a coating.

X-ray and thermal studies have shown that argentite has an isometric structure only at temperatures above 180° C. The dimensions of its cubic cell are different from those of galena, and the two cannot be considered isomorphous. At ordinary temperatures, argentite shows an orthorhombic structure which is identical with that of acanthite. Argentite and acanthite, therefore, represent respectively the high and low temperature forms of Ag_2S.

Cleavage: $a(100)$, $d(110)$ in traces. Fracture small subconchoidal. Perfectly sectile. H. = 2–2·5. G. = 7·20–7·36. Luster metallic. Color and streak blackish lead-gray; streak shining. Opaque. Often alters on the surface to a black earthy sulphide.

Comp. — Silver sulphide, Ag_2S = Sulphur 12·9, silver 87·1 = 100.

Pyr., etc. — In the open tube gives off sulphurous fumes. B.B. on charcoal fuses with intumescence in O.F., emitting sulphurous fumes, and yielding a globule of silver.

Diff. — Distinguished from other sulphides by being readily cut with a knife; also by yielding metallic silver on charcoal.

Alter. — Alters to native silver, silver sulpho-salts, etc.

Artif. — Argentite is very easily prepared artificially and in numerous ways. Sulphur, sulphur dioxide or hydrogen sulphide will act upon metallic silver or any of its common compounds, either in solution or as solids, to produce silver sulphide.

Obs. — Argentite is the most important primary mineral of silver. It occurs at times in large masses and also as microscopic inclusions in galena and probably in other sulphide ores. It also may occur as a secondary mineral. It is associated with other silver minerals and also with galena, tetrahedrite, chalcopyrite, bornite, pyrite, cobalt and nickel ores, limonite, calcite, quartz, etc.

Notable occurrences of argentite are as follows: in Czechoslovakia at Schemnitz (Selmeczbánya); in Bohemia at Joachimstal (Jáchymov); in Saxony in the Freiberg district in fine crystals, and associated with the cobalt and nickel minerals of Schneeberg, Annaberg, Marienberg, and Johanngeorgenstadt; in the Harz Mts., especially at Andreasberg. On Sardinia near Sarrabus and with native silver at Kongsberg, Norway. In Cornwall, especially at Liskeard. Found in Bolivia at Colquechaca, etc., in Peru; in Chile at Chañarcillo, Atacama, etc. Very common in the silver mines of Mexico, especially in Guanajuato and Zacatecas, and at Arizpe, Sonora. In the United States argentite is found with the copper ores of Butte, Montana; frequently observed in the silver districts of Colorado, as at Aspen, Leadville, and in the mines of the San Juan district; in Nevada was found in large amounts at the Comstock Lode, and at Tonopah.

Use. — An important ore of silver.

JALPAITE. Originally found at Jalpa, Mexico, and considered to be a cupriferous argentite. Material from the Altai Mts. yields the formula $3Ag_2S.Cu_2S$, and is birefringent and therefore pseudo-isometric.

Hessite. — Silver telluride, Ag_2Te. Isometric. Usually massive, compact, or fine-grained. Cleavage indistinct. Somewhat sectile. H. = 2·5–3. G. = 8·31–8·45. Color between lead-gray and steel-gray. From the Altai Mts.; in Rumania at Nagy-Ág, Rézbánya, and at Botés, near Zalatna; Chile near Arqueros, Coquimbo. In Mexico at San Sebastian, Jalisco. In the United States, Calaveras and Nevada counties, California; Boulder, Eagle and San Juan counties, Colorado. This species also often contains gold and thus graduates toward petzite.

Petzite. — $(Ag,Au)_2Te$ with $Ag : Au = 3 : 1$. Massive; granular to compact. Slightly sectile to brittle. H. = 2·5–3. G. = 8·7–9·02. Color steel-gray to iron-black; tarnishing.

From Nagy-Ág, Rumania; Kalgoorlie, West Australia. At various localities in Boulder Co., Colorado; from Calaveras and Tuolumne counties, California.

Aguilarite. — Silver selenide, Ag_2S and $Ag_2(S,Se)$.   In skeleton dodecahedral crystals. Sectile.   G. =· 7·586.   Color iron-black.   From Guanajuato, Mexico.

Naumannite. — Silver-lead selenide $(Ag_2,Pb)Se$.   In cubic crystals; also massive, granular, in thin plates.   Cleavage: cubic.   G. = 8·0.   Color and streak iron-black. From Tilkerode in the Harz Mts., Germany.   Formerly found in quantity at De Lamar, Owyhee Co., Idaho, and considered to be argentite.   Very sectile and malleable.   H. = 2·5. G. = 6·5–7.

Berzelianite. — Copper selenide, Cu_2Se.   Isometric atomic structure.   In thin dendritic crusts and disseminated.   G. = 6·71.   Color silver-white, tarnishing.   From Skrikerum, near Tryserum, Kalmar, Sweden; Lerbach, in the Harz Mts., Germany.

Eucairite. — $Cu_2Se.Ag_2Se$.   Isometric.   Massive, granular.   G. = 7·50.   Color between silver-white and lead-gray.   From the Skrikerum copper mine, near Tryserum, Kalmar, Sweden; from near Copiapo, Atacama, Chile; in the Sierra de Umango, La Rioja, Argentina.

Crookesite. — Selenide of copper and thallium, also silver (1–5 per cent), $(Cu,Tl,Ag)_2Se$. Massive, compact.   G. = 6·9.   Luster metallic.   Color lead-gray.   From the mine of Skrikerum, near Tryserum, Kalmar, Sweden.

Umangite. — $CuSe.Cu_2Se$.   Massive, fine-granular to compact.   H. = 3.   G. = 5·620. Color dark cherry-red.   From La Rioja, Argentina.

KLOCKMANNITE.   CuSe.   Perhaps hexagonal and isomorphous with covellite.   Granular.   Basal cleavage.   Slate-gray color.   From Sierra de Umango, Argentina, and the Harz Mts., Germany.   Stated to be distinct from umangite.

PENROSEITE.   Perhaps $3CuSe.2PbSe_2.5(Ni,Co)Se_2$ Orthorhombic.   Perfect cleavages || (001), (100), (010); distinct (110).   Radiating columnar structure.   Lead-gray color. G. = 6·93.   H. = 3.   Probably from Colquechaca, Bolivia.

2. *Chalcocite Group*

		$a : b : c$
Chalcocite	Cu_2S	$0·5822 : 1 : 0·9701$
Stromeyerite	$Ag_2S.Cu_2S$	$0·5822 : 1 : 0·9668$
Sternbergite	$Ag_2S.Fe_4S_5$	$0·5832 : 1 : 0·8391$
Acanthite	Ag_2S	$0·6886 : 1 : 0·9944$

The species of the CHALCOCITE GROUP crystallize in the orthorhombic system with a prismatic angle approximating to 60°; they are hence pseudo-hexagonal in form, especially when twinned.   The group is parallel to the Argentite Group, since Cu_2S like Ag_2S has a high temperature isometric form and a low temperature orthorhombic form.   Some authors include dyscrasite here (see p. 414).

CHALCOCITE.   Copper Glance.   Redruthite.

Orthorhombic.   Axes $a : b : c = 0·5822 : 1 : 0·9701$.

mm''', 110 ∧ 1$\bar{1}$0 = 60° 25'. cp, 001 ∧ 111 = 62° 35$\frac{1}{2}$'.
dd', 021 ∧ 0$\bar{2}$1 = 125° 28'. pp''', 111 ∧ 1$\bar{1}$1 = 53° 3$\frac{1}{2}$'.

Crystals pseudo-hexagonal in angle, also by twinning (tw. pl. $m(110)$). Rarely twinned on (032) or (112).   Often massive, structure granular to compact and impalpable.

Cu_2S is dimorphous.   Above 91° C. the orthorhombic form changes to an isometric modification.   Artif. crystals of the latter show cube and octahedron with twinning on (111).

Cleavage: $m(110)$ indistinct; etching of orientated crystals develops cleavages parallel to the three pinacoids.   Fracture conchoidal.   Rather sectile.   H. = 2·5–3.   G. = 5·5–5·8 (5·785 for artif. mineral).   Luster metallic.   Color and streak blackish lead-gray, often tarnished blue or green, dull. Opaque.

Comp. — Cuprous sulphide, Cu_2S = Sulphur $20 \cdot 2$, copper $79 \cdot 8 = 100$. Sometimes iron in small amount is present, also silver.

It has been shown experimentally that Cu_2S may take up in solid solution as much as 8 per cent CuS. Such absorption is accompanied by a lowering of the gravity, a darkening of the color, and a raising of the temperature of transformation to the isometric structure.

Pyr., etc. — In the open tube gives sulphurous fumes. B.B. on charcoal melts to a globule, which boils with spurting; the fine powder roasted at a low temperature on charcoal, then heated in R.F., yields a globule of metallic copper. Soluble in nitric acid.

Diff. — Resembles argentite but much more brittle; bornite has a different color on the fresh fracture and becomes magnetic B.B.

Artif. — Chalcocite has been prepared artificially by heating the vapors of cuprous chloride and hydrogen sulphide or by the treatment of cupric oxide with hydrogen sulphide; also by the heating of cupric solutions with ammonium sulphocyanate in a sealed tube.

Alter. — Alters to native copper, chalcopyrite, bornite, covellite, malachite, and azurite. Pseudomorphic after chalcopyrite, bornite, pyrite, galena and millerite.

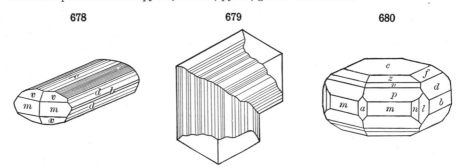

678   679   680

Obs. — Chalcocite is a very valuable ore of copper and is widespread in its occurrence. The origin of chalcocite has been the subject of many investigations. The problem is complex; it is apparent that the deposits differ widely and the reactions involved may be of many kinds. In the majority of cases it is to be found in the enriched sulphide zone of copper veins and is clearly secondary in origin. It has been formed in various ways by the action of descending copper-bearing solutions upon the original sulphide minerals of the veins. In other cases, however, chalcocite is primary and has been deposited as a direct precipitation from ascending waters. A third mode has been described in which ascending alkaline solutions have acted upon bornite to form chalcocite. It is associated commonly with chalcopyrite, bornite, covellite, tetrahedrite, and enargite, also with pyrite, sphalerite, galena, siderite, cuprite, malachite, azurite, etc. It sometimes carries valuable amounts of silver.

Cornwall affords splendid crystals, especially the districts of St. Just, St. Ives, Camborne, and Redruth (*redruthite*). Also in crystals from near Bogolovsk, Ural Mts., Russia. From Dognacska, Rumania. It is found in the copper ores of Montecatini, Tuscany. Occurs in large amounts at Tsumeb, near Otavi, South West Africa. From Mindouli, French Congo. From Chile, Peru, and Mexico. In the United States exceptional crystals came from Bristol, Connecticut. Also found in large amounts at Butte, Montana, and at Kennecott, in the Copper River district, Alaska. Occurs also in various copper mines, as at Ducktown, Tennessee, at Miami, etc., Arizona, and in Nevada and California.

Use. — An important ore of copper.

Stromeyerite. — $(Ag,Cu)_2S$, or $Ag_2S.Cu_2S$. Rarely in orthorhombic crystals, often twinned. Commonly massive, compact. H. $= 2 \cdot 5$–3. G. $= 6 \cdot 15$–$6 \cdot 3$. Luster metallic. Color and streak dark steel-gray. Stromeyerite is found in copper-silver veins and has apparently been usually formed by the reaction of descending silver solutions upon bornite, with which it is often intimately intergrown. The associated minerals include chalcopyrite, bornite, chalcocite, tetrahedrite, galena, sphalerite, pyrite, barite, and calcite. Found at Zmeyewskaja-Gora (Schlangén-Berg) near Zmyeinogorsk, Altai, Siberia; at Rudelstadt and Kupferberg, Silesia. From Mt. Lyell, Tasmania. From various localities in Chile. In Arizona at the Heintzelman and Silver King mines, Pinal Co.; at Butte, Montana. In Colorado in Gilpin and Ouray counties. In British Columbia at the Silver King mine, Nelson district, and from Cobalt, etc., in Ontario.

Cubanite. Chalmersite. — $Cu_2S.Fe_4S_5$. Orthorhombic. Axial ratio near that of chalcocite. In thin elongated prisms, vertically striated. Twins common with (110) as tw. pl., resembling chalcocite. H. = 3·5. G. = 4·7. Color brass- to bronze-yellow. Strongly magnetic. *Cubanite* was originally described from Barracanao, Cuba, later from Tunaberg and Kafveltorp in Sweden. *Chalmersite* came from the Morro Velho gold mine, Minas Geraes, Brazil and Prince William Sound, Alaska. The two appear to be identical.

STERNBERGITE.

Orthorhombic. Crystals tabular ‖ c(001). Commonly in fan-like aggregations; twins, tw. pl. m(110). Cleavage: c(001), highly perfect. Thin laminæ flexible, like tin-foil. H. = 1–1·5. G. = 4·215. Luster metallic. Color pinchbeck-brown. Streak black. Opaque.

Comp. — $AgFe_2S_3$ or $Ag_2S.Fe_4S_5$ = Sulphur 30·4, silver 34·2, iron 35·4 = 100.

It appears probable that small and varying amounts of FeS and S may be held in solid solution, and that *frieseite* and *argentopyrite* from Joachimstal and *argyropyrite* from Freiberg are varieties of sternbergite.

Obs. — Occurs with pyrargyrite and stephanite at Joachimstal, Bohemia, and Johanngeorgenstadt, Saxony.

Acanthite. — Silver sulphide, Ag_2S, like argentite. For relations to argentite, see under that mineral, p. 418. Orthorhombic. In slender prismatic crystals. Sectile. G. = 7·2–7·3. Color iron-black. Found at Joachimstal (Jáchymov), Bohemia; in Saxony near Freiberg and elsewhere. In Chihuahua and Zacatecas, Mexico. Reported in Colorado from Georgetown and Rico.

3. *Sphalerite Group.* RS. Isometric-tetrahedral

Sphalerite	ZnS	Onofrite	Hg(S,Se)	
Metacinnabarite	HgS	Coloradoite	HgTe	Massive
Guadalcazarite	(Hg,Zn)S			
Tiemannite	HgSo			

The SPHALERITE GROUP embraces a number of sulphides, selenides, etc., of zinc, mercury, and manganese. These are isometric-tetrahedral in crystallization. X-ray study shows the similarity in atomic structure of the above members of the Sphalerite Group.

SPHALERITE, ZINC BLENDE or BLENDE. Black-Jack, Mock-Lead, False Galena.

Isometric-tetrahedral. Atomic structure similar to that of diamond (cf. p. 40), with Zn and S atoms alternately taking the position of the C

681

682

683

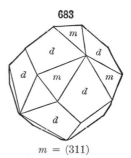

$m = (311)$

atoms. Often in tetrahedrons, the positive and negative forms frequently showing differences in luster, etching lines, etc. Cube, dodecahedron, and tristetrahedron forms also present at times.   Twins common: tw. pl. $o(111)$; twinning often repeated, sometimes as polysynthetic lamellæ.   Crystals often distorted or rounded.   Commonly massive cleavable, coarse to fine granular and compact; also foliated, sometimes fibrous and radiated or plumose; also botryoidal and other imitative shapes.   Cryptocrystalline to amorphous, the latter sometimes as a powder.

Cleavage: dodecahedral, highly perfect.   Fracture conchoidal.   Brittle. H. = 3·5–4.   G. = 3·9–4·1; 4·063 white, New Jersey.   Luster resinous to adamantine.   Color commonly yellow, brown, black; also red, green to white, and when pure nearly colorless.   Streak brownish to light yellow and white. Transparent to translucent.   Refractive index high: $n = 2·370$–$2·428$, rising with increase of iron contents.   Some varieties will phosphoresce when scratched, etc.

Comp. — Zinc sulphide, ZnS = Sulphur 33, zinc 67 = 100.   Often containing iron and manganese, and sometimes cadmium, mercury and rarely lead and tin.   Also sometimes contains traces of indium, gallium and thallium; may be argentiferous and auriferous.

Var. — 1.   *Ordinary.*   Containing little or no iron; from colorless white to yellowish brown, sometimes green; G. = 4·0–4·1.   The red or reddish brown transparent crystallized kinds are sometimes called *ruby blende* or *ruby zinc.*   The massive cleavable forms are the most common, varying from coarse to fine granular; also cryptocrystalline.   *Schalenblende* is a closely compact variety, of a pale liver-brown color, in concentric layers with reniform surface; galena and marcasite are often interstratified.   The fibrous forms are chiefly wurtzite.   A soft white amorphous form of zinc sulphide occurs in Cherokee Co., Kansas.

2.   *Ferriferous*: *Marmatite.*   Containing up to 20 per cent of iron; dark-brown to black; G. = 3·9–4·05.   The proportion of FeS to ZnS varies from 1 : 5 to 1 : 2, and the last ratio is that of the *christophite* of Breithaupt, a brilliant black sphalerite from St. Christophe mine, at Breitenbrunn, having G. = 3·91–3·923.

3.   *Cadmiferous*: *Přibramite.*   The amount of cadmium present in any sphalerite thus far analyzed is less than 5 per cent.

Pyr., etc. — Difficultly fusible.   In the open tube yields sulphurous fumes, and generally changes color.   B.B. on charcoal, in R.F., gives a coating of zinc oxide, which is yellow while hot and white after cooling.   If cadmium is present a reddish brown coating of cadmium oxide will form first.   With cobalt solution the zinc oxide coating gives a green color when heated in O.F.   Most varieties, after roasting, give with borax a reaction for iron.   Dissolves in hydrochloric acid with evolution of hydrogen sulphide.

Diff. — Varies widely in color and appearance, but distinguished by the resinous luster in all but deep black varieties; usually exhibits distinct cleavage; nearly infusible B.B.; yields a zinc oxide coating on charcoal.

Artif. — Sphalerite has been artificially formed by heating zinc solutions in hydrogen sulphide inclosed in a sealed tube; also by passing hydrogen sulphide over heated zinc chloride.

Obs. — Sphalerite is the most common mineral of zinc; in its occurrence and mode of origin it is closely allied with galena.   It is found in contact metamorphic deposits where the sulphide ores have been derived from some igneous intrusion and deposited usually by replacement in some adjacent rock, especially in the sedimentary types.   A very important kind of deposit is to be found in limestone or dolomitic rocks where the ore bodies have been formed by replacement processes.   It is found in veins in eruptive, sedimentary, and metamorphic rocks, associated with pyrite, chalcopyrite, galena, tetrahedrite, silver ores, calcite, dolomite, barite, fluorite, etc.   Of the two forms of zinc sulphide, sphalerite is the one which is stable below 1020° C., whereas wurtzite is the stable form at higher temperatures.   ZnS is deposited from alkaline solutions as sphalerite; from acid solutions both forms are deposited, the amount of sphalerite increasing with the temperature whereas that of wurtzite increases with the acidity of the solution.

Only those occurrences can be mentioned here that are notable as deposits of zinc ore or for the fine specimens that they furnish.   Some of the chief localities for crystallized sphalerite are: in Czechoslovakia in Bohemia at Příbram and at Schlaggenwald and Schemnitz (Selmeczbánya); in Rumania at Kapnikbánya and Felsöbánya (Baia Sprie); in Transylvania, at O-Rodna; at Neudorf in the Harz Mts.; at Ems in Hessen-Nassau; at Aacher

(Aix-la-Chapelle), Rhenish Prussia. In fine specimens at Bottino near Seravezza, Tuscany; in the Binnental, Switzerland, in fine isolated crystals in cavities in dolomite. Fine specimens come from France, at Laffrey, Isère, and Pont-Péan near Bruz, Ille-et-Vilaine. A beautiful transparent variety occurs at Picos de Europa, Santander, Spain. From Alston Moor, Cumberland; from St. Agnes and elsewhere in Cornwall; from Laxey, Isle of Man; from Weardale in Durham; in Scotland at Wanlockhead, Dumfries. Large beds occur at Åmmeberg, on Lake Vetter, Örebo, Sweden. In Japan large crystals are found at the Ani mine, Kayakusa, Ugo, and at Shiraita, Echigo. Transparent material of fine color comes from the Chivera mine, Cananea, Sonora, Mexico. In the United States it is found in many localities, the important ore districts being in Missouri, Colorado, Montana, Wisconsin, Idaho, and Kansas. Some localities noteworthy for the specimens they have produced are as follows: at Roxbury, Connecticut, brownish black, sometimes finely crystallized; in New Jersey a white variety (cleiophane) at Franklin; in Pennsylvania at the Wheatley and other mines at Phœnixville, in crystals; near Friedensville, Lehigh Co., a grayish wax-like variety; at Tiffin, Ohio; in Illinois at Marsden's diggings near Galena, in stalactites, some 6 inches or more in diameter and covered with crystallized marcasite and galena; in Wisconsin at Mineral Point in fine crystals often altered to smithsonite; in Missouri in beautiful crystallizations with galena, marcasite, and calcite at Joplin and other points in the southwestern part of the state and in the adjoining part of Kansas, the deposits in this section being of great extent and value. Sphalerite is found in many of the various ore veins of Colorado, etc.

Named *blende* because, while often resembling galena, it yielded no lead, the word in German meaning *blind* or *deceiving*. *Sphalerite* is from σφαλερός, *treacherous*.

Use. — The most important ore of zinc.

Metacinnabarite. — Mercuric sulphide, HgS. In composition like cinnabar, but occurs in *black* tetrahedral crystals; also massive. G. = 7·7. Metacinnabarite is a secondary mineral in origin and is found in the upper portions of mercury deposits. Usually associated with marcasite. From Idria in Gorizia, Italy (formerly in Carniola, Austria), in hemispherical aggregates. From Felsöbánya, Hungary. In California at the Redington and other mines near Knoxville, Napa Co.; in the Baker mine, etc., Lake Co.; at New Almaden, Santa Clara Co.; at New Idria, San Benito Co., etc.

Guadalcazarite. — Near metacinnabarite, but contains zinc (up to 4 per cent). Guadalcazar, San Luis Potosi, Mexico.

Tiemannite. — Mercuric selenide, HgSe. Isometric tetrahedral. Commonly massive; compact H. = 2·5. G. = 8·19 Utah; 8·30–8·47 Klaustal. Luster metallic. Color steel-gray to blackish lead gray. Streak nearly black. Occurs in the Harz Mts., near Zorge and at Tilkerode and Klaustal. Near Marysvale, Piute Co., Utah, in considerable amount.

ONOFRITE. Hg(S,Se) with Se = 4·5 to 8·4 per cent. San Onofre, near Plateros, northwest of Zacatecas, Mexico; with tiemannite from Marysvale, Piute Co., Utah. This last occurrence has been shown to be a mixture of tiemannite and sphalerite.

Coloradoite. — Mercuric telluride, HgTe. Massive. Cleavage, (111). Conchoidal fracture. H. = 2·5. G. = 8·07 (Kalgoorlie). Color iron-black. Originally found sparingly in Boulder Co., Colorado. Rather abundant at the Kalgoorlie district, West Australia. Material called *kalgoorlite* is a mixture of coloradoite and petzite.

Alabandite. — Manganese sulphide, MnS. Isometric. X-ray study shows an atomic structure either the same as that of halite or one closely similar. The symmetry might be tetrahedral or tetartohedral. The structure is distinct from that of sphalerite. Usually granular massive. Cleavage: cubic, perfect. G. = 3·95–4·04. Luster submetallic. Color iron-black. Streak green. Alabandite occurs in ore veins associated with sphalerite and other sulphides, rhodochrosite, etc. It has apparently been deposited from hot solutions by the action of hydrogen sulphide upon manganese-bearing waters. Originally from Alabanda in Aidin in Asia Minor. Occurs in Rumania at Nagy-Ág and elsewhere. In France at Adervielle, Hautes-Pyrénées. In Peru at Morococha, Province of Junin. At the Preciosa mine, on Cerro Tlachiaque, Puebla, Mexico. In Japan at Saimyoji, Ugo. In the United States fairly abundant at the Lucky Cuss mine, Tombstone, Arizona, also at the Queen of the West mine on Snake River, Summit Co., Colorado.

Oldhamite. — Calcium sulphide, CaS. Isometric. In pale brown spherules with cubic cleavage in the Busti meteorite. Also noted in Allegan meteorite.

PENTLANDITE.

Isometric.   Massive, granular.   Cleavage: octahedral.   Fracture uneven. Brittle.   H. = 3·5–4.   G. = 5·0.   Luster metallic.   Color light bronze-yellow.   Streak light bronze-brown.   Opaque.   Not magnetic.

Comp. — A sulphide of iron and nickel, (Fe,Ni)S.   In part, 2FeS.NiS = Sulphur 36·0, iron 42·0, nickel 22·0 = 100.

Obs. — Pentlandite commonly occurs intergrown with pyrrhotite and associated also with millerite, niccolite, gersdorffite, pyrite, marcasite and chalcopyrite.   In Norway it occurs with chalcopyrite at Espedalen, northwest of Lillehammer, Opland, and elsewhere. Occurs with pyrrhotite in the Sudbury district, Ontario, Canada, intimately associated with nickeliferous pyrrhotite.   It can be distinguished from the latter by its cleavage.

4. *Rhombohedral or Hexagonal Group*

			c	
Cinnabar	HgS	Rhombohedral-Trapezohedral	1·1453	
			c	c
Greenockite	CdS	Hexagonal-Hemimorphic	0·8109 or	0·9364
Wurtzite	ZnS	"	0·8175	0·9440
Millerite	NiS	Rhombohedral		0·9883
Niccolite	NiAs	Hexagonal?	0·8194	0·9462
Breithauptite	NiSb	"	0·8586	0·9915
Arite	Ni(Sb,As)			
Pyrrhotite	$Fe_{11}S_{12}$, etc.	Hexagonal	0·8701	1·0047
Troilite	FeS			
Covellite	CuS		1·1466	

This fourth group among the monosulphides includes several subdivisions, as shown in the scheme above, and the relations of the species are not in all cases perfectly clear.   It is to be noted that the sulphides of mercury and zinc, already represented in the sphalerite group, appear here again.   X-ray study has revealed new relationships between these minerals but has not yet clearly defined the whole group.   Covellite has been shown to have a structure similar to pyrrhotite, although its axial ratio is more nearly related to that of cinnabar.   Niccolite and breithauptite differ in their structure from millerite and show relationships to greenockite and wurtzite.   In some cases the present accepted value for c differs from that given above, but usually only in being a multiple or simple fraction of that quoted.   The former values are retained in order to show that whatever uncertainties exist in the relations between the members of the group they do have striking crystallographic characters in common.

If, as suggested by Groth, the prominent pyramids of wurtzite, greenockite, etc., be made pyramids of the second series (*e.g.,* $x = 11\bar{2}2$, instead of $10\bar{1}1$), then the values of c in the second column are obtained, which correspond to millerite.   The form of several of these species, however, is only imperfectly known.   A rhombohedral form for greenockite has been suggested.   Zincite (ZnO) shows close structural relations to greenockite and wurtzite.

CINNABAR.

Rhombohedral-trapezohedral. Axis $c = 1{\cdot}1453$. The atomic structure as given by X-ray study gives double this value for c.

$$rr',\ 10\bar{1}1 \wedge \bar{1}011 = 87° 23'.$$
$$ii',\ 40\bar{4}5 \wedge \bar{4}045 = 78° 0\tfrac{1}{2}'.$$
$$cr,\ 0001 \wedge 10\bar{1}1 = 52° 54'.$$

Crystals usually rhombohedral or thick tabular in habit, rarely showing trapezohedral faces; in rhombohedral penetration twins; also acicular prismatic. As in quartz, crystals may show a right- or left-handed development and turn the plane of polarized light. In crystalline incrustations, granular, massive; sometimes as an earthy coating.

Cleavage: $m(10\bar{1}0)$ perfect. Fracture subconchoidal, uneven. Somewhat sectile. H. = 2–2·5. G. = 8·0–8·2. Luster adamantine, inclining to metallic when dark-colored, and to dull in friable varieties. Color cochineal-red, often inclining to brownish red and lead-gray. Streak scarlet. Transparent to opaque. Optically +. High indices of refraction and strong birefringence. $\omega = 2{\cdot}91$, $\epsilon = 3{\cdot}27$. Shows strong circular polarization, being about fifteen times as strong as that exhibited by quartz. See Art. **402.**

Var. — 1. *Ordinary:* either (*a*) *crystallized;* (*b*) *massive,* granular embedded or compact; bright red to reddish brown in color; (*c*) *earthy* and bright red. 2. *Hepatic.* Of a liver-brown color, with sometimes a brownish streak, occasionally slaty in structure, though commonly granular or compact.

Comp. — Mercuric sulphide, HgS = Sulphur 13·8, mercury 86·2 = 100. Usually impure from the admixture of clay, iron oxide, bitumen.

Pyr. — In the closed tube alone a black sublimate of mercuric sulphide, but with sodium carbonate one of metallic mercury. Carefully heated in the open tube gives sulphurous fumes and metallic mercury, which condenses in minute globules on the cold walls of the tube. B.B. on charcoal wholly volatile, but only when quite free from gangue.

Diff. — Characterized by its color and vermilion streak, high specific gravity (reduced, however, by the gangue usually present), softness; also by the blowpipe characters (*e.g.*, in the closed tube). Resembles some varieties of hematite and cuprite.

Artif. — Cinnabar has been produced artificially by several methods which are, however, in general modifications of the two following types: (1) When the black mercury sulphide formed by the direct union of mercury and sulphur is sublimed, cinnabar is the product; (2) the black sulphide when treated with solutions of alkaline sulphides is converted into cinnabar. In general cinnabar is formed under alkaline conditions and metacinnabarite under acidic conditions. Three modifications of HgS have been prepared artificially. Cinnabar is formed from hydrothermal alkaline solutions and is stable up to its sublimation temperature, 580° C. Black metacinnabarite is precipitated from weakly acid solutions by sodium thiosulphate and a scarlet-red modification in a similar manner from concentrated neutral solutions.

Obs. — Cinnabar is the only common mineral of mercury and with rare exceptions constitutes the ore of the metal. It occurs in veins in rocks which have commonly a sedimentary origin, as slates, shales, sandstones, or limestones. It also frequently occurs in these rocks as impregnations and replacement deposits. Although it is found infrequently in igneous rocks, such rocks are commonly near by and are thought to have been the source of the metal. Cinnabar is almost always a primary mineral, which has been deposited from ascending alkaline solutions or as a result of solfataric action. Pyrite and marcasite, the sulphides of copper, stibnite, realgar, gold, etc., calcite, quartz or opal, barite, and fluorite are associated minerals; a bituminous mineral is also common.

Cinnabar was found in considerable amounts in the Bakhmut district in Ekaterinoslav, southern Russia, where it occurred as an impregnation in sandstone. Small twinned crystals come from Nikitovka in this district. In fine crystals from Mt. Avala near Belgrad, Yugoslavia; in Czechoslovakia at Szlana, near Rosenau (Rožňava), Gömör; at Horowitz (Hořovice) in Bohemia. It occurs in good crystals at Moschellandsberg in the Palatinate,

Bavaria. Important deposits are at Idria in Gorizia, Italy (formerly in Carniola, Austria). In fine crystals at Ripa near Seravezza and elsewhere in Tuscany. The most important mercury deposit of the world is at Almaden, Ciudad Real, Spain. An important deposit was located at Huancavelica, Peru. Found in water-worn pebbles in placer deposits in the Marowyne district, near the headwaters of the Tempati River, seventy-five miles southeast of Paramaribo, Dutch Guiana. In China from the provinces of Kweichow and Hunan, at times in large twinned rhombohedra. In various places in Mexico. In the United States the most important deposits are in California. Cinnabar has been mined in the Coast Range from Del Norte Co. in the north to San Diego Co. in the south. The most productive counties have been Lake, Napa, Santa Clara (New Almaden) and San Benito (New Idria). The deposits have been formed by the action of solfataric waters which have deposited the cinnabar along the contact between serpentine and metamorphic sandstones and shales. Cinnabar has also been found in Nevada, Utah, Oregon, and at Terlingua, Brewster Co., Texas.

The name cinnabar is supposed to come from India, where it is applied to the red resin, dragon's blood. The native cinnabar of Theophrastus is true cinnabar; he speaks of its affording quicksilver. The Latin name of cinnabar, *minium*, is now given to *red lead*, a substance which was early used for adulterating cinnabar, and so got at last the name.

Only comparatively few localities have furnished the mineral in quantity.

Use. — The most important ore of mercury.

GREENOCKITE.

Hexagonal-hemimorphic. Rarely in hemimorphic crystals; also as a coating.

Cleavage: $a(11\bar{2}0)$ distinct, $c(0001)$ imperfect. Fracture conchoidal. Brittle. H. = 3–3·5. G. = 4·9–5·0. Luster adamantine to resinous. Color honey-, citron-, or orange-yellow. Streak between orange-yellow and brick-red. Nearly transparent. Optically +. $\omega = 2\cdot506$, $\epsilon = 2\cdot529$.

684

Comp. — Cadmium sulphide, CdS = Sulphur 22·3, cadmium 77·7 = 100.

An artificial form ($\beta - CdS$) has been shown to have a structure similar to that of sphalerite.

Amorphous CdS, probably containing absorbed water, that occurs as a thin coating on sphalerite has been named *xanthochroite*.

Pyr., etc. — In the closed tube assumes a carmine-red color while hot, fading to the original yellow on cooling. In the open tube gives sulphurous fumes. B.B. on charcoal, either alone or with soda, gives in R.F. a reddish brown coating. Soluble in hydrochloric acid, affording hydrogen sulphide.

Artif. — Greenockite has been prepared artificially in several ways. Precipitated cadmium sulphide when fused with potassium carbonate and sulphur produced greenockite crystals; also when cadmium sulphate, calcium fluoride and barium sulphide were fused together. Greenockite is formed when cadmium oxide is heated in sulphur vapor.

Obs. — Occurs in short hexagonal crystals at Bishopton, Renfrew, Scotland; also at Wanlockhead, Dumfries. Usually occurs as an earthy coating associated with zinc minerals, commonly sphalerite. Such occurrences are noted at Přibram, Bohemia; at Bleiberg in Carinthia; at Pierrefitte, Hautes-Pyrénées, France. In the United States it is found at Franklin, New Jersey; at Freidensville, Pennsylvania; at Granby and elsewhere in Missouri; in Marion Co., Arkansas, it occurs coloring smithsonite bright yellow; from near Topaz, Mono Co., California. Not uncommon as a furnace product.

Use. — An ore of cadmium.

Wurtzite. — Zinc sulphide, ZnS, like sphalerite, but in hemimorphic hexagonal crystals closely related in structure, etc., to greenockite; also fibrous and massive. Cleavage as in greenockite. G. = 3·98. Optically +. $\omega = 2\cdot5$. Color brownish black. Wurtzite is the rare form of zinc sulphide. It is stable at temperatures above 1020° C. For its relations to sphalerite see under the latter, p. 421. Found at Mies, Bohemia; at Felsöbánya (Baia Sprie), Rumania; in crystals from Oruro and Chocaya, Potosi, Bolivia; at times in large tabular crystals from Quispisiza near Castro Virreyna,

Peru. In the United States in fine pyramidal crystals from Butte, Montana; in crystals from Joplin, Missouri, and from near Frisco, Beaver Co., Utah.

The massive, fibrous forms, *schalenblende*, occur at Přibram, Bohemia; Liskeard, Cornwall; etc. Many schalenblendes are mixtures of sphalerite and wurtzite.

MILLERITE. Capillary Pyrites.

Rhombohedral. Usually in very slender to capillary crystals, often in delicate radiating groups; sometimes interwoven like a wad of hair. Also in columnar tufted coatings, partly semi-globular and radiated. The rhombohedron $(01\bar{1}2)$ is a gliding plane and artificial twins may be formed. X-ray study shows that the unit cell contains three molecules of NiS.

Cleavage perfect parallel to $(10\bar{1}1)$ and $(01\bar{1}2)$. Fracture uneven. Brittle; capillary crystals elastic. H. = 3–3·5. G. = 5·3–5·65. Luster metallic. Color brass-yellow, inclining to bronze-yellow, with often a gray iridescent tarnish. Streak greenish black.

Comp. — Nickel sulphide, NiS = Sulphur 35·3, nickel 64·7 = 100.

Pyr., etc. — In the open tube sulphurous fumes. B.B. on charcoal fuses to a globule. When roasted, gives with borax and salt of phosphorus a violet bead in O.F., becoming gray in R.F. from reduced metallic nickel. On charcoal in R.F. the roasted mineral gives a coherent metallic and magnetic mass. Most varieties also show traces of copper and iron with the fluxes, rarely cobalt.

Artif. — Crystals of millerite have been formed artificially by treating under pressure a solution of nickel sulphate with hydrogen sulphide.

Obs. — Millerite commonly occurs in capillary crystals in cavities among the crystals of other minerals. Frequently found in the rocks of the coal measures associated with other nickel minerals, various sulphides, iron ores, etc. Also often associated with serpentine or with chrysolite-bearing rocks. Rarely as a sublimation product, as at Vesuvius. Has been noted in meteoric irons.

Occurs with nickel, cobalt, and silver ores at Joachimstal (Jáchymov), Bohemia; with iron ores at Dillenburg, Hessen-Nassau; at Müsen, Littfeld and Siegen, Westphalia. Found at Wissen on the Sieg and at Saarbrücken, Rhineland. In very fine hair-like crystals in cavities in siderite at Merthyr-Tydfil, Glamorgan, Wales. Occurs at the Sterling mine, Antwerp, New York in radiating groups of capillary crystals with ankerite in cavities in hematite; in Lancaster Co., Pennsylvania, at Gap mine, with pyrrhotite, in thin coatings of a radiated fibrous structure. With calcite, dolomite, and fluorite, forming delicate hair-like tufts in geodes in limestone, often penetrating the calcite crystals, at St. Louis, Missouri, at Keokuk, Iowa, and Milwaukee, Wisconsin. With a green chromiferous garnet in Orford township, Quebec; and from the Sudbury district, Ontario.

Use. — An ore of nickel.

BEYRICHITE. NiS like millerite, but with lower specific gravity (4·7) and slight differences in color. Laspeyres considers all millerite as formed by paramorphism from beyrichite. Found very rarely at Oberlahr near Altenkirchen, Rhineland.

NICCOLITE. Copper Nickel.

Hexagonal. Crystals rare. X-ray studies of the structure of niccolite and breithauptite show hexagonal symmetry with close relationships between the two. Usually massive, structure nearly impalpable; also reniform, columnar; reticulated, arborescent. Fracture uneven. Brittle. H. = 5–5·5. G. = 7·33–7·67. Luster metallic. Color pale copper-red. Streak pale brownish black. Opaque.

Comp. — Nickel arsenide, NiAs = Arsenic 56·1, nickel 43·9 = 100. Usually contains a little iron and cobalt, also sulphur; sometimes part of the arsenic is replaced by antimony, and then it grades toward breithauptite. The intermediate varieties have been called *arite*.

Pyr, etc. — In the closed tube on intense ignition gives a faint sublimate of arsenic. In the open tube a sublimate of arsenic trioxide, with a trace of sulphurous fumes, the assay becoming yellowish green.   On charcoal gives arsenical fumes and fuses to a globule, which, treated with borax glass, affords, by successive oxidation, reactions for iron, cobalt, and nickel; the antimonial varieties give also reactions for antimony.   Soluble in aqua regia.

Obs. — Niccolite is commonly associated with smaltite, chloanthite, annabergite, with native silver and the silver-arsenic minerals, pyrite, chalcopyrite and other sulphides, quartz, barite, etc.   It is found at Schladming, Styria, Austria; in Bohemia at Joachimstal (Jáchymov); at many points in Saxony as Schneeberg, Mansfield, Sangerhausen, etc.; in Hessen-Nassau at Richelsdorf in crystals; at Eisleben in Thuringia in twin crystals.   In France at Les Chalantes near Allemont, Isère, and at the mine of Ar near Eaux-Bonnes, Basses-Pyrénées (*arite*).   Found sparingly at Franklin, New Jersey; at Silver Cliff, Colorado. In Ontario with the silver and cobalt ores at Cobalt and on Silver Islet, Thunder Bay.   At Tilt Cove, Newfoundland.

Use. — An ore of nickel.

TEMISKAMITE.   Described as having composition Ni_4As_3, has been shown to be a mixture of *niccolite, maucherite* and a little *cobaltite*.   From Elk Lake, Ontario.

Breithauptite. — Nickel antimonide, $NiSb$.   Rarely in hexagonal crystals; usually massive, arborescent, disseminated.   G. = 7·54.   Color light copper-red.   Breithauptite is similar to niccolite in its occurrence.   Found at Andreasberg, in the Harz Mts.; at Mte. Narbo, near Sarrabus and elsewhere in Sardinia.   At Cobalt, Ontario.

PYRRHOTITE.   Magnetic Pyrites.

Hexagonal. $c = 0.8701$.

$$cs, 0001 \wedge 10\bar{1}1 \qquad = 45° \ 8'.$$
$$cu, 0001 \wedge 40\bar{4}1 \qquad = 76° \ 0'.$$

Twins: tw. pl. $s(10\bar{1}1)$, with vertical axes nearly at right angles (Fig. 444, p. 189).   Distinct crystals rare, commonly tabular; also acute pyramidal with faces striated horizontally.   Usually massive, with granular structure.

685

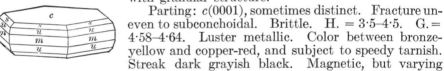

Parting: $c(0001)$, sometimes distinct.   Fracture uneven to subconchoidal.   Brittle.   H. = 3·5–4·5.   G. = 4·58–4·64.   Luster metallic.   Color between bronze-yellow and copper-red, and subject to speedy tarnish. Streak dark grayish black.   Magnetic, but varying much in intensity; sometimes possessing polarity.

Comp. — Ferrous sulphide containing variable amounts of dissolved sulphur.   In synthetic pyrrhotites the maximum of excess sulphur that could be obtained was 6 per cent.   Analyses show variation from Fe_5S_6 to $Fe_{16}S_{17}$. Often also contains nickel due probably to enclosed grains of pentlandite. Fe_7S_8 = Sulphur 39·6, iron 60·4 = 100.   (Cf. Art. **485**, p. 354.)

Pyrrhotite differs from troilite in containing more or less dissolved sulphur, while troilite, occurring in meteorites where there is always an excess of iron, may form the pure monosulphide.

Pyr, etc. — Unchanged in the closed tube.   In the open tube gives sulphurous fumes. On charcoal in R.F. fuses to a black magnetic mass; in O.F. is converted into red oxide, which with fluxes gives only an iron reaction when pure, but many varieties yield small amounts of nickel and cobalt.   Decomposed by hydrochloric acid, with evolution of hydrogen sulphide.

Diff. — Distinguished by its peculiar reddish bronze color; also by its magnetic properties.

Artif. — Pyrrhotite has been synthesized by the direct union of iron and sulphur and also when pyrite is heated in an atmosphere of hydrogen sulphide at 550°.   Pyrrhotite exists in two crystalline modifications, hexagonal at ordinary temperatures and orthorhombic above 138°.

Obs. — Pyrrhotite occurs at times in large amounts associated with basic igneous rocks, such as gabbro, norite, hornblende and augite rocks, etc., from which it has been segregated by some form of magmatic differentiation.   It is commonly associated with chalcopyrite, pyrite, pentlandite, magnetite, etc. It is also found in contact metamorphic deposits, in vein deposits, in pegmatites. Has been noted at fumeroles of Vesuvius. At times found in metamorphic limestones with garnet. Occurs in large crystal groups at Loben near St. Leonhard in the Lavant-Tal, Carinthia, Austria; in Trentino, Italy, in Val Passiria (Passeier Tal, Tyrol) in fine crystals. In Saxony at Freiberg and Schneeberg; in crystals at Bodenmais, Bavaria; at Andreasberg in the Harz Mts. In Switzerland in the Bristenstock tunnel, Uri. In Tuscany at Bottino near Serravezza. In fine crystal groups at Pont-Péan, near Bruz, Ille-et-Vilaine, France. Occurs in considerable amounts in Norway and Sweden as at Kongsberg, Modum, Snarum, etc. in Norway and at Falun and Klefva, Småland, Sweden. In very large crystals from Morro Velho, on Rio das Velhas, northwest of Ouro Preto, Minas Geraes, Brazil. In the United States from Standish, Maine, in crystals with andalusite; at Brewster, Putnam Co., New York; at the Gap mine, Lancaster Co., Pennsylvania, with chalcopyrite and millerite. In considerable quantity at Ducktown, Tennessee. Very large deposits of nickeliferous pyrrhotite occur at Sudbury, Ontario. In crystals from Elizabethtown, Ontario.

Named from πυρρότης, *reddish*.

Use. — Often becomes a valuable ore of nickel.

Troilite. — Ferrous sulphide, FeS. G. = 4·75–4·82. Color tombac-brown. Considered to be the end member of the pyrrhotite series. See above. Occurs commonly in iron meteorites, disseminated more or less sparingly through the mass; also in narrow veins usually separated from the iron by a thin layer of graphite. Also found in Del Norte Co., California, northeast of Cresent City, in massive form in serpentine with magnetite.

COVELLITE.

Hexagonal. X-ray study shows a complicated structure similar to that of pyrrhotite. The axial ratio is close to that of cinnabar, and the two minerals have often been grouped together. Crystals usually thin hexagonal plates. Often massive.

Cleavage: basal, perfect. Flexible in thin laminæ. H. = 1·5–2. G. = 4·6. Luster submetallic to resinous. Color indigo-blue or darker. Sometimes has a purple tarnish. Often shows fine purple color when moistened with water. Streak lead-gray to black. Opaque. In very thin plates translucent and shows pleochroism in light green colors. Optically +. $\omega = 1\cdot45_{Na}$.

Comp. — Cupric sulphide, CuS = Sulphur 33·6, copper 66·4 = 100.

Pyr., etc. — Fusible at 2·5 yielding sulphurous fumes. After roasting and moistening with hydrochloric acid gives azure-blue flame. Much sulphur in C.T.

Artif. — Covellite has been prepared artificially by heating in sealed tubes a cupric solution with ammonium sulphocyanate and by heating sphalerite in a solution of copper sulphate.

Obs. — Covellite occurs in copper veins associated commonly with chalcopyrite, chalcocite, bornite, enargite, etc., and it is in most cases secondary in its origin, formed during the alteration and enrichment of the copper minerals of a vein by the action of descending solutions. Frequently in intimate intergrowth with chalcocite. Also rarely observed as a sublimation product, as at Vesuvius and on the Island of Vulcano.

Covellite is found at Bor, northwest of Zaječar, Serbia, Yugoslavia; at Sangerhausen, Saxony. In large crystals from the Calabona mine, Alghero, Sardinia. In various localities in Chile, Bolivia, Argentina, Peru. In the United States covellite occurs at Butte, Montana, with chalcocite and enargite; in Colorado at Summitville, Rio Grande Co., and at Wagon Wheel Gap, Mineral Co.; near Laramie, Wyoming; in the La Sal district, San Juan Co., Utah. At Kennecott, Alaska.

HAUCHECORNITE. Described as Ni(Bi,Sb,S) but shown by microscopic tests to be a mixture of two unknown constituents. In tabular tetragonal crystals. H. = 5. G. = 6·4. Color light bronze-yellow. From Hamm a. d. Sieg, Rhineland, Germany.

C. Intermediate Division

The following species are sometimes regarded as Sulpho-salts, namely, Sulpho-ferrites, etc.

BORNITE. Peacock Ore. Purple Copper Ore. Variegated Copper Ore. Erubescite.

Isometric. Habit cubic, faces often rough or curved. Twins: tw. pl. $o(111)$, often penetration-twins. Crystals rare. Usually massive, structure granular or compact.

Cleavage: $o(111)$, in traces. Fracture small conchoidal, uneven. Brittle. H. = 3. G. = 4·9–5·4. Luster metallic. Color between copper-red and pinchbeck-brown on fresh fracture, speedily iridescent from tarnish. Streak pale grayish black. Opaque.

Comp. — A sulphide of copper and iron. Cu_5FeS_4. Copper 63·3, iron 11·1, sulphur 25·6 = 100.

The mineral often contains small amounts of chalcocite, etc., and therefore shows considerable variation in its percentage composition, giving from 50 to 70 per cent of copper and 15 to 6·5 per cent of iron.

Pyr., etc. — In the closed tube gives a faint sublimate of sulphur. In the open tube yields sulphurous fumes. B.B. on charcoal fuses in R.F. to a brittle magnetic globule. The roasted mineral gives with the fluxes the reactions of iron and copper, and with soda a metallic globule. Soluble in nitric acid with separation of sulphur.

Diff. — Distinguished (*e.g.*, from chalcocite) by the peculiar reddish color on the fresh fracture and by its brilliant tarnish; B.B. becomes strongly magnetic.

Artif. — Bornite has been obtained by fusing pyrite, copper and sulphur together; by heating a mixture of cuprous, cupric and ferric oxides in hydrogen sulphide at 100° to 200°.

Obs. — Bornite has been noted as a primary mineral of magmatic origin, being found in igneous rocks and pegmatite veins. It is also a common mineral of copper veins, being usually primary but occasionally secondary in origin, and occurring with chalcocite, etc., in the enriched portions of the veins. Bornite is a common and widespread mineral associated with chalcopyrite, chalcocite, covellite, tetrahedrite, pyrite, pyrrhotite, marcasite, arsenopyrite, etc. Has been noted with galena, sphalerite, magnetite, garnet, calcite, serpentine, etc. Common in quartz veins. It is usually massive in form, crystals being found at only a few localities.

From near Prägratten and Windisch-Matrei in the Tyrol, Austria, in crystals. In the cupriferous shales of the Mansfield district, Harz Mts. In the copper ores of Montecatini, Tuscany. In crystals from Redruth, Cornwall. In specimens of fine color from Androta, southwest of Vohémar, Madagascar. An important ore at Mt. Lyell, Tasmania. It is the principal ore of some Chilean mines, also common in Peru, Bolivia, and Mexico. Was abundant at Bristol, Connecticut, in fine crystals. Occurs in quantity at Butte, Montana. Frequently found in eastern Quebec, at the Acton mine, Bagot Co., etc.

Named after the mineralogist Ignatius von Born (1742–1791).

Use. — An ore of copper.

Germanite. — $Cu_3(Fe,Ge)S_4$? Isometric, probably tetrahedral. Massive. H. = 4. G. = 4·46–4·59. Color dark reddish gray. Metallic luster. Intimately intergrown with tennantite, pyrite, etc. at Tsumeb near Otavi, South West Africa.

Linnæite. — Isometric. A sulphide of cobalt, $Co_3S_4 = CoS.Co_2S_3$, analogous to the spinel group. Also contains nickel (var. *siegenite*). Commonly in octahedrons; also massive. X-ray study shows practically identical structure in linnæite, polydymite and sychnodymite. Structure is face-centered cubic. H. = 5·5. G. = 4·8–5. Color pale steel-gray, tarnishing copper-red. Occurs at Müsen and elsewhere in the Siegen district, Westphalia, associated with chalcopyrite, pyrite, galena, sphalerite, siderite, quartz, etc. With chalcopyrite at the Bastnaes mine, near Riddarhyttan, Vastmanland, Sweden. In large octahedral crystals from Katanga, Belgian Congo. At Mineral Hill, Carroll Co., Maryland, with chalcopyrite, bornite, sphalerite, pyrite, etc., and at Mine la Motte, in Missouri, mostly massive, sometimes in octahedral and cubo-octahedral crystals. *Carrollite*, from Finksburg, Carroll Co., Maryland, has been shown to be a mixture of linnæite in which small amounts of copper, iron, and nickel may be isomorphous with cobalt and intimately intergrown copper sulphides. *Sychnodymite*, cobalt-copper sulphide in small steel-gray octahedrons from the Siegen district, Westphalia, is probably also a variety of linnæite.

Violarite. — A nickel sulphide. Comp. given either as NiS_2 and then assumed to have same structure as artif. material which has been shown to belong to Pyrite Group, or more probably as $(Ni,Fe)_3S_4$ and then belonging in group with linnæite. Perfect cubic cleavage. Color, violet-gray. Tarnishes readily on exposure. Found with pyrite, chalcopyrite and pentlandite at the Key West mine, Clark Co., Nevada, as a replacement of pentlandite. Same mineral noted in southeastern Alaska. Reported from Sudbury, Ontario (see under polydymite), and Julian, California.

Polydymite. Nickel-linnæite. — Ni_3S_4. In octahedral crystals; frequently twinned. Structure same as that of linnæite. G. = 4·54–4·81. Color gray. From the Grüneau (Grünau) mine, southwest of Schutzback and southeast of Wissen on the Sieg, Rhineland. The polydymite reported from the Sudbury district, Ontario, is of doubtful character. It has been stated to be a mixture of pentlandite and violarite, but this conclusion has been disputed.

Daubréelite. — An iron-chromium sulphide, $FeS.Cr_2S_3$, occurring with troilite in some meteoric irons. Color black. G. = 5·01.

Badenite. — $(Co,Ni,Fe)_3(As,Bi)_4$? Massive granular to fibrous. G. = 7·1. Metallic. Color steel-gray. Fusible. From near Badeni-Ungureni, district of Muscel, Rumania, associated with erythrite, annabergite, malachite, siderite.

CHALCOPYRITE. — Copper Pyrites. Yellow Copper Ore.

Tetragonal-sphenoidal. Axis $c = 0.98525$.

pp', $111 \wedge \bar{1}\bar{1}1 = 108° 40'$. $pp_{,}$, $111 \wedge 1\bar{1}1 = 70° 7\frac{1}{2}'$. ce, $001 \wedge 101 = 44° 34\frac{1}{2}'$.

Crystals commonly tetrahedral in aspect, the sphenoidal faces $p(111)$ large, dull or oxidized; $p_{,}(1\bar{1}1)$ small and brilliant. Sometimes both forms equally developed, and then octahedral in form. Twins: (1) tw. pl. $p(111)$,

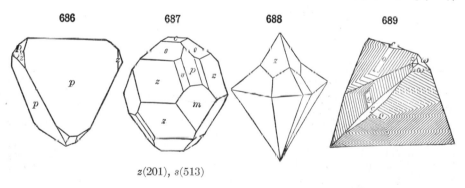

686   687   688   689

$z(201)$, $s(513)$

resembling spinel-twins (Fig. 443, p. 188); sometimes repeated as a fiveling (Fig. 688). (2) Tw. pl. and comp.-face $e(101)$ (Fig. 689), often in repeated twins. (3) Tw. pl. $m(110)$, tw. axis c, complementary penetration twins. X-ray analysis shows that the structure is similar to that of sphalerite, the layers of zinc atoms in that mineral being replaced by alternate layers of copper and iron atoms. Often massive, compact, sometimes botryoidal or reniform.

Cleavage: $z(201)$, sometimes distinct. Fracture uneven. Brittle. H. = 3·5–4. G. = 4·1–4·3. Luster metallic. Color brass-yellow; often tarnished or iridescent. Streak greenish black. Opaque.

Comp. — A sulphide of copper and iron, $CuFeS_2$ = Sulphur 35·0, copper 34·5, iron 30·5 = 100. Analyses often show variations from this formula, often due to mechanical admixture of pyrite.

Sometimes auriferous and argentiferous; also contains traces of selenium and thallium.

Pyr., etc. — In the closed tube often decrepitates, and gives a sulphur sublimate, in the open tube sulphurous fumes. On charcoal fuses to a magnetic globule; the residue moistened with hydrochloric acid and then touched with blowpipe flame gives intense blue flame color. Decomposed by nitric acid giving free sulphur and a green solution; ammonia in excess changes the green color to a deep blue, and precipitates red ferric hydroxide.

Diff. — Distinguished from pyrite by its inferior hardness and deeper yellow color. Resembles gold when disseminated in minute grains in quartz, but differs in being brittle and in having a black streak; further it is soluble in nitric acid.

Artif. — Chalcopyrite has been artificially prepared (1) by fusing pyrite and copper sulphide together; (2) by gently heating cupric and ferric oxides in an atmosphere of hydrogen sulphide.

Obs. — Chalcopyrite is the most common and important mineral containing copper. It is commonly of primary origin and from it, by various alteration processes, many other copper minerals are derived. It has been repeatedly observed as an original constituent of igneous rocks, and the ultimate source of the copper of ore deposits is to be found in rocks of this type. Noted in pegmatite veins. It occurs widely in metallic veins and in nests in gneiss and crystalline schists, also in serpentine rocks. The deposits are frequently near igneous rocks. In contact metamorphic deposits; at times broadly disseminated in schistose rocks. Often intimately associated with pyrite, also bornite, chalcocite, tetrahedrite, malachite, azurite, cuprite, siderite, quartz, etc.; sometimes with nickel and cobalt sulphides, pyrrhotite, etc.; at times with galena, sphalerite, arsenopyrite, cassiterite, etc. Observed coated with tetrahedrite crystals in parallel position, also as a coating over the latter. Frequently associated with sphalerite, its crystals often lying with parallel orientation upon the latter mineral.

Chalcopyrite is such a common mineral that only its outstanding occurrences can be mentioned. In Czechoslovakia at Schlaggenwald, Bohemia, and at Schemnitz (Selmeczbánya). In Saxony in well-defined crystals at Freiberg, and elsewhere. In the Harz Mts. at Rammelsberg near Goslar with pyrite and at Neudorf and Klaustal, and disseminated in the schists of the Mansfield district. In Westphalia in the region of Siegen, Müsen, and Littfeld, often in fine crystals. In Tuscany at the copper deposits of Montecatini and at Bottino near Seravezza. In France in unusual crystals at Ste. Marie aux Mines (Markirch), Alsace, and at the mine of Gardette near Bourg d'Oisans, Isère; also from Baigorry, Basses-Pyrénées. With pyrite at Rio Tinto, Spain. At Falun, Sweden, it occurs in large masses. In crystals at St. Agnes and numerous other localities in Cornwall. From Laxey on the Isle of Man and from Tavistock, Devonshire. From South Australia and New South Wales; various localities in Chile, Peru, and Bolivia. In crystals from Ani and Arakawa, Province of Ugo, Japan.

In the United States in very large crystals and massive at Ellenville, Ulster Co., New York. In Pennsylvania in exceptional crystals at the French Creek mines, Chester Co., associated with pyrite, magnetite, etc. In Missouri at Joplin in crystals with sphalerite, galena, dolomite, etc. In crystals at Central City and elsewhere in Gilpin Co., Colorado. The most important sulphide deposits of copper, in many of which chalcopyrite is the chief ore, are found in the states of Arizona, Montana, Utah, Nevada, New Mexico, California, Tennessee, and in Alaska. In Canada there are important deposits in British Columbia, Ontario, and Quebec.

Use. — The most important ore of copper.

Named from χαλκός, *brass*, and *pyrites*, by Henckel (1725).

D. Disulphides, Diarsenides, etc.

The disulphides, diarsenides, etc., embrace two distinct groups. The prominent metals included are the same in both, viz.: iron, cobalt and nickel. The groups present, therefore, several cases of isodimorphism, as is shown in the lists of species below. These sulphides are all relatively *hard*, H. = 5–6; they hence strike fire with a steel, and this has given the familiar name *pyrites* applied to most of them. The color varies between pale brass-yellow and tin-white.

Pyrite Group. RS_2, RAs_2, RSb_2.   Isometric-pyritohedral or -tetartohedral

Pyritohedral		Tetartohedral	
Pyrite	FeS_2	Cobaltite	CoAsS
Bravoite	$(Fe,Ni)S_2$	Gersdorffite	NiAsS
Cobaltnickelpyrite	$(Co,Ni,Fe)S_2$	Ullmannite	NiSbS
Hauerite	MnS_2		
Arsenoferrite	$FeAs_2$	Laurite	RuS_2
Sperrylite	$PtAs_2$	{ Smaltite	$CoAs_3$, also $(Co,Ni)As_3$
		{ Chloanthite	$NiAs_3$, also $(Ni,Co)As_3$
		Skutterudite	$CoAs_3$

Marcasite Group. RS_2, RAS_2, etc.   Orthorhombic

		$a:b:c$	$110 \wedge 1\bar{1}0$	$101 \wedge \bar{1}01$
Marcasite	FeS_2	$0.7662 : 1 : 1.2342$	74° 55′	116° 20′
Löllingite	$FeAs_2$	$0.6689 : 1 : 1.2331$	67° 33′	123° 3′
Safflorite	$CoAs_2$			
Rammelsbergite	$NiAs_2$			
Arsenopyrite	$FeS_2.FeAs_2$	$0.6773 : 1 : 1.1882$	68° 13′	120° 38′
Danaite	$(Fe,Co)S_2.(Fe,Co)As_2$			
Glaucodot	$(Co,Fe)S_2.(Co,Fe)As_2$	$0.6942 : 1 : 1.1925$	69° 32′	119° 35′
Wolfachite	$NiS_2.Ni(As,Sb)_2$			

The PYRITE GROUP includes, besides, the compounds of Fe, Co, Ni, also others of the related metals Mn and Pt. The crystallization is isometric-pyritohedral. X-ray analysis shows clearly the same structure for pyrite, sperrylite, cobaltite, gersdorffite and ullmannite, the last three, however, showing tetartohedral symmetry. Smaltite-chloanthite and skutterudite though probably pyritohedral and closely related to each other do show structural differences from pyrite.

The species of the MARCASITE GROUP crystallize in the orthorhombic system with prismatic angles of about 70° and 110° and a prominent macro-dome of about 60° and 120°. Hence fivefold and sixfold repeated twins are common with several species, in the one case the prism and in the other the macrodome named being the twinning-plane. X-ray analysis shows closely related structures for the members of this group; löllingite, safflorite and rammelsbergite have identical structures as also have arsenopyrite and glaucodot. Further the structure is closely related both to isometric and tetragonal (rutile) structures.

Pyrite Group

PYRITE. Iron Pyrites.

Isometric-pyritohedral. Cube and pyritohedron $e(210)$ the common forms, the faces of both often with striations ∥ edge $a(100)/e(210)$, due to oscillatory combination of these forms and tending to produce rounded faces; pyritohedral faces also striated ⊥ to this edge; octahedron also common. See Figs. 690–695, also Figs. 151–156, pp. 81, 82. Twins: tw. pl. = (110) and tw. ax. normal to this face, usually penetration-twins (Fig. 433, p. 187); rarely contact-twins. For illustration of atomic structure, see p. 41. Frequently massive, fine granular; sometimes subfibrous radiated; reniform, globular, stalactitic.

Cleavage: $a(100)$, $o(111)$, indistinct. Fracture conchoidal to uneven. Brittle. H. = 6–6·5. G. = 4·95–5·10; 4·967 Traversella, 5·027 Elba. Luster metallic, splendent to glistening. Color a pale brass-yellow, nearly uniform. Streak greenish black or brownish black. Opaque.

Comp. — Iron disulphide, FeS_2 = Sulphur 53·4, iron 46·6 = 100.

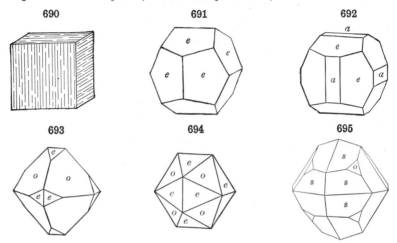

Nickel, cobalt, and thallium, and also copper in small quantities, sometimes replace part of the iron, or else occur as mixtures; selenium is sometimes present in traces. Gold is sometimes distributed invisibly through it, auriferous pyrite being an important source of gold. Arsenic is rarely present, as in octahedral crystals from French Creek, Pennsylvania. (0·2 per cent As.) A cobaltiferous pyrite from Falun, Sweden, has been named *cobaltpyrite*.

Pyr., etc. — Easily fusible, (2·5–3). Becomes magnetic on heating and yields sulphur dioxide. Gives an abundant sublimate of sulphur in the closed tube. Insoluble in hydrochloric acid. The fine powder is completely soluble in strong nitric acid.

Diff. — Distinguished from chalcopyrite by its greater hardness and paler color; in form and specific gravity different from marcasite, which has also a whiter color.

Alt. — Pyrite readily changes by oxidation to an iron sulphate or to the hydrated oxide, limonite, with sulphuric acid set free. Crystals of pyrite which have been changed on their surfaces to limonite are common. This change may continue until the original mineral has completely disappeared. Large masses of pyrite lying near the surface may be altered to a cellular mass of limonite — the *iron gossan* of the miners — while the sulphuric acid set free travels downward and enters into various important reactions with the unaltered minerals below. The alteration of pyrite to limonite may be continued until hematite is formed.

Obs. — Experiments show that pyrite is formed in neutral or alkaline solutions and at high temperatures. Marcasite, on the other hand, is deposited from acid solutions and is stable only at temperatures below 450° C. These sulphides can be formed through the action of hydrogen sulphide, although the reducing action of carbonaceous materials may also at times be of importance. Pyrite occurs in rocks of all ages and types, being most common in the metamorphic and sedimentary rocks. In the sedimentary rocks it occurs as either an accessory mineral, contemporaneous in origin with the other minerals, or as a fine and widespread impregnation of subsequent origin. It is also frequently found as a minor accessory constituent of igneous rocks. It is often associated with coal formations. At times it replaces the organic material of fossils. When disseminated in the rocks it usually occurs in small crystals, but in veins it may occur in crystals or with a granular or radiating structure. At times it is in nodular or concretionary forms.

Pyrite is very widespread in its occurrence, being the most common sulphide. At times it is found in very large amounts and is mined for its sulphur content or because it contains small amounts of some valuable metal. It is almost invariably to be found as a gangue mineral in those ore veins, the minerals of which have been deposited by ascending hot solu-- tions. The minerals that are associated with pyrite are many and of diverse characters. Among the sulphides are most commonly chalcopyrite, chalcocite, bornite, arsenopyrite, tetrahedrite, galena, sphalerite, marcasite, millerite, etc. The associated oxidized minerals include magnetite, hematite, siderite, cassiterite, quartz, etc. Some of the more notable localities for its occurrence are given below.

In Czechoslovakia pyrite occurs in crystals at Schemnitz (Selmeczbánya) and in a large deposit at Schmöllnitz (Smolnik); in Bohemia in crystals at Přibram; in Rumania at Facebay near Zalatna. From Waldenstein south of St. Leonhard in the Lavant-Tal, Carinthia, Austria, in fine crystals. In Saxony at Freiberg and elsewhere; in Westphalia in crystals in the Siegen district; in remarkable compound crystals from the clay at Vlotho southwest of Minden and in a considerable deposit at Meggen on the Lenne. In fine specimens from St. Gotthard, Switzerland. In Italy in exceptional crystals with magnetite, garnet, pyroxene (fassaite), hornblende, and epidote in Piedmont at Brosso and Traversella; at Gavorrano, Tuscany; pyritohedrons and other forms in great variety occur with hematite at Rio Marina and elsewhere on the island of Elba, the crystals being sometimes 5 to 6 inches in diameter. In France pyrite occurs in fine crystals at Saint-Pierre-de-Mesage near Vizille, Isère, and in large crystals altered to limonite from Estrèmes de Salles in Agos (vallée d'Argelès), Hautes-Pyrénées, and elsewhere. In southwestern Spain pyrite occurs in large deposits, as in the copper mine at Rio Tinto and at Tharsis, Province Huelva and in Portugal. In Sweden at Falun and at Klefva, Småland, and in Norway at Rorös, Sör-Tröndelag and in the Sulitelma district, Nordland In fine crystals from Liskeard, St. Ives, St. Just, and many other localities in Cornwall; at Tavistock in Devonshire. Notable crystals are found in various localities in Peru, Bolivia, Chile, Brazil, Japan, Mexico.

In the United States formerly at Roxbury, Connecticut, finely crystallized; in New York in St. Lawrence Co., at Rossie, fine crystals at the lead mine, and at Scoharie, in single and compound crystals; at Franklin, New Jersey in highly modified crystals; in Pennsylvania at the French Creek mines, octahedrons, and other forms, sometimes through distortion having a pseudo-tetragonal or orthorhombic symmetry, at Cornwall, Lebanon Co., in lustrous cubo-octahedrons; has been mined in Louisa and Prince William counties, Virginia, and at the large copper-bearing deposit at Ducktown, Tennessee; in flat circular concretions, " pyrite suns," from Sparta, Randolph Co., Illinois; in Colorado in fine crystals at Central City, Russell Gulch, and elsewhere in Gilpin Co., in Summit Co., at Leadville and Montezuma; in Arizona in good crystals from near Tucson; in Utah from Bingham Canyon, American Fork, Utah Co., and the Park City district; in various districts in California.

The name pyrite is derived from πῦρ, fire, and alludes to the sparks formed when the mineral is struck with a hammer; hence the early name pyrites, p. 432.

Use. — Pyrite often carries small amounts of copper or gold and becomes an important ore of these metals. It is also mined for its sulphur content which is used in the form of sulphur dioxide (used in the preparation of wood pulp for manufacture into paper), as sulphuric acid (used for many purposes, especially in the purification of kerosene and in the preparation of mineral fertilizers), and as the ferrous sulphate, copperas (used in dyeing, in inks, as a wood preservative, and as a disinfectant).

Bravoite. — (Fe,Ni)S$_2$. Contains nearly 20 per cent nickel In small grains and crystal fragments, apparently octahedral. Pale yellow with a faint reddish tarnish. Occurs disseminated through the vanadium ores at Minasragra, near Cerro de Pasco, Peru.

Cobaltnickelpyrite. *Hengleinite.* Iron sulphide with about 20 per cent cobalt and nickel, (Co,Ni,Fe)S$_2$. In minute pyritohedral crystals. Steel-gray color. Gray-black streak. H. = 5. G. = 4·716. Found at Müsen, Westphalia, Germany. Microscopic tests show that it is probably a mixture of siegenite and pyrite.

Hauerite. — Manganese disulphide, MnS$_2$. In octahedral or pyritohedral crystals; also massive. Shows differences in manner of oxidation, solution, etc. from pyrite but X-ray analysis indicates a similar structure. G. = 3·46. A rare mineral which occurs in association with gypsum and sulphur and has apparently been deposited from waters in which there was a small amount of manganese and a great excess of sulphur. In Czechoslovakia at Kalinka, near Neusohl (Ban Bystrica), and at Schemnitz (Selmeczbánya). In Sicily at Raddusa, west of Catania. Also in the Lake Wakatipu district, Collingwood, New Zealand.

Arsenoferrite. — Iron arsenide, probably FeAs$_2$. Isometric-pyritohedral. In small crystals. Color dark brown. H. = 5·5. G. = 6·42. Fine splinters transparent with ruby-red color. From the Binnental, Switzerland; also at Joachimstal (Jáchymov), Czechoslovakia.

Sperrylite. — Platinum diarsenide, PtAs$_2$. Isometric-pyritohedral. In minute cubes, or cubo-octahedrons with at times small pyritohedral or diploid faces. H. = 6–7. G. =

10·6. Luster metallic. Color tin-white. Streak black. Found originally in gold-quartz at the Vermilion mine, Algoma district, 22 miles west of Sudbury, Ontario. A few crystal grains were identified in the sands of Macon Co., Georgia. Found associated with covellite at the Rambler mine, Medicine Bow Mts., Wyoming. Recently found in large cubooctahedral crystals from northwest of Potgietersrust, Waterberg district, Transvaal, embedded in limonite. Noted in gold washings in eastern Siberia.

COBALTITE.

Isometric-tetartohedral. Commonly in cubes, or pyritohedrons, or combinations resembling common forms of pyrite. Also granular massive to compact.

Many cobalt crystals are paramorphs after an orthorhombic form since polished sections examined in reflected polarized light show anisotropic characters with twin lamellæ (probably orthorhombic). When heated to 850° C. such sections become permanently isotropic. X-ray study, however, indicates an isometric lattice or one with very slight divergences. The substitution of AsS for SS in the structure appears to account for the change from pyritohedral to tetartohedral symmetry. The evidences of tetartohedral symmetry are rare in crystals of cobaltite.

Cleavage: cubic, rather perfect. Fracture uneven. Brittle. H. = 5·5. G. = 6–6·3. Luster metallic. Color silver-white, inclined to red; also steel-gray, with a violet tinge, or grayish black when containing much iron. Streak grayish black.

Comp. — Sulpharsenide of cobalt, CoAsS or $CoS_2.CoAs_2$ = Sulphur 19·3, arsenic 45·2, cobalt 35·5 = 100.

Iron is present, and in the variety *ferrocobaltite* in large amount.

Pyr., etc. — Unaltered in the closed tube. In the open tube gives sulphurous fumes, and a crystalline sublimate of arsenic trioxide. B.B. on charcoal gives off sulphur and arsenic oxides, and fuses to a magnetic globule; with borax a cobalt-blue color. Soluble in warm nitric acid, with the separation of sulphur.

Obs. — Most commonly found in metasomatic contact deposits, often in quartz-rich gneiss, also in mica and hornblende schists and in diopside rocks. More rarely in typical veins. Occurs with other cobalt and nickel minerals, with copper and silver ores, with pyrite, pyrrhotite, molybdenite, siderite, calcite, quartz, tourmaline, etc.

Occurs in considerable amount at Dashkesan near Ellsavetpol, Azerbaijan, Trans-Caucasia. In Sweden localities notable for large, splendent, well-defined crystals are at Tunaberg, Södermanland; at Riddarhyttan and Hakansbö in Vastmanland; in Norway at Skutterud near Modum, Buskerud. From the Botallack mine, near St. Just, Cornwall. In the silver mines at Cobalt, Coleman township, Ontario.

Use. — An ore of cobalt.

Gersdorffite. — Sulpharsenide of nickel, NiAsS or $NiS_2.NiAs_2$. Iron, and sometimes cobalt, replace more or less of the nickel. Isometric-tetartohedral; usually massive. H. = 5·5. G. = 5·6–6·2. Color silver-white to steel-gray. Often in veins with siderite and other iron minerals, with niccolite, ullmannite, etc., with sphalerite, galena, chalcopyrite, etc. Occurs in Czechoslovakia at Dobsina (Dobschau); from Schladming in Styria, Austria. From Lobenstein, Thuringia; from Friedrichssegen and elsewhere near Ems, Hessen-Nassau; from Müsen, Westphalia. From the Sudbury district, Ontario. *Corynite* from Olsa, Carinthia, described as being near gersdorffite but containing antimony, has been shown to be a mixture.

Modderite. A cobalt arsenide. Bluish white color. Occurs with niccolite on the Witwatersrand, Transvaal.

Ullmannite. — Sulphantimonide of nickel, NiSbS or $NiS_2.NiSb_2$; arsenic is usually present in small amount. Isometric-tetartohedral; both pyritohedral and tetrahedral forms occur. Usually massive, granular. H. = 5–5·5. G. = 6·2–6·7. Color steel-gray to silver-white. Occurs in veins, frequently with siderite and other iron minerals; also at times with niccolite, gersdorffite, sphalerite, galena, and other sulphides. In Carinthia, Austria, at Lölling near Hüttenberg. In Westphalia in the district about Siegen and Müsen and at Petersbach near Hamm; from Lobenstein, Thuringia. In Sardinia at Monte Narba near Sarrabus. In France from the mine of Ar near Eaux-Bonnes, Basses-Pyrénées. In England at Brancepeth colliery, Durham.

VILLAMANINITE. Sulphide of Cu,Ni with smaller amounts of Co,Fe. Microscopic study indicates that it is a mixture. H. = 4·5. G. = 4·4–4·5. Color, iron-black. In irregular groups of cubo-octahedral crystals and in radiating nodular masses. In dolomite from Cármenes district, near Villamanín, León, Spain.

WILLYAMITE. Described as a cobalt-nickel sulphide-antimonide from Broken Hill, New South Wales, has been shown to be a mixture.

KALLILITE. Described as a nickel-antimony-bismuth sulphide from near Schönstein a.-d.-Sieg, Germany, but shown by mineralographic study to be a mixture.

Laurite. — Sulphide of ruthenium and osmium, probably essentially RuS$_2$. In minute octahedrons; in grains. Octahedral cleavage. H. = 7·5. G. = 6·99. Luster metallic. Color dark iron-black. From the platinum washings of Borneo. Also reported from Colombia and Oregon.

SMALTITE-CHLOANTHITE.

Isometric-pyritohedral. X-ray study shows that smaltite-chloanthite and skutterudite have similar atomic structures but show differences from the pyrite structure. Commonly massive; in reticulated and other imitative shapes.

Cleavage indistinct. Fracture granular and uneven. Brittle. H. = 5·5–6. G. = 5·7–6·8. Luster metallic. Color tin-white, inclining, when massive, to steel-gray, sometimes iridescent, or grayish from tarnish. Streak grayish black. Opaque.

Comp. — SMALTITE is essentially cobalt arsenide. CHLOANTHITE is nickel arsenide. Analyses show considerable variations, RAs$_3$, representing most nearly the composition.

Cobalt and nickel are usually both present, and thus these two species graduate into each other, and no sharp line can be drawn between them. Iron is also present in varying amount; the variety of chloanthite containing much iron has been called *chathamite*. Further sulphur is usually present, but only in small quantities. Microscopic examination of polished specimens shows probable zoning of different members of the group. It has been suggested that RAs$_2$, R$_2$As$_5$, and RAs$_3$, are all present. Material known as *keweenawite* is a mixture of smaltite, niccolite and domeykite.

Much that has been called smaltite is shown by the high specific gravity to belong to the orthorhombic species saflorite.

Pyr., etc. — In the closed tube gives a sublimate of metallic arsenic; in the open tube a white sublimate of arsenic trioxide, and sometimes traces of sulphur dioxide. B.B. on charcoal gives a coating of As$_2$O$_3$, the arsenical odor, and fuses to a globule, which, treated with successive portions of borax-glass, affords reactions for iron, cobalt, and nickel.

Obs. — Usually occurs in veins, accompanying other minerals of cobalt and nickel, frequently with ores of silver and copper; with sphalerite and galena, with arsenopyrite and native arsenic; sometimes with bismuth. The origin is probably pneumatolytic and deposition occurs at temperatures between 385° C. and 450° C.

Occurs at Joachimstal (Jáchymov), Bohemia in a reticulated variety embedded in calcite; at Dobsina (Dobschau), Czechoslovakia. With silver ores etc., at Schneeberg, Annaberg, Freiberg, etc., Saxony; in veins in the copper schists at Richelsdorf and Bieber, Hessen-Nassau; at Wittichen, Black Forest, Baden. Found in crystals in France at Les Chalantes, near Allemont, Isère, and at Ste. Marie aux Mines (Markirch), Alsace. From Province Huelva and elsewhere in Spain. At Wheal Sparnon in Cornwall. At Broken Hill, New South Wales. At Chatham, Connecticut, chloanthite (*chathamite*) associated generally with arsenopyrite and sometimes with niccolite. Rare at Franklin, New Jersey, in crystals. Found in considerable amount in the Cobalt district, Coleman township, Ontario.

Use. — Ores of cobalt and nickel.

Skutterudite. — Cobalt arsenide, CoAs$_3$. Isometric-pyritohedral. Also massive granular. Cleavage: $a(100)$, distinct. H. = 6. G. = 6·5–6·9. Color between tin-white and pale lead-gray. Found at Skutterud, near Modum, Buskerud, Norway. From Crete d'Omberenza, in Turtmann Tal, Valais, Switzerland. From Franklin, New Jersey; in Grant Co., New Mexico. From the Cobalt district, Ontario.

NICKEL-SKUTTERUDITE. (Ni,Co,Fe)As$_3$. Massive, granular. Color gray. From near Silver City, New Mexico.

BISMUTO-SMALTITE. $Co(As,Bi)_3$. A skutterudite containing bismuth. Color tin-white. G. = 6·92. Zschorlau, near Schneeberg, Saxony.

Marcasite Group

For the list of species and their relations, see p. 433.

MARCASITE. White iron pyrites.

Orthorhombic. Axes $a : b : c = 0·7662 : 1 : 1·2342$.

mm''', 110 $\wedge$ $1\overline{1}0$ = 74° 55'. ll', 011 $\wedge$ $0\overline{1}1$ = 101° 58'.
ee',   101 $\wedge$ $\overline{1}01$ = 116° 20'. cs, 001 $\wedge$ 111 = 63° 46'.

Twins: tw. pl. $m(110)$, sometimes in stellate fivelings (Fig. 462, p. 191, cf. Fig. 697); also tw. pl. $e(101)$, less common, the crystals crossing at angles

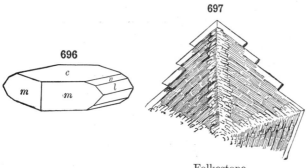

696

697

Folkestone

of nearly 60°. Crystals commonly tabular ‖ $c(001)$, also pyramidal; the brachy-domes striated ‖ edge $b(010)/c(001)$. Often massive; radiating fibrous; in stalactites; also globular, reniform, and other imitative shapes.

Cleavage: $m(110)$ poor; $l(011)$ in traces. Fracture uneven. Brittle. **H.** = 6–6·5. **G.** = 4·85–4·90. Luster metallic. Color pale bronze-yellow, deepening on exposure. Streak grayish or brownish black. Opaque.

Comp. — Iron disulphide, like pyrite, FeS_2 = Sulphur 53·4, iron 46·6 = 100. Arsenic is sometimes present in small amount.

Var. — The varieties named depend mainly on state of crystallization. *Radiated. Cockscomb Pyrite*: Aggregations of flattened twin crystals in crest-like forms. *Spear Pyrite*: Twin crystals, with re-entering angles a little like the head of a spear in form (Fig. 697). *Capillary*: In capillary crystallizations.

Pyr., etc. — Like pyrite. Very liable to decomposition, more so than pyrite.

Diff. — Resembles pyrite, but has a lower specific gravity, and the color when fresh (*e.g.*, after treatment with acid) is paler; when crystallized easily distinguished by the forms. More subject to tarnish and final decomposition than pyrite.

Marcasite can be distinguished chemically from pyrite by the following methods. When both minerals are finely powdered and treated with a little concentrated nitric acid, first in the cold and later, after vigorous action has ceased, by warming, it will be found that in the case of pyrite the greater part of the sulphur of the mineral has been oxidized and taken into solution as sulphuric acid, while in the case of marcasite most of the sulphur has separated in a free state. The Stokes method, which can be used quantitatively to determine the amounts of the two minerals in a mixture, depends upon the difference in their behavior when boiled with a standard solution of ferric sulphate. In the case of pyrite about 52 per cent of the sulphur is oxidized to sulphuric acid, while with marcasite only about 12 per cent is oxidized. Another method consists in boiling the minerals in a 3 per cent solution of $AgNO_3$, when marcasite will be tarnished to a tobacco-brown color, then red and finally blue, whereas pyrite becomes only slightly brownish.

Alt. — Marcasite being relatively unstable is easily altered. Specimens often disintegrate with the formation of ferrous sulphate and sulphuric acid. It also alters to pyrite, limonite, etc.

Obs. — Marcasite is a much less stable compound than pyrite and is formed under comparatively limited conditions. Experiments have shown that it is deposited at temperatures below 450° C. and from acid solutions. The higher the temperature of deposition the more acid must the solution contain; at ordinary temperatures marcasite may be deposited from nearly neutral solutions. Marcasite is formed in general under surface conditions; in deep veins where the minerals are deposited from ascending, hot, and usually alkaline waters, only pyrite will be formed.

It most commonly occurs, associated with galena, sphalerite, calcite and dolomite, etc., in replacement deposits in limestones, also in druses of ore veins where it is deposited upon other minerals and is clearly later in origin. At times is found in concretions, sometimes of considerable size, embedded in sedimentary rocks, clays, chalks, marls, etc. At times associated with lignites. Often forms the material in which fossils are preserved.

The spear-shaped variety occurs abundantly in clay in Bohemia of Czechoslovakia at Littmitz, west of Carlsbad (Karl Vary), at Altsattel, southeast of Přibram, and at Teplice (Teplitz); from Saxony at Freiberg and elsewhere. In France it occurs as spear-shaped twins at Cap Blanc Nez, Pas de Calais; also similarly in the chalk marl in Kent between Folkestone and Dover. In its cockscomb form it is found at Tavistock, Devonshire. Occurs at Weardale, Durham. From Guanajuato, Mexico. Notable localities in the United States are at Galena, Illinois, where it occurs in stalactites, with concentric layers of sphalerite and galena; from Wisconsin at Mineral Point, Iowa Co., and in crystals altered to limonite from Richland Co.; from the Joplin district, Missouri, associated with galena, sphalerite, dolomite.

The word *marcasite*, of Arabic or Moorish origin (and variously used by old writers, for bismuth, antimony), was the name of common crystallized pyrite among miners and mineralogists in later centuries, until near the close of the eighteenth. It was first given to this species by Haidinger in 1845.

Löllingite. — Essentially iron diarsenide, FeAs$_2$, but passing into Fe$_3$As$_4$ (*leucopyrite*); also tending toward arsenopyrite (FeAsS) and safflorite (CoAs$_2$). Bismuth and antimony are sometimes present. Usually massive. H. = 5–5·5. G. = 7·0–7·4 chiefly, also 6·8. Luster metallic. Color between silver-white and steel-gray. Streak grayish black.

Occurs in veins where it has been deposited from solutions at medium temperatures. Most commonly associated with calcite, limonite, siderite, various sulphides, and with cobalt, silver and gold ores. Occurs at Lölling near Hüttenberg, Carinthia, Austria. The arsenical iron from Reichenstein, Silesia, is in part löllingite, but mostly leucopyrite. From Andreasberg, in the Harz Mts.; from Breitenbrunn and from Geyer (*geyerite*) near Ehrenfriedersdorf in Saxony. *Glaucopyrite* is a variety from Guadalcanal in the Sierra Morena, Prov. Sevilla, Andalusia, Spain. In the United States has been found at Auburn and Paris, Maine; in Orange Co., New York, at Monroe and Edenville (*leucopyrite*); from Gunnison and San Juan counties, Colorado.

Safflorite. — Like smaltite, essentially cobalt diarsenide, CoAs$_2$, but with usually considerable amounts of iron and more rarely small amounts of nickel. Orthorhombic. Form near that of arsenopyrite. Usually massive. H. = 4·5–5. G. = 6·9–7·3. Color tin-white, soon tarnishing. Occurs in veins with other cobalt and nickel minerals, with bismuth, chalcopyrite, sphalerite, calcite, dolomite, etc. Found at Schneeberg, Saxony; at Bieber, in Hessen-Nassau; at Reinerzau near Wittichen, Baden. In Sweden at the Ko mine, Nordmark, and at Tunaberg in Södermanland. Reported from South Lorrain, Quebec.

Rammelsbergite. — Essentially nickel diarsenide, NiAs$_2$, but with isomorphous FeAs$_2$ and CoAs$_2$. Orthorhombic. Crystals resembling arsenopyrite; also massive. G. = 6·9–7·2. Color tin-white with tinge of red. Frequently found associated with quartz and horn stone, also with niccolite, chloanthite, pyrite, arsenopyrite, etc. Occurs at Lölling near Hüttenberg, Carinthia, Austria. At Schneeberg in Saxony; Richelsdorf, in Hessen-Nassau; at Eisleben in Thuringia. In Alsace at Ste. Marie aux Mines (Markirch). Found at Franklin, New Jersey, and in the silver veins of Cobalt, Ontario.

ARSENOPYRITE, or MISPICKEL.

Orthorhombic. Axes $a : b : c = 0{\cdot}6773 : 1 : 1{\cdot}1882$. (X-ray study shows that the unit cell has dimensions that correspond to this axial ratio except that the length of a should be doubled.)

$$
\begin{aligned}
&mm''', && 110 \wedge 1\bar{1}0 &&= 68° 13'. \\
&ee', && 101 \wedge \bar{1}01 &&= 120° 38'. \\
&uu', && 014 \wedge 0\bar{1}4 &&= 33° 5'. \\
&nn', && 012 \wedge 0\bar{1}2 &&= 61° 26'. \\
&qq', && 011 \wedge 0\bar{1}1 &&= 99° 50'.
\end{aligned}
$$

Twins: tw. pl. m(110), sometimes repeated like marcasite (Figs. 700 and 463, p. 191); e(101) cruciform twins, also trillings (Figs. 458, 459, p. 191). Crystals prismatic m(110) or flattened vertically by the oscillatory combination of brachydomes. Also columnar, straight, and divergent; granular, or compact.

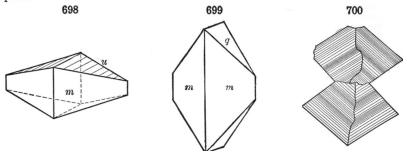

Cleavage: m(110) rather distinct; c(001) in faint traces. Fracture uneven. Brittle. H. = 5·5–6. G. = 5·9–6·2. Luster metallic. Color silver-white, inclining to steel-gray. Streak dark grayish black. Opaque.

Comp. — Sulpharsenide of iron, FeAsS or $FeS_2.FeAs_2$ = Arsenic 46·0, sulphur 19·7, iron 34·3 = 100. Part of the iron is sometimes replaced by cobalt, as in the variety *danaite* (3 to 9 per cent Co).

Pyr., etc. — In the closed tube may give at first a little yellow sulphide of arsenic and then a conspicuous sublimate of metallic arsenic which is of bright gray crystals near the heated end and of a brilliant black amorphous deposit farther away. In the open tube gives sulphurous fumes and a white sublimate of arsenic trioxide. B.B. on charcoal gives arsenical fumes and a magnetic globule. Decomposed by nitric acid with the separation of sulphur.

Diff. — Characterized by its hardness and tin-white color; closely resembles some of the sulphides and arsenides of cobalt and nickel, but identified, in most cases easily, by its blowpipe characters. Löllingite does not give a decided sulphur reaction.

Obs. — Arsenopyrite is the most common mineral containing arsenic, is widespread in its occurrence and is to be found under varying conditions. It is found with tin and tungsten ores in pneumatolytic deposits; with silver ores, galena, sphalerite, pyrite, chalcopyrite, tetrahedrite and with calcite, siderite, and quartz in veins formed by deposition from hot waters; with cobalt and nickel ores. It is also found in what are considered to be contact deposits. Frequently associated with gold. At times occurs disseminated in crystalline rocks, as gneiss, etc., in limestones and dolomites, in serpentines.

Notable localities for the occurrence of arsenopyrite are given below. In Salzburg, Austria, at Mitter-Berg, north of Mühlbach. In the silver mines of Saxony at Freiberg, Munzig near Meissen, Hohenstein near Glauchau, Altenberg, etc. In crystals from the Binnental, Valais, Switzerland. In Sweden at Sala in Vastmanland and at Tunaberg (*danaite*) in Södermanland; in Norway at Skutterud near Modum, Buskerud, and in crystals of danaite from Sulitelma, Nordland. In Cornwall at Redruth, St. Agnes, St. Just, etc.; in large crystals from Tavistock, Devonshire. In Bolivia at the San Baldomero mine, near Sorata, La Paz. In the United States occurs in fine crystals at Franconia, New Hampshire (*danaite*); from Roxbury, Connecticut; from Franklin, New Jersey; Emery, Montana, in crystals; from Leadville, Colorado. Large beds occur in quartz ore veins at Deloro, Marmora township, Hastings Co., Ontario.

The name mispickel is an old German term of doubtful origin. *Danaite* is from J. Freeman Dana of Boston (1793–1827), who made known the Franconia locality.

Use. — An ore of arsenic.

Gudmundite. FeSbS. Orthorhombic. $a : b : c$ = 0·6729 : 1 : 1·1868. Crystals elongated parallel to a axis. Twinned on (110). Silver-white to steel-gray. H. = 6. Occurs in a calcite vein associated with lead and zinc ores at Gudmundstrop, near Sala, Sweden.

Glaucodot. Sulpharsenide of cobalt and iron, (Co,Fe)AsS. In orthorhombic crystals (axes, etc., p. 433). Also massive. H. = 5. G. = 5·90–6·01. Luster metallic. Color grayish tin-white.   Occurs commonly with other cobalt minerals, particularly with cobaltite, with which it is often intergrown.   Also associated with chalcopyrite, pyrrhotite, galena, quartz, etc.   Found in fine crystals, often twins, at Hakansbö, Vastmanland, Sweden. From Skutterud near Modum, Buskerud, Norway.   From near Huasco, Atacama, Chile. Reported from Grant Co., Oregon.   Found in the silver veins of Cobalt, Ontario.   The danaite of Franconia, New Hampshire, approaches glaucodot in composition.   *Alloclasite*, which appears to be a mixture of glaucodot with some other mineral, occurs at Oravicza, Rumania.   Named from γλαυκος, *blue*, because used for making smalt.

Cooperite. $Pt(As,S)_2$.   Orthorhombic.   In minute and irregular crystal grains. Conchoidal fracture.   Color, steel-gray.   H. = 4–5·5.   G. = 9.   From platiniferous norite in the Rustenburg district, Transvaal, and from dunite of Lydenburg district, South Africa.

Wolfachite. — Probably $Ni(As,Sb)S$, near corynite.   In small crystals resembling arsenopyrite; also columnar radiated.   H. = 4·5–5.   G. = 6·372.   Color silver-white to tin-white.   From Wolfach, on the Kinzig, Baden, Germany.

Melonite. — A nickel telluride, $NiTe_2$.   One perfect cleavage.   In indistinct granular and foliated particles.   Color reddish white, with metallic luster.   H. = 1–1·5.   G. = 7·3. Reported from Magnolia, Boulder Co., Colorado.   Originally from Melones mine, and also from Stanislaus mine, Calaveras Co., California.   Found at Worturpa, New South Wales.

The following species are tellurides of gold, silver, etc.

SYLVANITE.   Graphic Tellurium.

Monoclinic. $a : b : c = 1·6339 : 1 : 1·1265$; $\beta = 89° 35'$.   Twins: tw. pl. $m(110)$ giving rise to branching arborescent forms resembling written characters;   also bladed and imperfectly columnar to granular.

Cleavage: $b(010)$ perfect.   Fracture uneven.   Brittle.   H. = 1·5–2.   G. = 7·9–8·3.   Luster metallic, brilliant.   Color and streak pure steel-gray to silver-white, inclining to yellow.

Comp. — Telluride of gold and silver $(Au,Ag)Te_2$; with Au : Ag = 1 : 1; Tellurium 62·1, gold 24·5, silver 13·4 = 100.

Pyr., etc. — When a little of the powdered mineral is heated in concentrated sulphuric acid a reddish violet color is given to the solution.   When treated with nitric acid is decomposed leaving residue of rusty colored gold.   A few drops of hydrochloric acid added to this solution yield an abundant precipitate of silver chloride.   In the open tube gives a white sublimate of tellurium dioxide which near the assay is gray; when treated with the blowpipe flame the sublimate fuses to clear transparent drops.   R.R. on charcoal fuses to a dark gray globule, covering the coal with a white coating, which treated in R.F. disappears, giving a bluish green color to the flame; after long blowing a yellow, malleable metallic globule is obtained.

Obs. — Sylvanite occurs in veins associated with native gold, other gold and silver tellurides, more rarely with tellurides of other metals, with pyrite, and other sulphides, with quartz, calcite, fluorite, etc.   Usually associated with igneous rocks and commonly in deposits formed in the upper zone of the earth's crust, but in Australia the deposits belong to a deeper-seated type.   Occurs in Transylvania of Rumania at Offenbánya and at Nagy-Ág. At Kalgoorlie, West Australia.   In Colorado has been found in Boulder Co., at Gold Hill, Sunshine, Magnolia, etc.; and from the gold-telluride district of Cripple Creek, El Paso Co. In California was found in Calaveras Co., in the Melones and Stanislaus mines at Carson Hill.   Named from Transylvania, where first found, and in allusion to *sylvanium*, one of the names at first proposed for the metal tellurium.

Use. — An ore of gold.

Krennerite. — A telluride of gold and silver $(Au,Ag)Te_2$ like sylvanite.   In prismatic crystals (orthorhombic), vertically striated.   Cleavage (001), perfect.   H. = 2·5.   G. = 8·353.   Color silver-white to brass-yellow.   Rarer than sylvanite but having same mode of occurrence, associations, etc.   From Nagy-Ág in Transylvania, Rumania, and Cripple Creek, El Paso Co., Colorado.

Calaverite. — A gold telluride, $AuTe_2$ with small amounts of silver.   Monoclinic.   In small lath-shaped crystals striated parallel to their length.   Massive granular to crystalline.

H. = 2·5. G. = 9·043. Color silver-white with often a faint yellow tinge. Tests similar to those for sylvanite with smaller amount of silver showing. A rare mineral but in Australia and Colorado becoming an important source of gold. Association and occurrence the same as with sylvanite. Found at Kalgoorlie in West Australia; at the Cripple Creek district, El Paso Co., Colorado, and at the Melones and Stanislaus mines at Carson Hill, Calaveras Co., California. A similar gold telluride but differing in its microchemical reactions has been described under the name *antamokite*. It occurs, associated with calaverite, at Antamok, Mountain Province, Philippine Islands.

Muthmannite. — $(Ag,Au)Te$. In tabular crystals usually elongated in one direction. One perfect cleavage parallel to elongation. H. = 2·5. Color bright brass-yellow, on fresh fracture gray-white. Probably from Nagy-Ág, Transylvania, Rumania. *Empressite*, $AgTe$, from the Empress-Josephine mine, in the Kerber Creek District, Colorado, is probably a gold-free variety, although it differs from muthmannite in not showing a cleavage. Massive. H. = 3–3·5. G. = 7·5. Color pale bronze.

Nagyagite. — A sulpho-telluride of lead and gold; some analyses show also about 7 per cent of antimony which was probably due to impurities. Orthorhombic. Crystals tabular ‖ $b(010)$; also granular massive, foliated. Cleavage: b perfect; flexible. H. = 1–1·5. G. = 6·85–7·2. Luster metallic, splendent. Streak and color blackish lead-gray. Opaque. From Nagy-Ág, and Offenbánya, Transylvania, Rumania. Reported from Colorado and Tararu Creek, New Zealand.

Oxysulphides

Here are included Kermesite, Sb_2S_2O, and Voltzite, Zn_5S_4O.

Kermesite. Pyrostibite. — Antimony oxysulphide, Sb_2S_2O or $2Sb_2S_3.Sb_2O_3$. Orthorhombic or monoclinic. Usually in tufts of capillary crystals. Cleavage: $a(100)$ perfect. H. = 1–1·5. G. = 4·5–4·6. Luster adamantine. Color cherry-red. $n = 2·75$.

Kermesite is a secondary mineral occurring as the alteration product of stibnite and associated with native antimony, stibnite, valentinite, senarmontite, etc. It is found in Czechoslovakia at Pernek near Bösing, northwest of Bratislava (Pressburg) (this locality is often given as Malaczka (Malacka)); also at Příbram, Bohemia. From Bräunsdorf near Freiberg, Saxony. Occurs at Sarrabus on Sardinia. In France at Les Chalantes near Allemont, Isère. In Algeria at Djebel Haminate southeast of Constantine. At South Ham, Wolfe Co., Quebec; near Lake George, southwest of Fredericton, Prince William parish, York Co., New Brunswick; and at the West Gore mines, Rawdon, Hants Co., Nova Scotia.

Named from *kermes*, a name given (from the Persian *qurmizq*, crimson) in the older chemistry to red amorphous antimony trisulphide, often mixed with antimony trioxide.

Voltzite. — Zinc oxysulphide, Zn_5S_4O or $4ZnS.ZnO$? In implanted spherical globules. H. = 4–4·5. G. = 3·66–3·80. Color dirty rose-red, yellowish. $n = 2·03$. Of secondary origin, commonly associated with various zinc minerals, sulphides, etc. From near Joachimstal (Jáchymov), Bohemia; from Geroldseck near Lahr, Baden. Was first found near Pontgibaud, Puy-de-Dôme, France. Reported from Djebel, Reças, Tunis.

III. SULPHO-SALTS

I. Sulpharsenites, Sulphantimonites, Sulphobismuthites.

II. Sulpharsenates, etc.

III. Sulphostannates, etc.

I. Sulpharsenites, Sulphantimonites, etc.

In these sulpho-salts, as further explained on p. 349, sulphur takes the place of the oxygen in the commoner and better understood oxygen acids (as carbonic acid, H_2CO_3, sulphuric acid, H_4SO_4, phosphoric acid, H_3PO_4, etc.).

The species included are salts of the sulpho-acids of trivalent *arsenic*, *antimony* and *bismuth*. The most important acids are the ortho-acids,

H_3AsS_3, etc., and the meta-acids, H_2AsS_2, etc.; but $H_4As_2S_5$, etc., and a series of others are included. The metals present as bases are chiefly *copper*, *silver*, *lead*; also *zinc, mercury, iron*, rarely others (as *nickel, cobalt*) in small amount. In view of the hypothetical character of many of the acids whose salts are here represented, there is a certain advantage, for the sake of comparison, in writing the composition after the dualistic method, $RS.As_2S_3$, $2RS.As_2S_3$, etc.

Many of the species of the sulpho-salt group are rare and can be only briefly mentioned. Further, as many of them are intimately associated with other similar minerals and are frequently imperfectly crystallized, good analyses are often rare. Therefore the interpretation of the available data and the proper classification of many species present difficulties. Fortunately several general studies of the group have been made recently, and these (especially those made by Wherry and Foshag and by Cesaro) have been largely followed in the classification given below.

It should be further mentioned that many authors include as part of the sulpho-salt group many of the species classified in this book as belonging to the intermediate division of the sulphides, interpreting their composition as sulpho-ferrites, etc.; see pp. 430–432.

A. Acidic Division RS : $(As,Sb,Bi)_2S_3 = 1:3, 1:2, 2:3, 3:4, 4:5$

Eichbergite	$(Cu,Fe)_2S.3(Bi,Sb)_2S_3.$	Chiviatite	$PbS.2Bi_2S_3$ or $2PbS.3Bi_2S_3$
Vrbaite	$Tl_2S.3(As,Sb)_2S_3$	Gladite	$2PbS.Cu_2S.5Bi_2S_3$
Livingstonite	$HgS.2Sb_2S_3$	Rezbanyite	$3PbS.Cu_2S.5Bi_2S_3$
Histrixite	$5CuFeS.2Sb_2S_3 7Bi_2S_3$		

B. Meta- Division RS : $(As,Sb,Bi)_2S_3 = 1:1$

Trechmannite	$Ag_2S.As_2S_3$	Rhombohedral
Platynite	$PbS.Bi_2S_3$	Rhombohedral

Zinkenite Group. Orthorhombic

Zinkenite	$PbS.Sb_2S_3$	Hutchinsonite	$PbS.(Tl,Ag)_2S.2Sb_2S_3$
Andorite	$2PbS.Ag_2S.3Sb_2S_3$	Chalcostibite	$Cu_2S.Sb_2S_3$
Lindströmite	$2PbS.Cu_2S.3Bi_2S_3$	Emplectite	$Cu_2S.Bi_2S_3$

Miargyrite Group. Monoclinic

Miargyrite	$Ag_2S.Sb_2S_3$	Sartorite	$PbS.As_2S_3$
Smithite	$Ag_2S.As_2S_3$	Lorandite	$Tl_2S.As_2S_3$

Matildite	$Ag_2S.Bi_2S_3$	GALENOBISMUTHITE	$PbS.Bi_2S_3$
Aramayoite	$Ag_2S.(Sb,Bi)_2S_3$	Berthierite	$FeS.Sb_2S_3$

C. Intermediate Division. RS : $(As,Sb,Bi)_2S_3 = 5:4, 3:2, 2:1, 5:2$

Fülöppite	$2PbS.3Sb_2S_3$	Monoclinic
Plagionite	$5PbS.4Sb_2S_3$	Monoclinic
Bismuthoplagionite	$5PbS.4Bi_2S_3$	Orthorhombic?

Baumhauerite	$4PbS.3As_2S_3$	Monoclinic
Fizelyite	$5PbS.Ag_2S.4Sb_2S_3$	Monoclinic
Ramdohrite	$3PbS.Ag_2S.3Sb_2S_3$	

Heteromorphite	$7PbS.4Sb_2S_3$	Monoclinic

Rathite	$3PbS.2As_2S_3$	Orthorhombic
Schirmerite	$3(Ag_2,Pb)S.2Bi_2S_3$	
Hammarite	$5PbS.3Bi_2S_3$	
Wittite	$5PbS.3Bi_2(S,Se)_3$	Orthorhombic
Benjaminite	$(Cu,Ag)_2S.2PbS.2Bi_2S_3$	
Klaprothite	$3Cu_2S.2Bi_2S_3$	Orthorhombic

Jamesonite Group.   Monoclinic or Orthorhombic

Jamesonite	$2PbS.Sb_2S_3$	Monoclinic
Dufrenoysite	$2PbS.As_2S_3$	Monoclinic
Owyheeite	$8PbS.2Ag_2S.5Sb_2S_3$	Orthorombic?
Cosalite	$2PbS.Bi_2S_3$	Orthorhombic

Kobellite	$2PbS.(Bi,Sb)_2S_3$	SCHAPBACHITE	$PbS.Ag_2S.Bi_2S_3$?
Berthonite	$5PbS.9Cu_2S.7Sb_2S_3$		

Semseyite	$9PbS.4Sb_2S_3$	Monoclinic

Boulangerite	$5PbS.2Sb_2S_3$	Orthorhombic
Freieslebenite	$5(Pb,Ag_2)S.2Sb_2S_3$	Monoclinic
Diaphorite	$5(Pb,Ag_2)S.2Sb_2S_3$	Orthorhombic

D.   Ortho- Division. $RS : (As,Sb,Bi)_2S_3 = 3 : 1$

Bournonite Group.   Orthorhombic

Bournonite	$2PbS.Cu_2S.Sb_2S_3$	Lillianite	$3PbS.Bi_2S_3$
Seligmannite	$2PbS.Cu_2S.As_2S_3$	Wittichenite	$3Cu_2S.Bi_2S_3$
Aikinite	$2PbS.Cu_2S.Bi_2S_3$		

Pyrargyrite Group.   Rhombohedral-hemimorphic

Pyrargyrite	$3Ag_2S.Sb_2S_3$	Proustite	$3Ag_2S.As_2S_3$

Pyrostilpnite	$3Ag_2S.Sb_2S_3$	Monoclinic
Samsonite	$2Ag_2S.MnS.Sb_2S_3$	Monoclinic
Xanthoconite	$3Ag_2S.As_2S_3$	Monoclinic
SANGUINITE	$3Ag_2S.As_2S_3$?	
Guitermanite	$3PbS.As_2S_3$?	
Stylotypite	$3Cu_2S.Sb_2S_3$?	
Falkenhaynite	$3Cu_2S.Sb_2S_3$?	Monoclinic
TAPALPITE	$3Ag_2(S,Te).Bi_2(S,Te)_3$	

Tetrahedrite Group. Isometric-tetrahedral

Tetrahedrite $3Cu_2S.Sb_2S_3$ **Tennantite** $3Cu_2S.As_2S_3$

E. Basic Division. RS : $(As,Sb,Bi)_2S_3 = 4 : 1, 5 : 1, 6 : 1, 9 : 1, 12 : 1$

Lengenbachite	$6PbS.Ag_2S.2As_2S_3$	Triclinic?

Meneghinite	$4PbS.Sb_2S_3$	Orthorhombic
Jordanite	$4PbS.As_2S_3$	Monoclinic
Goongarrite	$4PbS.Bi_2S_3$	Monoclinic?

Stephanite	$5Ag_2S.Sb_2S_3$	Orthorhombic
Geocronite	$5PbS.Sb_2S_3$	Orthorhombic

Goldfieldite	$5Cu_2S.(Sb,As,Bi)_2(S,Te)_3$?	

Beegerite	$6PbS.Bi_2S_3$	Isometric?

Polybasite Group. Monoclinic

Polybasite $9Ag_2S.Sb_2S_3$ **Pearccite** $9Ag_2S.As_2S_3$

Polyargyrite	$12Ag_2S.Sb_2S_3$	Isometric

Ultrabasite	$28PbS.11Ag_2S.3GeS_2.2Sb_2S_3$	Orthorhombic

A. Acidic Division

Eichbergite. — $(Cu,Fe)_2S.3(Bi,Sb)_2S_3$. Color iron-gray. H. > 0. G. = 5·36. Known in a single specimen from the Eichberg, Semmering Mts., Austria.

Vrbaite. — $Tl_2S.3(As,Sb)_2S_3$. Orthorhombic. H. = 3·5. G. = 5·3. Color gray-black to dark red in thin splinters. Streak light red. Occurs intergrown with realgar and orpiment at Allchar, near Rozsdan, northwest of Salonika, Macedonia, Greece.

Livingstonite. — $HgS.2Sb_2S_3$. Resembles stibnite in form. Color lead-gray; streak red. Biaxial, probably . $n > 2.72$. Birefringence strong. Elongation || Z H. = 2. G. = 4·81. From Mexico at Huitzuco, Guerrero and at Guadalcázar, San Luis Potosi.

Histrixite. — $5CuFeS_4.2Sb_2S_3.7Bi_2S_3$. Orthorhombic. In radiating groups of prismatic crystals. H. = 2. Color and streak steel-gray. Found at Ringville, Tasmania.

Chiviatite. $PbS.2Bi_2S_3$ or $2PbS.3Bi_2S_3$. Foliated massive. Color lead-gray. From Chiviato, Peru. Microscopic study of material from this locality showed it to be a mixture of bismuthinite and various copper minerals. *Cannizarrite* from fumaroles on Vulcano, Lipari Islands, is closely similar to chiviatite if not identical with it.

Gladite. — $2PbS.Cu_2S.5Bi_2S_3$. Prismatic crystals. Cleavages: (010) good, (100) imperfect. Color lead-gray. Black streak. H. = 2–3. G. = 6·96. Occurs on quartz with rezbanyite and galenobismutite at Gladhammar, Province of Kalmar, Sweden.

Rezbanyite. — $Cu_2S.3PbS.5Bi_2S_3$. Fine-granular, massive. Color lead-gray. G. = 6·1–6·4. From Rézbánya and Vaskö, Rumania.

B. Meta- Division. $RS.As_2S_3$, $RS.Sb_2S_3$, etc.

Trechmannite. — $Ag_2S.As_2S_3$. Rhombohedral. Crystals minute with prismatic habit. Good rhombohedral cleavage. Uniaxial, −. On heating becomes biaxial. $\omega = 2.6$.

High birefringence. H. = 1·5–2. Color and streak scarlet-vermilion. From the Binnental, Switzerland. α-*trechmanite* from the same locality has closely similar crystals but slightly different physical properties.

Platynite. — $PbS.Bi_2Se_3$. Rhombohedral. Basal and rhombohedral cleavages. H. = 2–3. G. = 7·98. Color like graphite. Streak shining. In small lamellæ in quartz at Falun, Sweden.

Zinkenite Group. Orthorhombic

ZINKENITE. Zinckenite.

Orthorhombic. Axes $a : b : c = 0·5575 : 1 : 0·6353$. Crystals seldom distinct; sometimes in nearly hexagonal forms through twinning. Lateral faces longitudinally striated. Also columnar, fibrous, massive.

Cleavage not distinct. Fracture slightly uneven. H. = 3–3·5. G. = 5·12–5·35. Luster metallic. Color and streak steel-gray. Opaque.

Comp. — $PbS.Sb_2S_3$ = Sulphur 22·3, antimony 41·8, lead 35·9 = 100. Arsenic sometimes replaces part of the antimony.

Pyr., etc. — Decrepitates and fuses very easily; in the closed tube gives a faint sublimate of sulphur, and antimony trisulphide. In the open tube sulphurous fumes and a white sublimate of oxide of antimony; the arsenical variety gives also arsenical fumes. On charcoal is almost entirely volatilized, giving a coating which on the outer edge is white, and near the assay dark yellow; with soda in R.F. yields globules of lead. Soluble in hot hydrochloric acid with evolution of hydrogen sulphide and separation of lead chloride on cooling.

Obs. — Zinkenite usually occurs in quartz associated with various other members of the sulpho-salt group. Found at Wolfsberg in the Harz Mts. in groups of slender columnar crystals. From the Ludwig mine near Hausach on the Kinzig, Black Forest, Baden. In fibrous masses from near Pontgibaud, Puy-de-Dôme, France. An argentiferous variety occurs in considerable amount at Dundas, Tasmania. Found with andorite at Oruro, Bolivia. In the United States at the antimony mines of Sevier Co., Arkansas; in Colorado in the Red Mountain district, San Juan Co.; in Nevada reported from Morey, Nye Co., and from Eureka, Eureka Co. A mineral closely similar to, if not identical with, zinkenite from Oruro, Bolivia, has been described under the name *keeleyite*.

Andorite. — $2PbS.Ag_2S.3Sb_2S_3$. In prismatic, orthorhombic crystals. H. = 3–3·5. G. = 5·5. Color dark gray to black. From Rumania at Felsöbánya (Baia Sprie); Oruro, Bolivia; Morey, Nye Co., Nevada. *Webnerite* and *sundtite* are identical with andorite.

LINDSTRÖMITE. $2PbS.Cu_2S.3Bi_2S_3$. In prismatic crystals, vertically striated. Cleavages parallel to (100), (010), and (110). Color lead-gray. H. = 3–3·5. G. = 7·01. Occurs on quartz at Gladhammar, Province of Kalmar, Sweden.

Hutchinsonite. — $PbS.(Tl,Ag)_2S.2As_2S_3$. Orthorhombic. In flattened rhombic prisms. Cleavage $a(100)$ good. Optically—. Ax. pl. (100). $X = b$ axis. $ρ < υ$ extreme. $β = 3·176$. High birefringence. H. = 1·5–2. G. = 4·6. Color scarlet to red. From the Binnental, Valais, Switzerland.

Chalcostibite. Wolfsbergite. — $Cu_2S.Sb_2S_3$. In small aggregated orthorhombic prisms; also fine granular, massive. G. = 4·75–5·0. Color between lead-gray and iron-gray. From Wolfsberg in the Harz Mts.; from Huanchaca, Potosi, and elsewhere in Bolivia. Occurs in quantity as aggregates of large rectangular crystals at Rar-el-Auz, in the wadi of Cherrat, east of Casablanca, Morocco. *Guejarite* from Spain is the same species.

Emplectite. — $Cu_2S.Bi_2S_3$. In thin striated orthorhombic prisms. Perfect basal cleavage. G. = 6·3–6·5. H. = 2. Color grayish white to tin-white. Black streak. From Czechoslovakia at Schlaggenwald, Bohemia; from Saxony near Schwarzenberg, at Annaberg, Johanngeorgenstadt, etc.; reported from Norway and Chile. *Cuprobismutite* described as argentiferous $3Cu_2S.4Bi_2S_3$, from Hall Valley, Park Co., Colorado, is probably identical with emplectite.

Miargyrite Group. Monoclinic

Miargyrite. — $Ag_2S.Sb_2S_3$. In complex monoclinic crystals, also massive. H. = 2–2·5. G. = 5·1–5·30. Luster metallic-adamantine. Color iron-black to steel-gray, in thin splinters deep blood-red. Streak cherry-red. Optically +. $α > 2·72$. Birefringence strong.

From Felsöbánya (Baia Sprie), Rumania; Přibram, Bohemia. In complex crystals from Braünsdorf near Freiberg, Saxony. From Chile and Bolivia. From the Sombrerete mine, Zacatecas, and elsewhere in Mexico. From the Flint and Silver City districts in Idaho and the Randsburg district, California.

Smithite. — $Ag_2S.As_2S_3$. Monoclinic. Crystals resemble a flattened hexagonal pyramid. One perfect cleavage (100). Optically −. Ax. pl. (010). $Z \wedge c$ axis $= +4°-6°$. $n = 3.27$? H. $= 1·5-2$. G. $= 4·9$. Color light red changing to orange-red on exposure to light. Streak vermilion. From the Binnental, Valais, Switzerland.

Sartorite. Skleroclase. — $PbS.As_2S_3$. In slender, striated crystals, probably monoclinic. G. $= 5·4$. Color dark lead-gray. Occurs in the dolomite of the Binnental, Valais, Switzerland. α-sartorite is similar but triclinic.

Lorandite. — $Tl_2S.As_2S_3$. Monoclinic. Highly modified tabular or prismatic crystals. Cleavage, perfect (100); distinct (001). Ax. pl. $\perp$ (010); $Z = b$ axis, Y nearly $\| a$ axis. 2V large. Probably optically $+$. $\alpha > 2·72$. Extreme birefringence. Color cochineal-red. From Allchar, near Rozsdan, northwest of Salonika, Macedonia; Rambler mine, Encampment, Wyoming.

Matildite. — $Ag_2S.Bi_2S_3$. In slender, prismatic crystals. G. $= 6·9$. Color gray. Found at the Matilda mine, near Morococha, Prov. Junin, Peru. From near Nikko, Shimotsuke, Japan. Occurs at Lake City, Hinsdale Co., Colorado, and at Cobalt, Ontario.

Aramayoite. — $Ag_2S.(Sb,Bi)_2S_3$. Triclinic, pseudo-tetragonal. Perfect cleavage in one plane with traces of other cleavages making 90° and 45° angles with each other. Color, iron-black, deep blood-red on very thin plates. Black streak. H. $= 2·5$. G. $= 5·6$. Associated with pyrite and tetrahedrite from Chocoya, Province of Sud-Chichas, Potosi, Bolivia.

GALENOBISMUTITE. $PbS.Bi_2S_3$; also with Ag,Cu. Crystalline columnar to compact. Color lead-gray to tin-white. G. $= 6·9$. Occurs with bismutite at Nordmark, Vermland, Sweden, where it sometimes carries gold. The seleniferous variety (weibullite — possibly a mixture of cosalite and guanajuatite) is from Falun, Sweden. The argentiferous variety, alaskaite, is from the Alaska mine, Poughkeepsie Gulch, San Juan Co., Colorado; also from near Cerro Bonete, Prov. Sur-Lipez, Bolivia. Material from Quartzburg district, Boise Co., Idaho, yields on analysis the simple formula. However, much that has been called galenobismutite is shown on microscopic study to be a mixture of cosalite, sphalerite, and bismuth.

Berthierite. — $FeS.Sb_2S_3$. Fibrous massive, granular. H. $= 2-3$. G. $- 4·0$. Color dark steel-gray. Commonly intimately associated with stibnite, the presence of which probably accounts for the variations observed in its analyses. Found at Arany-Idka near Košice (Kaschau or Kassa) in Czechoslovakia. In Saxony from Bräunsdorf and elsewhere in the Freiberg district. Originally described from the district of Pontgibaud, Puy-de-Dôme, France. The varieties murtourite and chazellite were named from localities no longer to be identified with certainty; the variety anglarite came from Anglar. Occurs rather abundantly near Charbes, Val de Villé, Alsace. From the antimony mine near Lake George, southwest of Fredericton, Prince William parish, York Co., New Brunswick.

C. Intermediate Division

Fülöppite. — $2PbS.3Sb_2S_3$. Monoclinic. Crystals small, usually prismatic or rhombohedral in habit. Lead-gray to steel-gray color. H. $=$ slightly greater than 2. G. $= 5·23$. Found at Baia Mare, Satul-Mare (formerly Nagybánya), Rumania.

Plagionite. — $5PbS.4Sb_2S_3$. (Plagionite, heteromorphite, and semseyite are all lead-antimony sulpho-salts with compositions ranging from $5PbS.4Sb_2S_3$ to $5PbS.2Sb_2S_3$. Analyses show variations from formulas given here. All are monoclinic. It has been suggested that they together form a morphotropic series with the vertical crystallographic axis increasing in length with increase in the percentage of lead. However, until the question can be definitely settled, it seems better to treat them as separate species.) Monoclinic. In thick tabular crystals, granular or massive. H. $= 2-3$. G. $= 5·5$. From Wolfsberg, in the Harz Mts., from Goldkronach, Fichtelgebirge, Bavaria. Noted at Leyvaux, Cantal, France. Liveingite from the Binnental, Valais, Switzerland, is the corresponding arsenic compound, $5PbS.4As_2S_3$.

Bismutoplagionite. — $5PbS.4Bi_2S_3$. In needle-like crystals and possibly orthorhombic. Color, faintly bluish lead-gray. H. $= 2·8$. G. $= 5·35$. Found probably near Wickes, Jefferson Co., Montana.

Baumhauerite. — $4PbS.3As_2S_3$. Monoclinic. In complex crystals with varied habit. One perfect cleavage. H. = 3. G. = 3·3. Metallic. Color lead to steel-gray. Streak, chocolate-brown. From the Binnental, Valais, Switzerland.

Fizelyite. — $5PbS.Ag_2S.4Sb_2S_3$. Monoclinic. In deeply striated prisms. Cleavage (010). Dark lead- or steel-gray color. Streak dark gray. H. = 2. Occurs with semseyite at Kisbánya, Comitat Szatmár, Rumania.

RAMDOHRITE. $3PbS.Ag_2S.3Sb_2S_3$. In prismatic or thick lance-shaped crystals. H. = 2. G. = **5·33**. Color gray-black with bluish tinge. Streak gray-black. Found with quartz and pyrite in the mine Chocaya la vieja, in Potosi, Bolivia.

Heteromorphite. — A lead-antimony sulpho-salt intermediate between plagionite and semseyite, see under plagionite. Approximately, $7Pbs.4Sb_2S_3$. Monoclinic. G. = 5·73. H. = 2–3. From Arnsberg, Westphalia; Wolfsberg in the Harz Mts.; Bottino, Tuscany.

Rathite. — $3PbS.2As_2S_3$. Orthorhombic, in prismatic crystals. Cleavage, b(010). Color, lead-gray. H. = 3. G. = 5·41. From the Binnental, Switzerland. *Wiltshireite* is apparently the same species.

Schirmerite. — $3(Ag_2,Pb)S.2Bi_2S_3$. Massive, granular. G. = 6·74. Color lead-gray. Treasury lode, Geneva district, Park Co., Colorado.

HAMMARITE. $5PbS.3Bi_2S_3$ or perhaps $2PbS.Cu_2S.2Bi_2S_3$. Monoclinic? In short prisms or needles. Cleavage (010), good. H. = 3–4. Color steel-gray with red tone. Streak black. Found on quartz at Gladhammar, Province of Kalmar, Sweden.

WITTITE. $5PbS.3Bi_2(S,Se)_3$. Orthorhombic or monoclinic. Good cleavage. Color light lead-gray. Streak black. H. = 2–2·5. G. = 7·12. Occurs at Falun, Kopparberg, Sweden, with quartz and magnetite in an iolite-bearing amphibole rock.

BENJAMINITE. $(Cu,Ag)_2S.2PbS.2Bi_2S_3$. One good cleavage. Doubly refracting. Color gray on fresh fracture, readily tarnishes. Metallic luster. H. = 3·3–3·5. Occurs with chalcopyrite, pyrite, covellite, muscovite, molybdenite, and fluorite, in quartz at the Outlaw Mine, 12 miles north of Manhattan, Nye Co., Nevada.

Klaprothite. Klaprotholite. — $3Cu_2S.2Bi_2S_3$. In furrowed prismatic orthorhombic crystals. G. = 4·6. H. = 2·5. Color steel-gray. Wittichen, Baden, and elsewhere in Germany.

Jamesonite Group. $2RS.As_2S_3$, $2RS.Sb_2S_3$, etc.   Monoclinic or Orthorhombic

JAMESONITE.

Monoclinic. Axes: $a : b : c = 0.8316 : 1 : 0.4260$. $\beta = 88° 36'$. mm''' $110 \wedge 1\bar{1}0 = 79° 28'$. In acicular crystals; common in capillary forms; also fibrous massive, parallel or divergent; compact massive.

Cleavage: basal, perfect. Fracture uneven to conchoidal. Brittle. H. = 2–3. G. = 5·5–6·0. Luster metallic. Color steel-gray to dark lead-gray. Streak grayish black. Opaque.

Comp. — $2PbS.Sb_2S_3$ = Sulphur 19·7, antimony 29·5, lead 50·8 = 100. Most varieties show a little iron (1 to 3 per cent), and some contain also silver, copper, and zinc.

It has been suggested that the iron shown by the analyses is an integral part of the mineral and that the formula should be $4PbS.FeS.3Sb_2S_3$ and that the usual jamesonite formula, $2PbS.Sb_2S_3$, belongs to the material commonly called *plumosite*. It has, however, been shown that by taking the length of the c axis of dufrenoysite at one-third the length commonly assigned to it an isomorphous relationship between that mineral and jamesonite is shown.

Pyr. — Same as for zinkenite, p. 446.

Obs. — Jamesonite is to be found in ore veins associated with other lead sulpho-salts, galena, stibnite, tetrahedrite, sphalerite, pyrite, etc., commonly in quartz with siderite, dolomite, rhodochrosite, and calcite. It is probable that many occurrences that have been credited to jamesonite are of some other lead sulpho-salt occurring in fibrous form. The most important localities for its occurrence are: Arany-Idka near Kosice (Kaschau or Kassa), Czechoslovakia; in Cornwall, at Endellion, etc. In Tasmania it occurs in quantity at Mt. Bischoff, at Dundas and at Zeehan. At the antimony mines in Sevier Co., Arkansas, and at Silver City, Pennington Co., South Dakota.

The *feather ore* occurs at Felsöbánya (Baia Sprie), Rumania; Schemnitz (Selmeczbánya) and Přibram, Czechoslovakia; in the Harz Mts. at Wolfsberg, etc. At Freiberg and Braünsdorf, Saxony; in Tuscany at Bottino near Seravezza; at Oruro and elsewhere in Bolivia. These so-called feather ores may be divided into flexible and brittle types, all the latter being referred to jamesonite and the former either to zinkenite, plumosite, boulangerite, or meneghinite.

Warrenite has been shown to be probably a mixture of jamesonite and zinkenite. *Comuccite* from St. Georgis, Sardinia, is probably to be referred to jamesonite.

Dufrenoysite. — $2PbS.As_2S_3$. Monoclinic. In highly modified crystals; also massive. Cleavage: $b(010)$ perfect. $n > 2.71$. Strong birefringence. H. = 3. G. = 5.55–5.57. Color blackish lead-gray. From the Binnental, Valais, Switzerland, in dolomite.

Owyheeite. — $2Ag_2S.8PbS.5Sb_2S_3$ or $Ag_2S.5PbS.3Sb_2S_3$. Probably orthorhombic. In acicular crystals or massive with an indistinct fibrous structure. Cleavage perpendicular to elongation of needles. Color light steel-gray to silver-white, yellowish tarnish. Reddish brown streak. H. = 2.5. Occurs in a quartz vein in the Poorman mine, Silver City district, Owyhee Co., Idaho. Originally described as a silver-jamesonite but later considered to be a distinct species.

Cosalite. — $2PbS.Bi_2S_3$. Orthorhombic. Usually massive, fibrous or radiated. H. = 2.5–3. G. = 6.39–6.75. Color lead- or steel-gray. From Rezbánya, Rumania. From the Bjelke mine at Nordmark, Vermland, Sweden (*bjelkite*). Reported from various localities in the New England Range, New South Wales. Originally found in a silver mine at Cosala, Sinaloa, Mexico. An argentiferous variety occurs at Candameña, Chihuahua, Mexico. In the United States in Colorado from near Parrott City, La Plata Co., and from Red Mountain and Poughkeepsie Gulch, San Juan Co. From Deerpark, Spokane Co., Washington. In Ontario from McElroy township (originally reported as galenobismutite) and from Cobalt district.

Kobellite. — $2PbS.(Bi,Sb)_2S_3$. Fibrous radiated or granular massive. G. = 6.3. Color lead-gray to steel-gray. From Hvena, Sweden; Ouray, Colorado.

Berthonite. — $5PbS.9Cu_2S.7Sb_2S_3$. Fine granular. Color lead-gray. Black streak. G. = 5.49. H. = 4–5. Associated with galena in small veins in the iron mines at Slata, Tunisia. Has been examined mineralographically and appears to be a definite species.

SCHAPBACHITE. $PbS.Ag_2S.Bi_2S_3$. From Schapbach, Baden. A doubtful species, probably a mixture.

Semseyite. — $9PbS.4Sb_2S_3$. For relations to plagionite and heteromorphite see under plagionite. Monoclinic. In tabular crystals and rounded aggregates. One perfect cleavage. Gray to black color. G. = 5.8–5.95. From Felsöbánya (Baia Sprie), Rumania. In the Harz Mts. at Wolfsberg. From Oruro, Bolivia.

BOULANGERITE.

Orthorhombic. Axes $a : b : c = 0.5527 : 1 : 0.7478$. In prismatic or tabular crystals or crystalline plumose masses; granular, compact. Cleavages (001) and (010). H. = 2.5–3. G. = 5.7–6.3. Luster metallic. Color bluish lead-gray; often covered with yellow spots from oxidation. Opaque. Streak red-brown.

Comp. — $5PbS.2Sb_2S_3 =$ Sulphur 18.9, antimony 25.7, lead 55.4 = 100.

Pyr. — Same as for zinkenite, p. 446.
Obs. — Found in ore veins associated with other lead sulpho-salts, galena, stibnite, tetrahedrite, sphalerite, pyrite, etc. Very commonly with siderite, quartz, dolomite, calcite, etc. From Nerchinsk, Transbaikalia; Přibram, Bohemia; Wolfsberg in the Harz Mts.; in Rhineland at Mayen in the Eifel, at Oberlahr near Altenkirchen, and at Horhausen; from Bottino near Seravezza, Tuscany. In France was originally found in considerable quantity at Molières, Gard, also at Ally, Haute-Loire. In massive and fibrous forms from Guerrouma, Alger, Algeria. In good crystals from Sala, Vastmanland, Sweden. In the United States from the Echo district, Nevada, and from Stevens Co., Washington.

Embrithite and *plumbostibite* are from Nerchinsk; they are probably impure varieties of boulangerite. *Mullanite*, from near Mullan, Idaho, is apparently identical with boulangerite. *Epiboulangerite*, from Altenberg, Saxony, described as having formula $3PbS.Sb_2S_5$, is probably only a mixture of boulangerite and galena.

FREIESLEBENITE.

Monoclinic. Axes $a : b : c = 0.5871 : 1 : 0.9277$; $\beta = 87° 46'$. Habit prismatic. H. = 2. G. = 6.2–6.4. Luster metallic. Color and streak light steel-gray inclining to silver-white, also to blackish lead-gray.

Comp. — $5(Pb,Ag_2)S.2Sb_2S_3$, or $2Ag_2S.3PbS.2Sb_2S_3$.

Obs. — From Felsöbánya and Kapnikbánya in Rumania. At Freiberg, Saxony. From Hiendelaencina, northeast of Guadalajara in the province of the same name, Spain. From Augusta Mt., Gunnison Co., Colorado.

Diaphorite. — Like freieslebenite in composition but orthorhombic in form. H. = 2.5. G. = 6.0–6.2. From Příbram, Bohemia; Bräunsdorf, near Freiberg, Saxony. Reported from Zancudo, Colombia, and from near Catorce, San Luis Potosi, Mexico. Found in the Lake Chelan district, Washington. Material originally described as *brongniardite* from Bolivia is apparently diaphorite.

D.   Ortho- Division. $3RS.As_2S_3$, $3RS.Sb_2S_3$, etc.

Bournonite Group.   Orthorhombic.   Prismatic angle 86° to 87°

BOURNONITE.   Wheel Ore.

Orthorhombic.   Axes: $a : b : c = 0.9380 : 1 : 0.8969$.

mm''', 110 $\wedge$ 1$\bar{1}$0 = 86° 20'	cn, 001 $\wedge$ 011 = 41° 53'
co, 001 $\wedge$ 101 = 43° 43'	cu, 001 $\wedge$ 112 = 33° 15'

Twins: tw. pl. $m(110)$, often repeated, forming cruciform and wheel shaped crystals.   Also massive; granular, compact.

Cleavage: $b(010)$ imperfect; $a(100)$, $c(001)$ less distinct.   Fracture subconchoidal to uneven.   Rather brittle.   H. = 2.5–3.   G. = 5.7–5.9.   Luster metallic, brilliant.   Color and streak steel-gray, inclining to blackish lead-gray or iron-black.   Opaque.

Comp. — $2PbS.Cu_2S.$ Sb_2S_3 = Sulphur 19.8, antimony 24.7, lead 42.5, copper 13.0 = 100.

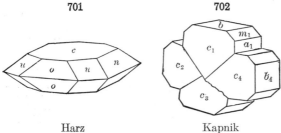

701   702

Harz   Kapnik

Pyr., etc. — In the closed tube decrepitates, and gives a dark red sublimate.   In the open tube gives sulphur dioxide, and a white sublimate of oxide of antimony.   B.B. on charcoal fuses easily, and at first coats the coal white; continued blowing gives a yellow coating of lead oxide; the residue, treated with soda in R.F., gives a globule of copper.   Decomposed by nitric acid, affording a blue solution, and leaving a residue of sulphur, and a white powder containing antimony and lead.

Obs. — Bournonite is one of the more common members of the sulpho-salt group.   It occurs in veins with galena, tetrahedrite, stibnite, chalcopyrite, sphalerite, etc.   Some of the more important localities, particularly those noted for their crystals follow.   From Příbram, Bohemia; in Rumania at Felsöbánya (Baia Sprie), and at Kapnikbánya; in Carinthia, Austria, as *wölchite*, which is bournonite altered on the surface, at Wölch near St. Gertraud, etc.; in the Harz Mts. at Neudorf where the crystals occasionally exceed an inch in diameter, and at Wolfsberg, Klaustal, and Andreasberg.   At Horhausen in the Rhineland.   In exceptional crystals from the mines of Pontgibaud, Puy-de-Dôme, France, and from Cornwall, particularly at Liskeard and at Wheal Boys, Endellion, where it was first found and called *endellione* by Count Bournon, after whom it was subsequently named. In Bolivia from the Pulacayo mine near Huanchaca, Potosi, and from Machacamarca. From Casapalca, Peru. In the United States from the Boggs mine, Big Bug district, Yavapai Co., Arizona; from Emery, Montana; in large crystals from Park City district, Utah; from Austin, Nevada.

Seligmannite. — $2PbS.Cu_2S.As_2S_3$, isomorphous with bournonite. Orthorhombic. $a : b : c = 0.9233 : 1 : 0.8734$. In small complex crystals. Commonly twinned with $m(110)$ as tw. pl. Color lead-gray. Chocolate streak. H. = 3. Found at Lengenbach quarry, Binnental, Valais, Switzerland; reported from Emery, Montana, and from Bingham, Utah.

Aikinite. — $2PbS.Cu_2S.Bi_2S_3$. Acicular crystals; also massive. G. = 6.1–6.8. Color blackish lead-gray. From Berezovsk near Ekaterinburg, Ural Mts.

Lillianite. — $3PbS.Bi_2S_3$. Orthorhombic. Crystalline and massive. Cleavage (100) good, (010) poor. Color steel-gray. H. = 2–3. G. = 7.0. In Sweden from Gladhammar, Kalmar, and from Vena in Nerike near Askersund, Orebo. An argentiferous variety from the mines of the Lillian Mining Co., near Leadville, Colorado.

Wittichenite. — $3Cu_2S.Bi_2S_3$. Rarely in crystals resembling bournonite; also massive. G. = 4.5. Color steel-gray or tin-white. Wittichen, Baden.

Pyrargyrite Group.   Rhombohedral-hemimorphic

PYRARGYRITE. Ruby Silver Ore.   Dark Red Silver Ore.

Rhombohedral-hemimorphic.   Axis: $c = 0.7892$; $0001 \wedge 10\bar{1}1 = 42° 20\frac{1}{2}'$.

$ee', 01\bar{1}2 \wedge \bar{1}012 = 42° 5'$
$rr', 10\bar{1}1 \wedge \bar{1}101 = 71° 22'$

$vv', 21\bar{3}1 \wedge \bar{2}3\bar{1}1 = 74° 25'$
$vv^v, 21\bar{3}1 \wedge 3\bar{1}\bar{2}1 = 35° 12'$

Crystals commonly prismatic.   Twins: tw. pl. $(10\bar{1}4)$ often in multiple twins; also $a(11\bar{2}0)$.   Also massive, compact.

Cleavage: $r(10\bar{1}1)$ distinct; $e(01\bar{1}2)$ imperfect.   Fracture conchoidal to uneven.   Brittle.   H. = 2.5.   G. = 5.77–5.86; 5.85 if pure.   Luster metallic-adamantine.   Color black to grayish black, by transmitted light deep red.   Streak purplish red.   Nearly opaque, but transparent in very thin splinters.   Optically $-$.   Refractive indices, $\omega = 3.084$, $\epsilon = 2.881$.

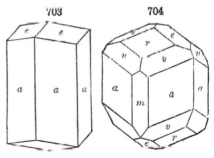

Comp. — $3Ag_2S.Sb_2S_3$ = Sulphur 17.8, antimony 22.3, silver 59.9 = 100. Some varieties contain small amounts of arsenic.

Pyr., etc. — In the closed tube fuses and gives a reddish sublimate of antimony oxysulphide; in the open tube sulphurous fumes and a white sublimate of oxide of antimony. B.B. on charcoal fuses with spurting to a globule, coats the coal white, and the assay is converted into silver sulphide, which, treated in O.F., or with soda in R.F., gives a globule of silver. In case arsenic is present it may be detected by fusing the pulverized mineral with soda on charcoal in R.F. Decomposed by nitric acid with the separation of sulphur and of antimony trioxide.

Obs. — Pyrargyrite and proustite are commonly primary minerals which are characteristically found in the upper portions of silver veins, having been deposited from ascending warm solutions but late in their origin. They are commonly associated with other silver minerals, galena, tetrahedrite, sphalerite, etc. The more important localities for the occurrence of pyrargyrite are as follows. In Czechoslovakia from Schemnitz (Selmecsbánya) and from Příbram. In the Harz Mts. at Andreasberg; in Saxony at Freiberg. In Bolivia at Colquechaca and in crystals with proustite at Chañarcillo, Atacama, Chile. In Mexico it serves as an ore of silver in Guanajuato and Zacatecas. In the United States it occurs in many of the silver localities of the Rocky Mountain states but usually in small amounts and seldom in crystals. In Colorado in the Ruby district, Gunnison Co., etc. In Nevada abundant about Austin and at the Comstock Lode. In considerable amounts at the Poorman mine, etc. in the Silver City district, Owyhee Co., Idaho. From the silver district of Cobalt, Ontario.

Named from πῦρ, *fire*, and ἄργυρος, *silver*, in allusion to the color.

PROUSTITE. Ruby Silver Ore. Light Red Silver Ore.

Rhombohedral-hemimorphic. Axis $c = 0.8039$; $0001 \wedge 10\bar{1}1 = 42° 52'$.

ee', $01\bar{1}2 \wedge \bar{1}012 = 42° 46'$ vv', $21\bar{3}1 \wedge 23\bar{1}1 = 74° 39'$
rr', $10\bar{1}1 \wedge \bar{1}101 = 72° 12'$ vvv, $21\bar{3}1 \wedge 3\bar{1}\bar{2}1 = 35° 18'$

Crystals often acute rhombohedral or scalenohedral. Twins: tw. pl. $u(10\bar{1}4)$ and $r(10\bar{1}1)$. Also massive, compact.

Cleavage: $r(10\bar{1}1)$ distinct. Fracture conchoidal to uneven. Brittle. H. = 2–2.5. G. = 5.57–5.64; 5.57 if pure. Luster adamantine. Color scarlet-vermilion; streak same, also inclined to aurora-red. Transparent to translucent. Optically negative. $\omega = 3.0$. $\epsilon = 2.7$.

Comp. — $3Ag_2S.As_2S_3$ = Sulphur 19.4, arsenic 15.2, silver 65.4 = 100.

Pyr., etc. — In the closed tube fuses easily, and gives a faint sublimate of arsenic trisulphide; in the open tube sulphurous fumes and a white crystalline sublimate of arsenic trioxide. B.B. on charcoal fuses and emits odors of sulphur and arsenic; with soda in R.F. gives a globule of silver. Decomposed by nitric acid, with separation of sulphur.

Obs. — See under pyrargyrite for mode of occurrence, etc. Found at Joachimstal in Bohemia, Czechoslovakia. In Saxony at Freiberg, etc. Was found in fine crystals at Ste. Marie aux Mines (Markirch), Alsace, France. At Chañarcillo, Atacama, Chile, in magnificent crystals. In Mexico at Batopilas, Chihuahua, etc. In the United States found rarely in any amount or with fine quality but noted from many silver districts of the western states. In Colorado at Red Mountain, San Juan district; Georgetown, Clear Creek Co., etc. In the mines of the Silver City district, Owyhee Co., Idaho. In various localities in Nevada. Noted at Cobalt, Ontario.

Named after the French chemist, J. L. Proust (1755–1826).

Pyrostilpnite. — Like pyrargyrite, $3Ag_2S.Sb_2S_3$. In tufts of slender (monoclinic) crystals. Cleavage (010). H. = 2. G. = 4.25. Color hyacinth-red. From Andreasberg in the Harz Mts.; Freiberg, Saxony; Přibram, Bohemia; Heazlewood, Tasmania; Chañarcillo, Atacama, Chile.

Samsonite. — $2Ag_2S.MnS.Sb_2S_3$. Monoclinic. Habit prismatic. Color, steel-black, red in transmitted light. H. = 2–3. Occurs in Samson vein of Andreasberg silver mines, Harz Mts., Germany.

Xanthoconite. — $3Ag_2S.As_2S_3$. Monoclinic. In tabular pseudo-orthorhombic crystals; also massive, reniform. Basal cleavage. H. = 2–3. G. = 5.5. Color orange-yellow. High indices and strong birefringence. Optically –. X nearly $\perp$ (001). 2E. = 125°. From Freiberg, Germany. Noted also in Idaho and at Cobalt, Ontario. *Rittingerite* is the same species.

SANGUINITE. Near proustite in composition. In glittering scales, hexagonal or rhombohedral. From Chañarcillo, Chile.

Guitermanite. — Perhaps $3PbS.As_2S_3$. Massive, compact. H. = 3. G. = 5.94. Color bluish gray. Zuñi mine, Silverton, Colorado.

Stylotypite. — $3Cu_2S.Sb_2S_3$? Monoclinic. G. = 4.7–5.2. Color iron-black. From Copiapo, Atacama, Chile; from Costrovirroyna, Peru.

FALKENHAYNITE. Perhaps $3Cu_2S.Sb_2S_3$. Massive, resembling galena. From Joachimstal, Bohemia. Perhaps identical with stylotypite.

TAPALPITE. A sulpho-telluride of bismuth and silver, perhaps $3Ag_2(S,Te).Bi_2(S,Te)_3$. Study of polished specimen shows it to be a mixture of tetradymite and argentite. Massive, granular. G. = 7.80. Sierra de Tapalpa, Jalisco, Mexico.

Tetrahedrite Group. Isometric-tetrahedral

TETRAHEDRITE. Gray Copper Ore. Fahlerz.

Isometric-tetrahedral. Habit tetrahedral. Twins: tw. pl. $o(111)$; also with parallel axes (Fig. 418, p. 184, Fig. 434, p. 187). Also massive; granular, coarse or fine; compact. X-ray study shows an atomic structure closely similar to that of sphalerite.

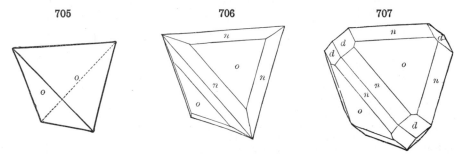

Cleavage none. Fracture subconchoidal to uneven. Rather brittle. H. = 3–4. G. = 4·4–5·1. Luster metallic, often splendent. Color between flint-gray and iron-black. Streak like color, sometimes inclining to brown and cherry-red. $n_{Li} > 2·72$. Opaque; sometimes subtranslucent (cherry-red) in very thin splinters.

Comp. — Essentially a copper-antimony sulphide. The fundamental formula is probably $3Cu_2S.Sb_2S_3$, which would satisfy the structure as derived from X-ray study. The analyses, however, show wide variations, and the mineral evidently contains several isomorphous molecules. Prior and Spencer suggest $3Cu_2S.Sb_2S_3$ with minor and varying amounts of $6(Zn,Fe)S.Sb_2S_3$. Wherry and Foshag write the formula as follows, $5Cu_2S.2(Cu,Fe,Zn)S.2Sb_2S_3$.

Antimony and arsenic are usually both present and thus tetrahedrite grades into the allied species tennantite. There are also varieties containing bismuth, chiefly at the arsenical end of the series, rarely selenium. Further the copper may be replaced by iron, zinc, silver, mercury, lead, manganese, and rarely cobalt and nickel.

Var. — *Ordinary.* Contains little or no silver. Color steel-gray to dark gray and iron-black. G. = 4·75–4·9.

Argentiferous; Freibergite. Contains 3 to 30 per cent of silver. Color usually steel gray, lighter than the ordinary varieties; sometimes iron-black; streak often reddish. G. = 4·85–5·0.

Mercurial; Schwazite. Contains 6 to 17 per cent of mercury. Color dark gray to iron-black. Luster often dull. G. = 5·10 chiefly.

Malinowskite, from Peru and a similar variety from Arizona, contain 13–16 per cent of lead.

Pyr., etc. — Differ in the different varieties. In the closed tube all the antimonial kinds fuse and give a dark red sublimate of antimony oxysulphide; if much arsenic is present, a sublimate of arsenic trisulphide first forms. In the open tube fuses, gives sulphurous fumes and a white sublimate of antimony oxide; if arsenic is present, a crystalline volatile sublimate condenses with the antimony; if the ore contains mercury it condenses in minute metallic globules. B.B. on charcoal fuses, gives a coating of the oxides of antimony and sometimes arsenic, zinc, and lead; arsenic is detected by the odor when the coating is treated in R.F. The roasted mineral gives with the fluxes reactions for iron and copper; with soda yields a globule of metallic copper. Decomposed by nitric acid, with separation of sulphur and antimony trioxide.

Diff. — Distinguished by its form, when crystallized, by its deep black color on fracture and brilliant metallic luster. It is harder than bournonite and much softer than magnetite; the blowpipe characters are usually distinctive.

Obs. — Tetrahedrite is the most common member of the sulpho-salt group; it is widespread in its occurrence and varied in its association. It is commonly found in copper or silver veins, usually as a primary constituent but at times it may be secondary in origin. It may occur in contact metamorphic deposits. It is commonly associated with chalcopyrite, bornite, pyrite, sphalerite, galena, argentite, with the ruby silvers, etc., with calcite, dolomite, siderite, barite, fluorite, and quartz. It frequently becomes an important ore of copper or because of its silver content an ore of that metal.

The following are some of the more important or interesting occurrences of tetrahedrite. In Rumania in exceptional crystals at Botés near Zalatna, at Kapnikbánya and Felsöbánya (Baia Sprie). In Czechoslovakia at Schemnitz (Selmeczbánya) and at Přibram, Bohemia (argentiferous). From Dillenburg in Hessen-Nassau and Horhausen in Rhineland. From Müsen in Westphalia; argentiferous in the Harz Mts. at Andreasberg, Klaustal, Neudorf, etc. In dodecahedral crystals from Brixlegg, Tyrol, Austria. In France exceptional crystals have been obtained from Alsace at Ste. Marie aux Mines (Markirch), at Framont and Urbeis; in the mines of Pontgibaud, Puy-de-Dôme; and in Hérault at Cabrières. Argentiferous tetrahedrite comes from a number of places in Algeria. Occurs in fine specimens in the copper-tin veins of Cornwall, especially at the Herodsfoot mine at Liskeard. From Bolivia at the Pulacayo mine, Huanchaca, Potosi.

The argentiferous variety, *freibergite*, occurs especially in the Freiberg district, Saxony, in addition to the localities already given. The mercurial variety, *schwazite*, is found in the neighborhood of Schwaz, Tyrol, Austria.

In the United States tetrahedrite occurs in Colorado near Central City, Gilpin Co., in fine crystals, and in many other localities. Argentiferous varieties occur in Pinal Co., Arizona, and in a number of districts in Nevada. Found in large amounts in the copper deposits of Bingham Canyon, Utah.

Use. — An ore of copper and frequently ore of the other metals, like silver, etc., that it may contain.

TENNANTITE.

Isometric-tetrahedral. Crystals often dodecahedral. Also massive, compact. H. = 3–4. G. = 4·37–4·49. Color blackish lead-gray to iron-black.

Comp. — Isomorphous with tetrahedrite, with arsenic replacing antimony. See under tetrahedrite.

Var. — Often contains antimony and thus grades into tetrahedrite. The original tennantite from Cornwall contains only copper and iron. In crystals, habit dodecahedral.

Tennantite is less widely distributed than tetrahedrite. Found in complex crystals from the Binnental, Valais, Switzerland (*binnite*). From Freiberg, Saxony. The argentiferous variety, *fredricite*, comes from Falun, Sweden. From the Cornish mines, particularly at Wheal Jewel at Gwennap, at St. Just, etc. From Morococha, Prov. Junin, Peru, the zinciferous variety, *sandbergerite*. Another similar variety, *miedziankite*, occurs as granular masses at Miedzianka, Poland. From other localities in Bolivia and Chile. In the United States, from Butte, Montana, from Clear Creek and Gilpin counties, etc., Colorado.

E. Basic Division

Lengenbachite. — 6PbS.Ag$_2$S.2As$_2$S$_3$. Probably triclinic. In thin blade-shaped crystals. One perfect cleavage. Soft. G. = 5·8. Color steel-gray. Streak black. From the Lengenbach quarry, Binnental, Valais, Switzerland.

Meneghinite. — 4PbS.Sb$_2$S$_3$. Orthorhombic. In slender prismatic crystals; also massive. G. = 6·34–6·43. Color blackish lead-gray. From Bottino, near Seravezza, and from near Pietrasanta, Tuscany, Italy; from near Schwarzenberg, Saxony, and from Goldkronach, in the Fichtelgebirge, Bavaria. At Marble Lake, Barrie Township, Frontenac Co., Ontario.

Jordanite. — 4PbS.As$_2$S$_3$. Monoclinic; often pseudohexagonal by twinning. Cleavage (010). H. = 3. G. = 5·5–6·4. Color lead-gray. From the Binnental, Valais, Switzerland; Nagy-Ág in Transylvania, Rumania; near Beuthen, Upper Silesia. A mineral closely similar to jordanite from Yunosawe mine in the Province of Mutsu, Japan, has been called *reniforite*.

Goongarrite. — $4PbS.Bi_2S_3$. Probably monoclinic. Fibrous to platy. Two good cleavages. H. = 3. G. = 7·23. From Lake Goongarrie, Western Australia. *Warthaite* from Hungary is apparently the same mineral.

STEPHANITE.   Brittle Silver Ore.

Orthorhombic.   Axes $a : b : c = 0·6292 : 1 : 0·6851$.

mm''', $110 \wedge 1\bar{1}0 = 64° 21'$ cd, $001 \wedge 021 = 53° 52'$
$c\beta$, $001 \wedge 101 = 47° 26'$ ch, $001 \wedge 112 = 32° 45'$
ck, $001 \wedge 011 = 34° 25'$ cP, $001 \wedge 111 = 52° 9'$

Crystals usually short prismatic or tabular $\| c(001)$. Twins: tw. pl. $m(110)$, often repeated, pseudo-hexagonal. Crystals show at times oblique striations on the prism faces, and because of this the mineral is often placed in the orthorhombic-hemimorphic class. Also massive, compact and disseminated.

708

Cleavage: $b(010)$, $d(021)$ imperfect. Fracture subconchoidal to uneven. Brittle. H. = 2–2·5. G. = 6·2–6·3. Luster metallic. Color and streak ironblack. Opaque.

Comp. — $5Ag_2S.Sb_2S_3$ = Sulphur 16·3, antimony 15·2, silver 68·5 = 100.

Pyr. — In the closed tube decrepitates, fuses, and after long heating gives a faint sublimate of antimony oxysulphide. In the open tube fuses, giving off antimonial and sulphurous fumes. B.B. on charcoal fuses with projection of small particles, coats the coal with oxide of antimony, which after long blowing is colored red from oxidized silver, and a globule of metallic silver is obtained. Soluble in dilute heated nitric acid, sulphur and antimony trioxide being deposited.

Obs. — Stephanite occurs in many silver deposits associated with other silver minerals, galena, tetrahedrite, sphalerite, etc. It is usually of primary origin but characteristically found in the upper portions of the veins, where it has been deposited from warm ascending solutions rather late in the formation of the deposit. From Czechoslovakia in crystals at Schemnitz (Selmeczbánya) and Příbram, etc. In Saxony at Freiberg and elsewhere. In unusual crystals from Andreasberg, Harz Mts. In twin crystals from Mte. Narba near Sarrabus, Sardinia. At various localities in Cornwall. From Chañarcillo, Atacama, Chile. In Mexico from Zacatecas in Zacatecas; in crystals from Arizpe, Sonora; and from Guanajuato. In the United States it was an important ore in the Comstock Lode, near Virginia City, Nevada; also from the Austin, Reese River, and Humboldt districts in Nevada.

Named after the Archduke Stephen, Mining Director of Austria.

Geocronite. — $5PbS.Sb_2S_3$. Rarely in orthorhombic crystals closely resembling those of stephanite; usually massive, granular. H. = 2–3. G. = 6·4. Color lead-gray. From Sala, Vastmanland, and Björkskogonäs, Örebo in Sweden; from Meredo, Galicia, Spain; Val di Castello, near Pietrasanta, Tuscany. *Kilbrickenite* from Kilbricken, Co. Clare, Ireland, is the same species.

GOLDFIELDITE. $5Cu_2S.(Sb,As,Bi)_2(S,Te)_3$. As a crust. Color, dark lead-gray. Conchoidal fracture. H. = 3–3·5. At Mohawk mine, Goldfield, Nevada. Probably a mixture.

Beegerite. — $6PbS.Bi_2S_3$. Massive, indistinctly crystallized. Cubic cleavage and therefore probably isometric. G. = 7·27. Color light to dark gray. In Colorado at two localities in Park Co., and from Poughkeepsie Gulch, Ouray Co. Noted also from Minusinsk district, Yeniseisk, Siberia.

Polybasite Group. $9R_2S.As_2S_3$, $9R_2S.Sb_2S_3$. Monoclinic, pseudo-rhombohedral

POLYBASITE.

Monoclinic. Axes $a : b : c = 1.7309 : 1 : 1.5796$, $\beta = 90° 0'$. Prismatic angle $60° 2'$. In short six-sided tabular prisms, with beveled edges; $c(001)$ faces with triangular striations; in part repeated twins, tw. pl. $m(110)$.

Cleavage: $c(001)$ imperfect. Fracture uneven. H. = 2–3. G. = 6.0–6.2. Luster metallic. Color iron-black, in thin splinters cherry-red. Streak black. Nearly opaque. Optically −. $n > 2.72$. Strong birefringence.

Comp. — $9Ag_2S.Sb_2S_3$ = Sulphur 15.0, antimony 9.4, silver 75.6 = 100. Part of the silver is replaced by copper; also the antimony by arsenic.

Pyr., etc. — In the open tube fuses, gives sulphurous and antimonial fumes, the latter forming a white sublimate, sometimes mixed with crystalline arsenic trioxide. B.B. fuses with spurting to a globule, gives off sulphurous (sometimes arsenical) fumes, and coats the coal with antimony trioxide; with long-continued blowing some varieties give a faint yellowish white coating of zinc oxide, and a metallic globule, which with salt of phosphorus reacts for copper, and cupelled with lead gives pure silver. Decomposed by nitric acid.

Obs. — Polybasite is found in many silver veins, at times in considerable amounts. It is associated with other silver and lead sulpho-salts, with various sulphides and with such non-metallic minerals as quartz, calcite, dolomite, barite. From Czechoslovakia at Schemnitz (Selmeczbánya) and at Přibram in Bohemia. At Freiberg in Saxony and at Andreasberg, Harz Mts. In Chile at Tres Puntas, Atacama. At many localities in Mexico, especially in Zacatecas, Guanajuato, and Durango; also from Arizpe, Sonora. In the United States common in small amounts in the silver mines of Colorado, from near Ouray, at Aspen, in Gilpin and Clear Creek counties, etc. In Nevada at Austin, and the Comstock Lode, etc. In Idaho at the silver mines of Owyhee Co., and elsewhere.

Named from πολύς, *many*, and βάσις, *base*, in allusion to the basic character of the compound.

Pearceite. — $9Ag_2S.As_2S_3$. Monoclinic, pseudo-rhombohedral. The arsenical variety of polybasite. From Aspen, Colorado; in Montana at Marysville, Lewis and Clarke Co., and at Neihart, Cascade Co.; in the Tintic district, Utah.

POLYARGYRITE. $12Ag_2S.Sb_2S_3$. In distinct isometric crystals. H. = 2–3. G. = 6.97. Color iron-black. Wolfach, Baden, Germany. Microscopic tests show that it is a mixture of argentite and tetrahedrite.

Ultrabasite. — $28PbS.11Ag_2S.3GeS_2.2Sb_2S_3$. Orthorhombic. Color and streak gray-black. H. = 5. G. = 6. From Freiberg, Saxony.

II. Sulpharsenates, Sulphantimonates

Here are included a few minerals, chiefly sulpho-salts of quintivalent arsenic and antimony. The analyses have been variously interpreted, and some authors write the formulas as sulpharsenites, etc., and include these minerals among the groups already described.

ENARGITE.

Orthorhombic. Axes: $a : b : c = 0.8711 : 1 : 0.8248$.

Crystals usually small; prismatic faces vertically striated. Twins: tw. pl. $x(320)$ in star-shaped trillings. Also massive, granular, or columnar.

Cleavage: $m(110)$ perfect; $a(100)$, $b(010)$ distinct; $c(001)$ indistinct. Fracture uneven. Brittle. H. = 3. G. = 4.43–4.45. Luster metallic. Color grayish black to iron-black. Streak grayish black. Opaque.

Comp. — $3Cu_2S.As_2S_5$ = Sulphur 32·6, arsenic 19·1, copper 48·3 = 100. Antimony is often present, cf. famatinite.

The formula has also been written $Cu_2S.4CuS.As_2S_3$.

Pyr. — In the closed tube decrepitates, and gives a sublimate of sulphur; at a higher temperature fuses, and gives a sublimate of sulphide of arsenic. In the open tube, heated gently, the powdered mineral gives off sulphurous and arsenical fumes, the latter condensing to a sublimate containing some antimony oxide. B.B. on charcoal fuses, and gives a faint coating of the oxides of arsenic, antimony, and zinc; the roasted mineral with the fluxes gives a globule of metallic copper. Soluble in aqua regia.

Obs. — Enargite is uncommon in its occurrence but in a few localities it is found in large amounts and becomes an important ore of copper and even of arsenic. It is in most instances of primary origin, favoring deposits that have been formed near the surface. It is associated with bornite, chalcocite, tetrahedrite, covellite, pyrite, sphalerite, quartz, barite, etc. Enargite is rare in European copper deposits but is found in an important occurrence at Bor, northwest of Zaječar, Serbia, Yugoslavia; also occurs at Parád, Comitat Heves, Hungary, and in twin crystals from Matzenköpfl, Brixlegg, Tyrol, Austria. From Morococha, Prov. Junin, Peru, in large masses. From Guayacana in the Cordilleras of Chile (*guayacanite*) and at the mine of Hediondas, Prov. of Coquimbo. From the Sierra de Famatina, La Rioja, Argentina. From Mancayan, island of Luzon, Philippines. In the United States enargite is an important ore at Butte, Montana. In small amounts in various districts in Colorado as at Red Mountain in the San Juan district, near Georgetown, Clear Creek Co., in Gilpin Co., etc. In Utah in the Tintic district and also from Bingham Canyon. From the Mogol district, Alpine Co., California. Found in the copper deposits at Kennecott, Copper River district, Alaska. *Clarite*, from the Clara mine, Schapbach, Baden, is the same species. *Luzonite* from the island of Luzon, Philippines, has the same composition as enargite but shows certain physical differences and has been thought to represent a dimorphous form. X-ray study, however, has shown a structure identical with that of enargite.

Use. — Serves as an ore of copper and arsenic.

Famatinite. — $3Cu_2S.Sb_2S_5$, isomorphous with enargite. X-ray study shows a structure identical with that of enargite. H. = 3–4. G. = 4·57. Color gray with tinge of copper-red. Rarer in its occurrence than enargite but found under the same conditions, the two minerals frequently occurring together. Occurs in the Sierra de Famatina, La Rioja, Argentina; from Cerro de Pasco, Peru, and from near Santiago, Chile. Found at Goldfield, Esmeralda Co., Nevada.

Sulvanite. – $3Cu_2S.V_2S_5$. Isometric; in cubic crystals from Utah. Also massive. Cubic cleavage. X-ray study shows an isometric structure. H. = 3·5. G. = 4·0. Color bronze-yellow. Streak nearly black. From near Burra, South Australia. Also found near Mercur, Utah.

Epigenite. — Perhaps $4Cu_2S.3FeS.As_2S_5$. Orthorhombic. In short prisms resembling arsenopyrite. Color steel-gray. From near Wittichen, Baden, Germany.

III. Sulphostannates, etc.

STANNITE. Tin Pyrites. Bell-metal Ore.

Tetragonal-sphenoidal. Pseudo-isometric-tetrahedral through twinning. Twinning, (1) always interpenetrant with $e(101)$ as tw. pl., (2) interpenetrant with twin axis $\perp$ to $p(111)$. Also massive, granular, and disseminated.

Cleavage: cubic, indistinct. Fracture uneven. Brittle. H. = 3·5. G. = 4·3–4·522; 4·506 Zinnwald. Luster metallic. Streak blackish. Color steel-gray to iron-black, the former when pure; sometimes a bluish tarnish; often yellowish from the presence of chalcopyrite. Opaque.

Comp. — A sulphostannate of copper, iron and sometimes zinc, $Cu_2S.FeS.SnS_2$ = Sulphur 29·9, tin 27·5, copper 29·5, iron 13·1 = 100.

Pyr., etc. — In the closed tube decrepitates, and gives a faint sublimate; in the open tube sulphurous fumes. B.B. on charcoal fuses to a globule, which in O.F. gives off sulphur dioxide and coats the coal with tin dioxide; the roasted mineral treated with borax gives reactions for iron and copper. Decomposed by nitric acid, affording a blue solution, with separation of sulphur and tin dioxide.

Obs. — Stannite occurs in tin-bearing veins associated with cassiterite, chalcopyrite, tetrahedrite, and pyrite, often containing various of these and other minerals as inclusions. It has been formed late in the period of deposition through the agency of ascending hot, alkaline solutions. From Zinnwald, Bohemia, in Czechoslovakia. In Cornwall at Wheal Rock near St. Agnes, at Carn Brea, more recently in considerable amount at St. Michael's Mount, also at Redruth, at Stenna Gwynne near St. Austell and at Camborne, etc. In important amounts at Zeehan, Tasmania. In Bolivia in crystals from Oruro; also from Potosi. Found sparingly in the Black Hills, South Dakota. *Bolivianite* is the name given to a supposedly new mineral, closely agreeing with stannite in character, from the Bolivian tin veins.

Argyrodite. — A silver sulpho-germanate, $4Ag_2S.GeS_2$. Isometric, crystals usually indistinct; at times they show octahedral and dodecahedral forms with frequent twinning according to the Spinel Law; also massive, compact. H. = 2·5. G. = 6·1–6·3. Luster metallic. Color steel-gray on a fresh fracture, with a tinge of red turning to violet. From the Himmelsfürst mine, Freiberg, Saxony; from Aullagas near Chayanta, northwest of Colquechaca, Bolivia.

Canfieldite. — $4Ag_2S.SnS_2$, the tin in part replaced by germanium. Isometric, in octahedrons with $d(110)$. Twins according to Spinel Law. H. = 2·5. G. = 6·28. Luster metallic. Color black. Associated with argyrodite at Aullagas near Chayanta, northwest of Colquechaca, Bolivia. Some of the mineral called *brongniartite* has been shown to be canfieldite.

Teallite. — $PbS.SnS_2$. Orthorhombic? In thin flexible folia. Perfect basal cleavage. H. = 1–2. G. = 6·4. Color blackish gray. Streak black. Reported from the Himmelsfürst mine at Freiberg, Saxony. Found in the silver mines of Santa Rosa, Antequera, Bolivia; also reported from Montserrat, Pazña, Bolivia. *Pufahlite*, or *zinc-teallite*, from Pazña, is probably a zinc-bearing variety.

Franckeite. — $5PbS.2SnS_2.Sb_2S_3$. Hexagonal or orthorhombic? In tabular crystals, and massive. Perfect basal cleavage. H. = 2·5. G. = 3·5–5·5. Color blackish gray to black. Found in Bolivia at Poopó in Oruro and at Las Animas, southeast of Chocaya, in Potosi.

Cylindrite. — $6PbS.6SnS_2.Sb_2S_3$. H. = 2·5–3. G. = 5·42. Luster metallic. Color blackish lead-gray. In cylindrical forms separating under pressure into distinct shells or folia. Poopó, in Oruro, Bolivia.

IV. HALOIDS. — CHLORIDES, BROMIDES, IODIDES; FLUORIDES

I. Anyhdrous Chlorides, Bromides, Iodides; Fluorides.
II. Oxychlorides; Oxyfluorides.
III. Hydrous Chlorides; Hydrous Fluorides.

The Fourth Class includes the haloids, that is, the compounds with the halogen elements, chlorine, bromine, iodine, and also the less closely related fluorine.

I. Anhydrous Chlorides, Bromides, Iodides; Fluorides

CALOMEL. Horn Quicksilver.

Tetragonal. Axis $c = 1·723$; 001 $\wedge$ 101 = 59° 52′. Crystals sometimes tabular $\parallel c(001)$; also pyramidal; often highly complex.

Cleavage: $a(100)$ rather distinct; also $r(111)$. Fracture conchoidal. Sectile. H. = 1–2. G. = 6·482. Luster adamantine. Color white, yellowish gray, or ash-gray, also grayish, and yellowish white, brown. Streak pale yellowish white. Translucent — subtranslucent. Optically +. $\omega = 1 \cdot 97$. $\epsilon = 2 \cdot 65$.

Comp. — Mercurous chloride, HgCl = Chlorine 15, mercury 85 = 100.

Pyr., etc. — In the closed tube volatilizes without fusion, condensing in the cold part of the tube as a white sublimate; with soda gives a sublimate of metallic mercury. B.B. on charcoal volatilizes, coating the coal white. Insoluble in water, but dissolved by aqua regia; blackens when treated with alkalies.

Obs. — Calomel occurs most commonly as a coating of minute crystals upon other minerals It has been deposited usually from hot solutions; at times probably formed by sublimation; also occurs as a mineral of secondary origin. Frequently associated with cinnabar and native mercury. Found at Moschellandsberg in the Palatinate, Bavaria; from Horowitz northwest of Přibram in Bohemia of Czechoslovakia; at Mt. Avala near Belgrad in Yugoslavia; also at Idria in Gorizia, Italy (formerly in Carniola, Austria), and from Almaden in Ciudad Real, Spain. In the United States in complex crystals at Terlingua, Brewster Co., Texas.

Calomel is an old term of uncertain origin and meaning, perhaps from καλός, *beautiful*, and μέλι, *honey*, the taste being sweet, and the compound the *Mercurius dulcis* of early chemistry; or from καλός and μέλας, *black*.

Kleinite. Mercurammonite. — A mercury ammonium chloride of uncertain composition. Hexagonal. Crystals short prismatic. Basal cleavage. H. = 3·5. G. = 8·0. Color yellow to orange, darkening on exposure. Uniaxial +, above 130° C. with $\omega = 2 \cdot 19$, $\epsilon = 2 \cdot 21$. At ordinary temperatures is biaxial −, probably triclinic with $\beta = 2 \cdot 18$ and strong dispersion, $\rho < v$. Volatile. From Terlingua, Brewster Co., Texas.

Mosesite. — A mercury-ammonium compound containing chlorine, sulphur trioxide and water. Near *kleinite* in composition. Isometric. Minute octahedrons. Spinel twins. H. = 3+. Color yellow. Doubly refracting at ordinary temperatures. $n = 2 \cdot 06$. Found sparingly at Terlingua, Texas.

Nantokite. — Cuprous chloride, CuCl. Isometric; tetrahedral (artif.) Granular, massive. Cleavage cubic. H. = 2–2·5. G. = 3·93. Luster adamantine. $n = 1 \cdot 93$. Colorless to white or grayish. From Nantoko, Chile; Broken Hill, New South Wales.

Marshite. — Cuprous iodide, CuI. Isometric-tetrahedral. Structure similar to that of sphalerite. Cleavage dodecahedral. H. = 2·5. G. = 5·6–5·9 Color oil-brown. $n = 2 \cdot 346$. Broken Hill, New South Wales,

Miersite. — Silver, copper iodide, 4AgI.CuI. Isometric; tetrahedral. H. = 2–3. G. = 5·64. Atomic structure similar to that of sphalerite. Dodecahedral cleavage. $n = 2 \cdot 2$. In bright yellow crystals from Broken Hill, New South Wales. *Cuproiodargyrite* from Huantajayu, Peru, belongs here also.

Halite Group. $\overset{I}{R}Cl$, $\overset{I}{R}Br$, $\overset{I}{R}I$. Isometric

Halite	NaCl	**Cerargyrite**	AgCl
Sylvite	KCl	**Embolite**	Ag(Cl,Br)
Sal Ammoniac	(NH₄)Cl	**Bromyrite**	AgBr
Huantajayite	20NaCl.AgCl	**Iodobromite**	Ag(Cl,Br,I)
Villiaumite	NaF		

The HALITE GROUP includes the halogen compounds of the closely related metals, sodium, potassium, and silver, also ammonium (NH₄). They crystallize in the isometric system, the cubic form being the most common.

HALITE. COMMON or ROCK SALT.

Isometric. Usually in cubes; crystals sometimes distorted, or with cavernous faces. The atomic structure shows normal symmetry (for details see p. 36), but certain irregularities in pits produced by etching have been thought to indicate plagiohedral symmetry. Also massive, granular to compact; less often columnar.

Cleavage: cubic, perfect. Fracture conchoidal. Rather brittle. H. = 2·5. G. = 2·1–2·6; pure crystals 2·164. Luster vitreous. Colorless or white also yellowish, reddish, bluish, purplish. At times shows a deep blue color which is usually localized in irregular spots. This has been variously explained as due to the presence of colloidal sodium, to a lower chloride of sodium, to the presence of organic matter, etc. Transparent to translucent. Soluble; taste saline. $n = 1·5442$. Highly diathermanous.

709   710

Comp. — Sodium chloride, NaCl = Chlorine 60·6, sodium 39·4 = 100. Commonly mixed with calcium sulphate, calcium chloride, magnesium chloride, and sometimes magnesium sulphate, which render it liable to deliquesce.

Pyr., etc. — In the closed tube fuses, often with decrepitation; when fused on the platinum wire colors the flame deep yellow. After intense ignition the residue gives an alkaline reaction upon moistened test paper. Nitric acid solution gives precipitate of silver chloride upon addition of silver nitrate. Dissolves readily in three parts of water.

Diff. — Distinguished by its solubility (taste), softness, perfect cubic cleavage.

Obs. — Common salt occurs in extensive but irregular beds in rocks of various ages, associated with gypsum, polyhalite, anhydrite, carnallite, clay, sandstone, and calcite; also in solution forming salt springs; similarly in the water of the ocean and salt seas. The deposits of salt have been formed by the gradual evaporation and ultimate drying up of enclosed bodies of salt water. Salt beds formed in this way are subsequently covered by other sedimentary deposits and gradually buried beneath the rock strata thus formed. The salt strata range from a few feet up to more than one hundred feet in thickness and have been found at depths of two thousand feet and more beneath the earth's surface. Also occurs as a sublimation product in volcanic regions and in arid districts as an efflorescence upon the surface of the ground.

Salt is generally granular in structure and commonly variously colored by impurities. Crystals occur in druses in salt beds where there has been solution and recrystallization, and also embedded in salt clays. Openings in salt clays are often filled with fine columnar salt with the columnar structure perpendicular to the walls of the cavities.

Halite is a very common mineral and is found in sedimentary rocks of all ages and widely distributed throughout the world. Only those localities that are most important either for scientific or commercial reasons can be mentioned here. In Europe a large deposit is found at Ipetsk near Orenburg in southeastern Russia. In Galicia, Poland, at Wieliczka, at Bochnia to the east and at Kalusz in large deposits. In Austria in Salzburg at Hallein, Hallstadt and Ischl, in Styria at Aussee and in Tyrol at Hall near Innsbruck. In Switzerland at Bex in Vaud. Famous deposits are found at Stassfurt, Saxony, and at the neighboring localities of Leopoldshall in Anhalt, etc., and at Wilhelmsglück in Würtemberg. These deposits contain many other saline minerals which vary at different depths in accord with the order of their solubilities. Crystallized halite is found on the Island of Sicily at Girgenti, Racalmuto, Castrigiovanni, etc. In France large bodies occur in the neighborhood of Dax in Landes, and from Vic and Dieuze in Lorraine. In Spain near Cardona in Barcelona. At Northwich in Cheshire, England. Salt is found in quantity in Algeria and Abyssinia; in India in the Punjab in enormous deposits in the Salt Range, at Khewra, etc. From China, Peru, Colombia, Santo Domingo, etc.

In the United States, salt has been found in large amount in central and western New York. Salt wells had long been worked in this region, but rock salt is now known to exist over a large area from Ithaca at the head of Cayuga Lake, Tompkins Co., and Canadaigua Lake, Ontario Co., through Livingston Co., also Genesee, Wyoming, and Erie counties. The salt is found in beds with an average thickness of 75 feet, but sometimes much thicker (in one instance 325 feet), and at varying depths from 1000 to 2000 feet and more; the depth increases southward with the dip of the strata. The rocks belong to the Salina period of the Upper Silurian. Extensive deposits of salt occur in Michigan, chiefly in Saginaw, Bay,

Midland, Isabella, Wayne, Manistee and Mason counties. Salt has also been found near Cleveland, Ohio, associated with gypsum. In Louisiana extensive beds occur in the southern portion of the state at and in the neighborhood of Petite Anse island. In Kansas in beds from ten to one hundred feet or more in thickness, in Ellsworth, Rice, Reno, Kingman, and Harper counties. In Arizona in fine transparent masses in the Verde Valley, Yavapai Co. In Nevada along the Virgin River in Clark Co., in distorted crystals from Humboldt Co. From Borax Lake, San Bernardino Co., California. Obtained from the waters of Great Salt Lake, Utah. In Ontario along the eastern shore of Lake Huron in Bruce, Huron, and Lambton counties.

Use. — The chief uses of salt are for culinary and preservative purposes. Soda ash is also made from it, being employed in the manufacture of glass, soap, bleaching, preparation of other sodium compounds, etc.

Huantajayite. — $20NaCl.AgCl$. In cubic crystals and as an incrustation. H. = 2. Not sectile. Color white. From Huantajaya, Tarapaca, Chile.

Villiaumite. — NaF. Isometric. In small carmine colored grains. Abnormally birefringent and pleochroic. Heated to 300° C. becomes isotropic. Soft. G. = 2·8. Refractive index = 1·33. Found in nephelite-syenite from the Islands of Los, French Guinea.

SYLVITE.

Isometric. The atomic structure as determined by X-ray study shows normal symmetry but because of etching figures sylvite has long been considered as belonging to the plagiohedral class. Cubes, often with octahedral truncations. Also in granular crystalline masses; compact.

Cleavage: cubic, perfect. Fracture uneven. Brittle. H. = 2. G. — 1·97–1·99. Luster vitreous. Colorless, white, bluish or yellowish red from inclusions. Soluble; taste resembling that of common salt, but bitter. $n = 1·490$. Diathermanous.

Comp. — Potassium chloride, KCl = Chlorine 47·6, potassium 52·4 = 100. Sometimes contains sodium chloride.

Pyr., etc. — B.B. in the platinum loop fuses, and gives a violet color to the outer flame; the color is often masked by the yellow due to the presence of sodium, in which case a blue filter must be used to eliminate the sodium flame. Dissolves completely in water (saline taste). After ignition residue reacts alkaline upon moistened test paper. Solution in nitric acid gives precipitate of silver chloride with silver nitrate.

Obs. — Sylvite is like halite in its mode of origin but much rarer in its occurrence. Occurs as a sublimation product at Vesuvius, about the fumaroles of the volcano. Also in Germany at Stassfurt, Saxony; and at Leopoldshall (*leopoldite*), Anhalt; at Kalusz in Galicia, Poland.

Use. A source of potash compounds used as fertilizers.

Sal Ammoniac. — Ammonium chloride, NH_4Cl. Isometric. $n = 1·639$. Commonly a sublimation product. Observed as a white incrustation about volcanic fumaroles. as at Etna, Vesuvius, etc.

Cerargyrite Group. Isometric-Normal

An isomorphous series of silver haloids in which silver chloride, bromide and iodide may mix in varying proportions. The suggestion has been made that the name cerargyrite be kept as the group name and that the different sub-species be named as follows: *chlorargyrite*, AgCl; *bromargyrite*, AgBr; *embolite*, Ag(Cl,Br); *iodembolite*, Ag(Cl,Br,I).

CERARGYRITE. Horn Silver.

Isometric. Habit cubic. The atomic structure is similar to that of halite. Twins: tw. pl. o(111). Usually massive and resembling wax or horn; sometimes columnar; often in crusts.

Cleavage: none. Fracture somewhat conchoidal. Highly sectile. H. =

1–1·5. G. = 5·552. Luster resinous to adamantine. Color pearl-gray, grayish green, whitish to colorless, rarely violet-blue; on exposure to the light turns violet-brown. Transparent to translucent. $n = 2·0611$.

Comp. — Silver chloride, AgCl = Chlorine, 24·7, silver 75·3 = 100. Some varieties contain mercury.

Pyr., etc. — In the closed tube fuses without decomposition. B.B. on charcoal gives a globule of metallic silver. Added to a bead of salt of phosphorus, previously saturated with oxide of copper and heated in O.F., imparts an intense azure-blue to the flame. Insoluble in nitric acid, but soluble in ammonia.

Obs. — Cerargyrite and the related minerals are usually the products of secondary action and are commonly found in the upper parts of silver deposits. Descending waters containing chlorine, bromine or iodine act upon the oxidation products of the primary silver minerals and so precipitate these relatively insoluble compounds. Commonly associated with other silver minerals, with lead, copper and zinc ores and their usual alteration products. At times as pseudomorphs after native silver, etc.

Cerargyrite has been found in the silver deposits of Saxony at Johanngeorgenstadt, Freiberg, etc.; from Andreasberg in the Harz Mts. In the Altai Mts., Tomsk, Asiatic Russia. In notable amounts at the Broken Hill district, New South Wales. The most important deposits are in various districts in Atacama, Chile; also at Potosi and elsewhere in Bolivia; in Mexico. In the United States in Colorado at Leadville, Lake Co., and elsewhere. In crystals from the Poorman mine, Silver City district, Owyhee Co., Idaho. Occurs also in various silver districts in Utah, Nevada, New Mexico, and Arizona.

Named from κέρας, *horn*, and ἄργυρος, *silver*.

Use. — An ore of silver.

Embolite. — Silver chloro-bromide Ag(Br,Cl) the ratio of chlorine to bromine varying widely. Isometric. Usually massive. Resembles cerargyrite, but color grayish green to yellowish green and yellow. $n = 2·15$. Abundant in Chile at Chañarcillo, etc. Found also at Broken Hill, New South Wales, and at Silver Reef, St. Arnaud, Victoria. From Tonopah, Nevada; Leadville, Colorado; Yuma County, Arizona; Georgetown and Silver City, New Mexico.

Bromyrite. — Silver bromide, AgBr. Isometric. G. = 5·8–6. Color bright yellow to amber-yellow; slightly greenish. $n = 2·25$. From Mexico; Chile.

Iodobromite. — 2AgCl.2AgBr.AgI. Isometric. G. = 5·713. Color sulphur-yellow, greenish. $n = 2·2$. From near Dernbach, Nassau; Broken Hill, New South Wales; Chile.

Iodyrite. — Silver iodide, AgI. Hexagonal-hemimorphic; usually in thin plates; pale yellow or green. Silver iodide has been shown to be dimorphous, changing to an isometric form at 146° C. G. = 5·5–5·7. Optically +. $\omega = 2·182$. From Durango, Mexico; Chañarcillo, Chile, etc.; Lake Valley, Sierra Co., New Mexico. In crystals from Broken Hill, New South Wales, and Tonopah, Nevada.

Fluorite Group

The species here included are Fluorite, CaF_2, and the rare yttrofluorite, and yttrocerite.

FLUORITE or FLUOR SPAR.

Isometric. Habit cubic; less frequently octahedral or dodecahedral; forms $f(310)$, $e(210)$ (fluoroids) common; also the vicinal form $\zeta(32·1·0\ ?)$, producing striations on $a(100)$ (Fig. 715); hexoctahedron $t(421)$ also common with the cube (Fig. 714). Cubic crystals sometimes grouped in parallel position to form a pseudo-octahedron. Because of this tendency to form crystals by the parallel grouping of minute cubes, the faces of the octahedron and some other forms may have a drusy character. Twins: tw. pl. $o(111)$, commonly penetration-twins (Fig. 715). Also massive; granular, coarse or fine; rarely columnar; compact. The atomic structure of fluorite has the calcium atoms arranged on a face-centered cubic lattice with the fluorine atoms lying at the centers of the eight small cubes of which the unit cell is composed, see p. 41.

Cleavage: $o(111)$ perfect. Fracture flat-conchoidal; of compact kinds, splintery. Brittle. H. = 4. G. = 3·01–3·25; 3·18 crystallized. Luster vit-

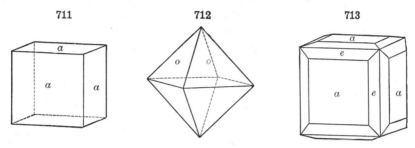

reous. Color white, yellow, green, rose- and crimson-red, violet-blue, sky-blue, and brown; wine-yellow, greenish blue, violet-blue, most common; red, rare. Color often varies in different portions of the same specimen, showing bands of different shades which lie parallel to the cubic planes.

Further, the color may be modified by various means, such as heat, X-rays, radium rays, ultra-violet light, pressure, etc. Streak white. Transparent — subtranslucent. Sometimes shows a bluish fluorescence. (See p. 275.) Some deeply colored specimens appear blue by reflected light and green by transmitted light. Some varieties phosphoresce when heated (p. 275) and others when scratched with a knife blade. $n = 1·4339$. Often shows abnormal birefringence which varies in bands that lie parallel to the cubic planes. This is probably due to internal tension.

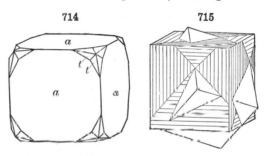

Comp. — Calcium fluoride, CaF_2 = Fluorine 48·9, calcium 51·1 = 100. Chlorine is sometimes present in minute quantities.

Var. — 1. *Ordinary*; (a) cleavable or crystallized, very various in colors; (b) fibrous to columnar, as the Derbyshire blue-john used for vases and other ornaments; (c) coarse to fine granular; (d) earthy, dull, and sometimes very soft. *Chlorophane* yields a green phosphorescent light when heated.

Pyr., etc. — In the closed tube decrepitates and sometimes phosphoresces. B.B. in the forceps and on charcoal fuses, coloring the flame orange, to an enamel which reacts alkaline on test paper. Fused in a closed tube with potassium bisulphate gives reaction for fluorine.

Diff. — Distinguished by its crystalline form, octahedral cleavage, relative softness (as compared with certain precious stones, also with the feldspars); etching power when treated with sulphuric acid. Does not effervesce with acid like calcite.

Obs. — Fluorite may be formed under widely varying conditions. It occurs most commonly as a vein mineral either in deposits in which it is the chief constituent or as a gangue mineral with various metallic ores, especially those of lead, silver, and zinc. Often associated with quartz, calcite, barite, etc. It is characteristic of pneumatolytic deposits, especially those carrying tin, and here is associated with tourmaline, topaz, lepidolite, apatite, etc. It is found in sedimentary rocks, as in dolomites and limestones. It is also found as a minor accessory mineral in granite and other acid igneous rocks. It occurs as a sublimation product in connection with volcanic rocks.

Fluorite is a common mineral and only the most important localities for its occurrence can be mentioned here. Famous localities are in the north of England, where it occurs as a gangue mineral in the lead veins which intersect the coal formation in Northumberland, at Alston Moor, Cleator Moor, etc., in Cumberland, at Weardale in Durham, and in Yorkshire. It is abundant in Derbyshire and also in Cornwall, at Redruth, Liskeard, St. Agnes, Menhenniot, and many other points. At Beer Alston in Devonshire. Common in the mining districts of Saxony at Freiberg, Annaberg, Gersdorf, etc.; from Andreasberg, Stolberg, etc., Harz Mts.; Wölsendorf, Bavaria; at Münstertal, Baden. In Austria from Rauris and Pinzgau in Salzburg. From Czechoslovakia at Gräben near Striegau in Silesia and at Zinnwald, Schlaggenwald, etc., in Bohemia. In Switzerland in pink octahedrons from the dolomite of the St. Gotthard region, at Göschenen-Alp in Uri, at Chamounix, and at Brienz, Bern. From near Baveno, Piedmont, Italy; with the lavas of Vesuvius. From near Kongsberg, Norway.

Some localities in the United States are Westmoreland, New Hampshire; Trumbull, Connecticut (*chlorophane*); in New York at Muscolonge Lake, Jefferson Co., at Macomb, St. Lawrence Co., both in very large sea-green cubes; at Phœnixville, Pennsylvania; at Amelia Court House, Virginia. Fluorite has been mined in the United States chiefly from western Kentucky and adjacent sections in Hardin (Rosiclare) and Pope counties, Illinois. Occurs at Tiffin, Ohio; from near St. Louis, Missouri; at Crystal Peak, Teller Co., and many other localities in Colorado. In colorless transparent crystals from Madoc, Hastings Co., Ontario.

Use. — As a flux in the making of steel; in the manufacture of opalescent glass; in enameling cooking utensils; the preparation of hydrofluoric acid; sometimes as an ornamental material.

Yttrofluorite. — CaF_2 with varying amounts of YF_3. (It has been shown experimentally that calcium fluoride can take up about 50 per cent YF_3.) Isometric. Atomic structure similar to that of fluorite. In granular masses. Imperfect octahedral cleavage. H. = 4·5. G. = 3·55. Color yellow, also with brown or green shades. $n = 1·46$. Found in pegmatite at Hundholmen in northern Norway.

Yttrocerite. — CaF_2 with varying amounts of $(Y,Ce)F_3$. (It has been shown experimentally that fluorite can take up about 55 per cent CeF_3. See further under yttrofluorite above.) The analyses of yttrocerite show small amounts of water. Massive cleavable to granular and earthy. H. = 4–5. G. = 3–4. Color, violet-blue, gray, reddish brown. $n = 1·434$. From near Falun, Sweden. From Sussex Co., New Jersey; from Warwick and Edenville, New York.

Zamboninite. — $CaF_2.2MgF_2$. Orthorhombic? In fine radiating fibers. White. Biaxial, +. Indices, 1·405–1·411. Fibers show parallel extinction. G. = 2·98–3·00. Formed by fumarole action in crevices of lava at Monti Rossi, Etna, Sicily.

Sellaite. — Magnesium fluoride, MgF_2. In prismatic tetragonal crystals. Cleavages (100), (110). Atomic structure similar to that of rutile. H. = 5–6. G. = 2·97–3·15. Colorless. Optically +. $\omega = 1·378$. From the moraine of the Gebroulaz glacier near Montiers, in Savoie, France, and from Vesuvius. *Belonesite* is the same species.

Cotunnite. — Lead chloride, $PbCl_2$. In acicular crystals (orthorhombic) and in semicrystalline masses. Cleavage: (001), perfect. Soft. G. = 5·3–5·8. Color white, yellowish. Optically +. $\beta = 2·217$. From Vesuvius; also Tarapaca, Chile.

Lawrencite. — Ferrous chloride, $FeCl_2$. Occurs in meteoric iron.

Molysite. — Ferric chloride, $FeCl_3$. Hexagonal. In brownish red to yellow incrustations. Easily decomposed. Found rarely about volcanic fumaroles, Vesuvius, etc.

Fluocerite. Tysonite. — Fluoride of the cerium metals, $(Ce,La,Di)F_3$. In thick hexagonal prisms, and massive. Cleavage: $c(001)$, perfect. H. = 4·5–5. G. = 6·13. Color pale wax-yellow, changing to yellowish and reddish brown. $n = 1·61$. From the granite of Pike's Peak, El Paso Co., Colorado (*tysonite*). *Fluocerite*, from near Falun and Österby, Sweden.

Hydrophilite. Chlorocalcite. — Calcium chloride, $KCaCl_3$. Orthorhombic? Pseudoisometric, usually intimately twinned. Biaxial, +. $\beta = 1·605$. In white cube-like crystals or as an incrustation at Vesuvius. *Bœumlerite* is same material intergrown with halite and tachhydrite from Leinetal, Germany. The original hydrophilite was described as $CaCl_2$. There is some doubt if this substance occurs in nature and further confusion concerning its crystal character.

Chlormanganokalite. — $4KCl.MnCl_2$. Rhombohedral. Yellow color. Optically +. $n = 1·59$. Easily decomposed. Found about fumaroles of Vesuvius.

Rinneite. — $FeCl_2.3KCl.NaCl$. Rhombohedral. In coarse granular masses. Prismatic cleavage. H. = 3. G. = 2·3. Colorless, rose, violet or yellow when fresh, becomes brown on exposure due to oxidation. $\omega = 1\cdot59$. Easily fusible. Astringent taste. Found in Germany at Wolkramshausen near Nordhausen and elsewhere in Saxony and near Diekholzen, Hanover.

CRYOLITE.

Monoclinic.   Axes $a : b : c = 0\cdot9663 : 1 : 1\cdot3882$; $\beta = 89° 49'$.

mm''', $110 \wedge 1\bar{1}0 = 88° \ 2'$.	ck, $001 \wedge \bar{1}01 = 55° 17'$.
cm, $001 \wedge 110 = 89° 52'$.	cr, $001 \wedge 011 = 54° 14'$.
cv, $001 \wedge 101 = 55° \ 2'$.	cp, $001 \wedge 111 = 63° 18'$.

Crystals often cubic in aspect and grouped in parallel position; often with twin lamellæ. Massive. The atomic structure has been shown to be similar to that of garnet. At about 570° C. changes to an isometric modification.

Parting at times due to twinning lamellæ parallel to $c(001)$, $m(110)$ and $k(\bar{1}01)$. Fracture uneven. Brittle. H. = 2·5. G. = 2·95–3·0. Luster vitreous to greasy; somewhat pearly on $c(001)$. Colorless to snow-white, sometimes reddish or brownish to brick-red or even black. Transparent to translucent. Optically +. Ax. pl. ⊥ (010). $Z \wedge c$ axis = $-44°$. $2V = 43°$. $\alpha = 1\cdot3385$, $\beta = 1\cdot3389$, $\gamma = 1\cdot3306$.

716

717

Comp. — A fluoride of sodium and aluminum, Na_3AlF_6 or $3NaF.AlF_3 =$ Fluorine 54·4, aluminum 12·8, sodium 32·8 = 100. A little iron sesquioxide is sometimes present as impurity.

Pyr., etc. — Fusible in small fragments in the flame of a candle. Heated in C. T. with potassium bisulphate gives fluorine reaction. In the forceps fuses very easily, coloring the flame yellow. On charcoal fuses easily to a clear bead, which on cooling becomes opaque; after long blowing, the assay spreads out, the fluoride of sodium is absorbed by the coal, a suffocating odor of fluorine is given off, and a crust of alumina remains, which, when heated with cobalt solution in O.F., gives a blue color. Soluble in sulphuric acid, with evolution of hydrofluoric acid.

Diff. — Distinguished by its extreme fusibility. Because of its low index of refraction the powdered mineral becomes almost invisible when placed in water. Its planes of parting (resembling cubic cleavage) and softness are characteristic.

Obs. — Occurs in a bay in Arksukfiord, in West Greenland, at Ivigtut (or Evigtok), about 12 m. from the Danish settlement of Arksuk, where it constitutes a large bed in a granitic vein in a gray gneiss. Cryolite and its alteration products, pachnolite, thomsenolite, prosopite, etc., also occur in limited quantity at the southern base of Pike's Peak, El Paso county, Colorado, north and west of Saint Peter's Dome. Also from near Miask, Ilmen Mts., Russia.

Named from $\kappa\rho\acute{v}os$, *frost*, $\lambda\acute{\iota}\theta os$, *stone*, hence meaning *ice-stone*, in allusion to the translucency of the white masses.

Use. — In the manufacture of sodium salts, certain kinds of glass and porcelain, and as a flux in the electrolytic process for the production of aluminum.

Cryolithionite. — $3NaF.3LiF.2AlF_3$. Isometric. In dodecahedral crystals. Has body-centered cubic structure, similar to that of garnet. Dodecahedral cleavage. H. = 2·5. G. = 2·78. Colorless or white. $n = 1\cdot34$. Occurs associated with cryolite both at Ivigtut and in the Ilmen Mts.

Chiolite. — $5NaF.3AlF_3$. In small pyramidal crystals (tetragonal); also massive granular. Cleavages: $c(001)$ perfect, $p(111)$ distinct. H. = 3·5–4. G. = 2·84–2·90. Color snow-white or colorless. Optically −. $\omega = 1·349$. $\epsilon = 1·342$. From near Miask in the Ilmen Mts., Russia; also with the Greenland cryolite.

Avogadrite. — $(K,Cs)BF_4$. Orthorhombic. In eight-sided crystals, tabular || (001). G. = 2·62. Optically −. Ax. pl. || (010). $X = c$ axis. $\beta = 1·3245$. 2V very large. Found as a sublimate at Vesuvius.

Hieratite. — K_2SiF_6. Isometric. Octahedral cleavage. Colorless, white, gray. $n = 1·340$. In stalactitic concretions. From the fumaroles of the crater of Vulcano, Lipari Islands; also from Vesuvius.

Malladrite. — Na_2SiF_6. Hexagonal. Optically −. $n =$ approx. 1·3. Found as minute prisms sometimes terminated by a pyramid, associated with avogadrite, and hieratite, among the sublimation products of Vesuvius.

II. Oxychlorides, Oxyfluorides

ATACAMITE.

Orthorhombic. Axes $a : b : c = 0·6613 : 1 : 0·7515$.

mm''', $110 \wedge \bar{1}10 = 66° 57'$. rr''', $111 \wedge 1\bar{1}1 = 52° 48'$.
ee', $011 \wedge 0\bar{1}1 = 73° 51'$. mr, $110 \wedge 111 = 36° 16\frac{1}{2}'$.

Commonly in slender prismatic crystals, vertically striated. Twins according to a complex law. (*Paratacamite* is twinned atacamite.) In confused crystalline aggregates; also massive, fibrous or granular to compact; as sand.

718   **719**

Cleavage: $b(010)$ highly perfect. Fracture conchoidal. Brittle. H. = 3–3·5. G. = 3·76–3·78. Luster adamantine to vitreous. Color bright green of various shades, dark emerald-green to blackish green. Streak apple-green. Transparent to translucent. Optically −. Ax. pl. (100). $X = b$ axis. 2V = 75°. $\rho < v$, strong. $\alpha = 1·831$. $\beta = 1·861$. $\gamma = 1·880$.

Comp. — $CuCl_2.3Cu(OH)_2$ = Chlorine 16·6, copper 14·9, cupric oxide 55·8, water 12·7 = 100.

The analyses show considerable variations, and the exact formula is uncertain.

Pyr., etc. — In the closed tube gives off much water, and forms a gray sublimate. B.B. on charcoal fuses, coloring the O.F. azure-blue, with a green edge, and giving two coatings, one brownish and the other grayish white; continued blowing yields a globule of metallic copper; the coatings, touched with the R.F., volatilize, coloring the flame azure-blue. In acids easily soluble.

Obs. — Always of secondary origin, commonly derived from malachite and cuprite and usually associated with these minerals. With other copper minerals, limonite, gypsum, etc. Originally found in Atacama, Chile, at Copiapo, etc. Also in Antofagasta at Taltal and in the Cerro Gordo. Also from Peru and Bolivia. In South Australia at Wallaroo, in the Burra district, etc. From Vesuvius. In the United States rarely from copper districts in Arizona, Wyoming and Utah.

Tallingite. — A hydrated copper chloride from the Botallack mine, Cornwall; in blue globular crusts.

Percylite. — A lead-copper oxychloride, perhaps $PbCl_2.CuO.H_2O$. Isometric. In sky-blue cubes. Cleavage, cubic. H. = 2–3. $n = 2·05$. From Sonora, Mexico. In Chile from Caracoles, Cerro Gordo, Antofagasta; also from Atacama and Challocollo, Tarapaca. From South Africa and Broken Hill, New South Wales.

Cumengite. Cumengéite. — $4PbCl_2.4CuO.5H_2O$. (Or $PbCl_2.Cu(OH)_2$; a formula derived from the X-ray study of the structure.) Tetragonal. Cleavages: (101), distinct; (110), poor. Indigo-blue color. Optically —. $\omega = 2·041$, $\epsilon = 1·926$. H. = 2·5. G. = 4·67. Occurs intimately associated with boléite from Boléo, near Santa Rosalia, Lower California, Mexico, partially or wholly surrounding the latter. To be distinguished from boléite by its lighter color.

Pseudoboléite. — $5PbCl_2.4CuO.6H_2O$. Tetragonal. Cleavages: (001), perfect; also pyramidal. Indigo-blue color. Optically —. $\omega = 2·03$, $\epsilon = 2·00$. H. = 2·4. G. = 5·0. Found only in parallel growth with boléite and cumengite at Boléo. Recently it has been suggested that pseudoboléite is identical with boléite and that the mineral is isometric with the formula, $3PbCl_2.3Cu(OH)_2.AgCl$. This has been largely based upon X-ray study. Other similar investigations have shown contradictory results. At present the preponderance of evidence indicates the identity of the two species and their tetragonal crystal structure.

Boléite. — $9PbCl_2.8CuO.3AgCl.9H_2O$. Tetragonal, pseudo-isometric. Twinned to form pseudo-cubes. Cleavages: (001), perfect; (101), distinct. Indigo-blue color. Optically —. $\omega = 2·04$, $\epsilon = 2·03$. Crystals show an isotropic center, supposed to be due to intimate twinning. See further under pseudoboléite, above. Occurs associated with cumengite, pseudoboléite, and various other copper and lead minerals at Boléo, near Santa Rosalia, Lower California, Mexico. Also noted in Chile near Huantajaya, Tarapaca, and at Challocollo, Cerro Gordo, Antofagasta; and at Broken Hill, New South Wales.

Diaboléite. — $2Pb(OH)_2.CuCl_2$. Tetragonal. Tabular. Cleavage parallel to (001). H. = 2·5. G. = 6·412. Color, bright sky-blue. Optically —. $n = 1·98$. Dichroic, O = deep blue, E = pale blue to almost colorless. Found associated with chloroxiphite embedded in mendipite from Higher Pitts, Mendip Hills, Somersetshire, England.

Chloroxiphite. — $2PbO.Pb(OH)_2.CuCl_2$. Monoclinic. Crystals elongated parallel to b axis. Tabular ‖ (001). Cleavages: (001) perfect, (010) distinct. H. = 2·5. G. = 6·76. Color dull olive- or pistachio-green. Resinous to adamantine luster. Optically —. Ax. pl. ⊥ (010). X nearly parallel to c axis. Refraction and birefringence, both high. Cleavage plates show strong pleochroism, bright emerald-green parallel to b axis and yellowish brown normal to it. Found embedded in mendipite of Higher Pitts, Mendip Hills, Somersetshire, England.

Penfieldite. — $PbO.2PbCl_2$. Hexagonal, prismatic. Cleavage: (0001). Colorless to white. Optically +. $\omega = 2·13$, $\epsilon = 2·21$. Occurs with fiedlerite, etc., in the ancient lead slags at Laurium, Greece.

Fiedlerite. — $PbO.2PbCl_2.H_2O$. Monoclinic. Tabular ‖ (100). Cleavages: (001), (100). Colorless to white. Optically —. $\beta = 2·10$. H. = 3·5. G. = 5·88. From the ancient lead slags at Laurium, Greece.

Matlockite. — Lead chlorofluoride, PbFCl. In tabular tetragonal crystals. H. = 2–3. G. = 7.21. Luster adamantine to pearly. Color yellowish or slightly greenish. Optically —. $\omega = 2·15$. $\epsilon = 2·04$. From Cromford, near Matlock, Derbyshire. Also noted from Laurium, Greece, and with percylite at Challocollo, Tarapaca, Chile.

Laurionite. — $Pb(OH)_2.PbCl_2$. Orthorhombic. In minute prismatic, colorless crystals. Cleavage, (010). Optically —. $\alpha = 2·077$, $\beta = 2·116$, $\gamma = 2·158$. H. = 2·3. G. = 6·24. Found in ancient lead slags at Laurium, Greece, associated with paralaurionite, penfieldite, matlockite, fiedlerite, phosgenite, cerussite, anglesite, etc. Also noted from Wheal Rose near Sithney, Cornwall.

Paralaurionite. — $Pb(OH)_2.PbCl_2$. Monoclinic, prismatic. Cleavage, (001). Colorless to white. $n = 2·146$. G. = 6·05. It has been suggested that laurionite is the same as paralaurionite but that submicroscopic twinning gives it pseudo-orthorhombic symmetry. Occurs with laurionite, etc., in the ancient lead slags at Laurium, Greece. Also from Wheal Rose near Sithney. Cornwall. *Rafaelite*, from the San Rafael mine, Cerro Gordo, Antofagasta, Chile, is the same mineral.

Mendipite. — $2PbO.PbCl_2$. Orthorhombic. In fibrous or columnar masses; often radiated. Prismatic cleavage. H. = 2·5–3. G. = 7–7·2. Color white. Pearly to adamantine luster. Ax. pl. (010). 2V = 90°. $\alpha = 2·24$, $\beta = 2·27$, $\gamma = 2·31$. From the Mendip Hills, Somersetshire, England; near Brilon, Westphalia. Material thought to have the composition $3PbO.PbCl_2$, has been called *pseudomendipite*.

Daviesite. — A lead oxychloride of uncertain composition. Orthorhombic. In minute prisms. Optically +. $\beta = 1·752$. From Beatrix mine, Cerro Gordo, Antofagasta, Chile.

Schwartzembergite. — $6PbO.3PbCl_2.PbI_2O_6$. Pseudo-tetragonal. In druses of minute crystals and crusts. Cleavage, (001). H. = 2·5. G. = 6·5. Color, honey-yellow. Biaxial, —. $\beta = 2·35$. From Cerro Gordo and various other points in northern Chile.

Lorettoite. — $6PbO.PbCl_2$. Tetragonal? Massive. Perfect basal cleavage. G. = 7·6. H. = 3. Fusible at 1. Color honey-yellow. Uniaxial, −. Indices, 2·33–2·40; From Loretto, Tennessee. *Chubutite*, a reddish yellow, tetragonal (?), oxychloride of lead, $7PbO.PbCl_2$, from Chubut, Argentina, is probably identical with lorettoite.

Hämatophanite. — $Pb(Cl,OH)_2.4PbO.2Fe_2O_3$. Tetragonal. Lamellar aggregates of thin plates. Micaceous cleavage parallel to base. H. = 2–3. G. = 7·7. Color dark red-brown with submetallic luster. Streak yellowish red. Optically −. Occurs associated with plumboferrite at Jakobsberg, Vermland, Sweden.

Kempite. — $MnCl_2.3MnO_2.3H_2O$. Orthorhombic. In minute prismatic crystals. Color emerald-green. H. = 3·5. G. = 2·94. Optically −. $X \perp (001)$. Ax. pl. = (010). α = 1·684, β = 1·695, γ = 1·698. Found very sparingly associated with pyrochroite, hausmannite, and rhodochrosite in a large boulder of manganese ore that formerly existed near San José, Santa Clara Co., California.

Daubreeite. — An earthy yellowish oxychloride of bismuth. From Cerro de Tazna, Bolivia.

Nocerite. — $Ca_3Mg_3O_2F_8$. In white hexagonal prismatic to acicular crystals. Colorless, brownish, rarely greenish. Optically −. ω = 1·509, ϵ = 1·485. G. = 2·96. Occurs in limestone blocks in the tufa of Nocera and elsewhere in Campania, Italy.

Koenenite. — An oxychloride of aluminum and magnesium. Rhombohedral. Perfect basal cleavage yielding flexible folia. Very soft. G. = 2·0. Color red, due to included hematite. Optically +. ω = 1·52, ϵ = 1·55. From near Volpriehausen in the Solling, Hanover, Germany.

The following are oxychlorides of mercury from the mercury deposits at Terlingua Texas. Associated minerals are montroydite, calomel, native mercury and calcite.

Eglestonite. — Hg_4Cl_2O. Isometric in minute crystals of dodecahedral habit. Many forms observed. H. = 2–3. G. = 8·3. Luster adamantine to resinous. Color brownish yellow darkening on exposure to black. n = 2·49. Volatile.

Terlinguaite. — Hg_2ClO. Monoclinic. In small striated prismatic crystals elongated parallel to the b-axis. Many forms observed. Cleavage perfect, (101). H. = 2–3. G. = 8·7. Luster adamantine. Color sulphur-yellow changing to olive-green on exposure. Optically −. α = 2·35, β = 2·64, γ = 2·66. Strong dispersion, $\rho < v$.

III. Hydrous Chlorides, Hydrous Fluorides, etc.

CARNALLITE.

Orthorhombic. Crystals rare. Commonly massive, granular.

No distinct cleavage. Fracture conchoidal. Brittle. H. = 2·5. G. = 1·60. Luster shining, greasy. Color milk-white, often reddish. Transparent to translucent. Strongly phosphorescent. Optically +. 2V = 70°. α = 1·466. β = 1·475. γ = 1·494. Ax. pl. (010). $Z = a$ axis. Taste bitter. Deliquescent.

Comp. — $KMgCl_3.6H_2O$ or $KCl.MgCl_2.6H_2O$ = Chlorine 38·3, potassium 14·1, magnesium 8·7, water 39·0 = 100.

Obs. — Carnallite is found in saline deposits, especially those containing potassium minerals as a deposit from low temperature solutions. Occurs in the salt beds of northern Germany, alternating with thinner beds of common salt and kieserite. At Stassfurt, Province of Saxony, Leopoldshall in Anhalt, etc. In large crystals from Beienrode, near Königshütte, Upper Silesia.

Use. — Carnallite is a source of potash compounds used in fertilizers.

Douglasite, associated with carnallite, is said to be $2KCl.FeCl_2.2H_2O$.

Bischofite. — $MgCl_2.6H_2O$. Monoclinic. Crystalline-granular or fibrous; colorless to white. H. = 1·5. Optically +. β = 1·524. Decomposes in air. From Leopoldshall and Stassfurt, Prussia.

Kremersite. — $KCl.NH_4Cl.FeCl_3.H_2O$. Orthorhombic. In pseudo-octahedral crystals. Color red. Found at acid fumaroles on Vesuvius and on Mt. Etna, Sicily.

Erythrosiderite. — $2KCl.FeCl_3.H_2O$. Orthorhombic. In tabular or pseudo-octahedral crystals. Color red. Optically +. β = 1·75. Dispersion strong, $\rho > v$. From Vesuvius and Etna.

Mitscherlichite. — $2KCl.CuCl_2.2H_2O$. Minute tetragonal crystals, greenish blue color. Optically −. $\omega = 1·637$, $\epsilon = 1·615$. G. = 2·418. Occurs on saline sublimations in the crater of Vesuvius.

Tachhydrite. — $CaCl_2.2MgCl_2.12H_2O$. Rhombohedral. Cleavage rhombohedral. Optically −. $\omega = 1·520$, $\epsilon = 1·512$. H. = 2. G. = 1·66. In wax- to honey-yellow masses. Found in salt deposits of northern Germany, at Stassfurt, Province of Saxony, etc.

Fluellite. — AlF_3H_2O. Orthorhombic. In colorless or white pyramids. H. = 3. G. = 2·17. Colorless or white. Optically +. $\alpha = 1·473$, $\beta = 1·490$, $\gamma = 1·511$. From Stenna Gwyn, near St. Austell, Cornwall.

Chloralluminite. — $AlCl_3.6H_2O$. Rhombohedral. Colorless, white, or yellowish. Optically −. $n = 1·6$. Found about fumaroles of Vesuvius.

Prosopite. — $CaF_2.2Al(F,OH)_3$. In monoclinic crystals, or granular massive. H. = 4·5. G. = 2·88. Colorless, white, grayish. Optically +. $\alpha = 1·501$, $\beta = 1·503$, $\gamma = 1·510$. From Altenberg, Saxony; from Schlaggenwald, Bohemia; St. Peter's Dome near Pike's Peak, Colorado; Dugway mining district, Toelle Co., Utah.

Pachnolite. — $NaF.CaF_2.AlF_3.H_2O$. Monoclinic. Prismatic. Prism angle = 98° 36′. Twinned on (100), giving pseudo-orthorhombic crystals. Cleavage, (001), poor. Colorless. Optically +. $\beta = 1·413$. H. = 3. G. = 2·98. An alteration product of cryolite at the localities for the latter in Greenland, Colorado, and the Ural Mts.

Thomsenolite. — Comp. same as for pachnolite. Monoclinic. Prism angle = 89° 46′. Crystals often resemble cubes, also prismatic. Cleavages: (001) perfect, (110) imperfect. Colorless. Optically −. $\beta = 1·414$. H. = 2. G. = 2·98. Occurrence same as for pachnolite.

Gearksutite. — $CaF_2.Al(F,OH)_3.H_2O$. Monoclinic. In fine acicular crystals, earthy, clay-like. White. Optically −. $\beta = 1·454$. H. = 2. G. = 2·75. Found with cryolite at the localities for the latter in Greenland, Colorado, and the Ural Mts. Also noted from green sand at Gingin, West Australia; in clay between Warm Springs and Hot Springs, Virginia; at Wagon Wheel Gap, Colorado.

Ralstonite. — $(Na_2,Mg)F_2.3Al(F,OH)_3.2H_2O$. Isometric. Octahedral crystals. Cleavage, octahedral. Colorless or white. $n = 1·43$. H. = 4·5. G. = 2·55. Found with cryolite and thomsenolite at Ivigtut, Greenland. Also noted from St. Peter's Dome, Colorado, and from Tanokamiyama, Omi, Japan.

ZIRKLERITE. $9(Fe,Mg,Ca)Cl_2.2Al_2O_3.3H_2O$. Hexagonal, rhombohedral. Cleavage rhombohedral. Optically +. n about 1·552. Weak birefringence. H. = 3·5. G. = 2·6. Found at Hope in Hanover and elsewhere, forming the chief constituent of a light gray massive to fine-grained rock in breccia-like layers in halite or potash salts.

Creedite. — $2CaF_2.2Al(F,OH)_3.CaSO_4.2H_2O$. Monoclinic. In grains, prismatic crystals and radiating masses. Usually colorless, rarely purple. H. = 3·5. G. = 2·71. Perfect cleavage, (100). Indices, 1·46–1·49. $2V = 64°$. $Y = b$ axis. Fusible with intumescence. Soluble in acids. Found near Wagon Wheel Gap, Creed Quadrangle, Colorado.

Trudellite. — $4AlCl_3.Al_2(SO_4)_3.4Al(OH)_3.30H_2O$. Hexagonal, rhombohedral. Indistinct rhombohedral cleavage. In compact masses. G. = 1·93. H − 2·5. Amber yellow color. Optically . $\omega = 1·560$. $\epsilon = 1·495$. Occurs with pickeringite, anhydrite, and gypsum in a breccia near Pintados, Tarapaca, Chile.

V. OXIDES

I. Oxides of Silicon.

II. Oxides of the Semi-Metals; Tellurium, Arsenic, Antimony, Bismuth; also Molybdenum, Tungsten.

III. Oxides of the Metals.

The Fifth Class, that of the OXIDES, is subdivided into three sections, according to the positive element present. The oxides of the non-metal silicon are placed by themselves, but it will be noted that the compounds of the related element titanium are included with those of the metals proper. This last is made necessary by the fact that in one of its forms TiO_2 is isomorphous with MnO_2 and PbO_2.

A series of oxygen compounds which are properly to be viewed as salts, *e.g.*, the species of the Spinel Group and a few others, are for convenience also included in this class.

I. Oxides of Silicon

QUARTZ.

Rhombohedral-trapezohedral.   Axis: $c = 1{\cdot}09997$.

rr', $10\bar{1}1 \wedge \bar{1}101 = 85° 46'$. mz, $10\bar{1}0 \wedge 01\bar{1}1 = 66° 52'$.
rz, $10\bar{1}1 \wedge 01\bar{1}1 = 46° 16'$. ms, $10\bar{1}0 \wedge 11\bar{2}1 = 37° 58'$.
mr, $10\bar{1}0 \wedge 10\bar{1}1 = 38° 13'$. mx, $10\bar{1}0 \wedge 51\bar{6}1 = 12° 1'$.

Crystals commonly prismatic, with the $m(10\bar{1}0)$ faces horizontally striated; terminated commonly by the two rhombohedrons, $r(10\bar{1}1)$ and $z(01\bar{1}1)$, in nearly equal development, giving the appearance of a hexagonal pyra-

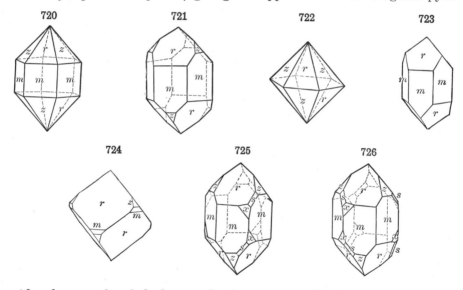

mid; when one rhombohedron predominates it is in almost all cases r. Often in double six-sided pyramids or *quartzoids* through the equal development of r and z; when r is relatively large the form then has a cubic aspect ($rr' = 85° 46'$). Crystals frequently distorted, when the correct orientation may be obscure except as shown by the striations on m. Crystals often elongated to acicular forms, and tapering through the oscillatory combination of successive rhombohedrons with the prism. Occasionally twisted or bent. Frequently in radiated masses with a surface of pyramids, or in druses. That quartz has a complicated atomic structure has been shown by X-ray study. The silicon atoms lie on three interpenetrating hexagonal lattices which have in the vertical direction a spiral arrangement in respect to each other. The oxygen atoms are apparently grouped in a tetrahedral manner about the silicon atoms. The unit cell contains three silicon atoms.

Quartz is enantiomorphous and simple crystals are either right- or left-handed. On a *right-handed* crystal (Fig. 725) the right trigonal pyramid $s(11\bar{2}1)$, if present, lies to the right of the *m* face, which is below the predominating positive rhombohedron *r*, and with this belong the positive right trapezohedrons, as $x(51\bar{6}1)$. On a *left-handed* crystal (Fig. 726), *s* lies to the left of the *m* below *r*. The right- and left-handed forms occur together only in twins. In the absence of trapezohedral faces the striations on *s* (∥ edge *r/m*), if distinct, serve to distinguish the faces *r* and *z*, and hence show the right- and left-handed character of the crystals. The right- and left-handed character is also revealed by etching (Art. **291**) and by pyroelectricity (Art. **450**).

Thermal study of quartz shows that it exists in two modifications, known as α- and β-quartz, or low and high temperature quartz. α-quartz is apparently hexagonal, trapezo-hedral-tetartohedral and is formed at temperatures below 573° while β-quartz is hexagonal, trapezohedral-hemihedral and forms at temperatures ranging from 573° to 870°. Above 870° tridymite is formed. β-quartz shows characteristically regular hexagonal bipyramids with or without a subordinate prism face. Twinning on $r(10\bar{1}1)$ is characteristic of β-quartz. Twinning with parallel *c* axes is very common in α-quartz but of no importance in β-quartz. X-ray study shows close similarity in structure between α- and β-quartz. The crystal angles of α-quartz change with increase of temperature up to 573°, the inversion point to β-quartz, while beyond this point they remain nearly constant. The temperature of inversion from α-quartz to β-quartz has been shown to rise with increase of pressure. In a similar manner at this point there is a sudden marked lowering of the refractive indices and birefringence. α-quartz occurs in veins and geodes and large pegmatites while the β modification is found in graphic granite, granite pegmatites, and porphyries. Tridymite when heated to about 1470° passes over into cristobalite. Quartz, tridymite and cristobalite are probably to be considered as polymers.

Twins: (1) tw. axis *c*, all axes parallel. (2) Tw. pl. *a*, sometimes called the *Brazil law*, usually as irregular penetration-twins (Fig. 727). (3) Tw. pl. $\xi(11\bar{2}2)$, contact-twins, the axes crossing at angles of 84° 33′ and with a prism face in common to the two individuals. (4) Tw. pl. $r(10\bar{1}1)$. See further p. 190 and Figs. 453–455. Massive forms common and in great variety, passing from the coarse or fine granular and crystalline kinds to those which are flint-like or cryptocrystalline. Sometimes mammillary, stalactitic, and in concretionary forms; as sand.

Cleavage not distinctly observed; sometimes fracture surfaces (∥ $r(10\bar{1}1)$, $z(01\bar{1}1)$ and $m(10\bar{1}0)$), developed by sudden cooling after being heated (see Art. 284). Fracture conchoidal to subconchoidal in crystallized forms, uneven to splintery in some massive kinds. Brittle to tough. H. = 7. G. = 2·653–2·660 in crystals; cryptocrystalline forms somewhat lower (to 2·60) if pure, but impure massive forms (*e.g.*, jasper) higher. Luster vitreous, sometimes greasy; splendent to nearly dull. Colorless when pure; often various shades of yellow, red, brown, green, blue, black. Streak white, of pure varieties; if impure, often the same as the color, but much paler. Transparent to opaque.

Optically +. Double refraction weak. Polarization circular; right-handed or left-handed, the optical character corresponding to right- and left-handed character of crystals, as defined above; in twins (law 2), both right and left forms sometimes united, sections then often

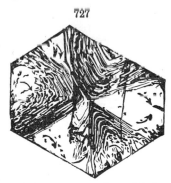

727

Basal section in polarized light, showing interpenetration of right- and left-handed portions. Des Cloizeaux.

showing Airy's spirals in the polariscope (cf. Art. **402**, p. 295, and Fig. 727). Rotatory power proportional to thickness of plate. Refractive indices for the D line, $\omega = 1\cdot54418$, $\epsilon = 1\cdot55328$; also rotatory power for section of 1^{mm} thickness, $\alpha = 21\cdot71°$ (D line). Pyroelectric; also electric by pressure or piezoelectric. See Arts. **450, 451.** On etching-figures, see Arts. **291, 292.**

Comp. — Silica, or silicon dioxide, SiO_2 = Oxygen 53·3, silicon 46·7 = 100.

In massive varieties often mixed with a little opal silica. Impure varieties contain iron oxide, calcium carbonate, clay, sand, and various minerals as inclusions.

Artif. — Quartz has been produced artificially in numerous ways. Recently crystals have been obtained at temperatures below 760° from melts containing dissolved silica which were composed of (1) a mixture of potassium and lithium chlorides, (2) vanadic acid, (3) sodium tungstate. At higher temperatures tridymite crystals formed.

Var. — A. PHENOCRYSTALLINE: Crystallized, vitreous in luster. B. CRYPTOCRYSTALLINE: Flint-like, massive.

The first division includes all ordinary vitreous quartz, whether having crystalline faces or not. The varieties under the second are in general acted upon somewhat more by attrition, and by chemical agents, as hydrofluoric acid, than those of the first. In all kinds made up of layers, as agate, successive layers are unequally eroded.

A. PHENOCRYSTALLINE OR VITREOUS VARIETIES

Ordinary Crystallized; Rock Crystal. — Colorless quartz, or nearly so, whether in distinct crystals or not. Here belong the Bristol diamonds, Lake George diamonds, Brazilian pebbles, etc. Some variations from the common type are: (*a*) cavernous crystals; (*b*) cap-quartz made up of separable layers or caps; (*c*) drusy quartz, a crust of small or minute quartz crystals; (*d*) radiated quartz, often separable into radiating parts, having pyramidal terminations; (*e*) fibrous, rarely delicately so, as a kind from Griqualand West, South Africa, altered from crocidolite (see *cat's-eye* below, also crocidolite, p. 578).

Asteriated; Star-quartz. — Containing within the crystal whitish or colored radiations along the diametral planes. Occasionally exhibits distinct asterism which is said to be due to the inclusion of submicroscopic needles of some other mineral arranged in parallel orientation.

Amethystine; Amethyst. — Clear purple, or bluish violet. Color perhaps due to manganese.

Rose. — Rose-red or pink, but becoming paler on exposure. Commonly massive. Luster sometimes a little greasy. Color perhaps due to titanium.

Yellow; False Topaz or *Citrine.* — Yellow and pellucid; resembling yellow topaz.

Smoky; Cairngorm Stone. — Smoky yellow to dark smoky brown, and often transparent; varying to brownish black. Color is probably due to some organic compound (Forster). Called *cairngorms* from the locality at Cairngorm, southwest of Banff, in Scotland. The name *morion* is given to nearly black varieties.

Milky. — Milk-white and nearly opaque. Luster often greasy.

Siderite, or *Sapphire-quartz.* — Of indigo or Berlin-blue color; a rare variety.

Sagenitic. — Inclosing acicular crystals of rutile. Other included minerals in acicular forms are: black tourmaline; goethite; stibnite; asbestos; actinolite; hornblende; epidote.

Cat's-eye exhibits opalescence, but without prismatic colors, especially when cut *en cabochon*, an effect sometimes due to fibers of asbestos. Also present in the siliceous pseudomorphs, after crocidolite, called *tiger-eye* (see crocidolite). The highly-prized Oriental cat's-eye is a variety of chrysoberyl.

Aventurine. — Spangled with scales of mica, hematite, or other mineral.

Impure from the presence of distinct minerals distributed densely through the mass. The more common kinds are those in which the impurities are: (*a*) *ferruginous*, either red or yellow, from anhydrous or hydrous iron sesquioxide; (*b*) *chloritic*, from some kind of chlorite; (*c*) *actinolitic*; (*d*) *micaceous*; (*e*) *arenaceous*, or sand.

Containing liquids in cavities. — The liquid, usually water (pure, or a mineral solution), or some petroleum-like compound. Quartz, especially smoky quartz, also often contains inclusions of both liquid and gaseous carbon dioxide.

B. CRYPTOCRYSTALLINE VARIETIES

Chalcedony. — Having the luster nearly of wax, and either transparent or translucent. G. = 2·6–2·64. Color white, grayish, blue, pale brown to dark brown, black. Also of other shades, and then having other names. Often mammillary, botryoidal, stalactitic, and occurring lining or filling cavities in rocks. It often contains some disseminated opal-silica. The thermal study of *chalcedony* has shown differences from quartz and it has been suggested that it constitutes a distinct species, but these differences may be due to the presence of a small amount of opal.   The name *enhydros* is given to nodules of chalcedony containing water, sometimes in large amount.   Embraced under the general name chalcedony is the crystalline form of silica which forms concretionary masses with radial-fibrous and concentric structure, and which, as shown by Rosenbusch, is optically *negative*, unlike true quartz.   It has $n = 1\cdot537$; G. $= 2\cdot59$–$2\cdot64$.   Often in spherulites, showing the spherulitic interference figure.   *Lussatite* of Mallard has a like structure, but is optically $+$ and has the specific gravity and refractive index of opal.   It may be a fibrous form of *tridymite*. *Quartzine* and *lutecite* are apparently fibrous chalcedony having the elongation of the fibers perpendicular to the *c*-axis (opposite to the usual orientation of chalcedony fibers) and showing optical anomalies that have led to their being considered distinct species.

   Carnelian.   Sard. — A clear red chalcedony, pale to deep in shade; also brownish red to brown.

   Chrysoprase. — An apple-green chalcedony, the color due to nickel oxide.

   Prase. — Translucent and dull leek-green.

   Plasma. — Rather bright green to leek-green, and also sometimes nearly emerald-green, and subtranslucent or feebly translucent.   *Heliotrope*, or *Blood-stone*, is the same stone essentially, with small spots of red jasper, looking like drops of blood.

   Agate. — A variegated chalcedony.   The colors are either (*a*) banded; or (*b*) irregularly clouded; or (*c*) due to visible impurities as in moss agate, which has brown moss-like or dendritic forms, as of manganese oxide, distributed through the mass.   The bands are delicate parallel lines, of white, pale and dark brown, bluish and other shades; they are sometimes straight, more often waving or zigzag, and occasionally concentric circular. The bands are the edges of layers of deposition, the agate having been formed by a deposit of silica from solutions intermittently supplied, in irregular cavities in rocks, and deriving their concentric waving courses from the irregularities of the walls of the cavity.   The layers differ in porosity, and therefore agates may be varied in color by artificial means, and this is done now to a large extent with the agates cut for ornament.   There is also *agatized wood*; wood petrified with clouded agate.

   Onyx. — Like agate in consisting of layers of different colors, white and black, white and red, etc., but the layers in even planes, and the banding straight, and hence its use for cameos.

   Sardonyx. — Like onyx in structure, but includes layers of carnelian (sard) along with others of white or whitish, and brown, and sometimes black colors.

   Agate-jasper. — An agate consisting of jasper with veinings of chalcedony.

   Siliceous sinter. — Irregularly cellular quartz, formed by deposition from waters containing silica or soluble silicates in solution.   See also under opal, p. 475.

   Flint. — Somewhat allied to chalcedony, but more opaque, and of dull colors, usually gray, smoky brown, and brownish black.   The exterior is often whitish, from mixture with lime or chalk, in which it is embedded.   Luster barely glistening, subvitreous.   Breaks with a deeply conchoidal fracture, and a sharp cutting edge.   The flint of the chalk formation consists largely of the remains of diatoms, sponges, and other marine productions. The coloring matter of the common kind is mostly carbonaceous matter.   Flint implements play an important part among the relics of early man.

   Hornstone. — Resembles flint, but is more brittle, the fracture more splintery.   *Chert* is a term often applied to hornstone, and to any impure, flinty rock, including the jaspers.

   Basanite, Lydian Stone, or *Touchstone.* — A velvet-black siliceous stone or flinty jasper, used on account of its hardness and black color for trying the purity of the precious metals. The color left on the stone after rubbing the metal across it indicates to the experienced eye the amount of alloy.   It is not splintery like hornstone.

   Jasper. — Impure opaque colored quartz; commonly red, also yellow, dark green and grayish blue.   *Striped* or *riband jasper* has the colors in broad stripes.   *Porcelain jasper* is nothing but baked clay, and differs from true jasper in being B.B. fusible on the edges.

C.   Besides the above there are also:

   Granular Quartz, Quartz-rock, or *Quartzite.* — A rock consisting of quartz grains very firmly compacted; the grains often hardly distinct.   *Quartzose Sandstone, Quartz-conglomerate.* — A rock made of pebbles of quartz with sand.   The pebbles sometimes are jasper and chalcedony, and make a beautiful stone when polished.   *Itacolumite,* or *Flexible Sandstone.* — A friable sand-rock, consisting mainly of quartz-sand, but containing a little mica, and possessing a degree of flexibility when in thin laminæ.   *Buhrstone,* or Burrstone. — A cellular, flinty rock, having the nature in part of coarse chalcedony.

   Pseudomorphous Quartz. — Quartz appears also under the forms of many of the mineral species, which it has taken through either the alteration or replacement of crystals of those species. The most common quartz pseudomorphs are those of calcite, barite, fluorite, and siderite. *Silicified wood* is quartz pseudomorph after wood (p. 358).

Lechateliérite is the name given to naturally occurring fused quartz as in fulgurites and inclusions in volcanic rocks. Noted also in the meteor crater west of Winslow, Arizona, produced apparently by the fusion of a fine-grained sandstone.

Pyr., etc. — B.B. unaltered; with borax dissolves slowly to a clear glass; with soda dissolves with effervescence; unacted upon by salt of phosphorus. Insoluble in hydrochloric acid, and only slightly acted upon by solutions of fixed caustic alkalies, the cryptocrystalline varieties to the greater extent. Soluble only in hydrofluoric acid. When fused and cooled it becomes opal-silica having G. = 2·2.

Diff. — Characterized in crystals by the form, glassy luster, and absence of cleavage; also in general by hardness and infusibility.

Micro. — Easily recognized in rock sections by its low refraction (" low relief," p. 235) and low birefringence ($\epsilon - \omega = 0·009$); the interference colors in good sections not rising above yellow of the first order; also by its limpidity and the positive uniaxial cross yielded by basal sections (p. 295, note), which remain dark when revolved between crossed nicols. Commonly in formless grains (granite), also with crystal outline (porphyry, etc.).

Obs. — Quartz is an essential component of certain igneous rocks, as granite, granite-porphyry, quartz-porphyry and rhyolite in the granite group; in such rocks it is commonly in formless grains or masses filling the interstices between the feldspar, as the last product of crystallization. Further it is an essential constituent in quartz-diorite, quartz-diorite porphyry and dacites in the diorite group; in the porphyries frequently in distinct crystals. It occurs also as an accessory in other feldspathic igneous rocks, such as syenite and trachyte. Among the metamorphic rocks it is an essential component of certain varieties of gneiss, of quartzite, etc. It forms the mass of common sandstone. It occurs as the vein-stone in various rocks, and forms a large part of mineral veins; as a foreign mineral in some limestones, etc., making geodes of crystals, or of chalcedony, agate, carnelian, etc.; as embedded nodules or masses in various limestones, constituting the flint of the Chalk formation, the hornstone of other limestones — these nodules sometimes becoming continuous layers; as masses of jasper occasionally in limestone. It is the principal material of the pebbles of gravel-beds, and of the sands of the seashore, and sandbeds everywhere. In graphic granite (*pegmatite*) the quartz individuals are arranged in parallel position in feldspar, the angular particles resembling written characters. The quartz grains in a fragmental sandstone are often found to have undergone a secondary growth by the deposition of crystallized silica with like orientation to the original nucleus. From a general study of the chemical and mineralogical character of the rocks of the earth's crust it has been estimated that quartz forms about twelve per cent of their constituents.

Quartz is the most common of minerals, and therefore only those localities that yield unusual crystals or exceptionally fine specimens can be mentioned here. Beautiful crystals, often of smoky quartz, come from many localities in the Swiss Alps, as at St. Gotthard, Ticino; in Val Tavetsch, Grisons; at Göschener Alp, Uri, etc. In Italy from Aosta, etc., in Piedmont; from the Carrara marble quarries in Tuscany; from Poretta, Emilia; from the Island of Elba. Crystals of amethyst are found at Zillertal, Tyrol, Austria, at Schemnitz (Selmecsbánya) in Slovakia, Czechoslovakia, and at Pokura near Nagy-Ág in Transylvania, Rumania. In France fine twin crystals come from the mine of Gardette, near Bourg d' Oisans, Isère, Dauphiné. In England fine crystals come from Frizington, Cleator Moor, and Alston Moor, Cumberland. Smoky quartz from Cairngorm, Banffshire, Scotland, and in the Mourne Mts., Ireland. Unusual chalcedony stalactites are found near Camborne in Cornwall. Large crystals come from Betafo and elsewhere in Madagascar. Famous localities for amethyst are at Mursinska, near Ekaterinburg, Ural Mts., Russia; in Madras, India; from Ceylon, Madagascar, and Uruguay; and especially in Rio do Sul and Minas Geraes, Brazil; also from Guanajuato, Mexico. Twin crystals are found in Kai, Japan. The finest carnelians and agates are found in India; Rio do Sul, Brazil; Uruguay; and formerly at Oberstein, Rhineland, Germany. Chrysoprase occurs at various localities in Silesia, Germany.

Notable localities in the United States include the following. In Maine rose quartz at Paris and smoky quartz at Auburn. In New York small and brilliant crystals are abundant in Herkimer Co., at Middleville, Little Falls, etc., loose in cavities in a quartzose rock, or embedded in loose earth; with hematite beds at Fowler and Edwards, etc., in St. Lawrence Co.; in fine groups at Ellenville, Ulster Co. In Virginia amethyst at Amherst in the county of the same name. In North Carolina fine specimens of both smoky and amethyst quartz occur in Alexander and Lincoln counties, etc. Beautiful colorless crystals occur at Hot Springs, Garland Co., Arkansas. Amethyst occurs on Keweenaw peninsula, Michigan. Rose quartz and chalcedony come from the Black Hills, South Dakota. Fine crystals of smoky quartz come from the granite of the Pike's Peak region in Colorado. Chalcedony occurs at Rico, Dolores Co., Colorado. In Montana smoky quartz from near Butte and both smoky and amethyst quartz on little Pipestone Creek, Jefferson Co. Amethyst crystals from Yellowstone Park, Wyoming. Fine amethyst also comes from the Thunder Bay district on Lake Superior, in Ontario.

The word quartz is of German provincial origin. Agate is from the name of the river Achates, in Sicily, whence specimens were brought, as stated by Theophrastus.

Use. — In its various colored forms as ornamental material; for abrading purposes; manufacture of porcelain, of glass; as wood filler; in paints, scouring soaps, etc.; as sand in mortars and cements; as quartzite, sandstone, etc., for building stone, etc.; as an acid flux in certain smelting operations.

TRIDYMITE.

Hexagonal or pseudo-hexagonal. Axis $c = 1.6530$. Crystals usually minute, thin tabular $\parallel c(0001)$; often in twins; also united in fan-shaped groups.

Cleavage: prismatic, not distinct; parting $\parallel c$, sometimes observed. Fracture conchoidal. Brittle. H. $= 7$. G. $= 2.28-2.33$. Luster vitreous, on c pearly. Colorless to white. Transparent. Optically $+$. Indices, $1.469-1.473$. Often exhibits anomalous refraction phenomena.

Comp. — Pure silica, SiO_2, like quartz.

Tridymite is formed above 800° C. See further under quartz, p. 471. There are two modifications of tridymite with inversion point between 117° C. and 163° C. The low temperature form, α-tridymite, has an orthorhombic structure; the high temperature form, β-tridymite, is hexagonal. The possibility of a third modification closely similar to α-tridymite and called α'-tridymite has been suggested as the result of X-ray study.

Pyr., etc. — Like quartz, but soluble in boiling sodium carbonate.

Obs. — Occurs chiefly in acidic volcanic rocks, rhyolite, trachyte, andesite, liparite, less often in dolerite; usually in cavities, often associated with sanidine, also hornblende, augite, hematite; sometimes in opal. First observed in crevices and druses in an augite-andesite from the Cerro San Cristobal, near Pachuca, Mexico; later proved to be rather generally distributed. Thus in trachyte of the Siebengebirge, Rhineland, Germany; of Euganean Hills in northern Italy; Puy Capucin (Mont-Dore), Puy-de-Dôme, France, etc. In the ejected masses from Vesuvius consisting chiefly of sanidine. In the lavas of Mt. Etna, Sicily, and Mt. Pelée, Martinique. From Kibōsan, Iigo, Japan. With quartz, feldspar, fayalite in lithophyses of Obsidian cliff, Yellowstone Park. In the andesite of Mt. Rainier, Washington.

Named from τριδυμος, *threefold*, in allusion to the common occurrence in trillings.

ASMANITE. A form of silica found in the meteoric iron of Breitenbach, in very minute grains, probably identical with tridymite; by some referred to the orthorhombic system.

Cristobalite. Christobalite. Silica in white octahedrons. Thermal study of the artificial mineral shows two modifications with inversion point lying between 198° C. and 275° C. The low-temperature form, α-cristobalite, is slightly birefringent, probably tetragonal, pseudo-isometric; the higher temperature form, β-cristobalite, is isometric with octahedral crystals which are often twinned according to the spinel law or occur in crystalline aggregates. G. $= 2.27$. $n = 1.486$. With tridymite in andesite of the Cerro San Cristobal, Pachuca, Mexico. In California from Inyo Co., and from Tuscan Springs, Tehama Co. Also noted in lava at Mayen, Rhineland, Germany, and in meteorites. For thermal relations to quartz and tridymite see under quartz, p. 471.

MELANOPHLOGITE. In minute cubes and spherical aggregates. Occurring with calcite and celestite implanted upon an incrustation of opaline silica over the sulphur crystals of Girgenti, Sicily. Consists of SiO_2 with 5 to 7 per cent of SO_3, perhaps SiO_2 with SiS_2. The mineral turns black superficially when heated B.B.

OPAL.

Amorphous (colloidal). Massive; sometimes small reniform, stalactitic, or large tuberose. Frequently pseudomorphous after other minerals. Also earthy.

Conchoidal fracture. H. $= 5.5-6.5$. G. $= 1.9-2.3$; when pure $2.1-2.2$.

Luster vitreous, frequently subvitreous; often inclining to resinous, and sometimes to pearly. Color white, yellow, red, brown, green, gray, blue, generally pale; dark colors arise from foreign admixtures; sometimes a rich play of colors, or different colors by refracted and reflected light. Streak white. Transparent to nearly opaque. $n = 1.43$.

Often shows double refraction similar to that observed in colloidal substances due to tension. The cause of the play of color in the precious opal was investigated by Brewster, who ascribed it to the presence of microscopic cavities. Behrends, however, has given a monograph on the subject (Ber. Ak. Wien, 64 (1), 1871), and has shown that this explanation is incorrect; he refers the colors to thin curved lamellæ of opal whose refractive power may differ by 0·1 from that of the mass. These are conceived to have been originally formed in parallel position, but have been changed, bent, and finally cracked and broken in the solidification of the groundmass.

Comp. — Silica, like quartz, with a varying amount of water, $SiO_2.nH_2O$. The percentage of water is usually small with a maximum about 10 per cent.

The opal condition is one of lower degrees of hardness and specific gravity, and, as generally believed, of incapability of crystallization. The water present varies from 2 to 13 per cent or more, but mostly from 3 to 9 per cent. Small quantities of ferric oxide, alumina lime, magnesia, and alkalies are usually present as impurities.

Var. — *Precious Opal.* — Exhibits a play of delicate colors.

Fire-opal. — Hyacinth-red to honey-yellow colors, with fire-like reflections, somewhat irised on turning.

Girasol. — Bluish white, translucent, with reddish reflections in a bright light.

Common Opal. — In part translucent; (a) *milk-opal*, milk-white to greenish, yellowish, bluish; (b) *Resin-opal*, wax-, honey- to ocher-yellow, with a resinous luster; (c) dull olive-green and mountain-green; (d) brick-red. Includes *Semiopal*; (e) *Hydrophane*, a variety which becomes more translucent or transparent in water.

Cacholong. — Opaque, bluish white, porcelain-white, pale yellowish or reddish.

Opal-agate. — Agate-like in structure, but consisting of *opal* of different shades of color.

Menilite. — In concretionary forms; opaque, dull grayish.

Jasp-opal. Opal-jasper. — Opal containing some yellow iron oxide and other impurities, and having the color of yellow jasper, with the luster of common opal.

Wood-opal. — Wood petrified by opal.

Hyalite. Muller's Glass. — Clear as glass and colorless, constituting globular concretions, and crusts with a globular or botryoidal surface; also passing into translucent, and whitish. Less readily dissolved in caustic alkalies than other varieties.

Schaumopal. — A porous variety from the Virunga district, East Africa.

Fiorite, Siliceous Sinter. — Includes translucent to opaque, grayish, whitish or brownish incrustations, porous, to firm in texture; sometimes fibrous-like or filamentous, and, when so, pearly in luster (then called *Pearl-sinter*); deposited from the siliceous waters of hot springs.

Geyserite. — Constitutes concretionary deposits about the geysers of the Yellowstone Park, Iceland, and New Zealand, presenting white or grayish, porous, stalactitic, filamentous, cauliflower-like forms, often of great beauty; also compact-massive, and scaly-massive.

Float-stone. — In light porous concretionary masses, white or grayish, sometimes cavernous, rough in fracture.

Tripolite. — Formed from the siliceous shells of diatoms (hence called *diatomite*) and other microscopic species, and occurring in extensive deposits. Includes *Infusorial Earth*, or *Earthy Tripolite*, a very fine-grained earth looking often like an earthy chalk, or a clay, but harsh to the feel, and scratching glass when rubbed on it.

Pyr., etc. — Yields water. B.B. infusible, but becomes opaque. Some yellow varieties, containing iron oxide, turn red. Soluble in hydrofluoric acid somewhat more readily than quartz; also soluble in caustic alkalies, but more readily in some varieties than in others.

Obs. — Opal is a non-crystalline, colloidal substance belonging to the group of the mineral-gels. It has been deposited at low temperatures from silica-bearing waters. It occurs in connection with many rock types, igneous, sedimentary, and metamorphic. It is found filling seams and fissures of igneous rocks, where it has been deposited during the last stages of the cooling of the rock. It occurs in some mineral veins and is deposited from many hot or warm springs. It is further formed during the weathering and alteration of many rocks. Its formation frequently accompanies the processes of serpentinization. It forms the siliceous skeletons of various sea organisms, as diatoms, sponges, etc., and ultimately may accumulate in extensive beds of diatomaceous earth, etc. It exists in most chalcedony and flint.

Precious opal occurs in an altered trachyte at Czerwenitza (Červenica) near Kaschau (Košice) in Slovakia; a beautiful blue and green opal comes from Barcoo River, etc., in Queensland; fine opal is found at White Cliffs, New South Wales, as filling in sandstone, in fossil wood, in the material of various fossil shells and bones, and in aggregates of radiating pseudomorphic crystals: gem opal is found in Mexico at Zimapan in Hidalgo (fire-opal) and in Queretaro. Gem opal, often of " black opal " type, comes from Humboldt Co., Nevada. *Hyalite* occurs in clinkstone at Waltsch (Valeč) in Bohemia, Czechoslovakia; from Gracias, Honduras; San Luis Potosi, Mexico; from Tate-yama, Etchu, Japan; at Kamloops, British Columbia. Siliceous sinter or geyserite is found in connection with geysers and hot springs at Yellowstone Park, Wyoming; at Steamboat Springs, Nevada; in Iceland and New Zealand. *Common opal* is very widespread in its occurrence.

Use. — In the colored varieties as a highly prized gem-stone.

II. Oxides of the Semi-Metals; also Molybdenum, Tungsten

Arsenolite. — Arsenic trioxide, As_2O_3. In isometric octahedrons; in crusts and earthy. Octahedral cleavage. Colorless or white. G. = 3·7. H. = 1·5. n = 1·755. A rare mineral of secondary origin associated with various other arsenical minerals. In Czechoslovakia at Schmöllnitz (Smolnik) intimately associated with claudetite; from Joachimstal (Jáchymov) in Bohemia; at various localities in Germany and France. In California occurred in large masses at the Amargosa mine, San Bernardino Co., and in crystals with enargite in Alpine Co.

Claudetite. — Also As_2O_3, but monoclinic in form. In thin plates, || (010). Cleavage (010), perfect. H. = 2·5. G. = 3·85–4·15. Colorless to white. Optically +. α = 1·87, β = 1·92, γ = 2·01. Strong dispersion, $\rho > v$. Of secondary origin, usually a sublimation product as a result of mine fires as at Schmöllnitz (Smolnik), Czechoslovakia and at San Domingos, Algarve, Portugal.

Senarmontite. — Antimony trioxide. Sb_2O_3. In isometric octahedrons, in crusts and granular massive. H. = 2. G. = 5·3. Colorless, grayish. n = 2·087. Formed by the oxidation of stibnite and other antimony minerals. Found in large amounts, either massive or in fine octahedral crystals, at the mine of Djebel-Haminate, northwest of Aïn-Beïda, district of Haracta, Constantine, Algeria. Also from Cornwall, Sardinia, and Borneo. From South Ham, Wolfe Co., Quebec.

Valentinite. — Sb_2O_3, in prismatic orthorhombic crystals. Cleavages (011), perfect, (010), distinct. H. = 2·5–3. G. = 5·76. Color white. Optically −. X = a axis. Ax. pl. || (001) for red light, || (010) for blue light, almost uniaxial for yellow light. α = 2·18, β and γ = 2·35. Occurs as an oxidation product of various antimony minerals. Occurs in Czechoslovakia at Přibram, Bohemia, and at Pernek near Bösing, northwest of Bratislava (Pressburg). In Rumania at Felsöbánya (Baia Spric). In Saxony at Bräunsdorf, near Freiberg. Found in large amount at the mine of Sensa northwest of Aïn-Beïda, Constantine, Algeria. Occurs at South Ham, Wolfe Co., Quebec.

Bismite. — Bismuth trioxide, Bi_2O_3. Commonly impure, often hydrous. Much so-called bismite is probably a bismuth hydroxide and it is possible that the anhydrous oxide does not occur as a mineral. Pulverulent, earthy; color straw-yellow. From Goldfield, Nevada, in minute silvery white, pearly scales that are hexagonal, rhombohedral; optically −. ω = 2·00, ϵ = 1·82. Artificial crystals of Bi_2O_3 are orthorhombic. Of secondary origin resulting from the oxidation of other bismuth compounds. Occurs with native gold at Beresovsk near Ekaterinburg, Ural Mts., Russia. In Saxony at Schneeberg and Johanngeorgenstadt. From Bolivia.

Tellurite. — Tellurium dioxide, TeO_2. In white to yellow slender orthorhombic, prismatic crystals. Cleavage (010), perfect. H. = 2. G. = 5·9. Ax. pl. (100). 2V nearly 90°. α = 2·00, β = 2·18, γ = 2·35. A product of oxidation of tellurides or native tellurium. Found in Rumania at Nagy-Ág, Hunyad, and at Facebay near Zalatna. Occurs in Colorado in Boulder Co.; also in Nye Co., Nevada.

Tungstite. — Tungsten trioxide, WO_3. Pulverulent, earthy; color yellow or yellowish green. Analyses of tungstic ocher from Salmo, B. C., prove it to have the composition

$WO_3.H_2O$; the same composition as the colloidal *meymacite* (a hydrated tungstic oxide from Meymac, Corrèze, France). The material from Salmo is composed of microscopic scales and is probably orthorhombic. Cleavage, (001), perfect. Optically $-$. $X \perp$ cleavage. $\alpha = 2.09$, $\beta = 2.24$, $\gamma = 2.26$. 2V small. A mineral of secondary origin, commonly associated with wolframite, etc. Noted from Cornwall and Cumberland, from Bolivia and at Monroe, Connecticut.

THOROTUNGSTITE. Essentially tungstic and thorium oxides with small amounts of water, rare earths, zirconia, alumina, etc. Formula doubtful. Massive and in microscopic crystals. Crystals are tabular prismatic; probably orthorhombic. Crystals show parallel extinction. $n > 1.74$. G. = 5.55. Occurs in shapeless blocks at the base of an eluvial deposit overlying granite at Pulai in the Kinta District of Perak, Malay States.

Cervantite. — $Sb_2O_3.Sb_2O_5$. Orthorhombic. In acicular crystals. Pulverulent or massive. H. = 4.5. G. = 4. Color yellow to white. Indices variable, 1.8–2.0. Occurs as an alteration product of stibnite, etc. In Spain at Cervantes in Lugo and elsewhere. Found in Cornwall at Endellion, etc. From various localities in Victoria, Australia. From Borneo. From Mexico in San Luis Potosi as pseudomorphs after stibnite, also from Durango and Sonora.

Stibiconite. — $H_2Sb_2O_5$. Mostly amorphous and variable in composition. Massive, compact. Color pale yellow to yellowish white. Indices 1.6–1.87. In part isotropic and in part birefringent. Associated with stibnite, etc., as an alteration product. From Goldkronach, Fichtelgebirge, Bavaria. With cervantite in Borneo. In extensive deposits at Altar, Sonora, Mexico. From the Empire district, Nevada.

III. Oxides of the Metals

A. ANHYDROUS OXIDES

I. Protoxides, R_2O and RO.

II. Sesquioxides, R_2O_3.

III. Intermediate, $\overset{\text{II III}}{R}R_2O_4$ or $RO.R_2O_3$, etc.

IV. Dioxides, RO_2.

The Anhydrous Oxides include, as shown above, three distinct divisions, the Protoxides, the Sesquioxides and the Dioxides. The remaining Intermediate Division embraces a number of oxygen compounds which are properly to be regarded chemically as salts of certain acids (aluminates, ferrates, etc.); here is included the well-characterized SPINEL GROUP.

Among the Protoxides the only distinct group is the PERICLASE GROUP, which includes the rare species Periclase, MgO, Manganosite, MnO, and Bunsenite, NiO. All of these are isometric in crystallization.

The Sesquioxides include the well-characterized HEMATITE GROUP, R_2O_3. The Dioxides include the prominent RUTILE GROUP, RO_2. Both of these groups are further defined later.

I. Protoxides, R_2O and RO

CUPRITE. Red Copper Ore.

Isometric-plagiohedral. Commonly in octahedrons; also in cubes and dodecahedrons, often highly modified. Plagiohedral faces sometimes distinct (see p. 87). X-ray study shows an atomic structure with normal symmetry. The copper atoms lie on a face-centered cubic lattice and the oxygen atoms on an interpenetrating body-centered cubic lattice. This structure does not explain the plagiohedral forms sometimes observed. Etching figures also indicate normal symmetry. At times in capillary crystals. Also massive, granular; sometimes earthy.

Cleavage: $o(111)$ interrupted. Fracture conchoidal, uneven. Brittle.
H. = 3·5–4. G. = 5·85–6·15. Luster adamantine or submetallic to earthy.
Color red, of various shades, particularly cochineal-red, sometimes almost
black; occasionally crimson-red by transmitted light.
Streak several shades of brownish red, shining. Subtransparent to subtranslucent. Refractive index, n
= 2·849.

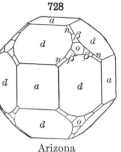

728

Arizona

Var. — 1. *Ordinary.* (a) Crystallized; commonly in octahedrons, dodecahedrons, cubes, and intermediate forms; the
crystals often with a crust of malachite; (b) massive.

2. *Capillary*; *Chalcotrichite.* Plush Copper Ore. In capillary or acicular crystallizations, which are sometimes cubes
elongated in the direction of the cubic axis.

3. *Earthy*; *Tile Ore.* Brick-red or reddish brown and
earthy, often mixed with red oxide of iron; sometimes nearly
black.

Comp. — Cuprous oxide, Cu_2O = Oxygen 11·2, copper 88·8 = 100.

Pyr., etc. — Unaltered in the closed tube. B.B. in the forceps fuses and colors the
flame emerald-green. On charcoal first blackens, then fuses, and is reduced to metallic
copper. With the fluxes gives reactions for copper. Soluble in concentrated hydrochloric
acid, and a strong solution when cooled and diluted with cold water yields a heavy, white
precipitate of cuprous chloride.

Diff. — Distinguished from hematite by inferior hardness, but is harder than cinnabar
and proustite and differs from them in the color of the streak; reactions for copper, B.B.,
are conclusive.

Obs. — Cuprite is of secondary origin, formed commonly during the oxidation and
alteration of copper sulphide deposits. Under such circumstances the solutions in contact
with the ore will contain copper salts which by reduction through the agency of organic
material or ferrous oxide will produce native copper or cuprite. Native copper through
reaction with such solutions will precipitate cuprite. Associated minerals are commonly
native copper, malachite, azurite, limonite, etc.

Cuprite is quite common and widespread in its occurrence. Some of its more notable
localities follow. From Nizhne-Tagilsk and Ekaterinburg, in Perm, Russia; from near
Rheinbreitbach (*chalcotrichite*), Rhineland; in isolated crystals, sometimes an inch in diameter, at Chessy near Lyon, Rhône, France. These crystals are generally altered in
part or wholly to malachite. In Cornwall in fine transparent crystals near St. Day, from
Liskeard, Redruth, and Truro, from St. Blazey (*chalcotrichite*), etc. In South Australia
from Burra district, and at Broken Hill, New South Wales. In Bolivia at Corocoro,
La Paz; in Chile in fine crystals from Andacollo, south of Coquimbo, and pseudomorphic
after brochantite from Chuquicamata, northeast of Antofagasta. In Arizona found at
the Copper Queen mine, Bisbee, in fine crystals and also as chalcotrichite; beautiful chalcotrichite from Morenci, also from the Globe district, from Jerome; and in crystals and
massive from Clifton, Greenlee Co. From Del Norte Co., California.

Use. — An ore of copper.

Ice. — H_2O. Hexagonal. Familiarly known in six-rayed snow crystals; also coating
ponds in winter, further as glaciers and icebergs.

Periclase Group

Periclase. — Magnesia, MgO. Isometric Structure similar to that of halite. In
cubes or octahedrons, and in grains. Cleavage cubic. H. = 6. G. = 3·67–3·90. n =
1·74. **Artif.** — Crystallized from a melt containing magnesium chloride and silica. A
contact mineral. Occurs in white limestone at Mte. Somma, Vesuvius; at the manganese
mines, Nordmark, Vermland, Sweden. Also at Riverside, California.

Manganosite. — Manganese protoxide, MnO. In isometric octahedrons. Cleavage
cubic. H. = 5·6. G. = 5·18. n = 2·16. Color emerald-green, becoming black on exposure. From Långbanshyttan and Nordmark, Vermland, Sweden; Franklin, New
Jersey.

Bunsenite. — Nickel protoxide, NiO. H. = 5·5. G. = 6·4–6·8. In green octahedrons. $n = 2·23$. From Johanngeorgenstadt, Saxony.

Cadmium oxide. — Isometric. In minute octahedrons. Forms a thin coating of black color and brilliant metallic luster upon calamine from Monte Poni, near Iglesias, Cagliari, Sardinia. Also formed artificially.

ZINCITE. Red Oxide of Zinc.

Hexagonal-hemimorphic. Axis $c = 1·5870$. Natural crystals rare (Fig. 44, p. 23); usually foliated massive, or in coarse particles and grains; also with granular structure. The atomic structure is closely related to that of greenockite and wurtzite.

Cleavage: $c(0001)$ perfect; prismatic, sometimes distinct. Fracture subconchoidal. Brittle. H. = 4–4·5. G. = 5·43–5·7. Luster subadamantine. Streak orange-yellow. Color deep red, also orange-yellow. The color is thought to be due to the presence of manganese oxide. Translucent to subtranslucent. Optically +. $\omega = 2·013$, $\epsilon = 2·029$.

Comp. — Zinc oxide, ZnO = Oxygen 19·7, zinc 80·3 = 100. Manganese protoxide is commonly present in varying small amounts.

Pyr., etc. — B.B. infusible; with the fluxes, on the platinum wire, gives reactions for manganese, and on charcoal in R.F. gives a coating of zinc oxide, yellow while hot, and white on cooling. The coating, moistened with cobalt solution and treated in O.F., assumes a green color. Soluble in acids.

Diff. — Characterized by its color, particularly that of the streak; by cleavage; by reactions B.B.

Artif. — Zincite is often formed as a furnace product. It is also produced when zinc chloride and water vapor act upon lime at red heat.

Obs. — Occurs in large amounts with franklinite and willemite, at Sterling Hill near Ogdensburg, and at Mine Hill, Franklin Furnace, Sussex Co., New Jersey, sometimes in lamellar masses in pink calcite. Has been reported from Olkusz in Kielce, Poland; from Bottino near Seravezza, Tuscany; near Paterna, Almeria, Spain; at Heazlewood mine, Tasmania. A not uncommon furnace product.

Use. — An ore of zinc.

Bromellite. — BeO. Hexagonal-hemimorphic. G. = 3·017. H. = 9. Optically +. $\omega = 1·719$, $\epsilon = 1·733$. White. Occurs with swedenborgite at Långbanshyttan, Vermland, Sweden.

Massicot. — Lead monoxide, PbO. Massive, scaly or earthy. Color yellow, reddish. Microscopic study of various specimens of massicot shows that the centers of the crystal plates correspond to massicot, and the borders to the artificial lead oxide litharge. The massicot is orthorhombic, nearly colorless, optically +. $\beta = 2·61$. Strong birefringence. The litharge is tetragonal, yellow-orange, optically −. $\omega = 2·65$. Strong birefringence. Optic axis is normal to plate. A rare mineral of secondary origin occurring associated with galena, etc.

Tenorite. — Cupric oxide, CuO. Triclinic, pseudo-monoclinic. In minute black scales with metallic luster. H. = 3–4. G. = 6·5. Occurs as a sublimation product in volcanic regions, as at Vesuvius and Etna; also as a black earthy material (*melaconite*) found in copper veins as an oxidation product, as at various localities in Central Europe, France, Spain, Cornwall, etc.; in the United States at Ducktown, Tennessee, formerly at Copper Harbor, Michigan; also from Globe and Bisbee, Arizona. Pitchy black material associated with cuprite, chrysocolla and malachite from Bisbee, Arizona, has been called *melanochalcite*. This has been shown to be a variable mixture of tenorite, chrysocolla, and malachite.

Paramelaconite is essentially cupric oxide, CuO, occurring in black pyramidal crystals referred to the tetragonal system. From the Copper Queen mine, Bisbee, Arizona.

Montroydite. — HgO. Orthorhombic. In minute highly modified crystals. Cleavage (010), perfect. H. = 1·5–2. Color and streak orange-red. Optically +. $\alpha = 2·37$, $\beta = 2·50$, $\gamma = 2·65$. Volatile. Found at Terlingua, Texas.

Hematite Group. R_2O_3. Rhombohedral

			c
Corundum	Al_2O_3		1·3630
Hematite	Fe_2O_3		1·3656
Ilmenite	$(Fe,Mg)O.TiO_2$ Tri-rhombohedral		1·3846
Senaite	$(Fe,Mn,Pb)O.TiO_2$	"	1·385
Pyrophanite	$MnO.TiO_2$	"	1·3692
Geikielite	$MgO.TiO_2$	"	1·370

The HEMATITE GROUP embraces the sesquioxides of aluminum and iron. These compounds crystallize in the rhombohedral class, hexagonal system, with a fundamental rhombohedron differing but little in angle from a cube. Both the minerals belonging here, Hematite and Corundum, are *hard*.

To these species the titanates of iron, magnesium, and manganese, Ilmenite, Senaite, Pyrophanite, and Geikielite, are closely related in form though belonging to the tri-rhombohedral class (phenacite type), and in atomic structure. It is to be noted, further, that hematite often contains titanium, and an artificial isomorphous compound, Ti_2O_3, has been described. Hence the ground for writing the formula of ilmenite $(Fe,Ti)_2O_3$, as is done by some authors. It is shown by Penfield, however, that the formula $(Fe,Mg)TiO_2$ is more correct. For other titanates see p. 688.

CORUNDUM.

Rhombohedral. Axis $c = 1\cdot3630$.

$cr,\ 0001 \wedge 10\bar{1}1 = 57° 34'$. $nn',\ 22\bar{4}3 \wedge \bar{2}4\bar{2}3 = 51° 58'$,
$cn,\ 0001 \wedge 22\bar{4}3 = 61° 11'$, $rr',\ 41\bar{8}3 \wedge \bar{4}8\bar{4}3 = 57° 38'$.
$rr'\ 10\bar{1}1 \wedge \bar{1}101 = 93° 56'$. $zz',\ 22\bar{4}1 \wedge \bar{2}4\bar{2}1 = 58° 55'$.

Twins: tw. pl. $r(10\bar{1}1)$, sometimes penetration-twins; often polysynthetic, and thus producing a laminated structure. Crystals usually rough and rounded. Also massive, with nearly rectangular parting or pseudo-cleavage;

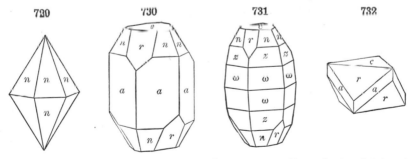

granular, coarse or fine. X-ray study shows a complicated atomic structure. The structure can be likened to a deformed halite structure in which the two different atoms are replaced by enantiomorphous Al_2O_3 groups.

Parting: $c(0001)$, sometimes perfect, but interrupted; also $r(10\bar{1}1)$ due to twinning, often prominent; $a(11\bar{2}0)$ less distinct. Fracture uneven to conchoidal. Brittle, when compact very tough. H. = 9. G. = 3·95–4·10. Luster adamantine to vitreous; on c sometimes pearly. Occasionally showing asterism. Color blue, red, yellow, brown, gray, and nearly white; streak uncolored. Dichroic in deeply colored varieties. Transparent to translucent. Normally uniaxial, negative; for sapphire $\omega = 1.7676$ to 1.7682 and $\epsilon = 1.7594$ to 1.7598. Often abnormally biaxial.

Var. — There are three subdivisions of the species prominently recognized in the arts, but differing only in purity and state of crystallization or structure.

1. SAPPHIRE, RUBY. — Includes the purer kinds of fine colors, transparent to translucent, useful as gems. Stones are named according to their colors: *Sapphire* blue; true *Ruby*, or *Oriental Ruby*, red; *Oriental Topaz*, yellow; *Oriental Emerald*, green; *Oriental Amethyst*, purple. The term *sapphire* is also often used as a general term to indicate corundum gems of any color except red. A variety having a stellate opalescence when viewed in the direction of the vertical axis of the crystal is the *Asteriated* Sapphire or Star Sapphire. This asterism is due to presence of minute cylindrical cavities which lie parallel to the prism planes.

2. CORUNDUM. — Includes the kinds of dark or dull colors and not transparent, colors light blue to gray, brown, and black. The original *adamantine spar* from India has a dark grayish smoky brown tint, but greenish or bluish by transmitted light, when translucent.

3. EMERY. — Includes granular corundum, of black or grayish black color, and contains magnetite or hematite intimately mixed. Sometimes associated with iron spinel or hercynite. Feels and looks much like a black fine-grained iron ore, which it was long considered to be. There are gradations from the evenly fine-grained emery to kinds in which the corundum is in distinct crystals.

Comp. — Alumina, $Al_2O_3 =$ Oxygen 47.1, aluminum $52.9 = 100$. The crystallized varieties are essentially pure containing only a little ferric oxide and silica. The variations in color are perhaps due to the presence of varying minute amounts of ferric, chromic, and manganese oxides. Analyses of emery show more or less impurity, chiefly magnetite.

Artif. — Crystallized corundum has been produced artificially in a number of different ways. Alumina dissolved in molten sodium sulphide, in a fused mixture of a fluoride and potassium carbonate or in fused lead oxide, will separate out as crystallized corundum. The melting point of artificial corundum is $2050°$ C. Gem material has been produced in this way, colored red, with a chromium salt, or blue by cobalt. Crystallized material can also be produced by fusing alumina in an electric arc. The artificial abrasive, *alundum*, is made by heating bauxite to $5000°$–$6000°$ in an electric furnace. Pear-shaped drops of gem material are made by fusing together small fragments of natural or artificial stones. Gems cut from them are known as " reconstructed " stones and have the crystalline and other physical properties of the natural mineral. Microscopic examination commonly shows minute bubbles and curved striae that serve to distinguish an artificial from a natural stone. Al_2O_3 has been shown to exist in four different forms, α, β, γ, and δ, Al_2O_3. The α form (corundum) is the only one that is stable at high temperatures.

Pyr., etc. — B.B. unaltered; slowly dissolved in borax and salt of phosphorus to a clear glass, which is colorless when free from iron; not acted upon by soda. The finely pulverized mineral, after long heating with cobalt solution, gives a beautiful blue color. Not acted upon by acids, but converted into a soluble compound by fusion with potassium bisulphate.

Diff. — Characterized by its hardness (scratching quartz and topaz), by its adamantine luster, high specific gravity and infusibility. The massive variety with rhombohedral parting resembles cleavable feldspar but is much harder and denser.

Micro. — In thin sections appears nearly colorless with high relief and low interference colors.

Obs. — Usually occurs as an accessory mineral in crystalline rocks, as granular limestone or dolomite, gneiss, mica slate, chlorite slate. The associated minerals often include some species of the chlorite group, as prochlorite, corundophilite, margarite, also tourmaline, spinel, kyanite, diaspore, and a series of aluminous minerals, in part produced from its alteration. Occasionally found as an original constituent of igneous rocks containing high percentages of alumina. In the Ural Mts. are found an anorthite rock containing nearly 60 per cent of corundum, a corundum syenite with 18 per cent, and a pegmatite with 35 per cent. A corundum anorthosite and corundum syenites are found in Canada. Important deposits of corundum in North Carolina and Georgia are associated with dunite rocks. Rarely observed as a contact-mineral. The fine sapphires are usually obtained from the beds of rivers, either in modified hexagonal prisms or in rolled masses, accompanied by grains of magnetite, and several kinds of gems, as spinel, etc. The emery of Asia Minor occurs in granular limestone.

The best rubies come from the mines in Upper Burma, north of Mandalay, in an area covering 25 to 30 square miles, of which Mogok is the center. The rubies occur *in situ* in crystalline limestone, also in the soil of the hillsides and in gem-bearing gravels of the Irrawaddy River. Blue sapphires as well as stones of many other colors, are brought from Ceylon from the Ratnapura and Rakwana districts, often as rolled pebbles, also as well-preserved crystals. Fine blue sapphires and also rubies from southeast of Bangkok near Battambang in Cambodia and Chantabun in Siam. Gem stones come from various parts of India and also large coarse crystals in the Carnatic district of southern Madras, also in Coimbatore and Mysore. Sapphires and pale blue opaque crystals come from Kashmir, northern India. Prismatic or tabular crystals from gray to beautiful blue from Takayama, Mino, Japan. From the Ilmen Mts., at Zlatoust and Miask. In Ticino, Switzerland, at Campolungo near St. Gotthard with a red or blue tinge in dolomite. In large opaque crystals and masses from various places in Madagascar, particularly in mica schist at Vatondrangy and elsewhere southeast of Antsirabe; also in crystals and rolled pebbles of gem quality in alluvial deposits. Large opaque crystals are found at Steinkopf, Namaqualand, Cape Province, South Africa, and in the Zoutpansberg and Pietersburg districts, Transvaal.

In the United States corundum occurs in Orange Co., New York, at Warwick, bluish and pink with spinel and, at Amity, white, blue, reddish crystals. In Sussex Co., New Jersey, at Franklin Furnace, at Newton in blue crystals, at Vernon in red crystals, at Sparta, etc. In Pennsylvania in Chester Co., at Corundum Hill near Unionville, Newlin township, was abundant in crystals and large masses, and in Delaware Co., in Aston township in large crystals, also in Lehigh Co., at Shimersville. Common at many points along a belt extending from Virginia across western North and South Carolina and Georgia to Dudleyville, Alabama. The localities at which most mining was done were Corundum Hill, near Franklin, Macon Co., North Carolina, and southeast at Laurel Creek, Georgia. Gem sapphires are found near Helena, Montana, in gold washings and in bars of the Missouri River, especially the Eldorado Bar; at Yogo Gulch on the Judith River, where they also occur in an andesite dike; also at other localities. In Ontario corundum occurs in many places in syenites and anorthosite in the south-central and eastern counties, especially in Hastings Co., in red and blue crystals at South Burgess in Leeds Co., and in many places in Renfrew Co.

Emery is found on the Island of Naxos and elsewhere in the Cyclades, Greece; also in Asia Minor, east of Ephesus in the mountains of Güme Dagh (Messogis), and in various districts north of Smyrna. In Massachusetts, at Chester, corundum and emery occur in a large vein.

Use. — Clear varieties of corundum form valuable gem stones as noted above. Also formerly largely used as an abrasive; at present various artificial abrasives are mostly used instead.

HEMATITE.

Rhombohedral. Axis $c = 1{\cdot}3656$.

cr, $0001 \wedge 10\bar{1}1 = 57° 37'$.
rr', $10\bar{1}1 \wedge \bar{1}10\bar{1} = 94° 0'$.
dd', $01\bar{1}2 \wedge \bar{1}012 = 64° 51'$.

uu', $10\bar{1}4 \wedge \bar{1}104 = 37° 2'$.
nn', $22\bar{4}3 \wedge \bar{2}4\bar{2}3 = 51° 59'$.
cn, $0001 \wedge 2\bar{2}43 = 61° 13'$.

Twins: tw. pl. (1) $c(0001)$, penetration-twins; (2) $r(01\bar{1}2)$, less common, usually as polysynthetic twinning lamellæ, producing a fine striation on $c(0001)$ and giving rise to a distinct parting or pseudo-cleavage $\parallel$ $r(10\bar{1}1)$. Crystals often thick to thin tabular $\parallel c$, and grouped in parallel position or in rosettes; c faces striated $\parallel$ edge c/d $(01\bar{1}2)$ and other forms due to oscillatory combination;

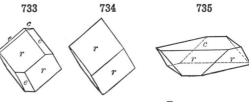

733   734   735

also in cube-like rhombohedrons; rhombohedral faces $u(10\bar{1}4)$ horizontally striated and often rounded over in convex forms. Also columnar to granular, botryoidal, and stalactitic shapes; also lamellar, laminæ joined parallel to c, and variously bent, thick or thin; also granular, friable, earthy or compact.

Parting: c(0001), due to lamellar structure; also r(10$\bar{1}$1), caused by twinning. Fracture subconchoidal to uneven. Brittle in compact forms; elastic

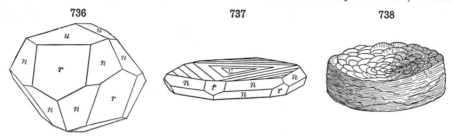

736   737   738

in thin laminæ; soft and unctuous in some loosely adherent scaly varieties. H. = 5·5–6·5. G. = 4·9–5·3; of crystals mostly 5·20–5·25; of some compact varieties, as low as 4·2. Luster metallic and occasionally splendent; sometimes dull. Color dark steel-gray or iron-black; in very thin particles blood-red by transmitted light; when earthy, red. Streak cherry-red or reddish brown. Opaque, except when in very thin laminæ.

Var. 1. *Specular.* Luster metallic, and crystals often splendent, whence the name *specular iron.* When the structure is foliated or micaceous, the ore is called *micaceous* hematite: some of the micaceous varieties are soft and unctuous. Some varieties are magnetic, but probably from admixed magnetite (Arts. **453, 455**).

2. *Compact Columnar*; or fibrous. The masses often long radiating; luster submetallic to metallic; color brownish red to iron-black. Sometimes called *red hematite*, to contrast it with limonite and turgite. Often in reniform masses with smooth fracture, called *kidney ore.*

3. *Red Ocherous.* Red and earthy. *Reddle* and *red chalk* are red ocher, mixed with more or less clay.

4. *Clay Iron-stone*; *Argillaceous hematite.* Hard, brownish black to reddish brown, often in part deep red; of submetallic to nonmetallic luster; and affording, like all the preceding, a red streak. It consists of oxide of iron with clay or sand, and sometimes other impurities.

Comp. — Iron sesquioxide, Fe_2O_3 = Oxygen 30, iron 70 = 100. Sometimes contains titanium and magnesium, and is thus closely related to ilmenite, p. 486.

Pyr., etc. — B.B. infusible; on charcoal in R.F. becomes magnetic; with borax gives the iron reactions. With soda on charcoal in R.F. is reduced to a gray magnetic powder. Slowly soluble in hydrochloric acid.

Diff. — Distinguished from magnetite by its *red streak*, also from limonite by the same means, as well as by its not containing water; from turgite by its greater hardness and by not decrepitating B.B. It is *hard* in all but some micaceous varieties (hence easily distinguished from the black sulphides); also *infusible*, and B.B. becomes strongly magnetic.

Artif. — Crystals of hematite have been made by decomposing ferric chloride by steam at a high temperature; also by the action of heated air and hydrochloric acid upon iron. Hematite has been crystallized from various artificial magmas, which must contain little or no ferrous iron.

Obs. — Hematite is formed in various ways and is found in rocks of all ages. It occurs in connection with volcanic activities as a sublimation product, usually as small thin crystal plates implanted upon lava. It is probably produced by the action of water vapor upon ferric chloride. Hematite is found also as a constituent of igneous rocks but only when the original magma was very deficient in ferrous oxide. It is characteristic therefore of the granites, syenites, trachytes, andesites, etc. At times it is of pneumatolytic origin, occurring in granites or pegmatites. It also occurs in the metamorphic rocks, at times in beds of great thickness, where it may have originated by the alteration of limonite, siderite, magnetite, etc. Apparently the hydrated iron oxides, when in the colloidal condition, may rather readily change into the anhydrous ferric oxide. It is formed in beds and irregular masses as the result of the weathering of iron-bearing rocks. Such residual ores are more commonly composed of limonite but may consist of hematite or a mixture of the two minerals. In the case of the Lake Superior hematite deposits the ore bodies have been concentrated by the action of meteoric waters upon sedimentary formations rich in iron carbonate and silicate. The oölitic hematite ores consist of small rounded, concretionary grains formed about a grain of sand, etc., as a nucleus. They occur at times in considerable beds, are of sedimentary origin and commonly of Paleozoic age.

Hematite is a widely distributed species, and only those localities which are of considerable interest either because of their economic value or because of the exceptional character of their specimens can be mentioned here. Hematite occurs in crystals in various mines in the district of Ekaterinburg, Ural Mts., Russia; in Rumania in crystals from Dognacska and in very thin plates from Kakuk-Berg in the Hargitta Mts., district of Odorhei (Udvárhely); from the Zillertal, Tyrol, Austria; in Saxony at Altenberg and Johanngeorgenstadt. From Switzerland beautiful specimens, composed of crystal tables grouped in the form of rosettes, are found at St. Gotthard, Ticino, and at Cavradi, Val Tavetsch, Grisons, and in crystals from the Binnental, Valais. In Italy it occurs at Traversella, Piedmont; in fine crystals as a result of volcanic action at Vesuvius, at Etna on Sicily, and at Stromboli, Lipari Islands. Beautiful crystallizations of hematite have long come from the Island of Elba. In France fine crystals are found at Framont, Alsace, and in Puy-de-Dôme on Mont Dore. Splendid specimens come from Cleator Moor and elsewhere in Cumberland. Found in crystals on Ascension Island, in the south Atlantic. Large and fine crystals are found in various localities in Minas Geraes, Brazil.

In North America, widely distributed, and sometimes in beds of vast thickness in rocks of the Archæan age. Very extensive and important hematite deposits are found along the southern and northwestern shores of Lake Superior. The various districts are known as *ranges* and are located as follows: The Marquette and Menominee Ranges in northern Michigan, the Penokee-Gogebic Range in Northern Wisconsin, the Mesabi, Vermilion and Cuyuna Ranges in Minnesota. Another district, the Michipicopen, is farther north in Canada. The ore bodies are the results of the concentration in favorable localities of the iron content of the original sedimentary rocks. These rocks contained cherty iron carbonates, pyrite-bearing iron carbonates and ferrous silicates. The ore bodies vary widely in form, many of them lying in trough-like structures formed by the deformation of impervious rock strata. The character of the ores varies from hard specular hematites to soft earthy ores. The latter are often mined by the use of steam shovels. Hematite is found in Wyoming in schist formations in Laramie and Carbon counties.

In New York, in Oneida, Herkimer, Madison, Wayne counties, a lenticular argillaceous variety, constituting one or two beds in the Clinton group of the Upper Silurian; the same in Pennsylvania, and as far south as Alabama, and in Canada, and Wisconsin, to the west; in Alabama there are extensive beds; prominent mines are near Birmingham. Besides these regions of large beds, there are numerous others of workable value, either crystallized or argillaceous. Some localities, interesting for their specimens, are in northern New York, at Gouverneur, Antwerp, Hermon, Edwards, Fowler, Canton, in St. Lawrence Co., at Antwerp in Jefferson Co., etc.; at Hawley, Mass., a micaceous variety; at Franklin, New Jersey, showing rhombohedral parting, in North and South Carolina a micaceous variety in schistose rocks, constituting the so-called *specular schist*, or *itabirite*. In crystals from the Pike's Peak region in Colorado. Hematite is mined in Nova Scotia and Newfoundland.

Named *hematite* from ἁιμα, *blood*.

Use. — The most important iron ore. Used also in red paints, as polishing rouge, etc.

MARTITE. Iron sesquioxide under an isometric form, occurring in octahedrons or dodecahedrons like magnetite, and believed to be pseudomorphous after magnetite; perhaps in part also after pyrite. Parting octahedral like magnetite. Fracture conchoidal. H. = 6–7. G. = 4·8–5·3. Luster submetallic. Color iron-black, sometimes with a bronzed tarnish. Streak reddish brown or purplish brown. Not magnetic, or only feebly so. The crystals are sometimes embedded in the massive sesquioxide. They are distinguished from magnetite by the red streak, and very feeble, if any, action on the magnetic needle. Martite is found at Rittersgrün, south of Schwarzenberg, in the Erzgebirge, Saxony; on Mont Dore, Puy-de-Dôme, France. It occurs in the schists of Minas Geraes, Brazil, and in large octahedrons at the Cerro de Mercado, Durango, Mexico. In the United States it is found at Monroe, New York, in the Marquette iron district in Michigan; at Twin Peaks, Millard Co., Utah. Also from Digby Co., Nova Scotia.

ILMENITE or MENACCANITE.   Titanic Iron Ore.

Tri-rhombohedral;   Axis $c = 1.3846$.

$$cr, \ 0001 \wedge 10\overline{1}1 = 57° \ 58\tfrac{1}{2}'.$$
$$rr', \ 10\overline{1}1 \wedge \overline{1}101 = 94° \ 29'.$$
$$cn, \ 0001 \wedge 22\overline{4}3 = 61° \ 33'.$$

Crystals usually thick tabular; also acute rhombohedral.   Often in thin plates or laminæ.   Massive, compact; in embedded grains, also loose as sand.

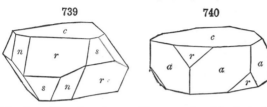

739   740

The atomic structure is similar to that of corundum, providing one-half of the latter's Al atoms be replaced by Fe and the other half by Ti.

Fracture conchoidal.   H. = 5–6.   G. = 4·5–5.   Luster submetallic.   Color iron-black.   Streak submetallic, powder black to brownish red.   Opaque.   Influences slightly the magnetic needle.

Comp. — If normal, $FeTiO_3$ or $FeO.TiO_2$ = Oxygen 31·6, titanium 31·6, iron 36·8 = 100.   Sometimes written $(Fe,Ti)_2O_3$, but probably to be regarded as an iron titanate.   Sometimes also contains magnesium (*picrotitanite*), replacing the ferrous iron; hence the general formula $(Fe,Mg)O.TiO_2$ (Penfield).   (Compare *geikielite*, p. 487.)   The variations in composition shown by analyses are in part, at least, explained by the fact that specimens often show a regular intergrowth of lamellæ of hematite or magnetite, in a manner analogous to the perthitic intergrowth of the feldspars.

Pyr., etc. — B.B. infusible in O.F., although slightly rounded on the edges in R.F. With borax and salt of phosphorus reacts for iron in O.F., and with the latter flux assumes a more or less intense brownish red color in R.F.; this treated with tin on charcoal changes to a violet-red color when the amount of titanium is not too small.   The pulverized mineral, heated with hydrochloric acid, is slowly dissolved to a yellow solution, which, filtered from the undecomposed mineral and boiled with the addition of tin-foil, assumes a beautiful blue or violet color.   Decomposed by fusion with bisulphate of sodium or potassium.

Diff. — Resembles hematite, but has a submetallic, nearly black, streak; not magnetic like magnetite.

Obs. — Occurs, as an accessory component, in many types of igneous rocks, assuming the place of magnetite, especially in gabbros and diorites.   It is one of the earliest constituents of the rock magma to crystallize.   It is often found in veins or large segregated masses near the borders of the igneous rock where it is supposed to have formed by local differentiation or fractional crystallization in the molten mass.   It is also found at times in metamorphic rocks.   Some principal European localities are near Miask in the Ilmen Mts., Urals (*ilmenite*); in the Binnental, Switzerland; in France as small crystals (*crichtonite*) from St. Cristophe, near Bourg d'Oisans, Isère.   One of the most remarkable is at Kragerö, Norway, where it occurs in veins or beds in diorite, at times in crystals weighing over 16 pounds.   Others are Egersund, Arendal, Snarum in Norway.   It is found in the form of sand at Menaccan, Cornwall (*menaccanite*).

In the United States it is found in crystals at Chester, Massachusetts; in Litchfield, Connecticut (*washingtonite*).   Fine crystals, sometimes an inch in diameter, occur in Warwick, Amity, and Monroe, Orange Co., New York.   Ilmenite is found with many of the magnetite deposits of the Adirondack region, New York.   It is also found in deposits of considerable size in the rocks of the Grenville series in Quebec, especially at Bay St. Paul, Charlevoix Co., and in the Seigniory of St. Francis, Beauce Co.

The titanic iron of massive rocks is extensively altered to a dull white opaque substance, called *leucoxene* by Gümbel.   This for the most part is to be identified with titanite.

Senaite. — $(Fe,Mn,Pb)O.TiO_2$.   Tri-rhombohedral.   Closely related crystallographically to ilmenite.   H. = 6.   G. = 5·3.   Color black.   Streak brownish black.   Optically —.   Found in the diamond-bearing sands of Diamantina, Minas Geraes, Brazil.

Pyrophanite. — Manganese titanate, $MnTiO_3$. In thin tabular rhombohedral crystals and scales, near ilmenite in form (p. 486). H. = 5. G. = 4.537. Luster vitreous to submetallic. Color deep blood-red. Streak ocher-yellow. Optically −. $\omega = 2\cdot481$, $\epsilon = 2\cdot210$. From the Harstig mine, Pajsberg, near Persberg, Vermland, Sweden; also at Queluz (in a rock called *queluzite*), southwest of Ouro Preto, Minas Geraes, Brazil.

Geikielite. — Magnesium iron titanate. $(Mg,Fe)TiO_3$. Hexagonal, rhombohedral. Structurally closely similar to ilmenite. Usually massive, as rolled pebbles. H. = 6. G. = 4. Color bluish or brownish black. Optically −. $\omega = 2\cdot31$, $\epsilon = 1\cdot95$. From the gem gravels of the Rakwana district, Ceylon.

Bixbyite. — $(Fe,Mn)_2O_3$. In black isometric crystals. X-ray study shows a body-centered cubic lattice with sixteen molecules to the unit cell. H. = 6–6·5. G. = 4·945. Occurs with topaz in cavities in rhyolite; from Utah. Noted at Ribes, Girona, Spain, and in Valle de las Plumas, northern Patagonia.

Högbomite. Hoegbomite. — An oxide corresponding to $RO.2R_2O_3$ with Al_2O_3,Fe_2O_3, MgO,TiO_2 chiefly. Rhombohedral. $c = 1.56$. Color, black. Cleavage imperfect, parallel to base and rhombohedron. Fracture conchoidal. H. = 6·5. G. = 3·81. Optically −. $\omega = 1\cdot853$. $\epsilon = 1\cdot803$. Pleochroic, E bright yellow-brown; O dark yellow-brown. Occurs as a rock-making mineral associated with iron ores, magnetite, ilmenite, pleonaste, corundum, etc., in the Ruoutevare district, Lapland. Also found in magnetite-corundum ore at Rödsand in Sondmore, Norway. Identified in microscopic form in a spinel emery near Whittles, Virginia.

SITAPARITE. $9Mn_2O_3.4Fe_2O_3.MnO_2.3CaO$. Not crystallized. Good cleavage. H. = 7. G. = 5·0. Color deep bronze. Streak black. Weakly magnetic. Found at Sitapar, District Chhindwara, Central Provinces, India.

VREDENBURGITE. $3Mn_3O_4.2Fe_2O_3$. Cleavage parallel to octahedron or tetragonal pyramid. H. = 6·5. G. = 4·8. Color bronze to dark steel-gray. Streak dark brown. Strongly magnetic. Completely soluble in acids. Found at Beldongri, District Nagpur, Central Provinces, and at Gravidi, District Vizagapatam, Madras, India.

III. Intermediate Oxides

The species here included are retained among the oxides, although chemically considered they are properly oxygen-salts, aluminates, ferrates, manganates, etc., and hence in a strict classification to be placed in section 5 of the Oxygen-salts. The one well-characterized group is the Spinel Group.

Spinel Group. $R\overset{II}{R}_2\overset{III}{O}_4$ or $\overset{II}{R}O.\overset{III}{R}_2O_3$. Isometric

Aluminum Spinels

Spinel	$MgO.Al_2O_3$
Ceylonite, pleonaste, ferro-picotite	$(Mg,Fe)O.(Al,Fe)_2O_3$
Magnesiochromite, picotite	$(Mg,Fe)O.(Al,Cr)_2O_3$
Hercynite	$FeO.Al_2O_3$
Picotite	$(Fe,Mg)O.(Al,Fe)_2O_3$
Chromohercynite	$FeO.(Al,Cr)_2O_3$
Gahnite (Automolite)	$ZnO.Al_2O_3$
Dysluite, Kreittonite	$(Zn,Mg,Fe,Mn)O.(Al,Fe)_2O_3$
Galaxite	$(Mn,Fe,Mg)O.(Al,Fe)_2O_3$

Iron Spinels

Magnetite	$FeO.Fe_2O_3$
Magnesioferrite	$(Mg,Fe)O.Fe_2O_3$
Franklinite	$(Fe,Zn,Mn)O.(Fe,Mn)_2O_3$
Jacobsite	$(Mn,Mg)O.(Fe,Mn)_2O_3$

Chromium Spinel

Chromite FeO.Cr$_2$O$_3$

(Fe,Mg)O.(Cr,Fe)$_2$O$_3$

The species of the Spinel Group are characterized by isometric crystallization, and, further, the octahedron is throughout the common form. X-ray studies show that the members of the Spinel Group have the same isometric atomic structure. This same structure is shown by a number of artificial compounds having the same type of chemical molecule as well as by the mineral linnæite (cf. p. 430). For spinel the structure may be described as having the O atoms arranged on a face-centered cubic lattice, with the Mg atoms lying in the center of tetrahedral groups of four O atoms while the Al atoms are each surrounded by a group of six O atoms. All of the species are *hard*; those with nonmetallic luster up to 7·5–8, the others from 5·5–6·5.

SPINEL.

Isometric. Usually in octahedrons, sometimes with dodecahedral truncations, rarely cubic. Twins: tw. pl. and comp. face *o*(111) common (Fig.

741   **742**

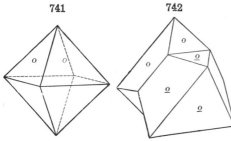

742), hence often called *spinel-twins*; also repeated and polysynthetic, producing tw. lamellæ.

Cleavage: *o*(111) imperfect. Fracture conchoidal. Brittle. H. = 8. G. = 3·5–4·1. Luster vitreous; splendent to nearly dull. Color red of various shades, passing into blue, green, yellow, brown and black; occasionally almost white. Streak white. Transparent to nearly opaque. Refractive index: *n* variable with composition, 1·7155–2·00.

Comp. — Magnesium aluminate, MgAl$_2$O$_4$ or MgO.Al$_2$O$_3$ = Alumina 71·8, magnesia 28·2 = 100. The magnesium may be in part replaced by ferrous iron or manganese, and the aluminum by ferric iron and chromium.

Var. — RUBY SPINEL or *Magnesia Spinel.* — Clear red or reddish; transparent to translucent; sometimes subtranslucent. G. = 3·63–3·71. Composition normal, with little or no iron, and sometimes chromium oxide to which the red color has been ascribed. The varieties are: (*a*) *Spinel-Ruby*, deep red; (*b*) *Balas-Ruby*, rose-red; (*c*) *Rubicelle*, yellow or orange-red; (*d*) *Almandine*, violet.

CEYLONITE or *Pleonaste, Iron-Magnesia Spinel.* — Color dark green, brown to black, mostly opaque or nearly so. G. = 3·5–3·6. Contains iron replacing the magnesium and perhaps also the aluminum, hence the formula (Mg,Fe)O.Al$_2$O$_3$ or (Mg,Fe)O.(Al,Fe)$_2$O$_8$, (this latter has been called *ferro-picotite*).

MAGNESIOCHROMITE, PICOTITE or *Chrome-Spinel.* — Contains chromium and also has the magnesium largely replaced by iron (Mg,Fe)O.(Al,Cr)$_2$O$_3$, hence lying between spinel proper and chromite. G. = 4·08. Color dark yellowish brown or greenish brown. Translucent to nearly opaque.

Pyr., etc. — B.B. alone infusible. Slowly soluble in borax, more readily in salt of phosphorus, with which it gives a reddish bead while hot, becoming faint chrome-green on cooling. Black varieties give reactions for iron with the fluxes. Soluble with difficulty in concentrated sulphuric acid. Decomposed by fusion with potassium bisulphate.

Diff. — Distinguished by its octahedral form, hardness, and infusibility; zircon has a higher specific gravity; the true ruby (p. 482) is harder and is distinguished optically; garnet is softer and fusible.

Micro. — In thin section shows light color and high relief. Isotropic.

Artif. — Artificial spinel crystals may be obtained by direct crystallization from the pure melt fused in the electric arc. They also form from melts of the oxides or fluorides of magnesium and aluminum dissolved in boric acid. The addition of chromium and iron oxides will produce various colors. Melting point about 2135° C. $n = 1\cdot723$. Artificial spinels are used as gem material. They often contain an excess of Al_2O_3 in solid solution.

Obs. — Spinel occurs frequently in crystals and rolled pebbles in the gem gravels, associated with the corundum stones, the spinel ruby often accompanying the true ruby. It occurs embedded in granular limestone or dolomite as a metamorphic mineral, frequently associated with chondrodite. Spinel (common spinel, also pleonaste, picotite, and chromite) occurs as an accessory constituent in many basic rocks, especially those of the peridodite group; its formation is the result of the crystallization of a magma very low in silica, high in magnesia, and containing alumina. Since, as in many peridotites, alkalies are absent, feldspars cannot form, and the Al_2O_3 present must form either spinel or corundum. The variety pleonaste occurs in rocks of this type that are high in their iron content and is commonly associated with magnetite, the two minerals being the first to crystallize from the magna. Spinel occurs in gneiss, amphibolite, etc., often as inclusions in garnet and iolite. It is frequently found in contact zones between eruptive rocks and carbonate rocks, occurring often in the cavities of the igneous rocks, owing its origin probably to pneumatolytic conditions or to deposition from superheated solutions.

Spinel occurs in Ceylon, Siam, and other eastern countries in beautiful colors, as rolled pebbles in the stream gravels. In Burma it is found in the Mogok district and at Ava near Mandalay. Small black splendent crystals occur in the ancient ejected masses of Mte. Somma, Vesuvius; in compact gehlenite at Monzoni, Val di Fassa, Trentino, Italy. At Åker, Södermanland, Sweden, is found a pale blue and pearl-gray variety in limestone. Occurs in large crystals near Ambatomainty, south of Betroka, and elsewhere in Madagascar.

From Amity, Orange Co., New York, to Andover, Sussex Co., New Jersey, a distance of about 30 miles, is a region of granular limestone and serpentine in which are many spinel localities; its colors are green, black, brown, and less commonly red; often associated with chondrodite. The more important localities in Orange Co., New York, are about Amity, and at Warwick, Edenville, Monroe, etc. In Sussex Co., New Jersey, Franklin affords crystals of various shades of black, blue, green, and red, which are sometimes transparent, and a bluish green *ceylonite*; Newton, pearl-gray crystals, with corundum, tourmaline, and rutile; Sparta, Hamburg, and Vernon are other localities; at Bryam, Hunterdon Co., New Jersey. Spinel occurs with the corundum of North Carolina, as at the Culsagee mines, near Franklin, Macon Co.

Good black spinel is found at South Burgess, Leeds Co., Ontario; in Quebec a bluish spinel, having a rough cubic form, occurs at Wakefield, Ottawa Co., and a blue spinel occurs in the Seigniory of Daillebout, Joliette Co.

Use. — The colored transparent varieties are used as gems.

Hercynite. — Iron Spinel, $FeAl_2O_4$. Isometric; massive, fine granular. H. $= 7\cdot5$–8. G. $= 3\cdot91$–$3\cdot95$. Color black. $n = 1\cdot800$. Hercynite occurs in Bohemia, Czechoslovakia, at Ronsperg (Ronšperk) in the Böhmer Wald, with corundum and iron hydroxide; also scattered through the granulites of Saxony. From a fine-grained gabbro at Le Prese on Lago di Poschiavo, Grisons, Switzerland. From near Erode, district of Coimbatore, Madras, India. From the tin drift near Moorina, Tasmania. A related iron-alumina spinel, with about 9 per cent MgO, occurs with magnetite and corundum in Cortlandt township, Westchester Co., New York. The name *chromohercynite* has been given to a spinel from the region between Farafagana and Vangaindrano, Madagascar, which consists of an isomorphous mixture in equal molecular proportions of the chromite and hercynite molecules. Color black. G. $= 4\cdot415$.

GAHNITE. — Zinc-Spinel.

Isometric. Habit octahedral, often with faces striated ‖ edge between dodecahedron and octahedron; also less commonly in dodecahedrons and modified cubes. Twins: tw. pl. $o(111)$.

Cleavage: $o(111)$ indistinct. Fracture conchoidal to uneven. Brittle. H. $= 7\cdot5$–8. G. $= 4\cdot0$–$4\cdot6$. $n_{gr} = 1\cdot82$ (Finland). Luster vitreous, or somewhat greasy. Color dark green, grayish green, deep leek-green, greenish black, bluish black, yellowish, or grayish brown; streak grayish. Subtransparent to nearly opaque.

Comp. — Zinc aluminate, $ZnAl_2O_4$ = Alumina 55·7, zinc oxide 44·3 = 100. The zinc is sometimes replaced by manganese or ferrous iron, the aluminum by ferric iron.

Var. — AUTOMOLITE, or *Zinc Gahnite.* — $ZnAl_2O_4$, with sometimes a little iron. G. = 4·1–4·6. Colors as above given.

DYSLUITE, or *Zinc-Manganese-Iron Gahnite.* — $(Zn,Fe,Mn)O.(Al,Fe)_2O_3$. Color yellowish brown or grayish brown. G. = 4–4·6.

KREITTONITE, or *Zinc-Iron Gahnite.* — $(Zn,Fe,Mg)O.(Al,Fe)_2O_3$. In crystals, and granular massive. H. = 7–8. G. = 4·48–4·89. Color velvet-black to greenish black; powder grayish green. Opaque.

Pyr., etc. — Gives a coating of zinc oxide when treated with a mixture of borax and soda on charcoal; otherwise like spinel.

Obs. — Gahnite occurs in zinc deposits and also, like pleonaste, as a contact mineral and in the crystalline schists. It occurs in large crystals at Bodenmais, Bavaria (*kreittonite*); in Sweden in a talcose schist near Falun, Kopparberg (*automolite*). In the United States at Rowe, Franklin Co., Massachusetts; in New Jersey at Franklin and Sterling Hill (*dysluite*). A cobaltiferous variety occurs at the copper mines of Carroll Co., Maryland. Sparingly at the Deake mica mine, Mitchell Co., North Carolina; at the Canton mine, Cherokee Co., Georgia; in large rough crystals at the Cotopaxi mine, Chaffee Co., Colorado, in part altered to a chloritic mineral.

Named after the Swedish chemist Gahn. The name *automolite*, of Ekeberg, is from αὐτόμολος, *a deserter*, alluding to the fact of the zinc occurring in an unexpected place.

Galaxite. — A manganese spinel. In minute grains. H = 7·5. G = 4·23. Brilliant black color with red-brown streak. $n = 1·923$. Occurs with alleghanyite and other manganese minerals in a vein at Bald Knob near town of Galax, Alleghany Co., North Carolina.

MAGNETITE. — Magnetic Iron Ore.

Isometric. Most commonly in octahedrons, also in dodecahedrons with faces striated ‖ edge between dodecahedron and octahedron (Fig. 745); in

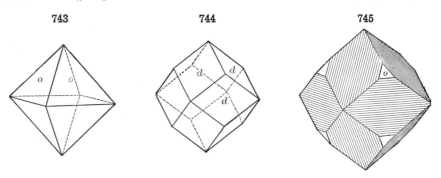

743   744   745

dendrites between plates of mica; crystals sometimes highly modified; cubic forms rare. Twins: tw. pl. *o*(111), sometimes as polysynthetic twinning lamellæ, producing striations on an octahedral face and often a pseudo-cleavage (Fig. 500, p. 198). Massive with laminated structure; granular, coarse or fine; impalpable.

Cleavage not distinct; parting octahedral, often highly developed. Fracture subconchoidal to uneven. Brittle. H. = 5·5–6·5. G. = 5·168–5·180, crystals. Luster metallic and splendent to submetallic and rather dull. Color iron-black. Streak black. Opaque, but in thin dendrites in mica nearly transparent and pale brown to black. Strongly magnetic; sometimes possessing polarity (lodestone).

Comp. — $\overset{\scriptstyle\text{II}}{\text{Fe}}\overset{\scriptstyle\text{III}}{\text{Fe}_2}\text{O}_4$ or $\text{FeO.Fe}_2\text{O}_3$ = Iron sesquioxide 69·0, iron protoxide 31·0 = 100; or, Oxygen 27·6, iron 72·4 = 100. The ferrous iron sometimes replaced by magnesium, and rarely nickel; also sometimes contains titanium (up to 6 per cent TiO_2).

746

Magnetite, when heated to 220° C. in oxygen, changes to red Fe_2O_3 without, however, noticeable change in magnetism or X-ray structure pattern, but when heated further to 550° C. the structure changes to that of hematite and magnetism disappears.

Var. — *Ordinary.* — (a) In crystals. (b) Massive, with pseudo-cleavage, also granular, coarse or fine. (c) As loose sand. (d) Ocherous: a black earthy kind. Ordinary magnetite is attracted by a magnet but has no power of attracting particles of iron itself. The property of polarity which distinguishes the *lodestone* is exceptional.

Magnesian. — G. = 4·41–4·42; luster submetallic; weak magnetic; in crystals from Sparta, New Jersey, and elsewhere.

Manganesian. — Containing 3·8 to 6·3 per cent manganese (*Manganmagnetite*). From Vester Silfberg, Sweden.

Pyr., etc. — B.B. very difficultly fusible. In O.F. loses its influence on the magnet. With the fluxes reacts like hematite. Soluble in hydrochloric acid and solution reacts for both ferrous and ferric iron.

Diff. — Distinguished from other members of the spinel group, as also from garnet, by its being attracted by the magnet, as well as by its high specific gravity; franklinite and chromite are only feebly magnetic (if at all), and have a brown or blackish brown streak; also, when massive, by its black streak from hematite and limonite; much harder than tetrahedrite.

Artif. Magnetite is frequently formed as a furnace product. It is easily formed in artificial magmas when they are low in the percentage of silica. It is formed by the breaking down of various minerals or by interreactions among minerals in processes similar to those of contact metamorphism.

Obs. — Magnetite is commonly found as a constituent of the crystalline rocks. It occurs widely distributed in grains in igneous rocks, being found most abundantly in the ferromagnesian types, in which at times it forms large segregated masses. These bodies have been formed through the processes of magmatic differentiation. They are often highly titaniferous, owing largely to an admixture of ilmenite, and then characteristic of the gabbros and similar rocks. Magnetite bodies are common in the Archæan rocks, where they are frequently of great extent and occur under the same conditions as those of hematite. Magnetite is abundant in metamorphic rocks, being formed either by contact or regional metamorphism. Magnetite is commonly, if not always, formed under conditions of high temperature. The theory has been advanced, however, that magnetite may be precipitated in the colloidal form at low temperatures and that later it may take on crystalline form. Magnetite is an ingredient in most of the massive variety of corundum, known as emery. It occurs in meteorites, and forms the crust on meteoric irons. It is present in dendrite-like forms in the mica of many localities and is perhaps of secondary origin. It is a common alteration product of minerals containing ferrous oxide, as in veins in serpentine that have resulted from the alteration of chrysolite. Common as the black constituent of beach sands, etc., sometimes in large amounts.

The largest deposits of magnetite in the world are found in Norrbotten, northern Sweden, at Kiruna and Gellivare. These bodies have been formed by magmatic segregation. Other important deposits in Sweden are in Vermland at Nordmark, Persberg, and Taberg, in Vastmanland at Norberg and in Upsala at Dannemora. These ores occur in irregular or lens-shaped masses in Archæan rocks, in or near crystalline limestones and intimately associated with masses of silicates like pyroxene, garnet, epidote, etc., known as *skarn*. They have been formed by the replacement of carbonate rocks through the action of hot solutions derived from intrusive magmas. Bedded metamorphic ores are to be found in the Varanger district in the extreme north of Norway. Also from Arendal, Norway. Octahedral crystals embedded in chloritic slate come from Falun, Sweden, and splendid dodecahedral crystals occur at Nordmark. Contact metamorphic deposits from Moravicza, Dognacska, etc., in the Banat district of Rumania have furnished magnetite ores for a long time. Large dodecahedral crystals come from Vaskö in the same district. Other localities for the crystallized mineral are Achmatovsk near Kussinsk in the Zlatoust district, Ural Mts.; in simple and twinned octahedrons in chlorite schist in the Zillertal, Tyrol, Austria, and in the Pfitschtal, Trentino (formerly in Tyrol), Italy; at Traversella in Piedmont; at Mte. Somma, Vesuvius; in the Binnental, Valais, Switzerland; Chillagoe, Queensland, Australia; from Minas Geraes, Brazil. The most powerful natural magnets are found in Siberia, in the Harz Mts., and on the Island of Elba.

In the United States magnetite occurs in large beds in the Archæan rocks of the Adirondack region in New York, especially in Warren, Essex and Clinton counties. The ores are in long lens- or pod-shaped bodies associated with an augite syenite and are thought to have had a magmatic origin. The chief district is near Port Henry and Mineville, Essex Co. Fine crystals and masses showing broad parting surfaces are found at Port Henry. It also occurs in New York in Saratoga, Herkimer, and Orange counties and at the Tilly Foster mine, Brewster, Putnam Co., in crystals and massive accompanied by chondrodite, etc. In New Jersey at Hamburg, and near Franklin, Sussex Co.; in Pennsylvania an important deposit is located at Cornwall, Lebanon Co.; in Chester Co., at the French Creek mines in brilliant octahedrons; in dendritic forms enclosed in mica from Pennsbury. From Magnet Cove, near Hot Springs, Arkansas, in crystals and as lodestone. In crystals from Millard Co., Utah. Fine crystals come from Fiormeza, Cuba. Magnetite is found in many districts of Ontario and Quebec.

Named from the locality *Magnesia* bordering on Macedonia. But Pliny favors Nicander's derivation from Magnes, who first discovered it, as the fable runs, by finding, on taking his herds to pasture, that the nails of his shoes and the iron ferrule of his staff adhered to the ground.

Use. — An important ore of iron.

Magnesioferrite. Magnoferrite. — $MgFe_2O_4$. In octahedrons. H. = 6–6·5 G. = 4·568–4·654. Luster, color, and streak as in magnetite. Strongly magnetic. Magnesioferrite is the product of the action of heated vapors containing water with magnesium and ferric chlorides and is to be found about volcanic fumaroles. Occurs at Vesuvius; from Stromboli, Lipari Islands; and from Sicily. Occurs on Puy de la Tache, on Mont Dore, Puy-de-Dôme, France.

FRANKLINITE.

Isometric. Habit octahedral; edges often rounded, and crystals passing into rounded grains. Massive, granular, coarse or fine to compact.

Pseudo-cleavage, or parting, octahedral, as in magnetite. Fracture conchoidal to uneven. Brittle. H. = 5·5–6·5. G. = 5·07–5·22. Luster metallic, sometimes dull. Color iron-black. Streak reddish brown or black. Opaque, except in thinnest section. $n = 2·36(Li)$. Slightly magnetic.

Comp. — $(Fe,Zn,Mn)O.(Fe,Mn)_2O_3$, but varying rather widely in the relative quantities of the different metals present, while conforming to the general formula of the spinel group.

Pyr., etc. — B.B. infusible. With borax in O.F. gives a reddish amethystine bead (manganese), and in R.F. this becomes bottle-green (iron). With soda gives a bluish green manganate, and on charcoal a faint coating of zinc oxide, which is much more marked when a mixture with borax and soda is used. Soluble in hydrochloric acid, sometimes with evolution of a small amount of chlorine.

Diff. — Resembles magnetite, but is only slightly attracted by the magnet, and has a dark brown streak; it also reacts for zinc on charcoal B.B.

Obs. — Franklinite is found in quantity only in the Franklin Furnace district, Sussex Co., New Jersey. It is the chief zinc mineral of two ore bodies, one at Mine Hill and the other at Stirling Hill, 3 miles apart. The ores lie in pre-Cambrian limestones associated with coarse gneisses of igneous origin and were presumably themselves of igneous metasomatic derivation. The chief associated minerals are willemite, zincite, and calcite. There are a large number of other minerals that may occur with franklinite, including many rare species some of which are to be found only at this locality. Some of these rarer associates are the zinc rhodonite (fowlerite), tephroite, gahnite, schefferite, axinite, manganese garnet, scapolite, apatite, sphalerite, galena, arsenopyrite. Franklinite has been noted in small amounts near Eibach, east of Dillenburg, Hessen-Nassau, in cubic crystals.

Use. — An ore of zinc.

Jacobsite. — $MgO.Fe_2O_3.(Mn,Fe)O.Fe_2O_3$. Isometric; in distorted octahedrons. Crystals show evidences of plagiohedral symmetry. Conchoidal fracture. H. = 6. G. =

4·75. Color deep black. Powder dark red-brown. Magnetic. In Vermland, Sweden, at Jacobsberg in the Nordmark district and at Långbanshyttan; at the Sjö and Glakärn mines in Örebro. Also from Debarstica near Tatar-Pazardžik, Bulgaria.

CHROMITE.

Isometric. In octahedrons. Commonly massive; fine granular to compact.

Fracture uneven. Brittle. H. = 5·5. G. = 4·1–4·9. Luster submetallic to metallic. Color between iron-black and brownish black, but sometimes yellowish red in very thin sections. Streak brown. Translucent to opaque. $n = 2·1$. Sometimes feebly magnetic.

Comp. — $FeCr_2O_4$ or $FeO.Cr_2O_3$ = Chromium sesquioxide 68·0, iron protoxide 32·0 = 100.

Shows considerable variation in composition and rarely conforms to the theoretical formula. The iron may be replaced by magnesium; also the chromium by aluminum and ferric iron. The varieties containing but little chromium (up to 10 per cent) are hardly more than varieties of spinel and are classed under picotite, p. 488.

Pyr., etc. — B.B. in O.F. infusible; in R.F. slightly rounded on the edges, and becomes magnetic. With borax and salt of phosphorus gives beads which, while hot, show only a reaction for iron, but on cooling become chrome-green; the green color is heightened by fusion on charcoal with metallic tin. Not acted upon by acids, but decomposed by fusion with potassium or sodium bisulphate.

Diff. — Distinguished from magnetite by feeble magnetic properties, streak and by yielding the reaction for chromic acid with the blowpipe.

Artif. — Chromite can be prepared artificially by fusing together chromic, ferric and boric oxides.

Obs. — Occurs in peridotite rocks and the serpentines derived from them, forming veins, or in embedded masses. It is one of the earliest minerals to crystallize in a cooling magma and its large ore bodies are probably formed during the solidification of the rock by the process of magmatic differentiation. When it occurs in serpentine it is commonly associated with magnetite. Not uncommon in meteoric irons, sometimes in nodules, less often in crystals.

Occurs in large deposits in Asia Minor near Brusa, Smyrna, and Antiochia. From various districts in the Ural Mts. Found in the Gulsen Mts. near Kraubat, southwest of Leoben, Styria, Austria. From Grachau near Frankenstein, Silesia. Was originally described from near Gassin, Var, France. In crystals on the islands of Unst and Fetlar in the Shetland Islands. A large deposit is at Selukwe, Southern Rhodesia. Has also been extensively mined in New Caledonia.

In the United States is found massive and in crystals at Hoboken, New Jersey, in serpentine and dolomite. In Pennsylvania near Unionville, Chester Co., and at Wood's Mine, near Texas, Lancaster Co. In Maryland, in the Bare Hills, near Baltimore, in veins or masses in serpentine. In various localities in North Carolina, the magnesian variety (*mitchellite*) coming from Webster in Jackson Co. From serpentine areas in many localities in California. From Port au Port, on the west coast of Newfoundland.

Use. — An ore of chromium. Used in certain hard steels, rustless steel, chromium plating, etc.; used in refractory bricks for metallurgical furnace linings; as source of certain red and yellow pigments and dyes.

CHROMITITE. Material in minute octahedral crystals occurring in sand at Zeljin Mt., Serbia, said to have composition, $FeCrO_3$.

TREVORITE. $NiO.Fe_2O_3$. Color black with greenish tint. Black streak. Strongly magnetic. H. = 5. G. = 5·165.

CHRYSOBERYL. Cymophane.

Orthorhombic. Axes $a : b : c = 0·4701 : 1 : 0·5800$.

mm''',	$110 \wedge 1\bar{1}0 =$	50° 21'.	pp',	$031 \wedge 0\bar{3}1 =$	120° 14'.
ss',	$120 \wedge \bar{1}20 =$	93° 32'.	oo',	$111 \wedge \bar{1}11 =$	90° 44'.
xx',	$101 \wedge \bar{1}01 =$	101° 57'.	oo''',	$111 \wedge 1\bar{1}1 =$	40° 7'.
ii',	$011 \wedge 0\bar{1}1 =$	60° 14'.	nn',	$121 \wedge \bar{1}21 =$	77° 43'.

Twins: tw. pl. $\rho(031)$, both contact- and penetration-twins; often repeated and forming pseudo-hexagonal crystals with or without reentrant angles (Fig. 421, p. 185). Crystals generally tabular $\parallel a(100)$. Face *a* striated vertically, in twins a feather-like striation (Fig. 748).

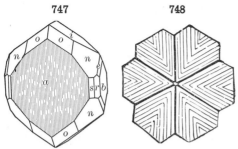

747   748

It is interesting to note that there is a close similarity between the crystal constants of chrysoberyl and those of chrysolite. The atomic structure of the two minerals as determined by X-ray study shows the same relationship.

Cleavage: $i(011)$ quite distinct; $b(010)$ imperfect; $a(100)$ more so. Fracture uneven to conchoidal. Brittle. H. = 8·5. G. = 3·5–3·84. Luster vitreous. Color asparagus-green, grass-green, emerald-green, greenish white, and yellowish green; greenish brown; yellow; sometimes raspberry- or columbine-red by transmitted light. Streak uncolored. Transparent to translucent. Sometimes a bluish opalescence or chatoyancy, and asteriated. Pleochroic, vibrations $\parallel Y$ (= *b* axis) orange-yellow, Z (= *c* axis) emerald-green, X (= *a* axis) columbine-red. Optically +. Ax. pl. $\parallel b(010)$. Bx. $\perp$ $c(001)$. $\alpha = 1\cdot747$. $\beta = 1\cdot748$. $\gamma = 1\cdot757$.   2E = 84° 43'. Indices and axial angle variable, even in different portions of a single specimen.

Var. — 1. *Ordinary.* — Color pale green, being colored by iron; also yellow and transparent and then used as a gem.

2. *Alexandrite.* — Color emerald-green, but columbine-red by artificial light; valued as a gem. G. = 3·644, mean of results. Supposed to be colored by chromium. Crystals often very large, and in twins, like Fig. 421, either six-sided or six-rayed.

3. *Cat's-eye.* — Color greenish and exhibiting a fine chatoyant effect; from Ceylon.

Comp. — Beryllium aluminate, $BeAl_2O_4$ or $BeO.Al_2O_3$ = Alumina 80·2, glucina 19·8 = 100.

Pyr., etc. — B.B. alone unaltered; with soda, the surface is merely rendered dull. With borax or salt of phosphorus fuses with great difficulty. Ignited with cobalt solution, the powdered mineral gives a bluish color. Not attacked by acids.

Diff. — Distinguished by its extreme hardness, greater than that of topaz; by its infusibility; also characterized by its tabular crystallization, in contrast with beryl.

Obs. — Chrysoberyl is to be found in granitic rocks and pegmatites and in mica schists. Also frequently in river sands and gravels. The variety *alexandrite* is found in mica schist on the Takowaja River, east of Ekaterinburg, Ural Mts. In the gold sands of the Orenburg district of the southern Urals, yellow in color. From a pegmatite at Marschendorf near Zöptau, Moravia, Czechoslovakia. In pebbles and crystals from the gem gravels of Ceylon. In diamond sands from near Somabula, south of Gwelo, Southern Rhodesia; similarly from Minas Geraes, Brazil. In crystals of gem quality from pegmatites in Madagascar.

In the United States found in pegmatites in Oxford Co., Maine, at Ragged Jack Mt., Hartford (twins), and at various other localities. At Haddam, Connecticut; at Greenfield near Saratoga, New York. Has been found in crystals in the rocks of New York City.

Chrysoberyl is from χρύσος, *golden*, βήρυλλος, *beryl. Cymophane*, from κύμα, *wave*, and φαίνω, *appear*, alludes to a peculiar opalescence the crystals sometimes exhibit. *Alexandrite* is after the Czar of Russia, Alexander I.

Use. — As a gem stone; see under **Var.** above.

Hausmannite. — Mn_3O_4 or $MnO.Mn_2O_3$. In tetragonal octahedrons and twins (Fig. 440, p. 188); also granular massive, particles strongly coherent. X-ray study shows an atomic structure closely related to that of magnetite. Cleavage, (001). H. = 5–5·5. G. = 4·856. Luster submetallic. Color brownish black. Streak chestnut-brown. Optically −. $\omega = 2\cdot46(Li)$. Hausmannite is one of the manganese minerals of primary origin; it occurs commonly in veins associated with acid igneous rocks. Found at Il-menau and near by at Örenstock, in Thuringia; at Ilfeld in the Harz Mts. In Vermland, Sweden, at Långbanshyttan north of Filipstadt. From near Ouro Preto, Minas Geraes, Brazil. Occurs intimately mixed with psilomelane in the Batesville district, Arkansas.

Hetærolite. Zinc-hausmannite. — $ZnO.Mn_2O_3$. Tetragonal. Octahedral habit. Per-fect basal cleavage. Black. Brown-black streak. Optically −. Indices, 2·10–2·34. H. = 6. G. = 4·85. B.B. infusible. Occurs at Franklin, New Jersey, and at the Wolf-tone mine (*wolftonite*) near Leadville, Colorado.

Hollandite. — A manganate of manganese, barium, and ferric iron. Tetragonal. Prismatic habit. Crystals up to 2 inches in length. Prism faces deeply striated. Ter-minated with low pyramid faces having dull luster. Has been considered to be the crystal-line form of psilomelane and suggestion made that it belongs in the scheelite group. H. = 4–6. G. = 4·7–5·0. Color, silvery-gray to black. Occurs in the manganese deposits of the Central Provinces of India; at Kajlidongri, Jhabua State; also at Sitapar and else-where in the Chhindwarah district, and in the Nagpur and Balaghat districts. *Coronadite*, described as a manganese-lead manganate from the Coronado vein of the Clifton-Morenci district, Arizona, has been shown to be a mixture of hollandite and some unidentified lead mineral. *Romanèchite* from Romanèche, Rhône, France, described as crystallized psilo-melane, has also been shown to be a mixture of hollandite and an unidentified mineral.

Cesàrolite. — $H_2PbMn_3O_8$. In cellular masses. Color, steel-gray. H. = 4·5. G. = 5·29. From Sidi-Amer-ben-Salem, Tunisia.

Minium. — Pb_3O_4 or $2PbO.PbO_2$. Pulverulent, as crystalline scales. H. = 2–3. G. = 4·6. Color vivid red, mixed with yellow; streak orange-yellow. n = 2·4. Strongly pleochroic. X = red-brown, Z = nearly colorless. Of secondary origin, derived usually from the alteration of galena or cerussite. Occurs in the Altai Mts., Siberia; from Bleialf in the Eifel, Rhineland; Badenweiler in Baden; from Broken Hill, New South Wales. In the United States it occurred in Idaho with native lead at the Jay Gould mine, Wood River district, Blaine Co., and with plattnerite and siderite near Gilmore, Lemhi Co.; also at the Rock mine, Leadville, Colorado.

Crednerite. — $CuO.Mn_2O_3$. Monoclinic? In pseudo-hexagonal plates, probably the result of twinning. Perfect cleavage parallel to tabular development. H. = 4·5. G. = 4·9–5·1. Luster metallic. Color iron-black to steel-gray. Streak black, brownish. From Friedrichroda, in Thuringia and the Mendip Hills, Somerset.

Delafossite. $Cu_2O.Fe_2O_3$. Rhombohedral. Crystals tabular parallel to base to equidimensional in habit. H. = 5·5. Easily fusible, becoming magnetic. Soluble in HCl. Color and streak, black. Occurs near Ekaterinburg, Ural Mts. At the Calumet and Arizona mine, Bisbee, Cochise Co., Arizona, with kaolin and ferruginous clay. Also in small spherical aggregates embedded in clay at Kimberly, Nevada.

Magnetoplumbite. An oxide of iron, manganese, and lead. $2RO.3R_2O_3$. Hex-agonal. Good basal cleavage. Color black. Streak dark brown. Strongly magnetic. G. = 5·517. Found associated with manganophyllite, kentrolite, etc., at Långbanshyttan, Vermland, Sweden.

Plumboferrite. — $PbO.2Fe_2O_3$. Rhombohedral, trapezohedral. Crystals thick tab-ular ‖ (001). Basal cleavage. In cleavable masses. H. = 5. G. = 6·0. Color, nearly black. Streak red, like hematite. Occurs with jacobsite at Jakobsberg, Nordmark, Sweden.

Zincdibraunite. $ZnO.2MnO_2.2H_2O$. Soft and earthy. Chocolate color. Occurs in cavities in calamine ore at Olkush, Russia.

BRAUNITE.

Tetragonal. Axis c = 0·9922. Commonly in octahedrons, nearly iso-metric in angle $(pp'\ 111 \wedge 1\bar{1}1 = 70°\ 7')$. Also massive. X-ray study shows eight molecules to the unit cell.

Cleavage: $p(111)$ perfect. Fracture uneven to subconchoidal. Brittle. H. = 6–6·5. G. = 4·75–4·82. Luster submetallic. Color and streak, dark brownish black to steel-gray.

Comp. — $3MnMnO_3.MnSiO_3$ = Silica 10·0, manganese protoxide 11·7, manganese sesquioxide 78·3 = 100.

Although the above formula agrees closely with the analyses, probably a better way of expressing the composition is to consider it as a mixture of the isomorphous molecules $MnMnO_3$ and $MnSiO_3$ in proportions nearly 3 : 1.

Pyr., etc. — B.B. infusible. With borax and salt of phosphorus gives an amethystine bead in O.F., becoming colorless in R.F. With soda gives a bluish green bead. Dissolves in hydrochloric acid leaving a residue of gelatinous silica. Marceline gelatinizes with acids.

Obs. — Is commonly of secondary but may be of primary origin. Occurs in veins traversing porphyry at Öhrenstock near Ilmenau, Thuringia; from near Ilfeld, Harz Mts.; at St. Marcel, Piedmont. At Långbanshyttan and elsewhere in Sweden. In India occurs in quantity at Kacharwaki, district of Nagpur, Central Provinces. From Miguel Burnier near Ouro Preto, Minas Geraes, Brazil. *Marceline* (heterocline) from St. Marcel, Piedmont, is impure braunite.

IV. Dioxides, RO_2

Rutile Group. Tetragonal

		c			c
Cassiterite	SnO_2	0·6723	**Rutile**	TiO_2	0·6442
Polianite	MnO_2	0·6647	**Plattnerite**	PbO_2	0·6764

The RUTILE GROUP includes the dioxides of the elements tin, manganese, titanium, and lead. These compounds crystallize in the tetragonal system with closely similar angles and axial ratio; furthermore in habit and method of twinning there is much similarity between the two best known species included here. Chemically these minerals are sometimes considered as salts of their respective acids, as stannyl metastannate, $(SnO)SnO_3$, for cassiterite and titanyl metatitanate, $(TiO)TiO_3$, for rutile.

X-ray study shows for cassiterite an atomic structure in which the tin atoms are arranged on a body-centered tetragonal lattice. The oxygen atoms lie in the same horizontal planes as the tin atoms and are grouped in pairs about each tin atom. Lines joining the pairs of oxygen atoms have the direction of one or the other of the prism diagonals and alternate in direction in each successive horizontal atomic layer. The other members of the group show closely similar structures.

With the Rutile Group is also sometimes included zircon, $ZrO_2.SiO_2$; $c = 0.6404$. The atomic structure is like that of this group with the zircon atoms placed at the corners of the tetragonal prism and the silicon atom at its center, or *vice versa*. In this work, however, zircon is classed among the silicates, with the allied species thorite, $ThO_2.SiO_2$; $c = 0.642$.

The phosphate xenotime, YPO_4, is also closely similar in crystal form and structure to this group; the same is true of sellaite, MgF_2 and tapiolite, $Fe(Ta,Nb)_2O_6$.

It may be added that ZrO_2, as the species baddeleyite, crystallizes in the monoclinic system.

CASSITERITE. Tin-stone, Tin Ore.

Tetragonal. Axis $c = 0.6723$.

ee', 101 $\wedge$ 011 = 46° 28'.	ms, 110 $\wedge$ 111 = 46° 27'.
ee'', 101 $\wedge$ $\bar{1}$01 = 67° 50'.	zz', 321 $\wedge$ 231 = 20° 53$\frac{1}{2}$'
ss', 111 $\wedge$ $\bar{1}$11 = 58° 19'.	zzvii, 321 $\wedge$ 3$\bar{2}$1 = 61° 42'.
ss'', 111 $\wedge$ $\bar{1}\bar{1}$1 = 87° 7'.	

Twins common: tw. pl. $e(101)$; both contact- and penetration-twins (Fig. 752); often repeated. Crystals low pyramidal; also prismatic and acutely terminated. Sometimes very slender (needle-tin). Prism zone vertically striated. For atomic structure see above. Often in reniform shapes, structure fibrous divergent; also massive, granular or impalpable; in rolled grains.

Cleavage: $a(100)$ imperfect; $s(111)$ more so; $m(110)$ hardly distinct. Fracture subconchoidal to uneven. Brittle. H. = 6–7. G. = 6·8–7·1. Luster adamantine, and crystals usually splendent. Color brown or black; sometimes red, gray, white, or yellow. Streak white, grayish, brownish.

Nearly transparent to opaque. Optically $+$. Indices: $\omega = 1\cdot9966$, $\epsilon = 2\cdot0934$.

Var. — *Ordinary.* Tin-stone. In crystals and massive.

Wood-tin. In botryoidal and reniform shapes, concentric in structure, and radiated fibrous internally, although very compact, with the color brownish, of mixed shades, looking somewhat like dry wood in its colors. *Toad's-eye tin* is the same, on a smaller scale. *Stream-tin* is the ore in the state of sand, as it occurs along the beds of streams or in gravel. *Silesite* was described as a silicate of tin occurring in chalcedony-like masses in the tin veins of Bolivia but is probably a mixture of wood-tin and silica.

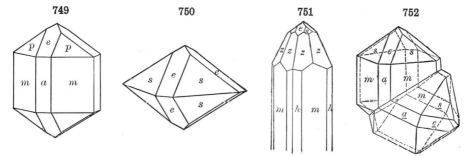

Comp. — Tin dioxide, SnO_2 = Oxygen 21·4, tin 78·6 = 100. A little Ta_2O_5 is sometimes present, also Fe_2O_3.

Pyr., etc. — B.B. alone unaltered. On charcoal with soda reduced to metallic tin, and gives a white coating. With the fluxes sometimes gives reactions for iron and manganese. Only slightly acted upon by acids. When fragments of cassiterite are placed in dilute hydrochloric acid together with a little metallic zinc the cassiterite becomes coated with a dull gray deposit of metallic zinc which becomes bright on friction.

Diff. — Distinguished by its high specific gravity, hardness, infusibility, and by its yielding metallic tin B.B.; resembles some varieties of garnet, sphalerite, and black tourmaline. Specific gravity (6·5) higher than that of rutile (4); wolframite is easily fusible.

Artif. — Cassiterite has been artificially prepared by the action of aqueous vapor upon tin tetrachloride in a heated tube and by other similar methods employing heated vapors.

Obs. — Cassiterite is the chief ore of tin and is found in many localities, although the important economic deposits are comparatively few in number. It has been noted as an original constituent of igneous rocks, but it commonly occurs in veins or stringers of quartz either lying in or close to granitic rocks or pegmatites. It has also been observed in quartz porphyry and rhyolite. It is commonly accompanied by minerals containing boron and fluorine, which indicates a pneumatolytic origin. The granites are frequently altered into material known as *greisen* in which the feldspars have been altered to mica, and topaz and tourmaline are present. The commonly associated minerals are quartz, wolframite, scheelite; also mica which is often a lithium-bearing species, topaz, tourmaline, axinite, apatite, fluorite; further pyrite, arsenopyrite, sphalerite, molybdenite, native bismuth, etc. Frequently occurs in sands and gravels.

Large deposits of vein tin ores occur in Cornwall. These were formerly very important but have been largely exhausted. The Dolcoath lode near Redruth is more than 3 miles long and has been worked for more than 3000 feet in depth. The cassiterite occurs often in fine crystals, and also as wood-tin and stream-tin. The more famous localities for specimens are St. Agnes, Callington, Camborne, St. Just, St. Austell. Pseudomorphs after feldspar are found at Wheal Coates near St. Agnes. On the continent of Europe unusual specimens are found at Schlaggenwald (Slavkov), Graupen (Krupka), and Zinnwald (Cinvald) in northwest Bohemia, Czechoslovakia, and in the neighboring district of Saxony at Altenberg and Ehrenfriedersdorf. Fine crystals, often twinned, are found at Villeder, Morbihan, France. From Pitkäranta on Lake Lagoda, Finland.

Important deposits, mostly placers, occur on the Malay Peninsula and the neighboring islands Banka and Billiton off the coast of Sumatra. Cassiterite is abundant in New South Wales over an area of 8500 square miles, fine crystals coming from Emmaville, Glen Innes (Glen Eden) and Elsmore, all in the New England Range. It also occurs in Victoria, in Queensland and at Mt. Bischoff, Tasmania. From South West Africa. Important deposits occur in Bolivia near La Paz, Oruro, and Potosi with associated minerals of lead, silver, and bismuth. In Mexico from the states of Durango, Guanajuato, Jalisco, etc.

Cassiterite does not occur in quantity in the United States. It is most commonly found in small amounts in the pegmatite veins. Such occurrences are found in New England as at Norway, Hebron, and elsewhere in Oxford Co., Maine; from New Hampshire, Massachusetts, and Connecticut. In Virginia, on Irish Creek, Rockbridge Co., with wolframite. In North and South Carolina and Alabama. In South Dakota near Harney Peak and near Custer City in the Black Hills, where it has been mined. Low-grade deposits occur at Temescal in the Santa Ana Mountains, Riverside Co., California. Has also been mined in the York district, Seward Peninsula, Alaska.

Use. — The most important ore of tin.

Polianite. — Manganese dioxide, MnO_2. Tetragonal. Atomic structure similar to that of cassiterite, see p. 496. In composite parallel groupings of minute crystals; also forming the outer shell of crystals having the form of manganite. H. = 6–6.5. G. = 4.992. Luster metallic. Color light steel-gray or iron-gray. Streak black. Definite and well-crystallized polianite is found only at Platten (Blatno) in Bohemia, Czechoslovakia, having been derived by the alteration of manganite. It is distinguished from pyrolusite by its hardness and its anhydrous character. Like pyrolusite it is often a pseudomorph after manganite.

RUTILE.

Tetragonal. Axis $c = 0.64415$.

$ll^{vii}, 310 \wedge 3\bar{1}0 = 36° 54'$.	$ss', 111 \wedge \bar{1}11 = 56 \ 25\frac{1}{2}'$.
$ee', 101 \wedge 011 = 45° 2'$.	$ss'', 111 \wedge \bar{1}\bar{1}1 = 84° 40'$.
$ee'', 101 \wedge \bar{1}01 = 65° 34\frac{1}{2}'$.	$tt', 313 \wedge 133 = 29° 6'$.

Twins: tw. pl. (1) $e(101)$; often geniculated (Figs. 755, 756); also contact-twins of very varied habit, sometimes sixlings and eightlings (Fig. 425, p. 185; Fig. 439, p. 188). (2) $v(301)$ rare, contact-twins (Fig. 441, p. 188). Crys-

753

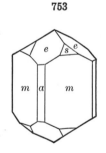

754

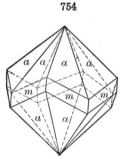

755

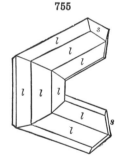

tals commonly prismatic, vertically striated or furrowed; often slender acicular. Occasionally compact, massive. For atomic structure, see p. 496.

756

Cleavage: $a(100)$ and $m(110)$ distinct; $s(111)$ in traces. Fracture subconchoidal to uneven. Brittle. H. = 6–6.5. G. = 4.18–4.25; also to 5.2. Luster metallic-adamantine. Color reddish brown, passing into red; sometimes yellowish, bluish, violet, black, rarely grass-green; by transmitted light deep red. Streak pale brown. Transparent to opaque. Optically +. Refractive indices high: $\omega = 2.6158$, $\epsilon = 2.9029$. Birefringence very high. Sometimes abnormally biaxial.

Comp. — Titanium dioxide, TiO_2 = Oxygen 40.0, titanium 60.0 = 100. A little iron is usually present, sometimes up to 10 per cent. While the iron present is often reported as ferric the probability is that in the unaltered mineral it existed in the ferrous state.

The formula for rutile may be written as a titanyl metatitanate $(TiO)TiO_3$. With this the ferrous titanate $FeTiO_3$ may be considered isomorphous and so account for the iron frequently present. It has been suggested that the tapiolite molecule, $FeO.Ta_2O_5$ is also isomorphous and that tapiolite belongs in the same group as rutile and cassiterite, see *ilmenorutile*, below. It has been noted that rutile and ilmenite may occur in intimate intergrowths, and the possible presence of the latter in this way could account for the iron content shown by many rutile analyses.

Var. — *Ordinary.* Brownish red and other shades, not black. G. = 4·18–4.25. Transparent quartz is sometimes penetrated thickly with acicular or capillary crystals. Dark smoky quartz penetrated with the acicular rutile or " rutilated quartz," is the *flèches d'amour* (or Venus hair-stone). Acicular crystals often implanted in parallel position on tabular crystals of hematite; also somewhat similarly on magnetite. Reticulated twin groups of slender crystals are known as *sagenite*, net-like.

Ferriferous. (a) *Nigrine* is black in color, whence the name; contains up to 30 per cent of ferrous titanate. (b) *Ilmenorutile* is a black variety, containing iron in the form of ferrous titanate, niobate and tantalate. G. = 5·14. *Strüverite* is the same mineral with greater amounts of the tantalate present. (c) *Iserine* from Iserweise, Bohemia, formerly considered to be a variety of ilmenite is probably also a ferriferous rutile. (d) *Paredrite* is one of the so-called " favas " found in the diamond sands of Minas Geraes, Brazil, which is composed of TiO_2 with a little water.

Pyr., etc. — B.B. infusible. With salt of phosphorus gives a colorless bead, which in R.F. assumes a violet color on cooling. Most varieties contain iron, and give a brownish yellow or red bead in R.F., the violet only appearing after treatment of the bead with metallic tin on charcoal. Insoluble in acids; made soluble by fusion with an alkali or alkaline carbonate. The solution containing an excess of acid, with the addition of tin, gives a beautiful violet color when concentrated.

Diff. — Characterized by its peculiar sub-adamantine luster and brownish red color. Differs from tourmaline, vesuvianite, augite in being entirely unaltered when heated alone B.B. Specific gravity about 4, of cassiterite 6·5.

Micro. — In thin sections shows red-brown to yellow color, very high relief and high order of interference color.

Artif. — Rutile has been formed artificially by heating titanic oxide with boric oxide, with sodium tungstate, etc. Rutile, octahedrite and brookite have all been formed by heating potassium titanate and calcium chloride in a current of hydrochloric acid gas and air. Rutile is formed at the highest temperature, brookite at lower temperatures, and octahedrite at the lowest of all. Rutile is apparently the most stable form of TiO_2, as crystals of both brookite and octahedrite have been found that have altered to rutile.

Obs. — Rutile occurs as an accessory mineral in igneous rocks as in hornblende diorites, syenites, granites, amphibolites; also in gneiss and mica schist and sometimes in granular limestone or dolomite. It frequently is secondary in origin, occurring as an alteration of mica, etc., in the igneous rocks or in the form of microlites in slates. Often in masses of quartz or feldspar and frequently in acicular crystals penetrating quartz. It has also been met with in hematite and ilmenite. It is common in grains or crystal fragments in many auriferous sands.

Rutile is widespread in small amounts, often in microscopic crystals. Only the most prominent localities can be mentioned here. Found in Austria at Modriach southwest of Graz, Styria; in the Sua-Alpe, Carinthia; at Rauris in Salzburg; at Prägratten and Windisch-Matrei in the Hohe Tauern, Tyrol; also in the Pfitschtal, Trentino (formerly Tyrol), Italy. In Switzerland in the Val Tavetsch, Grisons, at St. Gotthard, Ticino, and in the Binnental, Valais, and elsewhere. In France at Saint Yrieix south of Limoges, Haute-Vienne. In Norway at Snarum, Kragerö, Risör, etc. From Blumberg, Adelaide Co., South Australia. From Minas Geraes, Brazil.

In the United States rutile occurs in Vermont at Waterbury. At various localities in Massachusetts, Connecticut, and New York. In Pennsylvania in Chester Co. at Parksburg, Newlin, and at Sadsbury and the adjoining district in Lancaster Co. Important deposits lie in Amherst and Nelson counties, Virginia. In many localities in North Carolina, especially at Stony Point, Alexander Co., in splendent crystals of varied habit. In Georgia at Graves' Mountain, Lincoln Co., in large and lustrous crystals with lazulite. In Arkansas at Magnet Cove near Hot Springs.

Fine specimens of " rutilated quartz " come from Val Tavetsch, Grisons, and elsewhere in Switzerland; from Madagascar; Minas Geraes, Brazil; West Hartford, Vermont; Alexander Co., North Carolina.

Ilmenorutile was originally found in the Ilmen Mts., Russia; also noted from Ampangabé near Miandrarivo, Madagascar. *Strüverite* was originally described from Graveggia, Piedmont, Italy; has also been found in Madagascar at Tongafeno south of Betafo and at Fefena; also at the Etta mine, Keystone, Pennington Co., South Dakota.

Use. — A source of titanium.

Plattnerite. Lead dioxide, PbO_2. Tetragonal. Structure that of the Rutile Group, see p. 496. Rarely in prismatic crystals, usually massive. H. = 5–5·5. G. = 8·5. Luster submetallic. Color iron-black. Streak chestnut-brown. From Scotland at Leadhills, Lanark, and at Wanlockhead, Dumfries. Also in Idaho at Mullan, Coeur d'Alene district, and elsewhere in Shoshone Co. and in the Gilmore district, Lenhi Co.

Baddeleyite. — Zirconium dioxide, ZrO_2. Contains a small amount of hafnium oxide. In tabular monoclinic crystals. Cleavage, (001). H. = 6·5. G. = 5·5–6·0. Colorless to yellow, brown and black. Optically −. $\alpha = 2\cdot13$, $\beta = 2\cdot19$, $\gamma = 2\cdot20$. 2V. = 30°. Ax. pl. = (010). $X \wedge c$ axis = +12°. Dispersion strong, $\rho > v$. From Ceylon; from Brazil near Caldas, Minas Geraes and Jacupiranga, São Paulo (*brazilite*). From near Alnö, Västernorrland, Sweden. Noted at Mte. Somma, Vesuvius. Also near Bozeman, Montana. Various minerals occurring as rolled pebbles in the diamond sands of Brazil are known as *favas (beans)*. Some of them consist of nearly pure TiO_2, others of nearly pure ZrO_2, while others are various phosphates.

OCTAHEDRITE. Anatase.

Tetragonal. Axis $c = 1\cdot7771$.

Commonly octahedral in habit, either acute (p, 111), or obtuse (v, 117); also tabular, c(001) predominating; rarely prismatic crystals; frequently highly modified. X-ray analysis shows that the Ti atoms lie on a slightly deformed diamond lattice (see p. 40) with the O atoms arranged in pairs on either side of the Ti atoms in vertical lines.

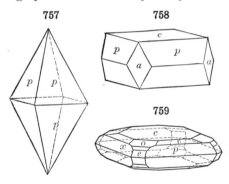

757 758 759

ee', 101 $\wedge$ 011 = 76° 5′.
ee'', 101 $\wedge$ $\bar{1}$01 = 121° 16′.
pp', 111 $\wedge$ $\bar{1}$11 = 82° 9′.
pp'', 111 $\wedge$ $\bar{1}\bar{1}$1 = 136° 36′.
zz', 113 $\wedge$ $\bar{1}$13 = 54° 1′.
vv', 117 $\wedge$ $\bar{1}$17 = 27° 39′.

Cleavage: c(001) and p(111) perfect. Fracture subconchoidal. Brittle. H. = 5·5–6. G. = 3·82–3·95; sometimes 4·11–4·16 after heating. Luster adamantine or metallic-adamantine. Color various shades of brown, passing into indigo-blue, and black; greenish yellow by transmitted light. Streak uncolored. Transparent to nearly opaque. Optically −. Birefringence rather high. Indices: $\omega = 2\cdot534$–$2\cdot564$, $\epsilon = 2\cdot488$–$2\cdot497$, varying probably with iron content. Sometimes abnormally biaxial.

Comp. — Titanium dioxide, TiO_2 = Oxygen 40·0, titanium 60·0 = 100. Commonly contains small amounts of iron oxide.

Pyr., etc. — Same as for rutile.

Artif. — See under rutile.

Obs. — Usually of secondary origin, derived from the alteration of other titanium-bearing minerals. Occurs in granite, quartz porphyry, gneiss, chlorite and mica schists associated with quartz, adular, hematite, apatite, and titanite, rutile, and brookite. From Switzerland in the Binnental, Valais, the variety *wiserine*, long supposed to be xenotime; at Cavradi, Val Tavetsch, Grisons. Abundant in fine crystals in the district of Bourg d'Oisans, Isère, France, associated with feldspar, axinite, and ilmenite. In chlorite near

Tavistock, Devonshire. In Minas Geraes, Brazil, in splendent crystals in quartz. In the United States octahedrite has been found in Massachusetts at Somerville. From Magnet Cove, near Hot Springs, Arkansas. Blue crystals with a large number of planes and reaching a centimeter in diameter have been found on Beaver Creek, Gunnison Co., Colorado.

BROOKITE.

Orthorhombic. Axes $a : b : c = 0.8416 : 1 : 0.9444$.

mm''', 110 ∧ 1$\bar{1}$0 = 80° 10'.		ee', 122 ∧ $\bar{1}$22 = 44° 23'.
zz', 112 ∧ $\bar{1}$12 = 53° 48'.		ee''', 122 ∧ 1$\bar{2}$2 = 78° 57'.
zz''', 112 ∧ 1$\bar{1}$2 = 44° 46'.		me, 110 ∧ 122 = 45° 42'.

Only in crystals, of varied habit. Atomic structure is similar to that of columbite, see p. 695.

Cleavage: m(110) indistinct; c(001) still more so. Fracture subconchoidal to uneven. Brittle. H. = 5·5–6. G. = 3·87–4·08. Luster metal-

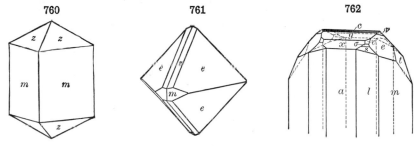

lic-adamantine to submetallic. Color hair-brown, yellowish, reddish, reddish brown, and translucent; also brown to iron-black, opaque. Streak uncolored to grayish or yellowish. $\alpha = 2.583$–2.584, $\beta = 2.585$–2.588, $\gamma = 2.705$–2.741. Optically +. Bx.$_{ac}$ always ⊥ (100). Ax. pl. ‖ (010) for blue and green light, ‖ (001) for red and yellow light. Uniaxial for a yellowish green color. Strong dispersion. Abnormal interference figure, see p. 326.

Comp. — Titanium dioxide, TiO_2 = Oxygen 40·0, titanium 60·0 = 100.

Pyr. — Same as for rutile.

Artif. — See under rutile.

Obs. – In mode of origin and associations brookite is similar to octahedrite. Found in the gold placers of the Ural Mts. at Zlatoust near Miask and on the river Sanarka, Orenburg district. From near Prägratten in the Virgen Tal, Tyrol, Austria. In Switzerland at St. Gotthard, Ticino, and at Bristenstock near Amsteg in the Maderaner Tal, Uri. From near Bourg d'Oisans, Isère, France. From near Tremadoc, Carnarvon, Wales.

In the United States found at Somerville, Massachusetts; at Ellenville, Ulster Co., New York, on quartz with chalcopyrite and galena; and in thick black crystals (*arkansite*) at Magnet Cove near Hot Springs, Arkansas.

Named after the English mineralogist, H. J. Brooke (1771–1857).

PYROLUSITE.

Orthorhombic, but pseudomorphous, commonly after manganite. X-ray study shows that pyrolusite has the same atomic structure as polianite, which is apparently the only crystalline form of MnO_2. Usually columnar, often divergent; also granular massive, and frequently in reniform coats.

Soft, often soiling the fingers. H. = 2–2·5. G. = 4·73–4·86. Luster metallic. Color iron-black, dark steel-gray, sometimes bluish. Streak black or bluish black, sometimes submetallic. Opaque.

Comp. — Manganese dioxide, MnO_2, like polianite (p. 498). Commonly contains a little water (2 per cent).

Pyr., etc. — Like polianite, but most varieties yield some water in the closed tube.

Diff. — Hardness less than that of psilomelane. Differs from iron ores in its reaction for manganese B.B. Easily distinguished from psilomelane by its inferior hardness, and usually by being crystalline. Its streak is black; that of manganite is more or less brown.

Obs. — Manganese ore deposits in general are secondary in origin, the manganese content of the rocks having been concentrated in favorable places. They often occur as irregular bodies in residual clays. It is thought that the manganese minerals were originally colloidal in character, having subsequent to deposition assumed a crystalline form. Pyrolusite and psilomelane are the two most common ore minerals.

Pyrolusite is a common mineral, widespread in its occurrence, and only a few of the more important localities can be given here. It is found at Platten (Blatno) and elsewhere in Bohemia, Czechoslovakia. Has been mined at Elgersburg, northwest of Ilmenau, Thuringia; near Johanngeorgenstadt, Saxony; from Rossbach, Nassau; with manganite at Ilfeld, Harz Mts.; from Horhausen, Rhineland. In Cornwall at Lanlivery. Large and important deposits occur in the Central Provinces, India, where pyrolusite occurs associated with psilomelane and braunite. High-grade manganese deposits also occur in Minas Geraes, Brazil, consisting chiefly of psilomelane but with pyrolusite also.

In the United States pyrolusite was found with psilomelane at various localities in Vermont. In the Appalachian region small bodies of manganese ores are to be found in the granites and schists of the Piedmont plateau and also in the area of residual iron ores of the Paleozoic sedimentary rocks. Has been mined at Batesville, Independence Co., and elsewhere in Arkansas. Frequently found in considerable quantity with the hematite ores of the Lake Superior region, as at Negaunee, Michigan. In New Brunswick in fine crystals near Bathurst, Gloucester Co., and in considerable amount near Upham, Kings Co. In Nova Scotia in Hants Co., at Tennycape, both crystalline and massive and in quantity at Cheverie; also from Bridgeville, Pictou Co.

The name is from πῦρ, *fire*, and λούειν, to *wash*, because used to discharge the brown and green (FeO) tints of glass; and for the same reason it is whimsically entitled by the French *le savon de verriers*.

Use. — An ore of manganese; as an oxidizing agent in manufacture of chlorine, bromine and oxygen; as a drier in paints, a decolorizer in glass and in electric batteries, as coloring material in bricks, pottery, glass, etc.

B. Hydrous Oxides

Among the hydrous oxides the Diaspore Group is well characterized. Here belong the hydroxides of aluminum, iron and manganese. The general formula is properly written $\overset{\text{III}}{R}O(OH)$. The three species here included are orthorhombic in crystallization with related angles and axial ratios; this relation is deviated from by manganite in the prismatic zone.

Another less prominent group is the Brucite Group, including the rhombohedral species Brucite, $Mg(OH)_2$, and Pyrochroite, $Mn(OH)_2$.

Gibbsite, $Al(OH)_3$, and Sassolite, $B(OH)_3$, are also related, and further Hydrotalcite and Pyroaurite.

Diaspore Group. $\overset{\text{III}}{R}O(OH)$ or $R_2O_3.H_2O$. Orthorhombic.

		$a : b : c$	$\dfrac{c}{a}$
Diaspore	$Al_2O_3.H_2O$	$0.9372 : 1 : 0.6039$	or 0.6443
Goethite	$Fe_2O_3.H_2O$	$0.9185 : 1 : 0.6068$	or 0.6606
Manganite	$Mn_2O_3.H_2O$	$0.8441 : 1 : 0.5448$	or 0.6463

DIASPORE.

Orthorhombic. Axes: $a : b : c = 0.9372 : 1 : 0.6039$. Crystals prismatic, mm''', $110 \wedge 1\bar{1}0, = 86° 17'$; usually thin, flattened $\parallel b(010)$; sometimes acicular. Also foliated massive and in thin scales; sometimes stalactitic.

Cleavage: $b(010)$ eminent; $h(210)$ less perfect. Fracture conchoidal, very brittle. H. = 6·5–7. G. = 3·3–3·5. Luster brilliant; pearly on cleavage-face, elsewhere vitreous. Color whitish, grayish white, greenish gray, hair-brown, yellowish, to colorless. At times pleochroic, X = dark violet or red-brown, Z = faint yellow. Transparent to subtranslucent. Optically +. Birefringence high. Ax. pl. ‖ $b(010)$. Bx. ⊥ $a(100)$. Dispersion $\rho < v$, feeble. 2V = 84°. α = 1·702. β = 1·722. γ = 1·750.

Comp. — AlO(OH) or $Al_2O_3.H_2O$ = Alumina 85·0, water 15·0 = 100.

Pyr., etc. — In the closed tube usually decrepitates strongly, separating into white pearly scales, and at a high temperature yields water. Infusible; ignited with cobalt solution gives a deep blue color. Not attacked by acids, but after ignition soluble in sulphuric acid.

Diff. — Distinguished by its hardness and pearly luster; also (B.B) by its decrepitation and yielding water; by the reaction for alumina with cobalt solution. Resembles some varieties of hornblende, but is harder.

Artif. — Diaspore crystals have been artificially formed by heating in a steel tube aluminum oxide in sodium hydroxide to temperatures less than 500°.

Obs. — Diaspore is commonly found associated with corundum or emery, probably as an alteration product of the oxide. It occurs similarly in bauxite deposits. Has been noted as an accessory mineral in metamorphic limestones. Noted as a microscopic constituent of certain igneous rocks, especially in southern Norway, apparently the result of the alteration of silicates like nephelite, sodalite, and natrolite.

Occurs near Kossoibrod, south of Ekaterinburg, Ural Mts., in a granular limestone with emery. Also with emery in the mountains of Güme Dagh (Messogis) east of Ephesus, Asia Minor, and on the Island of Naxos, Cyclades, Greece. At Schemnitz (Selmeczbánya), Czechoslovakia; in dolomite at Campolungo near St. Gotthard, Ticino, Switzerland. From Horrsjöberg near Ny, Vermland, Sweden (*empholite*). A manganiferous, rose to dark red variety from Postmasburg, Griqualand-West, South Africa, has been called *mangandiaspore*.

In the United States at the emery mines of Chester, Massachusetts, in large plates and crystals; with corundum and margarite at Newlin, near Unionville, Chester Co., Pennsylvania.

Named by Haüy from διασπείρειν, to scatter, alluding to the usual decrepitation before the blowpipe.

BOEHMITE. Probably $Al_2O_3.H_2O$. (Comp. deduced from similarity of X-ray pattern to that of lepidocrocite.) In microscopic orthorhombic plates. Prism angle 63°. Good basal cleavage. Ax. pl. ‖ (001). Z = b axis. Index slightly higher than for gibbsite. Found in the bauxites of Ariége and Var, France.

KAYSERITE. $Al_2O_3.H_2O$, same as diaspore. Monoclinic. Micaceous cleavage. H. = 5–6. Inclined extinction with angles as high as 46°. α = about 1·68. γ = about 1·74. Axial angle large. $\rho < v$. Of secondary origin, found on corundum crystals from Cerro Redondo, Minas, Uruguay. A hydrous oxide of aluminum with properties similar to those of kayserite, from the Ekaterinoslav mining district, Russia, has been called *tanatarite*.

GOETHITE.

763

Orthorhombic. Axes $a : b : c$ = 0·9185 : 1 : 0·6068.

mm''', 110 ∧ 1$\bar{1}$0 = 85° 8'. pp', 111 ∧ $\bar{1}$11 = 58° 55'.
ee', 011 ∧ 0$\bar{1}$1 = 62° 30'. pp''', 111 ∧ 1$\bar{1}$1 = 53° 42'.

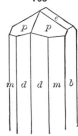

In prisms vertically striated, and often flattened into scales or tables ‖ $b(010)$. Also fibrous; foliated or in scales; massive, reniform and stalactitic, with concentric and radiated structure.

Cleavage: $b(010)$ very perfect. Fracture uneven. Brittle. H. = 5–5·5. G. = 4·28. Luster imperfect adamantine. Color yellowish, reddish, and blackish brown. Often blood-red by transmitted light. Streak brownish yellow to ocher-yellow. Optically −. Optical axes strongly dispersed. Bx.$_{ac}$ ⊥ (010) for all colors. Ax. pl. for red = (100);

for yellow = (001). $\alpha = 2\cdot26$. $\beta = 2\cdot39$. $\gamma = 2\cdot4$.   Only weakly pleochroic.

Var. — In thin scale-like or tabular crystals, usually attached by one edge. Also in acicular or capillary (not flexible) crystals, or slender prisms, often radiately grouped: the *Needle-Ironstone*. It passes into a variety with a velvety surface; the *pribramite* (*Sammetblende*) of Přibram, Bohemia, is of this kind. Also columnar, fibrous, etc., as above.

Comp. — FeO(OH) or $Fe_2O_3.H_2O$ = Oxygen 27·0, iron 62·9, water 10·1 = 100, or Iron sesquioxide 89·9, water 10·1 = 100.

Pyr., etc. — In the closed tube gives off water and is converted into red iron sesquioxide. With the fluxes like hematite; most varieties give a manganese reaction, and some, treated in the forceps in O.F., after moistening in sulphuric acid, impart a bluish green color to the flame (phosphoric acid). Soluble in hydrochloric acid.

Diff. — Distinguished from hematite by its yellow streak; from limonite by crystalline nature; it also contains less water than limonite.

Obs. — Goethite is most commonly found associated with limonite, more rarely with hematite. Also found with quartz, as an alteration product of a sulphide, especially pyrite; and as an inclusion, together with hematite and limonite in mica. In part, it is probable that it has been formed from limonite.

The variety *onegite* comes from Lake Onega, Olonetz, Russia. Occurs in a delicate fibrous variety with a velvety surface from Přibram, Bohemia, Czechoslovakia (*pribramite* or *sammetblende*). In lamelliform and foliated crystallizations of a hyacinth-red color with limonite from Eiserfeld near Siegen, Westphalia (*pyrrhosiderite*). From Altenberg, Saxony. In Cornwall from the Botallack mine at St. Just and at Lostwithiel in unusual crystals, also at Lanlivery. In the United States common in the Lake Superior hematite districts, notably in Michigan at the Jackson Iron mine, Negaunee, and elsewhere. The ocherous variety abundant in the Mesabi district, Minnesota, has been called *mesabite*. In Colorado at Florissant, Teller Co., in amethyst and at Crystal Peak, etc., in the Pike's Peak region, El Paso Co. Named *goethite* after the poet-philosopher Goethe (1749–1832).

A colloidal form of iron hydroxide having the composition of *goethite* and occurring as pseudomorphs after pyrite has been called *ehrenwerthite*.

Use. — An ore of iron.

Lepidocrocite. — A dimorphous form of goethite. Orthorhombic but with different axial ratio. Scaly. Cleavages, (010) perfect; (001) good; (100) fair. G. = 4.09. Optically +. $\beta = 2\cdot20$. Only weak dispersion (distinction from goethite). Pleochroic in thick sections. The occurrence of lepidocrocite is the same as goethite, in fact it is often difficult to distinguish between the two species. The following occurrences have been assigned to lepidocrocite. The *rubinglimmer* from Siegen, Westphalia; from the Eleonore mine on the Dünsberg near Giessen, Hessen-Nassau; the so-called *chileite*, an alteration product of pyrite, from near Coquimbo, Chile; from Easton, Pennsylvania.

HYDROGOETHITE. $3Fe_2O_3.4H_2O$.   Probably goethite with capillary water. Orthorhombic, radiating fibrous. H. = 4. G. = 3.7. Color and streak brick-red. With limonite at various localities in Tula, Russia.

MANGANITE.

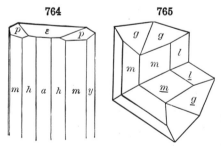

764   **765**

Orthorhombic.   Axes $a : b : c = 0\cdot8441 : 1 : 0\cdot5448$.

hh''', $410 \wedge 4\bar{1}0 = 23° 50'$.
mm''', $110 \wedge 1\bar{1}0 = 80° 20'$.
$\epsilon\epsilon'$, $205 \wedge \bar{2}05 = 28° 57'$.
ee', $011 \wedge 0\bar{1}1 = 57° 10'$.
pp', $111 \wedge \bar{1}11 = 59° 5\frac{1}{2}'$.

Crystals commonly prismatic, the faces deeply striated vertically; often grouped in bundles. Twins: tw. pl. $e(011)$. Also columnar; stalactitic.

Cleavage: $b(010)$ very perfect; $m(110)$ perfect. Fracture uneven. Brittle. H. = 4. G. = 4·2–4·4. Luster submetallic. Color dark steel-gray to iron-black. Streak reddish brown, sometimes nearly black. Opaque; in minute splinters sometimes brown by transmitted light.

Comp. — MnO(OH) or $Mn_2O_3.H_2O$ = Oxygen 27·3, manganese 62·4, water 10·3 = 100, or Manganese sesquioxide 89·7, water 10·3 = 100.

Pyr., etc. — In the closed tube yields water; manganese reactions with the fluxes, p. 371.

Obs. — Manganite is associated with the other manganese oxides and has a similar origin. It frequently alters to pyrolusite. Found often in veins associated with the acid igneous rocks, both as filling cavities and as a replacement of the neighboring rock. Barite and calcite are frequent associates.

Occurs in veins traversing a porphyry at Ilfeld, Harz Mts.; at Ilmenau in Thuringia; from Horhausen in Rhineland. Found at Bölet near Undenäs, Östergötland, Sweden. In Cornwall, at various places, occurring crystallized at the Botallack mine, St. Just. Also at Egremont, Cumberland; at Exeter, Devon; and from Granam near Towie, Aberdeen, Scotland.

In the Lake Superior iron district at the Jackson mine, Negaunee, Michigan, and elsewhere. In Nova Scotia at Bridgeville, Pictou Co., in Hants Co., etc.

Crystals having the composition of manganite but showing a sphenoidal habit have been found at Långbanshyttan, Vermland, Sweden, and have been called *sphenomanganite*.

Use. — An ore of manganese.

LIMONITE. Brown Hematite.

Not crystallized. A mineral colloid. Usually in stalactitic and botryoidal or mammillary forms, having a fibrous or subfibrous structure; also concretionary, massive; and occasionally earthy.

H. = 5–5·5. G. = 3·6–4·0. Luster silky, often submetallic; sometimes dull and earthy. Color of surface of fracture various shades of brown, commonly dark, and none bright; sometimes with a nearly black varnish-like exterior (var. *stilpnosiderite*); when earthy, brownish yellow, ocher-yellow. Streak yellowish brown. Opaque.

Var. — (1) *Compact.* Submetallic to silky in luster; often stalactitic, botryoidal, etc. (2) *Ocherous* or earthy, brownish yellow to ocher-yellow, often impure from the presence of clay, sand, etc. (3) *Bog ore.* The ore from marshy places, generally loose or porous in texture, often petrifying leaves, wood, nuts, etc. (4) *Brown clay-ironstone*, in compact masses, often in concretionary nodules.

Comp. — Approximately $2Fe_2O_3.3H_2O$ = Oxygen 25·7, iron 59·8, water 14·5 = 100, or Iron sesquioxide 85·5, water 14·5 = 100. The water content varies widely and it is probable that limonite is colloidal in character and may be considered as the amorphous form of goethite with adsorbed and capillary water. In the bog ores and ochers, sand, clay, phosphates, oxides of manganese, and humic or other acids of organic origin are very common impurities.

Pyr., etc. — Like goethite. Some varieties leave a siliceous skeleton in the salt of phosphorus bead, and a siliceous residue when dissolved in acids.

Diff. — Distinguished from hematite by its yellowish streak, inferior hardness, and its reaction for water. Does not decrepitate B.B., like turgite. Not crystallized like goethite and yields more water.

Obs. — Limonite is always of secondary origin, resulting from the alteration of other ores, or minerals, containing iron, through exposure to moisture, air, and carbonic or organic acids; it is derived largely from pyrite, magnetite, siderite, ferriferous dolomite, etc.; also from various silicates (as mica, pyroxene, hornblende, etc.), which contain iron in the ferrous state. It is widespread in its occurrence as it may be deposited at low pressures and temperatures from all solutions that contain iron. It may be formed in place by the direct alteration of some other iron mineral or it may be deposited from waters which have obtained their iron content from some distant source. Waters containing iron in solution when brought into marshy places deposit the metal usually in the form of limonite. The evaporation of the carbonic acid in the water is one cause for the separation of the iron oxide. This separation is also aided by the so-called "iron bacteria" which absorb the iron from the water and later deposit it as ferric hydroxide. Limonite consequently occupies, as a bog ore, marshy places, into which the iron has been carried by streams. It is also found in deposits associated with iron-bearing limestones where the original iron content of the rock has been largely dissolved and redeposited later in some favorable spot. Limonite forms the capping or *gossan*, iron hat, of many metallic veins. It is often associated with manganese ores.

Oölitic limonites or *minettes* occur in large quantities in Lorraine and Luxemburg and constitute the most important iron ore deposits of Europe. Limonite is also mined in Bavaria, the Harz Mts., and in Sweden. Some of the more important localities for the occurrence of fine specimens are as follows. From Rosenau (Rožňava), Czechoslovakia; Hüttenberg, Carinthia, Austria (pseudomorphous after siderite); in Saxony at Johann-georgenstadt and Freiberg; Siegen in Westphalia; Rossbach, Hessen-Nassau; from Elba, pseudomorphous after pyrite. In France from Framont, Alsace, and from Rancié near Vicdessos, Ariège. From Cornwall at Lanlivery and St. Just; in Gloucester at Frampton Cotterell near Bristol.

Abundant in the United States but in small deposits and of little economic value. Extensive beds existed at Salisbury and Kent, Litchfield Co., Connecticut, also in the neighboring towns of Dutchess Co., New York, and in a similar situation to the north in Berkshire Co., Massachusetts. Occurs in many localities in Pennsylvania, large pseudomorphs after pyrite crystals being found near Fruitville, Lancaster Co., and south of York, York Co. Many small residual deposits of limonite are to be found in Virginia, Tennessee, Georgia, and Alabama. Crystals pseudomorphic after pyrite are found at Pelican Point, Utah. From Londonderry, Colchester Co., Nova Scotia.

Named *Limonite* from λειμών, *meadow*.

Use. — An ore of iron; as a yellow pigment.

XANTHOSIDERITE. $Fe_2O_3.2H_2O$. Probably only a variety of limonite. In fine needles or fibers, stellate and concentric; also as an ocher. Color golden yellowish, brown to brownish red. Associated with manganese ores at Ilmenau, Thuringia, Germany, etc.

ESMERALDAITE. $Fe_2O_3.4H_2O$. Probably only a variety of limonite. In small pod-shaped masses enclosed in limonite. Conchoidal fracture. H. = 2·5. G. = 2·58. Color coal black. Yellow-brown streak. From Esmeralda Co., California.

Turgite. Hydrohematite. — Approximately $2Fe_2O_3.H_2O$. $Fe_2O_3 = 94·7$; $H_2O = 5·3$. Status as mineral species in doubt but probably to be considered as a solid solution of goethite with hematite together with enclosed and adsorbed water. Resembles limonite but has a red streak. H. = 6·5. G. = 4·14–4·6. Decrepitates B.B. Crimson in thin fibers. Strongly pleochroic. Parallel extinction. Turgite is rarer in its occurrence than limonite but is found associated with that mineral and occurring under the same conditions. It was originally described from the Ural Mts. at the Turnisk copper mine on the river Turya (the proper spelling of the mineral name should be *turite*) near Bogolovsk; also from near Bakal in the Zlatoust district. Hydrohematite was originally described from near Hof, Bavaria, and also at Siegen in Westphalia. Turgite is also found at Düsseldorf and at Horhausen, Rhineland.

In the United States it occurred abundantly in large botryoidal masses at the limonite ore bed of Salisbury, Connecticut, usually constituting the exterior layer of the limonite masses. Found in Pennsylvania at Chestnut Hill, Lancaster Co.

BAUXITE. Beauxite.

In round concretionary disseminated grains. Also massive, oölitic; and earthy, clay-like. G. = 2·55. Color whitish. grayish, to ocher-yellow, brown, and red. $n = 1·56$–$1·61$.

Var. — 1. In concretionary grains, or oölitic; *bauxite*. 2. Clay-like, *wocheinite*; the purer kind grayish, clay-like, containing very little iron oxide; also red from the iron oxide present.

Comp. — Essentially $Al_2O_3.2H_2O$ = Alumina 73·9, water 26·1 = 100; some analyses, however, give $Al_2O_3.H_2O$ like diaspore.

Bauxite is probably a mixture of varying character but containing large amounts of a colloidal form of $Al_2O_3.H_2O$. This substance has been called *sporogelite* or *diasporogelite*, *cliachite* and *alumogel*.

Iron sesquioxide is usually present, sometimes in large amount, in part replacing alumina, in part only an impurity. The name *hematogelite* has been suggested for this colloidal form of ferric oxide. Silica, phosphoric acid, carbonic acid, lime, magnesia are common impurities.

Obs. — Bauxite is of secondary origin and has commonly been formed under tropical climatic conditions by the prolonged weathering of aluminum-bearing rocks. In some localities it has also apparently been derived from the weathering of clay-bearing limestones. Its mode of occurrence and its physical characteristics indicate that it has originated as a colloidal precipitate, but the exact chemical reactions involved are not clear. In places the proximity of volcanic rocks suggests that the action of sulphuric acid derived from volcanic gases may have had an important part in its formation. Bauxite may occur in place as a direct derivative of the original rock and preserve some of the rock structure or it may have been transported and deposited in a sedimentary formation. In the tropics deposits consisting largely of aluminum and ferric hydroxides, with more or less free silica, are found in the residual soils. They are known as *laterites*. These vary widely in composition and purity but may at times become valuable as sources of aluminum or iron. It has been suggested that during the formation of laterites alkaline carbonates may have acted upon the aluminum silicates with the formation of an aluminum carbonate, which, because of its unstable character, has broken down and formed bauxite.

Bauxite occurs in quantity in the south of France over an area extending from Tarascon on the Rhône to Antibes near Nice, in the departments of Bouches-du-Rhône, Var, and Alpes Maritimes. An important district is about Baux (or Beaux) near St. Remy, Bouches-du-Rhône. Other deposits occur in Ariège and Hérault. The French bauxite varies in character, at times being massive, oölitic, or disseminated in grains. Found also in the Vogelsberg, Hessen-Nassau. From the Bihar Mts. in Rumania and from Croatia and Dalmatia in Yugoslavia, and in Istria, Italy. *Wocheinite* occurs in Carniola, Yugoslavia, between Feistritz and Lake Wochein.

In the United States bauxite deposits occur in a belt about 60 miles long between Jacksonville, Calhoun Co., Alabama, and Cartersville, Bartow Co., Georgia. Also from Pulaski and Saline counties, Arkansas.

Laterites are found in India, on Seycheele Island in the Indian Ocean, on Madagascar, in French Guiana, etc.

Use. — As an aluminum ore.

Brucite Group. R(OH)$_2$. Rhombohedral

BRUCITE.

Rhombohedral. Axis $c = 1.5208$; cr $0001 \wedge 10\bar{1}1 = 60° 20\frac{1}{2}'$, rr' $10\bar{1}1 \wedge \bar{1}101 = 97° 37\frac{1}{2}'$. X-ray study shows a structure in which the Mg and O atoms are arranged in horizontal planes in which they make hexagonal patterns. These horizontal planes are grouped into sets of three, the central plane containing the Mg atoms while the O atoms lie in the planes above and below. These groups of three structure planes are widely spaced from each other, thus accounting for the basal cleavage.

Crystals usually broad tabular. Also commonly foliated massive; fibrous, fibers separable and elastic.

H. = 2.5. G. = 2.38–2.4. Cleavage: $c(0001)$ eminent. Folia separable and flexible, nearly as in gypsum. Sectile. Luster ∥ c pearly, elsewhere waxy to vitreous. Color white, inclining to gray, blue, or green. Transparent to translucent. Optically +. Indices: $\omega = 1.5617$; $\epsilon = 1.5815$ (Li). Birefringence varies for different colors giving rise to an unusual succession of colors in the rings of the interference figure. Fibrous varieties show biaxial characters.

Comp. — Magnesium hydroxide, $Mg(OH)_2$ or $MgO.H_2O$ = Magnesia 69.0, water 31.0 = 100. Iron and manganese protoxide are sometimes present.

Var. — *Ordinary*, occurring in plates, white to pale greenish in color; strong pearly luster on the cleavage surface. *Nemalite* is a fibrous variety containing 4 to 5 per cent iron protoxide, with G. = 2·44. *Manganbrucite* contains manganese; occurs granular; color honey-yellow to brownish red. *Ferrobrucite* contains iron.

Pyr., etc. — In the closed tube gives off water, becoming opaque and friable, sometimes turning gray to brown; the manganesian variety becomes dark brown. B.B. infusible, glows with a bright light, and the ignited mineral reacts slightly alkaline to test-paper. Ignited with cobalt solution gives the pale pink color of magnesia. The pure mineral is soluble in acids without effervescence.

Diff. — Distinguished by its infusibility, softness, cleavage, and foliated structure. Is harder than talc and differs in its solubility in acids; the magnesia test and optical characters separate it from gypsum, which is also somewhat softer.

Obs. — Brucite is of secondary origin accompanying other magnesian minerals in serpentine, in metamorphic limestones, and with chlorite minerals. Found at Predazzo, Val di Flemm, Trentino, Italy, with calcite, hydromagnesite, and periclase (mixtures of which have been called *pencatite* and *predazzite*), and in Piedmont at the iron mine of Cogne, south of Aosta. Near Filipstadt in Vermland, Sweden, in rounded masses in limestone. Occurs in considerable veins traversing serpentine at Swenaness in Unst, Shetland Islands.

In the United States brucite occurs at the Tilly Foster iron mine, Brewster, New York, well crystallized. In Lancaster Co., Pennsylvania, at Wood's mine, Texas, in large plates or masses. From Crestmore, Riverside Co., California.

Nemalite, the fibrous variety, occurs at Hoboken, New Jersey. *Manganbrucite* occurs with hausmannite, etc., at Jacobsberg in the Nordmark district, Vermland, Sweden.

Named after the early American mineralogist, A. Bruce (1777–1818).

Pyrochroite. — Manganese hydroxide, $Mn(OH)_2$. Rhombohedral. Atomic structure similar to that of brucite. Usually foliated, like brucite. Perfect basal cleavage. Luster pearly. Color white, but growing dark on exposure. Optically −. $\omega = 1·723$. $\epsilon = 1·681$. H. = 2·5. G. = 3·26. B.B. infusible. Pyrochroite is found at various localities in Sweden, always associated with hausmannite; as in Vermland at Pajsberg near Persberg; at the Moss mine, Nordmark, and at Långbanshyttan (a variety from this locality containing iron has been called *iron-pyrochroite*). From near Prijedor, Bosnia, Yugoslavia. From Franklin, New Jersey.

Bäckströmite. Pseudopyrochroite. — Manganese hydroxide, $Mn(OH)_2$. Orthorhombic. Prismatic. Black color. From Långbanshyttan, Vermland, Sweden in intimate association with pyrochroite.

GIBBSITE. Hydrargillite.

Monoclinic. Axes $a : b : c = 1·7089 : 1 : 1·9184$; $\beta = 85° 29'$. Crystals tabular $\parallel c(001)$, hexagonal in aspect. Occasionally in spheroidal concretions. Also stalactitic, or small mammillary, incrusting, with smooth surface, and often a faint fibrous structure within.

Cleavage: $c(001)$ eminent. Tough. H. = 2·5–3·5. G. = 2·3–2·4. Color white, grayish, greenish, or reddish white. Luster of $c(001)$ pearly; of other faces vitreous; of surface of stalactites faint. Translucent; sometimes transparent in crystals. Optically +. 2V always small and may be 0°. Indices 1·554–1·589. Ax. pl. usually $\perp$ (010), at temperatures above 27° C. becomes $\parallel$ (010). 2V decreases with rise of temperature, being 0° at 27° C. A strong argillaceous odor when breathed on.

Comp. — Aluminum hydroxide, $Al(OH)_3$ or $Al_2O_3.3H_2O$ = Alumina 65·4, water 34·6 = 100.

Pyr., etc. — In the closed tube becomes white and opaque, and yields water. B.B. infusible, whitens, and does not impart a green color to the flame. Ignited with cobalt solution gives a deep blue color. Soluble in concentrated sulphuric acid.

Artif. — When solutions of sodium aluminate are slowly decomposed by carbon dioxide gibbsite is precipitated.

Obs. — Gibbsite is of secondary origin, resulting from the alteration of other aluminum-bearing minerals, especially the silicates. It occurs under the same conditions as bauxite and is present with that mineral in its characteristic deposits and in the laterites. At times it becomes the chief mineral of such occurrences.

The crystallized gibbsite (*hydrargillite*) occurred in the Schischimsk Mts. near Zlatoust in the Urals; also with natrolite on the small islands, Lille-Arö and Eikaholm, in the Langesundfiord, Southern Norway. Occurs in the laterites of India, Seychelle Island, and Madagascar. From Villa Rica near Ouro Preto and elsewhere in Minas Geraes, Brazil.

In the United States in stalactitic form at Richmond, Berkshire Co., Massachusetts, in a bed of limonite; at the Clove iron mine in Unionvale, Dutchess Co., New York; with corundum at Corundum Hill, near Unionville, Newlin township, Chester Co., Pennsylvania. Universally present in the bauxite deposits of Arkansas.

Named after Col. George Gibbs.

SHANYAVSKITE. Schanjawskite. $Al_2O_3.4H_2O$. Amorphous, transparent material found in dolomite, near Moscow, Russia.

Chalcophanite. Hydrofranklinite. — $(Mn,Zn)O.2MnO_2.2H_2O$. In druses of minute tabular rhombohedral crystals; sometimes octahedral in aspect. Also in foliated aggregates; stalactitic and plumose. Perfect basal cleavage. H. = 2·5. G. = 3·907. Luster metallic, brilliant. Color, bluish black to iron-black. Streak chocolate-brown, dull. Optically −. $n = 2·72$. Pleochroic, X = deep red, Z = nearly opaque. Occurs at Stirling Hill, near Ogdensburg, Sussex Co., New Jersey. From Leadville, Colorado.

HYDROHETÆROLITE. $2ZnO.2Mn_2O_3.1H_2O$. In radiating botryoidal masses. Black. An alteration product of hetærolite from Franklin, New Jersey, and Leadville, Colorado.

Quenselite. — $2PbO.Mn_2O_3.H_2O$. Monoclinic. Perfect basal cleavage. Color pitchblack. Streak dark brownish gray. H. − 2·5. G. = 6·812. With HCl evolves chlorine. Found at Långbanshyttan, Vermland, Sweden, in iron ore associated with calcite and barite.

PSILOMELANE.

Massive and botryoidal; reniform; stalactitic. H. = 5–7. G. = 3·3–4·7. Luster submetallic, dull. Streak brownish black, shining. Color iron-black, passing into dark steel-gray. Opaque.

Comp. — A manganese oxide containing varying amounts of barium, potassium and sodium oxides, and water. To be regarded as colloidal MnO_2 with various adsorbed impurities.

Pyr., etc. — In the closed tube most varieties yield water, and all lose oxygen on ignition; with the fluxes reacts for manganese. B.B. difficultly fusible. Soluble in hydrochloric acid, with evolution of chlorine.

Obs. — Psilomelane is a common mineral of manganese having a secondary origin with occurrence and associations like those of pyrolusite. It varies widely in composition and purity. It is probably always deposited in the colloidal state, which in some varieties has subsequently changed to a more or less crystalline condition. It is frequently intimately associated with pyrolusite, the two minerals often occurring in alternating layers. Often associated with limonite. A common constituent of laterite deposits. Barite and calcite are frequently associated minerals. Occurs as dendritic coatings upon other minerals.

Some of the more important localities for the occurrence of psilomelane follow. In Saxony at Schneeberg and Johanngeorgenstadt; in Thuringia at Ilmenau and Elgersburg near by; at Eiserfeld near Siegen, Westphalia (*calvonigrite*). The variety *romanèchite* in fibrous and botryoidal forms comes from Romanèche-Thorins, southwest of Macon, Saône-et-Loire, France. The variety *ranciéite* is found at Rancié near Vicdessos, Ariège, and in Pyrénées-Orientales from Fillols. Psilomelane is found in Cornwall at Lanlivery. It is found in the various manganese districts in the Central Provinces and elsewhere in India. Occurs at Miguel Burnier near Ouro Preto, Minas Geraes, Brazil. Occurs in the United States chiefly in small amounts associated with the Lake Superior hematite deposits, as at Negaunee, Michigan, etc.

Named from ψιλός, smooth or naked, and μέλας, black.

Use. — An ore of manganese.

The following mineral substances here included are mixtures of various oxides, chiefly of manganese (MnO_2, also MnO), cobalt, copper, with also iron, and from 10 to 20 per cent water. These are results of the decomposition of other ores — partly of oxides and sulphides, partly of manganesian carbonates, and can hardly be regarded as representing distinct mineral species.

WAD. In amorphous and reniform masses, either earthy or compact; also incrusting or as stains. Usually very soft, soiling the fingers; less often hard to H. = 6. G. = 3·0–

4·26; often loosely aggregated, and feeling very light to the hand. Color dull black, bluish or brownish black.

BOG MANGANESE consists mainly of oxide of manganese and water, with some oxide of iron, and often silica. alumina, baryta.

STAINIERITE. $(Co,Fe,Al)_2O_3.H_2O$. Orthorhombic? Compact. Mammillary. Color and streak, black. Birefringent. H. = 4·5. G. = 4·137. Possibly the crystalline equivalent of the colloidal heterogenite. X-ray study suggests goethite structure. From Mindingi, Katanga, Belgian Congo; also at Kadjilangue, Luamba, Congo.

HETEROGENITE. Approximately $CoO.2Co_2O_3.6H_2O$. Amorphous in globular or reniform masses. Found at Schneeberg, Saxony, and near Wittichen, Baden, as an alteration product of smaltite; a similar material, probably from northern Chile, with composition, $CuO.2CoO.Co_2O_3.4H_2O$, has been called *schulzenite*. Also another similar substance with composition, $CuO.CoO.3Co_2O_3.7H_2O$ has been found at Elizabethville, Katanga, Belgian Congo, filling cavities in malachite and chrysocolla. All three are probably colloidal without definite compositions.

ASBOLITE, or *Earthy Cobalt*, contains oxide of cobalt, which sometimes amounts to 32 per cent.

RABDIONITE. Contains copper, iron and manganese oxides.

LAMPADITE, or *Cupreous Manganese*, is a wad containing 4 to 18 per cent of oxide of copper, and often oxide of cobalt also.

LUBECKITE is a mixture of copper, cobalt and manganese oxides.

SKEMMATITE. $3MnO_2.2Fe_2O_3.6H_2O$. Color black. Streak dark brown. H. = 5·5–6. Fusible to magnetic globule. Alteration product of *pyroxmangite*. From Iva, Anderson Co., South Carolina.

BELDONGRITE. $6Mn_3O_5.Fe_2O_3.8H_2O$. Luster pitchy. Color black. From Beldóngri, District Nágpur, Central Provinces, India.

Sassolite. — Boric acid, $B(OH)_3$. Triclinic. In pseudo-hexagonal plates, tabular ‖ (001). Usually small, white, pearly scales. Perfect basal cleavage. Optically −. 2V very small. $\alpha = 1·340$, $\beta = 1·456$, $\gamma = 1·459$. Bx.$_{ac}$ nearly ⊥ (001). H. = 1. G. = 1·48. Very easily fusible. Found in the gases of certain fumaroles and in the waters of hot springs, accompanied by ammonium salts. The original source of the boric acid is under dispute. Occurs in Italy in the waters of the Tuscan lagoons, between Volterra and Massa Marittima. First found in the solid condition at Sasso on the Ombrone, northeast of Grosseto, Tuscany. Occurs also abundantly in the crater of Vulcano, Lipari Islands.

ALAÏTE. $V_2O_5.H_2O$. Rare. Found in dark bluish red moss-like masses in Alai Mts., Turkestan.

VANOXITE. $2V_2O_4.V_2O_5.8H_2O$. In minute crystals, with occasionally a rhombic outline. Compact. Color black. Opaque. Found as a replacement of wood and as a cement in sandstone in Paradox Valley, Montrose Co., Colorado.

Becquerelite. — $UO_3.2H_2O$. Orthorhombic. In minute crystals. Prism angles nearly 60°. Aragonite type of twinning frequent. Perfect cleavages ‖(001), (110). Color brownish yellow. Resinous luster. Optically −. X ⊥ (001). $\alpha = 1·75$, $\beta = 1·87$, $\gamma = 1·88$. 2V small. Heated to 100° C. becomes uniaxial. Pleochroic, deep yellow to nearly colorless. Radioactive. Found with curite, soddite, anglesite, and uraninite at Kasolo, Belgian Congo. *Schoepite* described from the same locality is closely similar if not identical with becquerelite. The chief differences are in the optical characters. $\alpha = 1·690$, $\beta = 1·714$, $\gamma = 1·735$. 2V large.

Ianthinite. — Perhaps $2UO_2.7H_2O$. Orthorhombic. Acicular crystals. Micaceous cleavage ‖ (100). Color, violet-black, altering on edges to yellow. Streak brown-violet. Optically −. Ax. pl. ‖(001). X ⊥ (100). $\alpha = 1·674$. $\beta = 1·90$. $\gamma = 1·92$. Strongly pleochroic, X = colorless, Y = violet, Z = dark violet. Found on uraninite with becquerelite and schoepite at Kasolo, Belgian Congo.

VI. OXYGEN-SALTS

The Sixth Class includes the salts of the various oxygen acids. These fall into the following seven sections: 1. Carbonates; 2. Silicates and Titanates; 3. Niobates and Tantalates; 4. Phosphates, Arsenates, etc.; also the

Nitrates; 5. Borates and Uranates; 6. Sulphates, Chromates and Tellurates; 7. Tungstates and Molybdates.

1. CARBONATES

A. Anhydrous Carbonates

The Anhydrous Carbonates include two distinct isomorphous groups, the CALCITE GROUP and the ARAGONITE GROUP. The metallic elements present in the former are calcium, magnesium, iron, manganese, zinc, and cobalt; in the latter, they are calcium, barium, strontium, and lead.

The species included are as follows:

1. *Calcite Group.* RCO₃. Rhombohedral

		rr'	c
Calcite	$CaCO_3$	74° 55′	0·8543
Dolomite	$CaCO_3.MgCO_3$ Tri-rhombohedral	73° 45′	0·8322
Ankerite	$CaCO_3.(Mg,Fe)CO_3$	73° 48′	0·8332
Magnesite	$MgCO_3$	72° 36′	0·8112
Siderite ·	$FeCO_3$	73° 0′	0·8184
Rhodochrosite	$MnCO_3$	73° 0′	0·8184
Smithsonite	$ZnCO_3$	72° 20′	0·8063
Sphærocobaltite	$CoCO_3$		

This list gives only the prominent species of this group; the names, etc. of the isomorphous intermediate compounds will be found under the description of the different members.

The CALCITE GROUP is characterized by rhombohedral crystallization. All the species show, when distinctly crystallized, perfect rhombohedral cleavage, the angle varying from 75° (and 105°) in calcite to 73° (and 107°) in siderite. This is exhibited in the table above. For the atomic structure of calcite as shown by X-ray study see p. 11. Similar studies for the other members of the group show the same type of structure for them all. Siderite and rhodochrosite have almost identical structures thus explaining the complete miscibility of these two molecules, while in the case of calcite and magnesite the dimensions of the unit cell differ enough to explain their very limited miscibility.

2. *Aragonite Group.* RCO₃. Orthorhombic

		mm'''	$a : b : c$
Aragonite	$CaCO_3$	63° 48′	0·0224 : 1 : 0·7206
Bromlite	$(Ca,Ba)CO_3$		
Witherite	$BaCO_3$	62° 12′	0·6032 : 1 : 0·7302
Strontianite	$SrCO_3$	62° 41′	0·6090 : 1 : 0·7239
Cerussite	$PbCO_3$	62° 46′	0·6100 : 1 : 0·7230

The species of the ARAGONITE GROUP crystallize in the orthorhombic system, but the relation to those of the Calcite Group is made more close by the fact that the prismatic angle varies only a few degrees from 60° (and 120°) and the twinned forms with the fundamental prism as twinning-plane are pseudo-hexagonal in character. X-ray study shows a complex structure which is orthorhombic but nearly hexagonal in character. The different molecules of the group replace each other to a certain degree but to a much less extent than in the Calcite Group.

1. *Calcite Group.* RCO₃. Rhombohedral

CALCITE. Calc Spar; Calcareous Spar.

Rhombohedral. Axis $c = 0.8543$.

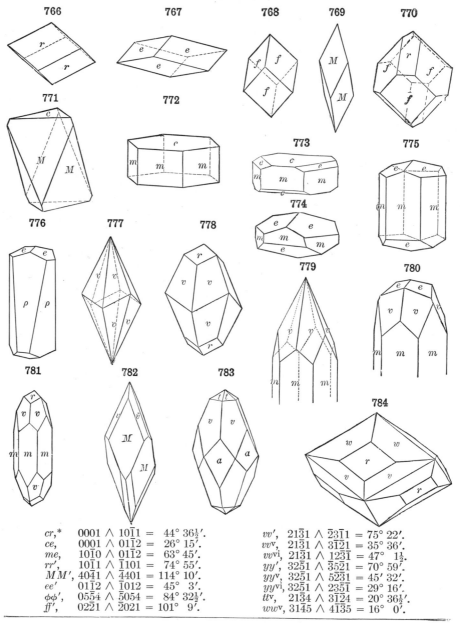

$$
\begin{array}{llll}
cr,^* & 0001 \wedge 10\bar{1}1 = & 44°\ 36\tfrac{1}{2}'. & \\
ce, & 0001 \wedge 01\bar{1}2 = & 26°\ 15'. & \\
me, & 10\bar{1}0 \wedge 01\bar{1}2 = & 63°\ 45'. & \\
rr', & 10\bar{1}1 \wedge \bar{1}101 = & 74°\ 55'. & \\
MM', & 40\bar{4}1 \wedge \bar{4}401 = & 114°\ 10'. & \\
ee', & 01\bar{1}2 \wedge \bar{1}012 = & 45°\ 3'. & \\
\phi\phi', & 05\bar{5}4 \wedge \bar{5}054 = & 84°\ 32\tfrac{1}{2}'. & \\
f\!f', & 02\bar{2}1 \wedge \bar{2}021 = & 101°\ 9'. & \\
\end{array}
$$

$$
\begin{array}{llll}
vv', & 21\bar{3}1 \wedge \bar{2}3\bar{1}1 = & 75°\ 22'. & \\
vv^{\text{v}}, & 21\bar{3}1 \wedge 3\bar{1}\bar{2}1 = & 35°\ 36'. & \\
vv^{\text{vi}}, & 21\bar{3}1 \wedge 1\bar{2}3\bar{1} = & 47°\ 1\tfrac{1}{2}. & \\
yy', & 32\bar{5}1 \wedge \bar{3}5\bar{2}1 = & 70°\ 59'. & \\
yy^{\text{v}}, & 32\bar{5}1 \wedge 5\bar{2}\bar{3}1 = & 45°\ 32'. & \\
yy^{\text{vi}}, & 32\bar{5}1 \wedge 23\bar{5}\bar{1} = & 29°\ 16'. & \\
tt^{\text{v}}, & 21\bar{3}4 \wedge 3\bar{1}\bar{2}4 = & 20°\ 36\tfrac{1}{2}'. & \\
ww^{\text{v}}, & 31\bar{4}5 \wedge 4\bar{1}\bar{3}5 = & 16°\ 0'. & \\
\end{array}
$$

* See the stereographic projection, Fig. 287, p. 125.

Habit of crystals very varied, as shown in the figures, from obtuse to acute rhombohedral; from thin tabular to long prismatic; and scalenohedral of many types, often highly modified. Calcite crystals exhibit a greater variety of forms and habits than any other mineral.

Twins (see Figs. 445–452, p. 189): (1) Tw. pl. $c(0001)$, common, the crystals having the same vertical axis. (2) Tw. pl. $e(01\bar{1}2)$, very common, the vertical axes inclined $127°\ 29\frac{1}{2}'$ and $52°\ 30\frac{1}{2}'$; often producing twinning lamellæ as in Iceland Spar, which are, in many cases, of secondary origin as in granular limestones (Fig. 785); this twinning may be produced artificially (see p. 210). (3) Tw. pl. $r(10\bar{1}1)$, not common; the vertical axes inclined $90°\ 46'$ and $89°\ 14'$. (4) Tw. pl. $f(02\bar{2}1)$, rare; the axes intersect at angles of $53°\ 46'$ and $126°\ 14'$.

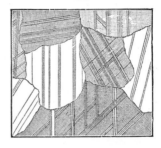

785

Section of crystalline limestone in polarized light.

Also fibrous, both coarse and fine; sometimes lamellar; often granular; from coarse to impalpable, and compact to earthy. Also stalactitic, tuberose, nodular, and other imitative forms.

Cleavage: $r(10\bar{1}1)$ highly perfect. Parting ‖ $e(01\bar{1}2)$ due to twinning. Fracture conchoidal, obtained with difficulty. H. = 3, but varying with the direction on the cleavage face; earthy kinds softer. G. = 2·710, in pure crystals, but varying somewhat widely in impure forms, as in those containing iron, manganese, etc. Luster vitreous to subvitreous to earthy. Color white or colorless; also various pale shades of gray, red, green, blue, violet, yellow; also brown and black when impure. Streak white or grayish. Transparent to opaque. Certain varieties show phosphorescence upon heating and others luminescence upon being exposed to sunlight or radium emanations.

Optically $-$. Birefringence very high. Refractive indices for the D line: $\omega = 1·65849$, $\epsilon = 1·48625$. Sometimes abnormally biaxial with $2V$ varying up to $30°$. Probably due to a molecular strain. The effect has been produced artificially.

Comp. — Calcium carbonate, $CaCO_3$ = Carbon dioxide 44·0, lime 56·0 = 100. Small quantities of magnesium, iron, manganese, zinc, and lead may be present replacing the calcium.

Calcite is the stable form of calcium carbonate at ordinary pressure and over a large temperature range. The occurrence of a form known as α-calcite or *clatolite* and supposed to be stable above 970° C. has been disproved. Aragonite and another form known as μ-CaCO$_3$ crystallize from solutions at temperatures below 90°. At higher temperatures they rapidly change to calcite. μ-CaCO$_3$ is very unstable; occurs in hexagonal plates; is optically $+$; $\omega = 1·550$, $\epsilon = 1.650$. *Vaterite* is a spherulitic variety of μ-CaCO$_3$.

Var. — The varieties are very numerous, and diverse in appearance. They depend mainly on the following points: differences in crystallization and structural condition, presence of impurities, etc., the extremes being perfect crystals and earthy massive forms; also on composition as affected by isomorphous replacement.

A. VARIETIES BASED CHIEFLY UPON CRYSTALLIZATION AND ACCIDENTAL IMPURITIES

1. *Ordinary.* — In crystals and cleavable masses, the crystals varying very widely in habit as already noted. *Dog-tooth Spar* is an acute scalenohedral form; *Nail-head Spar*, a composite variety having the form suggested by the name. The transparent variety from Iceland, used for polarizing prisms, etc., is called *Iceland Spar* or *Doubly-refracting Spar*. As regards *color*, crystallized calcite varies from the kinds which are perfectly

clear and colorless through yellow, pink, purple, blue, to brown and black.   The color is usually pale except as caused by impurities.   These impurities may be pyrite, native copper, malachite, sand, etc.; they are sometimes arranged in symmetrical form, as depending upon the growth of the crystals and hence produce many varieties.

Fontainebleau limestone, from Fontainebleau and Nemours, France, contains a large amount of sand, some 50 to 63 per cent.   Siliceous calcite crystals come from South Dakota, Wyoming, California, etc.

2. *Fibrous and lamellar kinds.* — *Satin Spar* is fine fibrous, with a silky luster; resembles fibrous gypsum, also called satin spar, but is much harder than gypsum and effervesces with acids.   *Lublinite* is a fibrous variety, originally supposed to be a new form of $CaCO_3$, but shown by X-ray analysis to be identical with calcite.

Argentine is a pearly lamellar calcite, the lamellæ more or less undulating; color white, grayish, yellowish.   *Aphrite*, in its harder and more sparry variety, is a foliated white pearly calcite, near argentine; in its softer kinds it approaches chalk, though lighter, pearly in luster, silvery white or yellowish in color, soft and greasy to the touch, and more or less scaly in structure.   Aphrite has been thought to be aragonite, pseudomorphous after gypsum.

3. *Granular massive to cryptocrystalline kinds*: Limestone, Marble, Chalk:

Granular limestone or *Saccharoidal limestone*, so named because like loaf sugar in fracture, varying from coarse to very fine granular, and hence to *compact* limestone; colors are various, as white, yellow, reddish, green; usually they are clouded and give a handsome effect when the material is polished.   When such limestones are fit for polishing, or for architectural or ornamental use, they are called *marbles*.   Many varieties have special names.   *Shell-marble* consists largely of fossil shells; *Lumachelle* or *fire-marble* is a dark brown shell-marble, with brilliant fire-like or chatoyant internal reflections.   *Ruin-marble* is a kind of a yellow to brown color, showing, when polished, figures bearing some resemblance to fortifications, temples, etc., in ruins, due to infiltration of iron oxide, etc.

Lithographic stone is a very even-grained compact limestone, of buff or drab color; as that of Solenhofen, Bavaria.   *Hydraulic limestone* is an impure limestone which after ignition sets; i.e., takes a solid form under water, due to the formation of a silicate.   The French varieties contain 2 or 3 per cent of magnesia, and 10 to 20 of silica and alumina (or clay).   The varieties in the United States contain 20 to 40 per cent of magnesia, and 12 to 30 per cent of silica and alumina.   *Hard compact limestone* varies from nearly pure white, through grayish, drab, buff, yellowish, and reddish shades, to bluish gray, dark brownish gray, and black, and sometimes variously veined.   Many kinds make beautiful marble when polished.   Red oxide of iron produces red of different shades.   Shades of green are due to iron protoxide, chromium oxide, iron silicate.

Chalk is white, grayish white, or yellowish, and soft enough to leave a trace on a board. It is composed of the shells of minute sea organisms.   *Calcareous marl* is a soft earthy deposit, with or without distinct fragments of shells; it contains much clay, and graduates into a calcareous clay.

Oölite is a granular limestone, its grains minute concretions, looking somewhat like the roe of fish, the name coming from ὠόν, egg.   *Pisolite* consists of concretions often as large as a small pea, or larger, having usually a distinct concentric structure.

Deposited from calcareous springs, streams, or in caverns, etc.   (a) *Stalactites* are calcareous cylinders or cones that hang from the roofs of limestone caverns, and which are formed from the waters that drip through the roof; these waters hold some calcium bicarbonate in solution, and leave calcium carbonate to form the stalactite when evaporation takes place.   Stalactites vary from transparent to nearly opaque; from a crystalline structure with single cleavage directions to coarse or fine granular cleavable and to radiating fibrous; from a white color and colorless to yellowish gray and brown.   (b) *Stalagmite* is the same material covering the floors of caverns, it being made from the waters that drop from the roofs, or from sources over the bottom or sides; cones of it sometimes rise from the floor to meet the stalactites above.   It consists of layers, irregularly curved, or bent. Stalagmite, or a solid kind of travertine (see below) when on a large scale, is the alabaster stone of ancient writers, that is, the stone of which ointment vases, of a certain form called *alabasters*, were made.   A locality near Thebes, now well known, was largely explored by the ancients, and the material has often been hence called *Egyptian alabaster*.   It was also formerly called *onyx* and *onychites* because of its beautiful banded structure.   In the arts it is often now called *Oriental alabaster* or *onyx marble*.   Very beautiful marble of this kind is obtained in Algeria.   *Mexican onyx* is a similar material obtained from Tecali, Puebla, Mexico; also in a beautiful brecciated form from the extinct crater of Zempoaltepec in southern Mexico.   Similar kinds occur in Missouri, Arizona, San Luis Obispo Co., California.   (c) *Calc-sinter, Travertine, Calc Tufa.*   Travertine is of essentially the same origin with stalagmite, but is distinctively a deposit from springs or rivers, especially where in large deposits, as along the river Anio, at Tivoli, near Rome, where the deposit is scores of feet in thickness. Similar material is being deposited at the Mammoth Hot Springs, Yellowstone Park. (d) *Agaric mineral.* Rock-milk is a very soft white material, breaking easily in the fingers, deposited sometimes in caverns, or about sources holding lime in solution. (e) *Rock-meal* is white and light, like cotton, becoming a powder on the slightest pressure.

B. Varieties based upon Composition

These include: *Dolomitic calcite.* Contains magnesium carbonate. Also *baricalcite* (which contains some $BaCO_3$); similarly, *strontianocalcite* ($SrCO_3$), *ferrocalcite* ($FeCO_3$), *manganocalcite* ($MnCO_3$) (see under *inesite*, p. 640), *zincocalcite* ($ZnCO_3$), *plumbocalcite* ($PbCO_3$), *cobaltocalcite* ($CoCO_3$). None of these various molecules appear in large amounts. The structure of calcite differs sufficiently from that of the other members of the group to prevent much isomorphous replacement.

Pyr., etc. — B.B. infusible, glows, and colors the flame reddish yellow; after ignition the assay reacts alkaline; moistened with hydrochloric acid imparts the characteristic lime color to the flame. In the solid mass effervesces when moistened with hydrochloric acid, and fragments dissolve with brisk effervescence even in cold acid. See further under *aragonite*, p. 521.

Diff. — Distinguishing characters: perfect rhombohedral cleavage; softness, can be scratched with a knife; effervescence in cold dilute acid; infusibility. Less hard and of lower specific gravity than aragonite (which see). Resembles in its different varieties the other rhombohedral carbonates, but is less hard, of lower specific gravity, and more readily attacked by acid. Also resembles some varieties of barite, but has lower specific gravity; it is less hard than feldspar and harder than gypsum.

Micro. — Recognized in thin sections by its low refraction and very high birefringence, the polarization colors in the thinnest sections attaining white of the highest order. The negative interference figure, with many closely crowded colored rings, is also characteristic. The rhombohedral cleavage is often shown in the fine fracture lines; systems of twinned lamellæ often conspicuous (Fig. 785), especially in crystalline limestone.

Artif. — Crystals of calcite are formed when a solution of calcium carbonate in dilute carbonic acid is evaporated slowly at ordinary temperatures. Calcite is formed when aragonite is heated, the transformation being complete at 470°.

Obs. — Calcite, in its various forms, is one of the most widely distributed of minerals. It is a minor constituent of secondary origin of the igneous rocks. It is the result of the action of carbonated waters upon calcium silicates. It is frequently observed in the amygdaloidal cavities of lavas, often in fine crystals. Calcite is a common and widespread constituent of the sedimentary rocks. It frequently occurs as the cementing material enclosing grains of other minerals. The percentages of calcite vary in such rocks until it may occur in great masses as the chief or sole constituent of the chalks, limestones, and the metamorphic marbles. Calcite is commonly deposited from lime-bearing waters under a wide range of conditions and with many varying associations. It is found in cavities and caves in limestones as crystalline encrustations and as stalactites and stalagmites; it is very common in many kinds of mineral veins, often in fine crystals. It is deposited in the form of calc-sinter or travertine, etc., especially in connection with hot springs, as at the Mammoth Hot Springs in the Yellowstone Park, Wyoming.

It is possible to give here only those localities which are especially noted for their fine crystals. These are to be found at Andreasberg in the Harz Mts.; in Saxony in the mines of Freiberg, Schneeberg, etc.; from Carinthia in Austria at Bleiberg and Hüttenberg; in Czechoslovakia at Schemnitz (Selmeczbánya) and at Příbram, Bohemia. Also from Maderaner Tal, Uri, Switzerland; from Poretta in Emilia, northern Italy; in France at Maronne between la Garde and Huez, Isère. In England famous localities for crystals are in Cumberland at Alston Moor and at Egremont (including nearby Pallaflat, etc.), at Frizington, etc.; at Matlock and Eyam in Derbyshire; at Weardale in Durham; at the Stank mine near Furness, Lancashire; in Cornwall at Liskeard and elsewhere. The Iceland spar has been obtained from Iceland near Helgustadir on the Eskefiord. It occurs in a large cavity in basalt. Twin crystals of great variety and beauty come from Guanajuato, Mexico.

In the United States crystallized calcite occurs in New York in St. Lawrence Co., at Rossie and vicinity, often in highly modified crystals, sometimes of great size; in Jefferson Co., near Oxbow, large, transparent crystals, sometimes with rose or purple color; in Niagara Co., near Lockport, scalenohedral crystals associated with dolomite, celestite, gypsum, and anhydrite. In New Jersey crystals of calcite are found associated with zeolites in the diabase rocks of West Paterson, Bergen Hill, and Great Notch.   In Ohio on Kelley's Island, Lake Erie, in pyramidal crystals.   In Michigan from the copper deposits splendid crystals of great variety and complexity of form, often containing scales of native copper.   In Illinois at Warsaw, in great variety of form, lining geodes and implanted on quartz crystals; also at Quincy and Galena.   In Missouri, with dolomite near St. Louis; also with sphalerite at Joplin, Granby, and other points in the zinc region in the southwestern part of the state, the crystals usually scalenohedral and of a wine-yellow color.   In South Dakota from the Bad Lands and the Black Hills.   Iceland spar occurs near Gray Cliff, Montana.   From Magdalena, New Mexico, and at the copper mines of Bisbee, Arizona.

Use. — In the manufacture of mortars and cements;   as a building and ornamental material;   as a flux in metallurgical operations;   Iceland spar is used to make polarizing prisms;   chalk as a fertilizer, in whitewash, etc.

THINOLITE.   A tufa deposit of calcium carbonate occurring on an enormous scale in northwestern Nevada;   also occurs about Mono Lake, California.   It forms layers of interlaced crystals of a pale yellow or light brown color and often skeleton structure except when covered by subsequent deposit of calcium carbonate.

DOLOMITE.   Pearl Spar pt.

Tri-rhombohedral.   Axis $c = 0.8322$.

cr, $0001 \wedge 10\bar{1}1 = 43° 52'$. MM', $40\bar{4}1 \wedge \bar{4}401 = 113° 53'$.
rr', $10\bar{1}1 \wedge \bar{1}101 = 73° 45'$.

Habit rhombohedral, usually $r(10\bar{1}1)$ or $M(40\bar{4}1)$;   the presence of rhombohedrons of the second or third series after the phenacite type very characteristic.   The r faces commonly curved or made up of sub-individuals, and thus

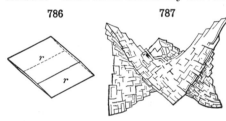

786   787

passing into saddle-shaped forms (Fig. 787).   Twinning on (0001) or (10$\bar{1}$1).   Sometimes shows polysynthetic twinning, tw. pl. (02$\bar{2}$1).   X-ray study shows a crystal symmetry lower than that of calcite.   A replacement of one-third of the Mg atoms by Fe does not change the character of the structure.

Also granular, coarse or fine, resembling ordinary marble.

Cleavage: $r(10\bar{1}1)$ perfect.   Fracture subconchoidal.   Brittle.   H. = 3·5–4.   G. = 2·8–2·9.   Luster vitreous, inclining to pearly in some varieties.   Color white, reddish, or greenish white;   also rose-red, green, brown, gray and black.   Transparent to translucent.   Optically −. $\omega = 1.68174$. $\epsilon = 1.50256$.   The indices and birefringence increase with increase of iron or manganese.

Comp. — Carbonate of calcium and magnesium, for normal dolomite $CaMg(CO_3)_2$ or $CaCO_3.MgCO_3$ = Carbon dioxide 47·9, lime 30·4, magnesia 21·7 = 100, or Calcium carbonate 54·35, magnesium carbonate 45·65 = 100. The carbonates of iron and manganese also frequently enter replacing the magnesium carbonate and grading to *ankerite*;   limited amounts of lime may replace the magnesia and vice versa;   rarely cobalt and zinc carbonates.

Pyr., etc. — B.B. acts like calcite.   In solution gives tests for magnesium and usually for iron.   Fragments thrown into cold acid, unlike calcite, are only very slowly acted upon, if at all, while in powder in warm acid the mineral is readily dissolved with effervescence. The ferriferous dolomites become brown on exposure.

Diff. — Resembles calcite (see p. 513), but generally to be distinguished in that it does not effervesce readily in the mass in cold acid.

Artif. — Artificial dolomite has been formed in several ways.   The results of many experiments would indicate that heat and pressure are favorable for its formation.   Sea water in contact with calcium carbonate when heated in a sealed tube produced dolomite. It has been observed that such reactions take place more readily with aragonite than with calcite, indicating the possibility of coral deposits (aragonite) being transformed into dolomite.

Micro. — Similar to calcite in thin sections except that it more often shows crystal outlines and less commonly polysynthetic twinning.

Obs. — Massive dolomite constitutes extensive strata in various regions, as in the dolomite district of the southern Tyrol. Crystalline and compact varieties are often associated with serpentine and other magnesian rocks, and with ordinary limestones. Dolomite rocks have been formed under different conditions and in different ways. It is generally agreed that dolomite, as a rock, is of secondary origin, having been transformed from ordinary limestone by the action of solutions containing magnesium. This change is commonly known as *dolomitization.* The waters containing magnesium are probably most commonly sea-waters, but also underground waters or thermal waters of deep-seated origin may have been active. The more favorable conditions for dolomitization would involve heat, pressure, high magnesium content of waters, and long periods of time. Consequently the older a limestone and the more deeply buried in the earth's crust the greater is the probability of its being converted into dolomite. Dolomite, frequently in crystals, is a common vein mineral, usually occurring with various metallic ores.

Only localities that are noted for their crystals of dolomite are given here. Such occurrences are at Schemnitz (Selmeczbánya) in Czechoslovakia; at Hall near Innsbruck in Tyrol. In Saxony at the mines of Freiberg and Schneeberg. In Switzerland crystals are found in Valais in the Binnental and the Simplon Tunnel, in Ticino at Campolungo near St. Gotthard and in Grisons at Tarasp (*taraspite*). In Italy from the Pfitschtal in Trentino (formerly Tyrol), from Traversella and Brosso in Piedmont, and from Miemo, Tuscany (*miemite*). In France near Vieillevigne, Loire-Inférieure. In Spain in unusual dark colored crystals from Teruel in the province of the same name. In England crystals are found at Frizington and elsewhere in Cumberland and at Laxey on the Isle of Man. Large clear crystals have been found near Djelfa in Algeria. At the Morro Velho gold mine on Rio das Velhas, northwest of Ouro Preto, Minas Geraes, Brazil. From Guanajuato, Mexico.

In the United States, in Vermont, at Roxbury, large, yellow, transparent rhombohedral crystals, in talc. In New York, at Lockport and Niagara Falls, with calcite, celestite and gypsum; at the Tilly Foster iron mine, Brewster, Putnam Co. In New Jersey at Hoboken, white hexagonal crystals and in rhombohedrons. From Phœnixville, Chester Co., Pennsylvania. In North Carolina at Hiddenite near Stony Point, Alexander Co., in fine rhombohedral crystals having nearly plane faces. At Warsaw, Hancock Co., Illinois; in Missouri near St. Louis and in saddle-shaped crystals in the zinc district of Joplin.

Named after Dolomieu (1750-1801), who announced some of the marked characteristics of the rock in 1791 — its not effervescing with acids, while burning like limestone, and solubility after heating in acids.

Use. — As a building and ornamental stone; for the manufacture of certain cements; for the production of magnesia used in the preparation of refractory linings in metallurgical furnaces.

Ankerite. — $CaCO_3,(Mg,Fe,Mn)CO_3$, or for normal ankerite $2CaCO_3.MgCO_3.FeCO_3$. In rhombohedral crystals; rr' $10\bar{1}1 \wedge \bar{1}101 = 73° 48'$; also crystalline massive, granular, compact. Color white, gray, reddish. Indices approximately $\omega = 1·727$, $\epsilon = 1·534$. $G. = 2·95-3·1$. $H. = 3·5$. Minerals that may be classed as ankerite are fairly common, having in general the same mode of occurrence and associations as dolomite; frequently associated with iron ores. Some of the better-known localities for ankerite follow. In the province of Styria in Austria associated with siderite, especially at Erzberg near Eisenerz. In Italy at Traversella in Piedmont. In Mexico at Guanajuato. In New York associated with hematite, at Antwerp, Jefferson Co. In Nova Scotia with the iron ores near Londonderry, Colchester Co.

MAGNESITE.

Rhombohedral. $c = 0·8112$. rr' $10\bar{1}1 \wedge \bar{1}101 = 72° 36'$. Crystals rare, usually rhombohedral, also prismatic. Commonly massive; granular cleavable to very compact; earthy.

Cleavage: $r(10\bar{1}1)$ perfect. Fracture flat conchoidal. Brittle. $H. = 3·5-4·5$. $G. = 3·0-3·12$, cryst. Luster vitreous; fibrous varieties sometimes silky. Color white, yellowish, or grayish white, brown. Transparent to opaque. Optically $-$. $\omega = 1\cdot717$. $\epsilon = 1\cdot515$. Indices and birefringence increase with amount of $FeCO_3$ present.

Comp. — Magnesium carbonate, $MgCO_3$ = Carbon dioxide $52\cdot4$, magnesia $47\cdot6 = 100$. Iron carbonate is often present.

Various compounds between magnesite and siderite have been described (see below), and it is thought that the two molecules form a continuous series and that these intermediate members do not exist as definite compounds. The following classification has been suggested.

Breunnerite with up to 30 per cent $FeCO_3$; white, yellowish, brownish, rarely black and bituminous; often becoming brown on exposure, and hence called *Brown Spar*.

Mesitite with between 30 and 50 per cent $FeCO_3$.

Pistomesite with between 50 and 70 per cent $FeCO_3$.

Sideroplesite with 70 to 95 per cent $FeCO_3$.

Pyr., etc. — B.B. resembles calcite and dolomite, and like the latter is but slightly acted upon by cold acids; in powder is readily dissolved with effervescence in warm hydrochloric acid. In solution gives strong test for magnesium with little or no calcium.

Obs. — Massive magnesite is commonly derived from the alteration of rocks rich in magnesium through the action of waters containing carbonic acid. Such magnesites are commonly colloidal in character and often contain opal silica. Crystallized magnesite is formed in various ways, as by the reaction of magnesium sulphate upon aragonite during the formation of saline deposits, as at Hall in the Tyrol. In other deposits it has apparently been formed through the metamorphism of original colloidal bodies and is to be found associated with talc, chlorite, serpentine, etc., as at Greiner in the Tyrol. It may also be formed through the gradual action of magnesium-bearing waters upon bodies of calcite, changing these first to dolomite and finally to pure magnesite. The crystalline magnesite deposits in the eastern Alps have been thought to have originated in this way. Large bodies of what is considered to be sedimentary magnesite have been found in the western United States. Magnesite is rarely to be found in the ore veins or as a pneumatolytic mineral.

Notable localities for the occurrence of magnesite are on the Island of Eubœa, Greece; at many localities in Styria, Austria, in the Mürztal near Kraubath, Bruck, Veitsch, Kathrein, and elsewhere; to the north at Mariazell and elsewhere; at Frankenstein in Silesia. From Snarum in Buskerud, Norway. Also found near Salem, Madras, India; near Platina (Fifield or Fiefield), New South Wales; in the Transvaal, South Africa, and in New Caledonia.

Commercial deposits of magnesite are found in California along the Coast Range from Mendocino Co. to south of Los Angeles, and on the western side of the Sierra Nevada from Placer Co. to Kern Co. Considerable bodies of sedimentary magnesite, in beds up to 300 feet in thickness, have been found in the Muddy Valley district, Clark Co., Nevada and at Bissell, Mohave Desert, California. In Canada magnesite occurs in Quebec at Grenville, Argenteuill Co., and at Sutton and Bolton in Brome Co. In small prismatic crystals from Orangedale, Nova Scotia.

Breunnerite is found at Greiner in the Zillertal, Tyrol, Austria, and from the Pfitschtal, Trentino, Italy (formerly the Tyrol). *Mesitite* is from Traversella, Piedmont, Italy, and with lazulite from Werfen in Salzburg, Austria. *Pistomesite* is from Thurnberg near Flauchau in Salzburg, Austria.

Use. — In the preparation of magnesite brick for the linings of metallurgical furnaces; in the manufacture of various chemical compounds, as epsom salts, magnesia, etc.

SIDERITE. Chalybite, Spathic Iron.

Rhombohedral. Axis $c = 0\cdot8184$.

cr,	$0001 \wedge 10\bar{1}1 = 43° 23'$.		rr',	$10\bar{1}1 \wedge \bar{1}101 = 73° 0'$.
cM,	$0001 \wedge 40\bar{4}1 = 75° 11'$.		MM',	$40\bar{4}1 \wedge \bar{4}401 = 113° 42'$.
cs,	$0001 \wedge 05\bar{5}1 = 78° 3'$.		ss',	$05\bar{5}1 \wedge \bar{5}051 = 115° 50'$.
cd,	$0001 \wedge 08\bar{8}1 = 82° 28'$.		dd',	$08\bar{8}1 \wedge \bar{8}081 = 118' 18\frac{1}{2}'$.

Crystals commonly rhombohedral $r(10\bar{1}1)$ or $e(01\bar{1}2)$, the faces often curved and built up of sub-individuals like dolomite. Often cleavable massive to coarse or fine granular. Also in botryoidal and globular forms, subfibrous within, occasionally silky fibrous; compact and earthy.

Cleavage: $r(10\bar{1}1)$ perfect. Twinning on $(01\bar{1}2)$, often in lamellæ. Fracture uneven or subconchoidal. Brittle. H. = 3·5–4. G. = 3·83–3·88. Luster vitreous, inclining to pearly. Color ash-gray, yellowish gray, greenish gray, also brown and brownish red, rarely green; and sometimes white. Streak white. Translucent to subtranslucent. Optically −. $\omega = 1·873$. $\epsilon = 1·633$. Indices and birefringence decrease as iron is replaced by magnesium or manganese.

Comp. — Iron protocarbonate, $FeCO_3$ = Carbon dioxide 37·9, iron protoxide 62·1 = 100 (Fe = 48·2 per cent). Manganese may be present (as in *oligonite, manganospherite*), also magnesium and calcium.

Apparently a complete isomorphous series may be formed between siderite and magnesite (see under magnesite, p. 517) and between siderite and rhodochrosite (see p. 520).

Pyr., etc. — In the closed tube decrepitates, gives off CO_2, blackens and becomes magnetic. B.B. blackens and fuses at 4·5–5. With the fluxes reacts for iron, and with soda and niter on platinum foil generally gives a manganese reaction. Only slowly acted upon by cold acid, but dissolves with brisk effervescence in hot hydrochloric acid. Exposure to the atmosphere darkens its color, rendering it often of a blackish brown or brownish red color.

Diff. — Characterized by rhombohedral form and cleavage. Specific gravity higher than that of calcite, dolomite and ankerite. Resembles some sphalerite but lacks the resinous luster, differs in cleavage angle and yields CO_2 (not H_2S) with hydrochloric acid.

Alt. — Readily alters on exposure to limonite, more rarely to hematite or magnetite.

Obs. — Siderite at times occurs in sedimentary deposits. The so-called *clay ironstones* and *black band ores* have this origin. The clay ironstone deposits are usually concretionary in structure, fine-grained or compact, and occur in clay or shale. The black band ores form dark colored beds in shale, closely associated with coal seams. Sedimentary siderites often have an origin similar to that of the limonite bog iron ores. They are precipitated by the action, in the absence of available oxygen, of organic matter upon a bicarbonate solution of ferrous iron. Siderite is also formed, at times in large amounts, by the replacement action of ferrous solutions upon limestones. It frequently occurs as a vein mineral, sometimes as the predominant mineral in veins of considerable size, but usually in subordinate amounts as a gangue mineral in metallic veins, especially with the lead-silver deposits.

Siderite is a common and widespread mineral, and only those deposits that are of commercial importance or of interest because of the quality of their specimens can be mentioned here. Important replacement deposits are found in the eastern Alps in Styria and Carinthia, Austria, as at Erzberg near Eisenerz in Styria and at Hüttenberg in Carinthia. Well-crystallized material occurs at Příbram in Bohemia, Czechoslovakia and at Felsöbánya (Baia Sprie), Rumania; also in the silver veins of Freiberg, Saxony; in the Harz Mts. at Neudorf near Harzgerode and at Stolberg; at Müsen near Siegen, Westphalia; at Lobenstein in Thuringia. Fine crystals occur also in Val Tavetsch, Grisons, Switzerland, and at Brosso and Traversella in Piedmont, Italy. In France large rhombohedrons are found at Allevard, Isère. In England exceptional specimens are found at Tavistock in Devonshire and in many of the mines of Cornwall, particularly those of Bodmin (prismatic), Camborne, Lanlivery, Redruth, and St. Austell (scalenohedral). Black band ores were formerly of considerable importance both in Germany and in England and Scotland. Siderite in fine cleavage masses occurs with cryolite at Ivigtut, Greenland. The crystallized mineral is found at Broken Hill, New South Wales; at Morro Velho gold mine on Rio das Velhas, northwest of Ouro Preto, Minas Geraes, Brazil; at Chorolque, Potosi, Bolivia.

In the United States it was found at Roxbury, Connecticut, in an extensive vein with sphalerite and pyrite; in Pennsylvania at the Gap mine, Lancaster Co. *Spherosiderite* occurs on trap rock at Weehawken, New Jersey, and Spokane, Washington. Clay ironstone is abundant in the coal regions of Pennsylvania, Ohio, etc.

Use. — An ore of iron.

RHODOCHROSITE. Dialogite.

Rhombohedral. Axis $c = 0·8184$, rr' $10\bar{1}1 \wedge \bar{1}101 = 73° 0'$. Distinct crystals not common; usually the rhombohedron $r(10\bar{1}1)$; also $e(01\bar{1}2)$, with rounded striated faces. Cleavable, massive to granular-massive and compact. Also globular and botryoidal, with columnar structure, sometimes indistinct; incrusting.

Cleavage: $r(10\bar{1}1)$ perfect. Fracture uneven. Brittle. H. = 3·5–4·5. G. = 3·45–3·60 and higher. Luster vitreous, inclining to pearly. Color shades of rose-red; yellowish gray, fawn-colored, dark red, brown. Streak white. Translucent to subtranslucent. Optically −. $\omega = 1\cdot820$. $\epsilon = 1\cdot600$. Indices and birefringence increase with increase in percentages of iron.

Comp. — Manganese protocarbonate, $MnCO_3$ = Carbon dioxide 38·3, manganese protoxide 61·7 = 100. Iron carbonate is usually present even up to 40 per cent, as in *manganosiderite*; the molecules of rhodochrosite and siderite are apparently completely miscible (see p. 519) and they may crystallize together in all proportions; sometimes the carbonate of calcium, as in *manganocalcite*, also magnesium, zinc, and rarely cobalt.

Pyr., etc. — B.B. changes to gray, brown, and black, and decrepitates strongly, but is infusible. With salt of phosphorus and borax in O.F. gives an amethystine-colored bead, in R.F. becomes colorless. With soda in O.F. a bluish green manganate. Dissolves with effervescence in warm hydrochloric acid. On exposure to the air changes to brown, and some bright rose-red varieties became paler.

Diff. — Characterized by its pink color, rhombohedral form and cleavage, effervescence in acids.

Obs. — Rhodochrosite is similar to siderite in its origin and mode of occurrence. It has been noted in deposits that are sedimentary in origin and in metamorphosed bodies that have been derived from sedimentary beds. It very commonly occurs as a gangue mineral of primary origin in vein deposits, especially associated with silver, lead, and copper ores and with other manganese minerals.

Some of the more noteworthy occurrences of rhodochrosite are given below. In Rumania at Nagy-Ág and Kapnikbánya and from various localities the ferriferous variety *ponite*; in Saxony at Freiberg; in Hessen-Nassau at Hainbach (botryoidal) near Diez and to the southeast at Oberneisen; at Siegen in Westphalia. In Cornwall at St. Just. A variety containing 45 per cent of zinc carbonate from Rosseto, Elba, has been called *zincorodochrosite*.

In the United States rhodochrosite occurs at Branchville, Connecticut; in New Jersey at Franklin; at Negaunee, Michigan; at Butte, Montana. In Colorado in Gilpin Co.; at Alma, Park Co.; in Lake Co., at the John Reed mine, Alicante, in beautiful clear rhombohedrons; in Chaffee, Saguache, and Ouray counties. Abundant in the silver mines at Austin, Lander Co., Nevada.

Named *rhodochrosite* from ρόδον, a *rose*, and χρώσις, *color*; and *dialogite*, from διαλογή, *doubt*.

Use. — A minor ore of manganese.

SMITHSONITE. Calamine pt. Dry-bone ore *Miners.*

Rhombohedral. Axis $c = 0\cdot8063$. rr' $10\bar{1}1 \wedge \bar{1}101 = 72° 20'$. Rarely well crystallized; faces $r(10\bar{1}1)$ generally curved and rough. Usually reniform, botryoidal, or stalactitic, and in crystalline incrustations; also granular, and sometimes impalpable, occasionally earthy and friable.

Cleavable: $r(10\bar{1}1)$ perfect. Fracture uneven to imperfectly conchoidal. Brittle. H. = 5·5. G. = 4·30–4·45. Luster vitreous, inclining to pearly. Streak white. Color white, often grayish, greenish, brownish white, sometimes green, blue and brown. Subtransparent to translucent. Optically −. $\omega = 1\cdot849$, $\epsilon = 1\cdot621$ (on a variety with 97.34 per cent $ZnCO_3$ from Broken Hill, Rhodesia).

Comp. — Zinc carbonate, $ZnCO_3$ = Carbon dioxide 35·2, zinc protoxide 64·8 = 100. Iron carbonate is often present (as in *monheimite*); also manganese and cobalt carbonates; further calcium and magnesium carbonates in traces; rarely cadmium and indium.

Pyr., etc. — In the closed tube loses carbon dioxide, and, if pure, is yellow while hot and white on cooling. B.B. infusible, giving characteristic zinc flame; moistened with cobalt solution and heated in O.F. gives a green color on cooling. With soda on charcoal coats the coal with the oxide, which is yellow while hot and white on cooling; this coating, moistened with cobalt solution, gives a green color after heating in O.F. Soluble in hydrochloric acid with effervescence.

Diff. — Distinguished from calamine, which it often closely resembles, by its effervescence in acids.

Obs. — Smithsonite is found both in veins and beds, especially in company with galena and sphalerite; also with copper and iron ores. It usually occurs in calcareous rocks and is commonly associated with calamine (the mixture of the two minerals is sometimes known as *galmei*), and sometimes with limonite. It frequently replaces limestone, pseudomorphs after calcite crystals being often observed. It is probably always of secondary origin and is commonly formed by the action of carbonated waters upon zinc sulphide. It is deposited under surface conditions of temperature and pressure. Often occurs as a porous honeycomb-like material, known commonly as *dry-bone ore*.

Notable occurrences of smithsonite are found at Nerchinsk, Transbaikalia, Siberia; at Laurium in Greece in great variety including a fine blue-green color; in Rumania at Dognacska and Rézbánya; at Bleiberg in Carinthia, Austria; in the district about Aix-la-Chapelle (Aachen), Rhineland, including Altenberg to the southwest and Moresnet in Belgium; at Chessy near Lyon, Rhône, France; on the island of Sardinia in the district to the northwest of Iglesias at Monteponi, Mte. Agruxiau, Mte. Malfidano (blue), Masua (in yellow stalactites with concentric banded structure), etc. In England smithsonite is found at Alston Moor in Cumberland and at Matlock in Derbyshire. Fine specimens come from Broken Hill, New South Wales, and from Tsumeb near Otavi, South West Africa, in clear green crystals. A pink cobaltiferous variety occurs at Boléo, Lower California, Mexico.

Among the occurrences of smithsonite in the United States may be listed; in Pennsylvania at Friedensville, Lehigh Co., and in Lancaster Co.; in Wisconsin at Mineral Point, Iowa Co., and at Shullsburg, Lafayette Co., constituting pseudomorphs after sphalerite and calcite. In Missouri at Granby, Newton Co., and elsewhere; in Marion Co., Arkansas, sometimes colored bright orange-yellow by greenockite and then locally known as *turkey-fat ore*; in New Mexico at Kelly near Magdalena, Socorro Co., in translucent green, botryoidal masses; in the Tintic district and elsewhere, Utah.

Named after James Smithson (1754–1829), who founded the Smithsonian Institution in Washington. The name calamine is frequently used in England, cf. calamine, p. 632.

Use. — An ore of zinc.

Sphærocobaltite. — Cobalt protocarbonate, $CoCO_3$. Rhombohedral. Perfect rhombohedral cleavage. In small spherical masses, with crystalline surface, rarely in crystals. H. = 3–4. G. = 4·1. Color rose-red. Optically —. $\omega = 1·855$, $\epsilon = 1·600$. From Schneeberg, Saxony. At Libiola near Casarza, in Liguria, Italy. From Boléo, Lower California, Mexico.

Codazzite. $(Ca,Mg,Fe,Ce)CO_3$. Rhombohedral. $r \wedge r' = 74°$. Color ashy brown. G. = 2·5. H. = 4. Abundant in the emerald mines of Colombia.

2. *Aragonite Group.* RCO_3. Orthorhombic

For list of species, see p. 511.

ARAGONITE.

Orthorhombic. Axes $a : b : c = 0.62244 . 1 . 0.72056$.

mm''', $110 \wedge 1\bar{1}0 = 63° 48'$. pp', $111 \wedge \bar{1}11 = 86° 24\frac{1}{2}'$.
kk', $011 \wedge 0\bar{1}1 = 71° 33'$. pp''', $111 \wedge 1\bar{1}1 = 50° 27'$.

Crystals often acicular, and characterized by the presence of acute domes or pyramids. Twins: tw. pl. $m(110)$ commonly repeated, producing pseudo-hexagonal forms (see Figs. 790–792). The twinned nature of these crystals is to be seen in the reentrant angles in the prism zone and by the striations on the composite basal plane (see Fig. 792) and by the varying optical orientation of the different sectors as seen in thin section under the microscope. Also globular, reniform, and coralloidal shapes; sometimes columnar, straight or divergent; also stalactitic; incrusting.

Cleavage: $b(010)$ distinct; also $m(110)$. Fracture subconchoidal. Brittle. H. = 3·5–4. G. = 2·93–2·95. Luster vitreous, inclining to resinous on surfaces of fracture. Color white; also gray, yellow, green, and violet; streak uncolored. Transparent to translucent. Optically $-$. Ax. pl. $\parallel$ $a(100)$. Bx. $\perp c(001)$. Dispersion $\rho < v$ small. $2V = 18°$. $\alpha = 1\cdot530$. $\beta = 1\cdot680$. $\gamma = 1\cdot685$.

Comp. — Calcium carbonate, $CaCO_3$ = Carbon dioxide 44·0, lime 56·0 = 100. Some varieties contain a little strontium, others lead, and rarely zinc.

Aragonite is under ordinary conditions of temperature and pressure relatively unstable, changing to calcite, although usually the rate of change is very slow. Crystals of aragonite have been observed which have completely changed to calcite; example of paramorphism (see p. 358).

Var. — *Ordinary.* (a) Crystallized in simple or compound crystals, the latter much the most common; often in radiating groups of acicular crystals. (b) Columnar; also fine fibrous with silky luster. (c) Massive.

Stalactitic or *stalagmitic*: Either compact or fibrous in structure, as with calcite; *Sprudelstein* is stalactitic from Karlsbad, Bohemia. *Coralloidal*: In groupings of delicate interlacing and coalescing stems, of a snow-white color, and looking a little like coral; often called *flos-ferri*; *Pisolitic*: Spherical concretions. *Tarnowitzite* is a kind containing lead carbonate (to 18 per cent), from Tarnowitz in Silesia; with G. = 2·99; also from Postenje, Serbia. *Zeyringite* is a calcareous sinter, probably aragonite, colored greenish white or sky-blue with nickel, from Zeyring, Styria. *Nicholsonite* is aragonite containing zinc from Leadville, Colorado, and the Tintic District, Utah. The substances called *conchite* and *ktypeite* are apparently porous forms of aragonite.

Pyr., etc. — B.B. whitens and falls to pieces, and sometimes, when containing strontia, imparts a more intensely red color to the flame than lime; otherwise reacts like calcite. When immersed in cobalt nitrate solution powder turns lilac and the color persists on boiling while calcite under like conditions remains uncolored or becomes blue on long boiling (Meigen's reaction). This test is not applicable when the minerals are already colored or when the two are intimately mixed. The distinction can, however, be emphasized by further treatment with ammonium sulphide, when the basic cobalt carbonate deposited on the grains of aragonite is converted into a black sulphide, calcite under the same conditions becoming only slightly gray.

Diff. — Distinguished from calcite by higher specific gravity and absence of rhombohedral cleavage; from the zeolites (*e.g.*, natrolite), etc., by effervescence in acid. Strontianite and witherite are fusible, higher in specific gravity and yield distinctive flames B.B. The resinous luster on fracture surfaces is to be noted.

Artif. — Aragonite will form when solutions of calcium carbonate are evaporated at temperatures from 80° to 100°; it will form at lower temperatures if the solution contains some sulphate or small amounts of the carbonates of strontium or lead.

Obs. — Aragonite is much less common in its occurrence than calcite. It is formed under a much narrower range of conditions than calcite and being the less stable form often changes to calcite with a change in surrounding conditions. It is formed through organic agencies, as a deposit from hot springs, and as a precipitate from saline solutions that contain a sulphate. The most common repositories for aragonite are beds of gypsum; also beds of iron ore, especially with siderite, as in the Styrian mines, where it occurs in coralloidal forms, and is called flos-ferri, " flowers of iron." It is found in cavities of basalts and other lavas; often associated with copper and iron sulphides, galena, and malachite. It constitutes the pearly layer of many shells and the skeleton material of corals.

Some of the more important localities for fine specimens of aragonite follow. In Czechoslovakia at Herrengrund (Urvölgy) north of Neusohl (Ban Bystrica) and Schemnitz (Selmeczbánya) in Slovakia and at Horschenz near Bilin in Bohemia in fine prismatic crystals. In Austria it occurs at Erzberg near Eisenerz, Styria, in fine examples of the flos-ferri variety; at Leogang in Salzburg and at Schwaz in Tyrol. In Italy on Mte. Somma, Vesuvius, and in fine twin crystals with the sulphur deposits in Sicily, especially at Girgenti and Cianciana. In France fine crystals are found at Framont in Alsace, at Vertaizon in Puy-de-Dôme, and in twins near Dax and Bastennes in Landes. The mineral was first described from Aragon, Spain, the most notable occurrence being in pseudo-hexagonal twin crystals at Molina, Province of Guadalajara. In England groups of acicular crystals are found in Cumberland at Alston Moor, Cleator Moor, and Frizington. At Corocoro, Bolivia, it occurs in twin crystals which are frequently replaced by native copper.

In the United States in Pennsylvania at Wood's mine, Texas, in Lancaster Co.; at Dubuque, Iowa, and at Mine-la-Motte, Missouri; in the Black Hills, South Dakota; as flos-ferri in the Organ Mts., New Mexico; in Arizona at Bisbee and near Tucson; in large six-sided twins at Fort Collins, Colorado.

Bromlite. Alstonite. — $(Ba,Ca)CO_3$, nearly $CaCO_3.BaCO_3$. May contain small amounts of $SrCO_3$. Orthorhombic. In pseudohexagonal pyramids (Figs. 640, 641, p. 328). Indices, 1·525–1·670. 2V small. Bromlite is a rare mineral, found at Bromley Hill, near Alston, Cumberland, associated with calcite and witherite; also in the galena veins at Fallowfield, near Hexham, Northumberland.

WITHERITE.

Orthorhombic. Axes $a : b : c = 0.6032 : 1 : 0.7302$. Crystals always repeated twins, simulating hexagonal pyramids, tw. pl. (110). Also massive, columnar or granular.

Cleavage: $b(010)$ distinct; $m(110)$ imperfect. Fracture uneven. Brittle. H. = 3–3·75. G. = 4·27–4·35. Luster vitreous, inclining to resinous on surfaces of fracture. Color white, yellowish, grayish. Streak white. Subtransparent to translucent. Optically −. Ax. pl. ∥ (010). Bx. ⊥ (001). 2V = 16°. $\alpha = 1.529$. $\beta = 1.676$. $\gamma = 1.677$.

Comp. — Barium carbonate, $BaCO_3$ = Carbon dioxide 22·3, baryta 77·7 = 100.

Inverts at 811° C. to a hexagonal form and at 982° C. to an isometric form.

Pyr., etc. — B.B. fuses at 2 to a bead, coloring the flame yellowish green; after fusion reacts alkaline. B.B. on charcoal with soda fuses easily, and is absorbed by the coal. Soluble in dilute hydrochloric acid; this solution, even when very much diluted, gives with sulphuric acid a white precipitate which is insoluble in acids.

Diff. — Distinguished by its high specific gravity; effervescence in acid; green coloration of the flame B.B. Barite is insoluble in hydrochloric acid.

Obs. — Witherite is of infrequent occurrence. It is most commonly found in veins associated with galena. It may have been formed by the direct crystallization from waters carrying barium carbonate or by the action of carbonated waters on other barium minerals.

In England important occurrences are at Alston Moor in Cumberland associated with galena and in Northumberland at Fallowfield near Hexham. At the latter locality it occurs in large quantities and often in splendid crystals. In Japan it occurs as radial aggregates at Tsubaki, Ugo. In the United States it has been found associated with barite near Lexington, Kentucky. Also reported from a silver-bearing vein in Gillies township, Thunder Bay district, Ontario, Canada.

Use. — A minor source of barium compounds.

STRONTIANITE.

Orthorhombic. Axes $a : b : c = 0.6090 : 1 : 0.7239$.

Crystals often acicular or acute spear-shaped, like aragonite. Twins: tw. pl. $m(110)$ common, giving pseudo-hexagonal types. Also columnar, fibrous and granular.

Cleavage: $m(110)$ nearly perfect; $b(010)$ in traces. Fracture uneven. Brittle. H. = 3·5–4. G. = 3·680–3·714. Luster vitreous; inclining to resinous on faces of fracture. Color pale asparagus-green, apple-green; also white, gray, yellow, and yellowish brown. Streak white. Transparent to translucent. Optically $-$. Ax. pl. $\parallel b(010)$. Bx. $\perp c(001)$. Dispersion $\rho < v$ small. 2V = 10° 36′. $\alpha = 1·516$, $\beta = 1·664$, $\gamma = 1·666$.

Comp. — Strontium carbonate, $SrCO_3$ = Carbon dioxide 29·9, strontia 70·1 = 100. A little calcium is sometimes present.

Inverts to a hexagonal form at 929° C.

Pyr., etc. — B.B. swells up, throws out minute sprouts, fuses only on the thin edges, and colors the flame strontia-red; the assay reacts alkaline after ignition. Moistened with hydrochloric acid and treated either B.B. or in the naked lamp gives an intense red color. Soluble in hydrochloric acid; the mediumly dilute solution when treated with sulphuric acid gives a white precipitate.

Diff. — Differs from related minerals, not carbonates, in effervescing with acids; has a higher specific gravity than aragonite and lower than witherite; colors the flame *red* B.B.

Obs. — Strontianite commonly occurs in veins in limestones or marls and also less frequently in eruptive rocks. It is found at times in small crystals in various metallic veins. It has probably been formed either by the direct crystallization from waters containing strontium carbonate or by the action of alkaline carbonated solutions upon celestite. It is commonly associated with celestite, calcite, etc.

Strontianite occurs in Austria at Leogang in Salzburg and at Brixlegg in Tyrol (*calciostrontianite*). In Germany it is found in the metallic veins of Klaustal in the Harz Mts., and at Bräunsdorf near Freiberg, Saxony. It occurs in large and commercially important amounts in the marls of Westphalia, both massive and in fine crystals. The important localities are at Hamm and at Drensteinfurt, a short distance to the north. It occurs at Strontian, in Argyll, Scotland.

In the United States it occurs at Schoharie, New York, in granular and columnar masses, and also in crystals, forming nests and geodes, often large, in the limestone, associated with barite, pyrite, and calcite.

Use. — A minor source of strontium compounds.

CERUSSITE. White Lead Ore.

Orthorhombic. Axes $a : b : c = 0·60997 : 1 : 0·72300$.

$$
\begin{aligned}
mm''', & \quad 110 \wedge 1\bar{1}0 = 62° 46'. \\
kk', & \quad 011 \wedge 0\bar{1}1 = 71° 44'. \\
ii', & \quad 021 \wedge 0\bar{2}1 = 110° 40'. \\
cp, & \quad 001 \wedge 111 = 54° 14'. \\
pp', & \quad 111 \wedge \bar{1}11 = 87° 42'. \\
pp''', & \quad 111 \wedge 1\bar{1}1 = 49° 59\tfrac{1}{2}'.
\end{aligned}
$$

Simple crystals often tabular $\parallel b(010)$, prismatic $\parallel c$ axis; also pyramidal. Twins: tw. pls. $m(110)$ and (130) very common, contact- and penetration-

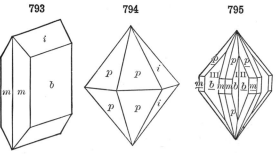

793 794 795

twins, often repeated yielding six-rayed stellate groups. Crystals grouped in clusters, and aggregates. Rarely fibrous, often granular massive and compact; earthy. Sometimes stalactitic.

Cleavage: $m(110)$ and $i(021)$ distinct. Fracture conchoidal. Very brittle. H. = 3–3·5. G. = 6·46–6·574. Luster adamantine, inclining to vitreous, resinous, or pearly; sometimes submetallic. Color white, gray, grayish black, sometimes tinged blue or green (copper); streak uncolored. Transparent to subtranslucent. Optically $-$. Ax. pl. $\parallel b(010)$. Bx. $\perp c(001)$. Dispersion $\rho > v$ large. $2V = 8° 14'$. The axial angle becomes smaller both with shorter wave-lengths of light and with the lowering of the temperature, becoming uniaxial in extreme violet at $15°$ C. $\alpha = 1\cdot804$. $\beta = 2\cdot076$. $\gamma = 2\cdot078$.

Comp. — Lead carbonate, $PbCO_3$ = Carbon dioxide 16·5, lead oxide 83·5 = 100.

Pyr., etc. — In the closed tube decrepitates, loses carbon dioxide, turns first yellow, and at a higher temperature dark red, but becomes again yellow on cooling. B.B. on charcoal fuses very easily, and in R.F. yields metallic lead. Soluble in dilute nitric acid with effervescence.

Diff. — Characterized by high specific gravity and adamantine luster; also by yielding lead B.B. Unlike anglesite, it effervesces with nitric acid.

Artif. — Cerussite has been produced artificially by the slow diffusion of a carbonate solution into a lead solution through a porous membrane; by the action of a carbonate solution upon a lead plate.

Obs. — Cerussite is a common lead mineral of secondary origin. It is found in the oxidized zones of lead veins and has been produced by the action of carbonated waters upon lead solutions. It has also been noted as a metasomatic replacement in limestones. It is commonly associated with galena, anglesite, and the other characteristic minerals of the oxidized lead veins.

Cerussite is a common mineral, and only those localities that are of note because of the exceptional character of the crystals that they contain can be mentioned here. Such localities are Mies and Přibram in Bohemia, Czechoslovakia; Bleiberg in Carinthia, Austria; in Germany at Zellerfeld in the Harz Mts., at Friedrichssegen near Ems in Hessen-Nassau, at Badenweiler, Baden. In France at Croix-aux-Mines, Vosges, and at Poullaouen and Huelgoat, Finistère. On Sardinia in the district about Iglesias and to the north, especially at Monteponi and Montevecchio. In Scotland at Leadhills, Lanark, formerly in fine crystals and in pseudomorphs after anglesite and leadhillite. In Tunis at Sidi-Amor-ben-Salem; at Tsumeb, near Otavi, in South West Africa. At Broken Hill, New South Wales; in New Caledonia at Mine Mérétrice on the Diachot River.

Among the notable localities in the United States are the Wheatley mines, Phœnixville, Pennsylvania, in fine crystals, often large; in Missouri at Joplin and Granby; in Colorado at Leadville and elsewhere; in Arizona at the Flux mine, Pima Co., in large crystalline masses; in crystals at the Red Cloud mine, Yuma Co.; in large twin crystals in the Organ Mts., near Las Cruces, Dona Ana Co., New Mexico; in Idaho in the Coeur d'Alene district at Wardner and elsewhere.

Use. — An ore of lead.

BARYTOCALCITE.

Monoclinic. Axes $a : b : c = 0\cdot7717 : 1 : 0\cdot6254$; $\beta = 73° 52'$. In crystals; also massive.

Cleavage: $m(110)$ perfect; $c(001)$ less so. The three cleavage planes make angles with each other that are near those of the rhombohedral cleavage planes of calcite. Structurally it shows relations to calcite, dolomite, and barite. The unit cell contains two molecules. Fracture uneven to subconchoidal. Brittle. H. = 4. G. = 3·64–3·66. Luster vitreous, inclining to resinous. Color white, grayish, greenish or yellowish. Streak white. Transparent to translucent. Optically $-$. $\alpha = 1\cdot525$. $\beta = 1\cdot684$. $\gamma = 1\cdot686$. Ax. pl. $\perp (010)$. $X \wedge c$ axis $= +64°$. $2V = 15°$.

Comp. — Carbonate of barium and calcium, $BaCO_3.CaCO_3$ = Carbon dioxide 29·6, baryta 51·5, lime 18·9 = 100.

Pyr., etc. — B.B. colors the flame yellowish green, and at a high temperature fuses on the thin edges and assumes a pale green color; the assay reacts alkaline after ignition. Soluble in dilute hydrochloric acid with effervescence. Dilute solution gives an abundant precipitate, $BaSO_4$, with a few drops of sulphuric acid.

Obs. — Occurs at Alston Moor in Cumberland, England, in limestone with barite and fluorite.

Bismutosphärite. — $Bi_2(CO_3)_3.2Bi_2O_3$. In spherical forms with radiated structure. Color yellow to gray or blackish brown. Uniaxial −. $\omega = 2.13$, $\epsilon = 1.94$. H. = 3–3·5. G. = 7·3–7·4. Easily fusible. Of secondary origin formed by the action of carbon dioxide upon either native bismuth or bismuthinite. Found at Neustädtel near Schneeberg, Saxony; in Madagascar at Ampangabe near Miandrarivo and elsewhere; from Guanajuato, Mexico, in tetragonal pseudomorphic crystals. Also sparingly in the United States at Willimantic and Portland, Connecticut. From the Stewart mine, Pala, San Diego Co., California.

Rutherfordine. — Uranyl carbonate, UO_2CO_3. Orthorhombic? A yellow ocher resulting from alteration of uraninite. Indices 1·72–1·80. G. = 4·8. From Uruguru Mts., Tanganyika Territory, East Africa.

Ancylite. — $4Ce(OH)CO_3.3SrCO_3.3H_2O$. Orthorhombic. In small pyramids with curved faces and edges. H. = 4·5. G. = 3·9. Color light yellow, orange, brown, gray. Optically −. $\alpha = 1.625$, $\beta = 1.700$, $\gamma = 1.735$. Infusible. From Narsarsuk, Greenland. A calcium variety, *calcio-ancylite*, is reported from Hibina-Toundra district, Kola peninsula, northern Russia. *Weibyeïte* from Langesundfiord, Norway, is a related mineral.

Ambatoarinite. — A carbonate of strontium and the rare earths. Orthorhombic? In crystals with parallel axes, forming skeleton-like groups. Color pink to black. Biaxial, −. Index, > 1·66. From Ambatoarina, near Ambositra, Madagascar.

Parisite. — A fluocarbonate of the cerium metals, $[(Ce,La,Di)F]_2Ca(CO_3)_3$. Rhombohedral. Crystals small and slender. Habit pyramidal or prismatic. Crystals horizontally grooved due to oscillatory combination of faces. Basal cleavage. H. = 4·5. G. = 4·358. Color brownish yellow. Optically +. $\omega = 1.676$. $\epsilon = 1.757$. From the emerald mines, Muso, Colombia; Montorfano, Lombardy, Italy; Narsarsuk, South Greenland (*synchisite*). In pegmatite veins of the granite at Quincy, Massachusetts, and from Ravalli, Montana.

Cordylite is a parisite containing barium from Narsarsuk, South Greenland. Other material from Narsarsuk thought to be a new species and named *synchisite* is probably parisite. X-ray study, however, is said to show that parisite and synchisite are distinct but closely related species. The formulas $2RFCO_3.CaCO_3$ and $RFCO_3.CaCO_3$ are suggested. Both are closely related structurally to bastnäsite. Cordylite is also said to show a structure differing from that of parisite.

Bastnäsite. Hamartite. — A fluocarbonate of the cerium metals $(RF)CO_3$. X-ray study indicates hexagonal, trigonal (benitoite class) symmetry. Prismatic crystals. Basal parting. H. = 4·5. G. = 4·948. Color wax-yellow to reddish brown. Uniaxial, +. $\omega = 1.717$, $\epsilon = 1.818$. Infusible B.B. From the Bastnäs mine, Riddarhyttan, Vastmanland, Sweden. From gold sands of the Barsovka River near Kyshtym, northeast of Zlatoust, Ural Mts., Russia. Found at various localities in Madagascar west of Ambositra. Occurs in parallel growth with tysonite in the Pike's Peak region in Colorado.

PHOSGENITE.

Tetragonal. Axis $c = 1.0876$. Crystals prismatic; sometimes tabular $\parallel c(001)$. Normal symmetry is shown by X-ray study.

Cleavage: $m(110)$, $a(100)$ distinct; also $c(001)$. Rather sectile. H. = 2·75–3. G. = 6·0–6·3. Luster adamantine. Color white, gray, and yellow. Streak white. Transparent to translucent. Optically +. $\omega = 2.114$. $\epsilon = 2.140$.

Comp. — Chlorocarbonate of lead, $(PbCl)_2CO_3$ or $PbCO_3.PbCl_2 = $ Lead carbonate 49·0, lead chloride 51·0 = 100.

Pyr., etc. — B.B. melts readily to a yellow globule, which on cooling becomes white and crystalline. On charcoal in R.F. gives metallic lead, with a white coating of lead chloride. Dissolves with effervescence in dilute nitric acid and solution reacts for chlorine with silver nitrate.

Obs. — Phosgenite is a rare mineral, found commonly in association with cerussite and apparently formed under the same conditions as that mineral. Occurs in fine crystals from Sardinia at Monteponi and Montevecchio near Iglesias and at Gibbas near Cagliari; in Derbyshire at Matlock and nearby Cromford. It occurs at Laurium, Greece, where it is the result of the action of sea-water upon ancient lead slags. In Tunis at Sidi-Amor-ben-

Salem and at Tsumeb near Otavi, South West Africa. In fine crystals from Dundas in Tasmania; also from Broken Hill, New South Wales.

Northupite. — $MgCO_3.Na_2CO_3.NaCl$. In isometric octahedrons. H. = 3·5–4. G. = 2·38. White to yellow or gray. $n = 1·514$. Easily fusible. Precipitated from saline waters containing a small amount of magnesium in solution. From Borax Lake, San Bernardino Co., California.

Tychite. — $2MgCO_3.2Na_2CO_3.Na_2SO_4$. Isometric. Octahedral habit. H. = 3·5. G. = 2·5. $n = 1·51$. Easily fusible. Very rare. From Borax Lake, San Bernardino Co., California, associated with northupite.

B. Acid, Basic, and Hydrous Carbonates

Teschemacherite. — Acid ammonium carbonate, HNH_4CO_3. Orthorhombic. In yellowish to white crystals. G. = 1·45. H. = 1·5. Indices, 1·423–1·536. From guano deposits on the coast of Africa (Saldanha Bay) and from the west coast of Patagonia, and on the Chincha Islands, Peru.

MALACHITE.

Monoclinic. Axes $a : b : c = 0·8809 : 1 : 0·4012$; $\beta = 61° 50'$.

Crystals rarely distinct, usually slender, acicular prisms (mm''' 110 $\wedge$ 1$\bar{1}$0 = 75° 40'), grouped in tufts and rosettes. Twins: tw. pl. $a(100)$ common. Commonly massive or incrusting, with surface botryoidal, or stalactitic, and structure divergent; often delicately compact fibrous, and banded in color; frequently granular or earthy.

Cleavage: $c(001)$ perfect; $b(010)$ less so. Fracture subconchoidal, uneven. Brittle. H. = 3·5–4. G. = 3·9–4·03. Luster of crystals adamantine, inclining to vitreous; of fibrous varieties more or less silky; often dull and earthy. Color bright green. Streak paler green. Translucent to subtranslucent to opaque. Optically —. $\alpha = 1·655$, $\beta = 1·875$, $\gamma = 1·909$. 2V = 43°. Ax. pl. $\parallel$ (010). $X \wedge c$ axis $= +23° 30'$. $X =$ nearly colorless, $Y =$ yellowish green, $Z =$ deep green

Comp. — Basic cupric carbonate, $CuCO_3.Cu(OH)_2 =$ Carbon dioxide 19·9, cupric oxide 71·9, water 8·2 = 100.

Cuprozincite and *paraurichalcite* are names given to zinc-bearing malachite-like minerals from Tsumeb, near Otavi, South West Africa. Cuprozincite has Cu:Zn = 9:2, and its optical properties are like those of malachite. The analyses of paraurichalcite vary, and it may only be an altered zinc-bearing malachite. See also rosasite below.

Pyr., etc. — In the closed tube blackens and yields water. B.B. fuses at 2, coloring the flame emerald-green; on charcoal is reduced to metallic copper; with the fluxes reacts like cuprite. Soluble in acids with effervescence.

Diff. — Characterized by green color and copper reactions B.B.; differs from other copper ores of a green color in its effervescence with acids.

Artif. — Malachite has been formed artificially by heating precipitated copper carbonate with a solution of ammonium carbonate for several days.

Obs. — Malachite is a common ore of copper, occurring typically in the oxidation zone of copper deposits, where it is associated with other ores of copper and is the product of their alteration. It is widespread in its occurrence, and only its more important localities can be mentioned here. It has been found in large amounts and in notable quality in the Ural Mts. near Nizhne-Taglisk. In these mines a bed of malachite was opened which yielded many tons of the mineral; one mass measured at top 9 by 18 feet; and the portion uncovered contained at least half a million pounds of the pure mineral. Much of this material was used for ornamental purposes. From Moldawa in the Banat, Rumania; in small crystals from Betzdorf near Siegen, Rhenish Prussia; from Chessy near Lyon, Rhône, France. In Cornwall at Liskeard and elsewhere. At Katanga in the Belgian Congo. In Northern Rhodesia at Bwana Mkubwa; at Tsumeb near Otavi, South West Africa; and in Namaqualand, Cape Province, South Africa. In South Australia in the Burra district and at Wallaroo.

In the United States the following localities may be mentioned. In Pennsylvania in Berks Co. in fibrous radiating masses; in the copper mines of Ducktown, Tennessee. In Arizona abundantly in fine masses and acicular crystals, with calcite at the Copper Queen mine Bisbee, Cochise Co.; also in Greenlee Co., at Morenci (6 miles from Clifton) in stalactitic forms with malachite and azurite in concentric bands; also in the Globe district, Gila Co., and elsewhere. From the Tintic district, Utah, and in pseudomorphs from Good Springs, Nevada.

Named from μαλαχή, *mallows*, in allusion to the green color.

Use. — An ore of copper; at times as an ornamental stone.

AZURITE. CHESSYLITE.

Monoclinic.   Axes $a : b : c = 0.8565 : 1 : 0.8844$; $\beta = 87° 35'$.

mm''', 110 ∧ 1$\bar{1}$0 = 81° 06'. ll', 023 ∧ 0$\bar{2}$3 = 61° 00'.
$c\sigma$,   001 ∧ 101 = 47° 06'. pp', 021 ∧ 0$\bar{2}$1 = 121° 00'.

Crystals varied in habit and highly modified.   Also massive, and presenting imitative shapes, having a columnar composition; also dull and earthy.

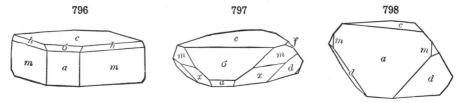

Cleavage: $p(021)$ perfect but interrupted; $a(100)$ less perfect. Fracture conchoidal. Brittle. H. = 3·5–4. G. = 3·77–3·89. Luster vitreous, almost adamantine. Color various shades of azure-blue, passing into Berlin-blue. Streak blue, lighter than the color. Transparent to subtranslucent. Optically +. $\alpha = 1.730$. $\beta = 1.758$. $\gamma = 1.838$. 2V = 68°. Ax. pl. ⊥ (010). $Z \wedge c$ axis = $-12° 30'$. Horizontal dispersion, $\rho > v$. $X = Y =$ clear blue, Z = dark to purple-blue.

Comp. — Basic cupric carbonate, $2CuCO_3.Cu(OH)_2$ = Carbon dioxide 25·6, cupric oxide 69·2, water 5·2 = 100.

Pyr., etc. — Same as in malachite.

Diff. — Characterized by its blue color; effervescence in nitric acid; copper reactions B.B.

Artif. — Azurite has been formed by allowing a solution of copper nitrate to lie in contact with fragments of calcite for several years.

Obs. — Azurite is much rarer in occurrence than malachite but, like the latter, is of secondary origin and found in the upper oxidized portions of ore deposits. It has been formed either by the action of carbonated waters upon copper compounds or by copper solutions upon limestones. Almost invariably associated with malachite; also other copper ores, various other minerals of secondary origin, etc. Occurs in fine crystals in Siberia at Tomsk and Solotuschinsk, Altai; in Greece at Laurium; at Moldawa in the Banat, Rumania; at Chessy near Lyon, Rhône, France, whence it derived its name *chessylite.* From Sardinia near Alghero; from Tsumeb near Otavi, South West Africa; in New South Wales at Broken Hill and elsewhere; at Mungana near Chillagoe, Queensland; in South Australia at Moonta, at Wallaroo, and near Adelaide.

Notable localities in the United States are few, the chief being in Arizona at Bisbee and Morenci and at Kelly, New Mexico.

Use. — An ore of copper.

Rosasite. — $(Cu,Zn)CO_3.(Cu,Zn)(OH)_2$. In mamillary spherules. These break up into rectangular plates with parallel extinction and positive elongation. Orthorhombic? Bright green to sky-blue color. Biaxial, −. 2V very small. $\alpha = 1.672$, β and $\gamma = 1.83$. Dispersion strong, $\rho < v$. From Rosas mine at Sulcis, Sardinia.

Aurichalcite. — A basic carbonate of zinc and copper, $2(Zn,Cu)CO_3.3(Zn,Cu)(OH)_2$. Monoclinic. In drusy incrustations. One perfect cleavage, (100)? G. = 3·64. Luster pearly. Color pale green to sky-blue. Optically −. $\alpha = 1·655$, $\beta = 1·740$, $\gamma = 1·744$. 2V very small. Y is nearly normal to cleavage. Elongated $\parallel Z$. Strong dispersion, $\rho < v$. X = nearly colorless, $Y = Z$ = pale green. Aurichalcite is a mineral of secondary origin, found in the oxidation zones of copper and zinc deposits. Often associated with malachite. The more important localities are at the zinc mines of Laurium, Greece; in Rumania at Moravicza and Rézbánya; near Campiglia in Tuscany; on Sardinia at Monteponi near Iglesias. In the United States at Salida, Chaffee Co., Colorado; in New Mexico at Kelly near Magdalena, Socorro Co.; in Utah at Big Cottonwood, Salt Lake Co., and at Stockton in Tooele Co.

Hydrozincite. — A basic zinc carbonate, $2ZnCO_3.3Zn(OH)_2$. Crystallographic and optical characters and the best analyses show that hydrozincite is to be regarded as a copper-free variety of aurichalcite. Monoclinic. Crystals are minute thin blades parallel to a perfect cleavage, either (100) or (001). Usually massive, fibrous, earthy or compact, as incrustations. G. = 3·58–3·8. H. = 2–2·5. Infusible B.B. Color white, grayish or yellowish. Optically −. Indices 1·63–1·75. 2V = 40°. Ax. pl. ⊥ (010). Z makes a small angle with direction of elongation of crystal blades. Hydrozincite is of secondary origin, formed usually by the alteration of sphalerite. It is commonly associated with smithsonite but is much rarer in its occurrence than that mineral. The more important localities for its occurrence include Bleiberg, etc., in Carinthia, Austria; on Sardinia at Monte Malfidano near Iglesias, etc.; was found in great quantities at the Dolores mine, Udias Valley, Province of Santander, Spain; from Llanidloes, Montgomery, Wales. In the United States has been found at Friedensville, Lehigh Co., Pennsylvania; in Marion Co., Arkansas (*marionite*); in Piute Co., Utah.

Loseyite. — $(Mn,Zn)CO_3.5(Mn,Zn)(OH)_2$. Contains also a small amount of magnesia. Monoclinic. Crystals elongated parallel to b axis. G. = 3·27. H. = 3. Color bluish white. Optically +. Ax. pl. $\parallel$ (010). $\alpha = 1·637$, $\beta = 1·648$, $\gamma = 1·676$. Found with pyrochroite, sussexite, chlorophœnicite, and calcite at Franklin, Sussex Co., New Jersey.

Otavite. A basic cadmium carbonate of uncertain composition. In crusts showing minute rhombohedral crystals. Color white to reddish. From Tsumeb, near Otavi, South West Africa.

Hydrocerussite. — A basic lead carbonate, $2PbCO_3.Pb(OH)_2$. Hexagonal (rhombohedral?). In thin colorless hexagonal plates. Basal cleavage. H. = 3·5. G. = 6·8. Optically −. $\omega = 2·09$, $\epsilon = 1·94$ (on artificial crystals). A rare mineral of secondary origin. Occurs on native lead at Långbanshyttan, Vermland, Sweden. Reported from Laurium, Greece. In Scotland at Wanlockhead, Dumfries, and from Leadhills, Lanark (*plumbonacrite*). In relative abundance and large crystals from the Mendip Hills, Somerset, England.

Dundasite. — A basic carbonate of lead and aluminum, $Pb(AlO)_2(CO_3)_2.4H_2O$. In small spherical aggregates of radiating acicular crystals. Color white. From Dundas and Mt. Read, Tasmania, and from near Trefriw, Carnarvonshire, Wales; Wensley, Derbyshire; near Maam, County Galway, Ireland.

Dawsonite. — A basic carbonate of aluminum and sodium, $Na_3Al(CO_3)_3.2Al(OH)_3$. Orthorhombic. In thin incrustations of white radiating acicular or bladed crystals. Perfect cleavage, (110). G. = 2·40. H. = 3. Fusible, 4·5. Optically −. $\alpha = 1·466$, $\beta = 1·542$, $\gamma = 1·596$. 2V = 77°. Ax. pl. $\parallel$ (001). $X = a$ axis. Formed probably through the decomposition of silicates containing sodium and aluminum. Found at Pian Castagnaio and Santa Fiora near Mte. Amiata, Province of Siena, Tuscany. Occurs east of Tenès, in Alger, Algeria. Originally found in a feldspathic dike near McGill University, Montreal, Canada.

Alumohydrocalcite. — A hydrated carbonate of lime and alumina. Monoclinic. In radiated spherulitic groups or as acicular crystals. Cleavages perfect $\parallel$ (100), imperfect $\parallel$ (010). Color, chalky white to pale blue, rarely violet, gray or light yellow. Optically −. $X \perp$ (010). $\alpha = 1·485$, $\beta = 1·553$, $\gamma = 1·570$. 2V = 50°–55°. From Khakassky district, Russia, associated with allophane, etc.

Thermonatrite. — Hydrous sodium carbonate, $Na_2CO_3.H_2O$. Orthorhombic. G. = 2·25. H. = 1–1·5. Optically −. $\alpha = 1·420$, $\beta = 1·506$, $\gamma = 1·524$. 2V = 48°. Ax. pl. $\parallel$ (001). $X = a$ axis. Occurs in various lakes, and as an efflorescence over the soil in many dry regions. Also about some mines and volcanoes (as at Vesuvius).

Nesquehonite. — Hydrous magnesium carbonate, $MgCO_3.3H_2O$. Orthorhombic. In radiating groups of prismatic crystals. Cleavages, (110) perfect, (001) poor. H. = 2·5. G. = 1·83–1·85. Colorless to white. Optically −. $\alpha = 1·412$, $\beta = 1·501$, $\gamma = 1·526$.

2V = 53°. Ax. pl. ∥ (001). $X = a$ axis. From a coal mine at Nesquehoning, Schuylkill Co., Pennsylvania. See lansfordite, p. 531.

Natron. — Hydrous sodium carbonate, $Na_2CO_3.10H_2O$. Occurring in nature only in solution, as in the soda lakes of Egypt, Nevada, California, etc., or mixed with the other sodium carbonates.

Pirssonite. — $CaCO_3.Na_2CO_3.2H_2O$. In prismatic crystals, orthorhombic-hemimorphic. H. = 3. G. = 2·35. Easily fusible. Colorless to white. Optically +. $α = 1·504$, $β = 1·509$, $γ = 1·575$. 2V = 33°. Ax. pl. ∥ (001). $Z = b$ axis. Borax Lake, San Bernardino Co., California.

GAY-LUSSITE.

Monoclinic.   Axes $a : b : c = 1·4897 : 1·4442$; $β = 78° 27'$.

$$mm''',\ 110 \wedge 1\bar{1}0 = 111° 10'.$$
$$ee',\ \ \ \ 011 \wedge 0\bar{1}1 = 109° 30'.$$
$$me,\ \ \ \ 110 \wedge 011 = \ \ 42° 21'.$$
$$rr',\ \ \ \ \bar{1}12 \wedge \bar{1}\bar{1}2 = \ \ 69° 29'.$$

Crystals often elongated ∥ a axis; also flattened wedge-shaped.   Cleavage: (110) perfect;   (001) rather difficult.   Fracture conchoidal.   Very brittle.

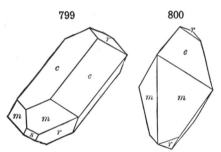

799   800

H. = 2–3.   G. = 1·93–1·95.   Luster vitreous.   Color white, yellowish white. Streak uncolored to grayish.   Translucent.   Optically −. $α = 1·444$, $β = 1·516$, $γ = 1·523$.   2V = 34°. Dispersion strong, $ρ < υ$. Ax. pl. ⊥ (010). $Z \wedge c$ axis = −14°.

Comp. — Hydrous carbonate of calcium and sodium, $CaCO_3.Na_2CO_3.$ $5H_2O$ = Calcium carbonate 33·8, sodium carbonate 35·8, water 30·4 = 100.

Pyr., etc. — Heated in a closed tube decrepitates and becomes opaque.   B.B. fuses easily to a white enamel, and colors the flame intensely yellow.   Dissolves in acids with a brisk effervescence; partly soluble in water, and reddens turmeric paper.

Obs. — Abundant at Lagunillas, near Merida, in Venezuela, in crystals disseminated at the bottom of a small lake, in a bed of clay, covering trona.   Also abundant in Little Salt Lake, or Soda Lake, in the Carson desert near Ragtown, Nevada, deposited upon the evaporation of the water.   From Sweetwater Valley, Wyoming.   Named after Gay-Lussac, the French chemist (1778–1850).

Lanthanite. — $(La,Di,Ce)_2(CO_3)_3.8H_2O$. In thin tabular orthorhombic crystals; also granular, earthy.   Cleavage (001), perfect.   H. = 3.   G. = 2·605.   Infusible B.B.   Color grayish white, pink, yellowish.   Optically −. $α = 1·52$, $β = 1·587$, $γ = 1·613$.   2V = 62°.   Ax. pl. ∥ (100). $X = c$ axis.   Found coating cerite at the Bastnäs mine near Riddarhyttan, Vastmanland, Sweden.   At the Sandford iron-ore bed, Moriah, Essex Co., New York, in delicate scales.   In Pennsylvania rarely in Silurian limestone with the zinc ores of the Saucon Valley, Lehigh Co., also as a mass of delicate pink crystals in the soil a few feet below the surface at Friedensville.

TRONA.   Urao.

Monoclinic.   Axes $a : b : c = 2·8460 : 1 : 2·9700$; $β = 77° 23'$.

$$ca,\ \ 001 \wedge 100 = 77° 23'.$$
$$co,\ \ 001 \wedge \bar{1}11 = 75° 53\tfrac{1}{2}'.$$
$$oo'',\ \bar{1}11 \wedge 11\bar{1} = 47° 35\tfrac{1}{4}'.$$

In plates ∥ (001) or elongated ∥ b axis, often fibrous or columnar massive. Cleavage: (100) perfect;   ($\bar{1}11$), (001) in traces.   Fracture uneven to subconchoidal.   H. = 2·5–3.   G. =·2·11–2·14.   Luster vitreous, glistening.

Color gray or yellowish white. Translucent. Taste alkaline. Optically $-$.
$\alpha = 1\cdot412$, $\beta = 1\cdot492$, $\gamma = 1\cdot540$. $2V = 72°$. Strong dispersion, $\rho < v$.
Ax. pl. $\perp$ (010). $X = b$ axis. $Y \wedge c$ axis $= +7°$.
    Comp. — $Na_2CO_3.HNaCO_3.2H_2O$ or $3Na_2O.4CO_2$.
$5H_2O$ = Carbon dioxide $38\cdot9$, soda $41\cdot2$, water $19\cdot9$
= 100.

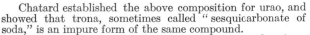

    Chatard established the above composition for urao, and
showed that trona, sometimes called " sesquicarbonate of
soda," is an impure form of the same compound.
    Pyr., etc. — In the closed tube yields water and carbon dioxide. B.B. easily fusible
and imparts an intensely yellow color to the flame. Soluble in water, and effervesces with
acids. Reacts alkaline with moistened test-paper.
    Obs. — Trona is found in the deposits of various saline lakes or is produced by the
evaporation of their waters. Occurs in the province of Fezzan, northern Africa, forming
superficial crusts. From the Soda Lakes, Wady Natrum, northwest of Cairo, Egypt, and
from the Oasis of Bilma, in the eastern Sahara Desert. Found at the bottom of a lake at
Lagunillas, Venezuela. In the United States has been found in connection with various
lake deposits in the western states, as in Churchill Co., Nevada, and in California at Borax
Lake, San Bernardino Co., in fine crystals, and at Owens Lake, Mono Co.
    NAHCOLITE. $NaHCO_3$. Occurs in small, white crystals mixed with trona, thermona-
trite, and thenardite lining the walls of an old tunnel about 9 miles west of Naples in the
Phlegrean Fields.
    Hydromagnesite. — Basic magnesium carbonate, $3MgCO_3.Mg(OH)_2.3H_2O$. Mono-
clinic. Twinned on (100) giving pseudo-orthorhombic symmetry. Perfect cleavage, (010);
distinct (100). Crystals small, tufted. Also amorphous; as chalky crusts. G. = $2\cdot16$.
H. = $3\cdot5$. B.B. infusible. Color and streak white. Optically $+$. $\alpha = 1\cdot523$, $\beta = 1\cdot527$,
$\gamma = 1\cdot545$. 2V medium. Ax. pl. $\perp$ (010). $Z = b$ axis. $X \wedge c$ axis $= +47°$. A rare
mineral commonly found in connection with serpentine and derived from the alteration of
magnesium-rich silicate rocks. Occurs in Czechoslovakia at Hrubschütz on the Iglawa,
north of Kromau, Moravia; at Kraubat in the Mürztal, Styria, Austria; from Limburg
near Sasbach, Baden. Occurs in Piedmont, Italy, at Emarese, Valle d'Aosta, and in Valle
Malenco. In the United States crystallized with serpentine and brucite, near Texas,
Lancaster Co., Pennsylvania; also similarly at Hoboken, New Jersey. In California at
Livermore, Alameda Co. Material closely similar from oalino crusts on lava at Alphar-
roëssa, Santorin Island, has been called giorgiosite.
    HYDROGIOBERTITE. $MgCO_3.Mg(OH)_2.2H_2O$. In light gray spherical forms. From the
neighborhood of Pollena, Vesuvius, Italy. Deposited from Phillips Springs, Napa Co.,
California. Both these occurrences have been shown to be mixtures of hydromagnesite
and some undetermined material, and it is probable that all hydrogiobertite is similar.
    Artinite. — $MgCO_3.Mg(OH)_2.3H_2O$. Orthorhombic. In minute groups of prismatic
crystals or radiating fibrous. H. = $2\cdot0$. G. = $2\cdot0$. White with silky luster. Opti-
cally $-$. $\alpha = 1\cdot489$, $\beta = 1\cdot534$, $\gamma = 1\cdot557$. 2V = $71°$. Crystals elongated $\parallel$ to Y.
Found in Lombardy, Italy, at Val Brutta and Franscia in the Val Lanterna, Valtellina, and
also at Emarese, Valle d'Aosta, Piedmont.
    Lansfordite. — $MgCO_3.5H_2O$. Monoclinic. Basal cleavage. H. = $2\cdot5$. G. = $1\cdot7$.
B.B. infusible. Optically $+$. $\alpha = 1\cdot456$, $\beta = 1\cdot468$, $\gamma = 1\cdot507$. 2V = $60°$. Ax. pl. $\perp$
(010). Z nearly parallel to c axis. Occurs as small stalactites in the anthracite mine at
Nesquehoning near Lansford, Schuylkill Co., Pennsylvania, changed on exposure to
nesquehonite. Noted from Atlin, British Columbia.
    GAJITE. A basic hydrous calcium, magnesium carbonate. Rhombohedral cleavage.
Granular structure. H. = $3\cdot5$. G. = $2\cdot62$. Color, white. Strong birefringence. Found
near Plešce, in the district Gorski-Kotar, Croatia, Yugoslavia.

    Hydrotalcite. — $MgCO_3.5Mg(OH)_2.2Al(OH)_3.4H_2O$. Hexagonal Lamellar, massive,
or foliated, somewhat fibrous. Perfect basal cleavage. Color white. Pearly luster.
Optically $-$. $\omega = 1\cdot512$, $\epsilon = 1\cdot489$. H. = 2. G. = $2\cdot06$. B.B. infusible. Occurs associated with magnetite in a talc schist from the Schischimsk mountains near Zlatoust, Urals
(völknerite). From Snarum, Buskerud, Norway in serpentine. Houghite is a hydrotalcite
derived from the alteration of spinel, from various localities in Jefferson and St. Lawrence
counties, New York.

Pyroaurite. — $MgCO_3.5Mg(OH)_2.2Fe(OH)_3.4H_2O$.   Hexagonal.   (X-ray studies indicate the possibility of two modifications, hexagonal normal and rhombohedral.)   In thin tabular crystals || to (0001).   Sometimes fibrous.   Perfect basal cleavage.   Color yellow to yellow-brown.   Luster pearly to greasy.   Optically −. $\omega = 1.565$.   H. = 2.5.   G. = 2.07.   B.B. infusible.   Occurs in Vermland, Sweden, at Långbanshyttan in gold-like submetallic scales (*pyroaurite*), and at the Moss mine at Nordmark.   In thin seams with a silvery white color in serpentine on Haaf-Grunay north of Fetlar, Shetland Islands, Scotland (*igelströmite*).

BRUGNATELLITE.   Comp. similar to pyroaurite above but contains less iron, the exact formula being in doubt.   Micaceous lamellar.   Perfect basal cleavage.   Color, flesh-pink.   Uniaxial, −. $\omega = 1.540$, $\epsilon = 1.510$.   Found in an old asbestos mine at Torre Santa Maria, Val Malenco, Lombardy, Italy.

Stichtite. — $MgCO_3.5Mg(OH)_2.2Cr(OH)_3.4H_2O$.   Hexagonal.   Micaceous.   In scales.   Perfect basal cleavage.   H. = 1.7.   G. = 2.16.   B.B. infusible.   Color lilac.   Optically −. $\omega = 1.542$, $\epsilon = 1.516$.   An alteration product of serpentine from Dundas, Tasmania.   Also noted from Black Lake area, Quebec.

Zaratite.   Emerald Nickel. — $NiCO_3.2Ni(OH)_2.4H_2O$?   Amorphous.   In mammillary incrustations; also massive compact.   H. = 3.   G. = 2.6.   B.B. infusible.   Color emerald-green. $n = 1.56$–1.61.   Occurs near Cape Ortegal, Galicia, Spain; from Swinaness, Unst, Shetland Islands.   In Tasmania from Heazlewood and also on the Whyte River.   From Dun Mt., Nelson, New Zealand, and Broken Hill, New South Wales.   Found at Igdlokunguak, Greenland.   In Pennsylvania at Texas, Lancaster Co., and at West Nottingham, Chester Co.

REMINGTONITE.   Originally described from a copper mine near Finksburg, Carroll Co., Maryland, as a hydrous cobalt carbonate.   It has been shown, however, that it is not a carbonate and is probably some fibrous mineral that has been colored by cobalt.   Also reported from Boléo, Lower California, and Mexico.

Tengerite. — A supposed yttrium carbonate.   In white pulverulent or fibrous coatings.   Optically +. $\alpha = 1.555$, $\beta = 1.57$, $\gamma = 1.585$.   2V large.   Crystals elongated parallel to X.   On gadolinite at Ytterby, northeast of Stockholm, Sweden.   A similar mineral is associated with the gadolinite of Llano Co., Texas.

Bismutite. — A basic bismuth carbonate, perhaps $Bi_2O_3.CO_2.H_2O$.   Incrusting fibrous, or earthy and pulverulent.   G. = 7.   H. = 4.   Easily fusible.   Color white, green, yellow and gray.   Biaxial, −. $n = 2.15$–2.28.   Elongated || Z.   Also isotropic in part.   Bismutite is of secondary origin, being derived chiefly by the alteration of bismuthinite and native bismuth.   Found in Bohemia at Joachimstal (Jáchymov) and at Ullersreuth near Hirschberg, Thuringia; in Saxony at Schneeberg and Johanngeorgenstadt.   Occurs in France at Meymac, Corrèze, and in Cornwall at St. Just and Redruth.   From Middlesex, Tasmania.   In Bolivia at Tazna, northwest of Chorolque and at the San Baldomero mine, near Sorata.   From Durango, Mexico.   In the United States in South Carolina at Brewer's mine, Chesterfield Co.; in Gaston Co., North Carolina.   From Mohave Co., Arizona.

BASOBISMUTITE.   A basic bismuth carbonate, $2Bi_2O_3.CO_2.H_2O$.   Occurs as a cement between crystals of beryl from the Sherlov Mts., Transbaikalia.

Uranothallite. — $2CaCO_3.U(CO_3)_2.10H_2O$.   Orthorhombic.   In scaly or granular crystalline aggregates.   Cleavage, (100), distinct.   H. = 2.5–3.   B.B. infusible.   Color siskin-green.   Optically +. $\alpha = 1.500$, $\beta = 1.503$, $\gamma = 1.539$.   2V = 40°. X || a axis.   Occurs on uraninite at Joachimstal, Bohemia.   *Liebigite* with a similar composition is probably the same species.   Occurs near Adrianople, Thrace, Greece; in Bohemia at Joachimstal; in Saxony at Schneeberg and Johanngeorgenstadt.

Voglite. — A hydrous carbonate of uranium, calcium and copper.   Triclinic?   In aggregations of crystalline scales.   Soft.   B.B. infusible.   Color emerald-green to bright grass-green.   Optically +. $\alpha = 1.541$, $\beta = 1.547$, $\gamma = 1.564$.   2V = 60°. X nearly ⊥ to crystal scales.   Pleochroic, $X = Y$ = dark blue-green, Z = pale yellow.   On uraninite, from the Elias mine near Joachimstal, Bohemia.

Oxygen Salts

2. SILICATES

The silicates comprise the largest chemical group among minerals.   They show a wide range in composition which is frequently very complex in character.   Recently, however, X-ray investigation has revealed important fundamental facts concerning their atomic structure and thrown much light upon the intricate problem of their composition. It is now established that the fundamental structural unit of all silicates has a tetrahedral shape with four oxygen atoms surrounding each atom of silicon. But these SiO_4 groups may be linked together in various ways to form indefinitely extended series. In the orthosilicates these SiO_4 groups occur independently with relations to the other elements present, as in an ordinary salt, such as a sulphate, a carbonate, etc. In other types of silicates, however, the SiO_4 groups are linked together by having one or more atoms of oxygen shared in common by neighboring groups. If two adjacent silicate groups have one atom of oxygen in common, individual groups with such formulas as Si_2O_7, Si_3O_9, Si_4O_{12}, Si_6O_{18} may occur. Further, the SiO_4 groups may form chains in which two oxygen atoms of two adjacent groups are held in common. Single chains of this type with a composition corresponding to SiO_3 occur in pyroxene. In amphibole two such chains are linked together, giving a composition represented by Si_4O_{11}. Such chains may be further linked together to form a silicon-oxygen sheet, thought to be characteristic of the micas, with a composition of Si_2O_5. Again the linking of the silicon-oxygen chains may take place so as to give rise to a three-dimensional network. Such a grouping is characteristic of the different forms of silica (quartz, tridymite, etc.), and by supposing a partial replacement of silicon by aluminum an extended acid network may be obtained into which basic atoms can be introduced. This latter structure is supposed to be characteristic of the zeolites, where metallic atoms may be artificially substituted for each other and the water content varied without disturbing the fundamental structure. In these various silicon-oxygen structures, metallic atoms are placed in such ways as to bind the whole structure together. It has been shown that where isomorphous replacement takes place the number of oxygen atoms (including F and OH) remains constant for each unit cell of the structure. Variations in composition may involve the substitution of Si by Al, of Al, Mg, and Fe by each other, of Na by Ca, etc. A general statement by Bragg[*] is " that a silicate should be regarded as a structure having a constant number of oxygen atoms in the unit, with a constant number of places for metal and silicon which can be filled by these elements in varying proportions consistent with a balance between valencies."

It can be seen that this view of the structure of the silicates largely does away with the customary assumption of the existence of distinctive silicate acid radicals and alters greatly the usual method of the classification of the silicates. Such a study of the silicates is, however, only just begun, and the time has not yet arrived when a fundamental reclassification can be attempted. Therefore, in this edition, the usual chemical classification of the silicates has been kept, but short statements concerning the new facts of structure and relations between species have been introduced in their approprate places.

The silicates are in part strictly anhydrous, in part hydrous, as the zeolites and the amorphous clays, etc. Furthermore, a large number of the silicates yield more or less water upon ignition, and in many cases it is known that they are, therefore, to be regarded as basic (or acid) silicates. The line, however, between the strictly anhydrous and hydrous silicates cannot be sharply drawn, since with many species which yield water upon ignition the part played by

---

[*] The substance of these paragraphs has been condensed from an article entitled, The Structure of Silicates, by W. L. Bragg, Zs. Kr., **74**, 237, 1930.

the elements forming the water is as yet uncertain. Furthermore, in the cases of several groups, the strict arrangement must be deviated from, since the relation of the species is best exhibited by introducing the related hydrous species immediately after the others.

This chapter closes with a section including the Titanates, Silico-titanates, Titano-niobates, etc., which connect the Silicates with the Niobates and Tantalates. Some Titanates have already been included among the Oxides.

Section A. Chiefly Anhydrous Silicates

I. Disilicates, Polysilicates
II. Metasilicates
III. Orthosilicates
IV. Subsilicates

The DISILICATES, RSi_2O_5, are salts of disilicic acid, $H_2Si_2O_5$, and have an oxygen ratio of silicon to bases of 4 : 1, as seen when the formula is written after the dualistic method, $RO.2SiO_2$.

The POLYSILICATES, $R_2Si_3O_8$, are salts of polysilicic acid, $H_4Si_3O_8$, and have an oxygen ratio of 3 : 1, as seen in $2RO.3SiO_2$.

The METASILICATES, $RSiO_3$, are salts of metasilicic acid, H_2SiO_3, and have an oxygen ratio of 2 : 1. They have hence been called *bisilicates*.

The ORTHOSILICATES, R_2SiO_4, are salts of orthosilicic acid, H_4SiO_4, and have an oxygen ratio of 1 : 1. They have hence been called *unisilicates*. The majority of the silicates fall into one of the last two groups.

Furthermore, there are a number of species characterized by an oxygen ratio of less than 1 : 1, *e.g.*, 3 : 4, 2 : 3, etc. These basic species are grouped as SUBSILICATES. Their true position is often in doubt; in most cases they are probably to be regarded as basic salts belonging to one of the other groups.

The above classification cannot, however, be carried through strictly, since there are many species which do not exactly conform to any one of the groups named, and often the true interpretation of the composition is doubtful. Furthermore, within the limits of a single group of species, connected closely in all essential characters, there may be a wide variation in the proportion of the acidic element. Thus the triclinic feldspars, placed among the polysilicates, range from the true polysilicate, $NaAlSi_3O_8$, to the orthosilicate, $CaAl_2Si_2O_8$, with many intermediate compounds, regarded as isomorphous mixtures of these extremes. Similarly of the scapolite group, which, however, is included among the orthosilicates, since the majority of the compounds observed approximate to that type. The micas form another example.

I. Disilicates, RSi_2O_5. Polysilicates, $R_2Si_3O_8$

PETALITE.

Monoclinic. $a : b : c = 1.153 : 1 : 0.744$. $\beta = 67° 34'$. Crystals rare (*castorite*). Usually massive, foliated cleavable (*petalite*).

Cleavage: $c(001)$ perfect; $o(\bar{2}01)$ easy; $z(\bar{9}05)$ difficult and imperfect. Fracture imperfectly conchoidal. Brittle. H. = 6–6.5. G. = 2.39–2.46. Luster vitreous, on $c(001)$ pearly. Colorless, white, gray, occasionally reddish or greenish white. Streak uncolored. Transparent to translucent. Optically +. $\alpha = 1.504$. $\beta = 1.510$. $\gamma = 1.516$. $2V = 83\frac{1}{2}°$. Ax. pl. ⊥ (010). $X \wedge a$ axis = 2°–8°.

Comp. — $LiAl(Si_2O_5)_2$ or $Li_2O.Al_2O_3.8SiO_2$ = Silica 78·4, alumina 16·7, lithia, 4·9 = 100.

When heated becomes uniaxial between 1000° and 1100° and isotropic at 1200° C.

Pyr., etc. — Gently heated emits a blue phosphorescent light. B.B. fuses quietly at 5 and gives the reaction for lithia. With borax it forms a clear, colorless glass. Not acted on by acids.

Obs. — Petalite occurs at the iron mine on the island of Utö, State of Stockholm, Sweden, with lepidolite, tourmaline, spodumene, and quartz; on Elba (*castorite*). In the United States, at Bolton, Worcester Co., Massachusetts, with scapolite; at Peru, Oxford Co., Maine, with spodumene in albite. The name *petalite* is from πέταλον, *a leaf*, alluding to the cleavage.

Milarite. — $K_2O.4CaO.4BeO.Al_2O_3.24SiO_2.H_2O$. In hexagonal prisms. Biaxial with small 2V, basal sections showing division into six biaxial sectors. On heating becomes at 750° C. uniaxial (*metamilarite*). Optically −. Indices, 1·529–1·532. H. = 5·5–6. G. = 2·55–2·59. Easily fusible. Colorless to pale green, glassy. From Val Giuf, northwest of Ruèras in Val Tavetsch, Grisons, Switzerland.

Eudidymite. — $HNaBeSi_3O_8$. Monoclinic. In white, glassy, twinned crystals, tabular in habit, ‖ (001). Lamellar twinning on (001). Cleavages (001), (551). X-ray study shows a similarity in structure between eudidymite and epididymite. H. = 6. G. = 2·553. Easily fusible. Optically +. α = 1·545, β = 1·546, γ = 1·551. 2V = 30°. Ax. pl. ‖ (010). $Z \wedge c$ axis = − 58°. Distinct dispersion, $\rho > v$. Occurs very sparingly in elæolite-syenite on the island Övre-Arö, Langesundfiord, Norway; from Narsarsuk, Greenland.

Epididymite. — Same composition as eudidymite. Orthorhombic. Tabular ‖ (001). Cleavages, (010) and (001), perfect. H. = 5·5. G. = 2·55. Easily fusible. Optically +. α = 1·544, β = 1·544, γ = 1·546. 2V = 22°. Ax. pl. ‖ (001). $Z = b$ axis. Occurs on the island of Arö, Langesundfiord, Norway and at Narsarsuk, Greenland.

Leifite. — $Na_4(AlF)_2Si_9O_{22}$. Colorless hexagonal prisms. Prismatic cleavage. H. = 6. G. = 2·57. Optically +. ω = 1·5177, ϵ = 1·5224. From drusy cavities in alkali-pegmatite veins at Narsarsuk, Greenland.

Feldspar Group

α. Monoclinic Section

		$a : b : c$	β'
Orthoclase	$KAlSi_3O_8$	0·6585 : 1 : 0·5554	116° 3'
Soda-Orthoclase	$\begin{cases} (K,Na)AlSi_3O_8 \\ (Na,K)AlSi_3O_8 \end{cases}$		
Hyalophane	$(K_2,Ba)Al_2Si_4O_{12}$	0·6584 : 1 : 0·5512	115° 35'
Celsian	$BaAl_2Si_2O_8$	0·657 : 1 : 0·554	115° 2'

β. Triclinic Section

Microcline	$KAlSi_3O_8$
Soda-microcline	$(K,Na)AlSi_3O_8$
Anorthoclase	$(Na,K)AlSi_3O_8$

Albite-anorthite Series. *Plagioclase Feldspars*

		$a : b : c$	α	β	γ
Albite	$NaAlSi_3O_8$	0·6335 : 1 : 0·5577	94° 3'	116° 29'	88° 9'
Oligoclase		0·6321 : 1 : 0·5524	93° 4'	116° 23'	90° 5'
Andesine	$\begin{cases} nNaAlSi_3O_8 \\ mCaAl_2Si_2O_8 \end{cases}$	0·6357 : 1 : 0·5521	93° 23'	116° 29'	89° 59'
Labradorite		0·6377 : 1 : 0·5547	93° 31'	116° 3'	89° 54½
Bytownite			93° 22'	116° 0'	90° 41'
Anorthite	$CaAl_2Si_2O_8$	0·6347 : 1 : 0·5501	93° 13'	115° 55'	91° 12'

The general characters of the species belonging in the FELDSPAR GROUP are as follows:

1, *Crystallization* in the monoclinic or triclinic systems, the crystals of the different species resembling each other closely in angle, in general habit, and in methods of twinning. The prismatic angle in all cases differs but a few degrees from 60° and 120°. X-ray study of the Feldspar Group shows the close similarity in structure that exists between its members. Four molecules are contained in the unit cell.

2, *Cleavage* in two similar directions parallel to the base c(001) and clino-pinacoid (or brachypinacoid), b(010), inclined at an angle of 90° or nearly 90°. 3, *Hardness* between 6 and 6·5. 4, *Specific Gravity* varying between 2·5 and 2·9, and mostly between 2·55 and 2·75. 5, *Color* white or pale shades of yellow, red or green, less commonly dark. 6, In composition silicates of aluminum with either potassium, sodium, or calcium, and rarely barium, while magnesium and iron are always absent. Furthermore, besides the several distinct species there are many intermediate compounds having a certain independence of character and yet connected with each other by insensible gradations; all the members of the series showing a close relationship not only in composition but also in crystalline form and optical characters.

The species of the Feldspar Group are classified, first as to form, and second with reference to composition. The *monoclinic* species include (see above): ORTHOCLASE, potassium feldspar, and SODA-ORTHOCLASE, potassium-sodium feldspar; also HYALOPHANE and CELSIAN, barium feldspars.

The *triclinic* species include: MICROCLINE and ANORTHOCLASE, potassium-sodium feldspars; ALBITE, sodium feldspar; ANORTHITE, calcium feldspar.

Also intermediate between albite and anorthite the isomorphous sub-species, sodium-calcium or calcium-sodium feldspars: OLIGOCLASE, ANDESINE, LABRADORITE, BYTOWNITE.

α. Monoclinic Section

ORTHOCLASE.

Monoclinic. Axes $a : b : c = 0.6585 : 1 : 0.5554$; $\beta = 63° 57'$.

mm''',	$110 \wedge 1\bar{1}0 = 61° 13'$.	cn,	$001 \wedge 021 = 44° 56\frac{1}{2}'$.
zz',	$130 \wedge \bar{1}30 = 58° 48'$.	nn',	$021 \wedge 0\bar{2}1 = 89° 53'$.
cx,	$001 \wedge \bar{1}01 = 50° 16\frac{1}{2}'$.	cm,	$001 \wedge 110 = 67° 47'$.
cy,	$001 \wedge \bar{2}01 = 80° 18'$.	co,	$001 \wedge \bar{1}11 = 55° 14\frac{1}{2}'$.

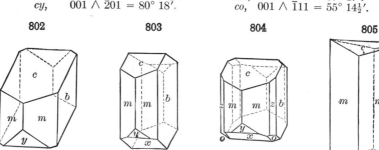

802   803   804   805

Twins: tw. pl. (1) a(100), or tw. axis c, the common *Carlsbad* twins, either of irregular penetration (Fig. 807) or contact type; the latter usually with b(010) as composition-face, often then (Fig. 808) with c(001) and x($\bar{1}$01) nearly in a plane, but to be distinguished by luster, cleavage, etc. (2) $n(021)$, the *Baveno* twins forming nearly square prisms (Fig. 809), since $cn = 44° 56\frac{1}{2}'$, and hence $cc = 89° 53'$; often repeated as fourlings (Fig. 473, p. 192), also in square prisms, elongated ∥ a axis. (3) $c(001)$, the *Manebach* twins (Fig. 810), usually contact-twins with c as composition face. Also other rarer laws.

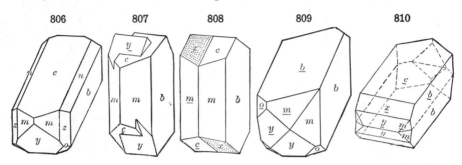

806   807   808   809   810

Crystals often prismatic ∥ c axis; sometimes orthorhombic in aspect (Fig. 805) since $c(001)$ and $x(\overline{1}01)$ are inclined at nearly equal angles to the vertical axis; also elongated ∥ a axis (Fig. 806) with $b(010)$ and $c(001)$ nearly equally developed; also thin tabular ∥ $b(010)$: rarely tabular ∥ $a(100)$, a face not often observed. Often massive, coarsely cleavable to granular; sometimes lamellar. Also compact crypto-crystalline, and flint-like or jasper-like.

Cleavage: (001) perfect; (010) somewhat less so; prismatic $m(110)$ imperfect, but usually more distinct parallel to one prismatic face than to the other. Parting sometimes distinct parallel to $a(100)$, also to a hemi-ortho-dome, inclined a few degrees to the orthopinacoid; this may produce a satin-like luster or schiller (p. 275), the latter also often present when the parting is not distinct. Fracture conchoidal to uneven. Brittle. H. = 6. G. = 2·56 (*adularia*), 2·57–2·58 (*sanidine*). Luster vitreous; on $c(001)$ often pearly. Colorless, white, pale yellow and flesh-red common, gray; rarely green. Streak uncolored.

For adularia, $\alpha = 1·519$, $\beta = 1·523$, $\gamma = 1·525$. Slightly different for sanidine. In adularia the ax. pl. is ⊥ (010) with $Z = b$ axis; X inclined only a few degrees to a axis (+3° to 7° (cf. Fig. 811) or in varieties rich in Na_2O from 10° to 12°). Axial angle variable. Usually 2V = about 70°. When adularia is heated the axial angle diminishes becoming 0° between 600° and 800° C. Beyond that point the ax. pl. becomes ∥ (010). Sanidine has this orientation for the axial plane with a small value for 2V. Dispersion $\rho > v$; horizontal, strongly marked, or inclined, according to position of axial plane. Optically −.

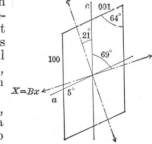

811

Comp. — A silicate of aluminum and potassium, $KAlSi_3O_8$ or $K_2O.Al_2O_3.6SiO_2$ − Silica 64·7, alumina 18·4, potash 16·9 = 100. Sodium is often also present, replacing part of the potassium, and sometimes exceeds it in amount; these varieties are embraced under the name soda-orthoclase (the name *barbierite* has been proposed for the monoclinic phase of $NaAlSi_3O_8$ whose existence is known only in crystal solution in orthoclase).

Ordinary orthoclase and the variety adularia are the low temperature forms of orthoclase. When crystallization takes place at high temperatures or orthoclase is heated to about 900° C. the variety sanidine is formed. Sanidine does not readily change to orthoclase on cooling.

Var. — The prominent varieties depend upon crystalline habit and method of occurrence more than upon difference of composition.

1. *Adularia.* The pure or nearly pure potassium silicate. Usually in crystals, like Fig. 805 in habit; often with vicinal planes; Baveno twins common. G. = 2·565. Transparent or nearly so. Often with a pearly opalescent reflection or schiller or a delicate play of colors; this variety is commonly known as *moonstone*, which has been shown by X-ray study to be a microperthitic to sub-microperthitic intergrowth of orthoclase and albite. The original adularia (Adular) is from the St. Gotthard region in Switzerland. *Valencianite*, from the silver mine of Valencia, Mexico, is adularia.

2. *Sanidine* or *glassy feldspar.* Occurs in crystals, often transparent and glassy, embedded in rhyolite, trachyte (as of the Siebengebirge, Germany), phonolite, etc. Habit often tabular || $b(010)$ (hence named from σανίς a *tablet*, or *board*); also in square prisms (b, c); Carlsbad twins common. Most varieties contain sodium as a prominent constituent, and hence belong to the soda-orthoclase. *Natronsanidine* is a sanidine-like soda-orthoclase from a soda liparite from Mitrowitza, Yugoslavia.

Rhyacolite. Occurs in glassy crystals at Monte Somma, Vesuvius; named from ρυαξ *stream* (lava stream).

3. *Isorthose* or *isorthoclase* is said to be a variety that is optically +, with 2V small, but otherwise like adularia.

4. *Ordinary.* In crystals, Carlsbad and other twins common; also massive or cleavable, varying in color from white to pale yellow, red or green, translucent; sometimes aventurine. Here belongs the common feldspar of granitoid rocks or granite veins. Usually contains a greater or less percentage of soda (soda-orthoclase). Compact cryptocrystalline orthoclase makes up the mass of much felsite, but to a greater or less degree admixed with quartz; of various colors, from white and brown to deep red. Much of what has been called orthoclase, or common potash feldspar, has proved to belong to the related triclinic species, microcline. Cf. p. 541, on the relations of the two species. Chesterlite and Amazon stone are microcline; also most aventurine orthoclase. *Loxoclase* contains sodium in considerable amount (7·6 Na_2O); from Hammond, St. Lawrence Co., New York. *Murchisonite* is a flesh-red feldspar similar to perthite (p. 540), with gold-yellow reflections in a direction ⊥ b (010) and nearly parallel to $\overline{7}01$ or $\overline{8}01$; from Dawlish and Exeter, England.

The spherulites noted in some volcanic rocks, as in the rhyolite of Obsidian Cliff in the Yellowstone Park, are believed to consist essentially of orthoclase needles with quartz. These are shown in Figs. 812 and 813 (from Iddings; much magnified) as they appear in polarized light (crossed nicols).

812 813

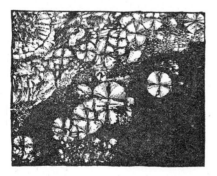

Pyr., etc. — B.B. fuses at 5; varieties containing much soda are more fusible. Loxoclase fuses at 4. Not acted upon by acids. Mixed with powdered gypsum and heated B.B. gives violet potassium flame visible through blue glass.

Diff. — Characterized by its crystalline form and the two cleavages at right angles to each other; harder than barite and calcite; not attacked by acids; difficultly fusible.   Massive corundum is much harder and has a higher specific gravity.

Micro. — Distinguished in rock sections by its low refraction (low relief) and low interference colors, which last scarcely rise to white of the first order — hence lower than those of quartz; also by its biaxial character in convergent light and by the distinct cleavages. It is colorless in ordinary light and may be limpid, but is frequently turbid and brownish from the presence of very minute scales of kaolin due to alteration from weathering; this change is especially common in the older granular rocks, as granite and gneiss.

Artif. — Orthoclase has not been produced artificially by the methods of dry fusion. It can, however, be crystallized from a dry melt when certain other substances, like tungstic acid, alkaline phosphates, etc., are added. The function of these additions in the reactions is not clear.   Orthoclase is more easily formed by hydrochemical methods.   It has been produced by heating gelatinous silica, alumina, caustic potash and water in a sealed tube. Orthoclase has also been formed by heating potassium silicate and water together with muscovite.   Pure synthetic orthoclase when heated to 1170° C. and above is converted into a fused glass and leucite.

Alter. — Orthoclase is frequently altered, especially through the action of carbonated or alkaline waters; the final result is often the removal of the potash and the formation of kaolin.   Steatite, talc, chlorite, leucite, mica, laumontite, occur as pseudomorphs after orthoclase; and cassiterite and calcite often replace these feldspars by some process of solution and substitution.

Obs. — Orthoclase is formed during the crystallization of igneous rocks; by pneumatolytic and hydrothermal agencies in pegmatite veins and in druses in the rocks, and more rarely by crystallization from aqueous solutions at low temperatures in veins.   It is a very common constituent of igneous rocks, being most frequently found in the more siliceous plutonic rocks, but it is also found in many eruptive and metamorphic rocks.   It occurs as an essential constituent in granite, gneiss, syenite; also in porphyries and as the variety sanidine in trachyte, phonolite, etc.   *Adularia* is characteristically found in the cavities of crystalline schists and also to a less extent in certain ore veins.   In the massive granitoid rocks orthoclase is seldom in distinct, well-formed, separable crystals, except as it occurs in veins and cavities; such crystals are more common, however, in volcanic rocks, like trachyte.   Large masses, frequently showing crystal forms, are common in the pegmatite veins.   It is also sometimes found as a vein constituent.   Fragments of orthoclase are found in the coarser sedimentary rocks, as arkose sandstone, graywacke, and conglomerates.

Orthoclase is perhaps the most common of all silicates, and therefore only the most important localities for its occurrence can be mentioned here.   The variety *adularia* occurs in the crystalline schists of the central and eastern Alps associated with smoky quartz, albite, etc.   Thus in Switzerland in the region of St. Gotthard, Ticino; in Uri in the Maderaner Tal, from near Realp in Urseren Tal, etc.; in Grisons in Val Tavetsch, and on Mt. Scopi east of Lukmanier in Val Medels; in Valais at Fiesch on the Rhône.   Further in the eastern Alps in Austria in the Zillertal, etc., of the Tyrol and in Salzburg in the Pinzgau district; also in the Pfitschtal, Trentino, Italy   From the island of Elba at La Colta between San Piero in Campo and San Ilario in Campo.   Fine crystals of orthoclase, often twinned, are obtained in Italy from Baveno on Lago Maggiore, Piedmont; from Val di Flemm, Trentino, a red variety.   In Bohemia of Czechoslovakia at Karlsbad (Karl Vary), Elbogen, etc.; from Bodenmais and Neubäu in Bavaria; in Silesia from Hirschberg, etc.; in France from Puy-de-Dôme; in Norway from the island of Stokö in the Langesundfiord and from Fredricksvärn near Larvik in Vestfold; from Cornwall at St. Agnes.   From Itrongahy near Betroka, Madagascar, in a clear yellow variety of gem quality.   In Japan from Tanokamiyama, Omi, mostly as Baveno twins and from Takayama, Mino, etc.   *Moonstone* is brought from Ceylon.   *Valencianite*, a variety of adularia, is from the Valencia silver mine in Guanajuato, Mexico.   Typical *sanidine* is prominent in the Rhineland in the trachyte of the Drachenfels in the Siebengebirge and at the Laacher See.

In the United States orthoclase is common in the pegmatite veins of the New England states.   In New York in St. Lawrence Co., at Rossie near Grasse Lake, and in the town of Hammond (*loxoclase*); in Lewis Co., near Natural Bridge and at Diana.   In Pennsylvania in Delaware Co., at Leiperville, Mineral Hill; the variety *lennilite* or *delawarite* comes from Lenni and *cassinite* from Blue Hill; in Chester Co., at French Creek.   *Adularia* is found at Hot Springs, Garland Co., Arkansas.   In Colorado, at the summit of Mt. Antero, Chaffee Co., in fine crystals; at Gunnison, Gunnison Co.; at Kokomo and Robinson, Summit Co. In Texas from Barringer Hill in Llano Co., in large crystals.   In Nevada from Goodsprings; in California from Mesa Grande, San Diego Co.

Use. — In the manufacture of porcelain, both in the body of the ware and in the glaze on its surface.

Perthite. As first described, a flesh-red aventurine feldspar from Perth, Ontario, Canada, called a soda-orthoclase, but shown by Gerhard to consist of interlaminated orthoclase and albite. Many similar occurrences have since been noted, as also those in which microcline and albite are similarly interlaminated, the latter called *microcline-perthite*, or microcline-albite-perthite; this is true in part of the original perthite. When the structure is discernible only with the help of the microscope it is called *microperthite*. Brögger has investigated not only the microperthites of Norway, but also other feldspars characterized by a marked schiller; he assumes the existence of an extremely fine interlamination of albite and orthoclase || ($\bar{8}01$), not discernible by the microscope (*cryptoperthite*), and connected with secondary planes of parting || (100) or ($\bar{8}01$), which is probably to be explained as due to incipient alteration. X-ray study has shown that orthoclase at high temperatures may hold in crystal solution other feldspar molecules (chiefly albite), which at these temperatures seem to possess the same monoclinic lattice as orthoclase. On cooling, however, unstable conditions may result, and the dissolved molecules may separate and show their normal triclinic lattices. This accounts for the intimate intergrowth of different feldspars shown in the perthites.

Hyalophane. — A barium-bearing adularia orthoclase, $KAlSi_3O_8$ with $BaAl_2Si_2O_8$ in which the orthoclase molecule is always dominant. In crystals, like adularia in habit (Fig. 805, p. 536); also massive. Cleavage: c(001) perfect; b(010) somewhat less so. H. = 6–6·5. G. = 2·805. Optically −. $\alpha = 1\cdot542$. $\beta = 1\cdot545$. $\gamma = 1\cdot547$ (for a mixture containing 70 per cent $KAlSi_3O_8$). Ax. pl. ⊥ (010). $Z = b$ axis. $X \wedge a$ axis from 0° to −25°. Occurs in a granular dolomite along with barite, etc., near Imfeld in the Binnental, Valais, Switzerland. Also from Sweden at the manganese mine at Jakobsberg in the Nordmark district, Vermland; also at the Sjö mines near Örebro, in Örebro. Noted at Franklin, New Jersey.

Celsian. — $BaAl_2Si_2O_8$, similar in composition to anorthite, but containing barium instead of calcium. Monoclinic. In crystals showing a number of forms; twinned according to Carlsbad, Manebach and Baveno laws. Usually cleavable massive. H. = 6–6·5. G. = 3·37. Colorless. Optically +. $\alpha = 1\cdot584$. $\beta = 1\cdot589$. $\gamma = 1\cdot594$. 2V = 86°. Ax. pl. || (010). $Z \wedge a$ axis = + 28°. From Jacobsberg, in the Nordmark district, Vermland, Sweden. Name *baryta-orthoclase* given to mixtures of celsian and orthoclase. *Paracelsian* from Candoglia, Toce valley, Piedmont, Italy, is the same species.

β. Triclinic Section

MICROCLINE.

Triclinic. Near orthoclase in angles and habit, but the angle bc(010 $\wedge$ 001) = about 89° 30′. Twins: like orthoclase, also polysynthetic twinning

814

according to the albite and pericline laws (p. 542), common, producing two series of fine tapering lamellæ nearly at right angles to each other, hence the characteristic grating-structure of a basal section in polarized light (Fig. 814). Also massive cleavable compact.

Cleavage: c(001) perfect; b(010) somewhat less so; M($1\bar{1}0$) sometimes distinct; m(110) also sometimes distinct, but less easy. Fracture uneven. Brittle. H. = 6–6·5. G. = 2·54–2·57. Luster vitreous, on c(001) sometimes pearly. Color white to pale cream-yellow, also red, green. Transparent to translucent. Optically −. Ax. pl. nearly perpendicular (82°–83°) to b(010). Z inclined 15° 26′ to a normal to b(010). Dispersion $\rho < v$ about Z. Extinction-angle on c(001), +15° 30′; on b(010), +5°–6° (Fig. 819, p. 542). $\alpha = 1\cdot518$, $\beta = 1\cdot522$, $\gamma = 1\cdot525$. 2V = 83°.

The essential identity of orthoclase and microcline has been urged by various authors on the ground that the properties of the former would belong to an aggregate of submicroscopic twinning lamellæ of the latter, according to the albite and pericline laws. However, slight but real differences in specific gravity, indices and axial angle point to different species. Further, it has been shown that orthoclase inverts from adularia to sanidine at 900° C., whereas microcline remains unchanged up to its melting point.

Comp. — Like orthoclase, $KAlSi_3O_8$ or $K_2O.Al_2O_3.6SiO_2$ = Silica 64·7, alumina 18·4, potash 16·9 = 100. Sodium is usually present in small amount: sometimes prominent, as in soda-microcline.

Pyr. — As for orthoclase.

Diff. — Resembles orthoclase but distinguished by optical characters (*e.g.*, the grating structure in polarized light, Fig. 814); also often shows fine twinning-striations on a basal surface (albite law).

Micro. — In thin sections like orthoclase but usually to be distinguished by the grating-like structure in polarized light due to triclinic twinning.

Obs. — Much of the potash feldspar commonly classed as orthoclase belongs here; in general, only an optical examination serves to establish the differences. It hence occurs under the same conditions as common orthoclase and is widespread. Only a few localities can be mentioned here. Beautiful *amazonstone* (the name given to the green variety) occurs in the Ilmen Mts. and at other points in the Urals. It is found at Baveno on Lago Maggiore, Piedmont, Italy. Occurs in Norway at Kragerö in Telemark; Arendal in Aust-Agder; near Frederiksvärn and Larvik in Vestfold; etc. Found in the gem-bearing pegmatites of Madagascar.

In the United States it occurs in Pennsylvania as the variety *chesterlite* at Poorhouse quarry, West Bradford, Chester Co., and as aventurine feldspar at Mineral Hill, Delaware Co. At Amelia Court House, Amelia Co., Virginia. In fine groups of large amazonstone crystals with deep color in the granite of the Pike's Peak district and at Florissant, Teller Co., Colorado.

Use. — Same as for orthoclase; sometimes as an ornamental material (amazonstone).

Anorthoclase. Soda-microcline. — A triclinic feldspar with a cleavage-angle, *bc* (010) $\wedge$ (001), varying but little from 90°. Form like that of the ordinary feldspars. Twinning as with orthoclase; also polysynthetic according to the albite and pericline laws; but in many cases the twinning laminæ very narrow and hence not distinct. Rhombic section (see p. 542) inclined on *b*(010) 4° to 6° to edge *b/c*. G. − 2 57–2 60. Cleavage, hardness, luster, and color as with other members of the group. Optically −. Extinction-angle on *c*(001) +1° to +6°; on *b*(010) 4° to 10°. Bx$_a$ nearly $\perp$ *y*($\overline{2}$01). Dispersion $\rho > v$; horizontal, distinct. α = 1·519. β = 1·55. γ = 1·527. Axial angle variable with temperature, becoming in part monoclinic in optical symmetry between 86° and 264° C., but again triclinic on cooling; this is true of those containing little calcium.

Chiefly a soda-potash, feldspar $NaAlSi_3O_8$ and $KAlSi_3O_8$, the sodium silicate usually in larger proportion (2 : 1, 3 : 1, etc.), as if consisting of albite and orthoclase molecules. Calcium ($CaAl_2Si_2O_8$) is also present in relatively very small amount.

These triclinic soda-potash feldspars are chiefly known from the andesitic lavas of Pantelleria, southwest of Sicily. Most of these feldspars come from a rock, called pantellerite. Also prominent from the augite-syenite of southern Norway and from the "Rhomben-porphyr" near Oslo. In East Africa the feldspars of Kilima Njaro in Tanganyika Territory, and of Mt. Kenya, Kenya Colony, belong here. Here is referred also a feldspar in crystals, tabular ‖ *c*(001), and twinned according to the Manebach and less often Baveno laws occurring in the lithophyses of the rhyolite of Obsidian Cliff, Yellowstone Park. It shows the blue opalescence in a direction parallel with a steep orthodome (cf. p. 537). *Sanidine-anorthoclase* is the name proposed for the triclinic sanidine occurring as inclusions in trachyte at Dächelsberg and Drachenfels, Bavaria.

Albite-anorthite Series. *Plagioclase Feldspars**

The albite ($NaAlSi_3O_8$) and anorthite ($CaAl_2Si_2O_8$) molecules are completely miscible and together form an isomorphous series ranging from the pure soda feldspar at one end to the pure lime feldspar at the other end.

---

* The triclinic feldspars of this series, in which the two cleavages *b*(010) and *c*(001) are oblique to each other, are often called in general *plagioclase* (from πλάγιος, *oblique*).

Tschermak first established the isomorphous relations between these two molecules. X-ray study has shown an identity of structure throughout the series. The sodium and calcium atoms, on the one hand, and the silica and aluminum atoms, on the other, may replace each other in the structure. Various names have been given to minerals falling in certain positions in the isomorphous series. They are given below with the approximate range in composition commonly assigned to each.

	Albite Molecule	Anorthite Molecule
ALBITE	100 to 90 per cent	0 to 10 per cent
OLIGOCLASE	90 to 70 per cent	10 to 30 per cent
ANDESINE	70 to 50 per cent	30 to 50 per cent
LABRADORITE	50 to 30 per cent	50 to 70 per cent
BYTOWNITE	30 to 10 per cent	70 to 90 per cent
ANORTHITE	10 to 0 per cent	90 to 100 per cent

815

816

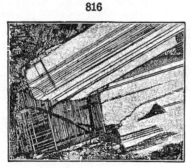

Plagioclase with twinning lamellæ. Fig. 815 section ‖ c(001) showing vibration-directions (cf. Fig. 819), ordinary light; Fig. 816 section in polarized light.

From albite to anorthite with the progressive change in composition (also specific gravity, melting points, etc.), there is also a corresponding change in crystallographic form, and in certain fundamental optical properties.

Crystalline Form. — The axial ratios and angles given on p. 535 show that these triclinic feldspars approach orthoclase closely in form, the most obvious difference being in the cleavage-angle bc (010) $\wedge$ (001), which is 90° in orthoclase, 86° 24′ in albite, and 85° 50′ in anorthite. There is also a change in the axial angle γ, which is 88° in albite, about 90° in oligoclase and andesine, and 91° in anorthite. This transition appears still more strikingly in the position of the "rhombic section," by which the twins according to the pericline law are united as explained below.

Twinning. — The plagioclase feldspars are often twinned in accordance with the Carlsbad,

817   818   819

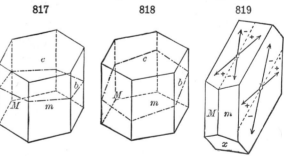

Fig. 817, rhombic section in albite. Fig. 818, same in anorthite. Fig. 819, typical form showing + and − extinction-directions on c(001) and b(010).

Baveno, and Manebach laws common with orthoclase (p. 536). Twinning is also almost universal according to the *albite law* — twinning plane the brachypinacoid; this is usually polysynthetic, i.e., repeated in the form of thin lamellæ, giving rise to fine striations on the basal cleavage surface (Figs. 815, 816). Twinning is also common according to the *pericline law* — twinning axis the crystal axis *b*; when polysynthetic this gives another series of fine striations seen on the brachypinacoid.

The composition-plane in this pericline twinning is a plane passing through the crystal in such a direction that its intersections with the prismatic faces and the brachypinacoid make equal plane angles with each other. The position of this rhombic section and the consequent direction of the striations on the brachypinacoid change rapidly with a small variation in the angle γ. In general it may be said to be approximately parallel to the base, but in albite it is inclined backward (+, Figs. 817 and 819) and in anorthite to the front (−, Fig. 818); for the intermediate species its position varies progressively with the composition. Thus for the average angle between the trace of this plane on the brachypinacoid and the edge b/c, we have for albite +21°; for oligoclase +6°; for andesine 0°; for labradorite −6°; for bytownite −12°; for anorthite −18°.

If the composition-plane is at right angles or nearly so to the basal plane, as happens in the case of microcline, the polysynthetic lamellæ then show prominently in a basal section, together with those due to the albite twinning. Hence the grating structure characteristic of microcline.

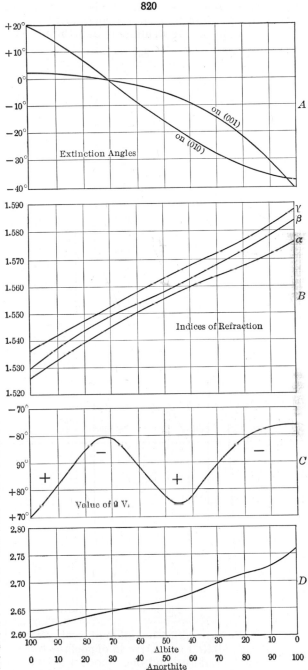

820

Optical Characters. — There is also a progressive change in the position of the X and Z directions and a consequent change in position of the optic axial plane in passing from albite to anorthite. This is most simply exhibited

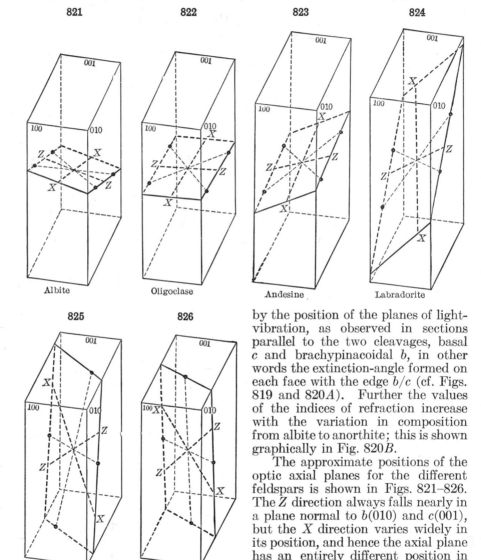

by the position of the planes of light-vibration, as observed in sections parallel to the two cleavages, basal c and brachypinacoidal b, in other words the extinction-angle formed on each face with the edge b/c (cf. Figs. 819 and 820A). Further the values of the indices of refraction increase with the variation in composition from albite to anorthite; this is shown graphically in Fig. 820B.

The approximate positions of the optic axial planes for the different feldspars is shown in Figs. 821–826. The Z direction always falls nearly in a plane normal to $b(010)$ and $c(001)$, but the X direction varies widely in its position, and hence the axial plane has an entirely different position in albite from what it has in anorthite.

The axial angle, 2V, is always large and its sign varies in different sections of the series, changing from positive with albite to negative in oligoclase-andesine, back again to positive in andesine-labradorite and once more negative in bytownite and anorthite. These variations are shown graphically in Fig. 820C. Another property of the plagioclase feldspars that varies with the composition is the specific gravity. This is illustrated in Fig. 820D.

Micro. — In rock sections the plagioclase feldspars are distinguished by their lack of color, low refractive relief, and low interference-colors, which in good sections are mainly dark gray and scarcely rise into white of the first order; also by their biaxial character in converging light. In the majority of cases they are easily told by the parallel bands or fine lamellæ which pass through them due to the multiple twinning according to the albite law; one set of bands or twin lamellæ exhibits in general a different interference-color from the other (cf. Figs. 815, 816). They are thus distinguished not only from quartz and orthoclase, with which they are often associated, but from all the common rock-making minerals. To distinguish the different species and sub-species from one another, as albite from labradorite or andesine, is more difficult. In sections having a definite orientation ($\parallel c(001)$ and $\parallel b(010)$) this can generally be done by determining the extinction angles (cf. p. 542 and Fig. 819). In general in rock sections special methods are required; these are discussed in the various texts devoted to this subject.

ALBITE.

Triclinic.   Axes $a : b : c = 0·6335 : 1 : 0·5577$; $\alpha = 94° 3'$, $\beta = 116° 29'$, $\gamma = 88° 9'$.

bc,	$010 \wedge 001$	$= 86° 24'$.
mM,	$110 \wedge 1\bar{1}0$	$= 59° 14'$.
bm,	$010 \wedge 110$	$= 60° 26'$.
cm,	$001 \wedge 110$	$= 65° 17'$.
cM,	$001 \wedge 1\bar{1}0$	$= 69° 10'$.
cx,	$001 \wedge \bar{1}01$	$= 52° 16'$.

Twins as with orthoclase; also very common, the tw. pl. $b(010)$, *albite law* (p. 542), usually contact-twins, and polysynthetic, consisting of thin lamellæ and with consequent fine striations on $c(001)$ (Fig. 830); tw. axis b axis, *pericline law*, contact-twins whose composition-face is the *rhombic section* (Figs. 817 and 832); often polysynthetic and showing fine striations which on $b(010)$ are inclined backward $+22°$ to the edge b/c.

827   828   829

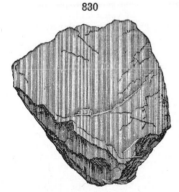

830

Crystals often tabular $\parallel b(010)$; also elongated $\parallel b$ axis as in the variety pericline. Also massive, either lamellar or granular; the laminæ often curved, sometimes divergent; granular varieties occasionally quite fine to impalpable.

Cleavage: $c(001)$ perfect; $b(010)$ somewhat less so; $m(110)$ imperfect. Fracture uneven to conchoidal. Brittle. H. = 6–6·5. G. = 2·60–2·62. Luster vitreous; on a cleavage surface often pearly. Color white; also occasionally bluish, gray, reddish, greenish, and green; sometimes having a bluish opalescence or play of colors on $c(001)$. Streak uncolored. Transparent to subtranslucent.

Optically $+$. Extinction-angle with edge $b/c = +3°$ to $2°$ on c, and $= +20°$ to $18°$ on b (Fig. 817). For position of axial plane, etc. see Fig. 821. Dispersion for Bx$_a$, $\rho < v$; also inclined, horizontal; Bx$_o$, $\rho > v$; inclined, crossed. α = 1·525. β = 1·529. γ = 1·536.   2V = 70°.   Birefringence weak, $\gamma - \alpha$ = 0·011.

Comp. — A silicate of aluminum and sodium, $NaAlSi_3O_8$ or $Na_2O.Al_2O_3$. $6SiO_2$ = Silica 68·7, alumina 19·5, soda 11·8 = 100. Calcium is usually present in small amount, as anorthite $(CaAl_2Si_2O_8)$, and as this increases it graduates through oligoclase-albite to oligoclase (cf. p. 547). Potassium may also be present, and it is then connected with anorthoclase and microcline.

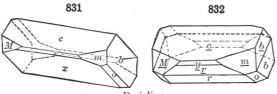

831   832

Pericline

Var. — *Ordinary.* In crystals and massive. The crystals often tabular ‖ b(010). The massive forms are usually nearly pure white, and often show wavy or curved laminæ. *Peristerite* is a whitish adularia-like albite, slightly iridescent, named from περιστερά, *pigeon.* *Aventurine* feldspar is a variety with reddish reflections from certain planes. These are thought to be due to thin lamellæ of hematite which lie along definite crystal planes of the feldspar. Their origin is probably due to an unmixing of an originally homogeneous feldspar that contained some iron oxide in solid solution. These lamellæ disappear when the feldspar is heated over 1200° C. A *moonstone* variety also occurs. *Pericline* from the chloritic schists of the Alps is in rather large opaque white crystals, with characteristic elongation in the direction of the *b* axis, as shown in Figs. 831 and 832, and commonly twinned with this as the twinning axis (pericline law).

Pyr., etc. — B.B. fuses at 4 to a colorless or white glass, imparting an intense yellow to the flame. Fusing point for pure albite = 1100° C. Not acted upon by acids.

Diff. — Resembles barite in some forms, but is harder and of lower specific gravity; does not effervesce with acid (like calcite). Distinguished optically and by the common twinning striations on c(001) from orthoclase; from the other triclinic feldspars partially by specific gravity and better by optical means (see p. 543).

Artif. — Albite acts, in regard to its artificial formation, like orthoclase, which see.

Alter. — While more stable than anorthite albite may change to kaolin, sericite, etc., under action of hot waters.

Obs. — Albite is a constituent of many igneous rocks, especially those of the alkaline type, as granite, syenite, diorite, etc.; also in the corresponding feldspathic lavas. It is more commonly found in the more acid rock types. In perthite (p. 540) it is interlaminated with orthoclase or microcline, and similar aggregations, often on a microscopic scale, are common in many rocks. Albite is common also in gneiss and occurs sometimes in the crystalline schists. It occurs in pegmatite veins, often associated with such rare minerals as beryl, tourmaline, allanite, columbite, etc. It may be deposited in veins and rock cavities from hot aqueous solutions. Also it is found in disseminated crystals in granular limestone and marble.

Albite is very common in its occurrence and only its most interesting localities are given here. Some of the most prominent European occurrences are in cavities and veins in the granite or gneiss rocks of the Swiss and Austrian Alps, etc. Thus in Switzerland in the St. Gotthard region in Ticino; in Valais in the Binnental, at Fiesch on the Rhône; in Grisons at Alp Ruschuna near Vals; in Uri at Bristenstock near Amsteg in the Maderaner Tal; on Mt. Scopi east of Lukmanier in the Val Medels (*pericline* with chlorite on the crystal surfaces and also partly within the crystals), etc. In Austria, in the Tyrol, at Schmirn in the druses of a dolomite and in the Zillertal; in Salzburg at Rauris and in the Habach Tal. In Trentino, Italy, it is found in the Pfitschtal. Other European localities are Mursinsk, north of Ekaterinburg, Ural Mts.; at Andreasberg in the Harz Mts. (*zygadite*); at Epprechstein west of Kirchenlamitz in the Fichtelgebirge, Bavaria. In France fine crystals come from Roc Tourné near Bourget, which is near Modane, Savoie; also from the district of Bourg d'Oisans, Isère. In Norway at Arendal in Aust-Agder (*hyposclerite*); Kragerö in Telemark; Snarum in Buskerud (*olafite*) and from Seiland in Finmarken. From the Mourne Mts., Co. Down, Ireland. Fine crystals come from the Morro Velho gold mine on Rio das Velhas, northwest of Ouro Preto, Minas Geraes, Brazil.

In the United States in the pegmatites of Maine at Auburn, Androscoggin Co.; Buckfield, Oxford Co., etc. At North Groton, Grafton Co., New Hampshire. In Massachusetts at Chesterfield, Hampshire Co., in lamellar masses (*cleavelandite*); in Connecticut at Haddam and Middletown, Middlesex Co.; at Branchville, Fairfield Co. In New York at Macomb and Pierrepont, St. Lawrence Co. In Virginia, at the mica mines near Amelia Court House, Amelia Co., in splendid crystallizations. In Colorado in the Pike's Peak region, Teller Co., with smoky quartz and amazonstone. From Barringer Hill, Llano Co., Texas. In Canada fine crystals at Wicklow, Hastings Co., Ontario. *Peristerite* occurs in Bathurst, Lanark Co., Ontario.

The name *albite* is derived from *albus*, white, in allusion to its common color.

Use. — Same as orthoclase but not so commonly employed; some varieties which show an opalescent play of colors when polished form a part of the ornamental material known as *moonstone*.

THE INTERMEDIATE PLAGIOCLASE FELDSPARS: OLIGOCLASE, ANDESINE, LABRADORITE, BYTOWNITE.

Since these do not constitute definite species but represent instead only certain sections of the albite-anorthite isomorphous series, it seems best to treat them together under one heading.

Triclinic. For crystal constants see table on p. 535. The cleavage angles, $bc(010) \wedge (001)$ are: oligoclase $= 86° 32'$; andesine $= 86° 14'$; labradorite $= 86° 4'$; bytownite $= 85° 56'$. Twinning as with albite. Crystals not common. Usually massive, cleavable, granular or compact.

Cleavages as in albite. H. $= 5$–6. G. $= 2\cdot60$–$2\cdot75$ (see Fig. 820D). Color white, gray, greenish, yellowish, brown, reddish; at times colorless. A play of colors is a common character, especially with certain labradorites. Blue and green are the predominant colors; but yellow, fire-red, and pearl-gray also occur. This effect has been shown to be, at least largely, due to the interference of light caused by reflection from thin lamellar inclusions of various minerals. These inclusions lie parallel to $b(010)$ or on a plane inclined at $15°$ to b. Varieties with aventurine reflections also occur. Transparent to subtranslucent. For the optical characters of the series see p. 544 and Figs. 822–825.

Comp. — For variations in composition see p. 542. A small percentage of the potash feldspar molecule is very commonly present; also rarely the barium feldspar molecule.

Pyr., etc. — Fusible in thin splinters B.B. with increasing difficulty toward the anorthite end of series. Insoluble or slightly attacked by HCl, the solubility increasing toward the anorthite end.

Obs. — *Oligoclase* is a common rock-making mineral, found especially in the more acid igneous rocks, as granite, gneiss, syenite, diorite, and the porphyries, and also in various effusive rocks, like andesite, trachyte, etc. It is sometimes associated with orthoclase in granite and similar rocks. Notable localities for its occurrence are in Aust-Agder, Norway at Arendal and nearby Tromö in crystals sometimes 2 to 3 inches long; and as the variety *sunstone* at Tvedestrand. In Sweden in the neighborhood of Stockholm at Ytterby, etc. In the United States in good crystals at Fine, St. Lawrence Co., New York. In Pennsylvania at Mineral Hill, Delaware Co., and at various points in Chester Co. At Bakersville, Mitchell Co., North Carolina, in clear glassy crystals, showing cleavage but no twinning. Named from ὀλίγος, *little*, and κλάσις, *fracture*.

Andesine occurs in many granular and volcanic rocks. It is less common than oligoclase and is especially characteristic of the rocks with medium silica content. Thus it is found in the Andes at Marmato, Colombia, as an ingredient of the rock called *andesite*, and in similar rocks elsewhere. From Bodenmais, Bavaria; in the rock known as *tonalite* from the Mte. Adamello district in Trentino, Italy. In France it occurs as crystals in a porphyry in the neighborhood of Saint-Raphaël in the Esterel, Var. Small crystals come from Mayeyama, Shinano, Japan.

Labradorite is an essential constituent of various igneous rocks, both plutonic and volcanic. It is especially characteristic of the more basic types and usually is associated with some member of the pyroxene or amphibole groups. Thus, it occurs in diorite, gabbro, norite, andesite, basalt, etc. At times it is found in amphibolites. Such rocks are most common among the formations of the Archæan era and are to be found in eastern Canada, northern New York, Greenland, Norway, Sweden, Finland, etc. It seldom occurs in distinct crystals. These are, however, found in the ashes of Monti Rossi on Etna in Sicily; in the quartz andesite of Verespátak, northeast of Abrudbánya, Siebenbürgen, Rumania. It is found in cleavable masses, frequently showing the characteristic play of colors, at various places, the most famous being in eastern Labrador where it occurs over a considerable area in an anorthosite associated with hornblende, hypersthene, and magnetite. Also found in Ontario and Quebec. It occurs abundantly through the central Adirondack region in northern New York.

Bytownite is uncommon, occurring rarely in certain basic plutonic and volcanic rocks. It was originally described from Bytown (now Ottawa) Ontario. This occurrence has been shown to be a mixture of anorthite with quartz, etc.

Use. — The varieties showing a play of colors are used as ornamental material.

MASKELYNITE. In colorless isotropic grains in meterorites; composition near labradorite. Probably represents a re-fused feldspar rather than an original glass.

ANORTHITE. Indianite.

Triclinic. Axes $a : b : c = 0.6347 : 1 : 0.5501$; $\alpha = 93° 13'$, $\beta = 115° 55\frac{1}{2}'$, $\gamma = 91° 12'$.

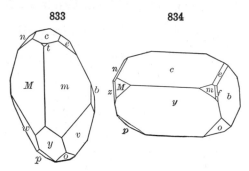

833   834

bc,	$010 \wedge 001$	$= 85° 50'$.
mM,	$110 \wedge \bar{1}10$	$= 59° 29'$.
bm,	$010 \wedge 110$	$= 58° 4'$.
cm,	$001 \wedge 110$	$= 65° 53'$.
cM,	$001 \wedge 1\bar{1}0$	$= 69° 20'$.
cy,	$001 \wedge \bar{2}01$	$= 81° 14'$.

Twins as with albite (p. 542 and p. 545). Crystals usually prismatic ‖ c axis (Fig. 833, also Fig. 384, p. 164), less often elongated ‖ b axis, like pericline (Fig. 834). Also massive, cleavable, with granular or coarse lamellar structure.

Cleavage: $c(001)$ perfect; $b(010)$ somewhat less so. Fracture conchoidal to uneven. Brittle. H. = 6–6.5. G. = 2.74–2.76. Color white, grayish, reddish. Streak uncolored. Transparent to translucent.

Optically —. For position of ax. pl. see Fig. 826. Extinction-angles on $c(001)$, −40° with edge b/c; on $b(010)$, −38° (Fig. 819, p. 542). Dispersion $\rho > v$. 2V = 77°. $\alpha = 1.575$. $\beta = 1.583$. $\gamma = 1.588$. Birefringence stronger than with albite.

Comp. — A silicate of aluminum and calcium, $CaAl_2Si_2O_8$ or $CaO.Al_2O_3$. $2SiO_2$ = Silica 43.2, alumina 36.7, lime 20.1 = 100. Soda (as $NaAlSi_3O_8$) is usually present in small amount, and as it increases there is a gradual transition through bytownite to labradorite. A small amount of $KAlSi_3O_8$ usually present.

Var. — *Anorthite* was described from the glassy crystals of Mte. Somma, Vesuvius; and *christianite* and *biotine* are the same mineral. *Thiorsauite* is the same from Iceland. *Indianite* is a white, grayish, or reddish granular anorthite from India, where it occurs as the gangue of corundum, first described in 1802 by Count Bournon. *Cyclopite* occurs in small, transparent, and glassy crystals, tabular ‖ $b(010)$, coating cavities in the dolerite of the Cyclopean Islands and near Trezza on Etna. *Lepolite, latrobite* also belong to anorthite. *Amphodelite* and *polyargite* are altered anorthite.

Pyr., etc. — B.B. fuses at 5 to a colorless glass. Artificial anorthite fuses at 1550° C. Anorthite from Mte. Somma, and indianite from the Carnatic, India, are decomposed by hydrochloric acid, with separation of gelatinous silica.

Artif. — Anorthite is the easiest of the feldspars to be formed artificially. Unlike the alkalic feldspars it can be easily formed in a dry fusion of its constituents. This method becomes progressively more difficult as the albite molecule is added to the composition. Anorthite is frequently observed in slags and is easily produced in artificial magmas. It further is often produced when more complex silicates are broken down by fusion.

Obs. — Anorthite is characteristically found in the basic igneous rocks, both volcanic and plutonic, as in andesite, basalt, diorite, gabbro, norite, etc. At times in rocks containing chrysolite; also in amphibolites. It occurs also in druses of ejected volcanic blocks and in the granular limestone of contact deposits. It is found as a constituent of some meteorites.

Anorthite occurs at Vesuvius in isolated blocks among the old lavas in the ravines of Mte. Somma, associated with sanidine, augite, mica, spinel, meionite, leucite. In amygdaloidal cavities in basalt on the Cyclopean Islands northeast of Catania, Sicily (*cyclopite*). In Trentino, Italy, on Pesmeda Alp at Monzoni, Val di Fassa, as translucent to cloudy crystals in calcite at contact deposits. In Södermanland, Sweden, at Tunaberg and at Åker (*rosite*). In Uudenmaan in Finland at Lojo (*amphodelite*) and near Orijärvi west of Lake Lojo (*lepodite* and *lindsayite*). In the lavas of Salem in the Carnatic district of Madras, India (*indianite*), associated with corundum. In the lavas of the island of Miyake, Japan, in large, white, porphyritic crystals covered with a black crust of lava. In crystals from Franklin, Sussex Co., New Jersey.

Anorthite was named in 1823 by Rose from ανορθός, *oblique*, the crystallization being triclinic.

Anemousite. — A feldspar having the composition, $Na_2O.2CaO.3Al_2O_3.9SiO_2$. This does not agree with any possible member of the albite-anorthite series. This is explained by assuming the presence in small amount of a sodium-anorthite molecule, $Na_2O.Al_2O_3.2SiO_2$, to which the name *carnegieite* has been given (see further under nephelite). Cleavage angle = 85° 59'. G. = 2·68. $\alpha = 1·555$. $\beta = 1·559$. $\gamma = 1·563$. 2V = 82° 48'. Found as loose crystals on Mte. Rosso, Island of Linosa, south of Sicily. Name derived from the ancient Greek name of the island. *Carnegieite* is named in honor of Andrew Carnegie.

II. Metasilicates. $RSiO_3$

Salts of Metasilicic Acid, H_3SiO_3; characterized by an oxygen ratio of 2 : 1 for silicon to bases. The Division closes with a number of species, in part of somewhat doubtful composition, forming a transition to the Orthosilicates.

The metasilicates include two prominent and well-characterized groups, viz., the Pyroxene Group and the Amphibole Group. There are also others less important.

Leucite Group. Isometric

In several respects leucite is allied to the species of the FELDSPAR GROUP, which immediately precede.

Leucite	$KAl(SiO_3)_2$	Isometric at 500°
	Pseudo-isometric at ordinary temperatures.	
Pollucite	$H_2Cs_4Al_4(SiO_3)_9$	Isometric

LEUCITE. Amphigène.

Isometric at about 600° C.; pseudo-isometric under ordinary conditions (see p. 331). Commonly in crystals varying in angle but little from the trapezohedron $n(211)$, sometimes with $a(100)$, and $d(110)$ as subordinate forms. Faces often showing fine striations due to twinning (Fig. 835). Also in disseminated grains; rarely massive granular.

Cleavage: $d(110)$ very imperfect. Fracture conchoidal. Brittle. H. = 5·5–6. G. = 2·45–2·50. Luster vitreous. Color white, ash-gray or smoke-gray. Streak uncolored. Translucent to opaque. Usually shows very feeble double refraction: $\alpha = 1.508$, $\gamma = 1.509$ (p. 331). Under crossed nicols shows weakly birefringent twinning bands.

Comp. — $KAl(SiO_3)_2$ or $K_2O.Al_2O_3.4SiO_2$ = Silica 55·0, alumina 23·5, potash 21·5 = 100.

Soda is present only in small quantities, unless as introduced by alteration; traces of lithium, also of rubidium and cæsium, have been detected. Leucite and analcite are closely related chemically as is shown by the fact that the two species can be converted into each other when heated with sodium or potassium chlorides or carbonates.

835

Pyr., etc. — B.B. infusible; with cobalt solution gives a blue color (aluminum). Decomposed by hydrochloric acid without gelatinization.

Diff. — Characterized by its trapezohedral form, absence of color, and infusibility. It is softer than garnet and harder than analcite; the latter yields water and fuses.

Micro. — Recognized in thin sections by its extremely low refraction, isotropic character, and the symmetrical arrangement of inclusions of glass, magnetite, etc. (Fig. 836; also Fig. 511, p. 202). Larger crystals are commonly not wholly isotropic and, further, show complicated systems of twinning-lines (Fig. 835); the birefringence is, however, very low, and the colors scarcely rise above dark gray; they are best seen by introduction of the quartz or gypsum plate yielding red of the first order. The smaller leucites, which lack this twinning or the inclusions, are only to be distinguished from sodalite or analcite by chemical tests.

836

Leucite crystals from the leucitite of the Bearpaw Mts., Montana (Pirsson). These show the progressive growth from skeleton forms to complete crystals with glass inclusions.

Artif. — Leucite is easily prepared artificially by simply fusing together its constituents in proper proportion and allowing the melt to crystallize slowly. The addition of potassium vanadate produces larger crystals. Leucite has been formed when microcline and biotite were fused together and also when muscovite was fused alone.

Obs. — Leucite occurs only in igneous rocks, and especially in the recent lavas, as one of the products of crystallization of magmas rich in potash and low in silica, and is associated with minerals having a high content of alumina and alkalies or with the ferromagnesian minerals. It is rarely observed in deep-seated rocks, evidently because it is easily altered and because under conditions of high pressure orthoclase is the more stable combination. Its former presence in such rocks is indicated by pseudomorphs, often of large size (*pseudo-leucite*), consisting of nephelite and orthoclase, of analcite, etc. It is found in leucitites, leucite-basalts, leucitophyres, leucite-phonolites, and leucite-tephrites; also in certain dike rocks. Leucite and a fused glass are formed when orthoclase is heated above 1170° C. This fact explains the unusual occurrences of leucite in rocks that contain enough potash ordinarily to form orthoclase and even in rocks where quartz is a product of a later crystallization. Leucite is found at times as loose crystals evidently ejected by explosive volcanic action.

Leucite is a comparatively rare mineral being prominent chiefly in the lavas of Italy and central Europe. Found in Italy on Vesuvius and Mte. Somma, where it is thickly disseminated through the lava in grains, and in large perfect crystals; also in the ejected masses. It occurs in central Italy from near Lago Bolsena to the north of Rome, where it occurs in a leucite-tephrite to Mti. Albani south of Rome, where some of the lavas appear to be almost entirely composed of it. Prominent localities are Capo di Bove, Albano, Ariccia, and Frascati. Further south it occurs in the lavas of the volcano Roccamonfina, east of Gaeta. In Rhineland leucite is found in the rocks of the Eifel district at Rieden near Andernach and at Laacher See. *Pseudoleucite* occurs in the rocks of the Kaiserstuhl in Baden and of Oberwiesental in the Erzgebirge, Saxony.

In the United States at Hamburg, Sussex Co., New Jersey (*analcite-pseudoleucite*); Magnet Cove near Hot Springs, Garland Co., Arkansas (*nephelite-orthoclase-pseudoleucite*);

in the Leucite Hills in the Green River Basin, Wyoming; in the Absaroka Range in the Yellowstone Park, Wyoming; in the Highwood and Bearpaw Mts., Montana (in part *pseudoleucite*). On the shores of Vancouver Island, British Columbia, where magnificent groups of crystals have been found as drift boulders.

Named from λευκός, *white*, in allusion to its color.

Pollucite. — Essentially $H_2O.2Cs_2O.2Al_2O_3.9SiO_2$. Isometric; often in cubes; also massive. H. = 6·5. G. = 2·901. Colorless. $n = 1·525$. Occurs very sparingly in the island of Elba, with petalite (castorite); also at Hebron and Rumford, Maine.

Ussingite. — $HNa_2Al(SiO_3)_3$. Triclinic. Three cleavages. G. = 2·5. H. = 6–7. Color reddish violet. Indices, 1·50–1·55. Easily fusible. Soluble in hydrochloric acid. Found in rolled masses from pegmatite at Kangerdluarsuk, Greenland; also from Kola Peninsula, Russian Lapland.

Pyroxene Group

Orthorhombic, Monoclinic, Triclinic

Composition for the most part that of a metasilicate, $RSiO_3$, with R = Ca,Mg,Fe chiefly, also Mn,Zn. Further $RSiO_3$ with $\overset{\text{II}}{R}(Fe,Al)_2SiO_6$, less often containing alkalies (Na,K), and then $RSiO_3$ with $\overset{\text{I}}{R}Al(SiO_3)_2$. Rarely including zirconium and titanium, also fluorine.

Orthorhombic Section

		$a : b : c$
Enstatite	$MgSiO_3$	1·0308 : 1 : 0·5885
Bronzite	$(Mg,Fe)SiO_3$	
Hypersthene	$(Fe,Mg)SiO_3$	1·0319 : 1 : 0·5868

The positions of the a and b axes are reversed from the usual orthorhombic orientation in order to emphasize the similarity of form between the orthorhombic and monoclinic pyroxenes.

Monoclinic Section

		$a : b : c$	β
Pyroxene		1·0921 : 1 : 0·5893	74° 10′

1. CLINOENSTATITE $MgSiO_3$
2. PIGEONITE   Intermediate between clinoenstatite and diopside
3. DIOPSIDE $CaMg(SiO_3)_2$
4. HEDENBERGITE $CaFe(SiO_3)_2$
5. AUGITE $CaMg(SiO_3)_2$ with $(Mg,Fe)(Al,Fe)_2SiO_6$

		$a : b : c$	β
Acmite (Ægirite)	$Na\overset{\text{III}}{Fe}(SiO_3)_2$	1·0996 : 1 : 0·6012	73° 11′
Jadeite	$NaAl(SiO_3)_2$	1·103 : 1 : 0·613	72° 44½′
Spodumene	$LiAl(SiO_3)_2$	1·1238 : 1 : 0·6355	69° 40′

Triclinic Section

		$a : b : c$	α	β	γ
Rhodonite	$MnSiO_3$	1·0729 : 1 : 0·6213	103° 18′	108° 44′	81° 39′
Babingtonite	$(Ca,Fe,Mn)SiO_3.Fe_2(SiO_3)_3$	1·0691 : 1 : 0·6308	104° 21½′	108° 31′	83° 34′

The PYROXENE GROUP embraces a number of species which, while falling in different systems — orthorhombic, monoclinic, and triclinic — are yet closely related in form. Thus all have a fundamental prism with an angle of 93° and 87°, parallel to which there is more or less distinct cleavage. Further, the angles in other prominent zones show a considerable degree of similarity. In composition the metasilicates of calcium, magnesium, and ferrous iron are most prominent, while compounds of the form $\overset{\text{II}}{R}(Al,\overset{\text{III}}{Fe})_2SiO_6$, $\overset{\text{I}}{R}Al(SiO_3)_2$ are also important.

The atomic structure of the monoclinic pyroxenes, as shown by X-ray analysis, has the following characteristics. Each silicon atom lies in the center of a tetrahedron with four oxygen atoms at its points. These tetrahedral groups are linked together into chains by the sharing of one oxygen atom between two adjacent groups, i.e., in each group two oxygen atoms also belong half to the groups on either side. This makes the silicon-oxygen ratio equal SiO_3. These chains of silicon-oxygen tetrahedra lie parallel to the vertical crystal axis and are bound together laterally by the calcium and magnesium atoms. The magnesium atoms lie within a group of six oxygen atoms while the calcium atoms are within a somewhat irregular group of eight oxygen atoms. The prismatic cleavage of pyroxene takes place between the silicon-oxygen chains.

The structure of the orthorhombic pyroxene, enstatite, is similar to that of the monoclinic pyroxene. The unit cell of enstatite corresponds very closely to two unit cells of pyroxene united by their (100) faces, one the reflection of the other. The unit cell gives an axial ratio in which the value of a (or b in the orientation used) is twice that usually given. Each cell contains sixteen molecules.

The species of the pyroxene group are closely related in composition to the corresponding species of the amphibole group, which also embraces members in the orthorhombic, monoclinic, and triclinic systems. In a number of cases the same chemical compound appears in each group; furthermore, a change by paramorphism of pyroxene to amphibole is often observed. In form also the two groups are related, as shown in the axial ratio; also in the parallel growth of crystals of monoclinic amphibole upon or about those of pyroxene (Fig. 487, p. 195). The axial ratios for the typical monoclinic species are:

Pyroxene $\quad a : b : c = 1\cdot0921 : 1 : 0\cdot5893 \quad \beta = 74° 10'$
Amphibole $\quad a : \frac{1}{2}b : c = 1\cdot1022 : 1 : 0\cdot5875 \quad \beta = 73° 58'$

See further on p. 569.

The optical relations of the prominent members of the Pyroxene Group, especially in regard to the connection between the position of the ether-axes and the crystallographic axes are exemplified in the following figures (Cross).

837

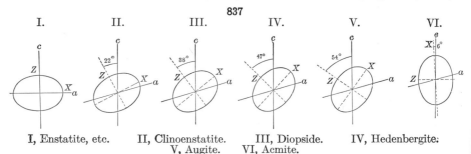

I, Enstatite, etc.   II, Clinoenstatite.   III, Diopside.   IV, Hedenbergite.
V, Augite.   VI, Acmite.

A corresponding exhibition of the prominent amphiboles is given under that group, Fig. 864, p. 570.

Orthorhombic Section

The orthorhombic pyroxenes enstatite and hypersthene form a continuous isomorphous series for which the name *enstenite* has been proposed. Commonly by enstatite is designated the $MgSiO_3$ end of the series including mixtures up to those containing 15 per cent of FeO. These are optically positive. The remainder of the series is optically negative and is known as hypersthene.

ENSTATITE.

Orthorhombic. Axes $a : b : c = 0.9702 : 1 : 0.5710$; (see also p. 551).

mm''', $110 \wedge 1\bar{1}0 = 88° 16'$.	rr', $223 \wedge \bar{2}23 = 40° 16\frac{1}{2}'$.
qq', $023 \wedge 0\bar{2}3 = 41° 41'$.	rr''', $223 \wedge 2\bar{2}3 = 39° 1\frac{1}{2}'$.

Twins rare: tw. pl. $h(014)$ as twinning lamellæ; also tw. pl. (101) as stellate twins crossing at angles of nearly 60°, sometimes six-rayed. Distinct crystals rare, habit prismatic. Usually massive, fibrous, or lamellar. See p. 552 for discussion of atomic structure and relation to monoclinic pyroxene.

Cleavage: $m(110)$ rather easy. Parting ∥ $b(010)$; also $a(100)$. Fracture uneven. Brittle. H. = 5·5. G. = 3·1–3·3. Luster, a little pearly on cleavage-surfaces to vitreous; often metalloidal in the bronzite variety. Color grayish, yellowish or greenish white, to olive-green and brown. Streak uncolored, grayish. Translucent to nearly opaque. Pleochroism weak, more marked in varieties relatively rich in iron. Optically +. Ax. pl. ∥ $b(010)$. $Bx_a \perp c(001)$. Cf. Fig. 839. Dispersion $\rho < v$ weak. Axial angle large and variable, increasing with the amount of iron, usually about 90° for FeO = 10 per cent. $\alpha = 1.650$, $\beta = 1.653$, $\gamma = 1.658$ (artif. $MgSiO_3$). $\alpha = 1.6607$, $\beta = 1.6658$, $\gamma = 1.6715$ (containing about 5 per cent FeO).

838

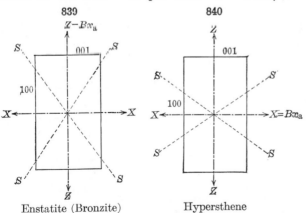

Bamle

Comp. — $MgSiO_3$ or $MgO.SiO_2$ = Silica 60, magnesia 40 = 100. Practically always contains iron giving the general formula, (Mg, Fe)SiO_3. The percentage of FeO present may vary up to about 15.

839

Z–Bx_a

Enstatite (Bronzite)

840

Hypersthene

Var. — 1. *With little or no iron.* Color white, yellowish, grayish, or greenish white; luster vitreous to pearly; G. = 3·10–3·13. *Chladnite* (*Shepardite* of Rose), which makes up 90 per cent of the Bishopville meteorite, belongs here and is the purest kind. *Victorite.* occurring in the Deesa meteoric iron in rosettes of acicular crystals, is similar.

2. *Ferriferous.* Color grayish green to olive-green and brown. Luster on cleavage surface often adamantine-pearly to submetallic or bronze-like (var. *bronzite*); this, however, is usually of secondary origin and is not essential. With the increase of iron passes to hypersthene, the optic axial angle changing so that in the latter $X = Bx_a \perp a$ (100). This is illustrated by Figs. 839, 840.

Pyr., etc. — B.B. almost infusible, being only slightly rounded on the thin edges; F. = 6. Insoluble in hydrochloric acid.

Artif. — Enstatite is formed from a melt having the proper composition at temperatures slightly under 1100°. At higher temperatures the monoclinic pyroxenes appear. Enstatite has also been formed by fusing olivine with silica. When serpentine is melted it breaks down into enstatite and olivine.

Micro. — In thin sections is colorless or light yellow or green; marked relief; prominent cleavage with parallel extinction; little pleochroism but becoming stronger with increase of iron; inclusions common lying parallel to brachypinacoid, producing characteristic schiller of mineral.

Obs. — Enstatite is a common constituent of those igneous rocks that are low in percentages of lime and to a much less extent of metamorphic rocks. It is common in pyroxenites, peridotites, gabbros, and norites, also in diorites. It likewise occurs in the porphyritic and extrusive equivalents of these rocks. Frequently found in serpentine, etc., that have been derived from their alteration. It is often associated in parallel growth with a monoclinic pyroxene, *e.g.*, diallage. A common mineral in both metallic and stony meteorites.

Occurs at the Zdjar-Berg in the Aloistal, Moravia of Czechoslovakia, in a serpentine-like rock called pseudophite; from Kraubat in the Mürztal, southwest of Leoben, Styria in Austria (*bronzite*); near Kupferberg, northwest of Marktschorgast in the Fichtelgebirge, Bavaria (*bronzite*); at Baste in the Radau-Tal, in the Harz Mts., (*protobastite*); in the so-called olivine bombs of the Dreiser Weiher near Daun in the Eifel, Rhineland (*bronzite*). In immense crystals at the apatite deposits of Kjörrestad near Bamble, Telemark, Norway, in part altered to steatite; also in steatite pseudomorphs at Snarum in Buskerud, Norway. In the peridotite associated with the diamond deposits of South Africa.

In the United States in New York at the Tilly Foster magnetite mine, Brewster, Putnam Co.; at Edwards, St. Lawrence Co. In Pennsylvania at Texas, Lancaster Co.; at Bare Hills near Baltimore, Maryland; in North Carolina at Webster, Jackson Co.

Named from ἐνσάτης, an *opponent*, because so refractory. The name *bronzite* has priority, but a bronze luster is not essential, and is far from universal.

HYPERSTHENE.

Orthorhombic. Axes $a : b : c = 0{\cdot}9713 : 1 : 0{\cdot}5704$; (see also p. 551).

mm''', 110 ∧ 1$\bar{1}$0 = 88° 20'. oo''', 111 ∧ 1$\bar{1}$1 = 52° 23'.
hh',   014 ∧ 0$\bar{1}$4 = 16° 14'. uu''', 232 ∧ 2$\bar{3}$2 = 72° 50'.

Crystals rare, habit prismatic, often tabular ‖ a(100), less often ‖ b(010). Usually foliated massive; sometimes in embedded spherical forms.

841   **842**   **843**

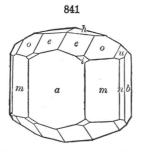 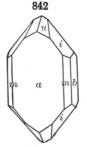 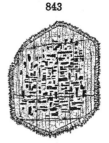

Figs. 841, *Amblystegite*, Laacher See; 842, Málnás; 843, Section ‖ b(010) showing inclusions; the exterior transformed to actinolite; from Lacroix.

Cleavage: m(110) distinct; parting b(010) good, a(100) rare. Fracture uneven. Brittle. H. = 5–6. G. = 3·40–3·50. Luster somewhat pearly on a cleavage-surface, and sometimes metalloidal. Color dark brownish green, grayish black, greenish black, pinchbeck-brown. Streak grayish, brownish gray. Translucent to nearly opaque. Pleochroism often strong, especially in the kinds with high iron percentage; thus ∥ X or a axis brownish red, Y or b axis reddish yellow, Z or c axis green. Optically −. Ax. pl. ∥ $b(010)$. Bx$_{ac}$ ⊥ $a(100)$. Dispersion $\rho > v$. Axial angle rather large and variable, diminishing with increase of iron, cf. enstatite, p. 553, and Figs. 839, 840, p. 553. Indices for variety with about 15 per cent FeO; $\alpha = 1.692$, $\beta = 1.702$, $\gamma = 1.705$.

Hypersthene often encloses minute tabular scales, usually of a brown color, arranged mostly parallel to the basal plane (Fig. 843); also less frequently vertical or inclined 30° to c axis; they may be brookite (goethite, hematite), but their true nature is doubtful. They are the cause of the peculiar metalloidal luster or schiller, and are often of secondary origin, being developed along the so-called " solution-planes " (p. 211).

Comp. — (Fe,Mg)SiO₃ with FeO greater than 15 per cent. Alumina is sometimes present (up to 10 per cent) and the composition then approximates to the aluminous pyroxenes.

Pyr., etc. — B.B. fuses to a black enamel, and on charcoal yields a magnetic mass; fuses more easily with increasing amount of iron. Partially decomposed by hydrochloric acid.

Micro. — In thin sections similar to enstatite except shows distinct reddish or greenish color with stronger pleochroism and is optically −.

Artif. — Similar to enstatite, which see.

Obs. — Hypersthene, associated with a triclinic feldspar (labradorite), is common in certain granular igneous rocks, as norite, hyperite, gabbro, and more commonly in the extrusive rocks, trachyte, and especially andesite, in which it occurs extensively in widely separated regions. It occurs in druses of eruptive rocks where it has evidently had a pneumatolytic origin. Also found in certain meteorites.

Hypersthene occurs in large crystals with pyrrhotite at Bodenmais, Bavaria; at various localities in Norway and Greenland. In large individuals with labradorite on the Isle St. Paul, Labrador, and in broad lamellar masses at Chateau Richer, Montmorency Co., Quebec. *Amblystegite* occurs in the sanidine bombs of the Laacher See district in the Eifel, Rhineland. *Szaboite* is from the andesite of Aranyer Berg near Piski on the Maros in Transylvania of Rumania.

In the United States it occurs in New York in the norites of the Cortlandt region on the Hudson River and in the Adirondack region. In the andesites of Buffalo Peaks, Park Co., Colorado, and of Mt. Shasta in northern California.

Hypersthene is named from ὑπέρ and σθένος, *very strong*, or *tough*.

BASTITE, or SCHILLER SPAR. An altered enstatite (or bronzite) having approximately the composition of serpentine. It occurs in foliated form in certain granular eruptive rocks and is characterized by a bronze-like metalloidal luster or schiller on the chief cleavage face $b(010)$, which " schillerization " (p. 275) is of secondary origin. H. = 3·5–4. G. = 2·5–2·7. Color leek-green to olive- and pistachio-green, and pinchbeck-brown. Pleochroism not marked. Optically −. Double refraction weak. Ax. pl. ∥ $a(010)$ (hence normal to that of enstatite). Bx$_a$ ⊥ $b(010)$. Dispersion $\rho > v$. The original bastite was from Baste near Harzburg in the Harz Mts., Germany; also from Todtmoos in the Schwarzwald, Germany.

PECKHAMITE, 2(Mg,Fe)SiO₃.(Mg,Fe)SiO₄. Occurs in rounded nodules in the meteorite of Estherville, Emmet Co., Iowa, May 10, 1879. G. = 3·23. Color light greenish yellow.

PYROXENE.

Monoclinic Section

Monoclinic. Axes $a : b : c = 1.0921 : 1 : 0.5893$; $\beta = 74° 10'$.

mm''',	$110 \wedge 1\bar{1}0$	$= 92° 50'$.
ca,	$001 \wedge 100$	$= 74° 10'$.
cp,	$001 \wedge \bar{1}01$	$= 31° 20'$.
ee',	$011 \wedge 0\bar{1}1$	$= 59° 6'$.
zz',	$021 \wedge 0\bar{2}1$	$= 97° 11'$.
cu,	$001 \wedge 111$	$= 33° 49\frac{1}{2}'$.

cv,	$001 \wedge 221$	$= 49° 54'$.
cm,	$001 \wedge 110$	$= 79° 9\frac{1}{2}'$.
cs,	$001 \wedge \bar{1}11$	$= 42° 2'$.
uu',	$111 \wedge 1\bar{1}1$	$= 48° 29'$.
ss',	$\bar{1}11 \wedge \bar{1}\bar{1}1$	$= 59° 11'$.
oo',	$\bar{2}21 \wedge \bar{2}\bar{2}1$	$= 84° 11'$.

Twins: tw. pl. (1) a(100), contact-twins, common (Fig. 850), sometimes polysynthetic. (2) c(001), as twinning lamellæ producing striations on the vertical faces and pseudocleavage or parting ‖ c(001) (Fig. 851); very common, often secondary. (3) y(101) cruciform-twins, not common (Fig. 477, p. 193). (4) W($\bar{1}$22) the vertical axes crossing at angles of nearly 60°; sometimes repeated as a six-rayed star (Fig. 476, p. 193). Crystals usually prismatic in habit, often short and thick, and either a square prism (a(100), b(010) prominent), or nearly square (93°, 87°) with m(110) predominating; some-

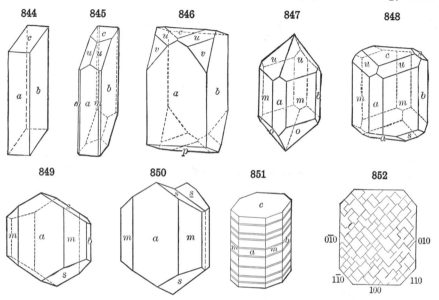

times a nearly symmetrical 8-sided prism with a, b, m (Fig. 851). Often coarsely lamellar, ‖ c(001) or a(100). Also granular, coarse or fine; rarely fibrous or columnar.

Cleavage: m(110) sometimes rather perfect, but interrupted, often only observed in thin sections ⊥ c axis (Fig. 852). Parting ‖ c(001), due to twinning, often prominent, especially in large crystals and lamellar masses (Fig. 851); also ‖ a(100) less distinct and not so common. Fracture uneven to conchoidal. Brittle. H. = 5–6. G. = 3·2–3·6, varying with the composition. Luster vitreous inclining to resinous; often dull; sometimes pearly ‖ c(001) in kinds showing parting. Color usually green of various dull shades, varying from nearly colorless, white, or grayish white to brown and black; rarely bright green, as in kinds containing chromium; also blue. Streak white to gray and grayish green. Transparent to opaque. Pleochroism usually weak, even in dark-colored varieties; sometimes marked, especially in violet-brown kinds containing titanium. (*Violaite* is name given to a highly pleochroic variety from the Caucasus Mts.)

Optically +. Birefringence strong, $(\gamma - \alpha) = 0·02 - 0·03$. Ax. pl. usually ‖ b(010). Bx_{ac} or $Z \wedge c$ axis = +36° in diopside, to +52° in augite (which see), or $Z \wedge c$(001) = 20° to 36°, the angle in general increasing with amount of iron. For diopside 2V = 59°. $\alpha = 1·664$, $\beta = 1·6715$, $\gamma = 1·694$.

Comp. — For the most part a normal metasilicate, $RSiO_3$, chiefly of calcium and magnesium, also iron, less often manganese and zinc. The alkali metals potassium and sodium present rarely, except in very small amount. Also in certain varieties containing the trivalent metals aluminum, ferric iron, and manganese. These last varieties may be most simply considered as molecular compounds of $Ca(Mg,Fe)Si_2O_6$ and $(Mg,Fe)(Al,Fe)_2SiO_6$, as suggested by Tschermak. Chromium is sometimes present in small amount; also titanium replacing silicon.

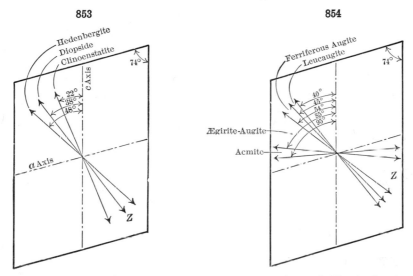

The name *pyroxene* is from πυρ, *fire* and ξενος, *stranger*, and records Haüy's idea that the mineral was, as he expresses it, " a stranger in the domain of fire," whereas, in fact, it is, next to the feldspars, the most universal constituent of igneous rocks.

The varieties are numerous and depend upon variations in composition chiefly; the more prominent of the varieties properly rank as sub-species.

1. CLINOENSTATITE. $MgSiO_3$. Study of artificial crystals show that they are monoclinic and either tabular after (100) or prismatic after (110) and always elongated parallel to the *c* axis. Polysynthetic twinning on *a*(100) very characteristic. Good prismatic cleavage at 88° 8′. Colorless to yellow. Ax. pl. ⊥ *b*(010). $Z \wedge c$ axis = 22°, see Fig. 837. $\alpha = 1·651$, $\beta = 1·654$, $\gamma = 1·660$. 2V = 53° 30′. Composition varies with iron replacing magnesium and grades into what has been called *clinohypersthene*. An increasing percentage of iron is accompanied with a rise in value of indices and of the extinction angle, $Z \wedge c$ axis.

$MgSiO_3$ can be crystallized from a melt having the theoretical composition at about 1500° C. or at a lower temperature from solution in molten calcium or magnesium vanadate. It is the most stable form of $MgSiO_3$. It has no true melting point but at 1557° C. breaks down into forsterite and silica.

Clinoenstatite occurs rarely in igneous rocks and meteorites. Clinohypersthene is known only in meteorites.

2. PIGEONITE. Intermediate between clinoenstatite and diopside, a mixture of the molecules $(Mg,Fe)SiO_3$ and $CaMg(SiO_3)_2$. A similar compound may occur in the series between clinoenstatite and hedenbergite. General physical properties like those of diopside. Ax. pl. $\perp$ (010) in varieties low in lime changing to $\parallel$ (010) at a content of 7 to 10 per cent CaO. Axial angle small and variable. Optically +. $Z \wedge c$ axis varies from 22° to 45°, increasing with lime content. Occurs at various points in diabase, basalt, gabbro, etc. Named from occurrence at Pigeon Point, Minnesota.

3. DIOPSIDE. Malacolite, Alalite. *Calcium-magnesium pyroxene.* Formula $CaMg(SiO_3)_2$ = Silica 55·6, lime 25·9, magnesia 18·5 = 100. Color white, yellowish, grayish white to pale green, and finally to dark green and nearly black; sometimes transparent and colorless, also rarely a fine blue. In prismatic crystals, often slender; also granular and columnar to lamellar massive. G. = 3·2–3·38. $Z \wedge c$ axis = $+36°$ to 40°. For indices see above. Iron is present usually in small amount as noted below, and the amount increases as it graduates toward true hedenbergite.

The following belong here: *Chrome-diopside*, contains chromium (1 to 2·8 per cent Cr_2O_3), often a bright green.

Malacolite, as originally described, was a pale-colored translucent variety from Sala, Sweden.

Alalite occurs in broad right-angled prisms, colorless to faint greenish or clear green, from the Mussa Alp in the Ala valley, Piedmont, Italy.

Traversellite, from Traversella, Piedmont, Italy, is similar.

Violan is a fine blue diopside from St. Marcel, Piedmont, Italy; occurring in prismatic crystals and massive.

Canaanite is a grayish white or bluish white pyroxene occurring with dolomite at Canaan, Connecticut.

Lavrovite is a pyroxene, colored green by vanadium, from the neighborhood of Lake Baikal, in eastern Siberia.

Mansjöite is a fluorine-bearing pyroxene composed largely of the diopside and hedenbergite molecules, occurring in contact metamorphic limestone from Mansjö Mountain, parish of Loos, Hälsingland, northern Sweden.

Diopside is named from δις, *twice* or *double*, and ὄψις, *appearance*. Malacolite is from μαλακός, *soft*, because softer than feldspar, with which it was associated.

4. HEDENBERGITE. *Calcium-iron pyroxene.* Formula $CaFe(SiO_3)_2$ = Silica 48·4, iron protoxide 29·4, lime 22·2 = 100. Color black. In crystals, and also lamellar massive. G. = 3·5–3·58. $Z \wedge c$ axis = $+46°$ to 48°. $\alpha = 1·739$, $\beta = 1·745$, $\gamma = 1·757$. Manganese is present in *manganhedenbergite* to 6·5 per cent. Color grayish green. G. = 3·55.

Between the two extremes, diopside and hedenbergite, there are numerous transitions conforming to the formula $Ca(Mg,Fe)Si_2O_6$. As the amount of iron increases the color changes from light to dark green to nearly black, the specific gravity increases from 3·2 to 3·6, and the angle $Z \wedge c$ axis also from 36° to 48°.

The following are varieties, coming under these two sub-species, based in part upon structure, in part on peculiarities of composition.

Salite (Sahlite), color grayish green to deep green and black; sometimes grayish and yellowish white; in crystals; also lamellar (parting $\parallel c(001)$), and granular massive; from Sala in Sweden. *Baikalite*, a dark, dingy green variety, in crystals, with parting $\parallel c(001)$, from Lake Baikal, in Siberia.

Coccolite is a granular variety, embedded in calcite, also forming loosely coherent to compact aggregates; color varying from white to pale green to dark green, and then containing considerable iron; the latter the original coccolite. Named from κόκκος, a *grain*.

DIALLAGE. A lamellar or thin-foliated pyroxene, characterized by a fine lamellar structure and parting $\parallel a(100)$, with also parting $\parallel b(010)$, and less often $\parallel c(001)$. Also a fibrous structure $\parallel c$ axis. Twinning $\parallel a(100)$, often polysynthetic; interlamination with an orthorhombic pyroxene common. Color grayish green to bright grass-green, and deep green; also brown. Luster of surface $a(100)$ often pearly, sometimes metalloidal or exhibiting schiller and resembling bronzite, from the presence of microscopic inclusions of secondary origin. Bx$_a$ $\wedge$ c axis = +39 to 40°; β = 1·681; $\gamma - \alpha$ = 0·024. H. = 4; G. = 3·2–3·35. In composition near diopside, but often containing alumina and sometimes in considerable amount, then properly to be classed with the augites. Often changed to amphibole, see smaragdite, and uralite, p. 574. Named from $\delta\iota\alpha\lambda\lambda\alpha\gamma\acute{\eta}$, *difference*, in allusion to the dissimilar planes of fracture. This is the characteristic pyroxene of gabbro, and other related rocks.

Omphacite. — The granular to foliated pyroxenic constituent of the garnet-rock called eclogite, often interlaminated with amphibole (smaragdite); color grass-green. Contains some Al_2O_3.

Schefferite. A manganese pyroxene, sometimes also containing much iron. Color brown to black.

In crystals, sometimes tabular $\parallel$ $c(001)$, also with $p(\bar{1}01)$ prominent, more often elongated in the direction of the zone $b(010)$: $p(\bar{1}01)$, rarely prismatic, $\parallel$ c axis. Twins, with $a(100)$ as tw. pl. very common. Also crystalline, massive. Cleavage prismatic, very distinct. Color yellowish brown to reddish brown; also black (*iron-scheferite*). Optically +. Z $\wedge$ c axis = 44° 25½′. The iron-scheferite from Pajsberg, Sweden, is black in color and has Z $\wedge$ c axis = +49° to 59° for different zones in the same crystal. The brown iron-scheferite (*urbanite*) from Långbanshyttan, Sweden, has Z $\wedge$ c axis = 69° 3′. It resembles garnet in appearance.

Jeffersonite is a manganese-zinc pyroxene from Franklin Furnace, New Jersey (but the zinc may be due to impurity). In large, coarse crystals with edges rounded and faces uneven. Color greenish black, on the exposed surface chocolate-brown.

Blanfordite. — A pyroxene containing some sodium, manganese and iron. Strongly pleochroic (rose-pink to sky-blue). Found with manganese ores in the Central Provinces, India.

5. Augite. *Aluminous pyroxene.* Composition chiefly $CaMgSi_2O_6$ with $(Mg,Fe)(Al,Fe)_2SiO_6$, and occasionally also containing alkalies and then graduating toward acmite. Titanium is also sometimes present. There are various explanations of the composition of the pyroxenes containing the sesquioxides; their exact character must still be considered as unsettled. Here belong:

a. Leucaugite. Color white or grayish. Contains alumina, with lime and magnesia, and little or no iron. Looks like diopside. Z $\wedge$ c axis = + 40° to 45°. H. − 6·5; G. = 3·19. Named from $\lambda\epsilon\upsilon\kappa\acute{o}s$, *white*.

b. Fassaite. Includes the pale to dark, sometimes deep-green crystals, or pistachio green and then resembling epidote. The aluminous kinds of diallage also belong here. Named from the locality in the Fassatal, Tyrol. *Pyrgom* is from $\pi\acute{\upsilon}\rho\gamma\omega\mu\alpha$, a *tower*.

c. Augite. Includes the greenish or brownish black and black kinds, occurring mostly in eruptive rocks. It is usually in short prismatic crystals, thick and stout, or tabular $\parallel$ $a(100)$; often twins (Figs. 849, 850). Ferric iron is here present in a relatively large amount and the angle Z $\wedge$ c axis ranges from + 45° to 54°. Indices variable, β = 1·69 − 1·71. When rich in iron or titanium becomes distinctly pleochroic; X = greenish yellow or reddish, Y = brownish red or violet, Z = greenish yellow, reddish, or violet. Named from $\alpha\upsilon\gamma\acute{\eta}$, *luster*.

d. Ægirite-augite. Here belong varieties of augite characterized by the presence of alkalies, especially soda, probably in the form of the acmite molecule, $NaFe(SiO_3)_2$. (Z $\wedge$ c axis = 54° to 85°, Fig. 854.) As the percentage of the acmite molecule increases there is an increase in value of indices, of 2V, and in the extinction angle Z $\wedge$ c axis. Strongly pleochroic, X and Y being shades of green, Z being yellow. Known chiefly from rocks rich in alkalies, as nephelite-syenite, phonolite, leucitite, etc.

Pyr., etc. — Varying widely, owing to the wide variations in composition in the different varieties, and often by insensible gradations. Fusibility, 3·75 in diopside; 3 in jeffersonite and augite; 2·5 in hedenbergite. Varieties rich in iron afford a magnetic globule when fused on charcoal, and in general the fusibility varies with the amount of iron. Many varieties give with the fluxes reactions for manganese. Most varieties are unacted upon by acids.

Diff. — Characterized by monoclinic crystallization and the prismatic angle of 87° and 93°, hence yielding nearly square prisms; these may be mistaken for scapolite if terminal faces are wanting or indistinct (but scapolite fuses easily B.B. with intumescence). The oblique parting (|| c(001), Fig. 851) often distinctive, also the common dull green to gray and brown colors. Amphibole differs in prismatic angle ($55\frac{1}{2}°$ and $124\frac{1}{2}°$) and cleavage, and in having common columnar to fibrous varieties, which are rare with pyroxene. (See also p. 571.)

Micro. — The common rock-forming pyroxenes are distinguished in thin sections by their high relief; usually greenish to olive tones of color; distinct system of interrupted cleavage-cracks crossing one another at nearly right angles in sections $\perp$ c axis (Fig. 852); high inter-ference-colors; general lack of pleochroism; large extinction-angle, 35° to 50° and higher, for sections || b(010). The last-named sections are easily recognized by showing the highest interference colors; yielding no optical figures in convergent light and having parallel cleavage-cracks, the latter in the direction of the vertical axis. See also ægirite, p. 561.

855

A zonal banding is common, the successive laminæ sometimes differing in extinction angle and pleochroism; also the hour-glass structure occasionally distinct (Fig. 855, from Lacroix).

Alter. — Pyroxene undergoes alteration in different ways. A change of molecular constitution without essential change of composition, *i.e.*, by *para-morphism* (using the word rather broadly), may result in the formation of some variety of amphibole. Thus, the white pyroxene crystals of Canaan, Connecticut, are often changed on the exterior to tremolite; similarly with other varieties at many localities. See *uralite*, p. 574. Also changed to steatite, serpentine, etc.

Obs. — Pyroxene is a very common mineral in igneous rocks, being the most important of the ferromagnesian minerals. Some rocks consist almost entirely of pyroxene. It most commonly occurs in volcanic rocks but is found also, but less abundantly, in connection with deep-seated rocks. It is a common mineral in crystalline limestone and dolomite, in serpentine and metamorphic schists; sometimes forms large beds or veins, especially in Archæan rocks. It occurs also in meteorites. The pyroxene of limestone is mostly white and light green or gray in color, falling under *diopside* (malacolite, salite, coccolite); that of most other metamorphic rocks is sometimes white or colorless, but usually green of different shades, from pale green to greenish black, and occasionally black; that of serpentine is sometimes in fine crystals, but often of the foliated green kind called *diallage*; that of eruptive rocks is usually the black to greenish black *augite*.

In limestone and other metamorphic rocks the associations are often amphibole, scapolite, vesuvianite, garnet, orthoclase, titanite, apatite, phlogopite, and sometimes brown tourmaline, chlorite, talc, zircon, spinel, rutile, etc. In eruptive rocks it may be in distinct embedded crystals, or in grains without external crystalline form; it often occurs with similarly disseminated chrysolite (olivine), crystals of orthoclase (sanidine), labradorite, leucite, etc.; also with an orthorhombic pyroxene, amphibole, etc.

Pyroxene, as an essential rock-making mineral, is especially common in basic eruptive rocks. Thus, as augite, with a triclinic feldspar (usually labradorite), magnetite, often chrysolite, in basalt, basaltic lavas and diabase; in andesite; also in trachyte; in peridotite; with nephelite in phonolite. Further with nephelite, orthoclase, etc., in nephelite-syenite and augite-syenite; also as diallage in gabbro; in many peridotites and the serpentines formed from them; as diopside (malacolite) in crystalline schists. In limburgite, augitite and pyroxenite, pyroxene is present as the prominent constituent, while feldspar is absent; it may also form rock masses alone nearly free from associated minerals.

It is possible here to note only a few of the most interesting occurrences. An attempt has been made in most cases to indicate to which variety each occurrence belongs. It is, however, not always possible to do this with accuracy. *Diopside* (*alalite, mussite*) occurs at Achmatovsk near Kussinsk in the Zlatoust district of the Ural Mts.; in the Austrian Tyrol in fine crystals from the Alpe Schwartzenstein in the Zillertal; from the Binnental in Valais, Switzerland. In Piedmont, Italy, fine crystals on the Mussa Alp in the Ala Valley, also from Traversella; *violan* is a blue diopside from St. Marcel. In Finland at Pargas, southeast of Åbo in Turun ja Porin, and elsewhere. In Sweden at Nordmark in Vermland fine crystals of varied habit ranging from diopside to hedenbergite in composition.

The more important districts for *hedenbergite* are in Sweden. It was originally described from Tunaberg, Södermanland; it also comes from Nordmark in Vermland, and *mangan-hedenbergite* is from Vester Silfberg, north of Norrbärke in Kopparberg. From Obira in Bungo, Japan. *Salite* comes from Sala in Vastmanland, Sweden; from near Arendal in Aust-Agder, Norway, etc. *Baikalite* is found near Lake Baikal, Siberia. *Coccolite* is also found at Arendal and at various places in New York state.

Diallage is a variety that is widespread in its occurrence. *Omphacite* comes from the

Sua-Alpe in Carinthia, Austria; from various points near Hof in the Fichtelgebirge, Bavaria.
Schefferite is from Långbanshyttan, Vermland, Sweden; also from Jacobsberg in the Nord-mark district, Vermland.

Augite occurs in Bohemia of Czechoslovakia in fine crystals in basalt tuff at Boreslau southeast of Teplice (Teplitz). In Italy on the Pesmeda Alp near Monzoni, Val di Fassa (*fassaite*) in Trentino (formerly in Tyrol); from Traversella in Piedmont; from the Albani Mts. south of Rome; from Vesuvius, green, brown, yellow to black, rarely white; from Stromboli on the Lipari Islands and from Etna on Sicily. In Finland from Ersby near Pargas in Turun ja Porin, and in Norway at Arendal in Aust-Agder.

In the United States the most important localities for the occurrence of pyroxene are at Canaan in Litchfield Co., Connecticut, and in the neighboring district in Massachusetts where white crystals of diopside are abundant in dolomite, often externally changed by uralitization to tremolite; in New York there are many localities among which are Ossining in Westchester Co. (white); in Orange Co., near Munroe, black coccolite and at Two Ponds near by, large crystals with scapolite, titanite, etc., in limestone; near Warwick in fine crystals; in Lewis Co., at Diana white and black crystals; in St. Lawrence Co., at De Kalb, fine diopside; also at Gouverneur, Rossie, Russell, Pitcairn, etc. In New Jersey at Franklin, Sussex Co., in good crystals and also the variety *jeffersonite*. At the Ducktown mines in Tennessee. Augite crystals occur at Twin Peaks, Millard Co., Utah. Augite crystals are common in the volcanic cinders on the Hawaiian Islands.

Pyroxene occurs in Canada at many points in the Archæan of Quebec and Ontario, espe-cially in connection with the apatite deposits. In Quebec at Hull and Wakefield in Ottawa Co., white crystals, and at Orford, Sherbrooke Co. In Ontario at various places in Renfrew Co., with apatite, titanite, etc.; in Lanark Co., at Bathurst, colorless or white crystals, and at North Burgess.

ACMITE. ÆGIRITE.

Monoclinic. Axes: $a : b : c = 1.0996 : 1 : 0.6012$; $\beta = 73° 11'$.

Twins: tw. pl. $a(100)$ very common; crystals often polysynthetic, with enclosed twinning lamellæ. Crystals long prismatic, vertically striated or channeled; acute terminations very characteristic.

The above applies to ordinary *acmite*. For *ægirite*, crystals prismatic, bluntly termi-nated; twins not common; also in groups or tufts of slender acicular to capillary crystals, and in fibrous forms.

Cleavage: $m(110)$ distinct; $b(010)$ less so. Fracture uneven. Brittle. H. = 6–6.5. G. = 3.50–3.55. Luster vitreous, inclining to resinous. Streak pale yellowish gray. Color brownish or reddish brown, green; in the fracture blackish green. Subtransparent to opaque. Optically −. Ax. pl. ‖ $b(010)$. $X \wedge c$ axis at various small angles, usually about 5° (see Fig. 837). $\alpha = 1.776$. $\beta = 1.819$. $\gamma = 1.836$ (on artificial material). Usually strongly pleochroic in colors of brown, yellow, and green with absorption $X > Y > Z$. 2V large. Inclined dispersion, $\rho > v$. Variation in composition by the introduction of the diopside and hedenbergite molecules modifies the optical character. With decreasing content of the ac-mite molecule the refractive indices, the disper-sion, and the birefringence decrease. Variations in the values of 2V and extinction angle also occur.

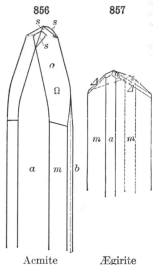

Acmite   Ægirite

Var. — Includes *acmite* in sharp-pointed crystals (Fig. 856) often twins. Also *ægirite* (Fig. 857) in crystals bluntly terminated, twins rare.

Crystals of acmite often show a marked zonal structure, green within and brown on the exterior, particularly || $a(100)$, $b(010)$, $p(\bar{1}01)$, $s(111)$. The brown portion (acmite) is feebly pleochroic, the green (ægirite) strongly pleochroic. Both have absorption $X > Y > Z$, but the former has X light brown with tinge of green, Y greenish yellow with tinge of brown, Z brownish yellow; the latter has X deep grass-green, Y lighter grass-green, Z yellowish brown to yellowish.

Comp. — Essentially $\overset{\text{III}}{\text{Na}}\text{Fe}(\text{SiO}_3)_2$ or $\text{Na}_2\text{O}.\text{Fe}_2\text{O}_3.4\text{SiO}_2$ = Silica 52·0, iron sesquioxide 34·6, soda 13·4 = 100.

The diopside-hedenbergite molecule $(\text{Ca}(\text{Mg},\text{Fe})(\text{SiO}_3)_2)$, can apparently mix in all proportions with the acmite molecule. Various hyphenated names have been proposed to describe the intermediate varieties, such as *ægirite-hedenbergite*, *jadeite-ægirite*. Acmite proper seems always to contain a notable amount of zirconia and rare earths.

Pyr., etc. — B.B. fuses at 2 to a lustrous black magnetic globule, coloring the flame deep yellow; with the fluxes reacts for iron and sometimes manganese. Slightly acted upon by acids.

Micro. — Ægirite is characterized in thin sections by its grass-green color; strong pleochroism in tones of green and yellow; the small extinction-angle in sections || $b(010)$. Distinguished from common green hornblende, with which it might be confounded, by the fact that in such sections the direction of extinction lying near the cleavage is negative (X), while the same direction in hornblende is positive (Z).

Artif. — Acmite can be produced artificially by fusing together its constituent oxides but usually under such conditions only a glass containing crystals of magnetite is formed.

Obs. — The original *acmite* occurs in a pegmatite vein, at Rundemyr, east of the little lake called Rokebergskjern, in the parish of Eker, near Kongsberg, Norway. It is in slender crystals, sometimes a foot long, embedded in feldspar and quartz.

Ægirite occurs especially in igneous rocks rich in soda and containing iron, commonly in rocks containing leucite or nephelite; thus in ægirite-granite, nephelite-syenite, and some varieties of phonolite; often in such cases iron-ore grains are wanting in the rocks, their place being taken by ægirite crystals. In the sub-variety of phonolite called tinguaite, the rock has often a deep greenish color due to the abundance of minute crystals of ægirite. It occurs with leucophanite, cancrinite, nephelite, etc., in the nephelite-syenite and augite-syenite of southern Norway, especially along the Langesundfiord in the Brevik district, etc. It occurs in large crystals in the pegmatite facies of the nephelite-syenites of southwest Greenland in the Julianehaab district at Narsarsuk, Kangerdluarsuk, Arsukfjord, etc.

In the United States it is found at Magnet Cove near Hot Springs, Garland Co., Arkansas, in fine prismatic crystals, up to 8 inches or more in length, often bent and twisted and with tapering terminations. A variety carrying vanadium occurs at Libby, Lincoln Co., Montana.

Acmite is named from ἄκμη, *point*, in allusion to the pointed extremities of the crystals; *Ægirite* is from Ægir, the Icelandic god of the sea.

JADEITE.

Monoclinic. Axes $a : b : c = 1\cdot103 : 1 : 0\cdot613$; $\beta = 72° 44\frac{1}{2}'$. Usually massive, with crystalline structure, sometimes granular, also obscurely columnar, fibrous foliated to closely compact.

Cleavage: prismatic, at angles of about 93° and 87°; also parting || $a(100)$ difficult. Fracture splintery. Extremely tough. H. = 6·5–7. G. = 3·3–3·5. Luster subvitreous, pearly on surfaces of cleavage. Color apple-green to nearly emerald-green, bluish green, leek-green, greenish white, and nearly white; sometimes white with spots of bright green. Translucent to subtranslucent. Optically +. Ax. pl. || $b(010)$. $Z \wedge c$ axis = 30° to 40°. 2V = large. For nearly pure $\text{NaAl}(\text{SiO}_3)_2$, $\alpha = 1\cdot654$, $\beta = 1\cdot659$, $\gamma = 1\cdot667$.

Comp. — Essentially a metasilicate of sodium and aluminum $\text{NaAl}(\text{SiO}_3)_2$ or $\text{Na}_2\text{O}.\text{Al}_2\text{O}_3.4\text{SiO}_2$ = Silica 59·4, alumina 25·2, soda 15·4 = 100. It is probable that both the acmite and diopside molecules are isomorphous with that of jadeite and that they may form more or less continuous series with each other.

Chloromelanite is a dark green to nearly black kind of jadeite (hence the name), containing iron sesquioxide and considerable amounts of the acmite and diopside molecules.

Pyr., etc. — B.B. fuses readily to a transparent blebby glass. Not attacked by acids after fusion, and thus differing from saussurite.

Obs. — The question of the origin and distribution of jadeite presents great difficulties. It has been suggested that it represents a deep-seated metamorphism of igneous rocks similar to the nephelite-syenites and phonolites. It occurs chiefly in eastern Asia; thus near Tawnaw in Myitkyina of Upper Burma, where it is associated with a serpentine that has been derived from an olivine rock. It also comes from Tibet and from the province of Yünnan in southern China. It is also reported from New Zealand and from Mexico. Much uncertainty prevails, however, as to the exact localities of occurrence, as most of its sources are in little-explored regions and because of the frequent confusion between jadeite and nephrite.

Jadeite has long been highly prized in the Orient, especially in China, where it is worked into ornaments and utensils of great variety and beauty. It is also found with the relics of early man, thus in the remains of the lake-dwellers of Switzerland, at various points in France, in Mexico, Greece, Egypt, and Asia Minor.

Use. — As the material jade, is used as an ornamental stone. See below.

JADE is a general term used to include various mineral substances of tough, compact texture and nearly white to dark green color used by early man for utensils and ornaments, and still highly valued in the East, especially in China. It includes properly two species only: *nephrite*, a variety of amphibole (p. 574), either tremolite or actinolite, with G. = 2·95–3·0, and *jadeite*, of the pyroxene group, with G. = 3·3–3·35; easily fusible.

The jade of China belongs to both species, so also that the Swiss lake-habitations and of Mexico. Of the two, however, the former, nephrite, is the more common and makes the jade (ax stone or Punamu stone) of the Maoris of New Zealand; also found in Alaska.

The name jade is also sometimes loosely used to embrace other minerals of more or less similar characters, and which have been or might be similarly used — thus sillimanite, pectolite, serpentine; also vesuvianite, garnet. Bowenite is a jade-like variety of serpentine. The "jade tenace" of de Saussure is now called saussurite.

859

SPODUMENE. Triphane.

Monoclinic. Axes $a : b : c = 1·1238 : 1 : 0·6355$; $\beta = 69° 40'$.

Twins: tw. pl. $a(100)$. Crystals prismatic $(mm''' 110 \wedge 1\bar{1}0 = 93° 0')$, often flattened $\parallel a(100)$; the vertical planes striated and furrowed; crystals sometimes very large. Also massive, cleavable.

858

Cleavage $m(110)$ perfect. A lamellar structure $\parallel a(100)$ sometimes very prominent, a crystal then separating into thin plates. Fracture uneven to subconchoidal. Brittle. H. = 6·5–7. G. = 3·13–3·20. Luster vitreous, on cleavage surfaces somewhat pearly. Color greenish white, grayish white, yellowish green, emerald-green, yellow, amethystine purple. Streak white. Transparent to translucent. Pleochroism strong in deep green varieties. Optically +. Ax. pl. $\parallel b(010)$. $Z \wedge c$ axis = $+26°$. Dispersion $\rho < v$. 2V = 54° to 69°. $\alpha = 1·65–1·66$; $\beta = 1·66–1·67$; $\gamma = 1·675–1·68$.

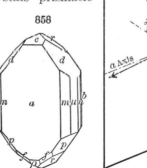

Huntington, Mass. Optical Orientation of Spodumene.

Hiddenite has a yellow-green to emerald-green color; the latter variety is used as a gem. In small ($\frac{1}{2}$ to 2 inches long) slender prismatic crystals, faces often etched.

Kunzite is a clear lilac-colored variety found near Pala, San Diego Co., California, and also at Vanakarata, Madagascar. The unaltered material from Branchville, Connecticut, shows the same color. Used as a gem stone.

Comp. — $LiAl(SiO_3)_2$ or $Li_2O.Al_2O_3.4SiO_2$ = Silica 64·5, alumina 27·4, lithia 8·4 = 100. Generally contains a little sodium; the variety hiddenite also chromium, to which the color may be due.

Pyr., etc. — B.B. becomes white and opaque, swells up, imparts a purple-red color (lithia) to the flame (sometimes obscured by sodium), and fuses at 3·5 to a clear or white glass. Not acted upon by acids. Kunzite shows strong phosphorescence with an orange-pink color when excited by an oscillating electric discharge, by ultra-violet rays, X-rays, or radium emanations.

Diff. — Characterized by its perfect parting $\parallel a(100)$ (in some varieties) as well as by prismatic cleavage; has a higher specific gravity and more pearly luster than feldspar or scapolite. Gives a red flame B.B. Less fusible than amblygonite.

Alter. — Spodumene very commonly undergoes alteration. First by the action of solutions containing soda it is changed to a mixture of eucryptite, $LiAlSiO_4$, and albite, $NaAlSi_3O_8$. Later through the influence of potash salts the eucryptite is changed to muscovite. This resulting mixture of albite and muscovite is known as *cymatolite*, having a wavy fibrous structure and silky luster. These alteration products are well shown in the specimens from Branchville, Connecticut.

Artif. — An artificial spodumene has been obtained together with other silicates by fusing together lithium carbonate, alumina and silica. This spodumene differs, however, from the natural mineral in its optical properties and has been called *β-spodumene*. The natural mineral, or *α-spodumene*, is transformed into the β modification on slow heating at 690° C.

Obs. — Spodumene occurs in granite pegmatites, sometimes in crystals of very great size. It is found on the island of Utö, state of Stockholm, Sweden; at Killiney, southeast of Dublin, Ireland. It is common in the soda-lithia pegmatites of Madagascar, in colors varying from yellowish green, yellow, colorless, to the rose of the *kunzite* variety. Notable localities are Anjanaboana, Maharitra on Mt. Bity, Tlapa, etc. In small transparent crystals of a pale yellow color from Minas Geraes, Brazil.

In the United States it occurs in the pegmatites of Oxford Co., Maine; in Massachusetts in Worcester Co., at Sterling and in Hampshire Co., at Goshen, Chesterfield, and Huntington (formerly Norwich). In Connecticut it was found at Branchville, Fairfield Co., in a pegmatite with lithiophilite, uraninite, and several rare manganesian phosphates, etc.; the crystals were often of considerable size and were commonly largely altered (see above). In North Carolina near Stony Point, Alexander Co., the variety *hiddenite*. In South Dakota at the Etta tin mine at Keystone in Pennington Co., in immense crystals up to 47 feet in length with diameters of from 3 to 5 feet. Such a crystal may weigh up to 90 tons. Gem crystals of the variety *kunzite* have been found in California in the lithia-pegmatites of San Diego Co., at Mesa Grande, Pala, and Rincon, and in Riverside Co., at Cahuilla, etc.

Triclinic Section

The following triclinic species are considered by some authors to be distinct from the pyroxenes and to form a separate group by themselves.

RHODONITE.

Triclinic. Axes $a : b : c = 1·07285 : 1 : 0·6213$; $\alpha = 103° 18'$; $\beta = 108° 44'$; $\gamma = 81° 39'$.

Crystals usually large and rough with rounded edges. Commonly tabular $\parallel c(001)$; sometimes resembling pyroxene in habit. Commonly massive, cleavable to compact; also in embedded grains. Crystals from Franklin, New Jersey, show polysynthetic twinning, with $c(001)$ as twinning plane.

ab,	$100 \wedge 010 = 94° 26'$.		mM,	$110 \wedge 1\bar{1}0 = 92° 28\frac{1}{2}'$.	
ac,	$100 \wedge 001 = 72° 36\frac{1}{2}'$.		cn,	$001 \wedge \bar{2}\bar{2}1 = 73° 52'$.	
bc,	$010 \wedge 001 = 78° 42\frac{1}{2}'$.		ck,	$001 \wedge \bar{2}21 = 62° 23'$.	
am,	$100 \wedge 110 = 48° 33'$.		kn,	$\bar{2}21 \wedge \bar{2}\bar{2}1 = 86° 5'$.	

Cleavage: $m(110)$, $M(1\bar{1}0)$ perfect; $c(001)$ less perfect. *Bustamite* shows perfect cleavage || (010). Fracture conchoidal to uneven; very tough when compact. H. = 5·5–6·5. G. = 3·4–3·68. Luster vitreous; on cleavage surfaces somewhat pearly. Color light brownish red, flesh-red, rose-pink; sometimes greenish or yellowish, when impure; often black outside from exposure. Streak white. Transparent to translucent. Optically + or −. 2V = 76°. α = 1·72–1·73, β = 1·73–1·74, γ = 1·73–1·744. *Bustamite* has 2V = 44°; α = 1·66, β = 1·67, γ = 1·676; ax. pl. and X nearly ⊥ to (010).

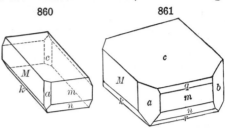

860   861

Franklin Furnace, N. J.

Comp. — Manganese metasilicate, $MnSiO_3$ or $MnO.SiO_2$ = Silica 45·9, manganese protoxide 54·1 = 100. Calcium appears to be always present and may be a necessary constituent. It has been suggested that the formula might be written $CaMn_5(SiO_3)_6$. *Bustamite* has much more calcium than ordinary rhodonite; the material from Franklin, New Jersey, has the composition $CaMn(SiO_3)_2$. It also differs from rhodonite in its optical properties. There is evidence that there is a continuous series from rhodonite to bustamite with a gradual change in the optical properties. *Fowlerite* is a variety with zinc replacing manganese.

Pyr., etc. — B.B. blackens and fuses with slight intumescence at 2·5; with the fluxes gives reactions for manganese; fowlerite gives with soda on charcoal a reaction for zinc. Slightly acted upon by acids. The calciferous varieties often effervesce from mechanical admixture of calcium carbonate. In powder, partly dissolves in hydrochloric acid, and the insoluble part becomes of a white color. Darkens on exposure to the air, and sometimes becomes nearly black.

Diff. — Characterized by its pink color; distinct cleavages; hardness; fusibility and manganese reactions B.B.

Obs. — Rhodonite occurs in various manganese ore bodies, frequently associated with rhodochrosite, etc. It occurs in the Ekaterinburg district of the Ural Mts., where it is found massive and is used for ornamental purposes; with tetrahedrite at Kapnikbánya in Rumania. In Italy at Campiglia Marittima in Tuscany. From Finland at Vittinge in Storkyre. It occurs in Vermland, Sweden, at Långbanshyttan in iron ore beds, and at the Pajsberg (near Persberg) iron mines (*paisbergite*). In crystals from Broken Hill, New South Wales. From Tetela di Xonotla in Puebla, Mexico (*bustamite*).

In the United States it occurs in Cummington, Hampshire Co., Massachusetts, and some of the neighboring towns. The variety *fowlerite* occurs in Sussex Co., New Jersey, at Mine Hill, Franklin and Sterling Hill, near Ogdensburgh, the two localities being only a short distance apart. It is usually embedded in calcite and is sometimes in splendid crystallizations. *Bustamite* is found at the same localities.

Named from ροδον, *a rose*, in allusion to the color.

Rhodonite is often altered chiefly by oxidation of the MnO (as in *marceline, dyssnite*); also by hydration (*stratopeite, neotocite*, etc.); further by introduction of CO_2 (*allagite, photicite*, etc.).

Use. — Rhodonite at times is used as an ornamental stone.

Pyroxmangite. — A triclinic, manganese-iron pyroxene. In cleavage masses. Cleavages (110) and (1$\bar{1}$0) and parting (010). Indices, 1·75–1·76. H. = 5·5–6. G. = 3·8. Color, amber to dark brown. Easily fusible to black magnetic globule. Alters to *skemmatite*. Found near Iva, Anderson Co., South Carolina.

Babingtonite. — $Ca_2Fe''Fe'''Si_5O_{14}(OH)$. In small black triclinic crystals, near rhodonite in angle (axes on p. 551). Cleavages, (110) and (1$\bar{1}$0), also (001) at times. H. = 5·5–6. G. = 3·35–3·37. Optically +. 2V = 60°–65°. α = 1·717, β = 1·730, γ = 1·752. Dispersion $\rho > v$, strong. Occurs at Herborn near Dillenburg, Hessen-Nassau; from Baveno on Lago Maggiore, Piedmont, Italy. Found in distinct crystals at

Arendal in Aust-Agder, Norway. In the United States in Massachusetts from Somerville, Middlesex Co., and at Athol, Worcester Co. In New Jersey in the zeolite deposits of Passaic Co.

Sobralite. — $(Mn,Fe,Ca,Mg)SiO_3$. A triclinic pyroxene. Prismatic cleavages. G. = 3·5. Brown color. $2V = 41°$. $n = 1·74$. Optically +. From eulysite rock in Södermanland, Sweden.

Wollastonite, pectolite, etc. have until recently been commonly placed in the Pyroxene Group to which they have certain relations, but because of definite crystallographic, structural, and optical differences it is better to separate them from the more typical members of the group.

WOLLASTONITE. Tabular Spar.

Monoclinic. Axes $a : b : c = 1·0531 : 1 : 0·9676$; $\beta = 84° 30'$.

mm''', 110 ∧ 1$\bar{1}$0 = 92° 42'.	cv, 001 ∧ 101 = 40° 3'.
hh''', 540 ∧ 5$\bar{4}$0 = 79° 58'.	cr, 001 ∧ $\bar{3}$01 = 74° 59'.
gg', 011 ∧ 0$\bar{1}$1 = 87° 51'.	ct, 001 ∧ $\bar{1}$01 = 45° 5'.

Twins: tw. pl. a(100). Crystals commonly tabular ‖ a(100) or c(001); also short prismatic. Usually cleavable massive to fibrous, fibers parallel or reticulated; also compact. X-ray study shows a structure unlike that of diopside but with a similarity to that of rhodonite.

862

Diana, N. Y.

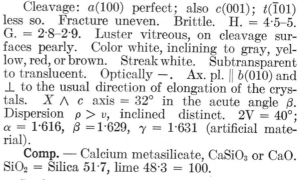

863

Optical Orientation of Wollastonite.

Cleavage: a(100) perfect; also c(001); t($\bar{1}$01) less so. Fracture uneven. Brittle. H. = 4·5–5. G. = 2·8–2·9. Luster vitreous, on cleavage surfaces pearly. Color white, inclining to gray, yellow, red, or brown. Streak white. Subtransparent to translucent. Optically −. Ax. pl. ‖ b(010) and ⊥ to the usual direction of elongation of the crystals. X ∧ c axis = 32° in the acute angle β. Dispersion $\rho > v$, inclined distinct. $2V = 40°$; $\alpha = 1·616$, $\beta = 1·629$, $\gamma = 1·631$ (artificial material).

Comp. — Calcium metasilicate, $CaSiO_3$ or CaO. SiO_2 = Silica 51·7, lime 48·3 = 100.

Synthetic experiments show that wollastonite can take into solid solution up to 17 per cent of the diopside molecule, to 67 per cent of the akermanite molecule and also a molecule having the composition, $5CaO.2MgO.6SiO_2$. It may also form solid solutions with other molecules.

When wollastonite is heated above 1200° C. it develops a basal cleavage, becomes pseudohexagonal, optically positive, nearly uniaxial but probably monoclinic. This material has been called *pseudowollastonite*.

Pyr., etc. — B.B. fuses quietly to a white, almost glassy globule. With hydrochloric acid decomposes with separation of silica; most varieties effervesce slightly from the presence of calcite. Often phosphoresces.

Micro. — In thin sections wollastonite is colorless with a moderate relief and medium birefringence. The plane of the optic axes is usually normal to the elongation of the crystals.

Artif. — Wollastonite may be obtained artificially by heating a glass of the composition $CaSiO_3$ to between 800° and 1000°. This is stable to 1200° C. At higher temperatures the pseudowollastonite modification is obtained.

Obs. — Wollastonite occurs chiefly as a contact mineral in crystalline limestones, formed by the action of silicic acid on the limestone, commonly in the presence of mineralizers and at a comparatively low temperature. Its occurrence in eruptive rocks is due to the inclusion in them of blocks of limestone. It is often associated with a lime garnet, diopside, etc.

It occurs in Rumania in the copper mines of Csiklova near Oravicza and at the latter place. In the ejected blocks of Mte. Somma, Vesuvius, rarely in fine crystals. In Finland in limestone at Pargas in Turun ja Porin and also at Perheniemi near Ihtis. In Chiapas, Mexico.

In the United States in New York at Willsboro in Essex Co.; in Lewis Co., at Diana near Natural Bridge in abundant large white crystals. On Isle Royal, Keweenaw Co., Michigan, a very compact variety. In California near Riverside, Riverside Co. In Quebec, Canada, at Grenville, Argenteuil Co., and at Wakefield, Ottawa Co.

Named after the English chemist, W. H. Wollaston (1766–1828).

Rivaite has been shown to consist of prisms of wollastonite embedded in a glass.

Alamosite. — Lead metasilicate, $PbSiO_3$. Monoclinic. Closely related to wollastonite in crystal forms. In radiating fibrous aggregates. Cleavage $\parallel b$ (010). G. = 6·5. H. = 4·5. Easily fusible. Colorless or white. Optically −. $2V = 65°$. Dispersion $\rho < v$ strong. $\alpha = 1·947, \beta = 1·961, \gamma = 1·968$. Found near Alamos, Sonora, Mexico.

PECTOLITE.

Monoclinic. Axes $a : b : c = 1·1140 : 1 : 0·9864;\ \beta = 84° 40'$.

Commonly in close aggregations of acicular crystals; elongated $\parallel b$ axis, but rarely terminated. Fibrous massive, radiated to stellate.

Cleavage: $a(100)$ and $c(001)$ perfect. Fracture uneven. Brittle. H. = 5. G. = 2·74–2·88. Luster of the surface of fracture silky or subvitreous. Color whitish or grayish. Subtranslucent to opaque. Optically +. Ax. pl. and $Z \perp b(010)$; X nearly $\perp a(100)$. $2V - 60°$. $\alpha - 1·595, \beta = 1·606, \gamma = 1·633$.

Comp. — $HNaCa_2(SiO_3)_3$ or $H_2O.Na_2O.4CaO.6SiO_2$ = Silica 54·2, lime 33·8, soda 9·3, water 2·7 = 100.

The amount of water present varies and is usually in excess of that required by the above formula. Pectolite is sometimes classed with the hydrous species allied to the zeolites. Magnesium may be present as in the variety, *walkerite*, and manganese, as in *manganpectolite*.

Pyr., etc. — In the closed tube yields water. B.B. fuses at 2 to a white enamel. Decomposed in part by hydrochloric acid with separation of silica as a jelly. Often gives out light when broken in the dark.

Obs. — Pectolite occurs like the zeolites, mostly in cavities or seams in basic eruptive rocks. It is one of the earliest of such minerals to form and is usually associated with the zeolitic minerals of low water content, as apophyllite, natrolite, and prehnite, and with calcite. It is occasionally observed in metamorphic rocks.

It was first observed in the Val di Fassa, Trentino, Italy, near Monzoni and on Mte. Baldo near Tiermo. From near Weardale, Durham, and in Scotland at Ratho, etc., near Edinburgh; from Corstorphine Hill (*walkerite*) near Kilsyth in Stirling.

In the United States common in the diabase rocks of northeastern New Jersey, as at Weehawken, Guttenberg, Bergen Hill, Great Notch, Paterson, and West Paterson, often in large and beautiful radiated groups. In a compact form on Isle Royal, Keweenaw Co., Michigan. From Magnet Cove, near Hot Springs, Garland Co., Arkansas, where it occurs in a nephelite-syenite (*manganpectolite*).

Yuksporite. Juksporite, Juxporite. Near pectolite in composition but containing more sodium and potassium. In platy or fibrous masses. Optically negative elongation. Color, pink, red. Occurs intergrown with sphene, with pectolite, astrophyllite, biotite, and ægirite from Kola peninsula, northern Russia.

Schizolite. — Like manganpectolite, $HNa(Ca,Mn)_2(SiO_3)_3$, but triclinic. In prismatic crystals similar to those of pectolite. Cleavages, (100) and (001). H. = 5–5·5. G. = 3·0–3·1. Color, light red to brown. Optically +. $\alpha = 1·631, \beta = 1·636, \gamma = 1·660$. From the nephelite syenite of Julianehaab, southern Greenland. Also noted on Kola peninsula, Russian Lapland.

Rosenbuschite. — Near pectolite, but contains zirconium. Monoclinic. In prismatic or acicular crystals. Cleavages, (001) perfect, (100), ($\overline{2}$01), distinct. H. = 5–6. G. = 3·3. Easily fusible. Optically +. $2V = 60°$. $\alpha = 1·682, \beta = 1·687, \gamma = 1·711$. Occurs very sparingly in the nephelite-syenite region of the Langesundfiord in southern

Norway. In nephelite-syenite-porphyry, Red Hill, Moultonboro, Carroll Co., New Hampshire.

The chemical relationships of the following rare minerals are uncertain; the first two have been considered as members of the Pyroxene Group.

Wöhlerite. — A zirconium-silicate and niobate of Ca, Na, etc. Monoclinic. In prismatic or tabular crystals. Cleavage ‖ (010). H. = 5·5–6. G. = 3·42. Color, yellow to brown. Optically −. Ax. pl. ⊥ (010). $X \wedge c$ axis = −45°. $\alpha = 1\cdot7$, $\beta = 1\cdot716$, $\gamma = 1\cdot726$. Occurs in pegmatite veins in zircon-syenite on several islands of the Langesundfiord, near Brevik, in Norway. Found on the Los Islands, French Guinea. In syenite from Red Hill, Moultonboro, Carroll Co., New Hampshire.

Låvenite. — A complex zirconium-silicate of Mn, Ca, etc., containing also F, Ti, Ta, etc. Monoclinic. In yellow to brown prismatic crystals. Cleavage, (100). G. = 3·5. H. = 6. Optically −. $n = 1\cdot70$–$1\cdot75$. Found on the island Låven in the Langesundfiord, southern Norway; also in nephelite-syenite from Brazil and the Los Islands, French Guinea. Also noted on the Azores.

Guarinite. Hiortdahlite. — Essentially $(Na_2,Ca)(Si,Zr)O_3$ with also fluorine. Triclinic. In pseudo-orthorhombic tabular (‖ 001) crystals. Cleavage (110). H. = 5·5. G. = 3·27. Color yellow to brown. Optically +. 2V nearly 90°. Strong dispersion. $n = 1\cdot652$–$1\cdot665$. *Guarinite* was found in small cavities in a grayish trachyte on Mte. Somma, Vesuvius. *Hiortdahlite* occurs sparingly on the island Mittel-Arö, in the Langesundfiord, southern Norway.

3. *Amphibole Group*

Orthorhombic, Monoclinic, Triclinic

Composition for the most part near that of a metasilicate, $RSiO_3$, with R = Ca,Mg,Fe chiefly, also Mn,Na_2,K_2,H_2. Further often containing aluminum and ferric iron. The composition in the case of some amphiboles is very complex with extensive replacements taking place between the different elements. It is impossible even to list here the different molecules that have been proposed to account for the variations in composition. Recently, however, the atomic structure of the amphiboles has been investigated by X-ray methods (see p. 533) and formulas suggested that conform both to these facts and to the chemical composition. The type molecule as derived by Warren is found to have a silicon-oxygen ratio of Si_4O_{11} instead of the metasilicate ratio of Si_4O_{12}. He finds further that in one-half of the unit cell the number of oxygen atoms or their equivalents is always twenty-four, that the number of the magnesium atoms, or their equivalents is always five, that the silicon atoms never exceed eight in number, and that a group containing calcium and alkali atoms may vary between two and three. His general formula conforming to these structural requirements is $(Ca,Na,K,Li)_{2-3}(Mg, Fe'', Fe''',Al,Ti)_5 (Fe''',Al,Si)_8(O,OH,F)_{24}$. This formula has been modified by Berman and Larsen to the following: $(Ca,Na)_2Na_{0-1}Mg(Mg,Al)_4(Al,Si)_2 Si_6O_{22}(O,OH,F)_2$, in which Fe ''and Fe''' may replace Mg and Al, respectively. In the list given immediately below, however, simplified metasilicate formulas have been used in most cases as giving the essential features of the compositions of the different members of the group. Fuller discussions of the compositions of each will be found in the succeeding pages.

Orthorhombic Section

		$a : b$
Anthophyllite	$(Mg.Fe)SiO_3$	0·5137 : 1

<table>
</table>

Monoclinic Section		$a : b : c$	β
Amphibole		0·5511 : 1 : 0·2938	73° 58'
1.	$MgSiO_3$		
2. CUMMINGTONITE	$(Mg,Fe)SiO_3$		
3. GRUNERITE	$(Fe,Mg)SiO_3$		
4. TREMOLITE	$CaMg_3(SiO_3)_4$		
5. ACTINOLITE	$Ca(Mg,Fe)_3(SiO_3)_4$		
6. RICHTERITE	$(K_2,Na_2,Mg,Ca,Mn)SiO_3$		
7. HORNBLENDE			

Edenite
Pargasite
Common hornblende
Basaltic hornblende

} Contain molecules approximating to the following in varying proportions.
$Ca(Mg,Fe)_3Si_4O_{12}$; $CaMg_2(Al,Fe)_2Si_3O_{12}$;
$NaAl(SiO_3)_2$

		$a : b : c$	β
Glaucophane	$Na(Al,Fe)(SiO_3)_2$ with $(Mg,Fe)SiO_3$	0.53 : 1 : 0.29	75°
Riebeckite	$NaFe(SiO_3)_2$ with $FeSiO_3$	0·5475 : 1 : 0·2295	76° 10'
Hastingsite	$Ca_2Na(Fe,Mg)_4(Al,Fe)(OH)_2(Al,Si)_8O_{22}$		
Arfvedsonite	$Na_3Fe_4Al(OH)_2Si_8O_{22}$	0·5496 : 1 : 0·2975	75° 44'

Triclinic Section

Ænigmatite $(Fe,Na_2)(Si,Ti)O_3.Na(Al,Fe)(SiO_3)_2$

The only species included under the triclinic section is the rare and imperfectly known ænigmatite (cossyrite).

The AMPHIBOLE GROUP embraces a number of species which, while falling in different systems, are yet closely related in form — as shown in the common prismatic cleavage of 54° to 56° — also in optical characters and chemical composition. As already noted (see p. 552), the species of this group form chemically a series parallel to that of the closely allied Pyroxene Group, and between them there is a close relationship in crystalline form and other characters. The Amphibole Group, however, is less fully developed, including fewer species, and those known show less variety in form.

The chief *distinctions* between pyroxene and amphibole proper are the following:
Prismatic angle with pyroxene 87° and 93°; with amphibole 56° and 124°; the prismatic cleavage being much more distinct in the latter.
With pyroxene, crystals usually short prismatic and often complex, structure of massive kinds mostly lamellar or granular; with amphibole, crystals chiefly long prismatic and simple, columnar and fibrous massive kinds the rule.
The specific gravity of most of the pyroxene varieties is higher than of the like varieties of amphibole. In composition of corresponding kinds, magnesium is present in larger amount in amphibole (Ca : Mg = 1 : 1 in diopside, = 1 : 3 in tremolite); alkalies more frequently play a prominent part in amphibole.

X-ray study has shown the structural relationships that exist between the pyroxene and amphibole groups. The amphibole structure can be derived by a reflection of the pyroxene structure over the plane (010). This results in giving a unit cell for amphibole which agrees with that of pyroxene except in the dimension parallel to the b axis. This is twice the value of the corresponding dimension of the pyroxene cell. In pyroxene, as stated on p. 552, the silicon-oxygen tetrahedral groups are arranged in single chains parallel to the c axis, whereas in amphibole the similar groups are united into double chains. These facts account for the crystallographic differences between the two groups, as for instance the difference between the respective cleavage angles. Further, the joining of two silicon-oxygen chains together reduces the silicon-oxygen ratio to Si_4O_{11} instead of Si_4O_{12} as in pyroxene. The positions of the calcium and magnesium atoms in the structure are similar to their positions in the pyroxene structure, see p. 552.

The optical relations of the prominent members of the group, as regards the position of the ether-axes, is exhibited by the following figures (Cross); compare Fig. 837, p. 552, for a similar representation for the corresponding members of the pyroxene group.

864

I.   II.   III.   IV.   V.   VI.

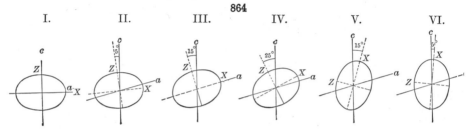

I. Anthophyllite.   II. Glaucophane.   III. Tremolite, etc.   IV. Hornblende.
V. Arfvedsonite.   VI. Riebeckite.

Orthorhombic Section

ANTHOPHYLLITE.

Orthorhombic. Axial ratio $a : b = 0.5137 : 1$. Crystals rare, habit prismatic ($mm''' \ 110 \wedge 1\bar{1}0 = 54° 23'$). Commonly lamellar, or fibrous massive; fibers often very slender; in aggregations of prisms.

Cleavage: prismatic, perfect; $b(010)$ less so; $a(100)$ sometimes distinct. H. = 5·5–6. G. = 2·85–3·2 (increasing with percentages of iron). Luster vitreous, somewhat pearly on the cleavage face. Color brownish gray, yellowish brown, clove-brown, brownish green, emerald-green, sometimes metalloidal. Streak uncolored or grayish. Transparent to subtranslucent. Sometimes pleochroic, $X = Y =$ brownish, $Z =$ pale yellow-green. Ax. pl. ‖ $b(010)$. $Z = c$ axis. 2V nearly 90°. Optically + in anthophyllite, − in gedrite. For artificial $MgSiO_3$, $\alpha = 1.584$, $\gamma = 1.597$; for variety with 27 per cent $FeSiO_3$, $\alpha = 1.633$, $\beta = 1.642$, $\gamma = 1.657$; for higher percentages of $FeSiO_3$ γ may rise to 1·698.

Comp. — $(Mg,Fe)SiO_3$, corresponding to enstatite-hypersthene in the pyroxene group. Aluminum is sometimes present in considerable amount. There is the same relation in optical character between anthophyllite (+) and gedrite (−) as between enstatite and hypersthene (cf. Figs. 839, 840, p. 553).

Var. — ANTHOPHYLLITE. Mg : Fe = 4 : 1, 3 : 1, etc. For 3 : 1 the percentage composition is: Silica 55·6, iron protoxide 16·6, magnesia 27·8 = 100. Anthophyllite sometimes occurs in forms resembling asbestos.

GEDRITE. Iron is present in larger amount, and also aluminum; it hence corresponds nearly to a hypersthene, some varieties of which are highly aluminous.

Amosite is a long-fibered ash-gray or greenish asbestos from South Africa intermediate between anthophyllite and gedrite in composition, and close to ferroanthophyllite.

Ferroanthophyllite is a name given to an iron anthophyllite from Idaho and elsewhere.

Hydrous anthophyllites have been repeatedly described, but in most cases they have been shown to be hydrated monoclinic amphiboles.

Pyr., etc. — B.B. fuses with difficulty to a black magnetic enamel; with the fluxes gives reactions for iron; unacted upon by acids.

Micro. — In sections colorless, non-pleochroic. Parallel extinction. Commonly fibrous.

Artif. — Anthophyllite is formed artificially when magnesium metasilicate is heated considerably above its melting point and then quickly cooled.

Obs. — Anthophyllite is found in the crystalline schists, at times becoming the chief constituent of the rock. It is thought to have usually been derived through the metamorphism of chrysolite. Occurs near Kongsberg, Norway; from Hermannschlag in Moravia of Czechoslovakia and at Dürrenstein near Krems on the Donau in Lower Austria. Found in many localities in southern Greenland. It occurs in the United States in Pennsylvania in Delaware Co., etc.; in North Carolina at Franklin, Macon Co.

The original *gedrite* is from the valley of Héas near Gèdres, Hautes Pyrénées, France. Similar anthophyllites have been observed in Norway at Bamble in Telemark, at Snarum in Buskerud, and in Greenland.

Named from *anthophyllum, clove*, in allusion to the clove-brown color.

Monoclinic Section

AMPHIBOLE. Hornblende.

Monoclinic. Axes $a : b : c = 0.5511 : 1 : 0.2938$; $\beta = 73° 58'$.

mm''', 110 $\wedge$ 1$\bar{1}$0 $= 55° 49'$.	rr', 011 $\wedge$ 0$\bar{1}$1 $= 31° 32'$.
ca, 001 $\wedge$ 100 $= 73° 58'$.	ii, 031 $\wedge$ 0$\bar{3}$1 $= 80° 32'$.
cp, 001 $\wedge$ $\bar{1}$01 $= 31° 0'$.	pr, $\bar{1}$01 $\wedge$ 011 $= 34° 25'$.

Twins: (1) tw. pl. a(100), common as contact-twins; rarely polysynthetic. (2) c(001), as tw. lamellæ, occasionally producing a parting analogous

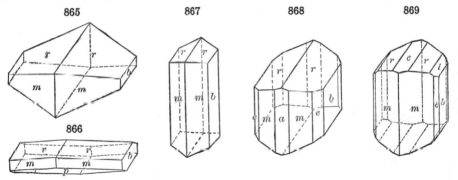

to that more common with pyroxene (Fig. 487, p. 195). Crystals commonly prismatic; usually terminated by the low clinodome, r(011), sometimes by r and p($\bar{1}$01) equally developed and then suggesting rhombohedral forms (as of tourmaline). Also columnar or fibrous, coarse or fine, fibers often like flax; rarely lamellar; also granular massive, coarse or fine, and usually strongly coherent, but sometimes friable.

The crystallographic position here adopted is that suggested by Tschermak, which best exhibits the relation between amphibole and pyroxene. Some authors retain the former position, according to which $p - (001)$, $r - (\bar{1}11)$, etc.

Cleavage: m(110) highly perfect; a(100), b(010) sometimes distinct. Fracture subconchoidal, uneven. Brittle. H. = 5–6. G. = 2.9–3.4, varying with the composition. Luster vitreous to pearly on cleavage faces;

fibrous varieties often silky. Color between black and white, through various shades of green, inclining to blackish green; also dark brown; rarely yellow, pink, rose-red. Streak uncolored, or paler than color. Sometimes nearly transparent; usually subtranslucent to opaque.

Pleochroism strongly marked in all the deeply colored varieties, as described beyond. Absorption usually $Z > Y > X$. Optically $-$, rarely $+$. Ax. pl. usually $\parallel b(010)$. Extinction-angle on $b(010)$, or $Z \wedge c$ axis $= +15°$ to $18°$ in most cases, but varying from about $1°$ up to $37°$; higher angles in

870   871   872

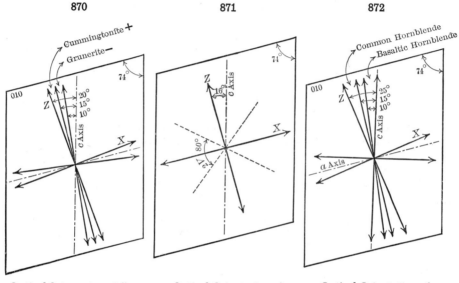

Optical Orientation of Cum-   Optical Orientation of   Optical Orientation of
mingtonite-Grunerite.   Tremolite-Actinolite.   Hornblendes.

rare instances. Hence also $Bx_{ac} \wedge c$ axis $= -75°$ to $-72°$, etc. See Figs. 870–872. Dispersion $\rho < v$. Axial angles variable; see beyond.

Optical characters, particularly indices of refraction, birefringence and extinction angles vary with change in composition, particularly with the total amount of iron present. In general the indices and extinction angles increase with increase of iron content while the birefringence decreases.

Comp. — In part near to a normal metasilicate of calcium and magnesium, $RSiO_3$, usually with iron, also manganese, and thus in general analogous to the pyroxenes. As stated on p. 568 the silicate oxygen ratio corresponds to Si_4O_{11}, instead of Si_4O_{12}. The alkali metals, sodium and potassium, also present, and more commonly so than with pyroxene. In part also aluminous, corresponding to the aluminous pyroxenes. Titanium sometimes is present and also rarely fluorine in small amount. Water is considered as an essential constituent.

The problem of the composition of the amphiboles is a complex one and has been the object of much study. Extensive replacements of one element by another may take place, giving rise to a wide variation in chemical composition. It is impossible to summarize here the various interpretations of the composition that have been offered in the past or even to indicate all the views that different mineralogists hold at present. See p. 568 and further under the different varieties described below.

1. MgSiO₃. An amphibole of this composition and corresponding to clinoenstatite of the Pyroxene Group is of doubtful natural occurrence. It has, however, been prepared artificially. It would form the theoretical end-member of the various amphibole series in which the molecules $FeSiO_3$, $MnSiO_3$, and $CaSiO_3$ appear in isomorphous relations.

The name *kupfferite* was originally given to what was thought to be a monoclinic amphibole largely composed of $MgSiO_3$; later it was assumed that the mineral in question was orthorhombic and it was included under anthophyllite. When the artificial monoclinic $MgSiO_3$ was described it was named *clinokupfferite*. Later the original kupfferite was shown to be monoclinic, as originally described, and the name was restored to the monoclinic $MgSiO_3$. However, new analyses of the original kupfferite show that it is more properly classed as an actinolite.

2. CUMMINGTONITE. $(Mg,Fe)SiO_3$. The magnesium and iron replace each other in varying amounts and the mineral grades into grunerite. Small amounts of manganese may be present. The range in composition of the cummingtonite-grunerite series is from about 70 per cent $MgSiO_3$ and 30 per cent $FeSiO_3$ to nearly 100 per cent $FeSiO_3$. The name cummingtonite is commonly restricted to that portion of the series containing between 50 and 70 per cent $MgSiO_3$. Usually fibrous or fibro-lamellar, often radiated. G. varies from about 3·1 to 3·6 for grunerite. Color brown. $Z \wedge c$ axis = 20° to 15° decreasing with increase in percentages of $FeSiO_3$. Optically +. 2V large. $\rho < v$. Indices for nearly equal mixtures of $MgSiO_3$ and $FeSiO_3$ are: $\alpha = 1·64$, $\beta = 1·65$, $\gamma = 1·67$. These values increase with increasing percentages of iron.

3. GRUNERITE. Grünerite. $(Fe,Mg)SiO_3$. The percentages of $FeSiO_3$ range from 50 to 100 per cent. $MnSiO_3$ may also be present in considerable amounts. See further under cummingtonite, above. Commonly fibrous or lamellar. G. = 3·4–3·6. Color brown. Silky luster. $Z \wedge c$ axis = 15° to 10°, decreasing with increase in percentage of $FeSiO_3$. Optically −. 2V large. $\rho > v$. $\alpha = 1·680$, $\beta = 1·707$, $\gamma = 1·726$ (for pure $FeSiO_3$). The values decrease regularly with decreasing percentages of the iron molecule; also the presence of manganese replacing the iron changes slightly the indices.

Dannemorite is a manganiferous member of the cummingtonite-grunerite series from Dannemora, Upsala, Sweden. Columnar or fibrous. Color yellowish brown to greenish gray. G. = 3·4–3·5. $Z \wedge c$ axis about 14°. Axial angle large.

4. TREMOLITE. Grammatite, nephrite in part. *Calcium-magnesium amphibole*. Formula has commonly been given as $CaMg_3Si_4O_{12}$; but from a study of the atomic structure by means of X-rays it is shown that the composition is more accurately given by the formula $Ca_2Mg_5(OH)_2(Si_4O_{11})_2$. Ferrous iron occurs replacing the magnesium, tremolite thus grading into actinolite. Small amounts of other molecules involving alkalies may also be present. Colors white to dark gray. In distinct crystals, either long-bladed or more rarely short and stout. In aggregates long and thin columnar, or fibrous; also compact granular massive (nephrite, below). G. = 2·9–3·2. Sometimes transparent and colorless. Optically −. Extinction-angle on $b(010)$, or $Z \wedge c$ axis = +16° to 18°, hence $Bx_a \wedge c$ axis = −74° to −72°. 2V = 80° to 88°. $\alpha = 1·60$, $\beta = 1·61$, $\gamma = 1·62$ for nearly pure tremolite, the values increasing with increase of the actinolite molecule.

Tremolite was named by Pini from the Tremola valley on the south side of the St. Gotthard.

Winchite is the name given to a blue amphibole near tremolite from the manganese mines of Central India.

5. ACTINOLITE.   *Calcium-magnesium-iron amphibole.*   Formula Ca_2 $(Mg,Fe)_5(OH)_2(Si_4O_{11})_2$. The amount of the iron molecule, $Ca_2Fe_5(OH)_2$ $(Si_4O_{11})_2$, varies up to between 40 and 50 per cent of the total. Alkalies also sometimes present in small amounts. Color bright green and grayish green. In crystals, either short- or long-bladed, as in tremolite; columnar or fibrous; granular massive. G. = 3–3.2. Sometimes transparent. The variety in long bright-green crystals is called *glassy actinolite*; the crystals break easily across the prism. The fibrous and radiated kinds are often called *asbestiform actinolite* and *radiated actinolite*. Actinolite owes its green color to the ferrous iron present.

Pleochroism distinct, increasing as the amount of iron increases, and hence the color becomes darker; Z emerald-green, Y yellow-green, X greenish yellow. Absorption $Z > Y > X$, Zillertal. Optically $-$. Extinction-angle on $b(010)$, $Z \wedge c$ axis = $+15°$ and $Bx_a \wedge c$ axis = $-75°$. $2V = 78°$; $\rho < v$; $\alpha = 1.614$, $\beta = 1.63$, $\gamma = 1.64$ (for variety with about 20 per cent of iron molecule).

Named actinolite from $\dot{\alpha}\kappa\tau\iota\varsigma$, *a ray*, and $\lambda\dot{\iota}\theta\varsigma$, *stone*, a translation of the German *Strahlstein* or *radiated stone*. Name changed to *actinote* by Haüy, without reason.

NEPHRITE. Jade in part. A tough, compact, fine-grained tremolite (or actinolite), breaking with a splintery fracture and glistening luster. H. = 6–6.5. G. = 2.96–3.1. Named from a supposed efficacy in diseases of the kidney, from $\nu\epsilon\phi\rho\dot{\varsigma}$, *kidney*. It varies in color from white (tremolite) to dark green (actinolite), in the latter, iron protoxide being present up to 6 or 7 per cent. The latter kind sometimes encloses distinct prismatic crystals of actinolite. A derivation from an original pyroxenic mineral has been suggested in some cases. Nephrite or jade was brought in the form of carved ornaments from Mexico or Peru soon after the discovery of America. A similar stone comes from Eastern Asia, New Zealand and Alaska. See jadeite, p. 562; jade, p. 563.

Széchenyiite is an amphibole occurring with jadeite from Central Asia.

ASBESTOS. Asbestus. Tremolite, actinolite, and other varieties of amphibole, excepting those containing much alumina, pass into fibrous varieties, the fibers of which are sometimes very long, fine, flexible, and easily separable by the fingers, and look like flax. These kinds are called *asbestos* (from the Greek for *incombustible*). The colors vary from white to green and wood-brown. The name *amianthus* is applied usually to the finer and more silky kinds. Much that is popularly called asbestos is *chrysotile*, or fibrous serpentine, containing 12 to 14 per cent of water. *Byssolite* is a stiff fibrous variety.

Mountain leather is in thin flexible sheets, made of interlaced fibers; and *mountain cork* the same in thicker pieces; both are so light as to float on water, and they are often hydrous, color white to gray or yellowish. *Mountain wood* is compact fibrous, and gray to brown in color, looking a little like dry wood.

SMARAGDITE. A thin-foliated variety of amphibole, near actinolite in composition but carrying some alumina. It has a light grass-green color, resembling much common green diallage. In many cases derived from pyroxene (diallage) by uralitization, see below. It retains much of the structure of the diallage and also often encloses remnants of the original mineral. It forms, along with whitish or greenish saussurite, a rock called saussurite-gabbro, the euphotide of the Alps. The original mineral is from Corsica, and the rock is the *verde di Corsica duro* of the arts.

URALITE. Pyroxene altered to amphibole. The crystals, when distinct, retain the form of the original mineral, but have the cleavage of amphibole. The change usually commences on the surface, transforming the outer layer into an aggregation of slender amphibole prisms, parallel in position to each other and to the parent pyroxene (cf. Fig. 843, p. 554). When the change is complete the entire crystal is made up of a bundle of amphibole needles or fibers. The color varies from white (tremolite) to pale or deep green, the latter the more common. In composition uralite appears to conform nearly to actinolite, as also in optical characters. The most prominent change in composition in passing from the original pyroxene is that corresponding to the difference existing between the two species in general, that is, an increase in the magnesium and decrease in calcium. The change, therefore, is not strictly a case of paramorphism, although usually so designated. Uralite was originally described by Rose in a rock from the Ural Mts. It has since been observed from many localities. The microscopic study of rocks has shown the process of " uralization " to be very common, and some authors regard many hornblendic rocks and schists to represent altered pyroxenic rocks on a large scale.

6. RICHTERITE. An amphibole containing MgO (18 to 21 per cent), CaO (5 to 8 per cent), MnO (5 to 12 per cent), alkali oxides (5 to 9 per cent).

In elongated crystals, seldom terminated. G. = 3·09. Color brown, yellow, rose-red. Transparent to translucent. $Z \wedge c$ axis = $+ 15°$–$20°$. Optically $-$. $2V = 68°$. $\alpha = 1·615$, $\beta = 1·629$, $\gamma = 1·636$. From Vermland, Sweden at Pajsberg near Persberg and at Långbanshyttan. Characterized by the presence of manganese and alkalies in relatively large amount.

Imerinite is a soda-amphibole, related to soda-richterite from the province Imerina, Madagascar.

7. ALUMINOUS AMPHIBOLE. Hornblende. Contains alumina or ferric iron, and usually both, with ferrous iron (sometimes manganese), magnesium, calcium, and alkalies. Hydroxyl and fluorine are commonly present in small amounts; also titanium in many varieties. The varieties of hornblende here included range from the light-colored *edenite*, containing but little iron, through the light to dark green *pargasite*, to the dark-colored or black *hornblende*, the color growing darker with increase in amount of iron. Extinction-angle variable, from 0° to 37°, see below. Pleochroism strong. Absorption usually $X < Y < Z$.

EDENITE. *Aluminous Magnesium-Calcium Amphibole.* Color white to gray and pale green, and also colorless; G. = 3·0–3·059. Resembles anthophyllite and tremolite. Named from the locality at Edenville, New York. To this variety belong various pale-colored amphiboles, having less than 5 per cent of iron oxides.

Koksharovite is a variety from the neighborhood of Lake Baikal, Siberia, named after the Russian mineralogist, von Koksharov.

COMMON HORNBLENDE, PARGASITE. Colors bright or dark green, and bluish green to grayish black and black. G. = 3·05–3·47. *Pargasite* is usually made to include green and bluish green kinds, occurring in stout lustrous crystals, or granular; and *common hornblende* the greenish black and black kinds, whether in stout crystals or long-bladed, columnar, fibrous, or massive granular. But no line can be drawn between them. The extinction-angle on $b(010)$, or $Z \wedge c$ axis = $+ 15°$ to $25°$ chiefly. Absorption, $X < Y < Z$.

Pargasite occurs at Pargas, Finland, in bluish green and grayish black crystals. $Z \wedge c$ axis = $+ 26°$; optically $-$. Dispersion inclined, $\rho > v$. $\beta = 1·64$; $\gamma - \alpha = 0·019$; $2V = 63°$. Pleochroism: Z greenish blue; Y emerald-green; X greenish yellow.

Hornblende is optically $-$; dispersion inclined, $\rho < v$; pleochroism, $X =$ pale yellow or brown, $Y =$ greenish, brownish yellow, brown, $Z =$ dark green, brown. From Edenville, New York. $\beta = 1·67$; $\gamma - \alpha = 0·0206$; $Z \wedge c$ axis = $23°$ 48'.

Basaltic hornblendes are brown to black hornblendes from basaltic and other igneous rocks. They commonly contain considerable titanium, with usually ferric iron and alkalies. They vary somewhat widely in optical characters. The angle $Z \wedge c$ axis = 0° to $+ 10°$ chiefly. Pleochroism, $X =$ pale brown or yellow, $Y =$ dark brown or brownish green, $Z =$ dark brown, dark blue, or dark olive-green. Hornblende from Linosa with 8·5 per cent TiO_2 has $\beta = 1·73$, $\gamma - \alpha = 0·068$.

Soretite is a hornblende from the anorthite-diorite rocks of Koswinsky in the northern Ural Mts.

Speziaite, from Traversella, Italy, is an iron amphibole with strong pleochroism; $X =$ green, $Y =$ yellow-brown, $Z =$ azure-blue; $Z \wedge c$ axis = 23°.

Syntagmatite is the black hornblende of Vesuvius.

Bergamaskite is an iron-amphibole containing almost no magnesia. From Monte Altino, Province of Bergamo, Italy.

Philipstadite, from Philipstad, Sweden, is an iron-magnesium amphibole showing unusual pleochroism.

Laneite is a dark-colored, strongly pleochroic amphibole occurring in riebeckite rocks. Axial angle very small. Optically −. Ax. pl. normal to (010). $Z \wedge c$ axis = 13°–28°.

Barroisite is described as an amphibole intermediate between hornblende and glaucophane.

Weinschenkite is a magnesium-calcium amphibole, poor in FeO, but rich in sesquioxides and water.

Rimpylite is an amphibole rich in sesquioxides, poor in magnesia.

Pyr. — Essentially the same as for the corresponding varieties of pyroxene, see p. 559.

Diff. — Distinguished from pyroxene (and tourmaline) by its distinct prismatic cleavage, yielding angles of 56° and 124°. Fibrous and columnar forms are much more common than with pyroxene, lamellar and foliated forms rare (see also pp. 560, 569). Crystals often long, slender, or bladed. Differs from the fibrous zeolites in not gelatinizing with acids. Epidote has a peculiar green color, is more fusible, and shows a different cleavage.

Micro. — In rock sections amphibole generally shows distinct colors, green, sometimes olive or brown, and is strongly pleochroic. Also recognized by its high relief; generally rather high interference-colors; by the very perfect system of cleavage-cracks crossing at angles of 56° and 124° in sections ⊥ *c* axis (Fig. 873). In sections ∥ *b*(010) (recognized by yielding no axial figure in convergent light, by showing the highest interference-colors, and by having parallel cleavage-cracks, ∥ *c* axis), the extinction-direction for common hornblendes makes a small angle (12°–15°) with the cleavage-cracks (*i.e.*, with *c* axis); further, this direction is positive, *Z* (different from common pyroxene and ægirite).

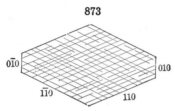

873

Artif. — Experiments on the artificial production of the amphiboles have shown that in general they are unstable at high temperatures and that their formation in igneous rocks is due either to the rapid cooling of the magma, to the presence of water or to some unusual conditions of pressure, etc. In general when the amphiboles are fused they are transformed into the corresponding pyroxenes.

Obs. — The amphiboles are common and widely distributed rock-making minerals. They are especially characteristic of the metamorphic rocks, although they may occur as important constituents of igneous rocks. The conditions under which they form are more restricted than is the case with the pyroxenes, mineralizing vapors apparently being necessary for their formation. *Tremolite* is always a product of metamorphism and is especially common in limestones, particularly the magnesian or dolomitic varieties. It also occurs as an alteration product of pyroxene or chrysolite in serpentines, etc. *Actinolite* (also nephrite) is common in schists, in steatitic rocks, and with serpentine; it also is found as a secondary mineral in certain igneous rocks, produced by the alteration of pyroxene, etc. Common dark green to black *hornblende* occurs in gneisses and schists, being the chief mineral of the amphibolites. It is a common constituent of igneous rocks, as granite, syenite, diorite, gabbro, etc. The variety known as basaltic hornblende is found only in igneous rocks, as in gabbro, basalt, etc.

Prominent foreign localities for amphibole are the following; others have been mentioned in connection with the descriptions of varieties. *Tremolite* is found in Austria in the Untersulzbachtal in Salzburg; in the Tyrol at Greiner in the Zillertal and near Prägratten. In dolomite at Campolungo near St. Gotthard in Ticino, Switzerland; in Italy in Val di Susa, Piedmont; on the island of Elba; etc. *Actinolite* is common in the crystalline schists of the central and eastern Alps, especially at Greiner in the Zillertal, Tyrol. Both tremolite and actinolite occur in fibrous form, constituting some of the asbestos of commerce. Thus in the Zillertal and elsewhere in Tyrol, Austria; at Zermatt in Valais, Switzerland; and on the island of Corsica. *Hornblende* is found in Bohemia of Czechoslovakia at Teplice (Teplitz) and at various points near Bilin. From Vesuvius; from Norway at Arendal in Aust-Agder; at Kragerö in Telemark; and at Snarum in Buskerud; from Sweden at Nordmark in Vermland.

The important localities in the United States follow. In Massachusetts, black crystals at Chester, Hampden Co.; *cummingtonite* at Cummington, Hampshire Co.; white crystals of tremolite at Lee, Berkshire Co., and at the neighboring locality of Canaan, Litchfield Co.; Connecticut. In New York in Orange Co., from Warwick; dark green crystals at Two Ponds near Monroe; from near Edenville, dark green, gray, or hair-brown crystals; in large and perfect crystals from near Amity. *Hudsonite* comes from Cornwall, Orange Co. In St. Lawrence Co. in fine crystals at Gouverneur and Russell, also at Rossie, Pierrepont, Macomb, Edwards, and De Kalb. In New Jersey at Franklin, Sussex Co.; actinolite and a zinc-manganese variety. In Pennsylvania actinolite at Mineral Hill in Delaware Co.; in Maryland asbestos at Pylesville, Harford Co.

In Canada tremolite is abundant in the limestones of Ontario and Quebec. Actinolite is found in Haliburton Co., Ontario. Black hornblende occurs at various localities in Quebec and Ontario, as in Renfrew Co., Ontario; at Grenville, Argenteuil Co., Quebec (*edenite*). Asbestos and mountain cork are found at Buckingham, Ottawa Co., Quebec.

GLAUCOPHANE.

Monoclinic; near amphibole in form. Crystals prismatic in habit, usually indistinct; commonly massive, fibrous, or columnar to granular.

Cleavage: $m(110)$ perfect. Fracture conchoidal to uneven. Brittle. H. = 6–6·5. G. = 3–3·15. Luster vitreous to pearly. Color azure-blue, lavender-blue, bluish black, grayish. Streak grayish blue. Translucent. Pleochroism strongly marked: Z sky-blue to ultramarine-blue, Y reddish or bluish violet, X yellowish green to colorless. Absorption $Z > Y > X$. Optically $-$. Ax. pl. || $b(010)$. $Z \wedge c$ axis = 4° to 6°, rarely higher values. 2V = 45°. Dispersion strong, $\rho < v$. $\alpha = 1·621$, $\beta = 1·638$, $\gamma = 1·639$.

Comp. — Essentially $Na(Al,Fe)(SiO_3)_2$ with $(Mg,Fe)SiO_3$ in varying proportions.

Pseudoglaucophane described as similar to glaucophane but with ax. pl. normal to (010). Occurs surrounding a nucleus of glaucophane in a quartzite from the Urals and also in a crystalline schist from Switzerland.

Obs. — Glaucophane is found only as a constituent of metamorphic rocks, as glaucophane schist, mica schist, eclogite, crystalline marble, etc. These may have been produced by the metamorphism of either igneous or sedimentary rocks. It is associated with such minerals as quartz, epidote, pyroxene, chlorite, garnet, etc.

Though comparatively rare in its occurrence, glaucophane is to be found in a number of widely separated places. It was first described from the island of Syra, one of the Cyclades. It occurs in Switzerland at Zermatt in Valais on the north side of the Matterhorn, as also on the south side in the Valtournanche, in Italy. Also *gastaldite* (a variety with a higher content of alumina) at other points in Italy on the south slope of the Alps in Piedmont as at Champ de Praz and elsewhere in the Valle d'Aosta; at La Beaume in the Val di Susa; etc. On the island of Corsica; from New Caledonia, and in Japan. *Rhodusit* is a fibrous variety from the island of Rhodes (Rhodus), Asia Minor; from Minusinsk in Yeniseisk, Siberia. *Holmquisite* is a lithium bearing variety from the island of Utö, State of Stockholm, Sweden. *Ternovskite* is a related alkalic amphibole from Ternovsky iron mine, Krivoy-Rog, Kherson, Russia.

In the United States glaucophane rocks are found over a large area in the Coast Range of California.

Glaucophane is named from γλαυκός, *bluish green*, and φαίνεσθαι, *to appear*.

Crossite. — An amphibole intermediate in composition between glaucophane and riebeckite, being optically more nearly related to the latter. Occurs in lath-shaped crystals. G. = 3·16. Color blue. Optically $-$. Ax. pl. normal to (010); also may be parallel to (010). $Z = b$ axis; $X \wedge c$ axis = 60°–80°. 2V small to 0°. Strong dispersion of axes and bisectrices. Strongly pleochroic with X = colorless or yellow, Y = blue, Z = violet. $\alpha = 1·657$, $\beta = 1·659$, $\gamma = 1·663$. Found in California near Berkeley, Contra Costa Co., and somewhat widely distributed in the crystalline schists of the Coast Ranges. Also from Venzolasca, Corsica.

Torendrikite. — An amphibole with composition approximating to $Na_2O.4MgO.CaO.FeO.Fe_2O_3.10SiO_2$. Probably chemically and optically intermediate between glaucophane and riebeckite. Also may be related to richterite and imerinite (p. 575). H. = 5. G. = 3·2. Color dark blue. $\beta = 1·665$. $X \wedge c$ axis = about 50°. Ax. pl. normal to (010). Axial angle large. Strong dispersion. Strongly pleochroic. Found in eruptive alkaline syenite rocks at Torendrika in the valley of the Imorona, Madagascar. Also reported from a nephelite-syenite, southeast of Tine, Wadai, French Equatorial Africa.

RIEBECKITE.

Monoclinic. Axes $a : b : c = 0·5475 : 1 : 0·2295$; $\beta = 76°\ 10'$. In embedded prismatic crystals, longitudinally striated. Cleavage: prismatic (56°)

perfect. H. = 4. G. = 3·4. Luster vitreous. Color dark blue to black. Optically − or +. Ax. pl. ‖ (010) or ⊥ (010). X or Z ∧ c axis = 0° to 5° 2V large. Strongly pleochroic; X = dark blue, Y = blue or greenish to brownish yellow, Z = yellow green to dark green. α = 1·693, β = 1·695, γ = 1·697. Dispersion marked.

Comp. — Essentially $\overset{\text{iii}}{\text{Na}}\text{Fe}(\text{SiO}_3)_2$ with FeSiO_3 in varying proportions. It corresponds closely to acmite (ægirite) among the pyroxenes.

Obs. — Originally described from the granite and syenite of the island of Socotra in the Indian Ocean, 120 m. N. E. of Cape Guardafui, the eastern extremity of Africa where it occurs in groups of prismatic crystals, often radiating and closely resembling tourmaline. Also from granulite at Piana on the Golfe de Porto, Corsica. In Portugal from Alter Predoso. Found at Narsarsuk, Julianehaab district, Greenland. It occurs in Madagascar (at times in large crystals) in the district about Ampasibitika. Reported also from the Province of Kano, Northern Nigeria.

In the United States it occurs in pegmatite at Quincy, Norfolk Co., Massachusetts. Reported from the San Francisco Mts., Coconino Co., Arizona. A mineral originally called arfvedsonite from St. Peter's Dome, Pike's Peak region, El Paso Co., Colorado, has been shown to be near riebeckite. A soda amphibole, related to riebeckite, from Bababudan Hills, Mysore, India, has been named *bababudanite*.

CROCIDOLITE. Blue Asbestos. — Probably to be considered as a fibrous variety of riebeckite. Fibrous, asbestos-like; fibers long but delicate, and easily separable. Also massive or earthy. H. = 4. G. = 3·2-3·3. Silky to dull luster. Color lavender-blue or leek-green. Strongly pleochroic; green to violet and indigo blue. X ∧ c axis = nearly 0°. α = 1·698, β = 1·699, γ = 1·706. Occurs in South Africa in Griqualand-West, north of the Orange River, and in the neighboring district of Prieska (often replaced by quartz, forming the ornamental material known as *tiger's-eye*, see p. 472). At Golling in Salzburg, Austria. From Wackenbach, near Schirmeck in the Vosges Mts., Bas-Rhin, France. In Bolivia from the province of Cochabamba. In the United States from West Quincy, Norfolk Co., Massachusetts; from near Cumberland Hill, Providence, Rhode Island. *Abriachanite* is an earthy amorphous form occurring in the Abriachan district, near Loch Ness in Inverness, Scotland.

Hastingsite. — A group of amphiboles low in silica with calcium, ferrous iron, magnesium, aluminum and smaller amounts of alkalies. The formula may be written, $\text{Ca}_2\text{Na}(\text{Fe,Mg})_4(\text{Al,Fe})(\text{OH})_2(\text{Al,Si})_8\text{O}_{22}$. Iron and magnesium replace each other in varying degrees. With the increase of magnesia the value of 2V increases and the indices of refraction decrease. Optically −. Ax. pl. in most cases ‖ (010). For a variety with approximately equal amounts of FeO and MgO, Z ∧ c axis = 16°; 2V = 80°; α = 1·676, β = 1·692, γ = 1·705. The original hastingsite was a variety rich in iron (*ferrohastingsite*) from nephelite-syenite at Dungannon, Hastings Co., Ontario. Such amphiboles have been found elsewhere in nephelite-syenites, nordmarkites, and certain granites. Varieties with high magnesia content (*magnesiohastingsite*) are characteristic of essexites and diorites. *Tamarite* is an amphibole similar to hastingsite from the alkali rocks of the Mariupol district, on the Sea of Azov in Ukraine, Russia. Y ∧ c axis = +14° to 18°. *Fluotamarite* is another variety from the same locality with strong pleochroism. X ∧ c axis = −12° to 16°.

OSANNITE. A soda-amphibole between riebeckite and arfvedsonite in which ax. pl. is normal to (010), and X is nearly parallel to c axis. 2V large. β = 1·69. Found in amphibole gneiss at Cervadaes and in pegmatitic patches of an alkali-syenite at Alter Predoso, Portugal. The riebeckite from Quincy, Massachusetts, may be the same mineral.

ARFVEDSONITE.

Monoclinic. Axes a : b : c = 0·5496 : 1 : 0·2975; β = 75° 44′.

Crystals long prisms, often tabular ‖ b(010), but seldom distinctly terminated; angles near those of amphibole; also in prismatic aggregates. Twins: tw. pl. a(100).

Cleavage: prismatic, perfect; b(010) less perfect. Fracture uneven. Brittle. H. = 6. G. = 3·44-3·45. Luster vitreous. Color pure black; in thin scales deep green. Streak deep bluish gray. Opaque except in thin splinters. Pleochroism strongly marked: Z = greenish yellow to blue-gray; Y = lavender-blue to brownish yellow; X = greenish blue. Optically $-$. Ax. pl. normal to (010). Axial angle, large. $\alpha = 1\cdot687$, $\beta = 1\cdot707$, $\gamma = 1\cdot708$. Extinction-angle on b(010), varies $X \wedge c$ axis from $-14°$ to $+20°$.

Comp. — A slightly basic metasilicate of sodium and ferrous iron chiefly with smaller amounts of aluminum and ferric iron. $Na_3Fe_4Al(OH)_2Si_8O_{22}$.

Pyr., etc. — B.B. fuses at 2 with intumescence to a black magnetic globule; colors the flame yellow (soda); with the fluxes gives reactions for iron and manganese. Not acted upon by acids.

Micro. — In thin sections shows brown- or gray-green or gray-violet colors; strongly pleochroic in blue and green tints; negative elongation.

Obs. — Arfvedsonite and amphiboles of similar character, containing much iron and soda, are constituents of certain igneous rocks which are rich in alkalies, as nephelite-syenite, certain porphyries, etc. Large and distinct crystals are found only in the pegmatite veins of such rocks. It occurs in black hornblende-like crystals in the Julianehaab district in southern Greenland at Kangerdluarsuk, Narsarsuk, etc., associated with sodalite, eudialyte, and feldspar; the Greenland crystals are sometimes 9 inches long. It also occurs in Norway in the nephelite-syenites and related rocks of the Langesundfiord district and in the region about Oslo.

In the United States arfvedsonite is found at Red Hill near Moultonboro, Carroll Co., New Hampshire; and at St. Peter's Dome in the Pike's Peak region in El Paso Co., Colorado.

Tschernichéwite from a magnetite bearing quartzite in the northern Ural Mts., is near arfvedsonite.

KATAPHORITE from Norway appears to be intermediate between arfvedsonite and barkevikite. $Z \wedge c$ axis = $+30°$ to $+60°$; absorption, $Y > Z > X$; pleochroism, X = yellow-brown, Y = reddish or greenish brown, Z = reddish or greenish yellow, pale reddish brown, or greenish blue. Optically $-$.

KAERSUTITE, a titaniferous amphibole from Kaersut, Umanaksfiord, North Greenland, has been shown by X-ray study to be related to arfvedsonite and barkevikite.

BARKEVIKITE. An amphibole near arfvedsonite but more basic. In prismatic crystals. Cleavage; prismatic $(55° 44\frac{1}{2}')$. H. = 5·5. G. = 3·428. Color deep velvet-black. Pleochroism marked, colors brownish. Extinction-angle with c axis on b (010) = 10°–14°. Optically $-$. $\alpha = 1\cdot68$, $\beta = 1\cdot69$, $\gamma = 1\cdot70$. 2V = 31°–52°. Occurs at the wöhlerite locality near Barkevik, on the Langesundfiord, and elsewhere in southern Norway. In large crystals at Tugar, Ayrshire, Scotland.

ÆNIGMATITE. *Cossyrite* — Essentially a titano-silicate of ferrous iron and sodium, but containing also aluminum and ferric iron. In prismatic triclinic crystals. Cleavage: prismatic, distinct (66°). Shows certain relations to the amphibole group in which it has commonly been included as a triclinic member. Structurally, however, it appears to be distinct and it is doubtful if it should be so classified. G. = 3·74–3·80. Color black. Optically $+$. Ax. pl. nearly parallel to (010). $Z \wedge c$ axis = 45°. 2V = 32° $\beta = 1\cdot80$. Strongly pleochroic in shades of brown with absorption, $X < Y < Z$. Ænigmatite is from the sodalite-syenite of the Julianehaab district in southern Greenland, at Kangerdluarsuk, Naujakasik, Siorarsuit, and Tupersiatsiak. *Cossyrite* occurs in minute crystals embedded in the liparite lavas of the island Pantellaria (ancient name Cossyra); also widespread in the rocks of East Africa. *Rhönite* is like ænigmatite but contains much less ferrous oxide and alkalies with increase in alumina, ferric oxide, etc. From basaltic rocks in the Rhön district in Hessen-Nassau and elsewhere in Germany and Bohemia.

WEINBERGERITE. Perhaps $NaAlSiO_4.3FeSiO_3$ Orthorhombic. In spherical aggregates of radiating fibers. Black color. From a meteoric iron at Codai Canal, Palni Hills, Madras, India.

BERYL.

Hexagonal. Axis $c = 0\cdot4989$.

X-ray study of the structure shows that the vertical axis of the unit cell has a value twice as great as above. The silicon atoms are at the centers of a group of four oxygen atoms lying at the points of tetrahedra. The tetrahedral groups are linked together by the sharing of oxygen atoms into rings, having the composition Si_6O_{18}. These silica rings are joined together by aluminum atoms lying in the center of a group of six oxygen atoms and by beryllium atoms in a similar group of four oxygen atoms. There are two molecules in each unit cell.

Crystals usually long prismatic, often striated vertically, rarely transversely; distinct terminations exceptional. Occasionally in large masses, coarse columnar or granular to compact.

$$cp, \ 0001 \wedge 10\bar{1}1 = 29° \ 56\tfrac{1}{2}'. \qquad cs, \ 0001 \wedge 11\bar{2}1 = 44° \ 56'.$$
$$co, \ 0001 \wedge 11\bar{2}2 = 26° \ 31'. \qquad pp', \ 10\bar{1}1 \wedge 01\bar{1}1 = 28° \ 54\tfrac{1}{2}'.$$

Cleavage: $c(0001)$, imperfect and indistinct; rarely shows imperfect prismatic cleavage. Fracture conchoidal to uneven. Brittle. H. = 7·5–8. G. = 2·63–2·80; usually 2·69–2·70. Luster vitreous, sometimes resinous. Colors emerald-green, pale green, passing into light blue, yellow and white; also pale rose-red. Streak white. Transparent to subtranslucent. Dichroism more or less distinct. Optically −. Birefringence low. Often abnormally biaxial. Indices vary with the presence of alkalies; ω = 1·568–1·602, ϵ = 1·564–1·595.

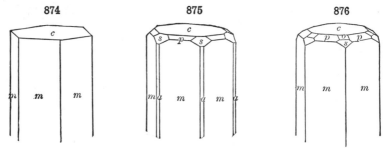

Var. — 1. *Emerald.* Color bright emerald-green, due to the presence of a little chromium; highly prized as a gem when clear and free from flaws.

2. *Ordinary; Beryl.* Generally in hexagonal prisms, often coarse and large; green the common color. The principal kinds are: (a) colorless; (b) bluish-green, called *aquamarine*; (c) apple-green; (d) greenish yellow to iron-yellow and honey-yellow; sometimes a clear bright yellow as in the *golden beryl* (a yellow gem variety from South West Africa has been called *heliodor,* containing a small amount of ferric oxide); (e) pale yellowish green; (f) clear sapphire blue; (g) pale sky-blue; (h) pale violet or reddish; (i) rose colored called *morganite* or *vorobyevite;* (j) opaque brownish yellow, of waxy or greasy luster. The *oriental emerald* of jewelry is emerald-colored sapphire.

Comp. — $Be_3Al_2(SiO_3)_6$ or $3BeO.Al_2O_3.6SiO_2$ = Silica 67·0, alumina 19·0, glucina 14·0 = 100.

Alkalies (Na_2O, Li_2O, Cs_2O) are sometimes present replacing the beryllium, from 0·25 to 5 per cent; also chemically combined water, including which the formula becomes $(R_2R)O.6BeO.2Al_2O_3.12SiO_2$.

Pyr., etc. — B.B. alone, unchanged or, if clear, becomes milky white and clouded; at a high temperature the edges are rounded, and ultimately a vesicular scoria is formed. Fusibility = 5·5, but somewhat lower for beryls rich in alkalies. Glass with borax, clear and colorless for beryl, a fine green for emerald. Unacted upon by acids.

Diff. — Characterized by its green or greenish blue color, glassy luster and hexagonal form; rarely massive, then easily mistaken for quartz. Distinguished from apatite by its hardness, not being scratched by a knife, also harder than green tourmaline: from chrysoberyl by its form; from euclase and topaz by its imperfect cleavage.

Artif. — Crystals of beryl have been produced artificially by fusing a mixture of silica, alumina and glucina with boric oxide as a flux.

Obs. — Beryl is most commonly found in granite rocks, either in druses in the granite or in pegmatite veins. It has also been noted in tin ores together with topaz and in mica schists. The emeralds found in Colombia occur in a bituminous limestone, a unique type of occurrence.

Beryl is widely distributed, and only the most important localities can be mentioned here. Foreign localities include the following: Aquamarine crystals of fine color and large size occur with topaz in the Adun-chilon Mts., south of Nerchinsk in Transbaikalia; also in the Ural Mts. at Miask with topaz; at Mursinsk north of Ekaterinburg and on the Takowaja River to the east of that place. The latter locality is also a famous source of emeralds, of large size, which are embedded in a mica schist. In Austria, small but beautiful green crystals are found in the Habachtal in Salzburg. Beryl is found in Bavaria at Bodenmais and Rabenstein; on Elba in beautiful crystals; from the Mourne Mts., Co. Down, Ireland.

In South West Africa at Rössing near Swakopmund. Emeralds are found in northeastern Transvaal. In Madagascar in many places both aquamarine and pink in color; as at Ampangabé near Miandrarivo, Tongafeno, Maharitra on Mt. Bity (rose-colored), Anjanaboana, Sahanivotry, etc. Beautiful aquamarines, at times of extraordinary size, come from the states of Bahia and Minas Geraes, Brazil. A crystal of aquamarine weighing more than 5 pounds was found near Santa Rita de Arassuahy, Minas Novas district, Minas Geraes. The world's most famous locality for emeralds is near Muso, 75 miles to the northeast of Bogota, Colombia.

In the United States, noteworthy localities for beryl are in Maine, at Albany, Oxford Co., where enormous crystals have been found, one being 18 feet long, 4 feet in diameter and weighing 18 tons; also at Paris, Oxford Co., and at Topsham, Sagadahoc Co. In New Hampshire beryl crystals of great size are found at Acworth, Sullivan Co., and at Grafton, Grafton Co. One crystal from the latter locality weighed nearly 3000 pounds and was over 4 feet long. In Massachusetts large beryls and also aquamarine crystals occur at Royalston, Worcester Co.; *goshenite* is found at Goshen, Hampshire Co. In Connecticut from the pegmatites in Middlesex Co., at Haddam and Haddam Neck, at Middletown and Portland; in large columnar masses at Branchville, Fairfield Co. In Pennsylvania in Delaware Co., at Leiperville, and Chester. In North Carolina in Mitchell and Yancey counties and also near Stony Point, Alexander Co., in imperfect emeralds associated with the green variety of spodumene, *hiddenite*. In Colorado near the summit of Mt. Antero, Chaffee Co., aquamarines. In California in rose and variously colored crystals at Cahuilla, Riverside Co., and in San Diego Co., at Pala, Mesa Grande, Ramona, etc.

Use. — The transparent mineral is used as a gem stone; see above under *Varieties*.

Eudialyte. — Essentially a metasilicate of Zr,Fe(Mn),Ca,Ce,Na,Cl, etc. The formula $(Na,Ca,Fe)_6Zr(OH,Cl)(SiO_3)_6$ has been proposed. In pink, red to brown tabular or rhombohedral crystals; also massive. Indistinct basal cleavage. H. = 5–5·5. G. = 2·9–3·0. Fusible at 2·5. Optically +. Indices variable. $\omega = 1·606–1·610$; $\epsilon = 1·610–1·613$. *Eucolite* is similar but optically −. $\omega = 1·620–1·643$; $\epsilon = 1·618–1·634$. Eudialyte and eucolite occur in nephelite-syenites and in their pegmatite facies. *Eudialyte* is found in the Julianehaab district of Greenland; on the island Sedlovatoi near Archangel on the White Sea, Russia, and at Lu Javr on the Kola peninsula, Russian Lapland. *Eucolite* is from the islands of the Langesundfiord in southern Norway. Also near Ampasibitika, Madagascar. Both occur at Magnet Cove, near Hot Springs in Garland Co., Arkansas, of a rich crimson to peach-blossom color. The name *mesodialyte* has been proposed for an intermediate member of the eudialyte-eucolite series.

Elpidite. — $H_6Na_2Zr(SiO_3)_6$. Orthorhombic. In fibrous crystals with prismatic cleavage or massive. H. = 7. G. = 2·54. Color white to brick-red. Optically +. $\alpha = 1·560$, $\beta = 1·565$, $\gamma = 1·674$. Ax. pl. || (010). $Z = a$ axis. From Narsarsuk in the Julianehaab district, southern Greenland. A variety containing much titanium from Mount Chibina, Russian Lapland, has been called *titanoelpidite*. Indices, $\alpha = 1·681$, $\beta = 1·686$, $\gamma = 1·698$.

Catapleiite. — $H_4(Na_2,Ca)ZrSi_3O_{11}$. Monoclinic; in thin pseudohexagonal plates. Becomes hexagonal at 140° C. Cleavages parallel to (110) and (010). H. = 6. G. = 2·75. Fusible at 3. Color light yellow to yellowish brown. Optically +. Ax. pl. || (010). Z nearly normal to (001). $\alpha = 1·591$, $\beta = 1·592$, $\gamma = 1·627$. From various islands in the Langesundfiord district of southern Norway. *Natron-catapleiite* contains only sodium and is found only on the island Lille-Arö. Catapleiite is reported from the Julianehaab district of southern Greenland, as being orthorhombic, optically −, and with lower indices of refraction. The mineral is also reported from Magnet Cove, near Hot Springs, Garland Co., Arkansas.

MURMANITE. A titanium-zirconium silicate of sodium, iron, manganese, and calcium. In tabular masses. Micaceous cleavage. Biaxial. Color violet. $\beta = 1\cdot735$. H. = 2–3. From Lu Javr and Hibina districts, Kola peninsula, northern Russia.

Cappelenite. — A boro-silicate of yttrium and barium. Hexagonal, rhombohedral. In greenish brown prismatic crystals. H. = 6. G. = 4·4. B.B. infusible. Optically −. $\omega = 1\cdot76$. Found on Lille-Arö in the Langesundfiord, southern Norway.

Melanocerite. — A fluo-silicate of the cerium and yttrium metals and calcium chiefly (also B, Ta, etc.). Hexagonal, rhombohedral. In brown to black tabular rhombohedral crystals. H. = 5–6. G. = 4·13. B.B. infusible. Optically −. $\omega = 1\cdot73$, $\epsilon = 1\cdot72$. Found very sparingly on the island Kjeö in the Langesundfiord, southern Norway.

Caryocerite. — Near melanocerite, containing ThO_2. Hexagonal, rhombohedral. Tabular crystal habit. H. = 5–6. G. = 4·29. Brown color. Isotropic because of alteration. $n = 1\cdot74$. Found rarely in Langesundfiord district, southern Norway.

Steenstrupine. — Is allied to this group. Rhombohedral. H. = 4. G. = 3·4. Color dark brown to nearly black. Optically −. From Kangerdluarsuk, Julianehaab district, Greenland.

Tritomite. — A fluo-silicate of thorium, the cerium and yttrium metals and calcium, with boron. Hexagonal, rhombohedral, hemimorphic. In dark brown crystals of acute triangular pyramidal form. H. = 5·5. G. = 4·2. Isotropic because of alteration and n about 1·75. Rare. From various places in the Langesundfiord district of southern Norway.

Leucophanite. — $(Ca,Na)_2BeSi_2(O,OH,F)_7$. Orthorhombic-sphenoidal. Crystals tabular || (001). Cleavages || to three pinacoids. Often twinned, tw. pls., (110), (001). Pseudo-tetragonal. H. = 4. G. = 2·96. Difficultly fusible with intumescence. Color greenish yellow. Optically −. Ax. pl. || (100). X normal to (001). $\alpha = 1\cdot571$, $\beta = 1\cdot595$, $\gamma = 1\cdot598$. Occurs sparingly on various islands in the Langesundfiord district, southern Norway.

Meliphanite. — A fluo-silicate of beryllium, calcium, and sodium, near leucophanite. $(Ca,Na)_2Be(Si,Al)_2(O,F)_7$. Tetragonal. In low square pyramids. Basal cleavage. H. = 5–5·5. G. = 3·0. Fuses with intumescence. Optically −. Sometimes slightly biaxial. Yellow color. $\omega = 1\cdot612$, $\epsilon = 1\cdot593$. From the Langesundfiord district of southern Norway.

Astrolite. $(Na,K)_2Fe(Al,Fe)_2(SiO_3)_5.H_2O$? Orthorhombic. In globular forms with radiating structure. One cleavage. H. = 3·5. G. = 2·8. Fusible, 3·5. Color green. Optically −. $\alpha = 1\cdot57$, $\beta = 1\cdot594$, $\gamma = 1\cdot597$. Found in a diabase tuff near Neumark, Vogtland district, Saxony.

The two following minerals are very similar to each other and should possibly be considered as one species.

Cuspidine. — $Ca_4Si_2O_7F_2$. Monoclinic. In minute spear-shaped crystals. Basal cleavage. H. = 5–6. G. = 2·8–2·9. Color pale rose-red. Optically +. 2V large. Ax. pl. || (010). $\alpha = 1\cdot590$, $\beta = 1\cdot595$, $\gamma = 1\cdot602$. Occurs sparingly in ejected masses from the tufa of Mte. Somma, Vesuvius. Also at Ariccia in the Albani Mts., south of Rome. In the United States from Franklin, Sussex Co., New Jersey.

Custerite. — $[Ca(F,OH)]_2SiO_3$. Monoclinic. In fine granular masses. Cleavages parallel to base and prism, all making nearly 90° with each other. Twinning plane $c(001)$, showing in twin lamellæ. H. = 5. G. = 2·91. Difficultly fusible. Color greenish gray. Transparent. Optically +. Ax. pl. normal to (010). Z nearly perpendicular to c (001). 2V large. $\alpha = 1\cdot586$, $\beta = 1\cdot59$, $\gamma = 1\cdot598$. Found in limestone contact zone at the Empire mine, Custer Co., Idaho. Also from Crestmore, Riverside Co., California.

Didymolite. — $2CaO.3Al_2O_3.9SiO_2$. Monoclinic. In small twinned crystals. H. = 4–5. G. = 2·71. Color dark gray. Opaque. Index 1·5. Difficultly fusible. Insoluble. Found as contact mineral in limestone from Tatarka River, Yeniseisk District, Siberia.

CORDIERITE. Iolite. Dichroite.

Orthorhombic. Axes $a : b : c = 0\cdot5871 : 1 : 0\cdot5585$. X-ray study of structure indicates possibility of hemimorphic symmetry.

Twins: tw. pl. $m(110)$, also $d(130)$, both yielding pseudo-hexagonal forms.

Habit short prismatic ($mm''' = 60°\,50'$) (Fig. 877). As embedded grains; also massive, compact.

Cleavage: b(010) distinct; a(100) and c(001) indistinct. Crystals often show a lamellar structure $\parallel c$(001), especially when slightly altered. Fracture subconchoidal. Brittle. H. = 7–7·5. G. = 2·60–2·66. Luster vitreous. Color various shades of blue, light or dark, smoky blue. Transparent to translucent. Pleochroism strongly marked except in thin sections. Absorption $Y > Z > X$. Y = dark violet, blue, or brown, Z = clear blue, X = clear yellow.

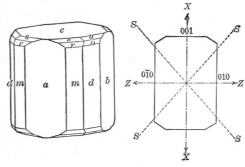

877 878

Pleochroic halos common, often bright yellow; best seen in sections $\parallel c$ axis. Exhibits idiophanous figures. Optically $-$. Has also been reported as optically $+$. Ax. pl. $\parallel a$(100). Bx. $\perp c$(001). Dispersion feeble, $\rho < v$. 2V = 40° to 84°. Indices variable, from 1·534 to 1·599.

Comp. — $Mg_2Al_4Si_5O_{18}$. This is the composition of artificial material that has the characters of cordierite. The natural mineral contains ferrous iron replacing a part of the magnesium; calcium and hydroxyl are also present.

The stable form of the artificial mineral crystallizes above 950° C. The natural mineral when heated to 1440° C. is changed into a mixture of sillimanite and glass.

Pyr., etc. — B.B. loses transparency and fuses at 5–5·5. Only partially decomposed by acids. Decomposed on fusion with alkaline carbonates.

Diff. — Characterized by its vitreous luster, color and pleochroism; fusible on the edges unlike quartz; less hard than sapphire.

Micro. — Recognized in thin sections by lack of color; low refraction and low interference-colors; it is very similar to quartz, but distinguished by its biaxial character; in volcanic rocks commonly shows distinct crystal outlines and a twinning of three individuals like aragonite. In the gneisses, etc., it is in formless grains, but the common occurrence of inclusions, especially of sillimanite needles, the pleochroic halos of a yellow color around small inclusions, particularly zircons, and the constant tendency to alteration to micaceous pinite seen along cleavages, help to distinguish it.

Obs. — Cordierite may be the direct product of igneous action, being stable over a considerable range of temperature and formed apparently without necessarily the aid of mineralizers. It is most commonly found in gneiss and crystalline schists and in contact metamorphic zones. It is also found in volcanic ejections, and in igneous rocks such as granite, andesite, dacite, rhyolite, kersantite, etc. It has been developed in sandstones that have been vitrified by contact with basalt and in sediments that have been altered by the slow burning of adjacent coal seams.

Only the more prominent localities are mentioned here; others will be found under the list of alteration products given below. At Bodenmais, Bavaria, in granite with pyrrhotite, etc.; in Finland at Orijärvi west of Lake Logo in Uudenmaan (*steinheilite*); in Norway at Arendal in Aust-Agder and at Kragerö in Telemark. In Greenland at Uiordlersuak in the Upernivik district. From the Mt. Bity district, Madagascar. Ceylon affords a transparent variety in small rolled masses of an intense blue color, the *saphir d'eau* of jewelers.

In the United States chiefly in Connecticut; in Middlesex Co. at Haddam, also in New London Co. to the north of Norwich, from near Guilford, New Haven Co., and in large altered crystals from Plymouth, Litchfield Co.

Named *Cordierite*, after Cordier, the French geologist (1777–1861). *Iolite* from *ἴον*, violet, and λίθος, stone; *Dichroite* (from δίχροος, *two-colored*), from its dichroism.

Alt. — The alteration of cordierite takes place so readily by ordinary exposure, that the mineral is most commonly found in an altered state, or enclosed in the altered cordierite. This change may be a simple hydration; or a removal of part of the protoxide bases by carbon dioxide; or the introduction of oxide of iron; or of alkalies, forming pinite and mica. The first step in the change consists in a division of the prisms of cordierite into plates parallel to the base, and a pearly foliation of the surfaces of these plates; with a change of color to grayish green and greenish gray, and sometimes brownish gray. As the alteration proceeds, the foliation becomes more complete; afterward it may be lost. The mineral in this altered condition has many names; as *hydrous cordierite* (including *bonsdorffite* and *auralite*) from Åbo, Turun ja Porin, Finland; *fahlunite* from Falun, Sweden, also *pyrargillite* from Helsingfors, Finland; *esmarkite* and *praseolite* from near Brevik, Langesund district, Norway, also *raumite* from Raumo, Turun ja Porin, Finland, and *peplolite* from Ramsberg, Orebrö, Sweden; *chlorophyllite* from Unity, Waldo Co., Maine; *aspasiolite* and *polychroilite* from Kragerö, Telemark, Norway. There are further alkaline kinds, as *pinite*, *cataspilite*, *gigantolite*, *iberite*, belonging to the Mica Group.

Use. — Cordierite is sometimes used as a gem.

Jurupaite. — $H_2(Ca,Mg)_2Si_2O_7$: Monoclinic? Radiating fibrous. White. H. = 4. G. = 2·75. Fusible at 2. $\alpha = 1·568$, $\gamma = 1·576$. $Z \wedge$ elongation = 31°. Found in Jurupa Mts. near Crestmore, Riverside Co., California.

The following are rare lead, zinc, and barium silicates:

Barysilite. — $Pb_3Si_2O_7$. Rhombohedral. In embedded masses with curved lamellar structure. Cleavage: basal. H. = 3. G. = 6·11–6·72. Color white; tarnishing on exposure. Optically −. $\omega = 2·033$, $\epsilon = 2·015$ (Franklin). Occurs in Vermland, Sweden, at the Harstig mine, Pajsberg near Persberg and at Långbanshyttan. Also at Franklin, Sussex Co., New Jersey.

Ganomalite. — $Pb_4(PbOH)_2Ca_4(Si_2O_7)_3$. Hexagonal. In prismatic crystals; also massive, granular. Prismatic cleavage. H. = 3. G. = 5·74. Easily fusible. Colorless to gray. Optically +. $\omega = 1·910$, $\epsilon = 1·945$. Occurs in Vermland, Sweden, at Långbanshyttan, and at Jacobsberg in Nordmark district.

Nasonite. — $Pb_4(PbCl)_2Ca_4(Si_2O_7)_3$. Closely related to ganomalite, and it is probable that the two minerals form one series. Massive, cleavable granular. H. = 4. G. = 5·4. Fusible. White. Optically +. $\omega = 1·913$, $\epsilon = 1·923$. From Franklin, Sussex Co., New Jersey, also from Långbanshyttan, Vermland, Sweden.

Margarosanite. — $Pb(Ca,Mn)_2(SiO_3)_3$. Triclinic. Slender prismatic crystals and cleavable granular. Three cleavages, one perfect. H. = 2·5–3. G. = 3·99. Easily fusible. Colorless and transparent with pearly luster. Optically −. $\alpha = 1·729$, $\beta = 1·773$, $\gamma = 1·807$. From Franklin, Sussex Co., New Jersey, and at Långbanshyttan, Vermland, Sweden.

Hyalotekite. — Approximately $(Pb,Ba,Ca)_9B_2(SiO_3)_{12}$. Orthorhombic? Massive; coarsely crystalline. H. = 5–5·5. G. = 3·81. Color white to pearly gray. Optically +. 2V small. Strong dispersion, $\rho < v$. $\beta = 1·96$. From Långbanshyttan, Vermland, Sweden.

Barylite. — $Be_2BaSi_2O_7$. Orthorhombic. Colorless, platy crystals. Cleavages (001) and (100). H. = 6–7. G. = 4·0. Optically + (Långbanshyttan), − (Franklin). Ax. pl. $\parallel$ (100). $X = b$ axis. $\alpha = 1·69$, $\beta = 1·70$, $\gamma = 1·705$. 2V large. From Långbanshyttan, Vermland, Sweden, and Franklin, Sussex Co., New Jersey.

Gillespite. — $FeO.BaO.4SiO_2$. Tetragonal or hexagonal. Cleavages (001), (100). H. = 4. G. = 3·33. Color, red. Uniaxial, −. $\omega = 1·621$. $\epsilon = 1·619$. Strongly pleochroic, O = colorless, E = red. Found in a rock specimen from a moraine near the head of Dry Delta, Alaska Range, Alaska.

Taramellite. — $Ba_4Fe\overset{II}{}Fe\overset{III}{}_4Si_{10}O_{31}$. Orthorhombic? Fibrous. H. = 5·5. G. = 3·9. Color reddish brown. Strong pleochroism, $X = Y$ = flesh-red, Z = almost black, $\beta = 1·77$. Found in limestone at Candoglia, in the Toce Valley, Piedmont, Italy.

Roeblingite. — $2PbSO_4.H_{10}Ca_7(SiO_4)_6$. Orthorhombic? In dense, white, compact, fibrous masses. H. = 3. G. = 3·433. Optically +. 2V small. $\beta = 1·64$. Fibers show parallel extinction and negative elongation. From Franklin, Sussex Co., New Jersey.

III. Orthosilicates. R_2SiO_4

Salts of Orthosilicic Acid, H_4SiO_4; characterized by an oxygen ratio of 1 : 1 for silicon to bases.

The following list includes the more prominent groups among the Orthosilicates.

A number of basic orthosilicates are here included, which yield water upon ignition; also others which are more or less basic than a normal orthosilicate, but which are of necessity introduced here in the classification, because of their relationship to other normal salts. The Mica Group is so closely related to many Hydrous Silicates that (with also Talc, Kaolinite, and some others) it is described later with them.

Nephelite Group. Hexagonal.
Sodalite Group. Isometric.
Helvite Group. Isometric-tetrahedral.
Garnet Group. Isometric.
Chrysolite Group. Orthorhombic.
Phenacite Group. Tri-rhombohedral.

Scapolite Group. Tetragonal-pyramidal.
Zircon Group. Tetragonal.
Danburite Group. Orthorhombic.
Datolite Group. Monoclinic.
Epidote Group. Monoclinic.

Nephelite Group. Hexagonal

Typical formula $\overset{I}{R}AlSiO_4$

Nephelite	$(Na,K)AlSiO_4$	$c = 0.8389$
Soda-nephelite (artif.)	$NaAlSiO_4$	
Eucryptite	$LiAlSiO_2$ **Kaliophilite**	$KAlSiO_4$

Cancrinite	$H_6Na_6Ca(NaCO_3)_2Al_8(SiO_4)_9$	$2c = 0.8448$
Microsommite	$(Na,K)_{10}Ca_4Al_{12}Si_{12}O_{52}SCl_4$	$2c = 0.8367$

The species of the NEPHELITE GROUP are hexagonal in crystallization and have in part the typical orthosilicate formula $\overset{I}{R}AlSiO_4$. From this formula nephelite itself deviates somewhat, though an artificial soda-nephelite, $NaAlSiO_4$, conforms to it. The species Cancrinite and Microsommite are related in form and also in composition, though in the latter respect somewhat complex. They serve to connect this group with the sodalite group following.

NEPHELITE. Nepheline. Elæolite.

Hexagonal-hemimorphic (p. 118). Axis $c = 0.83893$. The structure of nephelite corresponds to normal hexagonal symmetry; it is only in its etch figures that hemimorphic symmetry is suggested.

In thick six- or twelve-sided prisms with plane or modified summits. Also massive compact, and in embedded grains; structure sometimes thin columnar.

Cleavage: $m(10\bar{1}0)$ distinct; $c(0001)$ imperfect. Fracture subconchoidal. Brittle. H. = 5·5–6. G. = 2·55–2·65. Luster vitreous to greasy; a little opalescent in some varieties. Colorless, white, or yellowish; also, when massive, dark green, greenish or bluish gray, brownish red and brick-red. Transparent to opaque. Optically −. Indices low and variable: $\omega = 1.536$–1.549, $\epsilon = 1.532$–1.544.

Var. — 1. *Nephelite.* *Glassy.* — Usually in small glassy crystals or grains, transparent with vitreous luster, first found on Mte. Somma, Vesuvius. Characteristic particularly of younger eruptive rocks and lavas. 2. *Elæolite.* — In large coarse crystals, or more commonly massive, with a greasy luster, and reddish, greenish, brownish or gray in color. Usually clouded by minute inclusions. Characteristic of granular crystalline rocks, syenite, etc.

Comp. — $NaAlSiO_4$. This is the composition of the artificial mineral. Natural nephelite always contains silica in varying excess and also small amounts of potash. The composition usually approximates to $Na_8K_2Al_8Si_9O_{34}$.

Synthetic experiments, yielding crystals like nephelite with the composition $NaAlSiO_4$, lead to the conclusion that a natural soda-nephelite would be an orthosilicate with this formula. The presence of potassium is accounted for by the fact that $KAlSiO_4$ has been shown by experiment to be completely isomorphous with $NaAlSiO_4$. The presence of small amounts of calcium and excess silica has been assumed to be due to the albite and anorthite molecules in solid solution. The variation in composition may also be explained by considering normal nephelite, $NaAlSiO_4$, to take up in solid solution silica or other silicate molecules. The other species of the group are normal orthosilicates, viz., eucryptite $LiAlSiO_4$, and kaliophilite, $KAlSiO_4$.

Artificial nephelite on heating changes to an isometric form at 1248° C.; this inversion point is about 100° higher in the case of the natural mineral. This isometric form changes on rapid cooling to the triclinic substance to which the name *carnegieite* has been given (see p. 549).

Pyr., etc. — B.B. fuses quietly at 3·5 to a colorless glass, coloring the flame yellow. Gelatinizes with acids.

Diff. — Distinguished by its gelatinizing with acids from scapolite and feldspar, as also from apatite, from which it differs too in its greater hardness. Massive varieties have a characteristic greasy luster.

Micro. — Recognized in thin sections by its low refraction; very low interference-colors, which scarcely rise to gray; parallel extinction when in crystals; faint negative uniaxial cross yielded by basal sections in converging light. The negative character is best told by aid of the gypsum plate (see p. 291). Micro-chemical tests serve to distinguish non-characteristic particles from similar ones of alkali feldspar; the section is treated with dilute acid, and the resultant gelatinous silica, which coats the nephelite particles, stained with eosine or other dye.

Artif. — Nephelite is easily prepared artificially by fusing its constituents together in the proper proportions.

Obs. — Nephelite is rather widely distributed (as shown by the microscopic study of rocks) in igneous rocks as the product of crystallization of a magma rich in soda and at the same time low in silica (which last prevents the soda from being used up in the formation of albite). It is thus an essential component of the nephelite-syenites and phonolites where it is associated with alkali feldspars chiefly. It is also a constituent of more basic augitic rocks such as nephelinite, nephelite-basalts, nephelite-tephrites, theralite, etc., most of which are volcanic in origin. The variety *elæolite* is associated with the granular plutonic rocks, while the name *nephelite* was originally used for the fresh glassy crystals of the modern lavas; the terms have in this sense the same relative significance as orthoclase and sanidine. Modern usage, however, tends to drop the name *elæolite*.

Nephelite is a prominent constituent of a granitoid rock found near Miask in the Ilmen Mts., Russia. Also in a rock composed of orthoclase, nephelite, and sodalite, from Diträu (Ditro) in Transylvania, Rumania. From the basalt of the Katzenbuckel, near Heidelberg, Baden. In Italy it occurs in the older lavas of Mte. Somma, Vesuvius, and at Capo di Bove, near Rome. The syenites of southern Norway contain much nephelite; at Frederiksvärn and many other points in the Langesundfiord district. Similarly it is found in the sodalite-syenite of the Julianehaab district of southern Greenland, at Kangerdluarsuk, etc.

In the United States nephelite occurs massive and crystallized at Litchfield, Kennebec Co., Maine, with cancrinite; in the Ozark Mts., near Magnet Cove, Garland Co., Arkansas. Common in the syenites of Ontario; in large crystals from Dungannon, Hastings Co.

Named *nephelite* from νεφέλη, a cloud, in allusion to its becoming cloudy when immersed in strong acid; *elæolite* is from ἐλαίον, oil, in allusion to its greasy luster.

Alter. — Nephelite is easily altered and in various ways. It frequently changes to various zeolites, to analcite, sodalite, or cancrinite. Further, the alteration may produce micaceous material. The following are of such a character. *Gieseckite* is a pseudomorph after nephelite. It occurs in Greenland in six-sided greenish gray prisms of greasy luster; also at Diana in Lewis Co., New York. *Dysyntribite* from Diana is similar to gieseckite, as is also *liebenerite*, from the valley of Fleims, in Tyrol, Austria. See further PINITE under the MICA GROUP.

Eucryptite. — $LiAlSiO_4$. In symmetrically arranged crystals (hexagonal), embedded in albite and derived from the alteration of spodumene at Branchville, Connecticut (see Fig. 514, p. 203). G. = 2·667. Colorless or white. Optically −. $\omega = 1·545$.

Kaliophilite. Phacellite. Phacelite. Facellite. — KAlSiO$_4$. Commonly contains small amounts of isomorphous NaAlSiO$_4$ (see under nephelite, p. 585). In hexagonal prisms or bundles of slender acicular crystals, also in fine threads, cobweb-like. Basal cleavage. H. = 6. G. = 2·49–2·67. Colorless. Optically −. ω = 1·532, ϵ = 1·527. Occurs in ejected masses at Mte. Somma, Vesuvius.

CANCRINITE.

Hexagonal. Axis c = 0·4224; and mp $10\bar{1}0$ $\wedge$ $10\bar{1}1$ = 64°, pp' $10\bar{1}1$ $\wedge$ $01\bar{1}1$ = 25° 58′. Rarely in prismatic crystals with a low terminal pyramid. Usually massive. X-ray study shows a similarity in structure between cancrinite and the sodalite group.

Cleavage: prismatic, $m(10\bar{1}0)$ perfect; $a(11\bar{2}0)$ less so. H. = 5–6. G. = 2·42–2·5. Color white, gray, yellow, green, blue, reddish. Streak uncolored. Luster subvitreous, or a little pearly or greasy. Transparent to translucent. Optically −. ω = 1·515–1·524, ϵ = 1·491–1·502.

Comp. — Approximately 3H$_2$O.4Na$_2$O.CaO.4Al$_2$O$_3$.9SiO$_2$.2CO$_2$ = Silica 38·7, carbon dioxide 6·3, alumina 29·3, lime 4·0, soda 17·8, water 3·9 = 100.

The following type of formula has been suggested for the members of the cancrinite group: 3Na$_2$Al$_2$Si$_2$O$_8$.2CaCO$_3$, in which the first group may be replaced by K$_2$Al$_2$Si$_2$O$_8$ or CaAl$_2$Si$_2$O$_8$ and the second group by Na$_2$CO$_3$, Na$_2$CO$_3$.3H$_2$O, 2CaSO$_4$, or 2CaCl$_2$.

Pyr., etc. — In the closed tube gives water. B.B. loses color, and fuses (F. = 2) with intumescence to a white blebby glass, the very easy fusibility distinguishing it readily from nephelite. Effervesces with hydrochloric acid, and forms a jelly on heating, but not before.

Micro. — Recognized in thin sections by its low refraction, quite high interference-colors and negative uniaxial character. Its common association with nephelite, sodalite, etc., are valuable characteristics. Evolution of CO$_2$ with acid distinguishes it from all other minerals except the carbonates, which show much higher interference-colors.

Artif. — Cancrinite has been prepared artificially by heating under pressure a mixture of sodium silicate, alumina and sodium carbonate; also by the treatment of nephelite and labradorite by sodium carbonate at high temperatures.

Obs. — Cancrinite occurs only in igneous rocks of the nephelite-syenite and related groups. It is in part believed to be original, i.e., formed directly from the molten magma; in part held to be secondary and formed at the expense of nephelite by infiltrating waters holding calcium carbonate in solution. Occurs with a citron-yellow color at the graphite mine in the Tunkinsk Mts., west of Irkutsk in Siberia; at Miask in the Ilmen Mts., Russia. At Ditrău (Ditro) in Transylvania of Rumania, pale flesh red. Also at Barkevik and other points in the Langesundfiord district of southern Norway, nearly white, pale yellow, less often blue.

At Litchfield, Kennebec Co., Maine, with nephelite and blue sodalite. Occurs with nephelite in the syenites of Dungannon, Hastings Co., Ontario. Named after Count Cancrin, Russian Minister of Finance.

SULPHATIC CANCRINITE with nearly one-half the CO$_2$ replaced by SO$_3$ is found in an altered rock on Beaver Creek, Gunnison Co., Colorado. Has lower refractive indices and birefringence than *cancrinite*.

Microsommite. — Near cancrinite; contains the following molecules: Na$_2$Al$_2$Si$_2$O$_8$, K$_2$Al$_2$Si$_2$O$_8$, CaSO$_4$, CaCl$_2$. In minute colorless prismatic crystals (hexagonal. See Fig. 30, p. 20). From Vesuvius (Mte. Somma). H. = 6. G. = 2·42–2·53. Optically +. ω = 1·521, ϵ = 1·529.

Davyne. — Near microsommite. Contains the following molecules: Na$_2$Al$_2$Si$_2$O$_8$, K$_2$Al$_2$Si$_2$O$_8$, Na$_2$CO$_3$, CaSO$_4$, CaCl$_2$. Hexagonal. Colorless. Optically + (rarely −). ω = 1·518–1·522, ϵ = 1·517–1·527. Found at Mte. Somma, Vesuvius.

Sodalite Group. Isometric

The species of the Sodalite Group are isometric in crystallization and perhaps tetrahedral like the following group. X-ray study shows a close similarity in structure between the members of the Sodalite and Helvite Groups. In composition the minerals of the Sodalite Group are similar to the minerals of the Cancrinite Group and like it are peculiar in containing radicals with Cl, SO_4 and S, which are elements usually absent in the silicates. The molecules $K_2Al_2Si_2O_8$ and $CaAl_2Si_2O_8$ may occur in small amounts, replacing $Na_2Al_2Si_2O_8$.

Sodalite	$3Na_2Al_2Si_2O_8.2NaCl$
Haüynite	$3Na_2Al_2Si_2O_8.2CaSO_4$
Noselite	$3Na_2Al_2Si_2O_8.Na_2SO_4$
Lazurite	$3Na_2Al_2Si_2O_8.2Na_2S$

SODALITE.

Isometric. Study of the atomic structure shows a normal symmetry, although etch figures have suggested tetrahedral symmetry. The lattice is a simple cubic one, but near that of the body-centered cubic type. The unit cell contains two molecules. Common form the dodecahedron. Twins: tw. pl. $o(111)$, forming hexagonal prisms by elongation in the direction of an octahedral axis (Fig. 432, p. 187). Also massive, in embedded grains; in concentric nodules resembling chalcedony, formed from elæolite.

Cleavage: dodecahedral, more or less distinct. Fracture conchoidal to uneven. Brittle. H. = 5·5–6. G. = 2·14–2·30. Luster vitreous, sometimes inclining to greasy. Color gray, greenish, yellowish, white; sometimes blue, lavender-blue, light red. Transparent to translucent. Streak uncolored. $n = 1·4827$.

Comp. — $3NaAlSiO_4.NaCl$ = Silica 37·2, alumina 31·6, soda 25·6, chlorine 7·3 = 101·7, deduct (O = 2Cl) 1·7 = 100. Potassium replaces a small part of the sodium.

Pyr., etc. — In the closed tube the blue varieties become white and opaque. B.B. fuses with intumescence, at 3·5–4, to a colorless glass. Soluble in hydrochloric acid and yields gelatinous silica upon evaporation.

Diff. — Distinguished from much analcite, leucite, and haüynite by chemical tests alone; dissolving the mineral in dilute nitric acid and testing for chlorine is the simplest and best.

Micro. — Recognized in thin sections by its very low refraction, isotropic character and lack of good cleavage; also, in most cases, by its lack of color. In uncovered rock sections the minerals of this group may be distinguished from each other by covering them with a little nitric acid which is allowed to evaporate slowly. With sodalite crystals of sodium chloride will form; with haüynite crystals of gypsum; with noselite crystals of both compounds after the addition of calcium chloride; lazurite will evolve hydrogen sulphide which will blacken silver.

Artif. — Sodalite can be obtained by fusing nephelite with sodium chloride; also by the action of sodium carbonate and caustic soda upon muscovite at 500°. It has been produced also in various artificial magmas at temperatures below 700°.

Obs. — Sodalite occurs only in igneous rocks of the nephelite-syenite and related rock groups, as a product of the crystallization of a magma rich in soda; also as a product associated with enclosed masses and bombs ejected with such magmas in the form of lava, as at Vesuvius. Commonly associated with nephelite, cancrinite and eudialyte. In some instances it occurs as a secondary mineral, having been formed through the alteration of nephelite. Occurs at Miask in the Ilmen Mts., Russia, in a granite-like rock with nephelite and feldspar; at Ditrău (Ditro), Transylvania of Rumania, in a nephelite-syenite with cancrinite, etc. At Vesuvius on Mte. Somma in white translucent, dodecahedral crystals. A variety from Mte. Somma containing a small amount of molybdenum trioxide has been called *molybdosodalite*. At various points in the Langesundfiord district of southern Norway with a lavender-blue color. Further in West Greenland in the Julianehaab district in sodalite-syenite.

In the United States a blue variety occurs at Litchfield and West Gardiner, Kennebec Co., Maine. In a vein at Salem, Essex Co., Massachusetts, violet to azure-blue. In Canada, sodalite occurs in the nephelite-syenites of Quebec and at Dungannon, Hastings Co.,

Ontario. Also in fine large masses with blue color on Ice River, a tributary of the Beaver-foot, near Kicking Horse Pass, British Columbia.

Hackmanite. — A sodalite containing about 6 per cent of the molecule $3NaAlSiO_4.Na_2S$ from a rock called *tawite* from the Tawa valley on the Kola peninsula, Lapland. Color reddish violet which fades on exposure to light. $n = 1·487$.

AMELETITE. A silicate of soda and alumina with a small amount of chlorine, approximately $9Na_2O.\frac{1}{2}NaCl.6Al_2O_3.12SiO_2$. Refractive index lower than that for nephelite and higher than sodalite. Occurs as grains and minute crystals in phonolites of Dunedin, New Zealand.

HAUYNITE. Haüyne.

Isometric. Sometimes in dodecahedrons, octahedrons, etc.

Twins: tw. pl. $o(111)$; contact-twins, also polysynthetic; penetration-twins (Fig. 431, p. 187). Commonly in rounded grains, often looking like crystals with fused surfaces.

Cleavage: dodecahedral, rather distinct. Fracture flat conchoidal to uneven. Brittle. H. = 5·5–6. G. = 2·4–2·5. Luster vitreous, to somewhat greasy. Color bright blue, sky-blue, greenish blue; asparagus-green,

879

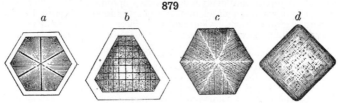

a   b   c   d

Sections of crystals of haüynite (after Möhl)

red, yellow. Streak slightly bluish to colorless. Subtransparent to translucent; often enclosing symmetrically arranged inclusions (Fig. 879). $n = 1·4961$.

Comp. — $3NaAlSiO_4.CaSO_4$. Silica 32·0, sulphur trioxide 14·2, alumina 27·2, lime 10·0, soda 16·6 = 100. The molecules $KAlSiO_4$ and $CaAl_2(SiO_4)_2$ may also be present in small amounts.

Pyr., etc. — In the closed tube retains its color. B.B. in the forceps fuses at 4·5 to a white glass. Soluble in hydrochloric acid and yields gelatinous silica upon evaporation. The solution gives a test for the sulphate radical with barium chloride.

Micro. — Similar to sodalite, which see.

Artif. — Has been produced artificially in the same ways as with sodalite with the use of a sulphate instead of a chloride.

Obs. — Found only in igneous rocks, especially in the extrusive types. Commonly in rocks that are low in silica and rich in alkalies and usually associated with nephelite or leucite.

In Italy haüynite occurs in the Vesuvian lavas, on Mte. Somma; on Mte. Vulture, south-west of Melfi, Basilikata, in various colors; in the lavas of the Campagna, near Rome. In the Eifel district, Rhineland, at Niedermendig, Laacher See, etc.; on the Kaiserstuhl in Baden.

Noselite. Nosean. — Near haüynite, but contains little or no lime. $3Na_2Al_2Si_2O_8.$ Na_2SO_4. Color grayish, bluish, brownish; sometimes nearly opaque from the presence of inclusions (cf. Fig. 879). H. = 5·5. G. = 2·25–2·4. $n = 1·48$–1·495. Noselite has the same mode of occurrence and associations as haüynite. Found at Laacher See in the Eifel district of Rhineland; from the Kaiserstuhl, Baden. From Albano south of Rome, Italy. From the phonolites of Cantal, France. Found in the lavas of the Canary and Cape Verde Islands.

LAZURITE. LAPIS LAZULI. Lasurite.

Isometric. In cubes and dodecahedrons. Commonly massive, compact. Cleavage: dodecahedral, imperfect. Fracture uneven. H. = 5–5·5.

G. = 2·38–2·45. Luster vitreous. Color rich Berlin-blue or azure-blue, violet-blue, greenish blue. Translucent. $n = 1·500$.

Comp. — Essentially $3NaAlSiO_4.Na_2S$, but containing also in isomorphous combination the haüynite and sodalite molecules. The character of the sulphide radical is uncertain since in artificial ultramarine it may be Na_2S, NaS, or NaS_2.

The heterogeneous character of what had long passed as a simple mineral under the name lapis lazuli was shown by Fischer (1869), Zirkel (1873), and more fully by Vogelsang (1873). The ordinary natural *lapis lazuli* is shown by Brögger and Bäckström to contain lazurite or haüynite (sometimes changed to a zeolite), a diopside free from iron, amphibole (koksharovite), mica (muscovite), calcite, pyrite; also in some varieties in relatively small amount scapolite, plagioclase, orthoclase (microperthite?), apatite, titanite, zircon, and an undetermined mineral optically + and probably uniaxial. Regarded by Brögger as a result of contact metamorphism in limestone.

Micro. — Similar to sodalite, which see.

Pyr., etc. — Heated in the closed tube gives off some moisture; the variety from Chile glows with a beetle-green light, but the color of the mineral remains blue on cooling. Fuses easily (3) with intumescence to a white glass. Soluble in hydrochloric acid and yields gelatinous silica upon evaporation and evolves hydrogen sulphide.

Obs. — Probably always a contact metamorphic mineral, occurring in limestones near their contact with a granitic rock. Commonly encloses small grains of pyrite.

Found in Transbaikalia at various points near the south end of Lake Baikal in a dolomitic limestone in connection with granite. From Afghanistan in the district of Badakhshan a few miles south of Firgamu in the valley of the Kokcha. From Bokhara, Turkestan, and from Persia. Rare in ejected masses at Vesuvius and in the Albani Mts., south of Rome. Further in Chile in the Andes of Ovalle, Coquimbo.

Use. — The richly colored varieties of lapis lazuli are highly esteemed for costly vases and ornamental furniture; also employed in the manufacture of mosaics; and when powdered constitutes the rich and durable paint called *ultramarine*. This has been replaced, however, by artificial ultramarine, now an important commercial product.

Helvite Group. Isometric-tetrahedral

Helvite	$3(Mn,Fe)BeSiO_4.MnS$
Danalite	$3(Fe,Zn,Mn)BeSiO_4.ZnS$
Eulytite	$Bi_4(SiO_4)_3$
Zunyite	$(Al(OH,F,Cl)_2)_6Al_2(SiO_4)_3$

The HELVITE GROUP includes several rare species, isometric-tetrahedral in crystallization and in composition and structure related to the species of the SODALITE GROUP.

HELVITE.

Isometric-tetrahedral. Commonly in tetrahedral crystals; also in spherical masses. The atomic structure is similar to that of sodalite. The unit cell contains two molecules.

Cleavage: octahedral in traces. Fracture uneven to conchoidal. Brittle. H. = 6–6·5. G. = 3·16–3·36. Luster vitreous, inclining to resinous. Color honey-yellow, inclining to yellowish brown, and siskin-green, reddish brown. Streak uncolored. Subtransparent. $n = 1·739$. Pyroelectric.

Comp. — $3(Mn,Fe)BeSiO_4.MnS$.

Pyr., etc. — Fuses at 3 in R.F. with intumescence to a yellowish brown opaque bead, becoming darker in R.F. With the fluxes gives the manganese reaction. Soluble in hydrochloric acid, giving hydrogen sulphide and yielding gelatinous silica upon evaporation.

Obs. — Helvite occurs in veins with quartz, hornblende, and iron ore and in pegmatite veins; also found at times in gneiss.

Found in the Ilmen Mts., Russia, near Miask, in large spherical masses in pegmatite. In Rumania at Kapnikbánya; in Saxony at Breitenbrunn and also near Schwarzenberg in gneiss. In Norway from the augite-syenites at various points in the Langesundfiord district. In the United States at the mica mines near Amelia Court House, Amelia Co., Virginia. Named by Werner, in allusion to its yellow color, from ἥλιος, *the sun.*

Danalite. — $3(Fe,Zn,Mn)BeSiO_4.ZnS.$ In octahedrons; usually massive. H. = 5·5–6. G. = 3·427. Color flesh-red to gray. $n = 1·754.$ Occurs in Essex Co., Massachusetts, in small grains disseminated in the Rockport granite, Cape Ann; with magnetite and quartz at the iron mine at Bartlett, Carroll Co., New Hampshire. Reported from West Cheyenne Cañon, El Paso Co., Colorado. Found also at Redruth, Cornwall.

Eulytite. — $Bi_4(SiO_4)_3.$ Usually in minute tetrahedral crystals; also in spherical forms. X-ray study shows four molecules to the unit cell. H. = 4·5. G. = 6·1. Color dark hairbrown to grayish, straw-yellow, or colorless. $n = 2·05.$ Found in Saxony with native bismuth near Schneeberg; also at Johanngeorgenstadt, in crystals on quartz.

Zunyite. — A highly basic orthosilicate of aluminum, approximately $(Al(OH,F,Cl)_2)_6$ $Al_2Si_3O_{12}.$ In minute transparent tetrahedrons. H. = 7. G. = 2·875. $n = 1·60.$ Occurs in highly aluminous shales near Postmasburg, South Africa. From the Zuñi mine, near Silverton, San Juan Co., and on Red Mountain, Ouray Co., Colorado.

Agricolite. — Same as for eulytite, $Bi_4(SiO_4)_3,$ but monoclinic. Fibrous. In globular or semi-globular forms. $n = 2·0.$ In Saxony at Johanngeorgenstadt and Schneeberg.

---

4. *Garnet Group.* Isometric

$$\overset{\text{II}}{R_3}\overset{\text{III}}{R_2}(SiO_4)_3 \quad \text{or} \quad 3RO.R_2O_3.3SiO_2$$

$$\overset{\text{II}}{R} = Ca,Mg,\overset{\text{II}}{Fe},Mn. \quad \overset{\text{III}}{R} = Al,\overset{\text{III}}{Fe},Cr,\overset{\text{III}}{Ti}.$$

Garnet

A. GROSSULARITE	$Ca_3Al_2(SiO_4)_3$		E. ANDRADITE	$Ca_3Fe_2(SiO_4)_3$
B. PYROPE	$Mg_3Al_2(SiO_4)_3$		Also	$(Ca,Mg)_3Fe_2(SiO_4)_3$
C. ALMANDITE	$Fe_3Al_2(SiO_4)_3$			$Ca_3Fe_2((Si,Ti)O_4)_3$
D. SPESSARTITE	$Mn_3Al_2(SiO_4)_3$		F. UVAROVITE	$Ca_3Cr_2(SiO_4)_3$

Schorlomite $Ca_3(Fe,Ti)_2((Si,Ti)O_4)_3$

The GARNET GROUP includes a series of important sub-species included under the same specific name. They all crystallize in the normal class of the isometric system and are alike in habit, the dodecahedron and trapezohedron being the common forms. They have also the same general formula, and while the elements present differ widely, there are many intermediate varieties. Some of the garnets include titanium, replacing silicon, and thus they are connected with the rare species schorlomite, which probably also has the same general formula.

GARNET.

Isometric. The dodecahedron and trapezohedron, $n(211),$ the common simple forms; also these in combination, or with the hexoctahedron $s(321).$ Cubic and octahedral faces rare. Often in irregular embedded grains. Also massive; granular, coarse or fine, and sometimes friable; lamellar, lamellæ thick and bent. Sometimes compact, cryptocrystalline like nephrite. X-ray study of the atomic structure shows a complicated unit cell containing eight molecules. The SiO_4 groups are independent of each other; the $\overset{\text{III}}{R}$ atoms lie in the center of a group of six oxygen atoms and the $\overset{\text{II}}{R}$ atoms in the center of a group of eight oxygen atoms.

Parting: d(110) sometimes rather distinct. Fracture subconchoidal to uneven. Brittle, sometimes friable when granular massive; very tough when compact cryptocrystalline. H. = 6·5–7·5. G. = 3·15–4·3, varying with the composition. Luster vitreous to resinous. Color red, brown, yellow, white, apple-green, black; some red and green, colors often bright. Streak white. Transparent to subtranslucent. Often exhibits anomalous double refraction,

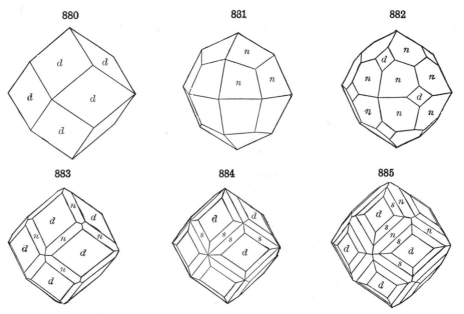

especially grossularite (also topazolite, etc.), see Art. **441**. Refractive index rather high, and varying directly with the composition. The different pure molecules have approximately the following indices.

Pyrope 1·705, Grossularite 1·735, Spessartite 1·800, Almandite 1·830, Uvarovite 1·870, Andradite 1·895.

Comp. — An orthosilicate having the general formula $\overset{\text{II}}{R_3}\overset{\text{III}}{R_2}(SiO_4)_3$, or $3RO.R_2O_3.3SiO_2$. The bivalent element may be calcium, magnesium, ferrous iron or manganese; the trivalent element, aluminum, ferric iron or chromium, rarely titanium; further, silicon is also sometimes replaced by titanium. The different garnet molecules are isomorphous with each other although there are apparently definite limits to their miscibility. The majority will be found to have two or three component molecules; in the case, however, where three are present one is commonly in subordinate amount. It has been pointed out that the garnets may be considered as consisting of two isomorphous series, a pyrope-almandite-spessartite series, and a grossularite-andradite series. These two have at least only a slight miscibility with each other. The index of refraction and specific gravity vary directly with the variation in composition.

Var. — There are three prominent groups, and various subdivisions under each, many of these blending into each other.

I. *Aluminum Garnet,* including

A.	GROSSULARITE	Calcium-Aluminum Garnet	$Ca_3Al_2(SiO_4)_3$
B.	PYROPE	Magnesium-Aluminum Garnet	$Mg_3Al_2(SiO_4)_3$
C.	ALMANDITE	Iron-Aluminum Garnet	$Fe_3Al_2(SiO_4)_3$
D.	SPESSARTITE	Manganese-Aluminum Garnet	$Mn_3Al_2(SiO_4)_3$

II. *Iron Garnet, including*

    E. ANDRADITE   Calcium-Iron Garnet $Ca_3Fe_2(SiO_4)_3$

    (1) Ordinary. (2) Magnesian. (3) Titaniferous. (4) Yttriferous.

III. *Chromium Garnet.*

    F. UVAROVITE   Calcium-Chromium Garnet $Ca_3Cr_2(SiO_4)_3$

The name Garnet is from the Latin *granatus,* meaning *like a grain,* and directly from *pomegranate,* the seeds of which are small, numerous, and red, in allusion to the aspect of the crystals.

A. GROSSULARITE. Essonite or Hessonite. Cinnamon-stone. *Calcium-aluminum Garnet.* Formula $3CaO.Al_2O_3.3SiO_2$ = Silica 40·0, alumina 22·7, lime 37·3 = 100. Often containing ferrous iron replacing the calcium, and ferric iron replacing aluminum, and hence graduating toward groups C and E. G. = 3·53. Color (*a*) colorless to white; (*b*) pale green; (*c*) amber- and honey-yellow; (*d*) wine yellow, brownish yellow, cinnamon-brown; (*e*) rose-red; rarely (*f*) emerald-green from the presence of chromium. Often shows optical anomalies (Art. **441**).

The original *grossularite* (*wiluite* in part) included the pale green from Siberia, and was so named from the botanical name for the gooseberry; G. = 3·42–3·72. *Cinnamon-stone,* or *essonite* (more properly *hessonite*), included a cinnamon-colored variety from Ceylon, there called *hyacinth;* but under this name the yellow and yellowish red kinds are usually included; named from ἥσσων, *inferior,* because of less hardness than the true hyacinth which it resembles. *Succinite* is an amber-colored kind from the Ala valley, Piedmont, Italy. *Romanzovite* is brown.

Pale green, yellowish, and yellow-brown garnets are not invariably grossularite; some (including topazolite, demantoid, etc.) belong to the group of Calcium-Iron Garnet; or Andradite.

B. PYROPE. Precious garnet in part. *Magnesium-aluminum Garnet.* Formula $3MgO.Al_2O_3.3SiO_2$ = Silica 44·8, alumina 25·4, magnesia 29·8 = 100. Magnesia predominates, but calcium and iron are also present; the original pyrope also contained chromium. G. = 3·51. Color deep red to nearly black. Often perfectly transparent and then prized as a gem. The name *pyrope* is from πυρωπός, *fire-like.*

Rhodolite, of delicate shades of pale rose-red and purple, brilliant by reflected light, corresponds in composition to two parts of pyrope and one of almandite; from Macon Co., North Carolina.

C. ALMANDITE. Almandine. Precious garnet in part. Common garnet in part. *Iron-aluminum Garnet.* Formula $3FeO.Al_2O_3.3SiO_2$ = Silica 36·2, alumina 20·5, iron protoxide 43·3 = 100. Ferric iron replaces the aluminum to a greater or less extent. Magnesium also replaces the ferrous iron, and thus it graduates toward pyrope, cf. rhodolite above. G. = 4·25. Color fine deep red, transparent, in *precious garnet;* brownish red, translucent or sub-translucent, in *common garnet;* black. Part of common garnet belongs to *Andradite.*

The Alabandic carbuncles of Pliny were so called because cut and polished at Alabanda. Hence the name almandine or almandite, now in use.

D. **Spessartite.** Spessartine. *Manganese-aluminum Garnet.* Formula $3MnO.Al_2O_3.3SiO_2$ = Silica 36·4, alumina 20·6, manganese protoxide 43·0 = 100. Ferrous iron replaces the manganese to a greater or less extent, and ferric iron also the aluminum. G. = 4·18. Color dark hyacinth-red, sometimes with a tinge of violet, to brownish red.

Partschinite, originally described as having the same composition as spessartite but monoclinic in crystallization, has been shown to be a garnet near spessartite. $n = 1·787$. From the auriferous sands of Olah-Pian near Sebes Sâsesc (Mühlbach) in Sibiu (Szeben), Transylvania, Rumania.

E. **Andradite.** Common Garnet, Black Garnet, etc. *Calcium-iron Garnet.* Formula $3CaO.Fe_2O_3.3SiO_2$ = Silica 35·5, iron sesquioxide 31·5, lime 33·0 = 100. Aluminum replaces the ferric iron; ferrous iron, manganese and sometimes magnesium replace the calcium. G. = 3·75. Colors various: wine-, topaz- and greenish yellow, apple-green to emerald-green; brownish red, brownish yellow; grayish green, dark green; brown; grayish black, black.

Named *Andradite* after the Portuguese mineralogist, d'Andrada, who in 1800 described and named one of the included subvarieties, *allochroite*. Chemically there are the following varieties:
1. Simple *Calcium-iron Garnet*, in which the protoxides are wholly or almost wholly lime. Includes: (*a*) *Topazolite*, having the color and transparency of topaz, and also sometimes green; crystals often showing a vicinal hexoctahedron. *Demantoid*, a grass-green to emerald-green variety with brilliant diamond-like luster, used as a gem. (*b*) *Colophonite*, a coarse granular kind, brownish yellow to dark reddish brown in color, resinous in luster, and usually with iridescent hues; named after the resin *colophony*. (*c*) *Melanite* (from μέλας, *black*), black, either dull or lustrous; but all black garnet is not here included. *Pyreneite* is grayish black melanite. (*d*) Dark green garnet, not distinguishable from some allochroite, except by chemical trials.
2. *Manganesian Calcium-iron Garnet.* (*a*) *Rothoffite*. The original *allochroite* was a manganesian iron-garnet of brown or reddish brown color, and of fine-grained massive structure. *Rothoffite*, from Långbanshyttan, Sweden, is similar, yellowish brown to liver-brown. Other common kinds of manganesian iron-garnet are light and dark, dusky green and black, and often in crystals. *Polyadelphite* is a massive brownish yellow kind, from Franklin Furnace, New Jersey. *Bredbergite*, from Sala, Sweden, contains a large amount of magnesia. (*b*) *Aplome* (properly haplome) has its dodecahedral faces striated parallel to the shorter diagonal, whence Haüy inferred that the fundamental form was the cube; and as this form is simpler than the dodecahedron, he gave it a name derived from ἁπλόος, simple. Color of the original aplome (of unknown locality) dark brown; also found yellowish green and brownish green at Schwarzenberg in Saxony, and on the Lena in Siberia.
3. *Titaniferous.* Contains titanium and probably both TiO_2 and Ti_2O_3; formula hence $3CaO.(Fe,Ti,Al)_2O_3.3(Si,Ti)O_2$. It thus graduates toward schorlomite. Color black.
4. *Yttriferous Calcium-iron Garnet.* Contains *yttria* in small amount; rare.

F. **Uvarovite.** Ouvarovite. Uwarowit. *Calcium-chromium Garnet.* Formula $3CaO.Cr_2O_3.3SiO_2$ = Silica 35·9, chromium sesquioxide 30·6, lime 33·5 = 100. Aluminum takes the place of the chromium in part. H. = 7·5. G. = 3·41–3·52. Color emerald-green.

Other varietal names have been proposed. *Skiagite* from Glen Skiag, Scotland, containing the ferrous-ferric garnet molecule. *Blythite* for garnets containing the manganous-manganic molecule. *Calderite* for a garnet containing the manganous-ferric molecule.

Pyr., etc. — Most varieties of garnet fuse easily to a light brown or black glass; F. = 3 in almandite, spessartite, and grossularite; 3·5 in andradite and pyrope; but uvarovite, the chrome-garnet, is almost infusible, F. = 6. Andradite and almandite fuse to a magnetic globule. Reactions with the fluxes vary with the bases. Almost all kinds react for iron; strong manganese reaction in spessartite, and less marked in other varieties; a chromium reaction in uvarovite and in most pyrope. Some varieties are partially decomposed by acids; all except uvarovite after ignition become soluble in hydrochloric acid, and generally yield gelatinous silica on evaporation. Decomposed on fusion with alkaline carbonates.

The density of garnets is largely diminished by fusion. Thus a Greenland garnet fell from 3·90 to 3·05 on fusion, and a Vilui grossularite from 3·63 to 2·95.

Diff. — Characterized by isometric crystallization, usually in isolated crystals, dodecahedrons or trapezohedrons; massive forms rare, then usually granular. Also distinguished by hardness, vitreous luster, and in the common kinds by the fusibility. Vesuvianite fuses more easily, zircon and quartz are infusible; the specific gravity is higher than for tourmaline, from which it differs in form; it is much harder than sphalerite.

Micro. — Distinguished in thin sections by its very high relief; lack of cleavage; isotropic character; usually shows a pale pink color; sometimes not readily told from some of the spinels.

Artif. — While members of the garnet group have been formed artificially their synthesis is difficult. Apparently they can be produced only under exact conditions of temperature and pressure that are difficult to reproduce. Natural garnets when fused break down into various other minerals.

Obs. — Garnet is common and widely distributed. It is usually an accessory rock-making mineral, being found in a great variety of rock types, more particularly in mica, hornblende, and chlorite schists and in gneiss; in crystalline limestones and in contact zones; also in granite, syenite, and occasionally in various volcanic rocks; sometimes in serpentine. It also at times is found in massive form as a prominent constituent of a rock. The particular mode of occurrence is largely dependent on the chemical composition of the garnet in question. *Grossularite* is especially characteristic of metamorphosed impure calcareous rocks, whether altered by local igneous or general metamorphic processes; it is thus commonly found in the contact zone of intruded igneous rocks and in the crystalline schists. *Almandite* is characteristic of the mica schists and metamorphic rocks containing alumina and iron; it occurs also in some igneous rocks as the result of later dynamic and metamorphic processes; it forms with the variety of amphibole called smaragdite, the rock eclogite. *Pyrope* is especially characteristic of such basic igneous rocks as are formed from magmas containing much magnesia and iron with little or no alkalies, as the peridotites, dunites, etc.; also found in the serpentines formed from these rocks; then often associated with spinel, chromite, etc. *Spessartite* occurs in granitic rocks, in quartzite, in whetstone schists (Belgium); it has been noted with topaz in lithophyses in rhyolite (Colorado). The black variety of andradite, *melanite*, is common in eruptive rocks, especially with nepheline and leucite, thus in phonolites, leucitophyres, nephelinites: in such cases often titaniferous or associated with a titaniferous garnet, sometimes in zonal intergrowth; it also occurs as a product of contact metamorphism. *Demantoid* occurs in serpentine. *Uvarovite* belongs particularly with chromite in serpentine; it occurs also in granular limestone.

Garnet crystals often contain inclusions of foreign matter, but only in part due to alteration; as, vesuvianite, calcite, epidote, quartz (Fig. 512, p. 202); at times the garnet is a mere shell, or perimorph, surrounding a nucleus of another species. A black garnet from Arendal, Norway, contains both calcite and epidote; crystals from Tvedestrand, Norway, are wholly calcite within, there being but a thin crust of garnet. Crystals from East Woodstock, Maine, are dodecahedrons with a thin shell of cinnamon-stone enclosing calcite; others from Raymond, Maine, show successive layers of garnet and calcite. Many such cases have been noted.

886

Garnets are often altered, thus to chlorite, serpentine; even to limonite. Crystals of pyrope are sometimes surrounded by a chloritic zone (*kelyphite* of Schrauf) not homogeneous. as shown in Fig. 886.

Only the most important localities for the occurrence of garnet can be mentioned here; others have been given in the preceding pages under the description of varieties. The foreign localities follow, arranged according to the different types. It is to be noted, however, that it is not always possible to definitely place a given occurrence in one particular variety.

Grossularite of a pale greenish color comes from the banks of the Vilui River in Yakutsk, Siberia; from Achmatovsk near Kussinsk in the Zlatoust district, Ural Mts., Russia. In Rumania in serpentine with vesuvianite from Oravicza and the nearby locality of Csiklova; also from Vaskö and Dognacska. In Italy from Monzoni, Val di Fassa, Trentino, and on the Mussa-Alp in the Ala valley, Piedmont; in honey-yellow octahedrons from Elba and with vesuvianite and wollastonite in ejected masses at Vesuvius. In Switzerland at Zermatt in Valais, etc. In Mexico in clear pink or rose-red dodecahedrons at Xalostoc,

Morelos, called variously *landerite*, *xalostocite*, and *rosolite*; also from Juarez, Lower California.

Pyrope occurs in Bohemia of Czechoslovakia in serpentine which has been derived from a peridotite from various points near Meronitz, where the variety used as a gem has been obtained; also from near Krems southwest of Budweis.

Almandite is probably the most common variety of garnet. It is found in the Tyrol, Austria, in hyacinth-red or brown crystals from the Zillertal and the Oetztal. At Bodö in Nordlund, Norway; in large dodecahedrons from Falun, Sweden. In gem quality from Ceylon, India, Brazil, southern Greenland, etc. Very large crystals are found in a chlorite schist on the Ampandramaika, a tributary of the Mangoky River, Madagascar. In well-defined individual crystals from pegmatite at Yamano, Hitachi Province, Japan.

Spessartite is from Aschaffenburg in the Spessart district of Bavaria; in Italy at St. Marcel in Piedmont and at San Piero in Campo, Alba. From Madagascar in transparent crystals with a clear orange color from Tsilaizina near Antsirabe and elsewhere.

Among the varieties of *andradite*, the beautiful green *demantoid* or " Uralian emerald " occurs in the gold washings of Nizhne-Tagilsk in the Ural Mts. In Rumania brown to green crystals come from Moravicza and Dognacska; emerald green at Dobsina (Dobschau) in Czechoslovakia; in the Ala Valley, Piedmont, Italy, the yellow to greenish *topazolite*, with its characteristic vicinal hexoctahedron. *Allochroite*, an apple-green and yellowish variety, occurs at Zermatt, Valais, Switzerland; brilliant black crystals (*melanite*) are found on Mte. Somma, Vesuvius, and from near Barèges, Hautes-Pyrénées, France (*pyreneite*). *Aplome* occurs in yellowish and brownish green crystals in Saxony at Breitenbrunn and Schwarzenberg; brown to black, highly modified crystals in the Pfitschtal, Trentino, Italy; other localities are Långbanshyttan, Vermland, Sweden (*rothoffite*); in large dodecahedrons at Arendal in Aust-Agder, Norway; from Pitkäranta on Lake Ladoga, Finland.

Uvarovite is found in the Ural Mts., Russia, at Saranovskaya near Bisersk; also in the vicinity of Kyshtymsk, northeast of Zlatoust. At Jordansmühl in Silesia.

In the United States, garnet is found in Maine at Phippsburg, Sagadahoc Co., in a manganesian variety as well as fine yellow cinnamon-stone; also from Raymond, Cumberland Co., in fine crystals. In Grafton Co., New Hampshire, at Hanover in small clear crystals, and at Warren in beautiful cinnamon garnets with green pyroxene, and at Grafton. At Newfane, Windham Co., Vermont, in large crystals in a chlorite slate. In Hampden Co., Massachusetts, at Russell in fine dark colored trapezohedral crystals, sometimes very large; from Westfield. In Connecticut at Haddam, Middlesex Co., a manganesian variety sometimes large; at Roxbury, Litchfield Co., in dodecahedral crystals in mica schist; in large trapezohedrons in mica schist at Redding, Fairfield Co. In New York at Willsboro, Essex Co., *colophonite* as a large vein in gneiss. In New Jersey at Franklin, Sussex Co., in various colors, also the variety *polyadelphite*. In Pennsylvania in Chester Co., at Pennsbury; at Keins' mine, near Knauertown; in Delaware Co., at Chester, Leiperville, Avondale, Middletown, etc.; uvarovite from Wood's chromite mine, Texas, Lancaster Co. In Amelia Co., Virginia, fine spessartite at Amelia Court House. *Rhodolite*, a rose-colored variety, is from Mason's Branch, Macon Co., North Carolina. Large dodecahedral crystals altered to chlorite are found near Marquette, Michigan. In Arkansas at Magnet Cove near Hot Springs, Garland Co., a titaniferous melanite. Fine pyrope comes from the northwestern part of New Mexico, and in northeastern Arizona about Fort Defiance, Apache Co. In Colorado in Chaffee Co., at Nathrop, fine spessartite in lithophyses in rhyolite with quartz and topaz; in large dodecahedral crystals, up to more than 14 pounds in weight, at Ruby Mt., near Salida, the exterior being usually green from a layer of chlorite due to alteration. In large crystals from Larimer Co.; topazolite from Crested Butte, Gunnison Co. In California, essonite, in fine transparent crystals, comes from the various tourmaline localities in San Diego Co.; fine garnets also from San Bernardino Co., and Tulare Co. (topazolite). Uvarovite from New Idria, San Benito Co. Fine crystals of ideal symmetry of a rich red color and an inch or more in diameter occur in the mica schists at Wrangell, at the mouth of the Stikine River, Alaska.

In Quebec an emerald-green chrome garnet at Orford, Sherbrooke, Co.; and in the same vicinity large cinnamon-red and yellowish garnets; fine colorless to pale olive-green, or brownish crystals at Wakefield, Ottawa Co., also others bright green carrying chromium. Also from Hull, Ottawa Co.

Use. — The various colored and transparent garnets are used as semiprecious gem stones. At times the mineral is also used as an abrasive.

Schorlomite. — Probably analogous to garnet, $3CaO.(Fe,Ti)_2O_3.3(Si,Ti)O_2$. Perhaps to be considered as a highly titaniferous variety of andradite. Usually massive, black, with conchoidal fracture and vitreous luster. H. = 7–7·5. G. = 3·81–3·88. *n* variable from 1·9 to 2·0. In small masses with nephelite and brookite from Magnet Cove, near Hot

Springs, Garland Co., Arkansas. In masses of considerable size in a nephelite-syenite from the Ice River, a tributary of the Beaverfoot near Kicking Horse Pass, British Columbia.

Chrysolite Group. R_2SiO_4. Orthorhombic

		mm'''	hh'	
		$110 \wedge 1\bar{1}0$	$011 \wedge 0\bar{1}1$	$a : b : c$
Monticellite	$CaMgSiO_4$	46° 54′	59° 52′	0·4337 : 1 : 0·5758
Glaucochroite	$CaMnSiO_4$	47° 36′	60° 18′	0·441 : 1 : 0·581
Forsterite	Mg_2SiO_4	49° 51′	60° 43′	0·4648 : 1 : 0·5857
Chrysolite	$(Mg,Fe)_2SiO_4$	49° 57′	60° 47′	0·4656 : 1 : 0·5865
Hortonolite	$(Fe,Mg,Mn)_2SiO_4$			
Fayalite	Fe_2SiO_4	49° 15′	60° 10′	0·4584 : 1 : 0·5793
Knebelite	$(Fe,Mn)_2SiO_4$			
Tephroite	Mn_2SiO_4	49° 24′	61° 25′	0·4600 : 1 : 0·5939
Larsenite	$PbZnSiO_4$			0·4339 : 1 : 0·5324

The CHRYSOLITE GROUP includes a series of orthosilicates of magnesium, calcium, iron and manganese. They all crystallize in the orthorhombic system with but little variation in axial ratio. The prismatic angle is about 50°, and that of the unit brachydome about 60°; corresponding to the latter three-fold twins are observed. The type species is chrysolite (or olivine), which contains both magnesium and iron in varying proportions and is hence intermediate between the comparatively rare magnesium and iron silicates. X-ray study of the atomic structure shows that the SiO_4 groups are independent of each other; that the magnesium atoms lie between irregular groups of six oxygen atoms and are of two kinds. In monticellite one set of magnesium atoms is replaced by calcium atoms. The oxygen atoms lie nearly in the positions required in a hexagonal close-packed structure.

In form, the species of the Chrysolite Group, R_2SiO_4, are closely related in angle to chrysoberyl, $BeAl_2O_4$; also somewhat less closely to the species of the Diaspore Group, $H_2Al_2O_4$, etc. There is also an interesting relation between the chrysolites and the humites (see p. 628).

CHRYSOLITE. Olivine. Peridot.

Orthorhombic. Axes $a : b : c = 0.46575 : 1 : 0.5865$.

mm''',	$110 \wedge 1\bar{1}0 =$	49° 57′.
ss',	$120 \wedge \bar{1}20 =$	94° 4′.
dd',	$101 \wedge \bar{1}01 =$	103° 6′.
kk',	$021 \wedge 0\bar{2}1 =$	99° 6′.
ee''',	$111 \wedge 1\bar{1}1 =$	40° 5′.
ff''',	$121 \wedge 1\bar{2}1 =$	72° 13′.

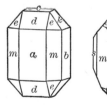

887   888

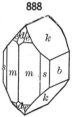

Twins rare: tw. pl. $h(011)$ with angle between basal planes of the two individuals = 60° 47′, penetration-twins, sometimes repeated; tw. pl. $w(012)$, the vertical axes crossing at an angle of about 30°. Crystals often flattened ∥ $a(100)$ or $b(010)$, less commonly elongated ∥ c axis. Massive, compact, or granular; in embedded grains.

Cleavage: $b(010)$ rather distinct; $a(100)$ less so. Fracture conchoidal. Brittle. H. = 6·5–7. G. = 3·27–3·37, increasing with the amount of iron; 3·57 for hyalosiderite (30 per cent FeO). Luster vitreous. Color green —

commonly olive-green, sometimes brownish, grayish red, grayish green, becoming yellowish brown or red by oxidation of the iron. Streak usually uncolored, rarely yellowish. Transparent to translucent. Optically $+$. Ax. pl. $\parallel c(001), Z \perp a(100)$. Dispersion $\rho < v$, weak. Axial angle large. 2V increases with rise in percentage of FeO; at about 13 per cent FeO the axial angle becomes 90°, and with a further increase in FeO content the sign becomes $-$. Indices variable, increasing with change in percentages of FeO; commonly $\alpha = 1 \cdot 635 - 1 \cdot 655$, $\beta = 1 \cdot 65 - 1 \cdot 67$, $\gamma = 1 \cdot 67 - 1 \cdot 69$.

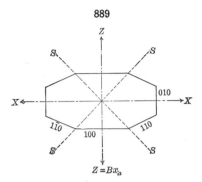

Var. *Precious.* — Of a pale yellowish green color, and transparent. G. = 3·441, 3·351. Occasionally seen in masses as large as "a turkey's egg," but usually much smaller. It has long been brought from the Levant for jewelry, but the exact locality is not known.

Common; *Olivine.* — Dark yellowish green to olive- or bottle-green. G. = 3·26-3·40. Disseminated in crystals or grains in basic igneous rocks, basalt and basaltic lavas, etc. *Hyalosiderite* is a highly ferruginous variety.

Comp. — $(Mg,Fe)_2SiO_4$ or $2(Mg,Fe)O.SiO_2$. The ratio of Mg : Fe varies widely, from 16 : 1, 12 : 1, etc., to 2 : 1 in hyalosiderite, and hence passing from forsterite on the one side to fayalite on the other. No sharp line can be drawn on either side. Titanium dioxide is sometimes present replacing silica; also tin and nickel in minute quantities.

Pyr., etc. — B.B. whitens, but is infusible in most cases; hyalosiderite and other varieties rich in iron fuse to a black magnetic globule; some kinds turn red upon heating. With the fluxes gives reactions for iron. Some varieties give reactions for titanium and manganese. Soluble in hydrochloric acid and yields gelatinous silica upon evaporation.

Diff. — Characterized by its infusibility, the yellow-green color, granular form and cleavage (quartz has none).

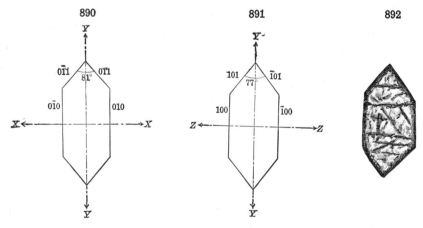

Micro. — Recognized in thin sections by its high relief; lack of color; its few but marked rough cleavage-cracks; high interference-colors, which are usually the brilliant and pronounced tones of the second order; parallel extinction; biaxial character; characteristic outlines (usually with acute terminations) when in distinct crystals (Figs. 889–891), its frequent association with iron ore and augite, and its very common alteration, in a greater or less degree, to serpentine, the first stages being marked by the separation of iron-ore grains along the lines of fracture (Fig. 892).

Artif. — The different members of the Chrysolite Group have been easily synthesized in various ways. They are often observed in slags.

Obs. — Chrysolite (olivine) has two distinct methods of occurrence: (*a*) in igneous rocks, as peridotite, norite, basalt, diabase and gabbro, formed by the crystallization of magmas low in silica and rich in magnesia; from an accessory component in such rocks the olivine may increase in amount until it is the main rock constituent as in the dunites; also (*b*) as the product of metamorphism of certain sedimentary rocks containing magnesia and silica, as in impure dolomites. In the dunites and peridotites the chrysolite is commonly associated with chromite, spinel, pyrope, etc., which are valuable indications also of the origin of serpentines derived from the olivine. In the metamorphic rocks the above are wanting, and carbonates, as dolomite, breunnerite, magnesite, etc., are the common associations; chrysolitic rocks of this latter kind may also occur altered to serpentine.

Chrysolite also occurs in grains, rarely in crystals, embedded in some meteoric irons. Also present in meteoric stones, frequently in spherical forms, or chondrules, sometimes made up of a multitude of grains with like (or unlike) optical orientation enclosing glass between.

Among the more prominent localities are: from Vesuvius in lava and on Mte. Somma in ejected masses where also more commonly forsterite occurs. Observed in the sanidine bombs at the Laacher See, Rhineland and elsewhere in the Eifel district and forming the mass of "olivine bombs" from the Dreiser Weiher near Daun in the same region. At Sasbach on the Kaiserstuhl in Baden, in basalt a variety containing much iron (*hyalosiderite*). In Austria from Kapfenstein, southeast of Graz, Styria. In serpentine at Snarum in Buskerud, Norway, in large crystals, themselves altered to serpentine. In Egypt in crystals of gem quality from St. John's Island in the Red Sea. Gem material also comes from the district of Mogok, north of Mandalay in Upper Burma; and from Minas Geraes, Brazil. Common in volcanic rocks from many localities.

In the United States at Thetford, Orange Co., Vermont, in boulders of basalt. In North Carolina at Webster, Jackson Co.; at Franklin, Macon Co., etc. In small, clear, olive-green grains with garnet at some points in Arizona and New Mexico.

Alteration of chrysolite often takes place through the oxidation of the iron; the mineral becomes brownish or reddish brown and iridescent. The process may end in leaving the cavity of the crystal filled with limonite or red oxide of iron. A very common kind of alteration is to the hydrous magnesium silicate, serpentine, with the partial removal of the iron or its separation in the form of grains of magnetite, also as iron sesquioxide; this change has often taken place on a large scale. See further under serpentine, p. 674.

Chrysolite is named from χρυσός, *gold*, and λίθος. The hyalosiderite, from ὕαλος, *glass*, and σίδηρος, *iron*. The *chrysolithus* of Pliny was probably our topaz; and his *topaz* our *chrysolite*.

Use. — The clear, fine green varieties are used as a gem stone; usually called *peridot*.

The axial ratios of the other members of the Chrysolite Group are given in the table on p. 597. The species are briefly characterized as follows:

Monticellite. — $CaMgSiO_4$. In small prismatic crystals or grains. H. = 5. G. = 3·2. Colorless to gray. Optical orientation as in chrysolite. Optically −. 2V = 75°. $\alpha = 1·65$, $\beta = 1·66$, $\gamma = 1·67$. Occurs in crystals embedded in limestone on Mte. Somma, Vesuvius. In small masses (*batrachite*) near Monzoni, Val di Fassa, Trentino, Italy. In crystals and grains in calcite at Magnet Cove near Hot Springs, Garland Co., Arkansas. Also found at Crestmore, Riverside Co., California. *Shannonite* from Shannon Tier, near Hobart, Tasmania, has been shown to be identical with monticellite.

Glaucochroite. — $CaMnSiO_4$. In prismatic crystals and grains. H. = 6. G. = 3·4. B.B. fusible at 3·5. Color, delicate bluish green. Optically −. 2V = 61°. Marked dispersion, $\rho > v$. $\alpha = 1·68$, $\beta = 1·72$, $\gamma = 1·73$. Found at Franklin Furnace, Sussex Co., New Jersey.

Forsterite. — Mg_2SiO_4. Orthorhombic. In small equidimensional or tabular crystals. Cleavages parallel to (001) and (010). H. = 6–7. G. = 3·21–3·33. Color white, greenish or yellow. Optically positive. 2V large. Ax. pl. || (001). Z ⊥ (100). $\alpha = 1·635$, $\beta = 1·651$, $\gamma = 1·670$. Occurs either in volcanic rocks or ejections or as a contact mineral in dolomite. Occurs in the ejected masses on Mte. Somma, Vesuvius; also in the Albani Mts., south of Rome; with serpentine at Snarum, Buskerud, Norway; north of Zlatoust, Ural Mts., and in limestone on the Kaiserstuhl, Baden. *Boltonite* is disseminated through a limestone at Bolton, Worcester Co., Massachusetts, in masses which are often over an inch through and rectangular in section.

Hortonolite. — $(Fe,Mg,Mn)_2SiO_4$. In rough dark-colored crystals or masses. G. = 3·91. Optically −. Indices, 1·768–1·803. Occurs at the iron mine of Monroe, Orange Co., New York; Iron Mine Hill, Cumberland, Rhode Island. Also reported from the Transvaal and Bohemia.

Fayalite. — Fe_2SiO_4. In crystals similar to those of chrysolite and massive. H. = 6·5. G. = 4·1. Optically −. Color light greenish yellow but brown to black on exposure. Ax. pl. || (001). $X \perp$ (010). 2V = 50°. $\alpha = 1·835$, $\beta = 1·877$, $\gamma = 1·886$. Fayalite is found in obsidian on the Lipari Islands, Sicily; from the island of Pantelleria; as nodules in volcanic rocks on Fayal, Azores. In France from Collobrières, Var. From the Mourne Mts., Co. Down, Ireland. *Manganfayalite* occurs at Tunaberg, Södermanland, Sweden. In the United States from Rockport near Gloucester, Essex Co., Massachusetts. In rhyolite at Obsidian Cliff in Yellowstone Park, Wyoming. Found at Cheyenne Canyon, Pike's Peak region, El Paso Co., Colorado.

Knebelite. — $(Fe,Mn)_2SiO_4$. Rarely in crystals with habit similar to those of chrysolite. H. = 6·5. G. = 4·0. Fusible at 3. Optical characters closely similar to those of fayalite. Found in the iron-manganese ore deposits of Sweden.

Tephroite. — Mn_2SiO_4; also with zinc, in the variety *roepperite*. Rarely in small crystals like those of chrysolite. H. = 6. G. = 4·1. Color flesh-red to ash-gray. Optically −. Ax. pl. || (100). $X = b$ axis. $\alpha = 1·77$, $\beta = 1·807$, $\gamma = 1·825$. 2V = 60°. Found at Sterling Hill and Franklin Furnace, Sussex Co., New Jersey. Also in Vermland, Sweden, at Pajsberg near Persberg and at Långbanshyttan. In France at Adervielle, Hautes-Pyrénées. *Picrotephroite* from Långbanshyttan is a variety intermediate between forsterite and tephroite.

Larsenite. — $PbZnSiO_4$. Orthorhombic. In slender striated prisms, occasionally tabular || (010). Prismatic cleavage, good. G. = 5·90. White and transparent. Adamantine luster. Optically −. Ax. pl. || (001). $X = a$ axis. $\alpha = 1·92$, $\beta = 1·95$, $\gamma = 1·96$. 2V = about 80°. Dispersion, $\rho > v$. Found at Franklin, Sussex Co., New Jersey, in veinlets associated with clinohedrite, hodgkinsonite, willemite, roeblingite, hardystonite, bementite, neotocite, calcium-larsenite, etc.

Calcium-larsenite. — $(Pb,Ca)ZnSiO_4$. Pb : Ca = about 1 : 1. Massive. Indistinct cleavage. G. = 4·421. White, opaque. Optically −. 2V = 5°. $\alpha = 1·760$, $\beta = \gamma = 1·769$. Another sample 2V = 40°. $\alpha = 1·760$, $\beta = 1·769$, $\gamma = 1·770$. Found in the same occurrence as larsenite at Franklin, Sussex Co., New Jersey.

Alleghanyite. — $5MnO.2SiO_2$. Orthorhombic. Lamellar twinning shown in thin section. In irregular grains. H. = 5·5. G. = 4·0. Color bright to grayish pink. Optically −. $\alpha = 1·756$, $\beta = 1·780$, $\gamma = 1·792$. 2V = 72°. $\rho > v$. Found in a vein with other manganese minerals (spessartite, rhodonite, manganiferous calcite, tephroite, galaxite, etc.), near Bald Knob, Alleghany Co., North Carolina.

Gosseletite. A manganiferous silicate (composition unknown). Orthorhombic. Prismatic cleavage. Green. Biaxial, +. Ax. pl. || (010). n high. Strong birefringence. 2V large. Pleochroic, X = clear green-yellow, Y = crude green, Z = warm olive-green. Found in a rock made up of hematite, phyllite and quartz, at Stavelot, southeastern Belgium.

Larnite. — Ca_2SiO_4. Monoclinic. In grains and imperfectly developed tabular crystals. Shows polysynthetic twinning || (100). Cleavage || (100), prominent. Color gray. Optically +. 2V moderately large. $\alpha = 1·707$, $\beta = 1·715$, $\gamma = 1·730$. Easily transformed by heating or shock to another substance probably identical with the artificial γ-Ca_2SiO_4. Occurs in a limestone contact zone at Scawt Hill, near Larne, Co. Antrim, Ireland, associated with spurrite, melilite, merwinite and spinel.

Merwinite. — $Ca_3Mg(SiO_4)_2$. Monoclinic. In grains, showing two laws of polysynthetic twinning. Cleavage || (010). H. = 6. G. = 3·150. Colorless to pale green. Vitreous luster. Optically +. $\alpha = 1·708$, $\beta = 1·711$, $\gamma = 1·718$. 2V = 66°. Dispersion $\rho > v$. $Z \perp$ (010). $X \wedge c$ axis = 36°. Occurs in intimate association with gehlenite, spurrite and monticellite in the contact metamorphic zone at Crestmore, near Riverside, California. Also noted with spurrite and larnite at Scawt Hill, Co. Antrim, Ireland, and at Velardeña, Durango, Mexico.

Trimerite. — $(Mn,Ca)_2SiO_4.Be_2SiO_4$. Monoclinic. Pseudo-hexagonal owing to twinning. In thick, tabular, prismatic crystals. Basal cleavage. H. = 6–7. G. = 3·474. Color salmon-pink to nearly colorless in small crystals. Optically −. Indices, 1·715–1·725. In Vermland, Sweden, at the Harstig mine at Pajsberg near Persberg and at Långbanshyttan.

Phenacite Group. R_2SiO_4. Tri-rhombohedral

		rr'	c
Willemite	Zn_2SiO_4	64° 30′	0·6775
Troostite	$(Zn,Mn)_2SiO_4$		
Phenacite	Be_2SiO_4	63° 24′	0·6611

The PHENACITE GROUP includes the above orthosilicates of zinc (manganese) and beryllium. Both belong to the tri-rhombohedral class of the trigonal division of the hexagonal system, and have nearly the same rhombohedral angle. X-ray study of phenacite and willemite shows a more complicated atomic structure than with the minerals of the chrysolite group. There are six molecules in the unit rhombohedral cell. The silicon and beryllium (or zinc) atoms are each surrounded by a tetrahedral group of oxygen atoms. Each oxygen atom is linked to one silicon and two beryllium (or zinc) atoms which lie nearly at the points of an equilateral triangle. The rare species trimerite, $Mn_2SiO_4.Be_2SiO_4$, which is pseudo-hexagonal (monoclinic) is perhaps to be regarded as connecting this group with the preceding Chrysolite Group.

Dioptase is probably related to the Phenacite Group with a similar structure. It also shows close crystallographic relations to the members of the Friedelite Group.

		rr'	c
Dioptase	H_2CuSiO_4 Tri-rhombohedral	54° 5′	0·5324

Friedelite Group

Friedelite	$6MnO.2Mn(OH,Cl)_2.6SiO_2.3H_2O$		
	Rhombohedral	56° 17′	0·5624
Pyrosmalite	$6(Fe,Mn)O.2(Fe,Mn)(OH,Cl)_2.6SiO_2.3H_2O$		
	Rhombohedral	53° 49′	0·5308
Schallerite	$6MnO.Mn_2((OH)_4(As_2O_3)).6SiO_2.3H_2O$?		
Molybdophyllite	$4(Mg,Pb)O.4(Mg,Pb)(OH)_2.4SiO_2.H_2O$		
	Rhombohedral		0·549

Various types of formulas have been assigned to the minerals of the Friedelite Group but the exact relations still seem uncertain. They show crystallographic and structural similarities. From the results of X-ray study on pyrosmalite it has been concluded that the formula should be written $3(Mn,Fe)(OH,Cl)_2.(Mn,Fe)Si_3O_7$. This, however, could be written similar to the formula given above and would differ only in having $4H_2O$ instead of $3H_2O$. Four such molecules are present in the structural unit cell. It has been suggested that the species bementite, dixenite, mcgovernite, hematolite, parsettensite and errite may also be related to this group.

WILLEMITE.

Tri-rhombohedral. Axis $c = 0·6775$; rr' $(10\bar{1}1) \wedge (\bar{1}101) = 64°$ 30′; ee' $(01\bar{1}2) \wedge (\bar{1}012) = 36°$ 17′.

In hexagonal prisms, sometimes long and slender, again short and stout; rarely showing subordinate faces distributed according to the phenacite type. Also massive and in disseminated grains; fibrous.

Cleavage: $c(0001)$ easy, Moresnet; difficult, New Jersey; $a(11\bar{2}0)$ easy,

New Jersey. Fracture conchoidal to uneven. Brittle. H. = 5·5. G. = 3·89–4·18. Luster vitreo-resinous, rather weak. Color white or greenish yellow, when purest; apple-green, flesh-red, grayish white, yellowish brown; often dark brown when impure. Streak uncolored. Transparent to opaque. Some varieties fluoresce strongly in ultra-violet rays, in green, yellow, etc. Artificial willemite fluoresces only when small amounts of manganese are present. Optically +. Refractive indices vary with the content of manganese; for pure zinc silicate, $\omega = 1·691$, $\epsilon = 1·719$.

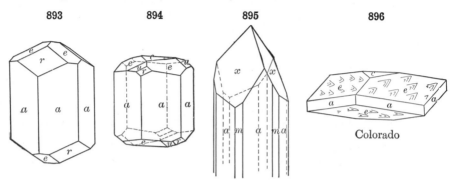

Figs. 893–895, New Jersey. $e(0\bar{1}12)$, $u(2\bar{1}\bar{1}3)$, $x(3\bar{1}\bar{2}1)$.

Comp. — Zinc orthosilicate, Zn_2SiO_4 or $2ZnO.SiO_2$ = Silica **27·0**, zinc oxide **73·0** = 100. Manganese often replaces a considerable part of the zinc (in *troostite*), and iron is also present in small amount.

Pyr., etc. — B.B. in the forceps glows and fuses with difficulty to a white enamel; the varieties from New Jersey fuse from 3·5 to 4. The powdered mineral on charcoal in R.F. gives a coating, yellow while hot and white on cooling, which, moistened with solution of cobalt, and treated in O.F., is colored bright green. With soda the coating is more readily obtained. Soluble in hydrochloric acid and yields gelatinous silica upon evaporation.

Obs. — Willemite occurs in zinc ore deposits associated with various other zinc minerals. Found at Altenberg near Moresnet in Belgium and between this point and Aachen (Aix-la-Chapelle) in Rhineland, massive or in small crystals. In Constantine, Algeria, at Bou Taleb, south of Sétif. From Mindouli in the French Congo; at Mumbwa, Northern Rhodesia; at Broken Hill, Rhodesia; and from Guchab near Otavi, in South West Africa. In Greenland from Musartut, in the Julianehaab district. In the United States in Sussex Co., New Jersey, at Mine Hill, Franklin Furnace, and at Sterling Hill, near Ogdensburg, 2 miles distant. Here it occurs in such quantity as to constitute an important ore of zinc. It occurs intimately mixed with zincite and franklinite, and is found massive in a great variety of colors; sometimes in reddish crystals (*troostite*) 6 inches long and an inch or more thick; rarely in slender transparent prisms of a delicate apple-green color. Rare at the Merritt mine, Socorro Co., New Mexico; at the Sedalia mine, Salida, Chaffee Co., Colorado; and from the Star district, Beaver Co., Utah. Named by Lévy after William I, King of the Netherlands.

Use. — An ore of zinc.

PHENACITE.

Tri-rhombohedral. Axis $c = 0·6611$; $rr'(10\bar{1}1) \wedge (\bar{1}101) = 63° 24'$.

Crystals commonly rhombohedral in habit, often lenticular in form, the prisms wanting; also prismatic, sometimes terminated by the rhombohedron of the third series, x (see further, pp. 127–129).

Cleavage: $a(11\bar{2}0)$ distinct; $r(10\bar{1}1)$ imperfect. Fracture conchoidal. Brittle. H. = 7·5–8. G. = 2·97–3·00. Luster vitreous. Colorless; also bright wine-yellow, pale rose-red; brown. Transparent to subtranslucent. Optically +. $\omega = 1.6540$; $\epsilon = 1.6697$.

Comp. — Beryllium orthosilicate, Be_2SiO_4 or $2BeO.SiO_2$ = Silica 54·45, glucina 45·55 = 100.

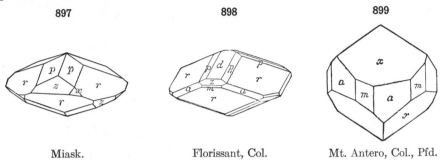

897　　　　　　　　898　　　　　　　　899

Miask.　　　　　Florissant, Col.　　　　Mt. Antero, Col., Pfd.

Pyr., etc. — Alone remains unaltered; with borax fuses with extreme slowness, unless pulverized, to a transparent glass. With soda affords a white enamel; with more, intumesces and becomes infusible. Dull blue with cobalt solution.

Obs. — Most commonly found in pegmatites as a pneumatolytic mineral, associated with microcline, topaz, and quartz; also in granite druses with adular, etc. Occurs (in prismatic crystals) in mica schist at the emerald and chrysoberyl mine on the Takowaja River, east of Ekaterinburg, in the Ural Mts., Russia, where the crystals are sometimes nearly 4 inches across, and one weighed more than 1 pound. Also in small rhombohedral crystals near Miask in the Ilmen Mts. In Switzerland at Reckingen, Valais; at Framont in the Vosges Mts., Alsace, France. Prismatic and twin crystals occur at Kragerö, Telemark, Norway. From the district of Usagara, Tanganyika Territory, East Africa. In exceptional crystals from San Miguel di Piracicaba, in Minas Geraes, Brazil. In the United States from Greenwood, Oxford Co., Maine; in New Hampshire on Bald Face Mt., North Chatham, Carroll Co. In Colorado on Topaz Butte, near Florissant, El Paso Co.; in flat rhombohedral crystals; also in prismatic crystals at Mt. Antero, Chaffee Co. Named from φέναξ, *a deceiver*, in allusion to its having been mistaken for quartz.

Dioptase. — H_2CuSiO_4, or $H_2O.CuO.SiO_2$. Tri-rhombohedral. Commonly in prismatic crystals ($ss'\ 02\bar{2}1 \wedge \bar{2}021 = 84°\ 33\frac{1}{2}'$). Also in crystalline aggregates; massive. Cleavage; $r(10\bar{1}1)$ perfect. Fracture conchoidal to uneven. H. = 5. G. = 3·28–3·35. Luster vitreous. Color emerald-green. Optically +. $\omega = 1.644–1.658$. $\epsilon = 1.697–1.709$.

Dioptase is a comparatively rare mineral, being found in only a few localities associated with other copper ores. Occurs in druses of well-defined crystals on quartz in a limestone west of the hill of Altyn-Tübe in the Kirghese Steppes, Russia. Fine specimens come from the copper deposits in the basin of the Niari River in southern French Congo at Mindouli, Pimbi, on the river Djoué, etc.; also in the Katanga district in the Belgian Congo; from Guchab near Otavi, South West Africa. From Copiapo in Atacama and elsewhere in Chile. In the United States dioptase occurs with the copper deposits of Arizona at Clifton, Greenlee Co., and at Florence and Riverside in Pinal Co.

900　　　　　901

Friedelite. — $H_7(MnCl)Mn_4Si_4O_{16}$, (see p. 601). Crystals commonly tabular || $c(0001)$; also massive, cleavable (|| to (0001)) to closely compact. H. = 4–5. G. = 3·07. Color rose-red. Optically −. $\omega = 1.66$, $\epsilon = 1.63$. From the manganese mine of Adervielle, Hautes Pyrénées, France; in Vermland, Sweden, at the Harstig mine, Pajsberg, near Persberg, and at the Sjö mines near Örebro in Örebro; from Franklin Furnace, New Jersey.

Pyrosmalite. — $H((Fe,Mn)Cl)(Fe,Mn)_4Si_4O_{16}$ (see p. 601). Crystals thick hexagonal prisms or tabular; also massive, foliated. Basal cleavage. H. = 4–4·5. G. = 3·06–3·19.

Color blackish green to pale liver-brown or gray. Optically $-$. $\omega = 1\cdot68$, $\epsilon = 1\cdot64$. From the iron mines of Nordmark in Vermland and at Dannemora, in Upsala, Sweden.

PARSETTENSITE. $3MnO.4SiO_2.4H_2O$? Massive, somewhat micaceous. G. $=2\cdot59$. Decomposed by HCl with separation of silica. Fusible with intumescence. Color copper-red. Uniaxial, $-$. $\omega = 1\cdot576$, $\epsilon = 1\cdot546$. Found in manganese deposits at Parsettens Alp, Val d'Err, Grisons, Switzerland. A mineral, supposed to differ in the amount of water present, occurring associated with parsettensite, and probably identical with it, has been named *errite*.

TINZENITE. $2CaO.Al_2O_3.Mn_2O_3.4SiO_2$? Good pinacoidal cleavage. Columnar structure. Color yellow to orange-red. Biaxial, $-$. $2V = 62°$. $\alpha = 1\cdot693$, $\beta = 1\cdot701$, $\gamma = 1\cdot704$. G. $= 3\cdot29$. Occurs in veins with quartz in the manganese deposits at Tinzen, Val d'Err, Grisons, Switzerland.

Schallerite. — $6MnO.Mn_2((OH)_4(As_2O_3)).6SiO_2.3H_2O$. Probably hexagonal. Massive. Basal cleavage. H. $= 4\cdot5–5$. G. $= 3\cdot37$. Difficultly fusible. Light brown color. Vitreous to waxy luster. Uniaxial, $-$. $\omega = 1\cdot704$, $\epsilon = 1\cdot679$. Occurs in veinlets in the ore body at Franklin, New Jersey. An iron-rich variety from the same locality has been called *ferroschallerite*.

Molybdophyllite. — $(Pb,Mg)SiO_4.H_2O$ (see also p. 601). Hexagonal, rhombohedral. In irregular foliated masses with perfect basal cleavage. H. $= 3–4$. G. $= 4\cdot7$. Difficultly fusible. Colorless to pale green. Optically $-$. $\omega = 1\cdot815$, $\epsilon = 1\cdot761$. Occurs at Långbanshyttan, Vermland, Sweden.

Scapolite Group. Tetragonal-tripyramidal

Meionite	$c = 0\cdot4393$	Mizzonite, Dipyre	$c = 0\cdot4424$
Wernerite	$c = 0\cdot4384$	Marialite	$c = 0\cdot4417$

The species of the SCAPOLITE GROUP crystallize in the pyramidal class of the tetragonal system with nearly the same axial ratio. They are white or grayish white in color, except when impure, and then rarely of dark color. Hardness $= 5–6\cdot5$; G. $= 2\cdot5–2\cdot8$. In composition they are essentially silicates of aluminum with calcium and sodium in varying amounts; radicals involving Cl, OH, CO_2, or SO_4 are also present in small amounts.

The Scapolites are analogous to the Feldspars in that they form a series with a gradual variation in composition, the amount of silica increasing with the increase of the alkali, soda, being 40 per cent in meionite and 64 per cent in marialite. A corresponding increase is observed also in the amount of chlorine, etc., present. Furthermore there is also a gradual change in specific gravity, in the value of ω, and in resistance to acids, from the easily decomposed meionite, with G. $= 2\cdot72$, to marialite, which is only slightly attacked and has G. $= 2\cdot63$. The variation in composition may be explained by the isomorphous mixtures of various molecules, the two most important being:

Meionite	$CaCO_3.3CaAl_2Si_2O_8$	Me
Marialite	$NaCl.3NaAlSi_3O_8$	Ma

Other molecules, which are present at times in important amounts, have been assumed to be as follows: $Na_2CO_3.3NaAlSi_3O_8$, $NaHCO_3.3NaAlSi_3O_8$, $Na_2SO_4.3NaAlSi_3O_8$, $NaHSO_4.3NaAlSi_3O_8$, $CaSO_4.3CaAl_2Si_2O_8$ (*silvialite*).

The composition therefore equals three plagioclase feldspar molecules (albite, anorthite, and probably carnegieite) plus one molecule of a salt which consists of a carbonate, sulphate, or chloride. It has also been suggested that, in the case of the molecules containing the sodium sulphate or carbonate radicals, these are present in one-half the amount given above, namely as $\frac{1}{2}Na_2CO_3.3NaAlSi_3O_8$ and $\frac{1}{2}Na_2SO_4.3NaAlSi_3O_8$.

A number of variety names have been applied to the scapolites, the significance of which is often in doubt. Winchell has proposed the following classification based upon the relative proportions of the two molecules, Me and Ma.

Marialite to include $Ma_{100}Me_0$ to $Ma_{80}Me_{20}$
Dipyre " $Ma_{80}Me_{20}$ to $Ma_{50}Me_{50}$
Mizzonite " $Ma_{50}Me_{50}$ to $Ma_{20}Me_{80}$
Meionite " $Ma_{20}Me_{80}$ to Ma_0Me_{100}

Much the same classification is used here, except that dipyre is used as a synonym for mizzonite and the central portion of the series which is the most common in occurrence is given the name wernerite.

The tetragonal species melilite and gehlenite are near the Scapolites in angle. The more common vesuvianite is also related.

MEIONITE.

Tetragonal. Axis $c = 0.439$. In prismatic crystals (Fig. 219, p. 102), either clear and glassy or milky white; also in crystalline grains and massive. Cleavage: $a(100)$ rather perfect, $m(110)$ somewhat less so. Fracture conchoidal. Brittle. H. = 5.5–6. G. = 2.70–2.74. Luster vitreous. Colorless to white. Transparent to translucent; often cracked within. Optically −. $\omega = 1.58$–1.60, $\epsilon = 1.55$–1.56.

Comp. — $CaCO_3.3CaAl_2Si_2O_8$.

The marialite molecule, $NaCl.3NaAlSi_3O_8$, may occur in varying amounts up to 20 per cent.

Obs. — Occurs in small crystals in cavities, usually in limestone blocks, on Mte. Somma, Vesuvius. Also in ejected masses at the Laacher See, Rhineland, Germany. A mineral in an amphibole-gneiss from the Black Forest, Germany, which is like meionite except for a basal cleavage has been called *pseudomeionite*.

WERNERITE. COMMON SCAPOLITE.

Tetragonal-tripyramidal. Axis $c = 0.4384$.

Crystals prismatic, usually coarse, with uneven faces and often large. The symmetry of the pyramidal class sometimes shown in the development of the faces $z(311)$ and $z_1(131)$. Also massive, granular, or with a faint fibrous appearance; sometimes columnar.

902   903

cc', $101 \wedge 011 = 32° 59'$.
rr', $111 \wedge \bar{1}11 = 43° 45'$.
mr, $110 \wedge 111 = 58° 12'$.
zz''', $311 \wedge 3\bar{1}1 = 29° 43'$.

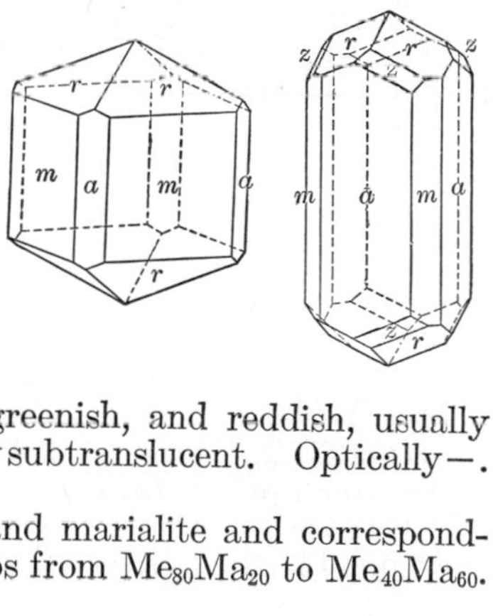

Cleavage: $a(100)$ and $m(110)$ rather distinct, but interrupted. Fracture subconchoidal. Brittle. H. = 5–6. G. = 2.66–2.73. Luster vitreous to pearly externally, inclining to resinous; cleavage and cross-fracture surface vitreous. Color white, gray, bluish, greenish, and reddish, usually light; streak uncolored. Transparent to faintly subtranslucent. Optically −. $\omega = 1.56$–1.58. $\epsilon = 1.54$–1.55.

Comp. — Intermediate between meionite and marialite and corresponding to a molecular combination of these in ratios from $Me_{80}Ma_{20}$ to $Me_{40}Ma_{60}$.

The silica varies from 46 to 54 per cent, and as its amount increases the soda and chlorine also increase.

Pyr., etc. — B.B. fuses easily with intumescence to a white blebby glass giving a strong sodium flame color. Imperfectly decomposed by hydrochloric acid.

Diff. — Characterized by its square form and prismatic cleavage (90°); resembles feldspar when massive, but has a characteristic fibrous appearance on the cleavage surface; it is also more fusible, and has a higher specific gravity; also distinguished by fusibility with intumescence from pyroxene (which see, p. 559).

Micro. — Recognized in thin sections by its low refraction; lack of color; rather high interference-colors reaching the yellows and reds of the first order, sections showing which extinguish parallel to the cleavage; by the distinct negative axial cross of basal sections which show the cleavage-cracks crossing at right angles.

Obs. — Common scapolite occurs in metamorphic rocks, as crystalline schists, gneisses, and amphibolites, especially those which are rich in lime or contain inclusions of limestone. It is most abundant as a contact metamorphic mineral in crystalline limestones near the contact with granitic or allied rocks; sometimes in beds of magnetite accompanying limestone. It is often associated with a light colored pyroxene, amphibole, garnet, and also with apatite, titanite, zircon. Scapolite has been shown also to be frequently a component of basic igneous rocks, especially those high in plagioclases containing much lime; it is regarded here as a secondary product through the alteration of the feldspar. The scapolites are easily altered; pseudomorphs of mica, more rarely other minerals, are common.

Prominent localities for the occurrence of scapolite are as follows: The pale blue or gray scapolite from Lake Baikal, Siberia, is the variety *glaucolite*. In Finland it occurs near Åbo in Turun ja Porin at Pargas and Laurinkari; at Hirvensalo, etc. From Arendal in Aust-Agder, Norway, and at Malsjö, Vermland, Sweden. *Passauite* is from Obernzell, east of Passau, Bavaria. Crystals of gem quality with a yellow color occur in pegmatite at Tsarasaotra east of Ankazobe, Madagascar.

In the United States the more important localities are as follows: In Massachusetts at Bolton in Worcester Co. In a great many places in New York, as in Orange Co., near Monroe at Two Ponds, sometimes in large crystals, and elsewhere; in Lewis Co., at Natural Bridge, Diana, etc., in fine crystals; in St. Lawrence Co. at Grasse Lake near Rossie in fine crystals, at Gouverneur in limestone; also at Macomb, Pierrepont, etc. At Franklin, Sussex Co., New Jersey.

In Canada scapolite is common in Ontario, as in Frontenac Co., at Bedford, and at Hinchinbrooke (Bobs Lake); in Lanark Co. at Bathurst; in Leeds Co., at Charleston Lake; also in Quebec at Grenville in Argenteuil Co., and at various points in Ottawa Co.

Mizzonite. Dipyre. — Here are included scapolites with 54 to 57 per cent SiO_2, corresponding to a molecular combination from $Me_{40}Ma_{60}$ to $Me_{20}Ma_{80}$. Indices, $\omega = 1\cdot53-1\cdot56$, $\epsilon = 1\cdot52-1\cdot54$. *Mizzonite* occurs in clear crystals in ejected masses on Mte. Somma, Vesuvius.

Dipyre occurs in elongated square prisms, often slender, sometimes large and coarse, in limestone and crystalline schists, chiefly from the Pyrenees; as at Seix in Ariège; Mauléon in Basses-Pyrénées; Ponzac in Hautes-Pyrénées; etc. *Couseranite* from the Pyrenees is a more or less altered form of dipyre.

Marialite. — Theoretically $NaCl.3NaAlSi_3O_8$, see p. 604. Indices, $\omega = 1\cdot539$, $\epsilon = 1\cdot537$. The actual mineral corresponds to $Ma_{85}Me_{15}$. It occurs in a basalt tuff, at Pianura, near Naples.

Sarcolite. — $(Ca,Na_2)_3Al_2(SiO_4)_3$. In small tetragonal crystals. X-ray study shows a structure more nearly related to that of the Scapolite Group than to that of the Melilite Group. H. = 6. G. = 2·545–2·932. Color flesh-red. Optically +. Indices, $\omega = 1\cdot604$, $\epsilon = 1\cdot615$. Sometimes abnormally biaxial with higher indices. From Mte. Somma, Vesuvius.

MELILITE.

Tetragonal. Axis $c = 0\cdot4548$. Usually in short square prisms ($a(100)$) or octagonal prisms (a, $m(110)$), also in tetragonal tables.

Cleavage: $c(001)$ distinct; $a(100)$ indistinct. Fracture conchoidal to uneven. Brittle. H. = 5. G. = 2·9–3·10. Luster vitreous, inclining to resinous. Color white, pale yellow, greenish, reddish, brown. Pleochroism distinct in yellow varieties. Sometimes exhibits optical anomalies. Optical characters vary with composition. Usually optically −, but + for åkermanite end of series. Indices for the chief component molecules; gehlenite, $\omega = 1 \cdot 67$, $\epsilon = 1 \cdot 66$; akermanite, $\omega = 1 \cdot 63$, $\epsilon = 1 \cdot 64$.

Comp. — The composition is complex but can usually be expressed as varying isomorphous mixtures of two molecules, that of gehlenite, $Ca_2Al_2SiO_7$, and that of åkermanite, $Ca_2MgSi_2O_7$. In addition the following hypothetical molecules have been assumed by various writers to be present in minor amounts; soda-melilite, $Na_2Si_3O_7$; sub-melilite, $CaSi_3O_7$; calcium-åkermanite, $Ca_3Si_2O_7$; iron-akermanite, $Ca_2\overset{\text{II}}{Fe}Si_2O_7$; iron-gehlenite, $Ca_2\overset{\text{III}}{Fe}_2SiO_7$.

Artif. — Melilite has been formed artificially by fusing together its constituent oxides. It is found in slags and has been produced in various artificial magmas.

Pyr., etc. — B.B. fuses at 3 to a yellowish or greenish glass. With the fluxes reacts for iron. Soluble in hydrochloric acid and yields gelatinous silica upon evaporation.

Micro. — Distinguished in thin sections by its moderate refraction; very low interference-colors, showing often the "ultra blue" (Capo di Bove); parallel extinction; negative character; usual development in tables parallel to the base and very common "peg structure" due to parallel rod-like inclusions penetrating the crystal from the basal planes inward: this, however, is not always easily seen.

Obs. — Melilite is a component of certain recent basic eruptive rocks formed from magmas very low in silica, rather deficient in alkalies, and containing considerable lime and alumina. In such cases melilite appears to crystallize in the place of the more acid plagioclase. Found in such rocks as basalts, leucite and nephelite rocks, etc., also in the intrusive rock called *alnöite*. Common in furnace slags and in Portland cement.

Melilite of yellow and brownish colors is found at Capo di Bove, near Rome, in leucitophyre with nephelite, augite, hornblende; at Vesuvius in dull yellow crystals (*somervillite*); not uncommon in certain basic eruptive rocks, as the *melilite-basalts* of Hochbohl near Owen in Württemberg; Hawaiian Islands; in Uvalde Co., Texas; etc. Occurs as chief constituent of rock on Beaver Creek, Gunnison Co., Colorado. Common in furnace slags. Melilite is named from μέλι, *honey*, in allusion to the color.

Humboldtilite occurs in cavernous blocks on Mte. Somma, Vesuvius, with greenish mica, also apatite, augite; the crystals are often rather large, and covered with a calcareous coating; less common in transparent lustrous crystals with nephelite, sarcolite, etc., in an augitic rock. *Zurlite* is impure humboldtilite. *Deeckeite* is a pseudomorph after *melilite*, found in a melilite basalt from the Kaiserstuhl, Baden, Germany. *Fuggerite* corresponds to a member of the gehlenite-akermanite series, 3 åk : 10 geh. Occurs in a contact zone adjoining the monzonite at Monzoni, Val di Fassa, Trentino, Italy.

Gehlenite. — $Ca_2Al_2SiO_7$. Crystals usually short square prisms. Axis $c = 0 \cdot 4001$. G. = $2 \cdot 9$–$3 \cdot 07$. Different shades of grayish green to liver-brown. See further under melilite above. Gehlenite is a contact mineral, occurring in limestone, usually at the contact with a monzonite-like rock. It is found in Trentino, Italy, at Monzoni, Val di Fassa; also in the Val di Fiemme. In rolled pebbles inclosing grains of vesuvianite at Oravicza, Rumania. Found in granular aggregates in the contact zone between limestone and diorite in the Velardeña mining district, Durango, Mexico. This last occurrence was described as containing a hypothetical molecule called *velardeñite* with the composition that is now assigned to gehlenite.

Åkermanite. — $Ca_2MgSi_2O_7$. Tetragonal. Isomorphous with gehlenite in melilite, which see. Found only in certain slags.

Hardystonite. — $Ca_2ZnSi_2O_7$. Tetragonal. A member of the Melilite Group. In granular masses. Basal cleavage and parting || (100) and (110). H. = 3–4. G. = 3·4. Color white. Optically −. $\omega = 1 \cdot 672$, $\epsilon = 1 \cdot 661$. From Franklin, Sussex Co., New Jersey.

Cebollite. — $H_2Ca_5Al_2Si_3O_{16}$. Orthorhombic (?). Fibrous. H. = 5. G. = 2·96. Color white to greenish gray. Indices, 1·59–1·63. Fusible at 5. Soluble in acids. Found as an alteration product of *melilite* near Cebolla Creek, Gunnison Co., Colorado.

VESUVIANITE. Idocrase.

Tetragonal. Axis $c = 0.5372$.

ce, $001 \wedge 101 = 28° 15'$.
cp, $001 \wedge 111 = 37° 13\frac{1}{2}'$.
ct, $001 \wedge 331 = 66° 18'$.

pp', $111 \wedge \bar{1}11 = 50° 39'$.
ss^{vii}, $311 \wedge 3\bar{1}1 = 31° 38'$.

Often in crystals, prismatic or pyramidal. Also massive; columnar, straight and divergent, or irregular; granular massive; cryptocrystalline. The atomic structure shows a similarity to that of garnet.

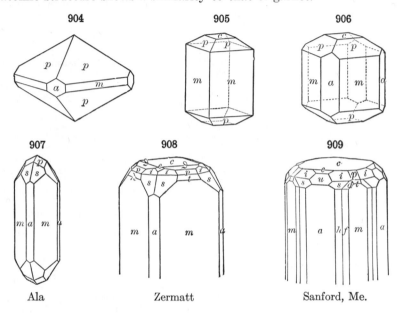

904   905   906

907   908   909

Ala   Zermatt   Sanford, Me.

Cleavage: $m(110)$ not very distinct; $a(100)$ and $c(001)$ still less so. Fracture subconchoidal to uneven. Brittle. H. = 6·5. G. = 3·35–3·45. Luster vitreous; often inclining to resinous. Color brown to green, and the latter frequently bright and clear; occasionally sulphur-yellow, and also pale blue. Streak white. Subtransparent to faintly subtranslucent. Dichroism not usually strong. Optically $-$; also $+$ rarely. Birefringence very low. Sometimes abnormally biaxial. Indices variable, from 1·701 to 1·736.

Comp. — A basic calcium-aluminum silicate, but of uncertain formula; perhaps $Ca_6[Al(OH,F)]Al_2(SiO_4)_5$. Ferric iron replaces part of the aluminum and magnesium the calcium. Fluorine, titanium, and boron may be present.

The suggestion has been made that a tetragonal variation of the garnet type of molecule is present in considerable amount with various other hypothetical molecules isomorphous with it. An analysis of vesuvianite from Franklin, New Jersey, shows a notable amount of beryllium oxide and yields the following formula: $2RO.6CaO.4BeO.Al_2O_3.6SiO_2$. The following formula has been derived from the results of X-ray study of structure; $Ca_{10}Al_4$ $(Mg,Fe)_2(OH)_4Si_9O_{34}$. There are four such molecules in the unit cell.

Pyr., etc. — B.B. fuses at 3 with intumescence to a greenish or brownish glass. With the fluxes gives reactions for iron, and some varieties a strong manganese reaction. *Cyprine*, a blue variety, gives a reaction for copper with salt of phosphorus. Partially decomposed by hydrochloric acid, and completely when the mineral has been previously ignited.

Diff. — Characterized by its tetragonal form and easy fusibility. Resembles some brown varieties of garnet, tourmaline, and epidote.

Micro. — Recognized in thin sections by its high refraction producing a very strong relief and its extremely low birefringence*; also in general by its color, pleochroism, and uniaxial negative character; the latter, on account of the low birefringence, being difficult to determine. The low birefringence, however, aids in distinguishing it from epidote, with which at times it may be confounded.

Obs. — Vesuvianite was first found among the ancient ejections of Vesuvius and the dolomitic blocks of Mte. Somma, whence its name. It commonly occurs as a contact mineral formed from the alteration of impure limestones, then usually associated with grossularite, phlogopite, diopside, wollastonite; also epidote; also in serpentine, chlorite schist, gneiss and related rocks. It is never a constituent of eruptive rocks.

It is found on the Achtaragda, a tributary of the Vilui River in Yakutsk, Siberia (sometimes called *wiluite* or *viluite*, like the grossular garnet from the same region; this variety contains boron, is optically $+$, and usually abnormally biaxial). At Achmatovsk near Kussinsk in the Zlatoust district in the Ural Mts. From Haslau, northwest of Eger in Bohemia of Czechoslovakia (*egeran*); in Rumania at Oravicza and neighboring Csiklova and at Moravicza. In Silesia of Czechoslovakia at Friedeberg and near Jordansmühl in German Silesia. In Italy vesuvianite occurs in Trentino on Canzocoli, southwest of Predazzo, Val di Fiemme; at Monzoni, Val di Fassa; in Piedmont on the Mussa Alp in the Ala Valley in transparent green or brown brilliant crystals; also from Locana northwest of Turin. From Ariccia and elsewhere in the Albani Mts., south of Rome. At Vesuvius it is hairbrown to olive-green, and sometimes in highly modified crystals. In Switzerland at Zermatt, Valais. In Norway at Arendal in Aust-Agder (*colophonite*); at Egg, near Christiansand in Vest-Agder; in the Eker parish, Buskerud, between Kongsberg and Drammen; in Telemarken the variety *cyprine*. From Finland on Lake Ladoga at Luppiko near Impilaks and at Pitkäranta. In well-defined crystals from Kiura, Bungo Province, Japan. In Mexico from the states of Morelos and Chiapas.

In the United States vesuvianite occurs in Maine at Auburn, Androscoggin Co.; at Sanford, York Co.; etc. In New York south of Amity, Orange Co., grayish and yellow-brown crystals. In New Jersey at Franklin, Sussex Co., the variety *cyprine*. From Magnet Cove near Hot Springs, Garland Co., Arkansas. In California it occurs at Crestmore, Riverside Co., and the closely compact variety, *californite*, with an olive-green to a grass-green color from Siskiyou and Tulare counties, etc.

In Quebec large brownish yellow crystals occur at Calumet Falls, Litchfield, Pontiac Co.; from Ottawa Co., at Templeton in brownish red crystals, and at Wakefield, green and bright yellow. It occurs in masses and minute crystals of a bright pink color from Black Lake, Megantic Co. (*mangan-vesuvianite*).

GENEVITE. Essentially a silicate of calcium and aluminum. Tetragonal, prismatic. Cleavages, (100) and (001). Color gray. Optically $-$. $\omega = 1.707$, $\epsilon = 1.698$. G. = 3.16. Occurs in limestone at Sidi bou Otmane, Morocco. Originally described as dipyre. Possibly identical with vesuvianite.

Zircon Group. $\overset{IV}{R}SiO_4$. Tetragonal

Zircon	$ZrSiO_4$	$c = 0.6404$
Thorite	$ThSiO_4$	$c = 0.6402$

This group includes the orthosilicates of zirconium and thorium, both alike in tetragonal crystallization, axial ratio, crystalline habit, and atomic structure.

These species are closely related to the members of the Rutile Group both as to crystal constants and atomic structure. See further on p. 496.

---

* Frequently minerals, which, like vesuvianite, melilite and zoisite, are doubly refracting but of extremely low birefringence and possibly (where they are positive for one color but negative for another), do not show a gray color between crossed nicols but a curious blue, at times an intense Berlin blue, which is quite distinct from the other blues of the color scale and is known as the " ultra blue."

ZIRCON.

Tetragonal. Axis $c = 0.64037$.

ee', $101 \wedge 011 = 44° 50'$.
ee'', $101 \wedge \bar{1}01 = 65° 16'$.
pp', $111 \wedge \bar{1}11 = 56° 40\frac{1}{2}'$.
uu', $331 \wedge \bar{3}31 = 83° 9'$.

mp, $110 \wedge 111 = 47° 50'$.
mu, $110 \wedge 331 = 20° 12\frac{1}{2}'$.
xx^{VII}, $311 \wedge 3\bar{1}1 = 32° 57'$.
ax, $100 \wedge 311 = 31° 43'$.

910   **911** 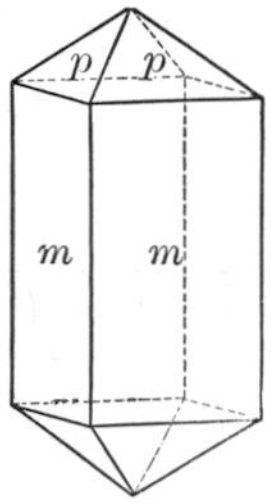   **912** 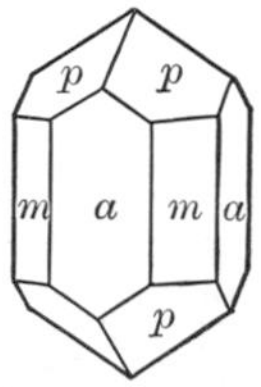   **913**

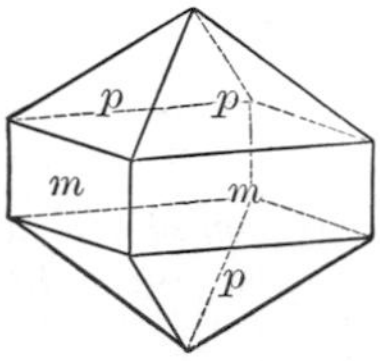

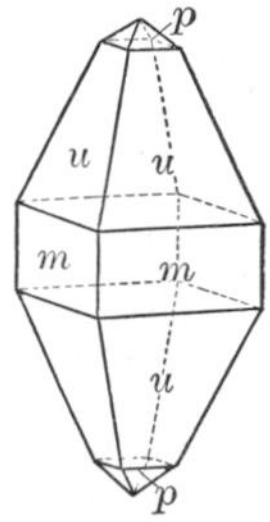

914 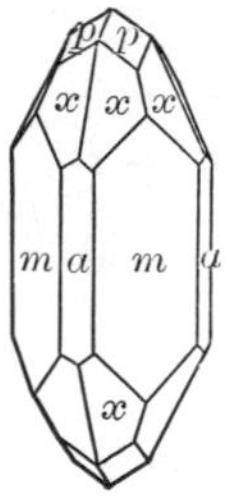   **915** 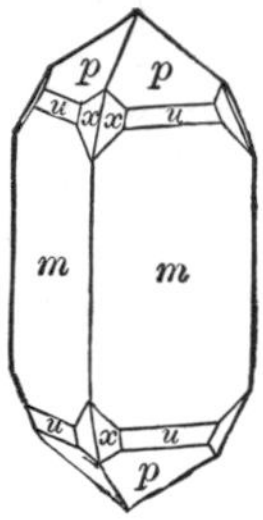   **916**

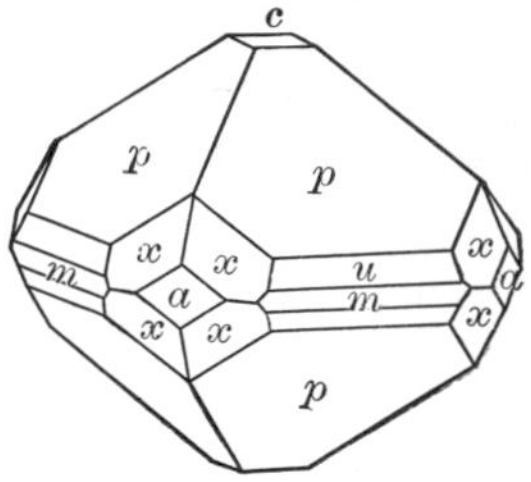

Colorado

Twins: tw. pl. $e(101)$, geniculated twins like rutile (Fig. 438, p. 188). Commonly in square prisms, sometimes pyramidal. Also in irregular forms and grains.

Cleavage: $m(110)$ imperfect; $p(111)$ less distinct. Fracture conchoidal. Brittle. H. = 7·5. G. = 4·68–4·70 most common, but varying widely from 4·2 to 4·86. Luster adamantine. Colorless, pale yellowish, grayish, yellowish green, brownish yellow, reddish brown. Streak uncolored. Transparent to subtranslucent and opaque. Optically +. Birefringence high. $\omega = 1.923$– 1.960, $\epsilon = 1.968$–2.015. Sometimes abnormally biaxial.

Hyacinth is the orange, reddish and brownish transparent kind used for gems. *Jargon* is a name given to the colorless or smoky zircons of Ceylon, in allusion to the fact that while resembling the diamond in luster, they are comparatively worthless; thence came the name *zircon*. A blue color has been given artificially to zircon by a special heat treatment in which the zircon is exposed to fumes liberated from a solution of cobalt nitrate and potassium ferrocyanide. Gem stones from this material have been called *starlite*.

Comp. — $ZrSiO_4$ or $ZrO_2.SiO_2$ = Silica 32·8, zirconia 67·2 = 100. A little iron (Fe_2O_3) is usually present.

Some varieties contain hafnium oxide up to 4 per cent. The rare earths occur in the variety *hagatalite* from Hagata, Iyo Province, Japan. *Oyamalite* is a variety containing considerable phosphorus from Oyama, Iyo Province, Japan.

Pyr., etc. — Infusible; the colorless varieties are unaltered, the red become colorless, while dark-colored varieties are made white; some varieties glow and increase in density by ignition. Not perceptibly acted upon by salt of phosphorus. In powder decomposed when fused with soda on the platinum wire, and if the product is dissolved in dilute hydrochloric acid it gives the orange color characteristic of zirconia when tested with turmeric paper. Not acted upon by acids except in fine powder with concentrated sulphuric acid. Decomposed by fusion with alkaline carbonates and bisulphates.

Diff. — Characterized by the prevailing square pyramid or square prism; also by its adamantine luster, hardness, high specific gravity, and infusibility; the diamond is optically isotropic.

Micro. — Recognized in thin sections by its very high relief; very high interference-colors, which approach white of the higher order except in very thin sections; positive uniaxial character. It is distinguished from cassiterite and rutile only by its lack of color, and from the latter also in many cases by method of occurrence.

Artif. — Zircon has been prepared artificially by heating zirconium oxide with quartz in gaseous silicon fluoride.

Obs. — A common accessory constituent of igneous rocks, especially those of the more acid feldspathic groups and particularly the kinds derived from magmas containing much soda, as granite, syenite, diorite, etc. It is one of the earliest minerals to crystallize from a cooling magma. Is generally present in minute crystals, but in pegmatitic facies often in large and well-formed individuals. Occurs more rarely elsewhere, as in granular limestone, chloritic and other schists; gneiss; sometimes in iron-ore beds. Crystals are common in most auriferous sands. Sometimes found in volcanic rocks, probably in part as inclusions derived from older rocks.

Zircon in distinct crystals is so common in the pegmatitic forms of the nephelite-syenite and augite-syenite of southern Norway (with ægirite, etc.) that this rock there and elsewhere has sometimes been called a " zircon-syenite."

Found at Miask, Ilmen Mts., and elsewhere in the gold regions of the Urals. Occurs as red crystals in lava at Niedermendig in the Eifel district of Rhineland. In Italy it is found in the Pfitschtal, Trentino, and at Mte. Somma, Vesuvius. In France it occurs in red crystals in river sand at Espaly near Le Puy, Haute-Loire. In Norway at Arendal in Aust-Agder, the altered variety *œrstedite*; at Hiterö, Vest-Agder; at Kragerö, Telemark, *tachyaphaltite*; at Fredriksvärn, Vestfold, in zircon-syenite and at many other points in the Langesundfiord region of southern Norway. In Sweden the altered zircon *cyrtolite* is found at Ytterby, northeast of Stockholm. Found in the alluvial sands in Ceylon. Of common occurrence in Madagascar as at Itrongahy near Betroka and on Mt. Ampanobe west of Fianarantsoa in large crystals.

In the United States zircon occurs at Litchfield, Kennebec Co., Maine; in Massachusetts at Chesterfield, Hampshire Co., the altered variety, *œrstedite*; in New York at Moriah, Essex Co., Two Ponds near Monroe, Orange Co., in St. Lawrence Co., in the town of Hammond near De Long's Mills; also at Rossie, Fine, Pitcairn, etc. In North Carolina abundant in the gold sands of Henderson Co., and elsewhere. In Colorado at various points in the Pike's Peak region. From Llano Co., Texas.

In Canada from Ontario in Renfrew Co., very large crystals, at Sebastopol, etc.; in Quebec at Grenville, Argenteuil Co., and in large crystals in the apatite deposits in Templeton and adjoining townships of Ottawa Co.

Malacon is an altered zircon. *Cyrtolite* is related but contains uranium, yttrium and other rare elements.

Naëgite is apparently zircon with yttrium, niobium-tantalum, thorium, and uranium oxides. Occurs in spheroidal aggregates near Takoyama, Mino, Japan. Color green, gray, brown. H. = 7·5. G. = 4·1.

Orvillite is an altered zircon from Caldas, Minas Geraes, Brazil.

Use. — Zircon in its transparent varieties serves frequently as a gem stone; also as a source of zirconium oxide used in the manufacture of the incandescent gas mantles.

Thorite. — Thorium silicate, $ThSiO_4$, like zircon in form and structure. Distinct prismatic cleavage. H. = 4·5–5. Usually hydrated, black in color, and then with G. = 4·5–5; also orange-yellow and with G. = 5·19–5·40 (*orangite*). Optically +. $\omega = 1·8$. Commonly altered and isotropic with $n = 1·68$–$1·72$. Found in Norway in the augite-syenite on the island of Lövö, opposite Brevik, and at other points in the Langesundfiord district. In large black crystals on the island of Landbö, and at other points near Arendal, Aust-Agder. In Sweden near Lindenäs in Kopparberg, both black thorite and orangite. Orangite and a variety containing uranium oxide, *uranothorite*, have been found at Ambatofotsy southwest of Soavinandriana and elsewhere in Madagascar. Uranothorite noted from Hybla, Ontario.

HYBLITE. Alteration materials of thorite occurring at Hybla, Ontario, in minute quantities have been called *alpha-* and *beta-hyblite*. From qualitative tests they appear to be hydrous basic sulpho-silicates of thorium with minor amounts of uranium, iron and lead. *Alpha-hyblite* is pearly or porcelain white. Isotropic. $n = 1.540$–1.545. *Beta-hyblite* is yellow-brown in color. Resinous luster. Isotropic. $n = 1.605$–1.610.

Auerlite. — Like zircon in form; supposed to be a silico-phosphate of thorium. H. = 2.5–3. G. = 4.1–4.8. Yellow color. Optically $+$. $\omega = 1.67$. Henderson Co., North Carolina.

Danburite-Topaz Group. Orthorhombic. $\overset{II}{R}\overset{III}{R}_2(SiO_4)_2$ or $(\overset{III}{R}O)\overset{III}{R}SiO_4$

Danburite	$CaB_2(SiO_4)_2$	$a : b : c = 0.5444 : 1 : 0.4807$
Topaz	$[Al(F,OH)_2]AlSiO_4$	$a : b : c = 0.5285 : 1 : 0.4770$
Andalusite	$(AlO)AlSiO_4$	$\frac{1}{2}b : a : \frac{2}{3}c = 0.5070 : 1 : 0.4749$
		or $a : b : c = 0.9861 : 1 : 0.7025$

Sillimanite	Al_2SiO_5	Orthorhombic	$a : b : c = 0.970 : 1 : 0.70$
Kyanite	Al_2SiO_5	Triclinic	

$a : b : c = 0.8994 : 1 : 0.7090;\ \alpha = 90° 5\frac{1}{2}',\ \beta = 101° 2',\ \gamma = 105° 44\frac{1}{2}'$.

Andalusite, Sillimanite, and Kyanite constitute trimorphous forms of Al_2SiO_5. The X-ray study of these minerals shows certain close relations in their atomic structures. The unit cells of all three have nearly the same dimensions, which agree in relative values with the axial ratios given above. Parallel to the vertical axes there are chains of aluminum atoms lying in the centers of groups of six oxygen atoms, each group sharing two oxygen atoms with each of the groups above and below. These aluminum-oxygen chains have therefore the composition of AlO_4. The positions of these chains are practically identical in andalusite and sillimanite and nearly the same in the case of kyanite. These vertical chains are linked together by the remaining Al, Si, and O atoms, and the variations in the structure of the three minerals are due to the different ways in which this linking takes place. In all cases the silicon lies between four oxygen atoms, giving in the structure independent SiO_4 groups. The remaining aluminum atom is supposed to lie between six oxygen atoms in kyanite, between five in andalusite, and four in sillimanite.

DANBURITE.

Orthorhombic. Axes $a : b : c = 0.5444 : 1 : 0.4807$.

917

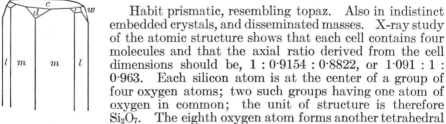

mm''', $110 \wedge 1\bar{1}0 = 57° 8'$. dd', $101 \wedge \bar{1}01 = 82° 53'$.
ll', $120 \wedge 1\bar{2}0 = 85' 8'$. ww', $041 \wedge 0\bar{4}1 = 125° 3'$.

Habit prismatic, resembling topaz. Also in indistinct embedded crystals, and disseminated masses. X-ray study of the atomic structure shows that each cell contains four molecules and that the axial ratio derived from the cell dimensions should be, $1 : 0.9154 : 0.8822$, or $1.091 : 1 : 0.963$. Each silicon atom is at the center of a group of four oxygen atoms; two such groups having one atom of oxygen in common; the unit of structure is therefore Si_2O_7. The eighth oxygen atom forms another tetrahedral group with one oxygen atom from each of three Si_2O_7 groups, and in the center of such groups lie the boron atoms.

Cleavage: $c(001)$ very indistinct. Fracture uneven to subconchoidal. Brittle. H. = 7–7·25. G. = 2·97–3·02. Color pale wine-yellow to colorless, yellowish white, dark wine-yellow, yellowish brown. Luster vitreous to greasy, on crystal surfaces brilliant. Transparent to translucent. Streak white. Strong dispersion, being optically − for red, yellow, and green light but + for blue light. Ax. pl. ∥ (001); $X = b$ axis. $2V = 88°$ to $90°$. $\alpha = 1·630$. $\beta = 1·633$. $\gamma = 1·636$.

Comp. — $CaB_2(SiO_4)_2$ or $CaO.B_2O_3.2SiO_2$ = Silica 48·8, boron trioxide 28·4, lime 22·8 = 100.

Pyr., etc. — B.B. fuses at 3·5 to a colorless glass, and imparts a green color to the O.F. (boron). Not decomposed by hydrochloric acid, but sufficiently attacked for the solution to give the reaction of boric acid with turmeric paper. When previously ignited gelatinizes with hydrochloric acid. Phosphoresces on heating, giving a reddish light.

Obs. — Occurs at Danbury, Fairfield Co., Connecticut, with microcline and oligoclase in dolomite. At Russell, St. Lawrence Co., New York, in fine crystals. On the Piz Valatscha, the northern spur of Mt. Scopi, east of Lukmanier in Val Medels, Uri, Switzerland, in slender prismatic crystals and elsewhere in Switzerland. In crystals from Obira, Bungo, Japan. In Madagascar at Maharitra on Mt. Bity; at Imalo near Mania, south of Betafo district; and in large crystals at Sahasonjo, northeast of Andina.

BARSOWITE. This doubtful species, occurring with blue corundum in the Ural Mts., is by some authors classed with danburite; composition $CaAl_2Si_2O_8$ like anorthite.

TOPAZ.

Orthorhombic. Axes $a : b : c = 0·52854 : 1 : 0·47698$.

mm''',	$110 \wedge 1\bar{1}0 =$	$55° 43'$.	yy',	$041 \wedge 0\bar{4}1 =$	$124° 41'$.	uu',	$111 \wedge \bar{1}11 =$	$78° 20'$.
ll',	$120 \wedge \bar{1}20 =$	$86° 49'$.	ci,	$001 \wedge 223 =$	$34° 14'$.	uu''',	$111 \wedge 1\bar{1}1 =$	$39° 0'$.
dd',	$201 \wedge \bar{2}01 =$	$122° 1'$.	cu,	$001 \wedge 111 =$	$45° 35'$.	oo',	$221 \wedge \bar{2}21 =$	$105° 7'$.
XX',	$043 \wedge 0\bar{4}3 =$	$64° 55'$.	co,	$001 \wedge 221 =$	$63° 54'$.	oo''',	$221 \wedge 2\bar{2}1 =$	$49° 37\frac{1}{2}'$.
ff',	$021 \wedge 0\bar{2}1 =$	$87° 18'$.						

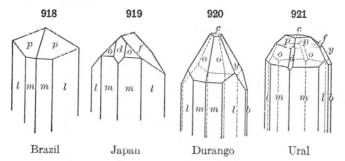

918   919   920   921

Brazil   Japan   Durango   Ural

Crystals commonly prismatic, $m(110)$ predominating; or $l(120)$ and the form then a nearly square prism resembling andalusite. Faces in the prismatic zone often vertically striated, and often showing vicinal planes. Also firm columnar; granular, coarse or fine. X-ray study of the atomic structure reveals the following facts: the SiO_4 groups are independent of each other; the aluminum atoms are linked to four oxygen atoms and two (F,OH) groups; each oxygen atom is linked to one silicon and two aluminum atoms; the (F,OH) groups lie between two aluminum atoms; there are four molecules in the unit cell, the dimensions of which are relatively 4·64, 8·78, and 8·38.

Cleavage: $c(001)$ highly perfect. Fracture subconchoidal to uneven. Brittle. H. = 8. G. = 3·4–3·6. Luster vitreous. Color straw-yellow, wine-yellow, white, grayish, greenish, bluish, reddish. Streak uncolored.

Transparent to subtranslucent. Optically +. Ax. pl. ‖ $b(010)$. Bx. ⊥ $c(001)$. Axial angles variable. $2V = 49°$ to $66°$. Dispersion, $\rho > v$, distinct. Indices, etc., vary with the varying amounts of hydroxyl and fluorine present. Refractive indices, Brazil:

For D $\qquad \alpha = 1.6294 \qquad \beta = 1.6308 \qquad \gamma = 1.6375$

Var. — *Ordinary.* In prismatic crystals usually colorless or pale yellow, less often pale blue, pink, etc. The yellow of the Brazilian crystals is changed by heating to a pale rose-pink. Often contains inclusions of liquid CO_2.

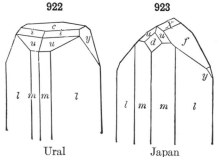

922   923

Ural   Japan

Physalite, or *pyrophysalite,* is a coarse nearly opaque variety, from Finbo, Sweden; intumesces when heated, hence its name from φυσαλίς, *bubble,* and πῦρ, *fire.* *Pycnite* has a columnar, very compact structure. Rose made out that the cleavage was the same, and the form probably the same; and Des Cloizeaux showed that the optical characters were those of topaz.

Comp. — $(AlF)_2SiO_4$; usually containing hydroxyl and then $[Al(F,OH)]_2$ SiO_4 or as given on p. 612. The former requires Silica 32·6, alumina 55·4, fluorine 20·7 = 108·7, deduct (O = 2F) 8·7 = 100.

Pyr., etc. — B.B. infusible. Fused in the closed tube, with potassium bisulphate gives the characteristic fluorine reactions. With cobalt solution the pulverized mineral gives a fine blue on heating. Only partially attacked by sulphuric acid. A variety of topaz from Brazil, when heated, assumes a pink or red hue, resembling the Balas ruby.

Diff. — Characterized by its prismatic crystals with angles of 56° (124°) or 87° (93°); also by the perfect basal cleavage; hardness; infusibility; yields fluorine B.B.

Artif. — Topaz has been made artificially by heating a mixture of silica and aluminum fluoride and then igniting this mixture in silicon fluoride gas.

Obs. — Topaz occurs especially in the highly acid igneous rocks of the granite family, as granite and rhyolite, in veins and cavities, where it appears to be the result of pneumatolytic action after the crystallization of the magma; sometimes also in the surrounding schists, gneisses, etc., as a result of such action. In these occurrences often accompanied by fluorite, cassiterite, tourmaline. Frequently occurs in tin-bearing pegmatites. Topaz alters easily into a compact mass of muscovite.

In magnificent specimens from the Adun-Chilon Mts., south of Nerchinsk in Transbaikalia in crystals of extraordinary size and perfection, transparent with a wine-yellow color. Fine topaz comes from the Ural Mts., Russia, at Alabaschka near Mursinsk north of Ekaterinburg; also from Miask in the Ilmen Mts.; and from the gold washings on the river Sanarka in the Orenburg district. *Pycnite* comes from the tin mines of Schlaggenwald (Slavkov) and Zinnwald (Cinvald) in Bohemia of Czechoslovakia and similarly in Saxony at Ehrenfriedersdorf and in smaller crystals at Altenberg and at Schneckenstein southeast of Muldenberg in Vogtland. Small limpid crystals found in the Mourne Mts., Co. Down, Ireland. *Physalite* occurs in crystals of great size, at Fossum near Modum, Norway, and in Sweden at Finbo near Falun, Kopparberg, and at nearby Broddbo. In Northern Nigeria at Jos in Bauchi; from Ceylon; from Mt. Bischoff, Tasmania, with tin ores; similarly at Emmaville, New South Wales. Fine crystals are found in Japan in Omi Province at Tanokamiyama and in Mino Province at Takayama. In the province of Minas Geraes, Brazil, at Ouro Preto and neighboring Villa Rica, of deep yellow color. Topaz also occurs in Mexico in Durango and San Luis Potosi.

In the United States at Stoneham, Oxford Co., Maine. In New Hampshire on Bald Face Mt., North Chatham, Carroll Co. In Connecticut at Trumbull, Fairfield Co. In Colorado in the Pike's Peak region, El Paso Co., at Crystal Peak, and Cheyenne Mt., etc., in fine colorless or pale blue crystals, and at Topaz Butte near Florissant; also at Nathrop, Chaffee Co. In Texas in fine crystals at Streeter, Mason Co. In the Thomas Mts., Juab Co., Utah, fine transparent crystals in rhyolite. In California from Ramona, San Diego Co.

The name topaz is from τσπάξιος, an island in the Red Sea, as stated by Pliny. But the topaz of Pliny was not the true topaz, as it " yielded to the file." Topaz was included by Pliny and earlier writers, as well as by many later, under the name *chrysolite*.

Use. — As a gem stone.

ANDALUSITE.

Orthorhombic. Axes $a : b : c = 0.9861 : 1 : 0.70245$.

$$mm''', 110 \wedge 1\bar{1}0 = 89° 12'.$$
$$ss', \quad 011 \wedge 0\bar{1}1 = 70° 10'.$$

924 **925**

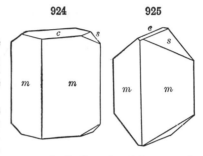

Usually in coarse prismatic forms, the prisms nearly square in form. Massive, imperfectly columnar; sometimes radiated and granular. For a discussion of the atomic structure see p. 612.

Cleavage: $m(110)$ distinct, sometimes perfect (Brazil); $a(100)$ less perfect; $b(010)$ in traces. Fracture uneven, subconchoidal. Brittle. H. = 7·5. G. = 3·16–3·20. Luster vitreous; often weak. Color whitish, rose-red, flesh-red, violet, pearl-gray, reddish brown, olive-green. Streak uncolored. Transparent to opaque, usually subtranslucent. Pleochroism strong in some colored varieties. Absorption strong, $X > Y > Z$. Sections normal to an optic axis are idiophanous or show the polarization-brushes distinctly (p. 317). Optically −. Ax. pl. ‖ $b(010)$. Bx. ⊥ $c(001)$. 2V = 85°. $\alpha = 1.634$. $\beta = 1.639$. $\gamma = 1.643$. *Manganandalusite* or *viridine* (considered by some authors to be a distinct species) is optically +; $X = a$ axis. $\alpha = 1.66, \beta = 1.67, \gamma = 1.69$. Dark green color and pleochroic yellow to green.

926

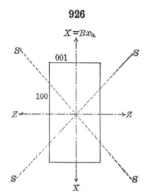

Var. — *Chiastolite*, or *Macle* is a variety in stout crystals having the axis and angles of a different color from the rest, owing to a regular arrangement of carbonaceous impurities through the interior, and hence exhibiting a colored cross, or a tesselated appearance in a transverse section. Fig. 927 shows sections of a crystal. *Viridine* is a green variety containing some iron and manganese from near Darmstadt, Germany.

Comp. — $Al_2SiO_5 = (AlO)AlSiO_4$ or $Al_2O_3.SiO_2 =$ Silica 36·8, alumina 63·2 = 100. Manganese is sometimes present, as in *manganandalusite* or *viridine*.

927

Andalusite when heated to 1400° C. changes to a mixture of *mullite*, $3Al_2O_3.2SiO_2$ and silica.

Pyr., etc. — B.B. infusible. With cobalt solution gives a blue color after ignition. Not decomposed by acids. Decomposed on fusion with caustic alkalies and alkaline carbonates.

Diff. — Characterized by the nearly square prism, pleochroism, hardness, infusibility; reaction for alumina B.B.

Micro. — Distinguished in thin sections by its high relief; low interference-colors, which are only slightly above those of quartz; negative biaxial character; negative extension of the crystals (diff. from sillimanite); rather distinct prismatic cleavage and the constant parallel extinction (diff. from pyroxenes, which have also greater birefringence); also by its characteristic arrangement of impurities when these are present (Fig. 927). The pleochroism, which is often lacking, is, when present, strong and characteristic.

Obs. — Andalusite is most common as a contact mineral in clay slates and argillaceous schists, especially in connection with granitic intrusions. It also occurs in gneiss, mica schist, and related rocks, associated with sillimanite and kyanite, with iolite, garnet, corundum, and tourmaline. Rarely in granite. Sometimes intergrown with sillimanite with parallel axes.

From Nankova in the Nerschinsk district in Transbaikalia (*chiastolite*); in the Ural Mts., at Mursinsk north of Ekaterinburg. In large crystals with kyanite at Lisens Alp in Selraintal, Tyrol, Austria. At Tillenburg in the Böhmer Wald near Neualbenreuth east of Waldsassen, Bavaria. In France andalusite is abundant in the rocks of Var; also in Ariège and elsewhere in the Pyrenees. Found in Andalusia in Spain. Remarkable crystals of chiastolite come from Mt. Howden, near Bimbowrie, South Australia. In Brazil, province of Minas Geraes, in fine crystals and rolled pebbles.

In the United States in fine pink crystals with pyrrhotite at Standish, Cumberland Co., Maine. In Massachusetts at Westford, Middlesex Co., abundant in crystals and in Worcester Co., at Sterling, chiastolite and at Lancaster both varieties. In Delaware Co., Pennsylvania, in large crystals at Leiperville and Upper Providence. Occurs in commercial quantity in the Inyo Range, Mono Co., California.

Named from Andalusia, the first locality noted. The name *macle* is from the Latin *macula*, a spot. Chiastolite is from χιάστος, *arranged diagonally*, and hence from *chi*, the Greek name for the letter X.

Use. — When clear and transparent may serve as a gem stone. Also used in the manufacture of the porcelain of spark plugs.

SILLIMANITE. Fibrolite.

Orthorhombic. Axes $a : b : c = 0.970 : 1 : 0.70$. $mm''' \, 110 \wedge 1\bar{1}0 = 88°$ 15', $hh' \, 230 \wedge \bar{2}30 = 69°$. Prismatic faces striated and rounded. Commonly in long slender crystals not distinctly terminated; often in close parallel groups, passing into fibrous and columnar massive forms; sometimes radiating. For atomic structure see p. 612.

Cleavage: $b(010)$ very perfect. Fracture uneven. H. = 6-7. G. = 3.23-3.24. Luster vitreous, approaching subadamantine. Color hair-brown, grayish brown, grayish white, grayish green, pale olive-green. Streak uncolored. Transparent to translucent. Pleochroism sometimes distinct. Optically +. Double refraction strong. Ax. pl. ∥ $b(010)$. Bx. ⊥ $a(100)$. Dispersion $\rho > v$. Axial angle and indices variable. 2V = 25° (approx.). $\alpha = 1.638-1.659$, $\beta = 1.642-1.660$, $\gamma = 1.653-1.680$.

Comp. — $Al_2SiO_5 = (AlO)AlSiO_4$, like andalusite. Silica 36.8, alumina 63.2 = 100.

Sillimanite is the most stable of the three aluminum silicates. But when heated to temperatures above 1600° C. changes to a mixture of mullite, $3Al_2O_3.2SiO_2$, and silica.

Pyr. — Same as andalusite.

Diff. — Characterized by its fibrous or columnar form; perfect cleavage; infusibility; reaction for alumina.

Micro. — In thin sections recognized by its form, usually with transverse fractures; parallel extinction; high interference-colors.

Artif. — Sillimanite has been made artificially by fusing its oxides together.

Obs. — Often present in the quartz of gneisses and sometimes of granites in very slender, minute prisms commonly aggregated together and sometimes intergrown with andalusite; cordierite is also a common associate; rarely as a contact mineral; often occurs with corundum.

Sillimanite has been observed in many localities. In the Tyrol on the Lisens Alp in Selraintal (*bucholzite*); from Matschendorf near Zöptau in Moravia of Czechoslovakia; from Bodenmais in Bavaria. Sillimanite is noted at many points in France, as in Ariège, and near Pontgibaud, Puy-de-Dôme, etc. From the Khasi Hills, Assam, India; at various points in Madagascar. Found at Mogok in Upper Burma and in Ceylon in gem quality with a pale sapphire-blue color.

In the United States sillimanite occurs in Connecticut at the falls of the Yantic, near Norwich, New London Co.; at Willimantic, Windham Co.; at Chester near Saybrook, Middlesex Co. In New York at Yorktown, Westchester Co., in distinct crystals; in Monroe, Orange Co. (*monroelite*). In Delaware Co., Pennsylvania, at Chester; at Mineral Hill, etc. In Delaware at Brandywine Springs. With corundum at the Culsagee mine, Macon Co., North Carolina.

Named *fibrolite* from the fibrous massive variety; *sillimanite*, after Prof. Benjamin Silliman of New Haven (1779–1864).

Bamlite, xenolite, wörthite probably belong to sillimanite; the last is altered.

Mullite. — $3Al_2O_3.2SiO_2$. Orthorhombic. In prisms, $m \wedge m''' = 89° 13'$. Cleavage $\|$ (010). Optically $+$. Ax. pl. $\|$ (010). $Z = c$ axis. $\alpha = 1.642$, $\gamma = 1.654$. $2V = 45°$–$50°$. Originally noted in artificial melts and in porcelains. Also formed when kyanite, andalusite, or sillimanite are heated to high temperatures. Found in fused argillaceous inclusions in Tertiary lavas on the Island of Mull, Scotland.

KYANITE. Cyanite. Disthene.

Triclinic. Axes $a : b : c = 0.8994 : 1 : 0.7090$; $\alpha = 90° 5\frac{1}{2}'$, $\beta = 101° 2'$, $\gamma = 105° 44\frac{1}{2}'$. $ac, 100 \wedge 001 = 78° 30'$; $bc, 010 \wedge 001 = 86° 45'$.

Usually in long bladed crystals, rarely terminated. Also coarsely bladed columnar to subfibrous. See p. 612 for discussion of atomic structure. See further under staurolite, p. 638, for structural relations to that mineral.

928

Cleavage: $a(100)$ very perfect; $b(010)$ less perfect; also parting $\| c(001)$. H. = 5–7.25; the least, 4–5, on $a(100) \| c$ axis; 6–7 on $a(100) \|$ edge $a(100)/c(001)$; 7 on $b(010)$. G. = 3.56–3.67. Luster vitreous to pearly. Color blue, white; blue along the center of the blades or crystals with white margins; also gray, green, black. Streak uncolored. Translucent to transparent. Pleochroism distinct in colored varieties. Optically $-$. Ax. pl. nearly $\perp a(100)$ and inclined to edge a/b on a about 30°, and about $7\frac{1}{2}°$ on $b(010)$, cf. Fig. 928. X nearly $\perp (100)$. $2V = 82°$. $\alpha = 1.717$. $\beta = 1.722$. $\gamma = 1.729$.

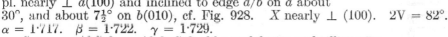

Comp. — Al_2SiO_5 or $Al_2O_3.SiO_2$, like andalusite and sillimanite.

When heated to above 1200° C. changes to a mixture of mullite, $3Al_2O_3.2SiO_2$, and silica.

Pyr., etc. — Same as for andalusite.

Diff. — Characterized by the bladed form; common blue color; varying hardness; infusibility; reaction for alumina.

Obs. — Occurs principally in gneiss and mica schist (both the ordinary variety with muscovite and also that with paragonite) often accompanied by garnet and sometimes by staurolite or corundum.

Some of the more important localities for its occurrence follow. From near Ekaterinburg, Ural Mts., Russia; Petschau, south of Karlsbad, Bohemia of Czechoslovakia; in Carinthia of Austria on the Sua-Alpe; in the Tyrol at Prägratten and on Mt. Greiner in the Zillertal and in Trentino, Italy, in the Pfitschtal (*rhœtizite*, a white variety). In Ticino, Switzerland, in the St. Gotthard region, especially on Pizzo Forno, Mte. Campione, in transparent crystals in a paragonite schist. In fine specimens from the neighborhood of Pontivy, Morbihan, France. In the United States at Chesterfield, Hampshire Co., Massachusetts; in Connecticut at Litchfield, Litchfield Co.; at Newtown, Fairfield Co., at Canton, Hartford Co., etc. In North Carolina at Crowder's Mountain, Gaston Co., and in beautiful clear green crystals in Yancey Co., etc. Named from κυανός, blue.

Datolite Group. Monoclinic

Basic Orthosilicates. $H\overset{\scriptscriptstyle II}{R}\overset{\scriptscriptstyle III}{R}SiO_5$ or $\overset{\scriptscriptstyle II}{R}_3\overset{\scriptscriptstyle III}{R}_2(SiO_5)_2$.　Oxygen ratio for R : Si $= 3 : 2$

$\overset{\scriptscriptstyle II}{R}$ = Ca,Be,Fe, chiefly;　$\overset{\scriptscriptstyle III}{R}$ = Boron, the yttrium (and cerium) metals, etc.

	$a : b : c$	β
Datolite	$0{\cdot}6345 : 1 : 1{\cdot}2657$	$89°\ 51'$
HCaBSiO$_5$ or Ca(BOH)SiO$_4$		
Homilite	$0{\cdot}6249 : 1 : 1{\cdot}2824$	$89°\ 21'$
Ca$_2$FeB$_2$Si$_2$O$_{10}$ or Ca$_2$Fe(BO)$_2$(SiO$_4$)$_2$		
	$2a : b : 4c$	
Euclase	$0{\cdot}6474 : 1 : 1{\cdot}3330$	$79°\ 44'$
HBeAlSiO$_5$ or Be(AlOH)SiO$_4$	a	
Gadolinite	$0{\cdot}6273 : 1 : 1{\cdot}3215$	$89°\ 26\frac{1}{2}'$
Be$_2$FeY$_2$Si$_2$O$_{10}$ or Be$_2$Fe(YO)$_2$(SiO$_4$)$_2$		

The species of the DATOLITE GROUP are usually regarded as basic ortho-silicates, the formulas being taken in the second form given above.　They all crystallize in the monoclinic system, and all but euclase conform closely in axial ratio; X-ray study, however, shows close structural relationship between datolite and euclase.

DATOLITE.

Monoclinic.　Axes $a : b : c = 0{\cdot}6345 : 1 : 1{\cdot}2657$; $\beta = 89°\ 51\frac{1}{3}'$.

mm''', $110 \wedge 1\bar{1}0 = 64°\ 47'$.	cn, $001 \wedge 111 = 66°\ 57'$.
ac,　$100 \wedge 001 = 89°\ 51'$.	cm, $001 \wedge 110 = 89°\ 53'$.
ax,　$100 \wedge 101 = 45°\ 0'$.	$c\epsilon$, $001 \wedge \bar{1}12 = 49°\ 49'$.
gg',　$012 \wedge 0\bar{1}2 = 64°\ 39\frac{1}{2}'$.	nn', $111 \wedge 1\bar{1}1 = 59°\ 4\frac{1}{2}'$.
m_xm_x', $011 \wedge 0\bar{1}1 = 103°\ 23'$.	$\epsilon\epsilon'$, $\bar{1}12 \wedge \bar{1}\bar{1}2 = 48°\ 19\frac{1}{2}'$.

Crystals varied in habit; usually short prismatic with either $m(110)$ or $m_x(011)$ predominating; sometimes tabular $\parallel$ $x(201)$; also of other types,

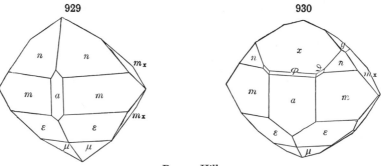

929　　　　　　　　　　　　　　　930

Bergen Hill

and often highly modified (Figs. 929–932).　Also botryoidal and globular, having a columnar structure; divergent and radiating; sometimes massive, granular to compact and crypto-crystalline.

Cleavage not observed.　Fracture conchoidal to uneven.　Brittle. H. $= 5$–$5{\cdot}5$.　G. $= 2{\cdot}9$–$3{\cdot}0$.　Luster vitreous, rarely subresinous on a surface of fracture.　Color white; sometimes grayish, pale green, yellow, red, or amethystine, rarely dirty olive-green or honey-yellow. Streak white. Transparent to translucent; rarely opaque white. Optically −. Ax. pl. ∥ (010) and Z nearly ∥ c axis. $2V = 74°$. $\alpha = 1·626$. $\beta = 1·654$. $\gamma = 1·670$.

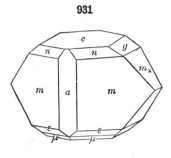

931

Bergen Hill

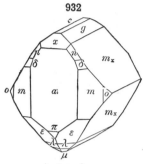

932

Andreasberg

Var. — 1. *Ordinary.* In glassy crystals of varied habit, usually with a greenish tinge. The angles in the prismatic and clinodome zones vary but little, *e.g.*, 110 ∧ 1̄10 = 64° 47′, while 011 ∧ 01̄1 = 66° 37′, etc. 2. *Compact massive.* Opaque, white, cream-colored, pink; breaking with the surface of porcelain or Wedgwood ware. From the Lake Superior region. 3. *Botryoidal; Botryolite.* Radiated columnar, having a botryoidal surface, and containing more water than the crystals, but optically identical.

Comp. — A basic orthosilicate of boron and calcium; empirically $HCaBSiO_5$ or $H_2O.2CaO.B_2O_3.2SiO_2$; this may be written $Ca(BOH)SiO_4 =$ Silica 37·6, boron trioxide 21·8, lime 35·0, water 5·6 = 100.

Pyr., etc. — In the closed tube gives off much water. B.B. fuses at 2 with intumescence to a clear glass, coloring the flame bright green. Gelatinizes with hydrochloric acid.
Diff. — Characterized by its glassy, greenish, complex crystals; easy fusibility and green flame, B.B.
Obs. — Datolite is found chiefly as a secondary mineral in veins and cavities in basic eruptive rocks, often associated with calcite, prehnite and various zeolites; sometimes associated with danburite; also in gneiss, diorite, and serpentine; in metallic veins; sometimes in beds of iron ore. Found at Andreasberg in the Harz Mts., in diabase and in veins of silver ore. In Italy from Trentino on the Seiser Alpe (Mont da Sous) and at Tiso (Theiss) northeast of Chiusa (Klausen) in geodes in amygdaloid; from Piedmont in granite at Baveno on Lago Maggiore; in Emilia at Toggiano near Modena in serpentine and from Serra del Zanchetti in the Apennines near Bologna in highly complex crystals. In Norway from Arendal in Aust-Agder in a bed of magnetite (*botryolite*). From Rosebery, Montagu Co., Tasmania.

In the United States not uncommon with the diabase of the Connecticut River valley as at Westfield, Hampden Co., Massachusetts; in Connecticut in Hartford Co., in large crystals at Tariffville and from near Hartford; at Meriden, New Haven Co., etc. In New Jersey in the trap rocks of Bergen Hill, Great Notch, Paterson, etc., in splendid crystals and associated with various zeolites, etc. In Michigan in both crystalline and compact opaque varieties associated with the trap rocks of the copper district of Keweenaw Co.

Named from δατὲισθαι, *to divide*, alluding to the granular structure of a massive variety.
Homilite. — $(Ca,Fe)_3B_2Si_2O_{10}$ or $(Ca,Fe)_3(BO)_2(SiO_4)_2$. Crystals often tabular ∥ $c(001)$; angles near those of datolite. H. = 5. G. = 3·38. Easily fusible. Color black, blackish brown. Optically +. Ax. pl. ⊥ (010). $Z = b$ axis; Y nearly ∥ c axis. $\alpha = 1·715, \beta = 1·725, \gamma = 1·738$. Altered material may be isotropic or anisotropic with variable optical properties. Found on Stokö and other islands, in the Langesundfiord, southern Norway.
Euclase. — $HBeAlSiO_5$ or $Be(AlOH)SiO_4$. In prismatic crystals. Cleavage ∥ $b(010)$ perfect. H. = 7·5. G. = 3·05–3·10. Luster vitreous. Colorless to pale green or blue. Optically +. Ax. pl. ∥ (010). $2V = 50°$. $\alpha = 1·652, \beta = 1·655, \gamma = 1·671$. Found in the auriferous sands in the Orenburg district of the southern Ural Mts., near the river San-

arka.   In Austria on Grieswies-Alpe in Rauris-Tal, Salzburg, and in Carinthia in the Gross-Glockner region, also from the Möll-Tal.   In pegmatite from Epprechstein, west Kirchen-lamitz, Fichtelgebirge, Bavaria.   From Minas Geraes, Brazil, at Boa Vista, etc., near Ouro Preto.

Gadolinite. — $Be_2FeY_2Si_2O_{10}$ or $Be_2Fe(YO)_2(SiO_4)_2$.   May contain considerable cerium oxide (*cergadolinite*).   The yttrium earths or " gadolinite-earths " (partly replaced by the oxides of cerium, lanthanum and didymium) form a complex group which contains consid-erable erbium, ytterbium, scandium, etc.   Crystals often prismatic, rough and coarse; commonly in masses.   Cleavage none.   Fracture conchoidal or splintery.   Brittle.   H. = 6·5–7.   G. = 4·0–4·5; normally 4·36–4·47 (anisotropic), 4·24–4·29 (isotropic and amorphous from alteration).   Luster vitreous to greasy.   Color black, greenish black, also brown.   Optically +.   Ax. pl. || (010). $Z \wedge c$ axis = 6° to 14°.   2V = 85°.   Indices variable, 1·77–1·82.   Becomes isotropic on alteration.   Occurs principally in pegmatite veins, often asso-ciated with allanite and other minerals containing rare elements, also with fluorine com-pounds.   From Kopparberg, Sweden, at the quarries of Finbo, Korarfvet, and Broddbo near Falun and also at Ytterby northeast of Stockholm.   In Norway on the island of Hiterö in Vest-Agder in large crystals and in the Sätersdal district of Vest-Agder.   In Llano Co., Texas, 5 miles south of Bluffton on the west bank of the Colorado River.   Here it occurs in nodular masses and rough crystals, sometimes up to 40 or 60 pounds in weight.   It is em-bedded in a quartzose pegmatite and is associated with allanite, yttrialite, nivenite, ferguson-ite, cyrtolite, gummite, etc.   Reported also from Mohave Co., Arizona, and from near Florissant, Teller Co., Colorado.

Yttrialite. — A silicate of thorium and the yttrium metals chiefly.   Massive; amor-phous.   G. = 4·575.   Color on the fresh fracture olive-green, changing to orange-yellow on surface.   Isotropic. $n = 1·758$.   Associated with the gadolinite of Llano Co., Texas.

Rowlandite. — An yttrium silicate, occurring massive with gadolinite of Llano Co., Texas; color drab-green.   H. = 6–7.   G. = 4·5.   Isotropic. $n = 1·725$.

Thalenite. — An yttrium silicate; $Y_2Si_2O_7$.   In tabular or prismatic monoclinic crystals.   H. = 6·5.   G. = 4·2.   Color flesh-red.   Optically −.   Ax. pl. ⊥ (010). Y nearly || c axis. $\alpha = 1·731$, $\beta = 1·738$, $\gamma = 1·744$.   Found in Sweden at Österby in Upsala and at Åskagen in Vermland.   Also from Hundholmen, northern Norway.

Thortveitite. — A silicate of the yttrium metals, $(Sc,Y)_2Si_2O_7$.   Monoclinic.   Habit prismatic.   Prismatic cleavage.   Usually twinned on $m(110)$.   H. = 6–7.   G. = 3·57.   Color grayish green to black.   Optically −. $\alpha = 1·756$, $\beta = 1·793$, $\gamma = 1·809$.   Ax. pl. (010). $X \wedge c$ axis = 5°.   2V = 65°.   Pleochroic; X = deep green, $Y = Z$ = brownish yellow.   In granite pegmatite in Iveland parish, Sätersdalen, Norway.   *Befanamite* is sim-ilar to thortveitite, except that it contains no yttrium and has a high content of Zr and Al.   From Befanamo, Madagascar.

Mackintoshite. — Silicate of uranium, thorium, cerium, etc.   Tetragonal.   Massive.   H. = 5·5.   G. = 5·4.   Color black.   Isotropic. $n = 1·77$.   Llano Co., Texas.

THOROGUMMITE.   Silicate of uranium, thorium, cerium, etc.   An alteration product of mackintoshite.   H. = 4.   G. = 4·5.   Color dark yellow-brown.   From Llano Co., Texas.   Two amorphous minerals from Wodgina, N. W. Division, West Australia, have been de-scribed as similar to mackintoshite and thorogummite but containing in addition some lead and calcium.   They have been named *maitlandite* (black) and *nicolayite* (yellow).

HYDROTHORITE.   A hydrous silicate of thorium.   Color pale pinkish buff.   Isotropic. $n = 1·638$.   Strongly radioactive.   H. = 1–2.   An alteration product of mackintoshite found in a tantalite-bearing pegmatite at Wodgina, Western Australia.

Epidote Group.   Orthorhombic and Monoclinic

$$\text{Basic Orthosilicates, } H\overset{\text{II III}}{R_2}R_3Si_3O_{13} \text{ or } \overset{\text{II}}{R_2}(ROH)\overset{\text{III}}{R_2}(SiO_4)_3$$

$$\overset{\text{II}}{R} = Ca,Fe; \quad \overset{\text{III}}{R} = Al,Fe,Mn,Ce, \text{ etc.}$$

α.   Orthorhombic Section

Zoisite	$Ca_2(AlOH)Al_2(SiO_4)_3$	$a : b : c$
		$0·6196 : 1 : 0·3429$

β. Monoclinic Section

		$a : b : c$	β
Clinozoisite	$Ca_2(AlOH)Al_2(SiO_4)_3$	1·583 : 1 : 1·814	64° 30′
Epidote	$\begin{cases} mCa_2(AlOH)Al_2(SiO_4)_3 \\ nCa_2(FeOH)Fe_2(SiO_4)_3 \end{cases}$	1·5787 : 1 : 1·8036	64° 37′
Piedmontite	$Ca_2(AlOH)(Al,Mn)_2(SiO_4)_3$	1·6100 : 1 : 1·8326	64° 39′
Allanite	$(Ca,Fe)_2(AlOH)(Al,Ce,Fe)_2(SiO_4)_3$	1·5509 : 1 : 1·7691	64° 59′

The EPIDOTE GROUP includes the above complex orthosilicates. The monoclinic species agree closely in form. To them the orthorhombic species zoisite is also related in angle, its prismatic zone corresponding to the monoclinic orthodomes, etc. Thus we have:

Zoisite mm''', 110 $\wedge$ 1$\bar{1}$0 = 63° 34′.   Epidote cr, 001 $\wedge$ $\bar{1}$01 = 63° 42′.
          uu', 021 $\wedge$ 0$\bar{2}$1 = 68° 54′. mm', 110 $\wedge$ $\bar{1}$10 = 70° 4′, etc.

X-ray studies of zoisite and epidote show close relationships between their atomic structures. The elementary cell of zoisite contains four molecules whereas that of epidote contains but two. The respective cells have two dimensions practically identical while the third dimension in zoisite is twice the corresponding dimension in epidote. This leads to a crystal orientation of zoisite in which axis b (as given above) becomes a, c becomes b, and a becomes c; so that the axial ratio equals 1 : 0·3429 : 0·6196 or 2·916 : 1 : 1·801. This last expression is practically that of epidote except that the value for a has been doubled.

ZOISITE.

Orthorhombic. Axes $a : b : c$ = 0·6196 : 1 : 0·34295.

    mm''', 110 $\wedge$ 1$\bar{1}$0 = 63° 34′. ff', 011 $\wedge$ 0$\bar{1}$1 = 37° 52′.
    dd', 101 $\wedge$ $\bar{1}$01 = 57° 56′. oo''', 111 $\wedge$ 1$\bar{1}$1 = 33° 24′.

Crystals prismatic, deeply striated or furrowed vertically, and seldom distinctly terminated. Also massive; columnar to compact.

Cleavage: b(010) very perfect. Fracture uneven to subconchoidal. Brittle. H. = 6–6·5. G. = 3·25–3·37. Luster vitreous; on the cleavage-face, b(010), pearly. Color grayish white, gray, yellowish brown, greenish gray, apple-green; also peach-blossom-red to rose-red. Streak uncolored. Transparent to subtranslucent.

Pleochroism strong in pink varieties. Optically +. Ax. pl. usually ‖ b(010) in iron-free zoisite; also ‖ c(001), with about 5 per cent Fe_2O_3. Bx. $\perp$ a(100). Dispersion strong, $\rho < v$ (iron-free); also $\rho > v$ (with 5 per cent Fe_2O_3). Axial angle variable even in the same crystal, increasing in value with increase in iron content. 2V varies widely from about 30° for varieties with little iron to 0° and then increasing again to about 60° with increase in iron content. α = 1·700. β = 1·703. γ = 1·706.

Var. — 1. *Ordinary.* Colors gray to white and brown; also green. Usually in indistinct prismatic or columnar forms; also in fibrous aggregates. G. = 3·226–3·381. *Unionite* is a very pure zoisite. 2. *Rose-red* or *Thulite.* Fragile; pleochroism strong. 3. *Compact, massive.* Includes the essential part of most of the mineral material known as *saussurite* (e.g., in saussurite-gabbro), which has arisen from the alteration of feldspar.

Comp. — $HCa_2Al_3Si_3O_{13}$ or $4CaO.3Al_2O_3.6SiO_2.H_2O$ = Silica 39·7, alumina 33·7, lime 24·6, water 2·0 = 100. The alumina is sometimes replaced by iron, thus graduating in composition toward epidote, which has the same general formula.

Pyr., etc. — B.B. swells and fuses at 3–3·5 to a white blebby mass. Not decomposed by acids; when previously ignited gelatinizes with hydrochloric acid. Gives off water when strongly ignited.

Diff. — Characterized by the columnar structure; fusibility with intumescence; resembles some amphibole.

Micro. — Distinguished in thin sections by its high relief and very low interference-colors; lack of color and biaxial character. From epidote it is distinguished by its lack of color and low birefringence; from vesuvianite by its color and biaxial character. Thin sections frequently show the " ultra blue " (p. 609) between crossed nicols.

Obs. — Probably the most common mode of occurrence of zoisite is in the material known as saussurite. Occurs especially in those crystalline schists which have been formed by the dynamic metamorphism of basic igneous rocks containing plagioclase rich in lime. Commonly accompanies some one of the amphiboles (actinolite, smaragdite, glaucophane, etc.); thus in amphibolite, glaucophane schist, eclogite; often associated with corundum.

The original zoisite is that of the eclogite of the Sau-Alpe in Carinthia (*saualpite*); other localities in Austria are at Rauris in Salzburg; in Tyrol in the Zillertal (*thulite*), near Prägratten and Windish-Matrei. Also in Trentino, Italy, in the Val Passiria (Passeier Tal) and at Sterzen (Sterzing). In Valais, Switzerland, at Zermatt. From Bavaria in the Fichtelgebirge at Gefrees, etc. From Loch Garve, Ross-shire, Scotland. From Juarez, Lower California, Mexico. *Thulite* occurs in Norway in the parish of Souland in Telemark, from Leksviken, Trondhjem, and also at an iron mine near Arendal in Aust-Agder. Further from Traversella in Piedmont, Italy.

In the United States, zoisite is found in Massachusetts at Goshen, Hampshire Co., at Conway, Franklin Co., etc. From Pennsylvania at Corundum Hill near Unionville, Chester Co., white (*unionite*), and at Leiperville, Delaware Co., etc. From the copper mines at Ducktown, Tennessee.

Clinozoisite. — $HCa_2Al_3Si_3O_{13}$, same as for zoisite. It forms a continuous series with epidote, $HCa_2(Al,Fe)_3Si_3O_{13}$. It is customary to separate under the name clinozoisite that portion of the series that contains less than 10 per cent of the iron molecule and is optically positive. Monoclinic, crystal characters like those of epidote. Colorless, light yellow, green, pink. Optical orientation same as for epidote. Indices and birefringence lower than for epidote. Indices for about 3 per cent Fe_2O_3; $\alpha = 1·724$, $\beta = 1·729$, $\gamma = 1·734$. $2V = 85°$. Originally described from the Goslerwand near Prägratten, Tyrol, Austria. Has been noted from other localities in the Tyrol, Piedmont, Bohemia, Switzerland, etc. From Lower California, Mexico. *Fouqueite* is probably the same as clinozoisite from an anorthite-gneiss in Ceylon.

EPIDOTE.   Pistacite.

Monoclinic.   Axes $a : b : c = 1·5787 : 1 : 1·8036$; $\beta = 64° 37'$.

mm''',	$110 \wedge 1\bar{1}0$ =	$109° 56'$.	cl,	$001 \wedge \bar{2}01$ =	$89° 26'$.
ca,	$001 \wedge 100$ =	$64° 37'$.	co,	$001 \wedge 011$ =	$58° 28'$.
ce,	$001 \wedge 101$ =	$34° 43'$.	cn,	$001 \wedge \bar{1}11$ =	$75° 11'$.
cr,	$001 \wedge \bar{1}01$ =	$63° 42'$.	an''',	$100 \wedge 11\bar{1}$ =	$69° 2'$.
ar',	$100 \wedge 10\bar{1}$ =	$51° 41'$.	nn''',	$\bar{1}11 \wedge 11\bar{1}$ =	$70° 29'$.

Twins: tw. pl. $a(100)$ common, often as embedded tw. lamellæ. Crystals usually prismatic ‖ the ortho-axis b and terminated at one extremity only; passing into acicular forms; the faces in the zone $a(100)/c(001)$ deeply striated. Also fibrous, divergent or parallel; granular, particles of various sizes, sometimes fine granular, and forming rock-masses.

Cleavage: $c(001)$ perfect; $a(100)$ imperfect. Fracture uneven. Brittle. H. = 6–7. G. = 3·25–3·5. Luster vitreous; on $c(001)$ inclining to pearly or resinous. Color pistachio-green or yellowish green to brownish green, greenish black, and black; sometimes clear red and yellow; also gray and grayish white, rarely colorless. Streak uncolored, grayish. Transparent to opaque: generally subtranslucent.

Pleochroism strong: vibrations ∥ Z green, Y brown and strongly absorbed, X yellow. Absorption usually $Y > Z > X$; but sometimes $Z > Y > X$ in the variety of epidote common in rocks. Often exhibits idiophanous figures; best in sections normal to an optic axis, but often to be observed in natural crystals (Sulzbach), especially if flattened ∥ $r(\bar{1}01)$. (See p. 317.) Op-

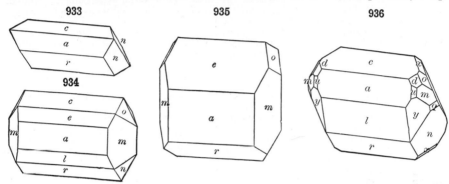

tically −. Ax. pl. ∥ $b(010)$. $X \wedge c$ axis = 0°–5°. Hence $Z \perp a(100)$ nearly. Dispersion inclined, strongly marked; of the axes feeble, $\rho > v$. Axial angle large. $\alpha = 1\cdot729$. $\beta = 1\cdot754$. $\gamma = 1\cdot768$. Indices and birefringence variable, rising in values with increase in iron content.

Var. — Epidote has ordinarily a peculiar yellowish green (pistachio) color, seldom found in other minerals. But this color passes into dark and light shades — black on one side and brown on the other; red, yellow and colorless varieties also occur.

Var. — 1. *Ordinary.* Color green of some shade, as described, the pistachio tint rarely absent. (*a*) In crystals. (*b*) Fibrous. (*c*) Granular massive. (*d*) *Scorza* is epidote sand from the gold washings in Transylvania. The Arendal, Norway, epidote (*arendalite*) is mostly in dark green crystals; that of Bourg d'Oisans, Isère, France (*thallite, delphinite, oisanite*), in yellowish green crystals, sometimes transparent. *Puschkinite* includes crystals from the auriferous sands of Ekaterinburg, Ural Mts. *Achmatite* is ordinary epidote from Achmatovsk, Ural Mts. A variety from Garda, Hoste Island, Terra del Fuego, is colorless and resembles zoisite.

2. The *bucklandite* from Achmatovsk, Ural Mts., described by Hermann, is black with a tinge of green, and differs from ordinary epidote in having the crystals nearly symmetrical and not, like other epidote, lengthened in the direction of the ortho-axis. G. = 3·51.

3. *Withamite.* — Carmine-red to straw-yellow, strongly pleochroic; deep crimson and straw-yellow. H. = 6–6·5; G. = 3·137; in small radiated groups. From Glencoe, in Argyllshire, Scotland. Sometimes referred to piedmontite, but contains little MnO.

4. *Tawmawite* or *chromepidote* is a chromium-bearing epidote from Tawmaw, Kachin Hills, Upper Burma. Deep green color and strong pleochroism, emerald-green to bright yellow.

Comp. — $HCa_2(Al,Fe)_3Si_3O_{13}$ or $H_2O.4CaO.3(Al,Fe)_2O_3.6SiO_2$, the ratio of aluminum to iron varies commonly from 6 : 1 to 3 : 2. Percentage composition:

For Al : Fe = 3 : 1 SiO_2 37·87, Al_2O_3 24·13, Fe_2O_3 12·60, CaO 23·51, H_2O 1·89 − 100. X-ray study shows two such molecules to the unit cell of the structure. *Picroepidote* is supposed to contain Mg in place of Ca.

Pyr., etc. — In the closed tube gives water on strong ignition. B.B. fuses with intumescence at 3–3·5 to a dark brown or black mass which is generally magnetic. Reacts for iron and sometimes for manganese with the fluxes. Partially decomposed by hydrochloric acid, but when previously ignited, gelatinizes with acid. Decomposed on fusion with alkaline carbonates.

Diff. — Characterized often by its peculiar yellowish green (pistachio) color; readily fusible and yields a magnetic globule B.B. Prismatic forms often longitudinally striated, but they have not the angle, cleavage or brittleness of tremolite; tourmaline has no distinct cleavage, is less fusible (in common forms) and usually shows its hexagonal form.

Micro. — Recognized in thin sections by its high refraction; strong interference-colors rising into those of the third order in ordinary sections; decided color and striking pleochroism; also by the fact that the plane of the optic axes lies transversely to the elongation of the crystals.

Obs. — Epidote is commonly formed by the metamorphism (both local igneous and of general dynamic character) of impure calcareous sedimentary rocks or igneous rocks containing much lime. It thus often occurs in gneissic rocks, mica schist, amphibole schist, serpentine; so also in quartzites, sandstones and limestones altered by neighboring igneous rocks. Often accompanies beds of magnetite or hematite in such rocks. It is often associated with quartz, feldspar, actinolite, axinite, chlorite, etc. It results from the alteration of various other minerals, as garnet, hornblende, augite, biotite, scapolite, plagioclase, chrysolite, etc. It has occasionally been observed in such conditions as to suggest its primary origin, as in a granite from Maryland, etc.

Only the more important occurrences of epidote can be given here. It is found at Achmatovsk near Kussinsk in the Zlatoust district of the Ural Mts. In Moravia of Czechoslovakia at Zöptau. In Austria it occurs at the Kappenwand in the Untersulzbachtal, Salzburg, in beautiful crystals with unusual complexity of form, luster, and transparency; in the Tyrol from the Zillertal, sometimes in small rose-red and greenish crystals and from near Prägratten. From Piedmont, Italy, in fine specimens from Traversella and the Ala valley. In Switzerland from the district of St. Gotthard, Ticino, and from Zermatt in Valais. In beautiful crystallizations from Bourg d'Oisans, Isère, France. In Norway from Arendal, Aust-Agder.

In the United States epidote is found in New Hampshire at Franconia and Warren. From Massachusetts at Huntington (formerly Norwich) in Hampshire Co., and elsewhere. In Connecticut near Willimantic, Windham Co., and in large splendid crystals at Haddam, Middlesex Co. In Colorado in Chaffee Co., at Calumet, and in fine specimens from near Salida. In good crystals from Riverside, Riverside Co., California. In extraordinary crystals from Sulzer, Prince of Wales Island, Alaska; also from Ketchikan.

Epidote was named by Haüy, from the Greek ἐπίδοσις, *increase*, translated by him, " qui à recu un accroissement," the base of the prism (rhomboidal prism) having one side longer than the other. *Pistacite*, from πιστακια, *the pistachio-nut*, refers to the color.

Piedmontite. — $HCa_2(Al,Mn)_3Si_3O_{13}$. Mn_2O_3 ranges from 5 to 15 per cent. Monoclinic. Crystal characters like those of epidote. H. = 6·5. G. = 3·404. Color reddish brown and reddish black. Pleochroism strong; X = yellow, Y = violet, Z = red. Optically +. Ax. pl. ‖ $b(010)$. $X \wedge c$ axis = $-6°$ to $-3°$. Indices variable, 1·73–1·82. Occurs with manganese ores at St. Marcel, Piedmont, Italy. In crystalline schists on Ile de Groix, Morbihan, France; from Jacobsberg in the Nordmark district, Vermland, Sweden; in glaucophane-schist in Japan. Occasionally in quartz porphyry, as in the antique red porphyry of Egypt, also that of Adams Co., Pennsylvania.

Sursassite. — $5MnO.2Al_2O_3.5SiO_2.3H_2O$. Monoclinic. A manganiferous member of the Epidote Group. In small radiated botryoidal masses. Crystals elongated ‖ b axis. Color deep reddish brown to copper-red. Ax. pl. ⊥ elongation of needles. Optically $-$. α = 1·736, β = 1·755, γ = 1·766. 2V medium. Strongly pleochroic. X = Z = light yellow, Y = reddish brown. From manganese deposits of the Val d'Err, Grisons, Switzerland.

Hancockite. — Belongs in Epidote Group containing PbO, MnO, CaO, SrO, MgO, Al_2O_3, Fe_2O_3, Mn_2O_3. Crystals which are very small and lath-shaped show characteristic epidote habit and closely related angles. H. = 6–7. G. = 4·0. Brownish red. Optically $-$. Ax. pl. ‖ (010). α = 1·79, β = 1·81, γ = 1·83. 2V = 50°. Found at Franklin, Sussex Co., New Jersey.

ALLANITE. Orthite.

Monoclinic. Axes, p. 621. In angle near epidote. Crystals often tabular ‖ $a(100)$; also long and slender to acicular prismatic by elongation ‖ axis b. Also massive and in embedded grains.

Cleavage: $a(100)$ and $c(001)$ in traces; also $m(110)$ sometimes observed. Fracture uneven or subconchoidal. Brittle. H. = 5·5–6. G. = 3·0–4·2. Luster submetallic, pitchy or resinous. Color brown to black. Subtranslucent to opaque. Pleochroism strong: Z brownish yellow, Y reddish brown, X greenish brown. Optically −. Ax. pl. usually $\parallel$ (010), sometimes $\parallel$ (100). $X \wedge c$ axis $= 32\frac{1}{2}°$ approx. 2V large. Indices vary widely, 1·64–1·80. Birefringence variable. Allanite is easily altered and then the mineral usually becomes isotropic with lower specific gravity and refractive index. Very commonly allanite shows a heterogeneous mixture of unaltered and altered material. Allanite embedded in biotite may produce a pleochroic halo in the latter mineral.

Var. — *Allanite.* The original mineral was from East Greenland, in tabular crystals or plates. Color black or brownish black. G. = 3·50–3·95. *Bucklandite* is anhydrous allanite in small black crystals from a magnetite mine near Arendal, Norway. *Bagrationite* occurs in black crystals which are like the bucklandite of Achmatovsk (epidote).

Orthite included, in its original use, the slender or acicular prismatic crystals, containing some water, from Finbo, near Falun, Sweden. But these graduate into massive forms, and some orthites are anhydrous, or as nearly so as most allanite. The name is from ὀρθός, *straight.*

Comp. — Like epidote $\overset{\text{ii iii}}{HRR_3Si_3O_{13}}$ or $H_2O.4RO.3R_2O_3.6SiO_2$ with $\overset{\text{ii}}{R}$ $=$ Ca and Fe, and $\overset{\text{iii}}{R}$ $=$ Al,Fe, the cerium metals Ce, Di, La, and in smaller amounts those of the yttrium group. Some varieties contain considerable water, but probably by alteration.

Pyr., etc. — Some varieties give much water in the closed tube, and all kinds yield a small amount on strong ignition. B.B. fuses easily and swells (F. = 2·5) to a dark, blebby, magnetic glass. With the fluxes reacts for iron. Most varieties gelatinize with hydrochloric acid, but if previously ignited are not decomposed by acid.

Obs. — Allanite is most characteristically found as an accessory mineral in the deep-seated igneous rocks, as granite, syenite, diorite; also in pegmatites. Further in metamorphic rocks that have been derived from igneous rocks, as gneiss, amphibolite, etc. Occurs in volcanic ejections and has been noted in limestone as a contact mineral. With magnetite bodies. Sometimes inclosed as a nucleus in crystals of the isomorphous species, epidote.

In the Ural Mts., the variety *bagrationite* from Achmatovsk near Kussinsk in the Zlatoust district and *uralorthite* from Miask in the Ilmen Mts. In Saxony allanite occurs at the Plauensche Grund near Dresden and the material called *bodenite* at Boden near Marienberg. In trachytic ejected masses at the Laacher See, Rhineland (*bucklandite* and *tautolite*); similarly at Vesuvius. In Norway *bucklandite* occurs in small black crystals in magnetite near Arendal, Aust-Agder; *orthite* is found at Kragerö in Telemark and at Hiterö, Vest-Agder, etc. *Orthite* occurs in acicular crystals sometimes a foot long at Finbo near Falun, Kopparberg, Sweden; also from Ytterby and Skeppsholm, near Stockholm, and the variety *cerine* at the Bastnaes mine near Riddarhyttan, Vastmanland. Allanite was first noted from eastern Greenland and has since been found at many localities in that country. *Orthite* occurs in large crystals at Sama near the junction of the rivers Sakay and Kitsamby, and elsewhere in Madagascar.

In the United States allanite occurs in New York in large crystals with magnetite at Moriah, Essex Co., and in Orange Co. at Monroe and at Edenville near Warwick. At Franklin, Sussex Co., New Jersey. In Pennsylvania in Chester Co., and elsewhere. In large masses in Amherst Co., Virginia, and also from Amelia Court House, Amelia Co. At the gadolinite locality in Llano Co., Texas.

NAGATELITE. A silicate and phosphate of aluminum, rare earths, calcium and iron. Monoclinic. In small prismatic crystals or tabular masses. H. = 5·5. G. = 3·91. Radioactive. Color black. Resinous luster. Ax. pl. $\parallel$ (010). $\alpha = 1·750, \beta = 1·760, \gamma = 1·765$. Pleochroic; X = brownish yellow, Y = reddish brown, Z = pale yellow. Found in pegmatite near Nagatejima, a small headland on the Noto peninsula, Japan. Considered to be a member of the Epidote Group, related to allanite.

AXINITE.

Triclinic.   Axes $a : b : c = 0.4921 : 1 : 0.4797$; $\alpha = 82° 54'$, $\beta = 91° 52'$, $\gamma = 131° 32'$.   (Other orientations are used by different authors.)

am, $100 \wedge 1\bar{1}0 = 15° 34'$.
aM, $100 \wedge 1\bar{1}0 = 28° 55'$.
as, $100 \wedge 201 = 21° 37'$.

Mr, $1\bar{1}0 \wedge 1\bar{1}1 = 45° 15'$.
mr, $110 \wedge 1\bar{1}1 = 64° 22'$.
ms, $110 \wedge 201 = 27° 57'$.

938

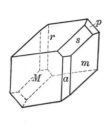

Dauphiné

939

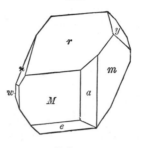

Poloma

940

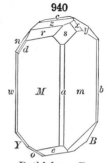

Bethlehem, Pa.

Crystals usually broad and acute-edged, but varied in habit.   Also massive, lamellar, lamellæ often curved; sometimes granular.

Cleavage: b(010) distinct.   Fracture conchoidal.   Brittle.   H. = 6·5–7.   G. = 3·27–3·29.   Luster highly glassy.   Color clove-brown, plum-blue, and pearl-gray; also honey-yellow, greenish yellow.   Streak uncolored.   Transparent to subtranslucent.   Pleochroism strong.   Optically −.   Ax. pl. and X approximately $\perp$ x(111).   Axial angles variable.   2V = 65°–70°.   Indices variable; $\alpha = 1.678$, $\beta = 1.685$, $\gamma = 1.688$.   Pyroelectric (p. 335).

Comp. — A boro-silicate of aluminum and calcium with varying amounts of iron and manganese.   Formula, $\overset{\text{II}\ \text{III}}{R_7R_4}B_2(SiO_4)_8$.   R = Calcium chiefly, sometimes in large excess, again in smaller amount and manganese prominent; iron is present in small quantity, also magnesium and basic hydrogen.

Pyr., etc. — B.B. fuses readily with intumescence, imparts a pale green color to the O.F., and fuses at 2 to a dark green to black glass; with borax in O.F. gives an amethystine bead (manganese), which in R.F. becomes yellow (iron).   Fused with a mixture of bisulphate of potash and fluorite on the platinum loop colors the flame green (boric acid).   Not decomposed by acids, but when previously ignited, gelatinizes with hydrochloric acid.

Obs. — Axinite occurs in cavities in granite or diabase and especially in the contact zones of these rocks.   Found in granite at Striegau, Silesia; in the Harz Mts., at Tressburg and Wormke-Tal; in Saxony at Thum near Ehrenfriedersdorf.   In Switzerland on Piz Valatscha, the northern spur of Mt. Scopi, east of Lukmanier in Val Medels in Uri, and at St. Gotthard in Ticino.   Axinite occurs in implanted glassy clove-brown crystals at St. Cristophe, near Bourg d'Oisans, Isère, France.   From Cornwall of a dark color, at the Botallack mine near St. Just; also at Lostwithiel.   From Roseberg, Montagu Co., Tasmania.   In Japan from near Obira, Bungo Province.

In the United States axinite is found at Franklin, Sussex Co., New Jersey in honey-yellow crystals.   At Bethlehem, Northampton Co., Pennsylvania.   In California from Riverside, Riverside Co., and from near Bonsall, south of Pala, San Diego Co.

Named from ἀξίνη, *an axe*, in allusion to the form of the crystals.

PREHNITE.

Orthorhombic-hemimorphic.   Axes $a : b : c = 0.8401 : 1 : 0.5549$.   X-ray study shows that the unit cell, which contains three molecules, has a vertical length corresponding to six times the value for the c axis here given.

Distinct individual crystals rare; usually tabular $\parallel$ $c(001)$; sometimes prismatic, mm''' (110) $\wedge$ (1$\bar{1}$0) = 80° 4′; again acute pyramidal. Commonly in groups of tabular crystals, united by $c(001)$ making broken forms, often barrel-shaped. Reniform, globular, and stalactitic with a crystalline surface. Structure imperfectly columnar or lamellar, strongly coherent; also compact granular or impalpable.

Cleavage: $c(001)$ distinct. Fracture uneven. Brittle. H. = 6–6·5. G. = 2·80–2·95. Luster vitreous; on base weak pearly. Color light green, oil-green, passing into white and gray; often fading on exposure. Subtransparent to translucent. Streak uncolored. Optically +. Ax. pl. $\parallel$ (010). $Z = c$ axis. 2V widely variable. $\alpha = 1·616$, $\beta = 1·626$, $\gamma = 1·649$. Frequently shows optical anomalies, perhaps caused by submicroscopic twinning.

Comp. — An acid orthosilicate, $H_2Ca_2Al_2(SiO_4)_3$ = Silica 43·7, alumina 24·8, lime 27·1, water 4·4 = 100. Ferric iron may replace the aluminum in small amounts.

Prehnite is sometimes classed with the zeolites, with which it is often associated; the water here, however, has been shown to go off only at a red heat, and hence plays a different part.

Pyr., etc. — In the closed tube yields water. B.B. fuses at 2 with intumescence to a blebby enamel-like glass. Decomposed slowly by hydrochloric acid without gelatinizing; after fusion dissolves readily with gelatinization.

Diff. — B.B. fuses readily, unlike beryl, green quartz, and chalcedony. Its hardness is greater than that of the zeolites.

Obs. — Occurs chiefly in basic eruptive rocks, basalt, diabase, etc., as a secondary mineral in veins and cavities, often associated with some of the zeolites, also datolite, pectolite, calcite, but commonly one of the first formed of the series; also less often in granite, gneiss, syenite, and then frequently associated with epidote. It sometimes occurs in amphibolites, augite-gneisses and in wollastonite rocks. May be a constituent of metamorphosed limestone and dolomite. Sometimes associated with native copper, as in the Lake Superior region.

In Austria in the Habachtal, Salzburg. In Italy in Trentino at Ratschinges near Sterzen (Sterzing) and in Val di Fassa; and in the Ala valley, Piedmont. In the Harz Mts., at Harzburg in the Radau-Tal; in Baden at Haslach in the Kinzig-Tal. In France from near Bourg d'Oisans, Isère. Occurs in Scotland in the Kilpatrick Hills, Dumbarton and the Campsie Hills, Stirling; also on Corstorphine Hill near Edinburgh. Found at Cradock, Cape Province, South Africa.

In the United States prehnite occurs in syenite at Somerville, Middlesex Co., Massachusetts; finely crystallized at Farmington, Hartford Co., Connecticut. From the trap quarries of New Jersey, at Paterson, Bergen Hill, Great Notch, etc. In Keweenaw Co., Michigan, from Isle Royale and from the copper deposits where it often occurs in intimate association with native copper.

Named (1790) after Col. Prehn, who brought the mineral from the Cape of Good Hope.

Harstigite. — An acid orthosilicate of manganese, calcium, and beryllium. Orthorhombic. In small colorless prismatic crystals. H. = 5·5. G. = 3·05. Optically +. Ax. pl. $\parallel$ (010). $Z = a$ axis. $\alpha = 1·678$, $\beta = 1·68$, $\gamma = 1·683$. 2V = 52°. Occurs with garnet and rhodonite at the Harstig mine, Pajsberg, near Persberg, Vermland, Sweden.

Grothine. — A silicate of calcium with aluminum and a little iron of uncertain composition. Probably related to harstigite. Orthorhombic. In small tabular crystals. H. = 5·5. G. = 3·09. Infusible. Decomposed by sulphuric acid. Colorless. Transparent. Optically +. Ax. pl. $\parallel$ (001). $Z = a$ axis. $\beta = 1·554$. Found with microsommite on limestone near Nocera and Sarno, Campania, Italy.

Fluosiderite. A silicate of lime, magnesia and small amount of alumina. Orthorhombic. Minute crystals. Granular crust of bright red color. In tuffs of Campania, Italy.

IV. Subsilicates

The species here included are basic salts, for the most part to be referred either to the metasilicates or orthosilicates, like many basic compounds already included in the preceding pages. Until their constitution is definitely settled, however, they are more conveniently grouped by themselves as SUBSILICATES. The only prominent group in this subdivision is the HUMITE GROUP.

Humite Group

			$a : b : c$	β
Norbergite				
	$Mg_2SiO_4.Mg(F,OH)_2$	Orthorhombic	1·086 : 1 : 1·887*	—
			(1·10 : 1 : 1·897)†	—
Chondrodite				
	$2Mg_2SiO_4.Mg(F,OH)_2$	Monoclinic	1·0863 : 1 : 3·1447	90°
Humite				
	$3Mg_2SiO_4.Mg(F,OH)_2$	Orthorhombic	1·0802 : 1 : 4·4033	—
Clinohumite				
	$4Mg_2SiO_4.Mg(F,OH)_2$	Monoclinic	1·0803 : 1 : 5·6588	90°

The species here included form a remarkable series both as regards crystalline form and chemical composition. In crystallization they have sensibly the same ratio for the lateral axes, while the vertical axes are almost exactly in the ratio of the numbers 3 : 5 : 7 : 9 (see also below). Furthermore, though two species are orthorhombic, and two are monoclinic, they here also correspond closely, since the axial angle β in the latter cases does not sensibly differ from 90°. Goldschmidt changes the crystal orientation of the members of this group by interchanging the a and c axes. This position is in accord with the results of the X-ray investigation of these minerals.

In composition, it was shown by Penfield and Howe (also Sjögren) that the last three species were basic orthosilicates in each of which the univalent group (MgF) or (MgOH) entered, while the Mg_2SiO_4 groups present were in the ratio of 2 : 3 : 4. From these facts it was predicted that a member of the group, then unknown, would be found in which the ratios would give only Mg_2SiO_4. This mineral, known as norbergite, has recently been discovered. In physical characters these species are very similar, and several of them may occur together at the same locality and even intercrystallized in parallel lamellæ.

The species of the group approximate closely in angle and structure to chrysolite. The axial ratios may be compared as follows:

Norbergite.........................a : b : $\frac{1}{3}c$ = 1·0803 : 1 : 0·6287
Chondrodite......................a : b : $\frac{1}{5}c$ = 1·0863 : 1 : 0·6289
Humite.............................a : b : $\frac{1}{7}c$ = 1·0802 : 1 : 0·6291
Clinohumite.......................a : b : $\frac{1}{9}c$ = 1·0803 : 1 : 0·6288
Chrysolite........................b : $2a$: c = 1·0735 : 1 : 0·6296

In the atomic structure there are independent SiO_4 groups. Each oxygen atom is linked to one silicon and three magnesium atoms. The latter lie within a group of six oxygen atoms. The humite structure is made up of atomic layers composed of magnesium silicate identical with the chrysolite structure. Between these layers lie other layers containing the (F,OH) groups. The different members of the group vary in their structure by the fre-

---

* Theoretical values as derived by Penfield and Howe.
† From measured crystals from Franklin, New Jersey.

quency in which these (F,OH) layers appear. It is to be seen that if in the formula nMg$_2$SiO$_4$.Mg(F,OH)$_2$, n is odd, the structure is orthorhombic, whereas if it is even, the structure becomes monoclinic.

NORBERGITE — CHONDRODITE — HUMITE — CLINOHUMITE.

Axial ratios as given above. Habit varied, Figs. 941 to 949. Twins common, the twinning planes inclined 60°, also 30°, to c(001) in the brachydome or clinodome zone, hence the axes crossing at angles near 60°; often repeated as trillings and as polysynthetic lamellæ (cf. Fig. 638, p. 327). Also twins, with c(001) as tw. plane. Two of the four species are often twinned together.

Cleavage: c(001) sometimes distinct. Fracture subconchoidal to uneven. Brittle. H. = 6–6·5. G. = 3·1–3·2. Luster vitreous to resinous. Color white, light yellow, honey-yellow to chestnut-brown and garnet- or hyacinth-red. Pleochroism sometimes distinct. Optically $+$. No consistent variation of optical constants throughout the group has been established.

Norbergite. — $\alpha = 1\cdot563$, $\beta = 1\cdot567$, $\gamma = 1\cdot590$. 2V = 49° 30′.
Chondrodite. — Absorption $X > Z > Y$. Ax. pl. and $Z \perp b$(010). $X \wedge a$ axis = 26° to 30°. $\alpha = 1\cdot59$–1·60, $\beta = 1\cdot60$–1·62, $\gamma = 1\cdot62$–1·64. 2V = approx. 80°.
Humite. — Ax. pl. $\parallel c$(001). $Z \perp b$(010). $\alpha = 1\cdot56$–1·62, $\beta = 1\cdot57$–1·63, $\gamma = 1\cdot58$–1·65. 2V = 68°.
Clinohumite. — Ax. pl. and $Z \perp b$(010). $X \wedge a$ axis = 9°. 2V = 76°. $\alpha = 1\cdot62$–1·66, $\beta = 1\cdot64$–1·67, $\gamma = 1\cdot65$–1·69.

Comp. — Basic fluosilicates of magnesium with related formulas as shown in the table above. Hydroxyl replaces part of the fluorine, and iron often takes the place of magnesium.

Pyr., etc. — B.B. infusible; some varieties blacken and then burn white. Fused with potassium bisulphate in the closed tube gives a reaction for fluorine. With the fluxes a reaction for iron. Gelatinizes with acids. Heated with sulphuric acid gives off silicon fluoride.

Obs. — These minerals most commonly occur in Archæan metamorphosed dolomitic limestones, and the usual presence of fluorine in their composition indicates pneumatolytic conditions for their formation. Similar conditions account for their occurrence in connection with iron ore bodies. In the case of the Vesuvian occurrence apparently their formation was due to the action of silicic acid upon limestone blocks in the presence of fluorine gas. The common associated minerals include phlogopite, spinel, pyroxene, chrysolite, and magnetite. Chondrodite, humite and clinohumite all occur at Vesuvius in the ejected masses both of limestone or feldspathic type found on Mte. Somma. They are associated with chrysolite, biotite, pyroxene, magnetite, spinel, vesuvianite, calcite, etc.; also less often with sanidine, meionite, nephelite. Of the three species, humite is the rarest and clinohumite of most frequent occurrence. They seldom all occur together in the same mass, and only rarely two of the species (as humite and clinohumite) appear together. Occasionally clinohumite interpenetrates crystals of humite, and parallel intergrowths with chrysolite have also been observed.

Norbergite was discovered at the Ostanmosoa iron mine, Norberg, Vastmanland, Sweden. Much that has been called chondrodite or humite from Franklin, Sussex Co., New Jersey, is norbergite.

Chondrodite at Mte. Somma, as noted above. From Finland at Pargas, southeast of Åbo, Turun ja Porin, of honey-yellow color in limestone; at Kafveltorp near Ljusnarsberg, Örebro, Sweden. In the United States abundant at the Tilly Foster magnetite deposit near Brewster, Putnam Co., New York, in garnet-red crystals of great beauty and variety of form. Material from Nordmarken originally described as *prolectite* and thought to be the first member of the Humite Group was later shown to be chondrodite.

Humite occurs at Mte. Somma. Also from the Ladu mine near Filipstad, Vermland, Sweden. In crystalline limestone with clinohumite in the Llanos de Juanar, Serrania de Ronda, Province of Malaga, Spain. In large, coarse, partly altered crystals at the Tilly Foster iron mine at Brewster, Putnam Co., New York.

Clinohumite occurs at Mte. Somma; in Llanos de Juanar, Malaga, Spain, as polysynthetic lamellæ in parallel intergrowth with humite; in crystalline limestone near Lake Baikal, Siberia. At Brewster, Putnam Co., New York, in rare but highly modified crystals. *Titanclinohumite* is a titaniferous variety (originally called *titanolivine*) from the Ala valley, Piedmont, Italy.

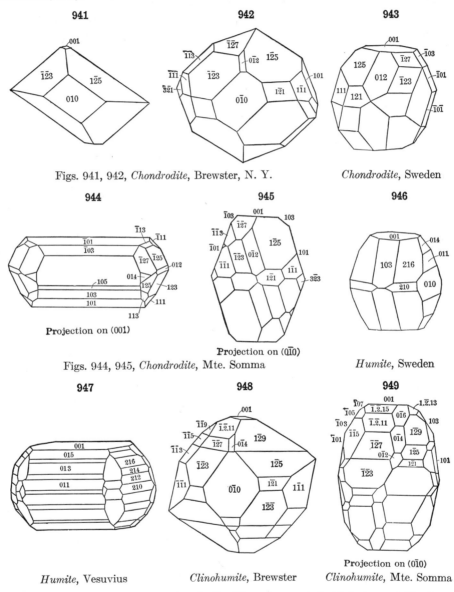

Figs. 941, 942, *Chondrodite*, Brewster, N. Y.　　　*Chondrodite*, Sweden

Projection on (001)
Figs. 944, 945, *Chondrodite*, Mte. Somma　　　*Humite*, Sweden

Humite, Vesuvius　　　*Clinohumite*, Brewster　　　*Clinohumite*, Mte. Somma

Numerous other localities of " chondrodite " have been noted, chiefly in crystalline limestone; most of them are probably to be referred to the species chondrodite, but the identity in many cases is not proved. At Brewster, Putnam Co., New York, large quantities of massive "chondrodite" occur associated with magnetite, enstatite, ripidolite, and from its extensive alteration serpentine has been formed on a large scale. In Orange Co., New York, at Warwick, Amity, and Edenville. In New Jersey, Sussex Co., a fine honey-yellow chondrodite from near Sparta; also from Vernon and Franklin.

The name chondrodite is from χόνδρος, *a grain*, alluding to the granular structure. Humite is from Sir Abraham Hume.

Leucophœnicite. — $Mn_5(MnOH)_2(SiO_4)_3$, similar to the humite type of formula. Monoclinic. In striated crystals elongated parallel to ortho-axis. Massive. H. = 5·5–6. G. = 3·8. Fusible. Color light purplish red. Optically −. $\alpha = 1·751$, $\beta = 1·771$, $\gamma = 1·782$. 2V = 74°. $X \perp$ to a cleavage. From Franklin, Sussex Co., New Jersey.

ILVAITE. Lievrite. Yenite.

Orthorhombic. Axes $a : b : c = 0·6665 : 1 : 0·4427$.   **950**

mm''', 110 $\wedge$ 1$\bar{1}$0 = 67° 22'. rr', 101 $\wedge$ $\bar{1}$01 = 67° 11'.
ss', 120 $\wedge$ 1$\bar{2}$0 = 73° 45'. oo', 111 $\wedge$ $\bar{1}$11 = 62° 33'.

Commonly in prisms, with prismatic faces vertically striated. Columnar or compact massive.

Cleavage: b(010), c(001) rather distinct. Fracture uneven. Brittle. H. = 5·5–6. G. = 3·99–4·05. Luster submetallic. Color iron-black or dark grayish black. Streak black, inclining to green or brown. Opaque. Optically −. Ax. pl. (100). $Z = c$ axis. Dispersion strong, $\rho < v$. $\alpha = 1·915$. Strongly pleochroic in very thin section.

Comp. — $CaFe_2(FeOH)(SiO_4)_2$ or $H_2O.2CaO.4FeO.Fe_2O_3.4SiO_2$ = Silica 29·3, iron sesquioxide 19·6, iron protoxide 35·2, lime 13·7, water 2·2 = 100. Manganese may replace part of the ferrous iron.

Pyr., etc. — B.B. fuses quietly at 2·5 to a black magnetic bead. With the fluxes reacts for iron. Some varieties give also a reaction for manganese. Gelatinizes with hydrochloric acid.

Obs. — Ilvaite occurs with magnetite ore bodies; with zinc and copper ores; in contact deposits; also with zeolites.

First found on the Rio Marina, and at Capo Calamita, on Elba where it occurs in large solitary crystals, and in aggregated crystallizations in dolomite with pyroxene, etc. In fine specimens from the island of Scriphos, Cyclades, Greece. From Hessen-Nassau near Herborn which is near Dillenburg. In Italy on Mte. Mulat near Predazzo in Val di Fiemme, Trentino, and in Tuscany at Campiglia Marittima. A fibrous variety from Vesuvius, Vulcano, etc., has been called *breislakite*. From Algeria at Cap Bougaroun near Collo, Constantine. From the sodalite-syenite of the Julianehaab district of southern Greenland at Kangerdluarsuk and Siorarsuit. Well-crystallized material has been found at various points in Japan. In the United States it occurs in fine crystals at the South Mountain mine, Owyhee Co., Idaho. Named *ilvaite* from the Latin name of the island of Elba.

Ardennite. Dewalquite. — A vanadio-silicate of aluminum and manganese; also containing arsenic. Orthorhombic. In prismatic crystals resembling ilvaite. Cleavages, (010) perfect, (110) distinct. H. = 6–7. G. = 3·620. Easily fusible. Yellow to yellowish brown. Optically +. Strong dispersion, $\rho > v$. Index about 1·79. Found at Salmchateau in the Ardennes, Belgium. Also in the Ala valley, Piedmont, Italy.

Chapmanite. — $5FeO.Sb_2O_5.5SiO_2.2H_2O$. Orthorhombic. In finely divided state. Lath-shaped crystals. Soft. G. = 3·58. Color green. Optically −. Positive elongation. $\alpha = 1·85$, $\gamma = 1·96$. 2V small. Found at the Keeley Mine in South Lorrain near Cobalt, Ontario, intimately associated with silver.

Långbanite. — Manganese silicate with ferrous antimonate; formula doubtful. Rhombohedral-tetartohedral. In iron-black hexagonal prismatic crystals. H. = 6·5. G. = 4·918. Luster metallic. Optically −. $\omega = 2·36$, $\epsilon = 2·31$. Occurs in Sweden at Långbanshyttan, Vermland and at the Sjö mines near Örebro in Örebro.

The following are rare lead silicates. See also p. 584.

Kentrolite. — Probably $3PbO.2Mn_2O_3.3SiO_2$. Orthorhombic. In minute prismatic crystals; often in sheaf-like forms; also massive. Cleavage, (110). H. = 5. G. = 6·19. Color dark reddish brown; black on the surface. Optically +. Ax. pl. || (010). $X = a$ axis. Strong dispersion, $\rho < v$. $\alpha = 2\cdot10$, $\beta = 2\cdot20$, $\gamma = 2\cdot31$. Originally found in southern Chile, exact locality unknown. Found in Vermland, Sweden, at Långbanshyttan and at Jacobsberg, Nordmark district. From Bena de Padru near Ozieri, province of Sassari, Sardinia.

Melanotekite. — $3PbO.2Fe_2O_3.3SiO_2$ or $(Fe_4O_3)Pb_3(SiO_4)_3$. Orthorhombic; prismatic. Massive; cleavable. H. = 6·5. G. = 5·73. Luster metallic to greasy. Color black to blackish gray. Optically +. $\alpha = 2\cdot12$, $\beta = 2\cdot17$, $\gamma = 2\cdot31$. Strongly pleochroic in thin section. Occurs with native lead at Långbanshyttan, Vermland, Sweden. Also in crystals resembling kentrolite at Hillsboro, Sierra Co., New Mexico.

Bertrandite. — $H_2Be_4Si_2O_9$ or $H_2O.4BeO.2SiO_2$. Orthorhombic-hemimorphic. In small tabular or prismatic crystals. Heart-shaped twins. Cleavage, (110) perfect; several other distinct cleavages. H. = 6–7. G. = 2·59–2·60. Colorless to pale yellow. Optically −. Ax. pl. || (010). $X = a$ axis. $\alpha = 1\cdot591$, $\beta = 1\cdot605$, $\gamma = 1\cdot614$. Bertrandite occurs most commonly in pegmatite in intimate association with beryl; sometimes in cavities formerly occupied by beryl crystals, and it has been noted as pseudomorphic after that mineral. It has, therefore, probably been derived by an alteration of beryl. From Irkutka Mt., Altai Mts., Siberia. At Pisek, Bohemia of Czechoslovakia. Originally discovered near Nantes, Loire-Inférieure, France. Also from la Villeder, Morbihan. From Iveland parish, Vest-Agder, Norway. At various points in Cornwall. In the United States from Maine in Oxford Co., at Albany, Greenwood, and Stoneham and at Brunswick, Sagadahoc Co., and Auburn, Androscoggin Co. At Amelia Court House, Amelia Co., Virginia. On Mt. Antero, Chaffee Co., Colorado with phenacite.

CALAMINE. Hemimorphite. Smithsonite.

Orthorhombic-hemimorphic. Axes $a : b : c = 0\cdot7834 : 1 : 0\cdot4778$.

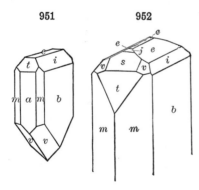

951   952

mm''',	$110 \wedge 1\bar{1}0 =$	$76°\ 9'$.
ss',	$101 \wedge \bar{1}01 =$	$62°\ 46'$.
tt',	$301 \wedge \bar{3}01 =$	$122°\ 41'$.
ee',	$011 \wedge 0\bar{1}1 =$	$51°\ 5'$.
ii',	$031 \wedge 0\bar{3}1 =$	$110°\ 12'$.
vv''',	$121 \wedge 1\bar{2}1 =$	$78°\ 26'$.

Crystals often tabular || $b(010)$; also prismatic; faces b vertically striated. Usually implanted and showing one extremity only. When doubly terminated crystals show hemimorphic development (cf. Fig. 951). Often grouped in sheaf-like forms and forming drusy surfaces in cavities. Also stalactitic, mammillary, botryoidal, and fibrous forms; massive and granular.

Cleavage: $m(110)$ perfect; $s(101)$ less sc; $c(001)$ in traces. Fracture uneven to subconchoidal. Brittle. H. = 4·5–5, the latter when crystallized. G. = 3·40–3·50. Luster vitreous; $c(001)$ subpearly, sometimes adamantine. Color white; sometimes with a delicate bluish or greenish shade; also yellowish to brown. Streak white. Transparent to translucent. Optically +. Ax. pl. || (100). $Z = c$ axis. 2V = 46°. $\alpha = 1\cdot614$. $\beta = 1\cdot617$. $\gamma = 1\cdot636$. Strongly pyroelectric.

Comp. — $H_2Zn_2SiO_5$ or $(ZnOH)_2SiO_3$ or $H_2O.2ZnO.SiO_2 =$ Silica 25·0, zinc oxide 67·5, water 7·5 = 100. The water goes off only at a red heat; unchanged at 340° C.

Pyr., etc. — In the closed tube decrepitates, whitens, and gives off water. B.B. almost infusible (F. = 6). On charcoal with soda gives a coating which is yellow while hot, and white on cooling. Moistened with cobalt solution, and heated in O.F., this coating assumes a bright green color, but the ignited mineral itself becomes blue. Gelatinizes with acids even when previously ignited.

Diff. — Characterized by its infusibility; reaction for zinc; gelatinization with acids. Resembles some smithsonite (which effervesces with acid), also prehnite.

Obs. — Calamine and smithsonite are usually found associated in veins or beds in stratified calcareous rocks accompanying sulphides of zinc, iron and lead. Some of the more important localities for its occurrence follow. Large crystals have been found at Nerchinsk in Transbaikalia, Siberia. In Rumania at Dognacska and in Carinthia of Austria at Bleiberg and Raibl. In Saxony at Altenberg. In fine specimens near Iglesias in Sardinia at Mte. Agruian, Monteponi, and La Duchessa near Domusnovas. From Spain in the provinces of Almeria and Santander. From the zinc district about Aachen (Aix-la-Chapelle) in Rhineland and Moresnet in Belgium. From Cumberland at Roughten Gill in acicular crystals and mammillary crusts, sky-blue and fine green; at Alston Moor, white; near Matlock in Derbyshire. Calamine occurs at various points in Algeria, in magnificent crystals at Djebel Guergour, near Lafayette, northwest of Setif, Constantine. In Mexico from Santa Eulalia, near Chihuahua.

In the United States was found at Sterling Hill, near Ogdensburg, Sussex Co., New Jersey, in fine clear crystalline masses. In Pennsylvania was extensively worked at Friedensville, Lehigh Co. Formerly abundant at Austin's mines in Wythe Co., Virginia. With the zinc deposits of southwestern Missouri, especially about Granby, Newton Co.; also from Aurora, Lawrence Co. In fine crystals from Elkhorn, Jefferson Co., Montana; at Leadville, Lake Co., Colorado; from the Organ Mts., Dona Ana Co., New Mexico. From the Emma mine, Cottonwood district, Salt Lake Co., Utah, greenish blue mammillary forms. From Eureka, Eureka Co., Nevada.

The name *Calamine* (with *Galmei* of the Germans) is commonly supposed to be a corruption of *Cadmia*. Agricola says it is from *calamus*, a *reed*, in allusion to the slender forms (stalactitic) common in the *cadmia fornacum*.

Use. — An ore of zinc.

Fraipontite. — $8ZnO.2Al_2O_3.5SiO_2.11H_2O$. In thin fibrous crusts. Color yellowish white. Silky luster. Biaxial, −. Ax. pl. ∥ length of fibers. Positive elongation. Soluble in HNO_3, gelatinizes upon evaporation. Easily fusible. Found on smithsonite at unknown locality, but believed to be Vielle-Montagne, Belgium.

Clinohedrite. — $H_2CaZnSiO_5$. Monoclinic-clinohedral. Crystals of varied habit. (See Figs. 372, 373, p. 157.) Cleavage ∥ (010), perfect. H. = 5·5. G. = 3·33. Fusible at 4. Soluble in HCl. Colorless or white to amethystine. Optically −. Ax. pl. ⊥ (010). $Z = b$ axis. Index, 1·67. From Franklin, Sussex Co., New Jersey.

Stokesite. — Perhaps $H_4CaSnSi_3O_{11}$. Orthorhombic. Prismatic cleavage. H. = 6. G. = 3·2. Fusibility, 4. Colorless. Optically +. Ax. pl. ∥ (010). $X = a$ axis. $α = 1·609, β = 1·613, γ = 1·619$. From Roscommon Cliff, St. Just, Cornwall.

Arandisite. — Basic silicate of tin. A mixture of two constituents, perhaps colloidal and crystalline phases. Massive, microscopically granular and fibrous. H. = 5. G. = 4. Varying shades of green. Waxy to resinous luster. Mostly isotropic with $n = 1·706$. In part weakly anisotropic with $n = 1·82$. Found with cassiterite, quartz, iron and copper sulphides in limestone, north of Arandis, South West Africa.

Carpholite. — $H_4MnAl_2Si_2O_{10}$. Orthorhombic. In radiated and stellated tufts. H. = 5. G. = 2·9. Color straw- to wax-yellow. Optically −. Ax. pl. ∥ (100). $X = b$ axis. $α = 1·61, β = 1·628, γ = 1·63$. $2V = 87°$. Occurs at the tin mines of Schlaggenwald (Slavkov), Bohemia; Wippra, in the Harz Mts., on quartz. From near Meuville, Ardennes, Belgium.

Lawsonite. — $H_4CaAl_2Si_2O_{10}$. In prismatic orthorhombic and tabular crystals; mm''', 110 ∧ 1$\bar{1}$0 = 67° 16'. Twinning on (110). Cleavages: (010) and (001) perfect, (110) poor. H. = 7–8. G. = 3·09. Fusibility, 4. Luster vitreous to greasy. Colorless, pale blue to grayish blue. Optically +. Ax. pl. ∥ (010). $Z = c$ axis. Strong dispersion, $ρ > v$. $α = 1·665, β = 1·674, γ = 1·684$. $2V = 84°$. Lawsonite characteristically occurs either as a secondary mineral in altered gabbros and diorites or as an accessory mineral in certain schists and gneisses which also commonly contain glaucophane. It has been suggested that it has been formed during metamorphism from the anorthite molecule of plagioclase feldspars. Originally described from a crystalline schist which is associated with serpentine on the Tiburon peninsula, Marin Co., California; also reported in the rocks of San Luis Obispo, Alameda, Santa Clara, and other counties in California. In Italy in the metamorphic rocks in Piedmont and in the massive rocks of the southern

Apennines. Also in France from Hautes-Alpes; from the island of Corsica; from New Caledonia.

Hibschite. — Same as for *lawsonite*, $H_4CaAl_2Si_2O_{10}$. In minute isometric crystals, usually octahedrons. H. = 6. G. = 3·0. Infusible. Colorless or pale yellow. $n = 1·67$. From the phonolite of Marienberg, near Aussig, Bohemia of Czechoslovakia; associated with melanite. Also in limestone inclusions from basalt of Aubenas, Ardéche, France.

Cerite. — A silicate of the cerium metals chiefly, with water. Tetragonal or orthorhombic. Crystals rare; commonly massive; granular. H. = 5·5. G. = 4·86. Color between clove-brown and cherry-red to gray. Uniaxial or biaxial with small optic angle. Optically +. $n = 1·82$. Occurs at Bastnaes mine near Riddarhyttan, Vastmanland, Sweden.

Törnebohmite. — A silicate of the cerium metals, chiefly $R_3(F,OH)(SiO_4)_2$. Monoclinic? H. = 4·5. G. = 4·9. Color, green to olive. $\beta = 1·85$. Biaxial, +. Strong dispersion, $\rho < v$. Pleochroic, rose to blue-green. Occurs at Bastnaes mine near Riddarhyttan, Vastmanland, Sweden.

Beckelite. — $Ca_3(Ce,La,Di)_4Si_3O_{15}$. Isometric. Crystals small, often microscopic. Cubic cleavage. H. = 5. G. = 4·1. Infusible. Color yellow. $n = 1·812$. Occurs with nepheline-syenite rocks near Mariupol, on the north shore of the Sea of Azov, Ukraine, Russia.

Lessingite. A silicate of calcium and the rare earths. H. = 4·5. G. = 4·7. Color greenish to reddish yellow; cerise-red on fresh fracture. Vitreous luster. Biaxial. 2V = 44°. $n = 1·78$. Found as rolled pebbles with bastnaesite, cerite, törnebohmite and allanite near Kyshtym, Ural Mts.

Bodenbenderite. A silicate and titanate of aluminum, yttrium, manganese, etc. Isometric. Dodecahedral habit. G. = 3·5. H. = 6–6·5. $n > 1·77$. Color, flesh-red; From albite-fluorite veins in the Sierra Chica, Sierra de Cordoba, Argentina.

Hellandite. — A basic silicate chiefly of the cerium metals, aluminum, manganese and calcium. Monoclinic. Prismatic habit. H. = 5·5. G. = 3·7. Fusible. Color brown. Optically +. Ax. pl. $\perp$ (010). $X = b$ axis. $Z \wedge c$ axis = $-43°$. $\beta = 1·65$. 2V = 80°. Found in pegmatite near Kragerö, Telemark, Norway.

Bazzite. — A silicate of scandium with other rare earth metals, iron and a little soda. Hexagonal. In minute prisms, often barrel shaped. H. = 6·5. G. = 2·8. Color azure-blue. Transparent in small individuals. Optically −. $\omega = 1·626$. $\epsilon = 1·605$. Strongly dichroic, O = pale greenish yellow, E = azure-blue. Infusible. Insoluble in ordinary acids. Found at Baveno, on Lago Maggiore, Piedmont, Italy.

Buszite. Silicate of the rare-earths, Nd, Pr, Er, and Eu. Hexagonal, trigonal. Yellowish red color. H. = 5·5. G. = 5. $n = 1·72$. A single short prismatic crystal found with beryl at Khan, South West Africa.

Angaralite. $2(Ca,Mg)0.5(Al,Fe)_2O_3.6SiO_2$. In thin tabular hexagonal (?) crystals: G. = 2·62. Color black from carbonaceous impurities. Uniaxial, +. In contact zone of limestone, near the Angara River, southern part of Yenisei district, Siberia.

TOURMALINE. Turmalin.

Rhombohedral-hemimorphic. Axis $c = 0·4477$.

$cr, 0001 \wedge 10\bar{1}1 = 27° 20'$. $rr', 10\bar{1}1 \wedge \bar{1}101 = 46° 52'$. $uu', 32\bar{5}1 \wedge \bar{3}5\bar{2}1 = 66° 1'$.
$co, 0001 \wedge 02\bar{2}1 = 45° 57'$. $oo', 02\bar{2}1 \wedge \bar{2}021 = 77° 0'$. $uu^v, 32\bar{5}1 \wedge 5\bar{3}\bar{2}1 = 42° 36'$.

Crystals usually prismatic in habit, often slender to acicular; rarely flattened, the prism nearly wanting. Prismatic faces strongly striated vertically, and the crystals hence often much rounded to barrel shaped. The cross section of the prism three-sided (m, Fig. 960), six-sided (a), or nine-sided (m and a). Crystals commonly hemimorphic. Sometimes isolated, but more commonly in parallel or radiating groups. Sometimes massive compact; also columnar, coarse or fine, parallel or divergent.

Cleavage: $a(11\bar{2}0)$, $r(10\bar{1}1)$ difficult. Fracture subconchoidal to uneven. Brittle and often rather friable. H. = 7–7·5. G. = 2·98–3·20. Luster vitreous to resinous. Color black, brownish black, bluish black, most common; blue, green, red, and sometimes of rich shades; rarely white or colorless; some specimens red internally and green externally; and others red at one extremity, and green, blue or black at the other; the zonal arrangement of different colors widely various both as to the colors and to crystallographic directions. Streak uncolored. Transparent to opaque.

Strongly dichroic, especially in deep-colored varieties; axial colors varying widely. Absorption for O much stronger than for E, thus sections $\parallel c$ axis transmit sensibly the extraordinary ray only, and hence their use (*e.g.*, in the tour-

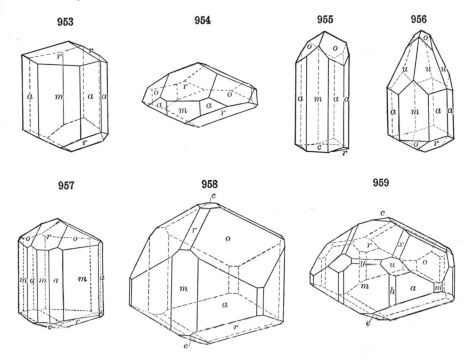

maline tongs (p. 267)) for giving polarized light. Exhibits idiophanous figures (p. 317). Optically $-$. Birefringence rather high, $\omega - \epsilon = 0.02$. Indices: $\omega_y = 1.6366$, $\epsilon_y = 1.6193$ colorless variety; $\omega_r = 1.6435$, $\epsilon_r = 1.6222$ blue-green. Sometimes abnormally biaxial. Becomes electric by friction; also strongly pyroelectric.

Var. — *Ordinary.* In crystals as above described; black much the most common. (*a*) *Rubellite;* the red, sometimes transparent; the Siberian is mostly violet-red (*siberite*), the Brazilian rose-red; that of Chesterfield and Goshen, Massachusetts, pale rose-red and opaque; that of Paris, Maine, fine ruby-red and transparent. (*b*) *Indicolite*, or *indigolite;* the blue, either pale or bluish black; named from the indigo-blue color. (*c*) *Brazilian Sapphire* (in jewelry); Berlin-blue and transparent. (*d*) *Brazilian Emerald, Chrysolite* (or *Peridot*) of *Brazil;* green and transparent. (*e*) *Peridot of Ceylon;* honey-yellow. (*f*) *Achroite;* colorless tourmaline, from Elba. (*g*) *Aphrizite;* black tourmaline, from Kragerö, Norway. (*h*) *Columnar* and *black;* coarse columnar. Resembles somewhat common hornblende, but has a more resinous fracture, and is without distinct cleavage or anything like a fibrous appearance in the texture; it often has the appearance on a broken surface of some kinds of soft coal. (*i*) *Dravite*, a brown tourmaline. (*j*) *Schorl* is black.

Comp. — A complex silicate of boron and aluminum, with also either magnesium, iron or the alkali metals prominent. A general formula may be written as $H_9Al_3(B.OH)_2Si_4O_{19}$ (Penfield and Foote) in which the hydrogen may be replaced by the alkalies and also the bivalent elements, Mg, Fe, Ca. Fluorine is commonly present in small amounts. Niggli writes the formula as $4R_6Al_2(SiO_4)_3.R_6Al_2(B_2O_5)_3$.

Many other formulas have been proposed by different authors but the composition must still be considered as uncertain. Various names have been given to hypothetical molecules assumed to be present, as: *kalbaite, elbaite, belbaite, dravite, schorl, rubellite, uvite, tsilaisite.*

The varieties based upon composition fall into three prominent groups, between which there are many gradations:

1. ALKALI TOURMALINE. Contains sodium or lithium, or both; also potassium. G. = 3·0–3·1. Color red to green; also colorless. From pegmatites.

2. IRON TOURMALINE. G. = 3·1–3·2. Color usually deep black. Accessory mineral in siliceous igneous rocks and in mica schists, etc.

3. MAGNESIUM TOURMALINE. G. = 3·0–3·09. Usually yellow-brown to brownish black; also colorless. From limestone or dolomite.

A *chromium tourmaline* also occurs. G. = 3·120. Color dark green.

Pyr., etc. — The magnesia varieties fuse rather easily to a white blebby glass or slag; the iron-magnesia varieties fuse with a strong heat to a blebby slag or enamel; the iron varieties fuse with difficulty, or, in some, only on the edges; the iron-magnesia-lithia varieties fuse on the edges, and often with great difficulty, and some are infusible; the lithia varieties are infusible. With the fluxes many varieties give reactions for iron and manganese. Fused with a mixture of potassium bisulphate and fluor-spar gives a distinct reaction for boric acid. Not decomposed by acids. Crystals, especially of the lighter colored varieties, show strong pyroelectricity.

Diff. — Characterized by its crystallization, prismatic forms usual, which are three-, six-, or nine-sided, and often with rhombohedral terminations; massive forms with columnar structure; also by absence of cleavage (unlike amphibole and epidote); in the common black kinds by the coal-like fracture; by hardness; by difficult fusibility (common kinds), compared with garnet and vesuvianite. The boron test is important.

Micro. — Readily distinguished in thin sections by its somewhat high relief; rather strong interference-colors; negative uniaxial character; decided colors in ordinary light in which basal sections often exhibit a zonal structure. Also, especially, by its remarkable absorption when the direction of crystal elongation is $\perp$ to the vibration-plane of the lower nicol; this with its lack of cleavage distinguishes it from biotite and amphibole, which alone among rock-making minerals show similar strong absorption.

Obs. — Commonly found in granite and gneisses as a result of fumarole action or of mineralizing gases in the fluid magma, especially in the pegmatite veins associated with such rocks; at the periphery of such masses or in the schists, or altered limestones, gneisses, etc., immediately adjoining them. It marks especially the boundaries of granitic masses, and its associate minerals are those characteristic of such occurrences; quartz, albite, microcline, muscovite, etc. Common in that variety of granite called greisen associated with cassiterite. It also occurs as an original constituent in granites and pegmatites, further to a less extent in their corresponding porphyries and rarely in lavas. The variety in granular limestone or dolomite is commonly brown; the bluish-black variety sometimes associated with tin ores; the brown with titanium; the lithium variety is often associated with lepidolite.

Tourmaline is so widespread in its occurrence that only the more important localities, noted either for their crystals or unusual colors, can be mentioned here. From Nerchinsk in Transbaikalia, Siberia. Black and variously colored tourmalines come from the Ekaterinburg district in the Ural Mts. In Czechoslovakia black crystals at Haslau northwest of Eger, Bohemia, and at Rozna near Neustadtl, Moravia. Brown crystals occur in Carniola of Yugoslavia at Dobrova near Unter Drauburg and to the southwest at Prävali and Gutenstein. In Saxony variously colored crystals occur at Wolkenburg and elsewhere near Penig. In Switzerland green at Campolungo near St. Gotthard, Ticino. Beautiful pink crystals come from San Piero in Campo, Elba. From Norway black crystals occur at Snarum in Buskerud and elsewhere. Fine black crystals have been obtained at Bovey Tracey, Devonshire; at various localities in Cornwall.

In Greenland black crystals are found at Karusulik in the Godthaab district. The pegmatites of Madagascar furnish exceptional crystals, both as to fine color and size; large pink crystals come from Anjanabonoina; from Maharitra on Mt. Bity; at Antandrokomby near Mt. Bity; from Ampantsikahitra. The crystals often show variation in color in the same specimen and at times are concentrically banded showing a triangular pattern of varying colors. Fine black crystals are found at various localities, as at Tongafeno, etc. Yellow and brown crystals come from Ceylon; also deep green and red from various localities in Minas Geraes, Brazil.

In the United States tourmaline is common in the pegmatites of southwestern Maine, the chief localities being in Oxford Co., at Paris and Hebron magnificent red and green crystals; pink crystals embedded in lepidolite from Rumford; pale green at Newry; and in Androscoggin Co., at Auburn and Poland. From New Hampshire in Grafton Co., at Orford, large brownish black crystals in steatite; at Grafton; in Sullivan Co., at Acworth and in large black crystals at Springfield. In Hampshire Co., Massachusetts, at Chesterfield and at Goshen. In Connecticut in Middlesex Co., at Haddam, and near Middletown, both black in mica slate and colored in pegmatite veins; in Fairfield Co., at Monroe, perfect brown crystals in mica slate. In New York in St. Lawrence Co., near Gouverneur, light and dark brown crystals, often highly modified; in splendent black crystals at Pierrepont; colorless and glassy at De Kalb; and at many other localities. In Sussex Co., New Jersey, at Hamburg, black and brown crystals in limestone and in grass-green crystals near Franklin. Black crystals occur at various points in Delaware Co., Pennsylvania. In beautiful colored crystals, red, pink, green, blue, etc., from the pegmatites of San Diego Co., California, at Mesa Grande, Pala, Ramona, Rincon, etc.

The name *turmalin* from *Turamali* in Cingalese (applied to zircon by jewelers of Ceylon) was introduced into Holland in 1703, with a lot of gems from Ceylon.

Use. — The variously colored and transparent varieties are used as gem stones; see under " Var." above.

Dumortierite. — A basic aluminum borosilicate, perhaps $8Al_2O_3.B_2O_3.6SiO_2.H_2O$ (Schaller). The water and boric oxide have been considered as variable in amount and basic in character with the general formula, $(AlO)_{16}Al_4(SiO_4)_7$ (Ford). On heating to 1500° C. it is converted into mullite $(3Al_2O_3.2SiO_2)$ with loss of boric oxide and water.

Orthorhombic. Prismatic angle approximately 60°. Usually in fibrous to columnar aggregates. Cleavage: $a(100)$, distinct; also prismatic, imperfect. H. = 7. G. = 3·26–3·36. Luster vitreous. Color bright smalt-blue to greenish blue. Transparent to translucent. Pleochroism very strong: X deep-blue or nearly colorless, Y yellow to red-violet or nearly colorless, Z colorless or pistachio-green. Exhibits idiophanous figures, analogous to andalusite. Optically −. Ax. pl. || $b(010)$. $X = c$ axis. $\alpha = 1·678$. $\beta = 1·686$. $\gamma = 1·689$.

Recognized in thin section by its rather high relief; low interference-colors (like those of quartz); occurrence in slender prisms, needles or fibers, with negative optical extension; parallel extinction; biaxial character and especially by its remarkable pleochroism.

Originally found near Lyon, Rhône, France, in fibrous forms embedded in feldspar in blocks of gneiss at Chaponost (the gneiss, however, originally came from near Beaunan); also at Brignais. In Silesia from Wolfschau near Schmiedeberg; in the cordierite of the gneiss of Tvedestrand in Aust-Agder, Norway. In Madagascar near Souvina north of Ambatofinandrahana, lavender-blue and violet-rose. From near Rio de Janeiro, Brazil, in pegmatite. From Nacozarl, Sonora, Mexico. In the United States it occurs near Harlem, New York Island; in Arizona at Clip, north of Yuma, Yuma Co., in a quartzose rock with kyanite. In Nevada from Oreana, Humboldt Co., and in a dumortierite schist over a large area in the Rochester mining district where it is quarried and used in the manufacture of porcelain for spark plugs. In California near Dehesa, San Diego Co., a violet-red variety; in Washington at Woodstock. In Canada it occurs in Ashby township, Addington and Lenox Co., Ontario.

STAUROLITE. Staurotide.

Orthorhombic. Axes $a : b : c = 0·4734 : 1 : 0·6828$.

mm''', 110 $\wedge$ 1$\bar{1}$0 = 50° 40'.	cr, 001 $\wedge$ 101 = 55° 16'.
rr', 101 $\wedge$ $\bar{1}$01 = 110° 32'.	mr, 110 $\wedge$ 101 = 42° 2'.

Twins cruciform: tw. pl. $x(032)$, the crystals crossing nearly at right angles (might also be explained as having tw. pl. $a(100)$ with a 90° angle between individuals); tw. pl. $z(232)$, crossing at an angle of 60° approximately (may also be explained as a rotation of 120° about the zone axis (101));

tw. pl. y(230) rare, also in repeated twins (cf. Figs. 423, p. 185; 465, 466, 467, p. 191). Crystals commonly prismatic and flattened ∥ b axis; often with rough surfaces.

X-ray structure of staurolite shows that it has in part an atomic arrangement identical with that of kyanite. It may be conceived as made up of

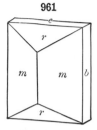

961

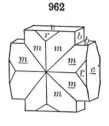

962

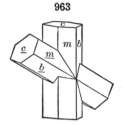

963

layers, lying perpendicular to the b axis, that show the structure of kyanite, alternating with layers that contain iron atoms and hydroxyl groups. The atomic arrangement on the (010) face of staurolite is identical with that on the (100) face of kyanite, thus explaining the parallel crystallization of the two minerals frequently observed. The unit cell contains four molecules. Its dimensions correspond to $1a$, $1b$, $\frac{1}{2}c$.

Cleavage: b(010) distinct, but interrupted; m(110) in traces. Fracture subconchoidal. Brittle. H. = 7–7·5. G. = 3·65–3·77. Subvitreous, in-

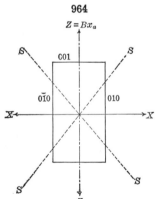

964

$Z = Bx_a$

clining to resinous. Color dark reddish brown to brownish black, and yellowish brown. Streak uncolored to grayish. Translucent to nearly or quite opaque. Pleochroism distinct: Z (= c axis) hyacinth-red to blood-red, X, Y yellowish red; or Z gold-yellow, X, Y light yellow to colorless. Optically +. Ax. pl. ∥ a(100). $Z \perp c$(001). 2V = 88° (approx.). $\alpha = 1\cdot736$. $\beta = 1\cdot741$. $\gamma = 1\cdot746$.

Comp. — $HFeAl_5Si_2O_{13}$, which may be written $(AlO)_4(AlOH)Fe(SiO_4)_2$ or $H_2O.2FeO.5Al_2O_3.4SiO_2$ = Silica 26·3, alumina 55·9, iron protoxide 15·8, water 2·0 = 100. Magnesium (also manganese) replaces a little of the ferrous iron; ferric iron, part of the aluminum.

As a result of X-ray study the following new formula has been proposed: $H_2FeAl_4Si_2O_{12}$. In order to show the relationship to kyanite this may be written $2Al_2SiO_5.Fe(OH)_2$. Variations in composition from this formula are explained as due to the intergrowth of small amounts of kyanite.

Nordmarkite from Nordmark, Sweden, contains manganese in large amounts.

Pyr., etc. — B.B. infusible, excepting the manganesian variety, which fuses easily to a black magnetic glass. With the fluxes gives reactions for iron, and sometimes for manganese. Imperfectly decomposed by sulphuric acid.

Diff. — Characterized by the obtuse prism (unlike andalusite, which is nearly square); by the frequency of twinning forms; by hardness and infusibility.

Micro. — Under the microscope, sections show a decided color (yellow to red or brown) and strong pleochroism (yellow and red); also characterized by strong refraction (high relief), rather bright interference-colors, parallel extinction and biaxial character (generally positive in the direction of elongation). Easily distinguished from rutile (p. 498) by its biaxial character and lower interference-colors.

Obs. — Usually found in crystalline schists, as mica schist, phyllite, and gneiss, as a result of regional or more rarely of contact metamorphism; often associated with garnet, sillimanite, kyanite, and tourmaline. Sometimes encloses symmetrically arranged carbonaceous impurities like andalusite (p. 615). Other impurities are also often present, especially silica, sometimes up to 30 to 40 per cent ; also garnet, mica, and perhaps magnetite, brookite.

Some notable localities for its occurrence follow. In Ticino, Switzerland, on Mte. Campione between Faido and Chironica with kyanite in a paragonite schist in lustrous, brown, translucent crystals; and at St. Gotthard.   In large twin crystals in the mica schists of Brittany at many places in Finistère and Morbihan.

In the United States abundant throughout the mica schists of New England; especially at Windham, Cumberland Co., Maine, where the mica slate is filled with large crystals. In New Hampshire, Grafton Co., in large brown crystals at Franconia, Sugar Hill, and Lisbon in mica slate; on the shores of Mink Pond, loose in the soil; in Sullivan Co., from Grantham and Claremont.   In Chesterfield, Hampshire Co., Massachusetts, in fine crystals.   In Litchfield Co., Connecticut, at Litchfield, black crystals.   At various points in the southern Allegheny Mts., as in Patrick Co., Virginia; in Cherokee Co., North Carolina, large coarse crystals; from Georgia in Fannin Co., loose in the soil in fine twinned crystals.

Named from σταυρός, *a cross*.

Use. — Occasionally a transparent stone is cut for a gem.

Kornerupine. — Near $MgAl_2SiO_6$. Contains also small amounts of alkalies, hydroxyl and boron. Orthorhombic.   In fibrous to columnar aggregates, resembling sillimanite. Prismatic cleavage. H. = 6·5. G. = 3·273 kornerupine; 3·341 prismatine. Colorless to white, or brown. Optically $-$. Ax. pl. || (100). $X = c$ axis. $\alpha = 1·665, \beta = 1·676, \gamma = 1·677$. 2V = 20°.

Kornerupine occurs at Fiskernæs, southwestern Greenland. Found in large clear crystals of a sea-green color and gem quality from Itrongahy, near Betroka, Madagascar. *Prismatine* is from Waldheim, Saxony.

Sapphirine. — $Mg_5Al_{12}Si_2O_{27}$. Monoclinic.   In indistinct tabular crystals. Usually in disseminated grains, or aggregations of grains. H. = 7·5. G. = 3·42–3·48. Color pale to dark blue or green. Optically $-$. Ax. pl. || (010). $Z \wedge c$ axis = 8°–15°. $\alpha = 1·704, \beta = 1·707, \gamma = 1·710$. 2V = 69°. From Fiskernæs, southwestern Greenland. Occurs near Betroka, Madagascar. From St. Urbain, Charlevoix Co., Quebec.

Grandidierite. — A basic silicate of aluminum, ferric iron, magnesium, ferrous iron, etc. Orthorhombic.   In elongated crystals. Two cleavages, (100) (010).   H. = 7·5.   G. = 3·0. Color bluish green. Optically $-$. Ax. pl. || (001). $X = a$ axis.   Strong dispersion, $\rho < v.$ $\alpha = 1·602,$ $\beta = 1·636,$ $\gamma = 1·639$. Strongly pleochroic; $X =$ dark blue-green, $Y =$ colorless, $Z =$ dark green.   Found in pegmatite at Andrahomana, southwest of Fort Dauphin, in southern Madagascar.

Serendibite. — $10(Ca,Mg)O.5Al_2O_3.B_2O_3.6SiO_2$. In irregular grains showing polysynthetic twinning; probably monoclinic or triclinic. H. = 6·7. G. = 3·4. Infusible. Color blue. Optically $+$. Strong dispersion, $\rho < v.$ $\alpha = 1·701,$ $\beta = 1·703,$ $\gamma = 1·706$. 2V nearly 90°. Pleochroism marked, $X = Y =$ yellow to green, $Z =$ blue.   From Gangapitiya near Ambakotte, Ceylon.

Silicomagnesiofluorite. — A fluosilicate of calcium and magnesium, perhaps, $H_2Ca_4Mg_3Si_2O_7F_{10}$. Radiating fibrous in spherical forms. H. = 2·5   G. = 2·9. Color ash gray, light greenish or bluish. Fusible. From Lupikko, near Impilaks on Lake Ladoga, Finland.

Aloisiite. Luigite. A basic silicate containing ferrous oxide, lime, magnesia, and soda. Amorphous. Color, brown to violet. Acts as a cement in a tuff found at Fort Portal, near the volcano Ruwenzori in the Western Province, Uganda, East Africa.

Poechite. $H_{16}Fe_8Mn_2Si_3O_{29}$. Colloidal. H. = 3·5–4. G. = 3·70. Color reddish brown to pitchy black. Opaque. Found in iron ore near Vareš, Bosnia, Yugloslavia.

SILICATES

Section B.   Chiefly Hydrous Species

The SILICATES of this second section include the true hydrous compounds, that is, those which contain water of crystallization, like the zeolites; also the hydrous amorphous species, as the clays, etc.   There are also included certain species — as the Micas, Talc, Kaolinite — which, while they yield water upon ignition, are without doubt to be taken as acid or basic metasilicates, orthosilicates, etc.   Their relation, however, is so close to other true hydrous species that it appears more natural to include them here than to have placed them in the preceding chapter with other acid and basic salts. Finally, some species are referred to here about whose chemical constitution and the part played by the water present there is still much doubt. The divisions recognized are as follows:

I. Zeolite Division

1. *Introductory Subdivision*. 2. *Zeolites*.

II. Mica Division

1. *Mica Group*. 2. *Clintonite Group*. 3. *Chlorite Group*.

III. Serpentine and Talc Division

Chiefly Silicates of Magnesium.

IV. Kaolin Division

Chiefly Silicates of Aluminum; for the most part belonging to the group of the clays.

V. Concluding Division

Species not included in the preceding divisions; chiefly silicates of the heavy metals, iron, manganese, etc.

I. Zeolite Division

1. *Introductory Subdivision*

Of the species here included, several, as Apophyllite, Okenite, etc., while not strictly ZEOLITES, are closely related to them in composition and method of occurrence. Pectolite (p. 567) and Prehnite (p. 626) are also sometimes classed here.

Inesite. — $H_2(Mn,Ca)_6Si_6O_{19}.3H_2O$. Triclinic. Crystals small, prismatic; also fibrous, radiated and spherulitic. Cleavages, (010) perfect, (100) good. H. = 6. G. = 3·029. Fusible at 3. Color rose- to flesh-red. Optically −. X nearly ⊥ (010). $\alpha = 1·609$, $\beta = 1·636, \gamma = 1·644$. 2V = 60°. Occurs at the manganese mines at Nanzenbach, northeast of Dillenburg, Hessen-Nassau. Noted at Nagybánya (Baia Mare), Rumania. *Rhodotilite* is the same species found at the Harstig mine, Pajsberg, near Persberg, Vermland, Sweden. Occurs also in Vermland at Jacobsberg in the Nordmark district and at Långbanshyttan. From Villa Carona, Durango, Mexico. *Agnolite* from Schemnitz (Selmecsbánya) in Slovakia of Czechoslovakia, described as a manganese silicate forming part of the material originally described as *manganocalcite*, has been shown to be identical with inesite.

Hillebrandite. — $Ca_2SiO_4.H_2O$. Orthorhombic; radiating fibrous. H. = 5·5. G. = 2·7. Fusible with difficulty. Color white. Optically −. $Z = c$ axis. Strong dispersion, $\rho < v$. $\alpha = 1·605$, $\beta = 1·61$, $\gamma = 1·612$. Found in contact zone between limestone and diorite in the Velardeña mining district, Durango, Mexico.

FOSHAGITE. $H_2Ca_5(SiO_4)_3.2H_2O$. Orthorhombic. Compact fibrous with fibers at times several inches long. G. = 2·36. Soluble in HCl with gelatinization. B.B. converted into a vitrified mass which is infusible. Snow-white color. Silky luster. Optically +. Positive elongation. $\alpha = \beta = 1·594, \gamma = 1·598$. Found in veins associated with vesuvianite, thaumasite and blue calcite at Crestmore, Riverside Co., California. Foshagite has the same crystal structure and same optical properties, except slightly lower indices, as hillebrandite and is probably an altered form of that mineral.

RADIOPHYLLITE. $CaSiO_3.H_2O$. In small white spheres, showing sometimes a radial-scaly structure. H. = 2–3. G. = 2·53. Uniaxial, −. Associated with phillipsite, calcite, and aragonite in fissures of nosean-phonolite from Schellkopf near Brenk, Rhineland.

Xonotlite. — $5CaSiO_3.H_2O$. Massive, in matted fibers, or from Virginia in needles. One good cleavage. Compact, very tough. H. = 6·5. G. = 2·7. Color white, gray, or pale pink. Fibers show parallel extinction and positive elongation. Biaxial, +. $X \perp$ cleavage. $\alpha = 1·583, \beta = 1·585, \gamma = 1·595$. 2V very small. Occurs at Tetela de Xonotla, Mexico. Also from Scotland and at Leesburg, Virginia. Materials from Santa Ynez, Santa Barbara Co., California, and Isle Royal, Michigan, described under the name *eakleite*, are xonotlite.

Crestmoreite. — Probably $4CaSiO_3.7H_2O$. Monoclinic. Compact. Color, snow-white. H. = 3. G. = 2·2. Easily fusible. Optically −. $\alpha = 1·593, \beta = 1·603, \gamma = 1·607$. 2V large. An alteration product of *wilkeite*. From Crestmore, Riverside Co., California.

Riversideite. — $2CaSiO_3.H_2O$. Orthorhombic? In compact fibrous veinlets. Silky luster. H. = 3. G. = 2·64. Easily fusible. Z = elongation. Indices 1·595–1·603. From Crestmore, Riverside Co., California.

Centrallasite. — $4CaO.7SiO_2.5H_2O$. Platy or lamellar to compact. H. = 2·5. G. = 2·51. Fuses with intumescence. Decomposed by HCl with separation of silica. White. Biaxial, −. $X \perp$ cleavage. $\alpha = 1·535, \beta = 1·548, \gamma = 1·549$. 2V small. Found in Bay of Fundy region, Nova Scotia and in pegmatite dike in limestone quarry at Crestmore, Riverside Co., California.

TRUSCOTTITE. $2(Ca,Mg)O.3SiO_2.3H_2O$. Color white. Pearly luster on cleavage faces. Biaxial, −. $\alpha = 1·528, \beta = 1·549, \gamma = 1·549$. 2V very small. As spherical aggregates from the Lelong Donok gold mine, Benkeelen, Sumatra.

Ganophyllite. — $7MnO.Al_2O_3.8SiO_2.6H_2O$. Monoclinic. In short prismatic crystals; also foliated, micaceous. Cleavage, (001) perfect. H. = 4–4·5. G. = 2·84. Easily fusible. Color brown. Optically −. Ax. pl. || (010). X nearly || c axis. Indices $\alpha = 1·573, \beta = 1·603, \gamma = 1·604$. From the Harstig mine, Pajsberg, near Persberg, Vermland, Sweden. Also at Franklin, Sussex Co., New Jersey.

RACEWINITE. A silicate of aluminum and calcium with a large amount of water which is given off readily and without materially changing the structure or optical properties. Coarsely crystalline but without crystal forms. Conchoidal fracture. H. = 2·5. G. = 1·94–1·98. Brown to yellow in thin splinters but nearly black in bulk. Bluish green on fresh fractures. Biaxial, −. $n =$ about 1.51. Large axial angle. From Highland Boy mine at Bingham, Utah, where it occurs in limestone with pyrite, etc.

Pumpellyite. — $6CaO.3Al_2O_3.7SiO_2.4H_2O$. Orthorhombic. In minute fibers or narrow plates, tabular || (001). Good cleavage || (001). H. = 5·5. G. = 3·2. Color, bluish green. Optically +. $Z \wedge c = 31°$. Y || length of fibers = b axis. $\alpha = 1·700, \beta = 1·707, \gamma = 1·718$. 2V large. Strong dispersion, $\rho < v$. $X = Z =$ colorless; Y − green. Found widely distributed in the amygdaloidal copper ores of Keweenaw Peninsula, Michigan. A supposed pumpellyite has been described from Limbe, northern Haiti. It agrees essentially with the above description but is considered monoclinic.

Lotrite. — $3(Ca,Mg)O.2(Al,Fe)_2O_3.4SiO_2.2H_2O$. Monoclinic? Massive, in an aggregate of small grains and leaves. One cleavage. H. = 7·5. G. = 3·2. Color green. Optically +. Ax. pl. normal to cleavage. $\beta = 1·67$. 2V small. Found in small veins in a chlorite schist in the valley of the Lotru, Transylvania, Rumania.

Okenite. — $H_2CaSi_2O_6.H_2O$. Triclinic. In minute blade-shaped crystals. Cleavage (010), perfect. Commonly fibrous; also compact. H. = 4·5–5. G. = 2·33. Fusible at 2·5. Color white, with a shade of yellow or blue. Optically −? Indices, 1·530–1·541. Occurs in basalt or related eruptive rocks; as in the Faroe Islands; Iceland; in Greenland at Kutdlisat in the Ritenbenk district. At Poona, near Bombay, India. On lava of Rio Putagan, a tributary of Rio Maule, Chile.

Gyrolite. — $H_2Ca_2Si_3O_9.H_2O$. Rhombohedral-tetartohedral. In white concretions, lamellar-radiate in structure. Perfect basal cleavage. H. = 3–4. G. = 2·34–2·45. Optically −. $\omega = 1·549$. At times abnormally biaxial. Found in Scotland on the Isle of Skye, Inverness, with stilbite, laumontite, etc. In Ireland at Collinward, near Belfast, Co. Antrim. Occurs on the Faroe Islands, and in Greenland. Found in various places in Bohemia, Czechoslovakia. From the railway cuts between Bombay and Poona, India. From Brazil in the Province of São Paula. In the United States at New Almaden, Santa Clara Co., California, with apophyllite. Also from Nova Scotia 25 miles southwest of Cape Blomidon, between Margaretville and Port George in Annapolis Co. *Reyerite* from Greenland is similar to gyrolite. *Zeophyllite* is a similar species which may be identical with gyrolite. Rhombohedral. In spherical forms with radiating foliated structure. Perfect basal cleavage. H. = 3. G. = 2·8. Color white. $\omega = 1·56$. From various localities in Bohemia; in the Salmon River district, near Higgins, Idaho.

APOPHYLLITE.

Tetragonal. Axis $c = 1\cdot2515$.

ay, $100 \wedge 310 = 18° 26'$.
cp, $001 \wedge 111 = 60° 32'$.

ap, $100 \wedge 111 = 52° 0'$.
pp', $111 \wedge \bar{1}11 = 76° 0'$.

965   **966**   **967**   **968**

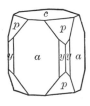 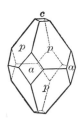 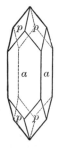 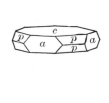

Habit varied; in square prisms (a (100)) usually short and terminated by c(001) or by c and p(111), and then resembling a cube or cubo-octahedron; also acute pyramidal (p (111)) with or without c and a; less often thin tabular ∥ c. Faces c often rough; a bright but vertically striated; p more or less uneven. Also massive and lamellar; rarely concentric radiated. X-ray study shows a tetragonal symmetry with two molecules contained in each unit cell. The silicon-oxygen tetrahedra are linked together in sheets having the composition Si_2O_5. These sheets lie parallel to (001) and their presence accounts for the perfect basal cleavage.

Cleavage: c(001) highly perfect; m(110) less so. Fracture uneven. Brittle. H. = $4\cdot5$–5. G. = $2\cdot3$–$2\cdot4$. Luster of c pearly; of other faces vitreous. Color white, or grayish; occasionally with a greenish, yellowish, or rose-red tint, flesh-red. Transparent; rarely nearly opaque. Birefringence low; usually +, also −. Often shows anomalous optical characters (Art. **441**, p. 329). Indices, $1\cdot535$–$1\cdot537$.

Comp. — $KF.Ca_4(Si_2O_5)_4.8H_2O$.

Various formulas for apophyllite have been suggested. The one given above represents quite closely the composition and further, fits the structure as determined by X-ray study.

Pyr., etc. — In the closed tube exfoliates, whitens, and yields water, which reacts acid. B.B. exfoliates, colors the flame violet (potash), and fuses to a white vesicular enamel. F. = $1\cdot5$. Decomposed by hydrochloric acid, with separation of slimy silica.

Diff. — Characterized by its tetragonal form, the square prism and pyramid the common habits; by the perfect basal cleavage and pearly luster on this surface.

Obs. — Occurs commonly as a secondary mineral in basalt and related rocks, with various zeolites, also datolite, pectolite, calcite; also occasionally in cavities in granite, gneiss, etc. Associated with certain ore veins and has been noted in cavities in limestone. The more important localities for the occurrence of apophyllite follow. In Bohemia of Czechoslovakia at several points near Aussig; in Trentino, Italy, on the Seiser Alpe (Mont da Sores). At Andreasberg in the Harz Mts., of a delicate pink color, in silver veins. On the Isle of Skye in Inverness, Scotland, at Talisker, etc. In Ireland at Collinward near Belfast, Co. Antrim. From the Faroe Islands and on Iceland at Berufjord, etc. From the De Beers diamond mines at Kimberley, Griqualand-West, South Africa. From India in crystals of unrivalled size and beauty in connection with trap rocks at Lonauli and Poona southeast of Bombay and in the Western Ghats or Sahyadri Mts., at Thul Ghat, Bhor Ghat, etc. From Guanajuato, Mexico, often of a beautiful pink color implanted upon amethyst.

In the United States with the trap rocks of southeastern New Jersey, at Bergen Hill, Paterson, West Paterson, etc., associated with various zeolites, etc. In Pennsylvania at the French Creek mines, Chester Co. With the copper deposits of Keweenaw Co., Michigan, at the Phœnix mine, etc. In Santa Clara Co., California, in large crystals at the mercury mines of New Almaden with bitumen and often stained brown by it.

In the Bay of Fundy region of Nova Scotia it occurs both massive and crystallized at Isle Haute, Cape d'Or, Cape Split, Cape Blomidon, etc.

Named by Haüy in allusion to its tendency to exfoliate under the blowpipe, from ἀπό and φύλλον, a leaf. Its whitish pearly aspect, resembling the eye of a fish after boiling, gave rise to the earlier name *Ichthyophthalmite*, from ἰχθύς, *fish*, ὀφθαλμός, *eye*.

2. *Zeolites*

The ZEOLITES form a family of well-defined hydrous silicates, closely related to each other in composition, in conditions of formation, and hence in mode of occurrence. They are often with right spoken of as analogous to the Feldspars, like which they are all silicates of aluminum with sodium and calcium chiefly, also rarely barium and strontium; magnesium, iron, etc., are absent or present only through impurity or alteration. Further, the composition in a number of cases corresponds to that of a hydrated feldspar; while fusion and slow recrystallization result in the formation from some of them of anorthite ($CaAl_2Si_2O_8$) or a calcium-albite ($CaAl_2Si_6O_{16}$) as shown by Doelter. The theory has been advanced by Winchell that in the zeolites the ratio $Al_2O_3 : CaO + Na_2O$ is always $1 : 1$; that the $Al + Si : O$ ratio is always $1 : 2$; and that in any given isomorphous part of the Zeolite Group the $Ca + Na : O$ ratio is constant. The first ratio is confirmed by G. Tschermak, who further concludes that all zeolites contain either $CaAl_2Si_2O_8$ or $Na_2Al_2Si_2O_8$ combined with a silicic acid, hydroxyl and water of crystallization. The Zeolites do not, however, form a single group of species related in crystallization, like the Feldspars, but include a number of independent groups widely diverse in form and distinct in composition; chief among these are the monoclinic PHILLIPSITE GROUP; the rhombohedral CHABAZITE GROUP, and the orthorhombic (and monoclinic) NATROLITE GROUP. A transition in composition between certain end compounds has been more or less well established in certain cases, but, unlike the Feldspars, with these species calcium and sodium seem to replace one another and an increase in alkali does not necessarily go with an increase in silica.

The water contained in the zeolites differs from the ordinary water of crystallization of other minerals. When the zeolites are heated, the water is given off readily and continuously and not in certain amounts at definite temperatures as is usually the case. Further, the partially dehydrated mineral can again take up an equal amount of water if exposed to water vapor. The optical characters change gradually on dehydration, but apparently the atomic structure (as shown by X-ray study) remains the same unless the process is carried nearly to completion. Further, the partially dehydrated mineral can absorb other materials in place of the water, such as air, ammonia, alcohol, hydrogen sulphide, iodine, etc. It would appear that the water occupies at least an unimportant position in the atomic structure of the zeolites, possibly being present as adsorbed water held in openings or channels of the structure (see further under analcite, p. 652). Another chemical peculiarity is the fact that the alkali metal present may be artificially replaced by silver and other metals.

Like other hydrous silicates they are characterized by inferior hardness, chiefly from 3·5 to 5·5, and the specific gravity is also lower than with corresponding anhydrous species, chiefly 2·0 to 2·4. Corresponding to these characters, they are rather readily decomposed by acids, many of them with gelatinization. The intumescence B.B., which gives the name to the family (from ζεῖν, *to boil*, and λίθος, *stone*) is characteristic of a large part of the species.

The Zeolites are all secondary minerals, occurring most commonly in cavities and veins in basic igneous rocks, as basalt, diabase, etc.; less frequently in granite, gneiss, etc. In these cases the lime and the soda in part have been chiefly yielded by the feldspar; the soda also by nephelite, sodalite, etc.; potash by leucite, etc. The different species of the family are often associated with each other; also with pectolite and apophyllite (sometimes included with the zeolites), datolite, prehnite and, further, calcite. Many of the zeolites have been produced synthetically by various hydrochemical reactions. In general they appear to have been formed in nature by reactions upon the feldspar or feldspathoid minerals.

Mordenite Group

Considerable confusion exists as to the relationships of the three following species. Schaller considers mordenite and flokite to be identical with the composition $(Ca,Na_2).Al_2O_3.9SiO_2.6H_2O$, monoclinic or triclinic crystallization with small inclined extinction (about 5°). He further considers ptilolite to have $10SiO_2$ and $7H_2O$, to be orthorhombic with parallel extinction and to be the most common member of the group. The mineral described by Pirsson from Wyoming and accepted by Dana as crystallized mordenite he considers to be a dimorphous form of ptilolite, monoclinic, with a large extinction angle, and gives it the name *clinoptilolite*. On the other hand, Bøggild considers flokite to be identical with ptilolite, and Walker and Parsons suggest that ptilolite is identical with mordenite.

Ptilolite. — $(Ca,K_2Na_2)O.Al_2O_3.10SiO_2.5H_2O$? Orthorhombic. In short capillary needles, aggregated in delicate tufts. Cleavages (100), (010). H. = 5. G. = 2·1. Easily fusible. Insoluble at HCl. Colorless, white. Optically −. Ax. pl. || (100). $X = c$ axis. $n = 1·475$. Occurs upon a bluish chalcedony in cavities in a vesicular augite-andesite found in fragments in the conglomerate beds of Green and Table mountains, Jefferson Co., and from Silver Cliff, Custer Co., Colorado; in Idaho at Challis, Custer Co. Found also at San Piero in Campo, Elba and at Berufjord, Iceland.

Flokite. — $(Ca,Na_2)O.Al_2O_3.9SiO_2.6H_2O$? Monoclinic. In slender prismatic crystals. Perfect cleavages parallel to (100) and (010). H. = 5. G. = 2·10. Fuses with intumescence. Colorless and transparent. Indices, 1·472–1·474. From Iceland.

Mordenite. — $(Ca,Na_2)O.Al_2O_3.9SiO_2.6H_2O$. Monoclinic. In minute crystals resembling heulandite in habit and angles (see below); also in small hemispherical or reniform concretions with fibrous structure. Cleavage (010), perfect. H. = 3–4. G. = 2·15. Fusible (4–5) with intumescence. Color white, yellowish or pinkish. Optically + or −. $Z = b$ axis, $X \wedge c$ axis = 4°. $\beta = 1·475$. Occurs near Morden, King's Co., Nova Scotia, in trap; also in western Wyoming near Hoodoo Peak; on the ridge forming the divide between Clark's Fork and the Lamar River (see *clinoptilolite* above). Also from Trentino, Italy, on the Seiser Alpe (Mont da Sores) and the Faroe Islands.

Ferrierite. — $R_2Al_2(Si_2O_5)_5.6H_2O$, $RO = MgO,Na_2O,H_2O$. Orthorhombic. In radiating groups of thin crystals, tabular || (100) and elongated || c axis. Cleavage, perfect || (100). H. = 3. G. = 2·150. Colorless or white. Vitreous to pearly luster. Optically +. Ax. pl. || (010). $X = a$ axis. $\alpha = 1·478$, $\beta = 1·479$, $\gamma = 1·482$. $2V = 50°$. Found in a railroad cut on north shore of Kamloops Lake, British Columbia, with chalcedony in seams in basalt.

Heulandite Group. Monoclinic

The following minerals show such close relationships in their crystallographic constants and similarities in their composition that it is probable that they form an isomorphous series.

		$a : b : c$	β
Mordenite	$(Ca,Na_2)O.Al_2O_3.9SiO_26H_2O$	$0{\cdot}401 : 1 : 0{\cdot}428$	$88°\ 31'$
Heulandite	$(Ca,Na_2)O.Al_2O_3.6SiO_2.5H_2O$	$0{\cdot}404 : 1 : 0{\cdot}429$	$88°\ 34'$
Epistilbite	$(Ca,Na_2)O.Al_2O_3.6SiO_2.5H_2O$	$0{\cdot}419 : 1 : 0{\cdot}432$	$89°\ 20'{}^*$
Brewsterite	$(Sr,Ba,Ca)O.Al_2O_3.6SiO_2.5H_2O$	$0{\cdot}405 : 1 : 0{\cdot}420$	$86°\ 20'$

HEULANDITE. Stilbite *some authors.*

Monoclinic. Axes $a : b : c = 0{\cdot}4035 : 1 : 0{\cdot}4293$; $\beta = 88°\ 34\frac{1}{2}'$.

mm''', $110 \wedge 1\bar{1}0 = 43°\ 56'$. cs, $001 \wedge \bar{2}01 = 66°\ 0'$.
ct, $001 \wedge 201 = 63°\ 40'$. cx, $001 \wedge 021 = 40°\ 38\frac{1}{2}'$.

Crystals sometimes flattened $\parallel b(010)$, the surface of pearly luster (Fig. 969; also Fig. 21, p. 13); form often suggestive of the orthorhombic system, since the angles cs and ct differ but little. Also in globular forms; granular. X-ray study gives a unit cell the dimensions of which are proportional in terms of the above axial ratio to $1a : 1b : 2c$.

969

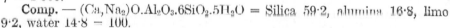

Cleavage: $b(010)$ perfect. Fracture subconchoidal to uneven. Brittle. H. = 3·5–4. G. = 2·18–2·22. Luster of b strong pearly; of other faces vitreous. Color various shades of white, passing into red, gray and brown. Streak white. Transparent to subtranslucent. Optically $+$. Ax. pl. and $Z \perp$ $b(010)$. Ax. pl. and X for some localities nearly $\parallel c(001)$; also for others nearly $\perp c$ in white light. Axial angle variable, from 0° to 92°; usually 2V nearly 34°. $\alpha = 1{\cdot}498$. $\beta = 1{\cdot}499$. $\gamma = 1{\cdot}505$.

Comp. — $(Ca,Na_2)O.Al_2O_3.6SiO_2.5H_2O$ = Silica 59·2, alumina 16·8, limo 9·2, water 14·8 = 100.

Strontia is usually present, sometimes up to 3·6 per cent.

When heulandite is heated from 25° to 190° C. there is a gradual and progressive rotation of the axial plane with a decrease in the size of the optic angle; when it is heated to higher temperatures the rotation of the axial plane is more rapid and the size of the optic angle increases. This change at 190° is due to the formation of the three molecule hydrate (*meta-heulandite*). It is always biaxial positive and the bisectrix is always parallel to the b axis. Above 177° the structure of the heulandite breaks down, although the effect of this is apparent in the optical tests only after the specimen has stood for a considerable period of time.

Pyr. — As with stilbite, p. 648.

Obs. — Heulandite occurs principally in cavities in basaltic and related rocks, associated with chabazite, stilbite and other zeolites; it may also occur in other volcanic rocks; in granites and pegmatites; more rarely in gneisses or crystalline schists. Occasionally in metalliferous veins.

The finest specimens of this species come from Iceland at Berufjord and Theigarhorn; in India on islands near Bombay and at Lonauli and Poona and in the Western Ghats or Sahyadri Mts. In Europe it occurs at Andreasberg in the Harz Mts.; in Val di Fassa, Trentino, Italy; in Switzerland at Giebelbach near Fiesch in the Rhône valley, Valais, and elsewhere. From Scotland at Campsie Hills, Stirling (a red variety with red stilbite) and at Kilpatrick Hills near Glasgow, Dumbarton. From the Faroe Islands.

---

* The original crystallographic orientation of epistilbite has been changed in order to show its relationship to this group.

In the United States with the trap rocks of northeastern New Jersey at Bergen Hill, West Paterson, and Great Notch. In minute crystals (*beaumontite*) at Jones's Falls, near Baltimore, Maryland. In the Bay of Fundy district, Nova Scotia, at Peter's Point; also at Cape Blomidon, Isle Haute, Wasson's Bluff, etc.

Named after the English mineralogical collector, H. Heuland, whose cabinet was the basis of the classical work (1837) of Lévy.

Epistilbite. — Like heulandite, $(Ca,Na_2)O.Al_2O_3.6SiO_2.5H_2O$. Monoclinic; uniformly twinned; habit prismatic. In radiated spherical aggregations; also granular. Cleavage || (010), perfect. H. = 4. G. = 2·25. Easily fusible. Color white. Optically −. Ax. pl. || (010). $Z \wedge c$ axis = 9°. Strong dispersion, $\rho < v$. $\alpha = 1·502, \beta = 1·510, \gamma = 1·512$. 2V = 44°. Occurs in Iceland at Berufjord with scolecite and at Theigarhorn, Djupivogr, and elsewhere; on the Faroe Islands; on the Isle of Skye, Inverness, Scotland, in small flesh-colored crystals. At Giebelbach near Fiesch in the Rhône valley, Valais, Switzerland. From Poona near Bombay, India. In small reddish crystals at Margaretsville, Annapolis Co., Nova Scotia. *Parastilbite* is from near Bogarfjord, Iceland and *reissite* is from Santorin (Thira) Island, Cyclades, Greece.

Brewsterite. — $(Sr,Ba,Ca)O.Al_2O_3.6SiO_2.5H_2O$. Monoclinic. In prismatic crystals. Cleavage || (010), perfect. H. = 5. G. = 2·45. Easily fusible. Color white, inclining to yellow and gray. Optically +. Ax. pl. ⊥ (010). $X \wedge c$ axis = 19°–34°. $\alpha = 1·510, \beta = 1·512, \gamma = 1·523$. First observed at Strontian in Argyll, Scotland. Occurs also at the Giant's Causeway, Antrim, Ireland. From the lead mines of St. Turpet, in the Black Forest, Baden. At the Col du Bonhomme, southwest of Mont Blanc, Savoie, France; near Barèges in Hautes-Pyrénées; and has been reported from St. Cristophe near Bourg d'Oisans, Isère.

Phillipsite Group. Monoclinic

		$a : b : c$	β
Wellsite	$(Ba,Ca,K_2)Al_2Si_3O_{10}.3H_2O$	0·768 : 1 : 1·245	53° 27′
Phillipsite	$(K_2,Ca)Al_2Si_4O_{12}.4\frac{1}{2}H_2O$	0·7095 : 1 : 1·2563	55° 37′
Harmotome	$(K_2,Ba)Al_2Si_5O_{14}.5H_2O$	0·7032 : 1 : 1·2310	55° 10′
Stilbite	$(Na_2,Ca)Al_2Si_6O_{16}.6H_2O$	0·7623 : 1 : 1·1940	50° 50′

The above species, while crystallizing in the monoclinic system, are remarkable for the pseudo-symmetry exhibited by their twinned forms. Certain of these twins are pseudo-orthorhombic, others pseudo-tetragonal and more complex twins even pseudo-isometric. Tschermak suggests another crystal orientation for the minerals of this group whereby β becomes nearly 90°. The orthorhombic minerals epidesmine and stellerite are probably closely related to the group.

970   971

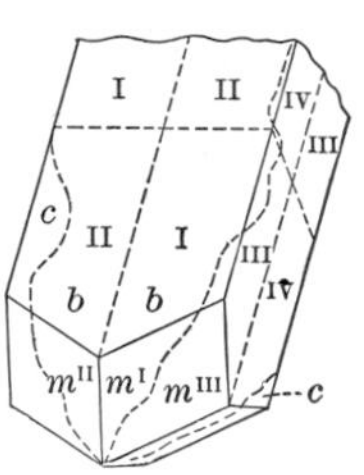

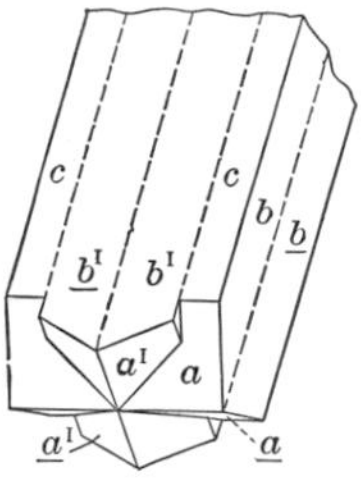

The chemical compositions of the different members of the group have been variously interpreted, and probably different isomorphous molecules are present. The above formulas are based upon reliable analyses of certain typical occurrences.

Wellsite. — $(Ba,Ca,K_2)Al_2Si_3O_{10}.3H_2O$. Sr and Na also present in small amount. Has been considered as a barium-bearing phillipsite. Monoclinic (axes above); in complex twins, analogous to those of phillipsite and harmotome (Figs. 970, 971). Brittle. No cleavage. H. = 4–4·5. G. = 2·278–2·366. Easily fusible. Luster vitreous. Colorless to white. Optically +. Bx ⊥ b(010). $\alpha = 1·498, \beta = 1·50, \gamma = 1·503$.

Occurs at the Buck Creek (Cullakanee) corundum mine in Clay Co., North Carolina; in isolated crystals attached to feldspar, also to hornblende and corundum; intimately associated with chabazite. Also found at Kurzy near Simferopol, Crimea, Russia (this occurrence stated to be a lime-rich harmotome).

PHILLIPSITE.

Monoclinic.   Axes $a : b : c = 0.7095 : 1 : 1.2563$; $\beta = 55° 37'$.

mm''', $110 \wedge 1\bar{1}0 = 60° 42'$. cm, $001 \wedge 110 = 60° 50'$.
$a'f$, $\bar{1}00 \wedge \bar{1}01 = 34° 23'$. ee', $011 \wedge 0\bar{1}1 = 92° 4'$.

Crystals uniformly penetration-twins, but often simulating orthorhombic or tetragonal forms.   Twins sometimes, but rarely, simple (1) with tw. pl. $c(001)$, and then cruciform so that diagonal parts on $b(010)$ belong together, hence a fourfold striation, ‖ edge b/m, may be often observed on b.   (2) Double twins, the simple twins just noted united with $e(011)$ as tw. pl., and, since ee' varies but little from 90°, the result is a nearly square prism, terminated by what appear to be pyramidal faces each with a double series of striations away from the medial line.   See Figs. 478–480, p. 193; also Fig. 426, p. 185.   Faces $b(010)$ often finely striated as just noted, but striations sometimes absent and in general not so distinct as with harmotome; also $m(110)$ striated ‖ edge b/m.   Crystals either isolated, or grouped in tufts or spheres, radiated within and bristled with angles at surface.

972

Cleavage: $c(001)$, $b(010)$, rather distinct.   Fracture uneven.   Brittle. H. = 4–4.5.   G. = 2.2.   Luster vitreous.   Color white, sometimes reddish. Streak uncolored.   Translucent to opaque.   Optically +.   Ax. pl. and $X \perp$ $b(010)$.   The position of Z varies widely.   Indices variable, $\alpha = 1.498$, $\beta = 1.500$, $\gamma = 1.503$.

Comp. — In some cases the formula is $(K_2,Ca)Al_2Si_4O_{12}.4\frac{1}{2}H_2O$.

Pyr., etc. — B.B. crumbles and fuses at 3 to a white enamel.   Gelatinizes with hydrochloric acid.

Obs. — Phillipsite occurs in cavities in basalts and phonolites.   It has been formed at comparatively low temperatures through the decomposition of feldspars by hot waters.

Found to the south of Rome, Italy, at Capo di Bove in small colorless crystals, and at Vallerano, etc.; among the lavas of Mte. Somma, Vesuvius; from Sicily.   Occurs in many places in Germany, as at the Kaiserstuhl in Baden with faujasite; in Rhineland in the basalt of the Limperich Kopf near Asbach and at Idar in Berkenfeld; in Hessen-Nassau at Stempel near Marburg; in Hessen at Annerod east of Giessen and at Nidda.   In translucent crystals in basalt, at the Giant's Causeway, Co. Antrim, Ireland.   In Victoria, Australia, at Richmond and Collingwood, parts of Melbourne.   *Pseudophillipsite*, found near Rome, Italy, differs from phillipsite only in the manner in which it loses water on heating.

973

HARMOTOME.

Monoclinic.   Axes $a : b : c = 0.7032 : 1 : 1.2310$; $\beta = 55° 10'$.

Crystals uniformly cruciform penetration-twins with $c(001)$ as tw. pl.; either (1) simple twins (Fig. 973) or (2) united as fourlings with tw. pl. $e(011)$.   These double twins often have the aspect of a square prism with diagonal pyramid, the latter with characteristic feather-like striations from the medial line.   Also in more complex groups analogous to those of phillipsite.

Cleavage: $b(010)$ easy, $c(001)$ less so.   Fracture uneven to subconchoidal.   Brittle.   H. = 4.5.   G. = 2.44–2.50.   Luster vitreous. Color white; passing into gray, yellow, red or brown.   Streak white.   Subtransparent to translucent. Ax. pl. and $Z \perp b(010)$. $X \wedge c$ axis $= 60°$. Optically $+$. $2V = 43°$. $\alpha = 1 \cdot 503$. $\beta = 1 \cdot 505$. $\gamma = 1 \cdot 508$.

Comp. — In part $(K_2,Ba)Al_2Si_5O_{14}.5H_2O$.

Pyr., etc. — B.B. whitens, then crumbles and fuses without intumescence at $3 \cdot 5$ to a white translucent glass. Some varieties phosphoresce when heated. Decomposed by hydrochloric acid without gelatinizing.

Obs. — Occurs in basalt and similar eruptive rocks, also phonolite, trachyte; not infrequently on gneiss, and in some metalliferous veins. Often associated with chabazite. Harmotome is found at Bodenmais in Bavaria; at Oberstein in Rhineland, implanted on agate in siliceous geodes; in a metalliferous vein at Andreasberg in the Harz Mts. At Kongsberg, Norway. In Scotland at Strontian in Argyll in fine crystals; in the amygdaloid of the Kilpatrick Hills, Dumbarton; and in the Campsie Hills, Stirling.

In the United States, in small brown crystals with stilbite on the gneiss of New York Island; near Port Arthur, Lake Superior.

Named from ἁρμός, *joint*, and τέμνειν, *to cut*, alluding to the fact that the pyramid (made by the prismatic faces in twinning position) divides parallel to the plane that passes through the terminal edges.

STILBITE. Desmine.

Monoclinic.   Axes $a : b : c = 0 \cdot 7623 : 1 : 1 \cdot 1940$; $\beta = 50° 50'$.

Crystals uniformly cruciform penetration-twins with tw. pl. $c(001)$, analogous to phillipsite and harmotome. The apparent form a rhombic pyramid

974

whose faces are in fact formed by the prism faces of the two individuals; the vertical faces being then the pinacoids $b(010)$ and $c(001)$ (cf. Figs. 642–644, p. 328). Usually thin tabular $\parallel b(010)$. These compound crystals are often grouped in nearly parallel position, forming sheaf-like aggregates with the side face (b), showing its characteristic pearly luster, often deeply depressed. Also divergent or radiated; sometimes globular and thin lamellar-columnar.

Cleavage: $b(010)$ perfect. Fracture uneven. Brittle. H. $= 3 \cdot 5$–4. G. $= 2 \cdot 094$–$2 \cdot 205$. Luster vitreous; of $b(010)$ pearly. Color white; occasionally yellow, brown or red, to brick-red. Streak uncolored. Transparent to translucent. Optically $-$. Ax. pl. $\parallel b(010)$. X inclined $5°$ to axis a. $2V = 33°$ (approx.). $\alpha = 1 \cdot 494$. $\beta = 1 \cdot 498$. $\gamma = 1 \cdot 500$.

Comp. — For most varieties $(Na_2,Ca)Al_2Si_6O_{16}.6H_2O$ or $(Na_2,Ca)O.Al_2O_3. 6SiO_2.6H_2O$.

Some kinds show a lower percentage of silica, and these have been called *hypostilbite*.

Pyr., etc. — B.B. exfoliates, swells up, curves into fan-like or vermicular forms, and fuses to a white enamel. F. $= 2$–$2 \cdot 5$. Decomposed by hydrochloric acid, without gelatinizing.

Diff. — Characterized by the frequency of radiating or sheaf-like forms; by the pearly luster on the clinopinacoid. Does not gelatinize with acids.

Obs. — Stilbite occurs mostly in amygdaloidal cavities in basalt, and similar rocks. It is also found in some metalliferous veins, and in granite and gneiss.

Stilbite occurs in Trentino, Italy, at the Pufler-loch on the Seiser Alpe (Mont da Sores) (*puflerite*); on the granite of Striegau, Silesia; at Andreasberg in the Harz Mts. From Giebelbach near Fiesch in the Rhône valley, Valais, Switzerland. In Aust-Agder, Norway, at Arendal with iron ore. In Scotland at Kilmacolm west of Glasgow in Renfrew and in red crystals at Old Kilpatrick, Dumbarton; from the Isle of Skye, Inverness. Abundant on the Faroe Islands; from Iceland at Berufjord, Rödefjord, etc. A common mineral in the Deccan trap area of India; fine crystals come from the Bhor and Thul Ghats in the Western Ghats or Sahyadri Mts., at Poona, etc. In fine specimens from Guanajuato and elsewhere in Mexico.

In the United States of fine quality in the trap quarries of northeastern New Jersey at

Bergen Hill, West Paterson, Great Notch, Upper Montclair, etc. In Pennsylvania at Frankford near Philadelphia. In Nova Scotia at the zeolite localities in the Bay of Fundy region; in Cumberland Co. at Partridge Island, Two Islands, etc.; at various points in Kings and Annapolis counties.

The name *stilbite* is from $\sigma\tau\iota\lambda\beta\eta$, *luster*, and *desmine* from $\delta\acute{\epsilon}\sigma\mu\eta$, *a bundle*.

Epidesmine. — Comp. same as for stilbite. Orthorhombic. In minute crystals, only the three pinacoids showing. Cleavages parallel to both vertical pinacoids. G. = 2·16. Easily fusible with intumescence. Colorless to yellow. Optically −. Ax. pl. || (100). $X \parallel c$ axis. $\alpha = 1\cdot485, \beta = 1\cdot495, \gamma = 1\cdot500$. Occurs as a crust on calcite from Schwarzenberg, Saxony. Found at Moore's Station, Mercer Co., New Jersey and south of Reading, Berks Co., Pennsylvania.

Stellerite. — $CaAl_2Si_7O_{18}.7H_2O$. Orthorhombic. Crystals tabular parallel to $b(010)$. Cleavage perfect parallel to $b(010)$, imperfect parallel to $a(100)$ and $c(001)$. H. = 3·5–4. G. = 2·12. Optically −. Ax. pl. || (010). $X = c$ axis. $\alpha = 1\cdot485, \beta = 1\cdot493, \gamma = 1\cdot495$. 2V = 44°. Found in cavity in a diabase tuff, Copper Island, Commander Islands, Behring Sea. Reported near Juneau, Alaska.

Gismondite. — Perhaps $(Ca,K_2)Al_2Si_2O_8.4H_2O$. Monoclinic. In twinned pyramidal crystals, pseudo-tetragonal. H. = 4·5. G. = 2·265. Easily fusible. Colorless or white, bluish white, grayish, reddish. Optically −. Ax. pl. ⊥ (010). Z nearly ⊥ (100). $\beta = 1\cdot539$. 2V large. Gismondite is of secondary origin formed at comparatively low temperatures by the alteration of plagioclase. It occurs in the leucitophyre of the Albani Mts., south of Rome, Italy, at Capo di Bove, and elsewhere; from Vesuvius. In Bohemia of Czechoslovakia at Salesel (Salesl), south of Aussig; from Schiffenberg near Giessen and elsewhere in Hessen; on the Hoher Berg, east of Bühne, Westphalia, in relatively large crystals.

LAUMONTITE. Leonhardite. Caporcianite.

Monoclinic. Axes $a : b : c = 1\cdot1451 : 1 : 0\cdot5906$; $\beta = 68° 46'$.

Twins: tw. pl. $a(100)$. Common form the prism $m(mm''' \ 110 \wedge 1\bar{1}0 = 93° 44')$ with oblique termination $e, \bar{2}01$ ($ce \ 001 \wedge \bar{2}01 = 56° 55'$). Also columnar, radiating and divergent.

Cleavage: $b(010)$ and $m(110)$ very perfect; $a(100)$ imperfect. Fracture uneven. Not very brittle. H. = 3·5–4. G. = 2·25–2·36. Luster vitreous, inclining to pearly upon the faces of cleavage. Color white, passing into yellow or gray, sometimes red. Streak uncolored. Transparent to translucent; becoming opaque and usually pulverulent on exposure. Optically . Ax. pl. || $b(010)$. $X \wedge c$ axis = +65° to 70°. Dispersion large, $\rho < v$; inclined, slight. 2V = 25°. $\alpha = 1\cdot513$. $\beta = 1\cdot524$. $\gamma = 1\cdot525$.

Comp. — $(Ca,Na_2)Al_2Si_4O_{12}.4H_2O$.

Var. — *β-Leonhardite* is a laumontite which has lost part of its water (to one molecule), and the same is probably true of *caporcianite*. A primary mineral with the same composition but with slightly different physical properties has been called *α-leonhardite*. *Schneiderite* is laumontite from the serpentine of Montecatini, Italy, which has undergone alteration through the action of magnesian solutions. *Vanadio-laumontite* is a yellow-red variety containing 2·5 per cent V_2O_5, from Ferghana, Turkestan.

Pyr., etc. — B.B. swells up and fuses at 2·5–3 to a white enamel. Gelatinizes with hydrochloric acid.

Obs. — Laumontite occurs in the cavities of many rock types; in basalt and similar eruptive rocks; also in the more acid rocks, as trachyte, andesite, granite, syenite, and in the metamorphic rocks, gneiss, mica schist, clay slate. Often in metalliferous veins, especially copper deposits.

Its principal occurrences follow. In Transylvania of Rumania at Nagy-Ág; in the Tyrol, Austria, in the Floitental of the Zillertal (*leonhardite*). From the Plauenscher Grund, near Dresden, Saxony. In Italy from Baveno on Lago Maggiore, Piedmont; *caporcianite* is from Mte. de Caporciano near Montecatini, Tuscany. From St. Gotthard, Ticino, Switzerland. In Scotland near Glasgow in the Kilpatrick Hills, Dumbarton, and at Kilmacolm, Renfrew. In India at Poona near Bombay and in the Western Ghats or Sahyadri Mts.

In the United States in northeastern New Jersey at Bergen Hill, West Paterson, Great Notch, etc. From near Philadelphia, Pennsylvania. Abundant in many places in the copper deposits in Keweenaw Co., Michigan. In Nova Scotia at various zeolite localities in the Bay of Fundy district.

Laubanite. — $Ca_2Al_2Si_5O_{15}.6H_2O$. Fibrous. H. = 4·5–5. G. = 2·23. Fusible. Color snow-white. Uniaxial, +. $\omega = 1·475$, $\epsilon = 1·486$. Occurs upon phillipsite in basalt at Lauban, Silesia.

Chabazite Group. Rhombohedral

		rr'	c		
Chabazite	$(Ca,Na_2)Al_2Si_4O_{12}.6H_2O$	85° 14′	1·0860		
Gmelinite	$(Na_2Ca)Al_2Si_4O_{12}.6H_2O$	68° 8′	0·7345	or	$\frac{3}{2}c = 1·1017$
Levynite	$CaAl_2Si_3O_{10}.5H_2O$	73° 56′	0·8357	or	$\frac{4}{3}c = 1·1143$

The Chabazite Group includes these three rhombohedral species. The fundamental rhombohedrons have different angles, but, as shown in the axial ratios above, they are closely related, since, taking the rhombohedron of Chabazite as the basis, that of Gmelinite has the symbol $(20\bar{2}3)$ and of Levynite $(30\bar{3}4)$. Because of optical evidence these minerals have been assumed to be triclinic or monoclinic and owe their hexagonal character to intimate twinning. The monoclinic crystal constants that have been derived on this assumption have close relations to those of the members of the Phillipsite Group.

The variation in composition often observed in the first two species has led to the rather plausible hypothesis that they are to be viewed as isomorphous mixtures of the feldspar-like compounds

$$(Ca,Na_2)Al_2Si_2O_8.4H_2O, \qquad (Ca,Na_2)Al_2Si_6O_{16}.8H_2O.$$

The exact formulas for these minerals are, however, still uncertain.

CHABAZITE.

Rhombohedral. Axis $c = 1·0860$; $0001 \wedge 10\bar{1}1 = 51° 25\frac{3}{4}′$.

Twins: (1) tw. axis c axis, penetration-twins common. (2) Tw. pl. $r(10\bar{1}1)$; contact-twins, rare. Form commonly the simple rhombohedron

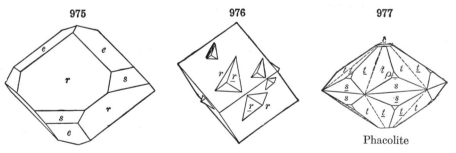

975   976   977

Phacolite

varying little in angle from a cube (rr' $10\bar{1}1 \wedge \bar{1}101 = 85° 14′$); also r and $e(01\bar{1}2)$. ($ee' = 54° 47′$). Also in complex twins. Also amorphous.

Cleavage: $r(10\bar{1}1)$ rather distinct. Fracture uneven. Brittle. H. = 4–5. G. = 2·08–2·16. Luster vitreous. Color white, flesh-red; streak uncolored. Transparent to translucent. Optically —; also + (Andreasberg, also *haydenite*). Birefringence low. The interference-figure usually confused; sometimes distinctly biaxial; basal sections then divided into sharply defined sectors with different optical orientation. These anomalous optical characters have been assumed to be due to the relative amounts of Ca and Na$_2$ present, to the amount of water, etc. Mean refractive index 1·48.

Var. — 1. *Ordinary.* The most common form is the fundamental rhombohedron, in which the angle is so near 90° that the crystals were at first mistaken for cubes. *Acadialite*, from Nova Scotia (*Acadia* of the French of 18th century), is a reddish chabazite; sometimes nearly colorless. *Haydenite* is a yellowish variety in small crystals from Jones's Falls, near Baltimore, Maryland. 2. *Phacolite* is a colorless variety occurring in twins of hexagonal form (Fig. 977), and lenticular in shape (whence the name from φακός, *a bean*); the original was from Leipa in Bohemia. Here belongs also *herschelite* (*seebachite*) from Richmond, Victoria; the composite twins of great variety and beauty. Probably also the original herschelite from Sicily. It occurs in flat, almost tabular, hexagonal prisms with rounded terminations divided into six sectors.

Comp. — Somewhat uncertain, since a rather wide variation is often noted even among specimens from the same locality. The composition usually corresponds to (Ca,Na$_2$)Al$_2$Si$_4$O$_{12}$.6H$_2$O.

Potassium is present in small amount, also sometimes, barium and strontium.

Pyr., etc. — B.B. intumesces and fuses to a blebby glass, nearly opaque. Decomposed by hydrochloric acid, with separation of slimy silica.

Diff. — Characterized by rhombohedral form (resembling a cube). It is harder than calcite and does not effervesce with acid; unlike calcite and fluorite in cleavage; fuses B.B. with intumescence unlike analcite.

Obs. — Chabazite occurs mostly in the amygdaloidal cavities of basalt and related rocks, and occasionally in gneiss, syenite, mica schist, hornblende schist, clay slate, etc. It has been found as a recent formation in connection with hot springs. It is a common zeolite, and only the most interesting occurrences can be listed here. In Bohemia of Czechoslovakia *phacolite* is found at various points near Böhmisch-Leipa; at Salesel (Salesl) and elsewhere near Aussig; Rübendörfel near Leitmeritz. In Hungary at Bogdány near Visegrad. In Rhineland at Oberstein with harmotome. In Hessen at various points near Giessen; at Striegau in Silesia. In Italy at Seiser Alpe (Mont da Sores) in Trentino; from Catania, Sicily, the original *herschelite* occurs at Aci Castello, south of Acireale. In Scotland at Kilmacolm, west of Glasgow, Renfrew, in crystals sometimes an inch across; in Ireland at the Giant's Causeway and elsewhere in Co. Antrim. On the Faroe Islands; from many points in Greenland. *Herschelite* (*seebachite*) occurs at Collingwood and Richmond, parts of Melbourne, Victoria, Australia.

In the United States chabazite occurs at West Paterson and elsewhere in the trap rocks of northeastern New Jersey; at Jones's Falls, near Baltimore, Maryland (*haydenite*). In Oregon at Goble, Columbia Co. In Nova Scotia wine-yellow or flesh-red (the latter being *acadialite*) associated with heulandite, analcite, and calcite, at Five Islands, Two Islands, Wasson's Bluff, Cape Blomidon, Cape d'Or, and elsewhere in the Bay of Fundy district.

The name *chabazite* is from χαβάζιος, an ancient name of a stone.

GMELINITE.

Rhombohedral. Axis c = 0·7345.

Crystals usually hexagonal in aspect; sometimes $\rho(01\bar{1}1)$ smaller than $r(10\bar{1}1)$, and habit rhombohedral; rr' $10\bar{1}1 \wedge \bar{1}101$ = 68° 8', $r\rho$ $10\bar{1}1 \wedge 01\bar{1}1$ = 37° 44'.

Cleavage: $m(10\bar{1}0)$ easy; $c(0001)$ sometimes distinct. Fracture uneven. Brittle. H. = 4·5. G. = 2·04–2·17. Luster vitreous. Colorless, yellowish white, greenish white, reddish white, flesh-red. Transparent to translucent. Optically positive, also negative. Birefringence very low. Interference-figure often disturbed, and basal sections divided optically into sections analogous to chabazite. Mean refractive index, 1·47.

Comp. — In part $(Na_2Ca)Al_2Si_4O_{12}.6H_2O$. See also p. 650.

Some authors class chabazite and gmelinite together as forming one isomorphous series; others, chiefly because of physical differences (cleavage, etc.), prefer to consider them as two species. The two minerals, in any case, are very closely related.

Pyr., etc. — B.B. fuses easily (F. = 2·5–3) to a white enamel. Decomposed by hydrochloric acid with separation of silica.

Obs. — Gmelinite occurs associated with other zeolites and was formed under the same conditions. It occurs on the Island of Cyprus near Pyrgos on Morphou Bay. At Andreasberg, Harz Mts. in argillaceous schist. In flesh-red crystals at Montecchio Maggiore, southwest of Vicenza in Venetia, Italy. In Co. Antrim, Ireland, at Glenarm and the island of Magee (some crystals ½ inch across), and elsewhere. At Talisker on the Isle of Skye, Inverness, Scotland, in large colorless crystals. In Australia from Flinders, Victoria.

In the United States from the trap rocks of northeastern New Jersey at Bergen Hill in fine white crystals, at Great Notch, Snake Hill, etc. From the Bay of Fundy district, Nova Scotia, at Cape Blomidon (*ledererite*); also at Two Islands and Five Islands.

Named *gmelinite* after Prof. Gmelin of Tübingen (1792–1860).

Levynite. — $CaAl_2Si_3O_{10}.5H_2O$. In rhombohedral crystals. H. = 4–4·5. G. = 2·09–2·16. Easily fusible. Colorless, white, grayish, reddish, yellowish. Optically −. ω = 1·496, ϵ = 1·491. In Ireland in Co. Antrim at Glenarm and Island Magee and in Co. Londonderry. On the Faroe Islands at Dalsnypen (the original locality) and elsewhere; in Iceland; and at various points about Disco Bay in Greenland. In the United States it has been found in the basalt of Table Mountain, near Golden, Jefferson Co., Colorado.

Arduinite. — A zeolite containing lime and soda. Probably orthorhombic. In radiating fibrous aggregates. G. = 2·26. Color red. Optically +. Ax. pl. ‖ cleavage. α = 1·474, β = 1·476, γ = 1·478. 2V nearly 90°. From Val dei Zuccanti, Venetia, Italy.

Offretite. — A potash zeolite, related to the species of the chabazite group. In basalt of Mont Simiouse, near Montbrison, Loire, France.

ANALCITE. Analcime.

Isometric. Usually in trapezohedrons; also cubes with faces $n(211)$; again the cubic faces replaced by a vicinal trisoctahedron. Sometimes in composite groups about a single crystal as nucleus (Fig. 411, p. 181). Also massive granular; compact with concentric structure.

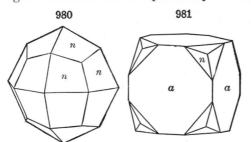

980   981

X-ray study shows that its atomic structure is made up of linked Al and Si tetrahedra grouped into fourfold and sixfold rings. Such a structure is pseudocubic. The silicon and aluminum atoms lie on the same series of points and presumably can replace each other, maintaining, however, the customary zeolite ratio of Si + Al : O = 1 : 2. Such a structure has comparatively large open spaces or channels in it which may account for the dehydration phenomena and for the ease by which silver and other metals may be substituted for the sodium atoms. The unit cell is complicated, containing sixteen molecules of $NaAlSi_2O_6.H_2O$.

Cleavage: cubic, in traces. Fracture subconchoidal. Brittle. H. = 5–5·5. G. = 2·22–2·29. Luster vitreous. Colorless, white; occasionally grayish, greenish, yellowish, or reddish white. Transparent to nearly opaque. Often shows weak double refraction, which is apparently connected with loss of water and consequent change in molecular structure (Art. **441**). $n =$ 1·4874.

Comp. — $NaAlSi_2O_6.H_2O = Na_2O.Al_2O_3.4SiO_2.2H_2O =$ Silica 54·5, alumina 23·2, soda 14·1, water 8·2 = 100.

Analyses show always a varying excess of silica and water above amounts required by the formula. It has been assumed that a molecule containing the acid $H_2Si_2O_5$ is present in solid solution in small amounts.

Pyr., etc. — Yields water in the closed tube. B.B. fuses at 2·5 to a colorless glass. Gelatinizes with hydrochloric acid.

Diff. — Characterized by trapezohedral form, but is softer than garnet, and yields water B.B., unlike leucite (which is also infusible); fuses without intumescence to a clear glass unlike chabazite. From leucite and sodalite surely distinguished only by chemical tests, *i.e.*, absence of chlorine in the nitric-acid test (see sodalite, p. 588), absence of much potash and abundance of soda in the solution, and evolution of much water from the powder in a closed glass tube below a red heat.

Micro. — Recognized in thin sections by its very low relief and isotropic character; often shows optical anomalies.

Obs. — Occurs frequently with other zeolites, also prehnite, calcite, etc., in cavities and seams in basic igneous rocks, as basalt, diabase, etc.; also in granite, gneiss, etc. It also may occur as replacing nephelite, leucite, sodalite, etc., and is found, therefore, as a constituent of rocks that originally contained these minerals. Recently shown to be also a rather widespread component of the groundmass of various basic igneous rocks, at times being the only alkali-alumina silicate present, as in the so-called analcite-basalts. Has been held in such cases to be a primary mineral produced by the crystallization of a magma containing considerable soda and water vapor held under pressure.

Analcite is one of the more common zeolites, and only its more important occurrences are listed below. It occurs near Aussig in Bohemia of Czechoslovakia. In the silver veins at Andreasberg in the Harz Mts. In Italy it occurs in Trentino at the Seiser Alpe (Mont da Sores) in large crystals and in the Val di Fassa; in Tuscany at Montecatini (*picranalcime*); from Vesuvius; in pellucid crystals from the Cyclopean Islands, northeast of Catania, Sicily. In Norway from Arendal in Aust-Agder in beds of iron ore; and on the islands of the Langesundfiord, in part as the result of the alteration of nephelite; varieties called *euthallite* and *eudnophite* are from this region. In Scotland at Talisker, Isle of Skye, Inverness; at Kilpatrick Hills, Dumbarton, etc. In Ireland from Co. Antrim. In fine crystals from Kerguelen Island in the Indian Ocean. In Australia at Flinders, Victoria. From Maze, Echigo Province, Japan.

In the United States it occurs in the trap quarries of northeastern New Jersey at Bergen Hill, West Paterson, etc. Abundant in fine crystals from the copper district in Keweenaw Co., Michigan. At Table Mountain, near Golden, Jefferson Co., Colorado. Fine specimens come from the zeolite districts on the Bay of Fundy, Nova Scotia, at Wasson's Bluff, Cape Blomidon, etc.

The name *analcime* is from ἄναλκις, *weak*, and alludes to its weak electric power when heated or rubbed. The correct derivative is *analcite*, as here adopted for the species.

Faujasite. — Perhaps $Na_2CaAl_4Si_{10}O_{28}.20H_2O$.

In isometric octahedrons. H. = 5. G. = 1·923. Colorless, white. $n = 1.48$. Occurs with augite in the limburgite of Sasbach in the Kaiserstuhl, Baden, Germany, etc.

Natrolite Group. Orthorhombic and Monoclinic

		$a : b : c$	β
Natrolite	$Na_2Al_2Si_3O_{10}.2H_2O$	0·9785 : 1 : 0·3536	
Scolecite	$CaAl_2Si_3O_{10}.3H_2O$	0·9764 : 1 : 0·3434	89° 18'
Mesolite	$\begin{cases} Na_2Al_2Si_3O_{10}.2H_2O \\ 2\ CaAl_2Si_3O_{10}.3H_2O \end{cases}$		

The three species of the NATROLITE GROUP agree closely in angle, though varying in crystalline system. Fibrous, radiating or divergent groups are common to all these species.

NATROLITE.

Orthorhombic. Axes $a : b : c = 0\cdot9785 : 1 : 0\cdot3536$.

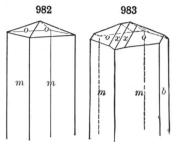

982   **983**

mm''', $110 \wedge 1\bar{1}0 = 88° 45\frac{1}{2}'$.
mo, $110 \wedge 111 = 63° 11'$.
oo', $111 \wedge \bar{1}11 = 37° 38'$.
oo''', $111 \wedge 1\bar{1}1 = 36° 47\frac{1}{2}'$.

Crystals prismatic, usually very slender to acicular; frequently divergent, or in stellate groups. Also fibrous, radiating, massive, granular, or compact. X-ray study confirms the orthorhombic symmetry. There are eight molecules in the unit cell.

Cleavage: $m(110)$ perfect; $b(010)$ imperfect, perhaps only a plane of parting. Fracture uneven. H. = 5–5·5. G. = 2·20–2·25. Luster vitreous, sometimes inclining to pearly, especially in fibrous varieties. Color white, or colorless; to grayish, yellowish, reddish to red. Transparent to translucent. Optically $+$. Ax. pl. $\parallel b(010)$. $Z \perp c(001)$. 2V = 63°. $\alpha = 1\cdot480$. $\beta = 1\cdot482$. $\gamma = 1\cdot493$.

Var. — *Ordinary.* Commonly either (a) in groups of slender colorless prismatic crystals, varying but little in angle from square prisms, often acicular, or (b) in fibrous divergent or radiated masses, vitreous in luster, or but slightly pearly (these radiated forms often resemble those of thomsonite and pectolite); often also (c) solid amygdules, usually radiated fibrous, and somewhat silky in luster within; (d) rarely compact massive. *Galactite* is ordinarily natrolite, in colorless needles from southern Scotland.

Bergmannite, spreustein, brevicite, are names which have been given to the natrolite from the augite-syenite of southern Norway, on the Langesundfiord, in the " Brevik " region, where it occurs fibrous, massive, and in long prismatic crystallizations, and from white to red in color. Derived in part from nephelite, in part from sodalite. *Iron-natrolite* is a dark green opaque variety, either crystalline or amorphous, from the Brevik region; the iron is due to inclusions.

Comp. — $Na_2Al_2Si_3O_{10}.2H_2O$ or $Na_2O.Al_2O_3.3SiO_2.2H_2O$ = Silica 47·4, alumina 26·8, Na_2O 16·3, water 9·5 = 100.

Pyr., etc. — In the closed tube whitens and becomes opaque. B.B. fuses quietly at 2 to a colorless glass. Fusible in the flame of an ordinary wax candle. Gelatinizes with acids.

Diff. — Distinguished from aragonite and pectolite by its easy fusibility and gelatinization with acid.

Obs. — Occurs in cavities in amygdaloidal basalt, and other related rocks. It is also found as the product of the alteration of nephelite, sodalite, plagioclase, in nephelite-syenite, etc. It is formed at low temperatures and is one of the latest of the zeolites to crystallize.

Some of the more important occurrences of natrolite are given below. In Bohemia of Czechoslovakia at various points near Aussig and Teplice (Teplitz); at Salesel (Salesl) south of Aussig; at Böhmish-Leipa and at Neubauer-Berg near Habichstein to the southeast and other nearby points. In Italy in the Val di Fassa, Trentino; in Venetia at Montecchio Maggiore, west of Vicenza; in Tuscany *savite* from Caporciano near Montecatini. In Württemberg at Hohentwiel in Hegau. In fine crystals from Puy de Marman and Puy de la Piquette, Puy-de-Dôme, France. Common in the augite-syenite of the Langesundfiord, southern Norway, near Brevik, etc., in fine crystallizations, also in radiated forms, etc. In Scotland in the amygdaloid of Bishopton, Renfrew (*galactite*), acicular crystals several inches long; at Glen Farg in Fife (*fargite*). In Ireland from Co. Antrim near Belfast; from Kinbane near Ballycastle. On the Faroe Islands and Iceland, and from various points in Greenland. From Kimberley in Griqualand-West, South Africa.

In the United States in fine specimens from the trap quarries of northeastern New Jersey, at Bergen Hill, West Paterson, Weehawken, Great Notch, etc. From Magnet Cove near Hot Springs, Garland Co., Arkansas. In Nova Scotia common in the zeolite localities of the Bay of Fundy district, at Cape Blomidon, Cape Split, etc. Large crystals come from the Ice Valley region of British Columbia.

Named *mesotype* by Haüy, from μέσος, *middle*, and τύρος, *type*, because the form of the crystal — in his view a square prism — was intermediate between the forms of stilbite and analcite. *Natrolite*, of Klaproth, is from *natron, soda;* it alludes to the presence of soda, whence also the name *soda-mesotype*, in contrast with scolecite, or *lime-mesotype*.

SCOLECITE.

Monoclinic. Axes $a : b : c = 0.9764 : 1 : 0.3434$; $\beta = 89° 18'$.

Crystals slender prismatic $(mm''' 110 \wedge 1\bar{1}0 = 88° 37\frac{1}{2}')$, twins showing a feather-like striation on $b(010)$, diverging upward; also as penetration-twins. Crystals in divergent groups. Also massive, fibrous and radiated, and in nodules. X-ray study confirms the monoclinic symmetry.

Cleavage: $m(110)$ nearly perfect. H. $= 5-5.5$. G. $= 2.16-2.4$. Luster vitreous, or silky when fibrous. Transparent to subtranslucent. Optically —. Ax. pl. and $Z \perp b(010)$. $X \wedge c$ axis $= 15°$ to $16°$. $2V = 36°$ (approx.). $\alpha = 1.512$. $\beta = 1.519$. $\gamma = 1.519$.

Comp. — $CaAl_2Si_3O_{10}.3H_2O$ or $CaO.Al_2O_3.3SiO_2.3H_2O$ = Silica 45.9, alumina 26.0, lime 14.3, water 13.8 — 100.

Metascolecite was the name originally given to an artificially dehydrated scolecite. It has recently been given to material which has been found on the Hegeberg, near Bodenbach on the Elbe. It has the composition of scolecite but the optical properties of the dehydrated material.

Pyr., etc. — B.B. sometimes curls up like a worm (whence the name from σκώληξ, a *worm*, which gives *scolecite*, and not *scolesite* or *scolezite*); other varieties intumesce but slightly, and all fuse at 2–2.2 to a white blebby enamel. Gelatinizes with acids like natrolite.

Obs. — Scolecite has the same mode of occurrence and origin as other zeolites; it is formed at low temperatures. It is found on the Kaiserstuhl in Baden; in the Maderaner Tal, Uri, Switzerland. In Iceland it occurs at Theigarhorn on the Berufjord on the east coast in large crystals. On the Faroe Islands. From Greenland. Common in fine crystals from the Deccan trap area in India, at Thul Ghat in the Western Ghats or Sahyadri Mts.; at Lonauli and Poona, southeast of Bombay, etc.

In the United States at Table Mountain, near Golden, Jefferson Co., Colorado. Reported from Paterson, New Jersey. From Nova Scotia in the Bay of Fundy district at Cape Blomidon, etc.; in Quebec from Black Lake, Megantic Co.

Mesolite. — Intermediate between natrolite and scolecite (see p. 653). Monoclinic (confirmed by X-ray study). In acicular and capillary crystals; delicate divergent tufts, etc. Cleavages prismatic (perfect) and basal. H. = 5. G. = 2.29. White or colorless. Optically +. Ax. pl. ⊥ to elongation. $\alpha - 1.505$, $\beta - 1.505$, $\gamma - 1.506$. Strong dispersion, $\rho > v$. Occurs in amygdaloidal basalt and similar rocks, associated with other zeolites. From the Cyclopean Islands northeast of Catania, Sicily. In Scotland at Talisker on the Isle of Skye, Inverness, in delicate interlacing crystals called *cotton-stone*, in feathery tufts and in solid masses of radiating crystals; at Kilmacolm west of Glasgow in Renfrew. In Co. Antrim, Ireland, at the Giant's Causeway and elsewhere (the varieties *antrimolite* and *harringtonite* occur in Co. Antrim). From the Faroe Islands; at Berufjord, Iceland, and at various points in Greenland. In India at the Bhor Ghat in the Western Ghats or Sahyadri Mts. From Richmond, a part of Melbourne, Victoria, Australia. In the United States with other zeolites on Fritz Island in the Schuylkill River, Berks Co., Pennsylvania; from Table Mountain, near Golden, Jefferson Co., Colorado. In Nova Scotia from the Bay of Fundy region. *Pseudomesolite* is the name given to a zeolite from Carlton Peak, Minnesota, like mesolite except for its optical characters. Two chemically identical but optically different materials from Ritter Hot Springs, Grant Co., Oregon; *mesolite* with elongation $\| Y$; *pseudomesolite* with elongation $\| Z$.

Edingtonite. — Perhaps $BaAl_2Si_3O_{10}.3H_2O$. In type of the formula and axial ratio shows relations to Natrolite Group. Crystals pyramidal in habit (orthorhombic, pseudo-tetragonal); also massive. H. = 4–4.5. G. = 2.694. Fusibility 5. White, grayish white, pink. Optically —. Ax. p. $\| (010)$. $X = c$ axis. $\alpha = 1.538$, $\beta = 1.549$, $\gamma = 1.554$. $2V = 50°$ approx. Occurs in the Kilpatrick Hills, near Glasgow, Dumbarton, Scotland, with harmotome. In Sweden from Bölet near Undenäs in Östergötland in large crystals.

Gonnardite. — $(Ca,Na_2)_2Al_2Si_5O_{15}.5\frac{1}{2}H_2O$. Orthorhombic. Fibrous. In spherules with radiating structure. H. = 4.5–5. G. = 2.3. Easily fusible. Color white. Optically +.

$X \perp$ to tabular development of crystals. $\alpha = 1\cdot514, \beta = 1\cdot515, \gamma = 1\cdot520$. 2V about 52°. From basalt of Gignat, Puy-de-Dôme, France.

THOMSONITE.

Orthorhombic. Axes $a : b : c = 0\cdot9932 : 1 : 1\cdot0066$.

Distinct crystals rare; in prisms, $mm''' \, 110 \wedge 1\bar{1}0 = 89° \, 37'$. Commonly columnar, structure radiated; in radiated spherical concretions; also closely compact.

Cleavage: $b(010)$ perfect; $a(100)$ less so; $c(001)$ in traces. Fracture uneven to subconchoidal. Brittle. H. = 5–5·5. G. = 2·3–2·4. Luster vitreous, more or less pearly. Snow-white; reddish green; impure varieties brown. Streak uncolored. Transparent to translucent. Pyroelectric. Optically +. Ax. pl. ∥ $c(001)$. $Z = b$ axis. 2V = 54° (approx.) Indices variable, 1·52–1·54.

Var. — 1. *Ordinary.* (*a*) In regular crystals, usually more or less rectangular in outline, prismatic in habit. (*b*) Prisms slender, often vesicular to radiated. (*c*) Radiated fibrous. (*d*) Spherical concretions, consisting of radiated fibers or slender crystals. Also massive, granular to impalpable, and white to reddish brown, less often green as in *lintonite*. The spherical massive forms also radiated with several centers and of varying colors, hence of much beauty when polished. *Ozarkite* is a white massive thomsonite from Arkansas. *Kalithomsonite* is a variety containing over 6 per cent K_2O, from Narsarsuk, Greenland. *Faroelite*, although by some authors considered as a distinct species, is apparently identical with thomsonite.

Comp. — $(Ca,Na_2)Al_2Si_2O_8.2\frac{1}{2}H_2O$. The ratio of Ca:Na varies from 3:1 to 1:1. Analyses also show variations in Al_2O_3 and SiO_2 percentages.

Pyr., etc. — B.B. fuses with intumescence at 2 to a white enamel. Gelatinizes with hydrochloric acid.

Diff. — Resembles some natrolite, but fuses to an opaque, not to a clear glass.

Obs. — Found in cavities in lava in amygdaloidal igneous rocks, sometimes with nephelite as a result of its alteration. It occurs in cavities in phonolite in Bohemia of Czechoslovakia at Seeberg near Kaaden; at various points near Aussig and also near Leitmeritz. In Trentino, Italy, at Tiso (Theiss) southeast of Chiusa (Klausen) and on the Seiser Alpe (Mont da Sores); from the Cyclopean Islands northeast of Catania, Sicily; in the lavas of Mte. Somma, Vesuvius (*comptonite*). In Scotland it occurs in amygdaloid in the Kilpatrick Hills, Dumbarton; at Bishopton in Renfrew; at Talisker on the Isle of Skye, Inverness, etc. In Ireland from Collinward near Belfast, Co. Antrim. On the Faroe Islands. From the Deccan trap area in India.

In the United States it is found at West Paterson and neighboring zeolite localities in northeastern New Jersey. From the amygdaloid of Grand Marais in Alger Co., Michigan, which yields the water-worn pebbles on the shore of Lake Superior resembling agate, in part green (*lintonite*). It occurs fibrous radiated and massive in the Ozark Mts., Arkansas (*ozarkite*). From Table Mountain, near Golden, Jefferson Co., Colorado. In Oregon from Goble, Columbia Co. Long slender crystallizations are obtained at Peter's Point, Nova Scotia; also from other points in the Bay of Fundy district.

HYDROTHOMSONITE. $(H_2,Na_2,Ca)Al_2Si_2O_8.5H_2O$. An alteration product of thomsonite or scolecite from Tschakwa near Batum on the Black Sea in Georgia.

Echellite. — $(Ca,Na_2)O.2Al_2O_3.3SiO_2.4H_2O$. Probably orthorhombic. In radiating, fibrous, spheroidal masses. Perfect cleavage ∥ elongation. White. H. = 5. Optically +. $\alpha = 1\cdot530$, $\beta = 1\cdot533$, $\gamma = 1\cdot545$. Elongated ∥ Y. From Sextant Portage, Abitibi River, Northern Ontario.

Erionite. — $H_2CaK_2Na_2Al_2Si_6O_{17}.5H_2O$. Orthorhombic. In aggregates of very slender fibers, resembling wool. G. = 1·997. White. Optically +. Z ∥ elongation. $\alpha = 1\cdot438$, $\beta = 1\cdot44$, $\gamma = 1\cdot452$. Occurs in cavities in rhyolite from Durkee, Baker Co., Oregon.

Bavenite. — $Ca_3Al_2(SiO_3)_6.H_2O$. Monoclinic. Fibrous-radiated groups of prismatic crystals. Cleavage (010). H. = 5·5. G. = 2·7. Color white. Optically +. Ax. pl. ⊥

(010). $Z = b$ axis. $X \wedge a$ axis $= 2°$. $\alpha = 1\cdot578$, $\beta = 1\cdot579$, $\gamma = 1\cdot583$. Occurs in pegmatitic druses in the granite of Baveno, Lago Maggiore, Piedmont, Italy.

Bityite. — A hydrous silicate of calcium and aluminum, with small amounts of the alkalies. Pseudo-hexagonal. In minute hexagonal plates which in polarized light show division into six biaxial sectors. Cleavage parallel to base. H. = 5·5. G. = 3·0. Easily fusible. Optically —. $X \perp$ cleavage. $\beta = 1\cdot652$. 2V small. Found as crystal crusts in pegmatite veins at Maharitra, on Mt. Bity, Madagascar.

Hydronephelite. $HNa_2Al_3Si_3O_{12}.3H_2O$. Massive, radiated. H. = 4·5–6. G. = 2·263. Color white; also dark gray. Index, 1·50. From Litchfield, Kennebec Co., Maine; said however to be a mixture of natrolite, hydrargillite and diaspore. *Ranite* from the Langesundfiord, Norway, is similar.

Dachiardite. — $(Na_2K_2,Ca)_3Al_4(Si_2O_5)_9.14H_2O$. Monoclinic. Occurring in small eight-sided prismatic twinned crystals, composed of eight sectors. Twinning plane, (110). Perfect cleavage || (100), (001). H. = 4–4·5. G. = 2·165. B.B. decrepitates, exfoliates and fuses to a white enamel. Decomposed by HCl. Transparent, colorless. Optically +. Ax. pl. $\perp$ (010). $X = b$ axis. $Z \wedge c$ axis $= 35°$. 2V = 65°. $\alpha = 1\cdot492$, $\beta = 1\cdot496$, $\gamma = 1\cdot500$. Found in granite pegmatite at San Piero in Campo, Elba.

II. Mica Division

The species embraced under this Division fall into three groups: 1, the Mica Group, including the Micas proper; 2, the Clintonite Group, or the Brittle Micas; 3, the Chlorite Group. Supplementary to these are the Vermiculites, hydrated compounds, chiefly results of the alteration of some one of the micas.

All of the above species have the characteristic micaceous structure, that is, they have highly perfect basal cleavage and yield easily thin laminæ. They belong to the monoclinic system, but the position of the bisectrix in general deviates but little from the normal to the plane of cleavage; all of them show on the basal section plane angles of 60° or 120°, marking the relative position of the chief zones of forms present, and giving them the appearance of hexagonal or rhombohedral symmetry; further, they are more or less closely related among themselves in the angles of prominent forms.

The species of this Division all yield water upon ignition, the micas mostly from 4 to 5 per cent, the chlorites from 10 to 13 per cent; this is probably to be regarded in all cases as water of constitution, and hence they are not properly *hydrous* silicates.

More or less closely related to these species are those of the Serpentine and Talc Division and the Kaolin Division following, many of which show distinctly a mica-like structure and cleavage and also pseudo-hexagonal symmetry.

1. *Mica Group.* Monoclinic

Muscovite	Potassium Mica	$H_2KAl_3(SiO_4)_3$

$a : b : c = 0\cdot57735 : 1 : 3\cdot3128 \qquad \beta = 89°\ 54'$

Paragonite	Sodium Mica	$H_2NaAl_3(SiO_4)_3$
Lepidolite	Lithium Mica	$(OH,F)_2KLiAl_2Si_3O_{10}$ in part.
Zinnwaldite	Lithium-iron Mica	
Biotite	Magnesium-iron Mica	$H_2K(Mg,Fe)_3(Al,Fe)(SiO_4)_3$ in part.

$a : b : c = 0\cdot57735 : 1 : 3\cdot2473 \qquad \beta = 90°\ 0'$

Phlogopite		$H_2KMg_3Al(SiO_4)_3$

Magnesium Mica; usually containing fluorine, nearly free from iron.

Lepidomelane   Annite.

Iron Micas. Contain ferric iron in large amount.

The species of the Mica Group crystallize in the monoclinic system, but with a close approximation to either rhombohedral or orthorhombic symmetry; the plane angles of the base are in all cases 60° or 120°. They are all characterized by highly perfect basal cleavage, yielding very thin, tough, and more or less elastic laminæ. The negative bisectrix, X, is very nearly normal to the basal plane, varying at most but a few degrees from this; hence a cleavage plate shows the axial interference-figure, which for the pseudo-rhombohedral kinds is often uniaxial or nearly uniaxial. Of the species named above, biotite has usually a very small axial angle, and is often sensibly uniaxial; the axial angle of phlogopite is also small, usually 10° to 12°; for muscovite, paragonite, lepidolite the angle is large, in air commonly from 50° to 70°.

The Micas may be referred to the same fundamental axial ratio with an angle of obliquity differing but little from 90°; they show to a considerable extent the same forms, and their isomorphism is further indicated by their not infrequent intercrystallization in parallel position, as biotite with muscovite, lepidolite with muscovite, etc.

A blow with a somewhat dull-pointed instrument on a cleavage plate of mica develops in all the species a six-rayed *percussion-figure* (Fig. 984, also

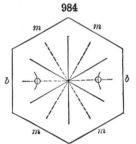

984

Fig. 517, p. 211), two lines of which are nearly parallel to the prismatic edges; the third, which is the most strongly characterized, is parallel to the clinopinacoid or plane of symmetry. The micas are often divided into two classes, according to the position of the plane of the optic axes. In the *first class* belong those kinds for which the optic axial plane is normal to $b(010)$, the plane of symmetry (Fig. 984); in the *second class* the axial plane is parallel to the plane of symmetry. The percussion figure serves to fix the crystallographic orientation when crystalline faces are wanting. A second series of lines at right angles to those mentioned may be more or less distinctly developed by pressure of a dull point on an elastic surface, forming the so-called *pressure-figure*; this is sometimes six-rayed, more often shows three branches only, and sometimes only two are developed. In Fig. 984 the position of the pressure-figure is indicated by the broken lines. These lines are connected with gliding-planes inclined some 67° to the plane of cleavage (see beyond).

The micas of the *first class* include: Muscovite, paragonite, lepidolite, also some rare varieties of biotite called anomite.

The *second class* embraces: Zinnwaldite and most biotite, including lepidomelane and phlogopite.

Sufficient work has been done in the X-ray examination of the micas to indicate the essentials of the atomic structure. The silicon atoms lie in the center of a tetrahedral arrangement of oxygen atoms, and these tetrahedral groups are linked together to form sheets in the atomic structure that lie parallel to the base and thus explain the mica cleavage. The Si–O tetrahedra are grouped together in a hexagonal-like ring which apparently accounts for the common pseudo-hexagonal character of the mica crystals. Such a Si–O sheet would have Si_2O_5 for its composition. This is the case in the related minerals talc and pyrophyllite. In the micas, however, one-fourth of the tetrahedral groups contains an aluminum atom instead of silicon, and in the so-called brittle micas one-half of the silicon atoms have been replaced by aluminum. On the basis of these observations, Pauling writes the following formulas:

Talc	$(OH)_2Mg_3(Si_4O_{10})$
Pyrophyllite	$(OH)_2Al_2(Si_4O_{10})$
Phlogopite	$(OH)_2KMg_3(AlSi_3O_{10})$
Muscovite	$(OH)_2KAl_2(AlSi_3O_{10})$
Margarite	$(OH)_2Ca_2Al_2(Al_2Si_2O_{10})$

Chemically considered, the micas are silicates, and in most cases orthosilicates, of aluminum with potassium and hydrogen, also often magnesium, ferrous iron, and in certain cases ferric iron, sodium, lithium (rarely rubidium and cæsium); further, rarely barium, manganese, chromium. Fluorine is prominent in some species, and titanium is also sometimes present. Other elements (boron, etc.) may be present in traces. All micas yield water upon ignition in consequence of the hydrogen (or hydroxyl) which they contain. The precise formulas to be given the different species and their isomorphous relations with each other, although they have been the subject of much study, must still be considered as uncertain.

MUSCOVITE. Common Mica. Potash Mica.

Monoclinic. Axes $a : b : c = 0.57735 : 1 : 3.3128$; $\beta = 89° 54'$.

Twins common according to the *mica-law*: tw. pl. a plane in the zone cM 001 $\wedge$ 221 normal to $c(001)$, the crystals often united by c. Crystals rhombic or hexagonal in outline with plane angles of 60° or 120°. Habit tabular, passing into tapering forms with planes more or less rough and strongly striated horizontally; vicinal forms common. Folia often very small and aggregated in stellate, plumose, or globular forms; or in scales, and scaly massive; also cryptocrystalline and compact massive. For structure, see above.

Cleavage: basal, eminent. Also planes of secondary cleavage as shown in the percussion-figure (see pp. 658 and 211); natural plates hence often yield

cM,	001 $\wedge$ 221	$= 85° 36'$.
$c\mu$,	001 $\wedge$ $\bar{1}11$	$= 81° 30'$.
MM',	221 $\wedge$ $2\bar{2}1$	$= 59° 48'$.
$\mu\mu'$,	$\bar{1}11 \wedge 1\bar{1}1$	$= 59° 16\frac{1}{2}'$.

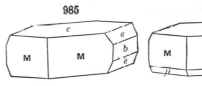

985 986

narrow strips or thin fibers || axis b, and less distinct in directions inclined 60° to this. Thin laminæ flexible and elastic when bent, very tough, harsh to the touch, passing into kinds which are less elastic and have a more or less unctuous or talc-like feel. Etching-figures on $c(001)$, monoclinic in symmetry (Fig. 521, p. 212).

H. $= 2$–2.25. G. $= 2.76$–3. Luster vitreous to more or less pearly or silky. Colorless, gray, brown, hair-brown, pale green, and violet, yellow, dark olive-green, rarely rose-red. Streak uncolored. Transparent to translucent.

Pleochroism usually feeble; distinct in some deep-colored varieties (see beyond). Absorption in the direction normal to the cleavage plane (vibrations || Y, Z) strong, much more so than transversely (vibrations || X); hence a crystal unless thin is nearly or quite opaque in the first direction though translucent through the prism. Optically $-$. Ax. pl. $\perp$ $b(010)$ and nearly

$\perp c(001)$. X inclined about $-1°$ (behind) to a normal to $c(001)$. Dispersion $\rho > v$. 2V variable, usually about $40°$, but diminishing in kinds (phengite) relatively high in silica. $\alpha = 1·552$. $\beta = 1·582$. $\gamma = 1·588$. Indices variable, increasing with the amount of iron present.

Var. — 1. *Ordinary Muscovite.* In crystals as above described, often tabular $\| c(001)$, also tapering with vertical faces rough and striated; the basal plane often rough unless as developed by cleavage. More commonly in plates without distinct outline, except as developed by pressure (see above); the plates sometimes very large, but passing into fine scales arranged in plumose or other forms. In normal muscovite the thin laminæ spring back with force when bent, the scales are more or less harsh to the touch, unless very small, and a pearly luster is seldom prominent.

2. DAMOURITE. Including *margarodite, gilbertite, hydro-muscovite*, and most HYDRO-MICA in general. Folia less elastic; luster somewhat pearly or silky and feel unctuous like talc. The scales are usually small and it passes into forms which are fine scaly or fibrous, as *sericite*, and finally into the compact cryptocrystalline kinds called *oncosine*, including much *pinite*. Often derived by alteration of kyanite, topaz, corundum, etc. Although often spoken of as *hydrous* micas, it does not appear that damourite and the allied varieties necessarily contain more water than ordinary muscovite; they may, however, give it off more readily.

Margarodite, as originally named, was the talc-like mica of Mt. Greiner in the Zillertal, Tyrol, Austria; granular to scaly in structure, luster pearly, color grayish white. *Gilbertite* occurs in whitish, silky forms from the tin mine of St. Austell, Cornwall. *Sericite* is a fine scaly muscovite united in fibrous aggregates and characterized by its silky luster (hence the name from $\sigma\eta\rho\iota\kappa\acute{o}s$, *silky*). It is a low temperature mineral.

Comp. — For the most part an orthosilicate of aluminum and potassium $(H,K)AlSiO_4$. If, as in the common kinds, $H : K = 2 : 1$, this becomes $H_2KAl_3(SiO_4)_3 = 2H_2O.K_2O.3Al_2O_3.6SiO_2 =$ Silica 45·2, alumina 38·5, potash 11·8, water 4·5 = 100.

Some kinds give a larger amount of silica (47 to 49 per cent) than corresponds to a normal orthosilicate, and they have been called *phengite*.

Iron is usually present in small amount only. Barium is rarely present, as in *œllacherite*. G. = 2·88–2·99. Chromium is also present in *fuchsite* from Schwarzenstein, Zillertal, Tyrol, and elsewhere.

Pyr., etc. — In the closed tube gives water. B.B. whitens and fuses on the thin edges (F. = 5·7) to a gray or yellow glass. With fluxes gives reactions for iron and sometimes manganese, rarely chromium. Not decomposed by acids. Decomposed on fusion with alkaline carbonates.

Diff. — Distinguished in normal kinds from all but the species of this division by the perfect basal cleavage and micaceous structure, the pale color separates it from most biotite; the laminæ are more flexible and elastic than those of phlogopite and still more than those of the brittle micas and the chlorites.

Micro. — In thin sections recognized by want of color and by the perfect cleavage shown by fine lines (as in Fig. 990, p. 664) in sections $\perp c(001)$, in a direction parallel to c. By reflected light under the microscope the same sections show a peculiar mottled surface with satin-like luster; birefringence rather high, hence interference-colors bright. Pleochroic halos are observed in mica about inclusions of certain radioactive minerals due to the effect of the alpha rays given off by the radioactive elements.

Obs. — Muscovite is the most common of the micas. It is an original constituent of the more crystalline rocks that are rich in potash and alumina, being most commonly found in granites and granitic pegmatites, but also in some syenites, nephelite-syenites, and their pegmatitic facies. In the volcanic rocks it is rare, appearing only as a secondary product. It is an essential constituent of the mica schists, phyllites, and related rocks, where it is often associated with biotite. The varieties with unctuous talc-like feel and pearly or silky luster are characteristic of much mica (" hydro-mica ") schist, which has often been erroneously called talcose schist. It occurs also in gneiss and may occur in the contact zones of various rocks; in crystalline limestone; in certain unaltered sedimentary rocks, etc. The largest and best-developed crystals occur in the miarolitic druses in granites and especially in the pegmatite dikes associated with granitic intrusions, either directly cutting the granite or in its vicinity. Often in such occurrences muscovite appears in enormous plates from which the mica or " isinglass " of commerce is obtained. It is then often associated with crystallized orthoclase, quartz, albite; also apatite, tourmaline, garnet, beryl, columbite, and other mineral species characteristic of granitic veins. Coarse lamellar aggregations often form the matrix of topaz, tourmaline, etc.

Muscovite is frequently of secondary origin, being derived from the alteration of other species, e.g., topaz, kyanite (damourite), feldspar (oncosine), etc.; cf. also pinite, beyond; muscovite forms with albite the mineral aggregate called cymatolite, derived from spodumene; cf., p. 564.

Muscovite often encloses flattened crystals of garnet, tourmaline, also quartz in thin plates between the sheets; further not infrequently magnetite in dendrite-like forms following in part the directions of the percussion-figure.

Muscovite is such a common species that only the best-known localities, noted for their unusual crystals or some especially interesting property, can be given here. From Russia in the region of Ekaterinburg, at Alabaschka near Mursinsk, Ural Mts., and in the Ilmen Mts. From Salzburg, Austria, in the Sulzbachtal; in Tyrol in the Zillertal, etc. In Switzerland in the Binnental, Valais, and at St. Gotthard in Ticino. From Norway at Bamble and Kragerö in Telemark and in the Langesundfiord district at Brevik. In Sweden at Falun in Kopparberg and on the Island of Utö in the State of Stockholm. In Turun ja Porin, Finland, at Pargas southeast of Åbo and at Skogböle near Kimito. From Cornwall, England, and in the Mourne Mts., Co. Down, Ireland. In large sheets from Greenland; also from the Uluguru Mts., district Morogoro, in Tanganyika Territory, East Africa; from Bengal in India and from Brazil.

In the United States fine specimens come from the pegmatite veins of southwestern Maine, in Androscoggin Co., at Auburn; in Oxford Co., at Buckfield in fine crystals, at Hebron, at Mt. Mica in the town of Paris, at Stoneham; and in Sagadohoc Co., at Topsham. In New Hampshire at Acworth in Sullivan Co., and at Grafton, Grafton Co., in large transparent plates. In Massachusetts at Chesterfield in Hampshire Co., sometimes pink in color, and at Goshen, rose-red; in Worcester Co., at South Royalston with beryl. In Connecticut in the pegmatites of Middlesex Co., at Middletown and Portland, at Haddam and Haddam Neck; at Branchville, Fairfield Co., both in large sheets and in aggregates with curved concentric structure. In Pennsylvania in Chester Co., at Pennsbury in fine hexagonal crystals of a brown color, weighing up to 100 pounds, often enclosing quartz, magnetite and hematite. In Maryland at Jones's Falls, near Baltimore, the plates showing by transmitted light a series of differently colored concentric hexagons. In Virginia at Amelia Court House, Amelia Co. In western North Carolina muscovite has been extensively mined in Mitchell, Yancey, Jackson, and Macon counties. Especially fine specimens come from Henry, Lincoln Co. From Habersham Co., Georgia. In crystals from Keystone, Pennington Co., Black Hills, South Dakota. Rose-colored muscovite is found at Taos, Taos Co., New Mexico.

Muscovite is named from *Vitrum Muscoviticum* or *Muscovy-glass*, formerly a popular name of the mineral.

Use. — As an insulating material in electrical apparatus; as a non-inflammable transparent material for furnace doors, etc.; in a finely divided form as a non-conductor of heat and fireproofing material; mixed with oil as a lubricant, etc.

Pinite. — A general term used to include a large number of alteration-products especially of cordierite, also spodumene, nephelite, scapolite, feldspar and other minerals. In composition essentially a hydrous silicate of aluminum and potassium corresponding more or less closely to muscovite, of which it is probably to be regarded as a massive, compact variety, usually very impure from the admixture of clay and other substances. Characters as follows: Amorphous; granular to cryptocrystalline. Rarely a submicaceous cleavage. H. = 2·5–3·5. G. = 2·6–2·85. Luster feeble, waxy. Color grayish white, grayish green, pea-green, dull green, brownish, reddish. Translucent to opaque. The following are some of the minerals also classed as pinite: *gigantolite*, *gieseckite* (see p. 586), *liebenerite*, *dysyntribite*, *parophite*, *rosite*, *wilsonite*, *killinite*.

Agalmatolite (pagodite) is like ordinary massive pinite in its amorphous compact texture, luster, and other physical characters, but contains more silica, which may be from free quartz or feldspar as impurity. The Chinese has H. = 2–2·5; G. = 2·785–2·815. Colors usually grayish, grayish green, brownish, yellowish. Named from ἄγαλμα, *an image; pagodite* is from *pagoda*, the Chinese carving the soft stone into miniature pagodas, images, etc. Part of the so-called agalmatolite of China is true pinite in composition, another part is compact pyrophyllite, and still another steatite (see these species).

Paragonite. — A sodium mica, corresponding to muscovite in composition; formula, $H_2NaAl_3(SiO_4)_3$. In fine pearly scales; also compact. G. = 2·78–2·90. Index, 1·60. Color yellowish, grayish, greenish; constitutes the mass of the rock at Mte. Campione between Faido and Chironica, Ticino, Switzerland, containing kyanite and staurolite; called paragonite-schist. Also from the Ochsenkopf, south of Schwarzenberg, Saxony. It forms the compact ground-mass, resembling soapstone, enclosing actinolite in the Zillertal, Tyrol, and the Pfitschtal in Trentino, Italy. From near Prägratten in the Virgen-Tal, Tyrol (*pregrattite*). From the Island of Syra in the Cyclades. *Cossaite* is a compact variety found at various points in Piedmont, Italy. *Hallerite*, considered to be a lithium-bearing paragonite, is found at Mesores, near Autun in Saône-et-Loire, France.

BADDECKITE, supposed to be an iron mica related to muscovite, is probably a mixture of hematite with clay. In small scales with a copper-red color. From near Baddeck, Victoria Co., Nova Scotia.

Roscoelite. — Essentially a muscovite in which vanadium has partly replaced the aluminum. In minute scales; micaceous structure and cleavage. H. = 2·5. G. = 2·97. Color clove-brown to greenish brown. Optically −. Ax. pl. ⊥ (010). X nearly ⊥ (001). $\alpha = 1\cdot610$, $\beta = 1\cdot685$, $\gamma = 1\cdot704$. Occurs in California at gold mines at different points to the northwest of Placerville, El Dorado Co. Also reported from Cripple Creek, Teller Co., Colorado, and at Kalgoorlie, Western Australia.

LEPIDOLITE. Lithia Mica.

In aggregates of short prisms, often with rounded terminal faces. Crystals sometimes twins or trillings according to the mica law. Also in cleavable plates, but commonly massive scaly-granular, coarse or fine.

Cleavage: basal, highly eminent. H. = 2·5–4. G. = 2·8–3·3. Luster pearly. Color rose-red, violet-gray or lilac, yellowish, grayish white, white. Translucent. Optically −. Ax. pl. usually ⊥ b(010). X nearly ⊥ (001). Axial angle variable 0° to 52°. $\alpha = 1\cdot530$. $\beta = 1\cdot553$. $\gamma = 1\cdot556$.

Comp. — Chiefly $(OH,F)_2KLiAl_2Si_3O_{10}$.

To account for variations from the above formula other isomorphous molecules have been assumed by Winchell as follows: lepidolite, $H_4K_2Li_2Al_4Si_6O_{22}$; polylithionite, $H_8K_2Li_4Al_2Si_6O_{22}$; protolithionite, $H_4K_2Fe_3Al_4Si_5O_{22}$; cryophyllite, $H_4K_2Li_2FeAl_2Si_7O_{22}$. Ordinary lepidolite usually contains the polylithionite molecule in varying amounts up to 50 per cent. Zinnwaldite is thought to be intermediate between lepidolite and protolithionite. The composition, however, must still be considered as uncertain.

Pyr., etc. — In the closed tube gives water and reaction for fluorine. B.B. fuses with intumescence at 2–2·5 to a white or grayish glass sometimes magnetic, coloring the flame purplish red at the moment of fusion (lithia). With the fluxes some varieties give reactions for iron and manganese. Attacked but not completely decomposed by acids. After fusion, gelatinizes with hydrochloric acid.

Obs. — Especially characteristic of granite pegmatite veins where it often occurs associated with the lithia-bearing tourmalines, amblygonite, spodumene, etc. Sometimes grown in parallel position with muscovite. More rarely in granite and gneiss.

Found in the Ural Mts., at Alabaschka near Mursinsk; lilac or reddish violet from Rožna near Neustadtl in Moravia of Czechoslovakia. In Saxony at Penig and nearby in large sheets. From the Island of Elba. Lepidolite is common in the soda-lithia pegmatites of Madagascar, often in large sheets and of a beautiful rose color. Important localities are Antsomgombato, south of Betafo, Ampangabé near Miandrarivo, Maharitra on Mt. Bity, etc.

In the United States common in the pegmatites of southwestern Maine; both granular and broad foliated varieties are found, associated with pink tourmaline, spodumene, amblygonite. Especially fine specimens come from Auburn, Androscoggin Co., and in Oxford Co., from Hebron, Buckfield, Paris. In Connecticut from Middlesex Co., at Haddam Neck and near Middletown. In California common in the tourmaline-bearing pegmatites of San Diego Co.

Named lepidolite from λεπίς, *scale*, after the earlier German name *Schuppenstein*, alluding to the scaly structure of the massive variety of Rožna.

Use. — As a source of lithium compounds.

Cookeite. — $(OH)_6LiAl_3Si_2O_6$. Monoclinic. In pseudo-hexagonal plates. In rounded aggregates. Micaceous cleavage. H. = 2·5. G. = 2·67. Folia flexible but not elastic. Color white, yellowish, pale pink, green, brown. A cleavage plate may be divided into six biaxial sectors with an uniaxial center. Optically +. $\alpha = 1\cdot576$, $\beta = 1\cdot579$, $\gamma = 1\cdot597$ (Buckfield). 2V = 0°–80°. Z ⊥ (0001). It often occurs as a pearly coating on crystals of pink tourmaline, of which it appears to be an alteration product. Occurs with tourmaline and lepidolite in Oxford Co., Maine, at Hebron, Mt. Mica near Paris, and at Buckfield. In Connecticut in Middlesex Co., at Haddam Neck.

Zinnwaldite. — An iron-lithia mica in form near biotite. See further under lepidolite above. Color pale violet, yellow to brown and dark gray. Optically −. $X \wedge c$ axis = 0°–4°. $\beta = 1\cdot57$. Zinnwaldite is most commonly a pneumatolytic mineral occurring in the cassiterite- and topaz-bearing pegmatites; also in granites. Found in the Erzgebirge of Saxony at Zinnwald, Altenberg, etc., in connection with tin veins; similarly in Cornwall, at St. Just, and elsewhere. From Narsarsuk, Greenland. In Alaska from the York district on Seward Peninsula.

Cryophyllite is a related lithium mica from the granite of Rockport on Cape Ann, Essex Co., Massachusetts, with danalite and annite. *Polylithionite* is a lithium mica from Kangerdluarsuk, in the Julianehaab district, Greenland. *Irvingite* is an alkali mica containing lithium found in pegmatite veins near Wausau, Marathon Co., Wisconsin.

Manandonite. — A basic boro-silicate of lithium and aluminum, $H_{24}Li_4Al_{14}B_4Si_6O_{53}$. Orthorhombic? Pseudohexagonal. Micaceous. In lamellar aggregates or mammillary crusts of hexagonal plates. Perfect basal cleavage. G. = 2·89. Easily fusible giving red flame. Color white. Luster pearly. Optically +. $Z \perp$ to cleavage. $\beta = 1\cdot6$. Axial angle small and variable. Found in pegmatite at Antandrokomby, near Mt. Bity on the Manandona River, Madagascar.

BIOTITE.

Monoclinic; pseudo-rhombohedral. Axes $a : b : c = 0\cdot57735 : 1 : 3\cdot2743$; $\beta = 90°$.

Habit tabular or short prismatic; the pyramidal faces often repeated in oscillatory combination. Crystals often apparently rhombohedral in symmetry since $r(101)$ and $z(132)$, $z'(1\bar{3}2)$, which are inclined to $c(001)$ at sensibly the same angle, often occur together; further, the zones to which these faces belong are inclined 120° to each other, hence the hexagonal outline of basal sections. Twins, according to the mica law, tw. pl. a plane in the prismatic zone $\perp c(001)$. Often in disseminated scales, sometimes in massive aggregations of cleavable scales.

co, 001 $\wedge$ 112 = 73° 1′.	cr, 001 $\wedge$ $\bar{1}$01 = 80° 0′.
cM, 001 $\wedge$ 221 = 85° 38′.	cz, 001 $\wedge$ 132 = 80° 0′.
$c\mu$, 001 $\wedge$ $\bar{1}$11 = 81° 19′.	MM', 221 $\wedge$ 221 = 59° 48½′.

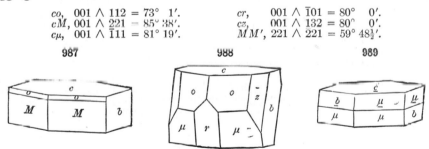

987   988   989

Cleavage: basal, highly perfect; planes of separation shown in the percussion-figure; also gliding-planes $\rho(\bar{2}05)$, $\varsigma(135)$ shown in the pressure-figure inclined about 66° to $c(001)$ and yielding pseudo-crystalline forms (Fig. 515, p. 210). H. = 2·5–3. G. = 2·7–3·1. Luster splendent, and more or less pearly on a cleavage surface, and sometimes submetallic when black; lateral surfaces vitreous when smooth and shining. Colors usually green to black, often deep black in thick crystals, and sometimes even in thin laminæ, unless the laminæ are very thin; such thin laminæ green, blood-red, or brown by transmitted light; also pale yellow to dark brown; rarely white. Streak uncolored. Transparent to opaque.

Pleochroism strong; $Y = Z =$ dark brown, reddish brown, green; $X =$

colorless, light yellow.   Absorption $Y = Z$ nearly, for X much less.   Hence sections ∥ $c(001)$ dark green or brown to opaque; those ⊥ (001) lighter and deep brown or green for vibrations ∥ (001), pale yellow, green or red for vibrations ⊥ (001).   Pleochroic halos often noted, particularly about microscopic inclusions.   Optically −.   Ax. pl. ∥ $b(010)$. X nearly coincident with the normal to $c(001)$, but inclined about half a degree, sometimes to the front, sometimes the reverse.   Indices varying with composition, $\alpha = 1\cdot58$–$1\cdot62$, $\gamma = 1\cdot63$–$1\cdot68$.   Axial angle usually very small, and often sensibly uniaxial; also up to 50°.   Birefringence high, $\gamma - \alpha = 0\cdot04$ to $0\cdot06$.

Comp. — Essentially $H_2K(Mg,Fe)_3Al(SiO_4)_3$.

Kunitz considers that biotite is composed of various mixtures of the phlogopite, $H_2KMg_3Al(SiO_4)_3$, and the lepidomelane, $H_2KFe_3(Al,Fe)(SiO_4)_3$ molecules.   Winchell gives the following members of the biotite system; phlogopite $H_4K_2Mg_6Al_2Si_6O_{24}$, eastonite $H_4K_2Mg_5Al_4Si_5O_{24}$, siderophyllite $H_4K_2Fe_5Al_4Si_5O_{24}$, annite $H_4K_2Fe_6Al_2Si_6O_{24}$.

Var. — Biotite was divided into two classes by Tschermak:
I. MEROXENE.   Axial plane ∥ $b(010)$.   II. ANOMITE.   Ax. pl. ⊥ $b(010)$.   Of these two kinds, meroxene includes nearly all ordinary biotite, while anomite is, so far as yet observed, of restricted occurrence, the typical localities being Greenwood Furnace, Orange Co., New York, and Lake Baikal in East Siberia.   *Meroxene* is a name early given to the Vesuvian biotite.   Anomite is from ανομος, contrary to law.   Anomite is considered by some to belong rather to the lepidolite series.

Haughtonite and *siderophyllite* are kinds of biotite containing much iron.

Manganophyllite is a manganesian biotite.   Occurs in aggregations of thin scales.   Color bronze- to copper-red.   Streak pale red.   From Vermland, Sweden, at Pajsberg near Persberg and at Långbanshyttan; Piedmont, Italy.   A titaniferous biotite from the Katzenbuckel, Baden, has been called *wodanite* and *titanbiotite*.   *Calciobiotite* is a brown to almost colorless variety of biotite, containing over 14 per cent CaO, from tuffs near Nocera, Naples, Italy.

Pyr., etc. — In the closed tube gives a little water.   Some varieties give the reaction for fluorine in the open tube; some kinds give little or no reaction for iron with the fluxes, while others give strong reactions for iron.   B.B. whitens and fuses on the thin edges.   Completely decomposed by sulphuric acid, leaving the silica in thin scales.

Diff. — Distinguished by its dark green to brown and black color and micaceous structure, usually nearly uniaxial.

Micro. — Recognized in thin sections by its brown (or green) color; strong pleochroism and strong absorption parallel to the elongation (unlike tourmaline).   Sections ∥ $c(001)$ are non-pleochroic, commonly exhibit more or less distinct hexagonal outlines and yield a negative sensibly uniaxial figure.   Sections ⊥ (001) are strongly pleochroic and are marked by fine parallel cleavage lines (Fig. 990); they also have nearly parallel extinction, and show high polarization colors; by reflected light they exhibit a peculiar mottled or watered sheen which is very characteristic and aids in distinguishing them from brown hornblende.

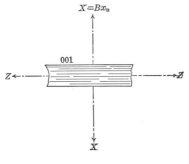

990

Obs. — Biotite is an important constituent of many different kinds of igneous rocks, especially those formed from magmas containing considerable potash and magnesia.   Common in certain varieties of granites, syenite, diorite, etc., of the massive granular type, and is also found, but more rarely, in their volcanic representatives at times in pegmatite veins in large sheets.   Its mode of occurrence suggests the presence of mineralizing vapors, especially those carrying fluorine, during its formation.   It occurs also as the product of metamorphic action in a variety of rocks, in gneisses, schists, etc.; and it is also to be found in contact metamorphic zones.   It occurs as an alteration product of various minerals, as augite, hornblende, scapolite, garnet, etc.   It is not infrequently associated in parallel position with muscovite, the latter, for example, forming the outer portions of plates having a nucleus of biotite.

Some of the prominent localities of crystallized biotite are as follows: Vesuvius, common particularly in ejected limestone masses on Mte. Somma, with augite, chrysolite, nephelite, humite, etc. The crystals are sometimes nearly colorless or yellow and then usually complex in form; also dark green to black. In Trentino at Monzoni, Val di Fassa. From the Zillertal, Austrian Tyrol. In Rhineland at Laacher See. In Norway at Arendal in Aust-Agder and elsewhere.

In the United States ordinary biotite is common in granite, gneiss, etc., but notable localities of distinct crystals are not numerous. It occurs with muscovite as a more or less prominent constituent of the pegmatite veins of Maine, New Hampshire, Massachusetts, Connecticut, Pennsylvania; also similarly in Virginia and North Carolina. Interesting specimens come from St. Lawrence Co., New York, at Rossie, etc., and at Monroe in Orange Co.; in North Carolina in crystals at the mica mines of Macon, Mitchell, Haywood, and Yancey counties. From the Pike's Peak region, El Paso Co., Colorado, where the variety *siderophyllite* was obtained.

CASWELLITE. An altered biotite from Franklin, Sussex Co., New Jersey.

PHLOGOPITE.

Monoclinic. In form and angles near biotite. Crystals prismatic, tapering; often large and coarse; in scales and plates.

Cleavage: basal, highly eminent. Thin laminæ tough and elastic. H. = 2·5–3. G. = 2·78–2·85. Luster pearly, often submetallic on cleavage surface. Color yellowish brown to brownish red, with often something of a copper-like reflection; also pale brownish yellow, green, white, colorless. Often exhibits asterism in transmitted light, due to regularly arranged inclusions. Pleochroism distinct in colored varieties: Z brownish red, Y brownish green, X yellow. Absorption $Z > Y > X$. Optically −. Ax. pl. $\parallel$ $b(010)$. X nearly $\perp$ $c(001)$. Axial angle small but variable even in the same specimen, from $0°$ to $50°$. Dispersion $\rho < v$. The axial angle appears to increase with the amount of iron. Indices variable, from 1·541–1·606.

The asterism of phlogopite, seen when a candle-flame is viewed through a thin sheet, is a common character, particularly prominent in the kinds from northern New York and Canada. It has been shown to be due to minute acicular inclusions, rutile or tourmaline, arranged chiefly in the direction of the rays of the pressure figure, producing a distinct six-rayed star; also parallel to the lines of the percussion-figure, giving a secondary star, usually less prominent than the other.

Comp. — A magnesium mica, near biotite, but containing little iron; potassium is prominent as in all the micas, and in most cases fluorine. Typical phlogopite is $H_2KMg_3Al(SiO_4)_3$. See further under biotite, p. 664.

Obs. — Phlogopite is especially characteristic of crystalline limestone or dolomite and is also found in serpentine. It is a product of both regional and contact metamorphism. It is often associated with pyroxene, amphibole, serpentine, etc. It more rarely occurs as a constituent of igneous rocks, being occasionally found in those that are rich in magnesia and poor in iron. Prominent localities are as follows: Almost colorless from Rézbánya in Rumania; in dolomite from Campolungo near St. Gotthard, Ticino, Switzerland; green and brown in Val di Fassa, Trentino, Italy; in crystalline limestone with diopside and pargasite at Pargas southeast of Åbo in Turun ja Porin, Finland; from Åker in Södermanland, Sweden. From the Island of Ceylon. Phlogopite is common in the crystalline limestones of Madagascar, at times in exceptionally large crystals.

In the United States phlogopite occurs in New York in St. Lawrence Co., at Edwards, colorless, at Clark's Hill in very large crystals; Rossie, Hammond, DeKalb, etc., and in Jefferson Co., near Antwerp at Oxbow and Vrooman's Lake; in Orange Co., at Monroe. In New Jersey at Franklin, Sussex Co., and elsewhere. In Canada it is common in the south central and eastern portions of Ontario; at North and South Burgess in fine crystals, sometimes very large; at Otter Lake, Eganville, Renfrew Co., etc. In Quebec in Ottawa Co., at Hull, Templeton, etc., and at Calumet Island, Pontiac Co.

Named from φλογωπός, *fire-like*, in allusion to the color.

Tæniolite. — Essentially a potassium-magnesium silicate. Monoclinic, belonging to the mica group. Perfect basal cleavage. Folia somewhat elastic. H. = 2·5–3. G. = 2·9. Colorless. Fusible. From Narsarsuk, southern Greenland and in the Hibina-Toundra district, Kola peninsula, northern Russia.

Lepidomelane. — Near biotite, but characterized by the large amount of ferric iron present. See further under biotite. H. = 3. G. = 3·1. Color black, brown. Optically —. Ax. pl. || (010). $X \perp$ (001). $\alpha = 1·586$, $\beta = 1·638$, $\gamma = 1·638$. 2V = 0°–40°. Lepidomelane occurs chiefly in feldspathic igneous rocks that are rich in iron content but low in magnesia, as certain granites, syenites and nephelite-syenites. It is commonly associated with ægirite or a soda amphibole. Found as a scaly mineral from Persberg, Vermland, Sweden. In the nephelite-syenites of the Langesundfiord district in southern Norway. It is found in granite rocks in Co. Carlow and Co. Donegal, Ireland. Similar iron micas occur in nephelite-syenite at Litchfield, Kennebec Co., Maine; at Haddam, Middlesex Co., Connecticut; near Baltimore, Maryland. *Annite* occurs in the Cape Ann granite, Essex Co., Massachusetts. In Ontario in Hastings Co. in large crystals in Faraday township, and at Dungannon in sodalite and at Marmora in arsenopyrite.

Alurgite. — A potassium-magnesium mica with small amounts of manganese; perhaps $6(H,K)_2O.2MgO.3Al_2O_3.12SiO_2$. Color copper-red. $\beta = 1·59$. From St. Marcel, Piedmont, Italy. *Mariposite* is similar. $\beta = 1·63$.

2. *Clintonite Group.* Monoclinic

The minerals here included are sometimes called the Brittle Micas. They are near the micas in cleavage, crystalline form and optical properties, but are marked physically by the brittleness of the laminæ, and chemically by their basic character.

In several respects they form a transition from the micas proper to the chlorites. Margarite, or calcium mica, is a basic silicate of aluminum and calcium, while Chloritoid is a basic silicate of aluminum and ferrous iron (with magnesium), like the chlorites.

MARGARITE.

Monoclinic. Rarely in distinct crystals. Usually in intersecting or aggregated laminæ; sometimes massive, with a scaly structure.

Cleavage: basal, perfect. Laminæ rather brittle. H. = 3·5–4·5. G. = 2·99–3·08. Luster of base pearly, of lateral faces vitreous. Color grayish, reddish white, pink, yellowish. Translucent, subtranslucent.

Optically —. Ax. pl. $\perp$ b(010). X approximately $\perp$ c(001), but varying more widely than the ordinary micas. Dispersion $\rho < v$. 2V = 0° to 67°. $\alpha = 1·632$, $\beta = 1·643$, $\gamma = 1·645$.

Comp. — $H_2CaAl_4Si_2O_{12}$ = Silica 30·2, alumina 51·3, lime 14·0, water 4·5 = 100.

A variety with most of the lime replaced by soda is called *soda-margarite* or *ephesite*. From near Ephesus, Asia Minor, and in the Postmasburg district, South Africa. *Lesleyite* from Unionville, Pennsylvania, is probably the corresponding *potash-margarite*.

Pyr., etc. — Yields water in the closed tube. B.B. whitens and fuses on the edges. Slowly and imperfectly decomposed by boiling hydrochloric acid.

Obs. — Associated commonly with corundum, and in many cases obviously formed directly from it. Rarely observed in chlorite schists. In the Ural Mts., it is found with the corundum of the Ekaterinburg district. *Diphanite* is from the emerald mines on the Takowaja River east of Ekaterinburg. Occurs at the emery deposits in the mountains of Güme Dagh (Messogis) east of Ephesus in Asia Minor. Similarly on the islands of Naxos, Nicaria, etc., in the Grecian archipelago. Occurs in chlorite from Mt. Greiner in the Zillertal, Tyrol, Austria, where it was first found. Also from near Sterzen (Sterzing) in the Pfitschtal, Trentino, Italy.

In the United States at the emery mine at Chester, Hampden Co., Massachusetts.

In Pennsylvania from Corundum Hill near Unionville in the township of Newlin, Chester Co., coating corundum crystals (*corundellite*); similarly at Morgan Station, Delaware Co. In North Carolina (*clingmanite*), in fine laminated crystals at the corundum mine near Buck Creek, Clay Co., and elsewhere.

Named *margarite* from μαργαρίτης, *pearl*.

SEYBERTITE. Clintonite. Brandisite.

Monoclinic, near biotite in form. Also foliated massive; sometimes lamellar radiate.

Cleavage: basal, perfect. Structure foliated, micaceous. Laminæ brittle. Percussion- and pressure-figures, as with mica but reversed in position. H. = 4–5. G. = 3–3·1. Luster pearly submetallic. Color reddish brown, yellowish, copper-red. Streak uncolored, or slightly yellowish or grayish. Pleochroism rather feeble. Optically −. Ax. pl. ⊥ $b(010)$ *seybertite*; ∥ $b(010)$ *brandisite*. X nearly ⊥ $c(001)$. Axial angles variable, but not large. $\alpha = 1{\cdot}646$. $\beta = 1{\cdot}657$. $\gamma = 1{\cdot}658$.

Var. — 1. The Amity *seybertite* (*clintonite*) is in reddish brown to copper-red brittle foliated masses; the surfaces of the folia often marked with equilateral triangles like some mica and chlorite. Axial angle 3°–13°.

2. *Brandisite* (*disterrite*), from the Val di Fassa, Trentino, is in hexagonal prisms of a yellowish green or leek-green color to reddish gray; H. = 5 of base; of sides, 6–6·5. Ax. pl. ∥ $b(010)$. Axial angle 15°–30°. Some of it pseudomorphous, after fassaite.

Comp. — In part $H_3(Mg,Ca)_5Al_5Si_2O_{18} = 3H_2O.10(Mg,Ca)O.5Al_2O_3.4SiO_2$.

Pyr., etc. — Yields water. B.B. infusible but whitens. In powder acted on by concentrated acids.

Obs. — *Seybertite* occurs at Amity, in the town of Warwick, Orange Co., New York, in limestone with serpentine. It has also been reported from Zlatoust, Ural Mts., Russia. *Brandisite* occurs on Mt. Monzoni, Val di Fassa, Trentino, Italy, in white limestone, with fassaite and black spinel.

Xanthophyllite. — Perhaps $H_8(Mg,Ca)_{14}Al_{16}Si_5O_{52}$. Monoclinic. Perfect basal cleavage. The original *xanthophyllite* is in crusts or in implanted globular forms. Leek-green. Optically negative. Ax. pl. ∥ (010). Axial angle usually very small, or sensibly uniaxial; sometimes 20°. Indices, 1·649–1·661. From near Zlatoust in the Ural Mts. Found at Crestmore, Riverside Co., California.

Waluewite (*Valuevite*) is the same species occurring in distinct pseudo-rhombohedral crystals. Folia brittle. H. = 4·6. G. = 3·093. Luster vitreous; on cleavage plane pearly. Color leek- to bottle-green. Transparent to translucent. Pleochroism rather feeble; ∥ c axis fine green; ⊥ c axis reddish brown. Optically −. Ax. pl. ∥ $b(010)$. X sensibly ⊥ $c(001)$. Axial angle 17° to 32°. Found with perovskite and other species in chloritic schists at Achmatovsk, near Kussinsk in the Zlatoust district in the southern Ural Mts.

CHLORITOID. Ottrelite.

Probably triclinic. Rarely in distinct tabular crystals, usually hexagonal in outline, often twinned with the individuals turned in azimuth 120° to each other. Crystals grouped in rosettes. Usually coarsely foliated massive; folia often curved or bent and brittle; also in thin scales or small plates disseminated through the containing rock.

Cleavage: basal, but less perfect than with the micas; also imperfect parallel to planes inclined to the base nearly 90° and to each other about 60°; $b(010)$ difficult. Laminæ brittle. H. = 6·5. G. = 3·52–3·57. Color dark gray, greenish gray, greenish black, grayish black, often grass-green in very thin plates. Streak uncolored, or grayish, or very slightly greenish. Luster of surface of cleavage somewhat pearly.

Pleochroism strong: Z yellow green, Y indigo-blue, X olive-green. Optically $+$.   Ax. pl. nearly $\|\, b(010)$. Z inclined about $12°$ or more to the normal to $c(001)$.   Dispersion $\rho > v$, large, also horizontal. $2V = 36°$ to $60°$. $\beta = 1\cdot72$.   Birefringence low, $\gamma - \alpha = 0\cdot007-0\cdot016$.

Comp. — For chloritoid $H_2(Fe,Mg)Al_2SiO_7$. If iron alone is present, this requires: Silica $23\cdot8$, alumina $40\cdot5$, iron protoxide $28\cdot5$, water $7\cdot2 = 100$.

The division into varieties has been made on the following basis: *Chloritoid* $H_2FeAl_2SiO_7$, *sismondine* $H_2(Fe,Mg)Al_2SiO_7$, *ottrelite* $H_2(Fe,Mn)Al_2SiO_7$.

Micro. — Recognized in thin sections by the crystal outlines and general micaceous appearance; high relief; green colors; distinct cleavage; frequent twinning; strong pleochroism and low interference-colors.   By the last character readily distinguished from the micas; also by the high relief and extinction oblique to the cleavage from the chlorites.

Obs. — Chloritoid (ottrelite, etc.) is characteristic of sedimentary rocks which have suffered dynamic metamorphism, especially in the earlier stages; it is thus found in phyllites, quartzites, mica schists, etc.   With more advanced degree of metamorphism it disappears.   Often grouped in fan-shaped, sheaf-like forms, also in irregular or rounded grains.

The original *chloritoid* from Kosoibrod, south of Ekaterinburg in the Ural Mts.   Other localities are in the mountains of Güme Dagh (Messogis) east of Ephesus in Asia Minor; from near Prägratten in the Tyrol, Austria; *sismondine* is from St. Marcel, Piedmont, Italy; from Ile de Groix in Morbihan, France; *ottrelite* occurs in small, oblong, shining scales or plates, more or less hexagonal in form and gray to black in color, in argillaceous schist near Ottrez, on the borders of Luxemburg, and from the Ardennes in France and Belgium.   *Salmite* is from Vielsalm, Luxemburg, Belgium; in large embedded crystals at Vanlup near Hillswick on Mainland, Shetland Islands, Scotland.   *Masonite* is from Natick, Kent Co., Rhode Island, in an argillaceous schist.   Also from the south shore of Michigamme Lake, Marquette Co., Michigan.   In Quebec at Brome and Sutton, Brome Co., and at Leeds, Megantic Co.

3. *Chlorite Group.*   Monoclinic

The CHLORITE GROUP takes its name from the fact that a large part of the minerals included in it are characterized by the *green* color common with silicates in which ferrous iron is prominent.   The species are in many respects closely related to the micas.   They crystallize in the monoclinic system, but in part with distinct monoclinic symmetry, in part with rhombohedral symmetry, with corresponding uniaxial optical character.   The plane angles of the base are also $60°$ or $120°$, marking the mutual inclinations of the chief zones of forms.   The mica-like basal cleavage is prominent in distinctly crystallized forms, but the laminæ are tough and comparatively inelastic.   Percussion- and pressure-figures may be obtained as with the micas and have the same orientation.   The etching-figures are in general monoclinic in symmetry, in part also asymmetric, suggesting a reference to the triclinic system.

Chemically considered the chlorites are silicates of aluminum with ferrous iron and magnesium and chemically combined water.   Ferric iron may be present replacing the aluminum in small amount; chromium enters similarly in some forms, which are then usually of a pink instead of the more common green color.   Manganese replaces the ferrous iron in a few cases.   Calcium and alkalies — characteristic of all the true micas — are conspicuously absent, or present only in small amount.   Various attempts have been made to determine the different isomorphous molecules that enter into the composition of the various members of the group.   Tschermak divided the chlorite minerals into two groups, the orthochlorites and the leptochlorites.   The composition of the orthochlorites has been explained by assuming isomorphous mixtures of $H_4Mg_3Si_2O_9$ (serpentine) and $H_4Mg_2Al_2SiO_9$ (amesite).   In both of these molecules ferrous iron may replace magnesium, and in the latter ferric iron (or chromium) may replace aluminum. On this basis the following classification of the orthochlorites has been made.

$$\text{Penninite} = H_4Mg_3Si_2O_9 > H_4Mg_2Al_2SiO_9$$
$$\text{Clinochlore} = 1H_4Mg_3Si_2O_9 .1H_4Mg_2Al_2SiO_9$$
$$\text{Prochlorite} = H_4Mg_3Si_2O_9 < H_4Mg_2Al_2SiO_9$$

The " leptochlorites " are more complicated in composition, have a higher iron-content, less water, and approach more nearly the composition of the " brittle micas." They commonly have higher refractive indices than the orthochlorites, and are in general optically negative and almost uniaxial in character. They commonly occur as scaly, dense, or earthy aggregates in rocks.

The only distinctly crystallized species of the Chlorite Group are clinochlore and penninite. These have similar compositions, but while the former is monoclinic in form and habit, the latter is pseudo-rhombohedral and usually uniaxial. Prochlorite (including some ripidolite) and corundophilite also occur in distinct cleavage masses.

Besides the species named there are other kinds less distinct in form, occurring in scales, also fibrous to massive or earthy; they are often of more or less undetermined composition, but in many cases, because of their extensive occurrence, of considerable geological importance. These latter forms occur as secondary minerals resulting from the alteration especially of ferro-magnesian silicates, such as biotite, pyroxene, amphibole; also garnet, vesuvianite, etc. They are often accompanied by other secondary minerals, as serpentine, limonite, calcite, etc., especially in the altered forms of basic rocks.

The rock-making chlorites are recognized in thin sections by their characteristic appearance in thin leaves, scales or fibers, sometimes aggregated into spherulites; by their greenish color; pleochroism; extinction parallel to the cleavage (unlike chloritoid and ottrelite); low relief and extremely low interference-colors, which frequently exhibit the " ultra-blue." By this latter character they are readily distinguished from the micas, which they strongly resemble and with which they are frequently associated.

PENNINITE. Pennine.

Apparently rhombohedral in form but strictly pseudo-rhombohedral and monoclinic.

Habit rhombohedral: sometimes thick tabular with $c(001)$ prominent, again steep rhombohedral; also in tapering six-sided pyramids. Rhombohedral faces often horizontally striated.

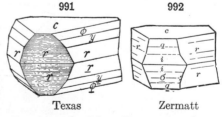

991   992

Twinning (1) *penninite law* with tw. pl. $c(001)$, contact-twins also united by $c(001)$; here corresponding faces differ 180° in position; (2) *mica law*, tw. pl. $\perp c(001)$ in the zone $001 \wedge 112$. Crystals often in crested groups. Also massive, consisting of an aggregation of scales; also compact cryptocrystalline.

Texas   Zermatt

Cleavage: $c(001)$ highly perfect. Laminæ flexible. Percussion-figure and pressure-figure as with clinochlore but less easy to obtain; not elastic. H. = 2–2·5. G. = 2·6–2·85. Luster of cleavage-surface pearly; of lateral plates vitreous, and sometimes brilliant. Color emerald- to olive-green; also violet, pink, rose-red, grayish red; occasionally yellowish and silver-white. Transparent to subtranslucent. Pleochroism distinct: usually $\parallel$ $c(001)$ green; $\perp$ c yellow. Optically $+$, also $-$, and sometimes both in adjacent laminæ of the same crystal. Usually sensibly uniaxial, but sometimes distinctly biaxial and both in the same section. Sometimes a uniaxial nucleus while the border is biaxial with $2E = 36°$, the latter probably to be referred to clinochlore. Ax. pl. $\parallel$ (010). Bx. $\perp$ (001). $\beta = 1·576$–$1·60$.

Var. — 1. *Penninite*, as first named, included a green crystallized chlorite from the Penninine Alps.

Kämmererite. — In hexagonal forms bounded by steep six-sided pyramids. Color kermes-red; peach-blossom-red. Pleochroism distinct. Optically $-$ from Lake Itkul near Miask and at Bisersk, Perm, Russia; $+$ from Texas, Pennsylvania. Uniaxial or biaxial with axial angle up to 20°. *Rhodophyllite* from Texas, Pennsylvania, and *rhodochrome* from Lake Itkul belong here.

Pseudophite is compact massive, without cleavage, and resembles serpentine.

Comp. — Essentially the same as for clinochlore, $H_8(Mg,Fe)_5Al_2Si_3O_{18}$. See further on p. 669.

Pyr., etc. — In the closed tube yields water. B.B. exfoliates somewhat and is difficultly fusible. With the fluxes all varieties give reactions for iron, and many varieties react for chromium. Partially decomposed by hydrochloric and completely by sulphuric acid.

Micro. — In thin sections shows pale green color and pleochroism; usually nearly uniaxial.

Obs. — To be found in chlorite schist and other crystalline schists. Occurs in Valais, Switzerland, with serpentine in the region of Zermatt; the crystals are sometimes 2 inches long and $1\frac{1}{2}$ inches thick; also from the Binnental. From the Zillertal in the Austrian Tyrol and in the Pfitschtal in Trentino, Italy, and in the green schists that lie between the two localities. From the Ala valley in Piedmont with clinochlore. In the United States from Snake Creek, Wasatch Co., Utah.

Kämmererite is found in Russia at the localities already mentioned, also in large crystals from Kraubat in the Mürztal, Styria, Austria. On the island of Unst, Shetland Islands, Scotland. Abundant at Texas, Lancaster Co., Pennsylvania. Also with chromite at various points in North Carolina. *Pseudophite* comes from Bernstein in Styria, Austria.

CLINOCHLORE. Ripidolite in part.

Monoclinic. Axes $a : b : c = 0·57735 : 1 : 2·2772$; $\beta = 89° 40'$.

Crystals usually hexagonal in form, often tabular $\parallel c(001)$. Plane angles of the basal section $= 60°$ or $120°$, and since closely similar angles are found in the zones which are separated by 60°, the symmetry approximates to that of the rhombohedral system.

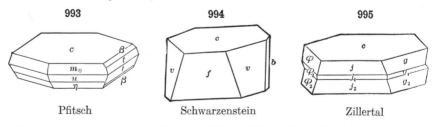

993	994	995
Pfitsch	Schwarzenstein	Zillertal

Twins: As with penninite, see Figs. 991, 992. Massive, coarse scaly granular to fine granular and earthy.

Cleavage: $c(001)$ highly perfect. Laminæ flexible tough, and but slightly elastic. Percussion- and pressure-figures orientated as with the micas (p. 658).

H. = 2–2·5.   G. = 2·65–2·78.   Luster of cleavage-face somewhat pearly. Color deep grass-green to olive-green; pale green to yellowish and white; also rose-red. Streak greenish white to uncolored. Transparent to translucent. Pleochroism not strong, for green varieties usually X green, Z yellow. Optically $+$. Ax. pl. $\parallel b(010)$. Z inclined somewhat to the normal to $c(001)$, forward, 2° to 7°. Dispersion $\rho < v$. Axial angles variable, even in the same crystal, 0° to 90°; sometimes sensibly uniaxial. Birefringence low. $\beta =$ 1·57–1·59.

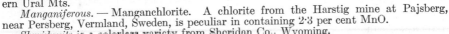

996

Achmatovsk

Var. — 1. *Ordinary;* green clinochlore, passing into bluish green; (*a*) in crystals, as described, usually with distinct monoclinic symmetry; (*b*) foliated; (*c*) massive.

Leuchtenbergite. — Contains usually little or no iron. Color white, pale green, yellowish; often resembles talc. From near Zlatoust in the Ural Mts.

Kotschubeite. — Contains several per cent of chromium oxide. Crystals rhombohedral in habit. Color rose-red. From the southern Ural Mts.

Manganiferous. — Manganchlorite. A chlorite from the Harstig mine at Pajsberg, near Persberg, Vermland, Sweden, is peculiar in containing 2·3 per cent MnO.

Sheridanite is a colorless variety from Sheridan Co., Wyoming.

Comp. — Normally $H_8Mg_5Al_2Si_3O_{18} = 4H_2O.5MgO.Al_2O_3.3SiO_2 =$ Silica 32·5, alumina 18·4, magnesia 36·1, water 13·0 = 100. Ferrous iron usually replaces a small part of the magnesia, and the same is true of manganese rarely; sometimes chromium replaces the aluminum. See further on p. 669.

Pyr., etc. — Yields water. B.B. in the platinum forceps whitens and fuses with difficulty on the edges to a grayish black glass. With borax, a clear glass colored by iron, and sometimes chromium. In sulphuric acid wholly decomposed.

Micro. — In thin sections characterized by pale green color and pleochroism; distinctly biaxial and optically $+$.

Obs. — Clinochlore is probably the most common of the chlorite minerals; it occurs in connection with chloritic and talcose rocks or schists and with serpentine. It is sometimes found in parallel position with biotite or phlogopite. It is usually of secondary origin, formed from the alteration of some other aluminous ferromagnesian mineral.

Prominent localities, especially those that furnish crystals, are as follows: From Achmatovsk near Kussinsk in the Zlatoust district, Ural Mts.; in the Zillertal, Tyrol, Austria, and in Italy in the Pfitschtal, Trentino, and in Piedmont on the Mussa Alp in the Ala valley. In Switzerland at Zermatt in Valais. In the United States at the Tilly Foster iron mine near Brewster, Putnam Co., New York, in part altered to serpentine. In Pennsylvania in Chester Co., at West Chester in large crystals and plates in talc; in Lancaster Co., at Wood's mine, Texas.

Leuchtenbergite comes from the Shiskimskaya Mts., near Zlatoust in the Urals; it is partly in large crystals, and partly quite small, embedded in steatite; the crystals are opaque and externally altered. Similar material occurs on Mt. Monzoni, Val di Fassa, Trentino, Italy. The white chlorite from Mauléon, Basses-Pyrénées, France, belongs here. A similar variety of clinochlore occurs at Amity, Orange Co., New York. *Kotschubeite* is from the southern Ural Mts., in the district of Ufaleisk, and elsewhere. Found also in California in Placer and Calaveras counties.

PROCHLORITE.   Ripidolite in part.

Monoclinic.   In six-sided tables or prisms, the side planes strongly furrowed and dull. Crystals often implanted by their sides, and in divergent groups, fan-shaped, vermicular, or spheroidal. Also in large folia. Massive, foliated, or granular.

H. = 1–2.   G. = 2·78–2·96.   Translucent to opaque; transparent only in very thin folia. Luster of cleavage surface feebly pearly. Color green, grass-green, olive-green, blackish green; across the axis by transmitted light sometimes red. Streak uncolored or greenish. Laminæ flexible, not elastic. Pleochroism distinct. Optically $+$, rarely $-$. Z only slightly inclined to the normal to $c(001)$. Axial angle small, often nearly uniaxial; again 2E = up to 50°. Dispersion $\rho < v$. Indices, 1·58–1·67.

Comp. — Lower in silicon than clinochlore, and with ferrous iron usually, but not always, in large amount. See further on p. 669.

Obs. — The occurrence of prochlorite is similar to other chlorite minerals; in chlorite schists and other metamorphic rocks; in serpentine; also in clefts of granite, gneiss, etc.; in smaller amounts with certain ore veins. At times as pseudomorphs after augite, hornblende, garnet, etc.

Common throughout the Alps. In Switzerland sometimes in implanted crystals, as at St. Gotthard, Ticino, often enveloping adularia, etc.; at Zermatt in Valais, etc.; at Traversella in Piedmont, Italy. In Austria on Mt. Greiner in the Zillertal and elsewhere in the Tyrol and from Rauris in Salzburg. Abundant in the rocks of Dauphiné, France, as at the mountains of Sept Lacs and St. Cristophe, Isère. Also massive in tin veins in Cornwall; at Arendal in Aust-Agder, Norway; at Sala, Vastmanland, Sweden.

In the United States from near Washington, District of Columbia; on Castle Mt., Batesville, Albemarle Co., Virginia, a massive form resembling soapstone. In North Carolina from the Culsagee mines in Macon Co., in broad plates of a dark green color and fine scaly. *Corundophilite* which occurs with corundum at Chester, Hampden Co., Massachusetts, is probably to be classed as prochlorite.

Rumpfite. — Probably a variety of prochlorite. Massive; granular, consisting of very fine scales. Color greenish white. Optically $+$. $X \perp (001)$. $\beta = 1.587$. $2V = 0°–10°$. Occurs in Styria, Austria, with talc in a magnesium carbonate rock near Sankt Michael near Leoben; at the Eichberg magnesite deposit in the Semmering district and similarly near Veitsch.

Amesite. — $H_4(Mg,Fe)_2Al_2SiO_9$. Monoclinic. In hexagonal plates, foliated, resembling the green talc from the Tyrol. Basal cleavage, perfect. $H. = 2.5–3$. $G. = 2.77$. Color apple-green. Luster pearly on cleavage face. Optically $+$, sensibly uniaxial. Ax. pl. || (010); Z nearly $\perp$ (001). $\alpha = \beta = 1.597, \gamma = 1.612$. Occurs with diaspore at Chester, Hampden Co., Massachusetts.

KOSSMATITE. Hydrous magnesium, aluminum, ferric iron silicate. In groups or rosettes of tables without crystal outline. Perfect basal cleavage. Folia non-elastic. Colorless, vitreous to pearly luster. Biaxial, $+$. $2E = 14°$. Occurs in a schistose zone in West Macedonia, associated with carbonates, corundum, diaspore, etc.

OTHER CHLORITES. Besides the chlorites already described which occur usually in distinct crystals or plates, there are, as noted on p. 669, forms varying from fine scaly to fibrous and earthy, which are prominent in rocks. In some cases they may belong to the species before described, but frequently the want of sufficient pure material has left their composition in doubt. These chlorites are commonly characterized by their green color, distinct pleochroism and low birefringence (p. 669).

The following are names which have been given particularly to the chlorites filling cavities or seams in basic igneous rocks: *aphrosiderite, diabantite, delessite, epichlorite, euralite, chlorophæite, hullite, pycnochlorite.*

The following are other related minerals:

Moravite. — $2FeO.2(Al,Fe)_2O_3.7SiO_2.2H_2O$. In lamellar, scaly and granular forms with perfect basal cleavage. $H. = 3.5$. $G. = 2.4$. Color iron-black. Fuses difficultly. Found at iron mines of Gobitschau near Sternberg, Moravia, Czechoslovakia, as an alteration product from thuringite.

Cronstedite. — $4FeO.2Fe_2O_3.3SiO_2.4H_2O$. Monoclinic. Occurs tapering in hexagonal pyramids; also in diverging groups; amorphous. Cleavage: basal, highly perfect. Thin laminæ elastic. $H. = 3.5$. $G. = 3.34–3.35$. Color coal-black to brownish black; by transmitted light in thin scales emerald-green. Streak dark olive-green. Optically $-$. Sensibly uniaxial. $X \perp$ cleavage. $\beta = 1.80$. In Bohemia of Czechoslovakia at Příbram and Kuttenberg. From Cornwall at Wheal Maudlin near Lostwithiel and at Wheal Jane at Truro, etc. In Brazil at Conghonas do Campo, southwest of Ouro Preto, Minas Geraes (*sideroschisolite*).

Thuringite. — $8FeO.4(Al,Fe)_2O_3.6SiO_2.9H_2O$. Monoclinic. Massive; an aggregation of minute pearly scales. Color olive-green to pistachio-green. $Y = Z =$ dark green, $X =$ nearly colorless. Optically $-$. $2V$ small. $\beta = 1.64–1.68$. Thuringite occurs with iron ores in various metamorphic rocks, frequently accompanied by calcite. Occurs in Thuringia at Schmiedefeld, east of Suhl, etc. At Zirm See northeast of Hellingenblut, Carinthia, Austria. In Moravia of Czechoslovakia at Gobitschau near Sternberg. In the United States from the metamorphic rocks on the Potomac River near Harper's Ferry, Jefferson Co., West Virginia (*owenite*). It forms the matrix enclosing garnet crystals that have been altered to a chlorite in the Lake Superior iron region, Michigan. From Hot Springs, Garland Co., Arkansas. *Stilpnochloran* is name given to an alteration product of thuringite from Gobitschau, near Sternberg, Moravia. In yellow to bronze-red scales.

Chamosite. — $15(Fe,Mg)O.5Al_2O_3.11SiO_2.16H_2O.$ Occurs compact or oölitic with H. about 3; G. = 3–3·4; color greenish gray to black. $n = 1.64$. It occurs with various ore deposits, as with sphalerite, galena, pyrite, siderite, etc. In Switzerland, chamosite forms thick beds of limited extent in limestone at Chamoson in the Rhône valley, Valais; also in the Windgälle, Maderaner Tal, Uri, etc. With iron carbonate and titanic iron at Schmiedefeld, east of Suhl, Thuringia; in Bohemia of Czechoslovakia, from the oölitic iron ore of Nučic near Unhoscht, southeast of Kladno. *Berthierine* constitutes a valuable bed of iron ore at Hayanes in Moselle. *Bavalite* is an oölitic mineral, apparently near chamosite, that forms beds in old schistose rocks in different parts of Brittany, in Côtes-du-Nord, etc.

Zebedassite. $5MgO.Al_2O_3.6SiO_2.4H_2O.$ Probably orthorhombic. Fibrous structure. Color, white. Silky luster. H. = 2. G. = 2·19. Indices 1·48–1·51. Strong birefringence. Positive elongation. Occurs in serpentine at Zebedassi in the Pavese Apennines, Italy.

Brunsvigite. — $9(Fe,Mg)O.2Al_2O_3.6SiO_2.8H_2O.$ In cryptocrystalline and foliated masses sometimes forming spherical radiated aggregates. Under microscope folia show hexagonal outline. Color olive-green to yellow-green. H. = 1·2. G. = 3·0. Optically −. Occurs in cavities in the gabbro of the Radau-Tal in the Harz Mts., Germany.

Griffithite. — $4(Mg,Fe,Ca)O.(Al,Fe)_2O_3.5SiO_2.7H_2O.$ Shown to have a variable water content with variation in atmospheric conditions. A member of the Chlorite Group. Color dark green. H. = 1. G. = 2·31. Fusible to magnetic slag. Pleochroic, X = pale yellow, Y = olive-green, Z = brown-green. Optically −. $X \perp (001).$ $\alpha = 1.485, \beta = 1.565, \gamma = 1.572.$ Occurs in amygdaloidal cavities in a basalt from Cahuenga Pass, Griffith Park, Los Angeles, California.

Stilpnomelane. — An iron silicate. In foliated plates; also fibrous, or as a velvety coating. G. = 2·77–2·96. Color black, greenish black. Optically −. $X \perp (001).$ 2V very small. Indices 1·54–1·78. Stilpnomelane is most commonly found with iron ores, magnetite, and limonite, also hematite and pyrite; frequently associated with calcite. Occurs in Moravia of Czechoslovakia at Sternberg in a bed of limonite. At Ober-Grund near Zuckmantel (Cukmantl) in Silesia of Czechoslovakia. From near Weilburg in Hessen-Nassau in a bed of iron ore. In Sweden near Nordmark, Vermland; also at Brunsjö near Grythyttan southwest of Ljusnarsberg, Örebro. In the United States the variety *chalcodite* occurs at the Sterling mine in Antwerp, Jefferson Co., New York, on hematite in a velvety coating of mica-like scales with a bronze color. From Rockyhill, Somerset Co., New Jersey.

Minguétite. — A member of Chlorite Group. A silicate of ferric and ferrous iron, intermediate between *stilpnomelane* and *lepidomelane*. G. = 2·86. Color blackish green. Strongly pleochroic, light yellow to opaque black. Optically −. Sensibly uniaxial. Fuses to a black magnetic enamel Decomposed by hydrochloric acid. From Minguet magnetite mine, near Segré, Maine-et-Loire, France.

Monrepite. A mica containing ferrous and ferric iron, found in the granite of the Rapakivi district, Finland.

Strigovite. — $H_4Fe_2(Al,Fe)_2Si_2O_{11}.$ In aggregations of minute hexagonal-like crystals. Color dark green. Optically −. Sensibly uniaxial. $X \perp (001).$ Indices, 1·65–1·67. Occurs as a fine coating over the minute crystals in cavities of the granite of Striegau in Silesia.

Spodiophyllite. — $(Na_2,K_2)_2(Mg,Fe)_3(Fe,Al)_2(SiO_3)_8.$ In rough hexagonal prisms. Micaceous cleavage. Laminæ brittle. H. = 3–3·2. G. = 2·6. Color ash-gray. Fusible. From Narsarsuk, southern Greenland.

Bardolite. A chlorite-like mineral but with higher potash content than usual with chlorites. Orthorhombic. Occurs in radially fibrous spherulites. G. = 2·47. Negative elongation. Birefringence higher than for chlorites. Occurs in a diabase dike near the village of Bardo, district Opatow, central Poland.

Viridite. Ferruginous chlorite, $4FeO.2SiO_2.3H_2O.$ Compact, composed of minute scales and needles. Micaceous cleavage. H. = 3. G. = 2·89. Color, leek-green. Occurs in dense chloritic iron ore between Sternberg, Moravia and Bennisch, Silesia.

Mackensite. $Fe_2O_3.SiO_2.2H_2O.$ Compact, composed of minute needles. Color, iron-black to greenish black. H. = 3. G. = 4·89. Occurs in a chloritic iron ore between Sternberg, Moravia, and Bennisch, Silesia.

APPENDIX TO THE MICA DIVISION — VERMICULITES

The VERMICULITE GROUP includes a number of micaceous minerals, all hydrated silicates, in part closely related to the chlorites, but varying somewhat widely in composition. They are alteration-products chiefly of the micas, biotite, phlogopite, etc., and retain more or less perfectly the micaceous cleavage, and often show the negative optical character and small axial angle of the original species. Many of them are of a more or less indefinite chemical nature, and the composition varies with that of the original mineral and with the degree of alteration.

The laminæ in general are soft, pliable, and inelastic; the luster pearly or bronze-like, and the color varies from white to yellow and brown. Heated to 100°–110° or dried over sulphuric acid most of the vermiculites lose considerable water, up to 10 per cent, which is probably hygroscopic; at 300° another portion is often given off; and at a red heat a somewhat larger amount is expelled. Connected with the loss of water upon ignition is the common physical character of exfoliation; some of the kinds especially show this to a marked degree, slowly opening out, when heated gradually, into long worm-like threads. This character has given the name to the group, from the Latin vermiculari, to breed worms. The minerals included can hardly rank as distinct species and only their names can be given here: Jefferisite, vermiculite, culsageeite, kerrite, lennilite, hallite, philadelphite, vaalite, maconite, dudleyite, pyrosclerite.

III. Serpentine and Talc Division

The leading species belonging here, Serpentine and Talc, are closely related to the Chlorite Group of the Mica Division preceding, as noted beyond. Some other magnesium silicates, in part amorphous, are included with them.

SERPENTINE.

Monoclinic. In distinct crystals, but only as pseudomorphs. Sometimes foliated, folia rarely separable; also delicately fibrous, the fibers often easily separable, and either flexible or brittle. Usually massive, but microscopically finely fibrous and felted, also fine granular to impalpable or cryptocrystalline; slaty. Crystalline in structure but often by compensation nearly isotropic; amorphous.

X-ray analysis of chrysotile shows that it is composed of bundles of parallel fibers, but although the individual fibers are elongated parallel to one crystallographic direction (c axis) their orientation otherwise varies. The structure is monoclinic. Four molecules of $H_4Mg_3Si_2O_9$ in the unit cell. Each silicon atom is surrounded by four oxygen atoms arranged at the points of a tetrahedron, a part of the oxygen atoms being shared by neighboring tetrahedral groups. Together these groups form chains extending parallel to the vertical axis. The chains are similar to those of the amphiboles with the composition Si_4O_{11}. The binding forces between these chains are weak thus accounting for the fibrous structure. The arrangement of the Mg and OH groups is similar to that in brucite.

Cleavage $b(010)$, sometimes distinct; also prismatic (50°) in chrysotile. Fracture usually conchoidal or splintery. Feel smooth, sometimes greasy. H. = 2·5–4, rarely 5·5. G. = 2·50–2·65; some fibrous varieties 2·2–2·3; retinalite, 2·36–2·55. Luster subresinous to greasy, pearly, earthy; resin-like, or wax-like; usually feeble. Color leek-green, blackish green; oil- and siskin-green; brownish red, brownish yellow; none bright; sometimes nearly white. On exposure, often becoming yellowish gray. Streak white, slightly shining. Translucent to opaque.

Pleochroism feeble. Optically −, perhaps also + in chrysotile. Double refraction weak. Ax. pl. ‖ $a(100)$. $X \perp b(010)$ the cleavage surface; Z ‖ elongation of fibers. 2V = 20° to 90°. Indices variable, from 1·490–1·571.

Var. — Many unsustained species have been made out of serpentine, differing in structure (massive, slaty, foliated, fibrous), or, as supposed, in chemical composition; and these now, in part, stand as varieties, along with some others based on variations in texture, etc. Some authors separate the mineral into two modifications: (*a*) *antigorite* (see below) with chemical and optical relations to the Chlorite Group, and (*b*) massive and fibrous *serpentine* and *chrysotile*.

A. IN CRYSTALS — PSEUDOMORPHS. The most common have the form of chrysolite. Other kinds are pseudomorphs after pyroxene, amphibole, spinel, chondrodite, garnet, phlogopite, etc. *Bastite* or *Schiller Spar* is enstatite (hypersthene) altered more or less completely to serpentine. See p. 555.

B. MASSIVE. 1. *Ordinary massive.* (*a*) *Precious or Noble Serpentine* is of a rich oil-green color, of pale or dark shades, and translucent even when in thick pieces. (*b*) *Common Serpentine* is of dark shades of color, and subtranslucent. The former has a hardness of 2·5–3; the latter often of 4 or beyond, owing to impurities.

Resinous. Retinalite. — Massive, honey-yellow to light oil-green, waxy or resin-like luster.

Bowenite (Nephrite *Bowen*). — Massive, of very fine granular texture, and much resembles nephrite, and was long so called. It is apple-green or greenish white in color; G. = 2·594–2·787; and it has the unusual hardness 5·5–6. From Smithfield, Rhode Island; also a similar kind from New Zealand, called *tangiwaite* or *tanyuwaite*.

Ricolite is a banded variety with a fine green color from Grant Co., New Mexico.

C. LAMELLAR. *Antigorite*, thin lamellar in structure, separating into translucent folia. H. = 2·5; G. = 2·622; color brownish green by reflected light; feel smooth, but not greasy. It shows a negative interference figure with X normal to cleavage. From Antigorio valley, Piedmont, Italy.

D. THIN FOLIATED. *Marmolite*, thin foliated; the laminæ brittle but separable. G. = 2·41; colors greenish white, bluish white to pale asparagus-green. From Hoboken, New Jersey.

E. FIBROUS. *Chrysotile.* Delicately fibrous, the fibers usually flexible and easily separating; luster silky, or silky metallic; color greenish white, green, olive-green, yellow and brownish; G. = 2·219. Often constitutes seams in serpentine. It includes most of the silky *amianthus* of serpentine rocks and much of what is popularly called *asbestus* (asbestus). Cf. p. 574.

Picrolite, columnar, but fibers or columns not easily flexible, and often not easily separable, or only affording a splintery fracture; color dark green to mountain-green, gray, brown. The original was from Taberg, Sweden. *Metaxite* is similar. *Baltimorite* is picrolite from Bare Hills, Maryland.

Radiotine is like serpentine except in regard to its solubility and specific gravity. In spherical aggregates of radiating fibers from near Dillenburg, Nassau.

F. SERPENTINE ROCKS. Serpentine often constitutes rock-masses. It frequently occurs mixed with more or less of dolomite, magnesite, or calcite, making a rock of clouded green, sometimes veined with white or pale green, called *verd*

997

998

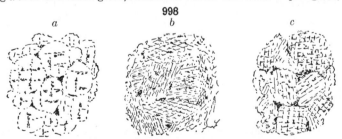

a, Serpentine derived from chrysolite;　*b*, from amphibole;　*c*, from pyroxene

antique, *ophiolite*, or *ophicalcite*. Serpentine rock is sometimes mottled with red, or has something of the aspect of a red porphyry; the reddish portions containing an unusual amount of oxide of iron. Any serpentine rock cut into slabs and polished is called *serpentine marble*.

Microscopic examination has established the fact that serpentine in rock-masses has been largely produced by the alteration of chrysolite, and many apparently homogeneous serpentines show more or less of this original mineral. In other cases it has resulted from the alteration of pyroxene or amphibole. Sections of the serpentine derived from chrysolite often show a peculiar structure, like the meshes of a net (Fig. 997); the lines marked by grains of magnetite also follow the original cracks and cleavage directions of the chrysolite (Fig. 998, a). The serpentine from amphibole and pyroxene commonly shows an analogous structure; the iron particles following the former cleavage lines. Hence the nature of the original mineral can often be inferred. Cf. Fig. 998, a, b, c (Pirsson).

Comp. — A magnesium silicate, $H_4Mg_3Si_2O_9$ or $3MgO.2SiO_2.2H_2O$ = Silica 44·1, magnesia 43·0, water 12·9 = 100. Iron protoxide often replaces a small part of the magnesium; nickel in small amount is sometimes present. The water is chiefly expelled at a red heat.

For chemical relations to the minerals of the Chlorite Group, see p. 668.

Pyr., etc. — In the closed tube yields water. B.B. fuses on the edges with difficulty. F. = 6. Gives usually an iron reaction. Decomposed by hydrochloric and sulphuric acids. From chrysotile the silica is left in fine fibers.

Diff. — Characterized by softness, absence of cleavage and feeble waxy or oily luster; low specific gravity; by yielding much water B.B.

Micro. — Readily recognized in thin sections by its greenish or yellowish green color; low relief and aggregate polarization due to its fibrous structure. When the fibers are parallel, the interference-colors are not very low, but the confused aggregates may show the "ultra blue" or even be isotropic. The constant association with other magnesia bearing minerals like chrysolite, pyroxene, hornblende, etc., is also characteristic. The presence of lines of iron oxide particles as noted above (Fig. 998) is characteristic.

Obs. — Serpentine is always a secondary mineral resulting. as noted above, from the alteration of non-aluminous silicates containing magnesia, particularly chrysolite, amphibole or pyroxene; more rarely from monticellite, chondrodite, vesuvianite, garnet, talc, etc. It may occur as a replacement of the original mineral, forming a pseudomorph, or it may be deposited elsewhere in the rock openings or as replacing some other mineral. It frequently forms large rock-masses, then being derived from the alteration of peridotites, dunites and other basic rocks of igneous origin; also of amphibolites or other similar metamorphic rocks which may have had a sedimentary origin. In the first case it is usually accompanied by spinel, garnet, chromite and sometimes nickel ores; in the second case by various carbonates such as dolomite, magnesite, breunnerite, etc. Serpentine is a mineral which occurs commonly, though rarely in masses of great size. In the metamorphic rocks it is commonly found as layers, lenticular bodies, etc.

Serpentine is so common a mineral and so widespread in its occurrence that only its most important and interesting localities can be mentioned here. In Saxony, massive ornamental material comes from Zöblitz; and the variety *metaxite* from Schwarzenberg; in Silesia at Reichenstein *picrolite* and *chrysotile*. A variety containing soda from the Zillertal, Tyrol, Austria, is called *nemaphyllite*. In Italy pseudomorphs after monticellite occur in the Val di Fassa, Trentino; the variety *antigorite* was originally described from Valle Antigorio, Piedmont. In Switzerland various types of serpentine are found near Zermatt, in Valais. *Thermophyllite* is found in Finland at Hoponsuo. In crystals pseudomorphous after chrysolite from Snarum in Buskerud, Norway. Fine serpentine from the Lizard, Cornwall. Asbestos occurs in the Transvaal in the Barberton district and at various points in Australia.

In the United States fine at Newburyport, Essex Co., Massachusetts. In Rhode Island *bowenite* at Smithfield, Providence Co. In New York from the Tilly Foster mine near Brewster, Putnam Co., pseudomorphic after chondrodite, also in cube-like cleavage forms, presumably pseudomorphous after some unknown mineral; in St. Lawrence Co., at Somerville and elsewhere. In New Jersey *marmolite*, etc., at Hoboken, associated with brucite, magnesite, etc.; at Montville, Morris Co., *chrysotile* and *retinalite* with common serpentine; in Sussex Co., at Vernon and Franklin. In Pennsylvania, at Easton, Northampton Co., pseudomorphous after pyroxene and amphibole; massive, fibrous and foliated of various colors at Texas, Lancaster Co.; at West Chester, Chester Co., *williamsite*; at Bare Hills near Baltimore in Maryland (*baltimorite*). Asbestos is found in Arizona from near Globe, Gila Co., from the Sierra Ancha, and the Grand Canyon.

In Canada abundant among the metamorphic rocks of Quebec between the Vermont boundary and Gaspe Co., at Danville, Richmond Co.; Thetford in Megantic Co.; Brough·

ton in Beauce Co., etc; in Ottawa Co., at Templeton. The asbestos often forms seams several inches in thickness in the massive mineral and has been extensively mined.

The names *Serpentine, Ophite, Lapis colubrinus,* allude to the green serpent-like cloudings of the serpentine marble. *Retinalite* is from ῥετινη, *resin; Picrolite,* from πικρός, *bitter,* in allusion to the magnesia (or bittererde) present; *Thermophyllite,* from θέρμη, *heat,* and φύλλον, *leaf,* on account of the exfoliation when heated; *Chrysotile,* from χρυσος, *golden,* and τιλος, *fibrous; Metaxite,* from μέταξα, *silk; Marmolite,* from μαρμαίρω, *to shine,* in allusion to its peculiar luster.

Use. — As an ornamental stone; the fibrous variety furnishes the greater part of the heat insulating material known as asbestos.

Deweylite. — A magnesian silicate near serpentine but with more water. Formula perhaps $4MgO.3SiO_2.6H_2O$. Amorphous, resembling gum arabic, or a resin. Microscopically fibrous. H. = 2–3·5. G. = 2·0–2·2. Color whitish, yellowish, reddish, brownish. Index, 1·55. Z parallel to fibers. Deweylite is usually found with serpentine as one of its decomposition products and commonly encloses a carbonate. It occurs as pseudomorphs after talc, etc. It occurs with serpentine at Kuttenberg in Bohemia of Czechoslovakia. From near Kraubat in the Mürztal in Styria, Austria, including the bright red variety called *eisengymnite.* With serpentine from the Val di Fiemme, Trentino, Italy; in granular limestone at Passau, Bavaria. In the United States it occurs at Middlefield, Hampshire Co., Massachusetts; in Pennsylvania at Texas, Lancaster Co., also in Berks, Chester, and Delaware counties. *Gymnite,* named from γυμνός, *naked,* in allusion to the locality at Bare Hills, Maryland, is the same species.

Genthite. Nickel Gymnite. — A gymnite with part of the magnesium replaced by nickel, $2NiO.2MgO.3SiO_2.6H_2O$. Probably should be included as a variety of garnierite. Amorphous, with a delicate stalactitic surface, incrusting. H. – 3–4; sometimes very soft. G. = 2·409. Luster resinous. Color pale apple-green, or yellowish. From Texas, Lancaster Co., Pennsylvania, in thin crusts on chromite. Also reported from near Prägratten in the Austrian Tyrol; from near Malaga, Spain; and on Michipicoten Island in Lake Superior, Ontario.

Garnierite. Noumeite. — An important ore of nickel, consisting essentially of a hydrated silicate of magnesium and nickel, perhaps $H_2(Ni,Mg)SiO_4$ + water, but very variable in composition, particularly as regards the nickel and magnesium; not always homogeneous. Amorphous. Soft and friable. G. = 2·3–2·8. Luster dull. Color bright apple-green, pale green to nearly white. Index, 1·59. In part unctuous; sometimes adheres to the tongue. Occurs in serpentine rock near Noumea, capital of New Caledonia, associated with chromic iron and steatite, where it is extensively mined. A similar ore occurs at Riddle in Douglas County, southern Oregon; also at Webster, Jackson Co., North Carolina. Reported also from the Transvaal in South Africa and from Madagascar. A nickel silicate found in peat in the Ural Mts. has been called *kerzinite.*

Nepouite. $3(Ni,Mg)O.2SiO_2.2H_2O$. Monoclinic. In microscopic crystal plates with hexagonal outline. Good cleavages; || (001) and (010). H. = 2–2·5. G. = 2·5–3·2. Color pale to deep green. Optically −. Ax. pl. || (010). X ⊥ (001). 2V very small. Indices vary with nickel content; 1·53–1·63. Occurs in the nickel deposits at Nepoui and elsewhere in New Caledonia.

Connarite. — A hydrous nickel silicate, perhaps $H_4Ni_2Si_3O_{10}$. Hexagonal. In small fragile grains. Perfect cleavage (0001). H. = 2·5–3. G. = 2·5. Color yellowish green. Optically −. ω = 1·59, ε = 1·56. From Röttis, north of Plauen in Saxony.

Maufite. Essentially an aluminum, nickel silicate, perhaps $(Mg,Ni,Fe)O.2Al_2O_3.3SiO_2.4H_2O$. In fibrous-like sheaves. Emerald-green color. H. = 3. G. = 2·27. Found in a small seam in serpentine near Umvukwe Geodetic Station, Umvukwe Mts., southern Rhodesia.

TALC.

Orthorhombic or monoclinic. Rarely in tabular crystals, hexagonal or rhombic with prismatic angle of 60°. Usually foliated massive; sometimes in globular and stellated groups; also granular massive, coarse or fine; fibrous (pseudomorphous); also compact or cryptocrystalline.

Cleavage: basal, perfect. Sectile. Flexible in thin laminæ, but not elastic. Percussion-figure a six-rayed star, as with the micas. Feel greasy. H. = 1–1·5. G. = 2·7–2·8. Luster pearly on cleavage surface. Color apple-green to white, or silvery white; also greenish gray and dark green; sometimes bright green perpendicular to cleavage surface, and brown and less translucent at right angles to this direction; brownish to blackish green and reddish when impure. Streak usually white; of dark green varieties lighter than the color. Subtransparent to translucent. Optically negative. Ax. pl. $\parallel$ $a(100)$. $X \perp c(001)$. Axial angle small, variable. Indices approx.; $\alpha = 1\cdot539$. $\beta = 1\cdot589$. $\gamma = 1\cdot589$.

Var. — *Foliated, Talc.* Consists of folia, usually easily separated, having a greasy feel, and presenting ordinarily light green, greenish white, and white colors. G. = $2\cdot55$–$2\cdot78$.

Massive, Steatite or *Soapstone.* a. Coarse granular, grayish green, and brownish gray in color; H. = 1–$2\cdot5$. *Pot-stone* is ordinary soapstone, more or less impure. b. Fine granular or cryptocrystalline, and soft enough to be used as chalk; as the *French chalk*, which is milk-white with a pearly luster. c. *Indurated talc.* An impure slaty talc, harder than ordinary talc.

Pseudomorphous. a. Fibrous, fine to coarse, altered from enstatite and tremolite. b. *Rensselaerite*, having the form of pyroxene from northern New York and Canada.

A peculiar form of clay which swells when wet, found at the Hacienda Santa Lucia near Mexico City, Mexico, and which corresponds in composition to a hydrated talc has been called *lucianite*.

Comp. — An acid metasilicate of magnesium, $H_2Mg_3(SiO_3)_4$ or H_2O. $3MgO.4SiO_2$ = Silica $63\cdot5$, magnesia $31\cdot7$, water $4\cdot8$ = 100. One-half the water is lost below dull red heat; the remainder goes off rapidly at about 900° C. Nickel is sometimes present in small amount.

Talc shows variations in composition, especially in the water content. A part of the water can be driven off by heat without any change in the crystalline or optical characters, suggesting that some of the water present is out of the crystal structure.

Pyr., etc. — In the closed tube B.B., when intensely ignited, most varieties yield water. In the platinum forceps whitens, exfoliates, and fuses with difficulty on the thin edges to a white enamel. Moistened with cobalt solution, assumes on ignition a pale red color. Not decomposed by acids. Rensselaerite is decomposed by concentrated sulphuric acid.

Diff. — Characterized by extreme softness, soapy feel; common foliated structure; pearly luster; it is flexible but inelastic. Yields water only on intense ignition.

Obs. — Talc or steatite is a very common mineral of secondary origin, and in the form of steatite constitutes extensive beds in some regions. It is often associated with serpentine, talcose or chloritic schist, and dolomite, and frequently contains crystals of dolomite, breunnerite, also asbestus, actinolite, tourmaline, magnetite. It has been formed by the alteration of non-aluminous magnesian silicates, like chrysolite, hypersthene, pyroxene, amphibole, etc. It is very common in pseudomorphs; see above.

Only a few of the most important localities for the occurrence of talc are given. Apple-green talc occurs at Mt. Greiner in the Zillertal, Austrian Tyrol; in Switzerland at St. Gotthard in Ticino and at Fiesch, etc., in Valais. Steatite comes from Göpfersgrün near Wunsiedel, Bavaria. From near Lizard Head in Cornwall with serpentine; from the Shetland Islands. In the Transvaal in the Barberton district. Steatite occurs in large amounts in India and China.

In the United States at various points in Vermont, New Hampshire, and Massachusetts. From Smithfield, Providence Co., Rhode Island; in New York at Edwards a fine fibrous talc (*agalite*); and elsewhere in St. Lawrence Co.; the so-called *rensselaerite* is common in northern New York. In Pennsylvania in Chester, Delaware, and Lancaster counties. In Maryland at Cooptown, Harford Co., with green, blue, and rose colors. Steatite occurs in Cherokee Co., North Carolina. In Canada talc and steatite are found in Brome Co., and elsewhere in Quebec and in Hastings Co., etc., in Ontario.

Use. — In the form of soapstone used for wash tubs, sinks, table tops, switchboards, hearth stones, furnace linings, etc.; the tips of gas burners, tailors' chalk, slate pencils, carved ornaments, etc.; in powdered form as filler in papers, as a lubricant, in toilet powders, etc.

GAVITE is apparently a variety of talc, differing in the amount of water present and in its solubility in acids. From Gava valley, Italy, occurring as a coating on garnet, chlorite, titanite, etc.

SEPIOLITE. Meerschaum.

Compact, with a smooth feel, and fine earthy texture, or clay-like. Microcroscopically is shown to be a mixture of fine fibrous material and an amorphous substance of apparently the same composition. The mixture forms the variety *meerschaum*. The fibrous mineral has been called *α-sepiolite* or *parasepiolite*; the amorphous material has been called *β-sepiolite*. H. = 2–2·5. G. = 2. Impressible by the finger nail. In dry masses floats on water. Color grayish white, white, or with a faint yellowish or reddish tinge, bluish green. Opaque. Biaxial, −. Indices, 1·52–1·53. 2V usually very small. *Parasepiolite* has a lower refraction index ($\beta = 1.506$), 2V = 50°.

Comp. — $H_4Mg_2Si_3O_{10}$ or $2H_2O.2MgO.3SiO_2$ = Silica 60·8, magnesia 27·1, water 12·1 = 100. Some analyses show more water, which is probably to be regarded as hygroscopic. Copper and nickel may replace part of the magnesium.

Pyr., etc. — In the closed tube yields first hygroscopic moisture, and at a higher temperature gives much water and a burnt smell. B.B. some varieties blacken, then burn white, and fuse with difficulty on the thin edges. With cobalt solution a pink color on ignition. Decomposed by hydrochloric acid with separation of silica. *Parasepiolite* gelatinizes with acid.

Obs. — Commonly found as a secondary mineral in connection with serpentine; also with magnesite and at times with opal. Occurs in Asia Minor, in masses in stratified earthy or alluvial deposits near Eski Shehr associated with magnesite. It is also found at Thebes in Bœotia, Greece, with opal and derived from serpentine. In Moravia of Czechoslovakia at Hrubschütz on the Iglawa north of Kronau. Formerly obtained from Morocco where it was used in place of soap and known as *Pierre de savon de Maroc.* In Spain at Vallecas southeast of Madrid. Found in the United States in Chester and Delaware counties, Pennsylvania; in New Mexico near Sapella, San Miguel Co., and in Grant Co.

A fibrous mineral, having the composition of sepiolite, occurs in Utah.

The word *meerschaum* is German for *sea-froth*, and alludes to its lightness and color. *Sepiolite* is from σηπια, *cuttlefish*, the bone of which is light and porous.

Spadaite. — Perhaps $MgSiO_3.H_2O$. Massive, amorphous. H. = 2·5. Fusible at 4. Color reddish. Optically −. Indices, 1·51–1·53. 2V small. From Capo di Bove, near Rome, Italy. Noted at Gold Hill, Nevada.

SAPONITE. Piotine.

Monoclinic or orthorhombic. Commonly massive, in nodules, or filling cavities. Soft, like butter or cheese, but brittle on drying. G. = 2·24–2·30. Luster greasy. Color white, yellowish, grayish green, bluish, reddish. Does not adhere to the tongue. Biaxial, −. $X \perp$ to a cleavage; $Z \parallel$ elongation. $\alpha = 1.479–1.490$, $\beta = 1.510–1.525$, $\gamma = 1.511–1.527$. 2V small (optical description of material from Michigan).

Comp. — A hydrous silicate of magnesium and aluminum; the mineral is usually amorphous and impure. Material from Michigan gave: $9MgO. Al_2O_3.10SiO_2.15–16H_2O$.

Pyr., etc. — B.B. gives out water very readily and blackens; thin splinters fuse with difficulty on the edge. Decomposed by sulphuric acid.

Obs. — Saponite characteristically occurs in the amygdaloidal openings of basic eruptive rocks, associated with calcite, zeolites, etc. It is commonly one of the last minerals to crystallize and apparently has been formed through the alteration of some magnesium silicate. Occurs at Lizard Head in Cornwall, in veins in serpentine; at various localities in Scotland; *bowlingite* is from near Bowling in Dumbarton, and *cathkinite* from the Cathkin Hills, near Glasgow. In Sweden at Svärdsjö in Kopparberg northeast of Falun (*piotine* and *saponite*). In the United States it occurs with the copper deposits of Keweenaw Peninsula, Michigan, and in Minnesota in amygdaloid on the north shore of Lake Superior (*thulite*), and in the Thunder Bay district of Ontario, Canada. Also occurs on St. George Island off the north shore of Prince Edward Island.

Saponite is from *sapo*, *soap;* and piotine from πιοτης, *fat.*

LASSALLITE. Composition perhaps $3MgO.2Al_2O_3.12SiO_2.8H_2O$. In snow-white fibrous masses. G. = 1·5. In France from the antimony mine at Meyssonial en Mercœur in Haute-Loire and at Can Pey near Arles-sur-Tech, Pyrénées-Orientales.

Iddingsite. — $MgO.Fe_2O_3.3SiO_2.4H_2O$. Orthorhombic. Foliated. Three pinacoidal cleavages. H. = 3. G. = 2·5–2·8. Color brown. Optically —, sometimes +. Ax. pl. ‖ (010). X = a axis. Indices very variable, from 1·608 to 1·765. Strong dispersion, $\rho < v$. Occurs as an alteration product of chrysolite at Carmelo Bay, Monterey Co., California; from a number of localities in Colorado, etc. Minerals identical with or closely resembling iddingsite have been noted as alteration products of chrysolite in the basalts of Traversa and Orosei, Sardinia, and have been given the names *traversite* and *oroseite.*

Celadonite. — A silicate of iron, magnesium and potassium. Earthy or in minute scales. Lamellar or fibrous. Perfect cleavage. Very soft. Color green. Optically —. 2V small. $\beta = 1·63$. From cavities in amygdaloid at Mte. Baldo near Verona, Venetia, Italy. Also from the Zillertal, Tyrol, Austria.

Glauconite. — Essentially a hydrous silicate of iron and potassium. Shows a considerable variation in composition and probably occurs in both colloidal and crystalline forms. Usually amorphous, and resembling earthy chlorite. Noted in small crystal grains in a dolomite from near Bonneterre, St. Francis Co., Missouri. Micaceous cleavage. Optically —. X nearly normal to cleavage. Indices, 1·60–1·63. 2V = 0°–20°. Color dull green. Occurs in rocks of nearly all geological ages; abundant in the " green sand," of the Chalk formation, sometimes constituting 75 to 90 per cent of the whole. Found abundantly in ocean sediments near the continental shores. It has been considered to have been formed by the alteration of augite, hornblende, mica, etc. A manganese glauconite from the Marsjat forest, Ural Mts., has been called *marsjatskite.* *Greenalite* is a green hydrated ferrous silicate found as granules in the cherty rock associated with iron ores of the Mesabi district, Minnesota. Resembles glauconite but contains no potash.

Pholidolite. — Corresponds approximately to $K_2O.12(Fe,Mg)O.Al_2O_3.13SiO_2.5H_2O$. Monoclinic. In minute pseudo-hexagonal crystalline scales. Perfect cleavage (001). H. = 4. G. = 2·408. Color grayish yellow. Optically —. X ⊥ (001). Indices 1·50–1·545. 2V small. From Taberg in Vermland, Sweden, with garnet, diopside, etc.

IV. KAOLIN DIVISION

KAOLIN MINERALS: KAOLINITE, NACRITE, DICKITE.

Monoclinic; in thin rhombic, rhomboidal or hexagonal scales or plates with angles of 60° and 120°. Usually constituting a clay-like mass, either compact, friable or mealy.

Cleavage: basal, perfect. Flexible, inelastic. H. = 2–2·5. G. = 2·6–2·63. Luster of plates, pearly; of mass, pearly to dull earthy. Color white, grayish white, yellowish, sometimes brownish, bluish or reddish. Scales transparent to translucent; usually unctuous and plastic.

The study of minerals commonly included under the name of kaolin shows three species with different optical characters. The suggested separation is into *nacrite* (from near Freiberg, Saxony), *dickite* (from island of Anglesey, Wales, and Red Mountain, near Ouray, Colorado), and *kaolinite.* All are monoclinic. Optical axial plane, ⊥ (010) and nearly ‖ (100). Indices $\alpha = 1·557$–$1·561$, $\beta = 1·562$–$1·565$, $\gamma = 1·563$–$1·566$. *Nacrite* and *dickite* are nearly or quite transparent and in tabular crystals. *Kaolinite* is translucent to opaque and in crystals elongated parallel to c axis. *Nacrite* is usually optically —, rarely +; *dickite* is +; *kaolinite* —. Z = b axis in all three. The angle between X and the normal to (001) = 10° to 12° in *nacrite*, 15° to 20° in *dickite*, 1° to $3\frac{1}{2}$° in *kaolinite*. Dispersion $\rho > v$, rarely $\rho < v$ in *nacrite*; $\rho < v$ in *dickite*; $\rho > v$ in *kaolinite*. *Kaolinite* absorbs dyes readily becoming pleochroic; the other two do not show this. Study by X-rays shows distinctly different molecular spacings in the three cases.

Var. — 1. *Kaolinite, Nacrite, Dickite.* In crystalline scales, pure white and with a satin luster in the mass. 2. *Ordinary.* Common kaolin, in part in crystalline scales but more or less impure including the compact *lithomarge*. 3. A kaolin-like mineral occurring in bluish or greenish crusts near Alushta in the Crimea has been named *alushtite*; a pink kaolin-like mineral from Yanokami Hill, Omi Province, Japan, has been called *takizolite*.

Comp. — $H_4Al_2Si_2O_9$, or $2H_2O.Al_2O_3.2SiO_2$ = Silica 46·5, alumina 39·5, water 14·0 = 100. The water goes off at a high temperature, above 330°.

Pyr., etc. — Yields water. B.B. infusible. Gives a blue color on ignition with cobalt solution. Insoluble in acids.

Diff. — Characterized by unctuous, soapy feel and the alumina reaction B.B. Resembles infusorial earth, but readily distinguished under the microscope.

Obs. — Ordinary kaolin is a result of the decomposition of aluminous minerals, especially the feldspar of granitic and gneissoid rocks and porphyries. In some regions where these rocks have decomposed on a large scale, the resulting clay remains in vast beds of *kaolin*, usually more or less mixed with free quartz, and sometimes with oxide of iron from some of the other minerals present. It may occur in bodies that have been formed in place by the direct decomposition of the original rock, or it may have been transported and occur as extensive sedimentary beds. Pure kaolinite in scales often occurs in connection with iron ores of the Coal formation and in various ore veins; at times in crystalline scales in cavities in quartz veins. Also met with accompanying diaspore and emery or corundum.

Kaolin is very widespread in its occurrence and only those localities that are important for the purity and size of the kaolin bodies or those in which it occurs in crystals can be given here. In Bohemia of Czechoslovakia at Schlan near Kladno and near Karlsbad. From Saxony at Brand near Freiberg; at Rocklitz in a porphyritic rock; near Zwickau; as bodies of kaolin near Meissen, Schneeberg, and in the Province of Saxony near Halle. At Saint Yrieix, south of Limoges, Haute-Vienne, France, is the best kaolin locality in Europe; it affords material for the famous Sèvres porcelain manufactory. Large quantities of kaolin are found at St. Austell and elsewhere in Cornwall and in Devonshire. Dickite occurs in crystalline plates near Amlwich on the island of Anglesey, Wales.

In the United States kaolin occurs in Pennsylvania at Summit Hill, Carbon Co., in the coal formation; at Pottsville, Schuylkill Co., etc. Near Richmond, Virginia, the mealy variety constitutes a bed of considerable extent in the Tertiary formation. At the National Bell mine, Red Mountain, Silverton, San Juan Co., Colorado (dickite), in very pure form in cavities of a quartz vein. Commercial kaolin is produced from various southern states, chiefly from Georgia and South Carolina.

The name *kaolin* is a corruption of the Chinese *Kauling*, meaning *high-ridge*, the name of a hill near Jauchau Fu, where the material is obtained.

Use. — The finer, purer grades used in the manufacture of porcelain, china, etc.; in the form of clay in pottery, stoneware, bricks, etc.

PHOLERITE. Near kaolinite, but some analyses give 15 per cent water. The original was from the coal mines of Fins, Dept. of Allier, France.

MILOSCHITE. A kaolin with varying amounts of chromium oxide (4 to 9 per cent) replacing aluminum oxide. H. = 2·5. G. = 2·1. Color pale indigo-blue, pale greenish blue, celandine-green. Luster waxy to dull. Optically −. $\beta = 1·56$. Originally from Rudniak in Serbia; also from Volterra, Tuscany, Italy. Recently from near Ely, Nevada.

VOLCHONSKOITE. Similar to miloschite with chromic and ferric oxides preponderating over aluminum oxide. Color bluish to grass-green. $\beta = 1·585$. In clay from Okhansk in Siberia.

Faratsihite. — $H_4(Al,Fe)_2Si_2O_9$. Intermediate between *kaolinite* and *chloropal*. Monoclinic. In microscopic hexagonal plates. Soft. G. = 2. Difficultly fusible. Decomposed by hydrochloric acid. Color pale yellow. Birefringence a little higher than that of kaolinite. Optically −. $n = 1·56$. From near Faratsiho, east of Mt. Ankaratra, Madagascar.

HYDROUS ALUMINUM SILICATES

HALLOYSITE.

Massive. Clay-like or earthy.

Fracture conchoidal. Hardly plastic. H. = 1–2. G. = 2·0–2·20. Luster somewhat pearly, or waxy, to dull. Color white, grayish, greenish, yellowish, bluish, reddish. Translucent to opaque, sometimes becoming translucent or even transparent in water, with an increase of one-fifth in weight. Isotropic. $n = 1.47–1.52$ varying with amount of water content.

Var. — *Ordinary.* Earthy or waxy in luster and opaque massive. *Galapectite* is halloysite of Angleur, Belgium. *Pseudosteatite* is an impure variety, dark green in color, with H. = 2·25. G. = 2·469. *Indianaite* is a white porcelain clay from Lawrence Co., Indiana, where it occurs with allophane in beds four to ten feet thick ($n = 1.538$).

Smectite is greenish, and in certain states of humidity appears transparent and almost gelatinous; it is from Condé, near Houdan, France.

Bole, in part, may belong here; that is, those colored, unctuous clays containing more or less iron oxide, which also have about 24 per cent of water; the iron gives them a brownish, yellowish or reddish color; but they may be mixtures. Here belongs *bergseife* (mountain-soap).

Comp. — A silicate of aluminum ($Al_2O_3.2SiO_2$) like kaolin, but amorphous and containing more water.

The water present is in part held mechanically since it is given off very readily. The resulting partially dehydrated material is near kaolin in composition. It is probable that halloysite is an amorphous mineral corresponding to kaolin but holding by capillarity or adsorption varying amounts of excess water.

Pyr., etc. — Yields water. B.B. infusible. A fine blue on ignition with cobalt solution. Decomposed by acids.

Obs. — Occurs often in veins or beds of ore, as a secondary product; also in granite and other rocks, being derived from the decomposition of some aluminous minerals. Found in many localities in small amounts.

TERMIERITE. A clay-like substance resembling halloysite of uncertain composition from the antimony mines of Meyssonial en Mercœur in Haute-Loire, France.

BATCHELORITE. $Al_2O_3.2SiO_2.H_2O$? A green foliated mineral from Mt. Lyell mine, Tasmania.

Leverrierite. — A hydrated silicate of aluminum. Monoclinic. Pseudo-hexagonal. In vermiculate aggregates. Perfect cleavage || (001). H. = 1·5. G. = 2·5–2·6. Difficultly fusible, usually with swelling. Colorless to brown. Biaxial with very small optical angle. Optically −. $\alpha = 1.558$, $\beta = \gamma = 1.602$. Occurs in schistose rocks at various points in France. In broad cleavage plates associated with quartz and manganese oxides at Beidell, Saguache Co., Colorado.

Anauxite. — $Al_2O_3.3SiO_2.2H_2O$. Monoclinic. In crystal plates, || (001), with hexagonal outline. Perfect basal cleavage. Shows same X-ray pattern as kaolinite and has nearly the same optical properties but differs in its composition. H. = 2·5. G. = 2·524. Color white. Pearly luster. Optically −. Ax. pl. ⊥ (010). $X = c$ axis. $\alpha = 1.559$. $\beta = 1.564$. $\gamma = 1.565$. 2V = 18°–31°. Dispersion $\rho > v$. From Bilin in Bohemia of Czechoslovakia, as alteration of augite and biotite. Occurs as scales in the Ione sandstone formation of California (*ionite*). Thought to be of wide occurrence in clays, especially in many so-called kaolins.

Beidellite. — $Al_2O_3.3SiO_2.4H_2O$. Probably orthorhombic. In thin crystal plates. Color white, reddish or brownish gray. Waxy to vitreous luster. $\alpha = 1.494$, $\beta = \gamma = 1.536$. 2V = 9°–16°. Occurs as a clay gouge made up of minute crystal plates at Beidell, Colorado. *Iron-beidellite* is a variety with considerable amount of Fe_2O_3.

ELBRUSSITE. A hydrous silicate of aluminum, ferric and ferrous iron, magnesium, etc. Massive. H. = 2. G. = 2·28. Fusible at 3. Dark chocolate-brown. Streak dark orange-yellow. Greasy luster. Uniaxial, −. $n = 1.56$. Found on the river Tschuhtschur, Karatschaev district, northern Caucasus.

KOCHITE. $2Al_2O_3.3SiO_2.5H_2O$? Isometric. Cubes, truncated by octahedrons. In granular aggregates of minute crystals. Color, white. $n = 1.590$. G. = 2·929. From Kochi-mura, Rikuchu Province, Japan.

KIMOLITE. Cimolite. A hydrous silicate of aluminum, $2Al_2O_3.9SiO_2.6H_2O$? Amorphous claylike, or chalky. Probably represents a mixture of minerals. Very soft. G. = 2·18–2·30. Color white, grayish white, reddish. From the island of Kimolos (Argentiera) of the Cyclades, Greece; from the province of Kiev, Russia (*pelikanite*); Berg Hradišt near Bilin in Bohemia of Czechoslovakia, pseudomorphic after augite (*anauxite* and *hunterite*) Similar material from Norway, Oxford Co., Maine.

Montmorillonite. — Perhaps $(Mg,Ca)O.Al_2O_3.5SiO_2.nH_2O$ with $n = 5$–7. Massive, clay-like. Shows the X-ray pattern of kaolinite and may be composed of that mineral enclosing colloidal particles within the structure. Very soft and tender. G. $= 2$. Luster feeble. Color white or grayish to rose-red, and bluish; also pistachio-green. Unctuous. Isotropic usually. $n = 1\cdot49$–$1\cdot56$. Always an alteration product of some aluminous mineral. From near Macska-Mező not far from Magyar-Lapos in Solnoc-Dobaca, Rumania. In France from Montmorillon in Vienne (*montmorillonite*); *confolensite* from Confolens in Charente and from Saint-Jean-de-Cole near Thiviers in Dordogne; *saponite* (or *smegmatite*) comes from Plombières in Vosges. From the basalt of Stolpen, Saxony (*stolpenite*). From the Giant's Causeway, Co. Antrim, Ireland (*erinite*).

In the United States montmorillonite is found in Maine at Auburn, Androscoggin Co., and in Oxford Co., at Stoneham and Paris; in Connecticut at Branchville, Fairfield Co.; and from California in San Diego Co. The clay known as *bentonite*, which occurs widely distributed in western United States and Canada and which has been derived from the alteration of volcanic ash or tuff, is usually largely composed of montmorillonite.

Collyrite. — $2Al_2O_3.SiO_2.9H_2O$. A clay-like mineral, white, with a glimmering luster, greasy feel, and adhering to the tongue. G. $= 2$–$2\cdot15$. Isotropic. $n = 1\cdot55$. From near Schemnitz (Selmecsbánya) in Slovakia of Czechoslovakia. In Bavaria at Bergensreuth east of Wunsiedel. From Haute-Garonne in the Pyrenees in the valley of Squiéry near the village of Oo (usually given as Ezquerra). At Hove, near Brighton, Sussex, in fissures in chalk, white and very soft.

PYROPHYLLITE.

Orthorhombic. Foliated, radiated lamellar or somewhat fibrous; also granular to compact or cryptocrystalline; the latter sometimes slaty.

Cleavage: basal, eminent. Laminæ flexible, not elastic. Feel greasy. H. $= 1$–2. G. $= 2\cdot8$–$2\cdot9$. Luster of folia pearly; of massive kinds dull and glistening. Color white, apple-green, grayish and brownish green, yellowish to ocher-yellow, grayish white. Subtransparent to opaque. Optically $-$. $X = c$ axis. $Z \parallel$ elongation. $\alpha = 1\cdot552$, $\beta = 1\cdot588$, $\gamma = 1\cdot600$. $2V = 57°$.

Var. — (1) *Foliated*, and often radiated, closely resembling talc in color, feel, luster and structure. (2) *Compact massive*, white, grayish and greenish, somewhat resembling compact steatite, or French chalk. This compact variety includes part of what has gone under the name of *agalmatolite* from China; it is used for slate-pencils, and is sometimes called *pencil-stone*.

Comp. — $H_2Al_2(SiO_3)_4$ or $H_2O.Al_2O_3.4SiO_2$ = Silica $66\cdot7$, alumina $28\cdot3$, water $5\cdot0 = 100$.

Pyr., etc. — Yields water, but only at a high temperature. B.B. whitens, and fuses with difficulty on the edges. The radiated varieties exfoliate in fan-like forms, swelling to many times the original volume of the assay. Moistened with cobalt solution and heated gives a deep blue color (alumina). Partially decomposed by sulphuric acid, and completely on fusion with alkaline carbonates.

Diff. — Resembles some talc, but distinguished by the reaction for alumina with cobalt solution.

Obs. — Compact pyrophyllite is the material or base of some schistose rocks. The foliated variety is often the gangue of kyanite. Pyrophyllite occurs in the Ural Mts., Russia, between Pyschminsk and Beresovsk in the Ekaterinburg district; in Valais, Switzerland, in the region between Visp, St. Niklaus and Zermatt. In Belgium at Spa and Malmédy in Liége and at Ottré in Luxembourg. From Sweden from the Westanå mine near Näsum west of Carlshamn in Kristianstad and with kyanite from Horrsjöberg in Vermland. In Brazil from Ouro Preto, Minas Geraes, in foliated masses of considerable extent.

In the United States in thin seams in coal slates at Mahoney City, Schuylkill Co., Pennsylvania. A compact pyrophyllite, resembling a slaty soapstone, is found in large beds in Deep River, Guilford Co., North Carolina; also in Orange Co., etc. From the Chesterfield district, Chesterfield Co., South Carolina, with lazulite and kyanite. On Graves Mt., in Lincoln Co., Georgia. Near Indian Gulch, Mariposa Co., California.

Use. — For the same purposes as talc, which see.

ALLOPHANE.

Amorphous. In incrustations, usually thin, with a mammillary surface, and resembling hyalite; sometimes stalactitic. Occasionally almost pulverulent.

Fracture imperfectly conchoidal and shining, to earthy. Very brittle. H. = 3. G. = 1·85–1·89. Luster vitreous to subresinous; bright and waxy internally. Color pale sky-blue, sometimes greenish to deep green, brown, yellow or colorless. Streak uncolored. Translucent. Isotropic. $n = 1·47$–1·49.

Comp. — Hydrous aluminum silicate, $Al_2SiO_5.nH_2O$.

Impurities are often present. The coloring matter of the blue variety is due to traces of chrysocolla, and substances intermediate between allophane and chrysocolla (mixtures) are not uncommon. The green variety is colored by malachite, and the yellowish and brown by iron. A highly ferriferous variety has been called *ferri-allophane*.

Pyr., etc. — Yields much water in the closed tube. B.B. crumbles but is infusible. Gives a blue color on ignition with cobalt solution. Gelatinizes with hydrochloric acid.

Obs. — Allophane is regarded as a result of the decomposition of some aluminous silicate (feldspar, etc.); and it often occurs incrusting fissures or cavities in ore veins, especially those of copper or zinc; often with limonite; at times in coal seams. First described from Gräfental, south of Saalfield, Thuringia (*riemannite*). At Friesdorf near Bonn, Rhineland, in lignite (*elhuyarite*). In Sardinia at the Rosas mine in Sulcis. From near Woolwich, Kent, in chalk. In the United States it occurs in Pennsylvania at the Friedensville zinc mines in Lehigh Co.; also at Cornwall, Lebanon Co.

Named from αλλός, *other*, and φαίνεσθαι, *to appear*, in allusion to its change of appearance under the blowpipe.

Viterbite is apparently a mixture of allophane and wavellite from Santa Rosa de Viterbo, state of Boyaca, Colombia.

MELITE. $2(Al,Fe)_2O_3.SiO_2.8H_2O$. In imperfect prisms. Stalactitic, massive. H. = 3. G. = 2·2. Color bluish brown. Infusible. From Saalfeld, Thuringia.

Schrötterite. — $8Al_2O_3.3SiO_2.30H_2O$? Perhaps a mixture. Resembles allophane; sometimes like gum in appearance. H. = 3–3·5. G. = 1·95–2·05. Color pale green or yellowish. Isotropic. $n = 1·58$. From Dollinger mountain, near Freienstein, northwest of Leoben, Styria, Austria; at the Falls of Little River, on the Sand Mt., Cherokee Co., Alabama.

The following are clay-like minerals or mineral substances: *Sinopite, smectite, catlinite*.

HYDROUS IRON SILICATES

CHLOROPAL. Nontronite.

Compact massive, with an opal-like appearance; earthy.

H. = 2·5–4·5. G. = 1·727–1·870, earthy varieties, the second a conchoidal specimen; 2·105, Ceylon. Color greenish yellow and pistachio-green. Opaque to subtranslucent. Fragile. Fracture conchoidal and splintery to earthy. Adheres feebly to the tongue. Optical characters vary. Biaxial, — (also reported as +). Indices 1·56–1·65.

Var. — *Chloropal* has the above-mentioned characters, and was named from the mineral occurring at Užhorod (Ungvar) in Ruthenia of Czechoslovakia.

Nontronite is pale straw-yellow or canary-yellow, and greenish, with an unctuous feel; flattens and grows lumpy under the pestle, and is polished by friction; from Nontron, Dordogne, France. *Pinguite* is siskin- and oil-green, extremely soft, like new-made soap, with a slightly resinous luster, not adhering to the tongue; from Wolkenstein and near Zwickau, in Saxony. *Graminite* has a grass-green color (whence the name), and occurs at Menzenberg, in the Siebengebirge, Rhineland; in thin fibrous seams, or as delicate lamellæ.

Comp. — A hydrated silicate of ferric iron, perhaps with the general formula $H_4Fe_2Si_2O_9$, analogous to kaolin and it is therefore frequently included as a member of the Kaolin Group. Alumina is present in some varieties.

The content of water is variable, depending upon atmospheric conditions. When heated the mineral loses water gradually and continuously without change in crystal character, even when heated to 300°. Indices of refraction and other optical properties undergo progressive change with the dehydration. On exposure to moist air most of the water is regained. It is probable that chloropal should include müllerite, morencite, and hœferite, mentioned below. Chloropal occurs mixed with opal, and graduates into it, and this accounts for the high silica of some of its analyses.

Pyr., etc. — Infusible. Gelatinizes with HCl.

Obs. — In addition to localities mentioned above, chloropal is found at Passau in Bavaria; from Meenser Steinberg near Göttingen in the Province of Hannover. In the United States it occurs at Lehigh Mt., south of Allentown, Lehigh Co., Pennsylvania. Reported from Palmetto Mts., Esmeralda Co., Nevada.

Hœferite. An iron silicate near chloropal. Color green. From Křitz, southwest of Rackovitz in Bohemia of Czechoslovakia.

Müllerite. *Zamboninite.* $Fe_2Si_3O_9.2H_2O$. Massive. Resembles nontronite. Soft. G. = 2·0. Color yellowish green. Infusible. From Nontron, Dordogne, France.

Morencite. A hydrated ferric silicate of uncertain composition. Fibrous. Color brownish yellow. From Morenci, Greenlee Co., Arizona.

Hisingerite. — A hydrated ferric silicate of uncertain composition. Amorphous, compact. Fracture conchoidal. H. ⇌ 3. G. = 2·5–3·0. Luster greasy. Color black to brownish black. Streak yellowish brown. Isotropic, $n = 1·49–1·66$. Also reported as finely crystalline, fibrous, biaxial −, with 2V very small. From Sweden at Riddarhyttan in Vastmanland; from Långbanshyttan in Vermland. *Degeröite* comes from Degerö near Helsingfors in Uudenmaan, Finland. Found on the west coast of Greenland, in the Godhavn district. In the United States found in Blaine Co., Idaho, and in Canada at Parry Sound, Ontario.

Canbyite. — $H_4Fe_2'''Si_2O_9.2H_2O$. In thin layers of crystalline flakes. One good cleavage. Optically −. $Z \perp$ cleavage. Indices variable, average being, $\alpha = 1·562, \beta = 1·580, \gamma = 1·582$. 2V very small. Occurs in thin layers lying between quartz and garnet in the Brandywine Quarry at Wilmington, Delaware. May be the crystalline phase of the amorphous hisingerite.

HYDROUS MANGANESE SILICATES

Bementite. — $H_{10}Mn_8Si_7O_{27}$. Orthorhombic. Cleavages || to three pinacoids, one perfect. In soft radiated masses resembling pyrophyllite. G. = 2·981. Color pale grayish yellow. Optically −. $X \perp$ to best cleavage. $\beta = 1·632–1·650$. 2V very small. From the zinc mines of Franklin, Sussex Co., New Jersey. Also noted at Pajsberg (*caryopilite*) near Persberg, Vermland, Sweden, and in the Olympic Mts., Washington.

Ectropite. — $Mn_{12}Si_8O_{28}.7H_2O$. Monoclinic(?). In thin tabular crystals. Good cleavage. H. = 4. G. = 2·46. Color brown. Opaque. Indices, 1·61–1·63. From Långbanshyttan, Vermland, Sweden. Shows close similarity to bementite in chemical and optical characters although it is apparently different crystallographically.

Sérandite. A hydrous silicate of manganese, calcium, and sodium. Monoclinic. Crystals elongated || b axis. Two cleavages || b axis. Color rose-red. Optically +. Ax. pl. $\perp$ (010). $\alpha = 1·660, \beta = 1·664, \gamma = 1·688$. From pegmatites in a soda syenite found on the island of Rouma, Los Islands, French Guinea.

Orientite. — $Ca_4\overset{\text{III}}{Mn}_4(SiO_4)_5.4H_2O$. Orthorhombic. Radiating prismatic. H. = 4·5–5. G. = 3. Brown to black. Transparent to opaque. Optically +. Ax. pl. || (001). $X = a$ axis. Strong dispersion, $\rho < v$. $\alpha = 1·758$. $\beta = 1·776$. $\gamma = 1·795$. From Oriente Province, Cuba.

Hodgkinsonite. — $3(Zn,Mn)O.SiO_2.H_2O$. Monoclinic. In acute pyramidal crystals; also at times prismatic || a axis or tabular || (001). Perfect basal cleavage. H. = 4·5–5. G. = 3·91. Decrepitates and then fuses readily. Soluble in acids. Color, bright pink to reddish brown. Optically −. Ax. pl. || (010). $Z \wedge c$ axis = 38°. $\alpha = 1·724, \beta = 1·742, \gamma = 1·746$. 2V = 50°–60°. From Franklin, Sussex Co., New Jersey.

Gageite. — A hydrous silicate of manganese, magnesium and zinc, $8RO.3SiO_2.3H_2O$. Probably orthorhombic. In radiating groups of needle-like crystals. G. = 3·58. Colorless and transparent. Biaxial, −. Z ‖ elongation. $\alpha = 1.723$, $\beta = 1.734$, $\gamma = 1.736$. 2V moderate. Strong dispersion, $\rho < v$. From Franklin, Sussex Co., New Jersey.

NEOTOCITE. A hydrated silicate of manganese and iron, of doubtful composition, usually derived from the alteration of rhodonite and other manganese silicates. Amorphous. Colloidal in character. Conchoidal fracture. H. = 3–4. G. about 2·8. Color black to dark brown and liver-brown.

Sturtite. — $6(Mn,Ca,Mg)O.Fe_2O_3.8SiO_2.23H_2O$. Compact. Brittle and friable. Black with yellow-brown streak. Isotropic. H. = 3. G. = 2·05. Difficultly fusible to magnetic mass. Decomposed by acids. From Broken Hill, New South Wales.

HYDROUS COPPER SILICATES

CHRYSOCOLLA.

Cryptocrystalline; often opal-like or enamel-like in texture; earthy. Incrusting or filling seams. Sometimes botryoidal. In microscopic acicular crystals from Mackay, Idaho.

Fracture conchoidal. Rather sectile; translucent varieties brittle. H. = 2·4. G. = 2–2·24. Luster vitreous, shining, earthy. Color mountain-green, bluish green, passing into sky-blue and turquois-blue; brown to black when impure. Streak, when pure, white. Translucent to opaque. Crystals from Idaho gave: Uniaxial, +; $\omega = 1.46$; $\epsilon = 1.57$; weakly pleochroic, O = colorless, E = pale blue-green. Indices variable and optical character different in different occurences.

Comp. — True chrysocolla appears to correspond to $CuSiO_3.2H_2O$ = Silica 34·3, copper oxide 45·2, water 20·5 = 100, the water being double that of dioptase.

Composition varies much through impurities; free silica, also alumina, black oxide of copper, oxide of iron (or limonite) and oxide of manganese may be present; the color consequently varies from bluish green to brown and black. It has been suggested that the composition of most chrysocolla is not definite but that it is usually in the form of a mineral gel with copper oxide, silica and water occurring in varying proportions according to the conditions of formation. The colloidal phase of chrysocolla has been called *cornuite* and *katangite*. A mixture of chrysocolla and gibbsite from Arenas, Sardinia, has been named *traversoite*.

Pyr., etc. — In the closed tube blackens and yields water. B.B. decrepitates, colors the flame emerald-green, but is infusible. With the fluxes gives the reactions for copper. With soda and charcoal a globule of metallic copper. Decomposed by acids without gelatinization.

Obs. — Chrysocolla is a mineral of secondary origin, commonly found associated with other secondary minerals of copper in the upper portions of copper veins. It is of frequent occurrence, only a few of the more significant localities being given here. In the Ural Mts., Russia, at Nizhne-Tagilsk (*demidovite*). Found in many of the copper mines of Cornwall, at Liskeard, etc.; in Cumberland at Roughten Gill, near Caldbeck. Fine specimens come from Katanga in the Belgian Congo. Common in the copper deposits of Chile. In the United States found in Pennsylvania at Cornwall, Lebanon Co., and in Berks Co. In Arizona in fine specimens, sometimes glassy green, at the Clifton-Morenci district, Greenlee Co.; in the Globe district, Gila Co.; at Bisbee, Cochise Co., etc.; from the Organ Mts., Dona Ana Co., New Mexico. From the Tintic district, Juab Co., Utah. From Mackay, Custer Co., Idaho, in microscopic acicular crystals in radiating groups or narrow bands of closely packed individuals oriented normal to sides of the bands.

Chrysocolla is derived from χρυσος, *gold*, and κόλλα, *glue*, and was formerly the name of a material used in soldering gold. The name is often applied now to borax, which is so employed.

Use. — Chrysocolla may serve as a minor ore of copper.

Bisbeeite. — $CuSiO_3.H_2O$. Orthorhombic, fibrous or in thin laths. Color pale blue to nearly white. Optically +. Ax. pl. ⊥ to laths. Z ‖ elongation. $\alpha = 1.615$, $\beta = 1.625$, $\gamma = 1.71$. 2V small. Pleochroic, very pale green to pale olive-brown. Found at Shattuck mine at Bisbee, Cochise Co., Arizona, resulting from the hydration of *shattuckite*.

Shattuckite. — $2CuSiO_3.H_2O$. Monoclinic. Compact, granular, fibrous. G. = 3·8. Color blue. Optically +. Ax. pl. ⊥ (010). Z nearly parallel to elongation. $\alpha = 1·752$, $\beta = 1·782, \gamma = 1·815$. 2V large. Pleochroic, dark to light blue. Found at Shattuck mine, Bisbee, Cochise Co., Arizona, forming pseudomorphs after *malachite*. Reported also at Tantara, Katanga, Belgian Congo.

Planchéite. — $6CuO.5SiO_2.2H_2O$, or perhaps $3CuSiO_3.H_2O$. Orthorhombic. Fibrous, often mammillary. H. = 5·5. G. = 3·3. Optically +. Z ‖ elongation. $\alpha = 1·645$, $\beta = 1·660$, $\gamma = 1·715$. Pleochroism, Z = blue, $X = Y$ = colorless. Found associated with dioptase, etc., in the French Congo at Mindouli, Pimbi, and on the river Djoué; from the Katanga district of the Belgian Congo; from South West Africa at Guchab near Otavi and from near Windhoek. Has been thought to be identical with shattuckite but though closely related the two species appear to be distinct.

Searlesite. — $NaB(SiO_3)_2.H_2O$. Monoclinic. Prismatic. In minute spherulites composed of radiating fibers. Perfect cleavage ‖ (010). G. = 2·45. Fusible. Decomposed by hydrochloric acid. Color white. Optically −. Ax. pl. ⊥ (010). $X \wedge c$ axis = +30°. $\alpha = 1·513$, $\beta = 1·533, \gamma = 1\,535$. Found at Searles Lake, San Bernardino Co., California. In crystals from near Coaldale, Esmeralda Co., Nevada.

Colerainite. — $4MgO.Al_2O_3.2SiO_2.5H_2O$. Hexagonal. In minute, thin, hexagonal plates, which are usually aggregated in rosettes or botryoidal forms. H. = 2·5–3. G. = 2·51. Colorless or white. Optically +. Index, 1·56. Occurs as veins in serpentine in Coleraine Township, Megantic Co., Quebec.

Tatarkaite. A complex hydrous silicate of aluminum, magnesium, etc. Tabular crystals. G. = 2·7. Color dark gray to black. Uniaxial, +. In limestone on the Tatarka river, Yeniseisk District, Siberia.

SILICATES CONTAINING VARIOUS OTHER ACID RADICALS

Cenosite. Kainosite. — $2CaO.(Ce,Y)_2O_3.CO_2.4SiO_2.H_2O$. Orthorhombic. Crystals short prismatic. One cleavage. H. = 5–6. G. = 3·4–3·6. Color yellowish brown. Optically −. $\alpha = 1·664$, $\beta = 1·689, \gamma = 1·692$ (North Burgess). 2V = 40°. Strong dispersion, $\rho < v$. From island of Hiterö, Vest-Agder, Norway, and at Nordmark, Vermland, Sweden. From North Burgess, Lanark Co., Ontario.

Britholite. — A complex silicate and phosphate of the cerium metals and calcium. Orthorhombic, in pseudo-hexagonal prisms. H. = 5·5. G. = 4·4. Color brown. Optically −. $X = c$ axis; $Y = a$ axis. $\alpha = 1·772$, $\beta = 1·775, \gamma = 1·777$. 2V small to medium. From nephelite-syenite region of Julianehaab, South Greenland.

Erikite. — Composition uncertain; essentially a silicate and phosphate of the cerium metals. Orthorhombic. In prismatic crystals. H. = 5·5. G. = 3·5. Color light yellow-brown to dark gray-brown. From nephelite-syenite at Nunarsiuatiak in the Julianehaab district, South Greenland.

Plazolite. — $3CaO.Al_2O_3.2(SiO_2,CO_2).2H_2O$. Isometric. In minute dodecahedrons. H. = 6·5. G. = 3·13. Colorless to light yellow. $n = 1·675$. From Crestmore, Riverside Co., California.

Thaumasite. — $CaSiO_3.CaCO_3.CaSO_4.15H_2O$. Hexagonal. Massive, compact, crystalline. Cleavage in traces. H. = 3·5. At first soft but hardens on exposure to the air. G. = 1·877. Color white. Uniaxial, −. $\omega = 1·507$. $\epsilon = 1·468$. Occurs filling cavities and crevices at the Bjelke mine, near Åreskutan, Jämtland. From Långbanshyttan, Vermland, Sweden. Also in fibrous crystalline masses at Paterson and West Paterson (crystals), New Jersey; from Beaver Co., Utah.

Spurrite. — $2Ca_2SiO_4.CaCO_3$. Probably monoclinic. In granular cleavable masses. H. = 5. G. = 3. Color pale gray. Infusible. Optically −. Ax. pl. ⊥ (010). Z nearly ‖ a axis. $\alpha = 1·64$, $\beta = 1·67$, $\gamma = 1·68$. 2V = 40°. From contact zone between limestone and diorite in Velardeña mining district, Mexico. Found in limestone at contact zone at Scawt Hill, Co. Antrim, Ireland.

Afwillite. — $3CaO.2SiO_2.3H_2O$. Monoclinic. Prismatic habit, elongated parallel to b axis. Perfect basal cleavage. Colorless or white. G. = 2·630. Optically +. Ax. pl. ‖ (010). $X \wedge c$ axis = −30°. $\alpha = 1·617$, $\beta = 1·620$, $\gamma = 1·634$. 2V = 58°. From the Dutoitspan diamond mine at Kimberley, South Africa. Also as an alteration product of spurrite at Scawt Hill, Larne, Co. Antrim, Ireland.

Scawtite. — A silicate and carbonate of lime, $6CaO.4SiO_2.3CO_2$. Monoclinic. Small plates in sub-parallel or slightly divergent groups. Cleavages (001) perfect, also (010). G. = 2·77. H. = 4·5–5. Colorless. Vitreous luster. Optically +. $Y = b$ axis.

$Z : (001) = 29°$. $\alpha = 1{\cdot}599$, $\beta = 1{\cdot}606$, $\gamma = 1{\cdot}621$.   Occurs as a rare constituent of a zone of hybrid rocks formed by the assimilation of limestone by dolerite at Scawt Hill, Co. Antrim, Ireland.

Uranophane.   Uranotil. — $CaO.2UO_3.2SiO_2.7H_2O$.   Orthorhombic.   Prismatic.   In radiated aggregations; massive, fibrous.   H. = 2–3.   G. = $3{\cdot}81$–$3{\cdot}90$.   Color yellow.   Biaxial, $-$. $X \perp$ to cleavage or tabular development. $Z \parallel$ elongation.   Strong dispersion, $\rho < v$. $\alpha = 1{\cdot}643$, $\beta = 1{\cdot}666$, $\gamma = 1{\cdot}669$.   Commonly occurs in granite.   Found at Kupferberg in Silesia; in the Oberpfalz, Bavaria, from Wölsendorf, south of Naabburg (*uranotil*) where it occurs in cavities in fluorite with uraninite.   From near Schneeberg, Saxony (*uranotil*).   From Mt. Painter in the Flinders Range, South Australia.   In the United States it is found at Avondale, Delaware Co., Pennsylvania.   At the mica mines of Mitchell Co., North Carolina, as an alteration product of gummite.

Sklodowskite.   Chinkolobwite. — $MgO.2UO_3.2SiO_2.7H_2O$.   Orthorhombic, prismatic.   Cleavage $\parallel$ (100), perfect.   Optically $+$ or $-$.   Ax. pl. $\parallel$ (001). $X = a$ axis. $\alpha = 1{\cdot}613$, $\beta = 1{\cdot}635$, $\gamma = 1{\cdot}657$.   Color, pale lemon-yellow.   Pleochrism: X = colorless, Y = pale yellow, Z = yellow.   Dispersion distinct.   G. = $3{\cdot}54$.   Radioactive.   Found in the Belgian Congo, associated with kasolite in cracks in a siliceous breccia at Kasolo, Katanga, and at Chinkolobwe.   *Droogmansite* is presumably a related mineral (no anal.) which occurs with sklodowskite and curite at Kasolo, Katanga, as small orange-yellow globules with a radially fibrous structure.   Perfect cleavage, with acute positive bisectrix perpendicular to it. $n > 1{\cdot}74$.

Kasolite. — $PbO.UO_3.SiO_2.H_2O$.   Monoclinic.   Minute prismatic crystals.   Cleavages: (001) perfect; (100), (010).   H. = 4–5.   G. = $5{\cdot}96$.   Soluble in acids with gelatinization.   Water in C. T.   Radioactive.   Color yellow to brown.   Luster resinous to greasy.   Optically $+$.   Ax. pl. $\perp$ (010). $\alpha = 1{\cdot}89$, $\beta = 1{\cdot}91$. $\gamma = 1{\cdot}95$.   Found associated with curite, torbernite, etc., at Kasolo, Katanga, Belgian Congo.

Soddyite.   Soddite. — $5UO_3.2SiO_2.6H_2O$?   Orthorhombic.   Occurs in minute prismatic crystals with prism faces vertically striated.   Also in minute pyramidal crystals with horizontal striations.   H. = 3–4.   G. = $4{\cdot}627$.   Radioactive.   Infusible.   In C. T. blackens, losing water and oxygen.   Soluble in HCl with gelatinization.   Yellow color.   Translucent to opaque.   Optically $-$.   Ax. pl. $\parallel$ (010). $Z = c$ axis. $\alpha = 1{\cdot}65$. $\beta = 1{\cdot}68$. $\gamma = 1{\cdot}71$.   Occurs intimately mixed with curite at Kasolo, Belgian Congo.

Dixenite. — $MnSiO_3.2Mn_2(OH)(AsO_3)_2$.   Hexagonal, rhombohedral.   In aggregates of thin folia.   Basal cleavage.   H. = 3–4.   G. = $4{\cdot}2$.   Color nearly black, red by transmitted light.   Optically $-$.   Indices, $1{\cdot}73$–$1{\cdot}97$.   From Långbanshyttan, Vermland, Sweden.

McGovernite.   Macgovernite. — $21(Mn,Mg,Zn)O.3SiO_2.\tfrac{1}{2}As_2O_3.As_2O_5.10H_2O$.   Probably hexagonal.   In coarse-grained granular masses.   Perfect almost micaceous cleavage parallel to base.   G. = $3{\cdot}719$.   Reddish, somewhat bronze-like color.   Deep red-brown in transmitted light.   Uniaxial, $+$. $\omega = 1{\cdot}754$.   From Sterling Hill, Franklin, Sussex Co., New Jersey.

TITANO-SILICATES, TITANATES

This section includes the common calcium titano-silicate, Titanite; also a number of silicates which contain titanium, but whose relations are not altogether clear; further the titanate, Perovskite, and niobo-titanate, Dysanalyte, which is intermediate between Perovskite and the species Pyrochlore, Microlite, Koppite of the following section.

In general the part played by titanium in the many silicates in which it enters is more or less uncertain.   It is probably in most cases, as shown in the preceding pages, to be taken as replacing the silicon; in others, however, it seems to play the part of a basic element; in schorlomite (p. 596) it may enter in both relations.

TITANITE.   Sphene.

Monoclinic.   Axes $a : b : c = 0{\cdot}7547 : 1 : 0{\cdot}8543$; $\beta = 60° 17'$.

mm''', $110 \wedge 1\bar{1}0 =$	$66° 29'$.	ll', $\bar{1}12 \wedge \bar{1}\bar{1}2 = 46°\ \ 7\tfrac{1}{2}'$.
cx,	$001 \wedge 102 = 21°\ \ 0'$.	cn, $001 \wedge 111 = 38° 16'$.
ss',	$021 \wedge 0\bar{2}1 = 112°\ \ 3'$.	cm, $001 \wedge 110 = 65° 30'$.
nn',	$111 \wedge 1\bar{1}1 = \ \ 43° 49'$.	cl, $001 \wedge \bar{1}12 = 40° 34'$.

Twins: tw. pl. $a(100)$ rather common, both contact-twins and cruciform penetration-twins. Crystals very varied in habit; often wedge-shaped and flattened ∥ $c(001)$; also prismatic. Sometimes massive, compact; rarely lamellar.

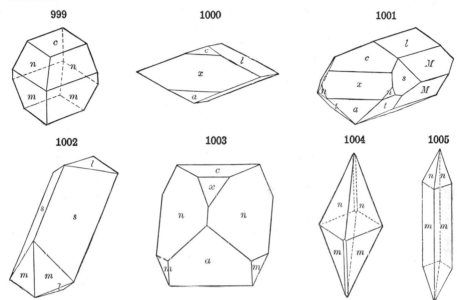

X-ray study shows a unit cell agreeing in its dimensions with the crystal constants given above and which contains four molecules. The SiO_4 groups present are independent of each other. The Ti atoms lie in the center of an octahedral group of six O atoms and the Ca atoms lie between seven O atoms.

Cleavage: $m(110)$ rather distinct; $a(100)$, $l(\bar{1}12)$ imperfect; in greenovite, $n(111)$ easy, $t(\bar{1}11)$ less so. Parting often easy ∥ $n(221)$ due to twinning lamellæ. H. = 5–5·5. G. = 3·4–3·56. Luster adamantine to resinous. Color brown, gray, yellow, green, rose-red and black. Streak white, slightly reddish in greenovite. Transparent to opaque.

Pleochroism in general rather feeble, but distinct in deep-colored kinds: Z, red with tinge of yellow; Y, yellow, often greenish; X, nearly colorless. Optically +. Ax. pl. ∥ $b(010)$. Z nearly ⊥ $x(102)$, i.e., $Z \wedge c$ axis = $+51°$. Dispersion $\rho > v$ very large, and hence the peculiarity of the axial interference-figure in white light. Axial angles variable but usually small. $\alpha = 1·900$. $\beta = 1·907$. $\gamma = 2·034$.

Var. — *Ordinary.* (a) *Titanite;* brown to black, the original being thus colored, also opaque or subtranslucent. (b) *Sphene* (named from σφήν, *a wedge*); of light shades, as yellow, greenish, etc., and often translucent; the original was yellow. *Ligurite* is an apple-green sphene. *Spinthère* (or Séméline) a greenish kind. *Lederite* is brown, opaque, or subtranslucent, of the form in Fig. 999.

Titanomorphite is a white, mostly granular alteration-product of rutile and ilmenite, not uncommon in certain crystalline rocks; here also belongs most leucoxene (see p. 486).

Manganesian; Greenovite. Red or rose-colored, owing to the presence of a little manganese; from St. Marcel, Piedmont, Italy; from Jothvád in Nárukot, India.

Containing yttrium or cerium. Here belong *grothite, alshedite, eucolite-titanite.* *Keilhauite* or *yttrotitanite* contains about 12 per cent $(Y,Ce)_2O_3$. Found near Arendal in Aust-

Agder, Norway and on the islands of Buö Askerö, Alve, and Narestö in a feldspathic rock. Also at Snarum in Buskerud.

Comp. — $CaTiSiO_5$ or $CaO.TiO_2.SiO_2$ = Silica 30·6, titanium dioxide 40·8, lime 28·6 = 100. Iron is present in varying amounts, sometimes manganese and also yttrium in some kinds.

Pyr., etc. — B.B. some varieties change color, becoming yellow, and fuse at 3 with intumescence, to a yellow, brown or black glass. With borax they afford a clear yellowish green glass. Imperfectly soluble in heated hydrochloric acid; and if the solution be concentrated along with tin, it assumes a fine violet color. With salt of phosphorus in R.F. gives a violet bead; varieties containing much iron require to be treated with the flux on charcoal with metallic tin. Completely decomposed by sulphuric and hydrofluoric acids.

Diff. — Characterized by its oblique crystallization, a wedge-shaped form common; by resinous (or adamantine) luster; hardness less than that of staurolite and greater than that of sphalerite. The reaction for titanium is distinctive, but less so in varieties containing much iron.

Micro. — Distinguished in thin sections by its acute-angled form, often lozenge-shaped; its generally pale brown tone; very high relief and remarkable birefringence, causing the section to show white of the higher order; by its biaxial character (showing many lemniscate curves); and by its great dispersion, which produces colored hyperbolas.

Artif. — Titanite is apparently produced artificially only with difficulty. It has been obtained by fusing together silica and titanic oxide with calcium chloride.

Obs. — Titanite, as an accessory component, is widespread as a rock-forming mineral, though confined mostly to the igneous rocks of intermediate composition; it is much more common in the plutonic granular types than in the volcanic forms. Thus it is found in the more basic hornblende granites, syenites, and diorites, and is especially common and characteristic in the nephelite-syenites. It occurs also in the metamorphic rocks and especially in the schists, gneisses, etc., rich in magnesia and iron and in certain granular limestones. It is also found in beds of iron ore. Commonly associated minerals are pyroxene, amphibole, chlorite, scapolite, zircon, apatite, etc. In cavities in gneiss and granite, it often accompanies adularia, smoky quartz, apatite, chlorite, etc., the crystals being sometimes coated with or penetrated by the chlorite.

The prominent localities for the occurrence of titanite include the following: At Schwarzenstein and Rotenkopf in the Zillertal, Tyrol, Austria. In Italy at Pfunders and in the Pfitschtal, Trentino; in Piedmont in the Ala valley in large, broad, yellowish or reddish crystals (*ligurite*); at St. Marcel the manganesian variety (*greenovite*); from Mte. Somma, Vesuvius. In Switzerland it occurs in Grisons in fine crystals of a pale green color and transparent in Val Tavetsch at Kreuzlital, and elsewhere near Sedrun, near Disentis, etc.; in Ticino in the St. Gotthard region at Val Sella, etc., and in Val Maggia; in Valais at Zermatt and in the Binnental. In France from Isère at Maronne (*spinthère*), and in fine crystals from near Bourg d'Oisans. In Norway in Aust-Agder at Arendal as pale yellowish green crystals (*aspidelite*); also at Risör; from Kragerö in Telemark. In Madagascar fine specimens have come from Midongy, from Ambalavaokely near Betroka, and elsewhere.

In the United States it occurs in New York at Diana near Natural Bridge, Lewis Co., in large dark brown crystals, among which is the variety *lederite*; at various points in St. Lawrence Co.; in Orange Co., in large crystals in the town of Monroe and near Edenville; in Putnam Co., at the Tilly Foster iron mine, Brewster, in very fine transparent greenish crystals. In New Jersey at Franklin, Sussex Co., of a honey-yellow color. From Magnet Cove near Hot Springs, Garland Co., Arkansas. In Canada titanite occurs in Renfrew Co., Ontario, at Eganville in very large dark brown crystals with apatite, etc.; similarly at other points where apatite is abundant. In Quebec at Litchfield in Pontiac Co. in fine black twinned crystals; also at various points in Ottawa Co.

Molengraaffite. — A titano-silicate of lime and soda. Monoclinic. In imperfect prismatic crystals. Cleavage (100) perfect. Color yellow-brown. Optically +. Ax. pl. ⊥ (010). Z nearly || c axis. $\alpha = 1·735$, $\beta = 1·74$, $\gamma = 1·77$. From a rock, "lujaurite," in the Pilands Berg, near Rustenberg, Transvaal, South Africa.

Fersmannite. — $2Na_2(O,F_2).4CaO.4TiO_2.3SiO_2$. Monoclinic. Crystals show a pseudo-tetragonal habit. H. = 5·5. G. = 3·44. Brown color. Vitreous luster. Optically −. Ax. pl. || (010). X nearly ⊥ (001). $\alpha = 1·886$, $\beta = 1·930$, $\gamma = 1·939$. $2V = 0°-7°$. Found associated with feldspar and ægirite in veins in nephelite-syenite at Chibines Mts., Kola peninsula, Russia.

Tscheffkinite. Chevkinite. — A titano-silicate of the cerium metals, iron, etc., but an alteration product, more or less heterogeneous, and the composition of the original mineral is very uncertain. Usually massive, amorphous. In orthorhombic or monoclinic (pseudo-orthorhombic) crystals from Madagascar. H. = 5–5·5. G. = 4·5. Color velvet black. Optically − (also reported as + at times). In part isotropic. Indices variable, 1·63–1·97. Pleochroic, X = nearly colorless, Y = pale reddish brown, Z = dark red-brown. Occurs in granite rocks and apparently as an alteration product of other minerals. Found sparingly near Miask in the Ilmen Mts., Russia. From Kanjamalai Hill, Salem district, Madras, India. From Madagascar in syenitic and granitic pegmatites from the district about Torendrika, west of Ambositra, and from near Betroka; at times in large crystals and masses. An isolated mass weighing 20 lbs. has been found on Hat Creek, near Massie's Mills, Nelson Co., Virginia; also found, south of this point, in Bedford Co.

Astrophyllite. — Probably $\overset{\text{I}}{R}_4\overset{\text{II}}{R}_4Ti(SiO_4)_4$ with $\overset{\text{I}}{R}$ = H, Na, K, and $\overset{\text{II}}{R}$ = Fe, Mn chiefly, including also Fe_2O_3. Orthorhombic. In elongated crystals; also in thin strips or blades; sometimes in stellate groups. Cleavage: $b(010)$ perfect like mica, but laminæ brittle. H. = 3. G. = 3·3–3·4. Luster submetallic, pearly. Color bronze-yellow to gold-yellow. Optically +; also reported as −. Ax. pl. ‖ (001). α = 1·68, β = 1·70, γ = 1·73. 2V large.

Occurs on the small islands in the Langesundfiord, near Brevik, Norway, in zircon-syenite, embedded in feldspar, with catapleiite, ægirite, black mica, etc. Similarly at Kangerdluarsuk and Narsarsuk, Greenland. Occurs in the nephelite-syenites on the Los Islands in French Guinea. Also with arfvedsonite and zircon at St. Peter's Dome, Pike's Peak, El Paso Co., Colorado.

Johnstrupite. — A silicate of the cerium metals, calcium and sodium chiefly, with titanium and fluorine. In prismatic monoclinic crystals. Cleavage ‖ (100). H. = 1. G. = 3·29. Easily fusible. Color brownish green. Optically +. Ax. pl. ‖ (010). X nearly ‖ c axis. Strong dispersion, $\rho > v$. α = 1·661, β = 1·666, γ = 1·673. 2V = 70°. From near Barkevik, in the Langesundfiord, southern Norway.

Mosandrite. — Near johnstrupite in form and composition and optical properties. α = 1·646, β = 1·649, γ = 1·658. Color reddish brown. Found in the Langesundfiord region of southern Norway, near Barkevik, and on the islands Lamöskjär (Låven) and Stokö.

Rinkite. — Near johnstrupite in composition. Monoclinic. Crystals prismatic or tabular ‖ (100). Cleavage ‖ (100). H. = 5. G. = 3·5. Infusible. Color yellow to brown. Optically +. Ax. pl. ⊥ (010). X = b axis. Strong dispersion, $\rho < v$. α = 1·665, β = 1·668, γ = 1·681. 2V = 43°. Occurs in sodalite-syenite in the Julianehaab district, Greenland, at Kangerdluarsuk, Naujakasik, and Narsarsuk.

Rinkolite. — Titano-silicate of cerium, calcium, strontium, and sodium. Monoclinic. Prismatic. Cleavages: (100) perfect, (010) good. H. = 5. G. = 3·4. Color brown, green, or yellow. Optically +. Y = b axis. X nearly parallel to c axis. β = 1·645–1·667. Occurs in large crystals in Hibina district, Kola peninsula, northern Russia. *Lovchorrite* or *lowtschorrite* is a colloidal material of similar character, found also here in large masses.

Narsarsukite. — A highly acidic titano-silicate of ferric iron and sodium. Tetragonal. In tabular crystals. Fine prismatic cleavage. H. = 7. G. = 2·7. Color honey-yellow, on weathering brownish gray or ocher-yellow. Optically +. ω = 1·609; ϵ = 1·630. Fusible. In pegmatite at Narsarsuk, southern Greenland.

Neptunite. — $(Na,K)_2(Fe,Mn)(Si,Ti)_5O_{12}$. In prismatic monoclinic crystals. Cleavage, (110). H. = 5–6. G. = 3·23. Color black. Streak, cinnamon-brown. Optically +. Ax. pl. ‖ (010). $Z \wedge c$ axis = +16°–20°. Strong dispersion, $\rho < v$. α = 1·69, β = 1·70, γ = 1·74. 2V = 49°. Pleochroic, yellow to deep-red. Found at Narsarsuk and elsewhere, southern Greenland, and at the benitoite locality in San Benito Co., California (originally called *carlosite*). *Mangan-neptunite* occurs on the Kola peninsula, northern Russia.

Benitoite. — $BaTiSi_3O_9$. Hexagonal, trigonal (ditrigonal-bipyramidal). In crystals with $\rho(10\bar{1}1)$ prominent. X-ray study of structure shows three SiO_4 groups are linked together in a ring giving the ratio Si_3O_9. H. = 6·2–6·5. G. = 3·6. Fusible at 3. Color sapphire-blue to light blue and colorless. Transparent. Strongly dichroic, O = colorless, E = deep blue. Optically +. ω = 1·757, ϵ = 1·804. Found associated with neptunite and natrolite near the headwaters of the San Benito River in San Benito Co., California.

Leucosphenite. — $Na_4Ba(TiO)_2(Si_2O_5)_5$. Monoclinic. In minute wedge-shaped or tabular ‖ (001) crystals. Distinct (010) cleavage. H. = 6·5. G. = 3·0. Difficultly fusible. Color white. Optically +. Ax. pl. ⊥ (010). X nearly ‖ a axis, and Y to c axis. α = 1·645, β = 1·661, γ = 1·688. 2V = 77°. From Narsarsuk, southern Greenland.

Lorenzenite. — $Na_2(TiO)_2Si_2O_7$. Contains considerable zirconia. Orthorhombic. In minute acicular crystals. Distinct cleavages || (100) and (110). H. = 6. G. = 3·4. Fusible. Optically −. Ax. pl. || (010). $X = a$ axis. $\alpha = 1·92$, $\beta = 2·01$, $\gamma = 2·02$. From Narsarsuk, southern Greenland. *Ramsayite* described from the Kola peninsula, northern Russia, is the same mineral without zirconia.

Joaquinite. — A titano-silicate of calcium and iron. Orthorhombic. Color, honey-yellow. Associated with *benitoite* from San Benito Co., California.

PEROVSKITE. Perofskite.

Isometric or orthorhombic pseudo-isometric. Crystals in general (Ural Mts., Zermatt, Switzerland) cubic in habit and often highly modified, but the faces often irregularly distributed. Cubic faces striated parallel to the edges and apparently penetration-twins, as if of pyritohedral individuals. Also in reniform masses showing small cubes. X-ray study of structure shows a cubic cell in which a Ca atom lies at the center, the Ti atoms at the corners and the O atoms at the middle points of the edges.

Cleavage: cubic, rather perfect. Fracture uneven to subconchoidal. Brittle. H. = 5·5. G. = 4. Luster adamantine to metallic-adamantine. Color pale yellow, honey-yellow, orange-yellow, reddish brown, grayish black. Streak colorless, grayish. Transparent to opaque. Usually exhibits double refraction in the larger crystals, with complex twinning of orthorhombic lamellæ. Mean index, about 2·38.

Geometrically considered, perovskite conforms to the isometric system; optically, however, it is uniformly biaxial and usually positive.

Comp. — Calcium titanate, $CaTiO_3$ = Titanium dioxide 58·9, lime 41·1 = 100. Iron is present in small amount replacing the calcium.

Pyr., etc. — In the forceps and on charcoal infusible. With salt of phosphorus in O.F. dissolves easily, giving a greenish bead while hot, which becomes colorless on cooling; in R.F. the bead changes to grayish green, and on cooling assumes a violet-blue color. Entirely decomposed by boiling sulphuric acid.

Obs. — Perovskite is most commonly found in chlorite, talc, or serpentine rocks. Also, usually as a microscopic constituent, in melilite-, nephelite-, and leucite-basalts. Occurs in small crystals or druses of crystals in a chlorite slate at Achmatovsk near Kussinsk in the Zlatoust district, Ural Mts., Russia. Found at Schelingen in the Kaiserstuhl, Baden. From Switzerland near the Findelen glacier in the valley of Zermatt, Valais, in large crystalline masses. In Italy at the Wildkreuzjoch and elsewhere in the Pfitschtal, Trentino; also in Piedmont at Emarese, Valle d'Aosta, etc.

Knopite. — Near perovskite but contains cerium. In black isometric crystals. G. = 4·2. Usually isotropic, $n = 2·30$. From near Alnö, Västernorrland, Sweden. Reported from near Leanchoil, British Columbia.

LOPARITE. A titanate of cerium, calcium and sodium, perhaps $(Na,Ce,Ca)_2(Ti,Nb)_2O_6$. In pseudo-cubic twins. Color black, streak red-brown. H. = 5·5. G. = 4·77. Member of the perovskite group. Occurs associated with eudialyte, ægirite, ramsayite, and sphene, on Kola peninsula, northern Russia.

Dysanalyte. — A titano-niobate of calcium and iron, like perovskite with lime replaced to some extent by iron, etc. Pseudo-isometric, probably orthorhombic. In cubic crystals. Cubic cleavage. H. = 5–6. G. = 4·1. Infusible. Color, iron-black. Isotropic, $n = 2·33$. Also at times biaxial, +. From the granular limestone of Vogtsburg, Kaiserstuhl, Baden, Germany. Has previously been called perovskite, but is in fact intermediate between the titanate, perovskite, and the niobates, pyrochlore and koppite. From Mte. Somma, Vesuvius.

A related mineral, which has also long passed as perovskite, occurs with magnetite, brookite, rutile, etc., at Magnet Cove, near Hot Springs, Garland Co., Arkansas. It is in octahedrons or cubo-octahedrons, black or brownish black in color and submetallic in luster.

Zirkelite. — $(Ca,Fe)(Zr,Ti,Th)_2O_5$. Isometric. Octahedral habit. In twins according to spinel law and in polysynthetic twinning. H. = 5·5. G. = 4·7. Color black.

Resinous luster. $n = 2 \cdot 19$. Found with baddeleyite in the decomposed magnetite-pyroxenite of Jacupiranga, São Paulo, Brazil. Also described from Ceylon.

Uhligite. — $Ca(Zr,Ti)_2O_5$ with Al_2TiO_5. Isometric. Octahedral. Twinned according to spinel law. Cubic cleavage. Color black. Brown and transparent on thin edges. Found in a nephelite-syenite on the shore of Lake Magad, Tanganyika Territory, East Africa.

Oliveiraite. A titanate of zirconium, $3ZrO_2.2TiO_2.2H_2O$. Amorphous. Color yellowish green. Associated with euxenite from near Pomba, Minas Geraes, Brazil.

Delorenzite. — A titanate of iron, uranium and yttrium of uncertain composition. Orthorhombic. Prismatic habit. H. $= 5-5 \cdot 5$. G. $= 4 \cdot 7$. Color black. Resinous luster. Found in pegmatite at Graveggia, Piedmont, Italy.

Yttrocrasite. — A hydrous titanate of the yttrium earths and thorium. Orthorhombic. H. $= 5 \cdot 5-6$. G. $= 4 \cdot 8$. Infusible. Black color with pitchy to resinous luster. Isotropic, $n = 2 \cdot 1$. Found in Burnet Co., three miles east of Barringer Hill, Llano Co., Texas.

Arizonite. $Fe_2O_3.3TiO_2$. Monoclinic? Crystal faces rough. H. $= 5 \cdot 5$. G. $= 4 \cdot 25$. Color dark steel-gray. Streak brown. Decomposed by hot concentrated sulphuric acid. Found with gadolinite, 25 miles southeast of Hackberry, Mohave Co., Arizona.

Pseudobrookite. — Fe_2TiO_5. Usually in minute orthorhombic crystals, tabular $\parallel$ (100) and often prismatic $\parallel$ the macro-axis. G. $= 4 \cdot 4-4 \cdot 98$. Color dark brown to black. Streak ocher-yellow. Found with hypersthene (*szaboite*) in cavities of the andesite of Aranyer Berg near Piski on the Maros, Transylvania, Rumania; in the nephelinite of the Katzen-buckel, Odenwald, Baden; on the lava of 1872 of Vesuvius; on Mont Dore, Puy-de-Dôme, France; with the apatite of Jumilla, Murcia, Spain. In large crystals at Havredal near Bamble in Telemark, Norway.

Kalkowskite. Kalkowskyn. $(Fe,Ce)_2O_3.4(Ti,Si)O_2$. In thin plates, showing fibrous structure. Conchoidal fracture. H. $= 3-4$. G. $= 4 \cdot 01$. Black, dark to light brown. Sub-metallic luster. Red-brown streak. $n > 1 \cdot 77$. From muscovite schist of Serra do Itacolumy, Minas Geraes, Brazil. May be identical with arizonite or pseudobrookite.

Brannerite. — Essentially $(UO,TiO,UO_2)TiO_3$. Prismatic crystals or granular. Black. Streak, dark greenish brown. H. $= 4 \cdot 5$. G. $= 4 \cdot 5-5 \cdot 4$. Isotropic, $n = 2 \cdot 3$. Found in gold placers, Stanley Basin, Idaho.

Oxygen Salts

3. NIOBATES, TANTALATES

The Niobates (Columbates) and Tantalates are chiefly salts of metaniobic and metatantalic acid, RNb_2O_6 and RTa_2O_6; also in part Pyroniobates, $R_2Nb_2O_7$, etc. Titanium is prominent in a number of the species, which are hence intermediate between the niobates and titanates. Niobium and tantalum are found also in a few rare silicates, as wöhlerite, lâvenite, etc. The following groups may be mentioned:

The isometric PYROCHLORE GROUP, the tetragonal FERGUSONITE GROUP, the orthorhombic COLUMBITE GROUP, the orthorhombic SAMARSKITE GROUP.

The species belonging in this class are for the most part rare, and are hence but briefly described.

PYROCHLORE.

Isometric. Commonly in octahedrons; also in grains. The natural mineral has a heterogeneous structure but recovers its normal structure on heating, which agrees with the structure of artificial pyrochlore. There are eight molecules in the elementary cell. Cleavage: octahedral, sometimes distinct. Fracture conchoidal. Brittle. H. $= 5-5 \cdot 5$. G. $= 4 \cdot 2-4 \cdot 36$. Infusible. Luster vitreous or resinous, the latter on fracture surfaces. Color brown, dark reddish or blackish brown. Streak light brown, yellowish brown. Subtranslucent to opaque. Isotropic, $n = 1 \cdot 96-2 \cdot 02$.

Comp. — Chiefly a niobate of the cerium metals, calcium and other bases, with also titanium, thorium, fluorine. Probably essentially a metaniobate with a titanate, $RNb_2O_6.R(Ti,Th)O_3$; fluorine is also present. The formula suggested by X-ray study is $(Na,Ca)_2(Nb,Ti)_2(O,F)_7$.

Obs. — Occurs in nephelite-syenite at Fredriksvärn and Larvik, Norway; on the island Lövö, opposite Brevik, and at several points in the Langesundfiord; near Miask in the Ilmen Mts., Russia. Named from πῦρ, *fire*, and χλωρός, *green*, because B.B. it becomes yellowish green. A variety of pyrochlore from near Wausau, Marathon Co., Wisconsin, has been called *marignacite*.

Koppite. — Essentially a pyroniobate of cerium, calcium, iron, etc., near pyrochlore. Has same atomic structure as pyrochlore. In minute brown dodecahedrons. H. = 5·5. G. = 4·5. Infusible. Isotropic, $n = 2\cdot15$. From Schelingen, Kaiserstuhl, Baden.

Hatchettolite. — A tantalo-niobate of uranium, near pyrochlore. Isometric, octahedral. H. = 5. G. = 4·8. Infusible. Color yellowish brown. Isotropic, $n = 1\cdot98$. Occurs with samarskite, in Mitchell Co., North Carolina. Found at Hybla, Hastings Co., Ontario.

Ishikawaite. A niobate and tantalate of uranium, with ferrous iron and rare earths. Orthorhombic. Color black. Opaque. H. = 5–6. G. = 6·2–6·4. Found at Ishikawa, Iwaki Province, Japan.

Ellsworthite. — Approximately $CaO.Nb_2O_5.2H_2O$. Contains also uranium oxides, titanium, etc. Isometric. In rounded crystals or massive. H. = 4. G. = 3·608–3·758. Isotropic. $n = 1\cdot89$. Color amber-yellow to dark chocolate-brown. Adamantine luster. Occurs in Ontario embedded in calcite and quartz from a pegmatite in Monteagle Township, Hastings Co., and in Haliburton Co.

Neotantalite. A niobate and tantalate of iron, manganese, etc. Isometric, in octahedrons. H. = 5–6. G. = 5·2. Color clear yellow. $n = 1\cdot96$. Found with kaolin at Echassières, Allier, France.

Chalcolamprite. $Na_4(CaF)_2Nb_2SiO_9$? Isometric. In small octahedrons. H. = 5·5. G. = 3·8. Color dark gray-brown. Crystal faces show a copper-red metallic iridescence. Isotropic, $n = 1\cdot87$. Occurs sparingly at Narsarsuk, South Greenland. *Endeiolite* is a similar mineral from the same locality supposed to have the same composition with the substitution of the hydroxyl group for the fluorine.

Samirésite. — A niobate and titanate of uranium, lead, etc. Isometric. In octahedrons. G. = 5·24. Color golden-yellow. Isotropic. $n = 1\cdot92$–$1\cdot96$. From near Samiresy, southeast of Antsirabe, Madagascar.

Microlite. — Essentially a calcium pyrotantalate, $Ca_2Ta_2O_7$, but containing also niobium, fluorine and a variety of bases in small amount. Isometric. Habit octahedral; crystals often very small and highly modified. H. = 5·5. G. = 5·5. Infusible. Color pale yellow to brown, rarely hyacinth-red. Isotropic, $n = 1\cdot93$. Microlite is found on the island of Utö, state of Stockholm, Sweden. In the granitic veins of Elba. Found in Greenland in the Narsarsuk district. In the United States occurs at Rumford, Oxford Co., Maine. It was originally found at Chesterfield, Hampshire Co., Massachusetts. Also in fine crystals and in large crystalline masses at Amelia Court House, Amelia Co., Virginia.

Pyrrhite. Probably a niobate related to pyrochlore. Has been classed variously with koppite and with microlite. Occurs in minute orange-yellow octahedrons. From Alabaschka, near Mursinsk in the Ural Mts. Noted from Mte. Somma, Vesuvius, and from San Miguel, Azores.

FERGUSONITE. Tyrite. Bragite.

Tetragonal-pyramidal. Axis $c = 1\cdot4643$. Crystals pyramidal or prismatic in habit. X-ray study shows that fergusonite normally has not a crystalline atomic structure but on heating to 400° C. it becomes crystalline, tetragonal.

Cleavage: $s(111)$ in traces. Fracture subconchoidal. Brittle. H. = 5·5–6. G. = 5·8, diminishing to 4·3 when largely hydrated. Luster externally dull, on the fracture brilliantly vitreous and submetallic. Color brownish black; in thin scales pale liver-brown. Streak pale brown. Subtranslucent to opaque. Index, 2·19.

Comp. — Essentially a metaniobate (and tantalate) of yttrium with erbium, cerium, uranium, etc., in varying amounts; also iron, calcium, etc.

General formula $\overset{\text{iii}}{R}(Nb,Ta)O_4$ with $\overset{\text{iii}}{R} = Y,Er,Ce$.

Water is usually present and sometimes in considerable amount, but probably not an original constituent; the specific gravity falls as the amount increases.

Obs. — Fergusonite was originally described from the Julianehaab district in Greenland. Also found in Sweden at Ytterby, northeast of Stockholm. In Norway it occurs in Aust-Agder at Helle near Arendal (*tyrite* and *bragite*), at Hampemyr on the island of Tromö (*tyrite*), at Evje in Satersdal and in Olstfold at Raade near Moss. From the Rakwana district in Ceylon. In Madagascar from near Ambatofotsikely west of Miandrarivo and to the southwest of Tananarive. It is found in Japan at Takayama, Mino.

In the United States fergusonite is found at Rockport, near Gloucester, Massachusetts; in the Brindletown gold district, Burke Co., North Carolina; also from near Spruce Pine, Mitchell Co., South Carolina. At the gadolinite locality in Llano Co., Texas, it occurs in considerable quantity and in masses sometimes weighing over a pound, also in large rough crystals. *Sipylite* from Amherst Co., Virginia, eroneously thought to contain erbium, has been shown to be identical with fergusonite.

RISÖRITE. A niobate of the yttrium metals; very similar to fergusonite but with a higher content of titanic acid, and probably to be classed as a variety of that mineral. X-ray study shows a tetragonal structure. H. = 5·5. G. = 4·18. Color yellow-brown. Isotropic, $n = 2\cdot0$. In pegmatite at Risör, Aust-Agder, Norway.

COLUMBITE-TANTALITE.

Orthorhombic. Axes $a : b : c = 0\cdot8285 : 1 : 0\cdot8898$.

yy''',	$210 \wedge 2\bar{1}0 = 45°\ 0'$.		$ee,$	$001 \wedge 021 = 60°\ 40'$;
mm''',	$110 \wedge 1\bar{1}0 = 79°\ 17'$.		$ao,$	$100 \wedge 111 = 51°\ 16'$;
gg',	$130 \wedge \bar{1}30 = 43°\ 50'$.		$cu,$	$001 \wedge 133 = 43°\ 48'$.
$ck,$	$001 \wedge 103 = 19°\ 42'$.		$uu',$	$133 \wedge 1\bar{3}3 = 29°\ 57'$.
$cq,$	$001 \wedge 023 = 30°\ 41'$.		$uu''',$	$133 \wedge 1\bar{3}3 = 79°\ 54'$.

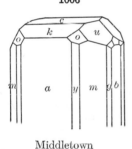

1006

Middletown

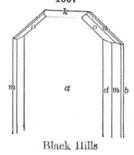

1007

Black Hills

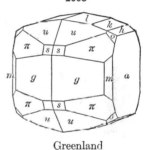

1008

Greenland

Twins: tw. pl. $e(021)$ common, usually contact-twins, heart-shaped (Fig. 407, p. 180), also penetration-twins; further tw. pl. $q(023)$ rare (Fig. 460, p. 191). Crystals short prismatic, often rectangular prisms with the three pinacoids prominent; also thin tabular || $a(100)$; the pyramids often but slightly developed, sometimes, however, acutely terminated by $u(133)$ alone. Also in large groups of parallel crystals, and massive.

X-ray study of the structure shows that the O atoms are arranged about the Nb and the Fe atoms at the points of octahedra. The Nb-O groups form a chain with opposite edges of the octahedra shared by the adjoining groups. The Fe-O groups have the same arrangement. These two different chains are connected by the sharing of octahedral corners with each other. The structure shows a close resemblance to that of brookite.

Cleavage: $a(100)$ rather distinct; $b(010)$ less so. Fracture subconchoidal to uneven. Brittle. H. = 6. G. = 5·3-7·3, varying with the composition (see below). Luster submetallic, often very brilliant, sub-resinous. Color iron-black, grayish and brownish black, opaque; rarely reddish brown and translucent; frequently iridescent. Streak dark red to black. Optically $+$ for tantalite; probably $-$ for columbite. $\beta = 2 \cdot 2 – 2 \cdot 4$. 2V large.

Comp. — Niobate and tantalate of iron and manganese, $(Fe,Mn)(Nb, Ta)_2O_6$, passing by insensible gradations from normal COLUMBITE, the nearly pure niobate, to normal TANTALITE, the nearly pure tantalate. The iron and manganese also vary widely. Tin and tungsten are present in small amounts. The percentage composition for $FeNb_2O_6 =$ Niobium pentoxide $82 \cdot 7$, iron protoxide $17 \cdot 3 = 100$; for $FeTa_2O_6 =$ Tantalum pentoxide $86 \cdot 1$, iron protoxide $13 \cdot 9 = 100$.

In some varieties, *manganocolumbite* or *manganotantalite*, the iron is largely replaced by manganese.

The connection between the specific gravity and the percentage of metallic acids is shown in the following table:

	G.	Ta$_2$O$_5$		G.	Ta$_2$O$_5$
Greenland	5·36	3·3	Bodenmais	5·92	27·1
Acworth, N. H.	5·65	15·8	Haddam	6·05	30·4
Limoges	5·70	13·8	Bodenmais	6·06	35·4
Bodenmais (*Dianite*)	5·74	13·4	Haddam	6·13	31·5
Haddam	5·85	10·0			
			Tantalite	7·03	65·6

Diff. — Distinguished (from black tourmaline, etc.) by orthorhombic crystallization, rectangular forms common; high specific gravity; submetallic luster, often with iridescent surface; cleavage much less distinct than for wolframite.

Pyr., etc. — For *tantalite*, B.B. alone unaltered. With salt of phosphorus dissolves slowly, giving an iron glass, which in R.F. is pale yellow on cooling; treated with tin on charcoal it becomes green. Decomposed on fusion with potassium bisulphate in the platinum spoon, and gives on treatment with dilute hydrochloric acid a yellow solution and a heavy white powder, which, on addition of metallic zinc, assumes a smalt-blue color; on dilution with water the blue color soon disappears. *Columbite*, when decomposed by fusion with caustic potash, and treated with hydrochloric and sulphuric acids, gives, on the addition of zinc, a blue color more lasting than with tantalite. Partially decomposed when the powdered mineral is evaporated to dryness with concentrated sulphuric acid, its color is changed to white, light gray, or yellow, and when boiled with hydrochloric acid and metallic zinc it gives a beautiful blue.

Obs. — Columbite-tantalite commonly occurs in pegmatite veins. It is found in many localities, the more important of which are given below. Wherever the occurrence is definitely known to belong to one or the other end of the series that fact is indicated by the appropriate initial following the name of the locality. From Miask in the Ilmen Mts., Russia, with samarskite, and from the gold-washings of the Sanarka river, Ural Mts. (*manganotantalite*). Occurs at Rabenstein and Bodenmais, Bavaria (C); in Finland at Skogböle (T) near Kimito in Turun ja Porin; at Tammela (C,T) in Hamern. In Norway at Raade and Anneröd (C) near Moss, Olstfold; near Kragerö in Telemark and at Broddho and Finbo (T) near Falun in Kopparberg. In France fine crystals are found in the district of Chanteloube (C,T) near Limoges, Haute-Vienne. It occurs in Greenland at Ivigtut (Evigtok) (C) in the Arsukfjord region, as brilliant crystals in cryolite. *Manganotantalite* occurs in Western Australia at Wodgina near Pilbarra and at Greenbushes near Bridgetown in Nelson. In Madagascar (C) from Andilana on Lake Alaotra and near Miandrarivo at Ampangabé and Ambatofotsikely. From Minas Geraes, Brazil (C).

In the United States, in Maine, at Standish (C), Cumberland Co., in splendent crystals; also at Stoneham, Oxford Co.; from Topsham, Sagadahoc Co., and *manganotantalite* from Auburn, Androscoggin Co., and Rumford, Oxford Co. In New Hampshire at Acworth (C), Sullivan Co. In Massachusetts, at Chesterfield, Hampshire Co., some fine crystals; from Northfield (T), Franklin Co. In Connecticut in the pegmatites of Haddam (C), Middletown, and Portland, Middlesex Co., some of the crystals very large; at Branchville (C,T), Fairfield Co., in fine crystals and aggregates of crystals, also in minute thin tabular crystals (*manganocolumbite*) on spodumene. In Virginia at Amelia Court House, Amelia Co. (C), in fine splendent crystals. In North Carolina, with samarskite crystals in parallel position in Mitchell Co. (C), also in Yancey Co. In Alabama in Coosa Co. (T). In South Dakota, in the Black Hills region, common in the pegmatite veins, the crystals and crystalline groups often large; one mass was estimated to have weighed 2000 pounds; most abundant at the Etta tin mine, Keystone, Pennington Co. (C,T). In Colorado on microcline in Cheyenne Canyon, Pike's Peak district.

Use. — Source of tantalum which has been used in making filaments for incandescent electric lights.

TODDITE. Similar to columbite in which some manganese and iron have been replaced by uranium. Pitch-black color. Brilliant submetallic luster. Fracture subconchoidal. Very thin grains showed isotropic character. H. = 6·5. G. = 5·041. From a pegmatite dike, in Dill township, Sudbury district, Ontario.

Tapiolite. — $(Fe,Mn)(Nb,Ta)_2O_6$. A variable series that may be considered as dimorphous with the columbite-tantalite series. The various molecules have been named as follows: tapiolite $FeTa_2O_6$, mossite $FeNb_2O_6$, ixiolite $MnTa_2O_6$. Tetragonal; in square pyramidal crystals. Crystallographically and structurally show close relations to the members of the Rutile Group, see p. 496. *Strüverite* and *ilmenorutile* may be considered as connecting the two groups. H. = 6. G. = 7·3–7·8. Infusible. Color pure black. Optically +. $\omega = 2·27$, $\epsilon = 2·42$ (Li). Occurs at Sukula, in the parish of Tammela, Hamern, Finland. In twin crystals from Topsham, Sagadahoc Co., Maine. *Mossite* is found at Berg, near Moss, Obstfold, Norway. *Skogbölite* and *ixiolite* come from Skogböle in Kimito, Turun ja Porin, Finland.

Stibiotantalite. — $(SbO)_2(Ta,Nb)_2O_6$. Orthorhombic, hemimorphic in direction of a axis. Polysynthetic twinning parallel to $a(100)$. Cleavage a (perfect). H. = 5·5. G. = 6·0 7·4 (varying with composition). Fusible. Color brown, reddish yellow, yellow. Luster adamantine to resinous. Optically +. Ax. pl. || (010). $Z = c$ axis. Strong dispersion, $\rho < v$. $\alpha = 2·39$, $\beta = 2·41$, $\gamma = 2·46$ (varying somewhat with composition). Originally found in tin-bearing sands of Greenbushes, near Bridgeton, Nelson, Western Australia. In crystals from Mesa Grande, San Diego Co., California.

Bismutotantalite. Ugandite. — A bismuth tantalate and niobate, probably $(BiO)_2$ $(Ta,Nb)_2O_6$. Orthorhombic. Crystals large and imperfect. Angles show similarity to those of stibiotantalite. G. = 8·15. H. = 5. Color black except on very thin edges. $n = 2·2$. From a pegmatite about 25 miles west northwest of Kampala, Uganda.

YTTROTANTALITE.

Orthorhombic. Axes $a : b : c = 0·5412 : 1 : 1·1330$. Crystals prismatic, $mm''' \, 110 \wedge 1\bar{1}0 = 56° 50'$.

Cleavage: $b(010)$ very indistinct. Fracture small conchoidal. H. = 5–5·5. G. = 5·5 5·9. Infusible. Luster submetallic to vitreous and greasy. Color black, brown, brownish yellow, straw-yellow. Streak gray to colorless. Opaque to subtranslucent. $n = 2·15$.

Comp. — Essentially $\overset{\text{II III}}{R}R_2(Ta,Nb)_4O_{15}.4H_2O$, with $\overset{\text{II}}{R} = Fe, Ca$, $\overset{\text{III}}{R} = Y$, Er, Ce, etc. The water may be secondary.

The so-called yellow yttrotantalite of Ytterby and Kårarfvet belongs to fergusonite.

Obs. — Occurs in Sweden at Ytterby, northeast of Stockholm, in red feldspar; at Finbo and Broddbo, near Falun, in Kopparberg.

1009

SAMARSKITE.

Orthorhombic. Axes $a : b : c = 0·5456 : 1 : 0·5178$. Crystals rectangular prisms $(a(100), b(010)$, with $e(101)$ prominent). Angles, $mm''' \, 110 \wedge 1\bar{1}0 = 57° 14'$; $ee' \, 101 \wedge \bar{1}01 = 87°$. Faces rough. Commonly massive, and in flattened embedded grains.

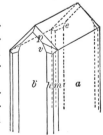

Cleavage: $b(010)$ imperfect. Fracture conchoidal. Brittle. H. = 5–6. G. = 5·6–5·8. Luster vitreous to resinous, splendent. Color velvet-black. Streak dark reddish brown. Nearly opaque. $n = 2·21$.

Comp. — $\overset{\text{II III}}{R_3}R_2(Nb,Ta)_6O_{21}$ with $\overset{\text{II}}{R} = Fe, Ca, UO_2$, etc.; $\overset{\text{III}}{R} = $ cerium and yttrium metals chiefly.

Pyr., etc. — In the closed tube decrepitates, glows, cracks open, and turns black. B.B. fuses on the edges to a black glass. With salt of phosphorus in both flames an emerald-green bead. With soda yields a manganese reaction. Decomposed on fusion with potassium bisulphate, yielding a yellow mass which on treatment with dilute hydrochloric acid separates white tantalic acid, and on boiling with metallic zinc gives a fine blue color. In powder sufficiently decomposed on boiling with concentrated sulphuric acid to give the blue reduction test when the acid fluid is treated with metallic zinc or tin.

Obs. — Occurs in reddish brown feldspar, with æschynite and columbite in the Ilmen mountains, near Miask, Russia; from Antanamalaza, southeast of Antsirabe, Madagascar. In the United States rather abundant and sometimes in large masses up to 20 lbs. at the mica mines in Mitchell Co., North Carolina, intimately associated with columbite; sparingly elsewhere. Varieties rich in lime from Hybla and Parry Sound, Ontario, have been called *calciosamarskite*. The mineral called *ånnerödite*, from a pegmatite vein at Ånneröd, near Moss, Olstfold, Norway, has been shown to be an intimate intergrowth of samarskite and columbite.

PLUMBONIOBITE. A niobate of yttrium, uranium, lead, iron, etc. Amorphous. Possibly identical with samarskite. H. = 5–5·5. G. = 4·81. Color dark brown to black. Found in mica mines at Morogoro, Tanganyika Territory, East Africa.

Ampangabéite. — A niobate and titanate of uranium, etc. In rectangular prisms, probably orthorhombic. Color brownish red. Luster greasy. H. = 4. G. = 3·97–4·29. Fuses to a black slag. Easily soluble in hydrochloric acid. Radioactive. Isotropic, n = 2·13. Found in parallel growth with columbite at Ampangabé and Ambatofotsikely, Madagascar. A similar mineral with a higher titanium content has been reported from Minas Geraes, Brazil.

Hielmite. — A stanno-tantalate (and niobate) of yttrium, iron, manganese, calcium. Crystals (pseudo-orthorhombic) usually rough; massive. G. = 5·82. Color pure black. Uniaxial, +. ω = 2·3, ϵ = 2·4. Strongly pleochroic, O = yellow-brown, E = nearly opaque. From the Kårarfvet mine, Falun, Kopparberg, Sweden.

Æschynite. — A niobate and titanate (thorate) of the cerium metals chiefly, also in small amount iron, calcium, etc. Crystals prismatic, pseudo-orthorhombic. Fracture small conchoidal. Brittle. H. = 5–6. G. = 4·93 Hitterö; 5·168 Miask. Infusible. Luster submetallic to resinous, nearly dull. Color nearly black, inclining to brownish yellow when translucent. Isotropic. n = 2·20–2·26. When heated shows birefringence.

From Miask in the Ilmen Mts., Russia, in feldspar with mica and zircon; also with euclase in the gold sands of the Orenburg District, Southern Ural Mts. From Hitterö, Vest-Agder, Norway. Named from αἰσχύνη, *shame*, by Berzelius, in allusion to the inability of chemical science, at the time of its discovery, to separate some of its constituents.

Polymignite. — A niobate and titanate (zirconate) of the cerium metals, iron, calcium. Pseudo-orthorhombic. Crystals slender prisms, vertically striated. H. = 6·5. G. = 4·8. Infusible. Color black. Isotropic, n = 2·2. Occurs at Frederiksvärn, near Larvik in Vestfold and elsewhere in Norway.

Euxenite. — A niobate and titanate of yttrium, erbium, cerium and uranium. Pseudo-orthorhombic. Crystals rare; commonly massive. H. = 6·5. G. = 4·7–5·0. Infusible. Color brownish black. Isotropic. n = 2·06–2·26.

Occurs in pegmatite veins at various localities in Norway as in Aust-Agder at Alve on Tromö near Arendal and at Tvedestrand; at Kragerö in Telemark; at Jolster in Sogn Og Fjordane, etc. In Finland at Impilaks (Impilahti) on Lake Ladoga. There are important occurrences in Madagascar; in considerable quantity near Antranotsiritra on the river Maevarano; fine crystals in the district of Ankazobe, especially on Mt. Volhambohitra; at various points to the southeast of Mandridrano; from Ambatofotsikely west of Miandrarivo; in the region about Betafo; at Samiresy southeast of Antsirabe. From Minas Geraes, Brazil. Found on the east coast of Greenland at Karra Akungnak. In the United States it has been found in Mitchell Co., North Carolina. A thorium-calcium euxenite from Lyndoch Township, Renfrew Co., Ontario, has been called *lyndochite*. An euxenite with preponderating tantalum from the Pilbara gold field, Western Australia has been called *tanteuxenite*.

LORANSKITE and **WIIKITE** are euxenite-like minerals from Impilaks, Finland. Usually in irregular masses but orthorhombic crystals are noted. H. = 6. G. = 3·8–4·8. Color black to brown and yellow. *Wiikite* has been shown to be a mixture. *Nuolaite* is the name given to a mixture of an opaque crystalline material and a colorless isotropic substance from the same locality.

Polycrase. — A niobate and titanate of yttrium, erbium, cerium, uranium, like euxenite. Crystals thin prismatic, orthorhombic. Fracture conchoidal. H. = 5–6. G. = 4·97–5·04. Luster vitreous to resinous. Color black, brownish in splinters. Isotropic. $n = 1·70$. Found in Norway as crystals in granite with gadolinite and allanite (*orthite*) on the island of Hiterö in Vest-Agder; near Arendal; etc. From Slättakra near Alseda in Jönköping, Sweden. In the United States, occurs in well-formed prismatic crystals in the gold sands of Henderson Co., North Carolina, the crystals being altered on the exterior to a yellow substance resembling gummite. Also in South Carolina, 4 miles from Marietta in Greenville Co. Named from πολυς, *many*, and κρᾱσις, *mixture*.

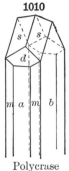

1010

Polycrase

Blomstrandine-Priorite. — Niobates and titanates of yttrium, erbium, cerium and uranium, similar to the *euxenite-polycrase* series. The two series may be dimorphous. The ratio of $Nb_2O_5 : TiO_2$ ranges from 1 : 2 in priorite; to 1 : 6 in blomstrandine. Orthorhombic. Crystals tabular parallel to $b(010)$. Most prominent forms are $b(010)$, $c(001)$ and $n(130)$. H. = 5·5. G. = 4·8–4·9. Color brownish black. Isotropic. $n = 2·14$. *Blomstrandine* was originally found in a pegmatite vein at Urstad on the island of Hiterö, Vest-Agder; also noted from near Arendal and at other points in southern Norway. *Priorite* was found in the Embabaan district of Swaziland, South Africa. Also from various localities in Madagascar.

Eschwegeite. A hydrous titanate, niobate and tantalate of yttrium and erbium. Conchoidal fracture. H. = 5·5. G. = 5·87. Color dark reddish gray. Isotropic. n between 2·15–2·20. Found as pebbles in the Upper Rio Doce, Minas Geraes, Brazil.

Betafite. — A niobate and titanate of uranium, etc. Isometric with octahedron and dodecahedron. H. = 5. G. = 4. Color, a greenish black. Opaque. Greasy luster. $n = 1·92$. Found in pegmatites from various localities in Madagascar, at Ambolotara, near Betafo; at Antanifotsy near Ampangabé; at Morafeno in large crystals; in crystals showing great variation in size from Ambatofotsky, southwest of Soavinandriana; at Andibakely and Sama near the junction of the rivers Sakay and Kitsamby; very large crystals at Ambatolampikely northwest of Mt. Sahapila; distorted crystals from northeast of Ankazobe. Reported from Sludianka, Transbaikalia.

Pisekite. A niobate and titanate of uranium and the rare earths. Shown by X-ray study to be amorphous. Occurs in prismatic crystals with habit similar to monazite. Presumably pseudomorphous. Color yellowish to black. Isotropic. Strongly radioactive. H. = 5·5–6. G. = 4·032. Found in beryl-bearing pegmatites at Pisek, Bohemia, Czechoslovakia.

Mendelyeevite. Mendelyevite. Calcium urano-titano-niobate. Isometric. Dodecahedral habit. Black color. Occurs in pegmatite near Sludianka, Transbaikalia.

Epistolite. — A niobate of uncertain composition. Analysis shows chiefly SiO_2, TiO_2, Na_2O, H_2O. Monoclinic. In rectangular plates, also in aggregates of curved folia. Basal cleavage perfect. H. = 1–1·5. G. = 2·9. Color white, grayish, brownish. Optically –. Ax. pl. = (010). $Z \wedge c$ axis = $| 7°$. $\alpha = 1·61$, $\beta = 1·65$, $\gamma = 1·68$. $2V = 80°$. Found in pegmatite veins or in massive albite from Julianehaab district, Greenland.

Oxygen Salts

4. PHOSPHATES, ARSENATES, VANADATES, ANTIMONATES

A. Anhydrous Phosphates, Arsenates, Vanadates, Antimonates

Normal phosphoric acid is H_3PO_4, and consequently normal phosphates have the formulas $\overset{\text{I}}{R}_3PO_4$, $\overset{\text{II}}{R}_3(PO_4)_2$ and $\overset{\text{III}}{R}PO_4$, and similarly for the arsenates, etc. Only a comparatively small number of species conform to this simple formula. Most species contain more than one metallic element, and in the prominent Apatite Group the radical (CaF), (CaCl) or (PbCl) enters; in the Wagnerite Group we have similarly $\overset{\text{II}}{(R}F)$ or (ROH).

XENOTIME.

Tetragonal.   Axis $c = 0.6187$, zz' ($111 \wedge \bar{1}11$) = $55° 30'$, zz'' ($111 \wedge \bar{1}\bar{1}1$) = $82° 22'$.   In crystals resembling zircon in habit;   sometimes compounded with zircon in parallel position (Fig. 488, p. 195).   In rolled grains.   X-ray study shows a close relationship in the atomic structure of xenotime, the Rutile Group, zircon, etc., see p. 496.

1011   **1012**

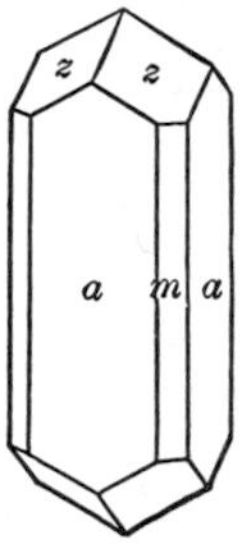

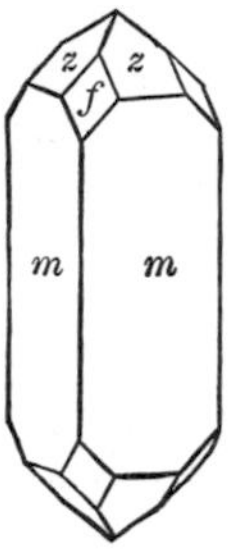

Cleavage: m(110) perfect.   Fracture uneven and splintery.   Brittle.   H. = 4–5.   G. = 4·45–4·56.   Luster resinous to vitreous.   Color yellowish brown, reddish brown, hair-brown, flesh-red, grayish white, wine-yellow, pale yellow;   streak pale brown, yellowish or reddish.   Opaque.   Optically +. $\omega = 1.72$. $\epsilon = 1.81$.

Comp. — Essentially yttrium phosphate, YPO_4 or $Y_2O_3 \cdot P_2O_5$ = Phosphorus pentoxide 38·6, yttria 61·4 = 100.   The yttrium metals may include erbium in large amount;   cerium is sometimes present;   also silicon and thorium as in monazite.

Pyr., etc. — B.B. infusible.   When moistened with sulphuric acid colors the flame bluish green.   Difficultly soluble in salt of phosphorus.   Insoluble in acids.

Diff. — Resembles zircon in its tetragonal form, but distinguished by inferior hardness and perfect prismatic cleavage.

Obs. — Occurs as an accessory mineral in pegmatite veins;   sometimes in minute embedded crystals generally distributed in granitic and gneissoid rocks.   Commonly associated with zircon and often encloses that mineral.   Of frequent occurrence in the pegmatite veins of Norway, as at Hiterö in Vest-Agder;   in Aust-Agder near Arendal and at Tvedestrand;   at Kragerö in Telemark;   at Raade near Moss in Olstfold;   etc.   It occurs in Sweden at Ytterby, northeast of Stockholm.   In Switzerland it is found on Mt. Fibia southwest of St. Gotthard, Ticino, and in the Binnental, Valais.   An accessory constituent in considerable quantity of the muscovite granites of Brazil, as in Minas Geraes from Dattas near Diamantina and from Pomba, etc., and from the states of Rio de Janeiro and São Paulo. *Hussakite* is a xenotime originally described from Brazil that contains sulphuric acid.

In the United States xenotime is observed rarely in the gneisses of New York City;   it is found in the gold sands of North Carolina in McDowell Co., near Dysortville;   at Brindletown in Burke Co., and near Green River, Henderson Co.;   in Mitchell Co., and in brilliant crystals with rutile in Alexander Co.   It occurs with tysonite in Cheyenne Canyon, Pike's Peak district, Colorado.

MONAZITE.

1013   **1014**

Monoclinic.   Axes $a : b : c = 0.9693 : 1 : 0.9256$; $\beta = 76° 20'$.

mm''',	$110 \wedge 1\bar{1}0 = 86° 34'$.
aw,	$100 \wedge 101 = 39° 12\frac{1}{2}'$.
$a'x$,	$\bar{1}00 \wedge \bar{1}01 = 53° 31'$.
ee',	$011 \wedge 0\bar{1}1 = 83° 56'$.
rr',	$111 \wedge 1\bar{1}1 = 60° 40'$.
vv',	$111 \wedge \bar{1}11 = 73° 19'$.

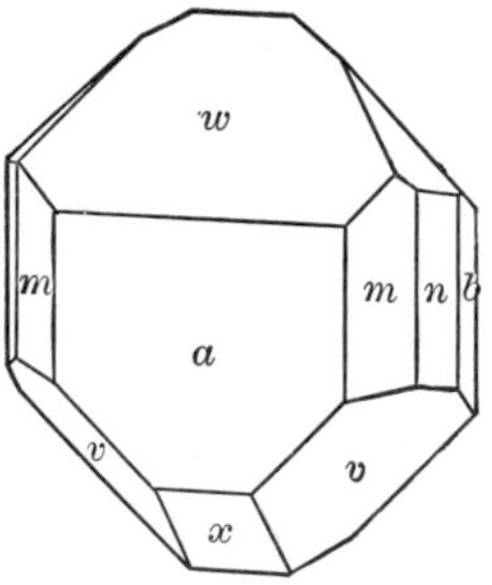

Crystals commonly small, often   flattened $\parallel$ a(100) or elongated $\parallel$ axis b;   sometimes

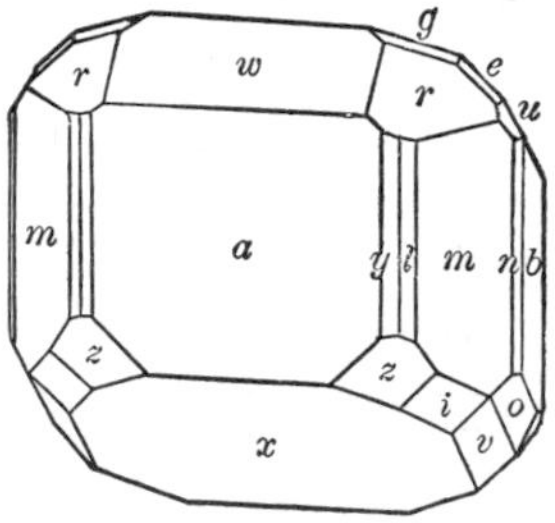

Norwich, Ct.   Switzerland

prismatic by extension of v(111);   also large and coarse.   In masses yielding angular fragments;   in rolled grains.

Cleavage: $c(001)$ sometimes perfect (parting?); also, $a(100)$ distinct; $b(010)$ difficult; sometimes showing parting $\parallel c(001)$, $m(110)$. Fracture conchoidal to uneven. Brittle. H. = 5–5·5. G. = 4·9–5·3; mostly 5·0 to 5·2. Luster inclining to resinous. Color hyacinth-red, clove-brown, reddish or yellowish brown. Subtransparent to subtranslucent. Optically +. Ax. pl. $\perp b(010)$ and nearly $\parallel a(100)$. $Z \wedge c$ axis = +1° to 4°. Dispersion $\rho < \upsilon$ weak; horizontal weak. 2V = 14°. $\alpha = 1·786$. $\beta = 1·788$. $\gamma = 1·837$.

Comp. — Phosphate of the cerium metals, essentially $(Ce,La,Di)PO_4$.

Most analyses show the presence of ThO_2 and SiO_2, usually, but not always, in the proper amount to form thorium silicate; that this is mechanically present is not certain but possible. It has been suggested that the thorium silicate is in solid solution with the cerium phosphate.

Pyr., etc. — B.B. infusible, turns gray, and when moistened with sulphuric acid colors the flame bluish green. With borax gives a bead yellow while hot and colorless on cooling; a saturated bead becomes enamel-white on flaming. Difficultly soluble in hydrochloric acid.

Obs. — Monazite occurs as an accessory mineral in granites, gneisses, aplites, and pegmatites; usually in small amount but in certain districts the rocks contain it abundantly and their detrital sands may carry it in commercial quantities.

It is found near Miask in the Ilmen Mts., Russia. The original *turnerite* came from Isère, France, probably from near St. Cristophe near Bourg d'Oisans, in small yellow or brown crystals. Similar specimens come from Switzerland in the Binnental, Valais, and in Grisons in Val Tavetsch and Val Cornera. Monazite occurs in Norway near Nöterö on Kristiania Fjord in Vestfold (*urdite*) in crystals sometimes 1 inch across; at various points near Arendal, Tvedestrand, and Risör in Aust-Agder; at Raade and other points near Moss in Olstfold. Monazite has been found in various places in India and the Malay States. It occurs, often in crystals of exceptional size and quality, in the potassic pegmatites of Madagascar, as at various places in the Ankazobe district, especially on Mt. Volhambohitra; at Ampangabe and Ambatofotsikely near Miandrarivo. In South Africa from Bronkhorstspruit east of Pretoria, Transvaal, and in the Embabaan district of Swaziland. Occurs in commercial deposits in the sands of Minas Geraes and Bahia, Brazil.

In the United States it was found in Connecticut in small crystals with the sillimanite from Norwich, also from Portland. Has been occasionally noted in the gneiss of New York City. In large coarse crystals and masses at Amelia Court House, Virginia. In North Carolina it occurs in Alexander Co., as splendid crystals at Millholland's Mill on Third Creek and also at Stony Point in large cruciform twins; in Madison Co., in considerable quantity; also in Mitchell and Burke counties; in the sands of Burke, Polk, McDowell, and Rutherford counties. Occurs also in the gold sands near Centerville, Boise Co., Idaho. Found in coarse pegmatite at Villeneuve, Ottawa Co., Quebec.

Monazite is named from μονάζειν, to be solitary, in allusion to its rare occurrence.

Cryptolite occurs in wine-yellow prisms and grains in the green and red apatite of Arondal, Norway, and is discovered on putting the apatite in dilute nitric acid. It is probably monazite.

Use. — Monazite is the chief source of thorium oxide which is used in the manufacture of incandescent gaslight mantles.

Berzeliite. — $(Ca,Na_2)(Mg,Mn)_2(AsO_4)_3$. Isometric, usually massive. Atomic structure is similar to that of garnet. H. = 5. Easily fusible. G. = 4·03. Color bright yellow. $n = 1·779$. In Vermland, Sweden, near Nordmark and Långbanshyttan. *Pyrrharsenite* from the Sjö mines, Sweden, contains also antimony; color yellowish red.

Caryinite. — Associated with berzeliite, is related to it, but contains lead. Orthorhombic. Usually massive. Cleavages (110), (010). H. = 3·5. G. = 4·25. Easily fusible. Color brown. Optically +. Ax. pl. $\parallel$ (100). $Z = b$ axis. $\alpha = 1·776$, $\beta = 1·780$, $\gamma = 1·805$. Alters to berzeliite.

Monimolite. — An antimonate of lead, iron and sometimes calcium; in part, $R_3Sb_2O_8$. Usually in octahedrons; massive, incrusting. H. = 5–6. G. = 6·58. Fusible. Color yellowish or brownish green. From the Harstig mine, Pajsberg, near Persberg, Vermland, Sweden.

Carminite. — Perhaps $Pb_3As_2O_8.10FeAsO_4$. Orthorhombic. In clusters of fine needles; also in spheroidal forms. H. = 2·5–3. G. = 4·1. Easily fusible. Color carmine to tile-red. Biaxial, +. $\beta = 2·06$. From the Luise mine at Horhausen, in Rhineland. From Calstock in Cornwall. Reported from Magnet, Tasmania.

Georgiadésite. — $Pb_3(AsO_4)_2.3PbCl_2$. Orthorhombic. In small crystals with hexagonal outline. H. = 3·5. G. = 7·1. Easily fusible. Resinous luster. Color white, brownish yellow. Optically +. Ax. pl. || (010). Z = c axis. β = 2·1. 2V large. Found on lead slags at Laurium, Greece.

Pucherite. — Bismuth vanadate, $BiVO_4$. In small orthorhombic crystals. Cleavage (001). H. = 4. G. = 6·25. Color reddish brown. Optically —. Ax. pl. || (100). X = c axis. α = 2·41, β = 2·50, γ = 2·51. Strong dispersion, $\rho < v$. 2V = 19°. From the Pucher mine, Schneeberg, Saxony; also from Eibenstock, Saxony; Ullersreuth near Hirschberg, Thuringia. From São José de Brejauba, Minas Geraes, Brazil; San Diego Co., California.

Armangite. — $Mn_3(AsO_3)_2$. Hexagonal-rhombohedral. Prismatic habit. H. = 4. G. = 4·23. Poor basal cleavage. Color black. streak brown. From Långbanshyttan, Vermland, Sweden.

Triphylite Group. Orthorhombic

		$a : b : c$
Triphylite	$Li(Fe,Mn)PO_4$	0·4348 : 1 : 0·5265
Lithiophilite	$Li(Mn,Fe)PO_4$	
Natrophilite	$NaMnPO_4$	

Orthophosphates of an alkali metal, lithium or sodium, with iron and manganese.

TRIPHYLITE-LITHIOPHILITE.

Orthorhombic. Axes $a : b : c$ = 0·4348 : 1 : 0·5265. Crystals rare, usually coarse and faces uneven. Commonly massive, cleavable to compact.

Cleavage: c(001) perfect; b(010) nearly perfect; m(110) interrupted. Fracture uneven to subconchoidal. H. = 4·5–5. G. = 3·42–3·56. Luster vitreous to resinous. Color greenish gray to bluish in triphylite; also pale pink to yellow and clove-brown in lithiophilite. Streak uncolored to grayish white. Transparent to translucent. Optical characters vary with the composition. Lithiophilite is optically +, with Ax. pl. || (001), β = 1·68, 2V = 63° (for variety with 9 per cent FeO); β = 1·69, 2V = 0° (for variety with 26·6 per cent FeO). Triphylite is optically —, with Ax. pl. || (100), β = 1·70, 2V = 0°–34°.

Comp. — A phosphate of iron, manganese and lithium, $Li(Fe,Mn)PO_4$, varying from the bluish gray TRIPHYLITE with little manganese to the salmon-pink or clove-brown LITHIOPHILITE with but little iron.

Typical *Triphylite* is $LiFePO_4$ = Phosphorus pentoxide 45·0, iron protoxide 45·5, lithia 9·5 = 100. Typical *Lithiophilite* is $LiMnPO_4$ = Phosphorus pentoxide 45·3, manganese protoxide, 45·1, lithia 9·6 = 100. Both Fe and Mn are always present.

Pyr., etc. — In the closed tube sometimes decrepitates, turns to a dark color, and gives off traces of water. B.B. fuses at 1·5, coloring the flame beautiful lithia-red in streaks, with a pale bluish green on the exterior of the cone of flame. With the fluxes reacts for iron and manganese; the iron reaction is feeble in pure lithiophilite. Soluble in hydrochloric acid.

Obs. — Found in pegmatites and pegmatitic granites associated with spodumene, beryl, tourmaline, garnet, micas, etc. *Triphylite* is found in Bavaria at the neighboring localities of Bodenmais and Rabenstein, and also at Hagendorf near Pleystein in the Oberpfalz. In Finland from Kietyö near Tammela, Hamern, and at Skogböle on Kimito, Turun ja Porin. In the United States it occurs in Maine at Peru, Oxford Co.; in New Hampshire at Grafton and North Groton, Grafton Co.; in Massachusetts at Huntington (formerly Norwich), Hampshire Co. Named from τρίς, *threefold*, and φυλή, *family*, in allusion to its containing three phosphates.

Lithiophilite occurs at Branchville, Fairfield Co., Connecticut. Also in Maine from Norway, Oxford Co., and Poland, Androscoggin Co.; also from Pala, San Diego Co., California. Named from *lithium* and φιλός, *friend*.

Natrophilite. — $NaMnPO_4$. Near triphylite in form. Chiefly massive, cleavable. H. = 4·5–5. G. = 3·41. Easily fusible. Color deep wine-yellow. Optically +. Ax. pl. || (001). $Z = b$ axis. $\alpha = 1\cdot671$, $\beta = 1\cdot674$, $\gamma = 1\cdot684$. 2V = 72°. Strong dispersion, $\rho < v$. Occurs sparingly at Branchville, Fairfield Co., Connecticut.

Graftonite. — $(Fe,Mn,Ca)_3P_2O_8$. Monoclinic. Lamellar. Basal cleavage. H. = 5. G. = 3·7. Fusible. Color when fresh salmon-pink, usually dark from alteration. Optically +. Ax. pl. ⊥ (010). $X = b$ axis; Y nearly || c axis. $\alpha = 1\cdot700$, $\beta = 1\cdot705$, $\gamma = 1\cdot724$. 2V medium. Dispersion, $\rho > v$. Occurs in laminated intergrowths with triphylite in a pegmatite from Grafton, and near North Groton, Grafton Co., New Hampshire.

Merrillite. — $3CaO.Na_2O.P_2O_5$. Uniaxial −. $\omega = 1\cdot623$. $\epsilon = 1\cdot620$. G. = 3·10. Colorless. Found in minute quantities in meteoric stones.

Dehrnite. — $6CaO.(Na,K)_2O.2P_2O_5.H_2O$. Hexagonal. Perfect basal cleavage. In grayish to greenish white crusts formed of fibrous to bladed crystals. Optically −. $\omega = 1\cdot60$–1·64, $\epsilon = 1\cdot59$–1·63. Crystals from Utah show a uniaxial hexagonal core surrounded by six biaxial segments. H. = 5. G. = 3·04. Found at Dehrn, Nassau, and from Cedar Valley, Utah, associated with wardite. May be a member of apatite group.

Beryllonite. — A phosphate of sodium and beryllium, $NaBePO_4$. Crystals short prismatic to tabular, orthorhombic. Cleavages, (001) perfect, (100) good, (010) poor. H. = 5·5–6. G. = 2·8. Luster vitreous; on $c(001)$ pearly. Colorless to white or pale yellowish. Optically −. Ax. pl. || (100). $X = c$ axis. $\alpha = 1\cdot552$, $\beta = 1\cdot558$, $\gamma = 1\cdot562$. 2V = 67°. In Oxford Co., Maine, from Stoneham, and in large crystals at Newry.

KOLBECKITE. A phosphate or silicophosphate of beryllium. Probably monoclinic. Twinned. Short prismatic crystals. Cleavage (010). H. = 3·5–4. G. = 2·39. Color blue to blue-gray. Strongly pleochroic. Found at Niederpöbel, near Schmiedeberg, Saxony.

Apatite Group

General formula	$R_5(F,Cl)[(P,As,V)O_4]_3 = (R(F,Cl))R_4[(P,As,V)O_4]_3$		
Apatite	$(CaF)Ca_4(PO_4)_3$	Fluor-apatite	$c = 0\cdot7346$
or	$(CaCl)Ca_4(PO_4)_3$	Chlor-apatite	
Pyromorphite	$(PbCl)Pb_4(PO_4)_3$		0·7362
Mimetite	$(PbCl)Pb_4(AsO_4)_3$		0·7224
Vanadinite	$(PbCl)Pb_4(VO_4)_3$		0·7122

In addition to the above species, there are also certain intermediate compounds containing lead and calcium; others with phosphorus and arsenic, or arsenic and vanadium, as noted beyond. Further the rare calcium arsenate, svabite, also seems to belong in this group. The radicals CaO, CaOH, may possibly replace the CaF radical in apatite. A probable member of the group, *wilkeite*, contains CO_3, SiO_2 and SO_4 in addition to usual radicals. *Fermorite* contains strontium.

The species of the APATITE GROUP crystallize in the hexagonal system, but all show, either by the subordinate faces, or in etching-figures, that they belong to the tripyramidal class (p. 116). They are chemically phosphates, arsenates, vanadates of calcium or lead (also manganese), with chlorine or fluorine. The latter element is probably present as a univalent radical CaF (or CaCl), etc., in general RF (or RCl), replacing one hydrogen atom in the acid $H_9(PO_4)_3$, so that the general formula is $(\overset{\text{II}}{R}F)\overset{\text{II}}{R}_4(PO_4)_3$, and similarly for the arsenates. This is a more correct way of viewing the composition than the other method sometimes adopted, viz., $3R_3(PO_4)_2.RF_2$, etc.

APATITE.

Hexagonal-tripyramidal.   Axis $c = 0.7346$.

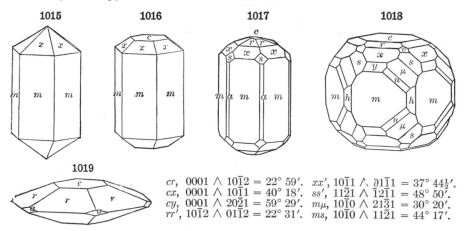

cr, $0001 \wedge 10\bar{1}2 = 22° 59'$.   xx', $10\bar{1}1 \wedge 0\bar{1}\bar{1}1 = 37° 44\frac{1}{2}'$.
cx, $0001 \wedge 10\bar{1}1 = 40° 18'$.   ss', $11\bar{2}1 \wedge \bar{1}2\bar{1}1 = 48° 50'$.
cy, $0001 \wedge 20\bar{2}1 = 59° 29'$.   mμ, $10\bar{1}0 \wedge 21\bar{3}1 = 30° 20'$.
rr', $10\bar{1}2 \wedge 01\bar{1}2 = 22° 31'$.   ms, $10\bar{1}0 \wedge 11\bar{2}1 = 44° 17'$.

Crystals varying from long prismatic to short prismatic and tabular. Also globular and reniform, with a fibrous or imperfectly columnar structure; massive, structure granular to compact. X-ray study shows two molecules in the unit cell the dimensions of which conform to the axial ratio given above.

Cleavage: $c(0001)$ imperfect; $m(10\bar{1}0)$ more so. Fracture conchoidal and uneven. Brittle. H. = 5, sometimes 4·5 when massive. G. = 3·17–3·23 crystals. Luster vitreous, inclining to subresinous. Streak white. Color usually sea-green, bluish green; often violet-blue; sometimes white; occasionally yellow, gray, red, flesh-red and brown. Transparent to opaque. Optically −. Birefringence low. $\omega = 1·632–1·648$, $\epsilon = 1·630–1·643$.

Var. — 1. *Ordinary.* Crystallized, or cleavable and granular massive. Colorless to green, blue, yellow, flesh-red. (*a*) The *asparagus-stone*, originally from Murcia, Spain, is yellowish green. *Moroxite*, from Arendal, Norway, is in greenish blue and bluish crystals. (*b*) *Lasurapatite* is a sky-blue variety with lapis-lazuli in Siberia. (*c*) *Francolite*, from Wheal Franco, near Tavistock, Devonshire, England, occurs in small crystalline stalactitic masses and in minute curving crystals.

Ordinary apatite is *fluor-apatite*, containing fluorine often with only a trace of chlorine, up to 0·5 per cent; rarely chlorine preponderates, and sometimes fluorine is entirely absent.

2. *Manganapatite* contains manganese replacing calcium to 10·5 per cent MnO; color dark bluish green. The presence of manganese causes an increase in refractive indices.

3. *Strontianapatite* or *fermorite* is a variety from India containing up to 10 per cent SrO, see p. 706.

4. *Vœlckerite* is name given to the possible isomorphous molecule, $Ca_4(CaO)(PO_4)_3$ and *hydroxyapatite* to $Ca_4(CaOH)(PO_4)_3$.

5. Varieties from the Laacker See district, Rhineland, that show small amounts of the sulphate and carbonate radicals have been called *sulphate-apatite* and *carbonate-apatite*.

6. *Fibrous, concretionary, stalactitic.* *Phosphorite* includes the fibrous concretionary and partly scaly mineral from Estremadura, Spain, and elsewhere. *Eupyrchroite*, from Crown Point, New York, belongs here; it is concentric in structure. *Staffelite* occurs incrusting the phosphorite of Staffel, Germany, in botryoidal, reniform, or stalactitic masses, fibrous and radiating. See p. 706.

7. *Earthy apatite; Osteolite.* Mostly altered apatite; *coprolites* are impure calcium phosphate.

Comp. — For *Fluor-apatite* $(CaF)Ca_4(PO_4)_3$; and for *Chlor-apatite* $(CaCl)Ca_4(PO_4)_3$; also written $3Ca_3P_2O_8.CaF_2$ and $3Ca_3P_2O_8.CaCl_2$. There are also intermediate compounds containing both fluorine and chlorine. The percentage composition for these normal varieties is as follows:

Fluor-apatite P_2O_5 42·3   CaO 55·5   F 3·8 = 101·6   or $Ca_3P_2O_8$ 92·25 CaF_2 **7·75** = 100
Chlor-apatite P_2O_5 41·0   CaO 53·8   Cl 6·8 = 101·6   or $Ca_3P_2O_8$ 89·4 $CaCl_2$ **10·6** = 100

Fluor-apatite is much more common than the other variety; here belongs the apatite of the Alps, Spain, St. Lawrence Co., New York, Canada.   Apatites in which chlorine is prominent are rare; this is true of some Norwegian kinds.

Pyr., etc. — B.B. in the forceps fuses with difficulty on the edges (F. = 4·5–5), coloring the flame reddish yellow; moistened with sulphuric acid and heated colors the flame pale bluish green (phosphoric acid).   Dissolves in hydrochloric and nitric acids, yielding with sulphuric acid a copious precipitate of calcium sulphate; the dilute nitric acid solution gives sometimes a precipitate of silver chloride on addition of silver nitrate.   Most varieties will give a slight test for fluorine, when heated with potassium bisulphate in a closed tube.

Diff. — Characterized by the common hexagonal form, but softer than beryl, being scratched by a knife; does not effervesce in acid (like calcite); difficultly fusible; yields a green flame B.B. after being moistened with sulphuric acid.

Micro. — Recognized in thin sections by its moderately high relief; extremely low birefringence (hence not often showing a distinct axial figure in basal sections), the interference colors in ordinary sections scarcely rising above gray of the first order; parallel extinction and negative extension; columnar form; lack of color and cleavage; and by the rude cross parting seen as occasional cracks crossing the prism.

Artif. — Apatite may be prepared artificially by fusing sodium phosphate with calcium fluoride or calcium chloride.

Obs. — Apatite is a widely distributed mineral, occurring in rocks of various kinds and ages, but is most common in metamorphic crystalline rocks, especially in granular limestone, in gneiss, syenite, hornblendic gneiss, mica schist, and occasionally in serpentine.   It occurs in many metalliferous veins especially those of pneumatolytic origin, as the tin veins.   It is often associated with beds of iron ore.   In the form of minute microscopic crystals it has an almost universal distribution as an accessory rock-forming mineral.   It is found in all kinds of igneous rocks and is one of the earliest products of the original crystallization.   In larger crystals it is especially characteristic of the pegmatite facies of igneous rocks, particularly the granites, and occurs there associated with quartz, feldspar, tourmaline, muscovite, beryl, etc.   It is sometimes present in ordinary stratified limestone, beds of sandstone or shale of the Silurian, Carboniferous, Jurassic, Cretaceous, or Tertiary.   It has been observed as the petrifying material of wood.

Only the most important localities for the occurrence of apatite can be mentioned here and especially those that afford crystals of exceptional quality or color.   It is found on the Takowaja River east of Ekaterinburg, Ural Mts., and in clear colorless crystals from Kiribinsk in the Orenburg district.   In Bohemia, Czechoslovakia, at Schlaggenwald (Slavkov) and Zinnwald (Cinvald).   In Salzburg, Austria, in complex small clear crystals with epidote on the Knappenwand, Untersulzbachtal, and in similar occurrences in the Zillertal, Tyrol, and the Pfitschtal in Trentino, Italy.   In Switzerland in Ticino at St. Gotthard, and in Val Maggia.   In Saxony purple crystals occur on Greifenstein near Ehrenfriedersdorf. In Norway the greenish blue variety, called *moroxite*, occurs at Arendal in Aust-Agder, at Snarum in Buskerud, and at Kragerö, Telemark.   In Sweden at Gellivare in Norrbotten and at Nordmark in Vermland.   The *asparagus-stone* or *spargelstein* of Jumilla, in Murcia, Spain, is pale yellow-green in color.   In England apatite is found with the tin ores of Cornwall; at Carrock Fell in Cumberland, in celandine-green crystals; in Devonshire, creamcolored at Bovey Tracey and at Wheal Franco (*francolite*) at Tavistock.   A variety similar to asparagus-stone comes from Cerro de Mercado, Durango, Mexico.   Apatite occurs in Japan in tabular colorless crystals at Ashio, Shimotsuke, and in prismatic, yellow crystals at Nosen, Kai.

In the United States crystals of apatite occur in the pegmatite veins of Maine at Mt. Apatite, Auburn, Androscoggin Co., in fine purple crystals and in Oxford Co., at Greenwood, Hebron, Paris, Stoneham, etc.; in New Hampshire at Grafton, Grafton Co.; in Massachusetts at Huntington (formerly Norwich) Hampshire Co.; in Connecticut at Haddam Neck in Middlesex Co., and at Branchville, Fairfield Co., dark greenish blue (*manganapatite*), also in white or colorless crystals, highly modified.   In New York large crystals are found in St. Lawrence Co., at Hammond, Gouverneur; in Rossie 2 miles north of the village of Oxbow (Jefferson Co.); also at Vrooman Lake, near Antwerp, Jefferson Co.; at Sandford mine, East Moriah, Essex Co., in magnetite; at the Tilly Foster iron mine near Brewster, Putnam

Co.; near Edenville, Orange Co., asparagus-green; fibrous mammillated (*eupyrchroite*) near Crown Point, Essex Co., about a mile south of Hammondville in large quantities. In New Jersey in Sussex Co., at Franklin and in large masses at Hurdstown. In North Carolina small highly modified yellow crystals occur in Alexander Co. In California lavender crystals come from Mesa Grande, San Diego Co.

Apatite is of common occurrence in the Laurentian gneiss of eastern Canada, usually associated with limestone and accompanied by pyroxene, amphibole, titanite, zircon, garnet, vesuvianite, etc. The largest deposits occur in the form of irregular veins or as basic intrusives cutting the gneiss. More rarely found in pegmatite veins. The chief localities are in central Ontario and in Ottawa Co., Quebec. Some of the more important localities in Ontario are Bedford in Frontenac Co., North Burgess in Lanark Co., and at Sebastopol and Eganville in Renfrew Co.; in Ottawa Co., Quebec, notable deposits are found in the townships of Buckingham, Templeton, Portland, Hull, and Wakefield. The crystals of apatite from this region are sometimes of enormous size; one from Buckingham weighed 550 pounds and measured 72·5 inches in circumference.

Apatite was named by Werner from ἀπατάειν, *to deceive*, older mineralogists having referred it to aquamarine, chrysolite, amethyst, fluorite, tourmaline, etc.

Besides the definite *mineral phosphates*, including normal apatite, phosphorite, etc., there are also extensive deposits of amorphous and colloidal phosphates, consisting largely of " bone phosphate " ($Ca_3P_2O_8$), of great economic importance, though not having a definite chemical composition and hence not strictly belonging to pure mineralogy. Here belong the phosphatic nodules, coprolites, bone beds, guano, etc. Extensive phosphatic deposits also occur in North Carolina, Alabama, Florida, Tennessee, and in the western states, Idaho, Utah, and Wyoming. Guano is bone phosphate of lime, mixed with the hydrous phosphates, and generally with some calcium carbonate, and often a little magnesia, alumina, iron, silica, gypsum, and other impurities. *Nauruite* is a resinous, colloidal calcium phosphate from Nauru or Pleasant Island in the Pacific Ocean. See also under collophanite, p. 718.

Use. — Apatite and phosphate rock are used chiefly as sources of mineral fertilizers. Some clear finely colored varieties of apatite may be used as gem stones. The mineral is too soft, however, to permit of extensive use for this purpose.

STAFFELITE. A carbonated calcium phosphate. Occurs incrusting the phosphorite of Staffel, Germany in botryoidal or stalactitic masses, fibrous and radiating; it is the result of the action of carbonated waters. H. = 4. G. = 3·128. Color leek- to dark green, greenish yellow. See under apatite, p. 704. *Francolite* (see p. 704) has a similar composition. *Dahllite*, from Bamble, Telemark, Norway, and elsewhere, is similar with small and varying amounts of water.

KURSKITE. $2Ca_3(PO_4)_2.CaF_2.CaCO_3$. In nodules. Cryptocrystalline. Probably colloidal. Color black. G. = 2·9. First found at Kursk, Russia; also widespread in the phosphorites of the Ukraine district, Russia, as a cement, as a replacement of organic remains, or in nodules.

Podolite. — $3Ca_3(PO_4)_2.CaCO_3$. Hexagonal. In microscopic prismatic crystals, also in spherulites. G. = 3·1. Color yellow. Shows division into biaxial sectors. $\beta = 1·63$. Occurs in cavities in the phosphorite nodules from near the Ushitsa River, Podolia, southern Russia. See also *staffelite* and *dahllite*, above.

Svabite. — A calcium arsenate, related to the species of the Apatite Group. Crystals hexagonal prisms; colorless; $c = 0·7143$. H. = 5. G. = 3·52. Optically −. Indices variable; 1·7–2·1. Occurs in Vermland, Sweden, at the Harstig mine, Pajsberg, near Persberg, and at Jacobsberg in the Nordmark district. Also at Franklin, Sussex Co., New Jersey.

Fermorite. — A member of the Apatite Group. $(Ca,Sr)_4[Ca(OH,F)][(P,As)O_4]_3$. H. = 5. G. = 3·52. Color pale pinkish white to white. Uniaxial, −. $\omega = 1·66$. Found with manganese ores at Sitapár, Chhindwára district, Central Provinces, India.

Wilkeite. — $3Ca_3(PO_4)_2.CaCO_3.3Ca_3[(SiO_4)(SO_4)].CaO$. Probably a member of *Apatite Group*. Hexagonal. H. = 5. G. = 3·23. Fusible at 5·5. Dissolves in acids with separation of silica. Color pale rose-red, yellow. Optically −: $\omega = 1·655$. In crystalline limestone at Crestmore, Riverside Co., California.

PYROMORPHITE. Green Lead Ore.

Hexagonal-tripyramidal. Axis $c = 0·7362$.

Crystals prismatic, often in rounded barrel-shaped forms; also in branching groups of prismatic crystals in nearly parallel position, tapering down to a slender point.   Often globular, reniform, and botryoidal or in wart-like shapes, with usually a subcolumnar structure; also fibrous, and granular.

Cleavage: $m(10\bar{1}0)$, $x(10\bar{1}1)$ in traces.   Fracture subconchoidal, uneven. Brittle. H. = 3·5–4. G. = 6·5–7·1 mostly, when pure; 5·9–6·5, when containing lime.   Luster resinous.   Color green, yellow, and brown, of different shades; sometimes wax-yellow and fine orange-yellow; also grayish white to milk-white.   Streak white, sometimes yellowish.   Subtransparent to subtranslucent.   Optically −. $\omega = 2\cdot050$. $\epsilon = 2\cdot042$.   At times when containing arsenic shows weak biaxial character.

1020

Var. — 1. *Ordinary.*   (a) *In crystals* as described; sometimes yellow and in rounded forms resembling campylite (*pseudo-campylite*).   (b) In *acicular* and *moss-like* aggregations.   (c) *Concretionary* groups or masses of crystals, having the surface angular.   (d) *Fibrous.*   (e) *Granular massive.*   (f) *Earthy;* incrusting.

2. *Polysphærite.*   Containing lime; color brown of different shades, yellowish gray, pale yellow to nearly white; streak white; G. = 5·89–6·44. Rarely in separate crystals; usually in groups, globular, mammillary. *Miesite,* from Mies in Bohemia, is a brown variety.   *Nussierite* is similar and impure, from Nussière, near Beaujeu, Rhône, France; color yellow, greenish or grayish; G. = 5·042.   3. *Chromiferous;* color brilliant red and orange.   4. *Arseniferous;* color green to white; G. = 5·5–6·6.   5. *Pseudomorphous;* (a) after galena; (b) cerussite.

Comp. — $(PbCl)Pb_4(PO_4)_3$ or also written $3Pb_3P_2O_8.PbCl_2$ = Phosphorus pentoxide 15·7, lead protoxide 82·2, chlorine 2·6 = 100·5, or Lead phosphate 89·7, lead chloride 10·3 = 100.

The phosphorus is often replaced by arsenic, and as the amount increases the species passes into mimetite.   Calcium also replaces the lead to a considerable extent.

Pyr., etc. — In the closed tube gives a white sublimate of lead chloride.   B.B. in the forceps fuses easily (F. = 1·5), coloring the flame bluish green; on charcoal fuses without reduction to a globule, which on cooling assumes a crystalline polyhedral form, while the coal is coated white from lead chloride and, nearer the assay, yellow from lead oxide.   With soda on charcoal yields metallic lead; some varieties contain arsenic, and give the odor of garlic in R.F. on charcoal.   Soluble in nitric acid.

Diff. — Distinguished by its hexagonal form; high specific gravity; resinous luster; blowpipe characters.

Obs. — Pyromorphite is a mineral of secondary origin, found frequently in lead deposits but seldom in large amounts   It has resulted through the action of phosphoric acid upon galena and cerussite.   The chief localities for the occurrence of pyromorphite, especially in crystals, follow.   From Beresovsk, near Ekaterinburg, Ural Mts., Russia.   In Czechoslovakia at Mies, Příbram, and Bleistadt, Bohemia, and Schemnitz (Selmecsbánya), Slovakia.   In Saxony at Zschopau and near Freiberg; in Hessen-Nassau in fine crystals at Friedrickssegen near Ems; in Baden at Hofsgrund in the Black Forest; at Klaustal in the Harz Mts.   In France the mines of Huelgoat and of Poullaouen in Finistère, Brittany, have furnished fine specimens of a brown color; also from Pontgibaud, Puy-de-Dôme, and from Croix-aux-Mines, Vosges.   Fine crystals come from Horcajo, Ciudad-Real, Spain. In England from Cornwall and from Roughten Gill, Cumberland, and in Scotland at Leadhills, Lanark.   From Mindouli, French Congo.   Occurs at Broken Hill, New South Wales.

In the United States pyromorphite occurs in Pennsylvania at the Wheatley mine, Phœnixville, Chester Co., and at Perkiomen, Montgomery Co.   In Davidson Co., North Carolina.   From Idaho in the Coeur d'Alene district in Shoshone Co., at Burke, Mace, Mullan, and Wardner.

Named from πῦρ, *fire*, μορφή, *form*, alluding to the crystalline form the globule assumes on cooling.   This species passes into mimetite.

Use. — A minor ore of lead.

MIMETITE.   Mimetesite.

Hexagonal-tripyramidal.   Axis $c = 0.7224$.

Habit of crystals like pyromorphite; sometimes rounded to globular forms. Also in mammillary crusts.

Cleavage: $x(10\bar{1}1)$ imperfect.   Fracture uneven.   Brittle.   H. = 3·5. G. = 7·0–7·25.   Luster resinous.   Color pale yellow, passing into brown; orange-yellow;   white or colorless.   Streak white or nearly so.   Subtransparent to translucent.   Optically –. $\omega = 2.135$. $\epsilon = 2.120$.   Sometimes shows biaxial character.

Var. — 1. *Ordinary.*   (a) *In crystals,* usually in rounded aggregates.   (b) *Capillary* or filamentous, especially marked in a variety from St. Prix-sous-Beuvray, France; somewhat like asbestos, and straw-yellow in color.   (c) *Concretionary.*

2. *Campylite,* from Drygill in Cumberland, England, has G. = 7·218, and is in barrel-shaped crystals (whence the name, from καμπυλος, *curved*), yellowish to brown and brownish red; contains 3 per cent P_2O_5.

Comp. — $(PbCl)Pb_4(AsO_4)_3$,   also   written $3Pb_3As_2O_8.PbCl_2$ = Arsenic pentoxide 23·2, lead protoxide 74·9, chlorine 2·4 = 100·5, or Lead arsenate 90·7, lead chloride 9·3 = 100.

Phosphorus replaces the arsenic in part, and calcium the lead.   *Endlichite* (p. 709) is intermediate between mimetite and vanadinite.

Pyr., etc. — In the closed tube like pyromorphite.   B.B. fuses at 1, and on charcoal gives in R.F. an arsenical odor, and is easily reduced to metallic lead, coating the coal at first with lead chloride, and later with arsenic trioxide and lead oxide.   Soluble in nitric acid.

Obs. — Mimetite is a fairly uncommon species to be found like pyromorphite in lead deposits that have undergone a secondary alteration.   Frequently associated with limonite and the carbonates.   It occurs in the Nerchinsk mining district, Transbaikalia, Siberia, in reniform masses and also in fine crystals.   From Příbram, in Bohemia, Czechoslovakia; from Johanngeorgenstadt, Saxony, in fine yellow crystals;   at Badenweiler, Baden.   In France in the Pontgibaud district, Puy-de-Dôme, *campylite.*   Mimetite occurs in various localities in Cornwall; in Cumberland from Roughten Gill, Drygill, etc.   In fine specimens from Tsumeb, near Otavi, South West Africa.   In Mexico at Santa Eulalia near Chihuahua in the state of the same name.

In the United States mimetite has been found at the Wheatley mine, Phœnixville, Chester Co., Pennsylvania.   At Eureka in the Tintic district, Utah.

Named from μιμητής, *imitator,* it closely resembling pyromorphite.

Use. — A minor ore of lead.

VANADINITE.

Hexagonal-tripyramidal.   Axis $c = 0.7122$.

Crystals prismatic, with smooth faces and sharp edges; sometimes cavernous, the crystals hollow prisms; also in rounded forms and in parallel groupings like pyromorphite.   In implanted globules or incrustations.

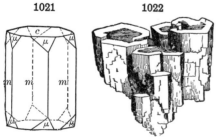

1021   1022

Fracture uneven, or flat conchoidal.   Brittle.   H. = 2·75–3.   G. = 6·66–7·10.   Luster of surface of fracture resinous.   Color deep ruby-red, light brownish yellow, straw-yellow, reddish brown.   Streak white or yellowish.   Subtranslucent to opaque.

Optically –. $\omega = 2.354$. $\epsilon = 2.299$.   For *endlichite,* $\omega = 2.25$, $\epsilon = 2.20$.

Comp. — $(PbCl)Pb_4(VO_4)_3$, also written $3Pb_3V_2O_8.PbCl_2 =$ Vanadium pentoxide 19·4, lead protoxide 78·7, chlorine 2·5 $= 100·6$, or Lead vanadate 90·2, lead chloride 9·8 $= 100$.

Phosphorus is sparingly present, also sometimes arsenic, both replacing vanadium. In *endlichite* the ratio of V : As $= 1 : 1$ nearly.

Pyr., etc. — In the closed tube decrepitates and yields a faint white sublimate. B.B. fuses easily, and on charcoal to a black lustrous mass, which in R.F. yields metallic lead and a coating of lead chloride; after completely oxidizing the lead in O.F. the black residue gives with salt of phosphorus an emerald-green bead in R.F., which becomes light yellow in O.F. Decomposed by hydrochloric acid.

Obs. — Vanadinite is an uncommon mineral usually found in altered lead deposits. Occurrences to be noted are: at Beresovsk near Ekaterinburg in the Ural Mts.; near Eisenkappel in Carinthia, Austria. From the Sierra Nevada, Andalusia, Spain. Found in Scotland at Wanlockhead, Dumfries, and from Leadhills, Lanark. Large crystals are found at Djebel Mahseur (Muhser) near Oudjda, Morocco. From Otteskoop in the Marico district, Transvaal. From the Serra de Cordoba, Argentina. Vanadinite was first noted from Zimapan, Hildago, Mexico.

In the United States frequently found in the mining regions of Arizona and New Mexico, often associated with wulfenite and descloizite. In Arizona the chief localities are at the Red Cloud mine, etc., in Yuma Co., in brilliant deep red crystals; from the Mammoth gold mine at Schultz near Oracle in the Catalina Mts., Pinal Co.; from the Old Yuma mine near Tucson, Pima Co.; from Yavapai Co., in brown barrel-shaped crystals; in the Globe district, Gila Co. In New Mexico in Sierra Co., at Lake Valley (*endlichite*) and Hillsboro, near Georgetown, Grant Co.; and at Kelly near Magdalena, Socorro Co. From the Black Hills, South Dakota.

Use. — A source of vanadium and a minor ore of lead.

HEDYPHANE. Has ordinarily been included as a calcium variety of mimetite. Massive cleavable. Color yellowish white. Uniaxial, $+$. $\omega = 1·948$, $\epsilon = 1·958$. (Franklin.) From Vermland, Sweden, at Långbanshyttan and Pajsberg near Persberg. Also from Franklin, Sussex Co., New Jersey.

Wagnerite Group. Monoclinic

		$a : b : c$	β
Wagnerite	$(MgF)MgPO_4$	$1·9145 : 1 : 1·5059;$	$71°\ 53'$
Triplite	$(RF)RPO_4$, $R = Fe : Mn - 2 : 1$, $1 : 1$, etc.		
Triploidite	$(ROH)RPO_4$, $R = Mn : Fe = 3 : 1$	$1·8572 : 1 : 1·4925;$	$71°\ 46'$
Adelite	$(MgOH)CaAsO_4$	$2·1978 : 1 : 1·5642;$	$73°\ 15'$
Tilasite	$(MgF)CaAsO_4$		
Sarkinite	$(MnOH)MnAsO_4$	$2·0017 : 1 : 1·5154;$	$62°\ 13\frac{1}{2}'$

Phosphates (and arsenates) of magnesium (calcium), iron and manganese containing fluorine (also hydroxyl). Formula R_2FPO_4 or $(RF)RPO_4$, etc.

WAGNERITE.

Monoclinic. Axes, see above. Crystals sometimes large and coarse. Also massive.

Cleavage: $a(100)$, $m(110)$ imperfect; $c(001)$ in traces. Fracture uneven and splintery. Brittle. H. $= 5–5·5$. G. $= 3·07–3·14$. Luster vitreous. Streak white. Color yellow, of different shades; often grayish, also flesh-red, greenish. Translucent. Optically $+$. Ax. pl. $\parallel$ (010). $Z \wedge c$ axis $= -21°$. $2V = 26°$ (approx.). $\alpha = 1·569$. $\beta = 1·570$. $\gamma = 1·582$.

Comp. — A fluo-phosphate of magnesium, $(MgF)MgPO_4$ or $Mg_3P_2O_8$. $MgF_2 =$ Phosphorus pentoxide 43·8, magnesia 49·3, fluorine 11·8 $= 104·9$, deduct $(O = 2F)$ 4·9 $= 100$. A little calcium replaces part of the magnesium.

Pyr., etc. — B.B. in the forceps fuses at 4 to a greenish gray glass; moistened with sulphuric acid colors the flame bluish green. With borax reacts for iron. On fusion with soda effervesces, but is not completely dissolved; gives a faint manganese reaction. Reacts for fluorine. Soluble in nitric and hydrochloric acids. With sulphuric acid evolves fumes of hydrofluoric acid.

Obs. — *Wagnerite* (in small highly modified crystals) occurs in the valley of Höllengraben, near Werfen, and at Raidelgraben near Bischofshofen in Salzburg, Austria. *Kjerulfine* (massive, cleavable; also in coarse crystals) is from Havredal and Kjörrestad, near Bamble, Telemark, Norway.

Spodiosite. — A calcium fluo-phosphate, perhaps $(CaF)CaPO_4$. In flattened prismatic monoclinic (?) crystals. Cleavages: (010) distinct, (001) poor. H. = 5. G. = 2·94. Difficultly fusible. Color ash-gray. Optically +. $\alpha = 1·663$, $\beta = 1·674$, $\gamma = 1·699$. 2V = 69°. Dispersion, $\rho > v$. Found in Vermland, Sweden, at the Nyttsta Kran mine, north of Filipstad, and from Nordmark.

TRIPLITE.

Monoclinic. Massive, imperfectly crystalline. Cleavage: unequal in two directions perpendicular to each other, one much the more distinct. Fracture small conchoidal. H. = 4–5·5. G. = 3·44–3·8. Luster resinous, inclining to adamantine. Color brown or blackish brown. Streak yellowish gray or brown. Subtranslucent to opaque. Optically +. Ax. pl. ‖ (010). $Z \wedge a$ axis = 42°. Indices vary with composition, 1·65–1·68. 2V large. Strong dispersion, $\rho > v$.

Comp. — $(RF)RPO_4$ or $R_3P_2O_8.RF_2$ with R = Fe and Mn, also Ca and Mg. The ratio varies widely from Fe : Mn = 1 : 1 to 2 : 1 (*zwieselite*); 1 : 2; 1 : 7.

Talktriplite is a variety from Horrsjöberg, near Ny, Vermland, Sweden; contains magnesium and calcium in large amount.

Pyr., etc. — B.B. fuses easily at 1·5 to a black magnetic globule; moistened with sulphuric acid colors the flame bluish green. With borax in O.F. gives an amethystine-colored glass (manganese); in R.F. a strong reaction for iron. With soda reacts for manganese. With sulphuric acid evolves hydrofluoric acid. Soluble in hydrochloric acid.

Obs. — Triplite occurs in pegmatite veins associated with quartz, feldspars, micas, beryl, apatite, fluorite, columbite, etc. It is found at Schlaggenwald (Slavkov) in Bohemia, Czechoslovakia. It was originally found at Chanteloube near Limoges, Haute-Vienne, France. *Zwieselite*, a clove-brown variety, is from Rabenstein, near Zwiesel, Bavaria. Found in the Serra de Cordoba, Argentina, in part altered to a mineral like heterosite. In the United States occurs at Stoneham, Oxford Co., Maine; at Branchville, Fairfield Co., Connecticut; from the Reagan mining district, White Pine Co., Nevada.

GRIPHITE. A problematical phosphate related to triplite occurring in embedded reniform masses. Probably amorphous. H. = 5·5. G. = 3·4. Easily fusible. Brown. Isotropic. $n = 1·63–1·65$. From the Riverton tin lode near Harney City, Pennington Co., South Dakota.

Triploidite. — Like triplite, but with the F replaced by (OH). Monoclinic. Commonly in crystalline aggregates. Fibrous to columnar. Cleavage (100) perfect. H. = 4·5–5. G. = 3·7. Easily fusible. Color yellowish to reddish brown. Optically +. Ax. pl. ‖ (010). Z nearly ‖ c axis. Y nearly ⊥ cleavage. $\alpha = 1·725$, $\beta = 1·726$, $\gamma = 1·730$. Strong dispersion, $\rho > v$. From Branchville, Fairfield Co., Connecticut. Also from Hagendorf, near Pleystein, Oberpfalz, Bavaria.

Adelite. — $(MgOH)CaAsO_4$. Monoclinic. Axes, see p. 709; also massive. H. = 5. G. = 3·74. Color gray or grayish yellow. Optically +. Ax. pl. ‖ (010). $Z \wedge c$ axis = +39°. $\alpha = 1·712$, $\beta = 1·721$, $\gamma = 1·731$. 2V large. With manganese ores in Vermland, Sweden; at Jacobsberg near Nordmark and at Långbanshyttan.

Tilasite. — Like adelite, but contains fluorine. Monoclinic. H. = 5. G. = 3·8. Optically −. $Z = b$ axis. $\alpha = 1·640$, $\beta = 1·660$, $\gamma = 1·675$. From Långbanshyttan, Vermland, Sweden. Also noted at Kajlidongri, Jhabua state, Central Provinces, India.

Sarkinite. — $(MnOH)MnAsO_4$. In monoclinic crystals; also in spherical forms. Cleavage (100) distinct. H. = 4–4·5. G. = 4·17. Fusible. Color rose-red, flesh-red, reddish yellow. Optically −. Ax. pl. ‖ (010). $Z \wedge c$ axis = +37°. $\alpha = 1·793$, $\beta = 1·806$, $\gamma = 1·808$. 2V large. From the iron-manganese mines of Pajsberg, near Persberg,

Vermland, Sweden; also at Långbanshyttan. *Polyarsenite* and *Xantharsenite* from the Sjö mine, Örebro, Sweden, and *Chondrarsenite* from Pajsberg, Sweden, are essentially the same.

Arsenoclasite. — $2Mn(OH)_2.Mn_3(AsO_4)_2$. Orthorhombic. Cleavage ∥ (010). Color red. Optically −. Ax. pl. ∥ (100). $X \perp$ (010). $\alpha = 1.787$, $\beta = 1.810$, $\gamma = 1.816$. From Långbanshyttan, Vermland, Sweden.

Arrojadite. — A phosphate of iron, manganese, etc. $4R'_3PO_4.9R''_3P_2O_8$. Monoclinic. Perfect cleavage. H. = 5+. Optically −. 2V = 71°. $\gamma = 1.70$. Color dark green. Pleochroic, $X = Y$ = colorless, Z = pale green. Found massive at Serro Branco, Picuhy, Parahybla, Brazil.

Sarcopside. — Perhaps $2R_3P_2O_8.RF_2$; R = Fe, Mn, Ca. Monoclinic? In irregular ellipsoids or distorted six-sided plates; or from Deering with fibrous appearance. Cleavage approximately perpendicular to fibers. H. = 4. G. = 3.64–3.73. Luster silky. Color flesh-red to lavender-blue; altering on exposure to blue, green, or brown. Optically−. $\beta = 1.728$. Occurs with vivianite and hureaulite in a pegmatite vein near Michelsdorf, Silesia. Also found in a small pegmatite vein in Deering, New Hampshire.

Trigonite. — $HPb_3Mn(AsO_3)_3$. Monoclinic-clinohedral. In small wedge-shaped crystals. Perfect cleavage ∥ (010). H. = 2–3. G. = 8.3. Color sulphur-yellow. Optically −. $\alpha = 2.07$. $\gamma = 2.12$. Ax. pl. ∥ (010). From Långbanshyttan, Vermland, Sweden.

Schultenite. — $HPbAsO_4$. Monoclinic. Thin tabular ∥ (010). Cleavage, good ∥ (010). Transparent, colorless. Brilliant vitreous to almost adamantine luster. Optically +. $X = b$ axis. $Z \wedge c$ axis = +66°. 2V = 58°. $\alpha = 1.89$, $\beta = 1.91$, $\gamma = 1.97$. Occurs on anglesite and bayldonite from Tsumeb, Otavi, South West Africa.

Herderite. — A fluo-phosphate of beryllium and calcium, $CaBe(F,OH)PO_4$. In prismatic crystals, monoclinic with complex twinning. H. = 5. G. = 3. Difficultly fusible. Luster vitreous. Color yellowish and greenish white. Optically −. Ax. pl. ∥ (010). Z nearly ∥ c axis. $\alpha = 1.592$, $\beta = 1.612$, $\gamma = 1.621$. 2V = 74°. From the tin mines of Ehrenfriedersdorf, Saxony; from Epprechtstein, Bavaria; also Oxford Co., Maine, at Stoneham, Buckfield, Greenwood, Hebron, Paris, and Newry (in radial fibrous aggregates), and in Androscoggin Co., at Auburn and Poland.

The following minerals are similar and are commonly placed together in the Hamlinite Group. They are closely related to the sulphates of the Alunite Group.

Goyazite. Hamlinite. — A basic phosphate of aluminum, and strontium. $SrAl_3(OH)_7 P_2O_8$. In colorless rhombohedral crystals. Perfect basal cleavage. H. = 4.5. G. = 3.16–3.28. Optically +. $\omega = 1.620$. $\epsilon = 1.630$. At times biaxial. Occurs with herderite, bertrandite, etc., at Stoneham, Oxford Co., Maine (*hamlinite*). In the diamond sands of Diamantina, Minas Geraes, Brazil (goyazite: originally described as a calcium phosphate but later shown to contain strontium and to be identical with hamlinite). Found also in Binnental, Valais, Switzerland (originally thought to be a new species and named *bowmanite*).

Gorceixite. A basic phosphate of aluminum and barium (with smaller amounts of calcium and cerium) $BaAl_3(OH)_7P_2O_7$? Microcrystalline, in rolled pebbles or "favas." H. = 6. G. = 3.1. Color brown and white. Uniaxial, +. $\omega = 1.625$. From the diamond sands of Minas Geraes, Brazil. *Geraesite* is similar but more acidic in composition.

Plumbogummite. — A basic phosphate of lead and aluminum. $PbAl_3(OH)_7P_2O_7$. Resembles drops or coatings of gum; as incrustations. Color yellowish, brownish. H. = 4–5. G. = 4–5. Uniaxial, +. $n = 1.65–1.67$. From Roughten Gill, Cumberland, England. *Hitchcockite* from Canton mine, Cherokee Co., Georgia, is closely identical. The material from Huelgoet, Brittany, France, is a mixture. *Schadeite* has been proposed as the name for the colloidal equivalent of plumbogummite.

Florencite. — A basic phosphate of aluminum and the cerium metals, closely analogous to goyazite to which it is related in form. $3Al_2O_3.Ce_2O_3.2P_2O_5.6H_2O$. Hexagonal, rhombohedral. Habit rhombohedral. Basal cleavage. H. = 5. G. = 3.58. Color pale yellow. Infusible. Optically +. $\omega = 1.680$, $\epsilon = 1.685$. Found in sands from near Ouro Preto and Diamantina, Minas Geraes, Brazil.

Crandallite. — $CaO.2Al_2O_3.P_2O_5.5H_2O$. Probably orthorhombic. In compact to cleavable masses. Microscopically fibrous. Cleavage ∥ elongation. H. = 4. Color white to light gray. Optically +. X ∥ elongation. $\alpha = 1.595$, $\gamma = 1.60$. Found at Brooklyn mine near Silver City, Tintic District, Utah. Also identified from Dehrn, Nassau.

Harttite. — A basic phosphate and sulphate of aluminum and strontium, $(Sr,Ca)O$: $2Al_2O_3.P_2O_5.SO_3.5H_2O$. Hexagonal. Usually microcrystalline as rolled pebbles. H. = 4·5–5. G. = 3·2. Color flesh-red. From the diamond sands of Bahia, Brazil.

Ježekite. — A fluo-phosphate of lime, soda, and alumina, $Na_4CaAl(AlO)(F,OH)_4(PO_4)_2$: Monoclinic. Cleavage perfect (100); imperfect (001). H. = 4·5. G. = 2·94. Colorless or white. Optically $-$. Ax. pl. $\parallel$ (010). $X \wedge a$ axis = $+29°$. Indices 1·55–1·56. From Ehrenfriedersdorf, Saxony.

Lacroixite. — A fluo-phosphate of soda, lime, manganese oxide, and alumina. $Na_4(Ca,Mn)_4Al_3(F,OH)_4P_3O_{16}.2H_2O$. Orthorhombic. Pyramidal cleavage. H. = 4·5. G. = 3·13. Color pale yellow or green. $\beta = 1·55$. Found at Ehrenfriedersdorf, Saxony.

Durangite. — A fluo-arsenate of sodium and aluminum, $Na(AlF)AsO_4$. In monoclinic crystals. Prismatic cleavage. H. = 5. G. = 3·94–4·07. Easily fusible. Color orange-red. Optically $-$. Ax. pl. $\parallel$ (010). $X \wedge c$ axis = $-25°$. $\alpha = 1·634$, $\beta = 1·673$, $\gamma = 1·685$. Pleochroic; X = orange-yellow, $Y = Z$ = almost colorless. From Durango, Mexico.

AMBLYGONITE. Hebronite.

Triclinic. Crystals large and coarse; forms rarely distinct. Usually cleavable to columnar and compact massive. Polysynthetic twinning lamellæ common.

Cleavage: $c(001)$ perfect, with pearly luster; $a(100)$ somewhat less so, vitreous; $e(0\bar{2}1)$ sometimes equally distinct; $M(1\bar{1}0)$ difficult; $ca(001) \wedge (100) = 75° 30'$, $ce(001) \wedge (0\bar{2}1) = 74° 40'$, $cM(001) \wedge (1\bar{1}0) = 92° 20'$. Fracture uneven to subconchoidal. Brittle. H. = 6. G. = 3·01–3·09. Luster vitreous to greasy, on $c(001)$ pearly. Color white to pale greenish, bluish, yellowish, grayish or brownish white. Streak white. Subtransparent to translucent. Optically $-$. $\alpha = 1·578$. $\beta = 1·595$. $\gamma = 1·598$. 2V = 50°. With increase in hydroxyl content becomes optically $+$ with higher indices.

Comp. — A fluo-phosphate of aluminum and lithium, $LiAl(F,OH)PO_4$. Sodium often replaces part of the lithium, and a small excess of water may be present.

Some authors give the name *montebrasite* to the hydroxyl end of the series and *amblygonite* to the fluorine end.

Pyr., etc. — In the closed tube yields water, which at a high heat is acid and corrodes the glass. B.B. fuses easily (at 2) with intumescence, and becomes opaque white on cooling. Colors the flame yellowish red with traces of green; the Hebron variety gives an intense lithia-red; moistened with sulphuric acid gives a bluish green to the flame. With borax and salt of phosphorus forms a transparent colorless glass. In fine powder dissolves easily in sulphuric acid, more slowly in hydrochloric acid.

Diff. — Distinguished by its easy fusibility and by yielding a red flame B.B., from feldspar, barite, calcite, etc.; also by the acid water in the tube from spodumene.

Obs. — Amblygonite occurs in granite pegmatite veins accompanied by various other lithium minerals. Occurs in Saxony at Penig and other nearby localities and near Geyer in the Erzgebirge. At Montebras, Creuze, France. In the Northern Territory, Australia, near Port Darwin. In the pegmatites of Oxford Co., Maine, at Hebron, Mt. Mica at Paris, Peru; also in Androscoggin Co., at Greenwood, Rumford, and Auburn. Sparingly at Branchville, Fairfield Co., Connecticut. From Pala, San Diego Co., California.

The name *amblygonite* is from ἀμβλύς, *blunt*, and γόνυ, *angle*.

Fremontite. Natramblygonite. Natromontebrasite. — $(Na,Li)Al(OH,F)PO_4$. Triclinic? Isomorphous with amblygonite. Crystals coarse with rough faces. Three cleavages. Usually in cleavage masses. Polysynthetic twinning shown under microscope. H. = 5·5. G. = 3·04. Easily fusible to a white enamel with strong sodium flame color. Luster vitreous to greasy. Color, grayish white to white. Translucent to opaque. Optically $+$. X nearly $\perp$ (001). $\alpha = 1·594$, $\beta = 1·603$, $\gamma = 1·615$. 2V large. From a pegmatite near Canon City, Fremont County, Colorado.

B. Basic Phosphates

This section includes a series of well-characterized basic phosphates, a number of which fall into the Olivenite Group. Acid phosphates are represented by one species only, the little known monetite, probably $HCaPO_4$, see p. 719.

Olivenite Group. Orthorhombic

		$a : b : c$
Olivenite	$Cu_2(OH)AsO_4$	$0\cdot9396 : 1 : 0\cdot6726$
Libethenite	$Cu_2(OH)PO_4$	$0\cdot9601 : 1 : 0\cdot7019$
Adamite	$Zn_2(OH)AsO_4$	$0\cdot9733 : 1 : 0\cdot7158$
Descloizite	$(Pb,Zn,Cu)_2(OH)VO_4$	

$$a : b : c = 0\cdot6368 : 1 : 0\cdot8045 \text{ or } \tfrac{3}{2}a : b : c = 0\cdot9552 : 1 : 0\cdot8045$$

The OLIVENITE GROUP includes several basic phosphates, arsenates, etc., of copper, zinc, and lead, with the general formula $(ROH)RPO_4,(ROH)RAsO_4$, etc. They crystallize in the orthorhombic system with similar form. It is to be noted that this group corresponds in a measure to the monoclinic Wagnerite Group, p. 709, which also includes basic members.

OLIVENITE.

Orthorhombic. Axes $a : b : c = 0\cdot9396 : 1 : 0\cdot6726$.

mm''', 110 ∧ 1$\bar{1}$0 = 86° 26'.　　　　ee', 011 ∧ 0$\bar{1}$1 = 67° 51'.
vv',　101 ∧ $\bar{1}$01 = 71° 11½'.　　　　ve, 101 ∧ 011 = 47° 34'.

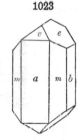

1023

Crystals prismatic, often acicular. Also globular and reniform, indistinctly fibrous, fibers straight and divergent, rarely irregular; also curved lamellar and granular.

Cleavage: m(110), b(010), e(011) in traces. Fracture conchoidal to uneven. Brittle. H. = 3. G. = 4·1–4·4. Luster adamantine to vitreous; of some fibrous varieties pearly. Color various shades of olive-green, passing into leek-, siskin-, pistachio-, and blackish green; also liver- and wood-brown; sometimes straw-yellow and grayish white. Streak olive-green to brown. Subtransparent to opaque. Optically + (perhaps sometimes −). Ax. pl. ‖ (001). $Z = b$ axis. Indices variable, $\beta = 1\cdot785$–$1\cdot795$. 2V nearly 90°. Strong dispersion, $\rho < v$ for optically +.

Var. — (a) *Crystallized.* (b) *Fibrous;* finely and divergently fibrous, of green, yellow, brown and gray, to white colors, with the surface sometimes velvety or acicular; found investing the common variety or passing into it; called *wood-copper* or *wood-arsenate*. (c) *Earthy;* nodular or massive; sometimes soft enough to soil the fingers.

Comp. — $Cu_3As_2O_8.Cu(OH)_2$ or $4CuO.As_2O_5.H_2O$ = Arsenic pentoxide 40·7, cupric oxide 56·1, water 3·2 = 100.

Pyr., etc. — In the closed tube gives water. B.B. fuses at 2, coloring the flame bluish green, and on cooling the fused mass appears crystalline. B.B. on charcoal fuses with deflagration, gives off arsenical fumes, and yields a metallic arsenide which with soda yields a globule of copper. With the fluxes reacts for copper. Soluble in nitric acid.

Obs. — Olivenite is a rare mineral of secondary origin found in copper deposits. Occurs in crystals with adamite from Cap Garonne, Var, France. Abundant in Cornwall as crystals with quartz at Liskeard, Gwennap, St. Day (Wheal Gorland), Redruth, etc. From near Tavistock in Devonshire. Found at Tsumeb, near Otavi, South West Africa. In the

United States at various mines in the neighborhood of Eureka, Tintic district, Utah, both in crystals and as *wood-copper*. The name olivenite alludes to the olive-green color.

Duftite. — Belongs to olivenite group with approximately half the copper replaced by lead, $(Pb,Cu)_3(AsO_4)_2.(Pb,Cu)(OH)_2$. In aggregates of minute curved crystals. H. = 3. G. = 6·19. Color bright olive-green to gray-green. Luster vitreous on fractured surfaces, dull on crystal faces. Optically −. $\alpha = 2·06$, $\beta = 2·08$, $\gamma = 2·09$. 2V large. Dispersion, $\rho > v$. Found associated with azurite at Tsumeb near Otavi, South West Africa.

Higginsite. — $CuCa(OH)AsO_4$. Orthorhombic. Small prismatic crystals. H. = 4·5. G. = 4·33. Easily fusible. Optically −. Ax. pl. || (010). $X = a$ axis. $\alpha = 1·800$, $\beta = 1·831$, $\gamma = 1·846$. 2V nearly 90°. Dispersion, $\rho > v$. Pleochroic, green, yellow-green, blue-green. From Higgins mine, Bisbee, Cochise Co., Arizona.

LIBETHENITE.

Orthorhombic. Axes $a : b : c = 0·9601 : 1 : 0·7019$.

mm''', 110 $\wedge$ 1$\bar{1}$0 = 87° 40'. ss''', 111 $\wedge$ 1$\bar{1}$1 = 59° 4$\frac{1}{2}$'.
ee',   011 $\wedge$ 0$\bar{1}$1 = 70° 8'. ss',   111 $\wedge$ $\bar{1}$11 = 61° 47$\frac{1}{2}$'.

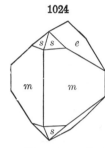

1024

In crystals usually small, short prismatic in habit; often united in druses. Also globular or reniform and compact. Cleavage: $a(100)$, $b(010)$ very indistinct. Fracture subconchoidal to uneven. Brittle. H. = 4. G. = 3·6–3·8. Luster resinous. Color olive-green, generally dark. Streak olive-green. Translucent to subtranslucent. Optically −. Ax. pl. || (001). $X = b$ axis. $\alpha = 1·702$, $\beta = 1·745$, $\gamma = 1·789$. 2V nearly 90°. Strong dispersion, $\rho > v$.

Comp. — $Cu_3P_2O_8.Cu(OH)_2$ or $4CuO.P_2O_5.H_2O$ = Phosphorus pentoxide 29·8, cupric oxide 66·4, water 3·8 = 100.

Pyr., etc. — In the closed tube yields water and turns black. B.B. fuses at 2 and colors the flame emerald-green. On charcoal with soda gives metallic copper, sometimes also an arsenical odor. Fused with metallic lead on charcoal is reduced to metallic copper, with the formation of lead phosphate, which treated in R.F. gives a crystalline polyhedral bead on cooling. With the fluxes reacts for copper. Soluble in nitric acid.

Obs. — Libethenite is a rare mineral occurring in certain copper deposits. It is found at Nizhne-Tagilsk in the Ural Mts.; in Slovakia of Czechoslovakia at Libethen (Libetbánya) near Neusohl (Ban Bystrica) in cavities with quartz associated with chalcopyrite; from Cornwall at Liskeard and Redruth. At the Mercedes mine, east of Coquimbo, Chile. In the United States from the Clifton-Morenci district, Greenlee, Co., Arizona.

Adamite. — $Zn_3As_2O_8.Zn(OH)_2$. In small orthorhombic crystals, often grouped in crusts and granular aggregations. H. = 3·5. G. = 4·34–4·35. Fusible. Color honeyyellow, violet, rose-red, green, colorless. Varies from optically + to −. Ax. pl. || (001). $X = a$ axis. $\alpha = 1·708$, $\beta = 1·744$, $\gamma = 1·773$. 2V nearly 90°. Strong dispersion, $\rho > v$ for optically −. A rare mineral occasionally found in zinc deposits. Occurs in Greece at the ancient zinc mines of Laurium. At Cap Garonne, Var, France, including the varieties that containing cobalt and copper, have been called *cobaltoadamite* and *cuproadamite*. From Aïn Achour, near Guelma, Constantine, Algeria. From Tsumeb, near Otavi, South West Africa. At Chañarcillo, Atacama, Chile.

Descloizite. — $R_3V_2O_8.R(OH)_2$ or $4RO.V_2O_5.H_2O$, also given as $2R_3V_2O_8.R(OH)_2$; R = Pb, Zn chiefly, and usually in the ratio 1 : 1 approx. *Cuprodescloizite* is a variety with about half of the zinc replaced by copper. *Mottramite* may be considered as another variety in which the zinc is almost entirely replaced by copper. *Psittacinite* belongs here also. In small orthorhombic crystals, often drusy; also massive, fibrous radiated with mammillary surface. H. = 3·5. G. = 5·9–6·2. Easily fusible. Color cherry-red and brownish red, to light or dark brown, black. Streak orange to brownish red or yellowish gray. Optically − or +. Ax. pl. || (010). $\alpha = 2·185$, $\beta = 2·265$, $\gamma = 2·35$. 2V about 90°. Strong dispersion, $\rho < v$. *Cuprodescloizite* has indices 2·17–2·33. 2V = 50°–70°. Color olive-green to brown.

Descloizite is a mineral of secondary origin found occasionally in lead-zinc deposits. Found near Eisenkappel, Carinthia, Austria. From Broken Hill, N. W. Rhodesia, and in fine specimens from Abenab near Grootfontein, South West Africa. In small crystals at various points in the Serra de Córdoba, Argentina. In the United States in New Mexico abundant at Lake Valley, Sierra Co.; near Georgetown, Grant Co., and at Magdalena, Socorro Co. In Arizona occurs at Schultz near Oracle, Pinal Co. *Cuprodescloizite* (also known as *tritochorite* and *ramirite*) occurs in crusts and reniform masses with radiated structure in Mexico in the states of San Luis Potosi, Zacatecas, and Guanajuato. A similar variety occurs in Arizona at Tombstone and Bisbee, in Cochise Co., and at Schultz, Pinal Co. *Mottramite* occurs in velvety black encrustations at Alderley Edge and at Mottram, south and southeast of Manchester, Cheshire, England. Also noted from Tsumeb, near Otavi, South West Africa.

EUSYNCHITE may be identical with descloizite. Massive: in nodular, stalactitic forms. G. = 5·596. Color yellowish red, reddish brown, greenish. From Hofsgrund near Freiburg in Baden, Germany. The same may be true of *arœoxene* from Dahn near Niederschlettenbach, Rhenish Bavaria, Germany.

Pyrobelonite. — $4(Mn,Pb)O.V_2O_5.H_2O$. Perhaps a descloizite in which all the zinc and a part of the lead have been replaced by manganese. Orthorhombic. In small acicular crystals. Fire-red color. H. = 3·5. G. = 5·377. Optically −. Ax. pl. || (001). $\beta =$ 2·36. From Långbanshyttan, Vermland, Sweden.

Tarbuttite. — $Zn_3P_2O_8.Zn(OH)_2$. Triclinic. Crystals striated and rounded, frequently in sheaf-like aggregates. Perfect basal cleavage. H. = 3·7. G. = 4·1. Easily fusible. Colorless to pale yellow, brown, red, or green. Optically −. $\alpha = 1·66$, $\beta = 1·70$, $\gamma =$ 1·71. 2V = 47°. From Broken Hill, N. W. Rhodesia.

Brackebuschite. — $(Pb,Mn,Fe)_3V_2O_8.H_2O$? Monoclinic? Prismatic crystals. Color black. Strongly pleochroic; X = nearly colorless, Y and Z = reddish brown. Optically +. $\beta = 2·36$. From the State of Córdoba, Argentina.

DECHENITE. Composition usually accepted as PbV_2O_6. Massive, botryoidal, nodular. G. = 5·6–5·81. Color deep red to yellowish red and brownish red. From near Eisenkappel, Carinthia, Austria; at Niederschlettenbach near Bundental on the Lauter, Bavarian Pfalz.

Calciovolborthite. — Probably $(Cu,Ca)_3V_2O_8.(Cu,Ca)(OH)_2$. Monoclinic? In pseudo-hexagonal scales, rosettes and gray, fine crystalline granular. H. = 3·5. G. = 3·5. Easily fusible. Color various shades of yellow-green. Optically +. $\beta = 2·05$. Dispersion very great. At times As_2O_5 replaces V_2O_5 with a consequent lowering of refractive indices. Occurs as scales or powder upon psilomelane at Friedricksroda in Thuringia. A similar mineral has been described from the Ferghana district of Turkestan. Also from southwest Colorado and adjoining portions of Utah, as near Telluride in San Miguel Co., and at Naturita and Paradox in Montrose Co., Colorado, and at Richardson in Grand Co., Utah.

TURANITE. A copper vanadate, $5CuO.V_2O_5.2H_2O$. Radiating fibrous. From Tyuya-Muyun, south of Andijan, in the Ferghana district, Turkestan.

Uzbekite. Usbekite. — $3CuO.V_2O_5.3H_2O$. In thin dark-green crusts which under the microscope are shown to be composed of fine aggregates of unoriented needles. Biaxial, −. $\alpha = 2·01$, $\beta = 2·04$, $\gamma = 2·07$. 2V large. Dispersion strong, $\rho < v$. From the Uzbekistana district in Ferghana, Russian Turkestan. This mineral has been described as consisting of two types: α-*uzbekite*, $2RO.V_2O_3.3H_2O$ with dark green color, and β-*uzbekite*, $3RO.V_2O_5.4H_2O$ with pale green color.

Fornacite. Furnacite. — A basic chrom-arsenate of lead and copper. In dark olive-green, small prismatic crystals. Biaxial. Optically +. Occurs with dioptase on the river Djoué, a tributary of the Congo, northwest of Brazzavillo, French Equatorial Africa.

Tsumebite. Preslite. — A basic lead and copper phosphate. Orthorhombic? In small tabular crystals. H. = 3·5. G. = 6·1. Easily fusible. Optically +. $\alpha = 1·885$, $\beta = 1·920$, $\gamma = 1·956$. 2V about 90°. Strong dispersion, $\rho < v$. Color emerald-green. From Tsumeb, near Otavi, South West Africa.

CLINOCLASITE. Aphanèse.

Monoclinic. Axes $a : b : c = 1·9069 : 1 : 3·8507$; $\beta = 80° 30'$.

Crystals prismatic ($m(110)$); also elongated || b axis; often grouped in nearly spherical forms. Also massive, hemispherical or reniform; structure radiated fibrous.

Cleavage: $c(001)$ highly perfect. Brittle. H. = 2·5–3. G. = 4·19–4·37. Luster: c pearly; elsewhere vitreous to resinous. Color internally dark verdigris-green; externally blackish blue-green. Streak bluish green. Subtransparent to translucent. Optically −. Ax. pl. ∥ (010). Z nearly ∥ a axis. $\alpha = 1.73$, $\beta = 1.87$, $\gamma = 1.91$. 2V medium. Strong dispersion $\rho < v$.

Comp. — $Cu_3As_2O_8.3Cu(OH)_2$ or $6CuO.As_2O_5.3H_2O$ = Arsenic pentoxide 30.3, cupric oxide 62.6, water 7.1 = 100.

Pyr., etc. — Same as for olivenite.

Obs. — Found in the mines of Cornwall, at Gwennap, St. Day, Redruth, etc.; also from near Tavistock, Devonshire. From Collahuasi, Tarapacá, Chile. In the United States from the Mammoth mine, near Eureka, Tintic district, Utah, in fine crystallizations. Named in allusion to the basal cleavage being oblique to the sides of the prism.

Erinite. — $Cu_3As_2O_8.2Cu(OH)_2$. Orthorhombic? In mammillated crystalline groups. One perfect cleavage. H. = 5. G. = 4. Easily fusible. Color fine emerald-green. Optically −. $Z \perp$ to cleavage. Y ∥ elongation. $\alpha = 1.82$. $\beta = 1.826$. $\gamma = 1.88$. 2V large. From Cornwall; also the Tintic district, Utah.

Dihydrite. — $Cu_3P_2O_8.2Cu(OH)_2$. Triclinic with monoclinic habit. In dark emerald-green crystals. Fibrous. Cleavage ∥ (010), imperfect. H. = 4.5–5. G. = 4–4.4. Easily fusible. Optically both + and −. Z nearly ∥ b axis. $\alpha = 1.719$, $\beta = 1.763$, $\gamma = 1.805$. 2V nearly 90°. Strong dispersion. From Nizhne-Tagilsk in the Ural Mts.; in Rhineland at Virneberg near Rheinbreitbach and at Ehl near Linz on the Rhine.

Pseudomalachite. — In part $Cu_3P_2O_8.3Cu(OH)_2$. Monoclinic? Fibrous to massive, resembling malachite in color and structure. H. = 4.5. G. = 3.6. Easily fusible. Optically −. Z nearly ∥ fibers. Indices = 1.73–1.88. From Rheinbreitbach, Germany; Nizhne-Tagilsk, Russia, etc. *Ehlite* from Ehl on the Rhine is closely allied. *Dihydrite, pseudomalachite* and *ehlite* are all closely similar if not identical. They have all been grouped under the general name of *lunnite*.

STASZICITE. Staszycyt. $5(Ca,Cu,Zn)O.As_2O_5.2H_2O$. Orthorhombic? Fibrous. Parallel extinction and positive elongation. Yellowish green. H. = 5.5–6. G. = 4.227. An alteration product of tennantite, associated with other copper ores at Miedzianka, west of Kielce, Poland.

Cornetite. — $Cu_3P_2O_8.3Cu(OH)_2$. Dimorphous with pseudomalachite. Orthorhombic. In minute crystals or crystalline crusts. H. = 4–5. G. = 4.10. Color blue. Optically −. Ax. pl. ∥ (001). $X = a$ axis. $\alpha = 1.765$, $\beta = 1.81$, $\gamma = 1.82$. 2E = 62°. From Bwana Mkubwa, northern Rhodesia, and Katanga, Belgian Congo.

DUFRENITE. Kraurite.

Monoclinic? Crystals rare, small, and indistinct. Usually massive, in nodules; radiated fibrous with drusy surface.

Cleavage: (010) perfect; (100) distinct. H. = 3.5–4. G. = 3.2–3.4. Luster silky, weak. Color dull leek-green, olive-green, or blackish green; alters on exposure to yellow and brown. Streak siskin-green. Subtranslucent to nearly opaque. Usually optically +. Z ∥ b axis. $\alpha = 1.830$, $\beta = 1.840$, $\gamma = 1.885$. 2V medium to 90°. Very strong dispersion. Strongly pleochroic; yellowish to green and dark red-brown. At times shows zonal structure with varying optical characters.

Comp. — Doubtful; in part $FePO_4.Fe(OH)_3 = 2Fe_2O_3.P_2O_5.3H_2O$ = Phosphorus pentoxide 27.5, iron sesquioxide 62.0, water 10.5 = 100.

Pyr., etc. — Same as for vivianite, but less water is given out in the closed tube. B.B. fuses easily to a slag.

Obs. — Dufrenite is most commonly found associated with limonite bodies, together with other phosphates. Frequently produced by the alteration of triplite. From Ullersreuth near Hirschberg, Thuringia; from Saxony at Hauptmannsgrün east of Reichenbach in Vogtland; from near Waldgrimes, northeast of Wetzlar in Hessen-Nassau; from Siegen, Westphalia. In France from Haute-Vienne, at Hureaux en Sylvestre, also from Vilate near Chanteloube; in Morbihan at Rochefort-en-Terre. In Belgium from the mine of Berneau near Visé, Liège. From Liskeard, etc., Cornwall. In the United States at Allentown, Monmouth Co., New Jersey. In Rockbridge Co., Virginia. From Sevier Co., Arkansas. *Dufreniberaunite* is the name given to a variety supposedly intermediate between dufrenite and beraunite from Hellertown, Northampton Co., Pennsylvania.

Delvauxite. A hydrated iron phosphate, perhaps $2Fe_2O_3.P_2O_5.9H_2O$ (a hydrated dufrenite?). Amorphous. H. = 2·5. G. = 1·8–2·0. Color yellowish brown to brownish black or reddish. Isotropic. $n = 1·72$. From Berneau, near Visé, Belgium, and in reniform concretions and compact masses from Litošice, Bohemia, Czechoslovakia.

LAZULITE.

Monoclinic: Axes $a : b : c = 0·9750 : 1 : 1·6483$; $\beta = 89° 14'$.

$at,\quad 100 \wedge 101 = 30° 24'.$
$pp',\ 111 \wedge 1\bar{1}1 = 79° 40'.$
$ee',\ \bar{1}11 \wedge \bar{1}\bar{1}1 = 80° 20'.$
$pe,\ 111 \wedge \bar{1}11 = 82° 30'.$

1025

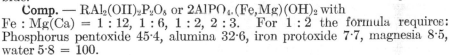

Crystals usually acute pyramidal in habit. Also massive, granular to compact.

Cleavage: prismatic, indistinct. Fracture uneven. Brittle. H. = 5–6. G. = 3·1. Luster vitreous. Color azure-blue; commonly a fine deep blue viewed along one axis, and a pale greenish blue along another. Streak white. Subtranslucent to opaque. Optically −. Ax. pl. ∥ (010). $X \wedge c$ axis = −9°. $\alpha = 1·612$. $\beta = 1·634$. $\gamma = 1·643$. 2V = 69°. Pleochroic; $X =$ colorless, $Y = Z =$ azureblue.

Comp. — $RAl_2(OH)_2P_2O_8$ or $2AlPO_4.(Fe,Mg)(OH)_2$ with Fe : Mg(Ca) = 1 : 12, 1 : 6, 1 : 2, 2 : 3. For 1 : 2 the formula requires: Phosphorus pentoxide 45·4, alumina 32·6, iron protoxide 7·7, magnesia 8·5, water 5·8 = 100.

Pyr., etc. — In the closed tube whitens and yields water. In the forceps whitens, cracks open, swells up, and without fusion falls to pieces, coloring the flame bluish green. B.B. with cobalt solution the blue color of the mineral is restored. The green color of the flame is made more intense by moistening the assay with sulphuric acid. With the fluxes gives an iron glass; with soda on charcoal an infusible mass. Unacted upon by acids, retaining perfectly its blue color.

Obs. — Lazulite is usually found in quartz or pegmatite veins. Occurs in Austria at various localities near Werfen, Salzburg; from near Krieglach in the Mürztal in Styria. In Switzerland from near Zermatt, Valais. In Vermland, Sweden, at Horrsjöberg near Ny, sometimes in large crystals. From central Madagascar, at times in large fragments; in crystals near Betafo. From Brazil in the diamond sands of Diamantina, Minas Geraes, and in Bahia. In the United States abundant with corundum at Crowder's Mt., Gaston Co., North Carolina; and in fine sky-blue crystals on Graves Mt., Lincoln Co., Georgia, with kyanite, rutile, pyrophyllite, etc.

The name lazulite is derived from an Arabic word, *azul*, meaning *heaven*, and alludes to the color of the mineral.

Tavistockite. — $Ca_3P_2O_8.2Al(OH)_3$. Orthorhombic. In microscopic acicular crystals, sometimes stellate groups. Cleavage (100), perfect. Color white. Optically +. Ax. pl. ∥ (100). Z ∥ c axis. $\alpha = 1·522$, $\beta = 1·530$, $\gamma = 1·544$. 2V = 71°. From Tavistock, Devonshire.

Bialite. Hydrous magnesium, calcium, aluminum phosphate. Perhaps a magnesian variety of tavistockite. Orthorhombic. In minute white needles. One cleavage which is ax. pl. Optically +. Indices 1·52–1·55. Occurs on compact brown phosphatic rock from Mushishimano, Katanga, Belgian Congo.

Kirrolite. Cirrolite. — Perhaps $Ca_3Al(PO_4)_3.Al(OH)_3$. Compact. H. = 5·5. G. = 3·08. Color pale yellow. Occurs at the iron mine at Westanå, near Näsum west of Carlshamn in Blekinge, Sweden.

Arseniosiderite. — $Ca_3Fe(AsO_4)_3.3Fe(OH)_3$. Tetragonal? Basal cleavage. In yellowish brown fibrous concretions. H. = 1·5. G. = 3·5–3·9. Easily fusible. Uniaxial (or biaxial with very small 2V). Optically −. Indices, 1·79–1·90. Strongly pleochroic, $O =$ dark red-brown, $E =$ almost colorless. From Schneeberg, Saxony; at Romanèche-Thorins, southwest of Macon, Saône-et-Loire, France. Reported from the Tintic district, Utah. A close similarity in optical characters suggests that mazapilite (p. 732) is identical with arseniosiderite.

Dussertite. — $Ca_3(AsO_4)_2.3Fe(OH)_3$. Hexagonal or hexagonal rhombohedral. In small crystals, tabular parallel to (0001). H. = 3·5. G. = 3·75. Color green. Uniaxial, —. $\omega = 1·87$, $\epsilon = 1·85$. Pleochroic in shades of greenish yellow. Occurs as crusts on quartz at Djedel Debar, northeast of Hammon Meskhoutine, Constantine, Algeria.

Allactite. — $Mn_3As_2O_8.4Mn(OH)_2$. Monoclinic. In small brownish red prismatic crystals or tabular || (100). Several indistinct cleavages. H. = 4·5. G. = 3·8. Fusible at 6. Optically —. Ax. pl. ⊥ (010) for blue, || (010) for red and yellow. $X \wedge c$ axis = —50°. $\alpha = 1·76$, $\beta = 1·78$, $\gamma = 1·78$. 2V very small. Dispersion very strong, $\rho > v$. In Vermland, Sweden, at the Moss mine, Nordmark, and at Långbanshyttan.

Synadelphite. — $2(Al,Mn)AsO_4.5Mn(OH)_2$. Monoclinic. In prismatic crystals; also in grains. H. = 4·5. G. = 3·5. Easily fusible. Color brownish black to black. Optically +. Ax. pl. ⊥ (010). $X \wedge c$ axis = 45°. $\alpha = 1·86$, $\beta = 1·87$, $\gamma = 1·90$. 2V small. From the Moss mine, Nordmark, Vermland, Sweden.

Allodelphite. — $5MnO.2(Mn,Al)_2O_3.As_2O_3.SiO_2.5H_2O$. Probably orthorhombic. In tabular elongated crystals. Conchoidal fracture. G. = 3·573. Color dark red-brown. Chocolate-brown streak. Optically +. $\alpha = 1·7244$, $\gamma = 1·7493$. 2V very small. From Långbanshyttan, Vermland, Sweden.

Flinkite. — $MnAsO_4.2Mn(OH)_2$. In minute orthorhombic crystals, tabular || $c(001)$; grouped in feather-like aggregates. H. = 4·5. G. = 3·87. Easily fusible. Color greenish brown. Optically +. Ax. pl. || (001). $Z = a$ axis. $\alpha = 1·783$, $\beta = 1·801$, $\gamma = 1·834$. 2V large. Pleochroic; X = pale brownish green, Y = yellowish green, Z = orange-brown. From the Harstig mine, Pajsberg, near Persberg, Vermland, Sweden.

Hematolite. — Perhaps $(Al,Mn)AsO_4.4Mn(OH)_2$. In rhombohedral crystals. Basal cleavage, perfect. H. = 3·5. G. = 3·4. Infusible. Color brownish red, black on the surface. Optically —. $\omega = 1·733$, $\epsilon = 1·714$. Also at times biaxial with small 2V. From the Moss mine, Nordmark, Vermland, Sweden.

Retzian. — A basic arsenate of the yttrium earths, manganese and calcium. In orthorhombic crystals. H. = 4. G. = 4·15. Infusible. Color chocolate- to chestnut-brown. Optically +. Ax. pl. || (010). $Z = a$ axis. $\alpha = 1·777$, $\beta = 1·788$, $\gamma = 1·800$. 2V large. Pleochroic; X = colorless, Y = pale yellow-brown, Z = red-brown. From the Moss mine, Nordmark, Vermland, Sweden.

Arseniopleite. — Perhaps $\overset{\text{II III}}{R_9}R_2(OH)_6(AsO_4)_6$; $\overset{\text{II}}{R}$ = Mn, Ca, also Pb, Mg; $\overset{\text{III}}{R}$ = Mn, also Fe. Rhombohedral with rhombohedral cleavage. Massive, cleavable. H. = 3·5. Easily fusible. Color brownish red. Optically +. $\omega = 1·794$, $\epsilon = 1·803$. Sometimes biaxial with small 2V. Occurs at the Sjö mine, near Örebro, Sweden, with rhodonite in crystalline limestone.

Manganostibiite. — $Mn_{10}Sb_2O_{15}$. Monoclinic? In small rods and fibers and embedded grains. Cleavage || (010), perfect. Infusible. Color black. Optically —. Ax. pl. ⊥ (010). Z = elongation. $\alpha = 1·92$, $\beta = 1·95$, $\gamma = 1·96$. 2V small. Strongly pleochroic; X = brown, $Y = Z$ = almost opaque. From Nordmark, Vermland, Sweden. *Hematostibiite* is closely related if not identical; from Sjö mine, near Örebro, Sweden.

Atelestite. — Basic bismuth arsenate, $H_2Bi_3AsO_8$. Monoclinic. In minute tabular crystals. Cleavage || (001), poor. H. = 4. G. = 6·4. Easily fusible. Color sulphur-yellow. Optically +. $\alpha = 2·14$, $\beta = 2·15$, $\gamma = 2·18$. 2V = 44°. Dispersion, $\rho < v$. From Schneeberg, Saxony.

C. Normal Hydrous Phosphates, etc.

The only important group among the normal hydrous phosphates is the monoclinic VIVIANITE GROUP.

Struvite. — $NH_4MgPO_4.6H_2O$. In orthorhombic-hemimorphic crystals (Fig. 342, p. 145). Cleavages, (001) perfect, (010) good. H. = 2. G. = 1·7. Fusible. White or yellowish. Optically +. Ax. pl. || (100). $Z = b$ axis. $\alpha = 1·495$, $\beta = 1·496$, $\gamma = 1·504$. 2V = 37°. Strong dispersion, $\rho < v$. Struvite is formed where magnesia solutions act upon a phosphate in the presence of ammonia. Such conditions may exist in bodies of guano, etc., which have remained undisturbed for a long period.

Collophanite. — Essentially $Ca_3P_2O_8.H_2O$. Composition variable, especially as to water content. Also contains small amounts of calcium carbonate, fluoride, sulphate, etc.

Amorphous. In layers resembling gymnite or opal. Conchoidal fracture. H. = 2–5. G. = 2·6–2·9. Difficultly fusible. Colorless or snow-white. Isotropic. $n = 1·57–1·63$. Collophanite is usually found in coral limestones which have been covered by beds of guano. From the island of Sombrero in the West Indies, having been formed in the elevated coral reef by infiltration of the salts from the overlying guano. *Monite* is similar from the islands Mona and Moneta in the West Indies, where it is associated with *monetite*, $HCaPO_4$, which occurs in yellowish white triclinic crystals; is optically + with $\beta = 1·518$. Also similarly found in various localities in the southern Pacific Ocean. Material apparently identical with collophanite, found in nodules in a sandy marl, near Grodno, Poland, has been named *grodnolite*. Collophanite is probably an important constituent of the phosphorites described under apatite, p. 704.

PYROPHOSPHORITE. $Mg_2P_2O_7.4(Ca_3P_2O_8.Ca_2P_2O_7)$? Massive, earthy. Color snow-white, dull. From the West Indies

---

Hopeite. — $Zn_3P_2O_8.4H_2O$. Orthorhombic. In minute prismatic crystals. Also in reniform masses. Three cleavages: $a(100)$, perfect; $b(010)$, good; $c(001)$, poor. H. = 3·2. G. = 3. Difficultly fusible. Color grayish white. Crystals from Broken Hill show interbanding of two modifications, α- and β-hopeite which have the same composition but differ in their optical characters. Both varieties are optically −, and $X = b$ axis. In α-hopeite $Z = a$ axis, while in β-hopeite $Z = a$ or c axes. Indices for α-hopeite are, $\alpha = 1·572$, $\beta = 1·591$, $\gamma = 1·592$. $2V = 36°$. In β-hopeite, $\alpha = 1·574$, $\beta = 1·582$, $\gamma = 1·582–1·59$. 2V very small. Found in cavities in calamine at the zinc mines of Moresnet, Belgium and at Altenberg, southwest of Aix-la-Chapelle, Rhineland; at the Broken Hill mines, Rhodesia. Also found at Salmo, Nelson mining district, British Columbia, (*hibbenite*).

Phosphophyllite. — Essentially a zinc phosphate with composition similar to hopeite, $R_3(PO_4)_2.4H_2O$. Monoclinic but with angles similar to those of hopeite. Cleavages: (100) perfect, (010), ($\bar{1}$02) distinct. H. = 3. G. = 3·1. Colorless to pale blue-green. Optically −. Ax. pl. ⊥ (010). $Z = b$ axis. $Y \wedge a$ axis = 50°. $\alpha = 1·595$, $\beta = 1·614$, $\gamma = 1·616$. 2V = 50°. From Hagendorf near Pleystein in the Oberpfalz, Bavaria.

Parahopeite. — $Zn_3P_2O_8.4H_2O$. Same as for hopeite. Triclinic. In minute tabular crystals, || (100) with deep striations; twinned on (100). Good cleavage || (010). H. = 3·7. G. = 3·3. Easily fusible. Colorless. Optically +. $\alpha = 1·614$, $\beta = 1·625$, $\gamma = 1·637$. 2V large. Found at Broken Hill, Rhodesia; also from Salmo, Nelson mining district, British Columbia.

Dickinsonite. — $3R_3P_2O_8.H_2O$ with R = Mn, Fe, Na_2 chiefly, also Ca, K_2, Li_2. Monoclinic. In tabular, pseudo-rhombohedral crystals; commonly foliated to micaceous. Cleavage (001), perfect. G. = 3·34. Easily fusible. Color olive- to oil-green, grass-green. Optically +. Ax. pl. ⊥ (010). $X = b$ axis. $\beta = 1·655–1·662$. 2V medium to large. Strong dispersion, $\rho > v$. From Branchville, Fairfield Co., Connecticut; also from Poland, Androscoggin Co., Maine.

Fillowite. — Formula as for dickinsonite, but differing in crystal angles. Monoclinic, pseudo-rhombohedral. Basal cleavage. In granular crystalline masses. H. = 4·5. G. = 3·43. Easily fusible. Color wax-yellow, yellowish to reddish brown, colorless. Optically +. $\alpha = 1·672$, $\beta = 1·672$, $\gamma = 1·676$. 2V small. Occurs sparingly at Branchville, Fairfield Co., Connecticut.

The three following species are related in composition and may be in crystalline form.

Roselite. — $(Ca,Co,Mg)_3As_2O_8.2H_2O$. Triclinic. In small crystals; often in druses and spherical aggregates. Lamellar twinning. One cleavage perfect. H. = 3·5. G. = 3·5–3·6. Easily fusible. Color light to dark rose-red. Optically +. $X \perp$ cleavage. $\beta = 1·725$. Strong dispersion, $\rho < v$. From Schneeberg, Saxony.

Brandtite. — $Ca_2MnAs_2O_8.2H_2O$. Monoclinic. In prismatic crystals; crystals often united in radiated groups. Cleavage (010). H. = 5·5. G. = 3·67. Fusible. Colorless to white. Optically +. Ax. pl. ⊥ (010). $\alpha = 1·709$, $\beta = 1·711$, $\gamma = 1·724$. From the Harstig mine, Pajsberg, near Persberg, Vermland, Sweden.

Fairfieldite. — A hydrous phosphate of calcium and manganese, $Ca_2MnP_2O_8.2H_2O$. Triclinic. In prismatic crystals; usually in foliated or fibrous crystalline aggregates. Cleavages: (010) perfect, (100) good. H. = 3·5. G. = 3·1. Fusible. Color white or greenish white to pale straw-yellow. Optically +. $\alpha = 1·636$, $\beta = 1·644$, $\gamma = 1·654$. 2V large. Dispersion, $\rho > v$. From Branchville, Fairfield Co., Connecticut; Rabenstein, near Zwiesel, Bavaria (*leucomanganite*). Recently found in Maine at Buckfield, Oxford Co., and at Poland, Androscoggin Co.

Messelite. — $(Ca,Fe)_3P_2O_8.2\frac{1}{2}H_2O$. Triclinic. In minute tabular crystals. One good cleavage. H. = 3–3·5. G. = 3. Colorless to brownish. Optically +. $\alpha = 1·644$, $\beta = 1·653$, $\gamma = 1·680$. From near Messel near Darmstadt, in Hesse, Germany. Perhaps an alteration of *anapaite* through loss of water.

Collinsite. — $2CaO.(Mg,Fe)O.P_2O_5.2\frac{1}{2}H_2O$. Triclinic. Four cleavages. H. = 3–3·5. G. = 2·95. Fusible at 3. Easily soluble in acids. Color light brown. Silky luster. Optically +. 2V = 80°. $\alpha = 1·632$, $\beta = 1·642$, $\gamma = 1·657$. Found as nodules associated with quercyite in a vein associated with asphalt and brecciated andesite on north side of Francois Lake, British Columbia.

Akrochordite. — Probably $Mn_4Mg(AsO_5)_2.6H_2O$. Monoclinic. In spherical aggregates of minute, nearly parallel crystals. H. = 4·5. G. = 3·2. Color red-brown with yellow tint. Optically +. Ax. pl. ⊥ (010). X = b axis. $Y \wedge c$ axis = 40°–45°. $\alpha = 1·672$, $\beta = 1·676$, $\gamma = 1·683$. 2V medium large.¶ Dispersion, $\rho < v$. Found in small amount with pyrochroite and barite at Långbanshyttan, Vermland, Sweden.

Anapaite. *Tamanite.* — $(Ca,Fe)_3P_2O_8.4H_2O$. Triclinic. In tabular crystals. One perfect cleavage. H. = 3·5. G. = 2·8. Color greenish white. Optically +. $\alpha = 1·602$, $\beta = 1·613$, $\gamma = 1·649$. From the limonite mines near Anapa on the Taman peninsula, Black Sea, Russia.

WENZELITE. Wentzelite. $(Mn,Fe,Mg)_3(PO_4)_2.5H_2O$? Monoclinic. In rosettes of small crystals. Pale rose-red color. Optically −. Ax. pl. ⊥ (010). X = b axis. $\beta = 1·655$. Occurs in cavities in dufrenite in pegmatite at Hagendorf, Bavaria. *Baldaufite* is isomorphous with wenzelite with greater amount of iron. Similar physical properties. From same locality.

Reddingite. — $Mn_3P_2O_8.3H_2O$. In orthorhombic crystals near scorodite in angle; also granular. Cleavage (010). H. = 3·5. G. = 3·102. Easily fusible. Color pinkish white to yellowish white. Optically +. Ax. pl. ‖ (010). Z = c axis. Indices variable with iron content, 1·643–1·683. Strong dispersion, $\rho > v$. From Branchville, in the town of Redding, Fairfield Co., Connecticut; also from Maine at Buckfield, Oxford Co., and at Poland, Androscoggin Co.

PHOSPHOFERRITE. An iron-rich variety of reddingite. Columnar. Cleavage (010) imperfect. White to yellow or pale green. H. = 4–5. G. = 3·16. Optically +. Ax. pl. ‖ (010). Z = c axis. From Hagendorf, near Pleystein, Oberpfalz, Bavaria.

Palaite. — Hydrous manganese phosphate, $5MnO.2P_2O_5.4H_2O$. Monoclinic? In crystalline masses. G. = 3·2. Easily fusible. Color, flesh-pink. Optically −. $\alpha = 1·652$, $\beta = 1·656$, $\gamma = 1·660$. 2V large. From Pala, San Diego Co., California. Derived by alteration from *lithiophilite* and alters into *hureaulite*.

Stewartite. — Hydrous manganese phosphate, $3MnO.P_2O_5.4H_2O$. Triclinic. In fibers or minute crystals. Cleavage (010). G. = 2·94. Optically −. $\alpha = 1·63$, $\beta = 1·66$, $\gamma = 1·69$. Pleochroic, colorless to yellow. Found as an alteration product of lithiophilite from Pala, San Diego Co., California.

Picropharmacolite. — $R_3As_2O_8.6H_2O$, with R = Ca : Mg = 5 : 1. Monoclinic? Cleavages (010), (100), perfect. In small spherical forms. Soft. G. = 2·58. Color white. Optically +. Ax. pl. ‖ (010). $\alpha = 1·631$, $\beta = 1·632$, $\gamma = 1·640$. From Richelsdorf in Hessen-Nassau and from Freiberg, Saxony. Also found at Joplin, Missouri.

Trichalcite. — $Cu_3As_2O_8.5H_2O$. Orthorhombic. In plates and radiated groups, columnar; dendritic. Color verdigris-green. Optically −. X ⊥ plates. $\alpha = 1·67$, $\beta = 1·686$, $\gamma = 1·698$. 2V large. From the Turnisk copper mine near Bogoslovsk, Ural Mts.; also in Pine Creek district, Shoshone Co., Idaho.

Vivianite Group.   Monoclinic

Vivianite	$Fe_3P_2O_8.8H_2O$	$a : b : c = 0·7498 : 1 : 0·7015$	$\beta = 75° 34'$
Symplesite	$Fe_3As_2O_8.8H_2O$	$0·7806 : 1 : 0·6812$	$72° 43'$
Bobierrite	$Mg_3P_2O_8.8H_2O$		
Hœrnesite	$Mg_3As_2O_8.8H_2O$		
Erythrite	$Co_3As_2O_8.8H_2O$	$0·75 : 1 : 0·70$	$75°$
Annabergite	$Ni_3As_2O_8.8H_2O$		
Cabrerite	$(Ni,Mg)_3As_2O_8.8H_2O$		
Köttigite	$Zn_3As_2O_8.8H_2O$		

The VIVIANITE GROUP includes hydrous phosphates of iron, magnesium, cobalt, nickel and zinc, all with eight molecules of water. The crystallization is monoclinic, and the angles, so far as known, correspond closely, as also does the optical orientation.

VIVIANITE.

Monoclinic. Crystals prismatic $(mm''' \; 110 \wedge 1\bar{1}0 = 71° 58')$; often in stellate groups. Also reniform and globular; structure divergent, fibrous, or earthy; also incrusting.

Cleavage: $b(010)$ highly perfect; $a(100)$ in traces; also fracture fibrous nearly $\perp c$ axis. Flexible in thin laminæ; sectile. H. $= 1\cdot5$–2. G. $= 2\cdot58$ –2$\cdot$68. Luster, $b(010)$ pearly or metallic pearly; other faces vitreous. Colorless when unaltered, blue to green, deepening on exposure. This change in color is due to a partial oxidation of ferrous to ferric iron on exposure to light. This change is also accompanied by a definite increase in the values for β and γ and a strong pleochroism. Streak colorless to bluish white, changing to indigo-blue and to liver-brown. Transparent to translucent; opaque after exposure. Pleochroism strong; $X = $ cobalt-blue, Y and $Z = $ pale greenish yellow. Optically $+$. Ax. pl. $\perp$ (010). $X = b$ axis. $Z \wedge c$ axis $= +28°$. $\alpha = 1\cdot580$. $\beta = 1\cdot598$. $\gamma = 1\cdot627$.

Comp. — Hydrous ferrous phosphate, $Fe_3P_2O_8.8H_2O = $ Phosphorus pentoxide 28$\cdot$3, iron protoxide 43$\cdot$0, water 28$\cdot$7 $= 100$.

Many analyses show the presence of iron sesquioxide due to alteration.

Pyr., etc. — In the closed tube yields neutral water, whitens, and exfoliates. B.B. fuses at 1$\cdot$5, coloring the flame bluish green, to a grayish black magnetic globule. With the fluxes reacts for iron. Soluble in hydrochloric acid.

Obs. — Occurs associated with pyrrhotite and pyrite in copper and tin veins; sometimes in narrow veins with gold, traversing graywacke; both friable and crystallized in beds of clay, and sometimes associated with limonite, or bog iron ore; often in cavities of fossils or buried bones.

From O-Rodna in Transylvania of Rumania; from Bodenmais in Bavaria in crystals. In France found in fine crystals where coal measures have been on fire, as at Commentry in Allier and at Cransac in Aveyron. At St Agnes in Cornwall large transparent indigo-blue crystals occur in pyrrhotite; also from other Cornish localities. A variety from the Taman and Kerch peninsulas on the Black Sea, Russia, that contains small quantities of manganese and magnesium, has been called *paravivianite*. From the Wannon River, Victoria, Australia. Found on Fernando Po island in the Bight of Biafra on the west coast of Africa. In Bolivia from Llallogua and Tazna. The earthy variety, sometimes called *blue iron-earth* or *native Prussian blue* (Fer azuré), occurs in Greenland, Carinthia, Guatemala, Bolivia, Victoria, etc.

In the United States vivianite is found in New Jersey at Allentown, and Shrewsburg, Monmouth Co.; at Mullica Hill, Gloucester Co. (*mullicite*). In Delaware at Middletown, New Castle Co., it occurs in the green sand in fine large crystals. In good crystals from Leadville, Lake Co., Colorado.

Symplesite. — Probably $Fe_3As_2O_8.8H_2O$. Monoclinic. In small prismatic crystals and in radiated spherical aggregates. Cleavage (010), perfect. H. $= 2\cdot5$. G. $= 2\cdot96$. Infusible. Color pale indigo, inclined to celandine-green. Optically $-$. Ax. pl. $\perp$ (010). $Z \wedge c$ axis $= +32°$. $\alpha = 1\cdot635$, $\beta = 1\cdot668$, $\gamma = 1\cdot702$. $2V = 87°$. Dispersion, $\rho > \upsilon$. Pleochroic; $X = $ deep blue, $Y = $ almost colorless, $Z = $ yellowish. Occurs at Felsöbánya (Baia Sprie), Rumania. From Lölling near Hüttenberg in Carinthia, Austria; from Lobenstein in Thuringia.

Ferrisymplesite. A hydrous arsenate of ferric iron, $3Fe_2O_3.2As_2O_5.16H_2O$. Fibrous. G. $= 2\cdot88$. Color deep amber-brown. Resinous luster. $n = 1\cdot650$. Birefringence strong. Found associated with erythrite and annabergite in the upper workings of the Hudson Bay mine, Cobalt, Ontario.

Bobierrite. — $Mg_3P_2O_8.8H_2O$. Monoclinic. In aggregates of minute crystals; also massive. Cleavage (010). H. $= 2$. G. $= 2\cdot4$. Fusible. Colorless to white. Optically $+$. Ax. pl. $\perp$ (010). $Z \wedge c$ axis $= +29°$. $\alpha = 1\cdot510$, $\beta = 1\cdot520$, $\gamma = 1\cdot543$. $2V = 71°$. From the guano of Mejillones in Tarapaca, Chile. *Hautefeuillite* is like bobierrite, but contains calcium. Monoclinic. $\beta = 1\cdot52$. From Bamble, Telemark, Norway.

Hoernesite. — $Mg_3As_2O_8.8H_2O$. Monoclinic. Perfect cleavage $\parallel$ (010). In crystals resembling gypsum; also columnar; stellar-foliated. H. = 1. G. = 2·6. Easily fusible. Color snow-white. Optically +. Ax. pl. $\perp$ (010). $Z \wedge c$ axis = +31°. α = 1·563, β = 1·571, γ = 1·596. From Rumania in the Banat and from Nagy-Ág in Hunyad; from Joachimstal, in Bohemia, Czechoslovakia. Also from Fiano, Italy.

ERYTHRITE. Cobalt bloom.

Monoclinic. Crystals prismatic and vertically striated. Also in globular and reniform shapes, having a drusy surface and a columnar structure; sometimes stellate. Also pulverulent and earthy, incrusting.

Cleavage: b(010) highly perfect. Sectile. H. = 1·5–2·5; least on b. G. = 2·95. Luster of b pearly; other faces adamantine to vitreous; also dull, earthy. Color crimson- and peach-red, sometimes gray. Streak a little paler than the color. Transparent to subtranslucent. Optically − or +. Ax. pl. $\perp$ (010). $Z \wedge c$ axis = +32°. α = 1·629. β = 1·663. γ = 1·701. 2V about 90°. Pleochroic; X = pale pink, Y = pale violet, Z = red.

Comp. — Hydrous cobalt arsenate, $Co_3As_2O_8.8H_2O$ = Arsenic pentoxide 38·4, cobalt protoxide 37·5, water 24·1 = 100. The cobalt is sometimes replaced by nickel, iron, and calcium.

Pyr., etc. — In the closed tube yields water at a gentle heat and turns bluish; at a higher heat gives off arsenic trioxide which condenses in crystals on the cool glass, and the residue has a dark gray or black color. B.B. in the forceps fuses at 2 to a gray bead, and colors the flame light blue (arsenic). B.B. on charcoal gives an arsenical odor, and fuses to a dark gray arsenide, which with borax gives the deep blue color characteristic of cobalt. Soluble in hydrochloric acid, giving a rose-red solution.

Obs. — Erythrite is a mineral of secondary origin, found commonly in the upper portions of cobalt mineral deposits. From Dognacska in Rumania and Joachimstal in Bohemia, Czechoslovakia. Occurs at Schneeberg, Saxony, in micaceous stellate aggregates; in Thuringia at Saalfield and at Glücksbrunn near Schweina. From Wittichen in the Black Forest, Baden; also from Richelsdorf in Hessen-Nassau. Earthy, peach-blossom-colored varieties occur at Chalantes, near Allemont, Isère, France. From Tunaberg in Södermanland, Sweden. At various localities in Cornwall and at Alston Moor in Cumberland. From several places in Chile. It is of rare occurrence in the United States. Reported from Lovelocks, Humboldt Co., Nevada, and at several points in California. With other cobalt minerals from Cobalt, Ontario. Named from ἐρυθρός, red.

Annabergite. — $Ni_3As_2O_8.8H_2O$. Monoclinic. In capillary crystals; also massive and disseminated. Cleavage (010), perfect. H. = 2·5–3. G. = 3. Fusible. Color fine apple-green. Optically −. Ax. pl. $\perp$ (010). $Z \wedge c$ axis = +35°. α = 1·622, β = 1·658, γ = 1·687. 2V = 84°. Dispersion, $\rho > v$. Annabergite is of secondary origin, occurring in the alteration zone near the surface of nickel deposits. From Laurium in Greece; in Saxony at Annaberg and Schneeberg; in Hessen-Nassau at Richelsdorf. Occurs on smaltite at Chalantes near Allemont, Isère, France. Rarely found in the United States. Occurs in the Cobalt district, Ontario.

Cabrerite. — $(Ni,Mg)_3As_2O_8.8H_2O$. May be considered as a magnesian variety of annabergite. Monoclinic. Like erythrite in habit. Also fibrous, radiated; reniform, granular. Cleavage (010), perfect. Color apple-green. Optically −. Ax. pl. $\perp$ (010). $Z \wedge c$ axis = 33°. α = 1·62, β = 1·654, γ = 1·689. 2V nearly 90°. From the Sierra Cabrera, Spain; at Laurium, Greece.

KOLOVRATITE. A vanadate of nickel of uncertain composition. Color yellow to greenish yellow. Found as thin botryoidal crusts on slates in Ferghana, Russian Turkestan.

Köttigite. — $Zn_3As_2O_8.8H_2O$. Monoclinic. Cleavage (010), perfect. Massive, or in crusts. H. = 2·5–3. G. = 3·1. Color light carmine- and peach-blossom-red. Optically +. Ax. pl. $\perp$ (010). $Z \wedge c$ axis = +37°. α = 1·662, β = 1·683, γ = 1·717. 2V = 77°. Dispersion, $\rho < v$. Occurs with smaltite at the cobalt mine Daniel, near Schneeberg, Germany.

Rhabdophanite. Scovillite. — A hydrous phosphate of the cerium and yttrium metals. Massive, small mammillary; as an incrustation. H. = 3·5. G. = 4. Color brown, pinkish or yellowish white. Uniaxial, +. ω = 1·654, ϵ = 1·703. Fibers show positive elongation. *Rhabdophanite* is from Cornwall; *Scovillite* is from the Scoville (limonite) ore bed in Salisbury, Connecticut.

WEINSCHENKITE. $(Y,Er)PO_4.2H_2O$. In globular masses and radiated needles. Closely resembles wavellite but said to be optically distinct. White. Infusible. Soluble in dilute acids. Occurs coating limonite ores in the Amberg-Auerbach mine, Bavaria. An associated mineral with much smaller amounts of the rare earths has been called *pseudo-wavellite*.

Churchite. — $(Ce,Ca)PO_4.2H_2O$. Orthorhombic? As a thin coating of minute tabular crystals. One perfect cleavage. H. = 3–3·5. G. = 3·14. Color pale smoke-gray tinged with flesh-red. Optically +. $Z \perp$ plates. $\beta = 1.62$. 2V very small. From Cornwall, England.

SCORODITE.

Orthorhombic. Axes $a : b : c = 0.8658 : 1 : 0.9541$.

dd', 120 $\wedge$ $\bar{1}20$ = 60° 1'. pp'', 111 $\wedge$ $\bar{1}\bar{1}1$ = 111° 6'.
pp', 111 $\wedge$ $\bar{1}11$ = 77° 8'. pp''', 111 $\wedge$ $1\bar{1}1$ = 65° 20'.

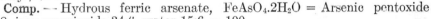

1026

Habit octahedral, also prismatic. Also earthy, amorphous. Cleavage: $d(120)$ imperfect; $a(100)$, $b(010)$ in traces. Fracture uneven. Brittle. H. = 3·5–4. G. = 3·1–3·3. Luster vitreous to subadamantine and subresinous. Color pale leek-green or liver-brown. Streak white. Subtransparent to translucent. Usually optically +, some varieties −. Ax. pl. || (100). $X = b$ axis. Indices variable, 1·74–1·92. Dispersion strong, usually $\rho > v$.

Comp. — Hydrous ferric arsenate, $FeAsO_4.2H_2O$ = Arsenic pentoxide 49·8, iron sesquioxide 34·6, water 15·6 = 100.

Pyr., etc. — In the closed tube yields neutral water and turns yellow. B.B. fuses easily, coloring the flame blue. B.B. on charcoal gives arsenical fumes, and with soda a black magnetic scoria. With the fluxes reacts for iron. Soluble in hydrochloric acid.

Obs. — Scorodite is an alteration product of arsenic minerals, especially arsenopyrite. It is also deposited from certain hot springs, as in Yellowstone Park, Wyoming. From the Adun-Chilon Mts., south of Nerchinsk, Transbaikalia, in fine crystals and also as an amorphous crust. From Laurium, Greece, and in Carinthia, Austria, at Lölling near Hüttenberg. Found in the granitic mountains on the Graul near Schwarzenberg, Saxony. Found at Dernbach northwest of Montabaur in Hessen-Nassau. From Vaulry, northwest of Limoges in Haute-Vienne, France. From various localities in Cornwall. From Djebel Debar, Constantine, Algeria. Occurs at Antonio Pereira, near Ouro Preto, Minas Geraes, Brazil. Found at Kiura, Bungo, Japan. In the United States scorodite occurs in Utah at the Mammoth mine near Eureka, Tintic district, and at Gold Hill, Tooele Co.; and as noted above in the Yellowstone Park, Wyoming.

Named from σκόροδον, *garlic*, alluding to the odor before the blowpipe.

Strengite. — $FePO_4.2H_2O$. Orthorhombic. Crystals rare; in habit and angle near scorodite; generally in spherical and botryoidal forms. Cleavages: (001) perfect, (100) imperfect. H. = 3–4. G. = 2·87. Fusible. Color pale red. Usually optically +, some varieties −. Ax. pl. || (100). $Z = c$ axis. Indices variable, 1·70–1·76. 2V varies greatly. Dispersion strong but variable. Occurs at the Eleonore iron mine on the Dünsberg, near Giessen in Hessen and at the Rothläufchen mine, near Waldgirmes in the same district. From the Kreuzberg, near Pleystein, Oberpfalz, Bavaria. From La Vilate near Chanteloube, Haute-Vienne, France. Occurs at Kiruna, Norbotten, Sweden. In the United States in distinct crystals with dufrenite in Rockbridge Co., Virginia. Noted from Pala, San Diego Co., California. *Globosite* from Ullersreuth near Hirschberg, Thuringia, and from Schneeberg, Saxony, may be identical with strengite.

Eggonite. — Hydrous aluminum phosphate, probably $AlPO_4.2H_2O$, because of crystallographic similarity to strengite. Orthorhombic. Cleavage || (100). Optically −. Ax. pl. || (010). $X \perp$ (100). $\beta = 1.59$. Originally found artificially attached to specimens of zinc ore from Altenberg, Rhineland. Later found in natural ocurrence on ores from Felsöbánya (Baia Sprie), Rumania.

Vilateite. — Hydrous iron phosphate with a little manganese. Perhaps a manganese strengite. Monoclinic. H. = 3–4. G. = 2·75. Color violet. Index, 1·74. Found in pegmatite at La Vilate near Chanteloube, Haute-Vienne, France.

Barrandite. — $(Al,Fe)PO_4.2H_2O$. Orthorhombic. Fibrous, in spheroidal concretions. Color pale shades of gray. H. = 4·5. G. = 2·6. Optically +. $Z \parallel c$ axis. β = 1·65–1·67. Intermediate in properties between strengite and variscite. Occurs at Cerhovic, southwest of Beraun in Bohemia of Czechoslovakia. From the island of Connétables in French Guiana. In the United States from Manhattan, Nye Co., Nevada.

Variscite. — $AlPO_4.2H_2O$. Orthorhombic. Octahedral habit. Commonly in crystalline aggregates and incrustations with reniform surface. H. = 4. G. = 2·5. Infusible. Color green. Optically –. β = 1·560–1·588. 2V moderately large. Positive elongation. From Messbach in Saxon Vogtland; Montgomery Co., Arkansas, on quartz; in nodular masses from Tooele Co., Utah (*utahlite*); crystallized from Lucin, Utah. *Peganite* from Striegis, near Freiberg, Saxony, and *lucinite* from Lucin, Utah, have been shown to be identical with variscite. *Bolivarite* from near Pontevedra, Spain, is probably also variscite.

Metavariscite. — $AlPO_4.2H_2O$. Orthorhombic. Tabular or prismatic. H. = 4. G. = 2·54. Optically +. Ax. pl. $\parallel$ (010). $X = c$ axis. β = 1·558. 2V = 55°. $\rho < v$. Found at Lucin, Utah.

Meyersite. A colloidal material, with composition $AlPO_4.2H_2O$, found on Necker Island (near the Hawaiian Islands) in the cavities of a lava flow, associated with guano and limestone.

Purpurite. — $2(Fe,Mn)PO_4.H_2O$. Orthorhombic. In small irregular masses. Cleavages: (001) perfect, (010) imperfect. H. = 4–4·5. G. = 3·4. Easily fusible. Color deep red or reddish purple. Optically +. Ax. pl. $\parallel$ (100). $Z = b$ axis. Indices, 1·85–1·92. 2V = 38°. Dispersion very strong. Pleochroic; X = gray, $Y = Z$ = deep blood-red. Found at several localities in Haute-Vienne, France. At Hagendorf near Pleystein, Bavaria. In the United States at Kings Mt., Gaston Co., North Carolina; sparingly from Pala and Rincon, San Diego Co., California; Hill City, Pennington Co., South Dakota, and Branchville, Fairfield Co., Connecticut.

Phosphosiderite. — $4FePO_4.7H_2O$. Monoclinic. Prismatic. Cleavages, (010) perfect, (001) distinct. H. = 3·5–4. G. = 2·76. Easily fusible. Color red. Optically –. Ax. pl. $\parallel$ (010). $X \wedge c$ axis = 3°–4°. α = 1·692, β = 1·725, γ = 1·738. 2V = 62°. Strong dispersion, $\rho > v$. In iron ore from Eiserfeld, southwest of Siegen, Westphalia; from the Kreuzberg near Pleystein, Oberpfalz, Bavaria. In crystals at San Giovanneddu near Gonnesa, Sardinia.

Callainite. — $AlPO_4.2\frac{1}{2}H_2O$. Massive; wax-like. Color apple- to emerald-green. From a Celtic grave in Lockmariaquer, Brittany.

Zepharovichite. $AlPO_4.3H_2O$. Perhaps impure wavellite. Cryptocrystalline fibrous. H. = 5·5. G. = 2·37. Infusible. Color yellowish or grayish white. β = 1·55. From Třenic near Cerhovic southwest of Beraun in Bohemia of Czechoslovakia.

Palmerite. $HK_2Al_2(PO_4)_3.7H_2O$. Amorphous, pulverulent. Color white. Occurs as a stratum in a guano deposit on Mte. Alburno, Salerno, Italy.

Rosiérésite. — A hydrous phosphate of aluminum with lead and copper. Amorphous. In stalactites. G. = 2·2. Infusible. Color yellow to brown. Isotropic. n = 1·5. Found in an abandoned copper mine at Rosiéres, near Carmaux, Tarn, France.

Koninckite. — $FePO_4.3H_2O$. In small spherical aggregates of radiating needles. H. = 3. G. = 2·4. Fusible. Color yellow. Shows variously biaxial and isotropic characters, n = 1·58–1·65. From Richelle, near Visé in Liége, Belgium.

Sicklerite. — A hydrous iron-manganese phosphate with lithia. $Fe_2O_3.6MnO.4P_2O_5.3(Li,H)_2O$. Orthorhombic? In cleavable masses. H. = 4. G. = 3·45. Easily fusible. Color dark brown. Optically –. $Z \perp$ to good cleavage. α = 1·715, β = 1·735, γ = 1·745. 2V large. Strong dispersion, $\rho > v$. Strongly pleochroic, X = deep red, Y = paler red, Z = very pale red. Crystals from Tintic district, Utah, are nearly uniaxial, optically +, indices 1·74–1·83, non-pleochroic. From Pala, San Diego Co., California.

Salmonsite. — A hydrous iron-manganese phosphate, $Fe_2O_3.9MnO.4P_2O_5.14H_2O$. Orthorhombic with 2 pinacoidal cleavages. Cleavable fibrous masses. H. = 4. G. = 2·88. Color buff. Optically +. $Z \parallel$ length of fibers. α = 1·655, β = 1·66, γ = 1·67. 2V large. Strong dispersion, $\rho < v$. Pleochroic; X = almost colorless, Y = yellow, Z = orange-yellow. Found at Pala, San Diego Co., California, as alteration of *hureaulite*.

Landesite. — $3Fe_2O_3.20MnO.8P_2O_5.27H_2O$. In rough octahedral-like crystals with one good cleavage and a poor cleavage at right angles to first. Biaxial, –. $Z \perp$ to best cleavage and $X \perp$ to poorer cleavage. α = 1·720, β = 1·728, γ = 1·735. 2V large. Pleochroic; X = dark brown, Y = light brown, Z = yellow. Found at the Berry quarry, Poland, Androscoggin Co., Maine, as an alteration of reddingite.

Acid Hydrous Phosphates, etc.

Haidingerite. — $HCaAsO_4.H_2O$. Orthorhombic. Cleavage (010), perfect. In minute crystal aggregates, botryoidal and drusy. H. = 2. G. = 2·85. Color white. Optically +. Ax. pl. || (100). $Z = c$ axis. $\alpha = 1·590$, $\beta = 1·602$, $\gamma = 1·638$. 2V = 58°. From Joachimstal, Bohemia, Czechoslovakia, with pharmacolite. Also reported from Wittichen near Wolfach, Black Forest, Baden.

PHARMACOLITE.

Monoclinic. Crystals rare. Commonly in delicate silky fibers; also botryoidal, stalactitic.

Cleavage: $b(010)$ perfect. Fracture uneven. Flexible in thin laminæ. H. = 2–2·5. G. = 2·64–2·73. Luster vitreous; on $b(010)$ inclining to pearly. Color white or grayish; frequently tinged red. Streak white. Translucent to opaque. Optically −. Ax. pl. ⊥ (010). $X \wedge c$ axis = +70°. $\alpha = 1·583$. $\beta = 1·589$. $\gamma = 1·594$. 2V = 77°.

Comp. — Probably $HCaAsO_4.2H_2O$ = Arsenic pentoxide 53·3, lime 25·9, water 20·8 = 100.

Obs. — Pharmacolite is a product of late alteration of mineral deposits carrying principally arsenopyrite and the arsenical ores of cobalt and silver. Occurs at Joachimstal in Bohemia, Czechoslovakia; from Schneeberg in Saxony; Andreasberg in the Harz Mts.; Richelsdorf in Hessen-Nassau; in crystals at Wittichen in the Black Forest, Baden; from Glücksbrunn near Schweina in Thuringia. In France it occurs at Ste. Marie aux Mines (Markirch) in Alsace, in botryoidal or globular groups. Named from φάρμακον, *poison*.

WAPPLERITE. $HCaAsO_4.3\frac{1}{2}H_2O$. Monoclinic or Triclinic. In minute crystals; also in incrustations. Cleavage (010), perfect. H. = 2–2·5. G. = 2·48. Colorless to white. Optically +. Z nearly || b axis. $\alpha = 1·525$, $\beta = 1·53$, $\gamma = 1·55$. Found with pharmacolite at Joachimstal, Bohemia, Czechoslovakia. Also from Schneeberg, Saxony.

Brushite. — $HCaPO_4.2H_2O$. In small slender monoclinic prisms; concretionary massive. Cleavages (010), (301) perfect. H. = 2. G. = 2·2. Easily fusible. Colorless to pale yellowish. Optically +. Ax. pl. ⊥ (010). $Z = b$ axis. $\alpha = 1·539$, $\beta = 1·545$, $\gamma = 1·551$. Occurs on the rock guano of Los Aves islands and the island of Sombrero in the West Indies. *Metabrushite* is apparently identical with brushite. *Stoffertite* is a mineral similar to brushite but said to contain a little more water. From guano deposits on the island of Mona, West Indies.

Martinite. — $H_2Ca_5(PO_4)_4.\frac{1}{2}H_2O$. Monoclinic? In rhombic shaped plates. G. = 2·9. Infusible. Colorless. Optically +. $\beta = 1·60$. From phosphorite deposits (from guano) on the island of Curacoa, West Indies. Also reported from Ascension Island, South Atlantic.

Hewettite. — $CaO.3V_2O_5.9H_2O$. Orthorhombic. In microscopic needles. G. = 2·5–2·6. Easily fusible. Color deep red. On heating loses water changing color through shades of brown to a bronze. Optically −. Z || elongation. $\alpha = 1·77$, $\beta = 2·18$, $\gamma = 2·35–2·4$. Pleochroic; Z = deep red, $X = Y$ = light orange. Found as an alteration of *patronite* at Minasragra, near Cerro de Pasco, Peru. Also observed from Paradox Valley, Montrose Co., Colorado.

Metahewettite. — Comp. same as for *hewettite*. In minute tabular orthorhombic crystals. On heating loses water changing from dark red to yellow-brown. G. = 2·5. Easily fusible. Optically −. X ⊥ to crystal tables, Z || elongation. $\alpha = 1·70$, $\beta = 2·10$, $\gamma = 2·23$. Pleochroic; X = light orange, Y = red, Z = deep red. Occurs as an impregnation in sandstone in southwestern Colorado and southeastern Utah; as at Paradox valley, Montrose Co., Colorado, and at Thompson's, Grand Co., Utah, etc.

Fervanite. — $2Fe_2O_3.2V_2O_5.5H_2O$. Probably monoclinic. Fibrous. Color, golden-brown. Brilliant luster. Biaxial, −. $\alpha = 2·186$, $\beta = 2·222$, $\gamma = 2·224$. 2V very small. Found in the carnotite district in Colorado and Utah.

Rossite. — $CaO.V_2O_5.4H_2O$. Triclinic. Prismatic habit. H. = 2–3. G. = 2·45. Yellow. Luster pearly to vitreous. 2V large. $\alpha = 1·710$, $\beta = 1·770$, $\gamma = 1·840$. Strong dispersion. Found as small glassy kernels embedded in metarossite at Bull Pen Canyon, San Miguel Co., Colorado.

Metarossite. — $CaO.V_2O_5.2H_2O$. A dehydration product of rossite. Yellow color. Luster pearly to dull. Biaxial. 2V large. Strong dispersion. $\alpha = 1·840$, β and γ greater than 1·85. Soft and friable. Found as small veinlets in sandstones at Bull Pen Canyon, San Miguel Co., Colorado.

Fernandinite. — $CaO.V_2O_4.5V_2O_5.14H_2O$. Massive. Fibrous. Color dull green. $n = 2·05$. Readily soluble in acids, partly soluble in water. Found at Minasragra, near Cerro de Pasco, Peru.

Melanovanadite. — $2CaO.3V_2O_5.2V_2O_4$. The mineral is very hydroscopic, taking up about 16·6 per cent H_2O on standing in a moist atmosphere. It is very possible that the naturally occurring mineral contained originally this amount of water. Monoclinic. In needle-like prismatic crystals. Color black, dark brown in very thin sections under the microscope. Streak very dark reddish brown. $H. = 2·5$. $G. = 3·477$. Easily fusible. Soluble in acids. Cleavage perfect $\parallel$ (010). Optically $-$. Ax. pl. $\perp$ (010). $Z = b$ axis. On (010) shows inclined extinction of 15°. $\alpha = 1·73$, $\beta = 1·96$, $\gamma = 1·98$. 2V small. Pleochroic; X = yellow-brown, $Y = Z$ = dark red-brown. From Minasragra, near Cerro de Pasco, Peru, associated with pascoite and a vanadium sulphide, perhaps patronite.

Pascoite. — Hydrous calcium vanadate, possibly $2CaO.3V_2O_5.11H_2O$. Monoclinic. Cleavage (010), poor. In grains. $H. = 2·5$. $G. = 2·46$. Easily fusible. Soluble in water. Color orange. Streak yellow. Optically $-$. Ax. pl. $\perp$ (010). $X = b$ axis. $\alpha = 1·775$, $\beta = 1·815$, $\gamma = 1·825$. Strong crossed dispersion. Found at Minasragra, near Cerro de Pasco, Peru.

Pintadoite. — Hydrous calcium vanadate, $2CaO.V_2O_5.9H_2O$. As an efflorescence. Color green. Found coating surfaces of sandstone in Canyon Pintado, San Juan Co., Utah.

Newberyite. — $HMgPO_4.3H_2O$. In white orthorhombic crystals. Cleavages: (010), perfect; (001), good. $H. = 3$. $G. = 2·1$. Optically $+$. Ax. pl. $\parallel$ (010). $Z = c$ axis. $\beta = 1·518$. From guano of Skipton Caves, southwest of Ballarat, Victoria, Australia; also from the guano of Mejillones, Tarapaca, Chile, and from Ascension Island, South Atlantic.

Hannayite. — $H_4(NH_4)_2Mg_3(PO_4)_4.8H_2O$. Triclinic or monoclinic. In prismatic or tabular crystals. Several cleavages. Soft. $G. = 1·9$. White or yellowish. Optically $-$. $X \perp$ to perfect cleavage. $\alpha = 1·555$, $\beta = 1·572$, $\gamma = 1·575$. 2V = 42°. Occurs with struvite and newberyite in the bat guano of the Skipton Caves, southwest of Ballarat, Victoria, Australia. **Schertelite**, $H_2(NH_4)_2Mg(PO_4)_2.4H_2O$, is a similar mineral occurring in small tabular crystals from the same locality.

Stercorite. Microcosmic salt. — $HNa(NH_4)PO_4.4H_2O$. Monoclinic? In white crystalline masses and nodules. $H. = 2$. $G. = 1·57$. Very easily fusible. Optically $+$. Ax. pl. $\perp$ (010). $Z = b$ axis. $Y \wedge c$ axis = 30°. $\alpha = 1·439$, $\beta = 1·441$, $\gamma = 1·469$. 2V = 36°. Dispersion, $\rho > v$. From guano on the island of Ichaboe, South West Africa, and on the Guanape Islands off the coast of Peru.

Hureaulite. — $H_2Mn_5(PO_4)_4.4H_2O$. In short prismatic crystals (monoclinic). Also massive, compact, or imperfectly fibrous. Cleavage (100) distinct. $H. = 5$. $G. = 3·18$. Easily fusible. Color yellowish, orange-red, rose, grayish. Optically $-$. Ax. pl. $\perp$ (010). $Z \wedge c$ axis = $+75°$. $\alpha = 1·647$, $\beta = 1·654$, $\gamma = 1·660$. 2V = 74°. Strong dispersion, $\rho < v$. From Hureaux in Sylvestre or at Vilate near Chanteloube, both north of Limoges, Haute Vienne, France. In the United States in Connecticut at Branchville, Fairfield Co., and reported from Portland, Middlesex Co. Also from Pala, San Diego Co., California.

Forbesite. — $H_2(Ni,Co)_2As_2O_8.8H_2O$. Structure fibro-crystalline. Color grayish white. Occurs in the desert of Atacama, Chile, in veins in a decomposed diorite. Appears to have been derived from chloanthite.

Ferrazite. $3(Ba,Pb)O.2P_2O_5.8H_2O$. A "fava" found in the diamond sands of Diamantina, Minas Geraes, Brazil. Color dark yellowish white. $G. = 3·0-3·3$.

Basic Hydrous Phosphates, etc.

Isoclasite. — $Ca_3P_2O_8.Ca(OH)_2.4H_2O$. Monoclinic. In minute white crystals; also columnar. Cleavage, (010), perfect. $H. = 1·5$. $G. = 2·9$. Fusible. Optically $+$. Ax. pl. $\perp$ (010). Z nearly $\parallel c$ axis. $\alpha = 1·565$, $\beta = 1·568$, $\gamma = 1·580$. 2V = 50°. From Joachimstal, Bohemia, Czechoslovakia.

Hemafibrite. — $Mn_3As_2O_8.3Mn(OH)_2.2H_2O$. Orthorhombic. Cleavage (010). Commonly in spherical radiated groups. $H. = 3$. $G. = 3·6$. Easily fusible. Color brownish red to garnet-red, becoming black. Optically $+$. Ax. pl. $\parallel$ (100). $Z = c$ axis. $\alpha = 1·87$, $\beta = 1·88$, $\gamma = 1·93$. From the Moss mine, Nordmark, Vermland, Sweden.

EUCHROITE.

Orthorhombic. Habit prismatic mm''' 110 $\wedge$ 1$\bar{1}$0 = 62° 40′. Cleavage: $m(110)$, $n(011)$ in traces. Fracture small conchoidal to uneven. Rather brittle. H. = 3·5–4. G. = 3·39. Easily fusible. Luster vitreous. Color bright emerald- or leek-green. Transparent to translucent. Optically +. Ax. pl. || (100). $Z = c$ axis. $\alpha = 1·695$, $\beta = 1·698$, $\gamma = 1·733$. 2V = 29°. Dispersion, $\rho > v$.

Comp. — $Cu_3As_2O_8.Cu(OH)_2.6H_2O$ = Arsenic pentoxide 34·2, cupric oxide 47·1, water 18·7 = 100.

Obs. — Occurs in quartzose mica slate at Libethen (Libetbánya) near Neusohl (Ban Bystrica) in Slovakia of Czechoslovakia, in crystals of considerable size, having much resemblance to dioptase. Named from ευχροα, *beautiful color*.

Conichalcite. — Perhaps $(Cu,Ca)_3As_2O_8.(Cu,Ca)(OH)_2.\frac{1}{2}H_2O$. Orthorhombic. Fibrous. Usually reniform and massive, resembling malachite. H. = 4·5. G. = 4·1. Fusible. Color pistachio-green to emerald-green. Described both as optically + with small 2V and as optically − with medium to large 2V. Z || elongation. Indices variable, 1·73–1·8. Found at the copper mine of Maya-Tass, province of Akmolinsk, western Siberia, in crystals. From Hinojosa de Cordoba in Andalusia, Spain. In the United States at the American Eagle mine, Tintic district, Utah, and in crystals from Lincoln Co., Nevada.

Freirinite. — $6(Cu,Ca)O.3Na_2O.2As_2O_5.6H_2O$. Probably tetragonal. Composed of fine flakes. Cleavages: basal, good; prismatic, imperfect. Uniaxial, −. $\omega = 1·748$, $\epsilon = 1·645$. Easily soluble in HCl. Fuses with intumescence. This is the mineral that was originally described as lavendulan, from San Juan, department of Freirini, Chile.

Bayldonite. — $(Pb,Cu)_3As_2O_8.(Pb,Cu)(OH)_2.H_2O$. Monoclinic? Fibrous. In mammillary concretions, drusy. H. = 4·5. G. = 4·35. Fusible. Color green. Optically +. Ax. pl. ⊥ (010). $Z \wedge c$ axis = 45°. $\alpha = 1·95$, $\beta = 1·97$, $\gamma = 1·99$. 2V large. Found in Cornwall at St. Hilary near Marazion, at St. Day, etc. Also occurs at Tsumeb, near Otavi, South West Africa. A molecule similar to that of bayldonite except that it has less water and which is supposed to be present with bayldonite at Tsumeb has been called *parabayldonite*. A similar compound found at the same locality has been named *cuproplumbite*.

Tagilite. — $Cu_3P_2O_8.Cu(OH)_2.2H_2O$. Monoclinic. Fibrous. In reniform or spheroidal concretions; earthy. Cleavage (010). H. = 3–4. G. = 4·1. Fusible. Color verdigris-to emerald green. Optically −. X nearly || elongation. $\alpha = 1·69$, $\beta = 1·84$, $\gamma = 1·85$. 2V small. Occurs at Nizhne-Tagilsk in the Ural Mts., Russia. From Ullersreuth near Hirschberg, Thuringia and in Chile at the Mercedes mine, east of Coquimbo.

Leucochalcite. — Probably $Cu_3As_2O_8.Cu(OH)_2.2H_2O$. Orthorhombic. In white, silky, acicular crystals. Easily fusible. Optically +. Ax. pl. || (001). $\alpha = 1·79$, $\beta = 1·807$, $\gamma = 1·84$. Strong dispersion, $\rho < v$. From the Wilhelmine mine near Schöllkripen in the Spessart, Bavaria.

Barthite. — $3ZnO.CuO.3As_2O_5.2H_2O$. In small monoclinic (?) crystals. H. = 3. G. = 4·10. Color grass-green. Optically +. Indices variable, 1·77–1·81. 2V large. Found in druses of a dolomite at Guchab, near Otavi, South West Africa.

Volborthite. — A hydrous vanadate of copper, barium, and calcium. Monoclinic? In small six-sided tables; in globular forms. One cleavage. H. = 3. G. = 3·5. Color olive-green, citron-yellow. $n = 2·02$. 2V variable. Strong dispersion, $\rho > v$. From Sissersk and Nizhne-Tagilsk near Ekaterinburg, Russia, and from several localities in the government of Perm. Occurs at Friedricksroda, Thuringia. Has been found in the United States at Richardson, Grand Co., Utah.

Tangeite. — Tangueite. — $2CaO.2CuO.V_2O_5.H_2O$. Found as fine fibrous and radiated botryoidal masses, etc. Color dark olive-green. Biaxial, −. Crystals positively elongated and show parallel extinction. $\alpha = 2·00$, $\beta = 2·01$, $\gamma = 2·02$. 2V large. Strong dispersion, $\rho > v$. Pleochroic; $X = Y$ = brown, Z = green. From Tange Ravine, Tyuya-Muyun, Alai Mts., Ferghana, Russian Turkestan. A greenish black mineral from Turkestan, differing slightly from volborthite in composition and considered to be the colloidal equivalent of tangeite, has been called *turkestan-volborthite*.

Hügelite. — A hydrous lead-zinc vanadate. Monoclinic. In microscopic hair-like crystals. G. = 5. Color orange-yellow to yellow-brown. Optically +. $Z = b$ axis. X nearly || c axis for red and Y nearly || c axis for blue. $\beta = 1·915$. 2V widely variable for different wave-lengths of light. Abnormal interference colors. From Reichenbach near Lahr, Baden.

Cornwallite. — $Cu_3As_2O_8.2Cu(OH)_2.H_2O.$ Orthorhombic? Massive, resembling malachite. H. = 4·5. G. = 4·16. Easily fusible. Color emerald-green. Optically +. $\alpha =$ 1·81, $\beta = 1·815$, $\gamma = 1·85$. 2V small. From Cornwall at St. Day, Gwennap, Liskeard, etc.

Tyrolite. Tirolit. — Perhaps $Cu_3As_2O_8.2Cu(OH)_2.7H_2O.$ Analyses show also the presence of CaO, CO_2, SO_3. It is impossible to say whether these radicals are an integral part of the composition or not. Orthorhombic. Usually in fan-shaped crystalline groups; in foliated aggregates; also massive. Cleavage (001), perfect, yielding soft thin flexible laminæ. H. = 1–1·5. G. = 3·1. Easily fusible. Color pale green inclining to sky-blue. Optically −. Ax. pl. || (010). $X = c$ axis. $\alpha = 1·694$, $\beta = 1·726$, $\gamma = 1·730$. 2V = 36°. Strong dispersion, $\rho > v$. Occurs frequently in altered copper deposits but always in small amounts. From Nerchinsk, Transbaikalia. Found in Slovakia of Czechoslovakia at Libethen (Libetbánya) and Pojnik near Neusohl (Ban Bystrica). In the Tyrol, Austria, at many localities, especially at Kogel near Brixlegg and on the Falkenstein near Schwaz. In France at Cap Garonne, Var. In the United States has been found at the Mammoth mine, etc., in the Tintic district, Utah.

Chlorophœnicite. — $10(Zn,Mn)O.As_2O_5.7H_2O.$ Monoclinic; crystals elongated parallel to b axis. Cleavage || (100) distinct. H. = 3–3·5. G. = 3·55. Color light green in daylight but light purple-red in artificial light. Optically + or −. Ax. pl. || (010). $\alpha =$ 1·682, $\beta = 1·690$, $\gamma = 1·697$. 2V large. Strong dispersion, $\rho > v$. With calcite, tephroite, and leucophœnicite in the franklinite-zincite ore at Franklin, Sussex Co., New Jersey.

Holdenite. — $8MnO.4ZnO.As_2O_5.5H_2O.$ Orthorhombic. Crystals tabular || (100). Cleavage || (010), poor. Color, clear pink to deep red and yellowish red. Optically +. Ax. pl. || (010). $Z \perp (100)$. 2V = 30°. Dispersion, $\rho > v$. $\alpha = 1·769$, $\beta = 1·770$, $\gamma =$ 1·785. From Franklin, Sussex Co., New Jersey.

Spencerite. — $Zn_3(PO_4)_2.Zn(OH)_2.3H_2O.$ When heated spencerite is converted into material with the composition of tarbutite, see p. 715. Monoclinic. In radiating and reticulated crystals. Twin crystals common. Cleavages: (100) very perfect, (010) perfect, (001) good; the three making nearly 90° angles with each other. G. = 3·12. H. = 2·7. Color white. Optically −. Ax. pl. ⊥ (010). X nearly || a axis. $\alpha = 1·586$, $\beta = 1·602$, $\gamma = 1·606$. 2V = 49°. From Hudson Bay Mine, Salmo, British Columbia.

SALMOITE. Probably a basic zinc phosphate. Biaxial, −. 2V moderately large. $\alpha = 1·645$, $\beta = 1·683$, $\gamma = 1·695$. Colorless. Occurs with spencerite at the Hudson Bay mine, Salmo, British Columbia.

CHALCOPHYLLITE.

Rhombohedral. Axis $c = 2·761$. $cr\ 0001 \wedge 10\bar{1}1 = 72°\ 2'$.

In tabular crystals; also foliated massive; in druses.

Cleavage: $c(0001)$ highly perfect; $r(10\bar{1}1)$ in traces. H. = 2. G. = 2·4–2·66. Luster of c pearly; of other faces vitreous or subadamantine.

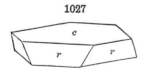

1027

Color emerald- or grass-green to verdigris-green. Streak somewhat paler than the color. Transparent to translucent. Optically −. $\omega = 1·632$. $\epsilon = 1·575$. Determinations made on material from Chile that had been preserved in a moist atmosphere gave $\omega = 1·618$, $\epsilon = 1·552$.

Comp. — A highly basic arsenate and sulphate of copper; formula uncertain, perhaps $20CuO.Al_2O_3.2As_2O_5.3SO_3.25H_2O.$

Pyr., etc. — In the closed tube decrepitates, yields much water, and gives a residue of olive-green scales. In other respects like olivenite. Soluble in nitric acid, and in ammonia.

Obs. — Chalcophyllite occurs in the zone of alteration of copper deposits. Found at Sayda in the Erzgebirge, Saxony; at Cap Garonne, Var, France. At various points in Cornwall, as at Redruth, St. Day, Gwennap, Liskeard, etc. From the Teniente mine near Rangagus, Chile. In the United States it occurs at Bisbee, Cochise Co., Arizona; at the Tintic district, Utah; from Sodaville, Mineral Co., Nevada.

Veszelyite. — $7(Zn,Cu)O.(P,As)_2O_5.9H_2O.$ Triclinic or monoclinic. H. = 3·5–4. G. = 3·5. Fusible. Color greenish blue. Optically +. $\alpha = 1·640$, $\beta = 1·658$, $\gamma = 1·695$. 2V = 71°. Strong dispersion, $\rho < v$. Occurs as a crystalline incrustation at Moravicza, in the Banat, Hungary; also reported from Dognacska and Vaskö. From Broken Hill.

Rhodesia, with monoclinic crystals; this occurrence may, however, be of arakawaite; see below.

Arakawaite. Kipushite. — $4CuO.2ZnO.P_2O_5.6\frac{1}{2}H_2O$. Monoclinic. Blue to bluish green. Optically +. Indices 1·63–1·74. From Arakawa mine, Province of Ugo, Japan. *Kipushite* is from Kipushi, Katanga, Belgian Congo.

WAVELLITE.

Orthorhombic. Axes $a : b : c = 0.5520 : 1 : 0.4067$. Crystals rare. Usually in aggregates, hemispherical or globular with crystalline surface, and radiated structure.

Cleavage: $p(101)$ and $b(010)$ rather perfect. Fracture uneven to sub-conchoidal. Brittle. H. = 3·25–4. G. = 2·316–2·337. Luster vitreous inclining to pearly and resinous. Color white, passing into yellow, green, gray, brown, and black. Streak white. Translucent. Optically +. Ax. pl. $\parallel$ (100). $Z = c$ axis. $\alpha = 1.525$, $\beta = 1.534$, $\gamma = 1.552$. 2V = 72°.

Comp. — $4AlPO_4.2Al(OH)_3.9H_2O$ = Phosphorus pentoxide 35·2, alumina 38·0, water 26·8 = 100. Fluorine is sometimes present, up to 2 per cent.

Pyr., etc. — In the closed tube gives off much water, the last portions of which may react acid (fluorine). B.B. in the forceps swells up and splits into fine infusible particles, coloring the flame pale green. Gives a blue color on ignition with cobalt solution. Soluble in hydrochloric acid, and also in caustic potash.

Obs. — Wavellite is commonly of secondary origin, associated with many rock types. It is frequently found in ore beds, especially those of limonite, and occurs with phosphorite deposits. Though occurring in many localities it is seldom found in quantity. It occurs in Bohemia of Czechoslovakia at the neighboring localities of Cerhovic and Zbirow, southwest of Beraun; at Kapnikbánya, Rumania (*kapnicite*); in Saxony from Frankenberg and Langenstriegis a short distance to the east; at Amberg in the Oberpfalz, Bavaria. In France at Montebras, Creuse, with amblygonite. Wavellite was originally described from a clay slate near Barnstable, Devonshire. From Co. Cork, Ireland.

In the United States wavellite occurs in a number of localities in Pennsylvania, especially at East Whitehead, Chester Co.; Moore's Mill, Cumberland Co.; in Juniata Co. In Arkansas at Magnet Cove near Hot Springs, Garland Co., in fine stellate radiations of a light to deep green color; also from Montgomery Co.

FISCHERITE. $AlPO_4.Al(OH)_3.2\frac{1}{2}H_2O$. In small prismatic crystals and in drusy crusts. Color green. Index, 1·55. From Nizhne-Tagilsk in the Ural Mts.; the other reported occurrences of fischerite at Roman Gladna, Hungary, has been proved to be vashegyite instead, and the original mineral may have been wavellite.

TURQUOIS. Turquoise.

Triclinic. Crystals minute and in angle near those of *chalcosiderite* with which it may be isomorphous. Usually massive; amorphous or cryptocrystalline. Reniform, stalactitic, or incrusting. In thin seams and disseminated grains. Also in rolled masses.

Cleavage in two directions in crystals; none in massive material. Fracture small conchoidal. Rather brittle. H. = 5–6. G. = 2·6–2·83. Luster somewhat waxy, feeble. Color sky-blue, bluish green to apple-green, and greenish gray. Streak white or greenish. Feebly subtranslucent to opaque. Optically +. $\alpha = 1.61$. $\beta = 1.62$. $\gamma = 1.65$. Strong dispersion, $\rho < v$.

Comp. — A hydrous phosphate of aluminum and copper $CuO.3Al_2O_3$. $2P_2O_5.9H_2O$ or perhaps $H_5(CuOH)[Al(OH)_2]_6(PO_4)_4$ = Phosphorus pentoxide 34·12, alumina 36·84, cupric oxide 9·57, water 19·47 = 100.

Penfield considered that the $H_2(CuOH)$ and $Al(OH)_2$ mutually replace each other in the orthophosphoric molecule.

Pyr., etc. — In the closed tube decrepitates, yields water, and turns brown or black. B.B. in the forceps becomes brown and assumes a glassy appearance, but does not fuse;

colors the flame green; moistened with hydrochloric acid the color is at first blue (copper chloride). With the fluxes reacts for copper. Soluble in hydrochloric acid.

Obs. — Turquois is a mineral of secondary origin, found in thin veins and small masses in various rock types which have undergone extensive alteration. In many cases the phosphoric acid has been derived from original apatite in the rocks, at times possibly from organic remains; the aluminum from the decomposition of the feldspars and the copper from small amounts of copper minerals present.

The highly prized oriental turquois occurs in narrow seams or in irregular patches in the brecciated portions of a porphyritic trachyte, associated with limonite, in Persia on the southern slopes of the Ali-Mirsai-Kuh Mts., between Nishapur and Meshed in Khorasan. From the Sinai Peninsula in the Wady Maghara. A greenish blue variety comes from Karkarallinsk in Semipalatinsk, Siberia. Also in the Kara-tjube Mts., south of Samarkand, Turkestan. An impure variety is found in Silesia at Jordansmühle and at Steine and at Ölnitz in Vogtland, Saxony. Occurs in porous masses at Montebras, Creuse, France.

In the United States turquois is found near Lynch Station, Campbell Co., Virginia, in minute crystals forming spherical groups. Occurs in the Los Cerillos Mts., southwest of Santa Fe, New Mexico, in a much altered trachytic rock; the deposit was early mined by the Mexicans and Indians and in recent years has been reopened and extensively worked. Also in New Mexico from the Jarilla Mts., Dona Ana Co.; in the Burro Mts., Grant Co. A pale green variety from Columbus in Esmeralda Co., Nevada.

Natural turquois of inferior color is often artificially treated to give it the tint desired. Moreover, many stones which are of a fine blue when first found retain the color only so long as they are kept moist, and when dry they fade, become a dirty green, and are of little value. Much of the turquois (not artificial) used in jewelry in former centuries, as well as the present, and that described in the early works on minerals, was *bone-turquois (called* also *odontolite*, from ὀδούς, *tooth*), which is fossil bone, or tooth, colored by a phosphate of iron. Its organic origin becomes manifest under a microscope. Moreover, true turquois, when decomposed by hydrochloric acid, gives a fine blue color with ammonia, which is not true of the odontolite.

Use. — As an ornamental material.

Vauxite. — $FeO.Al_2O_3.P_2O_5.6H_2O$. Triclinic. In aggregates of small tabular crystals. H. = 3·5. G. = 2·45. Color sky-blue to venetian-blue. Optically +. $\alpha = 1·551$, $\beta = 1·555$, $\gamma = 1·562$. 2V = 32°. Dispersion, $\rho > v$. Strongly pleochroic, colorless to blue. Occurs on wavellite at the tin mines of Llallagua, Bolivia.

Paravauxite. — $FeO.Al_2O_3.P_2O_5.5H_2O$. Triclinic. In small prismatic crystals. Perfect cleavage || (010). H. = 3. G. = 2·3. Colorless. Optically +. $\alpha = 1·554$, $\beta = 1·558$, $\gamma = 1·573$. 2V = 35°. Cleavage flakes show emergence of an optic axis. Occurs on wavellite at the tin mines of Llallagua, Bolivia.

Gordonite. — $MgO.Al_2O_3.P_2O_5.9H_2O$. Triclinic. Crystals apparently similar to those of paravauxite. In clear, glassy, lath-shaped crystals forming thin crusts. Perfect cleavage parallel to length of crystals. Biaxial, +. $\alpha = 1·534$, $\beta = 1·543$, $\gamma = 1·558$. 2V = 76°. Associated with variscite, etc. in the phosphate nodules from near Fairfield, Utah Co., Utah.

Pseudowavellite. Essentially a hydrous phosphate of aluminum and calcium; perhaps $5CaO.6Al_2O_3.4P_2O_5.18H_2O$. Rhombohedral. Optically +. $n = 1·63$. Occurs as white radiating incrustations on limonite and wavellite at Amberg, Bavaria. Also noted near Fairfield, Utah Co., Utah.

Wardite. — $2Na_2O.CaO.6Al_2O_3.4P_2O_5.17H_2O$. Tetragonal? Perfect basal cleavage. Forms light green or bluish green concretionary incrustations in cavities of nodular masses of variscite from Cedar Valley, Utah. H. = 5. G. = 2·81. Uniaxial, +. $\omega = 1·590$, $\epsilon = 1·599$. *Soumansite* from Montebras in Soumans, Creuse, France, agrees closely and is probably identical with wardite.

Sphærite. — Perhaps $4AlPO_4.6Al(OH)_3$. Orthorhombic. In globular drusy concretions with fibrous structure. One distinct cleavage. $H_1 = 4$. G. = 2·5. Color light gray, bluish. Optically −. $\alpha = 1·562$, $\beta = 1·576$, $\gamma = 1·588$. 2V large. Z || elongation. From Zaječow, north of St. Benigna, Bohemia, Czechoslovakia.

Liskeardite. — $(Al,Fe)AsO_4.2(Al,Fe)(OH)_3.5H_2O$. Orthorhombic? In thin incrusting layers, white or bluish. Cleavage || (010). H. = 4. G. = 3. Optically +. Ax. pl. $\perp$ (010). $Z = c$ axis. $\alpha = 1·661$, $\beta = 1·675$, $\gamma = 1·689$. 2V nearly 90°. From Liskeard, Cornwall, England.

Evansite. — $2AlPO_4.4Al(OH)_3.12H_2O$. Amorphous. Massive; reniform or botryoidal. H. = 4. G. = 2. Colorless, or milk-white. Isotropic. $n = 1·485$. Occurs in Czechoslovakia at Mt. Zceleznik near Szirk in Gömör of Slovakia, and at Gross-Tresny in Moravia. From the district of Vatomandry on the eastern coast of Madagascar. In the United

States has been found in the Coosa coal field of Alabama and from Goldburg, Custer Co., Idaho.

Cœruleolactite. — $3Al_2O_3.2P_2O_5.10H_2O$. Cryptocrystalline. In fibrous crusts. Milk-white to light copper-blue. H. = 5. G. = 2·5–2·7. Uniaxial, +. $\omega = 1·580$, $\epsilon = 1·588$. From near Katzenelnbogen, Hessen-Nassau; also East Whiteland Township, Chester Co., Pennsylvania. An earlier described mineral from Gumeahevsk, Ural Mts., under the name *planerite* is probably identical with cœruleolactite.

Augelite. — $2Al_2O_3.P_2O_5.3H_2O$. In tabular monoclinic crystals and massive. Cleavages || (101), ($\bar{1}$01). H. = 5. G. = 2·7. Colorless to white. Optically +. Ax. pl. || (010). $Z \wedge c$ axis = $-34°$. $\alpha = 1·574$, $\beta = 1·576$, $\gamma = 1·588$. $2V = 51°$. Found in Kristianstad, Sweden, at the Westanå mine near Näsum, west of Carlshamn, associated with other phosphates. Also from Bolivia at Machacamarca near Potosi and in Oruro.

The three following doubtful species all are found at the Westanå mine near Näsum west of Carlshamn in Kristianstad, Sweden.

Berlinite. $2Al_2O_3.2P_2O_5.H_2O$. Compact, massive. G. = 2·64. Colorless to grayish or rose-red.

Trolleite. $4Al_2O_3.3P_2O_5.3H_2O$. Compact, indistinctly cleavable. G. = 3·10. Color pale green. Possibly identical with lazulite.

Attacolite. $P_2O_5,Al_2O_3,MnO,CaO,H_2O$, etc.; formula doubtful. Massive. G. = 3·09. Color salmon-red.

Minervite. Under this name have been described a series of hydrated aluminum phosphates with potassium, derived from organic deposits, as guano, etc. From "Grotte de Minerve," valley of the Aude, France; caves in New South Wales; Réunion Island, etc.

Vashegyite. — $4Al_2O_3.3P_2O_5.30H_2O$. Fibrous. H. = 2–3. G. = 1·96. Color white or yellow to rust-brown when colored by iron oxide. $n = 1·50$. Fibers || Z. From iron mine Vashegy near Szirk in Gömör in Slovakia, Czechoslovakia. Reported from near Manhattan, Nye Co., Nevada.

PHARMACOSIDERITE.

Pseudo-isometric-tetrahedral. Commonly in cubes; also tetrahedral. Rarely granular.

Cleavage: $a(100)$ imperfect. Fracture uneven. Rather sectile. H. = 2·5. G. = 2·9–3. Luster adamantine to greasy, not very distinct. Color olive-, grass- or emerald-green, yellowish brown, honey-yellow. Streak green to brown, yellow, pale. Subtransparent to subtranslucent. Nearly isotropic, $n = 1·693$. At ordinary temperatures anisotropic, showing crystal divided into biaxial segments. Usually optically $-$. Very strong dispersion, $\rho > v$.

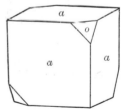

1028

Comp. — Perhaps $6FeAsO_4.2Fe(OH)_3.12H_2O$ = Arsenic pentoxide 43·1, iron sesquioxide 40·0, water 16·9 = 100. Some varieties contain K_2O.

Pyr., etc. — Same as for scorodite.

Obs. — Pharmacosiderite occurs associated with various ore bodies that contain arsenic minerals, as tetrahedrite, arsenopyrite, etc., and has been derived from their alteration. Occurs at Königsberg southwest of Schemnitz (Selmeczbánya) in Slovakia of Czechoslovakia; in Bavaria at Kahl and Aschaffenburg in the Spessart; from Saxony near Schwarzenberg. In France at Vaulry and at Puy-les-Vignes near Saint-Léonard, Haute Vienne. From mines in Cornwall, at St. Day, Liskeard, Redruth, Calstock, etc. In the United States found at the Mammoth mine, etc., in the Tintic district, Utah. Named from φαρμακον, *poison*, and σιδηρος, *iron*.

Ludlamite. — $2Fe_3P_2O_8.Fe(OH)_2.8H_2O$. Monoclinic. Occurs in small green tabular crystals. Cleavages, || (001) perfect, (100) distinct. H. = 3·5. G. = 3·7. Fusible. Optically +. Ax. pl. || (010). $Z \wedge c$ axis = $-67°$. $\alpha = 1·653$, $\beta = 1·675$, $\gamma = 1·697$. $2V$ large. From near Truro, Cornwall. *Lehnerite*, described from Hagendorf, Bavaria, is closely similar to ludlamite and probably identical with it.

Cacoxenite. — $FePO_4.Fe(OH)_3.4\frac{1}{2}H_2O$. Hexagonal. In radiated tufts of a yellow or brownish color. H. = 3·5. G. = 3·4. Fusible. Optically +. $\omega = 1·580$–$1·585$, $\epsilon = 1·64$–$1·65$. Occurs in Czechoslovakia in Bohemia near St. Benigna and at Zbirow near Cerhovic, which is southwest of Beraun. At the Eleonore mine on the Dünsberg, near Giessen in Hessen, and at Waldgrimes in the same district; also at Weilburg, Hessen-Nassau. In the United States from Lancaster and Cumberland counties, Pennsylvania.

XANTHOXENITE. Hydrous ferric phosphate with FeO, MnO, CaO, MgO, Al_2O_3. Monoclinic. In thin plates. Yellow. Pleochroic. G. = 2·84. From Hühnerkobel, Rabenstein, Bavaria. Near beraunite.

Beraunite. — Perhaps $2FePO_4.Fe(OH)_3.2\frac{1}{2}H_2O$. Monoclinic. Commonly in druses and in radiated globules and crusts. Cleavage || (100). H. = 2. G. = 2·9. Color reddish brown to dark hyacinth-red. Optically +. Ax. pl. ⊥ (010). $Z = b$ axis. Y nearly || c axis. $\alpha = 1·775$, $\beta = 1·786$, $\gamma = 1·815$. 2V medium. Dispersion, $\rho > v$. Beraunite is from near St. Benigna south of Beraun in Bohemia, Czechoslovakia. *Eleonorite* is from the Eleonore mine on the Dünsberg near Giessen, Hessen, and at the Rothläufchen mine near Waldgrimes in the same region. In the United States it is found in Pennsylvania at Hellertown, Northampton Co., and at Moore's Hill, Cumberland Co. Also from Sevier Co., Arkansas.

KERTSCHENITE, OXYKERTSCHENITE, are other hydrated ferric phosphates.

CHILDRENITE.

Orthorhombic. Axes $a : b : c = 0·7780 : 1 : 0·52575$.

mm''', 110 ∧ 1$\bar{1}$0 = 75° 46'. rr''', 131 ∧ 1$\bar{3}$1 = 105° 9'.
rr',   131 ∧ $\bar{1}$31 = 39° 47'. ss',   121 ∧ $\bar{1}$21 = 49° 56$\frac{1}{2}$'.

Only known in crystals. Cleavage: a(100) imperfect. Fracture uneven. H. = 4·5–5. G. = 3·18–3·24. Luster vitreous to resinous. Color yellowish white, pale yellowish brown, brownish black. Streak white to yellowish. Translucent. 2E = 74°. Optically −. Ax. pl. || (100). $X = b$ axis. $\alpha = 1·643$, $\beta = 1·678$, $\gamma = 1·684$. 2V = 45°. Dispersion strong, $\rho > v$.

Comp. — In general $AlPO_4.Fe(OH)_2.H_2O$. Phosphorus pentoxide 30·9, alumina 22·2, iron protoxide 31·3, water 15·6 = 100. Manganese replaces part of the iron and it hence graduates into eosphorite.

Pyr., etc. — In the closed tube gives off neutral water. B.B. swells up into ramifications, and fuses on the edges to a black mass, coloring the flame pale green. Heated on charcoal turns black and becomes magnetic. With soda gives a reaction for manganese. With borax and salt of phosphorus reacts for iron and manganese. Soluble in hydrochloric acid.

Obs. — Found at Greifenstein near Ehrenfriedersdorf, Saxony. From near Tavistock and elsewhere in Devonshire and in Cornwall at St. Austell. In the United States at Hebron, Oxford Co., Maine, in minute hair-brown prismatic crystals with amblygonite.

Kreuzbergite. Pleysteinite. — Aluminum phosphate with Fe,Mn,H_2O. Orthorhombic. Cleavage (010). White to yellow. Optically +. $X \perp$ (001). $n = 1·50$. 2V nearly 90°. From the Kreuzberg, Pleystein, Bavaria.

Eosphorite. — Form and composition as for childrenite, but containing chiefly manganese instead of iron. In prismatic crystals; also massive. Cleavage (100). H. = 5. G. = 3·1. Fusible. Color rose-pink, yellowish, etc. Optically −. Ax. pl. || (100). Z || c axis. $\alpha = 1·63$, $\beta = 1·65$, $\gamma = 1·66$. 2V = 40°–50°. Strong dispersion, $\rho < v$. From Branchville, Fairfield Co., Connecticut; at Newry and Buckfield, Oxford Co., Maine. Also noted at Hagendorf, near Pleystein, Oberpfalz, Bavaria.

Mazapilite. — $Ca_3Fe_2(AsO_4)_4.2FeO(OH).5H_2O$. Orthorhombic. In slender prismatic crystals. H. = 4·5. G. = 3·6. Easily fusible. Color black. Nearly uniaxial. Optically −. $\omega = 1·898$, $\epsilon = 1·815$. Pleochroic. X = nearly colorless, Z = dark red-brown. From Mazapil, Zacatecas, Mexico. May be identical with arseniosiderite, p. 717.

YUKONITE. $(Ca_3,Fe_2''')(AsO_4)_2.2Fe(OH)_3.5H_2O$. Amorphous. In irregular concretions. H. = 2–3. G. = 2·8. Color nearly black with brown tinge. Decrepitates at low heat, also when immersed in water. Easily fusible with intumescence. From Tagish Lake, Yukon Territory.

Calcioferrite. — $Ca_3Fe_2(PO_4)_4.Fe(OH)_3.8H_2O$? Monoclinic? Occurs in yellow to green nodules. Cleavage (001), perfect. H. = 2·5. G. = 2·5. Easily fusible. Nearly uniaxial. Optically −. $n = 1·56–1·58$. Found in clay at Battenberg, Hessen.

Borickite. — Perhaps $Ca_3Fe_2(PO_4)_4.12Fe(OH)_3.6H_2O$. Amorphous. Reniform massive; compact. H. = 3·5. G. = 2·7. Fusible. Color reddish brown. Isotropic. $n = 1·57–1·67$. In Bohemia of Czechoslovakia, at Nanačovic near Unhoscht, south of Kladno. Also from Leoben in Styria, Austria. Reported from the Tintic district, Utah. *Fouchérite*, probably identical with borickite, is from Fouchères, Aube, France. *Picite* described from St. Benigna, south of Beraun in Bohemia, may be the same species.

Eguëïite. A hydrous basic phosphate of ferric iron with calcium and aluminum. Amorphous. In small nodules with fibrous-lamellar structure. Index, 1·65. Fusibility 1. Easily soluble in hydrochloric acid. Found embedded in clay from Egai (Egueï) in the Military Territory of Chad in the Sudan.

Richellite. Perhaps $4FeP_2O_8.Fe_2OF_2(OH)_2.36H_2O$. Massive, compact or foliated. Color yellow. From Richelle, near Visé in Liége, Belgium.

LIROCONITE.

Monoclinic. Axes $a : b : c = 1·3191 : 1 : 1·6808$; $\beta = 88° 33'$.

mm''', $110 \wedge 1\bar{1}0 = 105° 39'$. me, $110 \wedge 011 = 46° 10'$.
ee', $011 \wedge 0\bar{1}1 = 118° 29'$. $m'e$, $\bar{1}10 \wedge 011 = 47° 24'$.

1029

Crystals resembling rhombic octahedrons. Rarely granular. Cleavage: $m(110)$, $e(011)$ indistinct. Fracture subconchoidal to uneven. Imperfectly sectile. H. = 2–2·5. G. = 2·9. Luster vitreous, inclining to resinous. Color and streak sky-blue to verdigris-green. Optically −. Ax. pl. $\perp$ (010). $Z \wedge c$ axis = −25°. $\alpha = 1·612$, $\beta = 1·652$, $\gamma = 1·675$. 2V = 67°.

Comp. − A hydrous arsenate of aluminum and copper, formula uncertain; analyses correspond nearly to $Cu_6Al(AsO_4)_5.3CuAl(OH)_5.20H_2O$ = Arsenic pentoxide 28·9, alumina 10·3, cupric oxide 35·9, water 24·9 = 100. Phosphorus replaces part of the arsenic.

Pyr., etc. — In the closed tube gives much water and turns olive-green. B.B. cracks open, but does not decrepitate; fuses less readily than olivenite to a dark gray slag; on charcoal cracks open, deflagrates, and gives reactions like olivenite. Soluble in nitric acid.

Obs. — Found as a mineral of secondary origin in copper deposits, associated with malachite, etc. In minute crystals from Herrengrund (Urvölgy), north of Neusohl (Ban Bystrica) in Slovakia of Czechoslovakia. With various ores of copper in Cornwall, especially at St. Day and Gwennap.

Chenevixite. — $2CuO.Fe_2O_3.As_2O_5.2H_2O$. Cryptocrystalline. Massive to compact. H. = 4. G. = 3·9. Easily fusible. Color dark green to greenish yellow. $n = 1·88$. From St. Day, Cornwall; in the Tintic district, Utah.

Henwoodite. A hydrated phosphate of aluminum and copper. In botryoidal globular masses. Color turquois-blue. From Liskeard, Cornwall.

Cœruleite. — $CuO.2Al_2O_3.As_2O_5.8H_2O$. Compact, made up of very minute crystals. G. = 2·8. Color, turquois-blue. Soluble in acids. From Huanaco, Taltal province, Chile.

Chalcosiderite. — $CuO.3Fe_2O_3.2P_2O_5.9H_2O$? Probably isomorphous with turquois. Triclinic. In sheaf-like crystalline groups, as incrustations. Cleavage (010). H. = 4·5. G. = 3·1. Fusible. Color light siskin-green. Optically −. X nearly || b axis. $\alpha = 1·77$, $\beta = 1·84$, $\gamma = 1·845$. 2V = 24°. Crossed dispersion very strong, $\rho > v$. Sections || (010) show abnormal interference colors. From Liskeard, Cornwall. Reported from Bisbee, Cochise Co., Arizona.

Andrewsite, also from Cornwall, is near chalcosiderite.

Kehoeite. − A hydrated phosphate of aluminum, zinc, etc. Amorphous. Massive. G. = 2·34. Isotropic. $n = 1·53$. From Galena, Lawrence Co., South Dakota.

Dennisonite. — $6CaO.Al_2O_3.2P_2O_5.5H_2O$. Probably hexagonal. In stout fibers. Perfect basal cleavage. Optically −. $\omega = 1·601$, $\epsilon = 1·591$. Occurs as thin white crusts associated with pseudo-wavellite, etc. from phosphate nodules found near Fairfield, Utah Co., Utah.

Deltaite. — 8CaO.5Al₂O₃.4P₂O₅.14H₂O. Hexagonal, rhombohedral? In minute prismatic crystals with triangular cross section (whence name from shape of Greek letter delta). Also fibrous, and in crusts composed of matted fibers. H. = 5. G. = 2·95. Gray color. Uniaxial, +. ω = 1·630, ε = 1·640. Occurs embedded in pseudo-wavellite and in chalky or cherty crusts in phosphate nodules from near Fairfield, Utah Co., Utah.

Millisite. — 2CaO.Na₂O.6Al₂O₃.4P₂O₅.17H₂O. In white fibers forming bands which resemble chalcedony. H. = 5·5. G. = 2·83. Biaxial, −. α = 1·584, β = 1·598, γ = 1·602. Intimately associated with wardite in the phosphate nodules from near Fairfield, Utah.

Englishite. — 4CaO.K₂O.4Al₂O₃.4P₂O₅.14H₂O. Probably orthorhombic. Highly perfect cleavage ‖ (001). Optically −. X ⊥ (001). α = 1·570, γ = 1·572. 2V small. Occurs in thin layers intimately associated with variscite in phosphate nodules from near Fairfield, Utah Co., Utah.

Lehiite. — 5CaO.(Na,K)₂O.4Al₂O₃.4P₂O₅.12H₂O. In white crusts composed of coarse fibers. H. = 5·5. G. = 2·89. Biaxial, −. α = 1·602, β = 1·616, γ = 1·629. Fibers show a large extinction angle. Occurs in phosphate nodules from near Fairfield, Utah Co., Utah.

Lewistonite. — 15CaO.(K,Na)₂O.4P₂O₅.8H₂O. Pseudo-hexagonal. In stout prisms which are shown optically to be composed of six biaxial segments with rarely an uniaxial core. Also in white powdery crusts. Perfect basal cleavage, probably also a prismatic cleavage. α = 1·613, β = 1·623, γ = 1·624. 2V = 42°. Uniaxial material is optically −, with ω = 1·624, ε = 1·613. H. = 5. G. = 3·06. Found with pesudo-wavellite, etc., in phosphate nodules from near Fairfield, Utah Co., Utah.

Roscherite. — (Mn,Fe,Ca)₂Al(OH)(PO₄)₂.2H₂O. Monoclinic. Crystals tabular. Cleavages (001) perfect, (010) good. H. = 4·5. G. = 2·9. Color brown. Optically −. Ax. pl. ⊥ (010). X = b axis. Y ∧ c axis = −15°. β = 1·639. 2V large. Strong dispersion, ρ > υ. A pneumatolytic mineral from Greifenstein, near Ehrenfriedersdorf, Saxony.

Uranite Group

TORBERNITE. Copper Uranite.

Orthorhombic, pseudo-tetragonal. Axis c = 2·974. Crystals usually square tables, sometimes very thin, again thick; less often pyramidal. Also foliated, micaceous.

Cleavages: c(001) perfect, micaceous; (100) distinct. Laminæ brittle. H. = 2–2·5. G. = 3·2. Luster of c pearly, other faces subadamantine. Color emerald- and grass-green, and sometimes leek-, apple-, and siskin-green. Streak paler than the color. Transparent to subtranslucent. Optically uniaxial; negative. ω = 1·592, ε = 1·582.

Comp. — A hydrous phosphate of uranium and copper, Cu(UO₂)₂P₂O₈. 12H₂O = Phosphorus pentoxide 14·1, uranium, trioxide 56·6, cupric oxide 7·9, water 21·4 = 100. Arsenic may replace part of the phosphorus.

In air at a temperature below 100° C. dehydration takes place with formation of *meta-torbernite I* (*metakupferuranite*, *metachalcolite*) with 8H₂O. G. = 3·68. ω = 1·623, ε = 1·625. Optically + for red light; − for blue light; isotropic for green light. At 130° C. another change takes place with further dehydration and formation of *meta-torbernite II.*

Pyr., etc. — In the closed tube yields water. Fuses at 2·5 to a blackish mass, and colors the flame green. With salt of phosphorus gives a green bead, which with tin on charcoal becomes on cooling opaque red (copper). With soda on charcoal gives a globule of copper. Soluble in nitric acid.

Obs. — Occurs associated with autunite and frequently in parallel growth with it; also with other uranium minerals. From Bohemia of Czechoslovakia at Joachimstal and Zinnwald (Cinvald); in Saxony at Johanngeorgenstadt, Eibenstock, and Schneeberg; splendid crystals from various localities in Cornwall, especially at Gunnis Lake near Calstock (*meta-torbernite I*), Redruth, etc. Occurs in South Australia at Mt. Painter, Flinders Range.

Zeunerite. — Cu(UO₂)₂As₂O₈.8H₂O. In tabular crystals resembling torbernite in form and color. Perfect cleavage ‖ (001). H. = 2–2·5. G. = 3·2. Fusible. Uniaxial? Optically −. ω = 1·643, ε = 1·623; (ω = 1·585, ε = 1·576, Schneeberg). Found in

Bohemia of Czechoslovakia at Joachimstal and Zinnwald (Cinvald). Originally found at Schneeberg, Saxony. Occurs in the Tintic district, Utah.

AUTUNITE. Lime Uranite.

Orthorhombic. In thin tabular crystals, nearly tetragonal in form and similar to torbernite in angle; also foliated, micaceous.

Cleavage: basal, eminent, also $\|$ (100), (010) and (110). Laminæ brittle. H. = 2–2·5. G. = 3·1. Luster of $c(001)$ pearly, elsewhere subadamantine. Color lemon- to sulphur-yellow. Streak yellowish. Transparent to translucent. Optically −. Ax. pl. $\| b(010)$. $X = c$ axis. $\alpha = 1\cdot553$. $\beta = 1\cdot575$. $\gamma = 1\cdot577$.

Comp. — A hydrous phosphate of uranium and calcium, probably analogous to meta-torbernite, $Ca(UO_2)_2P_2O_8.8H_2O$ or $CaO.2UO_3.P_2O_5.8H_2O =$ Phosphorus pentoxide 15·5, uranium trioxide 62·7, lime 6·1, water 15·7 = 100.

Some analyses give 10 and others 12 molecules of water, but it is not certain that the additional amount is essential.

Pyr., etc. — Same as for torbernite, but no reaction for copper.

Obs. — Autunite is commonly of secondary origin, usually associated with uraninite and other uranium minerals. Sometimes with silver, tin, and iron ores; occasionally in pegmatites. It is found in Saxony at Johanngeorgenstadt and at Falkenstein, and Schwarzenberg. In France it occurs at Saint-Symphorien and other localities near Autun in Saône-et-Loire. In northern Portugal at Sabugul, southeast of Guarda, and at Vizeu. In Cornwall, at Redruth, etc. Found on Mt. Painter, Flinders Range, South Australia. In the United States occurs in Connecticut sparingly in pegmatite near Middletown, Middlesex Co., and at Branchville, Fairfield Co.; in Pennsylvania at Philadelphia. Occurs in the mica mines of Mitchell Co., North Carolina. Found in the Black Hills, South Dakota.

Bassetite. — Composition probably the same as *autunite*. Monoclinic. $\beta = 89^\circ 17'$. Crystals tabular $\|$ (010). Twinned; tw. pl. $b(010)$. Cleavage parallel to three pinacoids. G. = 3·10. Color yellow. Transparent. Optically −. Ax. pl. $\perp$ (010). $Z \wedge c$ axis = −4°. $\alpha = 1\cdot558$, $\beta = 1\cdot574$, $\gamma = 1\cdot580$. From the Basset mines, Cornwall. Previously considered to be *autunite*.

Uranospinite. — Probably $Ca(UO_2)_2As_2O_8.8H_2O$. Orthorhombic, pseudo-tetragonal. In thin tabular crystals rectangular in outline. Basal cleavage, perfect. H. = 2–3. G. = 3·45. Fusible. Color siskin-green. Optically −. Sensibly uniaxial in outer zone of crystals to biaxial with $2V = 46^\circ$ toward the center of crystals. $X \perp$ (001). Indices, 1·560–1·587. From near Neustädtel, south of Schneeberg, Saxony.

Uranocircite. — $Ba(UO_2)_2P_2O_8.8H_2O$. In crystals similar to autunite. Cleavages $\|$ (001), (100), (010). H. = 2. G. = 3·5. Color yellow-green. Optically −. Ax. pl. $\| (010)$. $X \perp$ (001). $\alpha = 1\cdot610$, $\beta = 1\cdot623$, $\gamma = 1\cdot623$. 2V small. From near Falkenstein, Saxony. Reported from Vinanikarena, southeast of Antsirabe, Madagascar.

Uranospathite. — A hydrated uranyl phosphate. Orthorhombic, pseudo-tetragonal. In elongated tabular crystals. Cleavages parallel to the three pinacoids. G. = 2·5. Color yellow to pale green. Optically −. Ax. pl. $\| (010)$. $X = c$ axis. $\alpha = 1\cdot488$, $\beta = 1\cdot510$, $\gamma = 1\cdot521$. 2V = 69°. From Redruth, Cornwall. Previously considered to be *autunite*.

Carnotite. — Approximately, $K_2O.2UO_3.V_2O_5.2H_2O$. The amount of water is uncertain as the percentage present varies with the moisture content of the atmosphere. Orthorhombic. In the form of powder, sometimes in crystalline plates $\| c(001)$. Basal cleavage. Color yellow. Optically −. Ax. pl. $\| (100)$. $X = c$ axis. Indices 1·75–2·08. 2V = 40°–50°. Occurs as a yellow crystalline powder, or in loosely cohering masses, intimately mixed with quartzose material. It is found in large quantities in Mesa, San Miguel, and Montrose counties of southwestern Colorado and in adjoining districts in Utah. Is mined there not only for its uranium and vanadium content but also for the small amount of radium it contains. Noted also from Radium Hill, near Olary, South Australia, from Katanga, Belgian Congo, and from near Mauch Chunk, Carbon Co., Pennsylvania.

Tyuyamunite. — Calciocarnotite. — $CaO.2UO_3.V_2O_5.nH_2O$. The water present is commonly near 9 or 10 H_2O but varies in amount with the moisture content of the atmosphere. Orthorhombic. In scales. Or cryptocrystalline and earthy. Cleavages, (100) perfect, (010) and (100) distinct. Soft. G. = 3·7–4·3. Easily fusible. Optically −. Ax. pl. $\|$

(010). $X = c$ axis. $\alpha = 1.670$, $\beta = 1.870$, $\gamma = 1.895$. 2V = 36°. Found in limestone on Tyuya-Muyun hill, a northern spur of the Alai Mts., Ferghana, in Russian Turkestan. Similar material is reported from Colorado at Paradox Valley, Montrose Co., and from near Thompson's, Grand Co., and in the Henry Mts., Garfield Co., Utah.

Rauvite. A hydrous calcium uranium vanadate, $CaO.2UO_3.6V_2O_5.20H_2O$. Compact. Apparently metacolloidal. Color purplish black. Streak light brown. $n = 1.88$. Found in cracks in sandstone associated with carnotite, hewettite, etc., at Temple Mt., San Rafael Swell, Utah.

Sincosite. — $CaO.V_2O_4.P_2O_5.5H_2O$. Tetragonal. In small crystals tabular $\parallel$ (001). Cleavage $\parallel$ (001) good. Poorer cleavages $\parallel$ (100) and (110). G. = 2.84. Heated in C.T. changes from green to dark brown to black, decrepitates, yields water, and finally sinters to a dark-colored mass. Readily soluble in acids. Color leek-green. Mostly uniaxial but some crystals show biaxial figure with varying axial angle. Optically −. $\omega = 1.680$. $\epsilon = 1.655$. Indices slightly higher in biaxial material. Pleochroism strong. E nearly colorless to pale yellow and O gray-green. Occurs in a black carboniferous shale from Sincos, Peru.

Renardite. — $PbO.4UO_3.P_2O_5.9H_2O$. Orthorhombic. Minute tabular crystals $\parallel$ (100). Perfect cleavage $\parallel$ (100). G. = 4.0. Yellow color. Ax. pl. $\parallel$ (001). Optically −. $X \perp$ (100). $\alpha = 1.715$, $\beta = 1.736$, $\gamma = 1.739$. Occurs with dewindtite, dumontite, and torbernite in uranium ore at the Kasolo mine, Katanga, Belgian Congo.

Dewindtite. — $3PbO.5UO_3.2P_2O_5.12H_2O$. Orthorhombic. Crystals short prismatic or in minute plates $\parallel$ (010). G. = 4.08. Radioactive. Color, canary-yellow. Optically +. Ax. pl. $\parallel$ (001). $\alpha = 1.762$, $\beta = 1.763$, $\gamma = 1.765$. 2V large. Occurs intimately intergrown with torbernite at Kasolo, Katanga, Belgian Congo. *Stasite* is the same mineral.

Dumontite. — $2PbO.3UO_3.P_2O_5.5H_2O$. Probably orthorhombic. Prismatic habit, flattened parallel to (010). Optically +. 2V large. $n = 1.77$–1.89. Strong pleochroism; Y = deep yellow, X = pale yellow. Found in pockets in torbernite from Chinkolobwe, Belgian Congo.

Parsonite. — $2PbO.UO_3.P_2O_5.H_2O$. Monoclinic or triclinic. In minute tabular crystals. Structure earthy, minutely crystalline to compact. G. = 6.23. Radioactive. Fusible. Soluble in acids. Yields water in C. T. Color pale brown. Optically −. $\alpha = 1.85$, $\gamma = 1.86$. Associated with torbernite at Kasolo, Katanga, Belgian Congo.

Phosphuranylite. — $(UO_2)_3P_2O_8.6H_2O$. Probably monoclinic, pseudo-tetragonal. Tabular $\parallel$ (010). As a pulverulent incrustation. Color deep lemon-yellow. Optically −. $X = b$ axis. $\alpha = 1.691$, $\beta = 1.720$, $\gamma = 1.720$. 2V very small. Dispersion very strong, $\rho > v$. From Mitchell Co., North Carolina.

Trögerite. — $(UO_2)_3As_2O_8.12H_2O$. In thin druses of tabular crystals. Probably monoclinic, pseudo-tetragonal. Two cleavages at right angles. Soft. G. = 3.3. Easily fusible. Color lemon-yellow. Sensibly uniaxial. Optically −. $X \perp$ plates. $n = 1.58$–1.63. From Neustädtel, south of Schneeberg, Saxony.

Uvanite. — $2UO_3.3V_2O_5.15H_2O$. Orthorhombic. Fine granular. Two pinacoidal cleavages. Color brownish yellow. Optically +. $\alpha = 1.817$, $\beta = 1.879$, $\gamma = 2.057$. 2V = 52°. Found disseminated in rocks near Temple Rock, 45 miles southwest of Greenriver, Emery Co., Utah.

Ferghanite. — $U_3(VO_4)_2.6H_2O$. In scales. One perfect cleavage. H. = 2. G. = 3.3. Color sulphur-yellow. Biaxial. Indices and birefringence low. 2V large. From province of Ferghana, Russian Turkestan.

Walpurgite. — Probably $3UO_3.5Bi_2O_3.2As_2O_5.12H_2O$. Triclinic. In thin yellow crystals, $\parallel$ (010), resembling gypsum. H. = 3.5. G. = 5.8. Easily fusible. Color yellow. Optically −. X nearly $\perp$ (010). Indices, 1.87–2.05. Occurs at Neustädtel, south of Schneeberg, Saxony. Also reported from Joachimstal, Bohemia in Czechoslovakia.

Rhagite. — Perhaps $2BiAsO_4.3Bi(OH)_3$. In crystalline aggregates. Color yellowish green, wax-yellow. From Neustädtel, south of Schneeberg, Saxony.

Arseno-bismite. A hydrous bismuth arsenate. In cryptocrystalline aggregates. Color yellowish green, with tinge of brown. G. = 5.7. Index, 1.6. Found at Mammoth mine, Tintic district, Utah.

Mixite. — A hydrated basic arsenate of copper and bismuth, formula doubtful. In acicular crystals; as an incrustation. H. = 3–4. G. = 3.8. Easily fusible. Color green to whitish. Uniaxial or nearly so. Optically +. $\omega = 1.730$, $\epsilon = 1.810$. Dichroic, O = almost colorless, E = fine-green. From Joachimstal, Bohemia; Wittichen, Baden; Tintic district, Utah.

Antimonates; also Antimonites, Arsenites

A number of antimonates have been included in the preceding pages among the phosphates, arsenates, etc.

Bindheimite. — A hydrous antimonate of lead. Amorphous, reniform; also earthy or incrusting. Color gray, brownish, yellowish. $n = 1 \cdot 84$–$1 \cdot 87$. A result of the decomposition of other antimonial ores. Some of the more important localities for its occurrence are: Nerchinsk in Transbaikalia; in Spain in the Sierra Almagrera near Aquilas in Murcia; in Cornwall, especially near Endellion; from Djebel Nador, Constantine, Algeria. In the United States in Sevier Co., Arkansas; in South Dakota at Silver City, Pennington Co.; from the Coeur d'Alene district in Shoshone Co., Idaho; from Mineral, Eureka Co., and elsewhere in Nevada.

Romeite. — An antimonite of calcium, perhaps $Ca_5Sb_6O_{20}$. Pseudo-isometric. In groups of minute square octahedrons. H. above $5 \cdot 5$. G. $= 4 \cdot 7$–$5 \cdot 1$. Color hyacinth- or honey-yellow. $n = 1 \cdot 83$–$1 \cdot 87$. Romeite was found at St. Marcel in Piedmont, Italy. *Atopite* from Miguel Burnier, near Ouro Preto, Minas Geraes, Brazil, has been shown to be identical with romeite, as is probably also the case with the original atopite from Långbanshyttan, Vermland, Sweden.

Weslienite. — $Na_2FeCa_3Sb_4O_{15}$. Isometric. Octahedral habit. Crystals minute. Shows anomalous double refraction, being optically positive with large optical angle. $n = 2 \cdot 21$. Color honey-yellow to resinous brown. Vitreous to adamantine luster. Found at Långbanshyttan, Vermland, Sweden, in massive hematite, associated with manganophyllite, richterite, etc.

Schneebergite. — $(Ca,Fe)_2Sb_2O_6$. Isometric. Octahedral habit. Octahedral cleavage. H. $= 6 \cdot 5$. G. $= 5 \cdot 41$. Honey-yellow color. Adamantine luster. $n = 2 \cdot 09$. Associated with chalcopyrite, magnetite, calcite, etc., from Schneeberg, Trentino, Italy (formerly in Tyrol).

Swedenborgite. — $Na_2O.2Al_2O_3.Sb_2O_5$. Hexagonal. Basal cleavage. G. $= 4 \cdot 285$. H. $= 8$. Transparent, colorless to wine-yellow. Optically $-$. $\omega = 1 \cdot 772$, $\epsilon = 1 \cdot 770$. Occurs in calcite or granular hematite with richterite, manganophyllite, etc., at Långban shyttan, Vermland, Sweden.

Tripuhyite. — An iron antimonate. $Fe_2Sb_2O_7$. In microcrystalline aggregates of a dull greenish yellow color. G. $= 5 \cdot 8$. Fusible. Optically $+$. $\alpha = 2 \cdot 19$, $\beta = 2 \cdot 20$, $\gamma = 2 \cdot 33$. 2V small. Strong dispersion, $\rho < v$. From Tripuhy near Ouro Preto, Minas Geraes, Brazil.

Nadorite. — $PbSb_2O_4.PbCl_2$. In orthorhombic crystals. H. $= 3 \cdot 5$–4. G. $= 7$. Easily fusible. Color brownish yellow. Optically $+$. Ax. pl. $\parallel$ (010). $X - a$ axis. $\alpha = 2 \cdot 30$, $\beta = 2 \cdot 35$, $\gamma = 2 \cdot 40$. 2V very large. Dispersion, $\rho > v$, strong. From Djebel Nador, Constantine, Algeria.

Ecdemite. Heliophyllite. — Perhaps $Pb_4As_2O_7.2PbCl_2$. Tetragonal. In crystals, massive, and as an incrustation. H. $= 2 \cdot 5$. G. $= 6 \cdot 9$–$7 \cdot 1$. Easily fusible. Color bright yellow to green. Optically $-$. $\omega = 2 \cdot 32$, $\epsilon = 2 \cdot 25$. In part biaxial. Found at Långbanshyttan, Vermland, Sweden; also at the Harstig mine at Pajsberg, near Persberg, Vermland (*heliophyllite*). In the United States at Schultz, Pinal Co., Arizona.

Finnemanite. — $Pb_5Cl(AsO_3)_3$. Hexagonal. Prismatic. Distinct pyramidal cleavage. H. $= 2 \cdot 5$. G. $= 7 \cdot 26$. Color, gray to black, in thin flakes somewhat olive-green. Optically $-$. $\omega = 2 \cdot 295$. $\epsilon = 2 \cdot 285$. Found as a crystalline crust lining crevices in hematite at Långbanshyttan, Vermland, Sweden.

Ochrolite. — Probably $Pb_4Sb_2O_7.2PbCl_2$. In small crystals, united in diverging groups. Color sulphur-yellow. From Pajsberg, near Persberg, Vermland, Sweden.

Trippkeite. — $nCuO.As_2O_3$. In small bluish green, tetragonal crystals. Basal cleavage, perfect. Soft. Easily fusible. Optically $+$. $\omega = 1 \cdot 90$, $\epsilon = 2 \cdot 12$. From Copiapo, Atacama, Chile.

Schafarzikite is described as isomorphous with trippkeite with the formula, $nFeO.P_2O_3$. Prismatic tetragonal crystals. Cleavages, (110), (100). H. $= 3 \cdot 5$. G. $= 4 \cdot 3$. Color red to red-brown. Optically $+$. $n > 1 \cdot 74$. From Slovakia of Czechoslovakia at Pernek northwest of Bösing which is north of Bratislava (Pressburg).

Flajolotite. — $4FeSbO_4.3H_2O$. Compact or earthy. Color lemon-yellow. In nodular masses. From Hammam N'Bail, south of Djebel Nador, Constantine, Algeria.

Catoptrite. Katoptrite. — $14(Mn,Fe)O.2(Al,Fe)_2O_3.2SiO_2.Sb_2O_5$. Monoclinic. Crystals minute tabular parallel to $b(010)$. Perfect cleavage $\parallel$ (100). H. $= 5 \cdot 5$. G. $= 4 \cdot 5$.

Color black. In thin splinters, red. Optically −. Ax. pl. ‖ (010). $\beta = 1.95$. Pleochroic, red-brown to red-yellow. From Brattsfor mine, Nordmark, Vermland, Sweden.

Derbylite. — An antimo-titanate of iron. In prismatic, orthorhombic crystals. H. = 5. G. = 4·53. Color black. Sensibly uniaxial, +. Indices 2·45–2·51. From Tripuhy, near Ouro Preto, Minas Geraes, Brazil.

Lewisite. — $5CaO.2TiO_2.3Sb_2O_5$. In minute yellow to brown isometric octahedrons. Octahedral cleavage. H. = 5·5. G. = 4·9. Easily fusible. $n = 2.2$. From Tripuhy, near Ouro Preto, Minas Geraes, Brazil.

Mauzeliite. — A titano-antimonate of lead and calcium, related to lewisite. In dark brown isometric octahedrons. Jacobsberg, Nordmark district, Vermland, Sweden.

Ammiolite. A doubtful antimonite of mercury; forming a scarlet earthy mass. Chile.

Phosphates or Arsenates with Carbonates, Sulphates, Borates

Destinezite. — $2Fe_2O_3.P_2O_5.2SO_3.12H_2O$. Triclinic or monoclinic. Crystals hexagonal plates. H. = 3. G. = 2·1. White. Optically +. X nearly ⊥ (010). $\alpha = 1.615$, $\beta = 1.625$, $\gamma = 1.665$. 2V small. Dispersion, $\rho > v$. From Argenteau near Visé in Liège, Belgium and east of Kolin in Bohemia of Czechoslovakia. *Diadochite* is similar but with more water. Amorphous. Color brown or yellow. $n = 1.618$–1.70. From Thuringia near Saalfeld and at Arnsbach near Gräfental south of Saalfeld. In France in Isère near La Mure and in Finistère at Huelgoat.

Pitticite. — A hydrated arsenate and sulphate of ferric iron. Amorphous. Reniform and massive. Conchoidal fracture. Yellowish and reddish brown. H. = 2–3. G. = 2·5. Easily fusible. Index, 1·61–1·63. An alteration product formed from arsenopyrite. Found in Saxony at Freiberg and at Graul near Schwarzenberg and at Schneeberg. Occurs also in Cornwall at Redruth and St. Just. In the United States in the Tintic district, Utah, and at Manhattan, Nye Co., Nevada.

The following minerals form a group which shows certain relations to the Hamlinite Group, p. 711.

Svanbergite. — $2SrO.3Al_2O_3.2SO_3.P_2O_5.6H_2O$. In rhombohedral crystals. Basal cleavage, perfect. H. = 5. G. = 3·5. Color yellow to yellowish brown, rose-red. Optically +. $\omega = 1.63$. Also at times basal section divided into six biaxial segments. From Sweden at Horrsjöberg near Ny in Vermland and from Westanå mine near Näsum west of Carlshamn in Kristianstad. Sparingly at Chizeuil near Chalmoux, Saône-et-Loire, France.

Tikhvinite. — $2SrO.3Al_2O_3.P_2O_5.SO_3.6H_2O$. Found as microcrystalline masses, the size of peas. Color white. Anisotropic. $n = 1.62$. G. = 3·32. H. = 4·5. Occurs filling cavities in bauxite from Tikhvin district, Novgorod, Russia.

Hinsdalite. — $2PbO.3Al_2O_3.2SO_3.P_2O_5.6H_2O$. Pseudo-rhombohedral. In coarse, dull crystals. Cleavage, basal perfect. H. = 4·5. G. = 4·65. Colorless with greenish tone. Optically +. $\omega = 1.67$, $\epsilon = 1.70$. Usually biaxial with 2V up to 30°. Basal section divided into six segments. Found at Golden Fleece mine, Hinsdale Co., Colorado.

Beudantite. — A phosphate or arsenate with sulphate of ferric iron and lead; formula perhaps, $2PbO.3Fe_2O_3.2SO_3.As_2O_5.6H_2O$. In rhombohedral crystals. Basal cleavage. H. = 4. G. = 4·1. Fusible. Color green to brown and black. Optically −. $\omega = 1.96$. Also at times basal section shows six biaxial segments. Occurs in iron ore deposits especially with limonite as at Horhausen, Rhineland, and at Dernbach northwest of Montabaur, Hessen-Nassau. From Laurium, Greece. *Corkite* is the same mineral from near Cork, Ireland. Reported from Beaver Co., Utah.

Lossenite. — $4PbO.9Fe_2O_3.4SO_3.6As_2O_3.33H_2O$? Orthorhombic. Acute pyramidal crystals. H. = 3–4. Easily fusible. Color brownish red. Optically +. Ax. pl. ‖ (100). $Z = c$ axis. $\alpha = 1.783$, $\beta = 1.788$, $\gamma = 1.818$. 2V = 50°. Strong dispersion, $\rho > v$. From Laurium, Greece. Possibly identical with beudantite.

Lindackerite. — Perhaps $3NiO.6CuO.SO_3.2As_2O_5.7H_2O$. Probably monoclinic. In rosettes, and in reniform masses. Cleavage ‖ (010), perfect. H. = 2–2·5. G. = 2–2·5. Color verdigris- to apple-green. Optically +. Ax. pl. ‖ (010). $\alpha = 1.629$, $\beta = 1.662$, $\gamma = 1.727$. 2V = 73°. Strong dispersion, $\rho < v$. From Joachimstal, Bohemia, Czechoslovakia.

Lüneburgite. — $3MgO.B_2O_3.P_2O_5.8H_2O$. Probably monoclinic. Crystals in six-sided plates. Prismatic cleavage. In flattened masses, fibrous to earthy structure. G. = 2·05.

Fusible. Colorless. Optically $-$. Ax. pl. $\parallel$ (010). $\alpha = 1\cdot52$, $\beta = 1\cdot54$, $\gamma = 1\cdot545$. From Lüneburg, Hannover; also from Mejillones, Tarapaca, Chile.

Seamanite. — A hydrated borophosphate of manganese, perhaps $3MnO.(B_2O_5.P_2O_5).3H_2O$. Orthorhombic. In small acicular crystals. H. $= 4$. G. $= 3\cdot128$. Transparent. Pale yellow to wine-yellow. Optically $+$. Ax. pl. $\parallel$ (010). $Z \perp$ (001). $\alpha = 1\cdot640$, $\beta = 1\cdot663$, $\gamma = 1\cdot665$. $2V = $ about $40°$. Found in the Chicagon mine, Iron Co., Michigan.

Cahnite. — $4CaO.B_2O_3.As_2O_5.4H_2O$. Tetragonal, sphenoidal. Occurs in minute penetration twins. Crystal faces etched and rounded. Perfect cleavage $\parallel$ (110). H. $= 3$. G. $= 3\cdot156$. Easily fusible. Soluble in HCl. Color white. Vitreous luster. Optically positive. $\omega = 1\cdot662$, $\epsilon = 1\cdot663$. Found at Franklin, Sussex Co., New Jersey.

Nitrates

The Nitrates being largely soluble in water play but an unimportant rôle in Mineralogy.

SODA NITER. Chile saltpeter.

Rhombohedral. Axis $c = 0\cdot8276$; $rr'\ 10\bar{1}1 \wedge \bar{1}101 = 73° 30'$. Homœomorphous with calcite. Usually in massive form, as an incrustation or in beds. Cleavage: $r(10\bar{1}1)$ perfect. Fracture conchoidal, seldom observable. Rather sectile. H. $= 1\cdot5-2$. G. $= 2\cdot24-2\cdot29$. Luster vitreous. Color white; also reddish brown, gray and lemon-yellow. Transparent. Taste cooling. Optically $-$. $\omega = 1\cdot5874$, $\epsilon = 1\cdot3361$.

Comp. — Sodium nitrate, $NaNO_3$ = Nitrogen pentoxide $63\cdot5$, soda $36\cdot5$ = 100.

Pyr., etc. — Deflagrates on charcoal with less violence than niter, causing a yellow light, and also deliquesces. Colors the flame intensely yellow. Dissolves in three parts of water at 60° F.

Obs. — Found in great quantities in the desert region of northern Chile, in a narrow zone running north and south approximately 30 miles east of the coast through Tarapaca and Antofagasta and also in the neighboring portions of Bolivia. The ground is covered with beds of this salt, known as *caliche*, at times, several feet in thickness, along with gypsum, halite, glauber salt, etc. The *azufrado* or *caliche jaune* is a deposit rich in nitrates and colored yellow by alkaline iodides. Deposits also occur in Humboldt Co., Nevada, 25 miles east of Lovelock's Station; from near Calico, San Bernardino Co., and elsewhere in California.

Use. — A source of nitrates. The deposits in Chile are of great importance.

Niter. Saltpeter. Potassium nitrate, KNO_3. Orthorhombic. In thin white crusts and silky tufts. Cleavages, (011) perfect, (010) and (100) distinct. H. $- 2$. G. $= 2\cdot1$. Very easily fusible. Deflagrates violently on charcoal. Soluble in water. Colorless. Optically $-$. Ax. pl. $\parallel$ (100). $X = c$ axis. $\alpha - 1\cdot332$, $\beta = 1\cdot504$, $\gamma - 1\cdot504$. 2V very small. Occurs on the surface of the earth; on walls, rocks, etc. It forms abundantly in certain soils in Spain, Italy, Egypt, Arabia, Persia, and India, especially during hot weather succeeding rains. Its formation is aided by the action of bacteria upon animal remains, etc. It occurs in Chile in connection with the soda niter deposits. In the United States it has been found in caves in Kentucky, Tennessee, and the Mississippi valley in general.

Nitrocalcite. — Hydrous calcium nitrate, $Ca(NO_3)_2.nH_2O$. In efflorescent silky tufts and masses. One perfect cleavage. $\beta = 1\cdot50$. In many limestone caverns, as those of Kentucky.

Nitromagnesite. — $Mg(NO_3)_2.nH_2O$. Monoclinic? Fibrous. Easily fusible. Soluble in water. Colorless. Optically $-$. $\alpha = 1\cdot344$, $\beta = 1\cdot506$, $\gamma = 1\cdot506$. 2V very small. In efflorescences in limestone caves.

Nitrobarite. — Barium nitrate, $Ba(NO_3)_2$. Isometric-tetartohedral. Easily fusible. Soluble in water. Colorless. $n = 1\cdot57$. From Chile.

Gerhardtite. — Basic cupric nitrate, $Cu(NO_3)_2.3Cu(OH)_2$. In pyramidal orthorhombic crystals. Cleavages: (001) perfect, (100) less perfect. H. $= 2$. G. $= 3\cdot43$. Easily fusible. Color emerald-green. Ax. pl. $\parallel$ (010). $X = a$ axis. $\alpha = 1\cdot703$, $\beta = 1\cdot713$, $\gamma =$

1·722. 2V very large. Pleochroic; $X = Y$ = green, Z = blue. From the copper mines at Jerome, Yavapai Co., Arizona. Also from Katanga, Belgian Congo.

Buttgenbachite. — A hydrous chloro-nitrate of copper, probably analagous to connellite with the nitrate radical substituted for the sulphate group. Hexagonal. Prismatic; in minute flat needles. G. = 3·33. Color azure-blue. Parallel extinction with elongation || X. n = 1·747. Low birefringence. Found in a cavity of cuprite associated with native silver at Lihasi, Belgian Congo.

Julienite. — Hydrated nitrate of cobalt, isomorphous with buttgenbachite. In minute blue needles, probably tetragonal. Optically +. ω = 1·556, ϵ = 1·645. From Katanga, Belgian Congo.

Darapskite. — $NaNO_3.Na_2SO_4.H_2O$. Monoclinic. In square tabular crystals. Cleavages (100), (010), perfect. H. = 2–3. G. = 2·2. Soluble in water. Colorless. Optically —. Ax. pl. ⊥ (010). $Z \wedge c$ axis = 12°. α = 1·391, β = 1·481, γ = 1·486. Dispersion $\rho > v$. From near Taltal, Antofagasta, Chile.

NITROGLAUBERITE. $6NaNO_3.2Na_2SO_4.3H_2O$. Such a compound has been shown by synthetic studies not to be possible, the natural material being probably a mixture of darapskite and soda niter. From near Taltal, Antofagasta, Chile.

Lautarite. — Calcium iodate, $Ca(IO_3)_2$. In prismatic, monoclinic crystals, colorless to yellowish. Prismatic cleavage. H. = 4. G. = 4·6. Easily fusible. Optically +. Ax. pl. || (010). $X \wedge c$ axis = 25°. α = 1·792, β = 1·840, γ = 1·888. From the sodium nitrate deposits of Atacama, Chile.

Dietzeite. — $Ca(IO_3)_2.CaCrO_4$. Monoclinic; commonly fibrous or columnar. Shows definite crystallographic and structural relations to lautarite and crocoite. H. = 3–4. G. = 3·70. Easily fusible. Color dark gold-yellow. Optically —. Ax. pl. || (010). X nearly ⊥ cleavage. α = 1·825, β = 1·842, γ = 1·857. 2V nearly 90°. Strong dispersion, $\rho < v$. From the same region as lautarite.

Oxygen Salts

5. BORATES

The aluminates, ferrates, etc., allied chemically to the borates, have been already introduced among the oxides. They include the species of the Spinel Group, pp. 487–493, also Chrysoberyl, p. 493, etc.

SUSSEXITE.

Probably orthorhombic. In fibrous seams or veins. H. = 3. G. = 3·12. Luster silky to pearly. Color white with a tinge of pink or yellow. Translucent. Optically —. Fibers || X. α = 1·630, β = 1·709, γ = 1·712. 2V small.

Comp. — $HRBO_3$, where R = Mn, Zn and Mg = Boron trioxide 34·1, manganese protoxide, 41·5, magnesia 15·6, water 8·8 = 100. Here Mn (+Zn) : Mg = 3 : 2.

Pyr., etc. — In the closed tube darkens in color and yields neutral water. If turmeric paper is moistened with this water, and then with dilute hydrochloric acid, it assumes a red color (boric acid). In the forceps fuses in the flame of a candle (F. = 2), and B.B. in O.F. yields a black crystalline mass, coloring the flame intensely yellowish green. With the fluxes reacts for manganese. Soluble in hydrochloric acid.

Obs. — Found on Mine Hill, Franklin Furnace, Sussex Co., New Jersey, with franklinite, zincite, willemite, etc.

Ludwigite. Ferroludwigite. — Perhaps $3MgO.B_2O_3.FeO.Fe_2O_3$. Orthorhombic. In finely fibrous masses. H. = 5. G. = 4. Fusible. Color blackish green to nearly black. Optically +. Z || c axis. α = 1·84, β = 1·85, γ = 1·98. 2V very small. Strong dispersion, $\rho > v$. Pleochroic; $X = Y$ = bright green, Z = red-brown. From Rumania at Moravicza; also reported from Vaskö. In the United States found at Philipsburg, Granite Co., Montana. Also from the Cottonwood district, Salt Lake Co.. Utah. *Collbranite* from Korea is ludwigite.

VONSENITE. $3(Fe,Mg)O.B_2O_3.FeO.Fe_2O_3$. Similar to ludwigite with more ferrous iron. Riverside, Riverside Co., California.

Magnesioludwigite. — $3MgO.B_2O_3.MgO.Fe_2O_3$. The magnesium end of the ludwigite series. Similar to ludwigite in crystal and optical properties. From Mountain Lake mine, south of Brighton, Salt Lake Co., Utah.

Pinakiolite. Manganludwigite. — $3MgO.B_2O_3.MnO.Mn_2O_3$. Orthorhombic. In small rectangular tabular crystals. Cleavage $\parallel$ (010). H. = 6. G. = 3·88. Luster metallic. Color black. Optically −. Ax. pl. $\parallel$ (001). $X = b$ axis. $\alpha = 1\cdot908$, $\beta = 2\cdot05$, $\gamma = 2\cdot065$. 2V = 32°. From Långbanshyttan, Vermland, Sweden.

Nordenskiöldine. — A calcium-tin borate, $CaSn(BO_3)_2$. Rhombohedral. In basal plates. Perfect basal cleavage. H. = 5·5–6. G. = 4·2. Color sulphur-yellow. Optically −. $\omega = 1\cdot77$. From the island of Arö in the Langesundfiord, southern Norway.

Jeremejevite. Eremyeevite, Eremeyevite. Eichwaldite. — Aluminum borate, $AlBO_3$. Orthorhombic. In prismatic pseudo-hexagonal crystals. H. = 6·5. G. = 3·28. Colorless to pale yellow. Optically −. Crystals show both uniaxial and biaxial parts with complex relations to each other. $n = 1\cdot64$. From Mt. Soktuj, a northern extension of the Adun-Chilon Mts., south of Nerchinsk in Transbaikalia.

Hambergite. — $Be_2(OH)BO_3$. In grayish white orthorhombic prismatic crystals. X-ray study shows eight molecules to the unit cell. Cleavages: (010) perfect, (100) good. H. = 7·5. G. = 2·347. Optically +. Ax. pl. $\parallel$ (010). $Z = c$ axis. $\alpha = 1\cdot554$, $\beta = 1\cdot588$, $\gamma = 1\cdot628$. 2V = 87°. Occurs in a small pegmatite vein near Helgeraaen on the shore of the Langesundfiord, southern Norway. From Madagascar in notable crystals from south of Betafo, at Imalo near Mania, and also at Maharitra on Mt. Bity.

Szaibelyite. — $2Mg_5B_4O_{11}.3H_2O$. In small nodules; white outside, yellow within. Uniaxial, −. $\omega = 1\cdot65$, $\epsilon = 1\cdot58$. Other conflicting optical data are given. From Rézbánya, Rumania. Also with ludwigite west of Pioche, Lincoln Co., Nevada.

Camsellite. — $2MgO.B_2O_3.H_2O$. This formula, assigned to original material from British Columbia. Another occurrence later discovered in California was thought to have formula $2(Mg,Fe)O.(B_2O_3.SiO_2).H_2O$. Orthorhombic? Fibrous. H. < 3. G. = 2·60 (California). Readily fusible. Soluble in acids. White. Parallel extinction. Elongation negative. $\alpha = 1\cdot575$, $\beta = 1\cdot62$, $\gamma = 1\cdot65$. Found with dolomite in serpentine near Douglas Lake, British Columbia and associated with calcium and magnesium carbonates at Bolinas Bay, Marin Co., California.

BORACITE.

Isometric and tetrahedral in external form under ordinary conditions, but in molecular structure orthorhombic and pseudo-isometric; the structure becomes isotropic, as required by the form, only when heated to 265°. (See Art. **441**.)

1030   1031   1032

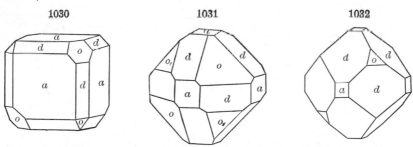

Habit cubic and tetrahedral or octahedral; also dodecahedral. Crystals usually isolated, embedded; less often in groups. Faces o(111) bright and smooth, $o_{\prime}(\bar{1}11)$ dull or uneven.

Cleavage: o, $o_{\prime}$ in traces. Fracture conchoidal, uneven. Brittle. H. = 7 in crystals. G. = 2·9. Luster vitreous, inclining to adamantine. Color white, inclining to gray, yellow and green. Streak white. Subtransparent to translucent. Commonly shows double refraction, which, however, disappears upon heating to 265°, when a section becomes isotropic. Optically +. Indices, 1·66–1·67. 2V large.

Strongly pyroelectric, the opposite polarity corresponding to the position of the + and − tetrahedral faces (see pp. 335, 336). The faces of the dull tetrahedron $o_l(1\bar{1}1)$ form the analogous pole, those of the polished form o (111) the antilogous pole.

Comp. — $Mg_6Cl_2B_{14}O_{26}$ or $5MgO.MgCl_2.7B_2O_3$.

Var. — 1. *Ordinary.* In crystals of varied habit. 2. *Massive,* with sometimes a subcolumnar structure; *stassfurtite* of Rose. It resembles a fine-grained white marble or granular limestone. *Parasite* of Volger is the plumose interior of some crystals of boracite. 3. *Eisenstassfurtite* contains some Fe.

Pyr., etc. — The massive variety gives water in the closed tube. B.B. both varieties fuse at 2 with intumescence to a white crystalline pearl, coloring the flame green; heated after moistening with cobalt solution assumes a deep pink color. Mixed with oxide of copper and heated on charcoal colors the flame deep azure-blue (copper chloride). Soluble in hydrochloric acid.

Alters very slowly on exposure, owing to the magnesium chloride present, which takes up water. It is the frequent presence of this deliquescent chloride in the massive mineral, thus originating, that led to the view that there was a hydrous boracite (stassfurtite). *Parasite* of Volger is a result of the same kind of alteration in the interior of crystals of boracite; this alteration giving it its somewhat plumose character, and introducing water.

Obs. — Observed in beds of anhydrite, gypsum or salt. Found in Hannover in crystals at Kalkberg and Schildstein in Lüneberg and at Hohenfels near Sehnde south of Lehrte. From Segeberg near Kiel in Schleswig-Holstein. Occurs massive, or as part of the rock, and in crystals, at Stassfurt in Province of Saxony. At Douglashall, Westeregeln, in Province of Saxony, pseudomorphous after quartz.

Fluoborite. — $3MgO.B_2O_3.3Mg(F,OH)_2$. Hexagonal. H. = probably 3·5. G. = 2·89. Colorless. Optically −. $\omega = 1·566$. $\epsilon = 1·528$. From Tallgruvan mine, east of Kallmora, Norberg, Sweden. Also from Franklin, Sussex Co., New Jersey.

Ascharite. $Mg_2B_2O_5.H_2O$. Orthorhombic? Fibrous. In white lumps with boracite. G. = 2·7. Optically −. $X \parallel$ fibers. Indices, 1·53–1·55. 2V small. From near Aschersleben and at Neu-Stassfurt, Province of Saxony.

Paternoite. — $MgB_8O_{13}.4H_2O$. Orthorhombic. Basal cleavage. In microscopic rhombic plates. Fine granular. G. = 2·11. Optically +. $X = b$ axis. $\alpha = 1·509$, $\beta = 1·528$, $\gamma = 1·548$. In salt beds at Monte Sambuco, Calascibetta, Sicily.

Rhodizite. — A hydrous borate of aluminum, beryllium, and alkalies. Pseudo-isometric-tetrahedral; in white, translucent dodecahedrons. H. = 8. G. = 3·4. Nearly isotropic. $n = 1·69$. Found on red tourmaline from near Ekaterinburg, Ural Mts.; from Madagascar at Antandrokomby near Mt. Bity and at Manjaka in the valley of Sahatany.

Warwickite. — $(Mg,Fe)_3TiB_2O_8$. Orthorhombic. In elongated prismatic crystals. Cleavage $\parallel$ (100). H. = 3–4. G. = 3·36. Color dark brown to dull black. Optically +. Ax. pl. $\parallel$ (010). $Z = a$ axis. $\alpha = 1·806$, $\beta = 1·809$, $\gamma = 1·830$. 2V small but variable. Occurs southwest of Edenville, in the town of Warwick, Orange Co., New York.

Howlite. — A silico-borate of calcium, $H_5Ca_2SiB_5O_{14}$. Monoclinic? Tabular $\parallel$ (100). In small white rounded nodules; also earthy. H. = 3·5. G. = 2·58. Easily fusible. Optically −. Ax. pl. $\perp$ (010). $Z \wedge c = 44°$. $\alpha = 1·586$, $\beta = 1·598$, $\gamma = 1·605$. 2V large. Occurs in Hants Co., Nova Scotia, at Brookville, about 3 miles south of Windsor, and also at Wentworth (formerly Winkworth). In California in large amounts in Los Angeles Co., near Lang, and in San Bernardino Co., in the Mohave desert, northeast of Daggett.

Bakerite. — A hydrated calcium borosilicate, $8CaO.5B_2O_3.6SiO_2.6H_2O$. In compact masses resembling unglazed porcelain. H. = 4·5. G. = 2·7–2·9. Fusible. Color white. $n = 1·64$. From borax deposits in Mohave desert, 16 miles northeast of Daggett, San Bernardino Co., California.

Lagonite. — $Fe_2O_3.3B_2O_3.3H_2O$. Amorphous. Color yellow. Isotropic. $n = 1·64$. An incrustation at the Tuscan lagoons, Italy.

Larderellite. — $(NH_4)_2B_{10}O_{16}.5H_2O$. Monoclinic. In minute plates with rhombic outline. Perfect cleavage $\perp$ plates. Soft. Easily fusible. White or yellowish. Optically +. $X = b$ axis. $Z \wedge c$ axis = 24°. $\alpha = 1·509$, $\beta = 1·52$, $\gamma = 1·561$. 2V = 60°. From Larderello in the Tuscan lagoons, Italy.

AMMONIOBORITE. $(NH_4)_2B_{10}O_{16}.5H_2O$. Differs from larderellite in crystal shape and refractive indices. $\alpha = 1.470$, $\beta = 1.487$, $\gamma = 1.540$. From the Tuscan lagoons, Italy.

COLEMANITE.

Monoclinic. Axes $a : b : c = 0.7748 : 1 : 0.5410$; $\beta = 69° 51'$.

Crystals usually short prismatic $(mm''' \ 110 \wedge 1\bar{1}0 = 72° 4')$. Massive cleavable to granular and compact.

Cleavable: $b(010)$ highly perfect; $c(001)$ distinct. Fracture uneven to subconchoidal. H. = 4–4.5. G. = 2.42. Luster vitreous to adamantine, brilliant. Colorless to milky white, yellowish white, gray. Transparent to translucent. Optically +. Ax. pl. $\perp$ (010). $Z \wedge c$ axis = 83°. $\alpha = 1.586$. $\beta = 1.592$. $\gamma = 1.614$. 2V = 55°.

Comp. — $Ca_2B_6O_{11}.5H_2O$, perhaps $HCa(BO_2)_3.2H_2O$ = Boron trioxide 50.9, lime 27.2, water 21.9 = 100.

Pyr. — B.B. decrepitates, exfoliates, sinters, and fuses imperfectly, coloring the flame yellowish green. Soluble in hot hydrochloric acid with separation of boric acid on cooling.

Obs. — First discovered in Death Valley, Inyo Co., California; later in Calico district, San Bernardino Co., in fine crystals. *Neocolemanite* from Lang, Los Angeles Co., California, is identical with colemanite. Found also in Kern, Riverside, and Ventura counties. Occurs in Clark Co., Nevada.

Priceite. — $5CaO.6B_2O_3.9H_2O$. Triclinic. In microscopic rhombic-shaped plates. Cleavages: (001) perfect, (110) and (010) poor. Crystalline to cryptocrystalline in compact masses. H. = 3–3.5. G. = 2.43. Color snow-white. Optically –. X inclined about 25° to normal to plates. $\alpha = 1.572$, $\beta = 1.591$, $\gamma = 1.594$. 2V = 32°. Dispersion, $\rho < v$. From Curry Co., Oregon, and Inyo Co., California. *Pandermite* from Panderma, Asia Minor, and Argentina is the same mineral.

Inyoite. — $2CaO.3B_2O_3.13H_2O$. Monoclinic. In large tabular crystals $\parallel$ (001). Cleavage, $c(001)$. H. = 2. G. = 1.87. Decrepitates and fuses with intumescence, giving green flame. Color white. Optically –. Ax. pl. $\parallel$ (010). $X \wedge c$ axis = +37°. $\alpha = 1.495$, $\beta = 1.51$, $\gamma = 1.52$. 2V = 70°–80°. Commonly alters to meyerhofferite. From Mt. Blanco district, on Furnace Creek, near Death Valley, Inyo Co., California, associated with colemanite. Also from Hillsborough, Albert Co., New Brunswick.

Meyerhofferite. — $2CaO.3B_2O_3.7H_2O$. Triclinic. Crystals prismatic, often tabular parallel to $a(100)$. Fibrous. Cleavage, $b(010)$. H. = 2. G. = 2.12. Fuses without decrepitation but with intumescence. Colorless to white. Optically –. Z makes small angle with normal to (001). $\alpha = 1.500$, $\beta = 1.535$, $\gamma = 1.560$. 2V = 80°. Found in California with inyoite (which see) as an alteration product.

Pinnoite. — $MgB_2O_4.3H_2O$. Tetragonal-pyramidal. Usually in nodules, radiated fibrous. H. = 3–4. G. = 2.29. Easily fusible. Color sulphur- or straw-yellow. Optically +. $\omega = 1.565$, $\epsilon = 1.575$. From Stassfurt, Province of Saxony.

Kaliborite. Heintzite. Hintzeite. — $K_2O.4MgO.11B_2O_6.18H_2O$. In small monoclinic crystals, sometimes aggregated together. Cleavages: (100) and (001) perfect. H. = 4–5. G. = 2.13. Very easily fusible. Colorless to white. Optically +. Ax. pl. $\perp$ (010). $Z \wedge c$ axis = +65°. $\alpha = 1.508$, $\beta = 1.525$, $\gamma = 1.550$. 2V = 81°. From Leopoldshall, in Anhalt near Stassfurt and at Schmidtmannshall near Aschersleben in the Province of Saxony.

Hulsite. — $12(Fe,Mg)O.2Fe_2O_3.1SnO_2.3B_2O_3.2H_2O$. Orthorhombic (?) as small crystals or tabular masses. H. = 3. G. = 4.3. Color and streak black. Fusible. Found in metamorphosed limestone at a granite contact at Brooks mountain, Seward Peninsula, Alaska. *Paigeite* is a similar mineral from the same locality with the composition, $30FeO.5Fe_2O_3.1SnO_2.6B_2O_3.5H_2O$.

BORAX.

Monoclinic. Axes $a : b : c = 1.0995 : 1 : 0.5632$; $\beta = 73° 25'$.

ca,	$001 \wedge 100 = 73° 25'$.	cz,	$001 \wedge \bar{2}21 = 64° 8'$.
mm''',	$110 \wedge 1\bar{1}0 = 93° 0'$.	oo',	$\bar{1}11 \wedge \bar{1}\bar{1}1 = 57° 27'$.
co,	$001 \wedge \bar{1}11 = 40° 31'$.	zz',	$\bar{2}21 \wedge \bar{2}\bar{2}1 = 83° 28'$.

Crystals prismatic, sometimes large; resembling pyroxene in habit and angles.

Cleavage: $a(100)$ perfect; $m(110)$ less so; $b(010)$ in traces. Fracture conchoidal. Rather brittle. H. = 2–2·5. G. = 1·69–1·72. Luster vitreous to resinous; sometimes earthy. Color white; sometimes grayish, bluish or greenish. Streak white. Translucent to opaque. Taste sweetish alkaline, feeble. Optically −. Ax. pl. ⊥ $b(010)$. $X = b$ axis. $Z \wedge c$ axis = −55°. $\alpha = 1\cdot447$. $\beta = 1\cdot470$. $\gamma = 1\cdot472$. 2V = 39°.

1033

Comp. — $Na_2B_4O_7.10H_2O$ or $Na_2O.2B_2O_3.10H_2O$ = Boron trioxide 36·6, soda 16·2, water 47·2 = 100.

Pyr., etc. — B.B. puffs up and afterward fuses to a transparent globule, called the glass of borax. Fused with fluorite and potassium bisulphate, it colors the flame around the assay a clear green. Soluble in water, yielding a faintly alkaline solution. Boiling water dissolves double its weight of this salt.

Obs. — Borax occurs in the waters of various saline lakes and in the salt deposits that have been formed through the evaporation of such lakes. It has been obtained since early times from the salt lakes of Kashmir and Tibet; it was brought to Europe as crude material called *tincal* and then purified. In the United States important deposits of borax are found in California; in Inyo Co., in the Death Valley region; in Lake Co., at Little Borax Lake, a few miles south of Clear Lake; in San Bernardino Co., at Searles Borax Lake, the most important deposit in the state, often in large crystals. In Nevada borax is obtained from Rhodes Marsh in Esmeralda Co.

Named borax from the Arabic *buraq*, which included also the *niter* (sodium carbonate) of ancient writers, the *natron* of the Egyptians. Borax was called chrysocolla by Agricola because used in soldering gold.

Use. — Borax is used for washing and cleansing; as an antiseptic and preservative; as a solvent for metallic oxides in soldering and welding; as a flux.

Tincalconite. — $Na_2O.2B_2O_3.5H_2O$. A dull white film of minutely crystallized material. Artificial product is rhombohedral. G. = 1·88. $\omega = 1\cdot460$, $\epsilon = 1\cdot473$. Occurs at Kramer district, Kern Co., California, as dehydration of borax or hydration of kernite.

Kernite. Rasorite. — $Na_2O.2B_2O_3.4H_2O$. Monoclinic. Cleavages: (001), (100), perfect; (101) distinct. H. = 3. G. = 1·953. Slowly soluble in cold water. B.B. swells and fuses to clear glass. Colorless to white. Transparent. Vitreous to pearly luster. Optically −. Ax. pl. ⊥ (010). $Z = b$ axis. $X \wedge c$ axis = 70°. $\alpha = 1\cdot454$, $\beta = 1\cdot472$, $\gamma = 1\cdot488$. 2V = 80°. Distinct dispersion, $\rho > v$. Found associated with ulexite in southeastern Kern Co., California, in sufficient amount to serve as a source of borates.

ULEXITE. Boronatrocalcite. Natroborocalcite.

Monoclinic. Usually in rounded masses, loose in texture, consisting of fine fibers, which are acicular or capillary crystals. H. = 1. G. = 1·65. Luster silky within. Color white. Tasteless. Optically +. Ax. pl. ⊥ (010). $Z \parallel b$ axis. $Y \wedge c$ axis 0° to 23°. $\alpha = 1\cdot500$. $\beta = 1\cdot508$. $\gamma = 1\cdot520$.

Comp. — A hydrous borate of sodium and calcium, probably $NaCaB_5O_9$. $8H_2O$ = Boron trioxide 43·0, lime 13·8, soda 7·7, water 35·5 = 100.

Pyr., etc. — Yields water. B.B. fuses at 1 with intumescence to a clear blebby glass, coloring the flame deep yellow. Moistened with sulphuric acid the color of the flame is momentarily changed to deep green. Not soluble in cold water, and but little so in hot; the solution alkaline in its reactions.

Obs. — Occurs in the dry plains of Iquique, Tarapaca, Chile (called *tiza*); in Argentina at Salinas de la Puna in Province Jujuy and also in Province Salta. In the United States it occurs in Nevada in large quantities in the salt marshes of the Columbus mining district in the western part of Esmeralda Co.; also from White Basin, Clarke Co. In California in Death Valley, Inyo Co.; from a few miles north of Desert Wells, Kern Co.; with colemanite at Lang, Los Angeles Co. In Nova Scotia in Hants Co., at Windsor and Brookville 3 miles to the south.

Named after the German chemist, G. L. Ulex.

Bechilite. $CaB_4O_7.4H_2O$. A doubtful species. In crusts, as a deposit from springs in Tuscany, Italy.

Hydroboracite. — $CaMgB_6O_{11}.6H_2O$. Monoclinic. Cleavages: (100), (010), perfect. Resembles fibrous and foliated gypsum. H. = 2. G. = 2. Easily fusible. Color white. Optically +. Ax. pl. || (010). $X \wedge c$ axis = 31°. $\alpha = 1.523$, $\beta = 1.534$, $\gamma = 1.570$. From the Caucasus Mts.; also from Stassfurt, Province of Saxony. Found near Ryan, Inyo Co., California.

Probertite. Kramerite. — $Na_2O.2CaO.5B_2O_3.10H_2O$. Monoclinic. Radiating prismatic habit. Cleavage (110) perfect. Colorless and transparent. Vitreous luster. H. = 3.5. G. = 2.141. Optically +. Ax. pl. || (010). $Z \wedge c$ axis = 12°. $\alpha = 1.515$, $\beta = 1.525$, $\gamma = 1.544$. Occurs in California in clay, borax or kernite at the Kramer district, Kern Co., and in considerable amount with ulexite at Ryan, Inyo Co.

Sulphoborite. — $2MgSO_4.4HMgBO_3.7H_2O$. In colorless prismatic orthorhombic crystals. Cleavages: (110), (001). H. = 4. G. = 2.4. Fusible. Optically −. Ax. pl. || (010). $X = c$ axis. $\alpha = 1.527$, $\beta = 1.540$, $\gamma = 1.544$. 2V = 70°. From the salt mines of Westeregeln, Province of Saxony.

Uranates

URANINITE. Cleveite. Bröggerite. Nivenite. Pitchblende.

Isometric. In octahedrons (o), also with dodecahedral faces (d); less often in cubes with o and d. Crystals rare. Usually massive and botryoidal; also in grains; structure sometimes columnar, or curved lamellar.

Fracture conchoidal to uneven. Brittle. H. = 5.5. G. = 9.0–9.7 of crystals; of massive altered forms from 6.4 upwards. Luster submetallic, to greasy or pitch-like, and dull. Color grayish, greenish, brownish, velvet-black. Streak brownish black, grayish, olive-green, a little shining. Opaque.

Comp. — A uranate of uranyl, lead, usually thorium (or zirconium), often the metals of the lanthanum and yttrium groups; also containing the gases nitrogen, helium and argon, in varying amounts up to 2.6 per cent. Calcium and water (essential?) are present in small quantities; iron also, but only as an impurity. The relation between the bases varies widely and no definite formula can be given. Radium was first discovered in this mineral and it has been shown that it and the helium present are products of the breaking down of the uranium.

It has been suggested that the UO_3 present in uraninite is due to oxidation and that the original isometric substance is UO_2 (ulrichite) with isomorphous replacements of ThO_2. This is considered to be isomorphous with thorianite.

Var. — The minerals provisionally included under the name uraninite are as follows:

1. *Crystallized. Uranniobite* from Norway. In crystals, usually octahedral, with G. varying for the most part from 9.0 to 9.7; occurs as an original constituent of coarse granites. The variety from Branchville, Connecticut, which is as free from alteration as any yet examined, contains chiefly UO_2 with a relatively small amount of UO_3. Thoria is prominent, while the earths of the lanthanum and yttrium groups are only sparingly represented.

Bröggerite, as analyzed by Hillebrand, gives the oxygen ratio of UO_3 to other bases of about 1 : 1; it occurs in octahedral crystals, also with $d(110)$ and $a(100)$. G. = 9.03.

Cleveite and *nivenite* contain UO_3 in larger amount than the other varieties mentioned, and are characterized by containing about 10 per cent of the yttrium earths. Cleveite is a variety from the Arendal, Norway, region occurring in cubic crystals modified by the dodecahedron and octahedron. G. = 7.49. It is particularly rich in the gas helium. Nivenite occurs massive, with indistinct crystallization. Color velvet-black. H. = 5.5. G. = 8.01. It is more soluble than other kinds of uraninite, being completely decomposed by the action for one hour of very dilute sulphuric acid at 100°. Material stated to be UO_3 in composition occurring in canary-yellow crystals with uraninite and uranophane in quartzite near Lusk, Wyoming, has been named *lambertite*.

2. *Massive*, amorphous or cryptocrystalline. Pitchblende. Contains no thoria; the rare earths also absent. Water is prominent and the specific gravity is much lower, in

some cases not above 6·5; these last differences are doubtless largely due to alteration. Here belong the kinds of pitchblende which occur in metalliferous veins, with sulphides of silver, lead, cobalt, nickel, iron, zinc, copper, as that from Johanngeorgenstadt, Germany; Přibram, Bohemia, etc.; probably also that from Black Hawk, Colorado.

Pyr., etc. — B.B. infusible, or only slightly rounded on the edges, sometimes coloring the outer flame green (copper). With borax and salt of phosphorus gives a yellow bead in O.F., becoming green in R.F. (uranium). With soda on charcoal gives a coating of lead oxide, and frequently the arsenic odor. Many specimens give reactions for sulphur and arsenic in the open tube. Soluble in nitric and sulphuric acids; the solubility differs widely in different varieties, being greater in those kinds containing the rare earths. Not attractable by the magnet. Strongly radioactive.

Obs. — As noted above, uraninite occurs either as a primary constituent of granitic pegmatites or as a secondary mineral with ores of silver, lead, copper, etc. It occurs associated with metallic minerals at Rézbánya in Rumania; at Přibram and Joachimstal in Bohemia of Czechoslovakia; in Saxony at Johanngeorgenstadt, Marienberg, Schneeberg, etc. Occurs in Norway near Moss in Olstfold at Ånneröd (*bröggerite*), Elvestad near Raade, etc.; also in Aust-Agder near Arendal (*bröggerite* and *cleveite*) and from Evje, Satersdal (*cleveite*). From various localities in Cornwall. From Mt. Lukwengule of the Uluguru Mts., in Morogoro, Tanganyika Territory, East Africa.

In the United States in Connecticut at feldspar quarries at Middletown and Portland, Middlesex Co., also in South Glastonbury, Hartford Co., and at Branchville, Fairfield Co. In Mitchell Co., North Carolina, at various mica mines. Often altered to gummite and uranophane. In Texas at the gadolinite locality in Llano Co. (*nivenite*). In quantity near Central City, Gilpin Co., Colorado. Important deposits of pitchblende are reported at Great Bear Lake, Mackenzie, Canada.

Use. — As a source of uranium and of radium salts.

Clarkeite. — $(Na_2,Pb)O.3UO_3.3H_2O$. Massive. Conchoidal fracture. H. = 4–4·5. G. = 6·39. Color very dark brown. Luster slightly waxy. Biaxial, −. $\alpha = 1·997$, $\beta = 2·098$, $\gamma = 2·108$. 2V = 30°–50°. Occurs surrounding uraninite from which it has been derived by alteration and is in turn enveloped by gummite. From Spruce Pine, Mitchell Co., North Carolina.

Gummite. — An alteration-product of uraninite of doubtful composition. In rounded or flattened pieces, looking much like gum. H. = 2·5. G. = 3·9–4·20. Luster greasy. Color reddish yellow to orange-red, reddish brown. Isotropic. $n = 1·61$. The varieties *eliasite* and *pittinite* are from Joachimstal, Bohemia in Czechoslovakia. In Saxony from Johanngeorgenstadt and Schneeberg. From Katanga, Belgian Congo. In the United States is abundant at the Flat Rock mine, Mitchell Co., North Carolina.

Yttrogummite. Occurs with cleveite in Norway as a decomposition-product.

Thorianite. — Chiefly $(Th,U)O_2$. Isometric, cubic habit. H. = 6·5. G. = 9·3. Color black. Radioactive. Isotropic, nearly opaque, $n = 2·2$. Obtained from gem gravels of Balangoda, Ceylon. Also noted from near Betroka, Madagascar.

Curite. — $2PbO.5UO_3.4H_2O$. Orthorhombic. In minute needles. H. = 4–5. G. = 7·192. Soluble in acids. Becomes brown and yields water in C. T. Strongly radioactive. Color reddish brown to deep yellow by transmitted light. Orange streak. Needles show parallel extinction with positive elongation. Optically −. $\alpha = 2·06$, $\beta = 2·11$, $\gamma = 2·15$. Found with many other rare minerals at Kasolo, Katanga, Belgian Congo.

Fourmarierite. A hydrous lead uranate. Orthorhombic. Crystals tabular || (100). Cleavage || (100). H. = 3–4. G. = 6·046. Color red. Adamantine luster. Optically −. Ax. pl. || (001). $X = a$ axis. $\alpha = 1·85$. $\beta = 1·92$, $\gamma = 1·94$. 2E large. Pleochroic in shades of yellow. An alteration product of uraninite found at Chinkolobwe, Katanga, Belgian Congo.

Uranosphærite. — $Bi_2O_3.2UO_3.3H_2O$. Orthorhombic? In half-globular aggregated forms. Cleavage || (100), perfect. H. = 2–3. G. = 6·36. Color orange-yellow, brick-red. Optically +. Ax. pl. || (010). $Z = c$ axis. $\alpha = 1·955$, $\beta = 1·985$, $\gamma = 2·05$. 2V very large. Strong dispersion, $\rho < v$. From Neustädtel south of Schneeberg, Saxony.

Oxygen Salts

6. SULPHATES, CHROMATES, TELLURATES

A. Anhydrous Sulphates, etc.

The important BARITE GROUP is the only one among the anhydrous sulphates and chromates.

Mascagnite. — Ammonium sulphate, $(NH_4)_2SO_4$. Orthorhombic. Usually in crusts and stalactitic forms. Cleavage, (001), distinct. H. = 2. G. = 1·76. Soluble in water. Colorless, yellowish, greenish. Optically +. Ax. pl. || (010). $Z = a$ axis. $\alpha = 1·521$, $\beta = 1·523$, $\gamma = 1·533$. 2V = 52°. Occurs about volcanoes, as at Etna, Vesuvius, etc., and has probably been formed by the action of sulphuric acid fumes upon sal ammoniac. Also occurs in guano deposits as on the Chincha and Guanape islands of Peru.

Taylorite. — $(NH_4)_2SO_4.5K_2SO_4$. Orthorhombic? In small compact lumps or concretions. H. = 2. Easily fusible. Colorless. Optically +. $\alpha = 1·447$, $\beta = 1·448$, $\gamma = 1·459$. 2V = 36°. Strong dispersion, $\rho > v$. From the guano of the Chincha Islands, Peru.

Thenardite. — Na_2SO_4. In orthorhombic crystals, pyramidal, short prismatic or tabular; also as twins (Fig. 406, p. 180). Basal cleavage. H. = 2·7. G. = 2·68. Easily fusible. Soluble in water. White to brownish. Optically +. Ax. pl. || (001). $Z = a$ axis. $\alpha = 1·471$, $\beta = 1·477$, $\gamma = 1·484$. 2V = 83°. Changes on heating at 235° C. to a negative uniaxial form, *metathenardite*. Often observed in connection with salt lakes, as on the shores of Lake Balkash in Semipalatinsk, Siberia, and at Shermakha northwest of Baku in Azerbaijan, Caucasus Mts. It occurs in Spain in the Provinces of Madrid and Toledo, especially at Espartinas near Aranjuez. Found in Chile near Tarapaca northeast of Iquique, Province of Tarapaca, and at Caracoles in Antofagasta; from near Aquas Blancas near Copiapo in Atacama, etc. In the United States in extensive beds on the Rio Verde in central Arizona. In California, at Borax Lake, San Bernardino Co. In Nevada at Rhodes Marsh, Mineral Co.

Aphthitalite. Glaserite. — $(K,Na)_2SO_4$. Rhombohedral; also massive, in crusts. Prismatic cleavage. H. = 3. G. = 2·7. Easily fusible. Soluble in water. Color white. Optically +. Indices 1·487–1·499. Occurs both in connection with volcanic lavas, where it has been formed by the action of sulphuric acid vapors upon the alkalies of the rocks or upon previously formed alkaline chlorides, and also in connection with saline deposits where it has been crystallized from solution. Found at Vesuvius upon lava; similarly at Etna on Sicily and from Kilauea, Hawaiian Islands. Occurs at Douglashall near Westeregeln, Province of Saxony. In the United States has been found at Borax Lake, San Bernardino Co., California. Artificial K_2SO_4, *arcanite*, has closely related optical and crystallographic properties to aphthitalite.

GLAUBERITE.

Monoclinic.   Axes $a : b : c = 1·2200 : 1 : 1·0275$; $\beta = 67° 49'$.

ca, $001 \wedge 100 = 67° 49'$. cs, $001 \wedge 111 = 43° 2'$.

mm''', $110 \wedge 1\bar{1}0 = 96° 58'$. cm, $001 \wedge 110 = 75° 30\frac{1}{2}'$.

1034

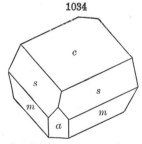

In crystals tabular || $c(001)$; also prismatic. Cleavage: c (001) perfect. Fracture conchoidal. Brittle. H. = 2·5–3. G. = 2·7–2·85. Luster vitreous. Color pale yellow or gray; sometimes brick-red. Streak white. Taste slightly saline. Optically −. Ax. pl. ⊥ (010). $X \wedge c$ axis = +30° and nearly ⊥ basal cleavage. 2V = 7°. $\alpha = 1·515$. $\beta = 1·535$. $\gamma = 1·536$. With rise in temperature 2V changes rapidly. The mineral becomes uniaxial for the D line at 40·6° C. See further p. 326.

Comp. — $Na_2SO_4.CaSO_4$ = Sulphur trioxide 57·6, lime 20·1, soda 22·3 = 100; or, Sodium sulphate 51·1, calcium sulphate 48·9 = 100.

Pyr., etc. — B.B. decrepitates, turns white, and fuses at 1·5 to a white enamel, coloring the flame intensely yellow. On charcoal fuses in O.F. to a clear bead; in R.F. a portion is absorbed by the charcoal, leaving an infusible hepatic residue. Soluble in hydrochloric acid. In water it loses its transparency, is partially dissolved, leaving a residue of calcium sulphate, and in a large excess this is completely dissolved.

Obs. — Occurs in salt deposits at various localities. Found in Salzburg, Austria, at Dürenberg near Hallein, at Ischl and Hallstadt, and in the neighboring locality of Berchtesgaden in Upper Bavaria. In the Province of Saxony at Douglashall near Westeregeln and at Leopoldshall in Anhalt near Stassfurt. In Spain at Villarrubia near Ocana in the Province of Toledo and at neighboring localities in the Province of Madrid. In India at the Mayo salt mines at Khewra in the Punjab. From Chile near Iquique in Tarapaca. In the United States in the Rio Verde valley, Yavapai Co., Arizona. In California, from Borax Lake, San Bernardino Co.

Langbeinite. — $K_2Mg_2(SO_4)_3$. Isometric-tetartohedral. In highly modified colorless crystals. G. = 2·83. $n = 1·535$. From Westeregeln and Stassfurt, Germany; Hall, Tyrol; Punjab, India.

MANGANOLANGBEINITE. $K_2Mn_2(SO_4)_3$. A sulphate of manganese and potassium, assumed to have this composition because of similarity of its physical characters with the artificial salt of this composition. Isometric, tetrahedral. In tetrahedrons. Rose-red color. $n = 1·572$. G. = 3·02. Found as small crystals in stalactites of thenardite and halite with sylvite and aphthitalite in a cavern of the recent lavas of Vesuvius.

Vanthoffite. — $3Na_2SO_4.MgSO_4$. Almost colorless crystalline material found at Wilhelmshall, near Stassfurt, Prussia.

Barite Group. RSO₄. Orthorhombic

		$m \wedge m'''$	dd'	oo'	
		$110 \wedge 1\bar{1}0$	$102 \wedge \bar{1}02$	$011 \wedge 0\bar{1}1$	$a : b : c$
Barite	BaSO₄	78° 22½′	77° 43′	105° 26′	0·8152 : 1 : 1·3136
Celestite	SrSO₄	75° 50′	78° 49′	104° 0′	0·7790 : 1 : 1·2801
Anglesite	PbSO₄	76° 16½′	78° 47′	104° 24½′	0·7852 : 1 : 1·2894

The BARITE GROUP includes the sulphates of barium, strontium, and lead, three species which are closely isomorphous, agreeing not only in axial ratio but also in crystalline habit and cleavage.

X-ray study of barite, celestite and anglesite shows that the atomic structure possesses a unit orthorhombic cell the dimensions of which correspond in terms of the crystal axial ratio to $2a$, $1b$, $1c$.

BARITE. Heavy Spar. Barytes.

Orthorhombic. Axes $a : b : c = 0·8152 : 1 : 1·3136$.

mm''',	$110 \wedge 1\bar{1}0 = 78° 22½′$.	dd''', $102 \wedge 10\bar{2} = 102° 17′$.
cd,	$001 \wedge 102 = 38° 51½′$.	oo''', $011 \wedge 01\bar{1} = 74° 34′$.
co,	$001 \wedge 011 = 52° 43′$.	cz, $001 \wedge 111 = 64° 19′$.

Crystals commonly tabular ∥ $c(001)$, and united in diverging groups having the axis b in common; also prismatic, most frequently ∥ axis b, $d(102)$ predominating; also ∥ axis c, $m(110)$ prominent; again ∥ axis a, with $o(011)$ prominent. Also in globular forms, fibrous or lamellar, crested; coarsely laminated, laminæ convergent and often curved; granular, resembling white marble, and earthy; colors sometimes banded as in stalagmite.

Cleavage: $c(001)$ perfect; $m(110)$ also perfect, Fig. 1035 the form yielded by cleavage; also $b(010)$ imperfect. Fracture uneven. Brittle. H. = 2·5–3·5. G. = 4·3–4·6. Luster vitreous, inclining to resinous; sometimes pearly on $c(001)$, less often on $m(110)$. Streak white. Color white; also inclining to yellow, gray, blue, red, or brown, dark brown. Transparent to translucent to opaque. Sometimes fetid, when rubbed. Optically +. Ax. pl. ‖ $b(010)$. $Z \perp a(100)$. $2V = 37° 30'$. $\alpha = 1\cdot636$. $\beta = 1\cdot637$. $\gamma = 1\cdot648$.

Var. — *Ordinary.* (*a*) Crystals usually broad or stout; sometimes very large; again in slender needles. (*b*) *Crested;* massive aggregations of tabular crystals, the crystals project-

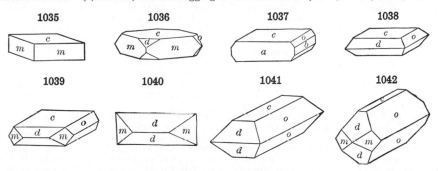

ing at surface into crest-like forms. (*c*) *Columnar;* the columns often coarse and loosely aggregated, and either radiated or par- allel; rarely fine fibrous. (*d*) In globular or nodular concre- tions, subfibrous or columnar within. *Bologna Stone* (from near Bologna) is here included; it was early a source of wonder be- cause of the phosphorescence it exhibited after heating with char- coal due to the barium sulphide formed. " Bologna phospho- rus " was made from it. (*e*) Lamellar, either straight or curved; the latter sometimes as aggregations of curved scale-like plates. (*f*) Granular. (*g*) Compact or cryptocrystalline. (*h*) Earthy. (*i*) Stalactitic and stalagmitic; similar in structure and origin to calcareous stalactites and stalagmites and of much beauty when polished. (*h*) Fetid; so called from the odor given off when struck or when two pieces are rubbed together, which odor may be due to carbonaceous matters present.

The barite of Muzsaj and of Betler, near Rosenau, Hungary, was early called *wolnyn. Cawk* is the ordinary barite of the Derbyshire lead mines. *Dreelite,* supposed to be rhom- bohedral, is simply barite. *Michel-lévyte* from Perkins' Mill, Templeton, Quebec (described as monoclinic), is peculiar in its pearly luster on *m,* twinning striations, etc.

Comp. — Barium sulphate, $BaSO_4$ = Sulphur trioxide 34·3, baryta 65·7 = 100.

Strontium sulphate is often present, (*barytocelestite*), also calcium sulphate; further, as impurities, silica, clay, bituminous or carbonaceous substances. A lead-bearing barite from Hokute in Formosa has been called *hokutolite.* It is a hot spring deposit. Similar material has been found at Shibukure near Akita, Japan. What is probably the same mineral from Chile has been named *weisbachite.*

Pyr., etc. — B.B. decrepitates and fuses at 3, coloring the flame yellowish green; the fused mass reacts alkaline with test paper. On charcoal reduced to a sulphide. With soda gives at first a clear pearl, but on continued blowing yields a hepatic mass, which spreads out and soaks into the coal. This reacts for sulphur (p. 372). Insoluble in acids.

Diff. — Characterized by high specific gravity (higher than celestite, aragonite, albite, calcite, gypsum, etc.); cleavage; insolubility; *green* coloration of the blowpipe flame. Albite is harder and calcite effervesces with acid.

Obs. — Barite is the most common mineral containing barium; it occurs at times in large amounts as veins or beds; also as a gangue mineral in various mineral veins in crystals and crystal groups. It is commonly associated with the ores of lead, also copper, iron, zinc, silver, nickel, cobalt, manganese, etc. It often accompanies stibnite, frequently associated with fluorite, quartz, calcite, dolomite, siderite, etc. Sometimes present in massive form with hematite deposits. It is found in limestones and sandstones, sometimes forming dis- tinct veins, and in the former often in crystals along with calcite and celestite; in the latter often with copper ores.   At times forms the cementing material in sandstones.   Sometimes occupies the cavities of amygdaloid, porphyry, etc.; forms earthy masses in beds of marl. Occurs as the petrifying material of fossils and occupies cavities in them.

Barite is of such frequent and wide occurrence that only the most important localities can be mentioned here, especially those which are noted for their unusual specimens.   In Rumania fine crystals occur in the mineral veins of Felsöbánya (Baia Sprie), Kapnikbánya, and Offenbánya, and in Czechoslovakia at Schemnitz (Selmeczbánya), Příbram, Gift-Berg northwest of Příbram, and Kremitz (Körmöczbánya).   It occurs at many points in Germany; especially to be noted are Freiberg and Marienberg in Saxony and in the Harz Mts. at Klaustal, etc.   There are many localities in France, especially in the district of Limagne in Puy-de-Dôme at Royat, etc.; also at Flaviac in Ardéche.   Barite occurs in large deposits in Spain.   In England there are many well-known localities; in Cornwall at Liskeard, etc.; in Westmorland at the Dufton lead mines in large transparent crystals, one of which weighed 100 pounds; in Cumberland at Alston Moor, Cleator Moor, Frizington, Egremont, and nearby Pallaflat, often in blue crystals with hematite and calcite; from Chirbury in Shropshire; in fine brown-colored stalactites from Newhaven near Youlgreave and from near Matlock.

In the United States the important localities include the following: in Connecticut at Cheshire, New Haven Co., large crystal groups; in New York at Pillarpoint opposite Sackett's Harbor in Jefferson Co., massive, affording large slabs, beautiful when polished; in St. Lawrence Co., fine tabular or prismatic crystals at Dekalb.   In Pennsylvania at the Perkiomen mine, Montgomery Co., in crystals; a fetid variety occurs in Berks Co.   In Virginia in Buckingham Co.   In large veins on Isle Royal, Michigan.   Common in Missouri.   Crystals enclosing quartz sand, "sand barite," from Norman, Cleveland Co., Oklahoma.   In distorted crystals, often occupying cavities in fossil bones, from the Bad Lands, South Dakota.   In fine crystals from near Fort Wallace, New Mexico.   From many places in Colorado, especially in light blue crystals from near Sterling, Logan Co.; on Apishapa River in Las Animas Co.; in Fremont Co., etc.; in Utah at Dugway, Tooele Co. In Canada occurs in Ontario in large veins on Jarvis, McKellars, and Pie islands in Lake Superior and in the Thunder Bay district.   In Nova Scotia in Colchester Co., in the Londonderry mines.

Named from βαρυς, *heavy*.

Use. — Source of barium hydroxide used in the refining of sugar; ground and used as a pigment, to give weight to paper, cloth, etc.

CELESTITE. Cœlestine.

Orthorhombic.   Axes $a : b : c = 0.7790 : 1 : 1.2800$.

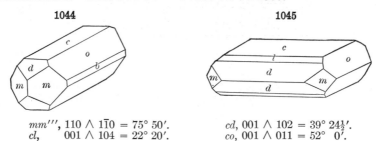

| 1044 | 1045 |

mm''', 110 ∧ 1\bar{1}0 = 75° 50'. cd, 001 ∧ 102 = 39° 24½'.
cl,   001 ∧ 104 = 22° 20'. co, 001 ∧ 011 = 52° 0'.

Crystals resembling those of barite in habit; commonly tabular ∥ c(001) or prismatic ∥ axis a or b; also more rarely pyramidal by the prominence of the forms ψ (133) or χ (144).   Also fibrous and radiated; sometimes globular; occasionally granular.

Cleavage: c(001) perfect; m(110) nearly perfect; b(010) less distinct. Fracture uneven.   H. = 3–3.5.   G. = 3.95–3.97.   Luster vitreous, sometimes inclining to pearly.   Streak white.   Color white, often faint bluish, and sometimes reddish.   Transparent to subtranslucent.   Optically +.   Ax. pl. ∥ (010).   Z ⊥ (100). $\alpha = 1.622$. $\beta = 1.624$. $\gamma = 1.631$.   2V = 51°.   Distinct dispersion, $\rho < v$.

Var. — 1. *Ordinary.* (*a*) In crystals of varied habit as noted above; a tinge of a delicate blue is very common and sometimes belongs to only a part of a crystal. The variety from Montmartre, near Paris, France, called *apotome*, is prismatic by extension of *o*(011) and doubly terminated by the pyramid ψ(133). (*b*) Fibrous, either parallel or radiated. (*c*) Lamellar; of rare occurrence. (*d*) Granular. (*e*) Concretionary. (*f*) Earthy; impure usually with carbonate of lime or clay.

Comp. — Strontium sulphate, $SrSO_4$ = Sulphur trioxide 43·6, strontia 56·4 = 100. Calcium and barium are sometimes present.

Pyr., etc. — B.B. frequently decrepitates, fuses at 3 to a white pearl, coloring the flame strontia-red; the fused mass reacts alkaline. On charcoal fuses, and in R.F. is converted into a difficultly fusible hepatic mass; this treated with hydrochloric acid and alcohol gives an intensely red flame. With soda on charcoal reacts like barite. Insoluble in acids.

Diff. — Characterized by form, cleavage, high specific gravity, *red* coloration of the blowpipe flame. Does not effervesce with acids like the carbonates (*e.g.*, strontianite); specific gravity lower than that of barite.

Obs. — Celestite is usually associated with limestone or sandstone, of various geological ages. At times it occurs in beds of gypsum, rock salt, etc. Frequently associated with sulphur in volcanic regions. Also in metallic veins with galena, sphalerite, etc. In amygdaloidal cavities of volcanic rocks. At times in large veins or beds, where it occurs on a commercial scale.

Found in Slovakia of Czechoslovakia at Herrengrund (Urvölgy) north of Neusohl (Ban Bystrica); at Leogang in Salzburg, Austria; at Scharfenberg, southeast of Meissen, Saxony. In fine specimens at Bex, Vaud, Switzerland. In Italy in the amygdules of a volcanic rock at Montecchio Maggiore, west of Vicenza in Venetia; near Bellisio in Ancona, Marche; with sulphur at Perticara, south of Cesena, Romagna; in Sicily in splendid groups of crystals with sulphur and gypsum at Girgenti and the nearby localities of Cattolica, Cianciana, Caltanissetta, etc. In large veins and frequently in fine crystals in the neighborhood of Bristol, Gloucester, at Clifton, Durdham Down, Yate, and Wickwar. Found in Egypt, southeast of Cairo at Djebel Mokattam in Wadi-el-Tih.

Celestite is found in the United States in New York at Lockport, Niagara Co.; at the lead mine at Rossie near Grasse Lake, St. Lawrence Co.; in Pennsylvania at Bellwood, Blair Co., in blue fibrous layers (this was the first celestite described). In Mineral Co., West Virginia, in pyramidal blue crystals. In cavities in limestone at Nashville, Tennessee. Finely crystallized specimens of a bluish tint are found at Put in Bay on Strontian Island in Lake Erie, and on Drummond Island in Lake Huron. In Brown Co., Kansas, a red variety in large crystals. At Glen Eyrie near Manitou, El Paso Co., Colorado, in fine clear crystals with the colemanite of Death Valley, Inyo Co., California.

Named from *cœlestis, celestial,* in allusion to the faint shades of blue often present.

Use. — Used in the preparation of strontium nitrate for fireworks; other salts used in the refining of sugar.

ANGLESITE.

Orthorhombic.   Axes $a : b : c = 0·7852 : 1 : 1·2894$.

mm''', 110 $\wedge$ 1$\bar{1}$0 = 76° 16½'. cd, 001 $\wedge$ 102 = 39° 23'.
cl,   001 $\wedge$ 104 = 22° 19'. co, 001 $\wedge$ 011 = 52° 12'.

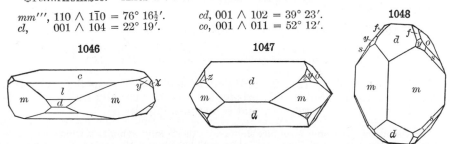

Crystals sometimes tabular ‖ *c*(001); more often prismatic in habit, and in all the three axial directions, *m*(110), *d*(102), *o*(011), predominating in the different cases; pyramidal of varied types. Also massive, granular to compact; stalactitic; nodular.

Cleavage: $c(001)$, $m(110)$ distinct, but interrupted. Fracture conchoidal. Very brittle. H. = 2·75–3. G. = 6·3–6·39. Luster highly adamantine in some specimens, in others inclining to resinous and vitreous. Color white, tinged yellow, gray, green, and sometimes blue. Streak uncolored. Transparent to opaque. Optically +. Ax. pl. || (010). $Z \perp$ (100). Dispersion strong, $\rho < v$. 2V = 60° to 75°. $\alpha = 1·877$. $\beta = 1·882$. $\gamma = 1·894$.

Comp. — Lead sulphate, $PbSO_4$ = Sulphur trioxide 26·4, lead oxide 73·6 = 100.

For varieties intermediate between anglesite and barite see under the latter, p. 749.

Pyr., etc. — B.B. decrepitates, fuses in the flame of a candle (F. = 1·5). On charcoal in O.F. fuses to a clear pearl, which on cooling becomes milk-white; in R.F. is reduced with effervescence to metallic lead. With soda on charcoal in R.F. gives metallic lead, and the soda is absorbed by the coal. Difficultly soluble in nitric acid.

Diff. — Characterized by high specific gravity; adamantine luster; cleavage; and by yielding lead B.B. Cerussite effervesces in nitric acid.

Obs. — Anglesite is commonly a mineral of secondary origin, having been formed usually by the oxidation of lead sulphide. It occurs in the oxidation zone of lead veins, its crystals often lying in cavities in galena, or it occurs in an earthy condition, forming concentric layers about a nucleus of unaltered galena.

It is of common occurrence, and only the more important localities, especially those for the crystallized mineral, can be mentioned here. Found in Carinthia, Austria, at Bleiberg. In Germany it is found at Badenweiler in Baden; at Müsen near Siegen in Westphalia. In fine crystals on the island of Sardinia in Cagliari, near Iglesias, at Monteponi and Montevecchio, and in the mines of San Giovanneddu near Gonnesa, etc. In Spain at Cartegena in Murcia and elsewhere. Originally found on the island of Anglesey, Wales. In fine specimens in Derbyshire at Matlock and the nearby locality of Cromford. In Scotland at Leadhills in Lanark, and formerly in large and fine crystals from Wanlockhead in Dumfries. In exceptional crystals from Sidi-Amor-ben-Salem in Tunis. In South West Africa at Tsumeb near Otavi. Occurs in notable specimens from Broken Hill, New South Wales, and from Dundas, Tasmania. From the mine Mérétrice in New Caledonia. In Mexico in large amounts, mostly massive, from Sierra Mojada, Coahuila, and in crystals from Sierra de los Lamentos in Chihuahua and at Boléo near Santa Rosalia in Lower California.

In the United States it was found in large crystals at the Wheatley mine, Phœnixville, Chester Co., Pennsylvania; less well crystallized from the Missouri lead mines. In crystals from Eureka in the Tintic district, Utah. From the Castle Dome district, Yuma Co., Arizona. From the Coeur d'Alene district in Shoshone Co., Idaho.

Named from the locality, Anglesey, where it was first found.

Use. — An ore of lead.

ANHYDRITE.

Orthorhombic. Axes $a : b : c = 0·8933 : 1 : 1·0008$.

mm''', 110 $\wedge$ $1\bar{1}0 = 83° 33'$. ss', 011 $\wedge$ $0\bar{1}1 = 90° 3'$.
rr',   101 $\wedge$ $\bar{1}01 = 96° 30'$. bo, 010 $\wedge$ 111 $= 56° 19'$.

Twins: 1, tw. pl. $d(012)$; 2, $r(101)$ occasionally as tw. lamellæ. Crystals not common, thick tabular, also prismatic || axis b. Usually massive, cleavable, fibrous, lamellar, granular, and sometimes impalpable.

X-ray study of the atomic structure shows an orthorhombic symmetry which is pseudo-tetragonal in the direction of the b axis. Each sulphur atom is the center of a tetrahedral group of four oxygen atoms and each calcium atom has eight oxygen atoms nearly equally spaced about it. The dimensions of the unit cell have the same relative values as the axial ratio given above.

Cleavage: in the three pinacoidal directions yielding rectangular fragments but with varying ease, thus, $c(001)$ very perfect; $b(010)$ also perfect; $a(100)$ somewhat less so. Fracture uneven, sometimes splintery. Brittle. H. = 3–3·5. G. = 2·899–2·985. Luster: c pearly, especially after heating in a closed tube; a somewhat greasy; b vitreous; in massive varieties, vitreous inclining to pearly. Color white, sometimes a grayish, bluish, or reddish tinge; also brickred. Streak grayish white. Optically +. Ax. pl. ‖ (010). Z ⊥ (100). $2V = 42°$. $\alpha = 1·571$. $\beta = 1·576$. $\gamma = 1·614$. Distinct dispersion, $\rho < v$.

1049

1050

1051

1049, 1050, Stassfurt   1051, Aussee

Var.—1. *Ordinary.* (a) Crystallized; crystals rare, more commonly massive and cleavable in its three rectangular directions. (b) Fibrous; either parallel, radiated or plumose. (c) Fine granular. (d) Scaly granular. *Vulpinite* is a scaly granular kind from Vulpino in Lombardy, Italy; it is cut and polished for ornamental purposes. A kind in contorted concretionary forms is the *tripestone*.

2. *Pseudomorphous;* in cubes after rock-salt.

Comp. — Anhydrous calcium sulphate, $CaSO_4$ = Sulphur trioxide, 58·8, lime 41·2 = 100.

Pyr., etc. — B.B. fuses at 3, coloring the flame reddish yellow, and yielding an enamel-like bead which reacts alkaline. On charcoal in R.F. reduced to a sulphide; with soda does not fuse to a clear globule, and is not absorbed by the coal like barite; is, however, decomposed, and yields a mass which blackens silver. Soluble in hydrochloric acid.

Diff. — Characterized by its cleavage in three rectangular directions (pseudo-cubic in aspect); harder than gypsum; does not effervesce with acids like the carbonates.

Obs. — Anhydrite occurs under much the same conditions as gypsum, the two minerals frequently being found together. It is chiefly found in sedimentary formations, with limestones and most commonly with salt beds. It also occurs in smaller amounts in connection with various metalliferous veins and at times in the amygdaloidal cavities of volcanic rocks.

The more important localities for the occurrence of anhydrite follow. From the salt beds of Wieliczka in Krakow, Poland. Found in Austria at Bleiberg in Carinthia; in salt deposits at Dürenberg near Hallein, at Ischl in Salzburg, and at Aussee in Styria; and from Hall near Innsbruck, Tyrol. At Stassfurt in the Province of Saxony in fine crystals embedded in kieserite and similarly at Leopoldshall in Anhalt. In Hannover at Lüneburg and at Wathlingen, south of Celle; in Upper Bavaria at Berchtesgaden; at Sulz on the Neckar in Würtemberg. Occurs in Switzerland at Bex, Vaud. From France at Arnave in Ariège.

In the United States at Lockport, Niagara Co., New York, fine blue with calcite and gypsum. From West Paterson, New Jersey, associated with zeolites. In cavities in limestone at Nashville, Tennessee. In Nova Scotia it occurs in considerable amounts at various localities. From Hillsborough, Albert Co., New Brunswick.

Anhydrite by absorption of moisture changes to gypsum. Extensive beds are sometimes thus altered in part or throughout, as at Bex, in Switzerland, where, by digging down 60 to 100 ft., the unaltered anhydrite may be found. Sometimes specimens of anhydrite are altered between the folia or over the exterior.

Bassanite. — $CaSO_4$. In white opaque crystals having form of gypsum but composed of slender needles in parallel arrangement. These show parallel extinction and positive elongation. G. = 2·69–2·76. Transformed into anhydrite at red heat. Found in blocks ejected from Vesuvius.

Zinkosite. — $ZnSO_4$. Reported as occurring at a mine in the Sierra Almagrera, near Aquilas in Murcia, Spain. The artificial zinc sulphate is probably orthorhombic. In crystals tabular ‖ (001). Optically −. $\beta = 1·67$. $2V$ small. Strong dispersion, $\rho < v$.

Hydrocyanite. — $CuSO_4$. Found at Vesuvius as a pale green to blue incrustation after the eruption of 1868.

Millosevichite. — Normal ferric and aluminum sulphate. As a violet incrustation, Alum Grotto, Island of Vulcano, Lipari Islands.

CROCOITE.

Monoclinic.   Axes $a : b : c = 0.9603 : 1 : 0.9159$; $\beta = 77° 33'$.

1052

mm''', 110 $\wedge$ 1$\bar{1}$0 = 86° 19'.
ck,   001 $\wedge$ $\bar{1}$01 = 49° 32'.

tt', 111 $\wedge$ 1$\bar{1}$1 = 60° 50'.
ct, 001 $\wedge$ 111 = 46° 58'.

Crystals usually prismatic, habit varied.   Also imperfectly columnar and granular.

Cleavage: m(110) rather distinct; c(001), a(100) less so. Fracture small conchoidal to uneven.   Sectile.   H. = 2·5–3.   G. = 5·9–6·1.   Luster adamantine to vitreous.   Color various shades of bright hyacinth-red.   Streak orange-yellow.   Translucent. Optically $+$.   Ax. pl. $\parallel$ (010). $Z \wedge c$ axis = $-5°$. $\alpha = 2.31$, $\beta = 2.37$, $\gamma = 2.66$.   2V = 57°.

Comp. — Lead chromate, $PbCrO_4$ = Chromium trioxide 31·1, lead protoxide 68·9 = 100.

Pyr., etc. — In the closed tube decrepitates, blackens, but recovers its original color on cooling.   B.B. fuses at 1·5, and on charcoal is reduced to metallic lead with deflagration, leaving a residue of chromium oxide, and giving a lead coating.   With salt of phosphorus gives an emerald-green bead in both flames.

Obs. — A mineral of secondary origin, deposited from hot solutions containing chromic acid that have acted upon the minerals of lead veins.   Occurs in the Ural Mts., Russia, at Beresovsk near Ekaterinburg, and elsewhere.   From Rézbánya in Rumania.   In Tasmania in fine crystals from near Dundas and from the Heazlewood River.   From Brazil in fine crystals in decomposed granite from Congonhas do Campo, southwest of Ouro Preto, Minas Geraes.   From the island of Luzon, Philippine Islands.   Occurs in limited quantities in the Vulture district, Maricopa Co., Arizona.

The name *crocoite* is from κρόκος, *saffron*.

Phœnicochroite. Phœnicite. — A basic lead chromate, 3PbO.2CrO₃.   In orthorhombic crystals and massive.   One perfect cleavage.   H. = 3.   G. = 5·75.   Easily fusible. Color between cochineal- and hyacinth-red.   Optically $+$. $\alpha = 2.34$, $\beta = 2.38$, $\gamma = 2.65$. 2V medium.   Strong dispersion, $\rho > v$.   From Beresovsk near Ekaterinburg, Ural Mts.

Vauquelinite. — A phospho-chromate of lead, perhaps 2(Pb,Cu)CrO₄.(Pb,Cu)₃P₂O₈. Monoclinic.   In crystals; also mammillary and reniform.   H. = 2·5–3.   G. = 6.   Color green to brown.   Optically $-$. X nearly $\parallel$ elongation. $\beta = 2.22$.   2V nearly 0°.   From Beresovsk near Ekaterinburg, Ural Mts.   Also with crocoite from Congonhas do Campo, Minas Geraes, Brazil.   In the United States reported from a lead mine near Ossining, Westchester Co., New York; in Lancaster Co., Pennsylvania; and probably at Vulture district, Maricopa Co., Arizona.

Bellite. — Lead chromate containing arsenious oxide.   Hexagonal.   In aggregates of delicate tufts.   H. = 2·5.   G. = 5·5.   Easily fusible.   Color crimson red, yellow to orange. Optically $-$. $\omega = 2.16$, $\epsilon = 2.14$.   From Magnet, Russell Co., Tasmania.

Tarapacaite. — K₂CrO₄.   Orthorhombic.   Sometimes twinned with pseudo-hexagonal habit.   In minute yellow crystals in cavities in crude soda niter (caliche) from Tarapaca, Chile.

Sulphates with Chlorides, Carbonates, etc. — In part hydrous

LEADHILLITE.

Monoclinic.   Axes $a : b : c = 1.7476 : 1 : 2.2154$; $\beta = 89° 48'$.

mm''', 110 $\wedge$ 1$\bar{1}$0 = 120° 27'.
cw,   001 $\wedge$ 101 =   51° 36'.

cx, 001 $\wedge$ 111 = 68° 31'.
cm, 001 $\wedge$ 110 = 89° 54'.

Twins: tw. pl. m(110), analogous to aragonite.   Crystals commonly tabular $\parallel$ c(001).

Cleavage: $c(001)$ very perfect; $a(100)$ in traces. Fracture conchoidal, scarcely observable. Rather sectile. H. $= 2\cdot5$. G. $= 6\cdot26–6\cdot44$. Luster of c pearly, other parts resinous, somewhat adamantine. Color white, passing into yellow, green, or gray. Streak uncolored. Transparent to translucent. Optically $-$. Ax. pl. $\perp$ (010). $X \wedge c$ axis $= 5°$. $\alpha = 1\cdot87$, $\beta = 2\cdot00$, $\gamma = 2\cdot01$. $2V = 10°$. Strong dispersion, $\rho < v$.

Comp. — Sulphato-carbonate of lead, $4PbO.SO_3.2CO_2.H_2O$ or $PbSO_4$. $2PbCO_3.Pb(OH)_2 =$ Sulphur trioxide $7\cdot4$, carbon dioxide $8\cdot2$, lead oxide $82\cdot7$, water $1\cdot7 = 100$.

Pyr., etc. — B.B. intumesces, fuses at $1\cdot5$, and turns yellow; but becomes white on cooling. Easily reduced on charcoal. With soda affords the reaction for sulphuric acid. Effervesces briskly in nitric acid, and leaves white lead sulphate undissolved. Yields water in the closed tube.

Obs. — Leadhillite is a mineral of secondary origin found in lead deposits as the result of the alteration of galena or cerussite. It occurs at Nerchinsk in Transbaikalia; from the Mala-Calzetta mine near Iglesias, Sardinia (*maxite*). From Leadhills, Lanark, Scotland; also in crystals from Red Gill, Cumberland, and near Taunton, Somerset; at Matlock in Derbyshire. In large crystals from Djebel Ressas, southeast of Tunis, Tunis. In the United States found at Granby, Newton Co., Missouri; from the Mammoth mine near Schultz, Pinal Co., Arizona. Also from the Tintic district, Utah.

SUSANNITE. Regarded at one time as rhombohedral and dimorphous with leadhillite, but probably only a modification of that species. From the Susanna mine, Leadhills, in Lanark, Scotland.

Schairerite. — $Na_2SO_4.Na(F,Cl)$. Rhombohedral. In minute crystals with a steep rhombohedron the predominating form. H. $= 3\cdot5$. G. $= 2\cdot612$. Easily fusible. Slowly but completely soluble in water. Colorless, transparent. Vitreous luster. Optically $+$. $\omega = 1\cdot440$, $\epsilon = 1\cdot445$. Found in small quantities with gaylussite, tychite, pirssonite, ther nardite, etc. At Searles Lake, San Bernardino Co., California.

Sulphohalite. — $2Na_2SO_4.NaCl.NaF$. In pale greenish yellow octahedrons and dodecahedrons. H. $= 3\cdot5$. G. $= 2\cdot43$. Very easily fusible. $n = 1\cdot455$. From Borax or Searles Lake, San Bernardino Co., California.

Caracolite. — Perhaps $Pb(OH)Cl.Na_2SO_4$. Orthorhombic. In pseudo-hexagonal pyramids. As a crystalline incrustation. H. $= 4\cdot5$. Easily fusible. Colorless. Optically $-$. $\alpha = 1\cdot743$, $\beta = 1\cdot754$, $\gamma = 1\cdot764$. $2V$ large. From Mina Beatriz, Sierra Gordo, 25 miles from Caracoles, Antofagasta, Chile.

Kainite. — $MgSO_4.KCl.3H_2O$. Monoclinic. Usually granular massive and in crusts. Cleavages $||$ (110), (100). H. $= 3$. G. $= 2\cdot1$. Easily fusible. Color white to dark flesh-red. Optically $-$. Ax. pl. $||$ (010). $X \wedge c$ axis $= -8°$. $\alpha = 1\cdot494$, $\beta = 1\cdot505$, $\gamma = 1\cdot516$. $2V = 85°$. Occurs in the upper portions of salt deposits, with halite, carnallite, kieserite, etc. It is found in beds of considerable thickness at Kalusz in Galicia, Poland. In the salt deposits of northern Germany in the Province of Saxony at Stassfurt, Aschersleben, etc., and from Asse near Wolfenbüttel, Brunswick.

Connellite. — Probably $CuSO_4.2CuCl_2.19Cu(OH)_2.H_2O$. Crystals slender, hexagonal prisms. H. $= 3$. G. $= 3\cdot4$. Easily fusible. Color fine blue. Optically $+$. $\omega = 1\cdot724$. Occurs at various localities in Cornwall. From the mines of Mouzaïa, Alger, Algeria, and from Namaqualand, Cape Province, South Africa. In the United States occurs in Arizona at Bisbee, Cochise Co., and from Eureka in the Tintic district, Utah. *Footeite* and *ceruleofibrite* are both identical with connellite.

Spangolite. — A highly basic sulphate of aluminum and copper, $Cu_6AlClSO_{10}.9H_2O$. In dark green hexagonal crystals (hemimorphic), tabular or short prismatic. Usually in very small crystals. Basal cleavage. H. $= 2\cdot5$. G. $= 3\cdot1$. Fusible. Optically $-$. $\omega = 1\cdot694$, $\epsilon = 1\cdot641$. From Arizona in Cochise Co. near Tombstone and at Bisbee and at Clifton, Greenlee Co.; in the Tintic district, Utah. Also at St. Day in Cornwall and from Arenas, south of Iglesias, Sardinia.

Hanksite. — $9Na_2SO_4.2Na_2CO_3.KCl$. In hexagonal prisms, short prismatic to tabular; also in quartzoids. Basal cleavage. H. $= 3$. G. $= 2\cdot5$. Easily fusible. Color white to yellow. Optically $-$. $\omega = 1\cdot481$, $\epsilon = 1\cdot461$. From Borax Lake, San Bernardino Co., California; also from Death Valley, Inyo Co.

B. Acid and Basic Sulphates

Misenite. — Probably acid potassium sulphate, $HKSO_4$. Probably monoclinic. In silky fibers of a white color. Easily fusible. Optically +. $Z \wedge$ elongation = $33°$. $\alpha = 1·475$, $\beta = 1·480$, $\gamma = 1·487$. 2V large. From Cape Miseno, near Naples, Italy.

BROCHANTITE.

Orthorhombic. Axes $a : b : c = 0·7739 : 1 : 0·4871$.

In groups of prismatic acicular crystals (mm''' 110 $\wedge$ 1$\bar{1}$0 = $75° 28'$) and drusy crusts; massive with reniform structure.

Cleavage: $b(010)$ very perfect; $m(110)$ in traces. Fracture uneven. H. = 3·5–4. G. = 3·9. Luster vitreous; a little pearly on the cleavage-face $b(010)$. Color emerald-green, blackish green. Streak paler green. Transparent to translucent. Optically —. Ax. pl. || (100). $X = b$ axis. $\alpha = 1·728$, $\beta = 1·771$, $\gamma = 1·800$. 2V = $72°$. Dispersion, $\rho < v$.

Comp. — A basic sulphate of copper, $CuSO_4.3Cu(OH)_2$ or $4CuO.SO_3.3H_2O$ = Sulphur trioxide 17·7, cupric oxide 70·3, water 12·0 = 100.

Pyr., etc. — Yields water, and at a higher temperature sulphuric acid, in the closed tube, and becomes black. B.B. fuses, and on charcoal affords metallic copper. With soda gives the reaction for sulphuric acid.

Obs. — Brochantite is of secondary origin, being found in the oxidation zones of copper deposits. Occurs in the Ural Mts., at Gumeshevsk, southwest of Ekaterinburg (*königine* or *königite* was from this locality); also from Nizhne-Tagilsk, etc. From Rézbánya in Rumania. From Sardinia at the Rosas mine in Sulcis. Found in various localities in Cornwall (in part *warringtonite*); in Cumberland near Roughten Gill. On Iceland at Krisuvig (*krisuvigite*). In brilliant crystals from Aïn-Barbar, Constantine, Algeria; in South West Africa at Tsumeb, near Otavi. Frequently observed in the Atacama desert region of northern Chile; at Chuquicamata, northeast of Antofagasta, Province of Antofagasta, in very large amounts; from Collahuasi, Tarapaca; also from Llai Llai in Aconcagua.

In the United States in Arizona at Bisbee, Cochise Co., at the Clifton-Morenci district in Greenlee Co., etc.; in Utah at the mines of Eureka and Mammoth, Tintic district, and at Frisco, Beaver Co. In California at the Cerro Gordo district, Inyo Co.

Antlerite. — Perhaps $CuSO_4.2Cu(OH)_2$. In light green soft lumps. Orthorhombic. Cleavage || (010), perfect. H. = 3. G. = 3·9. Optically +. Ax. pl. || (001). $Z = a$ axis. $\alpha = 1·726$, $\beta = 1·738$, $\gamma = 1·789$. 2V = $53°$. Very strong dispersion, $\rho < v$. Strongly pleochroic, X = yellowish green, $Y = Z$ = bluish green. Shows a close crystallographic and optical similarity to brochantite. From the Antler mine, Mohave Co., Arizona. Also from Chuquicamata, Chile. *Stelznerite* from Remolinos, Vallinar, Chile, is probably the same as antlerite. In prismatic crystals. G. = 3·9. *Heterobrochantite* from Chile has the same composition as antlerite but is reported to differ in its optical characters.

Lanarkite. — Basic lead sulphate, Pb_2SO_5. In monoclinic crystals. Basal cleavage perfect. H. = 2·5. G. = 6·4–6·8. Color greenish white, pale yellow or gray. Optically —. Ax. pl. || (010). $\alpha = 1·93$, $\beta = 1·99$, $\gamma = 2·02$. 2V = $47°$. Strong dispersion, $\rho > v$. Found at Leadhills, Lanark, Scotland. Reported from Siberia; Tanne in the Harz Mts.; from Bieberwier, northwest of Innsbruck, Tyrol, Austria; at Laquorre, Ariège, France.

Dolerophanite. — A basic cupric sulphate, Cu_2SO_5 (?) In small brown monoclinic crystals. From Vesuvius (eruption of 1868).

Caledonite. — A basic sulphate of lead and copper, perhaps $2(Pb,Cu)O.SO_3.H_2O$. Said at times to contain CO_2. In small prismatic orthorhombic crystals. Cleavages: (001) perfect, (100) distinct. H. = 2·5–3. G. = 6·4. Easily fusible. Color deep verdigris-green or bluish green. Optically —. Ax. pl. || (010). $X = a$ axis. $\alpha = 1·818$, $\beta = 1·866$, $\gamma = 1·909$. 2V = $85°$. A rare mineral of secondary origin. Found at Rézbánya in Rumania; from Sardinia near Iglesias. Occurs at Leadhills, Lanark, Scotland; at Red Gill in Cumberland. From Tsumeb, near Otavi, in South West Africa. An important occurrence is at Challocollo in the Sierra Gorda, desert of Atacama, Chile. In the United States it occurs in the Organ Mts., Dona Ana Co., New Mexico; Beaver Co., Utah; from the Cerro Gordo district, Inyo Co., California.

Linarite. — A basic sulphate of lead and copper, $(Pb,Cu)SO_4.(Pb,Cu)(OH)_2$: In deep blue monoclinic crystals. Cleavages: (100) perfect, (001) distinct. H. = 2·5. G. = 5·4. Easily fusible. Optically −. Ax. pl. ⊥ (010). $X \wedge c$ axis = −24°. α = 1·809, β = 1·838, γ = 1·859. 2V = 80°. Dispersion, $\rho < v$. Linarite occurs in certain lead-copper deposits as a rare mineral of secondary origin. Found at Nerchinsk in Transbaikalia, Siberia; from near Beresovsk near Ekaterinburg, Ural Mts. In Sardinia at the mines of Rosas in Sulcis, etc. Supposed to have been found formerly at Linares in Province Jaen, Spain. Formerly at Leadhills, Lanark, Scotland; in Cumberland at Roughten Gill, etc. Occurs at Tsumeb near Otavi, South West Africa; at Broken Hill, New South Wales. In Chile at Llai Llai in Aconcagua and elsewhere. In the United States from the Tintic district, Utah; from the Cerro Gordo district, Inyo Co., California.

Alumian. — Perhaps $Al_2O_3.2SO_3$. Rhombohedral. H. = 2–3. G. = 2·7. Crystalline or massive. White. Optically +. ω = 1·583, ϵ = 1·602. Sierra Almagrera, near Aquilas, Murcia, Spain.

C. Normal Hydrous Sulphates

Three well-characterized groups are included here. Two of these, the EPSOMITE GROUP and the MELANTERITE GROUP, have the same general formula, $RSO_4.7H_2O$, but in the first the crystallization is orthorhombic, in the second monoclinic. The species are best known from the artificial crystals of the laboratory; the native minerals are rarely crystallized. There is also the isometric ALUM GROUP, to which the same remark is applicable.

Lecontite. — $(Na,NH_4,K)_2SO_4.2H_2O$. Orthorhombic. Prismatic cleavage. H. = 2·5. Very easily fusible. Colorless. Optically −. Ax. pl. ∥ (001). $X = a$ axis. α = 1·440, β = 1·452, γ = 1·453. From bat guano in the cave of Las Piedras, near Comayagua, Honduras, Central America.

MIRABILITE. Glauber Salt.

Monoclinic. Crystals like pyroxene in habit and angle. Usually in efflorescent crusts.

Cleavage: $a(100)$, perfect; $c(001)$, $b(010)$ in traces. H. = 1·5–2. G. − 1·481. Luster vitreous. Color white. Transparent to opaque. Taste cool, then feebly saline and bitter. Optically −. Ax. pl. ⊥ (010). $Z \wedge c$ axis = +26° to 31°. β = 1·437. (On recrystallized material β = 1·395.)

Comp. — Hydrous sodium sulphate, $Na_2SO_4.10H_2O$ = Sulphur trioxide 24·8, soda 19·3, water 55·9 = 100.

Pyr., etc. — In the closed tube much water; gives an intense yellow color to the flame. Very soluble in water. Loses its water on exposure to dry air and falls to powder.

Obs. — Deposited from saline waters and occurs associated with halite and gypsum. Occurs in the salt deposits of Austria at Ischl, Hallein, and Hallstadt in Salzburg, and at Aussee in Styria. Abundantly in the hot springs at Karlsbad (Karl Vary) and at Sedlitz, south of Brüx in Bohemia of Czechoslovakia. In northern Spain from the district of La Rioja. From Kirkby Thore, Westmorland. Large quantities of this sulphate are obtained from the Great Salt Lake in Utah.

Kieserite. — $MgSO_4.H_2O$. Monoclinic. Usually massive, granular to compact. Several cleavages. H. = 3·5. G. = 2·57. Fusible. Color white, grayish, yellowish. Optically +. Ax. pl. ∥ (010). $Z \wedge c$ axis = +76°. β = 1·533. Dispersion, $\rho > v$. Found in large amounts in the salt deposits of Germany; at Stassfurt in the Province of Saxony and at Leopoldshall in Anhalt; from Wathlingen, south of Celle, Hannover. Found also at Hallstadt in Salzburg, Austria; at Kalusz in Galicia, Poland. In India from the Mayo mines at Kheura, Punjab.

Szomolnokite. — $FeSO_4.H_2O$. Monoclinic. Isomorphous with *kieserite*. In pyramids. G. = 3·08. Color yellow or brown. Found with other iron sulphates from Szomolnok, (Smolník), Slovakia of Czechoslovakia, Hungary. Apparently identical with *ferropallidite* from near Copiapo, Chile.

Szmikite. — $MnSO_4.H_2O$. Probably monoclinic. One cleavage. Stalactitic. H. = 1·5. G. = 3·15. Whitish, reddish. Indices, 1·57–1·62. From Felsöbánya, (Baia Sprie), Rumania.

GYPSUM.

Monoclinic. Axes $a : b : c = 0·6899 : 1 : 0·4124$; $\beta = 80° 42'$.

mm''',	$110 \wedge 1\bar{1}0 = 68° 30'$.		ll',	$111 \wedge 1\bar{1}1 = 36° 12'$.
cd,	$001 \wedge 101 = 28° 17'$.		nn',	$\bar{1}11 \wedge \bar{1}\bar{1}1 = 41° 20'$.
ct,	$001 \wedge \bar{1}01 = 33° 8\frac{1}{2}'$.		ml,	$110 \wedge 111 = 49° 9'$.
ce,	$001 \wedge \bar{1}03 = 11° 29'$.		$m'n$,	$\bar{1}10 \wedge \bar{1}11 = 59° 15'$.
vv',	$011 \wedge 0\bar{1}1 = 44° 17\frac{1}{2}'$.			

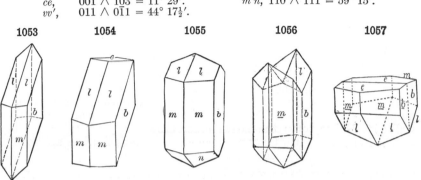

1053   1054   1055   1056   1057

Crystals usually simple in habit, common form flattened ∥ $b(010)$ or prismatic to acicular ∥ c axis; again prismatic by extension of $l(111)$. Also lenticular by rounding of $l(111)$ and $e(\bar{1}03)$. The form $e(\bar{1}03)$, whose faces are usually rough and convex, is nearly at right angles to the vertical axis (edge $m(110)/m'''$ $(1\bar{1}0)$), hence the apparent hemimorphic character of the twin (Fig. 1057). Simple crystals often with warped as well as curved surfaces. Also foliated massive; lamellar-stellate; often granular massive; and sometimes nearly impalpable. Twins: tw. pl. $a(100)$, very common, often the familiar swallow-tail twins. X-ray study of the crystal structure shows eight molecules in the unit cell. Each calcium atom lies between six complex atomic groups, consisting of four SO_4 tetrahedrons and two molecules of water.

1058

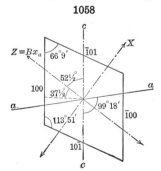

Cleavage: $b(010)$ eminent, yielding easily thin polished folia; $a(100)$, giving a surface with conchoidal fracture; $n(\bar{1}11)$, with a fibrous fracture ∥ $t(\bar{1}01)$; a cleavage fragment has the rhombic form of Fig. 1058, with plane angles of 66° and 114°. H. = 1·5–2. G. = 2·314–2·328, when in pure crystals. Luster of $b(010)$ pearly and shining, other faces subvitreous. Massive varieties often glistening, sometimes dull earthy. Color usually white; sometimes gray, flesh-red, honey-yellow, ocher-yellow, blue; impure varieties often black, brown, red, or reddish brown. Streak white. Transparent to opaque.

Optically +. Ax. pl. ∥ $b(010)$, and $Z \wedge c$ axis $= +52\frac{1}{2}°$ (at 9·4° C.), (cf. Fig. 1058). Dispersion $\rho > v$; also inclined strong. 2V = 58°. $\alpha = 1·520$. $\beta = 1·523$. $\gamma = 1·530$. The optic axial angle decreases rapidly with rise of temperature, becoming 0° for the D line at 91° C.; see further on p. 325.

Melanterite Group. $RSO_4.7H_2O$. Monoclinic

		$a : b : c$	
Melanterite	$FeSO_4.7H_2O$	$1.1828 : 1 : 1.5427$	$\beta = 75° 44'$
Mallardite	$MnSO_4.7H_2O$		
Pisanite	$(Fe,Cu)SO_4.7H_2O$	$1.1609 : 1 : 1.5110$	$74° 38'$
Bieberite	$CoSO_4.7H_2O$	$1.1815 : 1 : 1.5325$	$75° 20'$
Cupromagnesite	$(Cu,Mg)SO_4.7H_2O$		
Boothite	$CuSO_4.7H_2O$	$1.1622 : 1 : 1.500$	$74° 24'$

The species here included are the ordinary vitriols. They are identical in general formula with the species of the Epsomite group, and are regarded as essentially the same compounds under oblique crystallization.

MELANTERITE. Copperas.

Monoclinic. Usually capillary, fibrous, stalactitic, and concretionary; also massive, pulverulent. Cleavage: $c(001)$ perfect; $m(110)$ less so. Fracture conchoidal. Brittle. H. = 2. G. = 1.89–1.90. Easily fusible. Astringent taste. Luster vitreous. Color, various shades of green, passing into white; becoming yellowish on exposure. Streak uncolored. Subtransparent to translucent. Taste sweetish, astringent, and metallic. Optically+. Ax. pl $\parallel$ (010). $Z \wedge c$ axis = $-61°$. $\alpha = 1.471$. $\beta = 1.478$. $\gamma = 1.486$. 2V = $86°$.

Comp. — Hydrous ferrous sulphate, $FeSO_4.7H_2O$ = Sulphur trioxide 28.8, iron protoxide 25.9, water 45.3 = 100. Manganese and magnesium sometimes replace part of the iron.

Obs. — This salt usually results from the decomposition of pyrite or marcasite, which readily afford it, if occasionally moistened while exposed to the air. It is found in small amounts in many localities. Some of the more important are: Rammelsberg near Goslar, Harz Mts.; Bodenmais, Bavaria; in stalactites from the pyrite mine at Sain Bel, near Lyon, Rhône, France. From Falun, Kopparberg, Sweden. From Cornwall. In the United States found as small fibrous crystals at Leona Heights, Alameda Co., California. *Luckite* (1.9 per cent MnO) is from the Lucky Boy mine, Butterfield Cañon, Salt Lake Co., Utah.

Mallardite. $MnSO_4.7H_2O$. Fibrous, massive; colorless. From Lucky Boy mine, Butterfield Cañon, Salt Lake Co., Utah.

Pisanite. — $(Fe,Cu)SO_4.7H_2O$. CuO 10 to 18 per cent. Monoclinic. In concretionary and stalactitic forms. Cleavage $\parallel$ (001). H. = 2–3. G. = 2.15. Fusible. Color blue. Optically +. Ax. pl. $\parallel$ (010). X nearly $\parallel c$ axis. $\alpha = 1.472$, $\beta = 1.479$, $\gamma = 1.487$. 2V very large. Formed by the oxidation of cupriferous pyrite. Easily soluble and decomposable, consequently rarely observed. Originally described from a copper mine in the interior of Turkey, the exact locality unknown. Recently formed at the chalcopyrite and pyrite mines near Massa Marittima, Tuscany. Occurs associated with boothite at the pyrite mine at Sain Bel near Lyon, Rhône, France. From the mines of Rio Tinto, Huelva, Spain. In the United States occurs at Ducktown, Tennessee; Bingham Cañon, Utah; and from near Leona Heights, Alameda Co., California. *Iron-copper chalcanthite* is the name given to the product resulting from the natural dehydration of pisanite. Color, pale green. $\alpha = 1.517$, $\beta = 1.536$, $\gamma = 1.543$. Optically $-$.

Zinc-copper melanterite. — $(Zn,Cu,Fe)SO_4.7H_2O$, in which Zn : Cu : Fe = 100 : 98 : 19. Probably monoclinic. Consists of an aggregate of small rod-like forms. H. = 2. G. = 2.02. Pale greenish blue. Vitreous luster. Optically +. $\alpha = 1.479$, $\beta = 1.483$, $\gamma = 1.488$. 2V nearly $90°$. Readily fusible. Loses water in dry atmosphere. Readily soluble in water. Found on mine dumps at Vulcan, Gunnison Co., Colorado.

SALVADORITE. A copper-iron vitriol near pisanite. From the Salvador mine, Quetena, near Calama, Antofagasta, Chile.

Bieberite. — $CoSO_4.7H_2O$. Usually in stalactites and crusts. H. = 2. G. = 2.0. Color flesh- and rose-red. Optically $-$. $Z = b$ axis. $\alpha = 1.477$, $\beta = 1.483$, $\gamma = 1.489$.

2V nearly 90°. Formed by the oxidation of cobalt sulphide ores. It is very soluble and therefore of rare occurrence. Found as a recent formation in the old mines at Bieber, Hessen-Nassau, and elsewhere in Germany. From Leogang in Salzburg, Austria. *Cobalt-chalcanthite* is the name given to a natural dehydration product of bieberite.

Boothite. — $CuSO_4.7H_2O$. Usually massive. H. = 2–2·5. G. = 1·94. Color blue, paler than chalcanthite. Optically +. Ax. pl. ‖ (010). X nearly ‖ c axis. Indices near those of pisanite. Occurs as a mineral of secondary origin at the Alma pyrite mine, near Leona Heights, Alameda Co., and at a copper mine near Campo Seco, Calaveras Co., California. Noted from the pyrite mine at Sain Bel, near Lyon, Rhône, France.

Cupromagnesite. $(Cu,Mg)SO_4.7H_2O$. From Vesuvius.

CHALCANTHITE. Blue Vitriol.

Triclinic. $a : b : c = 0·5656 : 1 : 0·5507$; $\alpha = 82° 21'$, $\beta = 73° 11'$, $\gamma = 77° 37'$. Crystals commonly flattened ‖ $p(111)$. Occurs also massive, stalactitic, reniform, sometimes with fibrous structure.

Cleavage: $M(1\bar{1}0)$, $m(110)$, $p(111)$ imperfect. Fracture conchoidal. Brittle. H. = 2·5. G. = 2·12–2·30. Luster vitreous. Color Berlin-blue to sky-blue, of different shades; sometimes a little greenish. Streak uncolored. Subtransparent to translucent. Taste metallic and nauseous. Optically —. 2V = 56°. $\alpha = 1·516$. $\beta = 1·539$. $\gamma = 1·546$.

Comp. — Hydrous cupric sulphate, $CuSO_4.5H_2O$ = Sulphur trioxide 32·1, cupric oxide 31·8, water 36·1 = 100.

Artificial chalcanthites have been made in which a portion of the copper has been replaced by iron, manganese, zinc and cobalt.

Pyr., etc. — In the closed tube yields water, and at a higher temperature sulphur trioxide. Fusible at 3. B.B. with soda on charcoal yields metallic copper. With the fluxes reacts for copper. Soluble in water; a drop of the solution placed on a surface of iron coats it with metallic copper.

Obs. — Formed through the oxidation of chalcopyrite and other copper sulphides. Deposited from mine waters, often on the walls of abandoned workings. Since it is readily soluble in water it is to be found in quantity only in desert regions. Found at Herrengrund (Urvölgy) north of Neusohl (Ban Bystrica) in Slovakia of Czechoslovakia; from the Rio Tinto mines, Huelva, Spain. From various districts in Cornwall. Deposited from the fumaroles of Vesuvius and other volcanoes. In large amounts from the desert region of Chile, as at Chuquicamata, northeast of Antofagasta in the province of the same name; at Coquaquire, east of Challacollo, where it is the chief ore; at Copiapo, Atacama, etc. Also from Bolivia.

In the United States in the copper mines of Polk Co., Tennessee; from the mine waters of Butte, Montana; in Arizona in the Clifton-Morenci district, Greenlee Co.; in Nevada at Ely, Lincoln Co., and at Reno, Washoe Co.

Zinc-copper chalcanthite. — $(Zn,Cu,Fe)SO_4.5H_2O$. Name given to the product resulting from the natural dehydration of zinc-copper melanterite. Color pale blue. Optically —. $\alpha = 1·513$, $\beta = 1·533$, $\gamma = 1·540$.

Siderotil. — $FeSO_4.5H_2O$. Color white. Optically —. 2V = rather large. $\alpha = 1·526$, $\beta = 1·536$, $\gamma = 1·542$. Found at Idria, Gorizia, Italy.

Syngenite. Kaluzite. — $CaSO_4.K_2SO_4.H_2O$. In prismatic (monoclinic) crystals. Cleavages (110), (100) perfect. H. = 2·5. G. = 2·58. Easily fusible. Colorless or milky-white. Optically —. Ax. pl. ⊥ (010). $Z = b$ axis. $Y \wedge c$ axis = —3°. $\alpha = 1·501$, $\beta = 1·517$, $\gamma = 1·518$. 2V = 28°. Strong dispersion, $\rho < v$. Originally found at Kalusz in Galicia, Poland; also at Sondershausen, Thuringia. Also noted as a crust on lava from Maui, Hawaiian Islands.

Löweite. — $MgSO_4.Na_2SO_4.2\frac{1}{2}H_2O$. Tetragonal. Massive, cleavable, ‖ (001). H. = 3·5. G. = 2·4. Easily fusible. Color pale yellow. Optically —. $\omega = 1·490$, $\epsilon = 1·471$. Occurs at Ischl and Hallstadt, Salzburg, Austria. Reported from Joachimstal, Bohemia, Czechoslovakia. *Chile-löweite* from the saltpeter deposits in Chile may be the same material. From Ischl, Austria.

Blödite. — $MgSO_4.Na_2SO_4.4H_2O$. Crystals short prismatic, monoclinic; also massive granular or compact. H. = 3. G. = 2·23. Easily fusible. Colorless to greenish, yellowish, red. Optically —. Ax. pl. ‖ (010). $X \wedge c$ axis = 43°. $\alpha = 1·481$, $\beta = 1·483$,

$\gamma = 1.484$. $2V = 71°$. Strong dispersion, $\rho < v$. From the salt deposit at Hallstadt (*simonyite*), Salzburg, Austria; at Stassfurt, Province of Saxony, and at Leopoldshall in Anhalt. From the salt lakes of Astrakhan (*astrakanite*), Russia. From the Salt Range, Punjab, India. In Antofagasta, Chile, at Taltal and Chuquicamata. In New Mexico at Salt Lakes, near Estancia, Torrance Co.; in California at Soda Lake, San Luis Obispo Co.

Leonite. — $MgSO_4.K_2SO_4.4H_2O$. Monoclinic. H. $= 3$. G. $= 2.25$. Easily fusible. Colorless. Optically $-$. Ax. pl. $\parallel$ (010). Z nearly $\parallel c$ axis. $\alpha = 1.483$, $\beta = 1.487$, $\gamma = 1.490$. $2V = 86°$. From the salt deposits of Westeregeln, Province of Saxony and of Leopoldshall, Anhalt.

Boussingaultite. — $(NH_4)_2SO_4.MgSO_4.6H_2O$. Monoclinic. Cleavages, $\parallel$ ($\bar{2}01$) perfect; $\parallel$ (010) distinct. H. $= 2$. G. $= 1.7$. Very easily fusible. Optically $+$. Ax. pl. $\parallel$ (010). $Z \wedge c$ axis $= +95°$. $\alpha = 1.470$, $\beta = 1.472$, $\gamma = 1.479$. $2V = 51°$. From the boric acid lagoons, Tuscany, Italy. Also on South Mountain, near Santa Paula, Ventura Co., California.

Picromerite. — $MgSO_4.K_2SO_4.6H_2O$. As a white crystalline incrustation. Monoclinic. Cleavage $\parallel$ ($\bar{2}01$) perfect. H. $= 2.5$. G. $= 2.1$. Easily fusible. Colorless. Optically $+$. Ax. pl. $\parallel$ (010). $X \wedge c$ axis $= 14°$. $\alpha = 1.460$, $\beta = 1.462$, $\gamma = 1.475$. $2V = 48°$. From Vesuvius with *cyanochroite*, an isomorphous species in which copper replaces the magnesium. Also at Stassfurt (*schoenite*) and Aschersleben, Province of Saxony; at Leopoldshall in Anhalt. From Kalusz, Galicia, Poland.

Polyhalite. — $2CaSO_4.MgSO_4.K_2SO_4.2H_2O$. Triclinic. Usually in compact fibrous or lamellar masses. Cleavage $\parallel$ (100), distinct. H. $= 2.5$–3. G. $= 2.78$. Easily fusible. Color flesh- or brick-red. Optically $-$. $\alpha = 1.548$, $\beta = 1.562$, $\gamma = 1.567$. $2V = 70°$. Occurs at the mines of Ischl, Hallstadt, etc., in Salzburg, Austria; in Germany at Berchtesgaden, Bavaria; Stassfurt, Province of Saxony.

Hexahydrite. — $MgSO_4.6H_2O$. Monoclinic. Columnar to fibrous structure. Cleavage prismatic. G. $= 1.76$. Color, white with light green tone. Pearly luster. Opaque. Salty, bitter taste. B.B. exfoliates and yields water but does not fuse. Optically $-$. $\beta = 1.453$. Found in Lillooet district, British Columbia.

Mooreite. — $RSO_4.7R(OH)_2.4H_2O$. $R = Mg : Mn : Zn = 4 : 1 : 2$. Monoclinic. Crystals tabular $\parallel$ (010). Cleavage, (010), perfect. G. $- 2.47$. H. $= 3$. Color, clear, glassy-white. Optically $-$. $X \perp$ (010). $Z \wedge c$ axis $= 44°$. $\alpha = 1.533$, $\beta = 1.545$, $\gamma = 1.547$. Occurs with altered pyrochroite, rhodochrosite and zincite at Franklin, Sussex Co., New Jersey. A similar mineral called δ-*mooreite* occurs also with the same associations. Bluish white in color. Comp. — $RSO_4.6R(OH)_2.4H_2O$ with $R - Mg : Mn : Zn = 5 : 3 : 4$. $\alpha = 1.570$, $\beta = 1.584$, $\gamma = 1.585$.

Alum Group. Isometric

$$\overset{\text{I}}{R}Al(SO_4)_2.12H_2O \text{ or } \overset{\text{I}}{R}_2SO_4.Al_2(SO_4)_3.24H_2O$$

Potash Alum	$KAl(SO_4)_2.12H_2O$
Ammonia Alum, Tschermigite	$(NH_4)Al(SO_4)_2.12H_2O$
Soda Alum	$NaAl(SO_4)_2.12H_2O$

The ALUMS proper are isometric in crystallization and, chemically, are hydrous sulphates of aluminum with an alkali metal and 12 (i.e., if the formula is doubled, 24) molecules of water. The species listed above occur very sparingly in nature, and are best known in artificial form in the laboratory.

Potash Alum. — $KAl(SO_4)_2.12H_2O$. Isometric. Octahedral habit. H. $= 2$. G. $= 1.76$. Very easily fusible. Colorless. $n = 1.45$. Also *kalinite*, with the same composition. Probably monoclinic. Fibrous. Optically $-$. $Z = b$ axis. $Y \wedge c$ axis $= 13°$. $\alpha = 1.430$, $\beta = 1.452$, $\gamma = 1.458$. $2V$ very small to $52°$. These minerals are often formed by the weathering of schistose rocks, alum-slates, which contain disseminated pyrite and bituminous materials, and they may occur in large amounts in such formations. The action of ascending waters containing sulphuric acid which has been derived from either some volcanic source or through the oxidation of sulphides, upon rocks containing potassium and aluminum silicates will also form them. They occur frequently as an efflorescence upon argillaceous minerals. Found in connection with volcanoes, etc., as on the Island of Vulcano. Lipari Islands; at Vesuvius; at the solfatara at Cape Miseno, etc., near Naples.

Found in quantity in various places in Germany; in the department of Aveyron, France; in Spain; at Whitby, Yorkshire, and at Hurlet, near Glasgow, Scotland. Occurs in New South Wales; in South America in Chile, Bolivia, and Argentina. In the United States was found in quantity at Cape Sable, Maryland. In Nevada at Silver Peak, Esmeralda Co.

Tschermigite. Ammonia Alum; — $(NH_4)Al(SO_4)_2.12H_2O$. Isometric. Octahedral habit. H. = 2. G. = 1·64. Very easily fusible. Colorless or white. $n = 1·459$. Most commonly associated with lignite beds. Found in Bohemia of Czechoslovakia at Tschermig, east of Kaaden and from near Dux, etc. From the solfatara of Pozzuoli near Naples. In the United States from Tumucari, Quay Co., New Mexico, and from near Wamsutter, Sweetwater Co., Wyoming.

Soda Alum. — $NaAl(SO_4)_2.12H_2O$. Isometric. Known as artificial material but it is not certain that it occurs in nature. *Mendozite* has the same composition but is not isometric. Fibrous. H. = 3. G. = 1·73. Very easily fusible. Colorless. Optically —. Uniaxial, $\omega = 1·459$, $\epsilon = 1·431$. Also biaxial with very small axial angle. Formed by the action of sulphuric acid upon sodium-aluminum silicates. Found at the solfatara of Pozzuoli, near Naples. From Argentina at Mendoza and elsewhere and in Chile. Reported from Box Elder Co., Utah.

The HALOTRICHITES are oblique in crystallization, very commonly fibrous in structure, and are hydrous sulphates of aluminum with magnesium, manganese, etc.; the amount of water is given as 22 molecules, although some analyses would indicate $24H_2O$. Here belong:

Pickeringite. Magnesia Alum. — $MgSO_4.Al_2(SO_4)_3.22H_2O$. Monoclinic. In long fibrous masses; and in efflorescences. H. = 1. G. = 1·85. Easily fusible. Colorless, yellow to red. Optically —. Ax. pl. $\parallel$ (010). $Z \wedge c$ axis = 37°. $\alpha = 1·476$, $\beta = 1·480$, $\gamma = 1·483$. Commonly formed by the weathering of pyrite-bearing schists in the same manner as the other members of the alum group, with which it is frequently associated. Found in large amounts in the Cerros Pintados, near Iquique, Tarapaca, Chile, etc. In the United States from Tumucari, Quay Co., New Mexico. From Newport, Hants Co., Nova Scotia.

Halotrichite. Iron Alum. — $FeSO_4.Al_2(SO_4)_3.22H_2O$. Monoclinic. In yellowish silky fibrous forms. H. = 2. G. = 1·9. Fusible. $\beta = 1·49$. Formed in the same manner as the other alums. Occurs at Mörsfeld, east of Obermoschell, Bavarian Pfalz. At the solfatara of Pozzuoli, near Naples. From Björkbakkagård, Finland. Also at Urumiya, Persia. From Tierra Amarilla, near Copiapo, Chile. From Iceland (*hversalt*). A related mineral with other alums occurs in quantity in the Alum Mts., on the Gila River in Grant Co., Mexico.

Bilinite. — $FeSO_4.Fe_2(SO_4)_3.22H_2O$. Radiating fibrous. H. = 2. G. = 1·87. Color white to yellow. $n = 1·5$. From Schwaz, near Bilin, Bohemia, Czechoslovakia.

Apjohnite. Manganese Alum. — $MnSO_4.Al_2(SO_4)_3.22H_2O$. *Bushmanite* contains MgO. In fibrous or asbestiform masses; also as crusts and efflorescences. Monoclinic. H. = 1·5. G. = 1·8. White. Optically —. Ax. pl. $\parallel$ (010). $Z \wedge c$ axis = 29°. $\alpha = 1·478$, $\beta = 1·482$, $\gamma = 1·482$. 2V small. From Delagoa Bay, Portuguese East Africa. Found in Alum Cave, Sevier Co., Tennessee. *Bushmanite* occurs in a cave near the Bushman River, Cape Province, South Africa. A related alum is found in the Maderaner Tal, Uri, Switzerland (*keramohalite*) and at Alum Point, Salt Lake Co., Utah.

Dietrichite. — $(Zn,Fe,Mn)SO_4.Al_2(SO_4)_3.22H_2O$. Monoclinic. Fibrous. H. = 2. White. Optically +. Ax. pl. $\perp$ (010). $X = b$ axis. $Z \wedge c$ axis = 29°. $\alpha = 1·475$, $\beta = 1·480$, $\gamma = 1·488$. 2V large. A recent formation in an abandoned working at Felsöbánya (Baia Sprie), Rumania.

Kornelite. — $Fe_2(SO_4)_3.7\frac{1}{2}H_2O$. Orthorhombic. Prismatic. Cleavages $\parallel$ (100), (010). G. = 2·306. Color pale rose or violet. Silky luster. Optically +. Ax. pl. $\parallel$ (010). $Z \parallel b$ axis. $\beta = 1·59$. Dispersion, $\rho > v$. Found with voltaite and coquimbite in the pyrite mines of Szomolonok, Comitat Szepes, Hungary.

Coquimbite. — $Fe_2(SO_4)_3.9H_2O$. Rhombohedral. Granular massive. H. = 2. G. = 2·1. Fusible. Color white, yellowish, brownish. Optically +. $\omega = 1·550$, $\epsilon = 1·557$.

Strong dispersion with abnormal interference colors. From the Tierra Amarilla near Copiapo, Atacama, Chile. Has been noted from Vesuvius, from the Island of Vulcano, Spain, etc. In the United States has been found at the Redington mercury mine, Napa Co., California. Formed under fumarolic conditions during the burning of a portion of the United Verde copper mine, at Jerome, Arizona.

Alunogen. — $Al_2(SO_4)_3.16H_2O$. The water content is variable depending upon atmospheric conditions. Monoclinic or triclinic. Usually in delicate fibrous masses or crusts; massive. Sometimes in rosettes of thin crystals. H. = 1·5–2. G. = 1·65. Color white, or tinged with yellow or red. Optically $+$. $Z \wedge$ elongation of fibers = 42°. $\alpha = 1·473$, $\beta = 1·474$, $\gamma = 1·480$. Formed by volcanic action and the decomposition of pyrite in coal formations and alum slates, and is found at numerous localities. Among these are Königsberg in Slovakia and at Kollosoruk, southwest of Bilin in Bohemia. Has been observed at Vesuvius and at the solfatara of Pozzuoli, near Naples. From the Cerros Pintados near Iquique in Tarapaca, Chile. From the crater of the Pasto volcano in Dept. Cauca, Colombia. A white fibrous alunogen (?) occurs abundantly at Smoky Mt., Jackson Co., North Carolina. In Colorado at Doughty Springs and Alum Gulch. Extensive deposits occur in the Alum Mts., on the Gila River, Grant Co., New Mexico. Formed under fumarolic conditions during the burning of a portion of the United Verde copper mine at Jerome, Arizona.

DOUGHTYITE. A hydrated aluminum sulphate deposited by the alkaline waters of the Doughty Springs in Colorado.

Kröhnkite. — $CuSO_4.Na_2SO_4.2H_2O$. Monoclinic crystalline; massive, coarsely fibrous. Cleavages $\parallel$ (010) and (011). H. = 2·5. G. = 2·06. Very easily fusible. Color azure-blue. Optically $-$. Ax. pl. $\parallel$ (010). $X \wedge c$ axis = $-49°$. $\alpha = 1·544$, $\beta = 1·578$, $\gamma = 1·601$. 2V = 79°. Occurs in Chile at Chuquicamata, near Calama in Antofagasta, and at Collahuasi, Tarapaca, etc.

Natrochalcite. — $Cu_4(OH)_2(SO_4)_3.Na_2SO_4.2H_2O$. Monoclinic. Habit pyramidal. Perfect basal cleavage. H. = 4·5. G. = 2·3. Easily fusible. Color bright emerald-green. Optically $+$. Ax. pl. $\parallel$ (010). $Z \wedge c$ axis = 12°. $\alpha = 1·649$, $\beta = 1·656$, $\gamma = 1·714$. 2V = 37°. Strong dispersion, $\rho < v$. Found at Chuquicamata, near Calama, Antofagasta, Chile.

Ransomite. — $CuO.(Fe,Al)_2O_3.4SO_3.7H_2O$. Orthorhombic. Slender prisms. Perfect cleavage. H. = 2·5. G. = 2·632. Color bright sky-blue. Optically $+$. $\alpha = 1·631$, $\beta = 1·643$, $\gamma = 1·695$. Formed under fumarolic conditions during the burning of a portion of the United Verde copper mine, at Jerome, Arizona.

Guildite. — $3(Cu,Fe)O.2(Fe,Al)_2O_3.7SO_3.17H_2O$. Monoclinic. Habit cubic in appearance. Perfect cleavages $\parallel$ (100) and (001). H. = 2·5. G. = 2·725. Color, deep chestnut-brown. Optically $+$. $\alpha = 1·623$, $\beta = 1·630$, $\gamma = 1·684$. Pleochroic; $X = Y = $ pale yellow, $Z = $ greenish yellow. Formed under fumarolic conditions during the burning of a portion of the United Verde copper mine, at Jerome, Arizona. Associated with coquimbite and ransomite.

Louderbackite. — $2FeO.3(Fe,Al)_2O_3.10SO_3.35H_2O$. Orthorhombic. Two good cleavages. H. = 2·5 3. G. = 2·185. Color pale chestnut-brown. Optically $+$. $\alpha = 1·544$, $\beta = 1·558$, $\gamma = 1·581$. Found as a thin crystalline crust, coating pyrite and formed under fumarolic conditions during the burning of a portion of the United Verde copper mine, at Jerome, Arizona.

PHILLIPITE. Perhaps $CuSO_4.Fe_2(SO_4)_3.nH_2O$. In blue fibrous masses. Found at the copper mines in the Cordilleras of Condes, province of Santiago, Chile.

Krausite. — $K_2SO_4.Fe_2(SO_4)_3.2H_2O$. Monoclinic. In small crystals of varying habit, often prismatic. Cleavages, (001) perfect, (100) good. H. = 2·5. G. = 2·84. B.B. decrepitates and reduces to a black scoriaceous mass. Color pale yellowish green. Optically $+$. $Z = b$ axis. $X \wedge c$ axis = 35°. $\alpha = 1·588$, $\beta = 1·650$, $\gamma = 1·722$. 2V large. Found with alunite, coquimbite and other sulphates at Borate in the Calico Hills, San Bernardino Co., California. Noted also at Velardeña, Durango, Mexico.

Ferrinatrite. — $3Na_2SO_4.Fe_2(SO_4)_3.6H_2O$. Rhombohedral. Rarely in acicular crystals; usually in spherical forms. Cleavages: (10$\bar{1}$0) perfect, (0001) distinct. H. = 2. G. = 2·56. Easily fusible. Color greenish or gray to white. Optically $+$. $\omega = 1·559$, $\epsilon = 1·627$. From Sierra Gorda, near Caracoles, Antofagasta, Chile.

Sideronatrite. — $2Na_2O.Fe_2O_3.4SO_3.7H_2O$. Orthorhombic? Fibrous, massive. Perfect cleavage, (100). H. = 2–2·5. G. = 2·3. Easily fusible. Color yellow. Optically $+$. Ax. pl. $\parallel$ (010). $Z = c$ axis. $\alpha = 1·508$, $\beta = 1·525$, $\gamma = 1·586$. 2V = 58°. Strong dispersion, $\rho > v$. In Chile from near Huantajaya, Tarapaca, and from Sierra Gorda, Caracoles district, Antofagasta. Also on the Urus plateau, near Sarakaya, on the island Cheleken, in the Caspian Sea (*urusite*).

Voltaite. — Perhaps $3(K_2,Fe)O.(Al,Fe)_2O_3.6SO_3.9H_2O$. In octahedrons, etc. H. = 3–4. G. = 2.8. Color dull oil-green to brown or black. Isotropic. $n = 1.602$. Sometimes crystals are in part anisotropic. From the solfatara at Pozzuoli, near Naples. Occurs near Goslar, Harz Mts. Found at Kremnitz (Körmöczbanya), and Schmöllnitz, Czechoslovakia. From Chile in the Sierra Caparrosa between Sierra Gorda and Calama, Antofagasta. In the United States found in the Jerome district, Yavapai Co., Arizona. Similar material, but showing tetragonal crystals in complex twins, has been found in the region of Madeni Zakh, Persia.

Metavoltine. — $(H,K)_4(Fe'''OH)_2(SO_4)_4.5H_2O$. Hexagonal. In aggregates of minute yellow scales. H. = 2.5. G. = 2.5. Optically —. Indices, 1.57–1.59. Occurs with voltaite in Persia. From Vesuvius and from Cape Miseno, near Naples; also in the fumaroles on the islands of Vulcano, Lipari Islands and Milos in the Cyclades.

Slavikite. — $(Na,K)_2SO_4.Fe_{10}(OH)_6(SO_4)_{12}.63H_2O$. Hexagonal, rhombohedral. In minute crystals showing $(10\bar{1}1)$ and (0001). G. = 1.905. Color greenish yellow. Optically —. $\omega = 1.530$, $\epsilon = 1.506$. Pleochroic, O = lemon-yellow, E = colorless. Product of oxidation of pyrite, associated with other sulphates from Valachov hill, near Skřivan, Bohemia, Czechoslovakia.

Römerite. — $FeSO_4.Fe_2(SO_4)_3.14H_2O$. In tabular triclinic crystals; granular, massive. H. = 3–3.5. G. = 2.15. Fusible. Color chestnut-brown. Optically —. $\alpha = 1.524$, $\beta = 1.571$, $\gamma = 1.583$. 2V = 49°. Strong crossed dispersion, $\rho > v$. Abnormal interference colors. From Goslar in the Harz Mts.; Madeni Zakh, Persia. In Chile at Tierra Amarilla, east of Copiapo, Atacama. From Island Mountain, Trinity Co., California.

Basic Hydrous Sulphates

Langite. — Near brochantite. $CuSO_4.3Cu(OH)_2.H_2O$. Orthorhombic. Usually in fibro-lamellar, concretionary crusts. Cleavages: (001) perfect, (010) distinct. H. = 2.5–3. G. = 3.5. Fusible. Color blue to greenish blue. Optically —. Ax. pl. || (010). $X = c$ axis. $\alpha = 1.708$, $\beta = 1.760$, $\gamma = 1.798$. 2V = 81°. Originally described from Cornwall at St. Blazey and St. Just. Also from Herrengrund (Urvölgy) north of Neusohl (Ban Bystrica) in Slovakia of Czechoslovakia. From Cornwall.

Herrengrundite. — $2(CuOH)_2SO_4.Cu(OH)_2.3H_2O$ with one-fifth of the copper replaced by calcium. In thin tabular monoclinic crystals; usually in spherical groups. Cleavages: (001) perfect, (110) distinct. H. = 2.5. G. = 3.13. Fusible. Color emerald-green, bluish green. Optically —. Ax. pl. ⊥ (010). X nearly || c axis. $\alpha = 1.585$, $\beta = 1.649$, $\gamma = 1.660$. 2V = 39°. Strong dispersion, $\rho < v$. Pleochroic; X = pale green, Y and Z = deeper green. From Herrengrund (Urvölgy), north of Neusohl (Ban Bystrica) in Slovakia of Czechoslovakia.

Vernadskite. Vernadskyite. — $3CuSO_4.Cu(OH)_2.4H_2O$. In aggregates of minute crystals. H. = 3.5. Occurs as an alteration of *dolerophanite* at Vesuvius.

Kamarezite. — A hydrous basic copper sulphate from Kamareza, near Laurium, Greece.

Cyanotrichite. Lettsomite. — $4CuO.Al_2O_3.SO_3.8H_2O$. Orthorhombic. In velvet-like druses; in spherical forms. G. = 2.74. Fusible. Color bright blue. Optically +. Z || elongation. X ⊥ lath-shaped crystals. $\alpha = 1.588$, $\beta = 1.617$, $\gamma = 1.655$. 2V = 82°. Strong dispersion, $\rho < v$. Strongly pleochroic; X = nearly colorless, Y = pale blue, Z = bright blue. Occurs at Laurium, Greece. From Moldawa in the Banat, Rumania; Cap Garonne, Var. France. In Arizona at Morenci, Greenlee Co., and in the Grand Canyon; in the Tintic district, Utah.

Serpierite. — A basic sulphate of copper and zinc. Orthorhombic. In minute crystals, tabular, in tufts. Perfect (001) cleavage. G. = 2.52. Color bluish green. Optically —. Ax. pl. || (100). $X = c$ axis. $\alpha = 1.584$, $\beta = 1.642$, $\gamma = 1.647$. 2V = 35°. Strong dispersion, $\rho > v$. From Laurium, Greece. On Ross Island, Killarney, Co. Kerry, Ireland.

Beaverite. — $CuO.PbO.Fe_2O_3.2SO_3.4H_2O$. Hexagonal. In microscopic plates. G. = 4.36. Color, canary-yellow. Optically —. $\omega = 1.83–1.87$. Strong birefringence. From Horn Silver mine, Frisco, Beaver Co., Utah.

COPIAPITE.

Orthorhombic. Has been described as monoclinic but optical characters seem to conform to orthorhombic symmetry. Usually in loose aggregations of crystalline scales, or granular massive; incrusting.

Cleavage: $c(001)$. H. $= 2 \cdot 5$. G. $= 2 \cdot 103$. Luster pearly. Color sulphur-yellow, citron-yellow. Translucent. Optically $+$. Ax. pl. $\|$ (010). $X = c$ axis. $\alpha = 1 \cdot 506 - 1 \cdot 540$, $\beta = 1 \cdot 528 - 1 \cdot 550$, $\gamma = 1 \cdot 575 - 1 \cdot 600$. $2V = 45°$ to $74°$.

Comp. — A basic ferric sulphate, perhaps $Fe_4(OH)_2(SO_4)_5.18H_2O$.

Misy is an old term, which has been somewhat vaguely applied. It seems to belong in part here and in part also to other related species. *Janosite* is identical with copiapite.

Pyr., etc. — Yields water, and at a higher temperature sulphuric acid. On charcoal becomes magnetic, and with soda affords the reaction for sulphuric acid. With the fluxes reacts for iron. Soluble in water, and decomposed by boiling water.

Obs. — The most common ferric sulphate. Originally found near Copiapo, Atacama, Chile. Also from Potrerillos, Atacama, and in Coquimbo. From the Rammelsberg mine near Goslar, Harz Mts.; near Rio Marina, Elba; from Commentry, Allier, France; from Falun, Sweden. In the United States found in California at the Redington mine, Knoxville, Napa Co.; at Sulphur Bank, Lake Co.; also at the Alma mine, Leona Heights, Alameda Co. Formed under fumarolic conditions during the burning of a portion of the United Verde copper mine at Jerome, Arizona.

The following hydrous iron sulphates are of doubtful character and may all be identical with copiapite. *Quenstedtite*, originally described from Tierra Amarilla, east of Copiapo, Atacama, Chile. *Ihleite* occurs as efflorescences on graphite at Mugrau near Ober-Plau on the Moldau, southern Bohemia, Czechoslovakia. Also from near Rio Marina, Elba. *Knoxvillite* is found in greenish yellow masses at the Redington mine near Knoxville, Napa Co., California.

Other hydrated ferric sulphates:

Rhomboclase. — A hydrated acid ferric sulphate, $Fe_2O_3.4SO_3.9H_2O$. Orthorhombic or monoclinic but closely orthorhombic in character. In rhombic plates. Basal cleavage. H. $= 2$. Clear and colorless or gray and opaque. Optically $+$. Ax. pl. $\|$ (100). $X = c$ axis. $\beta = 1 \cdot 551$. Occurs at Smolnik (Szomolnok), Slovakia in Czechoslovakia.

Lausenite. Rogerside. Perhaps $Fe_2O_3.3SO_3.6H_2O$. Monoclinic. In aggregates of minute, silky fibers. $\alpha = 1 \cdot 598$, $\beta = 1 \cdot 628$, $\gamma = 1 \cdot 654$. Color white. Formed under fumarolic conditions during the burning of a portion of the United Verde copper mine, at Jerome, Arizona.

Castanite. $Fe_2O_3.2SO_3.8H_2O$. Triclinic, prismatic. Cleavages: (010) perfect, (110) (1$\bar{1}$0) imperfect. Color chestnut-brown. Optically $-$. X slightly inclined to normal to (010). $\alpha = 1 \cdot 553$, $\beta = 1 \cdot 643$, $\gamma = 1 \cdot 657$. From Sierra Gorda, near Caracoles, Antofagasta, Chile; also noted from Knoxville, Napa Co., California. Possibly identical with botryogen.

Utahite. $3Fe_2O_3.2SO_3.7H_2O$. In aggregates of fine scales. Uniaxial, $-$. $\omega = 1.82$. Color orange yellow. From the Tintic district, Utah; Guanaco, near Taltal, Antofagasta, Chile. Perhaps identical with *jarosite*.

Butlerite. — $(Fe,Al)_2O_3.2SO_3.5H_2O$. Orthorhombic. In minute pyramidal crystals. Cleavage $\|$ (010). Optically $-$. $\alpha = 1 \cdot 604$, $\beta = 1 \cdot 674$, $\gamma = 1 \cdot 731$. Formed under fumarolic conditions during the burning of a portion of the United Verde copper mine, Jerome, Arizona.

Amarantite. — $Fe_2O_3.2SO_3.7H_2O$. Triclinic. Usually in columnar or bladed masses, also radiated. Perfect cleavages, $\|$ (100), (010). H. $= 2 \cdot 5$. G. $= 2 \cdot 2$. Fusible. Color amaranth-red. Optically $-$. X nearly $\perp$ (100). $\alpha = 1 \cdot 51$, $\beta = 1 \cdot 605$, $\gamma = 1 \cdot 611$. Pleochroic; $X =$ nearly colorless. $Y =$ pale orange-yellow, $Z =$ orange-yellow. Occurs in Antofagasta, Chile, north of Sierra Gorda, near Caracoles; also from the Sierra Caparrosa between Calama and Sierra Gorda. *Hohmannite* is the same partially altered; this is probably also true of *paposite*.

Fibroferrite. — $Fe_2O_3.2SO_3.10H_2O$. Orthorhombic. In delicately fibrous aggregates. H. $= 2 - 2 \cdot 5$. G. $= 1 \cdot 9$. Fusible. Color pale yellow, nearly white. Optically $+$. $Z \|$ elongation of fibers. Indices variable, $\beta = 1 \cdot 518 - 1 \cdot 534$. From Pallières, Gard, France, and from the Tierra Amarilla, near Copiapo, Atacama, Chile.

Raimondite. $2Fe_2O_3.3SO_3.7H_2O$. In thin six-sided tables. Color between honey- and ocher-yellow. Uniaxial, $-$. From the tin mines of Ehrenfriedersdorf; mines of Bolivia. Probably identical with *jarosite*.

Carphosiderite. $3Fe_2O_3.4SO_3.7H_2O$? In reniform masses, and incrustations; also in micaceous lamellæ. H. $= 4$. G. $= 2 \cdot 6$. Color straw-yellow. Uniaxial, $-$. Indices, $1 \cdot 73 -$

1·82. From Greenland. Some carphosiderite has been shown to be jarosite and the species is doubtful.

BORGSTRÖMITE. $Fe_2O_3.SO_3.3H_2O$. Rhombohedral (artif.). Earthy. Yellow color. Formed by oxidation of pyrite and pyrrhotite. Occurs in Eno parish, Otravaara ore field, Finland.

PLANOFERRITE. $Fe_2O_3.SO_3.15H_2O$. Orthorhombic? In rhombic or hexagonal plates. Yellowish green to brown. From Morro Moreno, Antofagasta, Chile.

Glockerite. — $2Fe_2O_3.SO_3.6H_2O$? Massive, sparry or earthy; stalactitic. Color brown to ocher-yellow to pitch-black; dull green. Indices, 1·76–1·81. From Goslar, Harz Mts., Germany. Also from Obergrund, near Zuckmantel in Silesia of Czechoslovakia. From Falun, Sweden, and Modum, Norway.

REDINGTONITE. A hydrous chromium sulphate, in finely fibrous masses of a pale purple color. From Redington mercury mine, Knoxville, Napa Co., California.

CYPRUSITE. Perhaps $7Fe_2O_3.Al_2O_3.10SO_3.14H_2O$. An aggregation of microscopic crystals. Color yellowish. From the island of Cyprus.

Aluminite. Websterite. — $Al_2O_3.SO_3.9H_2O$. Orthorhombic? Usually in white earthy reniform masses, compact. H. = 1–2. G. = 1·66. Optically +. $X \parallel$ elongation. $\alpha = 1·459$, $\beta = 1·464$, $\gamma = 1·470$. 2V large. Occurs in clay where it has been formed by the action of sulphuric acid waters upon clay materials. From near Halle, Province of Saxony; also at Newhaven in Sussex. In France from Epernay, Marne, and at Auteuil, near Paris. Also at Mühlhausen, near Kralup on the Moldau in Bohemia of Czechoslovakia.

Paraluminite. — Near aluminite, but supposed to be $2Al_2O_3.SO_3.15H_2O$. Soft. White. Optically −. $X \parallel$ elongation. $\alpha = 1·462$, $\beta = 1·470$, $\gamma = 1·471$. 2V very small. Probably derived from alteration of aluminite. Found near Halle, Province of Saxony; also in France, from Huelgoat in Finistère and at Caden, Morhiban.

Felsöbányite. — $2Al_2O_3.SO_3.10H_2O$. Orthorhombic. Massive; in scaly concretions. Basal cleavage. H. = 1·5. G. = 2·3. Color snow-white. Optically +. $X \parallel$ elongation. $\alpha = 1·516$, $\beta = 1·518$, $\gamma = 1·533$. From near Felsöbánya (Baia Sprie), Rumania.

Botryogen. — Perhaps $2MgO.Fe_2O_3.4SO_3.15H_2O$. Monoclinic. Usually in reniform and botryoidal shapes. Cleavages: (010) perfect, (110) poor. H. = 2–2·5. G. = 2·1. Difficultly fusible. Color deep hyacinth-red, ocher-yellow. Optically +. Ax. pl. ⊥ (010). $Z \wedge c$ axis = −12°. $\alpha = 1·522$, $\beta = 1·529$, $\gamma = 1·577$. 2V = 41°. Pleochroic; X = yellow, Y = pale red, Z = deep orange-red. From Falun, Sweden. From the region of Medeni Zakh, Persia; from near Copiapo, Atacama, Chile. From the Redington mercury mine near Knoxville, Napa Co., California. *Quetenite* from Quetena, west of Calama, Antofagasta, Chile, is probably botryogen.

Alunite Group

ALUNITE. Alumstone.

Rhombohedral. Axis $c = 1·2520$. In rhombohedrons, resembling cubes (rr' $10\bar{1}1$ ∧ $\bar{1}101 = 90°$ 50'). Also massive, having a fibrous, granular, or impalpable texture.

Cleavage: $c(0001)$ distinct; $r(10\bar{1}1)$ in traces. Fracture flat conchoidal, uneven; of massive varieties splintery; and sometimes earthy. Brittle. H. = 3·5–4. G. = 2·58–2·752. Luster of r vitreous, basal plane somewhat pearly. Color white, sometimes grayish or reddish. Streak white. Transparent to subtranslucent. Optically +. $\omega = 1·572$. $\epsilon = 1·592$.

Comp. — Basic hydrous sulphate of aluminum and potassium, K_2Al_6 $(OH)_{12}(SO_4)_4$ = Sulphur trioxide 38·6, alumina 37·0, potash 11·4, water 13·0 = 100. Sometimes contains considerable soda, *natroalunite*.

Pyr., etc. — B.B. decrepitates, and is infusible. In the closed tube yields water, sometimes also ammonium sulphate, and at a higher temperature sulphurous and sulphuric oxides. Heated with cobalt solution affords a fine blue color. With soda and charcoal infusible, but yields a hepatic mass. Soluble in sulphuric acid.

Obs. — Alunite is most commonly associated with acid volcanic rocks where the rock has been largely altered and alunite formed owing to the presence of sulphuric acid solutions or vapors. The conditions of formation usually indicate high pressures and temperatures. Found in smaller amounts about volcanic fumaroles and also occurs in connection with sulphide ore bodies. Found in Ruthenia of Czechoslovakia at Beregszasz and Muzsay; in Italy at Tolfa, near Civitavecchia, Roma, in crystals. In France on Mont Dore, Puy-de-Dôme. With the hyalite and opal of Queretaro, Mexico. In the United States occurs in flattened rhombohedral crystals at Rosita Hills, Custer Co., Colorado. Also at Red Mountain in the San Juan district, Colorado. From Marysvale, Piute Co., Utah; from Goldfield, Esmeralda Co., Nevada. *Newtonite* from Sneed's Creek, Newton Co., Arkansas, originally described as an aluminum silicate near kaolin has been shown to be alunite. *Löwigite* is probably also to be classed with alunite.

JAROSITE.

Rhombohedral. Axis $c = 1 \cdot 2492$; $rr'\ 10\bar{1}1 \wedge \bar{1}101 = 90°\ 45'$, $cr\ 0001 \wedge 10\bar{1}1 = 55°\ 16'$. Often in druses of minute crystals; also fibrous, granular massive; in nodules, or as an incrustation.

Cleavage: $c(0001)$ distinct. Fracture uneven. Brittle. H. $= 2 \cdot 5$–$3 \cdot 5$. G. $= 3 \cdot 15$–$3 \cdot 26$. Luster vitreous to subadamantine: brilliant, also dull. Color ocher-yellow, yellowish brown, clove-brown. Streak yellow, shining. Optically $-$. $\omega = 1 \cdot 820$, $\epsilon = 1 \cdot 715$. Basal section may show division into six biaxial segments with 2V very small.

Comp. — $K_2Fe_6(OH)_{12}(SO_4)_4 =$ Sulphur trioxide $31 \cdot 9$, iron sesquioxide $47 \cdot 9$, potash $9 \cdot 4$, water $10 \cdot 8 = 100$.

Obs. — Jarosite has been formed probably in the same manner as alunite, under solfataric conditions with high temperature and pressure. The material known as *gelbeisenerz*, which should probably be included here, was originally found near Kollosoruk, southwest of Bilin in Bohemia of Czechoslovakia, and later at Modum, Norway. The original jarosite was from Barranco Jaroso, in the Sierra Almagrera, near Aquilas in Murcia, Spain. From near Laurium, Greece; from Schlaggenwald, Bohemia; from the Island of Elba. In France from Pallières in Gard (*pastreite*) and near Mâcon, Saône-et-Loire. From various mines in the Altai Mts., Siberia. From Chocaya in Potosi, Bolivia. In the United States in Arizona at the Vulture mine, Maricopa Co., and at Bisbee, Cochise Co.; at the Arrow mine in Chaffee Co., Colorado; in Brewster Co., Texas; in Dona Ana Co., New Mexico; in Lawrence Co., South Dakota. In Utah in the Tintic district and at Mercur, Tooele Co. It is probable that the minerals known as *cyprusite*, *carphosiderite*, *utahite*, are in part at least identical with jarosite.

Natrojarosite. — $Na_2Fe_6(OH)_{12}(SO_4)_4$. Rhombohedral. In minute tabular crystals. H. $= 3$. G. $= 3 \cdot 2$. Difficultly fusible. Color yellow-brown. Optically $-$. $\omega = 1 \cdot 832$, $\epsilon = 1 \cdot 750$. First described from near Sodaville, Mineral Co., Nevada, where it occurs as a glistening yellow-brown powder made up of minute crystals. Also from Cook's Peak, Luna Co., New Mexico. From Kingman, Mohave Co., Arizona. Also found at Cape Calamita, Elba; at Kundip, Western Australia.

Ammoniojarosite. — $(NH_4)_2Fe_6(OH)_{12}(SO_4)_4$. Found in small lumps and flattened nodules embedded in blackish lignite material associated with tschermigite from southern Utah. Under microscope shown to be composed of minute, transparent, pale yellow, tabular grains, some of which showed hexagonal outlines. Indices, $1 \cdot 750$–$1 \cdot 800$.

Plumbojarosite. — $PbFe_6(OH)_{12}(SO_4)_4$. Rhombohedral. In minute tabular crystals. Rhombohedral cleavage. G. $= 3 \cdot 67$. Color dark brown. Optically $-$. Indices, $1 \cdot 783$–$1 \cdot 878$. At times basal sections show a division into six biaxial segments with 2V very small. Found in the United States at Cook's Peak, Luna Co., New Mexico; in Utah in Beaver Co., in the Tintic district, etc.; from Mineral Co., Nevada. *Vegasite*, from Yellow Pine district, near Las Vegas, Clark Co., Nevada has been shown to be identical with plumbojarosite.

Argentojarosite. — $Ag_2Fe_6(OH)_{12}(SO_4)_4$. Hexagonal. In minute micaceous scales. Luster brilliant. Color yellow to brown. Optically $-$. $\omega = 1 \cdot 905$, $\epsilon = 1 \cdot 785$. Of secondary origin with silver ores in the Tintic Standard mine, Dividend, Utah.

Palmierite. — $(K,Na)_2Pb(SO_4)_2$. Hexagonal-rhombohedral. In microscopic plates, often hexagonal in outline. G. $= 4 \cdot 5$. Fusible. Decomposed by water. White. Luster somewhat pearly on base, otherwise vitreous. Optically negative. $\omega = 1.712$. Found in fumarole deposits at Vesuvius.

ALMERIITE. $Na_2SO_4.Al_2(SO_4)_3.5Al(OH)_3.H_2O$. Compact. White. From Almeria, Spain.

Ettringite. — Perhaps $6CaO.Al_2O_3.3SO_3.33H_2O$. Hexagonal. In minute colorless acicular crystals. Prismatic cleavage. H. = 2–2·5. G. = 1·8. Optically —. $\omega = 1.49$. From limestone-inclusions in lava, at Ettringen and Mayen in Rhineland. From Tombstone, Cochise Co., Arizona.

Zincaluminite. — $2ZnSO_4.4Zn(OH)_2.6Al(OH)_3.5H_2O$. In minute hexagonal plates. H. = 2·5–3. G. = 2·26. Color white, bluish. Uniaxial, —. $\omega = 1.534$, $\epsilon = 1.514$. From Laurium, Greece.

GLAUKOKERINITE. $Zn_{13}Al_8Cu_7(SO_4)_2O_{30}.34H_2O$. Radiating fibrous. G = 2·75 . H = 1. Blue color. Fibers show parallel extinction. $n = 1.542$. Strong birefringence. From Laurium, Greece. Name derived from γλαυχός, *blue* and χήρινος, *wax-like*.

Chalcoalumite. — $CuO.2Al_2O_3.SO_3.9H_2O$. Probably triclinic. In matted fibrous crusts. Under microscope shows lath-like structure with inclined extinction. Twinning, with twinning plane parallel to long edge of laths. Several cleavages. H. = 2·5. G. = 2·29. Difficultly fusible. Turquois-green color. Optically +. $\alpha = 1.523$, $\beta = 1.525$, $\gamma = 1.532$. 2V small. Strong dispersion, $\rho > v$. Found at Bisbee, Cochise Co., Arizona, with limonite and copper carbonates.

Johannite. — $(Cu,Fe,Na_2)O.UO_3.SO_3.4H_2O$. Triclinic. In druses or reniform masses or in minute lath-shaped crystals. Shows polysynthetic twinning in two directions. H. = 2. G. > 3·3. Infusible. Readily soluble in acids. Color greenish yellow to canary-yellow. Optically —. $\alpha = 1.57$, $\beta = 1.59$, $\gamma = 1.61$. 2V = nearly 90°. Dispersion strong. Pleochroism strong, X = colorless, Y = very pale yellow, Z = pale greenish yellow or canary-yellow. Found at Joachimstal, Bohemia, Czechoslovakia, and in Gilpin Co., Colorado (*gilpinite*).

URANOPILITE. Perhaps $CaO.8UO_3.2SO_3.25H_2O$. Triclinic? In velvety incrustations. Crystals small lath-shaped. G. = 3·9. Yellow. Optically +. $\alpha = 1.621$, $\beta = 1.623$, $\gamma = 1.631$. 2V large. Strong dispersion, $\rho < v$. From Johanngeorgenstadt, Saxony; Joachimstal, Bohemia in Czechoslovakia. Reported from Montrose Co., Colorado.

Zippeite. — Hydrous uranium sulphate. Monoclinic? In lath-shaped crystals with (010) cleavage. H. = 3. Yellow. Optically +. $\beta = 1.615$. Found at Joachimstal, Bohemia, and as a yellow crystalline powder in asphaltic sandstones at Fruita, Wayne Co., Utah. *Voglianite* and *uraconite* are similar uncertain uranium sulphates from Joachimstal.

Minasragrite. — An acid hydrous vanadyl sulphate, $V_2O_4.3SO_3.16H_2O$. Probably monoclinic. In granular aggregates, small mammillary masses, or in spherulites. Cleavages, (010) (110). Easily fusible. Soluble in cold water. Color blue. Optically —. Ax. pl. $\perp$ (010). $\alpha = 1.518$, $\beta = 1.530$, $\gamma = 1.542$. 2V large. Strongly pleochroic, X = deep blue, Y = pale blue, Z = colorless. Found as an efflorescence on patronite from Minasragra, near Cerro de Pasco, Peru.

KLEBELSBERGITE. Hydrous sulphate of antimony? Monoclinic. In aggregates of small, acicular or tabular crystals. Easily fusible. Color dark sulphur-yellow to orange-yellow. Optically —. X $\perp$ (010). $Y \wedge c$ axis slightly inclined. $n > 1.740$. From Felsöbánya (Baia Sprie), Rumania.

Tellurates; also Tellurites, Selenites

Montanite. — $Bi_2O_3.TeO_3.2H_2O$. Monoclinic? In earthy incrustations; yellowish to white. G. = 3·8. Easily fusible. Optically —. $n = 2.09$. 2V small. Very strong dispersion, $\rho < v$. Occurs incrusting tetradymite at Highland, Montana; in Davidson Co., North Carolina; at Norongo, Co. Murray, New South Wales.

Emmonsite. — Probably a hydrated ferric tellurite. Monoclinic. In thin yellow-green scales. Perfect (010) cleavage and two other imperfect cleavages. H. = 5. Optically —. Ax. pl. || (010). $\alpha = 1.95$, $\beta = 2.09$, $\gamma = 2.10$. 2V = 20°. Strong dispersion, $\rho > v$. From near Tombstone, Cochise Co., Arizona.

Durdenite. — Hydrous ferric tellurite, $Fe_2(TeO_3)_3.4H_2O$. Orthorhombic. In small mammillary forms. H. = 2–2·5. Greenish yellow. Optically —. Ax. pl. || (010). X = c axis. $\alpha = 1.70$, $\beta = 1.95$, $\gamma = 1.96$. 2V = 23°. Strong dispersion, $\rho > v$. From Ojojama district, Dept. Tegucigalpa, Honduras. Noted in Calaveras Co., California.

Chalcomenite. — $CuSeO_3.2H_2O$. In small blue monoclinic crystals. H. = 2·5–3. G. = 3·76. Easily fusible. Optically −. Ax. pl. ‖ (010) for red to green light, ⊥ (010) for blue light. $\alpha = 1·710$, $\beta = 1·731$, $\gamma = 1·732$. 2V = 0° for green light. From the Cerro de Cacheuta, near Mendoza, Argentina, with silver, copper selenides. Also reported from Sierra de Umango and Sierra Famatima, La Rioja, Argentina.

MOLYBDOMENITE is lead selenite and COBALTOMENITE probably cobalt selenite, from the same locality as chalcomenite.

Oxygen Salts

7. TUNGSTATES, MOLYBDATES

The monoclinic Wolframite Group and the tetragonal Scheelite Group are included here.

Wolframite Group

Ferberite	$FeWO_4$		
Wolframite	$(Fe,Mn)WO_4$	$a : b : c = 0·8255 : 1 : 0·8664$	$\beta = 89° 32'$
Hübnerite	$MnWO_4$		
Raspite	$PbWO_4$		

WOLFRAMITE.

Monoclinic. See above for axes.

Twins: (1) tw. axis c with $a(100)$ as comp.-face; (2) tw. pl. $k(023)$, Fig. 475, p. 193. Crystals commonly tabular ‖ $a(100)$; also prismatic. Faces in prismatic zone vertically striated. Often bladed, lamellar, coarse divergent columnar, granular.

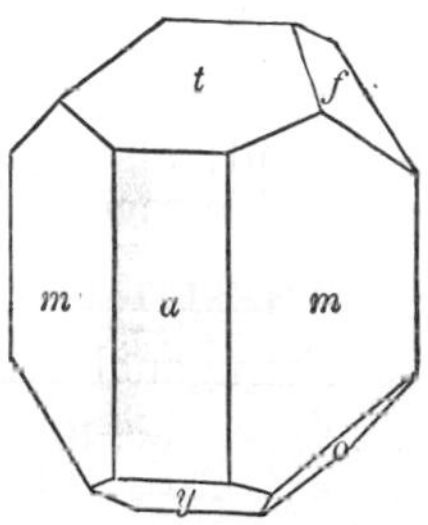

Cleavage: $b(010)$ very perfect; also parting ‖ $a(100)$, and ‖ $t(102)$. Fracture uneven. Brittle. H. = 5–5·5. G. = 7–7·5. Luster submetallic. Color dark grayish or brownish black. Streak nearly black. Opaque. Sometimes weakly magnetic. Optically +. Ax. pl. ⊥ (010). $Z \wedge c$ axis = +17° to 21°. $\beta = 2·4_{Li}$ for ferberite, 2·32 for wolframite, 2·22 for hübnerite.

1059

Comp. — Tungstate of iron and manganese (Fe,Mn) WO_4.

Forms a series from *ferberite*, $FeWO_4$ to *hübnerite*, $MnWO_4$. It has been suggested that ferberite should include that portion of the series containing up to 20 per cent $MnWO_4$, hübnerite the portion containing up to 20 per cent $FeWO_4$, and wolframite the remainder.

Pyr., etc. — Fuses B.B. easily (F. = 2–4) to a globule, which has a crystalline surface and is magnetic. The fusing point rises with increase in percentage of the hübnerite molecule. With salt of phosphorus gives a clear reddish yellow glass while hot which is paler on cooling; in R.F. becomes dark red; on charcoal with tin, if not too saturated, the bead assumes on cooling a green color, which continued treatment in R.F. changes to reddish yellow. With soda and niter on platinum foil fuses to a bluish green manganate. Decomposed by aqua regia with separation of tungstic acid as a yellow powder. Sufficiently decomposed by concentrated sulphuric acid, or even hydrochloric acid, to give a colorless solution, which, treated with metallic zinc, becomes intensely blue, but soon bleaches on dilution.

Obs. — Wolframite is commonly found in granite and pegmatite veins having been formed under pneumatolytic conditions. It is very commonly associated with cassiterite but occurs at times in veins that are free from that mineral. It is also deposited in veins with sulphide minerals that have been formed under conditions of less heat and pressure. Often a constituent of placer deposits.

Some of the more important localities for its occurrence are as follows: From the Adun-Chilon Mts., south of Nerchinsk in Transbaikalia; in Bohemia of Czechoslovakia at Schlaggenwald and Zinnwald in fine crystals; in Rumania at Felsöbánya (Baia Sprie); **in Saxony** at Ehrenfriedersdorf, etc.; in France in Haute Vienne, near Chanteloube, and at Puy-les-Vignes, near Saint-Léonard. In Spain from the Sierra Almagrera, near Aquilas in Murcia (*ferberite*) and elsewhere. With tin ores at various places in Cornwall. Important deposits are found in Lower Burma both in veins and placers. With cassiterite at various points in the New England Range, New South Wales. From Oruro in Bolivia. *Hübnerite* occurs in exceptional crystals at Cerro de Pasco, Peru.

In the United States in Connecticut at Long Hill in the town of Trumbull, Fairfield Co., often pseudomorphs after scheelite. From the St. Francis River in Madison Co., Missouri; in the Black Hills, South Dakota (*hübnerite*). In Colorado occurs at Nederland in Boulder Co. (*ferberite*), and near Silverton, San Juan Co. (*hübnerite*). In New Mexico it is found near Gage in Luna Co., and at Bonita Mt., Lincoln Co. *Hübnerite* comes from the Mammoth district, Nevada.

Reinite, described as a ferrous tungstate occurring in tetragonal crystals from various localities in Japan, is probably a pseudomorph after scheelite, similar to those found at Long Hill in Fairfield Co., Connecticut.

Use. — An ore of tungsten.

Raspite. — $PbWO_4$, same as stolzite. Monoclinic. In small crystals, tabular $\parallel$ (100). Cleavage $\parallel$ (100). H. = 2·5. Fusible. Color brownish yellow. Optically +. Ax. pl. $\parallel$ (010). $\beta = 2·27$. 2V almost 0°. From the Broken Hill mines, New South Wales; from the gold sands of Sumidouro, Minas Geraes, Brazil.

Scheelite Group.   Tetragonal-pyramidal

Scheelite	$CaWO_4$	pp' (111 $\wedge$ $\bar{1}$11) = 79° 55½'	$c = 1·5356$
Cuprotungstite	$CuWo_4$		
Cuproscheelite	$(Ca,Cu)WO_4$		
Powellite	$Ca(Mo,W)O_4$	80° 1'	$c = 1·5445$
Stolzite	$PbWO_4$	80° 15'	$c = 1·5667$
Wulfenite	$PbMoO_4$	80° 22'	$c = 1·5771$

The SCHEELITE GROUP includes the tungstates and molybdates of calcium and lead; also copper. In crystallization they belong to the Pyramidal Class of the Tetragonal System. Wulfenite is probably hemimorphic.

SCHEELITE.

Tetragonal-pyramidal.   Axis $c = 1·5356$.

ee', 101 $\wedge$ $\bar{1}$01 = 72° 40½'. pp', 111 $\wedge$ $\bar{1}$11 = 79° 55½'.
ce, 001 $\wedge$ 101 = 56° 56'. cp, 001 $\wedge$ 111 = 65° 16½'.

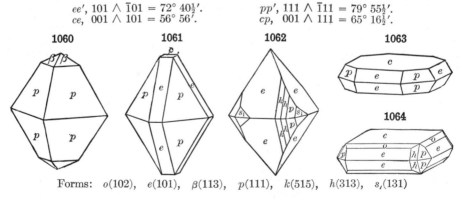

Forms: o(102), e(101), β(113), p(111), k(515), h(313), s_i(131)

Twins: (1) tw. pl. a(100), both contact- and penetration-twins (Fig. 442, p. 188). Habit octahedral, also tabular. Symmetry shown by faces k, h, s (Fig. 1062). Also reniform with columnar structure; massive granular.

Cleavage: p(111) most distinct; e(101) interrupted. Fracture uneven.

Brittle. **H.** = 4·5–5. **G.** = 5·9–6·1. Luster vitreous, inclining to adamantine. Color white, yellowish white, pale yellow, brownish, greenish, reddish. Streak white. Transparent to translucent. Optically +. $\omega = 1.918$. $\epsilon = 1.934$.

Comp. — Calcium tungstate, $CaWO_4$ = Tungsten trioxide 80·6, lime 19·4 = 100.

Molybdenum is usually present (to 8 per cent). Copper may replace calcium, see cuproscheelite.

Pyr., etc. — B.B. in the forceps fuses at 5 to a semi-transparent glass. Soluble with borax to a transparent glass, which afterward becomes opaque and crystalline. With salt of phosphorus forms a glass, colorless in outer flame, in inner, green when hot, and fine blue when cold; varieties containing iron require to be treated on charcoal with tin before the blue color appears. In hydrochloric or nitric acid decomposes, leaving a yellow powder soluble in ammonia. The hydrochloric acid solution treated with tin and boiled assumes a blue color, later changing to brown.

Obs. — Scheelite is formed under pneumatolytic or hydro-thermal conditions and is found in pegmatite veins or in ore veins which are associated with granites or gneiss. It has been observed in contact metamorphic deposits. It is found in Bohemia of Czechoslovakia at Schlaggenwald, Zinnwald, and Riesengrund in the Riesengebirge; in Saxony at Fürstenberg near Schwarzenberg and at Altenberg; in Italy at Mte. Mulat near Predazzo, Val di Flemm, Trentino, and at Traversella in Piedmont, in fine crystals. From near Guttannen in the Hasli Tal, Bern, Switzerland. In fine crystals from Framont in the Vosges, Alsace, France. At Pitkäranta on Lake Ladoga, Finland. It occurs at Carrock Fells, Cumberland, in fine crystals, and in the tin veins of Cornwall. From Sonora, Mexico.

In the United States at Long Hill, Trumbull, Fairfield Co., Connecticut; in South Dakota at Lead in Lawrence Co.; at Leadville, Lake Co., Colorado; at Dragoon, Cochise Co., Arizona; in California at Atolia, San Bernardino Co.

Use. — An ore of tungsten.

Cuprotungstite. — Cupric tungstate, $CuWO_4$. Cryptocrystalline, fibrous. H. = 4·5. Color green. $n = 2.15$. From the copper mines of Llamuco, near Santiago, Chile. *Cuproscheelite* is $(Ca,Cu)WO_4$, with 6·8 per cent CuO; color green. These minerals are apparently different stages in the alteration of scheelite through the replacement of calcium by copper. They occur in concentric layers about scheelite. *Cuprotungstite* was originally found at the copper mines of Llamuco, near Santiago, Chile. The original *cuproscheelite* was from the vicinity of La Paz, Lower California, Mexico. Also found near Montoro, Cordoba, Spain. Both substances are found in Georgiana Co., New South Wales. In the United States from Clifton district, Tooele Co., Utah; from Kern Co., California.

Powellite. — Calcium molybdate with calcium tungstate (10 per cent WO_3), $Ca(Mo,W)O_4$. In minute yellow tetragonal pyramids. Cleavage, (111). H. = 3·5. G. = 4·349. Fusible. Optically +. $\omega = 1.974$, $\epsilon = 1.984$. Formed through the oxidation of molybdenite; often associated with scheelite. Occurs in Siberia, near Minusinsk in Yeniseisk. In the United States from Houghton Co., Michigan. Originally described from Seven Devils district, Adams Co., Idaho. Also from Llano Co., Texas; the Clifton district, Tooele Co., Utah; Tonopah, Nye Co., Nevada; in California from the Black Mts., Kern Co.

Stolzite. — Lead tungstate, $PbWO_4$. In pyramidal tetragonal crystals. H. = 3. G. = 7·87–8·13. Easily fusible. Color green to gray or brown. Optically –. $\omega = 2.27$, $\epsilon = 2.19$. Zinnwald, Bohemia; Sardinia; Broken Hill, New South Wales; from Marianna, near Ouro Preto, Minas Geraes, Brazil.

Chillagite. — $3PbWO_4.PbMoO_4$. In tabular tetragonal crystals. Color yellow to brownish. H. = 3·5. G. = 7·5. Easily fusible. From Chillagoe, North Queensland.

WULFENITE.

Tetragonal-pyramidal; hemimorphic. Axis $c = 1.5771$.

cu, 001 ∧ 102 = 38° 15′.	uu', 102 ∧ 012 = 51° 56′.
ce, 001 ∧ 101 = 57° 37′.	ee', 101 ∧ 011 = 73° 20′.
cn, 001 ∧ 111 = 65° 51′.	nn', 111 ∧ 111 = 80° 22′.

Crystals commonly square tabular, sometimes extremely thin; less frequently octahedral; also prismatic. Hemimorphism sometimes distinct. Also granular massive, coarse or fine, firmly cohesive.

Cleavage: $n(111)$ very smooth; $c(001)$, $s(113)$ less distinct. Fracture subconchoidal. Brittle. H. = 2·75–3. G. = 6·7–7·0. Luster resinous or adamantine. Color wax- to orange-yellow, siskin- and olive-green, yellowish gray, grayish white to nearly colorless, brown; also orange to bright red.

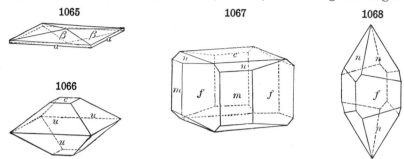

Streak white. Subtransparent to subtranslucent. Optically negative. Indices: $\omega_r = 2·402$, $\epsilon_r = 2·304$. At times shows biaxial characters.

Comp. — Lead molybdate, $PbMoO_4$ = Molybdenum trioxide 39·3, lead oxide 60·7 = 100. Calcium sometimes replaces the lead.

Pyr., etc. — B.B. decrepitates and fuses below 2. With salt of phosphorus in O.F. gives a yellowish green glass, which in R.F. becomes dark green. With soda on charcoal yields metallic lead. Decomposed on evaporation with hydrochloric acid, with the formation of lead chloride and molybdic oxide; on moistening the residue with water and adding metallic zinc, it gives an intense blue color, which does not fade on dilution of the liquid.

Obs. — Wulfenite is of secondary origin, being found in lead and zinc deposits in the oxidation zone. It has probably been formed subsequent to the first oxidation reactions by the action of molybdenum-bearing waters upon cerussite, etc. The more important localities for its occurrence are as follows: From Miess, near Prävali in Carniola, Yugoslavia, and the neighboring locality of Schwarzenbach. At Přibram in Bohemia of Czechoslovakia; at Bleiberg and elsewhere in Carinthia, Austria. In Morocco at Djebel Mahseur, near Oudjda; from Mindouli, French Congo. In New South Wales at Broken Hill. In Mexico from the Sierra de los Lamentos, Chihuahua, and from Sonora.

In the United States sparingly at the lead mine at Southampton, Hampshire Co., Massachusetts; in fine yellow and reddish orange crystals at Wheatley's mine, near Phœnixville, Chester Co., Pennsylvania. In New Mexico it occurs in fine crystals at the Organ Mts., near Las Cruces, Dona Ana Co.; at Hillsboro, Sierra Co.; and Silver City, Grant Co. In Arizona in fine, large, deep red crystals at the Hamburg and other mines in Yuma Co.; also at the Castle Dome district 30 miles distant; at the Mammoth mine near Schultz, which is near Oracle in the Catalina Mts., Pinal Co. In Utah in the Lucin district, Box Elder Co., and in the Little Cottonwood district, Salt Lake Co. In Nevada at Searchlight, Clark Co., and at Eureka in Eureka Co.

Named after the Austrian mineralogist Wülfen (1728–1805).

Use. — An ore of molybdenum.

Ferrimolybdite. Molybdite. — $Fe_2O_3.3MoO_3.8H_2O$. The material long known as molybdite and supposed to be molybdic oxide, MoO_3, has been shown to be rather a hydrous ferric molybdate. The existence of molybdic oxide as a mineral has not been established. The mineral is commonly impure and is frequently intimately associated with limonite. Orthorhombic. In crystal fibers elongated parallel to c axis. Cleavage (001), distinct. H. = 1·5. G. = 4·5. Sulphur-yellow color. Optically +. $Z \parallel$ to c axis. Dispersion distinct, $\rho < v$. $\alpha = 1·72$–$1·78$, $\beta = 1·73$–$1·79$, $\gamma = 1·87$–$2·05$. Pleochroic, X, Y = clear yellow, Z = dirty gray to canary-yellow. Occurs commonly in small amounts as an oxidation product of molybdenite.

Koechlinite. — A molybdate of bismuth, $Bi_2O_3.MoO_3$. Orthorhombic. In minute tabular crystals. Cleavage, (100). Easily fusible. Color, greenish yellow. Optically −. Ax. pl. $\parallel$ (001). $X = b$ axis. $\beta = 2·61$. From Schneeberg, Saxony.

Ferritungstite. — $Fe_2O_3.WO_3.6H_2O$. In microscopic hexagonal plates. Color pale yellow to brownish yellow. Decomposed by acids leaving yellow tungstic oxide. Optically —. $\omega = 1\cdot80$, $\epsilon = 1\cdot72$. Product of oxidation of wolframite from Germania Tungsten mine, Deer Trail district, Washington.

VII. SALTS OF ORGANIC ACIDS

Oxalates, Mellates

Whewellite. — Calcium oxalate, $CaC_2O_4.H_2O$. In small colorless monoclinic crystals. Several cleavages. H. $= 2\cdot5$. G. $= 2\cdot23$. Optically $+$. Ax. pl. $\perp$ (010). $Z \wedge c$ axis $= +29°$. $\alpha = 1\cdot490$, $\beta = 1\cdot555$, $\gamma = 1\cdot650$. Commonly occurs with carbonaceous materials. In Saxony associated with coal at Burgk, near Dresden, and at Zwickau. From Kladno and elsewhere in Bohemia of Czechoslovakia. In Alsace at the mine of Saint-Sylvestre at Urbeis.

Oxammite. — Ammonium oxalate, $(NH_4)_2C_2O_4.2H_2O$. Orthorhombic. Soft. G. $= 1\cdot48$. Easily fusible. White. Optically $-$. Ax. pl. $\parallel$ (100). $X = c$ axis. $\alpha = 1\cdot439$, $\beta = 1\cdot547$, $\gamma = 1\cdot595$. 2V $= 62°$. From the guano of the Guañape Islands, Peru.

Humboldtine. — Hydrous ferrous oxalate, $FeC_2O_4.2H_2O$. Orthorhombic. Crystals prismatic. Cleavages, (110), (100), (010). H. $= 2$. G. $= 2\cdot28$. Color yellow. Optically $+$. Ax. pl. $\parallel$ (010). $Z = c$ axis. $\alpha = 1\cdot494$, $\beta = 1\cdot561$, $\gamma = 1\cdot692$. 2V large. From near Bilin, Bohemia; at Grossalmerode, Hessen-Nassau; Capo d'Arco, Elba. At Kettle Point, Bosanquet township, Lambton Co., Ontario.

Mellite. — Hydrous aluminum mellate, $Al_2C_{12}O_{12}.18H_2O$. In square tetragonal pyramids; also massive, granular. H. $= 2–2\cdot5$. G. $= 1\cdot55–1\cdot65$. Color honey-yellow. Optically $-$. $\omega = 1\cdot539$, $\epsilon = 1\cdot511$. Occurs in brown coal at Artern in Saxony; from Czechoslovakia at Luschitz, south of Bilin in Bohemia; near Walchov, which is near Boskowitz in Moravia; in the Govt. of Tula in Russia; from Nerchinsk in Transbaikalia, Siberia. Reported from the Judith River, Montana.

VIII. HYDROCARBON COMPOUNDS

The Hydrocarbon compounds in general, with few exceptions, are *not homogeneous substances*, but mixtures, which by the action of solvents or by fractional distillation may be separated into two or more component parts. They are hence not definite mineral species and do not strictly belong to pure Mineralogy, rather, with the recent gums and resins, to Chemistry or, so far as they are of practical value, to Economic Geology. In the following pages they are treated for the most part with great brevity.

1. Simple Hydrocarbons. Chiefly members of the Paraffin Series C_nH_{2n+2}.

SCHEERERITE. In whitish monoclinic crystals. Perhaps a polymer of marsh-gas (CH_4). Found in brown coal at Uznach, Switzerland.

HATCHETTITE. Mountain Tallow. In thin plates, or massive. Like soft wax. Color yellowish. Indices $1\cdot47–1\cdot50$. Ratio of C to H $=$ nearly $1 : 1$. From the coal-measures near Merthyr-Tydvil in Glamorganshire, England; from Galicia.

PARAFFIN. A native crystallized paraffin has been described as occurring in cavities in basaltic lava near Paterno, Sicily. Indices, $1\cdot49–1\cdot52$.

OZOCERITE. Mineral wax in part. Like wax or spermaceti in appearance and consistency. Colorless to white when pure; often leek-green, yellowish, brownish yellow, brown. Indices, $1\cdot51–1\cdot54$. Essentially a paraffin, and consisting chiefly of one of the higher members of the series. Occurs in beds of coal, or associated bituminous deposits, as at Slanik, Moldavia; Rumania; Boryslaw in the Carpathians. Also occurs in southern Utah on a large scale.

Zietrisikite, Chrismatite, Urpethite are near ozocerite.

FICHTELITE. In white monoclinic tabular crystals. Perhaps C_5H_8. Occurs in thin layers of pine wood from peat-beds, near Redwitz, in the Fichtelgebirge, Bavaria; from Borkovic, Bohemia. *Hartite* has a similar occurrence.

NAPALITE. A yellow bituminous substance of the consistency of shoemaker's wax. C_3H_4. From the Phœnix mercury mine in Pope Valley, Napa county, California.

CURTISITE. Hydrocarbon, probably $C_{24}H_{18}$. Probably orthorhombic. In small irregular square or six-sided flakes. Perfect basal cleavage. Conchoidal fracture. H. < 2. G. = 1·236. Color, greenish yellow. Vitreous to adamantine luster. Optically +. Ax. pl. || (010). $Z \perp$ (001). $\alpha = 1·557$, $\beta = 1·734$, $\gamma = 2·07$. 2V = 83·5°. Pleochroism: X = pale yellow to nearly colorless, Y and Z = yellow. Occurs along cracks, etc. in a sandstone associated with opaline silica, realgar and metacinnabarite, in the hot spring area at Skaggs Springs, Sonoma Co., California, at a point where inflammable gases are given off in small quantity.

SIMONELLITE. Probably $C_{15}H_{28}$. Orthorhombic. Occurs in a white crystalline encrustation on lignite from Fognano, Montepulciano, Tuscany.

2. Oxygenated Hydrocarbons

AMBER. In irregular masses, with conchoidal fracture. H. = 2–2·5. G. = 1·096. Luster resinous. Color yellow, sometimes reddish, brownish, and whitish, often clouded, sometimes fluorescent. Transparent to translucent. Heated to 150° begins to soften, and finally melts at 250°–300°. Ratio for C : H : O = 40 : 64 : 4.

Part of the so-called amber is separated mineralogically as *succinite* (yielding succinic acid). Other related fossil resins from many other regions (*e.g.*, the Atlantic coast of the United States) have been noted. Some of them have been called *retinite, gedanite, glessite, rumänite, simetite, krantzite, chemawinite, delatynite, ajkaite*, etc.

Amber occurs abundantly on the Prussian coast of the Baltic from Dantzig to Memel; also on the coasts of Denmark, Sweden, and the Russian Baltic provinces. It is mined extensively, and is also found on the shores cast up by the waves after a heavy storm. Amber and the similar fossil resins are of vegetable origin, altered by fossilization; this is inferred both from its native situation with coal, or fossil wood, and from the occurrence of insects incased in it. Amber was early known to the ancients, and called ἤλεκτρον, *electrum*, whence, on account of its electrical susceptibilities, has been derived the word *electricity*.

COPALITE, or Highgate resin, is from the London blue clay. It is like the resin copal in hardness, color, luster, transparency, and difficult solubility in alcohol. Color clear pale yellow to dirty gray and dirty brown. Emits a resinous aromatic odor when broken.

TELEGDITE. A fossil resin from lignite at Száscsór, Transylvania.

The following are oxygenated hydrocarbons occurring with coal and peat deposits, etc.:

BATHVILLITE. Occurs in dull, brown, porous lumps in the torbanite or Boghead coal (of the Carboniferous formation) adjoining the lands of Torbane Hill, Bathville, Scotland. It may be an altered resin, or else material which has filtrated into the cavity from the surrounding torbanite.

TASMANITE. In minute reddish brown scales disseminated through a laminated shale; average diameter of scales about 0.03 in. Not dissolved at all by alcohol, ether, benzene, turpentine, or carbon disulphide, even when heated. Remarkable as yielding 5.3 per cent sulphur. From the river Mersey, north side of Tasmania; the rock is called *combustible shale*.

DYSODILE. In very thin folia, flexible, slightly elastic; yellow or greenish gray. Analysis gave 2.3 per cent sulphur and 1.7 per cent nitrogen. From lignite deposits at Melili, Sicily, and elsewhere.

GEOCERITE. A white, wax-like substance, separated from the brown coal of Gesterwitz, near Weissenfels. *Geomyricite* and *geocerellite* are other products from the same source.

LEUCOPETRITE. Also from the Gesterwitz brown coal. Between a resin and wax in physical characters.

PYRORETINITE. From brown coal near Aussig, Bohemia.

DOPPLERITE. In elastic or partly jelly-like masses; brownish black. An acid substance, or mixture of different acids, related to humic acid. Ratio for C, H, O, nearly 10 : 12 : 5. From peat beds near Aussee in Styria; Fichtelgebirge, Bavaria.

IDRIALITE. Occurs with the cinnabar of Idria. In the pure state white and crystalline in structure. In nature found only impure, being mixed with cinnabar, clay, and some pyrite and gypsum in a brownish black earthy material, called, from its combustibility and the presence of mercury, *inflammable cinnabar*.

POSEPNYTE. Occurs in hard, brittle plates or nodules, light green in color. From the Great Western mercury mine, Lake Co., California. See also napalite, p. 775.

FLAGSTAFFITE. $C_{10}H_{22}O_3$. Identical with terpin hydrate. Orthorhombic. In minute prisms. Colorless. $n = 1·51$. G. = 1·092. Found in cracks of buried tree trunks, near Flagstaff, Arizona.

HOELITE. $C_{14}H_8O_2$. Same as anthraquinone. As delicate needles. G. = 1·43. Negative elongation. $\beta = 1·75$. Found with sal-ammoniac and sulphur as crystallizations from gases due to a burning coal seam at Spitzbergen.

SCHARIZERITE. A nitrogenous carbon compound, found in the Drächenhöhle near Mixnitz in Styria, Austria.

The following are still more complex native hydrocarbon compounds of great importance from an economic standpoint.

Petroleum. — NAPHTHA; PETROLEUM. Mineral oil. Kerosene.
PITTASPHALT: Maltha. Mineral Tar.

Liquids or oils, in the crude state of disagreeable odor; varying widely in color, from colorless to dark yellow or brown and nearly black, the greenish brown color the most common; also in consistency from thin flowing kinds to those that are thick and viscous; and in specific gravity from 0·6 to 0·9. Petroleum, proper, passes by insensible gradations into *pittasphalt* or *maltha* (viscid bitumen); and the latter as insensibly into *asphalt* or solid bitumen.

Chemically, petroleum consists for the most part of members of the paraffin series, C_nH_{2n+2}, varying from marsh gas, CH_4, to the solid forms. The olefines, C_nH_{2n}, are also present in smaller amount. This is especially true of the American oils. Those of the Caucasus have a higher density, the volatile constituents are less prominent, they distill at about 150° and contain the benzenes, C_nH_{2n-6}, in considerable amount. There are present also members of the series C_nH_{2n-8}. The German petroleum is intermediate between the American and the Caucasian. The Canadian petroleum is especially rich in the solid paraffins.

Petroleum occurs in rocks or deposits of nearly all geological ages, from the Lower Silurian to the present epoch. It is associated most abundantly with argillaceous shales, sands and sandstones, but is found also permeating limestones, giving them a bituminous odor, and rendering them sometimes a considerable source of oil. From these oleiferous shales, sands and limestones the oil often exudes, and appears floating on the streams or lakes of the region, or rises in oil springs. It also exists collected in subterranean cavities in certain rocks, whence it issues in jets or fountains whenever an outlet is made by boring. The oil which fills the cavities has ordinarily been derived from the subjacent rocks; for the strata in which the cavities exist are frequently barren sandstones. The conditions required for the production of such subterranean accumulations would be therefore a bituminous oil-bearing or else oil-producing stratum at a greater or less depth below; cavities to receive the oil; an overlying stratum of close grained shale or limestone, not allowing of the easy escape of the naphtha vapors.

The important petroleum districts in the United States are: (1) The Appalachian including fields in New York, Pennsylvania, Ohio, West Virginia, Kentucky, Tennessee, (2) The Ohio-Indiana, (3) Illinois, (4) Kansas-Oklahoma, (5) Louisiana-Texas, (6) California, (7) Wyoming. In Canada oil chiefly produced in Ontario. Important fields in Mexico from Tampico to Tuxpam. The chief foreign districts are in the Baku region, Russia, in Galicia and Rumania, also in Borneo.

Asphaltum. Mineral Pitch. Asphalt.

Asphaltum, or mineral pitch, is a mixture of different hydrocarbons, part of which are oxygenated. Its ordinary characters are as follows: Amorphous. G. = 1–1·8; sometimes higher from impurities. Luster like that of black pitch. Color brownish black and black. Odor bituminous. Melts ordinarily at 90° to 100°, and burns with a bright flame. Soluble mostly or wholly in oil of turpentine, and partly or wholly in ether; commonly partly in alcohol. The more solid kinds graduate into the pittasphalts or mineral tar, and through these there is a gradation to petroleum. The fluid kinds change into the solid by the loss of a vaporizable portion on exposure, and also by a process of oxidation, which consists first in a loss of hydrogen, and finally in the oxygenation of a portion of the mass. The action of heat, alcohol, ether, naphtha and oil of turpentine, as well as direct analyses, show that the so-called asphaltum from different localities is very various in composition.

Asphaltum belongs to rocks of no particular age. The most abundant deposits are superficial. But these are generally, if not always, connected with rock deposits containing some kind of bituminous material or vegetable remains. Some of the noted localities of asphaltum are the region of the Dead Sea, or Lake Asphaltites, whence the most of the asphaltum of ancient writers; a lake on Trinidad, $1\frac{1}{2}$ m. in circuit, which is hot at the center, but is solid and cold toward the shores, and has its borders over a breadth of $\frac{3}{4}$ m. covered with the hardened pitch with trees flourishing over it; at various places in South America; in California, near the coast of St. Barbara; also in smaller quantities, elsewhere.

ELATERITE. Elastic Bitumen. Mineral Caoutchouc. Soft, elastic, sometimes much like india-rubber; occasionally hard and brittle. Color usually dark brown. Found at Castleton in Derbyshire, and elsewhere.

ALBERTITE. Differs from ordinary asphaltum in being only partially soluble in oil of turpentine, and in its very imperfect fusion when heated. H. = 1–2. G. = 1·097. Luster brilliant, pitch-like; color jet-black. Occurs filling an irregular fissure in rocks of the Lower Carboniferous in Nova Scotia. *Impsonite* from Impson valley, Indian Territory, is like albertite except that it is almost insoluble in turpentine.

GRAHAMITE. Resembles albertite in its pitch-black, lustrous appearance. H. = 2. G. = 1·145. Soluble mostly in oil of turpentine; partly in ether, naphtha or benzene; not at all in alcohol; wholly in chloroform and carbon disulphide. Melts only imperfectly, and with a decomposition of the surface. Occurs in masses several miles south of Parkersburg, filling a fissure in a Carboniferous sandstone; *kundaite* is a variety from Kunda, Esthonia, Russia.

GILSONITE, also called *Uintahite* or *Uintaite*. A variety of asphalt from near Ft. Duchesne, Utah, which has found many applications in the arts. Occurs in masses several inches in diameter, with conchoidal fracture; very brittle. H. = 2–2·5; G. = 1·065–1·070. Color black, brilliant and lustrous; streak and powder a rich brown. Fuses easily in the flame of a candle and burns with a brilliant flame, much like sealing-wax. Named after Mr. S. H. Gilson of Salt Lake City.

NIGRITE is a variety of asphaltum from Utah.

THUCHOLITE. Composition variable, showing large amounts of carbon, smaller amounts of water, uranium, thorium, and rare earth oxides and silica. Color jet-black. Luster brilliant. Conchoidal fracture. Opaque. H. = 3·5–4. G. = 1·78. Found as small irregularly rounded nodules embedded in feldspar, quartz or mica, often associated with uraninite and cyrtolite in a pegmatite vein near Parry Sound, Ontario. Also found under similar conditions near Buckingham, Quebec.

Mineral Coal. — Compact massive, without crystalline structure or cleavage; sometimes breaking with a degree of regularity, but from a jointed rather than a cleavage structure. Sometimes laminated; often faintly and delicately banded, successive layers differing slightly in luster. Fracture conchoidal to uneven. Brittle; rarely somewhat sectile. H. = 0·5–2·5. G. = 1–1·80. Luster dull to brilliant, and either earthy, resinous or submetallic. Color black, grayish black, brownish black, and occasionally iridescent; also sometimes dark brown. Opaque. Infusible to subfusible; but often becoming a soft, pliant or paste-like mass when heated. On distillation most kinds afford more or less of oily and tarry substances, which are mixtures of hydrocarbons and paraffin.

The varieties recognized depend partly (1) on the amount of the volatile ingredients afforded on destructive distillation; or (2) on the nature of these volatile compounds, for ingredients of similar composition may differ widely in volatility, etc.; (3) on structure, luster and other physical characters.

Coal is in general the result of the gradual change which has taken place in geological history in organic deposits, chiefly vegetable, and its form and composition depend upon the extent to which this change has gone on. Thus it passes from forms which still retain the original structure of the wood (peat, lignite) and through those with less of volatile or bituminous matter to anthracite and further to kinds which approach graphite.

1. **ANTHRACITE.** H. = 2–2·5. G. = 1·32–1·7. Luster bright, often submetallic, iron-black, and frequently iridescent. Fracture conchoidal. Volatile matter after drying 3–6 per cent. Burns with a feeble flame of a pale color. The anthracites of Pennsylvania contain ordinarily 85–93 per cent of carbon; those of South Wales, 88–95; of France, 80–38; of Saxony, 81; of southern Russia, sometimes 94 per cent. Anthracite graduates through semi-anthracite into bituminous coal, becoming less hard and containing more volatile matter; and an intermediate variety is called *free-burning* anthracite.

2. **BITUMINOUS COAL.** Burns in the fire with a yellow, smoky flame, and gives out on distillation hydrocarbon oils or tar; hence the name *bituminous.* The *ordinary* bituminous coals contain from 5–15 per cent (rarely 16 or 17) of oxygen (ash excluded); while the so-called *brown coal* or *lignite* contains from 20–36 per cent, after the expulsion, at 100°, of 15–36 per cent of water. The amount of hydrogen in each is from 4–7 per cent. Both have usually a bright, pitchy, greasy luster, a firm compact texture, are rather fragile compared with anthracite, and have G. = 1·14–1·40. The *brown* coals have often a brownish black color, whence the name, and more oxygen, but in these respects and others they shade into ordinary bituminous coals. The ordinary bituminous coal of Pennsylvania has G. = 1·26–1·37; of Newcastle, England, 1·27; of Scotland, 1·27–1·32; of France, 1·2–1·33; of Belgium, 1·27–1·3. The most prominent kinds are the following:

(a) *Caking or Coking Coal.* — A bituminous coal which softens and becomes pasty or semi-viscid in the fire. This softening takes place at the temperature of incipient decomposition, and is attended with the escape of bubbles of gas. On increasing the heat, the volatile products which result from the ultimate decomposition of the softened mass are driven off, and a coherent, grayish black, cellular or fritted mass (*coke*) is left. Amount of coke left (or part not volatile) varies from 50–85 per cent.

(b) *Non-Caking Coal.* — Like the preceding in all external characters, and often in ultimate composition; but burning freely without softening or any appearance of incipient fusion. There are all gradations between caking and non-caking bituminous coals.

(c) *Cannel Coal* (Parrot Coal). A variety of bituminous coal, and often caking; but differing from the preceding in texture, and to some extent in composition, as shown by its products on distillation. It is compact, with little or no luster, and without any appearance of a banded structure; and it breaks with a conchoidal fracture and smooth surface; color dull black or grayish black. On distillation it affords, after drying, 40 to 66 per cent of volatile matter, and the material volatilized includes a large proportion of burning and lubricating oils, much larger than the above kinds of bituminous coal; whence it is extensively used for the manufacture of such oils. It graduates into oil-producing coaly shales, the more compact of which it much resembles. *Torbanite* is a variety of cannel coal of a dark brown color, from Torbane Hill, near Bathville, Scotland; also called *Boghead Cannel.*

(d) *Brown Coal* (Lignite). The prominent characteristics of brown coal have already been mentioned. They are non-caking, but afford a large proportion of volatile matter; sometimes pitch-black, but often rather dull and brownish black. G. = 1·15–1·3. Brown coal is often called *lignite.* But this term is sometimes restricted to masses of coal which still retain the form of the original wood. *Jet* is a black variety of brown coal, compact in texture, and taking a good polish, whence its use in jewelry.

Coal occurs in beds, interstratified with shales, sandstones, and conglomerates, and sometimes limestones, forming distinct layers, which vary from a fraction of an inch to 30 feet or more in thickness. In the United States, the anthracites occur east of the Alleghany range, in rocks that have undergone great contortions and fracturings, while the bituminous coals are found extensively in many States farther west, in rocks that have been less disturbed; and this fact and other observations have led geologists to the view that the anthracites have lost their bitumen by the action of heat. The *origin* of coal is mainly vegetable, though animal life has contributed somewhat to the result. The beds were once beds of vegetation, analogous, in most respects, in mode of formation to the peat beds of modern times, yet in mode of burial often of a very different character. This vegetable origin is proved not only by the occurrence of the leaves, stems and logs of plants in the coal, but also by the presence throughout its texture, in many cases, of the forms of the original fibers; also by the direct observation that peat is a transition state between unaltered vegetable débris and brown coal, being sometimes found passing completely into true brown coal. *Peat* differs from true coal in want of homogeneity, it visibly containing vegetable fibers only partially altered; and wherever changed to a fine-textured homogeneous material, even though hardly consolidated, it may be true brown coal.

For an account of the chief coal fields, as also of the geological relations of the different coal deposits, reference is made to works on Economic Geology.

APPENDIX A

ON THE DRAWING OF CRYSTAL FIGURES

In the representation of crystals by figures it is customary to draw their edges as if they were projected upon some definite plane. Two sorts of projection are used: the *orthographic* in which the lines of projection fall at right angles and the *clinographic* where they fall at oblique angles upon the plane of projection. The second of these projections is the more important, and must be treated here in some detail. Two points are to be noted in regard to it. In the first place, in the drawings of crystals the point of view is supposed to be at an infinite distance, and it follows from this that all lines which are parallel on the crystal appear *parallel* in the drawing.

In the second place, in all ordinary cases, it is the complete ideal crystal which is represented, that is, the crystal with its full geometrical symmetry as explained on pp. 10 to 14 (cf. note on p. 13).

In general, drawings of crystals are made, either by constructing the figure upon a projection of its crystal axes, using the intercepts of the different faces upon the axes in order to determine the directions of the edges or by constructing the figure from the gnomonic (or stereographic) projection of the crystal forms. Both of these methods have their advantages and disadvantages. By drawing the crystal figure by the aid of a projection of its crystal axes the symmetry of the crystal and the relations of its faces to the axes are emphasized. In many cases, however, drawing from a projection of the poles of the crystal faces is simpler and takes less time. The student should be able to use both methods and consequently both are described below.

DRAWING OF CRYSTALS UPON PROJECTIONS OF THEIR CRYSTAL AXES

PROJECTION OF THE AXES

The projection of the particular axes required is obviously the first step in the process. These axes can be most easily obtained by making use of the Penfield Axial Protractor, illustrated in Fig. 1069.* The customary directions of the axes for the isometric, tetragonal, orthorhombic and hexagonal systems are given on the protractor and it is a simple matter, as explained below, to determine the directions of the inclined axes of the monoclinic and triclinic systems. Penfield drawing charts giving the projection of the isometric axes, which are easily modified for the tetragonal and orthorhombic systems, and of the hexagonal axes, (see Figs. 1070, 1071) are also quite convenient.

Isometric System. — The following explanation of the making of the projection of the isometric axes has been taken largely from Penfield's description.†

Fig. 1072 will make clear the principles upon which the projection of the isometric axes are based. Fig. 1072A is an orthographic projection (*a plan*, as seen from above) of a cube in two positions, one, *abcd*, in what may be called normal position, the other, *ABCD*, after a revolution of 18° 26′ to the left about its vertical axis. The broken-dashed lines throughout represent the axes. Fig. 1072B is likewise an orthographic projection of a cube in the position *ABCD* of A, when viewed from in front, the eye or point of vision being on a level with the crystal. In the position chosen, the apparent width of the side face *BCB'C'* is one-third that of the front face *ABA'B'*, this being de-

---

* The various Penfield crystal drawing apparatus may be obtained from the Mineralogical Laboratory of the Sheffield Scientific School of Yale University, New Haven, Connecticut.

† On Crystal Drawing; Am. J. Sc., **19**, 39, 1905.

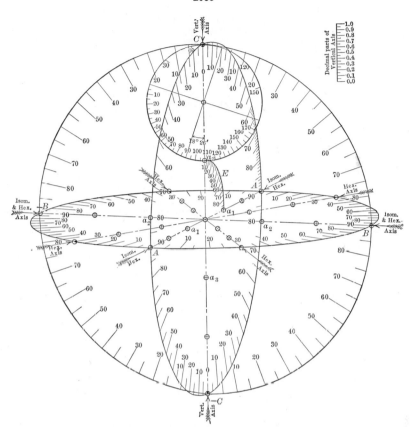

Protractor for plotting crystallographic axes; one-third natural size (after Penfield)

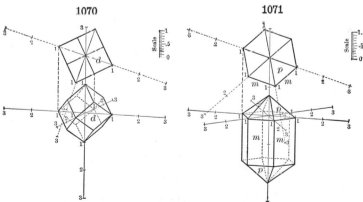

Scheme of the engraved axes of the isometric and hexagonal systems, one-sixth natural size (after Penfield)

pendent upon the angle of revolution 18° 26', the tangent of which is equal to one-third. To construct the angle 18° 26', draw a perpendicular at any point on the horizontal line, $X-Y$, Fig. 1072A as at o, make op equal one-third Oo, and join O and p. The next step in the construction is to change from orthographic to clinographic projection. In order to give crystal figures the appearance of solidity it is supposed that the eye or point of vision is raised, so that one looks down at an angle upon the crystal; thus, in the case under consideration, Fig. 1072C, the top face of the cube comes into view. The position of the crystal, however, is not changed, and the plane upon which the projection is made remains vertical.

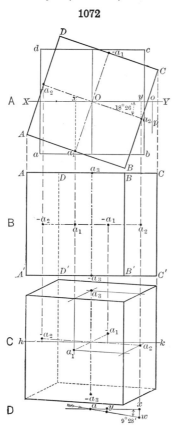

1072

From A it may be seen that the positive ends of the axes, a_1 and a_2 are forward of the line XY, the distances a_1x and a_2y being as 3 : 1. In B it must be imagined, and by the aid of a model it may easily be seen, that the extremities of these same axes are to the front of an imaginary vertical plane (the projection of XY above) passing through the center of the crystal, the distance being the same as a_1x and a_2y of the plan. In D the distance ax is drawn the same length as a_1x of the plan, and the amount to which it is supposed that the eye is raised, indicated by the arrow, is such that a, instead of being projected horizontally to x, is projected at an inclination of 9° 28' from the horizontal to w, the distance xw being one-sixth of ax; hence the angle 9° 28' is such that its tangent is one-sixth. Looking down upon a solid at an angle, and still making the projection on a vertical plane, may be designated as *clinographic projection;* accordingly, to plot the axes of a cube in clinographic projection in conformity with figures A, B and D draw the horizontal construction line hk, figure C, and cross it by four perpendiculars in vertical alignment with the points a_1, $-a_1$ and a_2, $-a_2$ of figures A and B. Then determine the extremities of the first, a_1, $-a_1$ axis by laying off distances equal to xw of figure D, or one-sixth a_1x of figure A, locating them below and above the horizontal line hk. The line a_1, $-a_1$ is thus the projection of the first, or front-to-back axis. In like manner determine the extremities of the second axis, a_2, $-a_2$, by laying off distances equal to xw of figure D, or one-sixth a_2y of figure A, plotted below and above the line hk. The line a_2, $-a_2$ is thus the projection of the second, or right-to-left axis. It is important to keep in mind that in clinographic projection there is no foreshortening of vertical distances. In figure C the axis a_2, $-a_2$ is somewhat, and a_1, $-a_1$ much foreshortened yet both represent axes of the same length as the vertical, a_3, $-a_3$.

It is wholly a matter of choice that the angle of revolution shown in Fig. 1072A is 18° 26', and that the eye is raised so as to look down upon a crystal at an angle of 9° 28' from the horizontal, as indicated by Fig. 1072D. Also it is evident that these angles may be varied to suit any special requirement. As a matter of fact, however, the angles 18° 26' and 9° 28' have been well chosen and

Development of the axes of the isometric system in orthographic and clinographic projection (after Penfield)

are established by long usage, and practically all the figures in clinographic projection, found in modern treatises on crystallography and mineralogy, have been drawn in accordance with them.

Tetragonal and Orthorhombic Systems. — The projection of tetragonal and orthorhombic axes can be easily obtained from the isometric axes by modifying the lengths of the various axes to conform to the axial ratio of the desired crystal. For instance with zircon the vertical axis has a relative length of $c = 0.64$ in respect to the equal lengths of the horizontal axes. By taking 0.64 of the unit length of the vertical axis of the isometric projection the crystal axes for a zircon figure are obtained. The Penfield axial charts all give decimal parts of the unit length of the isometric vertical axis, so that any proportion of this length can be found at once. In the orthorhombic system the lengths of both the a

and c axes must be modified. The desired point upon the c axis can be obtained as described above. In the case of the a axis the required point can be found by some simple method of construction. If, as is the case in the Penfield charts, a plan of the unforeshortened horizontal axes is given in a top view, the desired length can be laid off directly upon the a axis in this orthographic projection by means of the decimal scale and then projected vertically down upon its clinographic projection. Or the proper distance can be laid off on the vertical axis and then by means of a line drawn from this point parallel to a line joining the extremities of the c and a axes of the isometric projection the proper proportional part of the a axis can be determined by intersection.

Hexagonal System. — For projecting the hexagonal axes exactly the same principles may be made use of as were employed in the construction of the isometric axes. Fig. 1073A is an orthographic projection, a *plan*, of a hexagonal prism in two positions, one of them, a_1, a_2, etc., after a revolution of 18° 26' from what may be called normal position. In Fig. 1073B the extremities of the horizontal axes of A have been projected down upon the horizontal construction line hk, and a_1, a_2 and $-a_3$ which are forward of the line XY in A are located below the line hk in the clinographic projection, the distances from hk being one-sixth of a_1x, a_2y and $-a_3z$ of A. Fig. 1073C is a scheme for getting the distances which the extremities of the axes are dropped. The vertical axis in 1073B has been given the same length as the axes of the plan.

Monoclinic System. — In the case of the monoclinic axes the inclination and length of the a axis must be determined in each case. The axial chart, Fig. 1069, can be most conveniently used for this purpose. The ellipse in the figure, lettered A, C, $-A$, $-C$ gives the trace of the ends of the a and c axes as they are revolved in the A–C plane. To find, therefore, the inclination of the a axis it is only necessary to lay off the angle β by means of the graduation given on this ellipse. The unit length of the a axis may be determined in various ways. The plan of the axes given at the top of the chart may be used for this purpose. Fig. 1074 will illustrate the method of procedure as applied in the case of orthoclase, where $\beta = 64°$ and $a = 0.66$. The foreshortened length of the a axis is determined as indicated and then this length can be projected vertically downward upon the inclined a axis, the direction of which has been previously determined as described above.

Triclinic System. — In the construction of triclinic axes the inclination of the a axis and its length are determined in exactly the same manner as described in the preceding paragraph in the case of the monoclinic system. The direction of the b axis is determined as follows: The vertical plane of the b and c axes is revolved about the c axis through such an angle as will conform to the angle between the pinacoids 100 and 010. Care must be taken to note whether this plane is to be revolved toward the front or toward the back. If the angle between the normals to 100 and 010 is greater than 90° the right-hand end of this plane is to be revolved toward the front. Fig. 1075, which is a simplified portion of the axial chart, shows the necessary construction in order to obtain the direction of the b axis in the case of rhodonite in which 100 $\wedge$ 010 = 94° 26' and $\alpha = 103°$ 18'. The plane of the b–c axes will pass through the point p which is 94° 26' from $-a$. To locate the point b', which is the point where the b axis would emerge from the sphere, draw through the point p two or more chords from points where the vertical ellipses of the chart cross the horizontal ellipse, as lines a–p, $-a$–p, b–p, in Fig. 1075. Then

1073

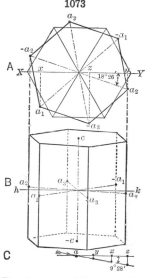

Development of the axes of the hexagonal system in orthographic and clinographic projection (after Penfield)

1074

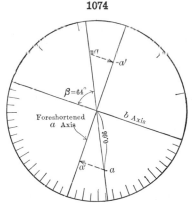

from points on these same vertical ellipses which are 13° 18′ below the horizontal plane draw chords parallel to the first series as x–x′, y–y′, z–z′. The point where these three chords meet determines the position of b′ and a line from this point drawn through the center of the chart determines the direction of the b axis, since it lies in the proper vertical plane and makes the angle α, 103° 18′, with the c axis. The foreshortened length of the b axis can

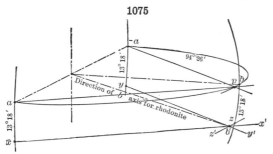

1075

be determined by the use of the orthographic projection of the a and b axes at the top of the chart in exactly the same manner as described under the monoclinic system and the point thus determined may be projected vertically downward upon the line of the b axis of the clinographic projection as already determined. It must be remembered, however, that the position of the b axis in the orthographic projection must conform to the position of the plane of the b and c axes or in the case of rhodonite have its right-hand end at an angle of 94° 26′ with the negative end of the projection of the a axis.

Drawing of Crystal Figures by Aid of Projections of their Axes. — In order to determine in the drawing the direction of any edge between two crystal faces it is necessary to establish two points, both of which shall be common to these two faces. A line connecting two such points will obviously have the desired direction. The positions of these points is commonly established by use of the linear or Quendstedt projection as explained in the following paragraphs, which have been taken almost verbatim from Penfield's description of the process.

The principle upon which the linear projection is based is very simple: *Every face of a crystal (shifted if necessary, but without change of direction) is made to intersect the vertical axis at* unity, *and then its intersection with the horizontal plane, or the plane of the a and b axis is indicated by a line.* For instance if a given face has the indices 111 it is clear that its linear projection would be a line passing through 1a and 1b, since the face under these conditions will also pass through 1c. If, however, the indices of the face are 112 it will only pass through 1/2 c when it passes through 1a and 1b. In order to fulfill, therefore, the requirements of the linear projection that the plane should pass through 1c the indices must be multiplied by two and then under these conditions the line in which the plane intercepts the horizontal plane, or in other words the linear projection of the face, will pass through 2a and 2b. In the case of a prism face with the indices 110, its linear projection will be a line having the same direction as a line joining 1a and 1b but passing through the point of intersection of these axes, since a vertical plane such as a prism can only pass through 1c when it also includes the c axis and so must have its linear projection pass through the point of intersection of the three axes.

When it is desired to find the direction of an edge made by the meeting of any two faces, the lines representing the linear projections of the faces are first drawn, and the point where they intersect is noted. Thus a point common to both faces is determined, which is located in the plane of the a and b axes. A second point common to the two faces is *unity* on the vertical axis, and a line from this point to where the lines of the linear projection intersect gives the desired direction.

1076

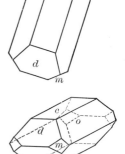

A simple illustration, chosen from the orthorhombic system, will serve to show how the linear projection may be employed in drawing. The example is a combination of barite, such as is shown in Fig. 1076. The axial ratio of barite is as follows:

$$a : b : c = 0.8152 : 1 : 1.3136$$

The forms shown in the figure and the symbols are: base c(001), prism m(110), brachydome o(011) and macrodome d(102).

Fig. 1077 represents the details of construction of the orthographic and clinographic projections shown in Fig. 1076. On the orthographic axes the axial lengths a and b are located, the vertical axis c being foreshortened to a point at the center. On the clinographic axes, centered at O, the ends of the axes a and b are located by dropping perpendiculars from corresponding points above, and the length of the vertical axis $1\cdot316$ is laid off above and below O by means of the scale of decimal parts, at points marked 1 and -1 in the figure. The lines of the linear projection needed for the two sets of axes are as follows: For the brachydome o, 011, the lines xz and $x'z'$, through b parallel to the a axis: For the macro-dome, d, 102 $(2a : \infty b : c)$, the lines xy and $x'y'$, through $2a$ parallel to the b axis: The prism $m(110)$ is parallel to the vertical axis, hence in order that such a plane shall satisfy the con-ditions of the linear projection and pass through *unity on the vertical axis*, it must be consid-ered as shifted (without change of direction) until it passes through the center: Its linear projection therefore is represented by the lines yz and $y'z'$, parallel to the directions $1a$

1077

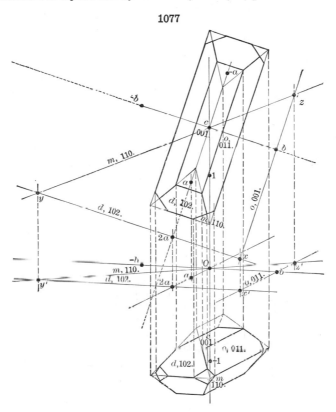

to $1b$ on the two sets of axes. Since a linear projection is made on the plane of the a and b axes, the intersection of any face with the base (001) has the same direction as the line rep-resenting its linear projection. It is well to note that the intersections x, y and z and x', y' and z' are in vertical alignment with one another.

Concerning the drawing of Fig. 1077, it is a simple matter to proportion the general outline of the barite crystal in orthographic projection. The direction of the edge between d, 102, and o, 011, is determined by finding the point x, where the lines of the linear projec-tion of d and o intersect, and drawing the edge parallel to the direction from x to the center c. The intersection of the prism m, 110, with d and o is a straight line, parallel to the direc-tion $1a$ to $1b$ or y to z. To construct the clinographic figure, at some convenient point be-neath the axes the horizontal middle edges of the crystal may be drawn parallel to the a and b axes, their lengths and intersections being determined by carrying down perpendicu-lars from the orthographic projection above. The intersection between d, 102, and o, 011, is determined by finding the point x' of the linear projection and drawing the edge parallel to the direction from x to 1 (*unity*) on the vertical axis, while the corresponding direction below is parallel to the direction x' to -1. The size of the prism m, 110, and its intersections with d and o may all be determined by carrying down perpendiculars from the orthographic projection above, but it is well to control the directions by means of the linear projection: The edges between m, 110, and d, 102; and m, 110, and o, 011, are parallel respectively to the directions y' to 1 and z' to 1. Having completed a figure, a copy free from construction lines may be had by placing the drawing over a clean sheet of paper and puncturing the intersections of all edges with a needle-point: An accurate tracing may then be made on the lower paper.

Should it happen that the linear projection made on the plane of the a and b axes gives intersections far removed from the center of the figure, a linear projection may be made on the clinographic axes either on the plane of the a and c axes or b and c axes, supposing that the faces pass, respectively, through *unity* on the b or the a axes.

1078

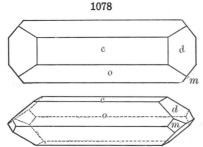

Importance of an Orthographic in connection with a Clinographic Projection. — Many students, on commencing the study of crystallography, fail to derive the benefit they should from the figures given in text-books. Generally clinographic projections are given almost exclusively, with perhaps occasional basal or orthographic projections, and beginners find it hard to reconcile many of the figures with the appearance of the models and crystals which they are intended to represent. For example, given only the clinographic projection of barite, Fig. 1076, it takes considerable training and knowledge of the projection employed to gain from the figure a correct idea of the proportions of the crystal which it actually represents. This may be shown by comparing Figs. 1076 and 1078, which represent the same crystal, drawn one with the a, the other with the b axis to the front. It is seen from Fig. 1078 that the crystal is far longer in the direction of the a axis than one would imagine from inspection of only the clinographic projection of Fig. 1076. The front or a axis is much foreshortened in clinographic projection, consequently by the use of only this one kind of projection there is a two-fold tendency to err; on the one hand, in drawing, one is inclined to represent those edges running parallel to the a axis by lines which are considerably too long, while, on the other hand, in studying figures there is a tendency to regard them as representing crystals which are too much compressed in the direction of the a axis. By using orthographic in connection with clinographic projections these tendencies are overcome. Having in mind the proportions of a certain crystal, or having at hand a model, it is easy to construct an orthographic projection in which the a and b axes are represented with their true proportions; then the construction of a clinographic projection of correct proportions follows as a comparatively simple matter. Without an orthographic projection it would have been a difficult task to have constructed the clinographic projection of Fig. 1078 with the proportions of the intercepts upon the a and b axes the same as in Fig. 1076, while with the orthographic projection orientated as in Fig. 1078 it was an easy matter. A combination of the two projections is preferable in many cases and from the two figures a proper conception of the development of the crystal may be had.

Drawing of Twin Crystals. — The axial protractor furnishes a convenient means for plotting the axes of twin crystals. The actual operation will differ with different problems but the general methods are the same. The two examples given will illustrate these methods.

1079

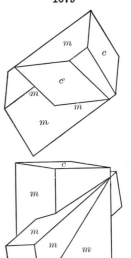

(1). *To plot the axes for the staurolite twin shown in Fig. 1079.* In this case the twinning plane is parallel to the crystal face $\overline{2}3\overline{2}$ which has the axial intercepts of $-3/2a$, b, $-3/2c$. For staurolite, $a : b : c = 0.473 : 1 : 0.683$, while the ϕ and ρ angles of the twinning plane are, $\phi = 010 \wedge \overline{2}30 = 54° 37'$ and $\rho = 001 \wedge \overline{2}3\overline{2} = 60° 31'$. To insure accuracy in plotting, the full lengths of the axes of the protractor have been regarded as unity. The first step is to locate on the clinographic projection the position of the twinning plane, $23\bar{2}$. This is shown in Fig. 1080 as the triangle from $-3/2a$ to b to $-3/2c$. The next step is to find the position of the twinning axis which will be normal to this plane. The coördinates of this twinning axis are given by the ϕ and ρ angles quoted above. The point p which is $54°\ 37'$ back from the pole to $0\bar{1}0$ or b marks the place where the normal to the prism face $\bar{2}30$ would emerge from the sphere. The normal to $\bar{2}3\bar{2}$, which is the twinning axis, will emerge on the meridian that runs through the point p and at such a distance below it that it will make the angle $60°\ 31'$ with the negative end of the c axis. Chords are drawn to p from the points where the a and b axes meet the equator of the sphere and then chords parallel to these are drawn from the points x, y and z which are in each case $60°\ 31'$ from the point where the negative end of the c axis cuts the spherical surface. The common meeting point of these chords T marks the place where the twinning axis pierces the spherical surface. The next step is to determine the point t at which the twinning axis cuts the twinning plane. The line OPp is by construction at right angles to the line connecting $-3/2a$ and $1b$. Therefore a vertical plane which is normal to the twinning plane would intersect that plane in the line connecting $-3/2c$ and P. The twinning axis OT would lie in this plane also. Consequently the point t, where OT and $-3/2c$-P intersect would lie both on the twinning axis and in the twinning plane. In order to make the method of construction clearer Fig. 1081 is given.

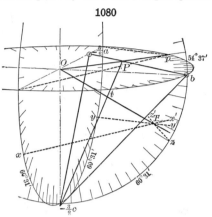

1080

Here the twinning axis is repeated from Fig. 1080. The twin position of the crystal is to be found by revolving it from its normal position through an arc of $180°$, using the twinning axis as the axis of revolution. This will turn the twinning plane about upon the point t as a pivot and so transpose the points $-3/2a$, b and $-3/2c$ to points equidistant from it in an opposite position. By drawing lines through t and laying off equal distances beyond that point the new points $-3/2A$, B and $-3/2C$ will be obtained. These points lie upon the three axes in their twin position and so determine their directions.

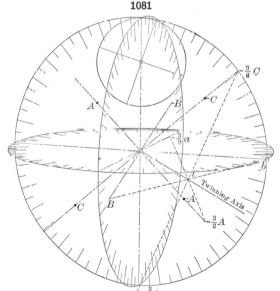

1081

The plotting of the twin axes in the top view follows similar methods. In order to make the construction clearer a separate figure, Fig. 1082, is given. The line O-t is laid off at an angle of $54°\ 37'$ to the b axis. Upon this line the point t is found by projection upward from the clinographic view below. This point t then becomes the point around which the axes are revolved $180°$ to their twin positions. The figure shows clearly the methods of construction and the directions of the axes for the twin.

Upon the twin axes found in this way the portion of the crystal in twin position is drawn in exactly the same manner as if it was in the normal position.

(2). *To plot the axes for the calcite twin shown in Fig. 1083.* In this case it was desired to represent a scalenohedron twinned upon the rhombohedron $f(02\bar{2}1)$ and so drawn that the twinning plane should be vertical and have the position of $b(010)$ of an orthorhombic crystal. The angle from $c(001)$ to $f(02\bar{2}1)$ equals 63° 7′. In order to make the face f vertical, the vertical axis must be inclined at an angle of 26° 53′, or the angle between the c axes of the two individuals composing the twin would be double this or 53° 46′. These relations are shown in Fig. 1084. As indicated in Fig. 1085 the position of these axes, c and C in the figure, are easily obtained at inclinations of 26° 53′ by use of the graduation of the vertical ellipse that passes through B and $-B$. The points X, X' and Y, Y' indicate the intersections with this same ellipse of the two planes containing the a_1, a_2 and a_3 axes in their respective inclined positions, the angles $-BX$, BX', and BY and $-BY'$ being in each case equal to 26° 53′. In order to have the twinning plane occupy a position parallel to the 010 plane of an orthorhombic crystal it is necessary to revolve the axes so that one of the a hexagonal axes shall coincide with the position of the a axis of the orthorhombic system, as $-a_3$, a_3 in Fig. 1085. The two other hexagonal axes corresponding to the axis c must therefore lie in a plane which includes $-a_3$, a_3 and the points X and X' and have such positions that they will make angles of 60° with $-a_3$, a_3. The construction necessary to determine the ends of these axes is as follows: Draw the two chords lettered x–x' through points that are 60° from $-a_3$ and a_3 and parallel to the direction of a

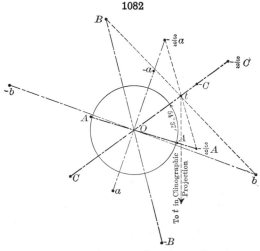

chord that would pass through $-B$ and X. In a similar way draw the two chords y–y' through the second pair of points that are 60° from $-a_3$ and a_3, parallel to the direction of a chord that would pass through the points B and X. The intersections of these two sets of chords determine the points a_1 and $-a_2$ which are the ends of these respective axes. The hexagon shown in the figure connects the ends of the a_1, a_2 and a_3 axes that lie in a

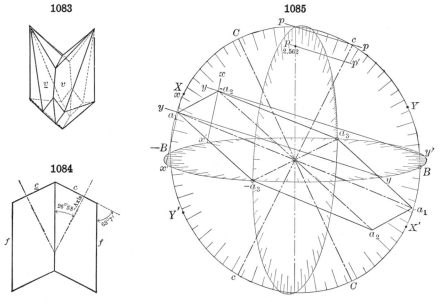

plane perpendicular to the axis c. The set of axes that belong to the axis C are to be found in a similar way. The length of the vertical axis is to be obtained by multiplying that of calcite, $c = 0.854$, by three and laying off on the vertical line the length obtained or 2.562. This is transferred to the twin axis c by drawing the line $p'-p'$ parallel to the line $p-p$. The desired figure of the calcite twin is to be drawn upon these two sets of inclined axes.

Drawing Crystals by use of the Stereographic and Gnomonic Projections

The following explanation of the methods of drawing crystals from the projections of their forms has been taken with only minor modifications from Penfield's description.[*]

1. Use of the Stereographic Projection

In explaining the method, a general example has been chosen; the construction of a drawing of a crystal of axinite, of the triclinic system. Fig. 1086A represents a stereographic projection of the ordinary forms of axinite, $m(110)$, $a(100)$, $M(1\bar{1}0)$, $p(111)$, $r(1\bar{1}1)$ and $s(201)$. As shown by the figure, the *first meridian*, locating the position of 010, has been chosen at 20° from the horizontal direction SS'.

Fig. 1086B is a *plan*, or an orthographic projection of an axinite crystal, as it appears when looked at in the direction of the vertical axis. It may be derived from the stereographic projection in a simple manner, as follows: The direction of the parallel edges made by the intersections of the faces in the zone m, s, r, m', A, is parallel to a tangent at either m or m', and this direction may be had most easily by laying a straight edge from m to m' and, by means of a 90° triangle, transposing the direction to B, as shown by the construction.

The construction of C, which may be called a *parallel-perspective* view, may next be explained: It is not a clinographic projection like the usual crystal drawings from axes, but an orthographic projection, made on a plane intersecting the sphere, represented by the stereographic projection, A, along the great circle SES'; the distance EC being 10°. The plane on which a drawing is to be made may, of course, have any desired inclination or position, but by making the distance CE equal 10° and taking the first meridian at 20° from S, almost the same effects of plan and parallel perspective are produced as in the conventional method of drawing from axes, where the eye is raised 9° 28′ and the crystal turned 18° 26′.

The easiest way to explain the construction of C from A is to imagine the sphere, represented by the stereographic projection, as revolved 80° about an axis joining S and S', or until the great circle SES' becomes horizontal. After such a revolution, the stereographic projection shown in A would appear as in D, and the parallel-perspective drawing, E, could then be derived from D in exactly the same manner as B was derived from A. This is, for example, because the great circle through m, s and r, D, intersects the graduated circle at x, where the pole of a vertical plane in the same zone would fall, provided one were present; hence the intersection of such a surface with the horizontal plane, and, consequently, the direction of the edges of the zone, would be parallel to a tangent at x: In other words, E is a *plan* of a crystal in the position represented by the stereographic projection, D. Although not a difficult matter to transpose the poles of a stereographic projection so as to derive D from A, it takes both time and skill to do the work with accuracy, and it is not at all necessary to go through the operation. To find the direction of the edges of any zone in C, for example msr, note first in A the point x, where the great circles msr and SES' cross. During the supposed revolution of 80° about the axis SS', the pole x follows the arc of a small circle and falls finally at x' (the same position as x of D) and a line at right angles to a diameter through x', as shown by the construction, is the desired direction for C. Similarly for the zones pr, MrM' and $MspM'$, their intersections with SES' at w, y and z are transposed by the revolution of 80° to w', y' and z'. The transposition of the poles w, x, y and z, A, to w', x', y' and z' may easily be accomplished in the following ways: (1) By means of the Penfield transparent, small-circle protractor (Fig. 86, p. 55) the distances of w, x, y and z from either S or S' may be determined and the corresponding number of degrees counted off on the graduated circle. (2) Find first the pole P of the great circle SES', where P is 90° from E or 80° from C, and is located by means of a stereographic scale or protractor (Fig. 80, p. 51): A straight line drawn through P and x will so intersect the graduated circle at x', that $S'x$ and $S'x'$ are equal in degrees. The

---

[*] Am. J. Sc., **21**, 206, 1906.

reason for this is not easily comprehended from A, but if it is imagined that the projection is revolved 90° about an axis AA', so as to bring S' at the center, the important poles and great circles to be considered will appear as in Fig. 1087, where P and C' are the poles, respectively, of the great circles $ES'E'$ and $AS'A'$, and x is $41\frac{1}{4}°$ from S' as in Fig. 1086A. It is evident from the symmetry of Fig. 1087 that a plane surface touching at C', P and x will so intersect the great circle $AS'A'$ that the distances $S'x$ and $S'x'$ are equal.   Now a plane passing through C', P, x and x', if extended, would intersect the sphere as a small

1086

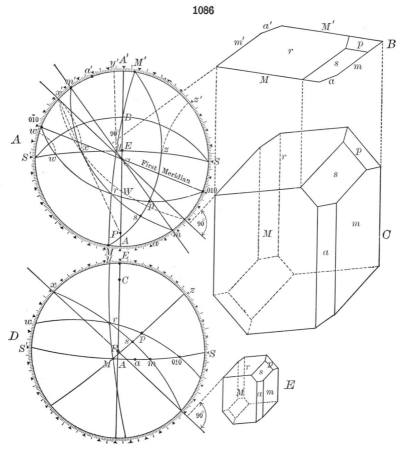

Development of a plan and parallel-perspective figure of axinite, triclinic system, from a stereographic projection (after Penfield)

circle, shown in the figure, but since this circle passes through C', which in Fig. 1086A is the pole of the stereographic projection (antipodal to C) it will be projected in Fig. 1086A as a straight line, drawn through P and x, since the intersections upon the plane of projection of all planes that pass through the point of vision of the projection will appear as straight lines.   (3) In Fig. 1087, B is located midway between E and A', $BS'B'$ is a great circle, and W, 40° from C, is its pole: It is now evident from the symmetry of the figure that a great circle through W and x so intersects the great circle $AS'A'$, that the distances $S'x$ and $S'x'$ are equal.   Transferring the foregoing relations to Fig. 1086A, W, 40° from C, is the pole of the great circle SBS', and a great circle drawn through W and x falls at x'.   However, it is not necessary to draw the great circle through W and x to locate the point x' on the graduated circle: By centering the Penfield transparent great circle protractor (Fig. 85, p. 55) at C, and turning it so that W and x fall on the same great circle, the point x may be transposed to x', and other points, w', y' and z', would be found in like manner.

The three foregoing methods of transposing x to x', z to z', etc., are about equally simple, and it may be pointed out that, supplied with transparent stereographic protractors, and having the poles of a crystal plotted in stereographic projection, it is only necessary to draw the great circle SES' and to locate one point, either W or P, in order to find the directions needed for a parallel-perspective drawing, corresponding to Fig. 1086C. Thus, with only a great circle protractor, the great circle through the poles of any zone may be traced, and its intersection with SES' noted and spaced off with dividers from either S or S'; then the great circle through the intersection just found and W is determined, and where it falls on the divided circle noted, when the desired direction may be had by means of a straight edge and 90° triangle, as already explained.

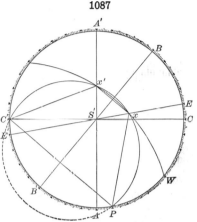

1087

2. Drawing of Twin Crystals by Use of the Stereographic Projection

In the great majority of cases the drawing of twin crystals can be most advantageously accomplished by the use of a stereographic projection of their forms. It is only necessary first to prepare a projection showing the poles of the faces in the normal and twin positions and then follow the methods outlined above. The preparation of the desired projection may, however, need some explanation. An illustrative example is given below taken from an article by Ford and Tillotson on some Baveno twins of orthoclase.*

According to the Baveno law of twinning the $n(021)$ face becomes the twinning plane and as the angle $c \wedge n = 44° 56\frac{1}{2}'$ the angle between c and c' (twin position) becomes 89° 53'. For the purposes of drawing it is quite accurate enough to assume that this angle is exactly 90° and that accordingly the c face of the twin will occupy a position parallel to that of the b face of the normal individual.

Fig. 1088 shows the forms observed on the crystal both in normal and in twin positions, the faces in twin position being indicated by open circles and a prime mark (') after their respective letters, while the zones in twin position are drawn in dashed lines. Starting out with the forms in normal position, the first face to transpose is the base c. This form, from the law of the twinning, will be transposed to c' where it occupies the same position as b of the normal individual, and it necessarily follows that b itself in being transposed will come to b' at the point where the normal c is located.

In turning therefore the crystal to the left from normal to twin position, the poles c and b travel along the great circle I through an arc of 90° until they reach their respective twin positions. We have, in other words, revolved the crystal 90° to the left about an axis which is parallel to the faces of the zone I. The pole of this axis is located on the stereographic projection at 90° from the great circle I and falls on the straight line II, another great circle which intersects zone I at right angles. This pole P is readily located by the stereographic protractor on the great circle II at 90° from c. The problem then is to revolve the poles of the faces from their normal positions about the point P to the left and through an arc of 90° in each case.

During the revolution the poles of the n faces remain on the great circle I and as the angle $n \wedge n = 90°$, the location of their poles when in twin position is identical with that of the normal position and n' falls on top of n. We can now transpose the great circle II from its normal to its twin position, since P remains stationary during the revolution and we have determined the twin position of c. The dashed arc II' gives the twin position of the great circle II. The twin position of y must lie on arc II' and can be readily located at y', the intersection of arc II' with a small circle about P having the radius $P \wedge y$. It is now possible to construct the arc of the zone III in its transposed position III', for we have two of the points, y' and n' of the latter, already located. By the aid of the Penfield transparent great circle protractor the position of the arc of the great circle on which these two points

---

* Am. J. Sc., **26**, 149, 1908.

lie can be determined.   On this arc, III', o' and m' must also lie.   Their positions are most easily determined by drawing arcs of small circles about b' with the required radii, $b \wedge o = 63° \ 8'$, $b \wedge m = 59° \ 22\frac{1}{2}'$ and the points at which they intersect arc III' locate the position of the poles o' and m'.   At the same time the corresponding points on IV' may be located, it being noted that IV' and III are the same arc.   But one other form remains to be transposed, the prism z.   We have already b' and m' located and it is a simple matter with the aid of the great circle protractor to determine the position of the great circle upon which they lie.   Then a small circle about b' with the proper radius, $b \wedge z = 29° \ 24'$, determines at once by its intersections with this arc the position of the poles of the z faces.

1088

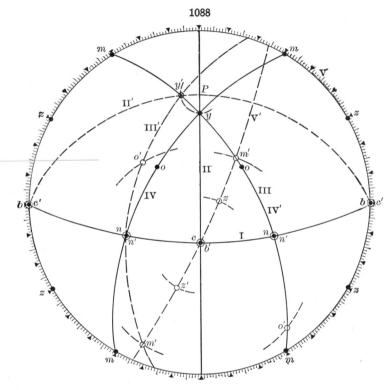

It may be pointed out that if it should be desired to make use of the methods of the gnomonic projection for the drawing of the figures as described below, the stereographic projection of the forms may be readily transformed into a gnomonic projection by doubling the angular distance from the center of the projection to each pole by the use of the stereographic protractor, Fig. 80, p. 51.

3. Use of the Gnomonic Projection

As an illustration, the method of drawing a simple combination of barite has been chosen. The forms shown in Fig. 1089 are $c(001)$, $m(110)$, $o(011)$ and $d(102)$.   The location of the poles in the gnomonic projection is shown in A, where, as in Fig. 1086A, the *first meridian* is taken at 20° from the horizontal direction SS'.   The poles of the prism m and the locations of S and S' (compare Fig. 1086A) fall in the gnomonic projection at infinity.   In any plan, such as Fig. 1089B, the direction of an edge made by the meeting of two faces is at right angles to a line joining the poles of the faces, shown in figures A and B by the direction at 90° to the line joining m'' and c.

The parallel-perspective view, 1089C, is an orthographic projection (compare Figs. 1086A and C) drawn on a plane passing through S and S', and intersecting the sphere on which the gnomonic projection is based as a great circle passing through E, Fig. 1089A, and drawn parallel to SS', the distance cE being 10°.   This great circle is called by Gold-schmidt the *Leitlinie*.   To find such intersections as between m''' and c, and m and d, figure C, note, as in Fig. 1086A, where the great circles through the poles of the faces intersect the *Leitlinie;* thus, the one through m''' and c at x, and that through m and d (through d parallel to mm'', since m and m'' are at infinity) at y.   Next imagine the points x and y transposed as in Fig. 1086A to x' and y', which latter points, however, are located at infinity: This transposition is done by locating first the so-called *Winkelpunkt*, W, of Goldschmidt, 40° from c in Fig. 1089A, and as in Fig. 1086A, 90° from a point B, which

1089

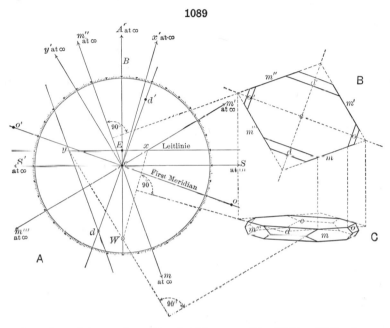

is an equal number of degrees from E and A' (compare Fig. 1087).   Of the three methods given above for transposing x and y to x' and y', the third may be easily applied in the gno-monic projection.   Great circles, or straight lines, through W and x and W and y, Fig. 1089A, if continued to infinity, would determine x' and y', which is accomplished by drawing lines parallel to Wx and Wy through the center.   It is not necessary, however, to draw the lines Wx and Wy, nor the parallel lines through the center; all that is needed to find the direc-tions of the edges $m''' \wedge c$ and $m \wedge d$ is to lay a straight edge from W to x, respectively W to y, and with a 90° triangle transpose the directions to C, as indicated in the drawings. The principles are exactly the same as worked out for the interrelations of Figs. 1086A and C.   As in the case of the stereographic projection, it is evident that, given the poles of a crystal plotted in the gnomonic projection, it would be necessary to draw only one line, the *Leitlinie*, and to locate one point, the *Winkelpunkt*, W, in order to find all pos-sible directions for a plan and parallel-perspective views, corresponding to Figs. 1089B and C.

APPENDIX B

TABLES USEFUL IN THE DETERMINATION OF MINERALS

This Appendix contains a series of tables, more or less complete, of minerals arranged according to chemical composition or to certain prominent crystallographic or physical characters. These, it is believed, will be of service not only to the student, but also to the skilled mineralogist.

The type used in the printing of the mineral names indicates their relative importance. Table I is a complete list of the species named in this book arranged first according to the prominent basic elements which they contain and secondly according to their acid radicals. Table II is of Minerals arranged according to their System of Crystallization. The other tables make no claim to completeness, being limited often to common and important species.

For an exhaustive system of Determinative Tables based particularly upon blowpipe and chemical characters, the student is referred to the work of Professors Brush and Penfield, mentioned on p. 361.

TABLE I. MINERALS ARRANGED ACCORDING TO CHEMICAL COMPOSITION

The following lists include all definitely described mineral species arranged first according to their important basic elements and secondly according to their acid radicals. If a given mineral contains two or more prominent bases its name is repeated in all the appropriate sections.

ALUMINUM

NOTE: — Aluminum is of such common occurrence among the silicate minerals that it is impracticable to list all of these minerals that contain it. Therefore only those silicates which are essentially aluminum minerals are included in the following list.

CRYOLITE, Na_3AlF_6.
Cryolithionite, $3NaF.3LiF.2AlF_3$.
Koenenite, Al,Mg, oxychloride.
Fluellite, $AlF_3.H_2O$.
Chloralluminite, $AlCl_3.6H_2O$.
Prosopite, $CaF_2.2Al(F,OH)_3$.
Pachnolite, Thomsenolite, $NaF.CaF_2.AlF_3.H_2O$.
Gearksutite, $CaF_2.Al(F,OH)_3.H_2O$.
Ralstonite, $(Na_2,Mg)F_2.3Al(F,OH)_3.2H_2O$.
Zirklerite, $9(Fe,Mg,Ca)Cl_2.2Al_2O_3.3H_2O$.
Creedite, $2CaF_2.2Al(F,OH)_3.CaSO_4.2H_2O$.
Trudellite, $4AlCl_3.Al_2(SO_4)_3.4Al(OH)_3.30H_2O$.
Corundum, Al_2O_3.
Högbomite, $Al_2O_3,Fe_2O_3,MgO,TiO_2$.
Spinel, $MgO.Al_2O_3$.
Hercynite, $FeO.Al_2O_3$.
Gahnite, $ZnO.Al_2O_3$.
Galaxite, $(Mn,Fe,Mg)O.(Al,Fe)_2O_3$.

Chrysoberyl, $BeO.Al_2O_3$.
DIASPORE, $Al_2O_3.H_2O$.
Boehmite, $Al_2O_3.H_2O$.
Kayserite, $Al_2O_3.H_2O$.
Bauxite, $Al_2O_3.2H_2O$.
GIBBSITE, $Al_2O_3.3H_2O$.
Shanyavskite, $Al_2O_3.4H_2O$.
Stainerite, $(Co,Fe,Al)_2O_3.H_2O$.
Dundasite, $Pb(AlO)_2(CO_3)_2$.
Dawsonite, $Na_3Al(CO_3)_3.2Al(OH)_3$.
Alumohydrocalcite, Hydrous Ca, Al, carbonate.
Hydrotalcite, $MgCO_3.5Mg(OH)_2.2Al(OH)_3.4H_2O$.
Zunyite, $[Al(OH,F,Cl)_2]_6Al_2Si_3O_{12}$.
Topaz, $[Al(F,OH)]_2SiO_4$.
ANDALUSITE, Al_2SiO_5.
SILLIMANITE, Al_2SiO_5.
Mullite, $3Al_2O_3.2SiO_2$.
Kyanite, Al_2SiO_5.

794

Dumortierite, $8Al_2O_3.B_2O_3.6SiO_2.H_2O$.
Staurolite, $(AlO)_4(AlOH)Fe(SiO_4)_2$.
Kaolin Minerals, $H_4Al_2Si_2O_9$.
Faratsihite, $H_4(Al,Fe)_2Si_2O_9$.
Halloysite, $H_4Al_2Si_2O_9.H_2O$.
Leverrierite, Hydrous Al silicate.
Anauxite, $Al_2O_3.3SiO_2.2H_2O$.
Beidellite, $Al_2O_3.3SiO_2.4H_2O$.
Elbrussite, Hydrous Al,Fe,Mg, silicate.
Cimolite, $2Al_2O_3.9SiO_2.6H_2O$.
Montmorillonite, $H_2Al_2Si_4O_{12}.nH_2O$.
Collyrite, $2Al_2O_3.SiO_2.9H_2O$.
PYROPHYLLITE, $H_2Al_2(SiO_3)_4$.
Allophane, $Al_2SiO_5.nH_2O$.
Melite, $2(Al,Fe)_2O_3.SiO_2.8H_2O$.
Schrötterite, $8Al_2O_3.3SiO_2.30H_2O$.
Goyazite, Al,Sr, phosphate.
Georceixite, $BaAl_3(OH)_7.P_2O_7$.
Plumbogummite, Pb,Al, phosphate.
Florencite, Al,Ce, phosphate.
Crandallite, $CaO.2Al_2O_3.P_2O_5.5H_2O$.
Harttite, $(Sr,Ca)O.2Al_2O_3.P_2O_5.SO_3.5H_2O$.
Durangite, $Na(AlF)AsO_4$.
Amblygonite, $LiAl(F,OH)PO_4$.
Fremontite, $(Na,Li)Al(OH,F)PO_4$.
Lazulite, $2AlPO_4.(Fe,Mg)(OH)_2$.
Tavistockite, $Ca_3P_2O_8.2Al(OH)_3$.
Bialite, Hydrous Mg,Ca,Al, phosphate.
Cirrolite, $Ca_3Al(PO_4)_3.Al(OH)_3$.
Synadelphite, $2(Al,Mn)AsO_4.5Mn(OH)_2$.
Hematolite, $(Al,Mn)AsO_4.4Mn(OH)_2$.
Eggonite, $AlPO_4.2H_2O$.
Barrandite, $(Al,Fe)PO_4.2H_2O$.
Variscite,
Metavariscite $\}$ $AlPO_4.2H_2O$.
Meyersite
Callainite, $AlPO_4.2\frac{1}{2}H_2O$.
Zepharovichite, $AlPO_4.3H_2O$.
Palmerite, $HK_2Al_2(PO_4)_3.7H_2O$.
Rosiérésite, Hydrous Al,Pb,Cu, phosphate.
WAVELLITE, $4AlPO_4.2Al(OH)_3.9H_2O$.
Fischerite, $AlPO_4.Al(OH)_7.2\frac{1}{2}H_2O$.
TURQUOIS, $CuO.3Al_2O_3.2P_2O_5.9H_2O$.
Vauxite, $FeO.Al_2O_3.P_2O_5.6H_2O$.
Paravauxite, $FeO.Al_2O_3.P_2O_5.5H_2O$.
Gordonite, $MgO.Al_2O_3.P_2O_5.9H_2O$.
Pseudowavellite, Hydrous Al,Ca, phosphate.
Wardite, $2Al_2O_3.P_2O_5.4H_2O$.
Sphærite, $4AlPO_4.6Al(OH)_3$.
Liskeardite, $(Al,Fe)AsO_4.2(Al,Fe)(OH)_3.$
  $5H_2O$.
Evansite, $2AlPO_4.4Al(OH)_3.12H_2O$.
Cœruleolactite, $3Al_2O_3.2P_2O_5.10 H_2O$.
Augelite, $2Al_2O_3.P_2O_5.3H_2O$.
Berlinite, Trolleite, $\}$ Hydrous
Attacolite, Vashegyite $\int$ Al phosphates
Childrenite, $AlPO_4.Fe(OH)_3.H_2O$.
Eosphorite, $2AlPO_4.2(Mn,Fe)(OH)_3.2H_2O$.
Egueiite, Hydrous Fe,Al,Ca, phosphate.
Liroconite, $Cu_6Al(AsO_4)_5.3CuAl(OH)_5$.
  $20H_2O$.
Henwoodite, Al,Cu, hydrous phosphate.
Cœruleite, $CuO.2Al_2O_3.As_2O_5.8H_2O$.

Kehoite, Hydrous Al,Zn, phosphate.
Dennisonite, $6CaO.Al_2O_3.2P_2O_5.5H_2O$.
Deltaite, $8CaO.5Al_2O_3.4P_2O_5.14H_2O$.
Millisite, $2CaO.Na_2O.6Al_2O_3.4P_2O_5.17H_2O$.
Englishite, $4CaO.K_2O.4Al_2O_3.4P_2O_5.14H_2O$.
Lehiite, $5CaO.(Na,K_2)O.4Al_2O_3.4P_2O_5.$
  $12H_2O$.
Roschérite,$(Mn,Fe,Ca)_2Al(OH)(PO_4)_2.2H_2O$.
Swedenborgite, $Na_2O.2Al_2O_3.Sb_2O_5$.
Svanbergite, Hydrous Al, Ca, phosphate
  and sulphate.
Tikhvinite, $2SrO.3Al_2O_3.P_2O_5.SO_3.6H_2O$.
Hinsdalite, $2PbO.3Al_2O_3.2SO_3.P_2O_5.6H_2O$.
Jeremejevite, $AlBO_3$.
Rhodizite, Al,K, borate.
Millosevichite, $(Fe,Al)_2(SO_4)_3$.
Spangolite, $Cu_6AlClSO_{10}.9H_2O$.
Alumian, $Al_2O_3.2SO_3$.
Potash Alum $\}$ $KAl(SO_4)_2.12H_2O$.
Kalinite
Tschermigite, $(NH_4)Al(SO_4)_2.12H_2O$.
Soda Alum $\}$ $NaAl(SO_4)_2.12H_2O$.
Mendozite
Pickeringite, $MgSO_4.Al_2(SO_4)_3.22H_2O$.
Halotrichite, $FeSO_4.Al_2(SO_4)_3.24H_2O$.
Apjohnite, $MnSO_4.Al_2(SO_4)_n.24H_2O$.
Dietrichite,$(Zn,Fe,Mn)SO_4.Al_2(SO_4)_3.22H_2O$.
Alunogen, $Al_2(SO_4)_3.18H_2O$.
Ransomite, $CuO.(Fe,Al)_2O_3.4SO_3.7H_2O$.
Guildite, $3(Cu,Fe)O.2(Fe,Al)_2O_3.7SO_3.$
  $17H_2O$.
Louderbackite, $2FeO.3(Fe,Al)_2O_3.10SO_3.$
  $35H_2O$.
Cyanotrichite, $4CuO.Al_2O_3.SO_3.8H_2O$.
Cyprusite, $7Fe_2O_3.Al_2O_3.10SO_3.14H_2O$.
Aluminite, $Al_2O_3.SO_3.9H_2O$.
Paraluminite, $2Al_2O_3.SO_3.10H_2O$.
Felsöbanyite, $2Al_2O_3.SO_3.10H_2O$.
Voltaite, $3(K_2,Fe)O.2(Al,Fe)_2O_3.6SO_3.9H_2O$.
Butlerite, $(Fe,Al)_2O_3.2SO_3.5H_2O$.
Alunite, $K_2Al_6(OH)_{12}.(SO_4)_4$.
Löwigite, $K_2O.3Al_2O_3.4SO_3.9H_2O$.
Almeriite, $Na_2SO_4.Al_2(SO_4)_3.5Al(OH)_3.H_2O$.
Ettringite, $6CaO.Al_2O_3.3SO_3.33H_2O$.
Zincaluminite, $2ZnSO_4.4Zn(OH)_2.6Al(OH)_2.$
  $5H_2O$.
Glaukokerinite, $Zn_{13}Al_8Cu_7(SO_4)_2O_{30}.34H_2O$.
Chalcoalumite, $CuO.2Al_2O_3.SO_3.9H_2O$.
Mellite, $Al_2C_{12}O_{12}.18H_2O$.

ANTIMONY

NOTE: — The antimonates are not in-
cluded in this list.
Allemontite, $SbAs_3$.
NATIVE ANTIMONY, Sb.
Stibnite, Sb_2S_3.
Kermesite, Sb_2S_2O.
Senarmontite, Valentinite, Sb_2O_3.
Cervantite, $Sb_2O_3.Sb_2O_5$.
Stibiconite, $H_2Sb_2O_5$.
Stibiotantalite, $(SbO)_2(Ta,Nb)_2O_6$.
Klebelsbergite, Hydrous Sb sulphate.

ARSENIC

NOTE: — The arsenates are not included in this list.

NATIVE ARSENIC, As.
Allemontite, $SbAs_3$.
REALGAR, AsS.
ORPIMENT, As_2S_3.
Jeromite, sulphide of arsenic.
Arsenopyrite, FeAsS.
Arsenolite, Claudetite, As_2O_3.

BARIUM

Hollandite, manganate of Mn,Ba,Fe'''.
Bromlite, $(Ba,Ca)CO_3$.
Witherite, $BaCO_3$.
Barytocalcite, $BaCO_3.CaCO_3$.
Hyalophane, a Ba-bearing feldspar.
Celsian, $BaAl_2Si_2O_8$.
Cappelenite, Y, Ba, boro-silicate.
Hyalotekite, $(Pb,Ba,Ca)_9B_2(SiO_3)_{12}$.
Barylite, $Be_2BaSi_2O_7$.
Gillespite, $FeO.BaO.4SiO_2$.
Taramellite, $Ba_4Fe'' Fe_4'''Si_{10}O_{31}$.
Brewsterite, $(Sr,Ba,Ca)O.Al_2O_3.6SiO_2.5H_2O$.
Wellsite, $(Ba,Ca,K_2)Al_2Si_3O_{10}.3H_2O$.
Harmotome, $(K_2,Ba)Al_2Si_5O_{14}.5H_2O$.
Edingtonite, $BaAl_2Si_3O_{10}.3H_2O$.
Benitoite, $BaTiSi_3O_9$.
Leucosphenite, $Na_4Ba(TiO)_2(Si_2O_5)_5$.
Georceixite, $BaO.2Al_2O_3.P_2O_5.5H_2O$.
Ferrazite, $3(Ba,Pb)O.2P_2O_5.8H_2O$.
Volborthite, Cu,Ba,Ca, vanadate.
Uranocircite, $Ba(UO_2)_2P_2O_8.8H_2O$.
Nitrobarite, $Ba(NO_3)_2$.
Barite, $BaSO_4$.

BERYLLIUM

Bromellite, BeO.
Chrysoberyl, $BeAl_2O_4$.
Milarite, $K_2O.4CaO.4BeO.Al_2O_3.24SiO_2.H_2O$.
Eudidymite, Epididymite, $HNaBeSi_3O_8$.
Beryl, $Be_3Al_2(SiO_3)_6$.
Leucophanite ⎱ Be,Ca,Na, silicates.
Meliphanite ⎰
Barylite, $Be_2BaSi_2O_7$.
Helvite, $3(Mn,Fe)BeSiO_4.MnS$.
Danalite, $3(Fe,Zn,Mn)BeSiO_4.ZnS$.
Phenacite, Be_2SiO_4.
Trimerite, $(Mn,Ca)_2SiO_4.Be_2SiO_4$.
Euclase, $HBeAlSiO_5$.
Gadolinite, $Be_2FeY_2Si_2O_{10}$.
Bertrandite, $H_2Be_4Si_2O_9$.
Beryllonite, $NaBePO_4$.
Kolbeckite, Be phosphate.
Herderite, $Ca[Be(F,OH)]PO_4$.
Hambergite, $Be_2(OH)BO_3$.

BISMUTH

NATIVE BISMUTH, Bi.
BISMUTHINITE, Bi_2S_3.
Guanajuatite, Bi_2Se_3.
Tetradymite, $Bi_2(Te,S)_3$.
Grünlingite, Bi_4TeS_3.
Josëite, Wehrlite, bismuth tellurides.
Daubréeite, Bi, oxychloride.
Bismite, Bi_2O_3.
Bismutospärite, $Bi_2(CO_3)_3.2Bi_2O_3$?
Bismutite, $Bi_2O_3.CO_2.H_2O$.
Basobismutite, $2Bi_2O_3.CO_2.H_2O$?
Eulytite, Agricolite, $Bi_4Si_3O_{12}$.
Bismutotantalite, Bi tantalate and niobate.
Pucherite, $BiVO_4$.
Atelestite, $H_2Bi_3AsO_8$.
Walpurgite, $Bi_{10}(UO_2)_3(OH)_{24}(AsO_4)_4$.
Rhagite, $2BiAsO_4.3Bi(OH)_3$.
Arseno-bismite, Hydrous Bi arsenate.
Mixite, Hydrous Cu, Bi, arsenate.
Uranosphærite, $Bi_2O_3.2UO_3.3H_2O$.
Montanite, $Bi_2O_3.TeO_3.2H_2O$.
Koechlinite, $Bi_2O_3.MoO_3$.

BORON

NOTE: — The borates are not included in this list.

Sassolite, $B(OH)_3$.
Cappelenite, Y,Ba, boro-silicate.
Hyalotekite, $(Pb,Ba,Ca)B_2(SiO_3)_{12}$.
DANBURITE, $CaB_2(SiO_4)_2$.
Datolite, $HCaBSiO_5$.
Homilite, $Ca_2FeB_2Si_2O_{10}$.
Axinite, Ca,Al, boro-silicate.
Tourmaline, complex boro-silicate.
Dumortierite, $8Al_2O_3.B_2O_3.6SiO_2.H_2O$.
Serendibite, $10(Ca,Mg)O.5Al_2O_3.B_2O_3.6SiO_2$.
Manandonite, $H_{24}Li_4Al_{14}B_4Si_6O_{53}$.
Bakerite, Hydrous Ca, boro-silicate.
Searlesite, $NaB(SiO_3)_2.H_2O$.
Lüneburgite, $3MgO.B_2O_3.P_2O_5.8H_2O$.

CADMIUM

Greenockite, CdS.
Cadmiumoxide, CdO.
Otavite, Cd carbonate.

CÆSIUM

Pollucite, $2Cs_2O.2Al_2O_3.9SiO_2.H_2O$.
Rhodizite, Al,K,Cs, borate.

CALCIUM

Oldhamite, CaS.
Fluorite, CaF_2.
Yttrofluorite, CaF_2 with YF_3.
Yttrocerite, CaF_2 with $(Y,Ce)F_3$.

Zamboninite, $CaF_2.2MgF_2$.
Hydrophilite, $KCaCl_2$.
Yttrofluorite, $(Ca_3,Y_2)F_6$.
Nocerite, $Ca_3Mg_3O_2F_8$.
Tachhydrite, $CaCl_2.2MgCl_2.12H_2O$.
Prosopite, $CaF_2.2Al(F,OH)_3$.
Pachnolite, Thomsenolite, $NaF.CaF_2.AlF_3.$
H_2O.
Gearksutite, $CaF_2.Al(F,OH)_3.H_2O$.
Zirklerite, $9(Fe,Mg,Ca)Cl_2.2Al_2O_3.3H_2O$.
Creedite, $2CaF_2.2Al(F,OH)_3.CaSO_4.2H_2O$.
Yttrocerite, $(Y,Er,Ce)F_3.5CaF_2.H_2O$.
Calcite, $CaCO_3$.
Dolomite, $CaCO_3.MgCO_3$.
Ankerite, $CaCO_3.(Mg,Fe,Mn)CO_3$.
Codazzite, $(Ca,Mg,Fe,Ce)CO_3$.
Aragonite, $CaCO_3$.
Bromlite, $(Ba,Ca)CO_3$.
Barytocalcite, $BaCO_3.CaCO_3$.
Parisite, $[(Ce,La,Di)F]_2Ca(CO_3)_2$.
Alumohydrocalcite, Hydrous Ca, Al, carbonate.
Pirssonite, $CaCO_3.Na_2CO_3.2H_2O$.
Gay-Lussite, $CaCO_3.Na_2CO_3.5H_2O$.
Gajite, Basic, hydrous Ca, Mg, carbonate.
Uranothallite, $2CaCO_3.U(CO_3)_2.10H_2O$.
Voglite, Hydrous U,Ca,Cu, carbonate.
Milarite, $K_2O.4CaO.4BeO.Al_2O_3.24SiO_2.H_2O$.
Oligoclase ⎫
Andesine ⎬ Mixtures of $NaAlSi_3O_8$ and $CaAl_2Si_2O_8$.
Labradorite ⎭
Anorthite, $CaAl_2Si_2O_8$.
Anemousite, $Na_2O.2CaO.3Al_2O_3.9SiO_2$.
Pyroxene, Ca, Mg, etc., silicate.
Babingtonite, $(Ca,Fe,Mn)SiO_3$ with $Fe_2(SiO_3)_3$.
Sobralite, $(Mn,Fe,Ca,Mg)SiO_3$.
Wollastonite, $CaSiO_3$.
PECTOLITE, $HNaCa_2(SiO_3)_3$.
Schizolite, $HNa(Ca,Mn)_2(SiO_3)_3$.
Rosenbuschite, near pectolite with Zr.
Wöhlerite, Zr-silicate and niobate of Ca,Na.
Lävenite, Zr-silicate of Mn,Ca.
Guarinite, $(Na_2,Ca)(Si,Zr)O_3$.
Amphibole Group, Ca, Mg, etc., silicates.
Torendrikite, $Na_2O.4MgO.CaO.FeO.Fe_2O_3.$
$10SiO_2$.
Leucophanite ⎫
Meliphanite ⎬ Na,Be,Ca, fluo-silicates.
Cuspidine, $(CaF)_2SiO_3$.
Custerite, $[Ca(F,OH)]_2SiO_3$.
Didymolite, $2CaO.3Al_2O_3.9SiO_2$.
Ganomalite, $Pb_4(PbOH)_2Ca_4(Si_2O_7)_3$.
Nasonite, $Pb_4(PbCl)_2Ca_4(Si_2O_7)_3$.
Margarosanite, $Pb(Ca,Mn)_2(SiO_3)_3$.
Roeblingite, $5(H_2CaSiO_4).2(CaPbSO_4)$.
Haüynite, $Na_2Ca(NaSO_4.Al)Al_2(SiO_4)_3$.
Grossularite, $Ca_3Al_2(SiO_4)_3$.
Andradite, $Ca_3Fe_2(SiO_4)_3$.
UVAROVITE, $Ca_3Cr_2(SiO_4)_3$.
Schorlomite, $Ca_3(Fe,Ti_2)(Si,Ti)O_{43}$.
Monticellite, $CaMgSiO_4$.
Glaucochroite, $CaMnSiO_4$.

Calcium-larsenite, $(Pb,Ca)ZnSiO_4$.
Larnite, Ca_2SiO_4.
Merwinite, $Ca_3Mg(SiO_4)_2$.
Trimerite, $(Mn,Ca)_2SiO_4.Be_2SiO_4$.
Tinzenite, $2CaO.Al_2O_3.Mn_2O_3.4SiO_2$.
SCAPOLITE GROUP, Mixtures of
$CaCO_3.3CaAl_2Si_2O_8$ and $NaCl.3NaAlSi_3O_8$.
Sarcolite, $(Ca,Na_2)_3Al_2(SiO_4)_3$.
Melilite, $Na_2(Ca,Mg)_{11}(Al,Fe)_4(SiO_4)_9$.
Cebollite, $H_2Ca_5Al_2Si_3O_{16}$.
Gehlenite, $Ca_3Al_2Si_2O_{10}$.
Hardystonite, $Ca_2ZnSi_2O_7$.
Vesuvianite, $Ca_6[Al(OH,F)]Al_2(SiO_4)_5$.
Genevite, Ca,Al, silicate.
DANBURITE, $CaB_2(SiO_4)_2$.
Datolite, $HCaBSiO_5$.
Homilite, $Ca_2FeB_2Si_2O_{10}$.
ZOISITE, $Ca_2(AlOH)Al_2(SiO_4)_3$.
Clinozoisite, $HCa_2Al_3Si_3O_{13}$.
Epidote, $Ca_2[(Al,Fe)OH](Al,Fe)_2(SiO_4)_3$.
Piedmontite, $HCa_2(Al,Mn)_3Si_3O_{13}$.
Allanite, $(Ca,Fe)_2(AlOH)(Al,Ce,Fe)_2(SiO_4)_3$.
Nagatelite, A silicate and phosphate of Al, Ce,Ca,Fe.
AXINITE, Ca,Al, boro-silicate.
PREHNITE, $H_2Ca_2Al_2(SiO_4)_3$.
Harstigite, Mn,Ca,Al, silicate.
Fluosiderite, Ca,Mg, silicate.
Grothine, Ca,Al, silicate.
ILVAITE, $CaFe_3(FeOH)(SiO_4)_2$.
Clinohedrite, $H_2CaZnSiO_4$.
Stokesite, $H_4CaSnSi_3O_{11}$.
Lawsonite, Hibschite, $H_4CaAl_2Si_2O_{10}$.
Beckelite, $Ca_3(Ce,La,Di)_4Si_3O_{15}$.
Lessingite, Ca,Ce, silicate.
Angaralite, $2(Ca,Mg)O.5(Al.Fe)_2O_3.6SiO_2$.
Serendibite, $10(Ca,Mg)O.5Al_2O_3.B_2O_3.6SiO_2$.
Silicomagnesiofluorite, $H_2Ca_4Mg_3Si_2O_7F_{10}$.
Aloisite, Fe,Ca,Mg,Na, silicate.
Inesite, $H_2(Mn,Ca)_6Si_6O_{19}.3H_2O$.
Hillebrandite, $Ca_2SiO_4.H_2O$.
Foshagite, $H_2Ca_5(SiO_4)_3.2H_2O$.
Radiophyllite, $CaSiO_3.H_2O$.
Xonotlite, $5CaSiO_3.H_2O$.
Crestmorcite, $4CaSiO_3.7H_2O$.
Riversideite, $2CaSiO_3.H_2O$.
Centrallassite, $4CaO.7SiO_2.5H_2O$.
Truscottite, $2(Ca,Mg)O.3SiO_2.3H_2O$.
Racewinite, Hydrous silicate of Al,Ca.
Pumpellyite, $6CaO.3Al_2O_3.7SiO_2.4H_2O$.
Lotrite, $3(Ca,Mg)O.2(Al,Fe)_2O_3.4SiO_2.2H_2O$.
Okenite, $H_2CaSi_2O_5.H_2O$.
Gyrolite, $H_2Ca_2Si_3O_9.H_2O$.
APOPHYLLITE, $KFCa_4(Si_2O_5)_4.8H_2O$.
Ptilolite, $(Ca,K_2,Na_2)Al_2Si_{10}O_{24}.5H_2O$.
Flokite, $(Ca,Na_2)O.Al_2O_3.9SiO_2.6H_2O$.
Mordenite, $(Ca,Na_2)O.Al_2O_3.9SiO_2.6H_2O$.
HEULANDITE, $(Ca,Na_2)O.Al_2O_3.6SiO_2.5H_2O$.
Brewsterite, $(Sr,Ba,Ca)O.Al_2O_3.6SiO_2.5H_2O$.
Epistilbite, $(Ca,Na_2)O.Al_2O_3.6SiO_2.5H_2O$.
Wellsite, $(Ba,Ca,K_2)Al_2Si_3O_{10}.3H_2O$.
Phillipsite, $(K_2,Ca)Al_2Si_4O_{12}.4\frac{1}{2}H_2O$.
Stilbite, $(Na_2Ca)Al_2Si_6O_{16}.6H_2O$.

Epidesmine, same as stilbite.
Stellerite, $CaAl_2Si_7O_{18}.7H_2O$.
Gismondite, $(Ca,K_2)Al_2Si_2O_8.4H_2O$.
Laumontite, $(Ca,Na_2)Al_2Si_4O_{12}.4H_2O$.
Laubanite, $Ca_2Al_2Si_5O_{15}.6H_2O$.
CHABAZITE, $(Ca,Na_2)Al_2Si_4O_{12}.6H_2O$.
Gmelinite, $(Na_2,Ca)Al_2Si_4O_{12}.6H_2O$.
Levynite, $CaAl_2Si_3O_{10}.5H_2O$.
Arduinite, Ca,Na, zeolite.
Faujasite, $Na_2CaAl_4Si_{10}O_{28}.20H_2O$.
Scolecite, $Ca(AlOH)_3(SiO_3)_3.2H_2O$.
Mesolite, $Na_2Al_2Si_3O_{10}.2H_2O+2[CaAl_2Si_3O_{10}$
    $3H_2O]$.
Gonnardite, $(Ca,Na_2)_2Al_2Si_5O_{15}.5\frac{1}{2}H_2O$.
Thomsonite, $(Na_2,Ca)Al_2Si_2O_8.2\frac{1}{2}H_2O$.
Hydrothomsonite, $(H_2,Na_2,Ca)Al_2Si_2O_8.$
    $5H_2O$.
Echellite, $(Ca,Na_2)O.2Al_2O_3.3SiO_2.4H_2O$.
Erionite, $H_2CaK_2Na_2Al_2Si_6O_{17}.5H_2O$.
Bavenite, $Ca_3Al_2(SiO_3)_6.H_2O$.
Bityite, Hydrous, Ca,Al, silicate.
Dachiardite, $(Na_2K_2,Ca)_3Al_4(Si_2O_5)_9.14H_2O$.
Margarite, $H_2CaAl_4Si_2O_{12}$.
Seybertite, $H_3(Mg,Ca)_5Al_5Si_2O_{18}$.
Xanthophyllite, $H_8(Mg,Ca)_{14}Al_{16}Si_5O_{52}$.
Griffithite, $4(Mg,Fe,Ca)O(Al,Fe)_2O_3.5SiO_2$.
    $7H_2O$.
Sérandite, Hydrous Mn,Ca,Na, silicate.
Sturtite, $6(Mn,Ca,Mg)O.Fe_2O_3.8SiO_2.23H_2O$.
Cenosite, $2CaO.(Ce,Y)_2O_3.CO_2.4SiO_2.H_2O$.
Plazolite, $3CaO.Al_2O_3.2(SiO_2,CO_2).2H_2O$.
Thaumasite, $CaSiO_3.CaCO_3.CaSO_4.15H_2O$.
Spurrite, $2Ca_2SiO_4.CaCO_3$.
Afwillite, $3CaO.2SiO_2.3H_2O$.
Scawtite, Ca silicate and carbonate.
Uranophane, $CaO.2UO_3.2SiO_2.7H_2O$.
TITANITE, $CaTiSiO_5$.
Molengraafite, Ca,Na, titano-silicate.
Fersmannite, $2Na_2(O,F_2).4CaO.4TiO_2.3SiO_2$.
Rinkolite, Ce,Ca,Sr,Na, titano-silicate.
Joaquinite, Ca,Fe, titano-silicate.
Perovskite, $CaTiO_3$.
Loparite, Ce,Ca,Na, titanate.
Dysanalyte, Ca,Fe, titano-niobate.
Zirkelite, $(Ca,Fe)(Zr,Ti,Th)_2O_5$.
Uhligite, $Ca(Zr,Ti)_2O_5$ with Al_2TiO_5.
Pyrochlore, Ca,Ce, niobate.
Koppite, Ca,Ce, niobate.
Ellsworthite, $CaO.Nb_2O_5.2H_2O$.
Chalcolamprite, $Na_4(CaF_2)Nb_2SiO_9$.
Microlite, $Ca_2Ta_2O_7$.
Mendelyeevite, Ca,U, titano-niobate.
Berzeliite, $(Ca,Na_2)(Mg,Mn)_2(AsO_4)_3$.
Graftonite, $(Fe,Mn,Ca)_3P_2O_8$.
Merrillite, $3CaO.Na_2O.P_2O_5$.
Dehrnite, $6CaO.(Na,K)_2O.2P_2O_5.H_2O$.
Apatite, $Ca_4(CaF)(PO_4)_3$.
Kurskite, $2Ca_3(PO_4)_2.CaF_2.CaCO_3$.
Podolite, $3Ca_3(PO_4)_2.CaCO_3$.
Svabite, Ca arsenate.
Fermorite, $(Ca,Sr)_4[Ca(OH,F)][(P,As)O_4]_3$.
Wilkeite, $3Ca_3(PO_4)_2.CaCO_3.3Ca_3[(SiO_4)$
    $(SO_4)].CaO$.

Spodiosite, $(CaF)CaPO_4$.
Adelite, $(MgOH)CaAsO_4$.
Tilasite, $(MgF)CaAsO_4$.
Herderite, $Ca[Be(F,OH)]PO_4$.
Ježekite, $Na_4CaAl(AlO)(F,OH)_4(PO_4)_2$.
Crandallite, $CaO.2Al_2O_3.P_2O_5.5H_2O$.
Lacroixite, $Na_4(Ca,Mn)_4Al_3(F,OH)_4P_3O_{16}.$
    $2H_2O$.
Higginsite, $CuCa(OH)AsO_4$.
Calciovolborthite, $(Cu,Ca)_3V_2O_8.(Cu,Ca)$
    $(OH)_2$.
Staszicite, $5(Ca,Cu,Zn)O.As_2O_5.2H_2O$.
Tavistockite, $Ca_3P_2O_8.2Al(OH)_2$.
Bialite, Hydrous Mg,Ca,Al, phosphate.
Cirrolite, $Ca_3Al(PO_4)_3.Al(OH)_3$.
Arseniosiderite, $Ca_3Fe(AsO_4)_3.3Fe(OH)_3$.
Dussertite, $Ca_3(AsO_4)_2.3Fe(OH)_3$.
Retzian, Y,Mn,Ca, arsenate.
Arseniopleite, $(Mn,Ca)_9(Mn,Fe)_2(OH)_6$
    $(AsO_4)_6$.
Collophanite, $Ca_3P_2O_8.H_2O$.
Pyrophosphorite, $Mg_2P_2O_7.4(Ca_3P_2O_8.$
    $Ca_2P_2O_7)$.
Roselite, $(Ca,Co,Mg)_3As_2O_8.2H_2O$.
Brandite, $Ca_2MnAs_2O_8.2H_2O$.
Fairfieldite, $Ca_2MnP_2O_8.2H_2O$.
Messelite, $(Ca,Fe)_3P_2O_8.2\frac{1}{2}H_2O$.
Collinsite, $2CaO.(Mg,Fe)O.P_2O_5.2\frac{1}{2}H_2O$.
Anapaite, $(Ca,Fe)_3P_2O_8.4H_2O$.
Picropharmacolite, $(Ca,Mg)As_2O_8.6H_2O$.
Churchite, Hydrous Ca,Ce, phosphate.
Haidingerite, $HCaAsO_4.H_2O$.
Pharmacolite, $HCaAsO_4.2H_2O$.
Wapplerite, $HCaAsO_4.3\frac{1}{2}H_2O$.
Brushite, $HCaPO_4.2H_2O$.
Martinite, $H_2Ca_5(PO_4)_4.\frac{1}{2}H_2O$.
Hewettite   } $CaO.3V_2O_5.9H_2O$.
Metahewettite }
Rossite, $CaO.V_2O_5.4H_2O$.
Metarossite, $CaO.V_2O_5.2H_2O$.
Fernandinite, $CaO.V_2O_4.5V_2O_5.14H_2O$.
Melanovanadite, $2CaO.3V_2O_5.2V_2O_4$.
Pascoite, $2CaO.3V_2O_5.11H_2O$.
Pintadoite, $2CaO.V_2O_5.9H_2O$.
Isoclasite, $Ca_3P_2O_8.Ca(OH)_2.4H_2O$.
Conichalcite, $(Cu,Ca)_3As_2O_8.(Cu,Ca)(OH)_2.$
    $\frac{1}{2}H_2O$.
Freirinite, $6(Cu,Ca)O.3Na_2O.2As_2O_5.6H_2O$.
Volborthite, Cu,Ba,Ca, vanadate.
Tangeite, $2CaO.2CuO.V_2O_5.H_2O$.
Pseudowavellite, Hydrous Al,Ca, phosphate.
Mazapilite, $Ca_3Fe_2(AsO_4)_4.2FeO(OH).5H_2O$.
Yukonite, $(Ca_3,Fe_2''')(AsO_4)_2.2Fe(OH)_3.$
    $5H_2O$.
Calcioferrite, $Ca_3Fe_2(PO_4)_4.Fe(OH)_3.8H_2O$.
Borickite, $Ca_3Fe_2(PO_4)_4.12Fe(OH)_3.6H_2O$.
Egueiite, Hydrous Fe,Al,Ca, phosphate.
Dennisonite, $6CaO.Al_2O_3.2P_2O_5.5H_2O$.
Deltaite, $8CaO.5Al_2O_3.4P_2O_5.14H_2O$.
Millisite, $2CaO.Na_2O.6Al_2O_3.4P_2O_5.17H_2O$.
Englishite, $4CaO.K_2O.4Al_2O_3.4P_2O_5.14H_2O$.
Lehiite, $5CaO.(Na,K)_2O.4Al_2O_3.4P_2O_5.12H_2O$.
Lewistonite, $15CaO.(K,Na)_2O.4P_2O_5.8H_2O$.

Roschérite, $(Mn,Fe,Ca)_2Al(OH)(PO_4)_2.$
$2H_2O.$

Autunite $\left.\right\}$ Bassetite $\left.\right\}$ $Ca(UO_2)_2P_2O_8.8H_2O.$

Uranospinite, $Ca(UO_2)_2As_2O_8.8H_2O.$
Tyuyamunite, $CaO.UO_3.V_2O_5.nH_2O.$
Rauvite, $CaO.2UO_3.6V_2O_5.20H_2O.$
Sincosite, $CaO.V_2O_4.P_2O_5.5H_2O.$
Romeite, $Ca_5Sb_6O_{20}.$
Weslienite, $Na_2FeCa_3Sb_4O_{15}.$
Schneebergite, $(Ca,Fe)_2Sb_2O_6.$
Lewisite, $5CaO.2TiO_2.3Sb_2O_5.$
Mauzeliite, Pb,Ca, titano-antimonate.
Svanbergite, Hydrous Al,Ca, phosphate and
sulphate.
Cahnite, $4CaO.B_2O_3.As_2O_5.4H_2O.$
Nitrocalcite, $Ca(NO_3)_2.nH_2O.$
Lautarite, $Ca(IO_3)_2.$
Dietzeite, Ca iodo-chromate.
Nordenskiöldine, $CaSn(BO_3)_2.$
Howlite, $H_5Ca_2B_5SiO_{14}.$
Bakerite, Hydrous Ca silico-borate.
COLEMANITE, $Ca_2B_6O_{11}.5H_2O.$
Priceite, $5CaO.6B_2O_3.9H_2O.$
Inyoite, $2CaO.3B_2O_3.13H_2O.$
Meyerhofferite, $2CaO.3B_2O_3.7H_2O.$
Ulexite, $NaCaB_5O_9.8H_2O.$
Probertite, $Na_2O.2CaO.5B_2O_3.10H_2O.$
Bechilite, $CaB_4O_7.4H_2O.$
Hydroboracite, $CaMgB_6O_{11}.6H_2O.$
GLAUBERITE, $Na_2SO_4.CaSO_4.$

Anhydrite $\left.\right\}$ Bassanite $\left.\right\}$ $CaSO_4.$

Gypsum, $CaSO_4.2H_2O.$
Syngenite, $CaSO_4.K_2SO_4.H_2O.$
Polyhalite, $2CaSO_4.MgSO_4.K_2SO_4.2H_2O.$
Ettringite, $6CaO.Al_2O_3.3SO_3.33H_2O.$
Uranopilite, $CaO.8UO_3.2SO_3.25H_2O.$
SCHEELITE, $CaWO_4.$
Powellite, $Ca(Mo,W)O_4.$
Whewellite, $CaC_2O_4.H_2O.$

CERIUM EARTHS

Yttrofluorite, CaF_2 with $YF_3.$
Yttrocerite, CaF_2 with $(Y,Ce)F_3.$
Fluocerite, $(Ce,La,Di)F_3.$
Codazzite, $(Ca,Mg,Fe,Ce)CO_3.$
Ancylite, $4Ce(OH)CO_3.3SrCO_3.3H_2O.$
Ambatoarinite, Rare earths, Sr, carbonate.
Parisite, $[(Ce,La,Di)F]_2.CaCO_3.$
Bastnäsite, $(CeF)CO_3.$
Lanthanite, $(La,Di,Ce)_2(CO_3)_3.8H_2O.$
Melanocerite $\left.\right\}$
Caryocerite $\left.\right\}$ Ca,Ce,Y, fluo-silicates.
Steenstrupine $\left.\right\}$
Tritomite, Th,Ce,Y,Ca, fluo-silicate.
Mackintoshite, U,Th,Ce, silicate.
Thorogummite, Silicate of U,Th,Ce.
Allanite, Ca,Fe,Ce,Al, silicate.
Nagatelite, A silicate and phosphate of Al,
Ce,Ca,Fe.
Cerite, Ce, etc., silicate.

Törnebohmite, Ce silicate.
Beckelite, $Ca_3(Ce,La,Di)_4Si_3O_{15}.$
Lessingite, Ca,Ce, silicate.
Buszite, Silicate of rare earths.
Hellandite, Ce, etc., Al,Mn,Ca, silicate.
Bazzite, Sc., etc., silicate.
Cenosite, $2CaO.(Ce,Y)_2O_3.CO_2.4SiO_2.H_2O.$
Britholite, Ce, etc., silicate and phosphate.
Erikite, Ce, etc., silicate and phosphate.
Tscheffkinite, Ce,Fe, titano-silicate.
Johnstrupite $\left.\right\}$
Mosandrite $\left.\right\}$
Rinkite $\left.\right\}$ Ce, etc., titano-silicates.
Rinkolite $\left.\right\}$
Knopite, Ca,Ce, titanate.
Loparite, Ce,Ca,Na, titanate.
Kalkowskite, $(Fe,Ce)_2O_3.4(Ti,Si)O_2.$
Pyrochlore, Ca,Ce, niobate.
Chalcolamprite, $R''Nb_2O_6F_2.R''SiO_3.$
Koppite, Ca,Ce, niobate.
Fergusonite, Y,Er,Ce,U, niobate.
Yttrotantalite, Fe,Ca,Y,Er,Ce, tantalate.
Samarskite, Fe,Ca,U,Ce,Y, niobate.
Aeschynite, Ce, niobate-titanate.
Polymignite, Ce,Fe,Ca, niobate-titanate.
Euxenite $\left.\right\}$
Polycrase $\left.\right\}$ Y,Ce,U, niobate-
Blomstrandine-Priorite $\left.\right\}$ titanates.
MONAZITE, $(Ce,La,Di)PO_4.$
Florencite, Ce,Al, phosphate.
Rhabdophanite, Hydrous Ce,Y, phosphate.
Churchite, Hydrous Ce,Ca, phosphate.

CHROMIUM

Daubréelite, $FeS.Cr_2S_3.$
Chromite, $FeO.Cr_2O_3.$
Stichtite, $MgCO_3.5Mg(OH)_2.2Cr(OH)_3.$
$4H_2O.$
Uvarovite, $Ca_3Cr_2(SiO_4)_3.$
Fornacite, Pb,Cu, chrom-arsenate.
Dietzite, Ca iodo-chromate.
CROCOITE, $PbCrO_4.$
Phoenicochroite, $3PbO.2CrO_3.$
Vauquelinite, $2(Pb,Cu)CrO_4.(Pb,Cu)_3P_2O_8.$
Bellite, Pb, arseno-chromate.
Redingtonite, Hydrous Cr sulphate.

COBALT

LINNÆITE, $Co_3S_4.$
Carrollite, $CuCo_2S_4.$
Badenite, $(Co,Ni,Fe)_2(As,Bi)_3.$
Cobaltnickelpyrite, $(Co,Ni,Fe)S_2.$
SMALTITE, $CoAs_2.$
COBALTITE, $CoAsS.$
Skutterudite, $CoAs_3.$
Safflorite, $CoAs_2.$
Glaucodot, $(Co,Fe)AsS.$
Stainerite, $(Co,Fe,Al)_2O_3.H_2O.$
Heterogenite, $CoO.2Co_2O_3.6H_2O.$
Sphærocobaltite, $CoCO_3.$
Roselite, $(Ca,Co,Mg)_3As_2O_8.2H_2O.$

Erythrite, $Co_3As_2O_8.8H_2O$.
Forbesite, $H_2(Ni,Co)_2As_2O_8.8H_2O$.
Julienite, Hydrous Co chloro-nitrate.
Bieberite, $CoSO_4.7H_2O$.

COPPER

Native Copper, Cu.
Horsfordite, Cu_6Sb.
Domeykite, Cu_3As.
Mohawkite, Cu_3As.
Algodonite, Cu_6As.
Whitneyite, Cu_9As.
Cocinerite, Cu_4AgS.
Rickardite, Cu_4Te_3.
Weissite, Cu_5Te_3.
Berzelianite, Cu_2Se.
Eucairite, $Cu_2Se.Ag_2Se$.
Crookesite, Cu,Tl, selenide.
Umangite, $CuSe.Cu_2Se$.
Klockmannite, CuSe.
Penroseite, $3CuSe.2PbSe_2.5(Ni,Co)Se_2$.
Chalcocite, Cu_2S.
Stromeyrite, $(Ag,Cu)_2S$.
Cubanite, $Cu_2S.Fe_4S_5$.
Covellite, CuS.
Bornite, Cu_5FeS_4.
Germanite, $Cu_3(Fe,Ge)S_4$.
Carrollite, $CuS.Co_2S_3$.
Chalcopyrite, $CuFeS_2$.
Villamaninite, Cu,Ni,Co,Fe, sulphide.
Eichbergite, $(Cu,Fe)_2S.3(Bi,Sb)_2S_3$.
Histrixite, $5CuFeS_2.2Sb_2S_3.7Bi_2S_3$.
Gladite, $2PbS.Cu_2S.5Bi_2S_3$.
Rezbanyite, $Cu_2S.3PbS.5Bi_2S_3$.
Lindströmite, $2PbS.Cu_2S.3Bi_2S_3$.
Chalcostibite, $Cu_2S.Sb_2S$
Emplectite, $Cu_2S.Bi_2S_3$.
Benjaminite, $(Cu,Ag)_2S.2PbS.2Bi_2S_3$.
Klaprotholite, $3Cu_2S.2Bi_2S_3$.
Berthonite, $5PbS.9Cu_2S.7Sb_2S_3$.
Bournonite, $2PbS.Cu_2S.Sb_2S_3$.
Seligmannite, $2PbS.Cu_2S.As_2S_3$.
Aikinite, $2PbS.Cu_2S.Bi_2S_3$.
Wittichenite, $3Cu_2S.Bi_2S_3$.
Stylotypite, $3Cu_2S.Sb_2S_3$.
Falkenhaynite, $3Cu_2S.Sb_2S_3$.
Tetrahedrite, $3Cu_2S.Sb_2S_3$.
Tennantite, $3Cu_2S.As_2S_3$.
Goldfieldite, $5Cu_2S.(Sb,As,Bi)_2(S,Te)_3$.
Enargite, $3Cu_2S.As_2S_5$.
Famatinite, $3Cu_2S.Sb_2S_5$.
Sulvanite, $3Cu_2S.V_2S_5$.
Epigenite, $4Cu_2S.3FeS.As_2S_5$.
Stannite, $Cu_2S.FeS.SnS_2$.
Nantokite, CuCl.
Marshite, CuI.
Miersite, $4AgI.CuI$.
Atacamite, $CuCl_2.3Cu(OH)_2$.
Tallingite, Hydrous Cu chloride.
Percylite, $PbCl_2.CuO.H_2O$.
Cumengite, $4PbCl_2.4CuO.5H_2O$.
Pseudoboléite, $5PbCl_2.4CuO.6H_2O$.

Boléite, $9PbCl_2.8CuO.3AgCl.9H_2O$.
Diaboléite, $2Pb(OH)_2.CuCl_2$.
Chloroxiphite, $2PbO.Pb(OH)_2, CuCl_2$.
Mitscherlichite, $2KCl.CuCl_2.2H_2O$.
Cuprite, Cu_2O.
Tenorite, Paramelaconite, CuO.
Crednerite, $CuO.Mn_2O_3$.
Delafossite, $Cu_2O.Fe_2O_3$.
Malachite, $CuCO_3.Cu(OH)_2$.
Azurite, $2CuCO_3.Cu(OH)_2$.
Rosasite, $(Cu,Zn)CO_3.(Cu,Zn)(OH)_2$.
Aurichalcite, $2(Zn,Cu)CO_3.3(Zn,Cu)(OH)_2$.
Voglite, Hydrous U,Ca,Cu, carbonate.
Dioptase, H_2CuSiO_4.
Chrysocolla, $CuSiO_3.2H_2O$.
Bisbeeite, $CuSiO_3.H_2O$.
Shattuckite, $2CuSiO_3.H_2O$.
Plancheite, $6CuO.5SiO_2.2H_2O$.
Olivenite, $Cu_2(OH)AsO_4$.
Duftite, $(Pb,Cu)_3(AsO_4)_2.(Pb,Cu)(OH)_2$.
Higginsite, $CuCa(OH)AsO_4$.
Libethenite, $Cu_2(OH)PO_4$.
Calciovolborthite, $(Cu,Ca)_3V_2O_8.$
  $(Cu,Ca)(OH)_2$.
Turanite, $5CuO.V_2O_5.2H_2O$.
Uzbekite, $3CuO.V_2O_5.3H_2O$.
Fornacite, Pb,Cu, chrom-arsenate.
Tsumebite, Pb,Cu, phosphate.
Clinoclasite, $Cu_3As_2O_8.3Cu(OH)_2$.
Erinite, $Cu_3As_2O_8.2Cu(OH)_2$.
Dihydrite, $Cu_3P_2O_8.2Cu(OH)_2$.
Pseudomalachite, $Cu_3P_2O_8.3Cu(OH)_2$.
Staszicite, $5(Ca,Cu,Zn)O.As_2O_5.2H_2O$.
Cornetite, $Cu_3P_2O_8.3Cu(OH)_2$.
Trichalcite, $Cu_3As_2O_8.5H_2O$.
Rosiérésite, Hydrous Al,Pb,Cu, phosphate.
Eucroite, $Cu_3As_2O_8.Cu(OH)_2.6H_2O$.
Conichalcite, $(Cu,Ca)_3As_2O_8.(Cu,Ca)(OH)_2.$
  $\frac{1}{2}H_2O$.
Freirinite, $6(Cu,Ca)O.3Na_2O.2As_2O_5.6H_2O$.
Bayldonite, $(Pb,Cu)_3As_2O_8.(Pb,Cu)(OH)_2.$
  H_2O.
Tagilite, $Cu_3P_2O_8.Cu(OH)_2.2H_2O$.
Leucochalcite, $Cu_3As_2O_8.Cu(OH)_2.2H_2O$.
Barthite, $3ZnO.CuO.3As_2O_5.2H_2O$.
Volborthite, Hydrous, Cu,Ba,Ca, vanadate.
Tangeite, $2CaO.2CuO.V_2O_5.H_2O$.
Cornwallite, $Cu_3As_2O_8.2Cu(OH)_2.H_2O$.
Tyrolite, $Cu_3As_2O_8.2Cu(OH)_2.7H_2O$.
Chalcophyllite, $7CuO.As_2O_5.14H_2O$.
Veszelyite, Hydrous Cu, Zn, phospho-
  arsenate.
Arakawaite, $4CuO.2ZnO.P_2O_5.6\frac{1}{2}H_2O$.
Turquois, $CuO.3Al_2O_3.2P_2O_5.9H_2O$.
Liroconite, $Cu_6Al(AsO_4)_5.3CuAl(OH)_5.$
  $20H_2O$.
Chenevixite, $2CuO.Fe_2O_3.As_2O_5.2H_2O$.
Henwoodite, Al,Cu, hydrous phosphate.
Cœruleite, $CuO.2Al_2O_3.As_2O_5.8H_2O$.
Chalcosiderite, $CuO.2Fe_2O_3.2P_2O_5.9H_2O$.
Torbernite, $Cu(UO_2)_2P_2O_8.8H_2O$.
Zeunerite, $Cu(UO_2)_2As_2O_8.8H_2O$.
Mixite, Hydrous Cu, Bi, arsenate.

Trippkeite, Cu, arsenite.
Lindackerite, $3NiO.6CuO.SO_3.2As_2O_5.7H_2O.$
Gerhardtite, $Cu(NO_3)_2.3Cu(OH)_2.$
Buttgenbachite, Hydrous Cu chloro-nitrate.
Hydrocyanite, $CuSO_4.$
Vauquelinite, $2(Pb,Cu)CrO_4.(Pb,Cu)_3P_2O_8.$
Connellite, $CuSO_4.2CuCl_2.19Cu(OH)_2.H_2O.$
Spangolite, $Cu_6AlClSO_{10}.9H_2O.$
BROCHANTITE, $CuSO_4.3Cu(OH)_2.$
Dolerophanite, $Cu_2SO_5.$
Antlerite, $CuSO_4.Cu(OH)_2.$
Caledonite, $2(Pb,Cu)O.SO_3.H_2O.$
Linarite, $(Pb,Cu)SO_4.(Pb,Cu)(OH)_2.$
Pisanite, $(Fe,Cu)SO_4.7H_2O.$
Zinc-copper melanterite, $(Zn,Cu,Fe)SO_4.$
    $7H_2O.$
Boothite, $CuSO_4.7H_2O.$
Cupromagnesite, $(Cu,Mg)SO_4.7H_2O.$
CHALCANTHITE, $CuSO_4.5H_2O.$
Zinc-copper chalcanthite, $(Zn,Cu,Fe)SO_4.$
    $5H_2O.$
Kröhnkite, $CuSO_4.Na_2SO_4.2H_2O.$
Natrochalcite, $Cu_4(OH)_2(SO_4)_2.Na_2SO_4.$
    $2H_2O.$
Ransomite, $CuO.(Fe,Al)_2O_3.4SO_3.7H_2O.$
Guildite, $3(Cu,Fe)O.2(Fe,Al)_2O_3.7SO_3.$
    $17H_2O.$
Phillipite, $CuSO_4.Fe_2(SO_4)_3.nH_2O.$
Langite, $CuSO_4.3Cu(OH)_2.H_2O.$
Herrengrundite, $2(CuOH)_2SO_4.Cu(OH)_2.$
Vernadskite, $3CuSO_4.Cu(OH)_2.4H_2O.$
Kamarezite, Hydrous basic Cu sulphate.
Kyanotrichite, $4CuO.Al_2O_3.SO_3.8H_2O.$
Serpierite, Hydrous basic Cu,Zn, sulphate.
Beaverite, $CuO.PbO.Fe_2O_3.2SO_3.4H_2O.$
Glaukokerinite, $Zn_{15}Al_5Cu_7(SO_4)_9O_{90}34H_2O.$
Chalcoalumite, $CuO.2Al_2O_3.SO_3.9H_2O.$
Johannite, Hydrous Cu,U, sulphate.
Chalcomenite, $CuSeO_3.2H_2O.$
Cuprotungstite, $CuWO_4.$

GOLD

Native Gold, $Au.$
Petzite, $(Ag,Au)_2Te.$
SYLVANITE, $(Au,Ag)Te_2.$
Krennerite, $(Au,Ag)Te_2.$
CALAVERITE, $AuTe_2.$
Muthmannite, $(Ag,Au)Te.$
Nagyagite, Au,Pb, sulpho-telluride.

IRON

Native Iron, Fe.
Awaruite, $FeNi_2.$
Josephinite, $FeNi_3.$
Cubanite, $Cu_2S.Fe_4S_5.$
Sternbergite, $Ag_2S.Fe_4S_5.$
Pentlandite, $(Fe,Ni)S.$
Pyrrhotite, $FeS.$
Troilite, $FeS.$
Daubréelite, $FeS.Cr_2S_3.$

Badenite, $(Co,Ni,Fe)_2(As,Bi)_3.$
Chalcopyrite, $CuFeS_2.$
Pyrite, $FeS_2.$
Bravoite, $(Fe,Ni)S_2.$
Cobaltnickelpyrite, $(Fe,Co,Ni)S_2.$
Arsenoferrite, $FeAs_2.$
Marcasite, $FeS_2.$
Löllingite, $FeAs_2.$
Arsenopyrite, $FeAsS.$
Gudmundite, $FeSbS.$
Eichbergite, $(Cu,Fe)_2S.3(Bi,Sb)_2S_3.$
Histrixite, $5CuFeS_2.2Sb_2S_3.7Bi_2S_3.$
Berthierite, $FeS.Sb_2S_3.$
Stylotypite, $3(Cu_2,Ag_2,Fe)S.Sb_2S_3.$
Molysite, $FeCl_3.$
Lawrencite, $FeCl_2.$
Rinneite, $FeCl_2.3KCl.NaCl.$
Hämatophanite, $Pb(Cl,OH)_2.4PbO.2Fe_2O_3.$
Kremersite, $KCl_2.NH_4Cl.FeCl_2.H_2O.$
Erythrosiderite, $2KCl.FeCl_3.H_2O.$
Zirklerite, $9(Fe,Mg,Ca)Cl_2.2Al_2O_3.3H_2O.$
Hematite, $Fe_2O_3.$
ILMENITE, $FeTiO_3.$
Senaite, $(Fe,Mn,Pb)O.TiO_2.$
Geikelite, $(Mg,Fe)O.TiO_2.$
Högbomite, $Al_2O_3,Fe_2O_3,MgO,TiO_2.$
Bixbyite, $(Fe,Mn)_2O_3.$
Sitaparite, $9Mn_2O_3.4Fe_2O_3.MnO_2.3CaO.$
Vredenburgite, $3Mn_3O_4.2Fe_2O_3.$
Hercynite, $FeO.Al_2O_3.$
Magnetite, $FeO.Fe_2O_3.$
Magnesioferrite, $MgO.Fe_2O_3.$
FRANKLINITE, $(Fe,Zn,Mn)O.$
    $(Fe,Mn)_2O_3.$
Jacobsite, $MgO.Fe_2O_3.(Mn,Fe)O.Fe_2O_3.$
Chromite, $FeO.Cr_2O_3$
Trevorite, $NiO.Fe_2O_3.$
Hollandite, manganate of Mn,Ba, Fe'''.
Delafossite, $Cu_2O.Fe_2O_3.$
Magnetoplumbite, Fe,Mn,Pb, oxides.
Plumboferrite, $PbO.2Fe_2O_3.$
Goethite, $Fe_2O_3.H_2O.$
Lepidocrocite, $Fe_2O_3.H_2O.$
Hydrogoethite, $3Fe_2O_3.4H_2O.$
Limonite, $2Fe_2O_3.3H_2O.$
Xanthosiderite, $Fe_2O_3.2H_2O.$
Esmeraldaite, $Fe_2O_3.4H_2O.$
Turgite, $2Fe_2O_3.H_2O.$
Stainerite, $(Co,Fe,Al)_2O_3.H_2O.$
Skemmatite, $3MnO_2.2Fe_2O_3.6H_2O.$
Beldongrite, $6Mn_2O_3.Fe_2O_3.8H_2O.$
Ankerite, $2CaCO_3.MgCO_3.FeCO_3.$
Siderite, $FeCO_3.$
Codazzite, $(Ca,Mg,Fe,Ce)CO_3.$
Pyroaurite, $MgCO_3.5Mg(OH)_2.2Fe(OH)_3.$
    $4H_2O.$
HYPERSTHENE, $(Fe,Mg)SiO_3.$
ACMITE, $NaFe(SiO_3)_2.$
Pyroxmangite, Mn,Fe, pyroxene.
Babingtonite, $(Ca,Fe,Mn)SiO_3,$ with
    $Fe_2(SiO_3)_3.$
Sobralite, $(Mn,Fe,Ca,Mg)SiO_3.$
ANTHOPHYLLITE, $(Mg,Fe)SiO_3.$

GLAUCOPHANE, $NaAl(SiO_3)_2$.
  $(Fe,Mg)SiO_3$.
Torendrikite, $Na_2O.4MgO.CaO.FeO.Fe_2O_3$.
  $10SiO_2$.
RIEBECKITE, $2NaFe(SiO_3)_2.FeSiO_3$.
CROCIDOLITE, $NaFe(SiO_3)_2.FeSiO_3$.
ARFVEDSONITE, Na,Ca,Fe, silicate.
Ænigmatite, Fe,Na,Ti-silicate.
Weinbergerite, $NaAlSiO_4.3FeSiO_3$.
Astrolite, $(Na,K)_2Fe(Al,Fe)_2(SiO_3)_5.H_2O$?
Gillespite, $FeO.BaO.4SiO_2$.
Taramellite, $Ba_4Fe''Fe_4'''Si_{10}O_{31}$.
Helvite, $3(Mn,Fe)BeSiO_4.MnS$.
Danalite, $3(Fe,Zn,Mn)BeSiO_4.ZnS$.
Almandite, $Fe_3Al_2(SiO_4)_3$.
Andradite, $Ca_3Fe_2(SiO_4)_3$.
Partschinite, $(Mn,Fe)_3Al_2Si_3O_{12}$.
Fayalite, Fe_2SiO_4.
Knebelite, $(Fe,Mn)_2SiO_4$.
Pyrosmalite, $6(Mn,Fe)O.2(Fe,Mn)$
  $(OH,Cl)_2.6SiO_2.3H_2O$.
Homilite, $(Ca,Fe)_3B_2Si_2O_{10}$.
Allanite, $(Ca,Fe)_2(AlOH)(Al,Ce,Fe)_2(SiO_4)_3$.
Nagatelite, A silicate and phosphate of
  Al,Ce,Ca,Fe.
ILVAITE, $CaFe_2(FeOH)(SiO_4)_2$.
Chapmanite, $5FeO.Sb_2O_5.5SiO_2.2H_2O$.
Melanotekite, $3PbO.2Fe_2O_3.3SiO_2$.
Angaralite, $2(Ca,Mg)O.5(Al,Fe)_2O_3.6SiO_2$.
STAUROLITE, $(AlO)_4(AlOH)Fe(SiO_4)_2$.
Grandidierite, Al,Fe,Mg, silicate.
Aloisite, Fe,Ca,Mg,Na, silicate.
Poechite, $H_{16}Fe_8Mn_2Si_9O_{29}$.
Lotrite, $3(Ca,Mg)O.2(Al,Fe)_2O_3.4SiO_2.2H_2O$.
Zinnwaldite, Li-Fe mica.
Biotite, Mg-Fe mica.
Lepidomelane, Iron mica.
Chloritoid, $H_2(Fe,Mg)Al_2SiO_7$.
Prochlorite, Fe,Mg, chlorite.
Amesite, $H_4(Mg,Fe)_2Al_2SiO_9$.
Kossmatite, Hydrous, Mg,Al,Fe''', silicate.
Moravite, $2FeO.2(Al,Fe)_2O_3.7SiO_2.2H_2O$.
Cronstedtite, $4FeO.2Fe_2O_3.3SiO_2.4H_2O$.
Thuringite, $8FeO.4(Al,Fe)_2O_3.6SiO_2.9H_2O$.
Brunsvigite, $9(Fe,Mg)O.2Al_2O_3.6SiO_2.8H_2O$.
Griffithite, $4(Mg,Fe,Ca)O.(Al,Fe)_2O_3.5SiO_2$.
  $7H_2O$.
Chamosite, Fe,Mg, silicate.
Stilpnomelane $\big\}$ Fe silicates.
Minguétite $\big\}$
Strigovite, $H_4Fe_2(Al,Fe)_2Si_2O_{11}$.
Spodiophyllite, $(Na_2,K_2)_2(Mg,Fe)_3(Fe,Al)_2$
  $(SiO_3)_8$.
Viridite, $4FeO.2SiO_2.3H_2O$.
Mackensite, $Fe_2O_3.SiO_2.2H_2O$.
Iddingsite, $MgO.Fe_2O_3.3SiO_2.4H_2O$.
Celadonite, Fe,Mg,K, silicate.
Glauconite, Hydrous Fe,K, silicate.
Pholidolite, $K_2O.12(Fe,Mg)O.Al_2O_3.13SiO_2$.
  $5H_2O$.
Faratsihite, $H_4(Al,Fe)_2Si_2O_9$.
Elbrussite, Hydrous Al,Fe,Mg, silicate.
Melite, $2(Al,Fe)_2O_3.SiO_2.8H_2O$.

Chloropal, $H_4Fe_2Si_2O_9$.
Müllerite, $Fe_2Si_3O_9.2H_2O$.
Morencite $\big\}$ Hydrous ferric silicates.
Hisingerite $\big\}$
Canbyite, $H_4Fe_2'''Si_2O_9.2H_2O$.
Sturtite, $6(Mn,Ca,Mg)O.Fe_2O_3.8SiO_2$.
  $23H_2O$.
Astrophyllite, Na,K,Fe,Mn, titano-silicate.
Narsarsukite, Fe,Na, titano-silicate.
Neptunite, Fe,Mn,Na,K, titano-silicate.
Joaquinite, Ca,Fe, titano-silicate.
Dysanalyte, Ca,Fe, titano-niobate.
Arizonite, $Fe_2O_3.3TiO_2$.
Zirkelite, $(Ca,Fe)(Zr,Ti,Th)_2O_5$.
Delorenzite, Fe,U,Y, titanate.
Pseudobrookite, Fe_2TiO_5.
Kalkowskite, $(Fe,Ce)_2O_3.4(Ti,Si)O_2$.
Neotantalite, Fe tantalate.
COLUMBITE, TANTALITE, (Fe,Mn)
  $(Nb,Ta)_2O_6$.
Tapiolite, $(Fe,Mn)(Ta,Nb)_2O_6$.
Yttrotantalite, Fe,Ca,Y,Er,Ce, tantalate.
Samarskite, Fe,U,Y, etc., niobate-tantalate.
Hielmite, Y,Fe,Mn,Ca, stanno-tantalate.
Monimolite, Pb,Fe, antimonate.
TRIPHYLITE, $Li(Fe,Mn)PO_4$.
Graftonite, $(Fe,Mn,Ca)_3P_2O_8$.
Triplite, $(RF)RPO_4$; $R = Fe,Mn$.
Triploidite $(ROH)RPO_4$; $R = Fe,Mn$.
Arrojadite, Fe,Mn, phosphate.
Sarcopside, Fe,Mn,Ca, phosphate.
Dufrenite, $FePO_4.Fe(OH)_3$.
Delvauxite, $2Fe_2O_3.P_2O_5.9H_2O$.
Lazulite, $2AlPO_4.(Fe,Mg)(OH)_2$.
Arseniosiderite, $Ca_3Fe(AsO_4)_3.3Fe(OH)_3$.
Dussertite, $Ca_3(AsO_4)_2.3Fe(OH)_3$.
Dickinsonite $\big\}$ Hydrous Mn,Fe,Na,
Fillowite $\big\}$ phosphates.
Messelite $(Ca,Fe)_3P_2O_8.2\frac{1}{2}H_2O$.
Collinsite, $2CaO.(Mg,Fe)O.P_2O_5.2\frac{1}{2}H_2O$.
Wenzelite, $(Mn,Fe,Mg)_3(PO_4)_2.5H_2O$.
Anapaite, $(Ca,Fe)_3P_2O_8.4H_2O$.
Vivianite, $Fe_3P_2O_8.8H_2O$.
Symplesite, $Fe_3As_2O_8.8H_2O$.
Ferrisymplesite, $3Fe_2O_3.2As_2O_5.16H_2O$.
Scorodite, $FeAsO_4.2H_2O$.
Strengite, $FePO_4.2H_2O$.
Vilateite, Hydrous Fe, Mn, phosphate.
Barrandite, $(Al,Fe)PO_4.2H_2O$.
Purpurite, $2(Fe,Mn)PO_4.H_2O$.
Phosphosiderite, $4FePO_4.7H_2O$.
Koninckite, $FePO_4.3H_2O$.
Sicklerite, $Fe_2O_3.6MnO.4P_2O_5.3(Li,H)_2O$.
Salmonsite, $Fe_2O_3.9MnO.4P_2O_5.14H_2O$.
Landesite, $3Fe_2O_3.20MnO.8P_2O_5.27H_2O$.
Fervanite, $2Fe_2O_3.2V_2O_5.5H_2O$.
Vauxite, $FeO.Al_2O_3.P_2O_5.6H_2O$.
Paravauxite, $FeO.Al_2O_3.P_2O_5.5H_2O$.
Liskeardite, $(Al,Fe)AsO_4.2(Al,Fe)(OH)_3$.
  $5H_2O$.
Pharmacosiderite, $6FeAsO_4.2Fe(OH)_3$.
  $12H_2O$.
Ludlamite, $2Fe_3P_2O_8.Fe(OH)_2.8H_2O$.

Cacoxenite, $FePO_4.Fe(OH)_3.4\frac{1}{2}H_2O$.
Beraunite, $2FePO_4.Fe(OH)_3.2\frac{1}{2}H_2O$.
Childrenite, $AlPO_4.Fe(OH)_2.H_2O$.
Mazapilite, $Ca_3Fe_2(AsO_4)_4.2FeO(OH).5H_2O$.
Yukonite, $(Ca_3,Fe_2''')(AsO_4)_2.2Fe(OH)_3$.
  $5H_2O$.
Calcioferrite, $Ca_3Fe_2(PO_4)_4.Fe(OH)_3.8H_2O$.
Borickite, $Ca_3Fe_2(PO_4)_4.12Fe(OH)_3.6H_2O$.
Egueiite, Hydrous Fe,Al,Ca, phosphate.
Richelite, $4FeP_2O_8.Fe_2OF_2(OH)_2.36H_2O$.
Chenevixite, $2CuO.Fe_2O_3.As_2O_5.2H_2O$.
Chalcosiderite, $CuO.3Fe_2O_3.2P_2O_5.9H_2O$.
Roschérite, $(Mn,Fe,Ca)_2Al(OH)(PO_4)_2$.
  $2H_2O$.
Tripuhyite, $Fe_2Sb_2O_7$.
Flajolotite, $4FeSbO_4.3H_2O$.
Catoptrite, $14(Mn,Fe)O.2(Al,Fe)_2O_3.2SiO_2$.
  Sb_2O_5.
Derbylite, Fe antimo-titanate.
Destinezite, $2Fe_2O_3.P_2O_5.2SO_3.2H_2O$.
Pitticite, Hydrous Fe arsenate and sulphate.
Beudantite, $3Fe_2O_3.2PbO.2SO_3.As_2O_5.6H_2O$.
Lossenite, Hydrous Fe,Pb, arsenate and sulphate.
Weslienite, $Na_2FeCa_3Sb_4O_{15}$.
Schneebergite, $(Ca,Fe)_2Sb_2O_6$.
Ludwigite, $3MgO.B_2O_3.FeO.Fe_2O_3$.
Vonsenite, $3(Fe,Mg)O.B_2O_3.FeO.Fe_2O_3$.
Magnesioludwigite, $3MgO.B_2O_3.MgO.Fe_2O_3$.
Warwickite, $(Mg,Fe)_3TiB_2O_8$.
Lagonite, $Fe_2O_3.3B_2O_3.3H_2O$.
Hulsite, $12(Fe,Mg)O.2Fe_2O_3.SnO_2.3B_2O_3$.
Millosevichite, $(Fe,Al)_2(SO_4)_3$.
Szomolnokite, $FeSO_4.H_2O$.
Ilesite, $(Mn,Zn,Fe)SO_4.4H_2O$.
Bianchite, $FeSO_4.2ZnSO_4.18H_2O$.
Melanterite, $FeSO_4.7H_2O$.
Pisanite, $(Fe,Cu)SO_4.7H_2O$.
Gidorotil, $FeSO_4.5H_2O$.
Halotrichite, $FeSO_4.Al_2(SO_4)_3.24H_2O$.
Bilinite, $FeSO_4.Fe_2(SO_4)_3.24H_2O$.
Dietrichite, $(Zn,Fe,Mn)SO_4.Al_2(SO_4)_2$.
  $22H_2O$.
Kornelite, $Fe_2(SO_4)_3.7\frac{1}{2}H_2O$.
Coquimbite, $Fe_2(SO_4)_3.9H_2O$.
Ransomite, $CuO.(Fe,Al)_2O_3.4SO_3.7H_2O$.
Guildite, $3(Cu,Fe)O.2(Fe,Al)_2O_3.7SO_3$.
  $17H_2O$.
Louderbackite, $2FeO.3(Fe,Al)_2O_3.10SO_3$.
  $35H_2O$.
Phillipite, $CuSO_4.Fe_2(SO_4)_3.nH_2O$.
Ferrinatrite, $3Na_2SO_4.Fe_2(SO_4)_3.6H_2O$.
Sideronatrite, $2Na_2O.Fe_2O_3.4SO_3.7H_2O$.
Voltaite, $3(K_2,Fe)O.2(Al,Fe)_2O_3.6SO_3.9H_2O$.
Metavoltine, $(H,K)_4(Fe'''OH)_2(SO_4)_4.5H_2O$.
Slavikite, $(Na,K)_2SO_4.Fe_{10}(OH)_6(SO_4)_{12}$.
  $63H_2O$.
Römerite, $FeSO_4.Fe_2(SO_4)_3.14H_2O$.
Beaverite, $CuO.PbO.Fe_2O_3.2SO_3.4H_2O$.
Vegasite, $PbO.3Fe_2O_3.3SO_3.6H_2O$.
Copiapite, $Fe_4(OH)_2(SO_4)_5$.

Rhomboclase, $Fe_2O_3.4SO_3.9H_2O$.
Lausenite, $Fe_2O_3.3SO_3.6H_2O$.
Castanite, $Fe_2O_3.2SO_3.8H_2O$.
Utahite, $3Fe_2O_3.2SO_3.7H_2O$.
Butlerite, $(Fe,Al)_2O_3.2SO_3.5H_2O$.
Amaranthite, $Fe_2O_3.2SO_3.7H_2O$.
Fibroferrite, $Fe_2O_3.2SO_3.10H_2O$.
Raimondite, $2Fe_2O_3.3SO_3.7H_2O$.
Carphosiderite, $3Fe_2O_3.4SO_3.7H_2O$.
Borgströmite, $Fe_2O_3.SO_3.3H_2O$.
Planoferrite, $Fe_2O_3.SO_3.15H_2O$.
Glockerite, $2Fe_2O_3.SO_3.6H_2O$.
Cyprusite, $7Fe_2O_3.Al_2O_3.10SO_3.14H_2O$.
Botryogen, $2MgO.Fe_2O_3.4SO_3.15H_2O$.
Sideronatrite, $2Na_2O.Fe_2O_3.4SO_3.7H_2O$.
Voltaite, $3(K_2,Fe)O.2(Al,Fe)_2O_3.6SO_3.9H_2O$.
Metavoltine, $5(K_2,Na_2,Fe)O.3Fe_2O_3.12SO_3$:
  $18H_2O$.
Jarosite, $K_2Fe_6(OH)_2(SO_4)_4$.
Natrojarosite, $Na_2Fe_6(OH)_{12}(SO_4)_4$.
Ammoniojarosite, $(NH_4)_2Fe_6(OH)_{12}(SO_4)_4$.
Plumbojarosite, $PbFe_6(OH)_{12}(SO_4)_4$.
Argentojarosite, $Ag_2Fe_6(OH)_{12}(SO_4)_4$.
Quetenite, $MgO.Fe_2O_3.3SO_3.13H_2O$.
Emmonsite, Hydrous Fe tellurate.
Durdenite, $Fe_2(TeO_3)_3.4H_2O$.
WOLFRAMITE, $(Fe,Mn)WO_4$.
Ferrimolybdite, $Fe_2O_3.3MoO_3.8H_2O$.
Ferritungstite, $Fe_2O_3.WO_3.6H_2O$.
Humboltine, $FeC_2O_4.2H_2O$.

LEAD

Native Lead, Pb.
Galena, PbS.
Altaite, Te.
Clausthalite, PbSe.
Naumannite, $(Ag_2,Pb)Se$.
Penroseite, $3CuSe.2PbSe_2.5(Ni,Co)Se_2$.
Chiviatite, $PbS.2Bi_2S_3$.
Gladite, $2PbS.Cu_2S.5Bi_2S_3$.
Rezbanyite, $3PbS.Cu_2S.5Bi_2S_3$.
Platynite, $PbS.Bi_2S_3$.
Zinkenite, $PbS.Gl_2S_3$.
Andorite, $2PbS.Ag_2S.3Sb_2S_3$.
Lindströmite, $2PbS.Cu_2S.3Bi_2S_3$.
Hutchinsonite, $PbS.(Tl,Ag)_2S.2Sb_2S_3$.
Sartorite, $PbS.As_2S_3$.
Galenobismuthite, $PbS.Bi_2S_3$.
Fülöppite, $2PbS.3Sb_2S_3$.
Plagionite, $5PbS.4Sb_2S_3$.
Bismutoplagionite, $5PbS.4Bi_2S_3$.
Baumhauerite, $4PbS.3As_2S_3$.
Fizelyite, $5PbS.Ag_2S.4Sb_2S_3$.
Ramdohrite, $3PbS.Ag_2S.3Sb_2S_3$.
Heteromorphite, $7PbS.4Sb_2S_3$.
Rathite, $3PbS.2As_2S_3$.
Schirmerite, $3(Ag_2,Pb)S.2Bi_2S_3$.
Hammarite, $5PbS.3Bi_2S_3$.
Wittite, $5PbS.3Bi_2(S,Se)_3$.
Benjaminite, $(Cu,Ag)_2S.2PbS.2Bi_2S_3$.

Jamesonite, $2PbS.Sb_2S_3$.
Dufrenoysite, $2PbS.As_2S_3$.
Ouyheeite, $8PbS.2Ag_2S.5Sb_2S_3$.
Cosalite, $2PbS.Bi_2S_3$.
Kobellite, $2PbS.(Bi,Sb)_2S_3$.
Berthonite, $5PbS.9Cu_2S.7Sb_2S_3$.
Schapbachite, $PbS.Ag_2S.Bi_2S_3$.
Semseyite, $9PbS.4Sb_2S_3$.
Boulangerite, $5PbS.2Sb_2S_3$.
Freieslebenite, $5(Pb,Ag_2)S.2Sb_2S_3$.
Diaphorite, $5(Pb,Ag_2)S.2Sb_2S_3$.
Bournonite, $2PbS.Cu_2S.Sb_2S_3$.
Seligmanite, $2PbS.Cu_2S.As_2S_3$.
Aikinite, $2PbS.Cu_2S.Bi_2S_3$.
Lillianite, $3PbS.Bi_2S_3$.
Guitermanite, $3PbS.As_2S_3$.
Lengenbachite, $6PbS.Ag_2S.2As_2S_3$.
Meneghinite, $4PbS.Sb_2S_3$.
Jordanite, $4PbS.As_2S_3$.
Coongarrite, $4PbS.Bi_2S_3$.
Geocronite, $5PbS.Sb_2S_3$.
Beegerite, $6PbS.Bi_2S_3$.
Ultrabasite, $28PbS.11Ag_2S.3GeS_2.2Sb_2S_3$.
Teallite, $PbSnS_2$.
Franckeite, $Pb_5Sn_3FeSb_2S_{14}$.
Cylindrite, $Pb_3Sn_4FeSb_2S_{14}$.
Cotunnite, $PbCl_2$.
Percylite, $PbCl_2.CuO.H_2O$.
Cumengite, $4PbCl_2.4CuO.5H_2O$.
Pseudo-boléite, $5PbCl_2.4CuO.6H_2O$.
Boléite, $9PbCl_2.8CuO.3AgCl.9H_2O$.
Diaboléite, $2Pb(OH)_2.CuCl_2$.
Chloroxiphite, $2PbO.Pb(OH)_2.CuCl_2$.
Matlockite, $PbCl_2.PbO$.
Mendipite, $PbCl_2.2PbO$.
Lorettoite, $PbCl_2.6PbO$.
Laurionite, $PbCl_2.Pb(OH)_2$.
Paralaurionite, $PbCl_2.Pb(OH)_2$.
Fiedlerite, $2PbCl_2.PbO.H_2O$.
Penfieldite, $2PbCl_2.PbO$.
Daviesite, Pb oxychloride.
Schwartzembergite, $6PbO.3PbCl_2.PbI_2O_5$.
Hämatophanite, $Pb(Cl,OH)_2.4PbO.2Fe_2O_3$.
Massicot, PbO.
Senaite, $(Fe,Mn,Pb)O.TiO_2$.
Minium, $2PbO.PbO_2$.
Magnetoplumbite, Fe,Mn,Pb oxides.
Plumboferrite, $PbO.2Fe_2O_3$.
Plattnerite, PbO_2.
Quenselite, $2PbO.Mn_2O_3.H_2O$.
Cerussite, $PbCO_3$.
PHOSGENITE, $PbCO_3.PbCl_2$.
Hydrocerussite, $2PbCO_3.Pb(OH)_2$.
Dundasite, $Pb(AlO)_2(CO_3)_2$.
Alamosite, $PbSiO_3$.
Barysilite, $Pb_3Si_2O_7$.
Ganomalite, $Pb_4(PbOH)_2Ca_4(Si_2O_7)_3$.
Nasonite, $Pb_4(PbCl)_2Ca_4(Si_2O_7)_3$.
Margarosanite, $Pb(Ca,Mn)_2(SiO_3)_3$.
Hyalotekite, $(Pb,Ba,Ca)B_2(SiO_3)_{12}$.
Roeblingite, $5(H_2CaSiO_4).2(CaPbSO_4)$.
Larsenite, $PbZnSiO_4$.
Calcium-larsenite, $(Pb,Ca)ZnSiO_4$.

Molybdophyllite, $4(Mg,Pb)O.4(Mg,Pb)$ $(OH)_2.4SiO_2.H_2O$.
Hancockite, Pb,Mn,Ca,Al, etc., silicate.
Kentrolite, $3PbO.2Mn_2O_3.3SiO_2$.
Melanotekite, $3PbO.2Fe_2O_3.3SiO_2$.
Kasolite, $PbO.UO_3.SiO_2.H_2O$.
Plumboniobite, Y,U,Pb,Fe, niobate.
Monimolite, Pb,Fe, antimonate.
Carminite, $Pb_3As_2O_8.10FeAsO_4$.
Georgiadésite, $Pb_3(AsO_4)_2.3PbCl_2$.
PYROMORPHITE, $Pb_4(PbCl)(PO_4)_3$.
Mimetite, $Pb_4(PbCl)(AsO_4)_3$.
Vanadinite, $Pb_4(PbCl)(VO_4)_3$.
Trigonite, $Pb_3MnH(AsO_3)_3$.
Schultenite, $HPbAsO_4$.
Plumbogummite, Pb,Al, phosphate.
Duftite, $(Pb,Cu)_3(AsO_4)_3.(Pb,Cu)(OH)_2$.
Descloizite, $(Pb,Zn)_2(OH)VO_4$.
Pyrobelonite, $4(Mn,Pb)O.V_2O_5.H_2O$.
Brackebuschite, $(Pb,Mn,Fe)_3V_2O_8.H_2O$.
Dechenite, PbV_2O_6.
Fornacite, Pb,Cu, chrom-arsenate.
Tsumebite, Pb,Cu, phosphate.
Rosiérésite, Hydrous Al,Pb,Cu, phosphate.
Ferrazite, $3(Ba,Pb)O.2P_2O_5.8H_2O$.
Bayldonite, $(Pb,Cu)_3As_2O_8.(Pb,Cu)(OH)_2$. H_2O.
Hügelite, Hydrous Pb,Zn, vanadate.
Renardite, $PbO.4UO_3.P_2O_5.9H_2O$.
Dewindtite, $3PbO.5UO_3.2P_2O_5.12H_2O$.
Dumontite, $2PbO.3UO_3.P_2O_5.5H_2O$.
Parsonsite, $2PbO.UO_3.P_2O_5.H_2O$.
Bindheimite, Hydrous Pb antimonate.
Nadorite, $PbSb_2O_4.PbCl_2$.
Ecdemite, $Pb_4As_2O_7.2PbCl_2$.
Finnemanite, $Pb_5Cl(AsO_3)_3$.
Ochrolite, $Pb_4Sb_2O_7.2PbCl_2$.
Mauzeliite, Pb,Ca, titano-antimonate.
Beudantite, $3Fe_2O_3.2PbO.2SO_3.As_2O_5.6H_2O$.
Hinsdalite, $3Al_2O_3.2PbO.2SO_3.P_2O_5.6H_2O$.
Lossenite, Hydrous Fe, Pb, arsenate and sulphate.
Clarkeite, $(Na_2Pb)O.3UO_3.3H_2O$.
Curite, $2PbO.5UO_3.4H_2O$.
Fourmarierite, Hydrous Pb uranate.
Anglesite, $PbSO_4$.
CROCOOITE, $PbCrO_4$.
Phœnicochroite, $3PbO.2CrO_3$.
Vauquelinite, $2(Pb,Cu)CrO_4.(Pb,Cu)_3P_2O_8$.
Bellite, Pb arseno-chromate.
Leadhillite, $PbSO_4.2PbCO_3.Pb(OH)_2$.
Caracolite, $Pb(OH)Cl.Na_2SO_4$.
Lanarkite, Pb_2SO_5.
Caledonite, $(Pb,Cu)SO_4.(Pb,Cu)(OH)_2$.
Linarite, $(Pb,Cu)SO_4.(Pb,Cu)(OH)_2$.
Beaverite, $CuO.PbO.Fe_2O_3.2SO_3.4H_2O$.
Vegasite, $PbO.3Fe_2O_3.3SO_3.6H_2O$.
Plumbojarosite, $PbFe_6(OH)_{12}(SO_4)_4$.
Palmierite, $(K,Na)_2Pb(SO_4)_2$.
Stolzite ⎱ $PbWO_4$.
Raspite ⎰
Chillagite, $3PbWO_4.PbMoO_4$.
WULFENITE, $PbMoO_4$.

LITHIUM

Cryolithionite, $3NaF.3LiF.2AlF_3$.
Petalite, $LiAl(Si_2O_5)_2$.
Spodumene, $LiAl(SiO_3)_2$.
Eucryptite, $LiAlSiO_4$.
Lepidolite, Lithium mica.
Cookeite, Lithium mica.
Zinnwaldite, Lithium-iron mica.
Manandonite, $H_{24}Li_4Al_{14}B_4Si_6O_{53}$.
Triphylite, $Li(Fe,Mn)PO_4$.
Lithiophilite, $Li(Mn,Fe)PO$.
Amblygonite, $LiAl(F,OH)PO_4$.
Fremontite, $(Na,Li)Al(OH,F)PO_4$;
Sicklerite, $Fe_2O_3.6MnO.4P_2O_5.3(Li,H)_2O$.

MAGNESIUM

Zamboninite, $CaF_2.2MgF_2$.
Sellaite, MgF_2.
Nocerite, $Ca_3Mg_3O_2F_8$.
Koenenite, Al,Mg, oxychloride.
Carnallite, $KCl.MgCl_2.6H_2O$.
Bischolite, $MgCl_2.6H_2O$.
Tachhydrite, $CaCl_2.2MgCl_2.12H_2O$.
Ralstonite, $(Na_2,Mg)F_2.3Al(F.OH)_3.2H_2O$.
Zirklerite, $9(Fe,Mg,Ca)Cl_2.2Al_2O_3.3H_2O$.
Periclase, MgO.
Geikielite, $(Mg,Fe)O.TiO_2$.
Hogbomite, $Al_2O_3,Fe_2O_3,MgO,TiO_2$.
Spinel, $MgO.Al_2O_3$.
Magnesioferrite, $MgO.Fe_2O_3$.
Jacobsite, $MgO.Fe_2O_3.(Fe,Mn)O.Fe_2O_3$.
Brucite, $Mg(OH)_2$.
Dolomite, $CaCO_3.MgCO_3$
Ankerite, $CaCO_3.(Mg,Fe,Mn)CO_3$.
Magnesite, $MgCO_3$.
Codazzite, $(Ca,Mg,Fe,Ce)CO_3$.
Northupite, $MgCO_3.Na_2CO_3.NaCl$.
Tychite, $2Mg CO_3.2Na_2CO_3.Na_2SO_4$.
Nesquehonite, $MgCO_3.3H_2O$.
Hydromagnesite, $3MgCO_3.Mg(OH)_2.3H_2O$.
Hydrogiobertite, $MgCO_3.Mg(OH)_2.2H_2O$.
Artinite, $MgCO_3.Mg(OH)_2.3H_2O$.
Lansfordite, $MgCO_3.5H_2O$.
Gajite, Basic, hydrous Ca,Mg, carbonate.
Stichtite, $MgCO_3.5Mg(OH)_2.2Cr(OH)_3$.
    $4H_2O$.
Hydrotalcite, $MgCO_3.5Mg(OH)_2.2Al(OH)_3$.
    $4H_2O$.
Pyroaurite, $MgCO_3.5Mg(OH)_2.2Fe(OH)_3$.
    $4H_2O$.
Enstatite, $MgSiO_3$.
Hypersthene, $(Fe,Mg)SiO_3$.
Pyroxene Group, Ca, Mg, etc., silicates.
Sobralite, $(Mn,Fe,Ca,Mg)SiO_3$.
Anthophyllite, $(Mg,Fe)SiO_3$.
Amphibole Group, Ca, Mg, etc., silicates
Glaucophane, $NaAl(SiO_3)_2.(Fe,Mg)SiO_3$.
Torendrikite, $Na_2O.4MgO.CaO.FeO.Fe_2O_3$.
    $10SiO_2$.
Cordierite, $Mg_2Al_4Si_5O_{18}$.
Pyrope, $Mg_3Al_2(SiO_4)_3$.

Chrysolite, $(Mg,Fe)_2SiO_4$.
Monticellite, $CaMgSiO_4$.
Fosterite, Mg_2SiO_4.
Hortonolite, $(Fe,Mg,Mn)_2SiO_4$.
Merwinite, $Ca_3Mg(SiO_4)_2$.
Molybdophyllite, $4(Mg,Pb)O.4(Mg,Pb)$
    $(OH)_2.4SiO_2.H_2O$.
Fluosiderite, Ca,Mg silicate.
Norbergite, $Mg_2SiO_4.Mg(F,OH)_2$.
Chondrodite, $2Mg_2SiO_4.Mg(F,OH)_2$.
Humite, $3Mg_2SiO_4.Mg(F,OH)_2$.
Clinohumite, $4Mg_2SiO_4Mg(F,OH)_2$.
Kornerupine, $MgAl_2SiO_6$.
Sapphirine, $Mg_5Al_{12}Si_2O_{27}$.
Serendibite, $10(Ca,Mg)O.5Al_2O_3.B_2O_3.6SiO_2$.
Silicomagnesiofluorite, $H_2Ca_4Mg_3Si_2O_7F_{10}$.
Truscottite, $2(Ca,Mg)O.3SiO_2.3H_2O$.
Lotrite, $3(Ca,Mg)O.2(Al,Fe)_2O_3.4SiO_2.2H_2O$.
Ferrierite, Hydrous Mg, Na, silicate.
Biotite, Magnesium-iron mica.
Phlogopite, Magnesium mica.
Tæniolite, K, Mg, silicate.
Alurgite, K,Mg,Mn mica.
Seybertite, $H_3(Mg,Ca)_5Al_5Si_2O_{18}$.
Xanthophyllite, $H_5(Mg,Ca)_{14}Al_{16}Si_5O_{52}$.
Chloritoid, $H_2(Fe,Mg)Al_2SiO_7$.
Clinochlore, Penninite, $H_8Mg_5Al_2Si_3O_{18}$.
Prochlorite, Fe,Mg, chlorite.
Amesite, $H_4(Mg,Fe)_2Al_2SiO_9$.
Kossmatite, Hydrous Mg,Al,Fe''', silicate.
Zebedassite, $5MgO.Al_2O_3.6SiO_2.4H_2O$.
Brunsvigite, $9(Fe,Mg)O.2Al_2O_3.6SiO_2.8H_2O$.
Griffithite, $4(Mg,Fe,Ca)O.(Al,Fe)_2O_3.5SiO_2$.
    $7H_2O$.
Spodiophyllite, $(Na_2,K_2)_2(Mg,Fe)_3(Fe,Al)_2$
    $(SiO_3)_8$.
Bardolite, Chlorite-like.
Serpentine, $H_4Mg_3Si_2O_9$.
Deweylite, $4MgO.3SiO_2.6H_2O$.
Genthite, $2NiO.2MgO.3SiO_2.6H_2O$.
Nopouite, $3(Ni,Mg)O.2SiO_2.2H_2O$.
Garnierite, $H_2(Ni,Mg)SiO_4.nH_2O$.
Talc, $H_2Mg_3(SiO_3)_4$.
Sepiolite, $H_4Mg_2Si_3O_{10}$.
Spadaite, $MgSiO_3.H_2O$.
Saponite, Hydrous Mg,Al, silicate.
Iddingsite, $MgO.Fe_2O_3.3SiO_2.4H_2O$.
Celadonite, Fe,Mg,K, silicate.
Pholidolite, $K_2O.12(Fe,Mg)O.Al_2O_3.13SiO_2$.
    $5H_2O$.
Elbrussite, Hydrous Al,Fe,Mg, silicate.
Sklodowskite, $MgO.2UO_3.2SiO_2.7H_2O$.
McGovernite, $21(Mn,Mg,Zn)O.\frac{1}{2}As_2O_3$.
    $As_2O_5.10H_2O$.
Sturtite, $6(Mn,Mg,Ca)O.Fe_2O_3.8SiO_2$.
    $23H_2O$.
Colerainite, $4MgO.Al_2O_3.2SiO_2.5H_2O$.
Tartarkaite, Al,Mg, hydrous silicate.
Berzeliite, $(Ca,Na_2)(Mg, Mn)_2(AsO_4)_3$;
Wagnerite, $(MgF)MgPO_4$.
Adelite, $(MgOH)CaAsO_4$.
Tilasite, $(MgF)CaAsO_4$.
Lazulite, $2AlPO_4.(Fe,Mg)(OH)_2$.

Bialite, Hydrous Mg,Ca,Al, phosphate.
Struvite, Hydrous, NH_4,Mg, phosphate.
Pyrophosphorite, $Mg_2P_2O_7.4(Ca_3P_2O_8.$
  $Ca_2P_2O_7)$.
Roselite, $(Ca,Co,Mg)_3As_2O_8.2H_2O$.
Collinsite, $2CaO.(Mg,Fe)O.P_2O_5.2\frac{1}{2}H_2O$.
Akrochordite, $Mn_4Mg(AsO_5)_2.6H_2O$.
Wenzelite, $(Mn,Fe,Mg)_3(PO_4)_2.5H_2O$.
Bobierrite, $Mg_3P_2O_8.8H_2O$.
Hoernesite, $Mg_3As_2O_8.8H_2O$.
Cabrerite, $(Ni,Mg)_3As_2O_8.8H_2O$.
Newberyite, $HMgPO_4.3H_2O$.
Hannayite, $H_4(NH_4)_2Mg(PO_4)_2.8H_2O$.
Schertelite, $H_2(NH_4)_2Mg(PO_4)_2.4H_2O$.
Gordonite, $MgO.Al_2O_3.P_2O_5.9H_2O$.
Lüneburgite, $3MgO.B_2O_3.P_2O_5.8H_2O$.
Nitromagnesite, $Mg(NO_3)_2.nH_2O$.
Sussexite, $H(Mn,Zn,Mg)BO_3$.
Ludwigite, $3MgO.B_2O_3.FeO.Fe_2O_3$.
Vonsenite, $3(Fe,Mg)O.B_2O_3.FeO.Fe_2O_3$.
Magnesioludwigite, $3MgO.B_2O_3.MgO.Fe_2O_3$.
Pinakiolite, $3MgO.B_2O_3.MnO.Mn_2O_3$.
Szaibelyite, $2Mg_5B_4O_{11}.3H_2O$.
Camsellite, $2MgO.B_2O_3.H_2O$.
BORACITE, $5MgO.MgCl_2.7B_2O_3$.
Fluoborite, $3MgO.3B_2O_3.3Mg(F,OH)_2$.
Ascharite, Hydrous Mg, borate.
Paternoite, $MgB_8O_{13}.4H_2O$.
Warwickite, $(Mg,Fe)_3TiB_2O_8$.
Pinnoite, $MgB_2O_4.3H_2O$.
Kaliborite, Hydrous Mg,K, borate.
Hulsite, $12(Fe,Mg)O.2Fe_2O_3.SnO_2.3B_2O_3.$
  $2H_2O$.
Hydroboracite, $CaMgB_6O_{11}.6H_2O$.
Sulphoborite, $2MgSO_4.4MgHBO_3.7H_2O$.
Langbeinite, $K_2Mg_2(SO_4)_3$.
Vanthoffite, $3Na_2SO_4.MgSO_4$.
Kainite, $MgSO_4.KCl.3H_2O$.
Kieserite, $MgSO_4.H_2O$.
Epsomite, $MgSO_4.7H_2O$.
Cupromagnesite, $(Cu,Mg)SO_4.7H_2O$.
Löweite, $MgSO_4.Na_2SO_4.2\frac{1}{2}H_2O$.
Blödite, $MgSO_4.Na_2SO_4.4H_2O$.
Leonite, $MgSO_4.K_2SO_4.4H_2O$.
Boussingaultite, $(NH_4)_2SO_4.MgSO_4.6H_2O$.
Picromerite, $MgSO_4.K_2SO_4.6H_2O$.
Polyhalite, $2CaSO_4.MgSO_4.K_2SO_4.2H_2O$.
Hexahydrite, $MgSO_4.6H_2O$.
Mooreite, Hydrous Mg,Mn,Zn, sulphate.
Pickeringite, $MgSO_4.Al_2(SO_4)_3.22H_2O$.
Botryogen, $2MgO.Fe_2O_3.4SO_3.15H_2O$.
Quetenite, $MgO.Fe_2O_3.3SO_3.13H_2O$.

MANGANESE

Alabandite, MnS.
Hauerite, MnS_2.
Samsonite, $2Ag_2S.MnS.Sb_2S_3$.
Sacchite, $MnCl_2$.
Chlormanganokalite, $4KCl.MnCl_2$.
Kempite, $MnCl_2.3MnO_2.3H_2O$.
Manganosite, MnO.
Senaite, $(Fe,Mn,Pb)O.TiO_2$.

Pyrophanite, $MnTiO_3$.
Bixbyite, $(Fe,Mn)_2O_3$.
Sitaparite, $9Mn_2O_3.4Fe_2O_3.MnO_2.3CaO$.
Vredenburgite, $3Mn_3O_4.2Fe_2O_3$.
Galaxite, $(Mn,Fe,Mg)O.(Al,Fe)_2O_3$.
FRANKLINITE, $(Fe,Zn,Mn)O.(Fe,Mn)_2O_3$.
Jacobsite, $MgO.Fe_2O_3.(Mn,Fe)O.Fe_2O_3$.
Hausmannite, $MnO.Mn_2O_3$.
Hetærolite, $ZnO.Mn_2O_3$.
Hollandite, Manganate of Mn,Ba, Fe'''.
Crednerite, $CuO.Mn_2O_3$.
Magnetoplumbite, Fe,Mn,Pb oxides.
Zincdibraunite, $ZnO.2MnO_2.2H_2O$.
BRAUNITE, $3Mn_2O_3.MnSiO_3$.
Polianite, MnO_2.
Pyrolusite, MnO_2.
Manganite, $Mn_2O_3.H_2O$.
Pyrochroite, $Mn(OH)_2$.
Bäckströmite, $Mn(OH)_2$.
Chalcophanite, $(Mn,Zn)O.2MnO_2.2H_2O$.
Hydrohetærolite, $2ZnO.2Mn_2O_3.H_2O$.
Quenselite, $2PbO.Mn_2O_3.H_2O$.
Psilomelane, Hydrous Mn manganate.
Wad, Mn oxides.
Skemmatite, $3MnO_2.2Fe_2O_3.6H_2O$.
Beldongrite, $6Mn_3O_5.Fe_2O_3.8H_2O$.
Rhodochrosite, $MnCO_3$.
Loseyite, $(Mn,Zn)CO_3.5(Mn,Zn)(OH)_2$.
Schizolite, $HNa(Ca,Mn)_2(SiO_3)_3$.
Rhodonite, $MnSiO_3$.
Pyroxmangite, Mn,Fe pyroxene.
Babingtonite, $(Ca,Fe,Mn)SiO_3$ with
  $Fe_2(SiO_3)_3$.
Sobralite, $(Mn,Fe,Ca,Mg)SiO_3$.
Låvenite, Zr-silicate of Mn, Ca.
Margarosanite, $Pb(Ca,Mn)_2(SiO_3)_3$.
Helvite, $3(Mn,Fe)BeSiO_4.MnS$.
Danalite, $3(Fe,Zn,Mn)BeSiO_4.ZnS$.
Spessartite, $Mn_3Al_2(SiO_4)_3$.
Partschinite, $(Mn,Fe)_3Al_2Si_3O_{12}$.
Glaucochroite, $CaMnSiO_4$.
Knebelite, $(Fe,Mn)_2SiO_4$.
Tephroite, Mn_2SiO_4.
Alleghanyite, $5MnO.2SiO_2$.
Gosseletite, Mn silicate.
Trimerite, $(Mn,Ca)_2SiO_4.Be_2SiO_4$.
Friedelite, $6MnO.2Mn(OH,Cl)_2.6SiO_2.3H_2O$.
Pyrosmalite, $6(Fe,Mn)O.2(Fe,Mn)$
  $(OH,Cl)_2.6SiO_2.3H_2O$.
Parsettensite, $3MnO.4SiO_2.4H_2O$.
Tinzenite, $2CaO.Al_2O_3.Mn_2O_3.4SiO_2$.
Shallerite, $6MnO.Mn_2(OH)_4.As_2O_3.6SiO_2.$
  $3H_2O$.
Piedmontite, $HCa_2(Al,Mn)_3Si_3O_{13}$.
Sursassite, $5MnO.2Al_2O_3.5SiO_2.3H_2O$.
Hancockite, Pb,Mn,Ca,Al, etc., silicate.
Harstigite, Mn,Ca,Al, silicate.
Leucophœnicite, $Mn_5(MnOH)_2(SiO_4)_3$.
Ardennite, Al,Mn,V, silicate.
Långbanite, Mn silicate with Fe antimo-
  nate.
Kentrolite, $3PbO.2Mn_2O_3.3SiO_2$.
Carpholite, $H_4MnAl_2Si_2O_{10}$.

Bodenbenderite, Silicate and titanate of Al, Y, Mn.

Poechite, $H_{16}Fe_3Mn_2Si_3O_{29}$.

Inesite, $H_2(Mn,Ca)_6Si_6O_{19}.3H_2O$.

Ganophyllite, $7MnO.Al_2O_3.8SiO_2.6H_2O$.

Alurgite, K,Mg,Mn, mica.

Dixenite, $MnSiO_3.2Mn_2(OH)AsO_3$.

McGovernite, $21(Mn,Mg,Zn)O.3SiO_2$. $\frac{1}{2}As_2O_3.As_2O_5.10H_2O$.

Bementite, $H_{10}Mn_8Si_7O_{27}$.

Ectropite, $Mn_2Si_3O_{28}.7H_2O$.

Hodgkinsonite, $3(Zn,Mn)O.SiO_2.H_2O$.

Gageite, Hydrous Mn,Mg,Zn, silicate.

Neotocite, Hydrous Mn, Fe, silicate.

Sérandite, Hydrous Mn,Ca,Na, silicate.

Sturtite, $6(Mn,Ca,Mg)O.Fe_2O_3.8SiO_2$. $23H_2O$.

Astrophyllite, Na,K,Fe,Mn,Ti-silicate.

Neptunite, Fe,Mn,K,Na, titano-silicate.

COLUMBITE-TANTALITE, (Fe,Mn) $(Nb,Ta)_2O_6$.

Tapiolite, $(Fe,Mn)(Nb,Ta)_2O_6$.

Hielmite, Y,Fe,Mn,Ca, stanno-tantalate.

Berzeliite, $(Ca,Na_2)(Mg,Mn)_2(AsO_4)_3$.

Lithiophilite, $Li(Mn,Fe)PO_4$.

Natrophilite, $NaMnPO_4$.

Graftonite, $(Fe,Mn,Ca)_3P_2O_8$.

Triplite, $(RF)RPO_4$; R = Fe,Mn.

Triploidite, $(ROH)RPO_4$; R = Mn,Fe.

Sarkinite, $(MnOH)MnAsO_4$.

Arsenoclasite, $2Mn(OH)_2.Mn_3(AsO_4)_2$.

Arrojadite, Fe,Mn, phosphate.

Sarcopside, Fe,Mn,Ca, phosphate.

Trigonite, $Pb_3MnH(AsO_3)_3$.

Lacroixite, $Na_4(Ca,Mn)_4Al_3(F,OH)_4P_3O_{10}$. $2H_2O$.

Pyrobelonite, $4(Mn,Pb)O.V_2O_5.H_2O$.

Brackebushite, $(Pb,Mn,Fe)_3V_2O_8.H_2O$.

Allactite, $Mn_3As_2O_8.4Mn(OH)_2$.

Synadelphite, $2(Al,Mn)AsO_4.5Mn(OH)_2$.

Allodelphite, $5MnO.2(Mn,Al)_2O_3.As_2O_3$. $SiO_2.5H_2O$.

Flinkite, $MnAsO_4.2Mn(OH)_2$.

Hematolite, $(Al,Mn)AsO_4.4Mn(OH)_2$.

Retzian, Y,Mn,Ca, phosphate.

Arseniopleite, $(Mn,Ca)_9(Mn,Fe)_2(OH)_6$ $(AsO_4)_6$.

Manganostibiite, $Mn_{10}Sb_2O_{15}$.

Dickinsonite ⎱ Hydrous Mn,Fe,Na,
Fillowite ⎰ phosphates.

Brandite, $Ca_2MnAs_2O_8.2H_2O$.

Fairfieldite, $Ca_3MnP_2O_8.2H_2O$.

Akrochordite, $Mn_4Mg(AsO_5)_2.6H_2O$.

Wenzelite, $(Mn,Fe,Mg)_3(PO_4)_2.5H_2O$.

Reddingite, $Mn_3P_2O_8.3H_2O$.

Palaite, $5MnO.2P_2O_5.4H_2O$.

Stewartite, $3MnO.P_2O_5.4H_2O$.

Purpurite, $2(Fe,Mn)PO_4.H_2O$.

Sicklerite, $Fe_2O_3.6MnO.4P_2O_5.3(Li,H)_2O$.

Salmonsite, $Fe_2O_3.9MnO.4P_2O_5.14H_2O$.

Landesite, $3Fe_2O_3.20MnO.8P_2O_5.27H_2O$.

Hureaulite, $H_2Mn_5(PO_4)_4.4H_2O$.

Hemafibrite, $Mn_3As_2O_8.3Mn(OH)_2.2H_2O$.

Chlorophœnicite, $10(Zn,Mn)O.As_2O_5 7H_2O$.

Holdenite, $8MnO.4ZnO.As_2O_5.5H_2O$.

Eosphorite, $2AlPO_4.2(Mn,Fe)(OH)_2.2H_2O$.

Roschérite, $(Mn,Fe,Ca)_2Al(OH)(PO_4)_2$. $2H_2O$.

Catoptrite, $14(Mn,Fe)O.2(Al,Fe)_2O_3.2SiO_2$. Sb_2O_5.

Seamanite, Hydrous Mn borophosphate.

Sussexite, $H(Mn,Zn,Mg)BO_3$.

Pinakiolite, $3MgO.B_2O_3.MnO.Mn_2O_3$.

Manganolangbeinite, $K_2Mn_2(SO_4)_3$.

Szmikite, $MnSO_4.H_2O$.

Ilesite, $(Mn,Zn,Fe)SO_4.4H_2O$.

Mallardite, $MnSO_4.7H_2O$.

Mooreite, Hydrous Mg,Mn,Zn, sulphate.

Apjohnite, $MnSO_4.Al_2(SO_4)_3.24H_2O$.

Dietrichite, $(Zn,Fe,Mn)SO_4.Al_2(SO_4)_3$. $22H_2O$.

MERCURY

Native Mercury, Hg.

Amalgam, (Ag,Hg).

Metacinnabarite, HgS.

Tiemannite, HgSe.

Onofrite, Hg(S,Se).

Coloradoite, HgTe.

Cinnabar, HgS.

Livingstonite, $HgS.2Sb_2S_3$.

Calomel, HgCl.

Kleinite, Hg,NH_4, chloride.

Eglestonite, Hg_4Cl_2O.

Terlinguaite, HgClO.

Mosesite, Hydrous Hg,NH_4, chloride.

Montroydite, HgO.

Ammiolite, Hg antimonite.

MOLYBDENUM

Molybdenite, MoS_2.

Molybdite, MoO_3.

Powellite, $Ca(Mo,W)O_4$.

Chillagite, $3PbWO_4.PbMoO_4$.

Wulfenite, $PbMoO_4$.

Ferrimolybdite, $Fe_2O_3.3MoO_3.8H_2O$.

Koechlinite, $Bi_2O_3.MoO_3$.

NICKEL

Awaruite, $FeNi_2$.

Josephinite, $FeNi_3$.

Maucherite, Ni_3As_2.

Penroseite, $3CuSe.2PbSe_2.5(Ni,Co)Se_2$.

PENTLANDITE, (Fe,Ni)S.

Millerite, NiS.

Beyrichite, NiS.

Hauchecornite, Ni(Bi,Sb,S)?

Niccolite, NiAs.

Breithauptite, NiSb.

Polydymite, Ni_3S_4.

Violarite, $(Ni,Fe)_3S_4$.

Badenite, $(Co,Ni,Fe)_3(As,Bi)_4$.

Bravoite, $(Fe,Ni)S_2$.

Cobaltnickelpyrite, $(Co,Ni,Fe)S_2$.
CHLOANTHITE, $NiAs_2$.
Gersdorffite, $NiAsS$.
Ullmanite, $NiSbS$.
Rammelsbergite, $NiAs_2$.
Wolfachite, $Ni(As,Sb)S$.
Melonite, $NiTe_2$.
Bunsenite, NiO.
Trevorite, $NiO.Fe_2O_3$.
Zaratite, $NiCO_3.2Ni(OH)_2.4H_2O$.
Genthite, $2NiO.2MgO.3SiO_2.6H_2O$.
Nepouite, $3(Ni,Mg)O.2SiO_2.2H_2O$.
Garnierite, $H_2(Ni,Mg)SiO_4.nH_2O$.
Maufite, Ni,Al, silicate.
Connarite, $H_4Ni_2Si_3O_{10}$.
Annabergite, $Ni_3As_2O_8.8H_2O$.
Cabrerite, $(Ni,Mg)_3As_2O_8.8H_2O$.
Kolovratite, Ni vanadate.
Forbesite, $H_2(Ni,Co)_2As_2O_8.8H_2O$.
Lindackerite, $3NiO.6CuO.SO_3.2As_2O_5.7H_2O$.
Morenosite, $NiSO_4.7H_2O$.

PLATINUM

Native Platinum, Pt.
Sperrylite, $PtAs_2$.
Cooperite, $Pt(As,S)_2$.

POTASSIUM

SYLVITE, KCl.
Chlormanganokalite, $4KCl.MnCl_2$.
Rinneite, $FeCl_2.3KCl.NaCl$.
Avogadrite, $(K,Cs)BF_4$.
Hieratite, K_2SiF_6.
CARNALLITE, $KCl.MgCl_2.6H_2O$.
Kremersite, $KCl,NH_4Cl.FeCl_2.H_2O$.
Erythrosiderite, $2KCl.FeCl_3.H_2O$.
Mitscherlichite, $2KCl.CuCl_2.2H_2O$.
Milarite, $K_2O.4CaO.4BeO.Al_2O_3.24SiO_2.H_2O$.
Orthoclase, Microcline, $KAlSi_3O_8$.
Hyalophane, A Ba-bearing feldspar.
Anorthoclase, $(Na,K)AlSi_3O_8$.
Leucite, $KAl(SiO_3)_2$.
Kaliophilite, $KAlSiO_4$.
Apophyllite, $KFCa_4(Si_2O_5)_4.8H_2O$.
Ptilolite, $(Ca,K_2,Na_2)Al_2Si_{10}O_{24}.5H_2O$.
Wellsite, $(Ba,Ca,K_2)Al_2Si_3O_{10}.3H_2O$.
Phillipsite, $(K_2,Ca)Al_2Si_4O_{12}.4\frac{1}{2}H_2O$.
Harmotome, $(K_2,Ba)Al_2Si_5O_{14}.5H_2O$.
Gismondite, $(Ca,K_2)Al_2Si_2O_8.4H_2O$.
Offretite, Potash zeolite.
Dachiardite, $(Na_2,K_2,Ca)_3Al_4(Si_2O_5)_9.$
$14H_2O$.
Muscovite, $H_2KAl_3(SiO_4)_3$.
Tæniolite, K, Mg, silicate.
Alurgite, K,Mg,Mn mica.
Spodiophyllite, $(Na_2K_2)_2(Mg,Fe)_3(Fe,Al)_2$
$(SiO_3)_8$.
Celadonite, Fe,Mg,K, silicate.
Glauconite, Hydrous Fe,K, silicate.
Astrophyllite, Na,K,Mn,Fe, titano-silicate.
Palmerite, $HK_2Al_2(PO_4)_3.7H_2O$.

Englishite, $4CaO.K_2O.4Al_2O_3.4P_2O_5.14H_2O$.
Lehiite, $5CaO.(Na,K)_2O.4Al_2O_3.4P_2O_5.$
$12H_2O$.
Lewistonite, $15CaO.(K,Na)_2O.4P_2O_5.8H_2O$.
Carnotite, $K_2O.2U_2O_3.V_2O_5.3H_2O$.
Niter, KNO_3.
Rhodizite, Al,K, borate.
Kaliborite, Hydrous Mg,K, borate.
Taylorite, $5K_2SO_4.(NH_4)_2SO_4$.
Aphthitalite, $(K,Na)_2SO_4$.
Langbeinite, $K_2Mg_2(SO_4)_3$.
Manganolangbeinite, $K_2Mn_2(SO_4)_3$.
Tarapacaite, K_2CrO_4.
Kainite, $MgSO_4.KCl.3H_2O$.
Hanksite, $9Na_2SO_4.2Na_2CO_3.KCl$.
Misenite, $HKSO_4$.
Lecontite, $(Na,NH_4,K)SO_4.2H_2O$.
Syngenite, $CaSO_4.K_2SO_4.H_2O$.
Leonite, $MgSO_4.K_2SO_4.4H_2O$.
Picromerite, $MgSO_4.K_2SO_4.6H_2O$.
Polyhalite, $2CaSO_4.MgSO_4.K_2SO_4.2H_2O$.
Potash Alum $\left.\begin{matrix} \\ \end{matrix}\right\}$ $KAl(SO_4)_2.12H_2O$.
Kalinite
Voltaite, $3(K_2,Fe)O.2(Al,Fe)_2O_3.6SO_3.9H_2O$.
Metavoltine, $(H,K)_4(Fe'''OH)_2(SO_4)_4.5H_2O$.
Slavikite, $(Na,K)_2SO_4.Fe_{10}(OH)_6(SO_4)_{12}.$
$63H_2O$.
ALUNITE, $K_2Al_6(OH)_{12}(SO_4)_4$.
Jarosite, $K_2Fe_6(OH)_{12}(SO_4)_4$.
Palmierite, $(K,Na)_2Pb(SO_4)_2$.
Löwigite, $K_2O.3Al_2O_3.4SO_3.9H_2O$.

SILVER

Native Silver, Ag.
Amalgam, (Ag,Hg).
Dyscrasite, Ag_3Sb.
Cocinerite, Cu_4AgS.
Stützite, Ag_4Te.
Naumannite, $(Ag_2,Pb)Se$.
Argentite, Ag_2S.
Hessite, Ag_2Te.
Petzite, $(Ag,Au)_2Te$.
Aguilarite, $Ag_2(S,Se)$.
Eucairite, $Cu_2Se.Ag_2Se$.
Crookesite, $(Cu,Tl,Ag)_2Se$.
Stromeyrite, $(Ag,Cu)_2S$.
Acanthite, Ag_2S.
Sternbergite, $Ag_2S.Fe_4S_5$.
Sylvanite, $(Au,Ag)Te_2$.
Krennerite, $(Au,Ag)Te_2$.
Muthmannite, $(Ag,Au)Te$.
Trechmannite, $Ag_2S.As_2S_3$.
Andorite, $2PbS.Ag_2S.3Sb_2S_3$.
Hutchinsonite, $PbS.(Tl,Ag)_2S.2Sb_2S_3$.
Miargyrite, $Ag_2S.Sb_2S_3$.
Smithite, $Ag_2S.As_2S_3$.
Matildite, $Ag_2S.Bi_2S_3$.
Aramayoite, $Ag_2S.(Sb,Bi)_2S_3$.
Fizelyite, $5PbS.Ag_2S.4Sb_2S_3$.
Ramdohrite, $3PbS.Ag_2S.3Sb_2S_3$.
Schirmerite, $3(Ag_2,Pb)S.2Bi_2S_3$.
Benjaminite $(Cu,Ag)_2S.2PbS.2Bi_2S_3$.

Owyheeite, $8PbS.2Ag_2S.5Sb_2S_3$.
Schapbachite, $PbS.Ag_2S.Bi_2S_3$.
Freieslebenite, $5(Pb,Ag_2)S.2Sb_2S_3$.
Diaphorite, $5(Pb,Ag_2)S.2Sb_2S_3$.
Pyrargyrite, $3Ag_2S.Sb_2S_3$.
Proustite, $3Ag_2S.As_2S_3$.
Pyrostilpnite, $3Ag_2S.Sb_2S_3$.
Samsonite, $2Ag_2S.MnS.Sb_2S_3$.
Xanthoconite, $3Ag_2S.As_2S_3$.
Sanguinite, $3Ag_2S.As_2S_3$.
Tapalpite, $3Ag_2(S,Te).Bi_2(S,Te)_3$.
Lengenbachite, $6PbS.Ag_2S.2As_2S_3$.
Stephanite $5Ag_2S.Sb_2S_3$.   =
Polybasite, $9Ag_2S.Sb_2S_3$.
Pearceite, $9Ag_2S.As_2S_3$.
Polyargyrite, $11Ag_2S.Sb_2S_3$.
Ultrabasite, $28PbS.11Ag_2S.3GeS_2.2Sb_2S_3$.
Argyrodite, $4Ag_2S.GeS_2$.
Canfieldite, $4Ag_2S.SnS_2$.
Cerargyrite, $AgCl$.
Embolite, $Ag(Br,Cl)$.
Bromyrite, $AgBr$.
Iodobromite, $2AgCl.2AgBr.AgI$.
Miersite, $4AgI.CuI$.
Iodyrite, AgI.
Argentojarosite, $Ag_2Fe_6(OH)_{12}(SO_4)_4$.

SODIUM

Halite, $NaCl$.
Villiaumite, NaF.
Huantajayite, $20NaCl.AgCl$.
Rinneite, $FeCl_3.3KCl.NaCl$.
CRYOLITE, Na_3AlF_6.
Cryolithionite, $3NaF.3LiF.2AlF_3$.
Chiolite, $5NaF.3AlF_3$.
Mallardite, Na_2SiF_6.
Ralstonite, $(Na_2Mg)F_2.3Al(F,OH)_3.2H_2O$.
Northupite, $MgCO_3.Na_2CO_3.NaCl$.
Tychite, $2MgCO_3.2Na_2CO_3.Na_2SO_4$.
Dawsonite, $Na_2Al(CO_3)_3.2Al(OH)_3$.
Thermonatrite, $Na_2CO_3.H_2O$.
Natron, $Na_2CO_3.10H_2O$.
Pirssonite, $CaCO_3.Na_2CO_3.2H_2O$.
Gay-Lussite, $CaCO_3.Na_2CO_3.5H_2O$.
Trona, $Na_2CO_3.HNaCO_3.2H_2O$.
Nahcolite, $NaHCO_3$.
Eudidymite, Epididymite, $HNaBeSi_3O_8$.
Leifite, $Na_4(AlF)_2Si_9O_{22}$.
Anorthoclase, $(Na,K)AlSi_3O_8$.
Albite, $NaAlSi_3O_8$.
Oligoclase ⎫
Andesine ⎬ Mixtures of $NaAlSi_3O_8$ and $CaAl_2Si_2O_8$.
Labradorite ⎭
Anemousite, $NaO.2CaO.3Al_2O_3.9SiO_2$.
Ussingite, $HNa_2Al(SiO_3)_3$.
ACMITE, $NaFe(SiO_3)_2$.
JADEITE, $NaAl(SiO_3)_2$.
PECTOLITE, $HNaCa_2(SiO_3)_3$.
Schizolite, $HNa(Ca,Mn)_2(SiO_3)_3$.
Rosenbuschite, near pectolite with Zr.
Wöhlerite, Zr-silicate and niobate of Ca,Na.
Guarinite, $(Na_2,Ca)(Si,Zr)O_3$.

GLAUCOPHANE, $NaAl(SiO_3)_2.(Fe,Mg)SiO_3$.
Torendrikite, $Na_2O.4MgO.CaO.FeO.Fe_2O_3.$ $10SiO_2$.
RIEBECKITE, $2NaFe(SiO_3)_2.FeSiO_3$.
CROCIDOLITE, $NaFe(SiO_3)_2.FeSiO_3$.
Arfvedsonite, Na,Ca,Fe'' silicate.
Ænigmatite, Fe,Na,Ti-silicate.
Weinbergerite, $NaAlSiO_4.3FeSiO_3$.
Elpidite, $H_6Na_2Zr(SiO_3)_6$.
Catapleiite, $H_4(Na_2,Ca)ZrSi_3O_{11}$.
Leucophanite ⎫
Meliphanite ⎬ Na,Be,Ca, fluo-silicates.
Nephelite, $NaAlSiO_4$.
CANCRINITE, $H_6Na_6Ca(NaCO_3)_2Al_8(SiO_4)_6$.
Microsommite, Davyne, near cancrinite.
SODALITE, $3Na_2Al_2Si_2O_8.2NaCl$.
Hackmanite, near sodalite.
Ameletite, Silicate of Na,Al,Cl.
HAUYNITE, $3Na_2Al_2Si_2O_8.2CaSO_4$.
Noselite, $3Na_2Al_2Si_2O_8.Na_2SO_4$.
LAZURITE, $3Na_2Al_2Si_2O_8.2Na_2S$.
SCAPOLITE GROUP, Mixtures of $CaCO_3.3CaAl_2Si_2O_8$ and $NaCl.3NaAlSi_3O_8$.
Sarcolite, $(Ca,Na_2)_3Al_2(SiO_4)_6$.
Melilite, $Na_2(Ca,Mg)_{11}(Al,Fe)_4(SiO_4)_9$.
Flokite, $(Ca,Na_2)O.Al_2O_3.9SiO_2.6H_2O$.
Mordenite, $(Ca,Na_2)O.Al_2O_3.9SiO_2.6H_2O$.
Ferrierite, Hydrous Mg,Na, silicate.
Heulandite, $(Ca,Na_2)O.Al_2O_3.6SiO_2.5H_2O$.
Epistilbite, $(Ca,Na_2)O.Al_2O_3.6SiO_2.5H_2O$.
Stilbite, $(Na_2,Ca)Al_2Si_6O_{16}.6H_2O$.
Epidesmine, Same as stilbite.
Laumontite, $(Ca,Na_2)Al_2Si_4O_{12}.4H_2O$.
CHABAZITE, $(Ca,Na_2)Al_2Si_4O_{12}.6H_2O$.
Gmelinite, $(Na_2,Ca)Al_2Si_4O_{12}.6H_2O$.
Arduinite, Ca,Na, zeolite.
Analcite, $NaAlSi_2O_6.H_2O$.
Faujasite, $Na_2CaAl_4Si_{10}O_{28}.20H_2O$.
Natrolite, $Na_2Al_2Si_3O_{10}.2H_2O$.
Mesolite, $Na_2Al_2Si_3O_{10}.2H_2O + 2[CaAl_2Si_3 O_{10}.3H_2O]$.
Gonnardite, $(Ca,Na_2)_2Al_2Si_3O_{10}.5\frac{1}{2}H_2O$.
Thomsonite, $(Na_2,Ca)Al_2Si_2O_8.2\frac{1}{2}H_2O$.
Hydrothomsonite, $(H_2,Na_2,Ca)Al_2Si_2O_8.$ $5H_2O$.
Echellite, $(Ca,Na_2)O.2Al_2O_3.3SiO_2.4H_2O$.
Erionite, $H_2CaK_2Na_2Al_5Si_6O_{17}.5H_2O$.
Hydronephelite, $HNa_2Al_3Si_3O_{12}.3H_2O$.
Dachiardite, $(Na_2,K_2,Ca)_3Al_4(Si_2O_5)_9.14H_2O$.
Paragonite, $H_2NaAl_3(SiO_4)_3$.
Spodiophyllite, $(Na_2,K_2)_2(Mg,Fe)_3(Fe,Al)_2$ $(SiO_3)_8$.
Sérandite, Hydrous Mn,Ca,Na, silicate.
Searlesite, $NaB(SiO_3)_2.H_2O$.
Molengraafite, Ca,Na, titano-silicate.
Astrophyllite, Na,K,Mn,Fe, titano-silicate.
Fersmannite, $2Na_2(O,F_2).4CaO.4TiO_2.3SiO_2$.
Rinkolite, Ce,Ca,Sr,Na, titano-silicate.
Narsarsukite, Fe,Na, titano-silicate.
Leucosphenite, $Na_4Ba(TiO)_2(Si_2O_5)_5$.
Lorenzenite, $Na_2(TiO)_2Si_2O_7$.
Loparite, Ce,Ca,Na, titanate.
Chalcolamprite, $Na_4(CaF_2)Nb_2SiO_9$.

Epistolite, Ti,Na, etc., niobate.
Berzeliite, $(Ca,Na_2)(Mg,Mn)_2(AsO_4)_3$.
Natrophilite, $NaMnPO_4$.
Merrillite, $3CaO.Na_2O.P_2O_5$.
Dehrnite, $6CaO.(Na_7K)_2O.2P_2O_5.H_2O$.
Beryllonite, $NaBePO_4$.
Ježekite, $Na_4CaAl(AlO)(F,OH)_4(PO_4)_2$.
Lacroixite, $Na_4(Ca,Mn)_4Al_3(F,OH)_4P_3O_{16}$.
  $2H_2O$.
Durangite, $Na(AlF)AsO_4$.
Fremontite, $(Na,Li)Al(OH,F)PO_4$.
Dickinsonite ⎫
Fillowite   ⎬ $3(Mn,Fe,Na_2)_3P_2O_8.H_2O$.
Stercorite, $HNa(NH_4)PO_4.4H_2O$.
Freirinite, $6(Cu,Ca)O.3Na_2O.2As_2O_5.6H_2O$.
Millisite, $2CaO.Na_2O.6Al_2O_3.4P_2O_5.17H_2O$.
Lehiite, $5CaO.(Na,K)_2O.4Al_2O_3.4P_2O_5$.
  $12H_2O$.
Lewistonite, $15CaO.(K,Na)_2O.4P_2O_5.8H_2O$.
Weslienite, $Na_2FeCa_3Sb_4O_{15}$.
Swedenborgite, $Na_2O.2Al_2O_3.Sb_2O_5$.
SODA NITER, $NaNO_3$.
Darapskite, $NaNO_3.Na_2SO_4.H_2O$.
Nitroglauberite, $6NaNO_3.2Na_2SO_4.3H_2O$.
Borax, $Na_2B_4O_7.10H_2O$.
Tincalconite, $Na_2O.2B_2O_3.5H_2O$.
Kernite, $Na_2O.2B_2O_3.4H_2O$.
Ulexite, $NaCaB_5O_9.8H_2O$.
Probertite, $Na_2O.2CaO.5B_2O_3.10H_2O$.
Clarkeite, $(Na_2,Pb)O.3UO_3.3H_2O$.
Thenardite, Na_2SO_4.
Aphthitalite, $(K,Na)_2SO_4$.
GLAUBERITE, $Na_2SO_4.CaSO_4$.
Vanthoffite, $3Na_2SO_4.MgSO_4$.
Schairerite, $Na_2SO_4.Na(F,Cl)$.
Sulphohalite, $3Na_2SO_4.NaCl.NaF$.
Caracolite, $Pb(OH)Cl.Na_2SO_4$.
Hanksite, $9Na_2SO_4.2Na_2CO_3.KCl$.
Lecontite, $(Na,NH_4,K)_2SO_4.2H_2O$.
Mirabilite, $Na_2SO_4.10H_2O$.
Löweite, $MgSO_4.Na_2SO_4.2\frac{1}{2}H_2O$.
Blödite, $MgSO_4.Na_2SO_4.4H_2O$.
Soda Alum ⎫
Mendozite ⎬ $NaAl(SO_4)_2.12H_2O$.
Kröhnkite, $CuSO_4.Na_2SO_4.2H_2O$.
Natrochalcite, $Cu_4(OH)_2(SO_4)_2.Na_2SO_4$.
  $2H_2O$.
Ferrinatrite, $3Na_2SO_4.Fe_2(SO_4)_3.6H_2O$.
Sideronatrite, $2Na_2O.Fe_2O_3.4SO_3.7H_2O$.
Metavoltine, $(H,K)_4(Fe'''OH)_2(SO_4)_4.5H_2O$.
Slavikite, $(Na,K)_2SO_4.Fe_{10}(OH)_6(SO_4)_{12}$.
  $63H_2O$.
Natrojarosite, $Na_2Fe_6(OH)_{12}(SO_4)_4$.
Palmierite, $(K,Na)_2Pb(SO_4)_2$.
Almeriite, $Na_2SO_4.Al_2(SO_4)_3.5Al(OH)_3.H_2O$.

STRONTIUM

Strontianite, $SrCO_3$.
Ancylite, $4Ce(OH)CO_3.3SrCO_3.3H_2O$.
Ambatoarinite, Rare earths, Sr, carbonate.
Brewsterite, $(Sr,Ba,Ca)O.Al_2O_3.6SiO_2.5H_2O$.
Rinkolite, Ce,Ca,Sr,Na, titano-silicate.

Fermorite, $(Ca,Sr)_4[Ca(OH,F)][(P,As)O_4]_3$.
Goyazite, Sr,Al, phosphate.
Harttite, Sr,Al, phosphate and sulphate.
Tikhvinite, $2SrO.3Al_2O_3.P_2O_5.SO_3.6H_2O$.
Celestite, $SrSO_4$.

THALLIUM

Vrbaite, $Tl_2S.3(As,Sb)_2S_3$.
Lorandite, $Tl_2S.As_2S_3$.
Hutchinsonite, $PbS.(Tl,Ag)_2S.2As_2S_3$.

THORIUM

Thorotungstite, Th and W oxides.
Caryocerite ⎫
Tritomite   ⎬ Ca,Ca,Y,Th, fluo-silicates.
Thorite, $ThSiO_4$.
Thorogummite, Silicate of U, Th, Ce.
Hydrothorite, Hydrous Th silicate.
Auerlite, Th silico-phosphate.
Yttrialite, Th, Y, silicate.
Mackintoshite, U, Th, Ce, silicate.
Zirkelite, $(Ca,Fe)(Zr,Ti,Th)_2O_5$.
Yttrocrasite, Hydrous Y, Th, titanate.
Pyrochlore, $RNb_2O_6.R(Ti,Th)O_3$.
MONAZITE, $(Ce,La,Di)PO_4$ with ThO_2.
Thorianite, Th and U oxides.

TIN

Stannite, $Cu_2S.FeS.SnS_2$.
Canfieldite, $4Ag_2S.SnS_2$.
Teallite, $PbSnS_2$.
Franckeite, $Pb_5Sn_3FeSb_2S_{14}$.
Cylindrite, $Pb_3Sn_4FeSb_2S_{14}$.
Cassiterite, SnO_2.
Stokesite, $H_4CaSnSi_3O_{11}$.
Arandisite, Basic Sn silicate.
Hielmite, Y,Fe,Mn,Ca, stanno-niobate.
Nordenskiöldine, $CaSn(BO_3)_2$.
Hulsite, $12(Fe,Mg)O.2Fe_2O_3.1SnO_2.3B_2O_3$
  $2H_2O$.

TITANIUM

ILMENITE, $FeTiO_3$.
Senaite, $(Fe,Mn,Pb)O.TiO_2$.
Högbomite, $Al_2O_3.Fe_2O_3.MgO.TiO_2$.
Geikielite, $(Mg,Fe)O.TiO_2$.
Pyrophanite, $MnTiO_3$.
Rutile, TiO_2.
Octahedrite, Brookite, TiO_2.
Schlorlomite, $Ca_3(Fe,Ti)_2((Si,Ti)O_4)_3$.
Titanite, $CaTiSiO_5$.
Molengraafite, Ca,Na, titano-silicate.
Fersmannite, $2Na_2(O,F_2).4CaO.4TiO_2.3SiO_2$
Tscheffkinite, Ce, etc., titano-silicate.
Astrophyllite, Na,K,Fe,Mn, titano-silicate
Johnstrupite ⎫
Mosandrite   ⎬ Ce, etc., titano-silicates.
Rinkite   ⎟
Rinkolite   ⎭

Narsarsukite, Fe,Na, titano-silicate.
Neptunite, Fe,Mn,Na,K, titano-silicate.
Benitoite, $BaTiSi_3O_9$.
Leucosphenite, $Na_4Ba(TiO)_2(Si_2O_5)_5$.
Lorenzenite, $Na_2(TiO)_2Si_2O_7$.
Joaquinite, Ca,Fe, titano-silicate.
PEROVSKITE, $CaTiO_3$.
Knopite, Ca,Ce, titanate.
Dysanalyte, Ca,Fe, titano-silicate.
Zirkelite, $(Ca,Fe)(Zr,Ti,Th)_2O_5$.
Uhligite, $Ca(Zr,Ti)_2O_5$ with Al_2TiO_5.
Oliveiraite, $3ZrO_2.2TiO_2.2H_2O$.
Delorenzite, Fe,U,Y, titanate.
Yttrocrasite, Hydrous Y,Th, titanate.
Arizonite, $Fe_2O_3.3TiO_2$.
Pseudobrookite, Fe_2TiO_5.
Kalkowskite, $(Fe,Ce)_2O_3.4(Ti,Si)O_2$.
Brannerite, $(UO,TiO,UO_2)'TiO_3$.
Pyrochlore, $RNb_2O_6.R(Ti,Th)O_3$.
Æschynite, Ce, niobate-titanate.
Polymignite, Ce,Fe,Ca, niobate-titanate.
Euxenite ⎫
Polycrase ⎬ Y,Ce,U, niobate-
Blomstrandine-Priorite ⎭　titanates.
Betafite, U, etc., niobate-titanate.
Epistolite, Na,Ti, etc., niobate.
Lewisite, $5CaO.2TiO_2.3Sb_2O_5$.
Mauzeliite, Pb,Ca, titano-antimonate.
Warwickite, $(Mg,Fe)_3TiB_2O_8$.

TUNGSTEN

Tungstenite, WS_2.
Tungstite, WO_3.
Thorotungstite, W and Th oxides.
WOLFRAMITE, $(Fe,Mn)WO_4$.
SCHEELITE, $CaWO_4$.
Cuprotungstite, $CuWO_4$.
Powellite, $Ca(Mo,W)O_4$.
Raspite ⎫
Stolzite ⎬ $PbWO_4$.
Chillagite, $3PbWO_4.PbMoO_4$.
Ferritungstite, $Fe_2O_3.WO_3.6H_2O$.

URANIUM

Becquerelite, $UO_3.2H_2O$.
Ianthinite, $2UO_2.7H_2O$.
Rutherfordine, $UO_2.CO_3$.
Uranothallite, $2CaCO_3.U(CO_3)_2.10H_2O$.
Voglite, Hydrous U,Ca,Cu, carbonate.
Mackintoshite, U,Th,Ce, silicate.
Thorogummite, Silicate of U,Th,Ce.
Uranophane, $CaO.2UO_3.2SiO_2.7H_2O$.
Sklodowskite, $MgO.2UO_3.2SiO_2.7H_2O$.
Kasolite, $PbO.UO_3.SiO_2.H_2O$.
Soddyite, $5UO_3.2SiO_2.6H_2O$.
Delorenzite, Fe,U,Y, titanate.
Brannerite, $(UO,TiO,UO_2)TiO_3$.
Hatchettolite, U, tantalo-niobate.
Ishikawaite, U niobate.
Samirésite, U, etc., niobate.
Fergusonite, Y,Er,U, niobate.

Samarskite, Fe,Ca,U,Ce,Y, niobate.
Ampangabéite, U, etc., niobate.
Euxenite ⎫ Y,Ce,U, niobate-
Polycrase ⎬ 　titanates.
Blomstrandine-Priorite ⎭
Betafite, U, niobate-titanate.
Pisekite, U,Ce, etc., niobate and titanate.
Mendelyeevite, Ca,U, niobate and titanate.
Plumboniobite, Y,U,Pb, niobate.
Torbernite, $Cu(UO_2)_2P_2O_8.8H_2O$.
Zeunerite, $Cu(UO_2)_2As_2O_8.8H_2O$.
Autunite ⎫ $Ca(UO_2)_2P_2O_8.8H_2O$.
Bassetite ⎬
Uranospinite, $Ca(UO_2)_2As_2O_8.8H_2O$.
Uranocircite, $Ba(UO_2)_2P_2O_8.8H_2O$.
Uranospathite, Hydrous uranyl phosphate.
CARNOTITE, $K_2O.2U_2O.V_2O_5.3H_2O$.
Tyuyamunite, $CaO.UO_3.V_2O_5.nH_2O$.
Rauvite, $CaO.2UO_3.6V_2O_5.20H_2O$.
Renardite, $PbO.4UO_3.P_2O_5.9H_2O$.
Dewindtite, $3PbO.5UO_3.2P_2O_5.12H_2O$.
Dumontite, $2PbO.3UO_3.P_2O_5.5H_2O$.
Parsonsite, $2PbO.UO_3.P_2O_5.H_2O$.
Phosphuranylite, $(UO_2)_3P_2O_8.6H_2O$.
Trögerite, $(UO_2)_3As_2O_8.12H_2O$.
Uvanite, $2UO_3.3V_2O_5.15H_2O$.
Ferghanite, $U_3(VO_4)_2.6H_2O$.
Walpurgite, $Bi_{10}(UO_2)_3(OH)_{24}(AsO_4)_4$.
URANINITE, Uranyl, etc., uranate.
Clarkeite, $(Na_2,Pb)O.3UO_3.3H_2O$.
Gummite, alteration of uraninite.
Thorianite, Th and U oxides.
Curite, $2PbO.5UO_3.4H_2O$.
Fourmarierite, Hydrous Pb uranate.
Uranosphærite, $Bi_2O_3.2UO_3.3H_2O$.
Johannite, Hydrous Cu,U sulphate.
Uranopilite, $CaO.8UO_3.2SO_3.25H_2O$.
Zippeite, Hydrous U sulphate.

VANADIUM

PATRONITE, VS_4.
Sulvanite, $3Cu_2S.V_2S_5$.
Alaïte, $V_2O_5.H_2O$.
Vanoxite, $2V_2O_4.V_2O_5.8H_2O$.
Ardennite, Al,Mn,V, silicate.
Roscoelite, Vanadium mica.
Puchcrite, $BiVO_4$.
Vanadinite, $Pb_4(PbCl)(VO_4)_3$.
Descloizite, $(Pb,Zn)_2(OH)VO_4$.
Pyrobelonite, $4PbO.7MnO.2V_2O_5.3H_2O$.
Dechenite, PbV_2O_6.
Calciovolborthite, $(Cu,Ca)_3V_2O_8.(Cu,Ca)(OH)_2$.
Turanite, $5CuO.V_2O_5.2H_2O$.
Usbekite, $3CuO.V_2O_5.3H_2O$.
Uvanite, $2UO_3.3V_2O_5.15H_2O$.
Ferghanite, $U_3(VO_4)_2.6H_2O$.
Hewettite ⎫ $CaO.3V_2O_5.9H_2O$.
Metahewettite ⎬
Fervanite, $2Fe_2O_3.2V_2O_5.5H_2O$.
Rossite, $CaO.V_2O_5.4H_2O$.
Metarossite, $CaO.V_2O_5.2H_2O$.

Fernandinite, $CaO.V_2O_4.5V_2O_5.14H_2O$.
Melanovanadite, $2CaO.3V_2O_5.2V_2O_4$.
Pascoite, $2CaO.3V_2O_5.11H_2O$.
Pintadoite, $2CaO.V_2O_5.9H_2O$.
Volborthite, Hydrous Cu,Ba,Ca, vanadate.
Tangeite, $2CaO.2CuO.V_2O_5.H_2O$.
Hügelite, Hydrous Pb,Zn, vanadate.
CARNOTITE, $K_2O.2U_2O_3.V_2O_5.3H_2O$.
Tyuyamunite, $CaO.UO_3.V_2O_5.nH_2O$.
Rauvite, $CaO.2UO_3.6V_2O_5.20H_2O$.
Sincosite, $CaO.V_2O_4.P_2O_5.5H_2O$.
Minasragrite, $V_2O_4.3SO_3.16H_2O$.

YTTRIUM, Etc.

Yttrofluorite, $(Ca_3,Y_2)F_6$.
Yttrocerite, $(Y,Er,Ce)F_3.5CaF_2.H_2O$.
Tengerite, Y carbonate.
Cappelenite, Y,Ba, boro-silicate.
Melanocerite ⎫
Caryocerite ⎬ Ca,Y,Ce, fluo-silicates.
Steenstrupine ⎭
Tritomite, Th,Ce,Y,Ca, fluo-silicate.
Gadolinite, $Be_2FeY_2Si_2O_{10}$.
Yttrialite, Th,Y, silicate.
Rowlandite, Y silicate.
Thalenite, $Y_2Si_2O_7$.
Thortveitite, $(Sc,Y)_2Si_2O_7$.
Bodenbenderite, Silicate and titanate of Al,Y,Mn.
Cenosite, $2CaO.(Ce,Y)_2O_3.CO_2.4SiO_2.H_2O$.
Delorenzite, Fe,U,Y, titanate.
Yttrocrasite, Hydrous, Y,Th, titanate.
Fergusonite, Y,Er, niobate.
Risörite, Y niobate.
Yttrotantalite, Y, etc., tantalate-niobate.
Samarskite, Fe,Ca,U,Ce,Y, niobate-tantalate.
Hielmite, Y,Fe,Mn,Ca, stanno-tantalate.
Euxenite ⎫
Polycrase ⎬ Y,Ce,U, niobate-titanates.
Blomstrandine-Priorite ⎭
Eschwegeite, Hydrous Y,Er, titano-niobate and tantalate.
Plumboniobite, Y,U,Pb,Fe, niobate.
XENOTIME, YPO_4.
Retzian, Y,Mn,Ca, arsenate.
Rhabdophanite, Hydrous Ce,Y, phosphate.
Weinschenkite, $(Y,Er)PO_4.2H_2O$.

ZINC

Sphalerite, ZnS.
Wurtzite, ZnS.
Voltzite, Zn_5S_4O.
ZINCITE, ZnO.
Gahnite, $ZnO.Al_2O_3$.
FRANKLINITE, $(Fe,Zn,Mn)O.(Fe,Mn)_2O_3$.
Hetærolite, $ZnO.Mn_2O_3$.
Zincdibraunite, $ZnO.2MnO_2.2H_2O$.
Chalcophanite, $(Mn,Zn)O.2MnO_2.2H_2O$.
Hydrohetærolite, $2ZnO.2Mn_2O_3.H_2O$.
Smithsonite, $ZnCO_3$.
Rosasite, $(Cu,Zn)CO_3.(Cu,Zn)(OH)_2$.

Aurichalcite, $2(Zn,Cu)CO_3.3(Zn,Cu)(OH)_2$.
Hydrozincite, $2ZnCO_3.3Zn(OH)_2$.
Loseyite, $(Mn,Zn)CO_3.5(Mn,Zn)(OH)_2$.
Danalite, $3(Fe,Zn,Mn)BeSiO_4.ZnS$.
Larsenite, $PbZnSiO_4$.
Calcium-larsenite, $(Pb,Ca)ZnSiO_4$.
Willemite, Zn_2SiO_4.
Hardystonite, $Ca_2ZnSi_2O_7$.
Calamine, H_2ZnSiO_5.
Fraipontite, $8ZnO.2Al_2O_3.5SiO_2.11H_2O$.
Clinohedrite, $H_2CaZnSiO_5$.
McGovernite, $21(Mn,Mg,Zn)O.3SiO_2.$
$\frac{1}{2}As_2O_3.As_2O_5.10H_2O$.
Hodgkinsonite, $3(Zn,Mn)O.SiO_2.H_2O$.
Gageite, Hydrous Mn,Mg,Zn, silicate.
Tarbuttite, $Zn_3P_2O_8.Zn(OH)_2$.
Adamite, $Zn_2(OH)AsO_4$.
Descloizite, $(Pb,Zn)_2(OH)VO_4$.
Staszicite, $5(Ca,Cu,Zn)O.As_2O_5.2H_2O$.
Hopeite ⎫
Parahopeite ⎬ $Zn_3P_2O_8.4H_2O$.
Phosphophyllite, Zn phosphate.
Köttigite, $Zn_3As_2O_8.8H_2O$.
Barthite, $3ZnO.CuO.3As_2O_5.2H_2O$.
Hügelite, Hydrous Pb,Zn, vanadate.
Chlorophœnicite, $10(Zn,Mn)O.As_2O_5.7H_2O$.
Holdenite, $8MnO.4ZnO.As_2O_5.5H_2O$.
Spencerite, $Zn_3(PO_4)_2.Zn(OH)_2.3H_2O$.
Salmoite, Zinc phosphate.
Veszelyite, Hydrous Cu,Zn, phospho-arsenate.
Arakawaite, $4CuO.2ZnO.P_2O_5.6\frac{1}{2}H_2O$.
Kehoeite, Hydrous Al,Zn, phosphate.
Sussexite, $H(Mn,Zn,Mg)BO_3$.
Zinkosite, $ZnSO_4$.
Ilesite, $(Mn,Zn,Fe)SO_4.4H_2O$.
Bianchite, $FeSO_4.2ZnSO_4.18H_2O$.
Goslarite, $ZnSO_4.7H_2O$.
Zinc-copper melanterite, $(Zn,Cu,Fe)SO_4.$
$7H_2O$.
Zinc-copper chalcanthite, $(Zn,Cu,Fe)SO_4.$
$5H_2O$.
Mooreite, Hydrous Mg,Mn,Zn, sulphate.
Dietrichite, $(Zn,Fe,Mn)SO_4.Al_2(SO_4)_3.$
$22H_2O$.
Serpierite, Hydrous Cu,Zn, sulphate.
Zincaluminite, $2ZnSO_4.4Zn(OH)_2.6Al(OH)_3.$
$5H_2O$.
Glaukokerinite, $Zn_{13}Al_8Cu_7(SO_4)_2O_{30}.34H_2O$.

ZIRCONIUM

Baddeleyite, ZrO_2.
Rosenbuschite, Na,Ca,Zr, silicate.
Wöhlerite, Na,Ca,Zr, silicate and niobate.
Låvenite, Mn,Ca,Zr, silicate.
Guarinite, $(Na_2,Ca)(Si,Zr)O_3$.
Eudialyte, Zr,Fe,Ca,Na, silicate.
Elpidite, $H_6Na_2Zr(SiO_3)_6$.
Catapleiite, $H_4(Na_2,Ca)ZrSi_3O_{11}$.
Zircon, Zr, SiO_4.
Zirkelite, $(Ca,Fe)(Zr,Ti,Th)_2O_5$.
Oliveiraite, $3ZrO_2.2TiO_2.2H_2O$.

TABLE II. MINERALS ARRANGED ACCORDING TO THEIR SYSTEM OF CRYSTALLIZATION

The following lists only include common or important species, whose crystallization is known, arranged according to the system to which they belong, and further classified by their luster and specific gravity; the hardness is also given in each case.*

I. CRYSTALLIZATION ISOMETRIC†

A. LUSTER NONMETALLIC

	Specific Gravity	Hardness		Specific Gravity	Hardness
Sylvite.........	1·98	2	Helvite.........	3·16–3·36	6–6·5
Halite.........	2·14	2·5	Garnet.........	3·3–4·3	6·5–7·5
Sodalite.......	2·14–2·30	5·5–6	Diamond.......	3·52	10
Analcite.......	2·2–2·3	5–5·5	Spinel.........	3·5–4·1	8
Noselite.......	2·25–2·4	5·5	Sphalerite.......	3·9–4·1	3·5–4
Haüynite.......	2·4–2·5	5·5–6	Gahnite........	4·0–4·6	7·5–8
Leucite.........	2·45–2·50	5·5–6	Cerargyrite......	5·55	1–1·5
Boracite.......	2·9–3	7	Cuprite.........	5·85–6·15	3·5–4
Pharmacosiderite	2·9–3	2·5	Eulytite........	6·11	4·5
Fluorite........	3·2	4			

B. LUSTER METALLIC (AND SUBMETALLIC)

	Specific Gravity	Hardness		Specific Gravity	Hardness
Chromite.......	4·3–4·57	5·5	Smaltite, Chloan-		
Tennantite......	4·4–4·49	3–4	thite.........	6·4–6·6	5·5–6
Tetrahedrite	4·4–5·1	3–4	Argentite.......	7·2–7·36	2–2·5
Pyrite..........	4·95–5·10	6–6·5	Galena.........	7·4–7·6	2·5–3
Franklinite......	5·07–5·22	6–6·5	Copper.........	8·8–8·9	2·5–3
Magnetite.......	5·18	6–6·5	Uraninite.......	9–9·7	5·5
Bornite.........	4·9–5·4	3	Silver..........	10·1–11·1	2·5–3
Cuprite........	5·85–6·15	3·5–4	Platinum.......	14–19	4–4·5
Cobaltite.......	6–6·3	5·5	Gold..........	15·6–19·3	2·5–3

* For complete determinative tables based on crystallization, see Crystallographic Tables for the Determination of Minerals by Goldschmidt and Gordon, 1928.

† Some pseudo-isometric species are here included. Species with submetallic luster are placed under B. but some species are included in both lists.

II. CRYSTALLIZATION TETRAGONAL

A. LUSTER NONMETALLIC

	Specific Gravity	Hardness		Specific Gravity	Hardness
Apophyllite.....	2·3–2·4	4·5–5	Xenotime.......	4·45–4·56	4–5
Wernerite			Thorite.........	4·4–5·4	4·5–5
(Scapolite) ...	2·66–2·73	5·5–6	Zircon..........	4·68–4·7	7·5
Meionite........	2·70–2·74	5·5–6	Scheelite.......	5·9–6·1	4·5–5
Vesuvianite.....	3·35–3·45	6·5	Phosgenite......	6–6·09	2·75–3
Torbernite......	3·4–3·6	2–2·5	Calomel........	6·48	1–2
Octahedrite.....	3·8–3·95	5·5–6	Wulfenite.......	6·7–7·0	2·75–3
Rutile..........	4·18–4·25	6–6·5	Cassiterite.....	6·8–7·1	6–7

B. LUSTER METALLIC (AND SUBMETALLIC)

	Specific Gravity	Hardness		Specific Gravity	Hardness
Chalcopyrite....	4·1–4·3	3·5–4	Hausmannite....	4·7–4·86	5–5·5
Rutile..........	4·18–4·25; 5·2	6–6·5	Braunite........	4·75–4·82	6–6·5

III. CRYSTALLIZATION HEXAGONAL*

Rhombohedral species are distinguished by a letter R

A. LUSTER NONMETALLIC

	Specific Gravity	Hardness		Specific Gravity	Hardness
Gmelinite, R....	2·04–2·17	4·5	Tourmaline, R...	2·98–3·20	7–7·5
Chabazite, R....	2·08–2·16	4–5	Apatite.........	3·17–3·23	5
Brucite, R......	2·38–2·4	2·5	Dioptase, R.....	3·28–3·35	5
Cancrinite......	2·42–2·5	5–6	Rhodochrosite, R.	3·45–3·60	3·5–4·5
Chalcophyllite, R	2·44–2·66	2	Siderite, R......	3·83–3·88	3·5–4
Nephelite......	2·55–2·65	5·5–6	Corundum, R....	3·95–4·10	9
Quartz, R......	2·65	7	Willemite, R.....	3·94–4·19	5·5
Beryl..........	2·64–2·7;2·80	7·5–8	Smithsonite, R...	4·30–4·45	5
Alunite, R......	2·67	3·5–4	Hematite, R.....	4·9–5·3	5·5–6·5
Penninite (pseu.)			Zincite.........	5·4–5·7	4–4·5
R............	2·6–2·85	2·25	Pyromorphite....	6·5–7·1	3·5–4
Calcite, R......	2·71	3	Vanadinite......	6·66–6·86	3
Dolomite, R....	2·8–2·9	3·5–4	Mimetite........	7·0–7·25	3·5
Ankerite, R.....	2·95–3·1	3·5–4	Cinnabar, R.....	8·08–8·2	2–2·5
Phenacite, R....	2·97–3·0	7·5–8			

* Some pseudo-hexagonal species are included.

B. Luster Metallic (and Submetallic)

	Specific Gravity	Hard-ness		Specific Gravity	Hard-ness
Graphite, R.....	2·1–2·2	1–1·5	Millerite, R.....	5·3–5·65	3–3·5
Ilmenite, R......	4·5–5	5–6	Pyrargyrite, R...	5·85	2·5
Pyrrhotite......	4·6	3·5–4·5	Niccolite........	7·3–7·67	5–5·5
Molybdenite.....	4·7–4·8	1–1·5	Cinnabar, R.....	8·0–8·2	2–2·5
Hematite, R....	5·2–5·3	5·5–6·5	Iridosmine, R....	19·3–21·1	6–7

IV. CRYSTALLIZATION ORTHORHOMBIC

A. Luster Nonmetallic

Struvite........	1·65–1·7	2	Dufrenite.......	3·23–3·4	3·5–4
Sulphur.........	2·07	1·5–2·5	Chrysolite.......	3·27–3·37	6·5–7
Natrolite.......	2·20–2·25	5–5·5	Diaspore	3·3–3·5	6·5–7
Thomsonite.....	2·3–2·4	5–5·5	Hypersthene.....	3·4–3·5	5·5
Wavellite.......	2·33	3·5–4	Calamine.......	3·4–3·5	4·5–5
Cordierite......	2·6–2·66	7–7·5	Topaz..........	3·4–3·65	8
Thenardite......	2·68–2·69	2–3	Triphylite.......	3·52–3·55	4·5–5
Talc...........	2·7–2·8	1–1·5	Chrysoberyl.....	3·5–3·8	8·5
Prehnite.......	2·8–2·95	6–6·5	Libethenite......	3·6–3·8	4
Anhydrite.......	2·90–2·98	3–3·5	Staurolite.......	3·65–3·75	7–7·5
Aragonite......	2·94	3·5–4	Strontianite.....	3·68–3·71	3·5–4
Danburite......	2·97–3·02	7–7·25	Brochantite.....	3·91	3·5–4
Humite........	3·1–3·2	6–6·5	Brookite........	3·87–4·07	5·5–6
Anthophyllite....	3·1–3·2	5·5–6	Celestite........	3·95–3·97	3–3·5
Andalusite......	3·16–3·2	7·5	Tephroite.......	4–4·12	5·5–6
Enstatite.......	3·15–3·3	5·5	Witherite.......	4·3–4·35	3–3·75
Autunite.......	3·05–3·19	2–2·5	Barite..........	4·5	2·5–3·5
Sillimanite.....	3·24	6–7	Samarskite......	5·6–5·8	5–6
Scorodite......	3·1–3·3	3·5–4	Anglesite.......	6·12–6·39	2·75–3
Forsterite.......	3·2–3·33	6–7	Caledonite......	6·4	2·5–3
Zoisite.........	3·25–3·37	6–6·5	Cerussite.......	6·46–6·57	3–3·5

B. Luster Metallic (and Submetallic)

Brookite........	3·87–4·07	5·5–6	Columbite......	5·36–6·0	6
Ilvaite..........	4·0–4·05	5·5–6	Chalcocite......	5·5–5·8	2·5–3
Manganite......	4·2–4·4	4	Bournonite......	5·7–5·9	2·5–3
Enargite........	4·43–4·45	3	Arsenopyrite....	5·9–6·2	5·5–6
Stibnite.........	4·5–4·6	2	Nagyagite......	6·85–7·2	1–1·5
Pyrolusite.......	4·73–4·86	2–2·5	Tantalite.......	7–7·3	6
Marcasite.......	4·85–4·9	6–6·5			

V. CRYSTALLIZATION MONOCLINIC

A. Luster Nonmetallic

	Specific Gravity	Hardness		Specific Gravity	Hardness
Mirabilite.......	1·48	1·5–2	Erythrite.......	2·95	1·5–2·5
Borax..........	1·69–1·72	2–2·5	Margarite.......	2·99–3·08	3·5–4·5
Melanterite.....	1·90	2	Amphibole......	2·9–3·4	5–6
Gay-Lussite.....	1·94	2–3	Lazulite........	3·06	5–6
Stilbite........	2·16–2·20	3·5–4	Euclase........	3·10	7·5
Scolecite.......	2·16–2·4	5–5·5	Glaucophane....	3·10–3·11	6–6·5
Heulandite......	2·18–2·22	3·5–4	Chondrodite.....	3·1–3·2	6–6·5
Phillipsite.......	2·2	4–4·5	Clinohumite.....	3·1–3·2	6–6·5
Blödite.........	2·25	2·5	Spodumene......	3·13–3·2	6·5–7
Laumontite.....	2·25–2·36	3·5–4	Pyroxene.......	3·3–3·6	5–6
Gypsum........	2·31–2·33	1·5–2	Epidote........	3·25–3·5	6–7
Petalite........	2·39–2·46	6–6·5	Jadeite.........	3·33–3·35	6·5–7
Harmotome.....	2·44–2·5	4·5	Piedmontite.....	3·40	6·5
Serpentine......	2·50–2·65	2·5–4	Arfvedsonite.....	3·44–3·45	6
Orthoclase......	2·57	6	Titanite........	3·4–3·65	5–5·5
Vivianite.......	2·58–2·68	1·5–2	Acmite.........	3·5–3·55	6–6·5
Kaolinite.......	2·6–2·63	2–2·5	Chloritoid?.....	3·52–3·57	6·5
Clinochlore.....	2·65–2·78	2–2·5	Azurite........	3·77–3·83	3·5–4
Pectolite........	2·68–2·78	5	Allanite........	3·5–4·2	5·5–6
Glauberite.....	2·7–2·85	2·5–3	Malachite.......	3·9–4·03	3·5–4
Muscovite......	2·76–3	2–2·5	Gadolinite......	4·0–4·5	6·5–7
Lepidolite......	2·8–2·9	2·5–4	Clinoclasite.....	4·19–4·36	2·5–3
Biotite.........	2·7–3·1	2·5–3	Monazite.......	4·9–5·3	5–5·5
Phlogopite......	2·78–2·85	2·5–3	Linarite......	5·3–5·45	2·5
Prochlorite.....	2·78–2·96	1–2	Crocoite........	5·9–6·1	2·5–3
Liroconite......	2·88	2–2·5	Tenorite........	5·8–6·25	3–4
Wollastonite.....	2·8–2·9	4·5–5	Leadhillite......	6·26–6·44	2·5
Pyrophyllite.....	2·8–2·9	1–2	Lanarkite.......	6·3–6·4	2–2·5
Datolite........	2·9–3·0	5–5·5	Hübnerite.......	7·2–7·5	5–5·5
Cryolite........	2·95–3	2·5			

B. Luster Metallic (and Submetallic)

	Specific Gravity	Hardness
Wolframite.................	7·2–7·5	5–5·5

VI. CRYSTALLIZATION TRICLINIC

A. Luster Nonmetallic

	Specific Gravity	Hardness		Specific Gravity	Hardness
Chalcanthite....	2·12–2·30	2·5	Anorthite.......	2·74–2·76	6–6·5
Microcline......	2·54–2·57	6–6·5	Amblygonite....	3·01–3·09	6
Albite..........	2·62–2·65	6–6·5	Axinite.........	3·27	6·5–7
Oligoclase.......	2·65–2·67	6–6·5	Rhodonite......	3·4–3·68	5·5–6·5
Andesine........	2·68–2·69	6–6·5	Kyanite........	3·56–3·67	5–7·25
Labradorite.....	2·70–2·72	6–6·5			

Octahedral. — Fluorite; Diamond. Magnetite (also Franklinite) has often distinct octahedral *parting.*

Dodecahedral. — Sphalerite. Also, imperfect, Sodalite; Hauynite.

Rhombohedral. — Calcite and other species of the same group (pp. 511–521) angles 75° and 105°.

Square Prismatic (90°). — Scapolite; Rutile; Xenotime.

Prismatic. — Barite (78½°, 101½°); Celestite; Amphibole (54° and 126°), etc.

Basal. — METALLIC LUSTER: Graphite; Molybdenite.

NONMETALLIC LUSTER: Apophyllite; Topaz; Talc; the Micas and Chlorites; Chalcophyllite, etc. Pyroxene often shows marked basal *parting.*

Pinacoidal. — METALLIC LUSTER: Stibnite.

NONMETALLIC LUSTER: Gypsum; Orpiment; Euclase; Diaspore; Sillimanite; Kyanite; Feldspars.

II. HARDNESS

1. **Soft Minerals.** — The following minerals are conspicuously *Soft*, that is, $H = 2$ or less; they hence have a *greasy* feel. (See further the Tables, pp. 813 to 816.)

METALLIC LUSTER: Graphite; Molybdenite; Tetradymite; Sternbergite; Argentite; Nagyagite; some of the Native Metals (Lead, etc.).

NONMETALLIC LUSTER: Talc; Pyrophyllite; Brucite; Tyrolite; Orpiment; Cerargyrite; Cinnabar; Sulphur; Gypsum.

Also Calomel, Arsenolite, and many hydrous sulphates, phosphates, etc.

2. **Hard Minerals.** — Minerals whose hardness is equal to or greater than 7 (Quartz $= 7$).

The following minerals are here included:

LUSTER NONMETALLIC

QUARTZ	7	Hambergite	7·5
Tridymite	7	ZIRCON	7·5
Barylite	7	ANDALUSITE	7·5
Dumortierite	7	BERYL	7·5–8
Danburite	7–7·25	Lawsonite	7·5–8
BORACITE	7	Phenacite	7·5–8
Zunyite	7	Gahnite	7·5–8
KYANITE	5–7·25	Hercynite	7·5–8
TOURMALINE	7–7·5	SPINEL	8
GARNET	6·5–7·5	TOPAZ	8
CORDIERITE	7–7·5	Rhodizite	8
STAUROLITE	7–7·5	CHRYSOBERYL	8·5
Schorlomite	7–7·5	CORUNDUM	9
Sapphirine	7·5	DIAMOND	10
Euclase	7·5		

The following minerals have hardness equal to 6 to 7, or 6.5–7.

LUSTER METALLIC: Iridosmine; Iridium; Sperrylite.

LUSTER NONMETALLIC: Ardennite; Axinite; Bertrandite; Cassiterite; Chrysolite; Diaspore; Elpidite; Epidote; Forsterite; Gadolinite; Jadeite; Partschinite; Sillimanite; Spodumene; Trimerite.

III. SPECIFIC GRAVITY

Attention is called to the remarks in Art. **307** (p. 221), on the relation of specific gravity to chemical composition. Also to the statements in Art. **308** as to the *average* specific gravity among minerals of metallic and nonmetallic luster respectively. The species in each of the separate lists of Table II of minerals classified with reference to crystallization are arranged according to ascending *specific gravities*. Hence the lists give at a glance minerals distinguished by both low and high density.

IV. LUSTER (See Art. **371**, p. 273)

Metallic. — Native metals; most Sulphides; some Oxides, those containing iron, manganese, lead, etc.

Submetallic. — Here belong chiefly certain iron and manganese compounds, as Ilmenite; Ilvaite; Columbite; Tantalite (and allied species); Wolframite; Braunite; Hausmannite. Also Brookite; Uraninite, etc.

Adamantine. — Here belong minerals of high refractive index: (*a*) Some *hard* minerals: Diamond; Corundum; Cassiterite; Zircon; Rutile. (*b*) Many species of *high density*, as compounds of lead, also of silver, copper, mercury. Thus, Cerussite, Anglesite, Phosgenite, etc.; Cerargyrite; Cuprite; some Cinnabar, etc. (*c*) Also certain varieties of Sphalerite, Titanite and Octahedrite.

Metallic-Adamantine. — Pyrargyrite; some varieties of the following: Cuprite; Cerussite; Octahedrite, Rutile, Brookite.

Resinous or Waxy. — Sphalerite; Sulphur; Elæolite; Serpentine; many Phosphates.

Vitreous. — Quartz and many Silicates, as Garnet, Beryl, etc.

Pearly. — The foliated species: Talc, Brucite, Pyrophyllite. Also (on cleavage surfaces) conspicuously the following: Apophyllite, Stilbite, Heulandite. Also, less prominent: Barite, Celestite; Diaspore; some Feldspar, and others.

Silky. — Some fibrous minerals, as Gypsum; Calcite; also Asbestus; Malachite.

V. COLOR

The following lists may be of some use in the way of suggestion. It is to be noted, however, that especially in the case of metallic minerals a slight surface change may alter the effect of color. Further, among minerals of nonmetallic luster particularly, no sharp line can be drawn between colors slightly different, and many variations of shade occur in the case of a single species. For these reasons no lists, unless inconveniently extended, could make any claim to completeness.

(*a*) METALLIC LUSTER

Silver-white, Tin-white. — Native Silver; Native Antimony, Arsenic and Tellurium; Amalgam; Arsenopyrite and Löllingite; several sulphides, arsenides, etc., of cobalt or nickel, as Cobaltite (reddish); some Tellurides; Bismuth (reddish). No sharp line can be drawn between these and the following group.

Steel-gray. — Platinum; Manganite; Chalcocite; Sylvanite; Bournonite.

Blue-gray. — Molybdenite; Galena.

Lead-gray. — Many sulphides, as Galena (bluish); Stibnite; many Sulpharsenites, etc., as Jamesonite, Dufrenoysite, etc.

Iron-black. — Graphite; Tetrahedrite; Polybasite; Stephanite; Enargite; Pyrolusite; Magnetite; Hematite; Franklinite.

Black (with submetallic luster). — Ilmenite; Limonite; Columbite; Tantalite, etc.; Wolframite; Ilvaite; Uraninite, etc. The following are usually brownish black: Braunite; Hausmannite.

Copper-red. — Native copper.

Bronze-red. — Bornite (quickly tarnished giving purplish tints); Niccolite.

Bronze-yellow. — Pyrrhotite; Pentlandite; Breithauptite.

Brass-yellow. — Chalcopyrite; Millerite (bronze). Pale brass-yellow: Pyrite; Marcasite (whiter than Pyrite).

Gold-yellow. — Native gold; chalcopyrite and pyrite sometimes are mistaken for gold.

Streak. — The following minerals of metallic luster are notable for the color of their *streak:*

Cochineal-red: Pyrargyrite.

Cherry-red: Miargyrite.

Dull Red: Hematite; Cuprite; some Cinnabar.

Scarlet: Cinnabar (usually nonmetallic).

Dark Brown: Manganite; Franklinite; Chromite.

Yellow: Limonite.

Tarnish. — The following are conspicuous for their bright or variegated tarnish: Chalcopyrite; Bornite (purplish tints); Tetrahedrite; some Limonite.

(b) Nonmetallic Luster

Colorless. — In Crystals: Quartz; Calcite; Aragonite; Gypsum; Cerussite; Anglesite; Albite; Barite; Adularia; Topaz; Apophyllite; Natrolite and other Zeolites; Celestite; Diaspore; Nephelite; Meionite; Calamine; Cryolite; Phenacite, etc.

Massive: Quartz; Calcite; Gypsum; Hyalite (botryoidal).

White. — Crystals: Amphibole (tremolite); Pyroxene (diopside, usually greenish).

Massive: Calcite; Milky Quartz; Feldspars, especially Albite; Barite; Cerussite, Scapolite; Talc; Meerschaum; Magnesite; Kaolinite; Amblygonite, etc.

Blue. — Blackish Blue: Azurite; Crocidolite.

Indigo-blue: Indicolite (Tourmaline); Vivianite.

Azure-blue: Lazulite; Azurite; Lapis Lazuli; Turquois.

Prussian-blue: Sapphire; Kyanite; Cordierite; Azurite; Chalcanthite and many copper compounds.

Sky-blue, Mountain-blue: Beryl; Celestite.

Violet-blue: Amethyst; Fluorite.

Greenish Blue: Amazon-stone; Chrysocolla; Calamine; Smithsonite; some Turquois; Beryl.

Green. — Blackish Green: Epidote; Serpentine; Pyroxene; Amphibole.

Emerald-green: Beryl (Emerald); Malachite; Dioptase; Atacamite; and many other copper compounds; Spodumene (hiddenite); Pyroxene (rare); Gahnite; Jadeite and Jade.

Bluish Green: Beryl; Apatite; Fluorite; Amazon-stone; Prehnite; Calamine; Smithsonite; Chrysocolla, Chlorite; some Turquoia.

Mountain Green: Beryl (aquamarine); Euclase.

Apple-green: Talc; Garnet; Chrysoprase; Willemite; Garnierite; Pyrophyllite; some Muscovite; Jadeite and Jade.

Pistachio-green: Epidote.

Grass-green: Pyromorphite; Wavellite; Variscite; Chrysoberyl.

Grayish Green: Amphibole and Pyroxene, many common kinds; Jasper; Jade.

Yellow-green to Olive-green: Beryl; Apatite; Chrysoberyl; Chrysolite (olive-green); Chlorite; Serpentine; Titanite; Datolite; Olivenite; Vesuvianite.

Yellow. — Sulphur-yellow: Sulphur; some Vesuvianite.

Orange-yellow: Orpiment; Wulfenite; Mimetite.

Straw-yellow, also Wine-yellow, Wax-yellow: Topaz; Sulphur; Fluorite; Cancrinite; Wulfenite; Vanadinite; Willemite; Calcite; Barite; Chrysolite; Chondrodite; Titanite; Datolite, etc.

Brownish Yellow: Much Sphalerite; Siderite; Goethite.

Ocher-yellow: Goethite: Yellow ocher (limonite).

Red. — Ruby-red: Ruby (corundum); Ruby spinel; much Garnet; Proustite; Vanadinite; Sphalerite; Chondrodite.

Cochineal-red: Cuprite; Cinnabar.

Hyacinth-red: Zircon; Crocoite.

Orange-red: Zincite; Realgar; Wulfenite.

Crimson-red: Tourmaline (rubellite); Spinel; Fluorite.

Scarlet-red: Cinnabar.

Brick-red: Some Hematite (red ocher).

Rose-red to Pink: Rose quartz; Rhodonite; Rhodochrosite; Erythrite; some Scapolite. Apophyllite and Zoisite; Eudialyte; Petalite; Margarite.

Peach-blossom Red to Lilac: Lepidolite; Rubellite.

Flesh-red: Some Orthoclase; Willemite (the variety troostite); some Chabazite; Stilbite and Heulandite; Apatite; rarely Calcite; Polyhalite.

Brownish Red: Jasper; Limonite; Garnet; Sphalerite; Siderite; Rutile.

Brown. — Reddish Brown: Some Garnet; some Sphalerite; Staurolite; Cassiterite; Rutile.

Clove-brown: Axinite; Zircon; Pyromorphite.

Yellowish Brown: Siderite and related carbonates; Sphalerite; Jasper; Limonite; Goethite; Tourmaline; Vesuvianite; Chondrodite; Staurolite.

Blackish Brown: Titanite; some Siderite; Sphalerite.

Smoky Brown: Quartz.

Black. — Tourmaline; black Garnet (melanite); some Mica (especially biotite); also some Amphibole, Pyroxene and Epidote (these are mostly greenish or brownish black); further, some Sphalerite and some kinds of Quartz (varying from smoky brown to black);

also Allanite; Samarskite.   Some black minerals with submetallic luster are mentioned on p. 820.

Streak. — The *streak* is to be noted in the case of some minerals with nonmetallic luster. By far the majority have, even when deeply colored in the mass (*e.g.* Tourmaline), a streak differing but little from white.   The following may be mentioned:

ORANGE-YELLOW: Zincite; Crocoite.

COCHINEAL-RED: Pyrargyrite and Proustite.

SCARLET-RED: Cinnabar.

BROWNISH RED: Cuprite; Hematite.

BROWN: Limonite.

The streak of the various copper, green and blue minerals, as Malachite, Azurite, etc., is about the same as the color of the mineral itself, though often a little paler.

GENERAL INDEX

A

Abbreviations, 5
Absorption of light, 246
    biaxial crystals, 316
    uniaxial crystals, 294
Accessory rock-making minerals, 380
Acicular structure, 205
Acid salts, 349
Acids, 348
Actinoelectricity, 336
Adamantine luster, 232, 273
Aggregate polarization, 329
Aggregates, crystalline, 204
    optical properties, 328
Airy's spirals, 295
Albite law (twinning), 193
Alkalies, test for, 349
Alkaline reaction, 363
    taste, 339
Alliaceous odor, 339
Alteration of minerals, 357
Aluminum tests for, 369
Amorphous structure, 8, 205
Amphibole schist, 386
Amplitude of vibration, 225
Amygdaloidal structure, 205
Analogous pole, 336
Analysis, blowpipe, 362
    chemical, 356
    microchemical, 356
Analyzer, 253
Andesite, 382
Angle, critical, 233
    of extinction, 304
Angles, measurement of, 171
    of isometric forms, 79, 82, 86
Anisometric crystals, 276
Anisotropic crystals, 276
Anomalies, optical, 329
Anorthic system, 162
Anorthosite, 382
Antilogous pole, 336
Antimony, tests for, 369
Aphanites, 381
Arborescent structure, see Dendritic, 205
Argillaceous odor, 339
Arkose, 384
Arsenic, tests for, 370
Artificial minerals, 1, 357
Association of minerals, 379
Asterism, 274
Astringent taste, 339
Asymmetric class, 165

A (continued, right column)

Athermanous, 334
Atom, 340
Atomic number, 340
    weight, 341
Axes, crystallographic, 18
    of symmetry, 11
    optic, 301
        dispersion of, 318, 321
Axial angle, optic, 302
    calculation, 303
    measurement of, 311
    plane, 43
    ratio, 43

B

Banded gneiss, 386
Barium, tests for, 370
Basal pinacoid, 94, 112, 139, 152
Basalt, 382
Bases, chemical, 348
Basic salts, 349
Baveno twins, 192
Becke test, 237, 239
Belonite, 202
Bertrand lens, 270
    ocular, 305
Berylloid, 114
Bevel, Bevelment, 74
Biaxial crystals, behavior of light in, 295
    positive and negative, 303
Biaxial examination in converging polarized
    light, 308
    in parallel polarized
    light, 304
    indicatrix, 299
    interference figure, 308
    optic axes, 308
Binary symmetry, 12
Biotite gneiss, 386
Bi-quartz wedge plate, 305
Birefringence, 247
    determination of, 261
Bisectrix, acute, 303
    obtuse, 303
Bismuth, tests for, 370
Bitter taste, 339
Bituminous odor, 339
Bivalent element, 346
Bladed structure, 204
Blebby bead, 363
Blowpipe, 361
    flame, 362
Borax bead tests, 368

INDEX TO SPECIES

A

Aarite, *v.* Arite, 427
Abriachanite, 578
Acadialite, 651
Acanthite, 421
Acerdese, *v.* Manganite
Achmatite, 623
Achroite, 635
Acmite, 561
Actinolite, Actinote, 574
Adamantine spar, 482
Adamine, 714
Adamite, 714
Adelite, 710
Adipocire, *v.* Hatchettite
Adular, Adularia, 538
Ædelite, *v.* Prehnite
Ægirine, 561
Ægirite, 561
Ægirite-augite, 559
Ænigmatite, 579
Æschynite, 698
Afwillite, 687
Agalite, 678
Agalmatolite, 661, 669
Agaric mineral, 515
Agate, 473
Agnolite, 640
Agricolite, 591
Aguilarite, 419
Aikinite, 451
Aikaite, 776
Åkermanite, 607
Akrochordite, 720
Alabandin, 423
Alabandite, 423
Alabaster, 759
     Oriental, 514
Alaïte, 510
Alalite, 558
Alamosite, 567
Alaskaite, 447
Alaun, *v.* Alum
Alaunstein, *v.* Alunite
Albertite, 778
Albite, 545
Alexandrite, 494
Algodonite, 415
Alisonite, 417
Allactite, Allaktit, 718
Allagite, 565
Allanite, 624
Alleghanyite, 600

Allemontite, 400
Allochroite, 594
Alloclasite, Alloklas, 441
Allodelphite, 718
Allopalladium, 407
Allophane, 684
Almandine, Almandite, 593, 488
Almeriite, 770
Aloisiite, 639
Alpha-calcite, 513
Alpha-hopeite, 719
Alpha-hyblite, 612
Alpha-leonhardite, 649
Alpha-quartz, 471
Alpha-sartorite, 447
Alpha-sepiolite, 679
Alpha-sulphur, 398
Alpha-trechmanite, 446
Alpha-uzbekite, 715
Alshedite, 689
Alstonite, 523
Altaite, 417
Alum, 763
Alumian, 756
Aluminite, 760
Alumohydrocalcite, 529
Alumstone, 768
Alundum, 482
Alunite, 768
Alumogel, 506
Alunogen, 765
Alurgite, 666
Alushtite, 681
Amalgam, 405
Amarantite, 767
Amazonite, 541
Amazonstone, 541
Ambatoarinite, 526
Amber, 776
Amblygonite, 712
Amblystegite, 555
Ameletite, 589
Amesite, 672
Amethyst, 472
     Oriental, 482
Amianthus, 574, 675
Ammiolite, 738
Ammonia alum, 764
Ammonioborite, 743
Ammoniojarosite, 769
Amosite, 570
Ampangabéite, 698
Amphibole, 571

AMPHIBOLE Group, 568
Amphigène, 549
Amphodelite, 548
Analcime, 652
Analcite, 652
Anapaite, 720
Anatase, 500
Anauxite, 682
Ancylite, 526
Andalusite, 615
Andesine, 547
Andorite, 446
Andradite, 594
Andrewsite, 733
Anemousite, 549
Angaralite, 634
Anglarite, 447
Anglesite, 751
Anhydrite, 752
Animikite, 415
Ankerite, 517
Annabergite, 722
Ännerödite, 697
Annite, 664, 666
Anomite, 664
Anorthite, 548
Anorthoclase, 541
Anthophyllite, 570
     Hydrous, 571
Anthracite, 778
Antigorite, 675
Antimonarsen, *v.* Allemontite
ANTIMONATES, 737
Antimonblende, *v.* Kermesite
Antimonglanz, *v.* Stibnite
Antimonite, 410
ANTIMONITES, 737
Antimonnickel, *v.* Breithauptite
Antimonsilber, *v.* Dyscrasite
Antimonsilberblende, *v.* Pyrargyrite
Antimony, 400
Antimony, Red, *v.* Kermesite
Antimony glance, 410
Antlerite, 756
Antrimolite, 655
Apatite, 704
Aphanèse, Aphanesite, 715
Aphrite, 514
Aphrizite, 635

833